T0391476

**Civil Engineering and Urban Research** collects papers resulting from the conference on Civil, Architecture and Urban Engineering (ICCAUE 2022), Xining, China, 24–26 June 2022. The primary goal is to promote research and developmental activities in civil engineering, architecture and urban research. Moreover, it aims to promote scientific information interchange between scholars from the top universities, business associations, research centers and high-tech enterprises working all around the world.

The conference conducts in-depth exchanges and discussions on relevant topics such as civil engineering and architecture, aiming to provide an academic and technical communication platform for scholars and engineers engaged in scientific research and engineering practice in the field of urban engineering, civil engineering and architecture design. By sharing the research status of scientific research achievements and cutting-edge technologies, it helps scholars and engineers all over the world comprehend the academic development trend and broaden research ideas. So as to strengthen international academic research, academic topics exchange and discussion, and promote the industrialization cooperation of academic achievements.

PROCEEDINGS OF THE 4TH INTERNATIONAL CONFERENCE ON CIVIL ARCHITECTURE AND URBAN ENGINEERING (ICCAUE 2022), XINING, CHINA, 24–26 JUNE 2022

# Civil Engineering and Urban Research

## Volume 2

*Edited by*

Hazem Samih Mohamed
*School of Civil Engineering and Geomatics, Southwest Petroleum University, China*

Jinfang Hou
*Tianjin Port Engineering Institute Co., Ltd. of CCCC First Harbor Engineering Co., Ltd., China*

CRC Press
Taylor & Francis Group
Boca Raton   London   New York   Leiden

CRC Press is an imprint of the
Taylor & Francis Group, an **informa** business

A BALKEMA BOOK

First published 2023
by CRC Press/Balkema
4 Park Square, Milton Park, Abingdon, Oxon, OX14 4RN
e-mail: enquiries@taylorandfrancis.com
www.routledge.com – www.taylorandfrancis.com

*CRC Press/Balkema is an imprint of the Taylor & Francis Group, an informa business*

SET
ISBN: 978-1-032-44484-0 (set hbk)
ISBN: 978-1-032-44485-7 (set pbk)

Volume 1
ISBN: 978-1-032-36838-2 (hbk)
ISBN: 978-1-032-36840-5 (pbk)
ISBN: 978-1-003-33406-4 (ebk)

DOI: 10.1201/9781003334064

Volume 2
ISBN: 978-1-032-44487-1 (hbk)
ISBN: 978-1-032-44489-5 (pbk)
ISBN: 978-1-003-37241-7 (ebk)

DOI: 10.1201/9781003372417

Typeset in Times New Roman
by MPS Limited, Chennai, India

# Table of contents

*Preface* xiii
*Committee members* xv

## VOLUME 2

### *Building material properties and building mechanics simulation experiment*

Effect of solar temperature field on effective stress in external tendons of externally prestressed concrete girder bridge 3
*Wei Feng, Henan Shi, Jianbao Miao & Jinsheng Du*

Effect of fine content on relative density of sand soils 11
*Qingsong Duan, Tao Han, Zenghui Yu & Haochen Wang*

The impact of overloaded vehicles on fatigue mechanical performance of steel-concrete composite beam-type bridges 18
*Jian-qing Bu, Zhi-bo Guo, Ji-ren Zhang & Jing-chuan Xun*

Finite element simulation of axial pressure test of UHPC-RC composite structure column 28
*Le Guo*

Research on the influence of emulsion powder on the shear mechanical property of polymer emulsion-based cement composite material 37
*Tengjiao Wang, Jinyu Xu, Erlei Bai, Biao Ren, Yipeng Ning & Ao Yao*

Research on fatigue performance of prestressed concrete girder bridge of heavy haul railway 44
*M.A. Qian, Si Jinyan, Zhao Jiacheng, Tang Qingchen, Zhu Li & Ren Xiaozuo*

Finite element analysis of fatigue performance of steel-concrete composite beams 51
*Yong Ma, Gesang Yundan, Shirap Tenzin, Jiacheng Zhao, Li Zhu, Jinyan Si & Yaoyu Zhu*

Study on low-temperature performance design parameters of highway asphalt binding materials in Liaoning Province 58
*Li Song*

Experimental study on bolted-cover plate connections under bending for modular steel buildings 63
*Chuchao Ren, Ke Cao & Siyuan Zhai*

Numerical simulation analysis of pulling mechanical behaviors of anti-floating anchor 69
*Zheng Xin, Zhang Shubo & Qiu Yang*

Seismic resistant design and performance of tall buildings 75
*Yi Lu*

Mechanical properties of basalt fiber-sisal fiber/fly ash regenerative concrete 84
*Zhiwei Jiang, Zhuo Li & Kai Xu*

Flexural strength analysis of steel-FRP hybrid-reinforced concrete beam with double reinforced rectangular section
*Yueqi Chen, Jia Tang & Jie Zhu*
92

Field test study about the behaviors of concrete sheet piles under different loading scenarios
*Lei Chen, Lei Zhang, Erlin Zhang & Yanyong An*
101

Research on physicochemical properties of gangue and analysis of its comprehensive utilization
*Bing Wang & Guoliang Bai*
111

Application of unified mechanics index of asphalt pavement failure in Xijing road
*Wei Tao, Su Yutao & Zhao Huaping*
117

Experimental study on mechanical properties of collapsible loess aggregate filling materials
*Yanpeng Zhu & Xiaotao Du*
122

Impact response of concrete-filled double steel tubular beams and its analytical evaluation
*Kailai Wang*
130

Experimental research on axial compression of regional confined concrete
*Jianying Zhou, Lidan Li & Kunhong Yan*
137

Experimental study on vertical in-situ blasting demolition of reinforced concrete water tower
*Xiaowu Huang, Xianqi Xie, Yongsheng Jia, Dongwang Zhong, Jinshan Sun & Yingkang Yao*
145

Shear resistance of asphalt mixture under pure shear of Xijing road
*Chunlei Wang, Huaping Zhao, Tao Wei & Junfeng Liu*
151

Analysis of bearing characteristics of rock-socketed pile based on laboratory model test
*Haijun Yu*
157

Optimal design of transverse and longitudinal sections of U-shaped cross-section seepage control canals
*Ling Liu, Shuhong Sun & Mengjiao Zhang*
163

Experimental study on the anti-horizontal force performance of flexural and non-flexural expanded stem piles with different strengths under horizontal loads
*Qian Yongmei, Li Huaqiang, Wang Ruozhu & Tian Wei*
169

Study on seismic behavior of composite hollow high pier with functionally recoverable reinforced concrete columns and corrugated steel plates
*Ziqi Li, Yanyan Fan, Li Wang & Jihua Zeng*
177

Development and application of prefabricated building construction simulation training system
*Jing Xiong & Kai Luo*
185

Experimental study on the dynamic characteristics of pile foundation model for saturated soil and swelling soil
*Jiaxing Li, Qiang Li, Xinyi Li & Qian Wang*
189

The application of high-density electrical method in karst roadbed investigation 195
*Xiaolu Yan, Yang Zhang, Lu Yang & Xiaolei Sun*

Bearing performance analysis of CFG piles in Tidal Zone 201
*Zhenxiang Shi, Wei Zhang, Whenzhen Li, Haotian Luo & Ke Wu*

Study on mechanical properties of construction waste subgrade filler under
long-term immersion condition 207
*Siyu Cao, Jiaqi Li, Weihao Wang, Xinhang Ren, Congye Shen,
Xiaoyong Lai & Qian Wu*

Research on the light prestressed hollow flue slab structure of super-large
diameter highway shield tunnel 215
*Jian Li, Lei Zhang, Yulin Yin, Bowen Tao, Zhanhu Yao & Yunli Li*

Research on pipeline deformation of foundation pit excavation 224
*Hao Hu, Jin Pang, LingChao Shou, Ting Bao & LiFeng Wang*

Study on properties of recycled aggregate and recycled concrete modified by
chemical modification reagent and biotechnology 233
*Chun-hui Lan & Nan-xing Wang*

Experimental investigation of mechanical properties of sleeve grouting materials 248
*Yuliang Qi, Keke Huang & Zhanzhong Li*

Research on influence of grouting defects on mechanical properties of sleeve joints 257
*Yuliang Qi, Keke Huang & Yequan Zhan*

Research on influencing factors on the mechanism of multi-ribbed composite wall 267
*Junbin Gu*

Numerical simulation of seismic performance for a light-weight steel frame with
steel tube column filled with aluminum foam under different structural parameters 273
*Qinglan Liu, Jianhua Shao, Zhanguang Wang & Jiangcheng Man*

Mechanical properties and numerical simulation analysis of v-shaped steel joints 279
*Wenting Liu, Xiaomeng Zhang, Qingying Ren & Shengze Li*

Research on strength and stability of underwater ring-ribbed cylindrical shell
structure considering corrosion effect 284
*Chuang You & Feiyu Chen*

Deformation and failure mechanism of surrounding rock considering the influence
of tensile yield on mechanical properties of rock mass 292
*Hai-feng Li, Biao Wang, Lei Chen, Yang Wang & Shuai Tao*

Design and verification of towing system of resistance model test for high-speed crafts 300
*Yongshun Wu, Sujun Yang, Panhao Shi, Pan Yan & Guozhao He*

Research and treatment measures on track deformation of bridge crane in
main powerhouse of LD hydropower station 306
*Zheng Si, Yanbing Wu, Bin Duan, Xingwei Hu, Rongping Tang & Huiyuan Zhang*

Experimental study on the load-bearing performance of extra-long piles in
saturated sandy soils 311
*Wang Yifei, Li Qiang, Hu Lin, Qiu Yiqin & Zhang Yongyuan*

Study on size effect of concrete porosity measured by water saturation method 318
*Xiaozhong Zhang & Guomin Sun*

Influence of slenderness ratio on eccentric compression performance for
GFRP confined reinforced concrete columns 323
*Jingshan Jiang, Xin Huang, Zhihua Wang, Chao Zhang & Youxin Wei*

Experiment on mechanical properties of sisal fiber-basalt fiber reinforced concrete 329
*He Xiang, Jingshan Jiang, Xin Huang, Zhaoyue Zhu, Yuan Zhang & Yuelai Qiao*

Development and performance evaluation of a new permeable pavement material 335
*Liu Yugui, Xu Jianhui, Yue Xiaowen, Zhao Yun & Wang Jie*

Preparation and performance evaluation of stain-resistant coating materials 340
*Liu Yugui, Xu Jianhui, Chen Cheng, Dai Jianfeng & Wang Jie*

Effects of grinding aids on properties of cement clinker 345
*Lang Du, Jing Peng, Xiang Zhou, Liang Li, Xiaomin Zhang, Shuangfu Zhou,
Zhao Shao, Honggen Chen, Shuang Li & Xiao Xiao*

Study on polyurethane thermal insulation materials for prefabricated buildings in
cold areas Based on the project research of "exploration and localization of passive
ultra-low energy consumption buildings with integrated assembly in
cold areas – Taking Shandong as an example" (No. 2020-r2-3) 351
*Wei Sisheng*

Study on the formation and transmission characteristics of soot particles in
Chinese cooking 356
*Yang Yuan, Neng Zhu, Jing Liu, Zhiqiang Li, Chunlong Li & Zhengzheng Zhang*

Effect of different admixtures on early static strength of geopolymer mortar 365
*Yipeng Ning, Ao Yao, Erlei Bai, Zhihang Wang & Biao Ren*

Example analysis of the virtual displacement principle of the rigid body
system in structural mechanics 371
*Ting Kang, Qiqi Sun, Erlei Bai & Huixiang Sun*

### *Urban planning and construction and environmental engineering analysis and management*

Remediation of different concentrations of cadmium and lead-polluted water by
*Spathiphyllum kochii* 377
*Mengjie Lou, Kaiweng Li, Shukang Liu, Changzhe Luo & Junying Zhao*

Park green space from the perspective of an aging society: Taking Dongguan
Botanical Garden Park as an example 383
*Rongbing Mu & Yuanlong Tan*

Developing a recycling logistics network for the disposal of urban construction waste 390
*Yunjin Yang & Yanlin Zhao*

Design and effect analysis of Southward External Shading of college
classroom in Guangzhou 396
*Yang Wang, Rui Hou & Xijia Sun*

Design of shallow foundation applied to large-area recent filling areas 403
*Yi Zhou & Feng Liang*

The configuration characteristics and spatial relationships of indoor plants in
large airports of China 411
*Yang Keran, Wuyun Bagen, Kato Shoko & Li Lu*

Research on the design of urban road systems under the concept of sponge cities  419
*Hongli Huang & Liangsong Li*

Development of urban public management platform based on geographic information technology  425
*Jing Wu*

Research on the renewal strategy of urban typical vitality space based on typology theory  431
*Jialei Li & Hongmei Li*

Research on integrated scheduling strategy of inter-terminal container transportation  436
*Sun Jiuzeng & Ding Yi*

Study on the control of contaminated gas in the negative pressure ward combined with local exhaust air  442
*Chenxu Zhou, Xiaoyong Peng & Hao Zhang*

Transformation of the old industrial zone in Dongguan City from the perspective of urban renewal  451
*Le Li*

The impact of urban green space on health at different spatial scales  458
*Tan Yuan Long*

Urban small house storage space design under the concept of multi-functional design  465
*Wei Dai & Xinru Mu*

Research on integration evaluation and development model of traffic and tourism in non-core urban areas: A case study of Pingyao Ancient City  477
*Shali Zhou & Weiyang Luo*

An investigation and quantitative analysis of color landscape of countryside residences in Northeastern China  484
*Zhihui Wang, Xu Lu & Shan Guan*

Research on urban color characteristics of Hailar District under the influence of regional culture  489
*Xuan Tang, Xu Lu & Jiayi Dai*

Inheritance of summer natural ventilation in traditional dwellings of Miao nationality in Qiandongnan, China  495
*Heyu Hao, Yuan Li & Yawei Yang*

Research on the forecasting settlement method for high-filled highway subgrade  505
*Dahai Zhang*

Integrated construction of a prefabricated residential building project from the perspective of high-quality  510
*Shengping Tang & Xiaoyi Hu*

A research of urban blue corridor planning from the perspective of public gealth – Case study of Stalin Park in Harbin  516
*Guanyan Xiao, Binxia Xue, Tongyu Li, Taorong Liu & Siyuan Guo*

Design of fire detection and early warning system for minority nationality buildings in Southwest China  527
*Mingxuan Li & Xiujuan Mei*

Research on green architecture in China based on sustainable development theory  535
*Yan Yu*

Research on influencing factors and strategies of historical and cultural
districts renewal—Take Yangzhou Nanhexia historic district for an example  543
*Xuyuan Zhang & Jianming Su*

Research on intelligent operation and platform of urban large bridge and
tunnel cluster project based on BIM technology  549
*Ying Gu, Youhui Yang, Renfu Li, Peng Zhang, Dayang Liu & Xinhua Si*

Application of implicative analysis principle in the design of artistic painting
retaining wall  557
*Yangyang Tan & Weiwei Zhu*

Analysis of the impact of urbanization on precipitation characteristics and
trends in Shanghai  562
*Sihui Dong & Tianya Xu*

Design of highway maintenance decision system based on Nifi  574
*Juming Hao & Xinxiu Zhang*

Influence on water environment improvement of diversion water from
Nanjiang sluice based on two-dimensional numerical simulation analysis  582
*Zhihao Fang & Dongfeng Li*

Research on three-station integration construction scheme  588
*Qifeng Zou, He Chen & Youchao Wu*

Study on treatment of printing and dyeing wastewater by different
adsorption combined processes  593
*Hongcui Li, Yue Wu, Chenjia Zang & Boran Xie*

## *Architectural model research and environmental numerical monitoring and analysis*

Power load forecasting based on canonical-correlation analysis and LSTM networks  601
*Duanxu Liu, Bin Wu & Li Sun*

Mechanism of traffic safety culture on traffic accidents based on a
structural equation model  609
*Zhenqing Hao*

Numerical simulation and analysis of propylene pipeline leakage accident
based on CFD  613
*Shilin Chen, Wei Ma, Gang Tao & Lijing Zhang*

Empowering the dual-carbon policy: A study on the development of new
information modeling of smart city in China  621
*Yunlong Li, Luge Xing & Tianxiang Zhang*

Numerical simulation of aerosol generation and distribution in sewage pipe networks  627
*Zixin Liu*

Research on discrimination model of seismic failure modes of RC
columns based on K-nearest neighbor algorithm  635
*Ao Lu, Chunhua Zhang, Fang Huang, Shuang Wang, Ying Liu & Zi Wu*

Identification and cracking strategies of the constraints on the promotion of prefabricated buildings under "double carbon" policy: An empirical study of the DEMATEL-ISM method 642
*Jingsheng Yang, Qiyuan Xu & Jing Gao*

Research on 3D building model construction based on UAV oblique photogrammetry 651
*Zengzeng Lian, Jingcheng Xu & Jiaqi Dong*

Research on concrete bridge crack safety evaluation system based on particle swarm optimization algorithm 658
*Yifan Gu*

Research on monitoring method of time-varying gravity dam deformation based on Bayesian dynamic linear model 664
*Lin Cheng, Jiamin Chen, Pengfei Xie, Chunhui Ma & Jie Yang*

Study on trajectory prediction, monitoring and early warning of disaster-causing objects in water intake area of coastal power plant 673
*Yunjia Sun, Chen Li & Baisu Zhu*

Statistical inference of calculated model uncertainty coefficient in structural performance modeling 680
*Cheng Kaikai, Hao Yun & Zhang Yumin*

Research on intelligent analysis system for aircraft-loaded bridge health monitoring 686
*Qijie Teng, Xuekui Gao, Fulai Wang, Shengping Ma & Rui Liu*

Non-parametric identification method of nonlinear rolling coefficients of floating structure based on Hilbert transform 692
*Minghao Yan, Weihao Sun & Qinglai Fan*

Risk factors identification in old residential areas renovation PPP projects based on hierarchical holographic modeling 698
*Bing Zhao & Shengyue Hao*

Research on deformation monitoring data processing of water conservancy and hydropower projects based on comprehensive analysis model of wavelet theory 706
*Huacheng Yang*

Parameter calculation and structural design of wave-maker in Harbor Basin 711
*Ping He, Tao Han & Chen Li*

Numerical simulation research on comparison of smoke exhaust means in road tunnels 718
*Mingxuan Li & Xiujuan Mei*

A model test study on vibration construction of U-shaped sheet pile for bank protection reinforcement 725
*Kexiong Wu, Pengyan Bi, Ling Zhang, Huimei Shi, Yang Ming & Guoping Xu*

Two-dimensional numerical simulation analysis of Qiliuqiu-Beitang river water environment improvement 733
*Zhihao Fang & Dongfeng Li*

Numerical simulation on the effects of temperature difference and heat loss on the oscillations of thermo-solutocapillary convection 739
*Jungeng Fan & Ruquan Liang*

Evaluation of internal environmental quality of civil air defense project based on matter-element extension evaluation model optimized by AHP entropy weight method 745
*Yipeng Ning, Dongmei Yao, Xin Luo, Yilun Zhang, Erlei Bai & Ao Yao*

Numerical simulation of size effect of progressive failure of circular tunnel
surrounding rock 753
*Hai-feng Li, Biao Wang, Lei Chen, Yang Wang & Shuai Tao*

Numerical analysis of the probability of blockage of slag discharge
wellbore based on the DDA method 760
*Shengquan Yung, Chengwu Peng, Zhongyu Lu & Ke Cao*

Remote monitoring and early warning system of shield tunneling 768
*Mao Hongmei & Nie Hongbin*

Numerical analysis of global-local buckling of sandwich structures 774
*Yongxiang Huang, Shun Liu, Ao Zhao, Di Wang, Minghao Gao,*
*Yanchuan Hui & Xiao Liu*

Parametric sensitivity analysis of the impact response of single-layer
spherical mesh shells 780
*Ruize Zhong, Chang Lin & Jun Huang*

Huantang West River water environment improvement analysis of diversion
water based on two-dimensional numerical simulation 789
*Donghui Hu, Haibiao Shen, Yishan Chen, Zihao Li & Mei Chen*

Hydraulic calculation and water hammer analysis of municipal water distribution
system with large drop 795
*Zongke Chen, Lishuang Yuan, Xiaowei Yang, Linyuan Li & Lingxuan Zou*

Author index 801

# Preface

The 2022 4th International Conference on Civil Architecture and Urban Engineering (ICCAUE 2022) was successfully held online through Zoom on June 24, 2022. Due to COVID-19 and the pandemic-related nationwide lockdowns and other coordinated restrictive measures, the organizers decided to hold a conference in a virtual format with the organization of access for all participants to the presented reports with comprehensive discussion for ensuring the event at a high scientific level.

The conference is an international conference for the presentation of technological advances and research results in the fields of civil architecture and urban engineering. The conference brings together leading researchers, engineers and scientists in the domain of interest from around the world. We warmly welcome previous and prospected authors submit your new research papers to ICCAUE 2022, and share the valuable experiences with the scientist and scholars around the world.

The safety and well-being of all conference participants is our priority. The COVID-19 is unpredictable, so conference postponement met uncertainty, while many scholars and researchers want to attend this long-waited conference and have academic exchanges with their peers. But I want to note that there are no barriers to science, and we continue to work on our research areas remotely, using modern technical means. Under this situation, the conference model was divided into three sessions, including oral presentations, keynote speeches, and online Q&A discussion. In the first part, some scholars, whose submissions were selected as the excellent papers, were given about 5–10 minutes to perform their oral presentations one by one. Then in the second part, keynote speakers were each allocated 30–45 minutes to hold their speeches.

More than 300 participants attended the meeting. We were greatly honored to have invited two professors as our Conference Chair. There were over 20 experts and scholars in the area of Civil Architecture and Urban Engineering representing different famous universities and institutes around the globe to form Conference Committees.

In the keynote presentation part, we invited three professors as our keynote speakers. The first keynote speakers, Assoc.Prof. Hazem Samih Mohamed, Egypt, Southwest Petroleum University was invited to present his talk *Rehabilitation of corroded offshore tubular joints with Carbon Fibre Reinforcement Polymers (CFRP) laminates.* Professorial Senior Engineer, Assistant Manager, Jinfang Hou, China, Tianjin Port Engineering Institute Co., Ltd. of CCCC First Harbor Engineering Co., Ltd. She was our second keynote speakers and presented a talk: *Research and application of self-propelled immersed tube transportation and installation integrated ship and complete construction technology.* Prof. Daxin Tian, Beihang University, China, our finale keynote speaker. He presented a talk: *Secure and Reliable Edge Computing for Cooperative Vehicle Infrastructure System (CVIS).*

We are glad to share with you that we received lots of submissions from the conference and we selected a bunch of high-quality papers and compiled them into the proceedings after rigorously reviewed them. These papers feature following topics but are not limited to: Civil Engineering, Architecture, Urban Engineering, Urban traffic management. All the papers have been through rigorous review and process to meet the requirements of international publication standard.

We are really grateful to the International/National advisory committee, keynote speakers, session chairs, organizing committee members, student volunteers and administrative assistance of the management section of University, including accounts section, digital media and publication house. Also, we are thankful to all the authors for contributing a large number of papers in the conference, because of which the conference became a story of success. It was the quality of their presentations and their passion to communicate with the other participants that really make this conference series a great success.

The Committee of ICCAUE 2022

# Committee members

**Conference Chair**
Professorial Senior Engineer Aimin Liu, *Tianjin Port Engineering Institute Co., Ltd. of CCCC First Harbor Engineering Co., Ltd., China*
Associate Professor Hazem Samih Mohamed, *Southwest Petroleum University, Egypt*

**Academic Committee Chair**
Senior Engineer Guangsi Chen, *Tianjin University, China*

**Academic Committee Members**
Prof. Xu Zhang, *Henan University of Technology, China*
Prof. Yuhang Wang, *Chongqing University, China*
Prof. Shuitao Gu, *Chongqing University, China*
Prof. Li Ma, *Southwest University of Science and Technology, China*
Prof. Lei Wang, *Changsha University of Science & Technology, China*
Assoc. Prof. Dr. Yang Wang, *Guangzhou Institute of Geography, China*
Assoc. Prof. Norzailawati Hj Mohd Noor, *International Islamic University of Malaysia, Malaysia*
Assoc. Prof. Meng Liu, *Shanghai Estuarine and Coastal Research Center, China*
Assoc. Prof. Chao Liu, *State Key Laboratory of Hydraulics and Mountain River Engineering, Sichuan University, China*
Assoc. Prof. He Zhang, *School of Architecture, Tian Jin University, China*
A. Prof. Norhisham Bakhary, *School of Civil Engineering, Universiti Teknologi Malaysia, Malaysia*
Dr. Zhongzheng Lyu, *Dalian University of Technology, China*

**Organizing Committee Chair**
Senior Engineer Bin Li, *Tianjin Port Engineering Institute Co., Ltd. of CCCC First Harbor Engineering Co., Ltd., China*

**Organizing Committee Members**
Professorial Senior Engineer Jinfang Hou, *Tianjin Port Engineering Institute Co., Ltd. of CCCC First Harbor Engineering Co., Ltd., China*
Assisstant Professor Ali Rahman, *School of Civil Engineering, Southwest Jiaotong University, China*
Dr. Binbin Xu, Geotechnical Engineering Department, *Tianjin Port Engineering Institute Co., Ltd. of CCCC First Harbor Engineering Co., Ltd., China*
Dr. Jingshuang Li, Geotechnical Engineering Department, *Tianjin Port Engineering Institute Co., Ltd. of CCCC First Harbor Engineering Co., Ltd., China*
Dr. Jianbao Fu, *Tianjin Port Engineering Institute Co., Ltd. of CCCC First Harbor Engineering Co., Ltd., China*
Dr. Yiteng Xu, *Tianjin Port Engineering Institute Co., Ltd. of CCCC First Harbor Engineering Co., Ltd., China*

Vice-Chief Engineer Zhifa Yu, *Tianjin Port Engineering Institute Co., Ltd. of CCCC First Harbor Engineering Co., Ltd., China*
Engineer Changyi Yu, *Tianjin Port Engineering Institute Co., Ltd. of CCCC First Harbor Engineering Co., Ltd., China*

**Publication Chair**
Senior Engineer Bin Li, *Tianjin Port Engineering Institute Co., Ltd. of CCCC First Harbor Engineering Co., Ltd., China*

*Building material properties and building mechanics simulation experiment*

*Civil Engineering and Urban Research – Mohamed & Hou (Eds)*
*© 2023 the Authors, ISBN 978-1-032-44487-1*

# Effect of solar temperature field on effective stress in external tendons of externally prestressed concrete girder bridge

Wei Feng
*Xi'an Highway Research Institute Co., Ltd., Xi'an, China*

Henan Shi
*School of Civil Engineering, Beijing Jiaotong University, Beijing, China*

Jianbao Miao
*Xi'an Highway Research Institute Co., Ltd., Xi'an, China*

Jinsheng Du*
*School of Civil Engineering, Beijing Jiaotong University, Beijing, China*

ABSTRACT: To study the influence of solar temperature field on effective stress in external tendons of externally prestressed concrete girder bridge and modify the prestress loss of external tendons in *Specifications for Strengthening Design of Highway Bridges* (Standard ID: JTG/TJ22-2008), the temperature-displacement coupling method of ABAQUS finite element software was used to calculate and compare with the measured effective stress in external tendons of an externally prestressed concrete girder bridge. Using the parameters of solar radiation, convective heat transfer and ambient temperature at the bridge site were determined by meteorological data. The results show that after considering the effect of temperature change on the effective stress in external tendons, the difference between the effective stress in external tendons and the measured value is less than 2.6%. By studying the relationship between the strain difference and temperature difference between the external tendons and the corresponding box girder floor, the relationship between the temperature change and the stress change of the external tendons is proposed.

## 1 INTRODUCTION

The external tendons of externally prestressed concrete girder bridge are generally arranged at the bottom of the beam or in the box girder, except for the anchorage blocks and steering blocks which are in contact with the girder, the rest are separated from the girder. The temperature linear expansion coefficient of external tendons and concrete girder is different, and the temperature change is not synchronized. The temperature deformation of the superstructure and the temperature change of external tendons will lead to the change of effective stress in external tendons.

Many scholars have studied the loss of external tendons prestress caused by temperature change. Xiong et al. (2004) concluded that the external prestressing tendons outside the concrete section of the girder will cause prestress loss due to the differences in external environmental factors such as day-night temperature difference, seasonal temperature difference, and the linear expansion coefficient between steel and concrete. Xu (2014) carried out the long-term test of reinforced concrete girder strengthened with CFRP bars. The results show that the factors affecting the long-term prestress loss include temperature change, concrete shrinkage, and creep, and CFRP relaxation. Cai

---

*Corresponding Authors: 306304417@qq.com, 1174015427@qq.com, 342838952@qq.com and jshdu@bjtu.edu.cn

DOI 10.1201/9781003372417-1

et al. (2011) derived the prestress loss formula of externally prestressed composite beam bridge caused by vertical temperature gradient change by force method principle. Lu (2021) proposed that the indoor temperature change and stress relaxation were the main factors affecting the effective stress in external tendons based on the external tendons stress monitoring during the reinforcement construction and operation of the Fuhe Bridge.

There is no provision in *Specifications for Strengthening Design of Highway Bridges* (Standard ID: JTG/TJ22-2008) for prestressing loss of external tendons caused by temperature change. The change of effective stress in external tendons caused by the solar temperature field is studied by finite element numerical calculation, the prestress loss of external tendons in the specification is corrected and compared with the measured effective stress in external tendons of an externally prestressed concrete girder bridge.

## 2 BASIC PRINCIPLE OF SOLAR TEMPERATURE FIELD TIME HISTORY ANALYSIS

Solar temperature field refers to the unevenly distributed temperature field formed by solar radiation and convective heat transfer outside the bridge due to the bridge structure being influenced by the intensity of insolation radiation, bridge orientation, geographical location, and other factors, and heat conduction occurring inside the bridge. Under the action of the solar temperature field, the bridge structure will produce temperature deformation, temperature secondary internal force, and temperature stress. Thus, the solar temperature field of an externally prestressed concrete girder bridge is studied from three aspects of heat conduction, solar radiation, and convective heat transfer.

### 2.1 *Heat conduction*

Under sunlight, the internal heat conduction of the bridge structure follows the partial differential equation of heat conduction proposed by Fourier, as given by Eq. (1).

$$\rho c \frac{\partial T}{\partial t} = \frac{\partial}{\partial x}\left(k\frac{\partial T}{\partial x}\right) + \frac{\partial}{\partial y}\left(k\frac{\partial T}{\partial y}\right) + \frac{\partial}{\partial z}\left(k\frac{\partial T}{\partial z}\right) + Q \tag{1}$$

where $t$, $x$, $y$, and $z$ are the time and space coordinates corresponding to the temperature field, respectively; $T$ is the temperature of the bridge structure at point $x$, $y$, and $z$ at time $t$; $p$, $c$, and $k$ refer to the density, specific heat capacity, and thermal conductivity of the material, respectively; $Q$ is the hydration heat effect of concrete.

The heat transfer along the length of the bridge members can be neglected based on many experiments. The analysis of the bridge solar temperature field is mostly for the long-term effect after the construction is completed and the effect of the heat of hydration of concrete can be ignored. Based on the above assumptions, Eq. (1) can be simplified to a two-dimensional form, which is mostly used in many numerical finite element calculations, as described in Eq. (2).

$$\rho c \frac{\partial T}{\partial \tau} = k\left(\frac{\partial^2 T}{\partial x^2} + \frac{\partial^2 T}{\partial y^2}\right) \tag{2}$$

### 2.2 *Solar radiation*

Solar radiation includes direct solar radiation, scattered radiation, and reflected radiation. Direct solar radiation refers to the radiation directly projected onto the ground by the sun in the form of parallel rays; scattered solar radiation refers to the radiation that reaches the surface of the structure from all angles of the sky when the solar radiation passes through the atmosphere and is scattered by atmospheric gases, dust, aerosols, reflected radiation mainly refers to the radiation that reaches the surface of the structure after the reflection of direct and scattered radiation by the surface objects. The bridge structure is mainly subjected to direct sunlight; thus, this research only considers the effect of direct solar radiation and ignores the effect of scattered and reflected radiation.

The intensity of direct solar radiation decays exponentially as sunlight passes through the atmosphere. Kehlbeck (1975), in a study of temperature stresses in concrete bridges, proposed a power exponential model to calculate the intensity of direct solar radiation reaching the surface based on the Bouguer-Lambert law. The expression is described in Eq. (3).

$$I_{bN} = 0.9^{mt_u} I_0 \tag{3}$$

where $m$ is the optical mass of the atmosphere corrected by barometric pressure, $m = 1/\sinh$; h is the solar altitude angle; $t_u$ is the Link's turbidity coefficient, which can be calculated according to the following empirical formula with the change of atmospheric conditions and seasons: $t_u = A_{tu} - B_u \cos(360°N/365)$; $N$ is the number of days since January 1, $A_{tu}, B_u$, respectively, indicating that the annual average value of Link's turbidity coefficient and the magnitude of change under different atmospheric conditions; $I_0$ is the intensity of solar radiation in the outer tangent plane of the atmosphere; $I_0 = 1367[1 + 0.033 \cos(360°N/365)]$.

The direct solar radiation intensity can be converted to the heat flow density of the concrete in the finite element numerical calculation. The heat flow density refers to the heat transferred per unit area in unit time. The actual heat flow density of solar shortwave radiation absorbed by the surface of the bridge structure can be expressed as $q_s = \alpha_s I_{bN}$, where $q_s$ is the total solar radiation heat flow density on the surface of the member; $\alpha_s$ is the solar radiation absorption rate on the surface of the member, and the absorption rate of ordinary concrete is 0.55–0.70.

### 2.3 Convection heat transfer

Convection heat transfer refers to the energy exchange between the fluid on the surface of the concrete box girder and the concrete box girder. The heat flux of convection heat transfer around the concrete box girder can be calculated by Eq. (4).

$$q_c = h_c(T_a - T) \tag{4}$$

where $T_a$ and $T$ are the external atmosphere and bridge structure temperature, respectively; for convective heat transfer coefficient $h_c$, the value in Mirambell (1990) is adopted. The internal convective heat transfer coefficient of box girder is $h_{cin} = 3.5$, the convective heat transfer coefficient of box girder roof is $h_{ctop} = 4.67 + 3.83v$, the convective heat transfer coefficient of box girder floor is $h_{cbot} = 2.17 + 3.83v$, and the convective heat transfer coefficient of box girder web is $h_{cweb} = 3.67 + 3.83v$, where $v$ is the wind speed.

### 2.4 Hourly wind speed

Unlike solar radiation and atmospheric temperature, the distribution of daily wind speed has no obvious regularity. Given the poor variation of wind speed and the most unfavorable sunshine temperature occurs on sunny days and the wind speed is small, the hourly wind speed is equal to the daily average wind speed according to Xiao (2010).

### 2.5 Hourly atmospheric temperature

In a year, the pattern of temperature change for each day, except for some days when sudden changes occur, is generally more consistent; in other words, the lowest temperature in a day occurs around sunrise and the highest temperature occurs around 2: 00 p.m. The calculation for the hourly temperature is shown in Eq. (5).

$$T(t) = \begin{cases} \bar{T} + \dfrac{\Delta T}{2} \sin \dfrac{(t+30)\pi}{24} & (0 \leq t < 6) \\[2ex] \bar{T} + \dfrac{\Delta T}{2} \sin \dfrac{(t-10)\pi}{16} & (6 \leq t < 14) \\[2ex] \bar{T} + \dfrac{\Delta T}{2} \sin \dfrac{(3t-22)\pi}{40} & (14 \leq t < 24) \end{cases} \tag{5}$$

where $T(t)$ is the atmospheric temperature at the time $t$ of the day; $\bar{T}$ is the daily average temperature, and $\bar{T} = (T_{max} + T_{min})/2$; $T_{max}$ is the daily maximum temperature; $T_{min}$ is the daily minimum temperature; $\Delta T$ is the daily temperature difference, and $\Delta T = T_{max} - T_{min}$.

## 3 CASE STUDY

### 3.1 *Case background*

The Panjiahe No. 2 Bridge is located in Hanzhong City, Shaanxi Province. The total length of the bridge is 79 m, with 3 × 25 m reinforced concrete continuous box girders for the superstructure and a single column pier for the substructure. The width of bridge deck is 0.5 m (crash barrier) + 8.0 m (traffic lane) + 0.5 m (crash barrier) = 9.0 m. The side span cross-section is shown in Figure 1. In 2015, the bridge was strengthened by thickening the webs and tensioning prestressing steel bundles in the negative moment zone at the top of each pier, tensioning in vitro prestressing steel bundles at the bottom of the girders in the positive moment zone of the side span. External tendons arrangements for side spans and negative moment zones at the pier top are shown in Figures 2 and 3. The effective stress in the external tendons of the bridge was tested in 2021.

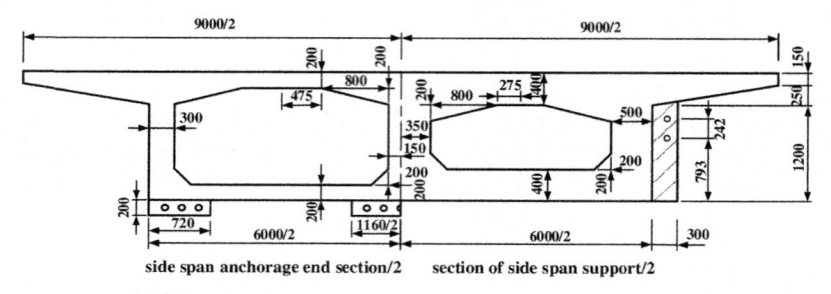

Figure 1. The side span cross section (mm).

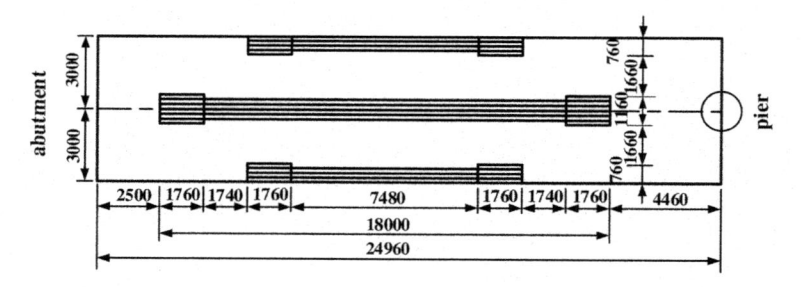

Figure 2. External tendons arrangements for side spans (mm).

The effective stress in external tendons was detected by the vibration frequency method. Using a high-sensitivity vibration pick-up sensor and its corresponding data acquisition equipment and analysis software, several order natural vibration frequencies are analyzed from the vibration of the structure, and then the tendons force is inversely calculated by the relationship between the tendons force and its natural vibration frequency, boundary conditions, stiffness and so on through frequency. When the bending stiffness of the tendons can be ignored, the tendon's force is calculated by Eq. (6).

$$T = \frac{4pL^2f_n^2}{n^2} \tag{6}$$

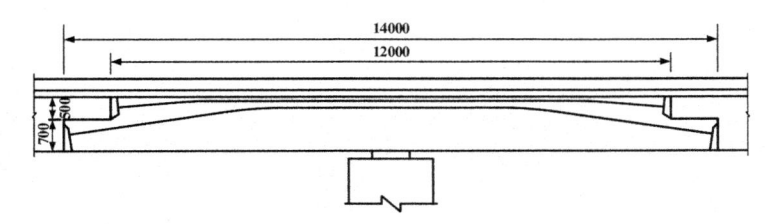

Figure 3.   External tendons arrangements for negative moment zones at pier top (mm).

where $T$ is the external tendons force; $p$ is the linear external tendons density, 1.101 kg/m; $L$ is the external tendons length, the length of two sides external tendons are both 11 m and the length of intermediate external tendons is 18 m; $f_n$ is the $n-$th self-vibration frequency of the external tendons, and the first order is taken for the calculation.

The effective stress test results and the effective stress are listed in Table 1.

Table 1.   Test results and code values of effective stress according to the specification.

| No. | Frequency 1 (Hz) | Frequency 2 (Hz) | Frequency 3 (Hz) | Mean (Hz) | Tendons force (kN) | Effective stress (MPa) | Code value (MPa) |
|---|---|---|---|---|---|---|---|
| L-T-1 | 4.896 | 4.898 | 4.896 | 4.897 | 127.8 | 919.2 | 945.0 |
| L-T-2 | 4.911 | 4.908 | 4.907 | 4.909 | 128.4 | 923.7 | 945.0 |
| L-T-3 | 4.925 | 4.919 | 4.919 | 4.921 | 129.0 | 928.4 | 945.0 |
| M-T-1 | 3.228 | 3.229 | 3.258 | 3.238 | 149.6 | 1076.3 | 1030.0 |
| M-T-2 | 3.139 | 3.136 | 3.137 | 3.137 | 140.4 | 1010.2 | 1030.0 |
| M-T-3 | 3.088 | 3.085 | 3.076 | 3.083 | 135.6 | 975.5 | 1030.0 |
| M-T-4 | 3.118 | 3.092 | 3.081 | 3.097 | 136.9 | 984.7 | 1030.0 |
| M-T-5 | 3.090 | 3.099 | 3.090 | 3.093 | 136.5 | 981.9 | 1030.0 |
| R-T-1 | 4.863 | 4.862 | 4.855 | 4.860 | 125.9 | 905.5 | 945.0 |
| R-T-2 | 4.987 | 4.985 | 5.062 | 5.011 | 133.8 | 962.8 | 945.0 |
| R-T-3 | 4.888 | 4.787 | 4.892 | 4.856 | 125.6 | 903.9 | 945.0 |

## 3.2   Numerical calculation

Based on ABAQUS finite element software, the model of externally prestressed concrete girder bridge is established, and the influence of solar temperature field on the effective stress in external tendons is calculated by the temperature-displacement coupling analysis step. Concrete adopts DC3D8 eight-node linear heat transfer hexahedron element, external tendons, and internal longitudinal reinforcement adopts DC1D2 two-node heat transfer connection element. Heat conduction parameters in material properties are defined as listed in Table 2. Surface radiation and surface heat exchange conditions are defined in the interaction module. The solar radiation absorption rate of ordinary concrete is 0.6, and the direct radiation intensity and solar radiation heat flux are shown

Table 2.   Material properties.

| Material | Coefficient of thermal conductivity (W/m·K) | Specific heat capacity (J/kg·K) | Density (kg/m$^3$) | Coefficient of linear expansion $\times 10^{-5}$ | Elastic modulus $\times 10^4$ MPa |
|---|---|---|---|---|---|
| Concrete | 1.5 | 850 | 2,400 | 1.0 | 3 |
| Steel | 49.8 | 465 | 7,800 | 1.2 | 21 |

in Figure 4. By querying the meteorological data, the maximum temperature, minimum temperature, and daily average wind speed of the day were determined. The hourly ambient temperature was determined by the method of Section 2.4 according to the daily maximum temperature and minimum temperature, as shown in Figure 5. MPC Beam constraint is used between the concrete anchorage block and the external tendons to ensure that they have the same curvature and rotation angle. The external prestress was simulated by the temperature reduction method, and the tension control stress was 1,209 MPa. The mesh size is 50 mm, with a total of 128,614 elements. The finite element model is shown in Figure 6.

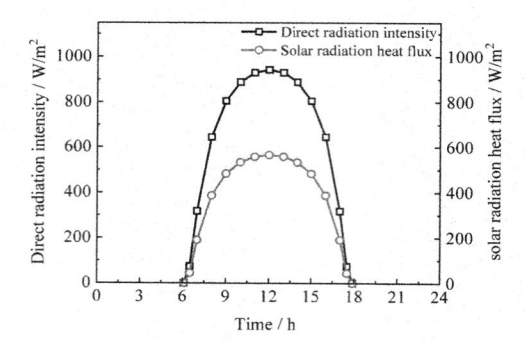 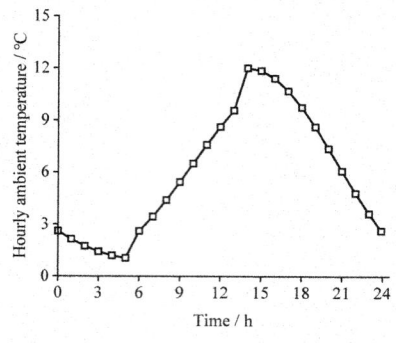

Figure 4. Solar direct radiation and heat flux density.

Figure 5. Ambient temperature.

Figure 6. Finite element model.

## 3.3 *Analysis and discussion*

After finite element calculation, the temperature change of the box girder bottom plate is shown in Figure 7, and the effective stress change in external tendons is shown in Figure 8.

Due to the slight difference in the linear expansion coefficient of temperature between the external tendons and the concrete box girder and the fact that the temperature change of the girders is not synchronized with the temperature change of the external tendons, it can be seen in Figures 7 and 8 that with the increase of temperature, the effective stress in external tendons shows a significant downward trend. On the contrary, with the decrease in temperature, the effective stress in external tendons presents a significant upward trend, which is consistent with the relationship between the measured effective stress in external tendons and the temperature change in Lu (2021).

It can be seen from Table 1 that the measured effective stresses in external tendons except M-T-1 and R-T-2 are smaller than those calculated according to the specification, and the test results of M-T-1 and R-T-2 can be regarded as errors. After considering the prestress loss caused by temperature change, the prestress loss value in the JTG/TJ22-2008 specification is corrected, as shown in Figure 9. The difference between the corrected value and the measured value is less than 2.6 %.

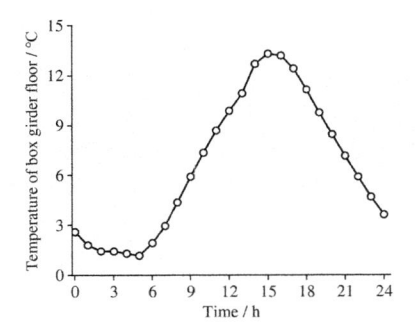
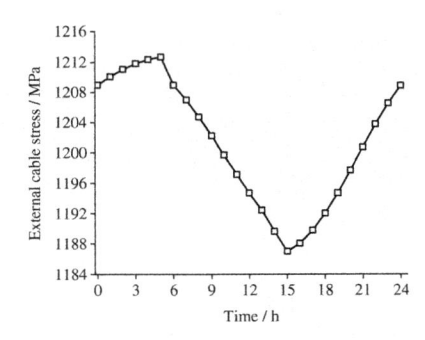

Figure 7. The temperature change of the box girder floor.

Figure 8. Variation of external tendons stresses with time.

The change of external tendons strain is inconsistent with the change of box girder strain. The relationship between external tendons strain and the strain difference and temperature difference of box girder bottom plate corresponding to the same section is shown in Figure 10. The relationship between strain difference and the temperature difference can be fitted as $\Delta\varepsilon = 2.0 \times 10^{-6}\Delta T + 3.7 \times 10^{-5}$, and the change of tendons stress in vitro can be expressed as $\Delta\sigma = \Delta\varepsilon E_p$.

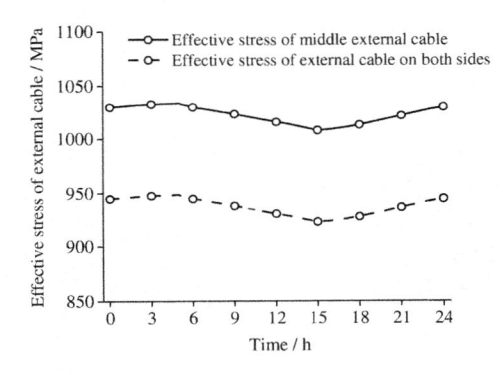
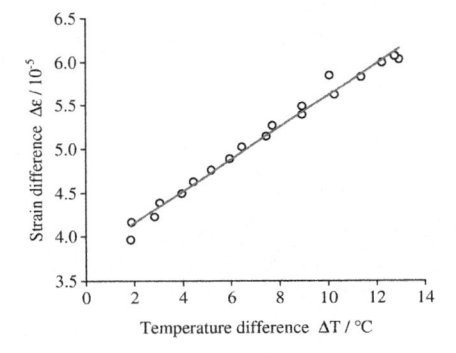

Figure 9. Variation of effective stress in external tendons with time.

Figure 10. Relationship between strain difference and temperature difference.

## 4 CONCLUSION

The influence of the solar temperature field on the effective stress in external tendons was studied by finite element numerical calculation. The internal heat conduction, solar radiation, convective heat transfer, and other parameters of concrete were determined. The effective stress change of external tendons was calculated by the ABAQUS temperature-displacement coupling analysis method, and the prestress loss in external tendons in the specification was corrected. The difference between the corrected value and the measured value was less than 2.6%. This method can get better results during calculate the variation law of effective stress in external tendons with time.

The formula of prestress loss caused by temperature change is proposed. The relationship between external tendons and strain difference and temperature difference of box girder bottom plate is determined by finite element numerical calculation, and the relationship between temperature difference and external tendons stress is determined.

ACKNOWLEDGMENT

This work was financially supported by the Transportation Science and Technology Project of the Shaanxi Transportation Department of China (20-10K).

REFERENCES

Cai J. J. & Tao M. X. (2011). Calculation of prestress loss of externally prestressed composite girder bridges. *Bridge Construction*. (06): 67–70.

Kehlbeck F. (1975). Einfluss der sonnenstrahlung bei bruckenbauwerken. *Technische Universitat Hannover*.

Lu F. L. (2021). Research on stress loss monitoring of external tendons in prestressed concrete box girders. *Journal of China & Foreign Highway*. 41 (02): 115–119.

Mirambell E. & Aguado A. (1990). Temperature and stress distributions in concrete box girder bridges. *Journal of Structural Engineering*. 116(9): 2388–2409.

*Specifications for Strengthening Design of Highway Bridges*: JTG/TJ22-2008. China Communications Press.

Xiao J. Z. & Song Z. W. (2010). Analysis of solar temperature action for concrete structure based on meteorological parameters. *China Civil Engineering Journal*. 43 (04): 30–36.

Xiong X. Y. & Gu W. (2004). The evaluation of prestressing loss in externally prestressed concrete structure. *Industrial Construction*. (07): 16–19.

Xu F. (2014). *Flexural Performance of Reinforced Concrete Beams Strengthend by External Prestressing CFRP Tendons*. Wu Han University.

*Civil Engineering and Urban Research – Mohamed & Hou (Eds)*
*© 2023 the Authors, ISBN 978-1-032-44487-1*

# Effect of fine content on relative density of sand soils

Qingsong Duan & Tao Han
*CCCC Road & Bridge Special Engineering Co., Ltd., Wuhan, Hubei, China*

Zenghui Yu* & Haochen Wang
*China Construction Eighth Bureau Second Engineering Co., Ltd., Jinan, China*

ABSTRACT: The relative density of sand soils varies with the fine content. The influence of non-plastic fine content on the maximum void ratio (minimum void ratio) of sand soils has been studied in the literature, but the influence of plastic fine content is less. Based on the dredger fill project of an industrial port abroad, the effect of plastic fine content on the relative density of sand soils is studied through a laboratory test. The test results show that (1) when the fine content is in the range of 0%–50%, the maximum dry density (minimum dry density) first increases and then decreases with the increase of fine content, and the ratio of minimum and maximum dry density decreases linearly with the increase of fine content; that (2) the relationship between relative compaction, relative density, and fine content is given; and that (3) the soil parameters of 20 boreholes are analyzed, and different empirical formulas are used to calculate the relative density. The results show that the modified Lunne's empirical formula for CPT-SPT using Kulhawy and Mayne's expression affected by fine content is the closest to calculated according to the definition, and can be used to estimate the $q_c$ or SPT-N value of hydraulic reclamation site.

## 1 INTRODUCTION

It has become an effective way to alleviate the shortage of land resources in many coastal cities at home and abroad to use the dredged sediment from harbor basins and waterways for land reclamation. The strength and compactness of the dredger fill foundation are very poor, so foundation treatment and reinforcement are needed to meet the needs of engineering construction. Chhanve Zohra and Abdul Hafeez (Chhanve 2019) analyzed the relationship between physical properties indexes of sand soils with non-plastic fine content less than 10% and provided a fitting relationship between the indexes. Munenori Hatanaka and Lei Feng (Munenori 2006) examined the method of estimating the relative compactness of sand and the errors associated with common empirical formulae. G. Agrawal and O. Pekin et al. (G. Agrawal 2010) studied the relative density of dredger fill estimated by the CPT test. There is a lognormal function relationship between compactness and normalized cone tip resistance. Jan van (Jan van 2012) given the effect of silt content on the maximum porosity (minimum porosity). However, the influence of fine content on the relative density of sand soils is mainly concentrated in the sandy soils with fine content of less than 20%, and there are few studies on the sand with fine content of more than 20%, and the calculation results of various empirical formulas are very different. Therefore, this paper studies the impact of fine content on the compaction characteristics of filled sand soils through laboratory tests and compares and analyzes the commonly used empirical formulas to provide a basis for foundation treatment of similar projects.

---

*Corresponding Author: 1976316086@qq.com

DOI 10.1201/9781003372417-2

## 2 ENGINEERING BACKGROUND

The foundation of an industrial port in a country is formed by the dredged soil of the harbor basin and channel, and the dredged sand of the harbor basin and channel is used as the blowing filler. The soil layer in the dredging area is mainly sandy soil from top to bottom, including the SP layer, SP-SM layer, SM layer, and SC layer (according to ASTM standards). It can be seen from Figure 1 that the grading curves of dredger fill mostly fall in areas C and D. The grading curve falls in different areas and different foundation treatment methods are adopted. Michelle (Mitchell 1970) gives the particle grading range suitable for the vibro-compaction method. The non-filler vibratory technique applies not only to the medium and coarse sand foundation with a fine content of less than 0.074 mm and no more than 10% but also to the silty sand and fine sand foundation with a fine content of no more than 10% (Lou 2012).

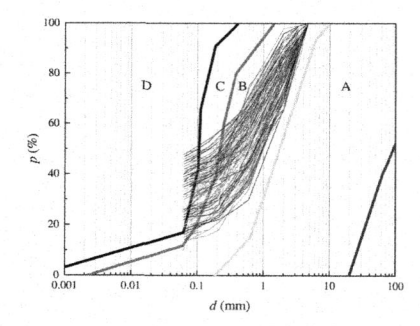 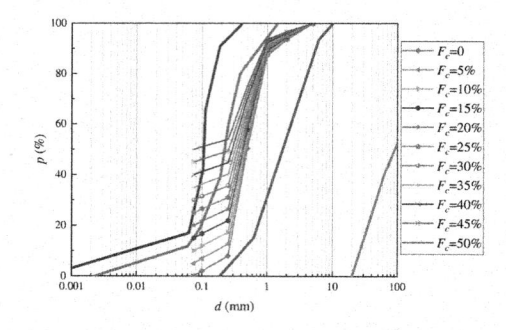

Figure 1.  Grading curve of SM and SC in site.       Figure 2.  Grading curve of laboratory test soil.

## 3 ANALYSIS OF INDOOR TEST RESULTS

According to the grading curve of sand fine-grained mixed soil from on-site dredging drilling (as shown in Figure 2), the test soil for the indoor compaction characteristic test is determined. According to the standard for geotechnical test methods (Standard ID: GB/T 50123-2019 2019), the specific gravity of sand soil with different fine content is measured. Figure 3 shows the comparison of specific gravity between on-site soil samples and indoor soil samples. The specific gravity of the indoor soil sample has a linear relationship with clay content. The $R^2$ of Equation (1) is greater than 0.97 and the $R^2$ of Equation (2) is 0.18.

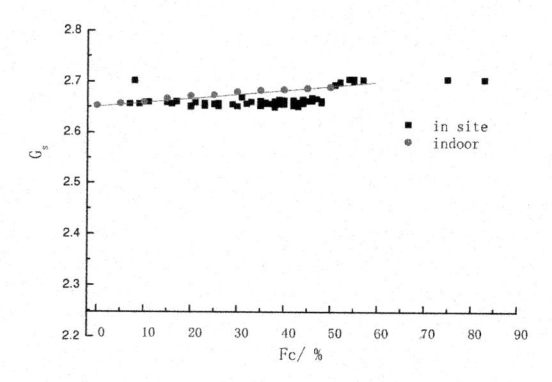

Figure 3.  Relationship between specific gravity and fine content.

It can be seen that $\rho_{d\,max}$ ($\rho_{d\,min}$) first increases and then decreases with the increase of $F_c$. The maximum value of $\rho_{d\,min}$ is 1.56 g/cm³, and the corresponding $F_c$ is 20%; the maximum value corresponding to $\rho_{d\,max}$ is 2.04 g/cm³, and the corresponding $F_c$ is 30%.

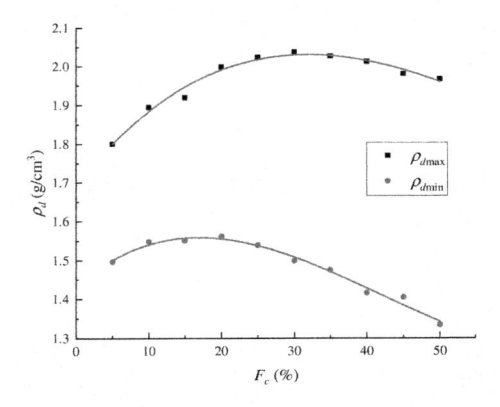

Figure 4.    Relationship between $\rho_d$ and $F_c$.

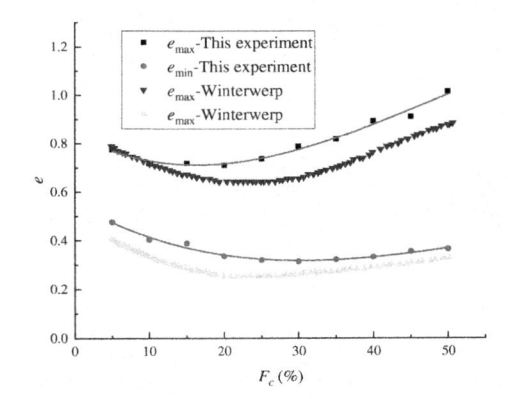

Figure 5.    Relationship between $e$ and $F_c$.

The relationship between the maximum void ratio $e_{max}$ (minimum void ratio $e_{min}$) is shown in Figure 5. $e_{max}$ ($e_{min}$) first decreases and then increases with the increase of $F_c$. Polynomial fitting is adopted for fitting. The comparison between the test results and the results of Winterwerp (1996) is also depicted. $e_{max}$ ($e_{min}$) of the test results is consistent with the law of fine particle content, and its value is slightly higher than the results of Winterwerp.

## 4   ANALYSES OF RELATIVE DENSITY AND RELATIVE COMPACTION

### 4.1   *Definition*

The relative compaction ($D_c$) is expressed as

$$D_c = \frac{\rho_d}{\rho_{d\,max}} \tag{1}$$

The relative density ($D_r$) is expressed as

$$D_r = \frac{e_{max} - e}{e_{max} - e_{min}} = \frac{(\rho_d - \rho_{d\,min})\rho_{d\,max}}{(\rho_{d\,max} - \rho_{d\,min})\rho_d} \tag{2}$$

The ratio of minimum density to maximum dry density ($R$) is expressed as

$$R_0 = \frac{\rho_{d\,min}}{\rho_{d\,max}} \tag{3}$$

Based on Equations (5) to (7), we can obtain the relationship between $D_c$ and $D_r$

$$D_r = \frac{D_c - R_0}{D_c(1 - R_0)} \times 100\% \text{ or } D_c = \frac{R_0}{1 - D_r(1 - R_0)} \tag{4}$$

The test results are sorted out to obtain the relationship between the maximum and minimum dry density ratio ($R$). Linear fitting is adopted, and the relationship formula is:

$$R_0 = 0.84091 - 0.31636F_c \ (R^2 > 0.95) \tag{5}$$

13

In Equations (1)–(5), $D_c$ is relative compaction; $D_r$ is relative density; $R$ is the ratio of minimum density to maximum dry density; $\rho_d$ is the dry density of soils; $\rho_{d\max}$ is maximum dry density; $\rho_{d\min}$ is minimum dry density; $e_{\max}$ is maximum void ratio; $e_{\min}$ is minimum void ratio; $e$ is void ratio; $F_c$ is fine content.

Substitute Equation (9) into Equation (8) to obtain the relationship between $D_c$, $D_r$, and $F_c$

$$D_c = \frac{0.84091 - 0.31636F_c}{1 - D_r(0.15909 + 0.31636F_c)} \tag{6}$$

### 4.2 Correction of relative compaction and relative density

(1) A Method Proposed by Lee and Singh (1971)

Based on the statistical analysis of 47 soil properties (from silty fine sand to coarse gravel, from uniform to well graded), given $R_0 = \rho_{d\min}/\rho_{d\max} = 0.80$, then the relationship between $D_c$ and $D_r$ is expressed as

$$D_c = 80 + 20D_r \tag{7}$$

Equation (7) does not affect the fine content of the sandy soils. Figure 6 presents the relationship between $D_c$ and $D_r$ when considering the fine content, and carries out linear fitting. The R-Squares are greater than 0.98.

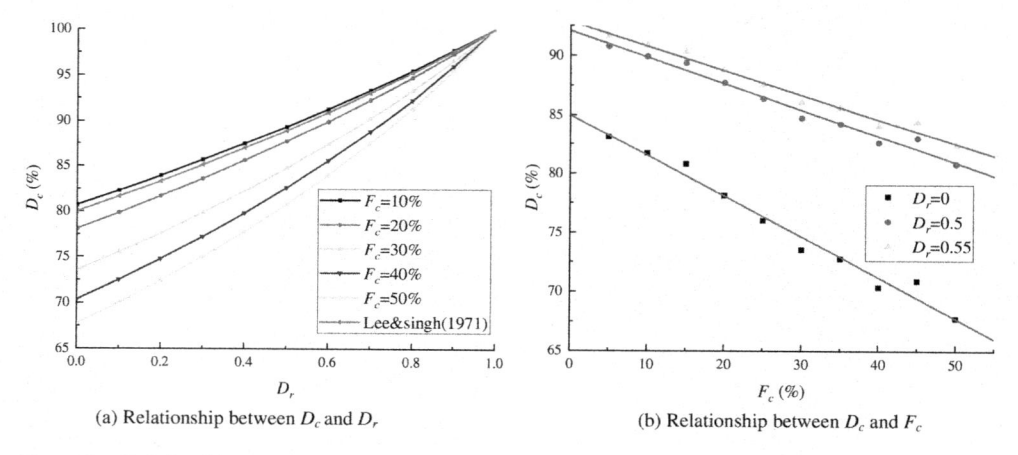

(a) Relationship between $D_c$ and $D_r$     (b) Relationship between $D_c$ and $F_c$

Figure 6.   Relationship curves.

As can be seen from Figure 6, the relationship between Lee and Singh (1971) is close to the $D_c - D_r$ curve of Equation (6) when $F_c = 10\%$. When $F_c = 0\%–15\%$, $R = 0.80$, which is consistent with Equation (6). When $F_c > 20\%$, as $R$ decreases with the increase of fine content, it is significantly smaller than the results obtained by Lee and Singh. When $D_r = 0$ and $F_c = 20\%–50\%$, $D_r$ in Lee and Singh's study is 80%, and $D_r$ obtained by Equation (6) is reduced by 8.04%–16.71% When $D_r = 0.5$ and $F_c = 20\%–50\%$, $D_c$ in Lee and Singh's study is 90%, and $D_c$ obtained by Equation (6) is reduced by 2.79%–9.84%. When $D_r = 0.55$ and $F_c = 20\%–50\%$, $D_c$ in Lee and Singh's study is 95%, and $D_c$ obtained by Equation (6) is reduced by 6.74%–12.94%.

(2) Meyerhof's method

Meyerhof sorted out the relationship according to the test results of Gibbs and Holtz (1957):

$$D_r = 21\sqrt{\frac{N}{0.7 + \sigma_v'/98}} \tag{8}$$

where $D_r$ is the relative compaction of sand soils (%); $N$ is the $N$-value of standard penetration test (SPT); $\sigma_v'$ is the effective overburden stress at the depth of SPT (kPa).

(3) A Method Proposed by Kokusho et al

Kokusho et al (1983) modified Equation (9) with the following relationship:

$$D_r = 36 \left( \frac{N}{1.5 + \sigma_v'/98} \right)^{0.37} \tag{9}$$

(4) Lunne's Method

According to the recommended equation of Lunne (1997), the relationship between relative compaction of sandy soils and CPT is:

$$D_r = \frac{1}{C_2} \ln \left[ \frac{q_c}{C_0(\sigma_v')^{C_1}} \right] \tag{10}$$

For normal sand soils, $C_0 = 157$, $C_1 = 0.55$, $C_2 = 2.41$, and $R = 0.96$; for over consolidated sand soils, $C_0 = 181$, $C_1 = 0.55$, $C_2 = 2.61$, and $R = 0.95$.

Empirical Equations (8)–(10) have certain accuracy for sand with fine content less than 20%. When the fine content is greater than 20%, the error is large, and the influence of fine content needs to be introduced. Kulhawy and Mayne (1991) gave the correction of fine content to CPT-SPT and obtained the following expression.

$$\frac{q_c/98}{N} = 4.25 - \frac{F_c}{41.3} \tag{11}$$

where n = 108; $r^2 = 0.414$; S. D = 0.89; Fc = 0–100%.

From Equation (11), find $q_c$ and substitute Equation (10) to obtain Equation (12).

$$D_r = \frac{1}{C_2} \ln \left[ \frac{98(4.25 - F_c/41.3)N}{C_0(\sigma_v')^{C_1}} \right] \tag{12}$$

### 4.3 *Analysis of calculation results*

Based on the data of 20 boreholes in the dredging area of the industrial port, the soil parameters with 50% less fine content are selected for analysis, and the scatter diagram is made after $D_r$ is calculated by the above method, as shown in Figure 7. Compared with the definition method, the Meyerhof formula applies to sand. The $D_r$ value calculated by SPT-N is significantly higher than the definition calculation value, especially when the SPT-N value is greater than 30. The empirical formula of Kokusho et al. is a modification of Meyerhof, and the area of $D_r$ value calculation results is reduced, but most calculation results are on the upper side of the formula calculation results. After the correction of CPT-SPT by Kulhawy and Mayne, the comparison between the calculation results of Lunne's empirical formula and definition method shows that the increase of fine content $\{(q_c/98)/N\}$ decreases, and when the fine content is 0%–50%, the $\{(q_c/98)/N\}$ is 4.25–3.04. Therefore, the $D_r$ value calculated by modified Lunne's empirical calculation is significantly lower than that of Meyerhof in the range of $F_c = 30\%–50\%$. As can be seen from Figures 7(c) and (d), when $F_c = 30\%–50\%$, the $D_r$ value calculated by the normal coefficient of consolidation is higher than the $D_r$ value calculated by the definition; the $D_r$ value calculated by the coefficient of over consolidation coincides is coincidence with the $D_r$ value calculated by the definition at most points. Therefore, for the influence of introducing fine content into Lunne's formula at the dredging site, the calculation result is closest to that calculated by the definition method.

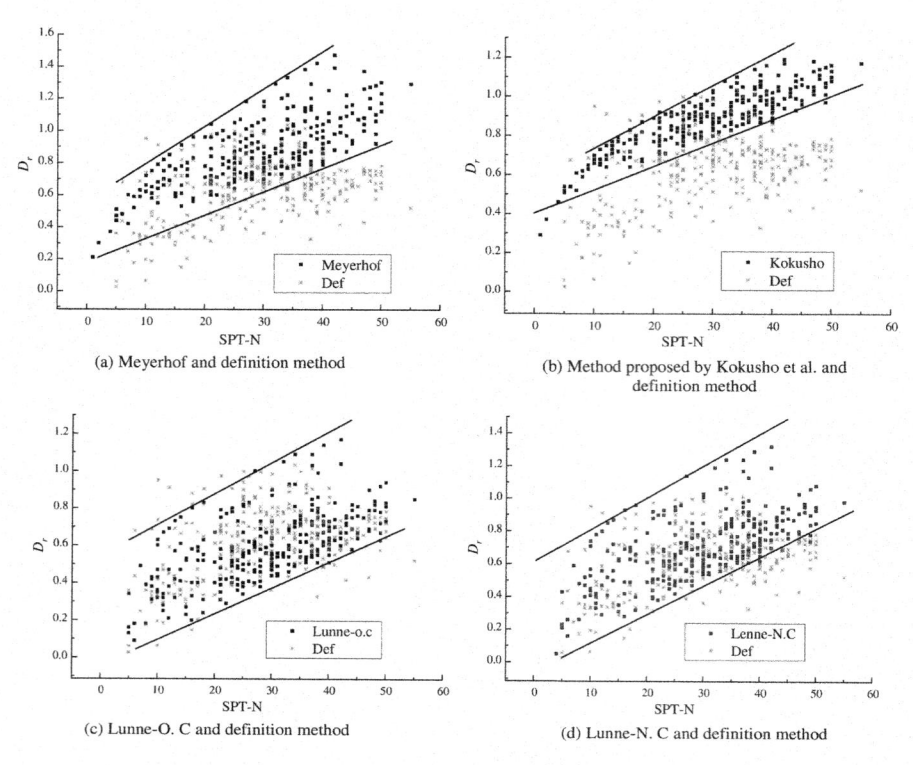

(a) Meyerhof and definition method

(b) Method proposed by Kokusho et al. and definition method

(c) Lunne-O. C and definition method

(d) Lunne-N. C and definition method

Figure 7. Comparison of the calculation results of $D_r$ at the dredging site.

## 5 CONCLUSIONS

(1) The fitting polynomials of maximum dry density (minimum dry density) and maximum void ratio (minimum void ratio) with fine content in the range of 0%–50% and with fine content are obtained through a laboratory test, and the goodness of fit is greater than 0.96. The maximum void ratio (minimum void ratio) is consistent with the results of Winterwerp (1996), and its value is slightly higher than that of Winterwerp.

(2) When $F_c$ = 0%–50%, the relationship between relative compaction, relative density, and fine content is given. The equation given by Lee and Singh is only applicable to sand soils with $F_c$ less than 10%.

(3) In the empirical method of estimating $D_r$, after modifying CPT-SPT by introducing Kulhawy and Mayne expression affected by fine content, Lunne's empirical formula is the closest to $D_r$ calculated according to the definition method. In practical applications, according to the relative compaction requirements, $D_r$ that meets the requirements can be determined according to the $D_r$-$D_c$-$F_c$ formula proposed in this paper, and then the $q_c$ value or SPT-N value can be determined according to the Lunne formula considering the influence of fine content.

## REFERENCES

Chhanve Zohra, Abdul Hafeez. *Correlations between Relative Density and Compaction Test Parameters-Part-b* [C]. 2 nd International Conference on Sustainable Development in Civil Engineering, MUET, Pakistan, 2019, 106–113.

G. Agrawal, O. Pekin etc. *Evaluating relative compaction of fills using CPT* [C]. 2nd International Symposium on Cone Penetration Testing, Huntington Beach, CA, USA, 2010.

Jan van't Hoff, Art Nooy van der Kolff. *Hydraulic Fill Manual For Dredging and Reclamation Works* [M]. CRC Press Taylor & Francis Group. 2012.

Lou Xioaming, Yu Zhiqiang. Review on the present situation of vibroflotation [J]. *Journal of Civil Engineering and Management*. 2012, 29 (1): 61–68.

Mitchell J K. In-place treatment of foundation soils [J]. *Journal of the Soil Mechanics and Foundations Division*, 1970, 96 (SM1): 73–110.

Munenori Hatanaka, Lei Feng. Estimating Relative Density of Sandy Soil [J]. *Soils and Foundations*. 2006, 46(1): 299–313.

Standard for geotechnical test methods (GB/T 50123-2019)[C]. 2019, Beijing.

*Civil Engineering and Urban Research – Mohamed & Hou (Eds)*
*© 2023 the Authors, ISBN 978-1-032-44487-1*

# The impact of overloaded vehicles on fatigue mechanical performance of steel-concrete composite beam-type bridges

Jian-qing Bu*
*School of Traffic and Transportation, Shijiazhuang Tiedao University, Shijiazhuang, Hebei, China*

Zhi-bo Guo*
*School of Civil Engineering, Shijiazhuang Tiedao University, Shijiazhuang, Hebei, China*

Ji-ren Zhang*
*Key Laboratory of Wind and Bridge Engineering of Hunan Province, Hunan University, Changsha, Hunan, China*

Jing-chuan Xun*
*China Construction Road and Bridge Group Co., Ltd., Shijiazhuang, Hebei, China*

ABSTRACT: Vehicle loads in service bridges are often higher than their design loads. The overload of vehicles will adversely affect mechanical performance such as bridge fatigue stiffness and residual bearing capacity. To further reveal the influence of vehicle overload on the mechanical performance of composite beam-type bridges, first, the existing theoretical calculation method of fatigue stiffness and residual bearing capacity of steel-concrete composite beams is improved. Second, the load amplification effect of overloaded vehicles is quantified based on the fatigue load calculation model. Finally, the degradation law of fatigue stiffness and residual bearing capacity of the bridge under different overload levels is analyzed by taking a real bridge as an example. The results show that with the increasing overload levels, the degradation rate, and the amount of degradation of fatigue stiffness and residual bearing capacity of the composite beam bridge both tend to increase, but the impact on the degradation of the residual bearing capacity is less.

## 1 INTRODUCTION

Steel-concrete composite beam-type bridges are composite structures by the exposed steel beam cross-section or steel truss beam cross-section through the connector and steel concrete bridge panel (Huang 2017). Many engineering practices show that the advantages of steel-concrete composite beam-type bridges have obvious technical advantages, rapid construction, good economic benefits, and beautiful shape (Huang 2017). However, the repeated effect of loading, especially the action of overloaded vehicles, will exacerbate the fatigue destruction of the structures, which will have a significant impact on the fatigue stiffness and the residual bearing capacity of the bridge structures (Liu et al. 2017). In the field of bridge fatigue stiffness research, Wang used the effective stiffness method to derive the composite coefficient calculated by the composite beam, the deflection calculation of composite beams considering different shear connection degrees is unified, and the calculation results of the example are in good agreement with the test results (Wang et al. 2005). Nie et al. (2009) put forward the calculation equation of mid-span residual deflection and verified the effectiveness of this method through the test results. In the research field

---

*Corresponding Authors: bujq@stdu.edu.cn, guozhibo19991010@163.com, 927288223@qq.com and 76201366@qq.com

  DOI 10.1201/9781003372417-3

of residual bearing capacity of steel-concrete composite beams, based on the residual strength and residual stiffness models of concrete slabs, steel beams, and studs, Wang proposed a calculation method of residual bearing capacity of stud and PBL composite beams considering the degradation of shear connection (Wang 2017). Since the existence of vehicle overload problem will accelerate the degradation of mechanical performance of steel-concrete composite beam-type bridges, this paper can provide theoretical support for the residual life prediction, performance assessment of steel-concrete composite beam-type bridges, and provide a reference for further improvement of fatigue damage theory of steel-concrete composite beam-type bridges.

## 2 FATIGUE STIFFNESS OF STEEL-CONCRETE COMPOSITE BEAMS

### 2.1 *Static load deflection of steel-concrete composite beam*

In this paper, the static load deflection $f_e$ of the composite beam is calculated by introducing the improved discounted stiffness method proposed in the literature reported by Xu et al. (2013), which takes the effective stiffness as the discounted stiffness considering the slip effect and introduces the calculated length factor in the calculation of the stiffness discount factor, as shown in Eqs. (1)–(8).

$$f_e = \frac{PL^3}{48B} \tag{1}$$

$$B = \frac{E_s I_0}{1 + \xi} \tag{2}$$

$$\xi = \eta \left[ \frac{1}{1 + (\mu/\pi)^2 (\alpha L)^2} \right] \tag{3}$$

$$\eta = \frac{d_{sc}^2 A_{00}}{I_{00}} \tag{4}$$

$$\alpha = \sqrt{\frac{n_s k A_{11}}{E_s I_{00} p}} \tag{5}$$

$$A_{00} = \frac{A_c A_s}{n_E A_s + A_c} \tag{6}$$

$$A_{11} = \frac{I_{00} + A_{00} d_{sc}^2}{A_{00}} \tag{7}$$

$$I_{00} = I_s + \frac{I_c}{n_E} \tag{8}$$

where $A_c$ is the area of the concrete slab, $A_s$ is the area of the steel beam, $I_s$ is the moment of inertia of the steel beam section, $I_c$ is the moment of inertia of the concrete slab section, $d_{sc}$ is the distance between the steel beam and the concrete slab at the section form center, $h$ is the section height, $k$ is the shear stiffness factor of the joint, $p$ is the longitudinal spacing of the joint, $n_s$ is the number of columns of the joint, $n_E$ is the ratio of steel to the concrete modulus of elasticity (Nie 2005), and $\mu$ is the calculated length factor and its value is related to the boundary conditions (Xu et al. 2013).

### 2.2 *Residual deflection of steel-concrete composite beams*

Residual deformation of the steel-concrete composite beams will be produced under fatigue loading, which is mainly caused by residual slip. Therefore, the residual slip at the steel-concrete interface after $n$ fatigue load cycles is the key to solving the residual deflection of the steel-concrete composite beam. The equations for the residual slip at the interface are given in the study of Hanswille et al. (2006)

$$\delta_{\mathrm{pl,N}} = C_1 - C_2 \ln\left(\frac{N_\mathrm{f} - n}{n}\right) \geq 0, 0 \leq \frac{n}{N_\mathrm{f}} \leq 0.9 \tag{9}$$

$$\delta_{\mathrm{pl,N}} = 0 \quad \frac{n}{N_f} = 0 \tag{10}$$

$$C_1 = 0.104 e^{3.95 \frac{P_{\mathrm{max}}}{P_{\mathrm{u,0}}}} \tag{11}$$

$$C_2 = 0.664 \frac{P_{\mathrm{min}}}{P_{\mathrm{u,0}}} + 0.029 \tag{12}$$

where $\delta_{\mathrm{pl,N}}$ is the interface residual slip (mm), $N_\mathrm{f}$ is the fatigue life of the studs, $P_{\mathrm{max}}$ is the upper load limit, $P_{\mathrm{min}}$ is the lower load limit, and $P_{\mathrm{u,0}}$ is the ultimate bearing capacity of a single stud. The calculation for the stud's ultimate bearing capacity is shown in Eq. (13).

$$V_{\mathrm{sud}} = \min\left\{0.43 A_\mathrm{s} \sqrt{E_\mathrm{c} f_{\mathrm{cd}}}, 0.7 A_\mathrm{s} f_{\mathrm{su}}\right\} \tag{13}$$

where $V_{\mathrm{sud}}$ is the design value of the shear bearing capacity of the stud, $E_\mathrm{c}$ is the modulus of elasticity of concrete, $f_{\mathrm{cd}}$ is the axial compressive strength of concrete, $A_\mathrm{s}$ is the cross-sectional area of the stud bar, and $f_{\mathrm{su}}$ is the ultimate strength of the material of the stud (Huang 2017).

Nie et al. (2009) derived an expression for the residual deflection in the span based on the residual slip at the interface of the composite beam, as shown in Eq. (14).

$$f_\mathrm{r} = \gamma \frac{\delta_{\mathrm{pl,N}} L}{12h} \tag{14}$$

where $\gamma$ is obtained by fitting the data and can be taken as 10.11 when no experimental data are available.

## 3   RESIDUAL BEARING CAPACITY OF STEEL-CONCRETE COMPOSITE BEAM

This section is based on the calculation method of the residual bearing capacity given in the literature reported by Wang (2017). Based on the degradation model of mechanical performance of each material and the degradation model of the stud bearing capacity with higher accuracy, the formulae for the residual bearing capacity of steel-concrete composite beams with different shear connection degrees are given.

### 3.1   *Material mechanical performance degradation model*

#### 3.1.1   *Studs bearing capacity degradation model*
Based on the basic theory of fracture mechanics, Rong et al. (2013) proposed an expression for calculating the residual bearing capacity considering the initial defects of the studs, as shown in Eq. (15).

$$\left(1 - \frac{P_{\mathrm{st}}(n)}{A_{\mathrm{st}} f_\mathrm{u}}\right)^M = \frac{N}{d_{\mathrm{st}}^M} \Delta\tau^{2-2M} n + \left(\frac{a_0}{d_{\mathrm{st}}}\right)^M \tag{15}$$

$$\Delta\tau = \frac{\Delta P S_0 \Delta l}{2 I_0 n_1 A_{\mathrm{st}}} \tag{16}$$

where $P_{\mathrm{st}}(n)$ is the residual bearing capacity of the studs after $n$ times of loading, $A_{\mathrm{st}}$ is the cross-sectional area of the stud, $f_\mathrm{u}$ is the ultimate strength of the stud, $d_{\mathrm{st}}$ is the diameter of the stud, $a_0$ is the initial crack length, and $\Delta\tau$ is the fatigue shear stress amplitude of the stud, which can be calculated according to Eq. (16). According to the research by Girhammar et al. (1993), we take $M = -1.05$ and $N = -6.19 \times 10^{-16}$.

### 3.1.2 Concrete degradation model

Concrete fatigue strength and stiffness can be calculated according to Eqs. (17)–(18).

$$f_c(n) = f_c - (f_c - \sigma_{c,max}) \left(\frac{n}{N_{c,f}}\right)^{c_1} \tag{17}$$

$$E_c(n) = \left(1 - 0.33\frac{n}{N_{c,f}}\right) E_{c,0} \tag{18}$$

where $n$ is the number of cycles of fatigue load, $N_{c,f}$ is the fatigue life of concrete, $f_c$ is the initial compressive strength of concrete, $\sigma_{c,max}$ is the maximum stress of concrete, $c_1$ is the test parameter of concrete material, which can be taken as 1 when no test is conducted, and $E_{c,0}$ is the initial modulus of elasticity of concrete.

### 3.1.3 Steel beam degradation model

Under fatigue loading, the residual strain of the steel is negligible, so it is considered that its fatigue modulus of elasticity and static modulus of elasticity remains the same. Its residual strength can be calculated according to Eq. (19).

$$f_s(n) = f_s - (f_s - \sigma_{s,max}) \left(\frac{n}{N_{s,f}}\right)^{c_2} \tag{19}$$

where $n$ is the number of fatigue load cycles, $N_{s,f}$ is the fatigue life of the steel beam, $f_s$ is the initial compressive strength of concrete, $\sigma_{s,max}$ is the maximum stress of the steel beam, $c_2$ is the material test parameter, which can be taken as 1 when no test is conducted.

### 3.2 The calculation for residual bearing capacity

The initial shear connection of the steel-composite composite beam is greater than 1, which is a fully shear connection. However, under the effect of fatigue load, the composite beam may change from full shear connection to partial shear connection. Therefore, when calculating the residual bearing capacity of the composite beam, the next calculation of the residual bearing capacity should be carried out according to the different shear connection degrees of the composite beam after the fatigue load. The shear connection degree $\eta$ of the steel-concrete composite beam under fatigue loading is defined as shown in Eq. (20).

$$\eta = \frac{n_s}{n_f'} \tag{20}$$

$$n_f' = \frac{\min\{A_c f_c(n), A_s f_s(n)\}}{P_{st}(n)} \tag{21}$$

where $n_s$ is the actual number of studs arranged within the shear span; $n_f'$ is the number of studs required to ensure complete shear connection within the shear span for $n$ times of fatigue loading; $P_{st}(n)$ is the residual bearing capacity of a single stud, which is calculated by Eq. (15); $f_c(n)$ and $f_s(n)$ are the residual strength of concrete and the material of steel beam, respectively, and they are calculated by Eq. (17) and Eq. (19); $A_c$ and $A_s$ are the cross-sectional areas of concrete slab and steel beam, respectively.

### 3.2.1 Residual bearing capacity of composite beam with partial shear connection

Part of the shear connection composite beam should consider the degradation of the mechanical performance of each member at the same time when calculating the residual bearing capacity. At this time it is assumed that the pressure of the concrete is equal to the composite force of the shear connections within a shear span, which can be divided into the following three cases according to the different locations of the plastic neutral axis of the section (as shown in Figure 1). Eqs. 22, 23, and 24 are the equations for calculating the residual bearing capacity of the composite beam in the three cases of partial shear connection, where the neutral axis is located within the upper flange of the steel beam, the concrete slab, and the web of the steel beam, respectively.

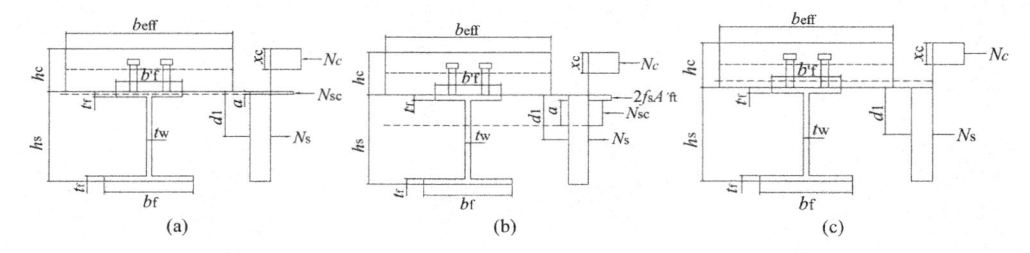

Figure 1.   Calculation of residual bearing capacity of partial shear connection composite beam.

$$M_{\mathrm{u}} = N_{\mathrm{s}}d_1 - \frac{N_{\mathrm{s}}^2}{4f_sb_f'} + \left(h_{\mathrm{c}} + \frac{N_{\mathrm{s}}}{2f_sb_f'}\right)N_{\mathrm{c}} - \left(\frac{1}{4f_sb_f'} + \frac{1}{2f_{\mathrm{c}}\left(n\right)b_{\mathrm{eff}}}\right)N_{\mathrm{c}}^2 \tag{22}$$

$$M_{\mathrm{u}} = N_{\mathrm{s}}\left(d_1 - t_f'\right) + f_sA_{\mathrm{ft}}'t_f' - \frac{1}{f_st_{\mathrm{w}}}\left(\frac{N_{\mathrm{s}}}{2} - f_sA_{\mathrm{ft}}'\right)^2 + \left(\frac{N_{\mathrm{s}} - 2f_sA_{\mathrm{ft}}'}{2f_st_{\mathrm{w}}} + h_{\mathrm{c}} + t_f'\right)N_{\mathrm{c}}$$
$$- \left(\frac{1}{4f_st_{\mathrm{w}}} + \frac{1}{2f_{\mathrm{c}}\left(n\right)b_{\mathrm{eff}}}\right)N_{\mathrm{c}}^2 \tag{23}$$

$$M_{\mathrm{u}} = N_{\mathrm{s}}\left(h_{\mathrm{c}} + d_1\right) - \left(\frac{1}{2f_{\mathrm{c}}\left(n\right)b_{\mathrm{eff}}}\right)N_{\mathrm{s}}^2 \tag{24}$$

where $N_{\mathrm{s}}$ is the steel beam composite force after considering the strength degradation of the lower flange plate of the steel beam, $N_{\mathrm{c}}$ is the concrete slab composite force after material strength degradation, $f_{\mathrm{s}}(n)$ is the residual steel strength, $x_{\mathrm{c}}$ is the concrete slab compressive height, $N_{\mathrm{sc}}$ is two times the steel beam compressive zone composite force, $a$ is the steel beam compressive zone height, $b_{\mathrm{eff}}$ is the concrete slab width, $b_f'$ is the steel beam upper flange width, $d_1$ is the steel beam composite force to the steel beam upper flange distance, $h_{\mathrm{c}}$ is the concrete slab height, $A_{\mathrm{ft}}'$ is the steel beam upper flange plate cross-sectional area, $t_{\mathrm{w}}$ is the steel beam web thickness, and $t_f'$ is the steel beam upper flange plate thickness. All other parameters are kept the same as above.

### 3.2.2   *Residual bearing capacity of fully shear-connected composite beam*
There is only one neutral axis in the composite beam in the case of a complete shear connection, and its bearing capacity can be calculated according to plasticity theory. At the same time, it is assumed that there is a sufficient connection between the steel beam and concrete to ensure their joint action, and only the degradation of the strength of the steel and concrete materials is considered. Without considering the degradation of the studs bearing capacity, the concrete in the tensile zone is not involved in the work. The calculation of the residual bearing capacity of the fully shear-connected composite beam is illustrated in Figure 2.

$$M_{\mathrm{u}} = N_{\mathrm{s}}d_1 - \frac{N_{\mathrm{s}}^2}{4f_sb_f'} + \left(h_{\mathrm{c}} + \frac{N_{\mathrm{s}}}{2f_sb_f'}\right)N_{\mathrm{c}} - \frac{1}{4f_sb_f'}N_{\mathrm{c}}^2 \tag{25}$$

$$M_{\mathrm{u}} = N_{\mathrm{s}}\left(d_1 - t_f'\right) + f_sA_{\mathrm{ft}}'t_f' - \frac{1}{f_st_{\mathrm{w}}}\left(\frac{N_{\mathrm{s}}}{2} - f_sA_{\mathrm{ft}}'\right)^2 + \left(\frac{N_{\mathrm{s}} - 2f_sA_{\mathrm{ft}}'}{2f_st_{\mathrm{w}}} + \frac{h_{\mathrm{c}}}{2} + t_f'\right)N_{\mathrm{c}} - \frac{1}{4f_st_{\mathrm{w}}}N_{\mathrm{c}}^2 \tag{26}$$

$$M_{\mathrm{u}} = N_{\mathrm{s}}\left(h_{\mathrm{c}} + d_1\right) - \left(\frac{1}{2f_{\mathrm{c}}\left(n\right)b_{\mathrm{eff}}}\right)N_{\mathrm{s}}^2 \tag{27}$$

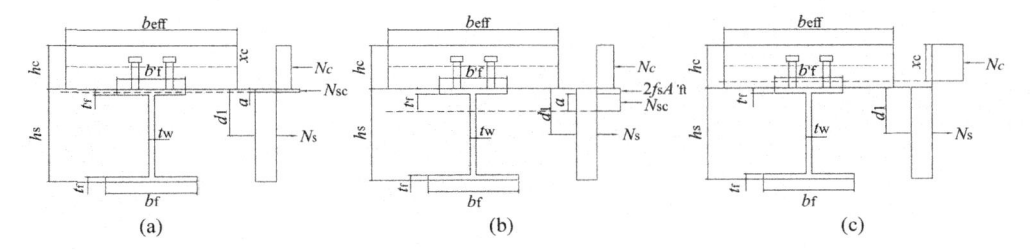

Figure 2. The calculation for the residual bearing capacity of a fully shear-connected composite beam.

Eqs. 25, 26, and 27 are for calculating the residual bearing capacity of the composite beam in the three cases of complete shear connection and are consistent with the above. The parameters in the equations remain the same as above.

## 4 MECHANICAL PERFORMANCE OF STEEL-CONCRETE COMPOSITE BEAM-TYPE BRIDGES UNDER OVERLOAD

### 4.1 Bridge parameters

Take a highway 30 m steel plate beam-concrete composite simple-supported beam bridge as an example. The specific design parameters of the structure are given by another researcher (Zhou et al. 2020). The limit state design method is adopted and the design is designed according to the design expression of the subfactor. The whole bridge elevation is arranged as shown in Figure 3. Among them, the steel beams are welded I-beams with 4 pieces and 3.2 m spacing. The cross-section of the middle beam is shown in Figure 4, and the characteristics of each material are shown in Table 1.

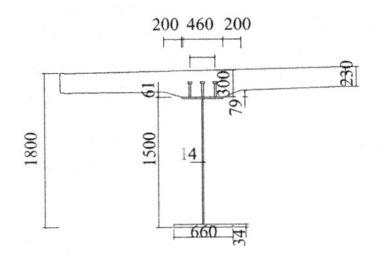
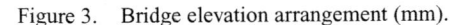

Figure 3. Bridge elevation arrangement (mm).

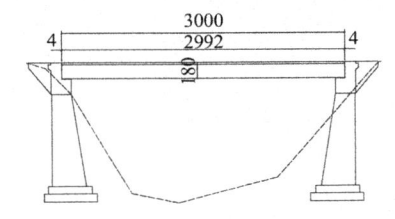

Figure 4. Mid-beam section sizes (mm).

Table 1. Properties of each material.

| Material | Elastic modulus (MPa) | Severe (kN/m³) | Coefficient of linear expansion | The design value of tensile strength (MPa) | The design value of compressive strength (MPa) |
|---|---|---|---|---|---|
| C50 concrete | $3.45 \times 10^4$ | 26 | $1.0 \times 10^{-5}$ | 1.83 | 22.8 |
| Q345qD Steel | $2.06 \times 10^5$ | 78.5 | $1.2 \times 10^{-5}$ | 270 | 270 |
| ML15 Steel | $2.00 \times 10^5$ | 78.5 | $1.2 \times 10^{-5}$ | 400 | – |

After calculation, the effective width of the concrete slab in the span of the center beam is 3.2 m and the effective width of the concrete slab at the pivot point of the center beam is also 3.2 m.

The geometric properties of the section are calculated using the equivalent conversion principle as shown in Table 2.

Table 2. Calculation results of cross-sectional geometric properties.

| Cross-section | Distance from the neutral axis to the beam bottom (mm) | Section moment of inertia ($mm^4$) | Sectional area ($mm^2$) |
|---|---|---|---|
| Concrete slab | 1,685 | $3.24 \times 10^9$ | 736,000 |
| Steel beam | 551 | $1.81 \times 10^{10}$ | 50,992 |
| Composite beam commutation | 1,352 | $6.50 \times 10^{10}$ | 173,658 |

### 4.2 Fatigue load amplitude

#### 4.2.1 Calculation model

The load amplitude borne by the bridge under the vehicle load is randomly varying and the random fatigue load can be converted to a constant amplitude fatigue load according to the equivalent damage theory. The overall design of the bridge can be calculated according to the lane load model given in *China's Design Code for Highway Steel Bridges* (Standard ID: JTG D64-2015) with a concentrated load of $0.7\,P_k$ and an average load of $0.3\,q_k$, which is taken according to the highway: Class I lane load standard. Its calculation is illustrated in Figure 5. The value of $P_k$ in Figure 5 is specified as shown in Table 3 and $q_k$ is taken as 10.5 kN/m.

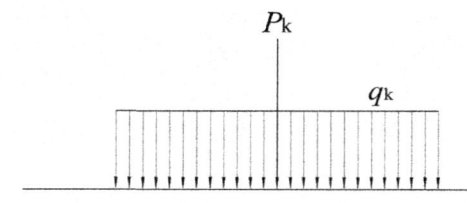

Figure 5. Calculation model for lane load.

Table 3. Value of concentrated load $P_K$.

| $L_e$ (m) | $L_e \leq 5$ | $5 < L_e < 50$ | $L_e \geq 50$ |
|---|---|---|---|
| $P_k$ (kN) | 270 | $2\,(L_e + 130)$ | 360 |

#### 4.2.2 Overloaded vehicle impact correction factor

According to China's *Highway Bridge Bearing Capacity Testing and Assessment Regulations* (Standard ID: JTG/T J21-2011), the live load influence correction coefficient is proposed to assess the bearing capacity of the bridge at the current stage of operation, as shown in Eq. (28).

$$\xi_q = \sqrt[3]{\xi_{q1}\xi_{q2}\xi_{q3}} \tag{28}$$

where $\xi_{q1}$ is the typical representative traffic impact correction factor; $\xi_{q2}$ is the large tonnage vehicle mixing impact correction factor; $\xi_{q3}$ is the axle load distribution impact correction factor, and the values are determined by the specification.

In this paper, the modified live load impact coefficient is defined as the overloaded vehicle impact correction coefficient. The bridge of the Gansu X245 line has been in overloaded operation for a long time. To take the effect of overloaded vehicles into account, Chen (2014) took the design traffic volume one level lower when calculating the typical representative traffic impact correction coefficient $\xi_{q1}$, thus raising the coefficient of $\xi_{q1}$ from 0.62 to 1.17, while $\xi_{q2}$ and $\xi_{q3}$ remained unchanged at 1.13 and 1.4, respectively. Finally, the calculation of the overloaded vehicle impact correction coefficient is 1.23 (Chen et al. 2014).

#### 4.2.3 Fatigue load amplitude under different overload degrees

Using the traffic volume information counted in the literature reported by Zhou et al. (2020), the degree of overload was distinguished by different overloaded vehicle impact correction coefficients.

The lane load calculation model was used to simulate the bridge under live load. The lower limitation of fatigue load is defined as the constant load borne by the composite beam bridge, and the upper limit of fatigue load is defined as the constant load and live load borne by the composite beam bridge, as shown in Table 4. The load magnitudes for the different working conditions are calculated without considering the crowd load and the results are shown in Table 4.

Table 4. Fatigue load amplitude at different overload levels.

| Operation status | Overloaded vehicle impact correction factor | The upper limit of load amplitude (kN) | The lower limit of load amplitude (kN) |
|---|---|---|---|
| Normal operation | 1.00 | 502 | 877 |
| General overload | 1.23 | 502 | 963 |
| Severe overload | 1.40 | 502 | 1027 |

## 4.3 Overload impact analysis

On the effect of overload on the fatigue stiffness of composite beam-type bridges, this paper takes the bridge design base period of 100 years as the end point of calculation. The loading interval is 5 years, using the daily traffic volume of 9,304 in the literature reported by Li (2015). The annual number of vehicle actions is 3,395,960 times and composite with the second section of this paper to calculate the fatigue stiffness value of the bridge in different service periods. The calculation results are organized into a fatigue stiffness degradation curve, as shown in Figure 6. With the increase in operating time, the fatigue stiffness degradation rate of the bridge in normal operation is slow. While the fatigue stiffness degradation rate of the bridge in general overload conditions is close to the fatigue stiffness degradation rate at the beginning and middle of loading and increases to a higher level at the end of loading. The fatigue stiffness degradation of the bridge in severe overload conditions is parabolic and the fatigue stiffness degradation rate increases with the operation time. It is shown that overloading has a significant effect on the degradation pattern of fatigue stiffness of composite beam-type bridges. Therefore, it is necessary to improve the fatigue resistance of the bridge while increasing the disposal of this overload problem to ensure the reliability of the bridge.

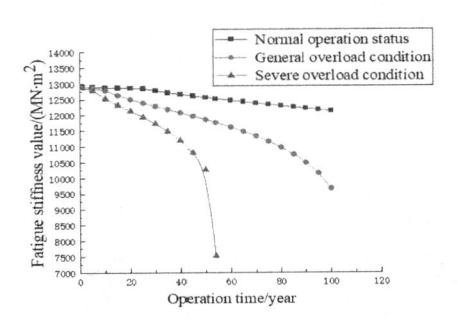

Figure 6. Fatigue stiffness values of composite beam-type bridges under different overload levels.

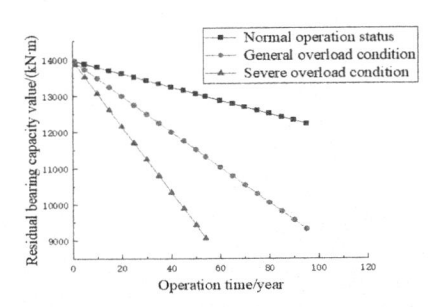

Figure 7. Residual bearing capacity values of composite beam-type bridges under different overload levels.

On the effect of overload on the residual bearing capacity of composite beam-type bridges, in this section, the shear connection degree in different operation conditions is kept in the range of complete shear connection. Using the calculation method of residual bearing capacity of the composite beam with complete shear connection given in Section 3.2, the load amplitude is taken according to Table 4 and the calculation results are organized into the residual bearing capacity degradation curve, as shown in Figure 7. It can be seen from the figure that the degradation of the residual bearing capacity of the composite beam bridge is roughly linear and the degradation rate is more stable, but the slope of the residual bearing capacity degradation curve of the composite beam bridge increases with the deepening of the overload degree, which indicates that the overload operation will aggravate and accelerate the degradation of the residual bearing capacity of the composite beam bridge.

## 5 CONCLUSIONS

The research results of this paper can provide theoretical support for the residual life prediction and performance evaluation of the steel-concrete combination beam-type bridges. It will further improve the fatigue and damage theory of the steel-concrete composite beam-type bridges. It has important guiding significance for bridge maintenance and reducing bridge collapse accidents.

(a) By introducing the improved discounted stiffness method, a fatigue deformation calculation method for composite beams with wider application and higher calculation accuracy is proposed. The validity of the method is proved by verifying the relevant test results.
(b) The influence of overloaded vehicles on the bridge load effect is considered. The correction coefficient for the influence of overloaded vehicles is proposed based on the existing correction coefficient for the influence of live load. The lane load calculation model is used to simulate the live load condition that the bridge is subjected to and the fatigue load amplitude of the bridge under different degrees of overload is calculated.
(c) The existence of overload vehicles has a significant effect on the fatigue stiffness and residual bearing capacity of the composite beam-type bridges. As the degree of overload increases, the degradation rate of the fatigue stiffness and the residual bearing capacity of the composite beam-type bridges are increasing, but the impact on the degradation of the residual bearing capacity is less. The fatigue stiffness degradation curve of the composite beam-type bridges changed from an S-shaped curve to a parabolic shape.
(d) The research in the paper is based on a load of frequent fatigue, which ignores the interaction between the load sequences. The impact of overload on the fatigue damage of the bridge may be weakened. Thus, further research is needed on the mechanical performance of steel-concrete composite under random loads so that the testing conditions of the test load are more suitable.

## ACKNOWLEDGMENTS

This work was supported by the National Key R&D Program of China (Grant Nos. 2021YFB2600600 and 2021YFB2600605) and the Key R&D Program of Hebei Province (Grant No. 19275405D).

## REFERENCES

Chen Junmin (2014). *Study on the load effect of concrete beam bridge under actual vehicle operation*. D. Fuzhou University.

Girhammar, U.A. & Gopu, V.K.A. (1993). Composite beam-columns with interlayer slip-exact analysis. *J. Journal of Structural Engineering*. 119 (4).

Hanswille, G., Porsch M. & Ustundag C. (2006). Resistance of headed studs subjected to fatigue loading part II: analytical study. *J. Journal of Constructional Steel Research*. 63 (4): 485–493.

Huang Qiao. (2017). *Design Principles of Steel-Concrete Composite Structures for Bridges*. M. People's Traffic Press.

Li Ming. (2015). *Study on the impact of overload operation on the safety of in-service reinforced concrete beam (slab) bridges*. D. Lanzhou Jiaotong University.

Liu Junli & Zhang Jinhao. (2017). Statistical analysis of bridge collapse cases caused by overload from 2007 to 2015. *J. Highway*. 62 (4): 5.

Nie Jianguo & Wang Yuhang. (2009). Calculation of deformation of steel-concrete composite beams under fatigue loading. *J. Journal of Tsinghua University: Natural Science Edition*. (12): 5.

Nie Jianguo. (2005). *Steel-concrete composite beam structures: tests, theory and applications*. M. Science Press.

Peng Guorong. (2003). *Experimental study on fatigue performance of steel-concrete composite beams*. D. Beijing Municipal Engineering Research Institute.

Rong Xueliang & Huang Qiao & Zhao Pin. (2013). Research on the anti-shear capacity of the nail connector considering fatigue damage. *J. China Highway Journal*. 2013, 26 (4): 7.

Wang Bing. (2017). *Study on the residual mechanical properties of steel-concrete composite beam type bridges based on fatigue cumulative damage effect*. D. Southeast University.

Wang Jingquan & Lv Zhitao & Liu Zhao. (2005). Composite coefficient method for deformation calculation of steel-concrete composite beams with partial shear connection. *J. Journal of Southeast University: Natural Science Edition*. 35 (A01): 6.

Xiang Yiqiang & He Baida. (2020). Calculation method of residual load capacity of pinned composite beams considering fatigue damage. *J. Journal of Hunan University: Natural Science Edition*. 47 (9): 7.

Xu Rongqiao & Chen Dequan. (2013). Improved discounted stiffness method for deflection calculation of composite beams. *J. Engineering Mechanics*. 30 (2): 7.

Zhou Xuhong & Liu Yongjian. (2020). *Steel bridge*. M. Beijing: People's Traffic Publishing House Co.

Civil Engineering and Urban Research – Mohamed & Hou (Eds)
© 2023 the Authors, ISBN 978-1-032-44487-1

# Finite element simulation of axial pressure test of UHPC-RC composite structure column

Le Guo*

*Chang'an University, Xi'an, China*

ABSTRACT: The calculation method of ordinary concrete (RC) and ultra-high-performance concrete (UHPC) plastic damage model is introduced and based on the RC column and UHPC-RC plastic (CDP) model, the parameter analysis, including the screw spacing of the UHPC layer, UHPC layer thickness, and internal concrete strength grade. The results show that the load-displacement curve of each specimen is gentle. In addition, some UHPC may be used instead of RC to increase the peak load of the column, increase the specimen thickness of the UHPC, and thus increase the peak load of the specimen.

## 1 INTRODUCTION

Ultra-high-performance concrete is a high-performance concrete composite designed based on the maximum packing density theory with super high strength, high durability, and high crack resistance. Due to its many properties, UHPC has become an ideal material for pier column members in extreme environments. Because the UHPC components contain many highly active materials, the maintenance conditions are relatively strict, and the price is relatively high, so it is not reasonable to make all the pier columns with UHPC. At present, the application of UHPC in pier columns is mainly to make prefabricated thin-wall members (Shan 2019), as the external reinforcement material of pier columns (Fu 2018).

To give full play to the ultra-high-pressure resistance and high durability of UHPC, it is proposed that UHPC is prefabricated into a thin-wall ring, and ordinary concrete is poured into it, and finally form a new column structure of UHPC-RC combination structure. At the same time, the prefabricated UHPC thin-walled ring can be used as the template of internal ordinary concrete, eliminating the formwork step in the construction of pier column, and reducing the amount of UHPC, which has a good development prospect.

In conclusion, the UHPC-RC combined structure column has better compression resistance and durability performance than the conventional concrete column. Based on Abaqus finite element software, this paper introduces the whole process of ultra-high performance concrete CDP model in detail, conducts axial pressure test simulation of UHPC-RC combined structure column, analyzes the influence of different parameters on the peak load of combined structure column, and provides some reference for subsequent tests.

## 2 TEST PIECE DESIGN AND TEST WORKING CONDITION

### 2.1 *Sample design*

The axial pressure test simulation adopts the UHPC-RC combination specimen of 600 mm long, an outer diameter of 240 mm, and a length ratio of 2.5. The spiral stirrup is arranged at the center of the pipe wall, and the specimen size and reinforcement are shown in Figure 1.

---

*Corresponding Author: 105655530@qq.com

  DOI 10.1201/9781003372417-4

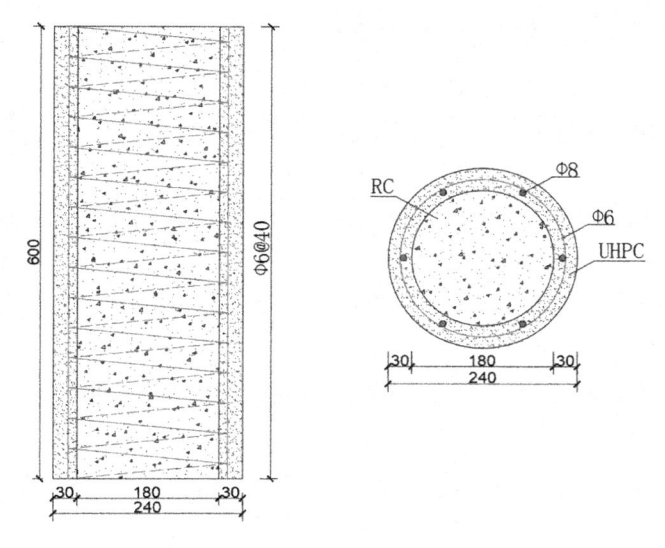

Figure 1.  Test piece size and reinforcement drawing.

## 2.2  *Operating condition of the test*

The axial pressure test is simulated on 7 UHPC-RC composite structure columns, and the test parameters include the thickness of the external UHPC layer, the strength grade of the internal ordinary concrete, and the spacing of the spiral stirrups in the UHPC layer. The test conditions are shown in Table 1.

Table 1.  Test conditions.

| Test number | Internal concrete strength (C) (MPa) | UHPC layer thickness (T) (mm) | Stirrup spacing (S) (mm) | Remarks |
|---|---|---|---|---|
| RC-T30-C40-S40 | C40 | 30 | 40 | Base group |
| U-R-T30-C30-S40 | C30 | 30 | 40 | The RC intensity |
| U-R-T30-C50-S40 | C50 | 30 | 40 | grade control group |
| U-R-T20-C40-S40 | C40 | 20 | 40 | The UHPC thickness |
| U-R-T25-C40-S40 | C40 | 25 | 40 | control group |
| U-R-T30-C40-S40 | C40 | 30 | 40 | |
| U-R-T35-C40-S40 | C40 | 35 | 40 | |
| U-R-T30-C40-S30 | C40 | 30 | 30 | The stirrup spacing |
| U-R-T30-C40-S40 | C40 | 30 | 40 | control group |
| U-R-T30-C40-S50 | C40 | 30 | 50 | |

## 3  FINITE ELEMENT MODEL ESTABLISHMENT

### 3.1  *Material constitutive relation selection*

#### 3.1.1  *Constitutive relation of ordinary concrete*

The concrete damage plasticity model is built in the Abaqus software, including an elastic rising section, inelastic rising section, and softening section after peak load. China's Code for Design of Concrete Structure (Standard code: GB/T 50010-2010) (referred to as the mixing gauge) gives the empirical constitutive relationship calculation formula for C20 to C80 concrete, so the finite element simulation adopts the empirical calculation formula in the mixing gauge to calculate the CDP model parameters of internal ordinary concrete.

To meet the requirements of CDP constitutive, the need to define the compression elastic limit stress and the peak stress and percentage, most European and American scholars think that the compression elastic limit point is 0.4 to 0.5, some scholars think that the small value will affect the convergence of the calculation results, so the compression elastic limit point is the peak stress point of 0.6 (Shi 2021). For the tensile elasticity limit point, the tensile peak stress is used as the tensile elasticity limit point (Zhang 2018).

The most critical problem in the CDP model is the calculation of the compression injury factor and pull damage factor. Using the damage factor calculation method (Birtel 2006; Mark 2006), as shown in Equations (1) and (2), the key is to determine the proportion factor. The two rated compression and stretching as 0.7 and 0.1.

$$d_c = 1 - \frac{\sigma/E}{\varepsilon_{pl}\,(1/r_c - 1) + \sigma/E} \tag{1}$$

$$d_t = 1 - \frac{\sigma/E}{\varepsilon_{pl}\,(1/r_t - 1) + \sigma/E} \tag{2}$$

The CDP model is also required to incorporate many plasticity parameters, including the viscosity parameters. The larger the calculation, the smaller the accuracy. At 0.0005, the requirements of accuracy and convergence can be met (Li 2021). The plastic parameters of the ordinary concrete are shown in Table 2.

Table 2.    Material parameters of ordinary concrete CDP model.

| Expansion angle $\Psi$ (°) | Eccentricity | $f_{b0}/f_{c0}$ | K | $\mu$ |
|---|---|---|---|---|
| 38 | 0.1 | 1.16 | 0.6667 | 0.0005 |

Considering the above analysis, taking C40 concrete as an example, the CDP tensile and compression configuration curves calculated according to the mixing gauge are shown in Figure 2.

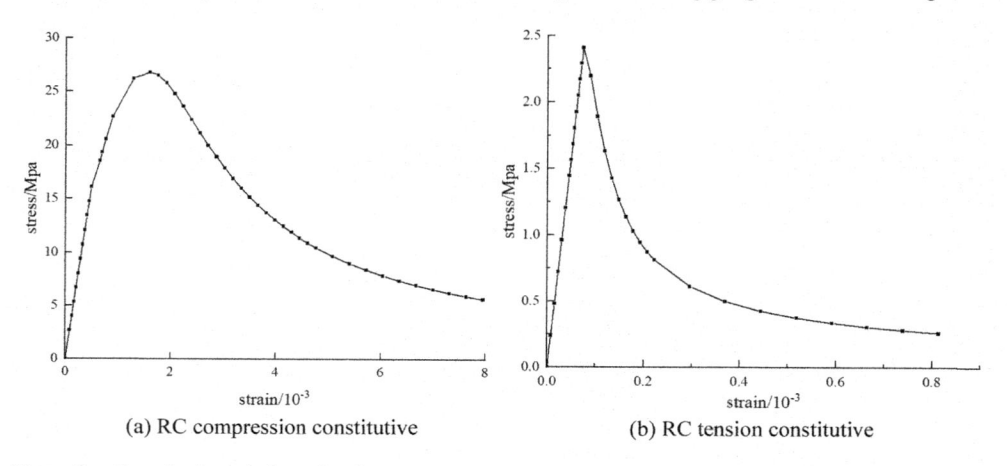

(a) RC compression constitutive                    (b) RC tension constitutive

Figure 2.    Constitutive relation of ordinary concrete.

### 3.1.2 *Constitutive relation of UHPC*

After years of research, domestic and foreign scholars have proposed various types of UHPC constitutive curves, this paper intends to select the data according to the constitutive curve from other researchers (Du 2014; Shan 2002). Its expression is shown in Equations (3) and (4). UHPC constitutive curve is shown in Figure 3. Meanwhile, the relevant mechanical parameters of UHPC refer to the reference literature (Zhang 2020), as detailed in Table 3.

$$y = \begin{cases} 1.117x + 0.415x^5 - 0.532x^6 (0 \leq x \leq 1) \\ \dfrac{x}{2.41(x-1)^2 + x}(x \geq 1) \end{cases} \tag{3}$$

$$y = \begin{cases} \dfrac{x}{0.92x^{1.09} + 0.08}(0 \leq x \leq 1) \\ \dfrac{x}{0.1(x-1)^{2.4} + x}(x \geq 1) \end{cases} \tag{4}$$

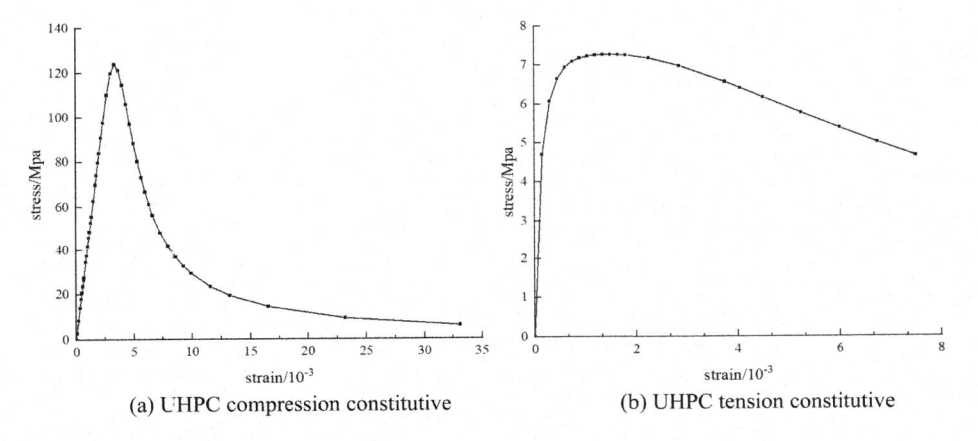

(a) UHPC compression constitutive      (b) UHPC tension constitutive

Figure 3. The UHPC constitutive relationship.

Table 3. UHPC basic mechanical parameters.

| Elastic modulus (MPa) | Axial pressure strength (MPa) | Peak pressure strain ($\mu\varepsilon$) | Ultimate tensile strength (MPa) | Peak pull strain ($\mu\varepsilon$) |
|---|---|---|---|---|
| $4.55 \times 10^4$ | 123.8 | 3309 | 7.27 | 1500 |

The UHPC plastic damage factor is calculated according to Najar's damage theory (Krajcinovic 1981), and according to Equation (5), the damage variable is defined as:

$$d = \frac{W_0 - W_\varepsilon}{W_0} \tag{5}$$

where
$W_0$—strain energy in the non-destructive state;
$W_\varepsilon$—strain energy under the damage state.
Figure 4 illustrates the linear damage plasticity model of Najar.

$$W_0 = \frac{E_0 \varepsilon^2}{2}$$

$$W_\varepsilon = \int \sigma \, d\varepsilon = \int f(\varepsilon) \, d\varepsilon$$

The calculation equation of the UHPC tension damage factor can be obtained by substituting Equation (5). As shown in Equation (6), the solution can be integrated by MATLAB to calculate the UHPC constitutive curve. This simulation uses the trapezoidal integral in the Gaussian integral.

$$d_c = \frac{\frac{E_0\varepsilon^2}{2} - \int f(\varepsilon)d\varepsilon}{\frac{E_0\varepsilon^2}{2}} \qquad (6)$$

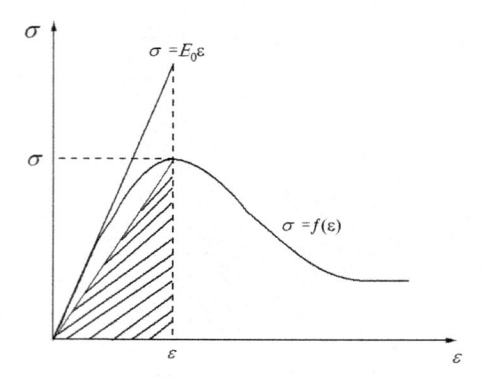

Figure 4.   A linear damage plasticity model diagram.

Table 4.   Material parameters of the UHPC CDP model.

| Expansion angle $\Psi$ (°) | Eccentricity | $f_{b0}/f_{c0}$ | K | $\mu$ |
| --- | --- | --- | --- | --- |
| 30 | 0.1 | 1.16 | 0.6667 | 0.005 |

### 3.1.3   Steel constitutive relationship

In finite element analysis, the stress-strain curve of steel usually includes (1) an ideal elastic-plastic model; (2) a double-fold line model; (3) a 3-fold line model. In this study, the HRB400 grade steel bar is used to model the double line, an ideal plastic model. This model incorporates the ideal elasticity before yield, the hardening stiffness before the limit strength, and Es = 0.01E0 of steel elastic modulus, where Es represents the hardening stiffness.

## 3.2   Finite element calculation method

### 3.2.1   Unit selection and grid division

UHPC-RC composite column finite element module uses separation unit, in which UHPC ring and internal ordinary concrete use C3D8R eight junction point linear hexahedral unit, spiral stirrup and longitudinal reinforcement use T3D2, while the test machine loads the actual loading process, R3D4 four junction point three-dimensional rigid quadrangle is used to simulate the upper and lower pressure plate, and establish the corresponding reference points in the center.

### 3.2.2   Interact

During modeling, the built-in (embedded) command is used to embed the steel cage in the external UHPC layer. The interface between the inner ordinary concrete and the outer UHPC, and the tangential friction coefficient between UHPC and the inner ordinary concrete is 0.3. The binding (Tie) command is used between the upper and lower plates and the combined components.

### 3.2.3 Boundary conditions and load application

In the modeling process, to facilitate calculation and conform to the actual test situation, the constraint form of the combined structure column is displacement constraint, and the bottom of the specimen is completely consolidated (U1 = U2 = U3 = U3 = UR1 = UR2 = UR3 = 0). To converge the calculation results, the loading method adopts displacement loading, which is 1/100 of the height of the test specimen.

## 4 PARAMETER ANALYSIS

### 4.1 Stirrup spacing

The stirrup spacing of each specimen is set as 30 mm, 40 mm, and 50 mm, the thickness of the UHPC layer is 30 mm, and the internal concrete strength grade is C40. If the remaining conditions remain unchanged, the load-displacement curve is obtained (see Figure 5).

In combining Figure 5 and Table 5, it is evident that specimens using the UHPC-RC combination structure have a greater peak load and peak displacement than those using ordinary concrete alone. When the stirrup spacing of the specimens is reduced from 40 mm to 30 mm, the peak load of the specimen increases and the peak displacement of the specimens also increases. When the stirrup spacing is reduced from 40 mm to 30 mm, the peak load increases by 8.74% of the specimens. The descending section of each specimen curve is slower, indicating that the thinner stirrup spacing has a better constraint effect on the concrete inside the composite structure.

The ductility coefficient of the specimen is expressed as $\mu = \Delta_u / \Delta_y$ where $\Delta_u$ refers to the specimen limit displacement and $\Delta_y$ represents the specimen yield displacement. This equation can reflect the inelastic deformation ability of the specimen under the load action (Liu 2019). There are clear yield points in the specimen load-displacement curve. The limit point of the curve

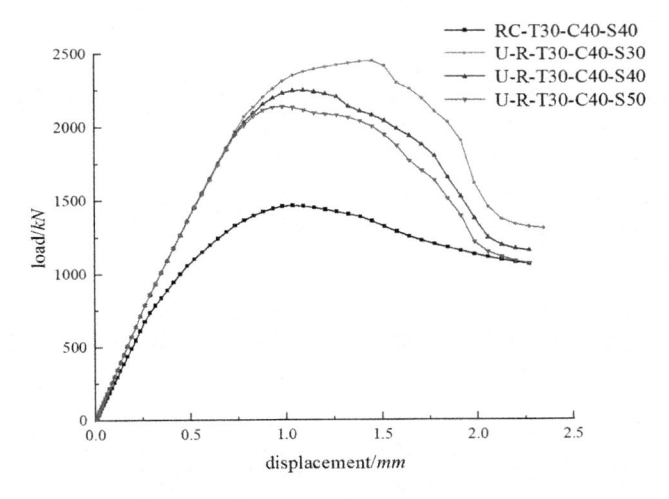

Figure 5. The load-displacement curve of different stirrup spacing.

Table 5. Peak load table of different stirrup spacing.

| Stirrup spacing | 40 (RC) | 30 | 40 | 50 |
| --- | --- | --- | --- | --- |
| Peak load | 1467.6 | 2449.7 | 2252.9 | 2140.7 |
| Load growth | 0 | 66.9% | 53.5% | 45.86 |
| Relative increase | / | / | 8%. 74 | 5%. 24 |

is selected by the general bending moment flexion method to select the corresponding displacement when the peak load of each specimen drops by 15%. For information regarding each specimen's ductility coefficient, please refer to Table 6. The higher the ductility coefficient, the greater the displacement when the specimen reaches the remaining 85% bearing capacity; the data in the table increases the ductility coefficient and the inelastic deformation ability of each specimen is gradually increased. When the stirrup spacing decreases from 40 mm to 30 mm, and the relative increase of the ductility coefficient reaches 7.53%.

Table 6.  Table of ductility coefficient of different stirrup spacing.

| Test number | UHPC-RC-50 | UHPC-RC-40 | UHPC-RC-30 |
| --- | --- | --- | --- |
| The yield displacement | 0.729 | 0.731 | 0.734 |
| Extreme displacement | 1.607 | 1.676 | 1.802 |
| Ductility factor | 2.189 | 2.283 | 2.455 |
| Relative increase | / | 4%. 29 | 7%. 53 |

### 4.2  UHPC thickness

The UHPC layer thickness of each specimen is set to 20 mm, 25 mm, 30 mm and 35 mm, the internal concrete strength grade is set to C40, the stirrup spacing is set to 40 mm, and the load-displacement curve in Figure 6 is unchanged.

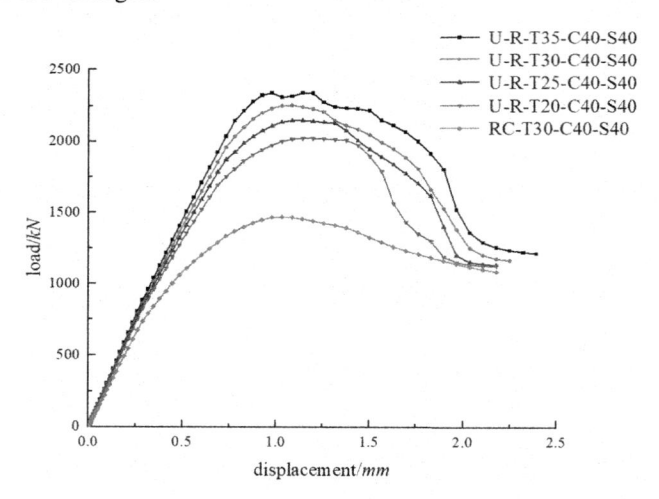

Figure 6.  The load-displacement curve of different UHPC thicknesses.

Combined with Figure 6 and Table 7, specimens with a UHPC-RC combination structure have a higher peak load than specimens with ordinary concrete, due to the thickness of the UHPC layer, the increase in peak load of the UHPC layer, and the application of the UHPC layer to the inner core concrete.

Table 7 shows that the thickness of the UHPC layer increases from 20 mm to 35 mm, and the peak load relative increase of each specimen decreases gradually from 30 mm to 35 mm, with only a 3 percent relative increase. Considering that the cost of UHPC is much higher than that of ordinary concrete when using the UHPC-RC combination structure test, the appropriate UHPC layer thickness should be selected according to the actual situation.

Table 7. Peak load table of different UHPC thickness.

| UHPC thickness | 0 | 20 | 25 | 30 | 35 |
|---|---|---|---|---|---|
| Peak load | 1467.6 | 2015.3 | 2147.2 | 2252.9 | 2338.6 |
| Load growth | 0 | 37.3% | 46.3% | 53.5% | 59.3% |
| Relative increase | / | / | 6%. 54 | 4%. 92 | 3%. 80 |

### 4.3 *Internal concrete strength grade*

The internal concrete strength grade of each specimen is set to C30, C40, and C50, the stirrup spacing is 40 mm, the UHPC layer thickness is 30 mm, and the load-displacement curve (Figure 7) is unchanged.

The UHPC-RC combination structure exhibits a greater peak load in comparison to ordinary concrete alone, as illustrated in Figure 7 and Table 8. The initial stiffness and a peak load of each composite structure increase with the increase of internal concrete strength grade, while the peak displacement decreases with the increase of internal concrete strength grade. The ductility coefficient of each specimen is shown in Table 9. As the strength grade of internal concrete increases, the ductility coefficient of each composite structure specimen decreases continuously. When the strength grade of internal concrete increases from C40 to C50, the ductility coefficient decreases by 12.48%.

The reason is that the higher the internal concrete strength, the smaller the strain of the stirrup, the smaller the constraint effect on the concrete, and the appropriate internal concrete strength grade should be selected according to the component environment in the project.

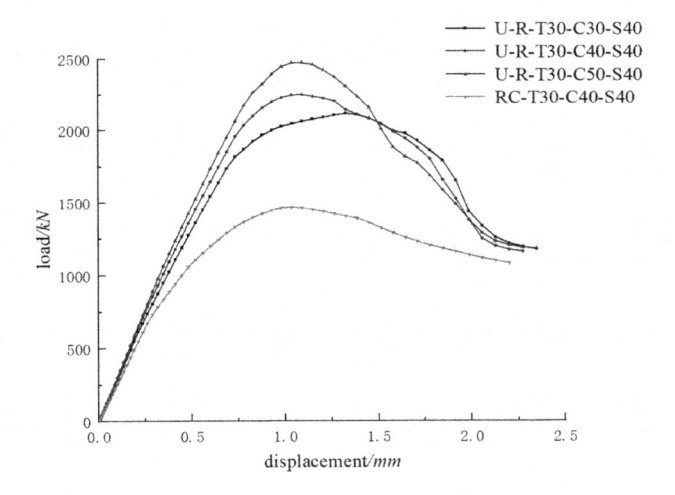

Figure 7. The load-displacement curve of different internal concrete strength.

Table 8. Peak load of different internal concrete strength.

| Internal concrete strength grade | C40 | C 30 | C40 | C50 |
|---|---|---|---|---|
| Peak load | 1467.6 | 2116.4 | 2252.9 | 2475.4 |
| Load growth | 0 | 44.2% | 53.5% | 68.7% |
| Relative increase | / | / | 6.44% | 9.87% |

Table 9. The ductility coefficient of different internal concrete strength.

| Internal concrete strength grade | C30 | C40 | C50 |
|---|---|---|---|
| The yield displacement | 0.731 | 0.729 | 0.732 |
| Extreme displacement | 1.840 | 1.676 | 1.473 |
| Ductility factor | 2.517 | 2.299 | 2.012 |
| Relative increase | / | −8.66% | −12.48% |

## 5 CONCLUSIONS

The simulation of the UHPC-RC column brings the following three conclusions.

1) The load-displacement curve of the UHPC-RC combined structure column is relatively full. Compared with the stirrup column using only ordinary concrete, the use of part of UHPC instead of ordinary concrete can make UHPC play the hoop effect and improve the peak load of the column.
2) With the decrease of stirrup spacing, the ductility coefficient and peak displacement of the UHPC-RC composite structure column are increased; the thicker the UHPC layer thickness, the smaller the relative increase of the peak load of the combined structure column; the higher the strength grade, the smaller the limit displacement of the combined structure specimen.
3) According to the above analysis, we can know that the stirrup spacing, UHPC layer thickness, and internal ordinary concrete strength are all important factors affecting the axial pressure performance of the UHPC-RC column. If the combined structure column is used in the actual project, the above parameters should be selected according to the environment of the structure and the relevant specifications to maximize its economic benefits.

## REFERENCES

Birtel V, Mark P. *Parameterised finite element modelling of RC beam shear failure J.* 2006 ABAQUS Users' Conference, 2006.

Du Renyuan. *Study on Limit Bearing Capacity of Active Powder Concrete Bch D.* Fuzhou University, 2014.

Fu Quanhong. *Experimental study on improving concrete durability of UHPC envelope D.* Tianjin University, 2018.

Krajcinovic D, Fonseka G U. The continuous damage theory of brittle materials, Part 1: General Theory *J. Journal of Applied Mechanics*, 1981, 48 (4): 809–815.

Li Qingfu, Kuang Yihang, Guo Wei. CDP model parameter calculation and value method validation *J. Journal of Zhengzhou University* (Engineering edition), 2021,42 (02): 43–48.

Liu chuanfei. *Study on mechanical properties of UHPC connection joints of fabricated piers D.* South China University of technology, 2019

Shan Bo *Test and Research on basic mechanical properties of reactive powder concrete D.* Hunan University, 2002

Shan Bo, Luo Xiaobing, Xiao Yan, Liu Fucai. Study on axial compression properties of short-concrete column *J. Journal of Xiangtan University* (Natural Science edition), 2019,41 (02): 85-93.

Shi Xinyu, Yao Yan, Wang Ling, Zhang Cheng. Impact of CDP model parameters based on uniaxial pull-press simulation *J. Building structure*, 2021,51 (S2): 999–1007.

Zhang Tian *Research on numerical simulation application of typical concrete model under monotonic and cyclic loading D.* Kunming University of technology, 2020.

*Civil Engineering and Urban Research – Mohamed & Hou (Eds)*

# Research on the influence of emulsion powder on the shear mechanical property of polymer emulsion-based cement composite material

Tengjiao Wang*
*Aviation Engineering School, Air Force Engineering University, Xi'an, China*

Jinyu Xu
*Aviation Engineering School, Air Force Engineering University, Xi'an, China*
*College of Mechanics and Civil Architecture, Northwest Polytechnic University, Xi'an, China*

Erlei Bai, Biao Ren, Yipeng Ning & Ao Yao
*Aviation Engineering School, Air Force Engineering University, Xi'an, China*

ABSTRACT: To study the effect of emulsion powder on the shear mechanical property of polymer emulsion-based cement composite material (PEBCC), five emulsion powder contents (0, 2.5%, 5%, 7.5%, and 10%) are subjected to shear testing. Based on the analysis of the test results, the modification mechanism of emulsion powder is further explored. The results show that the addition of emulsion powder can significantly improve the adhesion and cohesion among the components of PEBCC, thereby significantly improving the strength under shear load and significantly reducing the deformability. Currently, PEBCC tends to "harden". When PEBCC has both excellent strength and deformability, the best content of emulsion powder is 5%.

## 1 INTRODUCTION

The caulking material of airport cement concrete pavement is an important part of the airport pavement. The aging and failure of caulking material will cause rainwater, wind, and sand to penetrate the roadbed, which will further cause the pavement slab to appear dislocation, fracture, mud, and other diseases (Guo 2020; Hu 2021), which will eventually have a serious impact on flight safety. Therefore, the development and application of caulking material for airport cement concrete pavement have always attracted much attention. Currently commonly used caulking materials include organic polymer materials such as polysulfide (Teng 2013), silicone (Lago 2017; Wu 2011), polyurethane (Mao 2016), and polythiourethane (Xu 2015), with good tightness and deformation but poor strength and durability. Therefore, there is an urgent need for caulking material with high strength, deformability, and durability.

PEBCC is an organic or inorganic composite material that uses polymer emulsion as a matrix and cement and other inorganic powders as a reinforcement. Therefore, this type of material not only possesses the high elastoplasticity and strong deformability of polymer emulsion but also the high strength and great durability of cement hydration products to provide a new direction for the development and application of caulking material for cement concrete pavement. At present, domestic, and foreign studies on PEBCC mainly focus on modification optimization, adhesion, and waterproof performance (Fahad 2020; Xu 2017; Wang 2020; 2021). The initial studies (Ibraheim 2020) mainly discussed the effects of polymer emulsion content and cement content on

---

*Corresponding Author: wtengjiao83087@163.com

DOI 10.1201/9781003372417-5

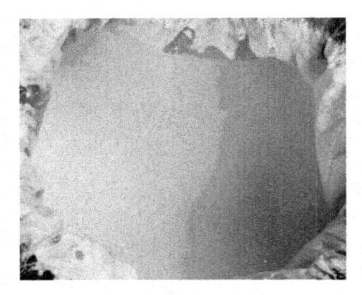

Figure 1. VAE emulsion.

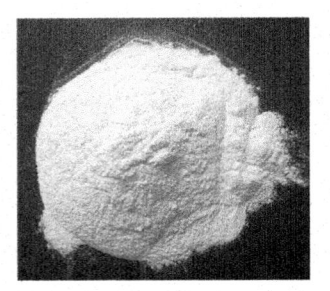

Figure 2. Dispersible emulsion powder.

the bonding performance of the composites to obtain the optimal content ratio of polymer emulsion and cement. Moreover, microscopic tests were carried out to analyze the influence of the change in the content ratio of polymer emulsion and cement on the microstructure of the composites (Onuaguluchi 2020; Pang 2021). As the research progresses, scholars now mainly study the optimization of different admixtures on the bonding performance of PEBCC and discuss the influence mechanism of admixtures to obtain the preparation method of composites with better performance (Wang 2020). However, these research works are mainly used to hole sealing and crack repair, and there are few studies have been done on its application in cement concrete pavement caulking.

In addition, when airplanes and vehicles pass cement concrete pavement, the two pavement slabs will have a displacement difference due to uneven force, which will produce a shearing effect on the caulking material (Wang 2020). Under the repeated action of the shear load caused by tire load, the caulking material is destroyed to failure, which adversely affects the safety of the pavement slab. Therefore, it is necessary to conduct an in-depth study on the shear mechanical property of the caulking material of cement concrete pavement.

Based on this, this paper develops a PEBCC for cement concrete pavement filling. Performance optimization is carried out by incorporating emulsion powder, and research on the effect of emulsion powder on mechanical properties such as PEBCC shear strength properties and deformability is carried out. Furthermore, the modification mechanism of emulsion powder is analyzed based on the microscopic test results.

## 2 TESTS

### 2.1 *Test materials*

As shown in Figure 1, the polymer emulsion uses vinyl acetate-ethylene copolymer emulsion (Celvolit 1350), namely VAE emulsion. Its technical indicators are shown in Table 1. The cement is 42.5 grade ordinary Portland cement (P•O 42.5), with technical indicators that meet the requirements of the Chinese national standard. As shown in Figure 2, quartz powder with 300 mesh and $SiO_2$ content larger than 99% and 5044N-type vinyl acetate/ethylene copolymer emulsion powder (re-dispersible emulsion powder) are selected. Its technical indicators are shown in Table 2. In addition, the test also selects SN-Dispersant 5040 type dispersant easily soluble in water, SN-345 type silicone defoamer suitable for high-viscosity elastic emulsion systems, and DN-12 type film-forming aids.

### 2.2 *Preparation of test piece*

Raw materials are prepared according to the five ratios listed in Table 3. The preparation of PEBCC follows the process shown in Figure 3. Among them, the content of emulsion powder is the percentage of emulsion powder mass to emulsion mass. The prepared PEBCC is injected into a cavity

Table 1. Technical indicators of VAE emulsion.

| Solid content | pH | Particle size ($\mu$m) | Brookfield viscosity (mPa•s) | Tg (°C) | MFT (°C) |
|---|---|---|---|---|---|
| 55±1% | 4.5–6.0 | 1.5 | 1500–5000 | −10 | 0 |

Table 2. Technical indicators of emulsion powder.

| Solid content | Particle size | Ash content | Water soluble | Corresponding emulsion particle size ($\mu$m) | $T_g$ of corresponding emulsion (°C) | MFT of corresponding emulsion (°C) |
|---|---|---|---|---|---|---|
| 99±1% | 400 $\mu$m sieve residue≤4% | 10±2% | 5% | 1–7 | 0 | 0 |

composed of cement mortar base material, anti-adhesive cushion, and anti-adhesive base film, and then cured for four days at room temperature, and for 24 days after removing the pad and bottom film. The test piece required for the test is obtained, as shown in Figure 4.

Table 3. The ratios of PEBCC (g).

| Test piece number | Emulsion powder content (%) | VAE emulsion | Cement | Quartz powder | Emulsion powder | SN-5040 dispersant | SN-345 defoamer | DN-12 filmforming aid |
|---|---|---|---|---|---|---|---|---|
| PEBCC01 | 0 | 100 | 15.8 | 29.2 | 0 | 1.02 | 0.73 | 6 |
| PEBCC02 | 2.5 | 100 | 15.8 | 29.2 | 2.5 | 1.02 | 0.73 | 6 |
| PEBCC03 | 5.0 | 100 | 15.8 | 29.2 | 5 | 1.02 | 0.73 | 6 |
| PEBCC04 | 7.5 | 100 | 15.8 | 29.2 | 7.5 | 1.02 | 0.73 | 6 |
| PEBCC05 | 10.0 | 100 | 15.8 | 29.2 | 10 | 1.02 | 0.73 | 6 |

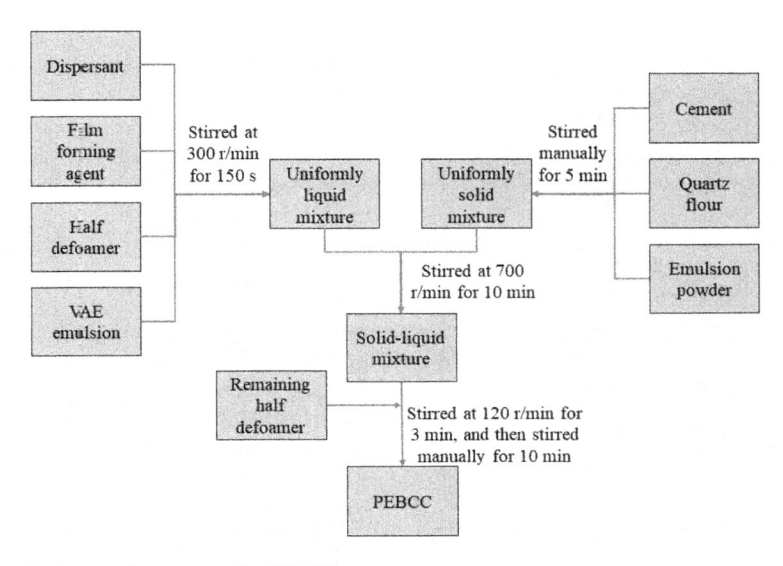

Figure 3. The preparation process for PEBCC.

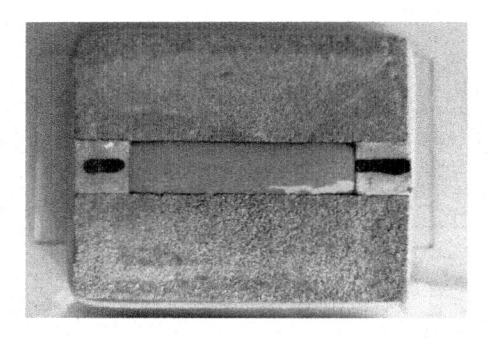

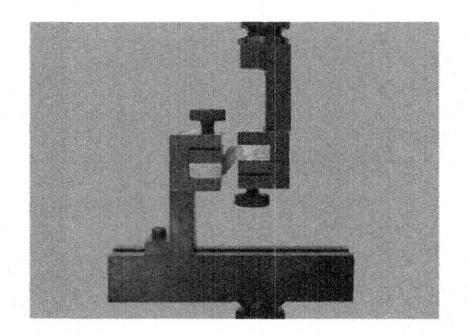

Figure 4.   PEBCC test piece.                    Figure 5.   Shearing fixture.

## 2.3   *Test method*

The HS-3001B electronic tensile testing machine is used to perform shear tests on PEBCC with five different emulsion powder contents. The test procedure is as follows. The shearing fixture is installed on the testing machine, as shown in Figure 5. The test piece is loaded into the shearing fixture. It should be noted that the cement mortar base material should be fixed and clamped. The right side of the fixture is pulled up at a rate of 5 mm/min until the specimen is broken. The load-displacement curve of the test is obtained. The result of each group is the average value of three repeated tests.

## 3   TEST RESULTS AND DISCUSSIONS

### 3.1   *Shear strength properties*

The shear strength properties of PEBCC are characterized by two indicators of shear strength and shear modulus. Figure 6 and Figure 7 show the influence of emulsion powder content on PEBCC shear strength and shear modulus. It can be seen from the figure that the shear strength and shear modulus of PEBCC both show an increasing trend with the increase in the content of emulsion powder. When the content of emulsion powder is greater than 2.5%, the shear strength of PEBCC increases significantly with the increase of the content of emulsion powder. When the emulsion powder content increases from 7.5% to 10%, the PEBCC shear modulus increases the most, reaching 11.24%. The addition of emulsion powder can significantly enhance the shear strength properties of PEBCC, and the enhancement effect increases with the increase of the amount.

### 3.2   *Shear deformability properties*

Shear deformability properties of PEBCC are characterized by two indicators of the shear peak strain and shear elongation at break. Figure 8 and Figure 9 show the influence of emulsion powder content on PEBCC shear peak strain and shear elongation at break, respectively. It can be seen from the figure that the shear peak strain of PEBCC increases with the increase in the content of emulsion powder, but the growth rate shows the slow-fast-slow law. When the amount of emulsion powder increases from 2.5% to 7.5%, the shear peak strain of PEBCC increases the fastest. The shear elongation at the break of PEBCC keeps decreasing as the amount of emulsion powder increases. When the emulsion powder content increases from 5% to 7.5%, the reduction rate of shear elongation at break is the fastest, with a decrease of 5.42%. The addition of emulsion powder can enhance the deformability of PEBCC when it reaches the ultimate bearing capacity, but ultimate deformability reduces.

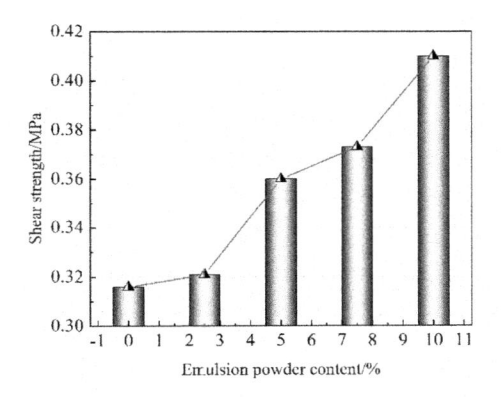

Figure 6. Effect of emulsion powder content on shear strength.

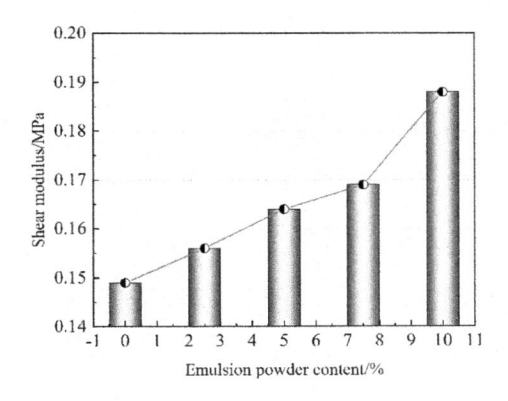

Figure 7. Effect of emulsion powder content on shear modulus.

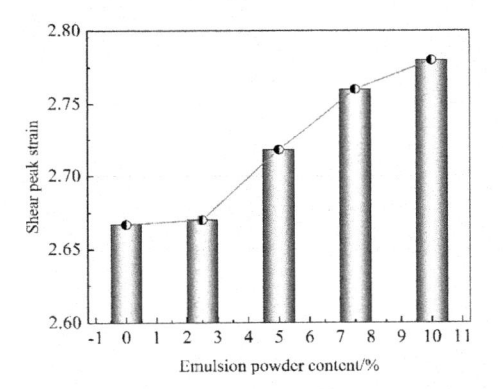

Figure 8. Effect of emulsion powder content on shear peak strain.

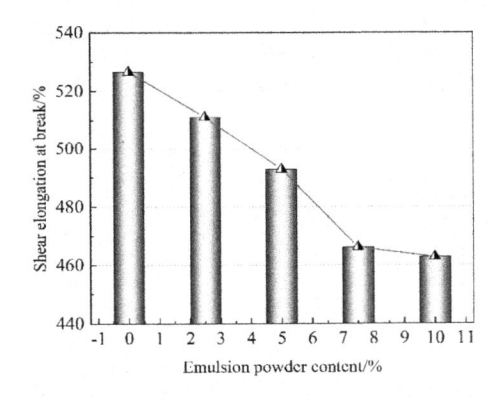

Figure 9. Effect of emulsion powder content on shear elongation at break.

Considering the influence of emulsion powder content on PEBCC strength and deformability and ensuring both good strength and deformability of the prepared PEBCC, the optimal content of emulsion powder is 5%.

## 4 MECHANISM ANALYSIS

The continuous film formed after the re-dispersible emulsion powder is dissolved in water can penetrate the gap between the polymer emulsion and the cement hydration product (see Figure 10), which further improves the density of PEBCC. At the same time, the high adhesion of the film can promote the adhesion of organic polymers and inorganic compounds, thereby enhancing the cohesion of PEBCC. Therefore, as the amount of emulsion powder increases, the shear strength and shear modulus of PEBCC increase significantly, which further enhances the shear peak strain and reduces elongation at break. At this time, the PEBCC tends to "harden".

## 5 CONCLUSIONS

Based on the results and discussions presented above, the conclusions are obtained as below.

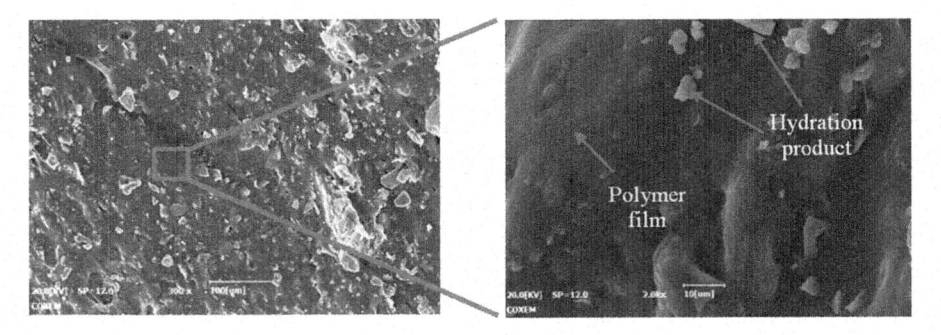

Figure 10. Microscopic morphology of PEBCC.

(1) The addition of emulsion powder can increase the cohesion of PEBCC and increase its bearing capacity under shear load, thereby improving the strength of PEBCC.
(2) The high cohesion caused by emulsion powder reduces the elasticity and plasticity of PEBCC and tends to "harden", resulting in a decrease in the overall deformability of PEBCC.
(3) Under the condition that PEBCC has both high strength and large deformation, the best content of emulsion powder is 5%.

## 6 LIMITATIONS

In this paper, the effect of emulsion powder on the shear mechanical properties of PEBCC was studied when the content of emulsion powder is in the range of 0–10%. Due to the small content range of emulsion powder, the research results have some limitations and it is necessary to increase the content range of emulsion powder and further study the effect of emulsion powder content on the shear mechanical properties of PEBCC in the next study. In addition, the study on the influence mechanism of emulsion powder is not enough only through the SEM test, so the MIP test and XRD test need to be further carried out to help discuss.

## REFERENCES

Fahad, B. M., & Matlik, B. M. (2020). Effect of Different Curing on Mechanical Properties of Polymer - Cement Composites. *Materials Science Forum* (Vol. 1002, pp. 627–635).
Guo, C., Cui, C., & Wang, F. (2020). Case study on quick treatment of voids under airport pavement by polymer grouting. *Journal of Materials in Civil Engineering*, 32(7), 05020006.
Hu, C., Weng, X., & Yan, X. (2021). A review of surface properties enhancement of pavement concrete. *IOP Conference Series: Earth and Environmental Science*, 638(1), 012101.
Ibraheim, S. M., & Hussein, S. I. (2020). Study on the contact angle, adhesion strength, and antibacterial activity of polymer/cement composites for waterproof coating. *Iraqi Journal of Science*, 1971–1977.
Lago, B. D., Biondini, F., Toniolo, G., & Tornaghi, M. L. (2017). Experimental investigation on the influence of silicone sealant on the seismic behaviour of precast façades. *Bulletin of Earthquake Engineering*, 15(4), 1771–1787.
Mao J., Pan L, Xiong T. (2016) Study on Adhesion of Silane Modified Polyurethane Sealant to PC . *Chemical Research and Application*, 11.
Onuaguluchi, O., & Banthia, N. (2020). Alkali–silica reaction resistance of cementitious material containing cacl2-blended acrylic polymer emulsion. *Journal of Materials in Civil Engineering*, 32(3), 04019378.
Pang, B., Jia, Y., Dai Pang, S., Zhang, Y., Du, H., Geng, G., & Liu, G. (2021). The interpenetration polymer network in a cement paste–waterborne epoxy system. *Cement and Concrete Research*, 139, 106236.
Teng, X., Wu, P., Sun, J., Xu, Y., & Yang, H. (2013). Polysulfide sealant modified with graphene oxide. Gaofenzi Cailiao Kexue Yu Gongcheng/*Polymeric Materials ence and Engineering*, 29(6), 54–57.

Wang T, Xu J, Zhu C, Ren W. (2020) Effects of cement-powder ratio on static mechanical properties and failure forms of styrene-acrylic emulsion-based cement composites. *Acta Materiae Compositae Sinica*, 37(9): 2324–2335.

Wang, T., Xu, J., Zhu, C., & Ren, W. (2020). Comparative study on the effects of various modified admixtures on the mechanical properties of styrene-acrylic emulsion-based cement composite materials. *Materials*, 13(1).

Wang, Z., Xu, J., Meng, X., Huang, Z., & Xia, W. (2021). Effect of VAE emulsion and inorganic fillers on the properties of styrene-acrylic emulsion based cement composite joint sealant. *Construction and Building Materials*, 286, 122976.

Wu, J. (2011). Preparation and properties of the silicone building sealant pfs. *New Chemical Materials*, 39(4), 71–73.

Xu, J., & Zhang, D. (2017). Pressure-sensitive properties of emulsion modified graphene nanoplatelets/cement composites. *Cement and Concrete Composites*, 84, 74–82.

Xu, S., Yang, H., Song, Q., Liu, Z., & Lei, Z. (2015). Preparation and performance study of fire-retardant two components polysulfide urethane sealant. *China Building Waterproofing*.

*Civil Engineering and Urban Research – Mohamed & Hou (Eds)*
*© 2023 the Authors, ISBN 978-1-032-44487-1*

# Research on fatigue performance of prestressed concrete girder bridge of heavy haul railway

M.A. Qian
*Shuohuang Railway Development Co., Ltd., Suning, China*

Si Jinyan
*Beijing Municipal Engineering Research Institute, Beijing, China*

Zhao Jiacheng, Tang Qingchen, Zhu Li* & Ren Xiaozuo
*School of Civil Engineering, Beijing Jiaotong University, Beijing, China*

ABSTRACT: With the development of the Shuohuang Railway's capacity expansion and reconstruction project, the axle load has increased and the operation density has increased, which makes the bridges along the line work under high load for a long time, and the fatigue deterioration accelerates, which seriously threatens the safety of vehicles and bridges. Taking the prestressed concrete T-beam of the Shuohuang heavy haul railway as the research object, based on the vehicle-bridge coupling dynamic analysis program, the S-N curve method is used to calculate the fatigue damage of the beam tensile steel and shear stirrup, to estimate their service life, and to further analyze the influence of track irregularity on the fatigue damage of the tensile steel and shear stirrup. The research results show that: the larger the axle load of the train, the longer the marshaling, and the more unfavorable the fatigue performance of the beam; the mid-span tensile steel bars can meet the 100-year service life, while the beam end shear stirrups are expected to have a service life of 66 years, which cannot meet the requirements of a 100-year operation period, considering the future growth in transportation volume; the worse the track irregularity, the smaller the fatigue life, and the more unfavorable the fatigue resistance.

## 1 INTRODUCTION

In the context of the rapid growth of transport volumes, bridges, as the most important component of transport infrastructure, have been one of the focal points of construction development. In recent years, to meet the growing demand for transport volume, the Shuohuang Railway has carried out capacity expansion and renovation projects with increased axle weight and higher operating density, causing bridges along the line to work under high loads for long periods and accelerating fatigue deterioration, which seriously threatens the safety of vehicles and bridges. Therefore, fatigue performance studies on heavy haul bridges are necessary.

Research into the fatigue of reinforced concrete has led to many mature conclusions since the early 20th century. Theoretically, relevant research has been carried out by scholars such as Basquin generalized the S-N curve (Basquin 1910). Miner formulated the P-M linear fatigue cumulative damage criterion (Miner 1945). Fangping Liu and Jianting Zhou proposed the S-level nonlinear fatigue strain model (Liu 2017). Many scholars have studied the fatigue problem of reinforced concrete through experiments and have concluded that after several repeated loading actions, the concrete in the compression zone of prestressed concrete beams under the fatigue upper limit load is still in the elastic stage, and the fatigue damage starts with ordinary reinforcement (ordinary tensile

---

*Corresponding Author: zhuli@bjtu.edu.cn

  DOI 10.1201/9781003372417-6

reinforcement or shear stirrups), while there is no fatigue damage of prestressed reinforcement (Che 1988; Li 1997, 2013; Luo 2007; 2003; Mohsen 1986; Zhong 1993). The fatigue damage of random variable amplitude fatigue loading on the specimen damage is greater than that caused by equal-amplitude fatigue loading (Naaman 1991); the development of cracks in bridges during fatigue damage (Feng 2006); the damage of stirrup can be measured by the stirrup strain under transient action (Yang 1994); and the trend of variation of relevant characteristic parameters in shear tests under fatigue loading (Xiao 2018). Chen Wan numerically analyzed the fatigue damage of a heavily loaded bridge using MATLAB simulations and gave fatigue damage values (Chen 2015). Based on these studies, we found that fatigue damage in prestressed concrete beams is controlled by the normal tensile steel bars and shear stirrups in the beams and that no fatigue damage occurs to the prestressed steel bundle and concrete, which can be measured by the strain in the reinforcement.

This paper takes a typical bridge type on the Shuohuang Heavy Haul Railway as the research object and uses the S-N curve method to calculate the fatigue damage of the tensile reinforcement and shear stirrups of the girder using the vehicle-bridge coupling calculation program to predict their service life, and further analyses the effect of track unevenness on the fatigue damage of the tensile reinforcement and shear stirrups.

## 2 STRESS-TIME CURVE FOR TRAINS CROSSING THE BRIDGE

In this study, a typical 32 m pre-stressed simple girder bridge of Shuohuang Railway is used as an example. The trains are mainly of C64K, C70, and C80 types, with axle weights of 21, 23, and 25 tons for the three types. Based on the vehicle-bridge coupling theory, a dynamic calculation program is developed to calculate the dynamic indexes of the bridge and extract the response times of the bridge under different train configurations for subsequent analysis and calculation.

According to previous research on fatigue problems of prestressed concrete, for bending fatigue problems, prestressed reinforced concrete is mainly controlled by the fatigue strength of the ordinary reinforcement at the span cross-section during fatigue damage; for shear fatigue problems, fatigue damage in diagonal sections is usually caused by fatigue fracture of stirrups. Therefore, the fatigue damage of the tensile reinforcement in the span and the shear stirrups at the end of the beam need to be analyzed.

### 2.1 *Calculation of tensile reinforcement stress in the span*

Considering that the object of study is a fully prestressed beam with high stiffness and cracks are not allowed to appear in the span during the normal service phase, the calculations are based on the following assumptions. (1) The beam is in a fully elastic working phase; (2) the reinforcement is well bonded to the concrete and the strain on the concrete at the same location as the reinforcement is the same. Based on these assumptions, and according to the basic finite element theory (Zeng 2008), the displacement-strain relationship exists as shown in the following equation.

$$\varepsilon_B^e = B\delta_B^e \tag{1}$$

where $\varepsilon_B^e$ is the strain in beam units; $\delta_B^e$ is the nodal displacements of beam units; $B$ is the displacement-strain matrix, represents the calculated transformation between strain and nodal displacement at a point, and can be obtained by deriving the shape function from the geometric coordinates of space. From the theoretical knowledge of elastic mechanics, the value of the positive stress can be determined by the following equation:

$$\sigma^e = EB\delta^e \tag{2}$$

where $E$ is the stress-strain matrix, which can generally use a constant matrix. Based on the above equation, the time course of the response of the bridge obtained from the coupled vehicle-bridge dynamic calculations can be calculated to obtain the results of the time course of the positive stress of the tensile reinforcement in the span of the bridge when each group of single trains passes through the bridge.

### 2.2 *Calculation of stirrup stress at beam ends*

Based on the bridge conditions and calculations, it was found that the stirrups in the range 1/16 L to 1/8 L are subjected to higher stresses and stress amplitudes and are relatively more dangerous. Therefore, fatigue damage calculations were carried out for stirrups in the range of 1/16 L to 1/8 L. Wen Yusong of Changsha Railway Institute (Wen 1997) gave a method for calculating stirrup stress in ordinary reinforced concrete beams, and the same can be deduced from the formula for calculating stirrup stress in prestressed concrete beams. The specific derivation process is as follows.

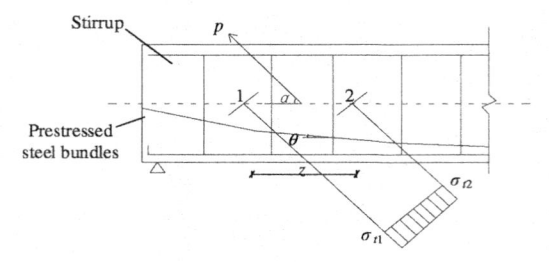

Figure 1. Schematic diagram of beam end stress analysis.

where $P$ is oblique tension in beam section between section 1 and section 2; $z$ is the length of beam section between section 1 and section 2; $\alpha$ is the angle of inclined main tension to horizontal; $\theta$ is the angle between prestressing beam and horizontal direction ($\theta = 5°$); $\sigma_{t1}$ is the principal tensile stress at section 1; $\sigma_{t2}$ is the principal tensile stress at section 2.

The beam section from point 1 to point 2 in the diagram is analyzed and the diagonal main tension $P$ on this beam section can be calculated from the following equation.

$$P = \frac{(\sigma_{t1} \cdot b \cdot c_1 + \sigma_{t2} \cdot b \cdot c_2)}{2} \tag{3}$$

where $b$ is the width of the web; $c_1$ and $c_2$ are internal force couple arms in sections 1 and 2, which can be calculated from the bending resistance of the beam.

For safety reasons, the shear resistance of the concrete and the dowel action of the longitudinal reinforcement is ignored, and the diagonal tension is borne entirely by the stresses and prestressing reinforcement. The stresses in the stirrups can be expressed as:

$$\sigma_{sv} = \frac{n_1 \cdot \sin \alpha \cdot P}{n_1 \cdot A_{sv} \cdot \frac{z}{S_v} + n_2 \cdot A_{sb} \cdot \cos (\alpha - \theta)} \tag{4}$$

where $n_1$ is the elastic modulus ratio of the stirrup to concrete; $n_2$ is the elastic modulus ratio of prestressed steel bundle to concrete; $A_{sv}$ is the cross-sectional area of stirrup; $A_{sp}$ is the cross-sectional area of prestressed steel bundle.

The externally applied shear timescale is obtained based on the vehicle-bridge coupling procedure and finally, the stress timescale of the shear stirrups under train load is found by Equation (4).

## 3 FATIGUE DAMAGE CALCULATION AND LIFE ASSESSMENT

Based on the calculated response of a single train crossing the bridge, the stress histories were counted statistically to obtain the fatigue stress spectrum. A rain flow counting program has been developed in MATLAB for the statistical counting of the stress time response of trains crossing the bridge to obtain the fatigue stress spectrum. After obtaining the stress spectra and the frequency of train passage, the fatigue damage, and the remaining life of the bridge under fatigue train loading during operation can be calculated.

### 3.1  Calculation of fatigue damage

In this paper, using the research results of Zeng and Li (Zeng 1999), this S-N curve is based on tests on plain reinforced concrete beams, the correctness of which has been verified in Zeng's study. The S-N fatigue curve equation is as follows.

$$\begin{cases} \log N = 15.1348 - 4.3827 \log \Delta\sigma \;\; (N < 10^7) \\ \log N = 18.8471 - 6.3827 \log \Delta\sigma \;\; (N > 10^7) \end{cases} \tag{5}$$

This paper follows the P-M linear cumulative damage principle when using the S-N curve for fatigue damage studies, with the following basic equation.

$$D = \sum_{i=1}^{k} \frac{n_i}{N_i} \tag{6}$$

where $D$ is the total fatigue damage level; $n_i$ is the number of cycles of stress amplitude $\Delta\sigma_i$; $N_i$ is the maximum number of cycles that the stress amplitude $\Delta\sigma_i$ can withstand; $k$ is the total number of stress amplitude categories. The extent of damage caused to the tensile reinforcement at the bottom of the span when a single ordinary train, a 10,000-ton train, and a 20,000-ton train cross the bridge can be calculated, as shown in Table 1.

Table 1.  Damage caused by a single train crossing the bridge.

|  | Ordinary marshaling trains | 10,000-ton marshaling trains | 20,000-ton marshaling trains |
|---|---|---|---|
| Mid-span tensile steel bar | $6.25 \times 10^{-8}$ | $1.27 \times 10^{-7}$ | $2.28 \times 10^{-7}$ |
| Shear stirrup at beam end | $4.17 \times 10^{-8}$ | $3.31 \times 10^{-7}$ | $6.23 \times 10^{-7}$ |

The cumulative fatigue damage since the commencement of train operation up to 2018 was calculated based on the extent of damage to the span mid-tension reinforcement and shear stirrup at the beam end by a single train and the frequency of each train formation, the results of which are shown in Table 2 below.

Table 2.  Calculated cumulative fatigue damage of reinforcement.

|  | Ordinary marshaling trains | 10,000-ton marshaling trains | 20,000-ton marshaling trains | Total |
|---|---|---|---|---|
| Mid-span tensile steel bar | $2.40 \times 10^{-2}$ | $1.26 \times 10^{-2}$ | $3.72 \times 10^{-3}$ | $4.03 \times 10^{-2}$ |
| Shear stirrup at beam end | $1.60 \times 10^{-2}$ | $3.36 \times 10^{-2}$ | $1.02 \times 10^{-2}$ | $5.98 \times 10^{-2}$ |

From the fatigue damage calculations in the table above, for the Shuohuang Railway, since the operation, the rate of damage accumulation at the beam ends is faster and the degree of damage is greater than the fatigue damage accumulation in the span, and the fatigue life control factor for the beams is the fatigue life of the shear stirrup.

### 3.2 *Lifetime assessment*

To estimate the service life of a bridge, a forecast of future annual traffic is required, here two working conditions are considered, one assuming no further growth in traffic from 2018 onwards and the second assessing the fatigue life by considering the trend of future traffic growth.

For the first condition, where the annual line traffic is maintained as is and does not increase, the fatigue life can be obtained by the following equation based on the fatigue damage accumulation calculation in Table 2.

$$A_r = \frac{1 - D_1}{D_0} \tag{7}$$

where $A_r$ is the fatigue life (years); $D_1$ is the accumulated fatigue damage to 2018; $D_0$ is the annual damage at current operating levels.

Under these conditions, the mid-span tensile steel bar is expected to have a service life of approximately 170 years and the shear stirrup at the beam end is expected to have a service life of approximately 79 years.

For the second working condition, according to the relevant information, Shuohuang Railway's long-term capacity target is to reach 500 million tons per year. According to the growing trend of the Shuohuang line in previous years, the set capacity is expected to grow at an annual rate of 10% per year, reaching 500 million tons of capacity in 2023. The fatigue damage accumulation from 2019 to 2023 can be calculated by Equation (6), and the results of the fatigue damage calculation for the calendar year during this period are shown in Table 3 below.

Table 3. Fatigue calculations for 2019 to 2023.

| | 2019 | 2020 | 2021 | 2022 | 2023 | Total |
|---|---|---|---|---|---|---|
| Mid-span tensile steel bar | $5.76 \times 10^{-3}$ | $5.79 \times 10^{-3}$ | $5.8 \times 10^{-3}$ | $5.88 \times 10^{-3}$ | $5.93 \times 10^{-3}$ | $2.92 \times 10^{-3}$ |
| Shear stirrup at beam end | $1.28 \times 10^{-2}$ | $1.29 \times 10^{-2}$ | $1.30 \times 10^{-2}$ | $1.31 \times 10^{-2}$ | $1.32 \times 10^{-2}$ | $6.49 \times 10^{-2}$ |

Considering the increase in freight traffic in recent years, the bridge life calculation can be calculated by the following equation.

$$A_r = \frac{1 - D_1 - D_2}{D_0} \tag{8}$$

where $D_2$ =estimated fatigue damage for 2019 to 2023.

Under these conditions, the mid-span tensile steel bar has a further service life of approximately 150 years and the shear stirrup at the beam end has a further service life of approximately 66 years.

### 3.3 *Summary*

(1) The fatigue damage calculation results show that a single 20,000 tons train will cause the most damage to the span center tensile reinforcement and girder end stirrup, followed by the 10,000 tons train and the smallest by the normal train, indicating that the larger the axle weight and the longer the group, the more unfavorable the fatigue performance of the girder.

(2) The life assessment results indicate that the span center tension reinforcement will not suffer fatigue damage during the operational period; whereas the shear stirrup at the girder end has insufficient fatigue resistance to meet the 100-year operational period requirement under both operating conditions, and fatigue damage may occur during the operational period.

# 4 THE EFFECT OF TRACK UNEVENNESS ON THE FATIGUE PERFORMANCE OF BRIDGES

Track irregularity is not constant during the bridge's operating period, and will change after being subjected to wear, corrosion, temperature, uneven settlement, and other factors, while the smoothness of the track is to some extent related to later maintenance work. Track irregularity, as one of the important excitations in the vehicle-bridge coupling system, has a large impact on the response of the girders. To facilitate the analysis and calculation of the effect of track irregularity on fatigue damage, the original track irregularity samples used in the previous section are directly multiplied by different scale factors, changing only the amplitude and not the wavelength.

The fatigue damage caused by a single train and the variation in the remaining life of the bridge for different levels of track irregularity excitation are shown in Table 4, respectively, and the trend is visualized in Figure 2.

Table 4. Calculated fatigue damage caused by a single train under different track irregularity excitation.

| Zoom factor | | 0.8 | 0.9 | 1.0 | 1.1 | 1.2 |
|---|---|---|---|---|---|---|
| Mid-span | Ordinary marshaling | $4.49 \times 10^{-8}$ | $4.47 \times 10^{-8}$ | $6.25 \times 10^{-8}$ | $1.03 \times 10^{-7}$ | $1.64 \times 10^{-7}$ |
| tensile | 10,000-ton marshaling | $5.48 \times 10^{-8}$ | $6.57 \times 10^{-8}$ | $1.27 \times 10^{-7}$ | $2.25 \times 10^{-7}$ | $3.89 \times 10^{-7}$ |
| steel bar | 20,000-ton marshaling | $9.94 \times 10^{-8}$ | $1.20 \times 10^{-7}$ | $2.28 \times 10^{-7}$ | $4.14 \times 10^{-7}$ | $7.17 \times 10^{-7}$ |
| Shear stirrup | Ordinary marshaling | $2.58 \times 10^{-8}$ | $3.25 \times 10^{-8}$ | $4.17 \times 10^{-8}$ | $5.33 \times 10^{-8}$ | $6.74 \times 10^{-8}$ |
| at beam end | 10,000-ton marshaling | $1.94 \times 10^{-7}$ | $2.48 \times 10^{-7}$ | $3.31 \times 10^{-7}$ | $4.38 \times 10^{-7}$ | $5.73 \times 10^{-7}$ |
| | 20,000-ton marshaling | $3.65 \times 10^{-7}$ | $4.68 \times 10^{-7}$ | $6.23 \times 10^{-7}$ | $8.26 \times 10^{-7}$ | $1.08 \times 10^{-6}$ |

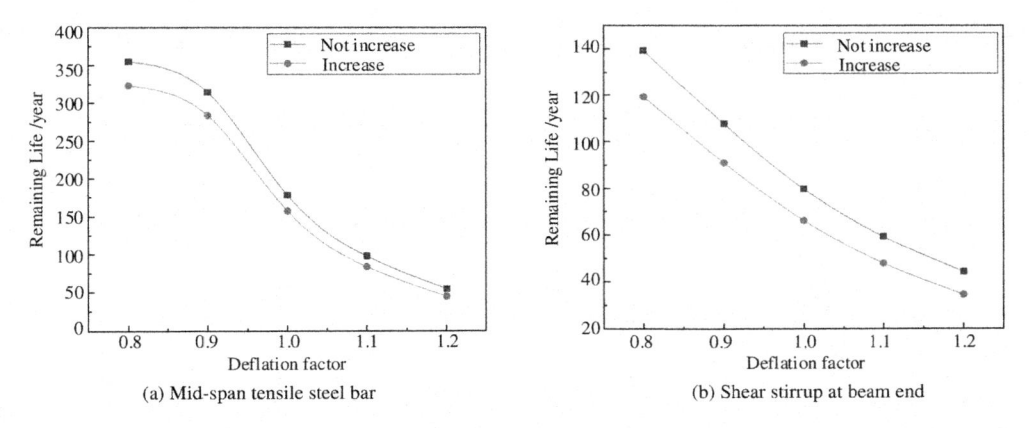

(a) Mid-span tensile steel bar  (b) Shear stirrup at beam end

Figure 2. The fatigue life of reinforcement bars as a function of track unevenness.

Table 4 shows that as the track irregularity gets worse, the damage caused to the tensile and shear reinforcement by the three formations of trains gets worse. The trend in the remaining fatigue life in Figure 2 also shows that the worse the track irregularity, the smaller the fatigue life and the more detrimental to fatigue resistance.

# 5 CONCLUSIONS

This paper analyses the damage caused to the span and girder ends of the bridge when trains cross the bridge and predict future traffic volumes to predict the fatigue service life of the bridge. The influence of the degree of track irregularity on fatigue damage and life is also analyzed, leading to the following conclusions.

(1) The greater the axle weight and the longer the grouping, the more unfavorable the fatigue performance of the beam. With the increase in the frequency of 10,000-ton trains and 20,000-ton trains, the accumulation of fatigue damage each year also shows a significantly accelerated trend.

(2) The mid-span tensile steel bar can meet the 100-year service life, whereas the shear stirrup at the beam end is expected to have a remaining service life of 66 years when future traffic growth is considered, and will not be able to meet the 100-year operational period requirement. The cumulative effect of heavy train operation on the damage at the beam ends is even greater.

(3) As track irregularities increase, fatigue life decreases, reducing fatigue resistance performance. Therefore, railway tracks require regular maintenance to improve fatigue resistance performance.

In this paper, only the fatigue damage and the remaining life of the bridge are studied and predicted, but no reasonable solutions are proposed. Therefore, the effectiveness of different strengthening methods for prestressed concrete girder bridges subjected to fatigue damage will be investigated subsequently and reasonable strengthening solutions will be proposed.

## REFERENCES

Basquin O. The exponential law of endurance test [J]. *Proc ASTM*, 1910, 10.

Che Huimin, He Guanghan. Fatigue testing of partially prestressed concrete slab girders [J]. *Bridge Construction*, 1988 (02): 19–26.

Chen Wan. *Numerical analysis of fatigue damage of bridge structures under heavy traffic* [D]. Hebei University of Technology, 2015.

Fangping Liu, Jianting Zhou. Research on Fatigue Strain and Fatigue Modulus of Concrete [J]. *Advances in Civil Engineering*, 2017, 2017: 1–7.

Feng Xiufeng. *Study on the fatigue performance of partially prestressed concrete beams with mixed reinforcement* [D]. Dalian University of Technology, 2006.

Li Chengming, Ma Chengli. Experimental study on the shear resistance of concrete beams with inclined sections under variable repetitive loads [J]. *Journal of Taiyuan University of Technology*, 1997 (S1): 36–39.

Li Jinzhou, Yu Zhiwu, Song Li. Study on the variation law of the neutral axis of heavily loaded railway bridges under fatigue repetitive loading [J]. *Journal of the China Railway Society*, 2013, 35 (06): 96–10.

Luo S. Y., Chen Y. K., Deng P. Q.. Experimental study on the fatigue performance of unbonded partially prestressed concrete beams [J]. *Journal of Building Structures*, 2007 (03): 98–10.

Luo S. Y., Yu Z. W., Nie J. G., et al. Experimental study on fatigue performance of self-compacting prestressed concrete beams [J]. *Journal of Building Structures*, 2003 (03): 76–81.

Miner MA. Cumulative damage in fatigue [J]. *Journal of Applied Mechanics*, 1945, 12 (3): 159–164.

Mohsen El Shahawi, Barrington deV. Batchelor. Fatigue of Partially Prestressed Concrete [J]. *Journal of Structural Engineering*, 1986, 112 (3).

Naaman, A E, Founas, M. Partially prestressed beams under random-amplitude fatigue loading [J]. *Journal of structural engineering New York*, N. Y., 1991, 117 (12): 3742–3761.

Wen Yusong, Zhang Qi. Fatigue damage of diagonal bars and hoop bars in concrete beams under the action of 25 T axle weight trains [J]. *Journal of Railway Science and Engineering*, 1997 (03): 15–20.

Xiao Weiqiang, Cui Lei. Experimental study on shear fatigue damage of reinforced concrete beams [J]. *Low Temperature Architecture Technology*, 2018, 40 (07): 31–36.

Yang DZ, Pu Qianhui, Yang Yq. Analysis and calculation of hoop behavior under fatigue loading [J]. *Journal of Southwest Jiaotong University*, 1994 (04): 391–397.

Zeng Pan. *A tutorial on the fundamentals of finite element analysis* [M]. Beijing: Higher Education Press, 2008.

Zeng Z-B, Li Z-Rong. Study of fatigue S-N curves of steel bars for ordinary concrete beams [J]. *China Civil Engineering Journal*, 1999 (05): 10–14.

*Civil Engineering and Urban Research – Mohamed & Hou (Eds)*
*© 2023 the Authors, ISBN 978-1-032-44487-1*

# Finite element analysis of fatigue performance of steel-concrete composite beams

Yong Ma, Gesang Yundan & Shirap Tenzin
*The Highway Construction Project Management Center, Transport Department of the Tibet Autonomous Region, Lhasa, Tibet, China*

Jiacheng Zhao & Li Zhu*
*School of Civil Engineering, Beijing Jiaotong University, Beijing, China*

Jinyan Si
*Beijing Municipal Engineering Research Institute, Beijing, China*

Yaoyu Zhu
*CCCC Highway Bridges National Engineering Research Centre Co. Ltd., Beijing, China*

ABSTRACT: In high-cold and high-altitude areas, both freeze-thaw cycles and chloride ions ($Cl^-$) in de-icing salts can adversely affect steel-concrete composite bridges. Therefore, the service performance of steel-concrete composite bridges under the coupling effect of freeze-thaw cycles and chloride ion corrosion has become one of the hot issues in the research of bridge structures in alpine and high-altitude regions. For the fatigue damage of steel-concrete composite beams, this paper investigates the problem by comparing numerical simulations with experiments. The final simulation results are in good agreement with the measured values and the following conclusions were obtained. The simulated stress amplitude ratios and simulated load differences for the specimens subjected to freeze-thaw cycles were smaller than those for the specimens not subjected to freeze-thaw cycles. The fatigue and static load properties of the steel-concrete composite beams are reduced after freeze-thaw testing. The simulated effect of cast-in-place slab composite beams is better than that of precast slab composite beams after freeze-thaw cycles.

## 1 INTRODUCTION

At present, the research on steel-concrete composite beams mainly lies in static performance. The research results are relatively remarkable and have been widely used in engineering practice. However, the fatigue problem of steel-concrete composite beams has not caused a large impact due to the short service time. The damage caused by fatigue loading on the composite beams has not been fully manifested. Therefore, the research on the fatigue mechanical properties of steel-concrete composite beams is still insufficient. And the fatigue problems of various forms of composite beams and those considering the fusion of various operating conditions (corrosion, freeze-thaw) have not been adequately studied. In conclusion, the fatigue problem of composite structures still needs further research (Nie 2012).

With the widespread use of steel-concrete composite structures, it has gradually become apparent that the forms of damage to steel-concrete composite beams are not only the shear damage of the studs but also the crushing of the concrete flange slab and the tensile fracture damage of the lower flange of the steel beam. Researchers have come to realize that fatigue problems also exist in the

---

*Corresponding Author: zhuli@bjtu.edu.cn

DOI 10.1201/9781003372417-7

steel beam tensioned flanges and concrete slabs in composite beams. The stress state of the steel beam also affects the fatigue performance of the studs (Fang 2007). Therefore, the fatigue problem of the composite beam structure should not only consider the force performance of the studs but should pay more attention to the overall force structure, from the concrete, steel beams, and shear connectors together.

Pedro Albrecht et al. (1995) derived the relationship between fatigue life and stress amplitude of the steel beam tensioned flange by fatigue tests on prestressed steel-concrete composite beams. Youn et al. (Youn 1998) determined the shear and fatigue strengths at different load loading positions out of static and fatigue tests on steel-concrete composite beams and fitted a relationship between fatigue life and interface shear based on the test results. Nie et al. (Nie 2012) summarized the research on the fatigue performance of steel-concrete composite beams, pointing out that the damage of steel-concrete composite beams is divided into two forms: the shear damage of the studs and the fracture damage of the lower flange of the steel beam. At the same time, he made a comparative analysis of the fatigue life formulae proposed by various countries for combined beams. Xinyuan Luo (Luo 2020) analyzed the deflection, slip, and shear force variation of studs by static load and fatigue tests of steel-concrete composite beams, and performed ABAQUS numerical simulations of the static load performance of the composite beams. The simulation results are in general agreement with the actual results. Guanling Ma (Ma 2021) analyzed the spanwise deflection and interfacial slip of a composite steel-concrete beam considering the effects of chloride salt corrosion by fatigue testing and ABAQUS modeling, with a high agreement between the test and simulation.

According to the above research on the fatigue performance of steel-concrete composite beams by domestic and foreign scholars, we can see that: the fatigue research of steel-concrete composite beams has begun to be combined with numerical simulation, and it is more inclined to derive the relationship formula with universal significance based on discrete test results, and it also saves more time and effort for the fatigue research of composite beams. However, due to the complexity of the fatigue test of composite beams, the numerical simulation needs to be further developed.

In this paper, we will use ABAQUS to establish solid cells to study the fatigue performance of steel-concrete composite beams and simulate the steel-concrete composite beams that have undergone corrosion test, freeze-thaw cycle test, and fatigue test. And the fatigue performance of the steel-concrete composite beam is further investigated by adjusting the parameters of the steel beam and concrete slab to simulate the effects of different working conditions.

## 2 CONSTRUCTION OF ABAQUS MODEL

The finite element model in this paper is designed as 6 pieces of I-beam-concrete combination beams, numbered SCB-1 to SCB-6 respectively. The fatigue tests of specimens SCB-1 and SCB-2 and SCB-3 are presented in the research of Luo (2020) and Ma (2021), and the test data of our group are used for specimens SCB-4 to SCB-6. The structural forms of the composite beams designed in this paper are divided into two types: cast-in-place concrete slab composite beams with a uniform arrangement of studs and precast concrete slab composite beams with a clustered arrangement of studs. All steel-concrete composite beam specimens met the design requirements for a fully shear connection. Figure 1 gives a schematic layout of the composite beam specimen. The steel beams are made of ordinary steel or weathering steel, and the parameters of the six specimens are shown in Table 1.

The steel-concrete composite beam consists of a steel beam, concrete slab, and studs, with the concrete slab containing the steel reinforcement. Because the focus of this paper is on the overall fatigue life and the overall force performance of the composite beam, the main components are modeled in solid cells. The structural reinforcement in the concrete wing slab is modeled by the beam cell model, which will improve the computational speed of the model and make the contact relations and later meshing simpler so that the convergence speed of the model increases rapidly. The overall model after assembly is shown in Figure 2.

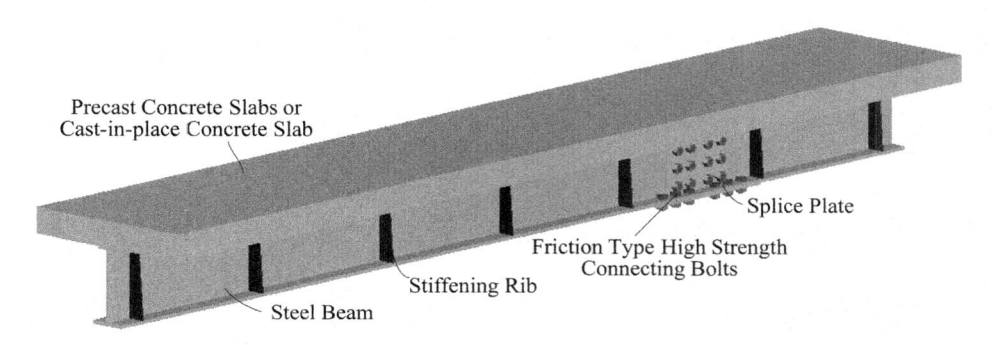

Figure 1. Schematic layout of the specimens.

Table 1. Specimen parameters.

| Number | Type of concrete slab | Arrangement of studs | Number of studs | Type of steel | Corrosion | Freeze-thaw |
|---|---|---|---|---|---|---|
| SCB-1 | Cast-in-place slab | Uniform distribution | 72 | Common steel | N | N |
| SCB-2 | Cast-in-place slab | Uniform distribution | 72 | Common steel | Y | N |
| SCB-3 | Precast slab | Cluster | 78 | Common steel | Y | N |
| SCB-4 | Cast-in-place slab | Uniform distribution | 72 | Common steel | Y | Y |
| SCB-5 | Cast-in-place slab | Uniform distribution | 72 | Weathering steel | Y | Y |
| SCB-6 | Precast slab | Cluster | 78 | Weathering steel | Y | Y |

For the established steel-concrete composite beam model, three different types of interaction constraints were used, the details of which are shown in Table 2.

Table 2. Contact relation of finite element model.

| Number | Contact relation pair | Interaction |
|---|---|---|
| 1 | The top surface of the upper flange of the steel beam and the bottom surface of the concrete slab | Surface-to-surface contact |
| 2 | The bottom surface of the upper flange plate of steel beam and the top surface of the web of intermediate steel beam | Tie |
| 3 | The top surface of the lower flange plate of steel beam and bottom surface of the web of intermediate steel beam | Tie |
| 4 | The contact surface between steel beam stiffening ribs and steel beam | Tie |
| 5 | The bottom surface of studs and the top surface of the upper flange of steel beam | Tie |
| 6 | The contact surface between the side of the stud and the concrete slab | Embedded region or surface-to-surface contact |
| 7 | Structural reinforcement and concrete slab | Embedded region |

## 3 FATIGUE STRESS AMPLITUDE SIMULATION COMPARISON

After the model is established, the lower and upper limits of the single fatigue load are applied to the model to obtain the stress values at the midspan section of the steel-concrete composite beam. Fatigue load loading is divided into two stages. In the first stage, the properties and strain conditions of the material are indexed to the initial state when calculating the stress range in the mid-span section of the composite beam. In the second stage, the steel-concrete composite beam has been

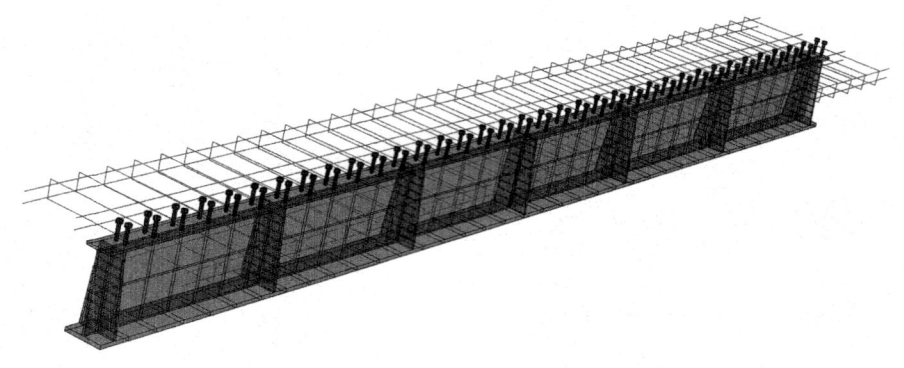

(a) Finite element model of cast-in-place concrete slab composite beam

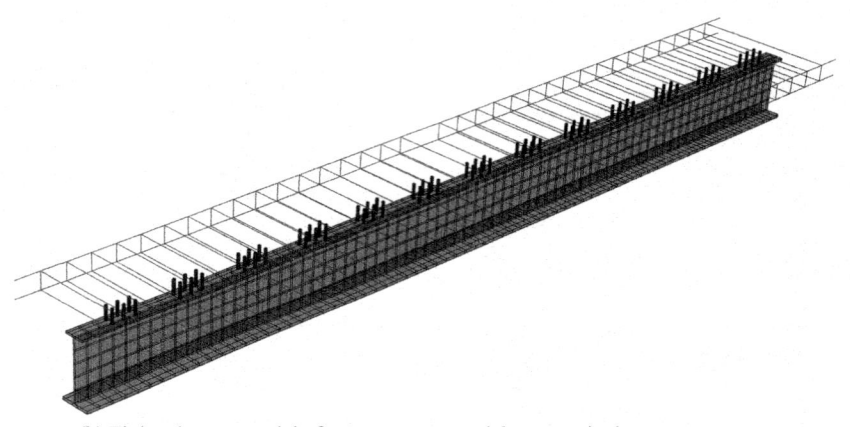

(b) Finite element model of precast concrete slab composite beam

Figure 2.    Finite element model of steel-concrete composite beam.

loaded with 2 million fatigue loads, and the internal fatigue damage will affect the stress-strain relationship of the material. At this point, the property state of the material needs to be corrected based on the strain gauge data of the mid-span section at the time of the test to ensure the accuracy of the simulation. After the submission job for the finite element model, the corresponding stress results are viewed in the visualization section, where the stress clouds of the specimens are shown in Figure 3 for the typical specimen precast concrete slab combination beam SCB-3 models.

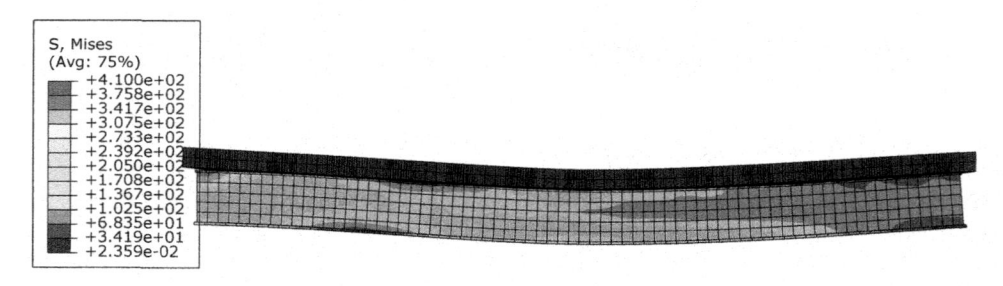

Figure 3.    Stress nephogram of specimen SCB-3.

It is well known that the magnitude of the fatigue stress amplitude determines the fatigue life of the steel-concrete composite beam specimen. From Table 3, we can see that the first stage ratios of SCB-1, SCB-3, and SCB-4 are 0.955, 0.934, and 0.932, and the second stage ratios are 0.915, 0.911, and 0.874. The simulation of the first stage stress amplitude of the actual test by the finite element model is more accurate than the second stage. This is due to the increase of the upper fatigue load limit in the second stage, which increases its fatigue stress amplitude and accelerates the fatigue damage of the actual specimen, causing the plastic strain of the overall structure to increase sharply, thus making the accuracy of the simulation results less accurate. It is also found that the ratio of simulated values to tested values of specimens SCB-4, SCB-5, and SCB-6 are smaller than those of SCB-1, SCB-2, and SCB-3. It was also found that the ratio of simulated to tested values of specimens SCB-4, SCB-5, and SCB-6 was smaller than that of SCB-1, SCB-2, and SCB-3. This indicates that the simulation of the overall structure of the steel-concrete composite beam after freeze-thaw cycles is not enough to rely on the values of the material properties test alone, and further research is needed. It also indicates that the performance of the steel-concrete composite beams will degrade and become more complex after freeze-thaw cycles, increasing the simulation error of the stress amplitude.

Table 3. Stress amplitude of finite element model and actual specimen.

| Number | First stage fatigue load | | Ratio | Second-stage fatigue loading | | Ratio |
| | Simulated value (MPa) | Test value (MPa) | Simulation test | Simulated value (MPa) | Test value (MPa) | Simulation test |
| --- | --- | --- | --- | --- | --- | --- |
| SCB-1 | 77.71 | 81.37 | 0.955 | 92.17 | 100.73 | 0.915 |
| SCB-2 | 78.22 | 83.66 | 0.935 | – | – | – |
| SCB-3 | 78.31 | 83.85 | 0.934 | 94.02 | 103.21 | 0.911 |
| SCB-4 | 82.14 | 88.16 | 0.932 | 131.38 | 150.29 | 0.874 |
| SCB-5 | 86.69 | 95.79 | 0.905 | – | – | – |
| SCB-6 | 93.50 | 105.65 | 0.885 | – | – | – |

## 4   LOAD-DEFLECTION CURVE COMPARISON

The upper limit of fatigue load was loaded onto the established finite element model and entered the post-processing program to extract the displacement variation at the midspan and compare it with the deflection of the actual test specimen, as shown in Figure 4. This part simulates the force performance of the combined beam under static load. The fatigue loading stage is divided into two parts, and the comparison is based on the actual loading conditions of the specimens. The first stage of the comparison of the static load performance is selected from the initial state of the steel-concrete composite beam, while the second stage of the comparison after 2 million fatigue loadings is selected from the state of the steel-concrete composite beam before the damage.

From Figure 4, it can be found that the load-deflection curves simulated by the finite element model and the actual load-deflection curves of the specimens generally agree well. The overall stiffness of the finite element model is slightly greater than the measured results. This is due to the initial defects in the specimens and the residual stresses during welding that leads to changes in the overall structural properties of the composite beam, resulting in a reduction in the stiffness of the composite beam when subjected to load, which leads to increase in deflection. From Figures 4 (a), (c), and (d), it can be seen that the load loading in the second stage will make the deflection increase rapidly, and the fatigue damage accumulated by the 2 million fatigue loads will appear in the second stage, resulting in a rapid decrease in the stiffness of the composite beam and increase in the difference between the finite element simulations. It is also found that the simulated differences of specimens SCB-4, SCB-5, and SCB-6 are larger than those of specimens SCB-1, SCB-2, and SCB-3, which further indicates that the rate of decrease of the overall structural stiffness will be accelerated and the deflection will increase rapidly after the steel-concrete composite beam undergoes freeze-thaw cycles under the same load. From Figures 4 (e) and (f), we can see that the simulated difference of precast concrete slab composite beam (SCB-6) is larger than that of cast-in-place concrete slab composite beam (SCB-5). On the one hand, it is because after the freeze-thaw cycle, the strength of the precast concrete slab decreases, which makes the grip on the studs decrease, resulting in a decrease in the connection performance of the composite beam, making its composite beam less stiff than the cast-in-place slab composite beam, which in turn leads to an increase in deflection. On the other hand, the constraint relationship between the studs and the concrete slab in the finite element model is surface-to-surface contact, and the constraint relationship is only tangential and normal, which is different from the actual stud constraint relationship, so it leads to the increase of the simulation difference.

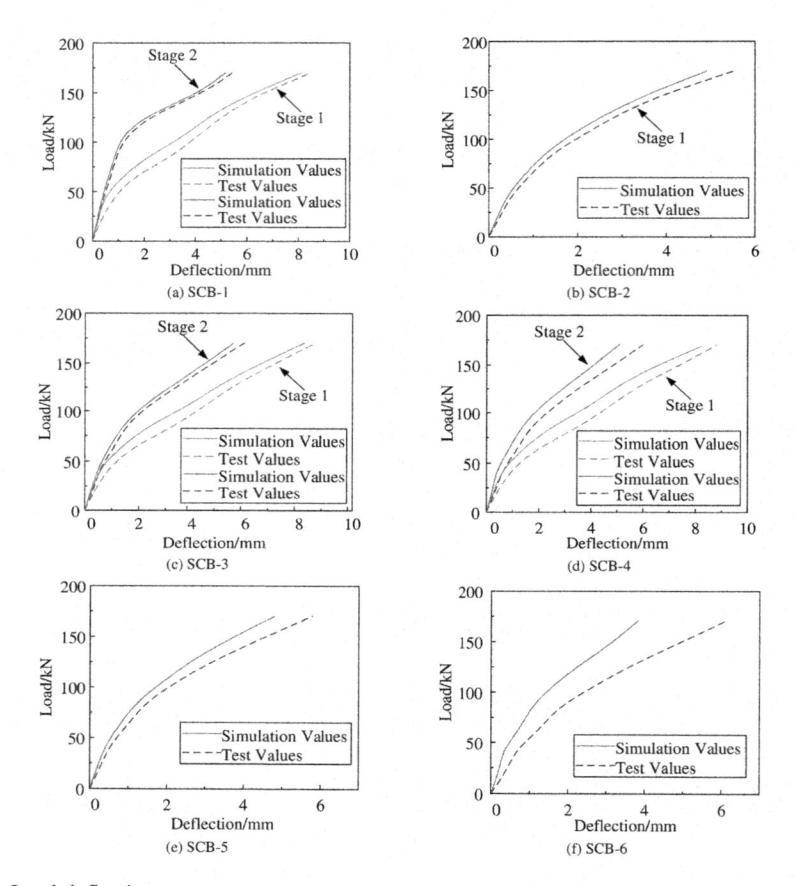

Figure 4. Load-deflection curves.

## 5 CONCLUSIONS

In this paper, we mainly used ABAQUS to simulate the composite beam specimens and conducted a comparative analysis of their fatigue performance. The following conclusions were mainly drawn.

(1) For steel-concrete composite beams, the fatigue loading in the second stage will have a greater impact on the composite beam specimens. Compared to the first stage, the steel-concrete composite beam in the second stage has reduced stiffness and increased deflection and slip.

This leads to an increase in the difference between measured and simulated values in the second stage of ABAQUS.

(2) The simulation results of the ABAQUS finite element model show that the fatigue performance and static load performance of the steel-concrete composite beam will be reduced after the freeze-thaw cycle test, and the deformation of the structure will be more complicated after the force is applied, which leads to an increase in the simulation error of the stress amplitude.

(3) After the freeze-thaw cycle, the simulated difference of the precast slab composite beam is larger than that of the cast-in-place slab composite beam, which is mainly related to the decrease of the concrete properties after the freeze-thaw cycle, i.e., the decrease of the bonding force of the concrete to the connection interface and the decrease of the grip force to the studs.

## ACKNOWLEDGMENT

The authors are very grateful for the financial support from the Science and Technology Project of Transportation of Tibet Autonomous Region. This work is under the project - Study on Key Technology of Construction of Weathering Steel-Concrete Composite Girder Bridges with Cast-In-Situ Deck in Tibet.

## REFERENCES

Fang Xing, Zhou Qihui. Fatigue assessment method for pinned weld toe in negative moment zone of combined beams [J]. *World Bridges*, 2007 (2): 75–77.

Luo Xinyuan. *Fatigue analysis and experimental study of steel-hybrid composite beams* [D]. Beijing Jiaotong University, 2020.

Ma Guanling. *Study on the fatigue performance of steel-concrete composite beams with different stud arrangement* [D]. Beijing Jiaotong University, 2021.

Nie JG, Wang YH. A review of fatigue performance of steel-concrete composite beams [J]. *Engineering Mechanics*, 2012, (06): 1–11.

Pedro Albrecht, Li Wulin, Hamid Saadatmanesh. Fatigue strength of prestressed composite steel-concrete beams [J]. *Journal of Structural Engineering*, 1995 (12): 1850–1856.

Youn Seok-Goo, Chang Sung-Pil. Behavior of composite bridge decks subjected to static and fatigue loading [J]. *ACI Structural Journal*, 1998, 95 (3): 249–258.

*Civil Engineering and Urban Research – Mohamed & Hou (Eds)*
*© 2023 the Authors, ISBN 978-1-032-44487-1*

# Study on low-temperature performance design parameters of highway asphalt binding materials in Liaoning Province

Li Song*
*Shenyang City Construction College, Shenyang, Liaoning, China*

ABSTRACT: According to the low-temperature cracking prediction model given in the current specification, the low-temperature cracking of asphalt pavement is controlled by controlling the design parameters of the asphalt binder. According to the low-temperature cracking model of asphalt pavement proposed in the current specification, the creep stiffness of asphalt mixtures in five climatic zones is investigated. The results show that temperature variables have a greater influence on the low-temperature cracking of asphalt pavement, and shows that the higher the highway grade, the lower the creep stiffness of asphalt. Therefore, in the design of highway asphalt pavement in seasonally frozen soil areas, the low-temperature cracking of asphalt pavement can be effectively controlled by selecting an asphalt binder whose creep stiffness meets the requirements of local climate, soil foundation type, and pavement grade design.

## 1 INTRODUCTION

Low-temperature cracking of asphalt pavement is one of the main problems that perplex the road engineering field at home and abroad. After temperature cracks appear on pavement, water will enter the pavement structure through cracks (Li 2010). Once water enters pavement surface cracks, it will stay inside the pavement structure, weakening the strength of the base and subgrade, accelerating road damage, and reducing driving quality (Li 2011).

As the main component of asphalt mixture, asphalt has a great influence on the low-temperature performance of asphalt mixture (Cao 2016). Therefore, the selection and evaluation basis of asphalt is very important. Zhang (2014) considered that the low-temperature crack resistance of asphalt material played a decisive role in the low-temperature shrinkage and fatigue cracking of asphalt mixture. Thus, the temperature crack of asphalt pavement can be reduced by improving the low-temperature crack resistance of asphalt material. The penetration degree, ductility, and wax content of asphalt are all related to low-temperature shrinkage cracking, but the creep stiffness of asphalt is the most fundamental factor to determine whether asphalt pavement cracks.

The low-temperature cracking prediction model is added in the current specification *Code for Design of Highway Asphalt Pavement* (JTG D50-2017), but the low-temperature cracking prediction model does not provide the sensitivity analysis of various factors affecting the low-temperature cracking of asphalt pavement, and the design requirements for asphalt creep stiffness are not clear. To solve the above problems, the creep stiffness of asphalt pavements at all levels of highways in five climatic zones of Liaoning Province is studied and analyzed, and the corresponding creep stiffness of asphalts is calculated, which can guide the selection of asphalt pavements at all levels of ordinary asphalt pavements in Liaoning Province.

---

*Corresponding Author: 117005312@qq.com

 DOI 10.1201/9781003372417-8

## 2 LIAONING PROVINCE CLIMATIC DIVISION

As a kind of linear structure, the road will inevitably pass-through regions with different climatic characteristics. To prolong the service life of pavement, the design of low-temperature performance of asphalt mixture must be adapted to the regional climate environment through which the road passes (Zhang 2014). Therefore, the climate division of Liaoning Province is carried out first. After the K-means clustering algorithm is used for the asphalt pavement climate partitioning, the partitioning is carried out according to different partitioning indexes, partitioning schemes, and partitioning methods. The process of the K-means clustering algorithm is described as follows:

1) Randomly select $k$ data objects as the center of mass of the initial cluster;

According to the distance between the object and the centroid of each cluster, allocating the object to the cluster represented by the centroid nearest to the object;

2) Recalculating the average value of data objects of each cluster, i.e., the center of mass;

$$\bar{x}_i = \sum_{x \in C_i} \frac{x}{|C_i|} \tag{1}$$

where $\bar{x}_i$ is the mean value of the cluster $C_i$, and $x$ is a point in space

3) If the center of mass of each cluster is not changed anymore or the specified convergence criterion is satisfied, the partition result is returned. Otherwise, go to step 2 to repeat the iteration.

The serviceability of asphalt concrete pavement in Liaoning Province is divided into 5 zones: Zone I—southeast near river sea and low mountain hilly area; Zone II—east low temperature and rain mountainous area; Zone III—middle river plain area; Zone IV—west dry hilly area; Zone V—south coastal hilly area with rain. The distribution of temperature index values in each region is shown in Table 1.

Table 1. Temperature index (°C).

| Index | $t_{I_{max}}$ | $t_{I_{min}}$ | $t_{II_{max}}$ | $t_{II_{min}}$ | $t_{III_{max}}$ | $t_{III_{min}}$ | $t_{IV_{max}}$ | $t_{IV_{min}}$ | $t_{V_{max}}$ | $t_{V_{min}}$ |
|---|---|---|---|---|---|---|---|---|---|---|
| Max. | 37.6 | −18.4 | 39.0 | −31.6 | 37.9 | −25.6 | 41.3 | −25.5 | 39.3 | −18.0 |
| Min. | 33.4 | −38.3 | 36.4 | −43.4 | 35.9 | −35.4 | 37.2 | −36.3 | 33.5 | −31.9 |
| Average | 35.5 | −28.4 | 37.7 | −37.5 | 36.9 | −30.5 | 39.3 | −30.9 | 36.4 | −25.0 |

## 3 SOIL FOUNDATION TYPES IN DIFFERENT NATURAL AREAS OF LIAONING PROVINCE

The results show that the low-temperature cracking of asphalt pavement is related to the type of soil foundation and humidity. Therefore, the properties and distribution of subgrade soil in different natural climate areas in Liaoning Province are analyzed.

The whole area of Liaoning Province can be divided into three parts: the eastern hilly hills, the western hilly hills, and the central plain. Hilly mountains are arranged on both sides of the east and west, inclined to the central plain, and the overall terrain is high in the east and west, and low in the middle. The north is high and the south is low, tilting from the land to the ocean. Therefore, there are great differences in engineering geological conditions in Liaoning Province, and there are obvious differences in geomorphological features, rock and soil types, soil strength, and moisture coefficient. To maintain consistency with the above climatic zones, the main distribution types, and characteristics of subgrade soil in each zone are introduced as follows according to the above climatic zones. Zone I: subgrade soil shall be considered as silty clay and sandy soil; Zone II: subgrade soil shall be considered as sandy soil; Zone III: subgrade soil is considered as sandy soil and silty clay; Zone IV: subgrade soil is considered as sandy soil; Zone V: subgrade soil is considered as clay and silty clay.

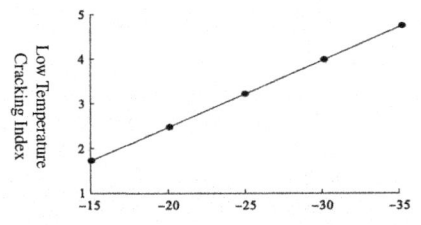
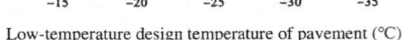
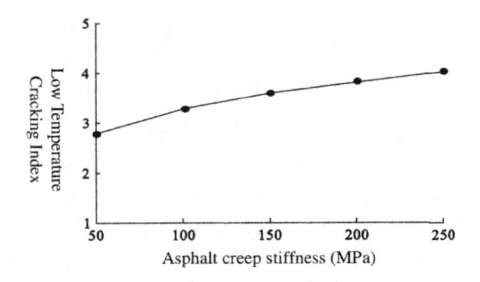

Figure 1. Influence of low-temperature design temperature of pavement on CI.

Figure 2. Influence of asphalt creep stiffness on CI.

## 4 SENSITIVITY ANALYSIS OF LOW TEMPERATURE CRACKING INDEX OF ASPHALT PAVEMENT

### 4.1 Cracking model

There are many factors affecting low-temperature cracking of asphalt pavement. According to the low temperature cracking prediction model given in the current specification *Code for Design of Highway Asphalt Pavement* (JTG D50-2017), it is known that the design temperature of pavement at low temperature, creep stiffness of asphalt, the thickness of asphalt mixture, and subgrade type all play an important role in the low-temperature cracking of asphalt pavement. The sensitivity analysis of the factors affecting low-temperature cracking of asphalt pavement is carried out to provide a reference for pavement design.

The low-temperature cracking prediction model of asphalt pavement is shown in Equation (2).

$$CI = 1.95 \times 10^{-3} S_t \lg b - 0.075(t + 0.07h_a) \lg S_t + 0.15 \tag{2}$$

where

$CI$ is the low-temperature cracking index of asphalt pavement;

$t$ is the designed low-temperature of pavement, which is the average value of annual minimum temperature for 10 consecutive years (°C);

$S_t$ is the creep stiffness of the surface layer asphalt bending beam under the condition of low-temperature design temperature plus 10°C test temperature, and the rheological test load of 180 seconds (MPa);

$h_a$ is the thickness of the asphalt mixture layer (mm);

$b$ is the subgrade type parameter, e.g., b(sand) = 5, b (silty clay) = 3, and b(clay) = 2.

### 4.2 Sensitivity analysis

To analyze the sensitivity of each factor to low-temperature cracking of asphalt pavement, five values of each calculation parameter except the subgrade soil type were determined respectively. When analyzing the influence of one parameter, the other three parameters remain unchanged. Draw the influence curve of the change of each parameter on the change of CI, and the results are shown in Figure 1–Figure 4.

From Figure 1 to Figure 4, we can draw the following conclusions:

1) Under the same other conditions, the lower the low-temperature design temperature of pavement, the higher the low-temperature cracking index CI, and the more the temperature shrinkage cracks of asphalt pavement. The CI increases by 0.15 for every 1°C decrease in temperature. The CI is linearly related to the design temperature $t$ of the pavement low temperature. The temperature change has a great influence on the low-temperature cracking index, while the CI has a high-temperature sensitivity.

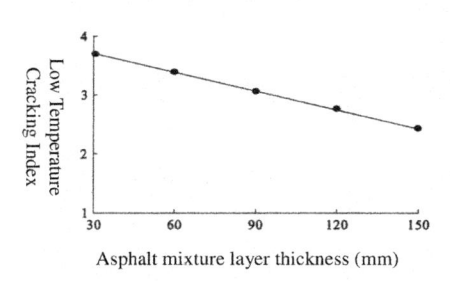

Asphalt mixture layer thickness (mm)

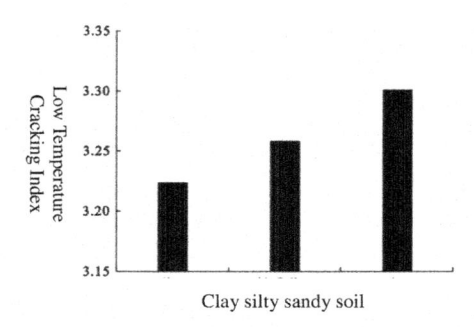

Clay silty sandy soil

Figure 3. Influence of asphalt mixture layer thickness on CI.

Figure 4. Influence of different types of subgrade soil on CI.

2) Under the same other conditions, the creep stiffness of asphalt increases, and the low-temperature cracking index CI increases. With the increase of creep stiffness of asphalt, the increased speed of CI gradually slows down. The relationship between CI and asphalt creep stiffness $St$ is nonlinear. Compared with pavement design temperature variation, the low temperature cracking index is affected by the low temperature cracking index, and the increase of asphalt creep stiffness is slightly slower.

3) Under the same other conditions, the thinner the asphalt surface, the larger the low temperature cracking index CI, and the more the temperature shrinkage cracks of asphalt pavement. The CI decreases by 0.32 every time the thickness of the asphalt mixture layer increases by 30 mm. The relationship between CI and the thickness ha of the asphalt mixture is linearly negative. The thickness of the asphalt mixture has a great influence on the low-temperature cracking index, and the sensitivity to the low-temperature cracking index CI is high.

4) Under the same other conditions, the low temperature cracking index CI of the asphalt pavement of clay subgrade is the lowest, followed by the asphalt pavement of silty clay subgrade, and the highest low temperature cracking index is the asphalt pavement of sandy subgrade. As shown in Figure 4, the change of subgrade soil type has little influence on low temperature cracking index CI, and the sensitivity of CI to subgrade type is low.

## 5 STUDY ON CREEP STIFFNESS OF ASPHALT BASED ON LOW-TEMPERATURE CRACKING MODEL

According to the current specification and the temperature parameters and soil foundation type parameters of five natural climatic zones in Liaoning Province, the stiffness modulus of asphalt is inversely calculated by using the low temperature cracking prediction model given in Equation (2) and the requirements of low temperature cracking index of different grades of highway given.

Table 2. Low-temperature cracking index requirements.

| Road hierarchy | Low temperature cracking index CI |
| --- | --- |
| First-class highway | $\leq 3$ |
| Second-class highway | $\leq 5$ |
| Third- and fourth-class highways | $\leq 7$ |

According to the design scope of asphalt pavement thickness of different grades of highways in Liaoning Province, the thickness of asphalt mixture of expressways and first-class highways is considered 15–20 cm, and the thickness of asphalt mixture of second-class highways is considered as 7–15 cm, and the thickness of asphalt mixture of third-class and fourth-class highways is considered as 3–7 cm.

The range of creep stiffness of asphalt to be reached by Equation (2) can be calculated. Figure 5 shows the variation of creep stiffness requirements of asphalt in different sub-areas and different grade highways in Liaoning Province drawn from the data in the above table.

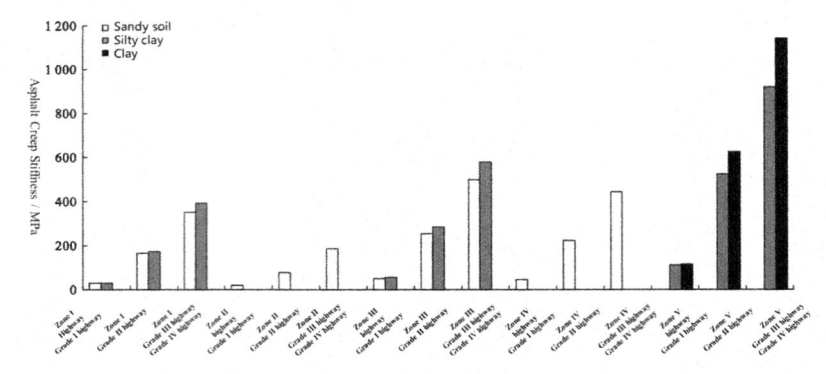

Figure 5. Variation law of asphalt creep modulus of highway asphalt pavement in different districts in Liaoning province.

## 6 CONCLUSIONS

1) The temperature variable has a great influence on the low-temperature cracking of asphalt pavement. The lower the temperature, the lower the asphalt pavement, and the lower the asphalt creep stiffness, and the decreasing speed of asphalt creep stiffness also decreases with the lower temperature. The creep stiffness of asphalt decreases by 44.97 MPa for every 1°C decrease in temperature.
2) High-class highways perform better at low temperatures, resulting in lower creep stiffness. Asphalt used on high-speed and first-class highways has a creep stiffness of 21.7%, while asphalt used on second-class highways has a creep stiffness of 46.3%.
3) The requirement of low-temperature performance of asphalt in sandy soil subgrade is higher than that of silty clay subgrade, and the requirement of low-temperature performance of asphalt in silty clay subgrade is higher than that of clay subgrade.

## REFERENCES

Cao Yu. *Research on low temperature cracking resistance of asphalt mixture* [D]. Chongqing: Chongqing Jiaotong University, 2016.
Chen Junqing. *Research on application technology of asphalt pavement based on low temperature performance* [D]. Xi'an: Chang'an University, 2013.
Gu Mingda. *Research on the low temperature performance evaluation method for asphalt and asphalt mixtures* [D]. Ji'nan: Shandong Jianzhu University, 2017.
Li Xiaojuan, Han Sen, Jia Zhiqing, et al. Low temperature bending creep test of asphalt mixture based on cracking resistance [J]. *Journal of Guangxi University*, 2011, 36 (1): 142–146.
Li Xiaojuan, Han Sen, Li Yuan, et al. Experimental study on low temperature crack resistance index of asphalt binder [J]. *Journal of Wuhan University of Technology* 2010, 32 (7): 81–84.
Li Xinjun, Marasteanu mo, Andreak, et al. Factors study in low -temperature fracture resistance of asphalt concrete [J]. *Journal of Materials in Civil Engineering*, 2010, 22 (22): 145–152.
Yao Zukang. *Asphalt pavement structure design* [M]. Beijing: People Communications Press, 2011.
Zhang Minjiang, Dong Shi. Research on comprehensive zoning technology of service performance of highway asphalt concrete pavement [J]. *Highway*, 2014, 8: 85–91.

*Civil Engineering and Urban Research – Mohamed & Hou (Eds)*
*© 2023 the Authors, ISBN 978-1-032-44487-1*

# Experimental study on bolted-cover plate connections under bending for modular steel buildings

Chuchao Ren*
*Guangdong Polytechnic of Water Resources and Electric Engineering, Guangdong, China*

Ke Cao* & Siyuan Zhai*
*Chongqing University, Chongqing, China*

ABSTRACT:   The module-to-module connections are critical to ensure the overall stress for modular steel buildings (MSBs). Using fully bolted assembly can better improve the efficiency of on-site construction. The bolted-cover plate connection was proposed to realize rapid splicing between multiple modules. Moreover, the connection with a dual-slot hole was designed to improve the problem that multi-module splicing is very sensitive to installation error. To proceed, two full-scale connections were performed for bending behavior. The test results presented failure mode, stiffness properties, ductile performance, and strain distribution of the specimens. The results indicated that the proposed bolted-cover plate connection has excellent bending behavior. The connection's stiffness is classified as semi-rigid. The design of dual-slot bolt holes hardly affects the connection's static performance and improves the construction error sensitivity of multi-module splicing.

## 1   INTRODUCTION

Modular steel building (MSB) is the construction form of prefabricated modular units on the construction site. The module unit is made in the factory, which can be a fully decorated room unit with wall panels, hydropower, and other systems (Ferdousa et al. 2019). MSB has outstanding advantages such as rapid construction, reducing carbon emissions, and safe construction, and it has been constructed in many regions (Srisangeerthanan et al. 2020). MSB is an integral building composed of multiple modules. Therefore, the reliable connection between modules is the key to ensuring the structure's integrity (Lacey et al. 2020). Various structural forms of connection between modules have been proposed, including plug-in self-lock connection type (Dai et al. 2018), rotary inter-module connection type (Chen et al. 2019), vertical post-tensioned connection type (Sanches et al. 2018), post-tensioned bolted steel connection type (Lacey et al. 2019), bolted-plug device connection type (Chen et al. 2017), etc. Accordingly, the mechanical performance of the above connections has been studied and evaluated.

Nevertheless, some problems in the construction form of connections proposed in the existing research, include low field splicing efficiency, difficulty in multi-module combination, and insufficient bending stiffness of connections. Additionally, due to uncertainties in the construction site and splicing under non-ideal conditions, modules may not be spliced due to installation errors. The installation errors may come from the initial defects of components, on-site assembly deviation, or stress deformation between modules, which are challenging to eliminate. Consequently, it is necessary to improve the error sensitivity of module installation. Figure 1 presents the details of the proposed connection. The connection between modules is completed through high-strength

---

*Corresponding Authors: 100828@gdsdxy.cn, caoke@cqu.edu.cn and 20191602034t@cqu.edu.cn

DOI 10.1201/9781003372417-9

bolts and cover plates. Stiffening ribs are arranged between the module column and the beam at the welding position. Two tests were conducted to analyze the bending behavior of the bolted-cover plate connection. Specifically, the bolt hole shape in the specimens is divided into two types: round and slotted. The research results could serve as an effective reference for the inter-module bolted connection design of MSBs.

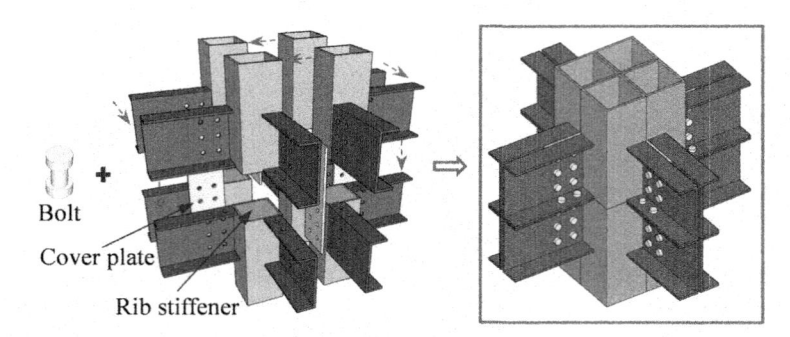

Figure 1.   Conceiving the bolted-cover plate connection for MSBs.

## 2   EXPERIMENTAL PROGRAM

### 2.1   *Details of the specimen*

Two full-scale specimens were subjected to study of the proposed module-to-module connection's working mechanism and bending behavior. The relationship between the specimens and the modular frame is shown in Figure 2. Each specimen contains four module corners, and the beam and column in each module were welded. Considering the test device, the force-length of the beam and column in the connection specimen is the distance between the bending points of the frame. The column, beam, and cover plate sections are identical for all the specimens to compare bearing behavior directly. Figure 3 shows different connection constructions. The bolt holes in specimen MS1 are all around. Correspondingly, the bolt holes in specimen MS2 are all slotted holes.

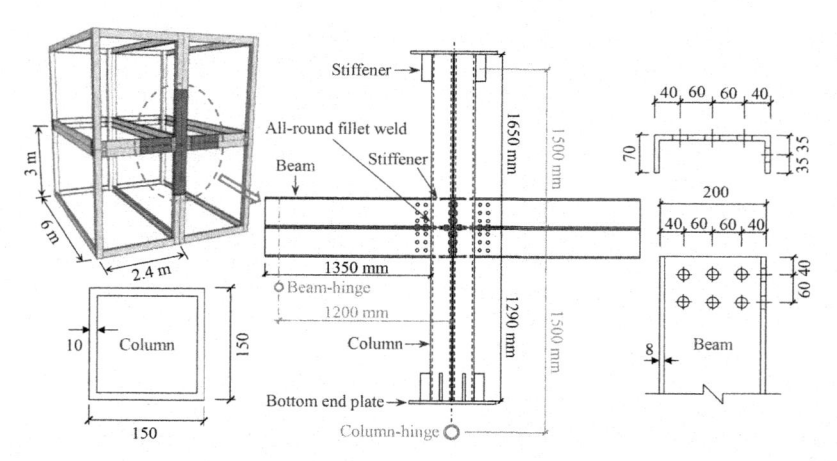

Figure 2.   Detailed information on specimens.

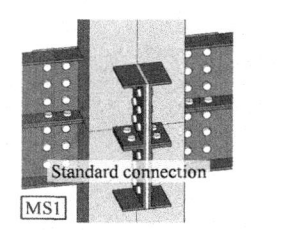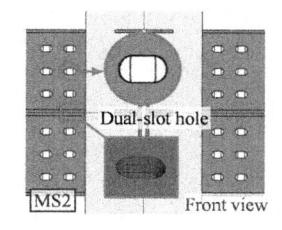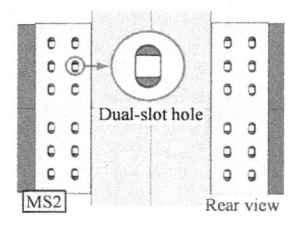

Figure 3.    Different connection constructions.

## 2.2  *Loading and measurement program*

The column-end loading method was used in the tests. As shown in Figure 4, the specimen was installed by a self-balancing device. The column bottom was hinged, and only in-plane rotation was allowed. The beam end is controlled by a chain rod that only enables the rotations and horizontal translations in-plane. The restraint member at the beam end was used to restrain the out-of-plane instability of beams during the test. All test specimens' horizontal and vertical forces are applied through the jack. In the loading stage, the vertical load was applied and maintained first. Then the horizontal load was applied. As suggested, the test will stop when the load at the column end drops to 85% of the ultimate strength.

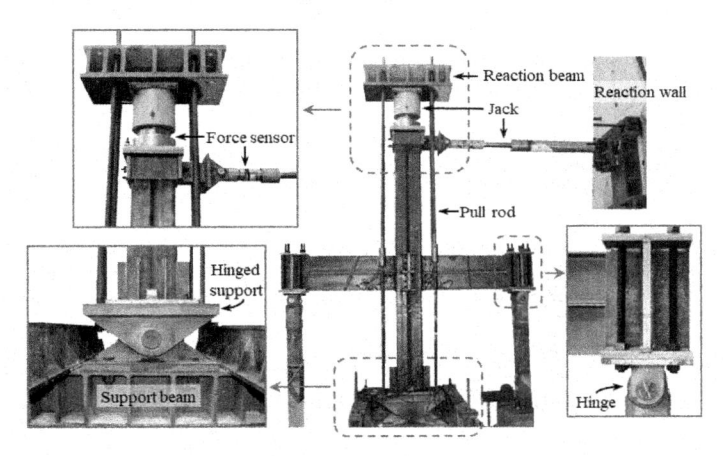

Figure 4.    Experimental setup photograph.

The primary measurement targets are force, displacement, and strain data at key positions. Force is measured by a dynamic force sensor connected to the jack. Four linear variable differential transformers (LVDTs) were placed to measure the displacement of the specimens. Through the measurement results of four LVDTs, the rotation angle can be converted and mutually verified. The specific measurement arrangement is shown in Figure 5.

## 3  EXPERIMENTAL RESULTS AND DISCUSSION

### 3.1  *Failure modes*

The failure characteristics of specimens are summarized as follows: (1) integral bending of module beam; (2) local buckling of beam flange; (3) fracture or necking of beam flange; (4) partial fracture of beam web; (5) module beam-ends dislocation. No obvious deformation of the module column

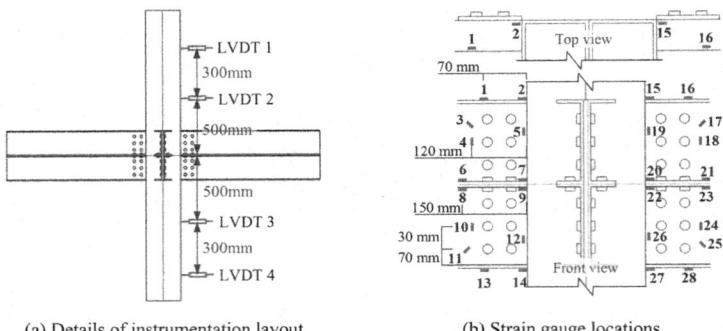

(a) Details of instrumentation layout     (b) Strain gauge locations

Figure 5.    Measurement arrangement.

and cover plate was found in the test. The final failure characteristics of the specimen MS1 are shown in Figure 6. When the rotation almost reached 0.11 rad, a complete fracture occurred at the bolt hole of the beam flange. Flange fracture is the control failure mode of the connection bearing capacity. The test stops due to global bending and severe local buckling of beams. The beam flange connection cracked at the bolt hole due to the large interlayer shear force. The dislocation deformation of the beam ends of the upper and lower modules is caused by the bending and local buckling of the beam. Figure 7 shows the failure characteristics of specimen MS2 when it was loaded to 0.09 rad. Similar to specimen MS1, flange buckling and module beam-end dislocation occurred. No significant dislocation was observed between adjacent modules in specimen MS2.

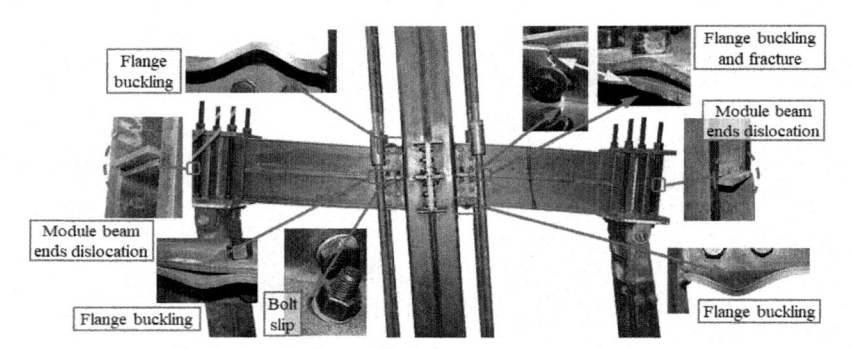

Figure 6.    Failure characteristics of the specimen MS1.

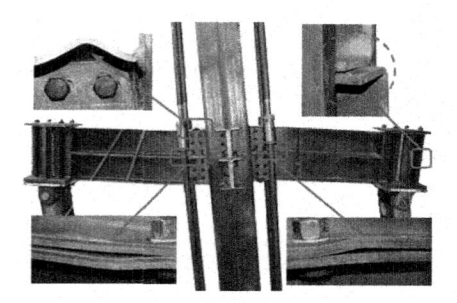
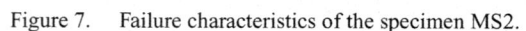
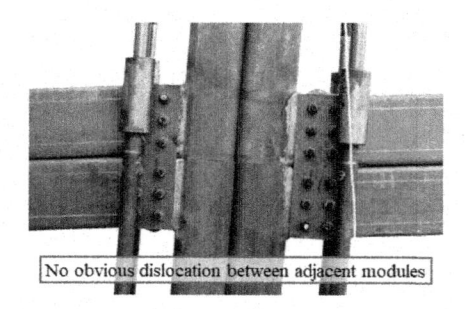

Figure 7.    Failure characteristics of the specimen MS2.

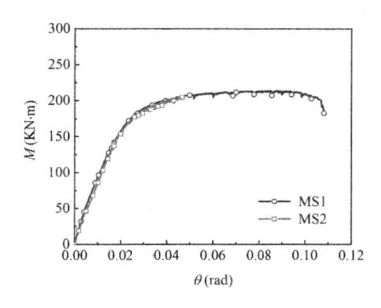
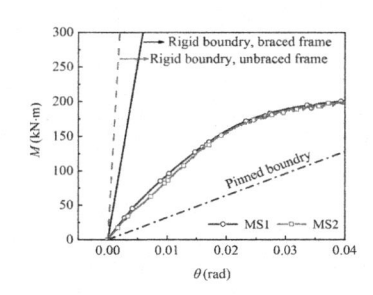

Figure 8.    Moment-rotation curves.          Figure 9.    Stiffness classification.

## 3.2  *Moment rotation response*

The moment and rotation angle are measured and calculated according to the measurement. Generally, the rotation of the connection is mainly composed of three parts: the bending of the column, the bending of the beam, and the shear deformation in the core area of the connection. By the calculation, the ratio obtained from the conversion of four LVDTs is approximately 10%, indicating the bending deformation of the column and the shear deformation of the connection domain are negligible. Hence, the bending of the beam mainly contributes to the rotation angle.

As shown in Figure 8, when the bearing capacity decreases to 85 % of the peak, the rotation of the specimen MS1 exceeds 0.1 rad, indicating that the connection has good rotational deformation. The moment-rotation curve shows prominent elastoplastic characteristics. Only part of the moment-rotation curve of specimen MS2 is recorded due to the device reason. However, it can be seen intuitively that its mechanical performance is similar to that of specimen MS1. This feature is favorable for promoting the application of bolted connections with dual-slot holes.

The calculation results of each eigenvalue of the moment-rotation curve are presented in Table 1. The difference between the service-level stiffness ($K_{se}$) and bearing capacity ($M_m$) of specimens MS1 and MS2 is within 3.6 %, reflecting that the design of slotted bolt holes has little effect on the mechanical properties of connections. In addition, the proposed connections meet the ductility requirement of the GB 50011-2010, including the elastic inter-story drift ratio limit of 0.004 rad and the elastic-plastic inter-story drift ratio limit of 0.02 rad. Furthermore, this paper adopted the relevant part of Eurocode 3 Part 1-8 for steel connections to classify the proposed bolted connection. The analysis results are presented in Figure 9, classified as semi-rigid connections.

Table 1.    Measured moments, stiffnesses, and rotations of connections.

| No. of specimen | $K_{ie}$ | $K_{se}$ | $M_y$ | $\theta_y$ | $M_m$ | $\theta_m$ | $M_u$ | $\theta_u$ | $\mu_\theta$ |
|---|---|---|---|---|---|---|---|---|---|
| MS1 | 11103.3 | 8655.0 | 185.4 | 0.028 | 213.7 | 0.088 | 181.6 | 0.11 | 3.9 |
| MS2 | 9909.5 | 8342.1 | 164.8 | 0.022 | 206.0 | 0.047 | – | – | – |

*Note: $K_{ie}$ (kN·m·rad$^{-1}$) denotes the initial stiffness of connection, defined as the scant flexural stiffness corresponding to 20% $M_m$ at the moment-rotation curve; $K_{se}$ (kN·m·rad$^{-1}$) denotes the service-level stiffness of connection, defined as scant flexural stiffness corresponding to 60% $M_m$ at the moment-rotation curve; $M_y$ (kN·m) and $\theta_y$ (rad) denotes the yield moment, and the yield angle, respectively; $M_m$ and $\theta_m$ denote the maximum moment and maximum rotation angle; $M_u$ and $\theta_u$ denote the moment and rotation angle at failure state; $\mu_\theta$ ($\theta_y/\theta_u$) denotes the angular displacement ductility coefficient.

## 3.3  *Moment strain response*

The moment-strain curve can reflect the mechanical behavior of specimens from the material level. Figure 10 shows the typical moment-strain curves of specimens, which has the following characteristics. (a) As the main bending component, the strain of the beam flange develops rapidly and significantly. The strain values for each component in the connection have significant changes near the connection's yield moment. (b) The typical strain trends in specimen MS1 and specimen MS2 are similar, reflecting the similar mechanical properties of both connections.

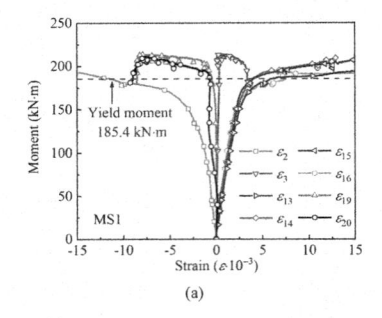 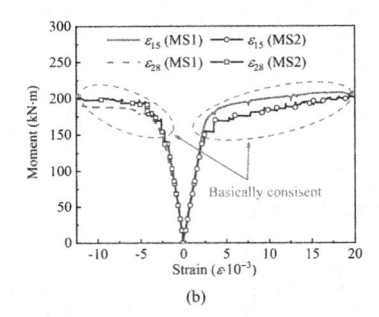

(a)          (b)

Figure 10.   Typical moment-strain curves of specimens.

## 4   CONCLUSIONS

The proposed bolted-cover plate connection has excellent bending behavior. Moreover, the design of slotted bolt holes hardly affects the mechanical performance of the connection, and it improves the error sensitivity of splicing between modules.

The overall bending and local buckling of the modular beam are the failure characteristic of all connections. Flange fracture is the control failure mode of the connection's bearing capacity.

Connection attributes of all specimens are classified as semi-rigid connections.

## ACKNOWLEDGMENTS

This work was financially supported by the Young Researcher Creative Foundation of Guangdong (Grant No. 2020KQNCX175) and the National Science Foundation for Young Scientists of China (Grant No. 51808068).

## REFERENCES

Chen, Z. & Liu, J. & Yu, Y. (2017). Experimental study on interior connections in modular steel buildings. *Eng. Struct.* 147, 625–638.

Chen, Z. & Liu, Y. & Zhong, X. & Liu, J. (2019). Rotational stiffness of inter-module connection in mid-rise modular steel buildings. *Eng. Struct.* 196, 109273.

Dai, X. M. & Zong, L. & Ding, Y. & Li, Z. X. (2019). Experimental study on seismic behavior of a novel plug-in self-lock joint for modular steel construction. *Eng. Struct.* 181, 143–164.

EN 1993-1-8: 2005 CEN. *Eurocode 3: Design of steel structures part 1-8: Design of joints*, European Committee for Standardization, Brussels, 2005.

Ferdousa, W. & Bai, Y. & Ngo, T. D. & Manalo, A. & Mendis P. (2019). New advancements, challenges and opportunities of multi-story modular buildings-A state-of-the-art review. *Eng. Struct.* 183, 883–893.

GB 50011-2010. *Code for Seismic Design of Buildings*, Architecture Industrial Press of China, Beijing, China, 2016.

Lacey, A. W. & Chen, W. & Hao, H. & Bi, K. & Tallowin, F. J. (2019). Shear behavior of post-tensioned inter-module connection for modular steel buildings. *J. Constr. Steel. Res.* 162, 105707.

Lacey, A. W. & Chen, W. & Hao, H. & Bi, K. (2019). Review of bolted inter-module connections in modular steel buildings. *J. Build. Eng.* 23, 207–219.

Sanches, R. & Mercan, O. & Roberts, B. (2018). Experimental investigations of vertical post-tensioned connection for modular steel structures. *Eng. Struct.* 175, 776–789.

Srisangeerthanan, S. & Hashemi, M. J. & Rajeev, P. & Gad, E. & Fernando, S. (2020). Review of performance requirements for inter-module connections in multi-story modular buildings. *J. Build. Eng.* 28, 101087.

*Civil Engineering and Urban Research – Mohamed & Hou (Eds)*
*© 2023 the Authors, ISBN 978-1-032-44487-1*

# Numerical simulation analysis of pulling mechanical behaviors of anti-floating anchor

Zheng Xin, Zhang Shubo* & Qiu Yang
*College of Civil Engineering and Architecture, Shandong University of Science and Technology,
Qingdao, China*

ABSTRACT:    This paper presents a new numerical simulation of the anti-floating anchor using finite element numerical simulation software. The built-in cohesive element is used to simulate the change of stiffness at the contact interface of anti-floating anchors, and the mechanical behaviors and damage mechanisms of anti-floating anchors are successfully simulated and analyzed. The results show that: the results of the numerical simulation were consistent with the testing information derived from the field anchor pullout test and could be explored as a new numerical simulation method. The ultimate deformation of the anchor rods during pulling was basically within 10 mm, and it complied with the engineering acceptance specifications. In the numerical simulation, it was found that the distribution of the shear stress and axial force of the anti-floating anchor is uneven and irregular. The distribution characteristics of the shear stress and axial force of the reinforced anti-floating anchor were that the stress on the head of the anchor was relatively high, while the stress on the bottom of the anchor was relatively low, which can be used as a reference for the design and optimization of the anti-floating anchor.

## 1    INTRODUCTION

To solve the contradiction between limited land resources and rapid urbanization, people pay increased attention to the development and utilization of underground space. When vigorously developing underground space, the anti-floating of building structure becomes the key problem to be solved (Kou et al. 2015).

At present, the main methods to solve the problem of anti-floating are ballast anti-floating, drainage, anti-floating piles, and anti-floating anchors. With its wide range of applications, convenient construction, flexible layout, low cost, and other advantages, the anti-floating anchor has been proved to be the most effective tool to offset the buoyancy of large structures (Yan et al. 2021). As early as the 1980s, anti-floating anchors were used in many deep foundation projects to counteract buoyancy forces (Peng et al. 2000). After extensive practical application, many scholars have carried out research on anti-floating anchors.

Under the joint action of the upper load and the buoyancy of the groundwater, the force situation of the foundation, anchor rod, and foundation is very complex, and the application of finite element analysis technology can simulate this complex problem well. Bai et al. (2019) investigated the pullout characteristics and deformation law of full-length bonded GFRP floating anchors by combining field pullout destructive tests with finite element numerical simulations. Zeng et al. (2004) used a combination of in-situ pullout test and ANSYS finite element software to obtain some basic mechanical behaviors and laws of floating anchor rods. The numerical model was used to calculate the stress distribution and the magnitude of stress value of the anti-floating anchors in

---

*Corresponding Author: zhangshubo_1993@sdust.edu.cn

the results so that the anti-floating anchors can be accurately deployed, the arrangement of anti-floating anchors can be strengthened in the area affected by underground buoyancy, the arrangement of anti-floating anchors can be weakened in the position less affected by underground water buoyancy, and the arrangement of anti-floating anchors can be eliminated in the position not affected by underground water buoyancy (Zhou 2014).

In this paper, we use ABAQUS software to build a three-dimensional finite element model for the Qingdao commercial center project, combine the in-situ anchor pullout experiments at the construction site to verify the correctness of the numerical simulation, and further investigate the force mechanism of the anti-floating anchors and find the pullout mechanical characteristics of the anti-floating anchors.

## 2 PROJECT PROFILE

The proposed commercial center project is located north of Binhai Avenue, south of Zhujiang Road, and on both sides of Emeishan Road in Huangdao District, Qingdao, with convenient traffic. The proposed commercial center project has a construction area of 141,904 $m^2$, and the multi-story frame building and pure basement have a weak buoyancy resistance, and the foundation can be designed according to the height of the buoyancy resistance level recommended in the geological survey report. In this project, especially the part where the podium and the main building are connected, the requirement of its anti-floating ability is high, and the deformation between them must be synergistic; otherwise, it will cause uneven uplift and cracks in the main part. Therefore, anti-floating anchor measures must be adopted. Anti-floating anchor grouting material adopts P.O42.5 ordinary silicate cement, the water-cement ratio is 0.5. The anchor rod is a full-length bonded anchor rod with a total length of 4.20 m and a reinforcement diameter of 32 mm, and the specific construction of the anchor rod is shown in Figure 1.

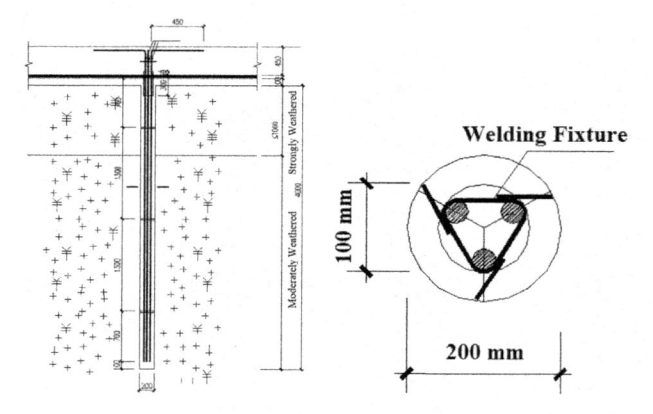

Figure 1. Structural design drawing of the anti-floating anchor rod.

## 3 PULLING TEST NUMERICAL SIMULATION

### 3.1 *Numerical model construction*

In this paper, to investigate the mechanical characteristics of floating anchor rods for optimizing the design and better serving the project, this model takes the floating support design project of the Qingdao commercial center project of Chengfa as a reference and establishes 3D numerical simulation analysis by combining with the anchor pull-out test process on-site, the construction of finite element numerical analysis is divided into three main parts.

## Part I: 3D numerical modeling and meshing

To avoid the influence of boundary conditions, we take the width of the model to be more than twice the length of the anti-floating anchor, and the depth of the model to be more than twice the length of the anti-floating anchor. The length of the floating anchor is 4.2 m. The grid is divided into three-dimensional hexahedral elements. The model and mesh division are shown in Figure 2.

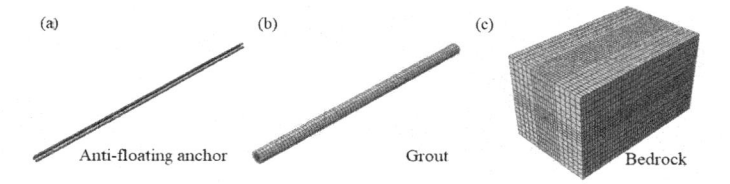

Figure 2.    Geometric model.

## Part II: Model parameter assignment

The anti-floating design is all at the bottom of the pit, mainly involving medium-weathering andesite. To ensure the accuracy of simulation results, all parameters of numerical simulation are taken from the geological survey report and design scheme. The bedrock part adopts the elastic-plastic model, and the plastic model adopts the Drucker plastic model. Due to the large rigidity of the anti-floating anchor rod and grouting body, we decided to use the complete elastomer. The specific material mechanical parameters are shown in Table 1.

Table 1.    Physical and mechanical parameters.

| Part | Density ($kg/m^3$) | Elastic Modulus (GPa) | Poisson's Ratio | Internal Friction Angle (°) |
| --- | --- | --- | --- | --- |
| Anchor rod | 7800 | 200 | 0.3 | / |
| Grout | 2300 | 20 | 0.2 | / |
| Bedrock | 2200 | 4 | 0.25 | 55 |

## Part III: Interaction Relationships

In this paper, the most important point different from previous numerical simulations is the addition of cohesive elements. In the interaction property setting, we add cohesive elements such as the contact relationship between anchor rod and grouting body, grouting force, and surrounding rock, which obeys the traction separation criterion. The traction separation criterion conforms to the rule-like bilinear stress-displacement law proposed by Xu et al. (1996), as shown in Figure 3. In this criterion, the interaction between elements is divided into two types: tensile behavior and shear behavior. According to the stress-displacement law, the area of the purple and yellow areas under the stress-displacement curve are the released energy in tensile and shear failure, respectively. Based on this property, the tensile and shear interaction relations at the interface in anchor pullout experiments can be simulated. The element stiffness of the cohesive force is 2,000 MPa/mm, the maximum nominal stress damage principle is used, the maximum damage stress is 60 MPa, and the friction coefficient is 0.2. The static general analysis step is used for the analysis step, and the boundary conditions are displacement constraints, which constrain the displacements in the bottom X, Y, and Z directions, and the displacements in the lateral constraints X and Y directions.

### 3.2  *Verification of model calculation results*

To test the correctness of our numerical simulation, the field experimental data were compared with the numerical simulation of the ultimate pullout force model to verify whether the numerical simulation is consistent with the real anchor rod force state. The applied external force was the

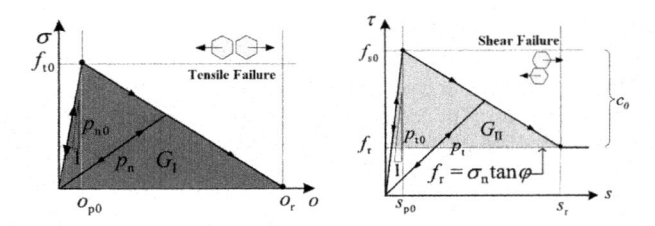

Figure 3. Traction separation criteria.

same as the field pullout monitoring test load, i.e., 60 kN, 180 kN, 300 kN, 420 kN, 500 kN, and 620 kN were applied to the end of the floating anchor to simulate the real anchor pullout test process. The test and simulation results are shown in Figure 4. The numerical simulation curve and the test curve are in very good agreement. Both the numerical simulation and the test show similar trends, indicating that this simulation can be used for the analysis of mechanical behaviors of floating anchor rods.

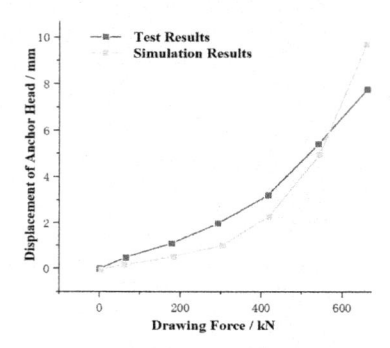

Figure 4. Comparison between test and numeric simulation.

## 4 ANALYSES OF PULLING MECHANICAL BEHAVIORS OF ANTI-FLOATING ANCHOR

### 4.1 *Axial stress analysis of anchor*

The study of the deformation of the surrounding rock and the anchor rod during the pulling process of the anti-floating anchor is very important for the optimization of design and construction. Since the anchor rod is cylindrical, its deformation and stresses are mainly in the axial and tangential directions. We investigate the axial stress and deformation of the surrounding rock, and the output numerical simulation results are shown in Figures 5 and 6.

Due to the cohesion, mechanical occlusion, and friction between the anti-floating anchor and the rock soil layer, the anti-floating anchor and the surrounding rock mass bear the pull-out load together. According to the numerical simulation results in Figure 7 and from the macro point of view, the stress of the anchor rod is not uniform during the pulling process, and the stress law of the whole anchor rod shows that the stress is gradually decreasing from shallow to deep. At the lowest stress at the bottom of the anchor rod, the stress at the head of the anchor rod reaches the maximum. At the same time, we can reflect the influence range of the anchor rod in the pulling process according to the range of stress change, and can further find out the stress range of the anchor rod.

The data are further extracted in Figure 8 to analyze the force of the anchor rod. Changes in the force of the anchor rod under different loads are the same as that reflected in the cloud diagram, and the force decreases gradually from the anchor head position downward. The force of the anchor rod

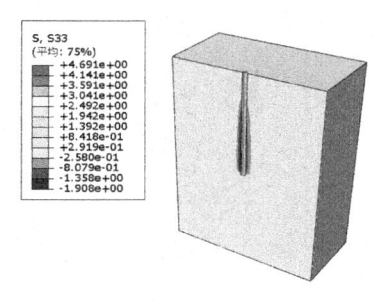

Figure 5. Stress nephogram of rock mass along the depth direction.

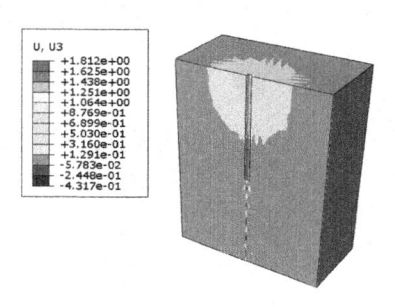

Figure 6. Nephogram of rock mass deformation along the depth direction.

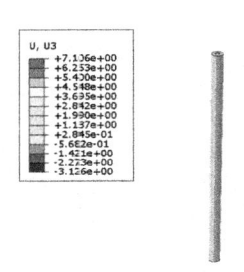

Figure 7. Bolt axial deformation.

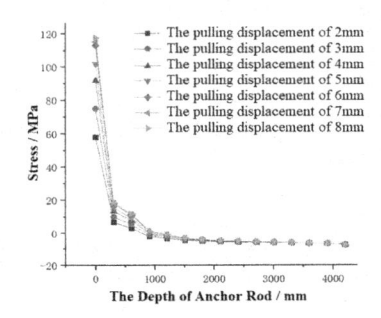

Figure 8. Stress variation along the depth of the anchor rod.

is also shown in the figure. The force of the anchor rod reinforcement decreases rapidly from the position of 1.5 mm, and the changes of the anchor rod tend to be stable with the increase in depth. Another point is that changes in force in the anchor head part are larger as the load increases, but the force change at the bottom of the anchor rod is smaller as the depth increases. This indicates that it is the anchorage of the anchor head part that is critical in the process of bearing the pullout load of the anti-floating anchor. Therefore, we can see that it is not better to make the anchor rods longer, but to selectively strengthen the segments, especially for the top of the anchor rods, and to selectively reduce the reinforcement or reduce the length of the rods for the bottom of the anchor rods. In this way, the effect of anti-floating anchors can be guaranteed while saving money, and the design of anti-floating anchors can be optimized to provide a reference for future engineering practice.

### 4.2 *Shear stress analysis of anchor*

We further investigate and analyze the shear stress and deformation of the surrounding rock and anchor rod, and the numerical simulation results are shown in Figure 9. From the numerical simulation, the cohesion, mechanical occlusion, and friction between the anti-floating anchor and the rock soil layer result in the anti-floating anchor being subjected to the pull-out load together with the surrounding rock mass. Thus, the shear force mainly exists on the contact surface between the anchor rod and the surrounding rock.

Figure 10 shows the shear stress distribution curve of the reinforced anti-floating anchor. The distribution of shear stress is irregular, and it decays exponentially with the increase in embedded depth of the anchor rod. It changes greatly within the depth of 1,000 mm, and the stress attenuation amplitude is inapparent when the depth exceeds 1,000 mm. As the head of the anchor rod is gradually pulled out, the stress of the anchor head changes greatly. When the pull-out displacement is 7 mm, the shear stress reaches 2.52 MPa. In summary, the shear stress of the steel anti-floating anchor is maximum at the head of the anchor and minimum at the bottom of the anchor, which

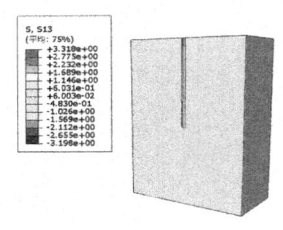
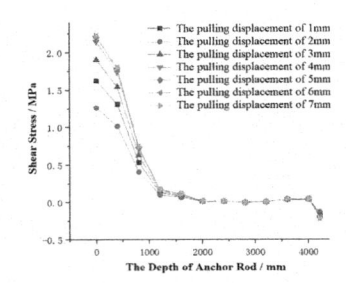

Figure 9. Nephogram of surrounding rock shear stress.

Figure 10. Anchor shear stress curve.

decreases with the increase of the anchor depth. Numerical simulations show that the main reason for this result is the bonding unit, which is suitable for simulating the contact between the anchor rod and the surrounding rock.

## 5 CONCLUSIONS

This paper presents a new numerical simulation of the anti-floating anchor using finite element numerical simulation software. The built-in cohesive element is used to simulate the change of stiffness of the contact interface of the anti-floating anchors, and the mechanical behaviors and damage mechanism of the anti-floating anchors are successfully simulated and analyzed. The main conclusions are as follows.

(1) The proposed numerical simulation method is consistent with the testing information derived from the field anchor pullout test and can be explored as a new numerical simulation method.
(2) The ultimate deformation of the designed anti-floating anchor is basically within 10 mm during the pulling process, which can resist the load of groundwater and is based on the engineering acceptance specifications.
(3) In the numerical simulation, it is found that the distribution of the shear stress and axial force of the anti-floating anchor is uneven and irregular. The distribution characteristics of the shear stress and axial force of the reinforced anti-floating anchor are that the stress on the head of the anchor is relatively high, while the stress on the bottom of the anchor is relatively low.

## REFERENCES

Bai X Y, Zhang M Y, Kuang Z, Wang Y H & Yan N. (2019). Analysis on pullout behaviors of full-bonding glass fiber reinforced polymer anti-floating anchor. *Journal of Central South University* (Science and Technology) (08), 1991–2000.

Kou, H. L., Guo, W., & Zhang, M. Y. (2015). Pullout performance of gfrp anti-floating anchor in weathered soil. *Tunnelling and Underground Space Technology*, 49 (jun.), 408–416.

Peng T, Wu W, Chen D B, Yang H M. (2000). Application of prestressed anti-floating anchor under complex geological conditions. *Geotechnical Investigation & Surveying* (02), 31–33.

Xu X-P, Needleman A. (1996). Numerical simulations of dynamic crack growth along an interface. *International Journal of Fracture*, 74 (4): 289–324.

Yan, N., Liu, X., Zhang, M., Bai, X., & Wang, Z. (2021). Analytical calculation of critical anchoring length of steel bar and GFRP anti-floating anchors in rock foundation. *Mathematical Problems in Engineering*, 2021 (1), 1–10.

Zeng G J, Wang X N, Hu D W. (2004). Analysis on Present Situation of Anti-floating Technology Application. *Underground Space* (01), 105–109+142.

Zhou, X. L. (2014). Simulation and numerical analysis on anti-float anchor by field rock and soil tests. *Applied Mechanics & Materials*, (686), 671–675.

Civil Engineering and Urban Research – Mohamed & Hou (Eds)
© 2023 the Authors, ISBN 978-1-032-44487-1

# Seismic resistant design and performance of tall buildings

Yi Lu*
*International Campus of Zhejiang University, Haining, China*

ABSTRACT: In modern society, tall buildings are more and more common, and gradually become a construction trend. And because earthquakes do great harm to tall buildings, it is important to consider the design and behavior of seismic-resistant characters of tall buildings. Based on many sources and research about the earthquake and seismic resistant structures of tall buildings, this article concludes and summarizes the characteristics and development of tall buildings, the characteristics of an earthquake, the design of tall building seismic resistant structures, and the analysis method of tall buildings seismic resistant technology. This article could provide the basic literature reference and introduction, characters, and basic analysis methods for the seismic resistance design of tall buildings.

## 1 INTRODUCTION

With the development of economics and technology, more and more high-rise buildings are erected in the city. The performance and design of tall buildings have always been a challenge for civil engineering engineers because they should guarantee both functionality and security. Earthquake, an inevitable natural disaster, has been challenging the safety performance of tall buildings for a long time. Its centrality and uncertainty also increase the difficulties of seismic resistance of tall buildings. Therefore, the design and behavior of seismic-resistant structures of tall buildings are very vital. The purpose of this paper focuses on the characteristics of an earthquake, the design of tall building seismic resistant structures, and the analysis method of tall buildings' seismic resistant technology.

## 2 CHARACTERISTICS OF TALL BUILDINGS

### 2.1 The frame of tall buildings

In history of modern tall buildings, modern tall buildings had emerged in the United States. In 1883, an insurance company building of 11 floors with the first brick stone self-supporting and steel frame structure appeared in Chicago. In 1931, the imperial building was built in New York with a height of 381 meters, and 102 floors (William 2001). After the 1960s, structure innovation greatly impacted the industry. With the strong help of the "non-structure" cladding and interior partitions, it is no longer a striped column and the interior of the frame. Instead, the frame tube, support tube, bundle, frame-shear wall interaction, leg system, and other new structural modules become the support elements. Due to various technologies, especially the emergence of computer and engineering pioneers, this new system could come true. The technologies for structure analysis that make these systems feasible can be proved by computers while performing important parameters to determine each applicable height range and safety performance of the system (William 2001).

*Corresponding Author: yil.20@intl.zju.edu.cn

DOI 10.1201/9781003372417-11

Therefore, the morphology and height of tall buildings had great breakthroughs, and the World Trade Center buildings in New York are typical examples.

## 2.2 *The materials of tall buildings*

For the materials, concrete is the first choice for medium-high residential buildings; but for high-level and ultra-high-level buildings, steel occupies the dominance. For instance, the Sears Building is a steel bundle, the World Trade Center and the Standard Petroleum Tower are steel frame tubes, the earlier Empire Building is a steel frame, and the John Hancock Center is a steel pipe (William 2001).

But from the mid-1980s to today, a series of innovations in materials have changed the tall building design industry once again. One of these innovations is a major shift from all-steel structures to systems that utilize the advantages of steel and concrete in structure. This transformation takes advantage of steel and concrete's strengths: steel is lightweight, ideal for building long-span office floors, and can be easily resized; concrete is very cost-effective in carrying the weight of the tower, and the heavy mass is conducive to reducing building movement due to earthquake and wind (William 2001).

## 2.3 *The morphology of tall buildings*

In terms of morphology, modern tall buildings break through the limitation of the aspect ratio. The modern tall buildings are characterized by their elongation and sensitivity to resonance effects and seismic waves (Spence & Kareem 2013). All newer towers have long, column-less, and leasing spans. Also, they have similar elongation, such as the John Hancock Center, which has an aspect ratio of 6.6:1 (William 2001).

By the end of the 20th century, the research and discussion of the theories and practical applications of structural seismic performance have been fully developed in the field of seismic engineering (Zhu et al. 2004). For the seismic performance of tall buildings, the support of these theories is more needed. Structural engineering innovation and its impact on skyscrapers continue play an important role in changing the characteristics of tall buildings. The current level of innovation is quite high and mainly focuses on composite systems, materials, wind engineering, motion control, and optimization methods (William 2001). Today, with the rapid development of the economy and the improvement of living, the morphology and system of building structures become more complex and diverse to meet the dual needs of the economy and society. The development and implementation of performance-based seismic design concepts have resulted in a new generation of super-tall buildings that are more flexible, slimmer, visually appealing, and, most importantly, function properly under a variety of load conditions (Pan et al. 2020). These buildings have brought us new problems. The newly revised specification puts forward specific requirements for these structures. They are of practical significance for structural engineers to design and analyze these structures correctly (Chen et al. 2004).

## 3 EARTHQUAKE CHARACTERS AND IMPACT

An earthquake is one of the inevitable natural disasters which affects the larger scope, concentration, and intensity (Joyner & Sasani 2020). Due to the uncertainty of the numerical model of earthquake disasters and structures, it is difficult to reduce the damage of earthquake disasters to a certain extent.

Seismic hazards are affected by multiple sources of uncertainty regarding the magnitude, location, and intensity of future earthquakes (Joyner & Sasani 2020). Probability Seismic Hazard Analysis (PSHA), a statistical model based on past seismic records and the calibration of earthquake prediction models, is the only readily available and widely used seismic hazard analysis method. It can be seen as a useful tool for disaster preparedness. The second source of perceived uncertainty is the uncertainty in structural models, which affects demand and capacity. Demand uncertainty can come from a variety of sources, including idealized constitutive material models, the size and distribution of gravity loads, construction errors, and basic modeling assumptions. However, in the current assessment, these sources of requirement uncertainty are considered small compared to the variability between records and are therefore ignored (Joyner & Sasani 2020).

According to Joyner & Sasani (2020), Models 4, 7, and 10 are a 15-story reinforced concrete special moment frame structure in San Francisco that receives a series of seismic ground motion records and use the result to assess deformation responses to a range of seismic intensities. From these analyses, demand statistics, as well as capacity models and seismic hazards, can be estimated and used to assess the expected cost of maintenance and loss of function for each building over the life cycle. Performing this analysis on a set of values for each design variable will demonstrate the impact of stiffness, strength, and deformation capacity on the seismic performance of the building, so this analysis can be used to assess and enhance community resilience (Joyner & Sasani 2020).

Seismic hazard is estimated based on the USGS's seismic hazard data, which is in terms of geometric mean spectral acceleration provided for a range of fundamental periods, soil types, and return periods. Figure 1 shows the hazard curves, presented in terms of annual frequency of exceedance (AFE), for each structure designed based on the minimum code requirements, obtained by fitting a 4th-order polynomial to the USGS data in log-space (Joyner & Sasani 2020).

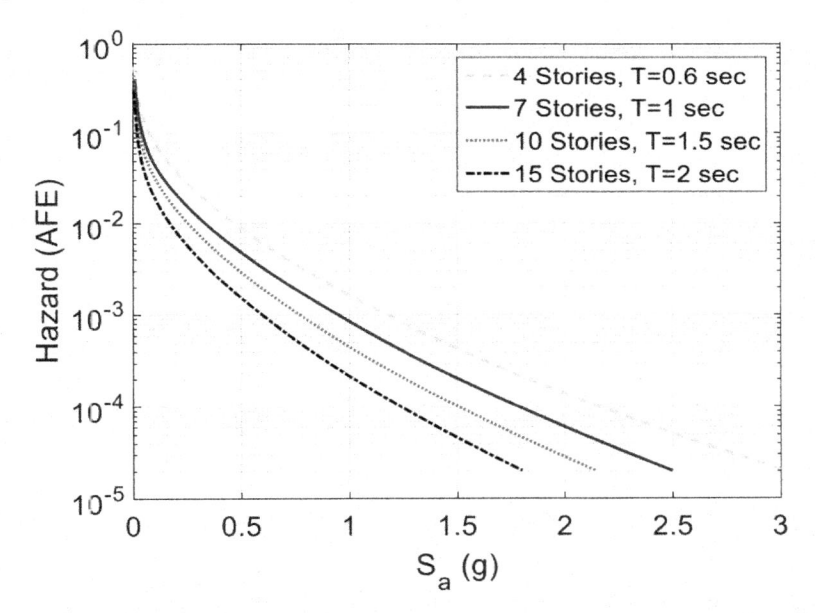

Figure 1. Hazard curves for geometric mean Sa for each structure (AFE stands for the annual frequency of exceedance). (Joyner & Sasani 2020).

The recent two decades of earthquake disasters in urban areas have shown that, although earthquake engineering research is progressing, the damage and economic losses caused by earthquakes are keep rising. For example, the earthquake in Christchurch of New Zealand on 2011.2.22 caused total economic losses in the range of $1.1–15 billion (Xiong, et al. 2016). For tall buildings, it is even more devastating, because collapsed tall buildings can cause multiple injuries on the ground.

Due to urbanization, the number of tall buildings in urban areas has increased significantly. These buildings usually contain many dwellings and many personal assets. In addition, tall buildings are widely used as offices, hospitals, commercial or financial centers, and communications or power hubs, which are essential to the function of the city. Modern tall buildings built over the past 30 years are often very strong and can withstand the collapse of a whole when an earthquake strikes.

Still, severe earthquake damage to tall buildings is common. Therefore, it is very important to accurately predict the earthquake damage to tall buildings and their related economic losses for the simulation of earthquake damage in urban areas (Xiong, et al. 2016).

## 4  SEISMIC RESISTANT FACTORS

### 4.1  *Safety factor*

Safety and collapse prevention has been and should remain, the primary concern of building codes. Earthquake risk now permeates many aspects of today's building codes. In fact, with the release of ASCE 7 in 2016, considerable progress has been made recently in developing these regulations. Among other things, ASCE 7-16 now specifically states that the Risk Class IV structure is intended to maintain functionality immediately after a design-level event (Joyner & Sasani 2020). The seismic performance of buildings plays an important role in seismic resistance. An important measure in building performance evaluation is the damage of structural and nonstructural elements when they deform beyond their bearing capacity during an earthquake. Through a risk-based performance assessment process, the expected impact of changes in stiffness, strength, and deformation ability on building damage can be quantified, and the results can be used to make targeted improvements to building codes to enhance seismic resistance (Joyner & Sasani 2020). Tall buildings are more flexible than short buildings, and they are sensitive to different frequency ranges of seismic excitation. The seismic success of tall buildings is enhanced by using the average characteristics of earthquakes and the research on typical characteristics of tall buildings, and there is no special seismic hazard due only to the height (Bhowmik 2015).

### 4.2  *Collapse prevention factor*

The seismic design of tall building structures is mainly accomplished through performance-based methods, followed by a series of response evaluations based on deformation and force. The performance under strong earthquakes, dynamic characteristics (e.g., mass, stiffness, and damping), and the response to environmental loads (e.g., wind and seismic events) should be carefully considered (Spence & Kareem 2013). To improve the performance of the structure under the action of a strong earthquake, composite structural members, including steel columns and steel sheet shear walls, are often considered. When subjected to a strong earthquake, the structure of a tall building will experience a great deal of nonlinear force (Ren, et al. 2018). And the structure may undergo serious degradation. Meanwhile, many structural elements may lose their bearing capacity due to excessive deformation. To understand the behavior of structures under strong earthquakes, a systematic approach should be developed for the nonlinear modeling of structures undergoing large deformation and severe degradation (Ren, et al. 2018).

To discuss the relation between the reinforcing steel and the structural response under severe earthquakes, Figure 2 presents the inter-story drift ratio curves for the different structural schemes under different earthquakes with the same PGA = 400 cm/s$^2$. We can find in Figure 2 that the structure model developed based on the RC scheme experiences serious deformation in lower levels under the excitation of the EL Centro record. There is even a maximum inter-story drift ratio exceeding 1/100. By using the steel-reinforced columns, the deformation of structures decreases but certain deformation concentration still appears in the bottom region. The steel plate reinforced shear walls placed in the bottom levels restrain the excessive inter-story deformations very effectively. But for the deformation of the upper levels, may be enlarged by the SPRSW. In Figures 2(b) and 2(c), the distributions of structural deformations are not very sensitive to the reinforcing steel. In Figure 2(c), the inter-story drift ratio of the upper levels is enlarged by the reinforcing steel. And in Figure 2(d), the steel plate reinforced shear wall reduces the inter-story deformation of the whole structure effectively (Ren et al. 2018).

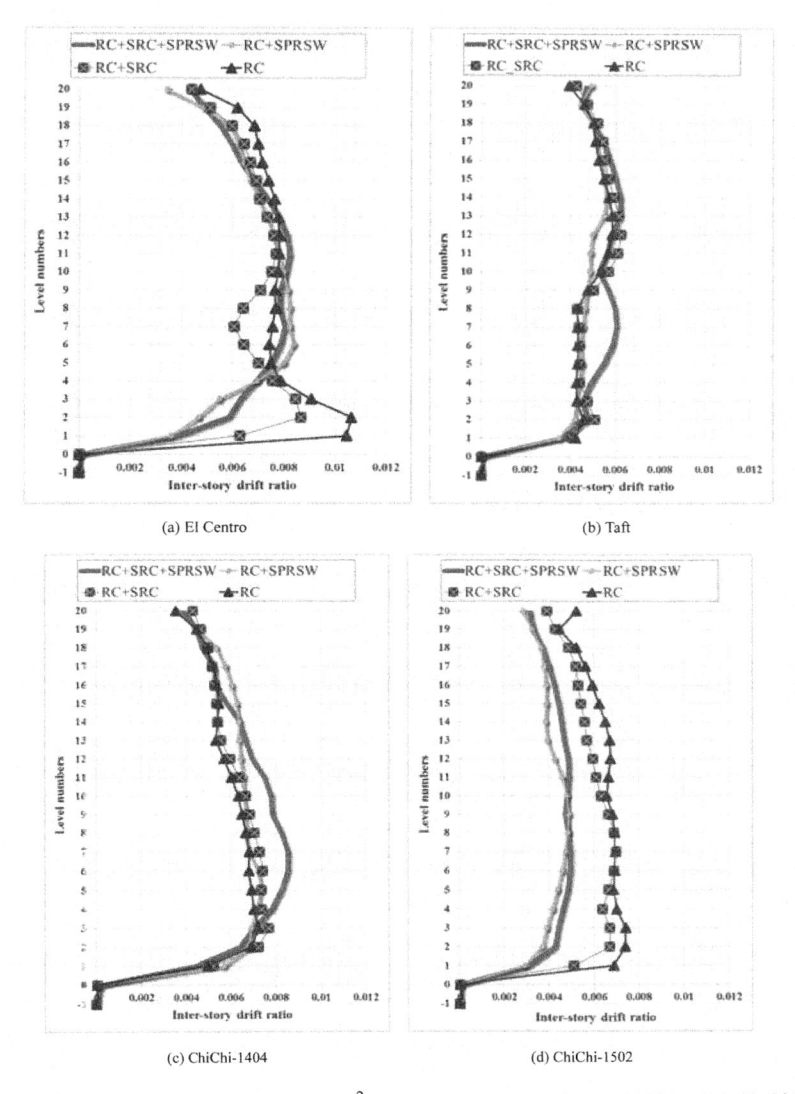

(a) El Centro

(b) Taft

(c) ChiChi-1404

(d) ChiChi-1502

Figure 2. Inter-story drift ratios (PGA = 400 cm/s$^2$). (a) El Centro. (b) Taft. (c) ChiChi-1404. (d) ChiChi-1502. (Ren et al. 2018).

### 4.3 *Damage resistance factor*

For the seismic characteristics of tall buildings, one of the basic goals of performance-based design is to prevent the overall collapse of the structure. Structures excited by earthquakes usually go through a series of stages, including the elastic stage, failure stage, partial collapse stage, and complete collapse stage. Therefore, we should use complex methods to evaluate structural damage and collapse under strong earthquakes to implement performance-based design procedures (Huang, Ren & Li, 2017). The start of structural failure depends on the weakness of the structure, including shear wall thickness, section shape, and transition layer of truss arrangement. When the ground beneath the foundations of buildings shakes, they vibrate like that of the ground around them. The inertial forces of the structure will shear the buildings, which makes stress concentrate in the structural joints and fragile walls. It can cause partial or complete failure of the structure. Compared with short structures, tall structures tend to amplify the amplitude of periodic motions over long periods. Every structure has a prevailing resonance, which is characteristic of the structure. Taller buildings tend to be used for longer periods than shorter buildings, which makes them relatively vulnerable to damage. Therefore, care must be taken in the analysis of tall structures (Bhowmik 2015). As for the horizontal vibration of an earthquake, it mainly causes translational vibration of the symmetrical structure. The horizontal seismic force is distributed according to the vertical stiffness distribution of the structure. Because the center of mass and the center of stiffness are not superimposed in the asymmetric structure, the torsional vibration can also be excited under horizontal earthquake action. The structural members far away from the rigid center often share a large horizontal shear force, which can easily cause serious damage and exceed the limit of translational displacement (Zhou & Yi 2008). Thus, the lateral stiffness of buildings plays an important role in limiting floor drift. At the same time, multiple collapse modes should be considered in the design of super-tall buildings, even for ground motion records with similar waveforms but different intensities (Huang, Ren & Li 2017).

## 5 SEISMIC-RESISTANT ANALYSIS METHODS

For most of the seismic resistant analyses, they would base on the building models. There are many methods to model the building dynamic. Modern finite element modeling capabilities allow for highly sophisticated models of complex frame structures with a wide range of precise structural details and capture subtle elements and system-level response characteristics that are useful to performance evaluation. Such models, however, tend to be computationally demanding. Thus, in selecting the appropriate method for the current evaluations, which entail over 60,000 nonlinear time history analyses, a balance between model sophistication and computational efficiency is needed to allow the completion of the analyses on a reasonable time scale (Joyner & Sasani 2020).

To analyze a tall structure, many analysis procedures are useful such as a) equivalent static analysis, b) response spectrum analysis, c) linear dynamic analysis, d) nonlinear static analysis or nonlinear pushover analysis, and e) nonlinear dynamic analysis.

### 5.1 *Equivalent static analysis*

Equivalent static analysis is a kind of response spectrum of seismic design. It can simulate and analyze the forces which act on buildings, and the ground motion effect due to earthquakes. During the analysis, it always sets the building to respond with fundamental mode. This type of analysis is used for estimating displacements of structures, and it is especially suitable for structures and individual frames. The earthquake load will be converted to an equivalent static and horizontal force, then applied to the individual frames. The given force is equal to the multiplication of the acceleration response spectrum and its weight. In this analysis, the response is studied from a response spectrum where the building's natural frequency is given by calculating the building design criteria. This analysis procedure is highly used for many building codes, including tall buildings and short buildings by set different modes in analysis (Bhowmik 2015).

### 5.2 *Response spectrum analysis*

Response spectrum analysis is a kind of statistical linear-dynamic analysis. It measures the mode of vibration and indicates the maximum seismic response of the elastic structure. It depends on the theory of structural dynamics and is derived from basic principles. This analysis contains a structural period function with several data of velocity, acceleration, displacement, the measurement for a given damping level, and time history. The response spectrum analysis combines dynamic performance with type selection of structure, which is greatly helpful for decision-making in design. This analysis includes the multiple modes of response of a building, so it is greatly required in many building codes. The response of a structure is also prescribed as a summation of many special modes that in a vibrating string correlate with the "harmonics". A response is studied from the spectrum design from each mode, depending on the modal mass and frequency. Then, an estimate can be calculated for all responses of structures by combining all data. The response spectrum analysis is not suitable for structures that are more irregular and taller. The more irregular, taller structures are preferred to use static analysis or nonlinear dynamic analysis. The response spectra have some limitations. These are only widely used for linear systems. Although nonlinear seismic spectra in design with broad structural applications have developed a lot, it is hard to associate the multimode response quickly and easily (Bhowmik 2015).

### 5.3 *Linear dynamic analysis*

For weaker seismic effects, the static analysis procedure might be useful. But for stronger seismic effects, much higher buildings, and buildings with irregularities or nonorthogonal systems, the dynamic analysis procedure is more likely to use. In this process of linear dynamic analysis, the structure is analyzed as multiple degrees of freedom system with a viscous damping matrix and elastic stiffness matrix. Always use time history analysis and modal special analysis when analyzing the seismic effects. The linear dynamic analysis considers higher modes and this is better than the linear static analysis. However, because all of these depend on linear elastic response, their application decreases as nonlinear behavior increases. In this analysis, the reaction of the structure's ground motion is deliberated in the domain time and all the phase information is sustained, and only linear properties are taken up (Bhowmik 2015).

### 5.4 *Nonlinear static analysis*

The nonlinear static analysis could also be called pushover analysis. It is an analysis for under-sustained vertical loads and gradually rising lateral loads. The forces induced by an earthquake are described by static lateral loads. The analysis could acquire a sketch of displacement versus total base shear in a structure. It would show any weakness and failure details. Nonlinear static analysis is impacted by force and displacement. This procedure is applied for the known loads.

A major purpose of the push-over analysis is to assess seismic performance and how much seismic intensity it can withstand. According to the push-over analysis, the structural reaction force under different lateral push-over loads can be obtained, and then the foundation shear force (total lateral push-over loads) under different free vibration periods can be obtained. Finally, the actual response curve structure (the ratio of foundation shear force to structure mass and free vibration period) can be achieved.

For the evaluation of seismic performance, pushover analysis is more suitable for analyzing seismic vulnerability. This analysis offers enough information on seismic demands decided by the ground's motion on the system.

The pushover analysis cannot describe the phenomena property and it depends on the static loading. So, there are some important records of deformation that might be tracked. To assess the structural system performance by estimating the strength, and deformation of the structure is the motive of this analysis. The assessment depends on some important parameters such as inter-story drift, global drift, inelastic element deformation, element connection forces, and deformations between elements. This analysis can prospect as a method to credit force and deformation, which calculates in an estimated manner for the new distribution of internal forces that cannot be resisted within the elastic range (Bhowmik 2015).

The process of push-over analysis for structures is as follows.

(1) Calculate the vertical load of the structure. The vertical loads applied to the structure in the early stage, including the dead weight and live load of the building, remain unchanged during the push-over process.
(2) Apply lateral load. The base shear force of the structure can be obtained by the response spectrum under low-intensity earthquakes, and the lateral load distribution with triangular

distribution can be determined by the response spectrum method with mode superposition. Then, the load is progressively applied to the structure until the structure is destroyed.

(3) Judge the structural performance level (including local and overall structural performance) (Zhu et al. 2004).

### 5.5 *Nonlinear dynamic analysis*

The nonlinear dynamic analysis gives results with low unpredictability because this analysis uses the summation of ground motion records with the details of the structural model. In this analysis, the structural model estimates the deformation for all the degrees of freedom. This analysis is necessary for important configuration according to building codes. The response calculation can catch and analyze the ground's motion and use this information as an input for the earthquake. The calculation of ground motion records and the estimation of structural response distributions require various analyses. As the characteristics of seismic response are based on intensity and earthquake characters, many measurements are needed to describe different earthquakes (Bhowmik 2015).

## 6  CONCLUSIONS

This paper focuses on the analysis of the characteristics of tall buildings, seismic impact on tall buildings, and the main ways of seismic design. Based on the history of morphology, structure, and material of tall buildings from ancient to modern, the paper analyses the influence of earthquakes on seismic resistance of tall buildings, and the main considerations of tall building designs. Some suggestions and analysis methods for seismic design of tall buildings are also provided.

Due to the strong uncertainty and concentration of earthquakes, probabilistic seismic hazard analysis (PSHA) is needed to evaluate and analyze seismic structures, quantify the expected impacts of changes in stiffness, strength, and deformation capacity on building damage, and results of the analysis can be used to make targeted improvements to building codes to improve seismic capacity. The seismic-resistant design of tall building structures is mainly accomplished through performance-based methods, followed by a series of response evaluations based on deformation and force. Careful consideration should be given to the performance under strong earthquakes, and the response of dynamic characteristics, such as mass, stiffness, and damping to environmental loads. Equivalent static analysis, response spectrum analysis, linear dynamic analysis, nonlinear static analysis, and nonlinear dynamic analysis are five kinds of analysis methods for seismic resistance. Each method has specific research objectives, advantages, and disadvantages.

This paper focuses on the review of the literature, and mainly on the qualitative analysis, design concept, and analysis method. Its weakness does not provide a very professional and data-oriented solution to analyze the tall building structure. At the same time, the coverage of seismic structure design for tall buildings is not very wide.

Various types of building materials can be developed in the future, and different combinations can be used to develop more innovative seismic structural units or structural cores. Similarly, we can develop structural analysis software with more analysis effects and unit-whole analysis to analyze the seismic structure of buildings in a more comprehensive manner.

## ACKNOWLEDGMENT

We are very grateful to Prof. Oral Buyukozturk for his inspiration and suggestions for the content, specific parts, and arrangement of this paper.

# REFERENCES

Bhowmik, K., Tarafder, N., and Kumar, N. "Earthquake Resistant Techniques and Analysis of Tall Buildings," *International Journal of Research in Engineering and Technology*, no. 25, 2015, 12.

Chen, J., Ding, J., Pan, Y., Luo, Z., & Xi, Z. (2004, January). The Holistic Design and Analysis for an Out-of-Codes Tall Building Structure. In IABSE Symposium Report (Vol. 88, No. 5, pp. 48–53). *International Association for Bridge and Structural Engineering.*

Huang, T., Ren, X., & Li, J. (2017). Incremental dynamic analysis of seismic collapse of super-tall building structures. *The Structural Design of Tall and Special Buildings*, 26 (16), e1370.

Joyner, M. D., & Sasani, M. (2020). Building performance for earthquake resilience. *Engineering Structures*, 210, 110371.

Pan, Y., Ventura, C. E., Xiong, H., & Zhang, F. L. (2020). Model updating and seismic response of a super tall building in Shanghai. *Computers & Structures*, 239, 106285.

Ren, X., Bai, Q., Yang, C., & Li, J. (2018). Seismic behavior of tall buildings using steel–concrete composite columns and shear walls. *The Structural Design of Tall and Special Buildings*, 27 (4), e1441.

Spence, S, and A Kareem. "Tall Buildings and Damping: A Concept-Based Data-Driven Model," 2013, 64.

William, Baker. "*Building Systems and Concepts-Structural Innovation*," CTBUH Research Paper, 2001.

Wu, J., Gan, G., and Xuan, J. "Seismic Analysis and Design of Twin Towers Conjoined Tall Building in Zhoushan Administrative Center," *Journal of Engineering Design* 11, no. 2, 2004, 4.

Xiong, C., Lu, X., Guan, H., & Xu, Z. (2016). A nonlinear computational model for regional seismic simulation of tall buildings. *Bulletin of Earthquake Engineering*, 14 (4), 1047–1069.

Zhou, Y., & Yi, W. J. (2008). *Ambient Vibration Measurement and Earthquake Resistant Behavior Analysis on a Twotower Tall Building with Enlarged Base*. In 14th World Conference on Earthquake Engineering.

Zhu, Jiejiang, Peijun Zhang, Xilin Lyu, and Baisheng Rong. "Push-Over Analysis for Concrete Structures of Tall Building." *Journal of Shanghai University* (EnglishEdition), 2004, 8.

*Civil Engineering and Urban Research – Mohamed & Hou (Eds)*
*© 2023 the Authors, ISBN 978-1-032-44487-1*

# Mechanical properties of basalt fiber-sisal fiber/fly ash regenerative concrete

Zhiwei Jiang*, Zhuo Li* & Kai Xu*
*School of Architectural Engineering, Nanjing Institute of Technology, Nanjing, China*

ABSTRACT:   The slump, cube compressive strength, and splitting tensile strength experiments of 16 groups of basalt fiber (BF)-sisal fiber (SF)/fly ash regenerative concrete and 1 group of C30 reference regenerative concrete were conducted by orthogonal experimental method. The results show that the cube compressive strength and splitting tensile strength of basalt fiber (BF)-sisal fiber (SF)/fly ash regenerative concrete is higher than that of C30 base regeneration concrete. The addition of basalt fiber and sisal fiber can reduce the slump of concrete, and basalt fiber can reduce the slump more significantly. Fly ash is a significant factor affecting the cube compressive strength of regenerative concrete. With the increase of fly ash substitute cement rate, the cube compressive strength of the cube increases first and then decreases. Basalt fiber is a highly significant factor affecting the splitting tensile strength of regenerative concrete, and sisal fiber is a significant factor, and the splitting tensile strength increases with the volume rate of the two fibers. At present, the mechanical properties of regenerative concrete are poor, and the incorporation of natural fiber into regenerative concrete is rarely studied. The purpose of this paper is to investigate the influence of sisal fiber, basalt fiber, and fly ash on the mechanical properties of regenerative concrete, then obtain the optimal combination under different applications, and promote the application of regenerative concrete.

## 1   INTRODUCTION

Regenerative concrete refers to the concrete that uses waste building materials as recycled coarse aggregate. The surface of recycled coarse aggregate is rough, with more edges and corners, many internal fine cracks, and high vorticity, which directly affects the working and mechanical properties of regenerative concrete. Topcu found that the cube compressive strength of regenerative concrete decreased with the increase of recycled coarse aggregate (Topcu 1997). Ravindrarajah (Ravindrarajah & Tam 1985) and Hansen's study found that the cube compressive strength and folding strength of regenerative concrete were lower than the reference concrete (Ananthi & Sakthieswaran 2015).

To improve the performance of regenerative concrete, Fang et al. found that polypropylene fiber and slag as external admixture concrete mechanical properties were slightly improved in the later period (Fang et al. 2019). Gu et al. found that when the amount of slag powder in regenerative concrete was 50%, its cube compressive strength increased with the substitution rate of waste concrete (Sun & Gu 2014). Zhang et al. studied the influence of activated cinder powder and fly ash blending on the performance of regenerative concrete, and found that it increased the cube compressive strength of regenerative concrete (Zhang et al. 2017).

Mixed fiber concrete is a new composite material, which combines two or more fiber-reinforced materials into concrete so that it cannot only play the advantages of their respective fibers but also reflect the synergistic working effect between fibers (Li 2015). Nicolas et al. believe that the

---

*Corresponding Authors: 3387367492@qq.com, lizhuo991208@gmail.com and 1192891355@qq.com

DOI 10.1201/9781003372417-12

dispersion of mixed fiber has an important impact on the performance of concrete, but excessive mixed fiber makes the performance of concrete reduced (Nicolas et al. 2010). Chi et al. studied that the incorporation of fiber into concrete can improve its cube compressive strength, and the ductility of fiber concrete is better than single fiber concrete. However, there are relatively few studies on natural fiber mixed with concrete (Chi et al. 2014).

To further study the influence of basalt fiber, sisal fiber, and fly ash substitute cement on the mechanical properties of basalt fiber (BF)-sisal fiber (SF)/fly ash regenerative concrete. Four horizontal L16 ($4^5$) orthogonal experiment of basalt fiber volume rate, sisal fiber volume rate, and fly ash cement rate is designed, and the slump experiment of basalt fiber (BF)-sisal fiber (SF)/fly ash regenerative concrete, cube compressive strength and splitting strength were carried out. The regenerative concrete prepared in this paper can improve the shortage of mechanical properties, expand the research of natural fiber in the concrete field, be low cost, be in line with the concept of green building, and has a certain reference value for promoting the application of regenerative concrete engineering.

## 2 EXPERIMENTAL STUDY

### 2.1 Materials

The P.O42.5 ordinary Portland cement is used. Fly ash is a grade fly ash, with a water demand ratio of 95.75%. Basalt fiber (BF) adopts short-cut basalt fiber, and sisal fiber (SF) uses sisal fiber produced by the Guangxi Sisal fiber Group. The main performance parameters of basalt and sword fibers are shown in Table 1.

Table 1. Main performance parameters of basalt fiber (BF) and sisal fiber (SF).

| Fiber | Basait fiber | Sisal fiber |
|---|---|---|
| Density (g/cm$^3$) | 2.64 | 1.45 |
| Splitting tensile strength (MPa) | 4,000 | 510 |
| Modulus of elasticity (GPa) | 100 | 26 |
| Monofilament diameter ($\mu$m) | 12 | 8 |
| Length (mm) | 12 | 15 |
| Elongation at break (%) | 3.1 | 6 |

The sand used in the experiment is the natural fine quartz sand in Lingshou County, Hebei Province, with a silicon content of 97.89% and a fineness modulus of 1.40. Fly ash replaces some cement by 0%, 10%, 20%, 30%, and 40% of the total quality of cement. Water reduction agent for Shaanxi Qinfen brand polycarboxylic acid efficient water reduction agent, water reduction rate is 37%. The recycled coarse aggregate is taken from waste concrete and processed by Yongfeng Factory of Wuhan Zhongjian Commercial Concrete Co., Ltd. with a particle size of 5 mm to 20 mm. The main performance parameters comparing the properties of regenerated crude aggregate and natural aggregate are shown in Table 2.

### 2.2 Experimental design

The orthogonal experiment method is a design method to effectively reduce the number of experiments when the number of total experiment groups is large. Specifically, the experiment (Fang & Ma 2001) will select the representative points from the whole experiment according to the orthogonality. To study the influence of fiber and fly ash on the mechanical properties of basalt fiber (BF)-sisal fiber (SF)/fly ash regenerative concrete, the volume rate $V_b$ of basalt fiber in concrete,

Table 2. Main performance parameters of regenerated crude aggregate.

| The aggregate type | Apparent density (kg/m$^3$) | Water absorption (%) | Component index (%) | Mortar content (%) |
|---|---|---|---|---|
| Natural aggregate | 2788 | 0.6 | 2.7 | 0 |
| Regenerative aggregate | 2482 | 4.8 | 8.7 | 34 |

$V_c$ of sisal fiber, and fly cement generation rate $R_s$ are considered, and four levels are set for each factor, which is listed in Table 3. Experiment factors and levels are three factors and four levels. L16 (4$^5$) orthogonal table is used for experiment design, two columns are empty columns, and C30 reference regenerative concrete are matched as shown in Table 4.

Table 3. Experiment factors and level of basalt fiber-sisal fiber/fly ash regenerative concrete.

| Level | Factor | | | | |
|---|---|---|---|---|---|
| | $V_b$ (A) (%) | $V_c$ (B) (%) | $R_s$ (C) (%) | Blank column (D) | Blank column (E) |
| 1 | 0.1 | 0.1 | 10 | 1 | 1 |
| 2 | 0.2 | 0.2 | 20 | 2 | 2 |
| 3 | 0.3 | 0.3 | 30 | 3 | 3 |
| 4 | 0.4 | 0.4 | 40 | 4 | 4 |

Notes: $V_b$ stands for the volume fraction of basalt fibers in concrete, $V_c$ stands for the volume fraction of sisal fibers in concrete, and $R_s$ stands for the mass fraction ratio of fly ash for cement.

Table 4. Proportion of C30 reference regenerative concrete (kg/m$^3$).

| Types | Cement | Fly ash | Sand | Stone | Water | Water reducer |
|---|---|---|---|---|---|---|
| S-1 | 350 | 0 | 640 | 1040 | 215 | 7 |

Notes: S refers to the reference regenerative concrete.

### 2.3 Experimental piece preparation and experimental method

Basalt fibers were modified with 10% glacial acetic acid solution before use. The incorporation of fiber by dry mixing method is easy to cause fiber formation, so sodium carboxymethyl cellulose (as a fiber dispersant) is added in the preparation process, which is 0.7% of the total mass of the cementitious material. Before use, the sword hemp fiber was soaked in 1% NaOH solution for 30min, then washed with clean water, and dried naturally.

Throughout the preparation process, the gel material, fiber, and sodium carboxymethyl cellulose were mixed dry for 1.5 min, then add sand and stone to the dry mix for 2 min, finally add water reducer and water mixed wet for 2.5 min, and finally make 100 mm × 100 mm × 100 mm cubic concrete experiment. The strength experiment is conducted after standard indoor curing for 28 d under natural conditions. The curing temperature varies from 16 to 24°C, and the humidity is greater than 95%. The concrete slump experiment shall be conducted according to the *Standard for Performance experiment Methods of Ordinary Concrete Mixes* (GB/T 50080-2016), the *Standard for experiment Methods of General Concrete* (GB/T 50081-2002), and the YAW-4206 pressure experimenter of *Metus Industrial System (China) Co., Ltd.* Since the experiment is 100 mm × 100 mm × 100 mm non-standard experiment, the cube compressive strength and splitting tensile strength are converted to the standard experiment block result by the size conversion coefficient. The size coefficient of the cube compressive strength is 0.95 and the splitting tensile strength is 0.85 (GB/T 50081-2002).

## 3 RESULTS AND DISCUSSIONS

The slump degree, cube compressive strength, splitting tensile strength, and tensile pressure ratio of 1 C30 reference regenerative concrete and 16 basalt fiber (BF)-sisal fiber (SF)/fly ash regenerative concrete are shown in Figure 1. It can be seen from the table that the cube compressive strength and splitting tensile strength of 16 basalt-sisal fiber/fly ash regenerative concrete are higher than that of C30 base regenerative concrete, and the maximum cube compressive strength of the cube appears in ZJ-3, which is 27.6% higher compared with the base regenerative concrete. The maximum splitting tensile strength of cracking appears in ZJ-15, which is 44.8% higher than that compared with the base regenerative concrete. To investigate the influence of fiber and fly ash on the slump degree, cube compressive strength, splitting tensile strength, and the tensile ratio of basalt fiber (BF)-sisal fiber (SF)/fly ash regenerative concrete, the data in Figure 1 were analyzed by the data processing and analysis software SPSS. The results of the analysis of slump, cube compressive strength, cube splitting strength, and the tensile ratio of $V_b$, $V_c$, and $R_s$ are listed in Figure 2 and Figure 3, respectively. $B_{ij}$ in Figure 2 represents the mean of experimental results for the $i$-th factor at the $j$-th level, which can be expressed as:

$$B_{ij} = \frac{\sum_{m=1}^{n} K_{ij,m}}{n} \tag{1}$$

where $K_{ij,m}$ is the $m$-th result of factor $i$ under level $j$; $n$ is the number of results of factor $i$ under level $j$; $R_i$ represents the difference between the $B_{ij}$ maximum and minimum values at each level, i. e., $R_i \cdot \max \cdot B_{ij} \cdot \min \cdot B_{ij}$.

As can be seen in Figure 2, the extreme difference between the three factors for the slump, cube compressive strength, splitting tensile strength, and tensile ratio are greater than the empty column, indicating that the orthogonal experiment results are credible (Chen, 2011), and the experiment results can accurately reflect the change law of mechanical properties of basalt-sisal fiber/fly ash regenerative concrete.

### 3.1 *Basalt fiber-sisal fiber/ fly ash regenerative concrete slump*

As can be seen from Figure 2, the influence degree of the three factors on the slump degree of new mixed concrete is basalt fiber (119.5 mm) > sisal fiber (68.25 mm) > fly ash (32.25 mm). As $V_b$, $V_c$, and $R_s$ increase, the concrete slump gradually. $V_b$ increased from 0.1% to 0.4% and concrete slump decreased by 63.6%. The $V_c$ increased from 0.1% to 0.4%, and the concrete slump degree fell by 37.7%. This shows that the concrete slump decreases. The main reason is that the fiber will form a non-directional support system inside the concrete, which hinders the separation of the concrete components; the second reason is that the incorporation of the fiber consumes the cement mortar originally used to wrap the coarse aggregate and the fine aggregate, which causes the cement mortar to weaken the sliding flow effect of the aggregate, and reduces the slump degree of the new concrete. $R_s$ increased from 10% to 40% and the slump decreased by 32%. The reason is that the fly ash particles are porous honeycomb structures, fly ash has strong water absorption, and water absorption is greater than cement. But the loss of cement is not as obvious as fiber.

Slump is a direct factor of the working performance of regenerative concrete, considering the best combination is $A_1B_1C_1$, namely 0.1% $V_b$, $V_c$ 0.1%, and $R_s$ 10%. According to the results of

Figure 1. Experiment results of slump and strength of basalt fiber-sisal fiber/fly ash regenerative concrete.

| Material types | Factor | | | Concrete slump (mm) | Cube compressive strength $f_{cu}$ (MPa) | Splitting tensile strength $f_{ts}$ (MPa) | Tension compression ratio $(f_{ts}/f_{cu})$ |
|---|---|---|---|---|---|---|---|
| | A (%) | B (%) | C (%) | | | | |
| S-1 | 0 | 0 | 0 | 175 | 34.3 | 2.61 | 0.076093 |
| ZJ-1 | 0.1 | 0.1 | 10 | 165 | 37.2 | 3 | 0.080645 |
| ZJ-2 | 0.1 | 0.2 | 20 | 144 | 38.9 | 3.09 | 0.079434 |
| ZJ-3 | 0.1 | 0.3 | 30 | 127 | 43.8 | 3.48 | 0.079452 |
| ZJ-4 | 0.1 | 0.4 | 40 | 83 | 37.8 | 3.47 | 0.091798 |
| ZJ-5 | 0.2 | 0.1 | 20 | 137 | 38.2 | 3.41 | 0.089267 |
| ZJ-6 | 0.2 | 0.2 | 10 | 122 | 37.9 | 3.29 | 0.086807 |
| ZJ-7 | 0.2 | 0.3 | 40 | 80 | 38.2 | 3.54 | 0.092670 |
| ZJ-8 | 0.2 | 0.4 | 30 | 93 | 40.3 | 3.6 | 0.089330 |
| ZJ-9 | 0.3 | 0.1 | 30 | 107 | 41.9 | 3.7 | 0.088305 |
| ZJ-10 | 0.3 | 0.2 | 10 | 91 | 37.3 | 3.56 | 0.095442 |
| ZJ-11 | 0.3 | 0.3 | 40 | 73 | 40.5 | 3.54 | 0.087407 |
| ZJ-12 | 0.3 | 0.4 | 20 | 65 | 42.3 | 3.68 | 0.086997 |
| ZJ-13 | 0.4 | 0.1 | 40 | 70 | 35.6 | 3.71 | 0.104213 |
| ZJ-14 | 0.4 | 0.2 | 30 | 44 | 40.5 | 3.75 | 0.092592 |
| ZJ-15 | 0.4 | 0.3 | 20 | 36 | 41.1 | 3.78 | 0.091970 |
| ZJ-16 | 0.4 | 0.4 | 10 | 20 | 39.7 | 3.63 | 0.091435 |

Notes: S stands for reference regenerative concrete; ZJ stands for basalt fiber-sisal fiber/fly ash regenerative concrete.

variance analysis of basalt-ma fiber/fly ash regenerative concrete slump in Figure 3, basalt fiber is a significant factor of basalt fiber (BF)-sisal fiber (SF)/fly ash regenerative concrete slump, followed by sisal fiber, and fly ash is a non-significant factor.

### 3.2 Cube compressive strength

As can be seen from Figure 2, the influence degree of the three factors on the cube compressive strength of concrete is fly ash (6.5 MPa) > basalt fiber (3.9 MPa) > sisal fiber (3.8 MPa). The effect of both fibers on the cube compressive strength was insignificant, and the overall change range was within 7.4%. $V_b$ and $V_c$ increased from 0.1% to 0.3%, the cube compressive strength increased; $V_b$ and $V_c$ increased from 0.3% to 0.4%, and the cube compressive strength decreased. The cube compressive strength of concrete cube increases with the incorporation of fiber and then decreases, which shows that the reasonable fiber volume rate is conducive to the improvement of cube compressive strength, and too high a fiber volume rate will appear negative confounding effect. The reason is that the appropriate amount of fiber is mixed, the fiber can be evenly dispersed in the concrete and form a certain bonding force between the mortar, producing a certain crack resistance, and the non-directional support system formed by the fiber can also bear the load together with the concrete, improve the cube compressive strength. However, too many fibers will make it gather in the concrete interior, forming a stress concentration point, reducing the effective utilization rate of the fiber, resulting in the reduction of the cube compressive strength of the concrete cube.

$R_s$ increased from 10% to 40% and the cube compressive strength increased by 12.3% but decreased as $R_s$ increased from 30% to 40%. It shows that the appropriate use of fly ash can enhance the cube compressive strength of concrete, but the high rate of fly cement is not conducive to improving the cube compressive strength of concrete. The analysis reason is that the fly ash material itself does not have the water hard cementing effect, but if the powder exists simultaneously with water, water thermalization phenomenon occurs, and calcium hydroxide or other alkaline oxide chemical reaction, the production of compounds with water hardening cementing, can improve the

Figure 2. Range analysis results of slump and strength of basalt fiber-sisal fiber/fly ash regenerative concrete.

| Examination index | Range calculation | Basalt fiber | Sisal fiber | Fly ash | Blank-column (D) | Blank-column (D) |
|---|---|---|---|---|---|---|
| Concrete slump (mm) | $B_{i1}$ | 187.75 | 181.25 | 99.75 | 90.00 | 97.75 |
| | $B_{i2}$ | 163.25 | 158.75 | 91.25 | 87.00 | 96.25 |
| | $B_{i3}$ | 138.50 | 132.50 | 82.25 | 98.25 | 88.00 |
| | $B_{i4}$ | 68.25 | 113.00 | 67.25 | 96.25 | 91.25 |
| | $R_i$ | 119.50 | 68.25 | 32.50 | 11.25 | 9.75 |
| Cube compressive strength $f_{cu}$(MPa) | $B_{i1}$ | 52.5 | 52.4 | 52.8 | 53.7 | 53.1 |
| | $B_{i2}$ | 54.0 | 53.0 | 54.3 | 53.7 | 53.6 |
| | $B_{i3}$ | 56.4 | 56.2 | 56.0 | 51.6 | 53.0 |
| | $B_{i4}$ | 54.8 | 55.8 | 49.5 | 52.6 | 53.8 |
| | $R_i$ | 3.9 | 3.8 | 6.5 | 2.1 | 0.8 |
| Splitting tensile strength $f_{ts}$(MPa) | $B_{i1}$ | 4.33 | 5.14 | 5.08 | 5.20 | 5.20 |
| | $B_{i2}$ | 5.18 | 5.10 | 5.20 | 5.33 | 5.30 |
| | $B_{i3}$ | 5.42 | 5.30 | 5.55 | 5.23 | 5.30 |
| | $B_{i4}$ | 5.83 | 5.45 | 5.34 | 5.27 | 5.20 |
| | $R_i$ | 1.5 | 0.35 | 0.47 | 0.13 | 0.1 |
| Tension compression ratio ($f_{ts}/f_{cu}$) | $B_{i1}$ | 0.120356 | 0.128470 | 0.131041 | 0.123307 | 0.127403 |
| | $B_{i2}$ | 0.125393 | 0.126300 | 0.129003 | 0.125001 | 0.128311 |
| | $B_{i3}$ | 0.125181 | 0.126008 | 0.130241 | 0.127348 | 0.129945 |
| | $B_{i4}$ | 0.160300 | 0.127130 | 0.139340 | 0.126318 | 0.126787 |
| | $R_i$ | 0.039944 | 0.002462 | 0.010337 | 0.004041 | 0.003158 |

Notes: $B_{ij}$ is the average of the experiment results of factor $i$ at level $j$; $R_i$ stands for the range value of factor $i$.

Figure 3. Variance analysis of slump and strength of basalt fiber-sisal fiber/fly ash regenerative concrete.

| Examination index | Sources of variation | $Ss$ | $D_f$ | $M_s$ | $F$ | Sig |
|---|---|---|---|---|---|---|
| Concrete slump (mm) | Basalt fiber | 23398.250 | 3 | 5849.563 | 38.718 | ** |
| | Sisal fiber | 13507.750 | 3 | 3376.938 | 12.259 | * |
| | Fly ash | 7178.250 | 3 | 1794.562 | 1.887 | – |
| | Error | 5343.235 | 6 | 1021.239 | | |
| Cube compressive strength $f_{cu}$(MPa) | Basalt fiber | 32.137 | 3 | 8.034 | 1.491 | – |
| | Sisal fiber | 43.257 | 3 | 10.814 | 2.425 | * |
| | Fly ash | 54.272 | 3 | 13.568 | 3.830 | ** |
| | Error | 8.543 | 6 | 0.943 | | |
| Splitting tensile strength $f_{ts}$ (MPa) | Basalt fiber | 1.250 | 3 | 0.313 | 13.438 | *** |
| | Sisal fiber | 0.864 | 3 | 0.216 | 3.893 | ** |
| | Fly ash | 0.792 | 3 | 0.198 | 3.223 | * |
| | Error | 0.097 | 6 | 0.088 | | |
| Tension compression ratio ($f_{ts}/f_{cu}$) | Basalt fiber | 0.000 | 3 | 2.379E-5 | 24.862 | ** |
| | Sisal fiber | 0.000 | 3 | 4.525E-5 | 1.957 | – |
| | Fly ash | 0.000 | 3 | 2.810E-5 | 13.655 | * |
| | Error | 1.810E-5 | 6 | 2.010E-6 | | |

Notes: $S_S$ stands for the sum of squares, $D_f$ stands for the degree of freedom, $M_S$ stands for mean square, and $F$ represents $F$-value. "***" represents highest marked, "**" represents highly marked, "*" represents marked, and "-" represents no marked. $F_{0.05}(3, 6) = 4.76$.

strength and hardness of concrete material. But when the fly ash is too much, the hydration reaction is difficult to proceed thoroughly, thus reducing the strength of the concrete.

Thus, the optimal combination is $A_3B_3C_3$, that is, 0.3% for $V_b$, $V_c$ 0.3%, and $R_s$ 30%. According to the results of variance analysis of cube compressive strength of regenerative concrete cube in Figure 3, fly ash is a highly significant factor of basalt fiber (BF)-sisal fiber (SF)/fly ash regenerative concrete cube, sisal fiber is a significant factor, and basalt fiber is a non-significant factor.

### 3.3 *Splitting tensile strength*

As can be seen from Figure 2, the influence degree of the three factors on the cracking splitting tensile strength of concrete is basalt fiber (1.5 MPa) > fly ash (0.47 MPa) > sisal fiber (0.35 MPa). Basalt fiber and Sisal fiber both increased the splitting tensile strength. $V_b$ and $V_c$ increased from 0.1% to 0.4%, and the splitting tensile strength increased by 34.6% and 6.9%, respectively; this indicates that the mixture of two fibers increased the splitting tensile strength of concrete. The effect of $R_s$ on cleavage strength tended to increase first and then decrease. The reason is that due to the incorporation of fly ash, the bonding effect between fine aggregate and cement slurry is enhanced, and the splitting tensile strength is somewhat improved. However, the excessive incorporation of fly ash is difficult to thoroughly hydrate and the splitting tensile strength is reduced.

Therefore, the optimal combination is $A_4B_4C_3$, that is, 0.4% for $V_b$, $V_c$ 0.4%, and $R_s$ 30%. According to the variance analysis of concrete splitting tensile strength in Figure 3, basalt fiber is the highly significant factor of concrete splitting tensile strength; sisal fiber and fly ash are the significant factors of splitting tensile strength of basalt fiber (BF)-sisal fiber (SF)/fly ash regenerative concrete.

### 3.4 *Tension compression ratio*

It can be seen from Figure 2 that the influence degree of the three factors on the concrete tensile pressure ratio is basalt fiber (0.0399) > fly ash (0.0103) > sisal fiber (0.0024). According to the variance analysis of concrete tensile pressure ratio in Figure 3, basalt fiber is a significant factor in concrete tensile pressure ratio; fly ash is a significant factor in concrete tension compression ratio, and sisal fiber is a non-significant factor.

## 4 CONCLUSIONS

1. The splitting tensile strength enhancement of basalt fiber and sisal fiber on basalt fiber (BF)-sisal fiber (SF)/fly ash regenerative concrete is greater than the cube compressive strength enhancement of the cube.
2. The significant factor in the tension ratio of basalt-sisal fiber/fly ash regenerative concrete is basalt fiber and fly ash, and sisal fiber has less influence on the tension ratio of basalt fiber (BF)-sisal fiber (SF)/fly ash regenerative concrete.
3. The incorporation of fiber and fly ash for cement will lead to reducing the slump of new basalt-sisal fiber/fly ash regenerative concrete. Basalt fiber (BF) and sisal fiber (SF) are the significant factors of a regenerative concrete slump, and fly ash is the non-significant factor of a regenerative concrete slump.
4. The cube compressive strength and splitting tensile strength of basalt fiber (BF)-sisal fiber (SF)/fly ash regenerative concrete are higher than those of C30 reference regenerative concrete.
5. The insufficient orthogonal experimental data used in this study is not conducive to establishing an accurate and complete strength prediction model between the mechanical properties of regenerative concrete and basalt fiber volume rate and fly ash replacement rate of cement.
6. The next research plan of this paper is to conduct regression analysis on the results of the orthogonal experiment, initially establish the intensity prediction model, and expand the experiment to obtain more experiment data to improve the model to meet the needs of engineering application.

## ACKNOWLEDGMENT

This research was funded by the Jiangsu 2020 Students' Innovation and Entrepreneurship Training Program (Grant Nos. 202011276001Z and 202111276022Z).

## REFERENCES

Ananthi P, Sakthieswaran N. (2015). Jute Fibre Wrapped RC Short Column. *International Journal for Innovative Research in Science and Technology*, 2015, 2(1):54–60.

Chen Y. (2011). A significance experiment in multi-level or- thogonal designs with only one replicate[J]. *Chinese Journal of Applied Probabil-ity*, 2011, 27(5):497–510 (in Chinese).

Chi Y, Xu L, Zhang Y. (2014). Experimental study on hybrid fiber-reinforced concrete subjected to uniaxial compression[J]. *Journal of Materials in Civil Engineering*, 2014, 26(2): 211–218.

Fang Jinmiao, Tu Jinsong, Li Wenli. (2019). Study on regenerative concrete properties of mixed propylene fiber and slag[J]. *Journal of Chengdu University* (Natural Science edition), 2019, 38 (02): 218–222.

Fang Kaitai, Ma Changxing. (2001). *Orthogonal and uniform experimental design* [M]. Beijing: Science press, 2001(in Chinese).

GB/T 50080-2016. Standardization Administration of the People's Republic of China. *Standard for experiment method of performance on ordinary fresh concrete: GB/T 50080-2016*[S], Beijing: China Standards Press, 2016 (in Chinese).

GB/T 50081-2002. Standardization Administration of the People's Republic of China. *Standard for experiment method of mechanical properties on ordinary concrete: GB/T 50081-2002*[S], Beijing: China Standards Press, 2002 (in Chinese).

Li Xibo. (2015). *Constitutive relationship of mixed fiber high strength concrete*[D]. Guangzhou: Guangzhou University, 2015.

Nicolas Ali Libre,Mohammad Shekarchi,Mehrdad Mahoutian,Parviz Soroushian. (2010). Mechanical prop- erties of hybrid fiber reinforced lightweight aggregate concrete made with natural pumice[J]. *Construction and Building Materials*, 2010, 25(5).

Ravindrarajah.R., Tam.C.T. (1985). Properties of Concrete Made with Crushed Concrete as Coarse Aggregate. *Magazine of Concrete Research*. 1985, 37(130):29~38.

Sun Jiaguo, Gu Yanling. (2014). Analysis[J]. *Journal of Chongqing University of Science and Technology* (Natural Science edition), 2014, 16 (03): 109-111 + 160.

Topcu I. B. (1997). Physical and mechanical properties of concrete produced with waste concrete[J].*Cement and Concrete Research*, 1997, 27 (12):1817–1823.

Zhang Liming, Li Jia, Liu Fuming. (2017). Effect of activated cinder powder blending on the properties of regenerative concrete[J]. *Concrete*, 2017 (05): 72–74.

*Civil Engineering and Urban Research – Mohamed & Hou (Eds)*
© 2023 the Authors, ISBN 978-1-032-44487-1

# Flexural strength analysis of steel-FRP hybrid-reinforced concrete beam with double reinforced rectangular section

Yueqi Chen*
*School of Transportation and Geomatics Engineering, Shenyang Jianzhu University, Shenyang, China*

Jia Tang
*School of Hydraulic and Electric Power, Heilongjiang University, Heilongjiang, China*

Jie Zhu
*School of Urban Construction, Zhejiang Shuren University, Zhejiang, China*

ABSTRACT: Fiber-reinforced polymer (FRP) has become a substitutable concrete reinforced material due to its high tensile strength, lightweight, and non-corrosion problem relative to the steel bars. However, the elastic modulus of FRP is many times smaller than the modulus of rebars. the modulus of GFRP in this paper is only about 20% of rebars. Therefore, the amount of deflection value is presented in FRP-reinforced concrete (FRP-RC) beams, and actual engineering use may be affected. Finite element (FE) modeling software ABAQUS is used to build the FE models including a steel-reinforced concrete beam, an FRP-reinforced concrete beam, and hybrid-reinforced concrete beams to simulate the numerical bending experiments. The simulation results demonstrate that the yield step appears more clearly with the ratio of rebars increasing in hybrid-reinforced concrete beams, but flexural capacity under the condition of high deflection deformation is reduced compared with FRP-RC beams. It can be concluded that by adjusting the appropriate proportion between FRP bars and rebars the requirements of bending bearing capacity can be satisfied in actual engineering.

## 1 INTRODUCTION

Fiber-reinforced polymer (FRP) is a kind of high-performance composite material formed consisting of mixing fiber material as loading components and matrix material into a certain proportion. FRP bars have become a substitute for steel in various aspects of engineering materials nowadays. For example, Ashrafi studied the fire resistance of FRP bars and Yi et al. conducted durability experiments on FRP bars and reinforced concrete (Ashrafi et al. 2017; Denvid & Hoat 2010; Yan & Zhang 2001; Yi et al. 2005). Compared with traditional steel bars, FRP is widely utilized in the aeronautic, mechanical, and chemical industries due to its superior properties including durability, fire resistance, strong chemical corrosion resistance, high tensile strength, low weight, and non-electrical conductivity as compared to traditional steel bars (Aiello & Ombres 2000; Hollaway 1978; Peece et al. 2000). With the development of research and discovery of FRP, it has been widely used in the field of civil engineering structure. In seasonal countries such as Switzerland, using FRP bars is one of the solutions to the problem of traditional steel bars being damaged using ice salts to prevent ice from forming on roads (Denvid & Hoat 2010). During the construction process, various engineering quality problems occasionally occur. Building collapse is one of the most serious structural damage accidents. That is a good and feasible method to reinforce FRP bars and mix

---

*Corresponding Author: 18407285@masu.edu.cn

DOI 10.1201/9781003372417-13

them with traditional steel bars in bridge buildings to avoid accidents and earthquakes. At the same time, FRP fiber material also has the shortcomings of poor plasticity and low thermal stability. Shanghai transit Line 15 is a successful application of glass fiber in bridge engineering, which is faced with the problems of complicated municipal lines and difficult construction, especially in the vicinity of the Bridge section of Zhangjiatang Port. During the construction, the experts involved in the project also need to consider the powerful influence of the river on the section and the corrosion of the steel bar itself (Zhou 2018). Hence, in the process of material selection, through the comparison of materials, experts took full advantage of the FRP bars. Therefore, the concrete material with FRP bars is used to replace some ordinary reinforced concrete materials, aim to save cost, shorten the construction period, and avoid a lot of safety risks.

In summary, with the wide application of new materials, a lot of analysis and experiments on the flexural and bonding properties of mixed FRP reinforced concrete beams have been carried out and corresponding results were obtained. Liu et al. studied the flexural strength, deflection, and crack behavior of high-strength concrete beams strengthened with GFRP bars according to experiments, and verified the influence of reinforcement layer layout and support design of mixed reinforced concrete structures (Liu & Yuan 2013). The research of Ge et al. determined the occurrence standard and flexural capacity of mixed reinforced concrete beams with ideal flexural failure modes according to the bending behavior of reinforced concrete beams erected basalt fiber reinforced polymer (BFRP) analysis (Ge, et al. 2015). Wu et al. (2013) conducted an experimental study on the bonding performance of glass fiber and concrete.

In this paper, ABAQUS finite element analysis software will be used to conduct numerical simulation experiments on the flexural capacity of four groups of concrete beams, including a pure reinforced concrete beam, a pure FRP reinforced concrete beam, and two mixed concrete beams with reinforcement and FRP reinforcement. By changing the proportion relationship between reinforcement and FRP reinforcement in the tensile area, the relationship between the flexural capacity of the hybrid-reinforced concrete beams and the number of reinforcement and FRP reinforcement will be explored.

## 2 THE BEAM MODEL

To make better use of FRP bars in the engineering field, taking advantage of the tensile zone reinforcement can bear a higher bearing capacity. Besides, the compression reinforcement reduces the height of the concrete in the compression zone. The ductility of the section is also improved, and the seismic effect is greatly strengthened, which is more in line with the characteristics of the current engineering needs. Therefore, this paper adopts two rectangular cross-section beams with double bars longitudinally arranged as the basic model, further following the needs of actual engineering characteristics.

In this paper, four specimens of FRP bars and traditional steel bars are adopted to simulate the failure of concrete structures under variable concentrated load. The specimens were designed as BS#. The stirrups in the model can make the tensile reinforcement fully play a role in the destruction. All specimens are shown in Table 1. The thickness of the concrete protective layer is 20 mm. The frame ribs and stirrups in this paper are HRB400 $\phi$10 mm and HRB300 $\phi$8 mm @100 respectively. The stirrup in the model can make the tensile reinforcement fully play a role in the destruction. In addition, in the reinforcement process, the size of the double-reinforced rectangular section beam is 2,400 mm long, 150 mm wide, and 300 mm high, which is shown in Figure 1. And the material characteristics of concrete are shown in Table 2, where $f_{cu}$ is the concrete strength, $f_c$ is the concrete compression strength, $f_t$ is the concrete tensile strength, $E_c$ is the elastic modulus of concrete, and $\upsilon_c$ is the concrete poison's ratio.

Concrete-concrete material nonlinearity is obtained by implementing the concrete plastic failure model, and concrete plastic parameters. The parameters in the concrete damage plastic model are as shown in Table 2, where $f_{b0}$, and $f_{c0}$ are the ratio of double-shaft compressive strength to the uniaxial compressive strength of concrete; K is the parameter that determines the yield surface

Table 1. Parameters of tensile bars.

| Specimens | Reinforcement material | Tensile region layout |
|---|---|---|
| BS1 | 3 $\times\phi$14 steel bars | o o o* |
| BS2 | 2 $\times\phi$14 steel bars + 1 $\times\phi$14 GFRP bar | o ● o |
| BS3 | 1 $\times\phi$14 steel bar + 2 $\times\phi$14 GFRP bars | ● o ● |
| BS4 | 3 $\times\phi$14 GFRP bars | ● ● ● |

*The hollow circle "o" are traditional steel bars and the red solid circle "●" are FRP.

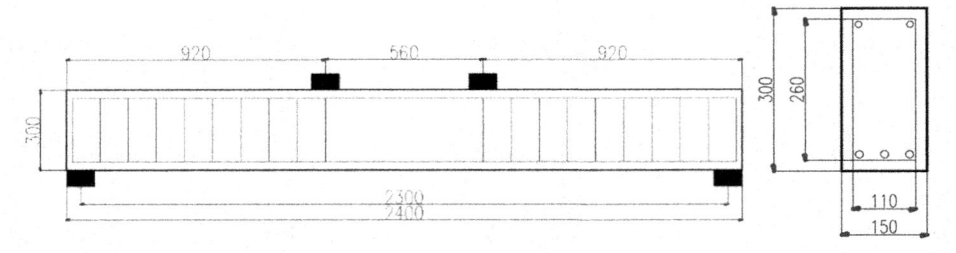

Figure 1. Details of the beam model.

Table 2. Concrete characteristics.

| The elastic properties | | | | | The shape features | | | | |
|---|---|---|---|---|---|---|---|---|---|
| $f_{cu}$ | $f_c$ | $f_t$ | $E_c$ | $\upsilon_c$ | Dilation angle | Eccentricity | $f_{b0}/f_{c0}$ | K | Viscosity parameter |
| MPa | MPa | MPa | GPa | / | ° | / | / | / | / |
| 40 | 26.8 | 2.51 | 32599.84 | 0.2 | 30 | 0.1 | 1.16 | 0.6667 | 0.0005 |

form. The parameters in Table 2 are taken from GB50010-2010. The stress of compressive and tensile concrete as shown in Figures 2 and 3.

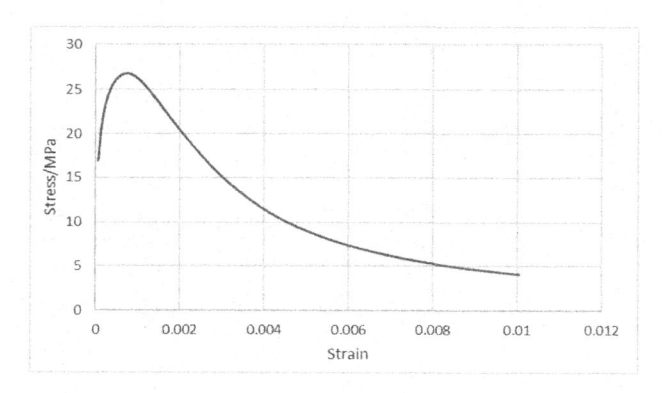

Figure 2. Concrete compressive stress-strain curve.

## 3 METHODOLOGIES

Engineering mechanics problems can be roughly divided into linear deformation and nonlinear deformation in two different scopes. This paper discusses the problems within the nonlinear scope

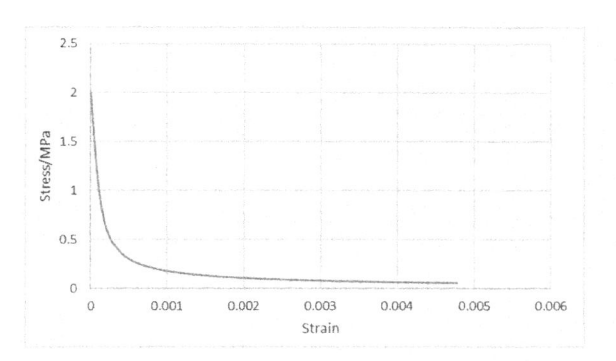

Figure 3.    Concrete tensile stress-strain curve.

and uses ABAQUS software to model and analyze them. It has the advantages of strong nonlinear analysis ability, fast numerical calculation speed, high precision of calculation results, and relatively low analysis cost. ABAQUS also has a user-friendly interface, which can make the results more intuitive visualization. The ABAQUS numerical simulation software is used to compare the BS1 and other four groups of experimental models and obtain the load-displacement curve image of the beam. The influence of the different amounts of FRP reinforcement on the beam is studied, and through the comparison with ordinary traditional steel, judge the advantages of FRP materials (Aiello 2000; Newhook 2000), the results of this numerical simulation test can also be used as the numerical basis for subsequent construction applications or laboratory experiments.

Numerical simulation analysis is the method used in this study. The used software is ABAQUS 2022. After the above model is put into the software through the assembly module in ABAQUS, the contact interaction is established. It should be emphasized that in this study, the rigid pads are given a higher elastic modulus to simulate their rigidity better. All rigid pads relate to the concrete by tie contact, and all bars are implanted into the concrete by embedded contact. At the same time, the reference points named RP-#, are set at the center points of the upper (lower) surfaces of four rigid pads. There are provided with coupling contact interaction on the reference points and the surface of the pads, corresponding boundary constraints are set at the RP of the lower pads, and displacement loading is set at the upper RP. The static and general analysis method is adopted in this finite element numerical simulation analysis, and the analysis step period is 1 s. In the mesh module, the mesh size of concrete and pads is 40 mm, and the element type is C3D8R, 8-node solid element. Because in RC-beam, the bars almost only bear axial force, so in this model, to simplify the calculation process and improve calculation speed, all kinds of bars are defined as T3D2, 2-node linear, 3-D truss element, and the element mesh size is 35 mm. The constitutive relations used in this study are shown in the above Figures 2 and 3. Thereinto, three models are given in ABAQUS to define the damage of concrete, namely the polished crack concrete model, brittle crack concrete model, and concrete damage plasticity (CDP) model. The CDP model allows to define the strain hardening during compression and can be defined as the corresponding strain rate sensitive, which is more consistent with the behavior of concrete. Therefore, this model uses the CDP model to define the damage to concrete. In the model, the elastic and plastic definitions are used to simulate the ideal elastic-plastic of reinforcement, and the elastic definition is used to simulate FRP reinforcement. The mesh structure and embedded reinforcement cage of the concrete beam are shown in Figure 4.

To intuitively reflect the failure state of concrete beams, the concept of damage factor is introduced in this paper to output the damage to concrete. As shown in Figure 5, the figure shows the cloud map of concrete beam compression damage in the BS4 group. It can be seen from the image that the concrete compression area is seriously damaged and finally crushed, while Figure 6 is the cloud map of reinforcement stress at the same analysis step time. Therefore, the FRP bars in the

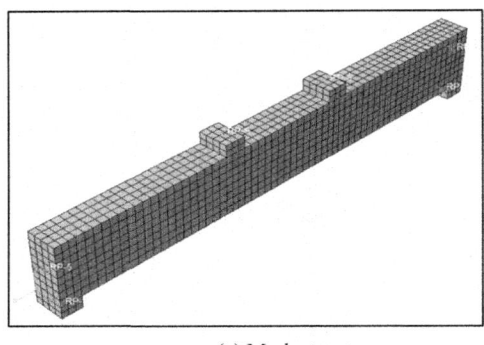 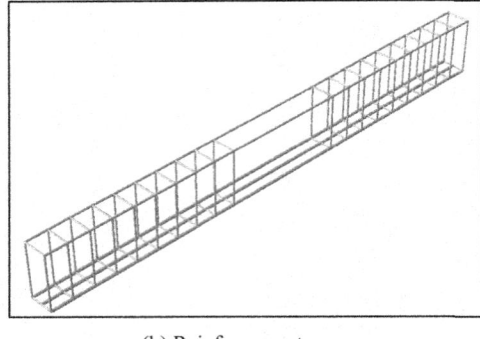

(a) Mesh structure　　　　　　　(b) Reinforcement cage

Figure 4.　The FE models.

tensile area do not reach the ultimate strength, which is the beam failure caused by the compressive failure of concrete. The failure forms of other groups of concrete beams are also the crushing of concrete.

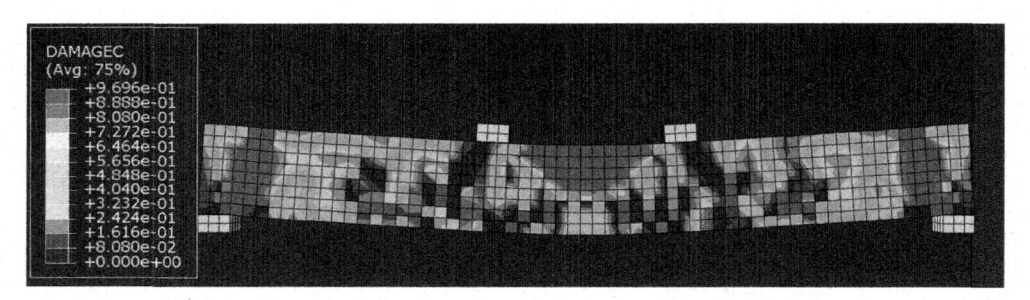

Figure 5.　The DAMAGE image of concrete in the BS4 group.

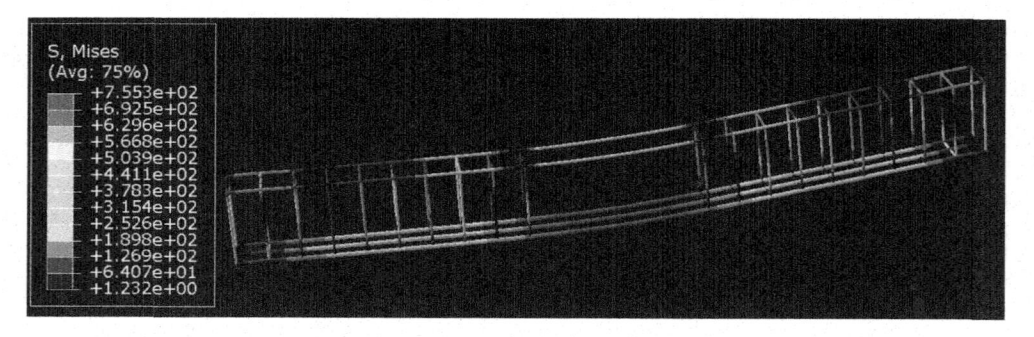

Figure 6.　The cloud map of reinforcement stress in the BS4 group.

## 4　RESULTS AND DISCUSSION

This work uses post-processing to extract the loading point's vertical displacement (U2) and the vertical response force (RF2), as well as the load-deflection curves for each group of models.

Figure 7(a) depicts the load-deflection curves of four groups of concrete beams. In the BS1 diagram, the obvious yield step in the load-deflection curve can be seen, and the rebars in the tensile area yield at a load of 130 kN, after which the curve is almost a flat straight line. The BS4 group is made entirely of FRP-reinforced concrete. It has a load-deflection curve with no yield step and is nearly straight. The concrete in the compression area gets crushed and damaged when the load exceeds roughly 178 kN. At the same time, the load-deflection curves of these two groups of pure steel reinforced concrete beams and pure FRP reinforced concrete beams are compared with those of pure reinforced concrete beams and pure FRP reinforced concrete beams obtained by Linh et al. (2018). The trend and shape of the curve are almost identical, which also verifies the accuracy of the model used in this paper. The curves mentioned in the paper are shown in Figure 7(b).

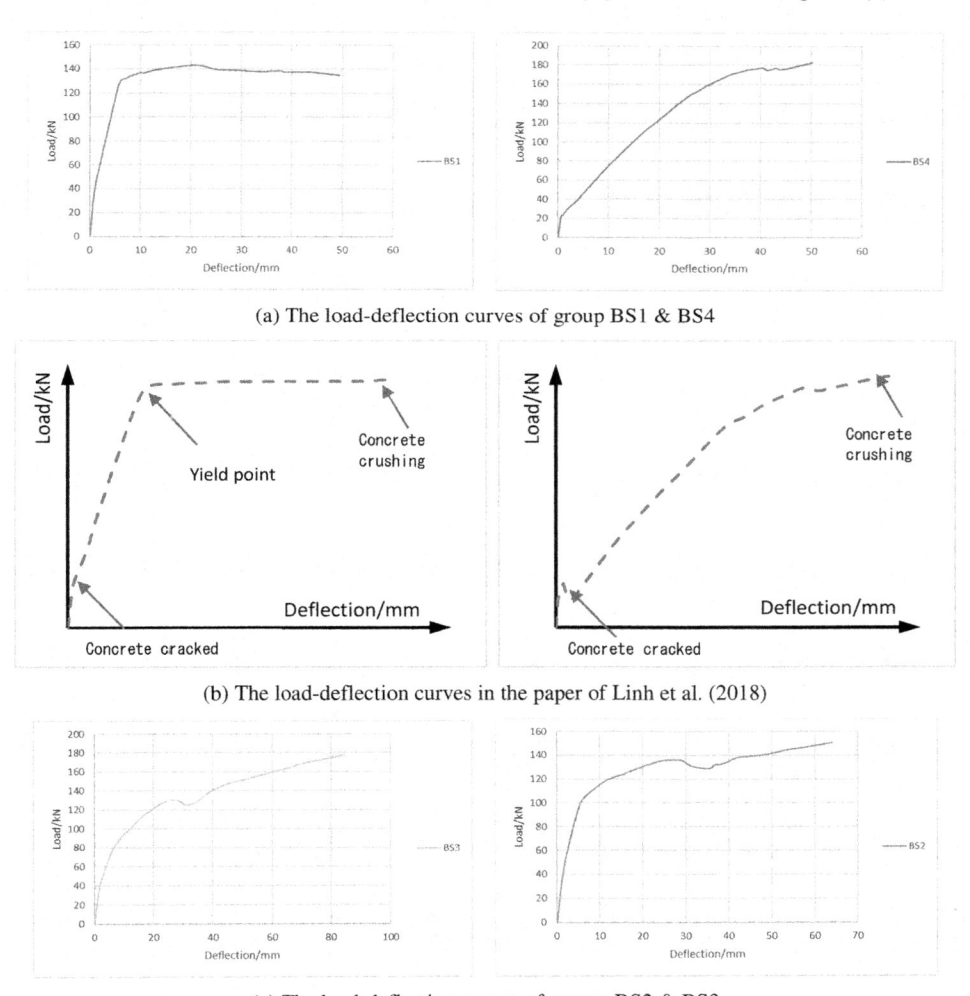

(a) The load-deflection curves of group BS1 & BS4

(b) The load-deflection curves in the paper of Linh et al. (2018)

(c) The load-deflection curves of groups BS2 & BS3

Figure 7.    The Load-Deflection curves of a single specimen.

In Figure 7(c), although the yield point is not as obvious as that of the BS1 group, the image can still be roughly divided into two sections: before and after the yield point. The values of load and deflection of the BS2 group relatively are small at the yield point, which is 100 kN and 5 mm respectively, and the values of load and deflection of the BS3 group are relatively large at the yield point, which are 130 kN and 25 mm, respectively. Both have the trend of load increasing with displacement, but the concrete strength in the compression area is not enough, so FRP bars are unable to give full play to their bending performance inadequately reinforced concrete beam.

According to Figure 8(a), the traditional steel-reinforced concrete beam (BS1) and FRP-reinforced concrete beam (BS4) show two distinct characteristics. Since the steel-reinforced concrete cracks, the rebars in the tensile area participate in the work, and its deflection development is not obvious. With the steady increase of load, the specimen has the ultimate load at the displacement of about 20 mm, while the ultimate load value is about 145 kN. However, once the FRP-reinforced concrete cracks, the displacement increases sharply with the load until the concrete is crushed and damaged. The deflection of this specimen is about 40 mm, while the ultimate load value is about 180 kN. Moreover, the deflection is about twice that of the steel-reinforced concrete beam, but the ultimate load value is about 24.1% larger than that of the steel-reinforced concrete beam.

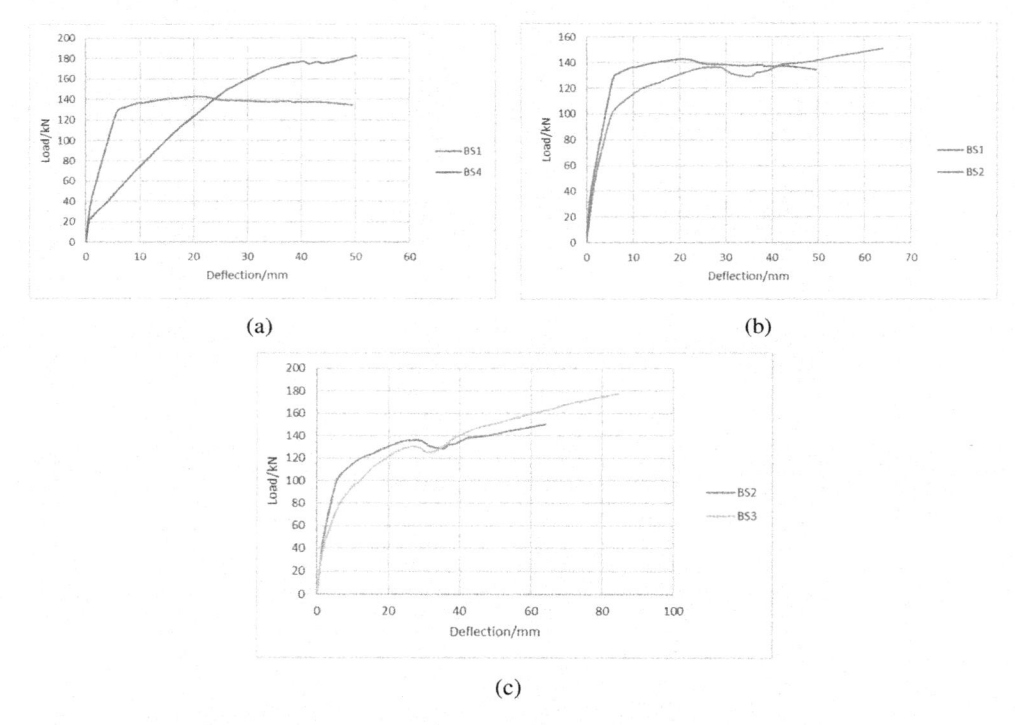

(a)

(b)

(c)

Figure 8.    Comparison diagram of load-deflection curves

Compared with the BS1 group, a load of tensile reinforcement in the BS2 group still increases uniformly after yielding. However, the steel rebars constitutive relation in this paper is regarded as a two-line model, which does not consider the stiffness hardening of steel rebars. Therefore, the load of the steel-reinforced concrete beam after rebars yielding hardly increases with the increase of deflection, but the displacement value where ultimate strength appeared is significantly smaller than that of the other three groups. In Figure 8(c), the comparison between the BS2 group and the BS3 group can also verify the above statement. By observing the load-displacement curve, the yield point of the BS2 group appears at the displacement of 5 mm, and the load value is 100 kN, while the yield point of the BS3 group appears at the displacement of 10mm, and the load value is also 100 kN, but the subsequent load growth value of BS3 group is significantly higher than that of BS2 group.

## 5 CONCLUSION AND FUTURE WORK

In this paper, the numerical simulation experiments of flexural skills of several specimens are proceeded by the usage of FE analysis software, and the mechanical property and ductility of a steel-reinforced concrete beam, an FRP-reinforced concrete beams, and two hybrid-reinforced concrete beams are compared. The conclusions can be drawn as follows.

(1) There is almost no fracture failure of FRP bars in the FRP-reinforced concrete beam and hybrid-reinforced beams, because of the characteristics of FRP with high tensile strength and low elastic modulus. The FRP-RC tends to compressive crushed but not FRP bars brittle failure. Therefore, when FRP reinforcement or hybrid reinforcement is used in actual engineering projects, over-reinforcement seems to be an effective way to give better play to its performance.
(2) With the decrease of the proportion of FRP bars and the increase of the proportion of steel rebars in the hybrid-reinforced concrete beams, the yield step becomes more obvious, and the development of deflection gradually mitigates from the rapid growth, which provides an important reference for the early warning of concrete structure destruction in practical engineering. Notwithstanding, with the appearance of the yield step and buffer segment, the high ultimate bending strength of FRP reinforcement under large deflection conditions is replaced.
(3) Through the above test results, the proportion between steel rebars and FRP bars needs to be adjusted according to the actual requirements of the engineering project. The over-reinforcement can be used appropriately on the premise of controlling the cost to make FRP bars more practical.

It should be emphasized that the embedded contact is used for the contact between bars and concrete in this research, so the bonding and sliding effects between them are not considered. However, the bond-slip capacity of FRP is different from that of steel bars in concrete, and this affects the actual flexural bearing capacity. Due to the difference in fiber types, FRP is classified as GFRP, BFRP, CFRP, etc. The mechanical properties of FRP will also be slightly different due to different production processes and different construction techniques. Because there is no experimental support for the research done in this paper, a simple finite element model is established only according to quoting the measured mechanical property parameters of GFRP in the existing papers, and the flexural bearing capacity of double-reinforced rectangular section is explored. In future studies, various FRP parameters measured in the laboratory and the model established in this paper will be used as the basis. Considering the bond slip effect, the difference in the flexural capacity of different FRP-steel mixed reinforcements will be compared.

## REFERENCES

Aiello, Antonietta, M., Ombres & Luciano. (2002). Structural performances of concrete beams with hybrid (fiber reinforced polymer-steel) reinforcements. *J. Compos. Constr.* 6(2):133–40.
Aiello, M.A. & Ombres L. (2000). Load-deflection analysis of FRP reinforced concrete flexural members. *J.J Compos Constr* 4(4):164.
Almansoor, A.A.A & Tu, J.W. (2017). Using ABAQUS finite element Analysis to investigate the influence of FRP types and Reinforcement ratio on Flexural capacity of beams reinforced with FRP Rod. *J.Architectural Engineering and Civil Engineering*:72.
Ashrafi, H., Bazli, M., Najafabadi, E.P. & Oskouei, A.V. (2017). The effect of the mechanical and thermal properties of FRP bars on their tensile performance under elevated temperatures. *J.Constr Build Mater* (157): 1001–1010.
Denvid, L. & Hoat, J.P. (2010). Experimental study of hybrid FRP reinforced concrete beams. *J.Eng.Struct* 32(2010):3857–3865.
Denvid, L. & Hoat. J.P. (2010). Experimental study of hybrid FRP reinforced concrete beams. *J.Eng. Struct* 32(2012): 3857–3865.

Ge, W., Zhang, J., Cao, D. & Tu, Y. (2015). Flexural behaviors of hybrid concrete beams reinforced with BFRP bars and steel bars. *J. Constr Build Mater* (87):28–37.

Hollaway, L. & Robinson, E.Y. (1981). Glass reinforced plastics in construction: engineering aspects.J.J Eng Mater- Technology, *T Asme* 103(1):78.

Linh, V.H.B., Boonchai, S. & Tamon, U. (2018). Ductility of Concrete Beams Reinforced with Both Fiber-Reinforced Polymer and Steel Tension Bars. *J.J Adv Concr Technol* 16 (2018):531–548.

Liu, Y.H. & Yong, Y. (2013). Arrangement of hybrid rebars on flexural behavior of HSC beams. *J. Compos.B. Eng.* 45(1):22–31.

Newhook, J.P. (2000). Design of under-reinforced concrete T-sections with GFRP reinforcement. *J.Bridges and Structures*: 153–60.

Peece, M., Manfredi, G. & Cosenza, E. (2002). Experimental response and code models of GFRP RC beams in bending. *J.J Compos Constr* 4(4):71.

Wu, B., Chu, F., Que, Kan, C. & Zhang, Z.Q. (2013). *Experimental study on bonding properties of glass fiber reinforced concrete*.C.The 12th Cross-Strait Tunnel and Underground Engineering Symposium Proceedings of Engineering Science Sichuan, (8)17.

Yan, Q.F. & Zhang, J.G. (2021). Application and development of fiber reinforced composites in civil engineering. *J.Secience Technology and Engineering* 21 (36): 15314–15322.

Yi, C., Julio, F., Davalos, Indrajit, R. & Hyeong, Y.K. (2005). Accelerated aging tests for evaluations of durability performance of FRP reinforcing bars for concrete structures. *J.Compos. Struct* 78 (1).

Zhou, J.S. (2018). Design and Application of Glass fiber Reinforcement in Bridge Engineering. *J.Tunnel and Rail Transit* (1).

*Civil Engineering and Urban Research – Mohamed & Hou (Eds)*
*© 2023 the Authors, ISBN 978-1-032-44487-1*

# Field test study about the behaviors of concrete sheet piles under different loading scenarios

Lei Chen*
*Tianjin Survey and Design Institute for Water Transport Engineering Co., Ltd., Tianjin, China*

Lei Zhang*
*Shandong University of Science and Technology, Qingdao, China*

Erlin Zhang* & Yanyong An*
*Tianjin Survey and Design Institute for Water Transport Engineering Co., Ltd., Tianjin, China*

ABSTRACT: Upgrading waterways with traditional shoring methods can be difficult and very costly, especially when facing a shortage of land. It is an effective solution to support the embankment with concrete sheet pile walls. However, few field tests were studied about the behaviors of concrete sheet pile walls in muddy clays under different loading scenarios (including embankment surface surcharges, and channel over-dredging). During the upgrading of the Hangzhou Shanghai Waterway of Beijing Hangzhou Grand Canal in China, four concrete sheet piles were instrumented and readings of earth pressure, settlement, and horizontal displacement of shoring structures were monitored over eight months. The field test results indicated that the deflections of non-anchored sheet piles were approximately corresponding to the typical results of cantilever sheet piles with line load. Increasing the sheet pile length could decrease the sheet pile deformation and increase the soil stability on buried parts of non-anchored sheet piles. The horizontal displacement of the sheet piles with anchors can be well controlled to a reasonable range even with 20 kPa surface surcharge loads performed behind them. The channel dredging depth should also be an important safety consideration. The research of field tests provides design experience and a theoretical basis for the application of vertical revetment sheet piles in inland upgrading waterways.

## 1 INTRODUCTION

With the rapid development of regional economic integration in the Yangtze River Delta and the increase in urbanization, the water channel has to carry more vessel traffic. There is an urgent need for the channel to be upgraded. However, due to the dense distribution of tourism areas, factories, and docks along the embankment, it is difficult to widen the channel. The possible solution is to dredge the channel and support the embankment with concrete sheet pile walls.

Sheet pile walls have been widely used and studied in excavation support projects and wharves (e.g., Briancon & Simon 2011; Byrum & Mcdevitt 2004; Cacoilo et al. 1998; Tan & Paikowsky 2004; Tan 2008; Tan & Lu 2009; Jamshidi et al. 2010; Wu et al. 1999). Recently, other supporting structures like deep foundations have also been applied and studied (e.g., Briancon & Siman 2011; Chen et al. 2010; Han et al. 2002; Liu et al. 2007; Li and Yang 2017; Yang et al. 2017), which can provide references for the application of concrete sheet piles in soft soils. However, limited information is available on the behaviors of sheet pile walls in muddy silty soil considering the

---

*Corresponding Authors: tkschenlei@126.com, chenlyrand@gmail.com, 153318537@qq.com and anyy@tiwte.ac.cn

DOI 10.1201/9781003372417-14

combination of large surcharge loadings on the embankment, embankment soil saturation, and channel dredge, considering the long-term impact of soil strength and creep, as discussed by Leshchinsky (1997).

This paper presents results from full-scaled field tests to investigate the application of the concrete sheet pile wall in soft soils under different loading scenarios to upgrade the water channel. In this paper, first, the geotechnical condition of the test site is presented. Then, the instrumentation of the sheet pile walls using inclinometers and stress sensors is described. Finally, the test results and discussion are included.

## 2 FIELD TEST PLAN

### 2.1 Field test overview

The field test site on the Hangzhou-Shanghai waterway is 200 m long and located at Xiuzhou of Jiaxing, China. When the field test was performed, the water surface width was kept at 65 m and the bottom channel width was dredged to 45 m. The cross-section of the waterway is shown in Figure 1. To investigate the soil engineering properties, a series of laboratory testing programs were performed as shown in Table 1.

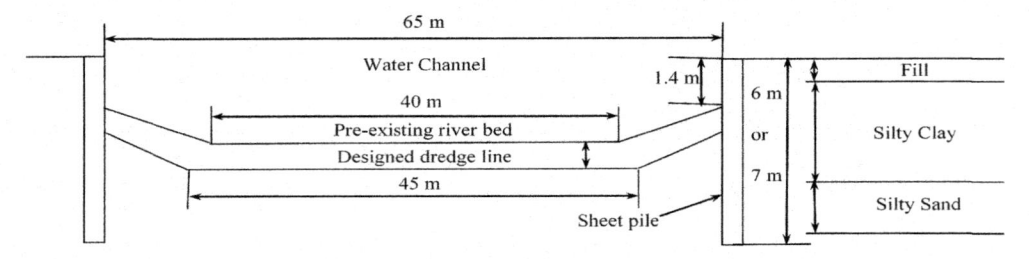

Figure 1. Profile view of the testing channel.

Table 1. Soil properties at test sections.

|  | $w$ (%) | $\gamma$ (kN/m³) | E (MPa) | $v$ | $c$ (kPa) | $\varphi$ (°) | Thickness (m) |
|---|---|---|---|---|---|---|---|
| Fill | 6.7-20.3 | 19.5 | 5 | 0.31 | 15.0 | 7.2 | 0.7-1.4 |
| Silty Clay | 26.1-39.9 | 20.0 | 11.53 | 0.32 | 28.4 | 25.3 | 2.1-4.5 |
| Silty Sand | 28.9 | 19.5 | 9.34 | 0.31 | 22.3 | 32.3 | 0.0-2.2 |
| Muddy Clay | 40.5 | 19.6 | 11.56 | 0.30 | 15.3 | 8.8 | >2.0 |

### 2.2 Field test setup

The plan view of the test site is shown in Figure 2. Field tests were performed on two adjacent areas (left side and right side) behind the concrete sheet pile wall. Outside the concrete wall is the water channel. On the left side of the embankment without vertical surcharge loads, 6 m and 7 m cantilever sheet piles were separately installed. On the right side of the embankment with vertical surcharge loads, anchors were installed connecting the concrete sheet pile wall to another concrete wall which was constructed to hold anchors 8m behind the concrete sheet pile wall, four check tests were performed to test the anchored concrete sheet pile wall under different loading scenarios.

The loading scenarios include 20 kPa surcharge loads at 2 m behind the sheet pile wall, and 0.5 m channel over-dredging. The two test areas were 100m away from each other to minimize interactions. The combinations of different loading scenarios in each check test are listed in Table 2.

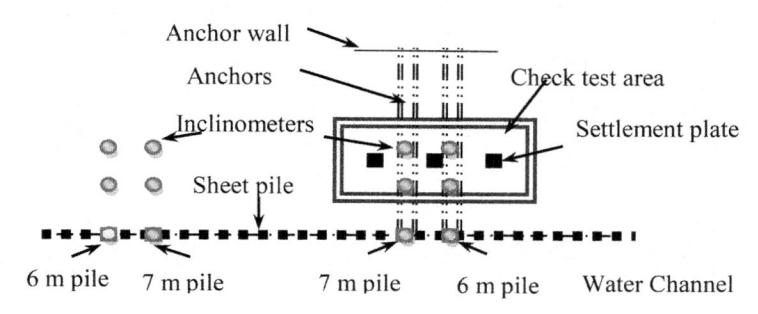

Figure 2. Plan view of the field test site.

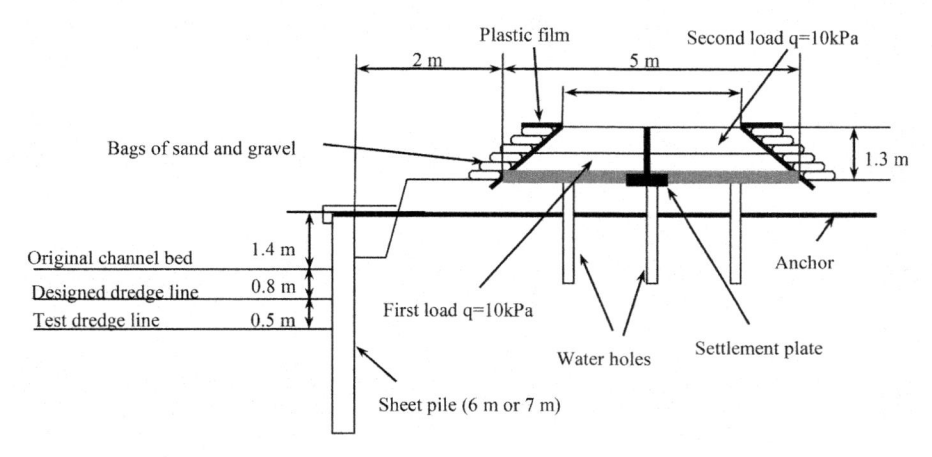

Figure 3. Profile of field test site.

Before the surcharge loads, settlement plates for soil surface settlement measurement were set on the embankment soil surface and inclinometers were installed to measure the soil and pile horizontal displacements. In each test, the settlements, horizontal displacements of the embankment soil and the sheet pile wall, and lateral pressures and body forces on the sheet pile wall were monitored.

Table 2. Check test item.

| Test | Over-dredging | Embankment surcharge | Start | End |
|---|---|---|---|---|
| Test 1 | 0.0 m | No | Feb. 01, 2009 | Mar. 25, 2009 |
| Test 2 | 0.5 m | No | Mar.25, 2009 | Apr. 17, 2009 |
| Test 3 | 0.5 m | 10 kPa | Apr. 17, 2009 | May 09, 2009 |
| Test 4 | 0.5 m | 20 kPa | May 09, 2009 | May 31, 2009 |

Note: The groundwater level was increased by 1m.

## 2.3 Design of surcharge loading

The profile of the embankment with sheet piles and surcharge loading setups is shown in Figure 3. In the surcharge loading zone, loadings were applied by the self-weight of filling materials. There were two loading steps with 10 kPa in each step. In the first step, a 25 cm thick layer of sand was put on the ground surface with a sheet of geotextile on top. Then, 40 cm thick soil with a bulk density of about 15 kN/m$^3$ was piled up on it. When the settlement of the ground surface was smaller than

0.5 mm/d for 7 days, the second loading step was to apply a 65 cm thick layer of sand. Then, two layers of sand were placed to apply a 20 kPa surcharge pressure. Finally, all four lateral sides of the surcharge loading soil were covered with two layers of plastic film to prevent water leakage and supported by sandbags, as shown in Figure 3.

### 2.4 *Manufacturing of concrete sheet piles*

The concrete sheet piles with sensors were manufactured before installation. Those sensors consisted of clusters of pressure cells, inclinometer casings, and water level casings. Earth pressure cells were installed on one side of the pile to measure the lateral earth pressure on the pile, the inclinometer casings were fixed to the center of the steel skeleton, and a steel pile cap was attached to the top of the pile to protect the pile during construction. After being manufactured, sheet piles were pressed into the clay in the mid-month of Jan. 2009.

## 3 FIELD TEST RESULTS

### 3.1 *Sheet Pile Wall Deflections*

Figure shows the deflections of all sheet piles including 6m-long sheet piles without anchors (designated 6SP), 7 m-long sheet piles without anchors (designated 7SP), 6 m-long sheet piles with anchors (designated 6SPA), and 7 m-long sheet piles with anchors (designated 7SPA).

Figure (a) and 4(b) indicate that the deflections of non-anchored sheet piles were approximately corresponding to the typical results of cantilever sheet piles with line load (e.g., Choudhury et al. 2006). Because of no surcharge loads on the embankment surface behind non-anchored sheet piles, deflections of those sheet piles remained smaller than 4.0 mm. From the comparison of 6SP and 7SP, it appears that the sheet pile length had a negligible effect on the sheet pile deflections for the depths from the ground surface to 3 m. However, increasing pile length had an obvious effect on the deflections at depths below 3 m. In addition, as expected, post dredging sheet pile deflection was observed after the completion of channel dredging. The 7-m sheet pile deflection reached its stabilization much faster than the 6-m sheet pile. The horizontal displacements of the 6SP gradually increased in the following five months after dredging, while the deflections of the 7SP below the depth of 3 m were completed only one month after dredging. Increasing sheet pile length had little influence on the deflection rate of the 7SP above 3 m.

To test the concrete sheet pile's ability to support embankment with surcharge loads, the sheet piles were anchored at the test section (i.e., the 6SPA and 7SPA). Figure (c) and 4(d) show the results from the beginning of the first check test. The horizontal displacement of the 6SPA piles was well controlled to a reasonable range by anchors even with surcharge loads performed behind them. However, the lateral movement for the 7SPA piles was as large as 12 mm after check tests, due to the breaking of 3 anchors on these piles during check tests.

### 3.2 *Soil horizontal movements*

The soil horizontal movement data were collected from the inclinometers behind the sheet pile wall. As shown in Figure 4, the maximum values of all sheet pile displacements were below 13 mm, which was relatively small and can fulfill the safety requirement. Due to the moment produced by the soil moving in the upper layers, inclinometers recorded negative displacements below 7 m. There was a trend of increasing displacements close to the lower ends due to the result of deep-seated movements.

Figure 4(a) and 5(c) indicate that soil behind 6SP reached its stability in the first month after dredging, and the soil behind 7SP stopped consolidation within 3 months after the channel dredging, corresponding to the larger lateral earth pressure on 7SP comparing to that on 6SP, as shown in Figure 5(a) and 5(c). Horizontal displacements on top of the anchored sheet piles were larger

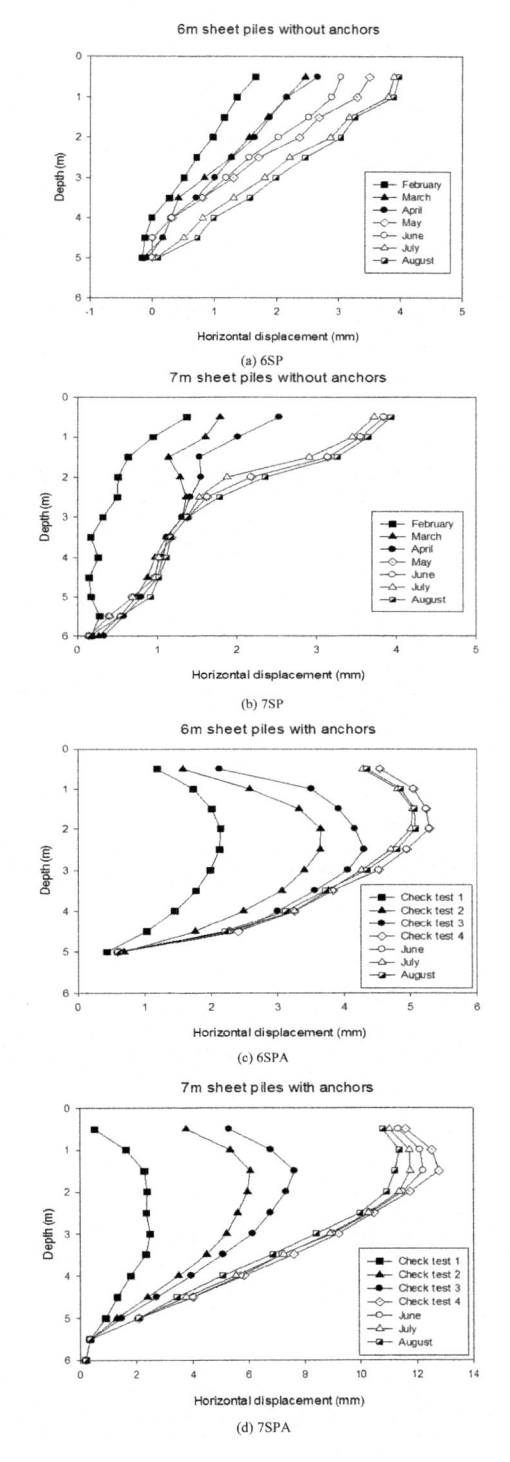

Figure 4. Measured lateral movement in the test section.

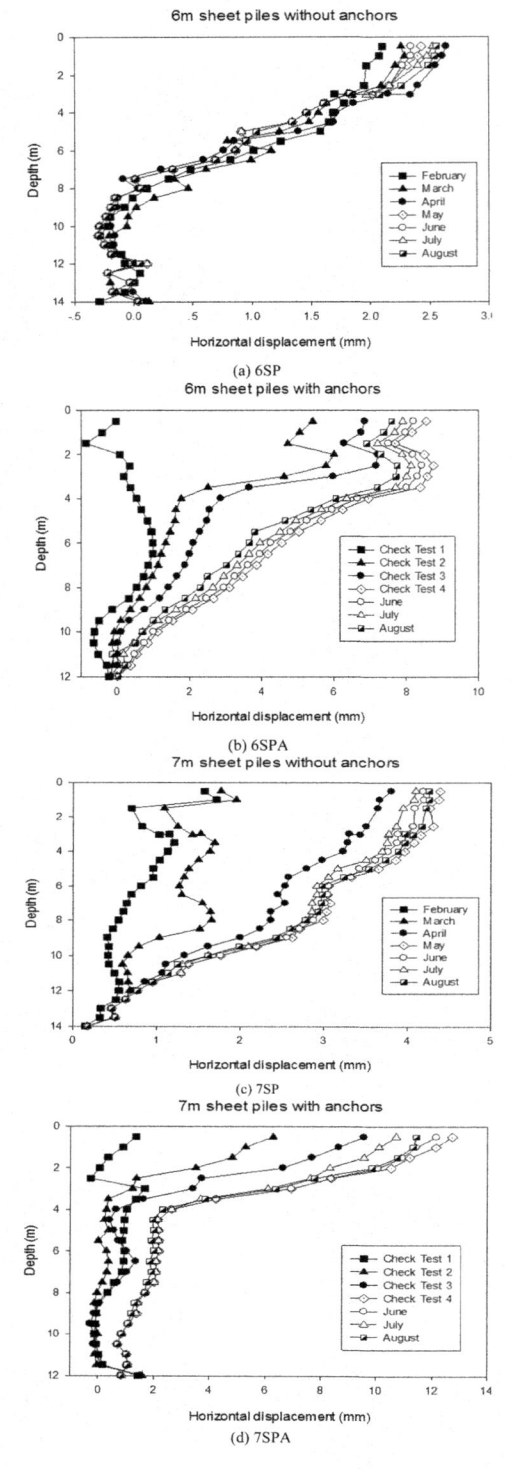

Figure 5. Horizontal displacement versus time.

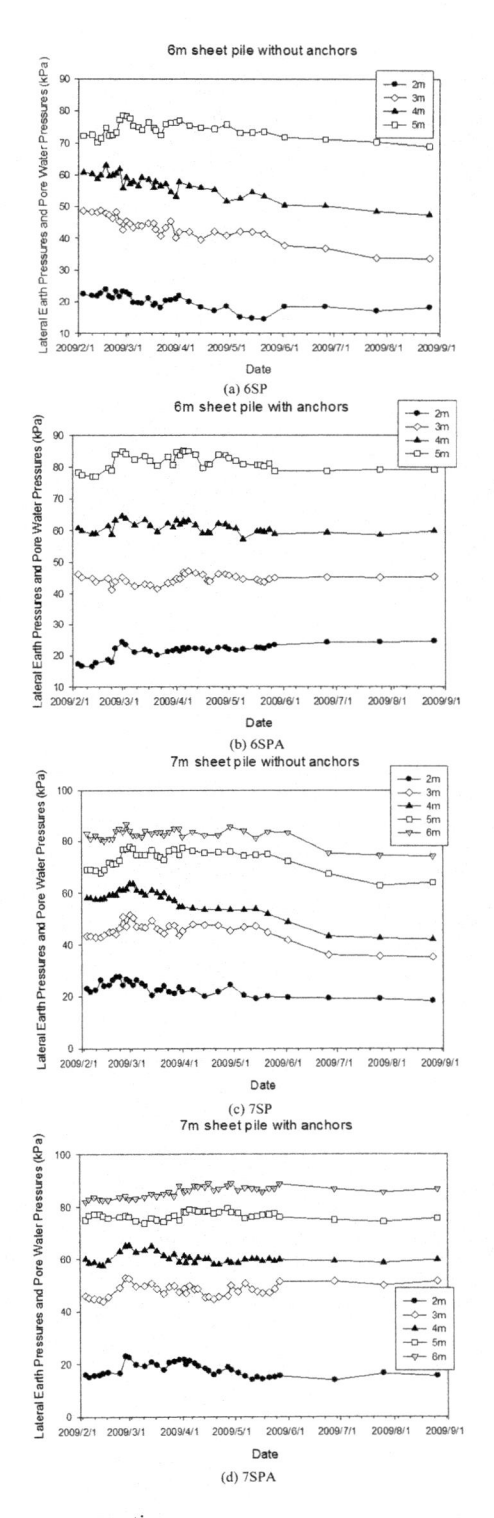

Figure 6.    Lateral earth pressure versus time.

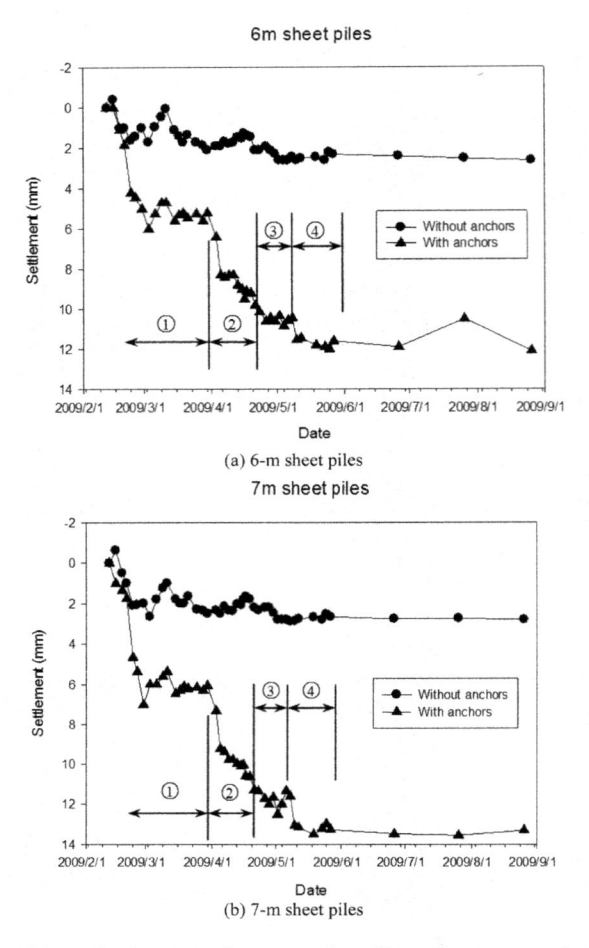

(a) 6-m sheet piles

(b) 7-m sheet piles

Figure 7. Settlement of the embankment surface versus time. Note: ① represents check test 1; ② represents check test 2; ③ represents check test 3; ④ represents check test 4.

than those on top of non-anchored sheet piles, which was the result of surcharge loads applied by check tests behind the anchored sheet pile wall. The maximum horizontal displacements of the soils behind the 7SPA were approximately 4 mm larger than the ones behind 6SPA, which was not expected. This phenomenon which was also found in Figure should be the result of an anchor being broken. However, total displacements were small, and this increase would not impact the embankment safety.

### 3.3 *Lateral earth pressure*

Figure 5 shows the change of measured lateral earth pressure from the February of 2009 upon the beginning of dredging to the end of August 2009 on all four sheet piles. From Figure 5, the pressure on the non-anchored sheet piles gradually decreased with time to a minimum value at which the consolidation process was completed. Similar observations were also reported by Hanna and Al-Romhein (2008) in a laboratory test. Because of surcharge loadings on top of embankment in check tests, the earth pressure on anchored sheet piles increased over time. The values of earth pressures were larger than those on non-anchored sheet piles. After dredging, earth pressures on non-anchored sheet piles decreased over a long time due to soil creep, but earth pressures on anchored sheet piles slightly increased because of complicated in-site conditions applied by check tests.

Earth pressure on 7SP was larger than that on 6SP, which is due to the larger restriction of the soil lateral movement behind 7SP. The same phenomena were also observed between 6-m and 7-m anchored sheet piles. From Figure 5, the two different types of 7-m sheet piles were close to each other, so the over-dredging in front of the 7-m sheet piles may affect the earth pressure and deformation for both sheet piles.

### 3.4 *Settlement of embankment surface*

Figure 7 shows that the soil settlements on the embankment surface had the same trends as soil horizontal movements. Like the horizontal displacements, soil settlements behind anchored sheet pile walls were larger than those behind the non-anchored sheet pile walls due to check tests. The soil settlements gradually stopped after check tests. The maximum value of soil settlement behind 7SPA was smaller than 14 mm, which was 1 mm larger than that behind 6SPA.

There were two sudden increases in settlements behind the anchored sheet pile walls. The first increase in February 2009 was the result of channel dredging from February 10th to 15th of 2009. There was twice more increase of settlements behind anchored sheet pile walls, which should be the result of 0.5 m over-dredging in front of sheet pile walls. The other sudden increase occurred at the beginning of check test 2. The main reason for this increase in settlement during check test 2 was that the water content increased lowered soil suction under the loading zero and soil was compressed. There was no sudden increase in check test 3 and check test 4, so an increase in groundwater level could significantly affect the embankment safety and its effect was relatively larger than that of a 20 kPa surcharge load on the embankment surface.

## 4  CONCLUSIONS

Through the field test results of sheet pile walls under different loads, the following conclusions are drawn:

1. The deflections of non-anchored sheet piles were approximately corresponding to the typical results of cantilever sheet piles with line load. It appears that the sheet pile length had a negligible effect on the sheet pile deflections for the depths from the ground surface to 3 m and increasing pile length reached its stabilization much faster.
2. The horizontal displacement of the sheet piles with anchors can be well controlled to a reasonable range even with 20 kPa surface surcharge loads performed behind them, but non-anchored sheet piles could not.
3. The channel dredging depth should also be an important safety consideration. 0.5 m over-dredging could cause twice more soil and sheet pile movements, which should be more dramatic when there are large soil movements.
4. The effects of the sheet pile length on soil and sheet pile performance were relatively smaller than the channel over-dredging. However, increasing the sheet pile length could decrease the sheet pile deformation and increase the soil stability on buried parts of non-anchored sheet piles.

The research of field test shows that the concrete sheet pile can meet the requirements of channel upgrading and has important practical value, especially in the coastal soft area, which provides a theoretical basis for the popularization and application of vertical revetment sheet pile in an inland waterway.

# REFERENCES

Briançon, L., and Simon, B. (2011). "Performance of pile-supported embankment over soft soil: full-scale experiment." *Journal of Geotechnical and Geoenvironmental Engineering*, 138(4), 551–561.

Byrum, C. R., Macdevitt, K. C., and Magnan, S. J. (2004). *"Instrumented Geofoam and Sheet Pile Wall for a Roadway Lane Addition in a Peat Marsh."* Geotechnical Special Publication, 1(126), pp. 600–608.

Cacoilo, D., Tamaro, G., and Edinger, P. (2004). *"Design and Performance of a Tied-Back Sheet Pile Wall in Soft Clay."* Geotechnical Special Publication, 83, pp. 14–21.

Cai, M. and Kaiser, P.K. (2007). "FLAC/PFC coupled numerical simulation of AE in large-scale underground excavations." *International Journal of Rock Mechanics & Mining Science*, 44(4), pp. 550–564.

Chen, R.P., Xu, Z. Z., Chen, Y. M., Ling, D. S., and Zhu B. (2010). "Field Tests on Pile-Supported Embankments over Soft Ground." *Journal of Geotechnical and Geoenvironmental Engineering*, 136(6), pp. 777–785.

Choudhury, D., Singh, S., and Goel, S. (2006). "New approach for analysis of cantilever sheet pile with line load." *Canadian geotechnical journal*, 43(5), 540–549.

Han, J., and Gabr, M. A. (2002). "Numerical analysis of geosynthetic-reinforced and pile-supported earth platforms over soft soil." *Journal of Geotechnical and Geoenvironmental Engineering*, 128(1), 44–53.

Hanna, A., and Al-Romhein, R. (2008). "At-Rest Earth Pressure of Overconsolidated Cohesionless Soil." *Journal of Geotechnical and Geoenvironmental Engineering*, 134(3), pp. 408–412.

Indraratna, B., , and Balachandran, S. (1992). "Performance of test embankment constructed to failure on soft marine clay." *Journal of Geotechnical Engineering*, 118(1), 12–33.

Wu, Z., Sasaki, Y., and Kusakabe, O. (1999). *"A Study on Lateral Bearing Capacity of Sheet Pile in Clay Slope."* Doboku Gakkai Ronbunshu, (631), 257–272.

Zhang, L., Zhang, F., and Hua, M. (2011). *"Application of Sheet Pile Wall in a Channel to Upgrade Waterways."* In Slope Stability and Earth Retaining Walls, pp. 164–171.

*Civil Engineering and Urban Research – Mohamed & Hou (Eds)*
© *2023 the Authors, ISBN 978-1-032-44487-1*

# Research on physicochemical properties of gangue and analysis of its comprehensive utilization

Bing Wang* & Guoliang Bai*

*School of Civil Engineering, Xi'an University of Architecture & Technology, Xi'an, PR China*
*Key Lab of Structural Engineering and Earthquake Resistance, Ministry of Education (XAUAT), Xi'an, China*

ABSTRACT: China is one of the world's largest coal producers. In the future, coal will still be the main part of primary energy production and consumption in China. Gangue is an appendage of coal mining and the coal washing process. To rationally utilize gangue resources, this paper selects gangue washed in mines and analyzes its chemical composition and mineral composition by X-ray fluorescence spectroscopy and X-ray diffraction. The test results show that: in terms of chemical composition, gangue is mainly composed of $SiO_2$ and $Al_2O_3$, of which $SiO_2$ accounts for about 54.44% and $Al_2O_3$ accounts for about 20.08%; in terms of mineral composition, gangue is mainly composed of quartz and clay minerals, of which quartz accounts for about 28%, clay minerals account for about 42%. Through the analysis of the test results, it is concluded that gangue can be used in the construction field to produce building materials and refractory materials in the chemical field, which provides a direction for the utilization of gangue resources.

## 1 INTRODUCTION

China is a resource-rich country, but due to the large population base in China, China is a country with a shortage of resources per capita. Today, with the improvement of people's living standards, China's demand for coal is also increasing (Miao 2020). In recent years, China's coal production is shown in Figure 1. China is rich in coal resources, but there are few high-quality resources. It faces severe challenges to ensure a stable supply of coal (Sun 2005).

Gangue is a kind of black-gray rock with lower carbon content and is harder than coal, which is associated with the coal seam in the process of coal formation. The gangue and the washing gangue picked out during the coal washing process account for about 15% to 25% of the coal output (Zhang 2018). Figure 1 shows the results of calculating the annual output of gangue in China with gangue accounting for 20% of coal production. The output of gangue in China is increasing every year, and the output of gangue in China will still increase year by year with an increase of 500 million tons to 800 million tons. Gangue is the largest solid waste discharged. The massive accumulation of gangue will occupy the land and block the river; long-term accumulation will also pollute the water body, soil, and air, and even cause safety accidents (Shang 2022). Studying the characteristics of gangue and rationally utilizing gangue as resources can reduce the discharge of industrial sewage and release a large amount of land (Sun 2016).

In recent years, many scholars have carried out research on the properties of gangue. In general, the chemical composition of gangue is mainly oxides of aluminum and silicon, but the chemical composition of different types of gangues is different. At present, there have been many studies on gangue (Jiao 2021; Li 2021; Ma 2018; Zhu 2021, 2021; Zhang 2021), but there is a lack of systematic research on the physical properties and chemical composition of gangue in different regions and strata. Therefore, in this paper, gangue was selected for gangue washing, and the

---

*Corresponding Authors: hqliu@xauat.edu.cn and guoliangbai@126.com

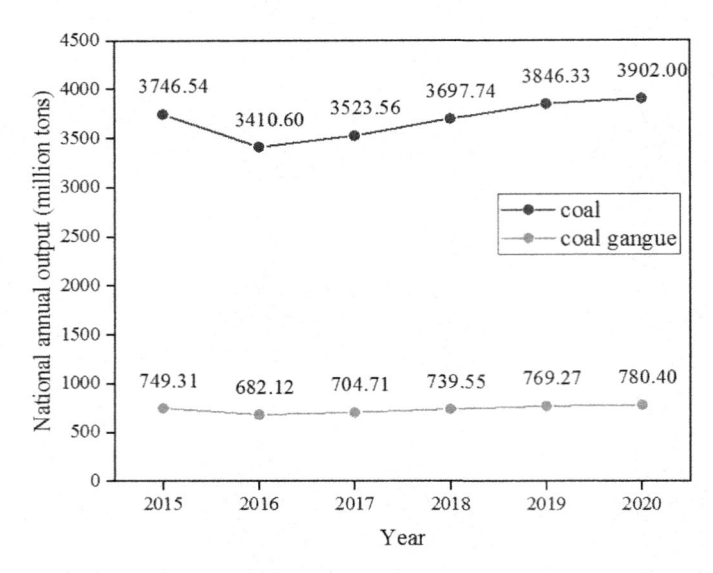

Figure 1. China's coal output and gangue output in the past six years.

chemical composition and mineral composition of gangue were analyzed by the XRF test and XRD test, which provided a reference for the classification of coal gangue and the improvement of its resource utilization rate.

## 2 TRIAL OVERVIEW

### 2.1 *Materials*

The physical and chemical properties of gangue are related to its rock type and mineral composition. Fully understanding the chemical composition of gangue and the mineral composition and characteristics of various types of gangues is the key to efficient resource utilization of gangue. The samples collected in gangue storage sites are generally larger in volume and weight, while the number of samples used in physical and chemical analysis is small (Wang 2015). The gangue selected in this paper, as shown in Figure 2, is broken into small pieces by the machine and made into fine powder by hand, and then the fine gangue is ground to 200 mesh powder with a ball mill (Figure 3), and the tablet method is used (Figure 4) to make test samples, as shown in Figure 5. Put it in a blast drying oven to dry to constant weight.

Figure 2. Gangue.

Figure 3. Ball mill.

Figure 4.  Tablet instrument.

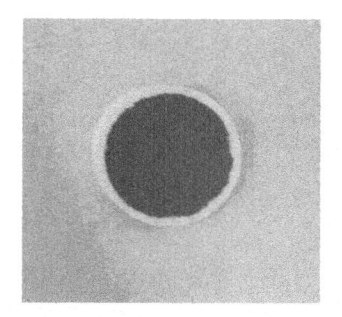

Figure 5.  Test sample.

## 2.2  *Chemical composition*

The chemical composition analysis method adopts X-ray fluorescence spectrometry (XRF), which has gradually become a commonly used instrumental analysis method due to its fast analysis speed, high precision, and simple operation. The principle is that when chemical elements in gangue are bombarded by accelerated particles such as electrons, protons, and ions, or excited by X-ray tubes, characteristic X-rays can be emitted, which are called elemental X-fluorescence rays (Wang 2015). The instrument used in the test is an S4 PIONEER X-rays fluorescence spectrometer from Bruker, Germany, as shown in Figure 6. After the gangue sample is processed by the tableting method, the content of each element in the sample is obtained according to the X-ray fluorescence intensity of the element to be measured and the calibration curve made by the standard sample.

Figure 6.  S4 PIONEER X-ray fluorescence spectrometer.

Figure 7.  D8 Advance X-ray diffractometer.

## 2.3  *Mineral composition*

The analysis method of the mineral composition is X-ray diffraction (XRD). The principle is that different crystals have specific lattice types and unit cell sizes, so they have a unique series of X-ray patterns that can be used for identification. The mineral composition of gangue can be well determined by the X-ray diffractometer test, which has no damage to the sample and has the advantages of no pollution, and fast and high measurement accuracy (Wang 2015). The experimental instrument is a Bruker D8 Advance X-ray diffractometer, as shown in Figure 7.

# 3  TEST RESULTS AND ANALYSIS

## 3.1  *Analysis of physical and chemical properties*

The test results of the chemical composition determination of gangue are shown in Table 1, and the test results of the mineral composition determination are shown in Figure 8.

Table 1.   Chemical composition of gangue (%).

| Test item | $SiO_2$ | $Al_2O_3$ | $Fe_2O_3$ | CaO | $K_2O$ | MgO | $TiO_2$ |
|---|---|---|---|---|---|---|---|
| Gangue | 54.44 | 20.08 | 3.79 | 1.38 | 2.74 | 1.25 | 0.97 |

It can be seen from Table 1 that gangue is mainly composed of $SiO_2$ and $Al_2O_3$, of which $SiO_2$ accounts for about 54.44% and $Al_2O_3$ accounts for about 20.08%. In addition, gangue also contains a small number of elements such as $Fe_2O_3$, CaO, $K_2O$, MgO, and $TiO_2$. Compared with Table 2 in the literature (Wang 2015), the mineral composition of gangue and rock shale gangue has a certain degree of similarity, so gangue can be divided into rock shale gangue.

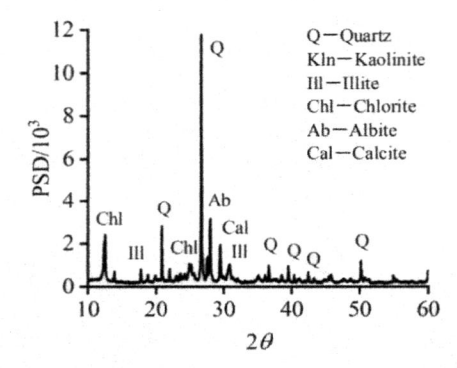

Figure 8.   XRD pattern of gangue.

It can be found from Figure 8 that the gangue is mainly composed of siliceous minerals (quartz) and aluminum clay minerals (kaolinite, illite, and chlorite). Quartz is geologically hard, but mostly appears in small particles, which are small fragments; clay minerals are softer and mostly appear in layers, which is the reason gangue is easily broken (Bai 2022). Quartz accounts for about 28%, and clay minerals account for about 42%, which is consistent with the literature (Shi 2008) that typical gangue contains about 30% to 50% clay minerals, 20% to 30% quartz and 10% to 20% other minerals are consistent with this conclusion.

## 3.2  *Way of use*

The Chinese standard *Code for Design of Highway Subgrade* (JTG D30-2015) proposed that the sum of Si, Al, and Fe minerals should not be less than 70% when gangue is used as embankment filler. The test result is about 93%, which fulfills the requirements. The ratio of $Al_2O_3/SiO_2$ in the gangue in this test is 0.37, which is between 0.3 and 0.7. According to the standard *Classification of Gangue* (GB/T 29162-2012), the gangue is classified by the ratio, and it is concluded that the gangue detected this time is medium aluminum gangue, which can be used as concrete aggregate. The content of $SiO_2$ in gangue is about 54.44%, the content of $Al_2O_3$ is about 20.08%, and the content of $Fe_2O_3$ is about 3.79%, which can be used as an effective resource for preparing refractory fibers (Wang 2015).

The content of illite in gangue is relatively high, accounting for about 27% of the total mass. Using the structural characteristics of illite, the activity of gangue is enhanced to release $Si^{4+}$ and $Al^{3+}$, which is also consistent with the conclusion of the literature (Hua 2015). Activation of gangue is one of the methods to achieve efficient utilization of gangue. Among the main components of gangue, $SiO_2$ and $Al_2O_3$ are both the main raw materials for the synthesis of nitrogen oxide refractories by carbothermic reduction and nitriding process - Sialon, so gangue can also prepare nitrogen oxide refractories (Wang 2015).

### 3.3 *Benefit analysis*

As a raw material for construction products, gangue can greatly improve the recycling and utilization of gangue resources, and can also greatly reduce the environmental damage caused by the accumulation of gangue (Liu 2019). A utilization approach that integrates social and environmental benefits (Bai 2022). The activated powder of gangue is also called active powder. It has economic and environmental benefits as a partial replacement of cement and silica fume as concrete active admixtures. Saving the amount of standard coal and reducing the use of cement can reduce the emissions of carbon dioxide and other harmful gases. At present, the global demand for refractory fiber materials is increasing (Li 2018). The rational use of gangue to replace coke gemstone and mullite to produce refractory fibers can also protect the ecological environment and efficiently recycle resources.

## 4 CONCLUSIONS

In this paper, gangue is used as the object to study its physical and chemical properties, and the following conclusions can be drawn mainly through experimental research:

(1) In terms of chemical composition, gangue is mainly composed of $SiO_2$ and $Al_2O_3$, of which the content of $SiO_2$ is about 54.44%, and the content of $Al_2O_3$ is about 20.08%. In terms of mineral composition, gangue is mainly composed of quartz and clay minerals, of which quartz accounts for about 28% and clay minerals account for about 42%.
(2) Combined with the analysis of the results of XRF and XRD tests, it shows that gangue can be used to produce building materials, as aggregates for concrete, as embankment fillers, to prepare reactive powder concrete, etc. In the chemical industry, it can be used to prepare nitrogen oxide refractories, refractory fiber, etc.
(3) The use of gangue in building materials can not only bring economic benefits but also avoid a series of environmental problems caused by the accumulation of gangue; the use of gangue in the preparation of new process products not only has huge market potential but also brings the rational use of solid waste gangue resources has considerable benefits in ecological environmental protection and efficient recycling of resources.
(4) For the gangue detected this time, we can continue to analyze its calorific value and carbon content, and other indicators, and then discuss the feasibility of gangue as a low calorific value material, gangue power generation, and other fields.

ACKNOWLEDGEMENTS

This work was supported by the National Natural Science Foundation of China (Grant No. 52078410); the Shaanxi Provincial Science and Technology Plan Achievement Promotion Project (Grant No. 2020CGHJ-017); and the Key Laboratory Project of Shaanxi Provincial Department of Education (Grant No. 20JS071).

## REFERENCES

Bai, G, L. (2022). Experimental study on compressive strength of coal gangue concrete from different mine sources in northern Shaanxi mining area. J/OL. *Chinese Journal of Civil Engineering*: 1–11.

Bai, G, L. (2022). Physical and chemical properties of coal gangue and its influence on concrete strength. J/OL. *Journal of Building Structures*: 1–12.

GB/T 29162-2012. (2012). Coal gangue classification, China Construction Industry Press, Beijing.

Hua, L. (2015). Activation and application of coal gangue activity. J. *Comprehensive utilization of fly ash*, (02):29–31.

Jiao, H. (2021). Variation of bacterial community in soil reclaimed by coal gangue filling. J. *Chinese Journal of Coal*, 46(10):3332–3341.

JTG D30-2015. (2015). Specification for Design of Highway Subgrade, China Communications Press, Beijing.

Li, H, X. (2018). An overview of the development of refractory materials. J. *Journal of Inorganic Materials*, 33(02):198–205.

Li, Z. (2021). Research progress on comprehensive utilization of coal gangue. J. *Mineral Protection and Utilization*, 41(06):165–178.

Liu, H, Q. (2019). Experimental study on bearing capacity of coal gangue concrete column. D. Xi'an: Xi'an University of Architecture and Technology.

Ma, H, Q. (2018). Compressive strength and durability of coal gangue aggregate concrete. J. *Material guide*, 32(14):2390–2395.

Miao, Q. (2020). Analysis and Guarantee Research of Coal Resources Availability in my country. J. *Energy and Environment*, (02):6–8+23.

Shang, Y, Sang N. (2022). Characteristics and phytotoxicity of heavy metal pollution in soil around coal gangue accumulation area. J/OL. *Environmental Science*:1–12.

Shi Tao, Xu Bi-wan, Shi Hui-sheng. (2008). The evolution of coal gangue(CG)-calcium hydroxide(CH)-gypsum-H2O system. J. *Materials and Structures*, 41(7).

Sun, C, B. (2016). Coal gangue and its comprehensive utilization at home and abroad. J. *Coal technology*, 35(03):286–288.

Sun, Y, B. (2005). Determination of special reserves of coal resources. J. *Chinese Journal of Coal*, (05): 119–122.

Wang, X, D. (2015). Comprehensive utilization technology and detection method of coal gangue. M, *Science Press*, Beijng.

Zhang, C, S. (2018). Coal gangue resource recycling technology. M. *Chemical Industry Press*, Beijing.

Zhang, W, Q. (2021). Preparation and Microstructure of Coal Gangue-Based Geopolymers. J. *Journal of China University of Mining and Technology*, 50(03):539–547.

Zhu, H, G. (2021). Research on compressive strength and frost resistance of coal gangue fine aggregate-slag concrete. J. *Material guide*, 35(22):22085–22091.

Zhu, Y, Y. (2021). Characteristics of calcined coal gangue fine aggregate and its effect on the improvement of mortar performance. J. *Chinese Journal of Coal*, 46(11):3657–3669.

Civil Engineering and Urban Research – Mohamed & Hou (Eds)
© 2023 the Authors, ISBN 978-1-032-44487-1

# Application of unified mechanics index of asphalt pavement failure in Xijing road

Wei Tao*, Su Yutao* & Zhao Huaping*
*Xingtai Road & Bridge Construction Group Corporation, Xingtai, China*

ABSTRACT: A new mechanical control index (APPDI) on asphalt pavement damage is proposed based on the Mohr-Coulomb criterion. Taking Xijing road in Hebei Province as an example, based on the finite element calculation, the mechanical model of pavement rutting failure is analyzed with APPDI. The results indicate that the composition of the principal stress has a great influence on asphalt pavement damage. The different components of the principle stress correspond to rutting under high-temperature conditions (60°C). Combined with the Mohr circle and the location in the pavement structure, APPDI can be used to judge rutting on asphalt pavement and optimize pavement and material design.

## 1 INTRODUCTION

Rutting is one of the main diseases of asphalt pavement, which seriously affects driving safety and reduces the performance of the road (Li 2018; Shen 2004; Sun 2005; Wen 2016). It is generally believed that the generation of rutting is mainly affected by both internal and external aspects. The main external factors refer to temperature, load, etc. (Barksdale 1977; Hitch 1976; Meier 1989; Oduroh 2000). The internal factors are mainly related to the material, design, and construction of the asphalt mixture.

Xijing Road in Hebei Province is an important traffic lane in Huanghua. A one-way driving road with four lanes was designed. Asphalt concrete pavement with a semi-rigid base structure was used.

Taking the Xijing road in Hebei Province as an example, damage mechanical modes were induced by combining APPDI on the basis of finite element calculation and were contrasted with real damage phenomena. Combining Mohr-circles and emerging positions, the results are used to define the material design to deduce the rutting damage of asphalt pavements.

## 2 UNIFIED MECHANICS MODEL OF PAVEMENT FAILURE

The M-C criterion is used to analyze and predict the failure mode and stress state of asphalt pavement. A damage control index of pavement structure, which is the ratio of the radius of the Mohr circle at any point to the distance from the center of the circle to the envelope, is defined as the unified mechanical evaluation index for the damage of asphalt pavement structure (Sousa 1995; Van 1976).

The damage envelope of the Mohr circle and Mohr-Coulomb is shown in Figure 1, and APPDI can be expressed as:

$$\text{APPDI} = \frac{\tau_{O_1}}{O_1 M} = \frac{(\sigma_1 - \sigma_3) \cdot \sqrt{\tan^2 \varphi + 1}}{|\tan \varphi \cdot (\sigma_1 + \sigma_3) + 2c|} \tag{1}$$

---

*Corresponding Authors: 198708268@qq.com, 898606781@qq.com and 250598901@qq.com

DOI 10.1201/9781003372417-16

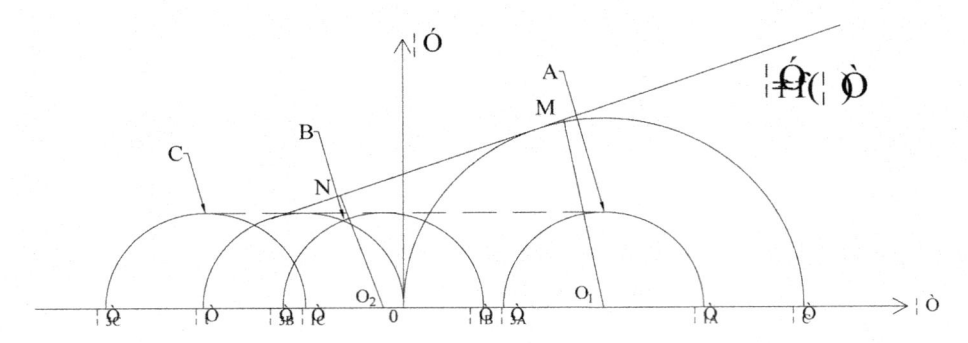

Figure 1.  Mohr circle and Mohr-Coulomb failure envelope.

It means that when APPDI < 1, it indicates that the stress Mohr circle of the point does not intersect with the envelope; that is, the point generates elastic deformation, and the deformation recovers after unloading. When APPDI > 1, there are three circumstances: 1) When $|\sigma 1| > |\sigma 3|$, the stress Mohr circle intersects with the envelope line, and the point is dominated by compressive stress; that is, the point produces tension-compression composite shear plastic deformation dominated by compression under external load. 2) When $|\sigma 1| < |\sigma 3|$, the point is dominated by tensile stress, and the point contacts the envelope in two ways under external load. When ① is tangent to the envelope, $|\sigma 3| < |\sigma t|$, tension-compression composite shear plastic deformation is mainly produced. When ② is intersected with the envelope, $|\sigma 3| > |\sigma t|$, the pull-based pull damage occurs.

## 3   INTEGRATED DESIGN OF PAVEMENT STRUCTURE MATERIALS

### 3.1   *Engineering overview*

This paper is based on the Xijing Road in Hebei Province. The upper layer of the main line is a 4 cm thick ARHM-13 rubber asphalt mixture, and the second layer is a 6 cm thick AC-16C modified asphalt concrete. The basement consists of 2 layers, the upper base layer is 18cm thick cement-stabilized stones, and the lower base layer is 30cm thick lime-treated soil.

### 3.2   *Preliminary pavement material parameters*

The calculated temperatures of the asphalt surface layer are 60°C (high temperature), and the following parameters can be obtained according to Formulas (2) and (3), as shown in Tables 1 and 2:

$$E_T = 10^{0.018(20-T)}E_{20} \tag{2}$$

$$c = 2.42037 * 10^{-8} * E^2 + 2.22716 * 10^{-4} * E \tag{3}$$

### 3.3   *Unified Mechanical Analysis of Pavement Structure Failure*

The pavement structure load distribution and model are constructed. The two-dimensional plane strain model refers to the "Highway Asphalt Pavement Design Specification (JTG D50-2017)." The standard axle load of the double-wheel group of 100 kN is used as the calculation axle load, and the elastic layered continuous theoretical system is used as the calculation theoretical basis. The road load and calculation points are shown in Figure 2, and the standard axial load calculation parameters are shown in Table 3.

The two-dimensional elastic finite element model adopts 4-node isoparametric elements. Through the convergence calculation of pavement structure stress, the calculation depth of the subgrade is 8 m, and the width is 6 m. It is assumed that the interlayer contact state is completely

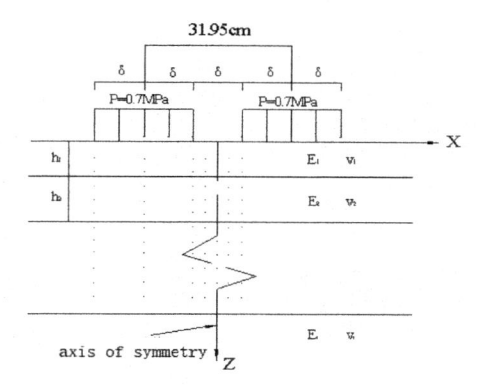

Figure 2. Road wheel load mode.

continuous. The boundary condition is assumed that the bottom is completely constrained, and there is no X-direction displacement in the left X direction. The finite element mesh model of the pavement structure is shown in Figure 3.

### 3.4 Calculation results and analysis

Under high-temperature conditions (when the road temperature is 60°C), the APPDI contours and the APPDI distribution diagrams of different depths and their typical stress Mohr circles are shown in Figure 3.

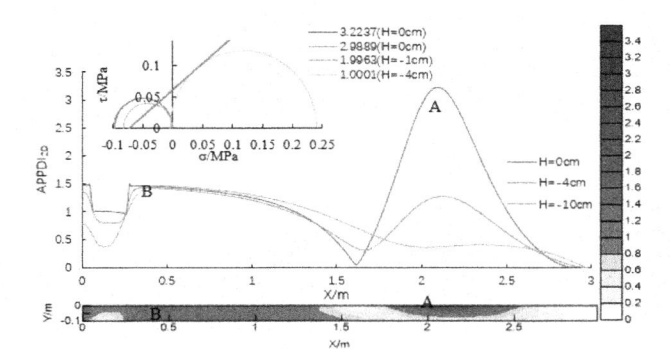

Figure 3. APPDI contours of asphalt surface under high-temperature conditions and APPDI distribution diagrams of different depths and their typical stress Mohr circles.

Figure 3 shows that there are 1461 potentially dangerous points under the plane strain model with APPDI greater than 1. The maximum value of APPDI in the upper layer is 3.224, in which $\sigma 1 = 0.000022558$ MPa, $\sigma 3 = -0.0984$ MPa. The node is located at the road surface of a 2 m lateral wheel load, $|\sigma 1| < |\sigma 3|$. From the calculation, we can conclude that the stress state is tensile-compression composite shear stress, and the failure mode is deduced as a Top-Down crack. The maximum value of APPDI in the lower layer is 1.4487, in which $\sigma 1 = 0.70626$ MPa, and $\sigma 3 = -0.04619$ MPa. The node is located at the bottom of the asphalt surface layer (B) of 0.48 m lateral wheel load. The stress state is tensile-compression composite shear stress with tensile-compression stress ratio $|\sigma 3|: |\sigma 1| \approx 1 : 15$, and the failure mode is deduced to be rutting.

## 3.5 Optimization of asphalt pavement material parameters

Based on the road performance and safety performance, the material parameters of the asphalt pavement surface are optimized. Taking the tensile-compression stress ratio and cohesion of the material as the measurement indexes, the cohesion of the upper layer APPDI2 D < 1 in the asphalt pavement structure at high temperatures is taken as the strength parameter.

Experimented at high temperature (60°C), the analysis results of the material parameter table of the upper layer of asphalt in Hebei West Road under the simplified plane strain model are shown in Table 1.

Table 1. Material parameter table of asphalt upper layer.

| Temperature | Optimizing Cohesive force c/MPa | H cm | $\varphi/°$ | Elastic Modulus E/MPa | APPDI maximum value | Number of dangerous points |
|---|---|---|---|---|---|---|
| 60°C | 0.0612(top layer) | 5 | 40 | 267 | 1.2116 | 143 |
| | 0.0523(lower layer) | 7 | 42 | 229 | 1.0072 | 262 |
| | 0.3885(top layer) | 6 | 42 | 1500 | 0.9982 | 0 |
| | 0.4183(lower layer) | 7 | 42 | 1600 | 0.8002 | 0 |

Table 1 shows the number of dangerous points under high-temperature conditions when the material cohesion (c) of the upper layer of the optimized asphalt pavement increases from 0.0612 MPa to 0.3885 MPa and the material cohesion (c) of the lower layer increases from 0.0523 MPa to 0.4183 MPa. The elastic modulus E increases from 267 MPa to 1500 MPa and from 229 MPa to 1600MPa. The upper layer thickness of structure 1 increases from 4 cm to 6 cm, and the $\varphi$ value increases from 40° to 43°; the thickness of the upper layer of the second structure increases from 5cm to 6cm, and the $\varphi$ value increases from 40° to 42°.

When the upper layer c = 0.3885 MPa and the lower layer c = 0.4183 MPa, the number of dangerous points in the model is 0. Therefpre, c = 0.3885 MPa and 0.4183 MPa are separately used as material design parameters at a high temperature (60°C) for two layers.

## 4 CONCLUSION

The results of pavement structure optimization of Hebei West Road are as follows: from the calculation, at high temperature, the cohesion of the upper layer is 0.3885 MPa, $\varphi$ is 43°, the thickness is 6 cm, and the cohesion of the lower layer is 0.4183 MPa, $\varphi$ is 42°. The program is supported by Xingtai scientific program 2020ZC033 and program 2020ZC029.

## REFERENCES

Barksdale Richard D. Performance of asphalt concrete pavements[1]. American Society of Civil Engineers, *Transportation Engineering Journal*, 1977, 103(1):55–73.

Hitch L.S., Russell R.B.C. *Sand-bitumen for road bases: an examination of five methods of measuring stability*[R]. Transport & Road Research Laboratory (Great Britain), TRRL Report,1976.

Jinan Shen., Fupu Li., Jing Chen. *Early damage and prevention countermeasures of highway asphalt pavement*[M]. Beijing: China Communications Press, 2004.12: 157–217.

Lijun Sun. *Behavior Theory of Asphalt Pavement Structure*[M].Beijing: China Communications Press, 2005.11:214–219.

Meier Jr. WR., Elnicky Edward J. Laboratory evaluation of shape and surface texture of fine aggregate for asphalt concrete[J]. *Transportation Research Record*, 1989, 1250:25–34.

Oduroh PaulK, Mahboub Kamyar C., Anderson P.E., et al. Flat and elongated aggregates in Superpave regime[J]. *Journal of Materials in Civil Engineering*, 2000, 12(2): 124–130.

Sousa Jorge B, Way George, Harvey John T, et al. Comparison of mix design concepts[J]. *Transportation Research Record*, 1995, 1492: 151–160.

Suwen Li. Problems and Measures in Asphalt Pavement Design[J]. *Communications Science and Technology Heilongjiang*, 2018, 41(09):64–65.

Van de Loo P.J. Practical approach to the prediction of rutting in asphalt pavements: the Shell method[J]. *Transportation Research Record*, 1976, 616:15–21.

Wen Tang., Xuewen Wu., Lijun Sun. Study on Multi-parameter Prediction Model of Asphalt Pavement Rutting[J]. *Journal of China & Foreign Highway*, 2016, 36(01):45–49.

*Civil Engineering and Urban Research – Mohamed & Hou (Eds)*
© 2023 the Authors, ISBN 978-1-032-44487-1

# Experimental study on mechanical properties of collapsible loess aggregate filling materials

Yanpeng Zhu* & Xiaotao Du*
*College of Civil Engineering Lanzhou University of Technology, Lanzhou, Gansu, China*

ABSTRACT:    In the collapsible loess area, the foundation trenches of new buildings and structures will be subject to the consolidation and settlement of the backfill soil itself after backfilling, which will cause a large area of uneven settlement in the backfill area of the foundation trenches, and the backfill soil will pose a safety hazard to the normal use of the building. Using collapsible loess and red sandstone, which are common in Northwest China, as raw materials, a kind of collapsible loess with low consolidation settlement, no collapsibility, low permeability, satisfactory strength, and low price is formed during the backfilling process. In the test, the fine aggregate is collapsible loess accounting for 70% of the total aggregate, the coarse aggregate is red sandstone accounting for 30% of the total aggregate, the content of soil solidifying agent is 15% of the aggregate, and the amount of water is 30%. The dosage is 25% of the total mass of aggregate and curing agent. The collapsible loess is set without a curing agent after stirring under the same conditions, and red sandstone without a curing agent after stirring under the same conditions. The collapsible loess with the curing agent added and the red sandstone with the curing agent added after stirring under the same conditions were the four control groups. The study found that the unconfined compressive strength of collapsible loess filling materials was better than that of the control group, the consolidation settlement and collapsibility were much smaller than those of the control group, and the cohesion and internal friction angle were larger than those of the control group.

## 1 INTRODUCTION

With the advancement of my country's urbanization process, the construction of infrastructure such as subway projects and high-rise building projects in China produces a huge amount of low-strength waste building materials such as silt, mud, loess, etc. during the excavation of foundation pits. Such waste improvement techniques have also been developed (Du & Wang 2018). According to engineering needs and geotechnical characteristics, the use of cement to add in-situ soil and water to mix evenly to form a filling material with sufficient strength to meet the requirements can solve the problem of backfilling the foundation groove of buildings.

Zhou Yongxiang and Liu Xudong (Zhou & Wang 2019) used foundation pit trench soil to add curing agent and water to prepare fluidized filling materials, which were successfully applied to foundation trench backfill projects such as integrated pipe corridors in Beijing and Chengdu. Yuan Qiu (Yuan et al. 2011, 2021) studied the influence of different cement content, water consumption, maintenance environment, and other factors on the flow properties and mechanical properties of fluidized fly ash materials through laboratory test methods. Zou Xianyun (Zou 2012) studied the applicability of high fluidity fly ash used in the back of abutments, culverts, and retaining walls of the roadbed. Wang Zhiyuan (Wang et al. 2015) proposed a fluidized backfill material with fly ash as the main component because of a large amount of pulverized coal waste in arid regions and

---

*Corresponding Author: 1412065336@qq.com

                                                    DOI 10.1201/9781003372417-17

the engineering problems of limited working surface or insufficient compaction in dead corners of structures in backfill projects.

Red sandstone is widely distributed in the Shaanxi-Gansu-Ningxia region and other basins in China (Zeng 2010). It is a special type of rock often encountered in road construction in my country. Due to its special physical and engineering properties, red sandstone directly affects as a filling material and greatly affects its durability of use. If you want to use local materials to improve the durability of red sandstone, it must be used as a filling material after necessary improvements. After the red sandstone is mixed with loess, cement, and water, it can become a geotechnical improvement material with certain mechanical strength and durability after an appropriate curing period and can be widely used in construction. Given the bad characteristics of red sandstone, many scholars have done research. Yu Zehong (Yu et al. 2005) discussed the shear strength and deformation characteristics of red sandstone under different compaction degrees and obtained the peak strength of red sandstone weathered soil within a certain range. Zhu Yanbo (Zhu et al. 2013) conducted laboratory tests after adding lime, cement, and fly ash to mudstone and studied the engineering characteristics and improvement mechanism of mudstone improved soil. Based on the concept of fully using resources and protecting the environment, this paper studied the properties of loess and red sandstone and thus concluded that the filling material can be used as a backfill for foundation trenches.

## 2 ANALYSIS OF THE MECHANISM OF ACTION

Loess and red sandstone, as common industrial raw materials in engineering, cannot be used alone because of their unique mechanical properties in practical applications. The two materials are mixed with an appropriate amount of curing agent to improve their mechanical properties and be used in practical engineering. When loess is added to the red sandstone soil, the loess itself has a larger particle contact area and tightly cements the red sandstone soil, thereby enhancing the strength of the soil, and enhancing its water stability and integrity.

When cement is mixed into red sandstone soil, two main chemical reactions will occur: cation exchange and pozzolanic reaction. The $Ca(OH)_2$ generated by the hydrolysis of cement creates an alkaline environment for the soil, and in this environment, $Ca^{2+}$ can replace the metal cations such as $K^+$ and $Fe^{2+}$ on the surface of the soil particles, so that the soil particles can quickly agglomerate. At the same time, the physical properties of the mixed soil are changed, and the initial strength of the soil is improved. The pozzolanic reaction is that in an alkaline environment, active $SiO_2$, $Al_2O_3$, and other colloid-forming ions in soil particles react with $Ca(OH)_2$ to form calcium silicate, calcium aluminate, and other substances, which mainly improve the later strength of the soil.

## 3 TEST PLAN

### 3.1 *Unconfined compressive strength test*

To explore the influence of collapsible loess and red sandstone-solidified soil on the mechanical properties, by referring to the relevant literature (Gan 2014; Zhu et al. 2017; Zhang 2018), the method of control experiment was used to conduct relevant experimental research. The composition materials of collapsible loess filling materials are: fine aggregate is collapsible loess accounting for 70% of the total aggregate, coarse aggregate is red sandstone accounting for 30% of the total aggregate, and the soil solidifying agent is mixed with The amount is 15% of the aggregate, and the amount of water is 25% of the total mass of the aggregate and the curing agent. Control group 1 is the collapsible loess without a curing agent after stirring under the same conditions, control group 2 is red sandstone without a curing agent after stirring under the same conditions, control group 3 is collapsible loess with a curing agent added after stirring under the same conditions, and control group 4 is red sandstone with a curing agent added after stirring under the same conditions.

The loess, red sandstone, and cement were mixed according to the proportion, the samples were prepared after fully stirring, and the unconfined compressive strength of each group of samples was measured to explore the strength of the filling material.

According to the Standard of Geotechnical Test Methods (Ministry of Water Resources of the People's Republic of China 2019), the collapsible loess and red sandstone were mixed with cement according to the mixing ratio requirements. After re-stirring evenly, the mixed soil samples were weighed. The other control groups were prepared in the same way. It is 150mm*150mm*150mm.

### 3.2 *Consolidation test*

In order to judge the consolidation settlement of the collapsible loess filling material, the standard consolidation compression test was carried out on the filling material. After preparing the test block and demoulding, researchers use a ring knife to select three groups of samples and a consolidation instrument to perform a standard consolidation test. Under the action of each level of pressure, the sample is consolidated for 24 hours, or the sample deformation does not change more than 0.01mm per hour, and the next level of pressure is applied to record the stable reading.

### 3.3 *Direct shear test*

The samples were demolded after vibrating and compacting with a shaking table. According to the "Standards for Geotechnical Test Methods"[13], four soil samples were taken for the direct shear test. The direct shear test was carried out with a type direct shearing instrument, and the normal stress of 100 kPa, 200 kPa, 300 kPa, and 400 kPa was applied for the fast-direct shear test. The shear speed was 0.8mm/min, and the sample was sheared within 3min-5min. The measured curve determines the corresponding shear strength value, thereby plotting the Coulomb strength envelope.

### 3.4 *Indoor penetration test*

The permeability coefficient of the sample can be obtained by performing the permeability test on the soil sample. During the variable water head test, the casing water head will gradually decrease with time. The initial and end water head readings were recorded in each time period. The following formula for calculating the permeability coefficient of variable water head is given in "Test Method Standard"[13].

$$k_T = 2.3 \frac{al}{A(t_1 - t_2)} \log \frac{H_1}{H_2} \qquad (1)$$

In formula (1): $a$ is the sectional area of the variable head pipe, $cm^2$; A is the sectional area of the sample, $cm^2$; 2.3 is the transformation factor of ln and log; L is the seepage path, cm; t is the height of the sample, cm; $t_1$, $t_2$ are the initial and end times of the reading head, s; $H_1$, $H_2$ are the starting and ending heads.

## 4 ANALYSIS OF TEST RESULTS

### 4.1 *Analysis of unconfined compressive strength results*

Through the unconfined compressive strength test of the collapsible loess filling material and other control groups, the failure diagram of the sample is shown in Figure 1. The compressive strength of the collapsible loess filling material and the control group test are analyzed. The specimen after compressive failure is shown in Figure 2.

It can be seen from Figure 2 that with the continuous increase of the curing age, the compressive strength of the collapsible loess filling material and each control group gradually increased, reaching the maximum value at 28d, and the collapsible loess filling material had no side. The limit compressive strength value is 3.319 Mpa; the compressive strength value of control group

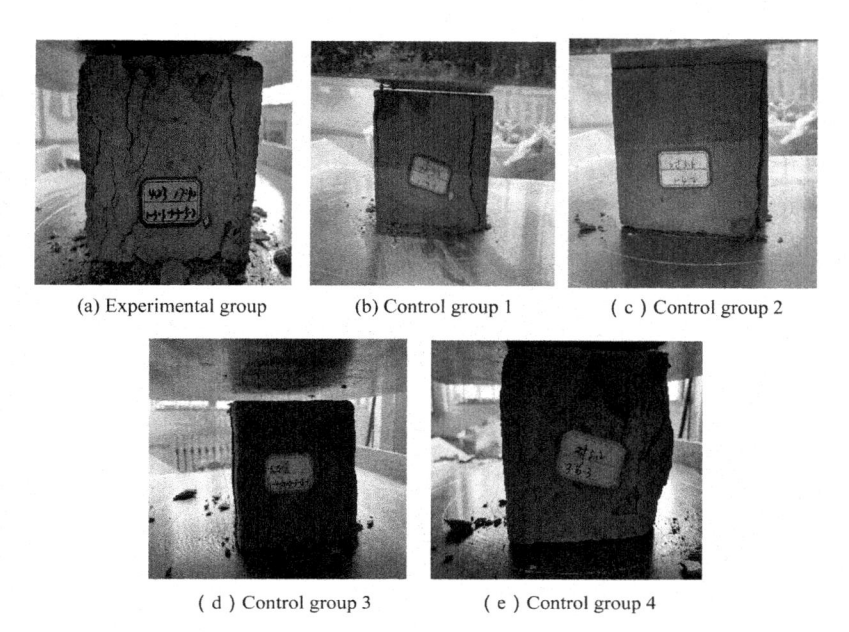

(a) Experimental group    (b) Control group 1    ( c ) Control group 2

( d ) Control group 3    ( e ) Control group 4

Figure 1.    Failure diagram of unconfined compressive strength specimen.

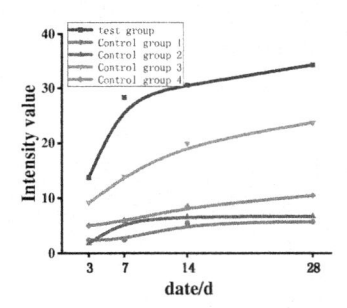

Figure 2.    Changes of compressive strength with age in test group and control group.

1 is 0.126Mpa; the compressive strength value of control group 2 is 0.086 Mpa; the compressive strength value of the control group 3 is 2.149Mpa; the compressive strength value is 0.192Mpa, which shows that the unconfined compressive strength value of the collapsible loess filling material is significantly better than that of other control groups.

## 4.2   *Analysis of direct shear test results*

Shear strength refers to the stress generated on the sliding surface when the shear failure occurs in the sliding surface inside the soil. At present, the measurement of shear strength parameters is mainly done with a shearing instrument. The internal friction angle and cohesion parameters of the sample are obtained through the shear test. The shear failure of the sample is shown in Figure 3. In engineering, it is an important mechanical index for soil measurement, and the shear strength is calculated according to the formula. It can be seen from the formula that the two important indicators of shear strength are cohesion and internal friction angle. The internal friction angle of soil reflects that the interaction between soil particles will produce certain friction characteristics. It mainly includes the surface friction force between soil particles and the occlusal force between particles, while soil cohesion mainly comes from the internal straightness of the particles. The stress is caused by the mutual attraction of shears.

Figure 3.   Shear failure specimen.

Table 1.   Direct shear test data table.

| Test type | Cohesion /KPa | Internal friction angle /° |
| --- | --- | --- |
| Test group | 83.65 | 54.25 |
| Control group 1 | 26.8 | 14.46 |
| Control group 2 | 34.1 | 35.9 |
| Control group 3 | 51.7 | 46.64 |
| Control group 4 | 60.6 | 51.58 |

It can be seen from Table 1 that the cohesion and internal friction angle of the collapsible loess filling materials are significantly improved compared with other control tests, which is due to the mixture of coarse and fine particles of the collapsible loess and the red sandstone soil during the sample preparation process, which improves the compactness of the sample and strengthens the frictional bite force between the particles. During the shearing process, the stronger the effect of adjacent particles on the relative movement, the greater the internal friction angle.

### 4.3   Soil Compression performance analysis

The compressibility of soil refers to the characteristic of the volume reduction of soil under the action of pressure. The compressive properties of the soil were analyzed by the consolidation test of the filling material and the soil of the control groups. The results are shown in Table 2. Compared with the compressive properties of the undisturbed red sandstone soil, the compressive properties of the collapsible loess filling material are greatly improved because the porosity of the soil sample is greatly reduced after the collapsible loess filling material is added with cement. From Table 2, it can be known that the compressive coefficient $\alpha$1-2<0.1 Mpa, the compressive index Cc<0.2, and the compressive modulus Es>15Mpa of the collapsible loess filling material, according to the "Geotechnical Test Method Standard"[13], the collapsibility can be judged the loess filling material belongs to low compressibility soil.

Table 2.   Consolidation test data.

| Test type | $\partial_{1-2}/Mpa$ | $E_{1-2}/Mpa$ | Compression index |
| --- | --- | --- | --- |
| Test group | 0.09 | 22.32 | 0.051 |

Table 3. Compression performance index of red sandstone undisturbed soil.

| Rock name | $\partial_{1-2}/Mpa$ | $E_{1-2}/Mpa$ |
|---|---|---|
| Strongly weathered sandstone | 0.16 | 28.7 |
| Strongly weathered sandstone | 0.21 | 30.1 |

Table 4. Collapse coefficient data.

| Test type | Collapsibility coefficient |
|---|---|
| Test group | 0.001 |
| Control group 1 | 0.04 |
| Control group 2 | 0.012 |
| Control group 3 | 0.003 |
| Control group 4 | 0.005 |

By comparing and analyzing the collapsible coefficient of the collapsible loess filling material soil sample and the control group soil sample, it can be seen from Table 4 that the collapsible coefficient of the filling material is 0.001, according to the "Geotechnical Test Method Standard" [13] knowing the collapsible coefficient, and it can be considered that the collapsible loess filling material has no collapsibility.

## 4.4 *Analysis of penetration test results*

The phenomenon that water seeps and moves in the pores of the soil is called the permeability of the soil, which is expressed by the permeability coefficient. When the hydraulic gradient is 1, the permeability rate is the permeability coefficient. The permeability coefficient is an index reflecting the infiltration capacity of the soil. The permeability coefficient of the soil body is determined by conducting a penetration test on the soil body, and the size of the permeability coefficient reflects the strength of the soil infiltration capacity. The penetration test process is shown in Figure 4.

(a) Ring knife cutting soil sample (b) Sharpened ring knife soil sample (c) TST-55 Permeameter

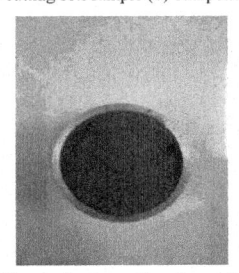 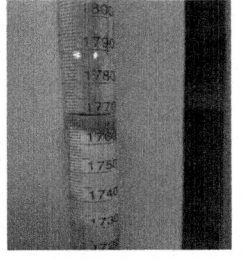

(d) Ring knife for cutting the soil sample (e) Variable head

Figure 4. Penetration test process.

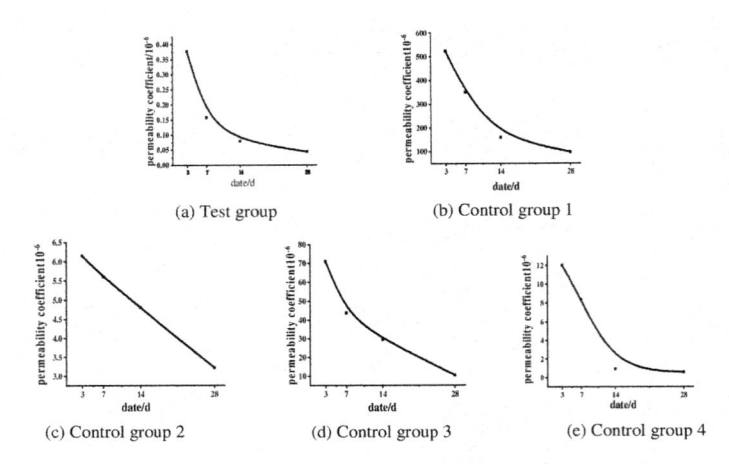

(a) Test group    (b) Control group 1

(c) Control group 2    (d) Control group 3    (e) Control group 4

Figure 5.    Changes in permeability coefficient between test group and control group.

Figure 5 shows the change curve of the permeability coefficient of the collapsible loess filling material and the control groups with the curing age, and it can be seen that the permeability coefficient of the sample shows a decreasing trend with the increase of age. With the increase of the period, the moisture content of the soil gradually decreases, the permeability of the soil to water weakens, and the permeability of the soil gradually weakens. The test results show that the permeability coefficient of the collapsible loess filling material is much lower than that of the control groups when the curing age is 28 days, and the filling material has good permeability.

## 5    CONCLUSION

In this paper, the collapsible loess accounts for 70% of the total aggregate, the coarse aggregate is red sandstone accounts for 30% of the total aggregate, the cement content is 15% of the aggregate, and the water content is 25% of the total mass of aggregate and cement. Through the unconfined compressive strength test, consolidation compression test, direct shear test, and indoor penetration test, it can be found that the performance of the collapsible loess filling material is better than the control groups. The test conclusions are as follows:

(1) The test uses collapsible loess and red sandstone as the main aggregates, adding cement as a curing agent to form a backfill material for foundation trenches, eliminating the collapsible of loess and the easy weathering and disintegration of red sandstone.
(2) The unconfined compressive strength ratio of collapsible loess filling material is compared with the collapsible loess without curing agent after stirring under the same conditions and the red sandstone without curing agent after stirring under the same conditions. The strength of the collapsible loess after adding the curing agent and the red sandstone adding the curing agent after stirring under the same conditions is higher in four cases, and with the increase of age, the compressive strength of the soil reaches the maximum at 28 d.
(3) Compared with the control groups, the collapsible loess filling material samples have lower consolidation settlement, no collapsibility, low permeability, and high strength, which can meet the requirements of using collapsible loess as the main raw material as the foundation trench filling.
(4) In the future, the flow properties of filling materials can be studied on this basis, and field tests can be carried out to apply them to actual projects and form a complete set of construction techniques to eliminate the backfilling of foundation trenches for new buildings and structures.

The consolidation and settlement of itself can improve the environment and economic benefits of collapsible loess areas.

## REFERENCES

Du Yanqing, Wang Xinqi. Experimental study on the treatment of waste mud in engineering[J]. *Tianjin Construction Technology*, 2018, 28(1); 47–50.

Gan Wenning. *Experimental study and application of red sandstone engineering characteristics* [D]. Hefei: Hefei University of Technology, 2014.

Ministry of Water Resources of the People's Republic of China. GB/T50123-2019, *Standards for Geotechnical Test Methods*[S], China Planning Press, 2019.

Wang Zhiyuan, Zhang Hong, Qian Jinsong, Ling Jianming, Yuan Qiu. Research on the flow properties of the backfill material mixture of industrial waste slag and fly ash in arid areas [J]. *Land Resources and Environment*,2015,29(04): 160–165.

Yu Zehong, Wei Hongwei, Zou Yinsheng. Strength and deformation characteristics of reinforced red sandstone weathered soil[J]. *Chinese Journal of Rock Mechanics and Engineering*,2005(15):2770–2779.

Yuan Qiu, Gao Dong, Qian Jinsong. Laboratory study on fluid fly ash[J]. *Shanghai Highway,* 2012(1): 61–65.

Yuan Qiu, Qian Jinsong, Yang Jianjiang. Laboratory study on fluid fly ash[J]. *Fly Ash*, 2011, 23(3): 1–3.

Zeng Qinglin. *Research on the influence of water on argillaceous red sandstone*[D]. Guangzhou: South China University of Technology, 2010.

Zhang Yixin. *Study on physical and mechanical properties and permeability of red sandstone in Lanzhou area*[D]. [Master's Thesis]. Lanzhou University of Technology, 2018.

Zhou Yongxiang, Wang Jizhong. Principle of ready-mixed solidified soil and its engineering application prospects[J]. *New Building Materials*,2019,46(10):117–120.

Zhu Yanbo, Yu Hongming, Yang Yanxia, et al. Laboratory test study on the properties of red-bed mudstone improved soil[J]. *Chinese Journal of Rock Mechanics and Engineering*, 2013, 32(2:): 425–432.

Zhu Yanpeng, Li Huijun, Yang Xiaohui, et al. Experimental study on improvement of red sandstone by loess and road performance[J]. *Journal of Water Resources and Architectural Engineering*, 2017,15(06):12–15.

Zou Xianyun. *Study on the applicability of high-fluid fly ash in the three backfills of subgrade*[D]. Chongqing: Chongqing Jiaotong University, 2012.

*Civil Engineering and Urban Research – Mohamed & Hou (Eds)*
*© 2023 the Authors, ISBN 978-1-032-44487-1*

# Impact response of concrete-filled double steel tubular beams and its analytical evaluation

Kailai Wang*

*Department of Civil Engineering, Yangzhou University, Yangzhou, China*

ABSTRACT: Concrete-filled double steel tubular (CFDST) members are commonly utilized in structures for their high strength and ductility. With service time, corrosion of the steel tubes, material aging, and live loads induced fatigue generated on the surface of steel tubes. The cracks threaten structural safety, especially when the damaged members are subjected to impact loads such as vessels, vehicles, and rocks. This paper investigates the influence of the initial crack length on CFDST beams under impact loads using the finite element approach. The paper indicates that initial crack length has a greater effect on the deflection of CFDST beams if it is greater than a critical value. With increased crack length, the overall trend is that peak impact force decreases for the damage softens the beam. It is recommended that once a fracture is formed on the surface, maintenance measures should be adopted to repair and enhance its impact resistance.

## 1 INTRODUCTION

Concrete-filled double steel tubular (CFDST) members, which are made up of two tubes with concrete sandwiched between them, are one of the most improved types of members (as shown in Figure 1). CFDST members have greater strength, advanced stiffness, and increased ductility when compared to regular reinforced concrete members, making them appropriate for earthquake-prone areas (Inai et al. 2004; Lee et al. 2011; Li et al. 2014; Skalomenos et al. 2016). CFDST is a composite construction that incorporates the benefits of both concrete and steel tubes (Brauns 1999).

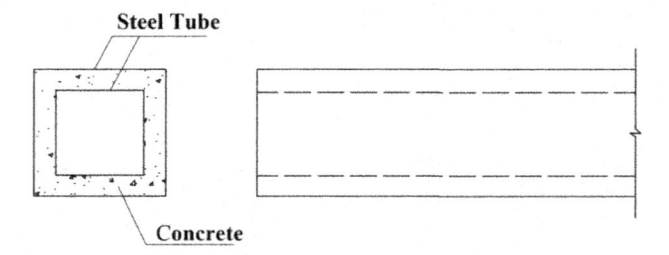

Figure 1. CFDST member.

With the development of different types of CFDST, many efforts have been made to investigate their behavior. Ci et al. proposed a theoretical model for simulating CFDST column test ultimate strengths and load-axial strain correlations (Ci et al. 2022). Ahmed et al. suggested a computer simulation approach for modeling the nonlinear behavior of eccentrically loaded short CFDST columns (Ahmed et al. 2020). Duan et al. developed a formula for calculating the shear capacity of

*Corresponding Author: 191404313@yzu.edu.cn

 DOI 10.1201/9781003372417-18

circular-in-square CFDST specimens that took both the shear-span ratio and the hollow ratio into account (Duan et al. 2022). The researchers above studied the static behavior of CFDST members.

Some CFDST structures, however, must be built to withstand impact loads, which can occur when comparably stiff large objects collide at moderate speeds, such as falling boulders in mountainous terrain or falling big loads in factories and warehouses due to accidents. Wang et al. summarized the current study methods, the range of the external impact energy, the material properties of the concrete and the steel, as well as various boundary conditions, etc., to investigate the impact performance of the CFDST members (Wang et al. 2017). Wang et al. also did a drop hammer impact test of CFDST members to study the impact performance of this kind of composite structure under multiple transverse impacts. The deformation mechanism of the composite tubes was analyzed, and the impact force-and-time history curves during each impact. The global and the local deformation data after each impact for the composite specimens were also obtained (Wang et al. 2017). According to the testing result, the impact resistance of the composite tube is considerably influenced by the thickness of the exterior steel tube.

In summary, researchers primarily concentrated on the performance of pristine CFDST parts under static and impact loads. However, with time, a fracture on the surface of a steel tube caused by erosion or loading-induced fatigue may develop, resulting in a reduction in structural capability. Few attempts have assessed the safety of such deteriorating structural parts or structures when subjected to impact loads.

This paper investigates the influence of the initial crack length on CFDST beams under impact loads using the finite element (FE) approach. The behavior of CFDST beams with various fracture lengths subjected to impact loads is thoroughly explored through numerical analysis. Some meaningful results are obtained, which can be used to guide the safety assessment and reinforcement of damaged CFDST members.

## 2 FINITE ELEMENT MODEL

The paper uses ABAQUS to develop a nonlinear FE model to investigate the dynamic behavior of cracked CFDST beams under impact loads. The length of the beam is 1.6m, the distance between two supports is 1.3m, and the thickness of the outer and inner layer is 4mm, as shown in Figure 2. The hammer was simulated as 500kg and dropped from the height of 1.2m. To simulate the initial crack, a cut was set as 1cm, 2cm, 3cm, and 4cm, respectively, on the outer tube at the middle of the span. Following the creation of the model, an explicit dynamic analysis was used to calculate the displacement and the impact force at the midpoint of the span. The FE model is presented in Figure 3.

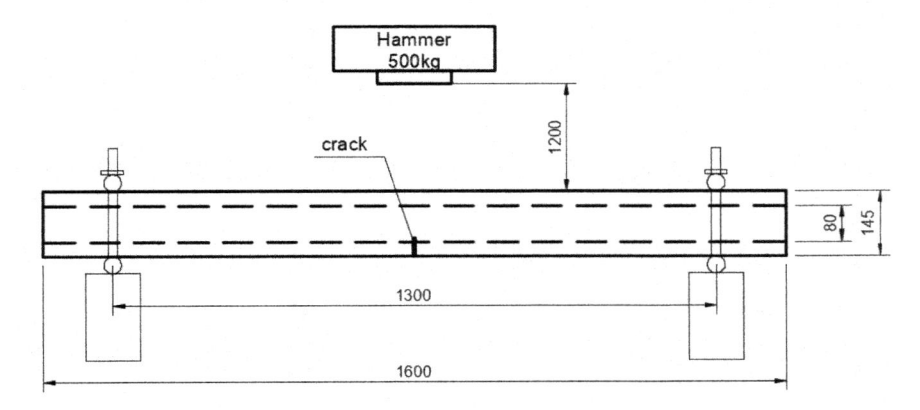

Figure 2.  Model setup. (unit:mm).

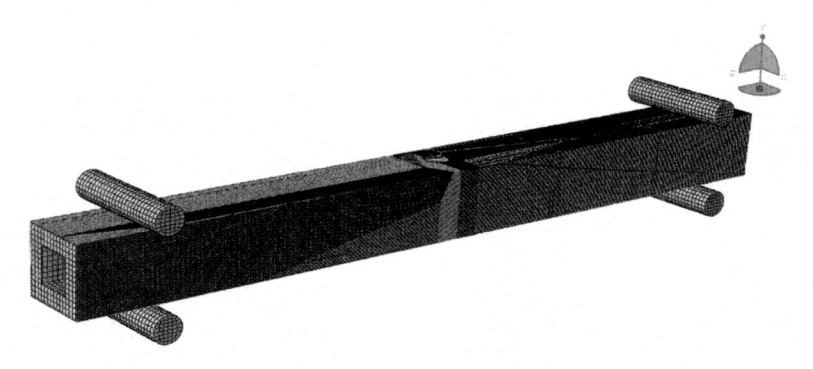

Figure 3.    Finite element model.

In the FE model, core concrete and steel tubes were represented with four-node shell elements with reduced integration (S4R) and eight-node brick elements with reduced integration (C3D8R). The corrosion modeling achieved by "Model change" in the tube thickness direction could not be easily performed with shell elements, which is why solid components were chosen for the steel tube. A total of 8 layers of elements are applied in the thickness direction of the steel tube to ensure that solid elements achieve equal accuracy to shell elements. To save time and money, the rigid shell element (R3D4) was used to replicate the drop hammer. The drop hammer was represented by a hard shell with the same mass and form as the hammer surface. The deformation of the drop hammer is quite minor compared to the deformation of steel tubes and concrete; therefore, its simplification could further ensure the accuracy of the results in this study. Four rigid cylinders were used to approximate the bearings. The degree of bearing freedom was set.

To achieve an evenly distributed mesh style, structural meshing is used. The impact area of the beam has a higher mesh density due to the complex contact behavior and substantial deformation. The beam's boundaries are designed to be simply supported for simplicity without sacrificing generality, and the impact is set to occur at the mid-span of the beam. The numerical simulation considered geometry nonlinearity, and the thickness of the steel tube was estimated using nine integrated points. In this paper, the ductile damage hypothesis is used to examine the failure of the steel tube element. The interaction between steel tube (out and inner layer) and concrete, drop weight, and steel tubular beam was simulated in the FE model using hard contact (normal direction) and the Coulomb friction model (tangential direction) in ABAQUS. In ABAQUS, general contact was defined in the contact surface, with a friction coefficient of 0.15. The density of the steel tube is 7850 kg/m$^3$, and the elastic modulus is $2.1 \times 10^{11}$ Pa. Also, it has a Poisson ratio of 0.3, a yield stress of 235 MPa, and a fracture strain of 0.26. The paper used the Cowper-Symonds model to calculate the yield strength of steel plates at various strain rates.

The concrete-damage-plastic model (CDP) was used to predict the nonlinear behavior of core concrete, with the effects of strain rate on concrete characteristics considered. The density of the concrete is 2360 kg/m$^3$, the elastic modulus is $2.2 \times 10^{10}$ Pa, and it has a Poisson ratio of 0.2. Concrete's dynamic compressive strength rises as the strain rate rises. The Comite Euro-International du Beton details the parameters for estimating the dynamic compressive strength of concrete (Beton 1990). Based on the model proposed by Malvar and Ross (Malvar & Ross 1998).

## 3   NUMERICAL SIMULATION RESULTS AND DISCUSSION

The dynamic behavior of damaged CFDST beams under drop weight impact is simulated. Figure 3 shows the fractures on the CFDST beams after impact under four working conditions. Table 1 corresponds to each working condition to its initial crack length on the surface of the CFDST beam.

It is known that under the same impact energy, the initial crack length significantly affects the final fracture.

Figure 4 presents the displacement at the middle of the span for each initial crack length. It can be known that with increased crack length, under the same impact energy, the deflection at the middle-of-the-span increases, especially when the crack length is larger than 2 cm.

Table 1. Length of the initial crack on each beam.

| Number | Beam | Length of the initial crack on the beam |
|---|---|---|
| (a) | Beam1 | 1cm |
| (b) | Beam4 | 4cm |

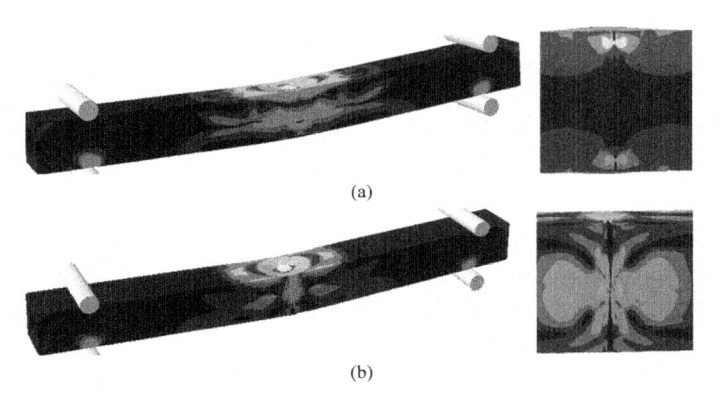

(a)

(b)

Figure 4. Fracture of CFDST beams after impact.

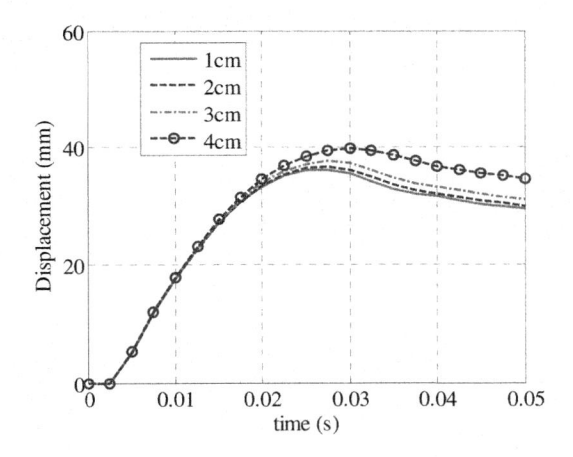

Figure 5. Displacement at middle-of-the-span of CFDST beams.

The impact force's time history is depicted in Figure 6. The peak impact force is affected by the severity of the damage. With increased crack length, the overall trend is that peak impact force decreases for the damage softens the beam.

The research of Hou & Han is used to prove the accuracy of the FE model in this paper (Hou & Han 2018). They studied the whole-life performance of concrete-filled steel tubular (CFST)

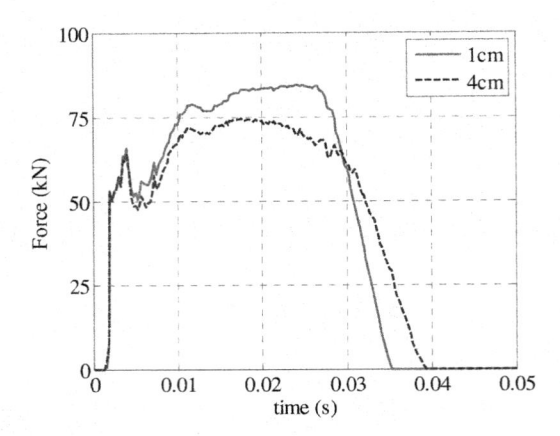

Figure 6. Impact force for the CFDST beams.

columns through numerical simulation. A finite element analysis (FEA) model is established to simulate the performance of degraded CFST structures under lateral impact.

Table 2 shows three accompanying load paths they designed for comparative analysis:

Table 2. Four simulation cases.

|  | Sustained long-term load | Corrosion | Impact load |
| --- | --- | --- | --- |
| Case 1 | × | × | × |
| Case 2 | ✓ | × | ✓ |
| Case 3 | ✓ | ✓ | × |
| Case 4 | ✓ | ✓ | ✓ |

Such four simulation cases are compared to highlight the influence of different actions on the performance of CFST columns. Figure 7 compares axial compression ($N$) versus axial displacement ($u_a$) curves in four simulation cases.

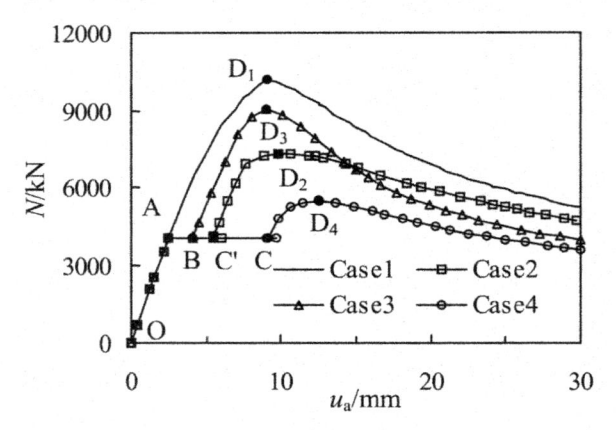

Figure 7. Axial compression ($N$) versus axial displacement ($u_a$) curves.

The analysis shows that the long-term impact, lateral corrosion, and residual compressive strength of steel tube columns are significantly reduced.

Corrosion will reduce the static and transverse impact resistance of concrete-filled steel tubular columns. When the impact resistance of the column is reduced, there will be greater residual transverse deflection in the column than without corrosion under the same impact energy. The results are similar to the analysis of this study.

## 4 CONCLUSION AND FURTHER WORK

The impact behavior of CFDST beams is investigated through numerical analysis. According to the report, the initial fracture length has a bigger effect on the deflection of CFDST beams if it is larger than a critical value. For example, in this paper, the deflection increased significantly when the crack length was greater than 2 cm. The results also indicate that damage severity affects the peak impact force. With increased crack length, the overall trend is that peak impact force decreases for the damage softens the beam.

Therefore, it is recommended that once a crack is formed on the surface, maintenance measures should be adopted to repair and enhance its impact resistance. The structural health monitoring system could monitor that the crack length does not exceed the critical value.

In this paper, the fracture increases significantly with initial crack length. The fracture length increment is also related to the strength of concrete, the thickness, or the inner-to-outer diameter ratio of CFDST beams. For further research, it is suggested that optional design be done to select reasonable sizes of inner and outer steel tubes by simulating the fracture of CFDST beams of different sizes under impact loads.

Due to some limitations in this paper, for further work, experiments should be designed to prove the accuracy of the FE model. It is also recommended that a formula could be summarized to express the critical value of the initial crack length of CFDST beams under impact loads.

## REFERENCES

Ahmed, M. et al. (2020). Experimental and numerical investigations of eccentrically loaded rectangular concrete-filled double steel tubular columns. *J. Constr. Steel. Res.* 105949.

Brauns, J. (1999). Analysis of stress state in concrete-filled steel column. *J. Constr. Steel. Res.* 49(2), 189–196.

Camargo, A.L. et al. (2019). Fire resistance of axially and rotationally restrained concrete-filled double-skin and double-tube hollow steel columns. *J. Struct. Eng.* 145(11), 04019128.

Ci, J. et al. (2022). Axial compressive behavior of circular concrete-filled double steel tubular short columns. *Adv. Struct. Eng.* 25(2), 259–276.

Comite Euro-International du Beton (1990). *CEB-FIP Model Code Trowbridge Wiltshire* UK: Redwood Books.

Duan, L.X. et al. (2022). Shear response of circular-in-square CFDST members: Experimental investigation and finite element analysis. *J. Constr. Steel. Res.* 190, 107160.

Ekmekyapar, T. & H.G. Hasan (2019). The influence of the inner steel tube on the compression behavior of the concrete-filled double skin steel tube (CFDST) columns. *Marine Structures.* 66, 197–212.

Fujikake, K., Li, B. & S. Soeun (2009). Impact response of reinforced concrete beam and its analytical evaluation. *J. Struct. Eng.* 135(8), 938–950.

Hou, C.C. et al. (2016). Flexural behavior of circular concrete-filled steel tubes (CFST) under sustained load and chloride corrosion. *Thin-Walled Structures.* 107, 182–196.

Inai, E. et al. (2004). Behavior of concrete-filled steel tube beam columns. *J. Struct. Eng.* 130(2), 189–202.

Lee, S.H. et al. (2011). Behavior of high-strength circular concrete-filled steel tubular (CFST) column under eccentric loading. *J. Constr. Steel. Res.* 67(1), 1–13.

Li, W., Han, L.H. & T.M. Chan (2014). Tensile behavior of concrete-filled double-skin steel tubular members. *J. Constr. Steel. Res.* 99, 35–46.

Malvar, L.J. & C.A. Ross(1998). Review of strain rate effects for concrete in tension. *ACI Mater. J.* 95, 735–739.

Skalomenos, K.A. et al. (2016). Experimental behavior of concrete-filled steel tube columns using ultrahigh-strength steel. *J. Struct. Eng.* 142(9), 04016057.

Thai, S. et al. (2019). Concrete-filled steel tubular columns: Test database, design, and calibration. *J. Constr. Steel. Res.* 157, 161–181.

Uenaka, K. & H.Kitoh (2011). Mechanical behavior of concrete-filled double skin tubular circular deep beams. *Thin-Walled Structures* 49(2), 256–263.

Wang, F., Young, B. & L. Gardner (2019). Experimental study of square and rectangular CFDST sections with stainless steel outer tubes under axial compression. *J. Struct. Eng.* 145(11), 04019139.

Wang, R., Han, L.H. & C.C. Hou (2013). Behavior of concrete-filled steel tubular (CFST) members under lateral impact: Experiment and FEA model. *J. Constr. Steel. Res.* 80, 188–201.

Wang, W. et al. (2019). Behavior of ultra-high performance fiber-reinforced concrete (UHPFRC) filled steel tubular members under lateral impact loading International. *Journal of Impact Engineering.* 132, 103314.

Wang, Y. & X.D. Qian (2017). Behavior of concrete-filled double skin steel tubes under multiple transverse impacts. *J. Shock. Vib.* 36(2), 1.

*Civil Engineering and Urban Research – Mohamed & Hou (Eds)*
*© 2023 the Authors, ISBN 978-1-032-44487-1*

# Experimental research on axial compression of regional confined concrete

Jianying Zhou* & Lidan Li
*School of Architecture and Engineering, Kaili University, Kaili, Guizhou, China*

Kunhong Yan
*Comprehensive Administrative Law Enforcement Bureau of Nanming, Guiyang, Guiyang, Guizhou, China*

ABSTRACT:   In this paper, the axial compression of 13 $250 \times 250 \times 500$ short columns under axial compression and three medium-long columns of $180 \times 180 \times 2400$ under axial compression are analyzed through experimental research. The research shows that the restraint effect of the form of regional restraint reinforcement is better than that of ordinary restraint reinforcement. Through the analysis, the author predicts that the high-strength steel bar in the regionally confined concrete column can fully exert its material strength, but the degree of its exertion still needs more test data to prove it. The test analysis shows that the reliability of the constraint coefficient K specified in the "Technical Code for Regional Confined Concrete Structures" still needs to be verified by test data.

## 1   INTRODUCTION

The most effective position of the traditional restraint method is located in the core area of the component. It is reasonable for axial compression members, but in actual engineering, most members are not axially compressed, so it is of little significance to constrain the core concrete. Therefore, in 2004, Cao Xinming (CAO 2004) proposed the concept of regional constraints based on the study of stirrup-confined concrete. Region-constrained concrete imposes constraints where constraints are required, so the constraints can be flexibly changed, overcoming the shortcomings of traditional constraints.

In region-constrained concrete structures, the section is divided into zones. Longitudinal reinforcement and transverse stirrups are arranged in these areas, and the connection between the various areas is achieved by concrete and restraint stirrups. Unlike traditional reinforced concrete, Region-constrained concrete not only constrains the full section, but also combines the way of area confinement. This way, the reinforced skeleton composed of stirrups and longitudinal reinforcements in each area effectively constrains the concrete. At the same time, as shown in Figure 1, due to the different amount of restraining stirrups in each restraint area, the cross-sectional area is divided into a strong restraint area and a weak restraint area.

The research on area-confined concrete started late, and the front-line engineering personnel did not have enough knowledge and understanding of region-constrained concrete, which limited the application and promotion of area-confined concrete technology. The research results of the past 19 years show that area-constrained concrete has good mechanical properties. Experimental studies (Cao 2008; Guo 2006; Wang 2007; Wang 2008; Zhu 2007) and practical engineering project applications found that regionally confined concrete columns have a higher bearing capacity and ductility than traditional reinforced concrete columns. Ordinary concrete compression members are limited by the ultimate strain of concrete, and high-strength steel bars cannot fully exert their material strength. Whether the regionally confined concrete compression members are subject to this limitation and the application degree of high-strength steel bars in the regionally confined

---

*Corresponding Author: Jiangong@klxy.com

DOI 10.1201/9781003372417-19

concrete compression members, have not been studied by scholars. According to the "Technical Code for Regional Confined Concrete Structures" (DBJ 52/T082-2016 2017), the compressive strength value of the restraint should meet the following requirements:

$$f_{cc} = (1 + k)f_c \tag{1}$$

Among them, the constraint coefficient is:

$$k = \left(f_y\rho_s + f_{yv}\rho_v\right)/f_c \tag{2}$$

And $k \geq 0.48$

$$\frac{\rho_v}{\rho_s} \approx 1 \sim 1.5 \tag{3}$$

If $\rho_v \geq 1.5\rho_s$, take $\rho_v = 1.5\rho_s$;
If $\rho_v \leq \rho_s$, take $\rho_v = \rho_s$.
$\rho_v$—Volume Stirrup Ratio of Constraint Stirrups, $\rho_v = \frac{A_{scl}\sum l_{cc}}{A_{cc}S}$
$\rho_s$—Longitudinal reinforcement ratio, $\rho_s = \frac{A_{sc}}{A_{cc}}$
For regional confined concrete beams, the cross-sectional area of the constrained area is $A_{cc} = b_1 x_{cc}$. However, scholars have not systematically studied the reliability of the formula. In this paper, the strength of the steel bars in the confined concrete column in the axial compression area and the restraint coefficient will be studied by experiments.

## 2 TEST OVERVIEW

In this paper, 13 short columns of $250\times250\times500\text{mm}^3$ and three medium and long columns of $180\times180\times2400\text{mm}^3$ were subjected to axial compression comparison tests. In the short column, seven ordinary confinement stirrup concrete, 6 area confinement concrete members, and the end of the stirrup are designed in two forms: welding and end hook. Among the three medium and long columns, one small square hoop, one ordinary well-shaped hoop, one area confinement concrete member, and the ends of the stirrups are all welded. The measured strength of the member materials and the reinforcement situation are shown in Table 1, and the cross-sectional form is shown in Figure 2.

Table 1. Strength and reinforcement of member materials.

| Specimen number | $f_{ck}$ (MPa) | $f_c$ (MPa) | $f_y$ (MPa) | Longitudinal stress reinforcement | Stirrup strength (MPa) | Stirrup Spacing (mm) | Sectional form |
|---|---|---|---|---|---|---|---|
| SNCC1 | 48.9 | 31.93 | 535 | 12Φ10 | 497 | 90 | d |
| SNCC2 | 48.9 | 31.93 | 535 | 12Φ10 | 497 | 90 | d |
| SNCC3 | 48.9 | 31.93 | 535 | 12Φ10 | 497 | 40 | d |
| SNCC4 | 48.9 | 31.93 | 535 | 12Φ10 | 497 | 65 | d |
| SNCC5 | 48.9 | 31.93 | 535 | 12Φ10 | 497 | 50 | d |
| SNCC6 | 48.9 | 31.93 | 535 | 12Φ10 | 497 | 50 | d |
| SNCC7 | 48.9 | 31.93 | 535 | 12Φ10 | 497 | 65 | d |
| SRCC1 | 48.9 | 31.93 | 535 | 16Φ10 | 497 | 40 | a |
| SRCC2 | 48.9 | 31.93 | 535 | 16Φ10 | 497 | 50 | a |
| SRCC3 | 48.9 | 31.93 | 535 | 16Φ10 | 497 | 50 | a |
| SRCC4 | 48.9 | 31.93 | 535 | 16Φ10 | 497 | 40 | a |
| SRCC5 | 48.9 | 31.93 | 535 | 16Φ10 | 497 | 65 | a |
| SRCC6 | 48.9 | 31.93 | 535 | 16Φ10 | 497 | 65 | a |
| MNCC1 | 25.9 | 17.48 | 400 | 12Φ12 | 235 | 50 | c |
| MRCC1 | 25.9 | 17.48 | 400 | 16Φ10 | 235 | 55 | b |
| MRCC2 | 25.9 | 17.48 | 400 | 16Φ10 | 235 | 60 | a |

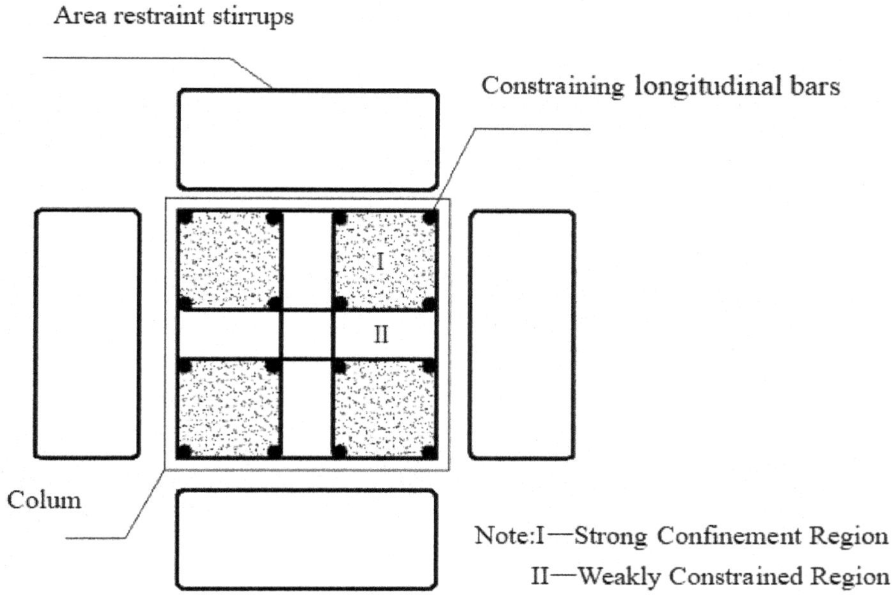

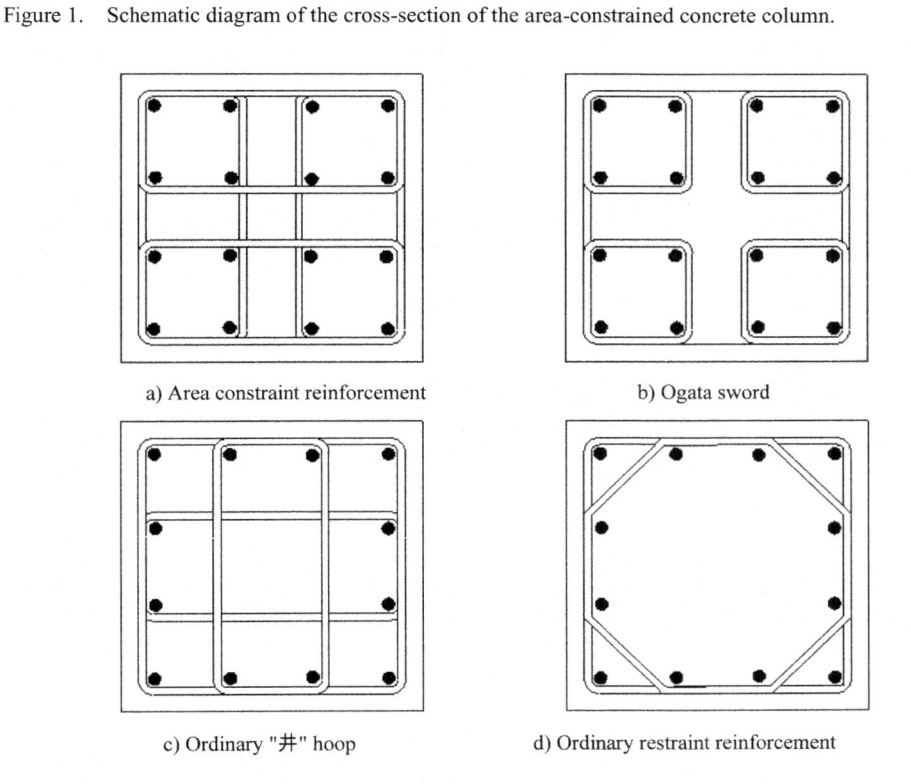

Figure 1.    Schematic diagram of the cross-section of the area-constrained concrete column.

a) Area constraint reinforcement

b) Ogata sword

c) Ordinary "井" hoop

d) Ordinary restraint reinforcement

Figure 2.    Member section and reinforcement.

# 3 STRAIN ANALYSIS OF SHORT COLUMN UNDER AXIAL COMPRESSION

In this test, resistance strain gauges were arranged on the longitudinal reinforcement, stirrup, and concrete surface to measure the strain of the reinforcement and concrete, and a displacement gauge was installed to measure the axial deformation of the components.

## 3.1 *Short column load-strain analysis*

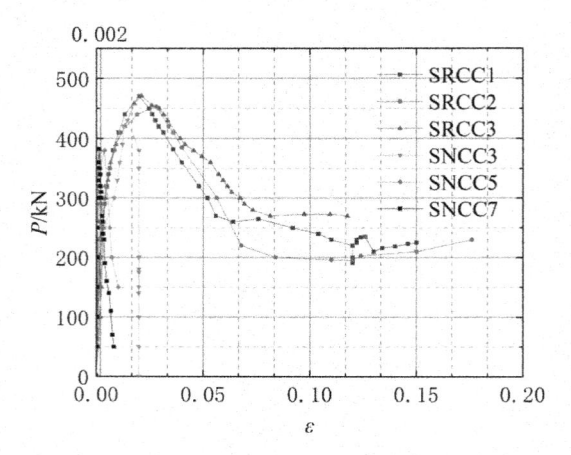

Figure 3.    Load-strain curve of the short column under axial compression.

The load-strain relationship curve can reflect the mechanical performance of the member. The load-strain relationship curve obtained from the test is shown in Figure 3. This paper analyzes three regionally restrained concrete short columns and three ordinary restrained stirrup concrete short columns with stirrup spacings of 40 mm, 50 mm, and 60 mm. It can be seen from Figure 3 that:

(1) The regional confinement axial compression short columns with different stirrup spacings have similar shapes, and the ascending section does not change much. When the strain is between 0 and 0.002, the load-strain curves of the zone-confined concrete column and the ordinary stirrup column almost coincide. When the strain is greater than 0.002, the load-strain curves of the area-confined concrete column and the ordinary stirrup column show obvious differences. It shows that the restraint effect came into play when the strain of the area restrained concrete column exceeds 0.002.

(2) Compared with the common restrained stirrup concrete short column, the load-strain relationship curve of the regional restrained stirrup concrete short column is fuller, the descending section is gentler, and the slope of the common restrained stirrup concrete column is steeper.

(3) Compared with the common confinement stirrup concrete short column, the peak strain of the regional confinement stirrup concrete short column increases significantly. The peak strains of the zone-confined concrete columns are 0.02, 0.028, and 0.0212. If the longitudinal reinforcement buckling is not considered, the calculated value of the corresponding reinforcement stress should reach 4000 MPa. The high-strength steel bar is predicted to fully exert its strength in the area-constrained concrete axial compression member. This conclusion needs to be verified experimentally.

## 3.2 *Load-strain analysis of longitudinal reinforcement*

The longitudinal reinforcement load-strain relationship curve obtained from the test is shown in Figure 4. In this paper, three regionally restrained concrete short columns (SRCC1, SRCC2, and SRCC3) and one ordinary restrained stirrup concrete short column (SNCC3) are used for analysis.

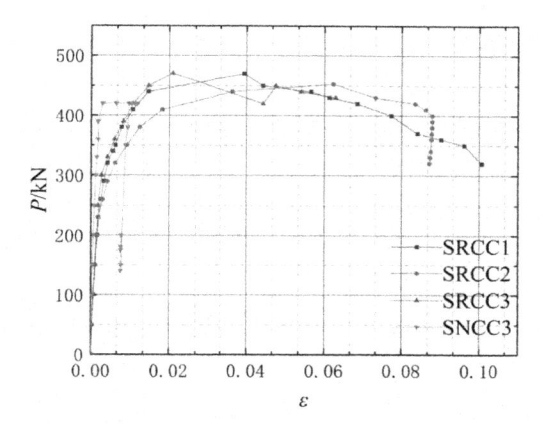

Figure 4.   Load-strain curve of longitudinal reinforcement.

It can be seen from Figure 4 that:

(1) The peak strain of the SNCC3 steel bar is 0.0028, and the corresponding calculated value of the steel bar stress is 560 MPa. The peak strain of SRCC1, SRCC2, and SRCC3 bars exceeded 0.02. Similar to the conclusion obtained in Section 2.1, if the longitudinal reinforcement buckling is not considered, the calculated value of the corresponding reinforcement stress should reach 4000 MPa. The longitudinal reinforcement stress is greater than its strength value of 535MPa, indicating that the reinforcement strength is fully utilized.

(2) Compared with the common restrained stirrup concrete short column, the longitudinal peak strain of the regional restrained stirrup concrete short column is significantly increased. It shows that the common constrained stirrup concrete short column can make full use of the strength of the steel bar. The author predicts that high-strength steel bars can give full play to their strength in regionally bound concrete axial compression members. However, this conclusion needs to be verified experimentally.

(3) The peak strain of SNCC3 steel bar is 0.0028, which exceeds the ultimate compressive strain of concrete when the axial center is compressed in the "Code for Design of Concrete Structures" of 0.002. The reason is that the spacing between the stirrups of SNCC3 is 40mm, and the stirrups are denser, which restrains the concrete and increases the ultimate compressive strain of the concrete.

(4) The stirrup spacing of SRCC1 and SNCC3 is 40mm, but the peak strain of SRCC1 is significantly higher than that of SNCC3. It shows that the confinement effect of the regional confinement stirrups is better than that of the ordinary confinement stirrups.

## 4   ANALYSIS OF BEARING CAPACITY OF AXIAL COMPRESSION COLUMN

### 4.1   *Calculation of Column Bearing Capacity in Axial Compression Zone*

Calculated value $P_0$ of bearing capacity without considering constraints

The calculated value $P_0$ of the bearing capacity without considering the restraint effect is calculated by the following formula.

$$p_0 = 0.9\varphi \left( f_c A + f_y' A_s' \right) \tag{4}$$

$\varphi$—Stability factor of reinforced concrete members;
$f_c$—Standard value of concrete axial compressive strength;
$A$—Member cross-sectional area;

$f'_y$—Longitudinal bar yield strength;

$A'_s$—Area of longitudinal reinforcement.

2. Calculated value $P_1$ of bearing capacity considering constraints

Considering the restraint effect, the bearing capacity calculation value P1 is calculated by the following formula.

$$p_1 = 0.9\varphi f_{cc} A_{cc} \tag{5}$$

The bearing capacity calculation results are shown in Table 2.

### 4.2 *Analysis of test results*

The bearing capacity test value P is shown in Table 2. It can be seen from Table 2 that:

(1) For ordinary restrained stirrup concrete columns, the test value of the bearing capacity is increased by 25%-40% for the short column, and the increase of the medium long column is 25%-40% compared with the calculation value of the bearing capacity considering the restraint, reaching 1.77%. It shows that there is still a certain margin for calculating the bearing capacity by considering the constraints. The calculated value of the bearing capacity considering the restraint is higher than that without considering the restraint, and the improvement range increases with the increase of the hoop ratio and the restraint coefficient. It shows that when the constraint coefficient is large, the bearing capacity of ordinary constrained concrete short columns can also be calculated by considering the constraint effect, and this method is more economical.

Table 2. Constraint coefficient and bearing capacity result table.

| Specimen number | $\rho_s$ | $\rho_v$ | $k$ | $f_{cc}$ (MPa) | $P_o$ (kN) | $P_1$ (kN) | $P$ (kN) | $\frac{(p_1-p_0)}{p_0}$ | $\frac{(p-p_1)}{p_1}$ |
|---|---|---|---|---|---|---|---|---|---|
| SNCC1 | 0.019 | 0.018 | 0.629 | 52.015 | 2249.6 | 2265.8 | 3800 | 0.72% | 40.37% |
| SNCC2 | 0.019 | 0.018 | 0.629 | 52.015 | 2249.6 | 2265.8 | 3135 | 0.72% | 27.73% |
| SNCC3 | 0.019 | 0.041 | 0.967 | 62.795 | 2249.6 | 2735.4 | 4300 | 21.59% | 36.39% |
| SNCC4 | 0.019 | 0.025 | 0.720 | 54.929 | 2249.6 | 2392.7 | 3800 | 6.36% | 37.03% |
| SNCC5 | 0.019 | 0.033 | 0.839 | 58.705 | 2249.6 | 2557.3 | 3800 | 13.67% | 32.71% |
| SNCC6 | 0.019 | 0.033 | 0.839 | 58.705 | 2249.6 | 2557.2 | 3800 | 13.67% | 32.71% |
| SNCC7 | 0.019 | 0.025 | 0.720 | 54.929 | 2249.6 | 2392.7 | 3820 | 6.36% | 37.36% |
| SRCC1 | 0.026 | 0.064 | 1.041 | 65.159 | 2400.8 | 2838.4 | 4700 | 18.22% | 39.61% |
| SRCC2 | 0.026 | 0.052 | 1.041 | 65.159 | 2400.8 | 2838.3 | 4530 | 18.22% | 37.34% |
| SRCC3 | 0.026 | 0.052 | 1.041 | 65.159 | 2400.8 | 2838.3 | 4710 | 18.22% | 39.74% |
| SRCC4 | 0.026 | 0.064 | 1.041 | 65.159 | 2400.8 | 2838.3 | 4700 | 18.22% | 39.61% |
| SRCC5 | 0.026 | 0.040 | 1.041 | 65.159 | 2400.8 | 2838.3 | 4100 | 18.22% | 30.77% |
| SRCC6 | 0.026 | 0.040 | 1.041 | 65.159 | 2400.8 | 2838.3 | 4030 | 18.22% | 29.57% |
| MNCC1 | 0.060 | 0.036 | 2.191 | 55.783 | 976.9 | 1129.6 | 1150 | 15.63% | 1.77% |
| MRCC1 | 0.056 | 0.036 | 2.028 | 52.927 | 942.1 | 1071.8 | 1200 | 13.76% | 10.69% |
| MRCC2 | 0.056 | 0.035 | 2.028 | 52.927 | 942.1 | 1071.8 | 1300 | 13.76% | 17.56% |

(2) For regionally restrained stirrup concrete short columns, the calculation value of the bearing capacity considering the restraint effect is the same. The reason is that the volume hoop ratio $\rho_v$ of the restraining stirrups of the regionally restrained concrete short columns is larger, which is greater than $1.5\rho_s = 0.39$. According to the "Technical Code for Regional Confined Concrete Structures," when calculating the constraint coefficient $k$, take $\rho_v = 1.5\rho_s$. Therefore, the restraint coefficients of the restrained concrete short columns in all regions are the same, so the calculation values of the bearing capacity considering the restraint effect are the same.

(3) For medium and long columns of regionally restrained stirrup concrete, the calculation values of the bearing capacity considering the restraint effect are the same. The reason is the $\rho_v \leq \rho_s$.

According to the "Technical Code for Regional Confined Concrete Structures," when calculating the constraint coefficient $k$, take $\rho_v = \rho_s = 0.056$. Therefore, the restraint coefficients of the restrained concrete short columns in all regions are the same. Therefore, the calculation value of the bearing capacity considering the restraint effect is the same.

(4) For regionally restrained concrete columns, compared with the calculation value of bearing capacity considering restraint, the test value of a short column is increased by about 30%–40%, and the increase of medium and long columns is about 10%–20%. On the one hand, the "Technical Code for Regional Confined Concrete Structures" stipulates "if $\rho_v \geq 1.5\rho_s$, take $\rho_v = 1.5\rho_s$" when calculating the constraint coefficient, which is more secure. However, its reliability and whether there is a more economic value still need to be determined through a large number of experiments. On the other hand, it is stated that the reliability of the "if $\rho_v \leq \rho_s$, take $\rho_v = \rho_s$" stipulated in the "Technical Code for Regional Confined Concrete Structures" when calculating the constraint coefficient still needs to be verified by tests. Thirdly, compared with the bearing capacity considering the restraint, there is no single reason why the increase in the test value of the medium and long columns is smaller than that of the short columns. Whether it is because the restraint effect in the middle and long columns cannot be fully exerted, or because the "Technical Code for Regional Confined Concrete Structures" stipulates that "if $\rho_v \leq \rho_s$, taking $\rho_v = \rho_s$" is unsafe when calculating the restraint coefficient, it is still uncertain. The reason still needs to be verified experimentally.

## 5 CONCLUSIONS

In this paper, 13 short columns of $250 \times 250 \times 500$ and three medium and long columns of $180 \times 180 \times 2400$ are respectively subjected to axial compression test analysis. The following conclusions are drawn from the analysis:

(1) The restraint effect of the regional restraint reinforcement is better than that of the common restraint.

(2) The stress of the steel bars in the area-confined concrete short column exceeds its material strength by 535 MPa. It shows that high-strength and high-rib can give full play to its material strength in area-confined concrete, but more experimental studies are needed to verify it.

(3) The "Technical Code for Regional Confined Concrete Structures" stipulates "if $\rho_v \geq 1.5\rho_s$, take $\rho_v = 1.5\rho_s$" when calculating the constraint coefficient, which is more secure. However, its reliability and economic value still need to be determined through many experiments.

(4) Compared with the bearing capacity considering the restraint, there is no single reason why the increase in the test value of the medium and long columns is smaller than that of the short columns. Whether it is because the restraint effect in the middle and long columns cannot be fully exerted, or because the "Technical Code for Regional Confined Concrete Structures" stipulates that "if $\rho_v \leq \rho_s$, taking $\rho_v = \rho_s$" is unsafe when calculating the restraint coefficient, it is still uncertain. The reason still needs to be verified experimentally.

## ACKNOWLEDGMENTS

We are grateful to the Youth Science and Technology Talents Growth Project of the Education Department of Guizhou Province (No.: QianJiaoHe KY Zi[2019]196) for sponsoring this paper.

## REFERENCES

Cao X M & Bai J (2004). *Confined concrete in moment element*[ISCC-2004].
Cao X M, Xiao C A, Xiao Jc & Ying L L (2008). Analysis of area confined concrete. *J. Engineering Seismic and Reinforcement Reconstruction*, (05), 112–115.

Cao X M, Yang L L & Zhu G L (2008). Research on the mechanical performance of area-constrained concrete axially compressed rectangular columns. *J. Journal of Chongqing Jianzhu University*. (03), 83–86.

DBJ 52/T082-2016 (2017). Technical Code for Regional Confined Concrete Structures. S. China Architecture & Building Press.

Guo X Z (2006). *Nonlinear analysis of regionally confined high-strength concrete short columns under axial compression*. D. Guizhou University.

Wang X B, Wang K, Luo X Y & Ma J (2007). Research on the axial compression ratio of area-constrained concrete columns. *J. Fujian Architecture*. (12), 46–48.

Wang X B. *Experimental research on regionally confined concrete axial compression column*. D. Guizhou University, 2008.

Zhu G L (2007). *Experimental study on zone-confined high-strength concrete short column under axial compression D*. Guizhou University.

*Civil Engineering and Urban Research – Mohamed & Hou (Eds)*
*© 2023 the Authors, ISBN 978-1-032-44487-1*

# Experimental study on vertical in-situ blasting demolition of reinforced concrete water tower

Xiaowu Huang*
*College of Science, Wuhan University of Science and Technology, Wuhan, China*
*Hubei Key Laboratory of Blasting Engineering, Wuhan, China*
*Wuhan Explosions & Blasting Co. Ltd., Wuhan, China*

Xianqi Xie & Yongsheng Jia
*Hubei Key Laboratory of Blasting Engineering, Wuhan, China*
*Wuhan Explosions & Blasting Co. Ltd., Wuhan, China*

Dongwang Zhong
*College of Science, Wuhan University of Science and Technology, Wuhan, China*

Jinshan Sun
*Hubei Key Laboratory of Blasting Engineering, Wuhan, China*

Yingkang Yao
*Hubei Key Laboratory of Blasting Engineering, Wuhan, China*
*Wuhan Explosions & Blasting Co. Ltd., Wuhan, China*

ABSTRACT:    In this study, we developed an in-situ vertical blasting demolition technology for reinforced concrete water towers contained in limited spaces. Using high-speed photography and numerical simulations, we experimented on and modeled the in-situ collapse of a vertical water tower. We conducted a comprehensive analysis of the impact failure mechanism and collapse process. Regression analysis for the water tower revealed that it fell at an acceleration of 9.4 m/s$^2$. The "separated" finite element model was used to approximately restore the collapse process of the water tower and accurately capture the impact and moment of collision of each section of the cylinder. The process by which damage was caused to the water tower cylinder was complex, as the damage was accumulated from multiple impacts. Not only did our method control the collapse accumulation range of the water tower, but it also controlled the damage of the blasting dust.

## 1    INTRODUCTION

As a safe and efficient demolition technology, blasting demolition is mostly used to demolish tall buildings and high-rise structures such as chimneys, cooling towers, and water towers. It plays an essential role in urban building renewal and upgrading industrial facilities (Xie 2008; Xu 2003; Wang 2000).

In recent years, the environment for blasting demolition has become increasingly complicated, and the size of the site required for directional collapse has become increasingly smaller. To shorten the collapse distance of the water tower, scholars have proposed designs to increase the elevation of the blasting cut (high cut) based on the bottom cut. For example, Peng Bingyang (Peng 2007) placed a blast cut at a distance of 1.2 m above the ground to demolish a 40-m high cylindrical thin-walled water tower. Zhang Songfeng (Zhang 2018) applied a blasting cut at 1.6 m above the ground and demolished a 35-m high umbrella-shaped reinforced concrete water tower, which resulted in a collapsed length of approximately 34 m. Xianqi et al. (Xie 2009) positioned a blasting cut at 4.0

---

*Corresponding Author: blasting_huang@163.com

DOI 10.1201/9781003372417-20

m above the ground to demolish a 38-m high inverted conical reinforced concrete water tower, for which the collapsed length was about 30 m. Ren Zhiyuan et al. (Ren 2008) situated a blasting cut 15 m away from the bottom of a 28.6-m high inverted conical water tower in a limited surrounding space, and the collapsed length was approximately 12.5 m.

Based on the blasting demolition project of a 38-m high reinforced concrete water tower in Huangpi District, Wuhan, this study utilized a high-speed camera to observe the process of the instability and collapse of the water tower demolition and used a dynamic strain gauge to test the dynamic response of the bottom cylinder. A comparative analysis was carried out using a numerical simulation. In addition, the vibration effect on the area surrounding the water tower was monitored using a blasting vibration meter. We comprehensively studied the mathematical law of the collapsed movement of the water tower, the vertical in-situ blasting demolition, the failure characteristics of the cylinder, and the vibration effect on the ground upon contact.

## 2   CASES OF IN-SITU BLASTING DEMOLITION OF WATER TOWER

The height of the water tower to be demolished was 38.1 m; the height of the tower structure itself was 28.6 m, and the height of the water tank at the top was 9.5 m. The lower support tube of the water tower was cylindrical, with an inner diameter of 2.0 m and a wall thickness of 0.18 m. There were five equipment maintenance platforms in the tube, which were 4.9 m, 9.8 m, 14.7 m, 19.8 m, and 24.7 m above the ground, respectively. The water tank was an inverted cone structure, and its maximum outer diameter was 10.64 m. The water tower barrel and water tank were equipped with a single-layer steel mesh. The main reinforcement was $\varphi16mm@100mm$, and the stirrup was $\varphi8mm@100mm$. On the north side of the water tower, there was an inspection door with a height of 2.0 m and a width of 0.6 m. On the east and west sides of the tower, there were circular observation windows with a diameter of 66 cm.

The overall blasting demolition plan, according to the structural characteristics of the water tower and the surrounding environment, was to collapse the cylinder vertically in situ and break the water tank by hydraulic blasting. For the convenience of construction, the blasting incisions were arranged at the bottom of the water tower and next to the maintenance platforms on each floor, for a total of six blasting cuts (see Table 1 and Figure 1).

Table 1.   Blasting cut height table.

| Blasting cut | 1 | 2 | 3 | 4 | 5 | 6 |
|---|---|---|---|---|---|---|
| Cut height (m) | 3.0 | 1.5 | 1.5 | 1.5 | 1.5 | 0.5 |

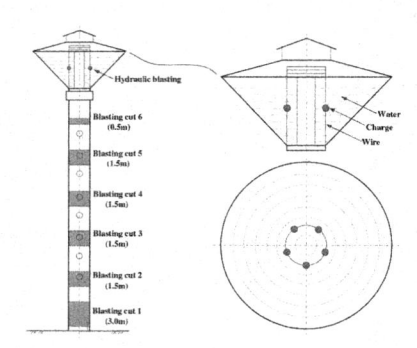

Figure 1.   Blasting diagram of cylinder and water tank.

Drilling and blasting were used for the six blasting cuts in the water tower cylinder, with a hole a distance of a=25 cm, a row spacing of b=25 cm, a wall thickness of $\delta$=18 cm, an explosive unit consumption of k=3000 g / m$^3$, a single hole charge quantity of q=40 g, and a total charge quantity of 20 kg. Five charges were fixed with an iron wire for the hydraulic blasting of the water tank. The weight of a single charge was 2 kg, and the depth the charge was placed into the water was 4 m. The Ms19 detonator (nominal delay time 1700 ± 150 ms) was used as the detonator on all the charges, and all the detonators were connected in series outside the holes. The cylinder blasting cut and water tank hydraulic blasting were detonated from bottom to top.

## 3 HIGH-SPEED PHTOTGRAGHY OBSERVATION AND ANALYSIS

### 3.1 *Photographic observation plan*

The high-speed photography observation point was 80 m away from the south side of the water tower. We used the Y7-S2 high-speed camera produced by the American IDT company. The frame resolution was 1920×1080. Considering the actual light conditions on the scene, the sampling rate was set to 5000 fps (i.e., the interval time of each frame was 0.2 ms). To clearly show the trajectory of the collapsed water tower, a red protective cloth was wrapped around each blasting cut.

### 3.2 *Observation results and analysis*

The times at which the high-speed camera captured each frame were read to obtain the blasting time of each layer of the water tower cylinder and the water tank hydraulic blasting. These results are shown in Table 2.

Table 2. Time table of water tower collapse.

| Blasting area | Frames | Design initiation time (ms) | Observation initiation time (ms) | Detonation error (ms) |
| --- | --- | --- | --- | --- |
| Cut 1 | 5810 | 0 | 0 | 0 |
| Cut 2 | 7340 | 310 | 306 | −4 |
| Cut 3 | 7990 | 420 | 436 | +16 |
| Cut 4 | 7720 | 530 | 382 | −148 |
| Cut 5 | 9240 | 640 | 686 | +46 |
| Cut 6 | 9360 | 640 | 710 | +70 |
| Water tank blasting | 12820 | 1430 | 1402 | −28 |

Note: the observed initiation time was the initiation time of the first detonating charge that can be observed.

The blasting was performed in the following order: cut 1, cut 2, cut 4, cut 3, cut 5, cut 6, and water tank blasting, as shown in Figure 2. Because the delay time of the detonator was 148 ms earlier, cut 4 detonated 54 ms earlier than cut 3. Although the delay error of the initiating detonator affected the sequence of initiation of the blasting cuts, it had little effect on the overall collapse of the water tower. At frame 12820, water was observed to overflow from the top and bottom of the water tank. At this moment, the water tower had fallen 9.1 m. According to the observations, the blasting dust was mainly generated at the moment of initial detonation. In this case, the charge for the water pressure blasting was too small to break the water tank, resulting in no obvious water mist. In addition, the initial time of hydraulic blasting was slightly later, and there was no time to cover the blasting dust. To enhance the effect of water pressure blasting while considering the dust reduction, we recommend that water pressure blasting be started first. The apex of the water tower was selected as the observation point, and the distance it moved was read at different times. It took about 2.4 seconds from the detonation of blasting cut 1 for the water tank to reach the ground. Due

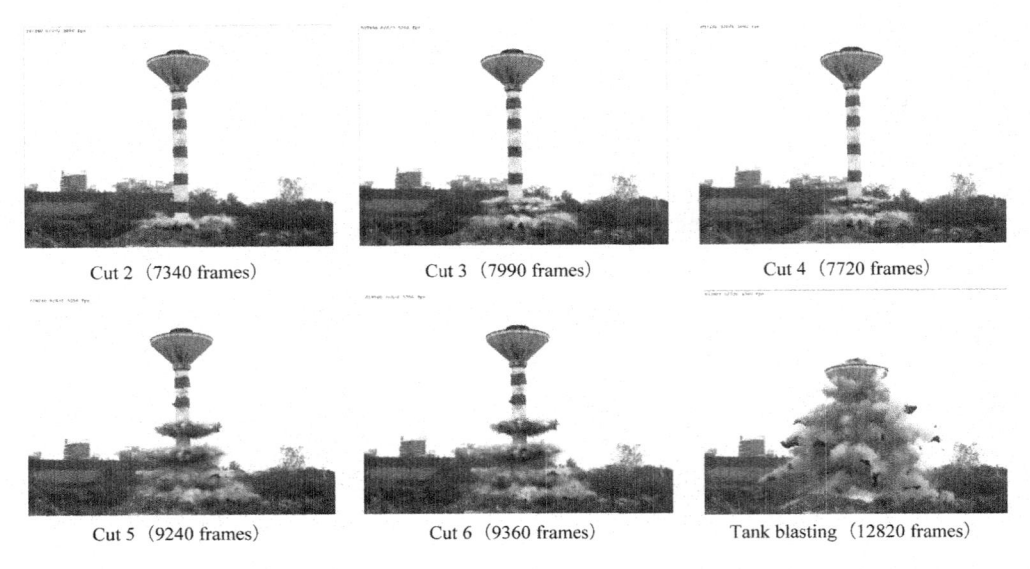

Cut 2 (7340 frames)  Cut 3 (7990 frames)  Cut 4 (7720 frames)

Cut 5 (9240 frames)  Cut 6 (9360 frames)  Tank blasting (12820 frames)

Figure 2. Blasting moment.

to the impact of blasting dust, the movement of the water tower can be seen clearly in only the first 2 s. A regression analysis revealed that the water tower's motion during collapse was similar to a free fall motion. The acceleration of the water tower's collapse was approximately 9.4 m/s$^2$, slightly less than the acceleration due to gravity (9.8 m/s$^2$). The discrepancy was due to the damping effect of the steel mesh and slag block.

## 4  FINITE ELEMENT NUMERICAL SIMULATION

### 4.1  *Model building*

Numerical simulations can continuously, dynamically, and repeatedly reproduce the entire process of instability, dumping, ground contact, and structure stacking. In order to accurately describe the geometric shape of the object, a 1:1 scale "separated" model of the reinforced concrete water tower was developed in the LS-DYNA dynamic finite element program. A Lagrange finite element mesh with an explicit integral was adopted to consider the mechanical properties of the concrete and reinforcement independently.

The concrete used SOLID164 solid units with a unit size of 0.1 cm × 0.1 cm × 0.1 cm, and the material used the 159*MAT_CSCM_CONCRETE model (Murray 2007). The continuous surface cap model (CSCM) material can capture the nonlinear behavior of concrete. It can exhibit different inelastic responses during tension and compression, soft plastic deformation during compression, crack propagation damage, and the strain rate effect during tension. The longitudinal bars and stirrups were made of BEAM161 elements with an element length of 0.1 cm. Each steel bar was made of Hughes-Liu beam elements and was located in accurate positions in the concrete grid. A 2×2 Gauss point integration was used in the section. The material used the No. 3 *MAT_PLASTIC_KINEMATIC plastic dynamic hardening model (Hallquist 2013), which fails when the effective plastic strain reaches the limit strain.

The concrete and steel bars used solid and beam elements. In LS-DYNA, the *CON-STRAINED_LAGRANGE_IN_SOLID option based on penalty function constraints was used to achieve the deformation coordination between the two. The bottom of the water tower barrel was fully fixed, and the ground was a rigid board (*MAT_RIGID). Automatic single-sided contact

148

was used between different parts of the structure, and automatic point-to-face contact was used between the reinforcement and the ground. The quality of the water injection was equivalent to the density of the concrete of the water tank structure. In addition, the structure model was simplified, and the water pressure blasting process of the water tank was ignored.

Each blasting cut of the water tower was performed by *MAT_ADD_EROSION to control the failure of the cylinder element. In order to balance the initial stress state, the cylinder element was deleted after 0.3 s. The failure time refers to the observation results of high-speed photography.

### 4.2 *Simulation results and analysis*

A comparison of the displacement curve (in the collapse direction) of node 48185 at the top of the water tower to the observation results and the movement of a free-falling body is shown in Figure 3.

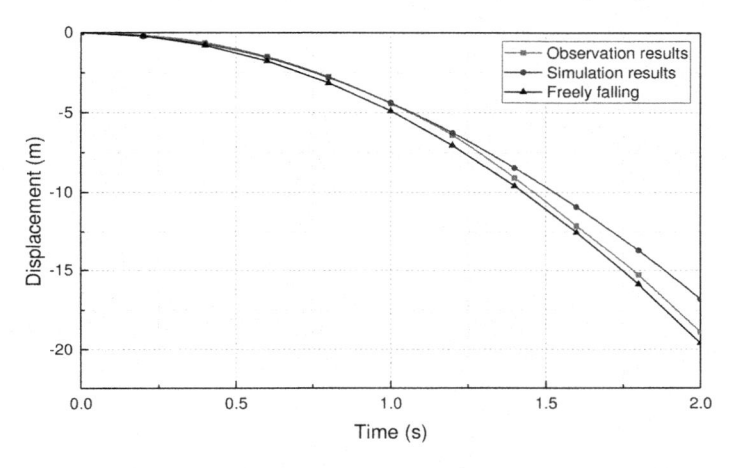

Figure 3.    The trajectory of the water tower blasting collapse.

A comparison of the simulation results of the collapse of the water tower with the actual blasting result is shown in Figure 4. The simulation result of the barrel part was closer to the actual value. However, part of the water tank remained intact after striking the ground due to the residue of the cylinder generating a buffer effect in the process of collapse (the water pressure explosion load was not considered). In practical engineering, water pressure blasting is used in the water tank. The energy of the explosion acted uniformly on the water tank wall through the water, causing the water

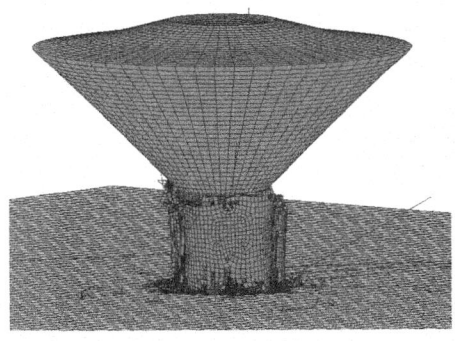

a. Simulation result.                                            b. Actual result.

Figure 4.    Blasting heap of the water tower.

tank to disintegrate fully. The blasting heap of the whole water tower, which was 2.4 m high and 11.8 m wide, was concentrated, effectively shortening the collapse distance of the water tower and reducing the impact on the surrounding environment. This was conducive to subsequent residue removal and transportation.

## 5 CONCLUSION

To solve the problem of the demolition of towering structures in confined spaces, innovative technology for the vertical in-situ collapse demolition of water towers was proposed. By ordering the vertical collapse and touchdown in sequence and timing the simultaneous water pressure blasting off the top water tank, our method not only controlled the collapse accumulation range of a high-rise water tower, but also reduced the vibration effect of touchdown and controlled the damage of the blasting dust.

By combining actual engineering cases, high-speed photography, and numerical simulations, a comprehensive analysis of the impact failure mechanism and collapse process was carried out. The following conclusions can be drawn.

(1) The collapse process of the vertical in-situ blasting demolition of the water tower was similar to free fall motion. Regression analysis showed that the downward acceleration of the water tower was about 9.4 m/s$^2$, slightly less than the acceleration due to gravity. The damping effect of the steel mesh and the concrete slag, which touched the ground first, hindered the downward speed of the water tower.

(2) Using the "separated" finite element model, we approximated the collapse process of the water tower and accurately captured the impact and moment of collision of each section of the cylinder. The No. 159 concrete material model can adequately simulate the collapse and impact damage of the water tower.

## ACKNOWLEDGMENTS

All authors contributed to the design and implementation of the research and to the analysis of the results. Huang X.W. wrote the paper. Xie X.Q. and Zhong D.W. supervised the project. All authors have read and agreed to the published version of the manuscript.

This research was partly funded by the Natural Science Foundation of Hubei Province under Grant 2020CFA043 and the Key R&D Project of Hubei Province under Grant 2020BCA084.

## REFERENCES

J. O. Hallquist et al., *LS-DYNA Keyword User's Manual*, Livermore Software Technology Corporation, 2013.

Peng Bin-yang, Wang Jiang-ping. Blasting demolition of a thin-wall reinforced concrete water tower [J]. *Blasting*, 2007(04):63–64+76.

Ren Zhi-yuan, Cheng Gui-hai, Jin Yang, et al. Demolition of a 150-ton cone water tower under complicated circumstances by directional blasting [J]. *Blasting*, 2008,25(04):62–66.

Wang Xu-guang, Yu Ya-lun. Demolition blasting technology faced in the 21 century [J]. *Engineering Blasting*, 2000(01):32–35.

Xie Xian-qi, Liu Chang-bang, Jia Yong-sheng, et al. Controlled blasting demolition of 38m high reverse cone shape RC water tower [J]. *Blasting*, 2009,26(02):61–63.

Xie Xian-qi, Lu Wen-bo. 3P (Precise, Punctilious and Perfect) blasting, *Engineering Blasting*, 2008(03):1–7.

Xu Shu-lei, Zheng Xue-zhao, Wang Xiao-lin, et al. Brief description about the current situation of demolition blasting at home and abroad [J]. *Blasting*, 2003(02):20–23.

Y. D. Murray, A. Abu-Odeh, and R. Bligh, *Evaluation of LSD-YNA Concrete Material Model*, vol. 159, 2007.

Zhang Song-feng, Lei Zhen, Gao Wen-jiao, et al. Explosive demolition of umbrella reinforced concrete water tower [J]. *Blasting*, 2018,35(04):90–93.

*Civil Engineering and Urban Research – Mohamed & Hou (Eds)*
*© 2023 the Authors, ISBN 978-1-032-44487-1*

# Shear resistance of asphalt mixture under pure shear of Xijing road

Chunlei Wang*, Huaping Zhao*, Tao Wei* & Junfeng Liu*
*Xingtai Road & Bridge Construction Group Corporation, Xingtai, China*

ABSTRACT: To study the shear characteristics of asphalt mixture under pure shear load, a pure shear test device that cooperates with the Marshall stability tester was developed. The stress of asphalt mixture under pure shear load was analyzed by the finite element method, and the stress state of the device was verified. The effectiveness of the test device and method was verified by laboratory tests. The correlation between shear strength and compressive resilient modulus obtained from the pure shear test was analyzed. The results show that: in the process of the pure shear test, the shear stress in the shear plane of asphalt mixture can be divided into three stages, and the peak value of shear stress is obvious. The shear strength of ARHM-13, AC-13, and AC-16 asphalt mixture decreases with the increase or decrease of test temperature, and the order of shear strength at each temperature is AC-16 > ARHM-13 > AC-13. ARHM-13, AC-13, and AC-16, with different nominal particle sizes, gradation, materials of asphalt mixture shear strength, and compressive modulus of resilience, showed a significant linear correlation. The developed pure shear test device and method can be used to evaluate the shear characteristics of asphalt mixture, and it is simple and effective.

## 1 GENERAL INSTRUCTIONS

Rutting is one of the main diseases of asphalt pavement. The shear resistance of the asphalt mixture is an important index to characterize its rutting resistance (Hao 2020; Li 2019). Domestic experts and scholars have done a lot of research on the shear test method, shear performance, and shear evaluation index of asphalt mixture (Han 2014; Li 2016; Wang 2009). At present, the test methods for the shear performance of asphalt mixtures mainly include triaxial test, uniaxial penetration test, direct shear or oblique shear test, and Superpave shear test (Bi 2005; Wang 2012, 2010; Wu 2019; Xie 2016). The stress state of the triaxial compression test is clear, and the test accuracy is high. However, the triaxial compression test at high temperatures is expensive, and the operation is more complicated. The stress state of the triaxial compression test is clear, and the test accuracy is high. However, the triaxial compression test at high temperatures is expensive, and the operation is more complicated. At present, it is only used for indoor tests and research work, and there are few engineering applications. The uniaxial penetration test can perform shear tests on indoor forming and on-site cored samples. For non-standard tests, correction coefficients are needed to correct the test results (Wang 2012). The direct shear test is similar to the oblique shear test, and the test is simple and easy to operate. Among them, the stress concentration is easy to occur when the shear tool of the direct shear test is in contact with the large particle aggregate of the test piece and the discreteness of the derived test data is large. In the oblique shear test, the height of the test piece greatly impacts the test results (Wu 2019), and the time of the direct shear test and the oblique shear test is not conducive to use on the construction site. The Superpave shear test (SST) is similar to pure shear, and the test data are accurate and reliable. However, it requires high accuracy of

---

*Corresponding Authors: 80249347@qq.com, 250598901@qq.com, 198708268@qq.com and 395542095@qq.com

loading and measuring equipment, and the equipment is expensive, which is not convenient for popularization and application (Wang 2010; Wu 2019).

This paper develops a pure shear test device that can be used together with the Marshall stability instrument. Numerical analysis and laboratory tests verify the effectiveness of the test device and method. The results show that the mechanical mechanism of the device is simple and clear, the test operation is simple, and it has the potential for large-scale popularization and application.

## 2 NUMERICAL ANALYSIS OF ASPHALT MIXTURE STRESS UNDER PURE SHEAR LOADING

Generalized shear includes pure shear state (Lord angle $\theta=0o$), tension and compression combined shear state (Lord angle $-30o<\theta<0o$), compression and tension combined shear state (Lord angle $0o<\theta<30o$).

The developed double-sided shear test device is shown in Figure 1 to realize the pure shear state of asphalt mixture specimen and the length of shear sleeve × wide × thick = 140 × 120 × 30mm, the diameter of internal boring round hole $\Phi= 102$mm, the horizontal limit plate is 120mm long, 90mm wide and 10mm thick, and the center of the circle is 20mm upward eccentric of the sleeve axis. The finite element method is used to analyze the internal stress change state of the mixture specimen during the double-sided shearing process. The finite element method simulates the stress state, and the 8-node hexahedral solid element is used to simulate the asphalt mixture specimens and fixtures. In the simulation, the elastic modulus of the asphalt mixture is 1000 MPa, the Poisson's ratio is 0.4, the sleeve and horizontal limit plate modulus is 200,000 MPa, and the Poisson's ratio is 0.3. The finite element model of the shear test is shown in Figure 2, and the boundary condition setting is shown in Figure 2(b). From the shear surface stress cloud diagram (as shown in Figure 3), it can be seen that the first principal stress and the third principal stress are minimal except for the outer ring part of the shear surface, which can be simplified to a pure shear state.

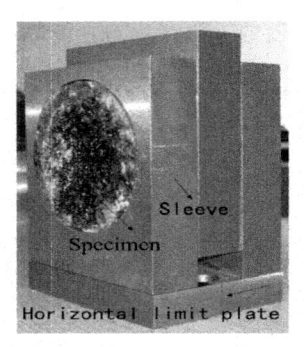

Figure 1.    Shear test device.

## 3   PURE SHEAR TESTS OF ASPHALT MIXTURE

### 3.1   *Raw materials*

ARHM-13, AC-13, and AC-16 asphalt mixtures are used in the test. The grading adopts the median value of AC-13 and AC-16 in the specification. ARHM-13 uses rubber asphalt, AC-13 and AC-16 use matrix asphalt, aggregates, and mineral powder made of limestone, and all performance indexes of asphalt and asphalt mixture meet the specification requirements. Marshall compacts the asphalt mixture specimen, and the formed specimen is shown in Figure 8.

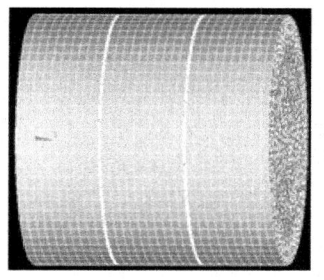 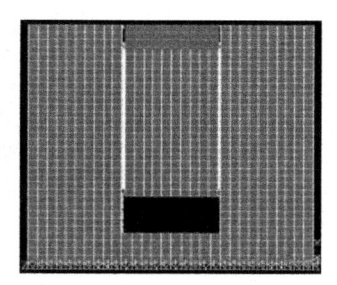

(a) Finite element model of the specimen  (b) Boundary constraints

Figure 2.    Finite element model of shear test.

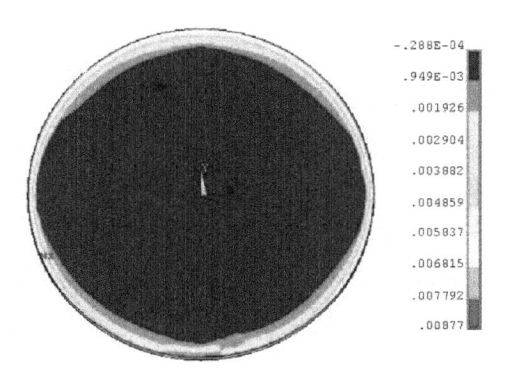 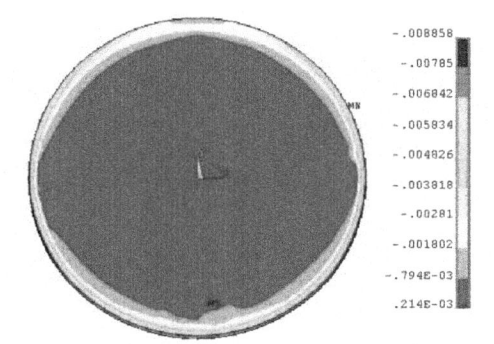

(a) First principal stress nephogram of shear plane  (b) Third principal stress nephogram of shear plane

Figure 3.    Stress cloud diagram of shear plane.

Figure 4.    Formed asphalt mixture sample.

### 3.2  *Pure shear test results*

The pure shear test failure specimen is shown in Figure 5. It can be seen from the figure that the shear failure interface is clear, and the failure cracks are distributed in a straight line along the shear plane. Taking the pure shear test results of AC-13 asphalt mixture specimens as an example, the shear stress-strain curve is shown in Figure 6. The comparison of shear strength test results is shown in Figure 7.

Figure 5.   Destroy specimens in a pure shear test.

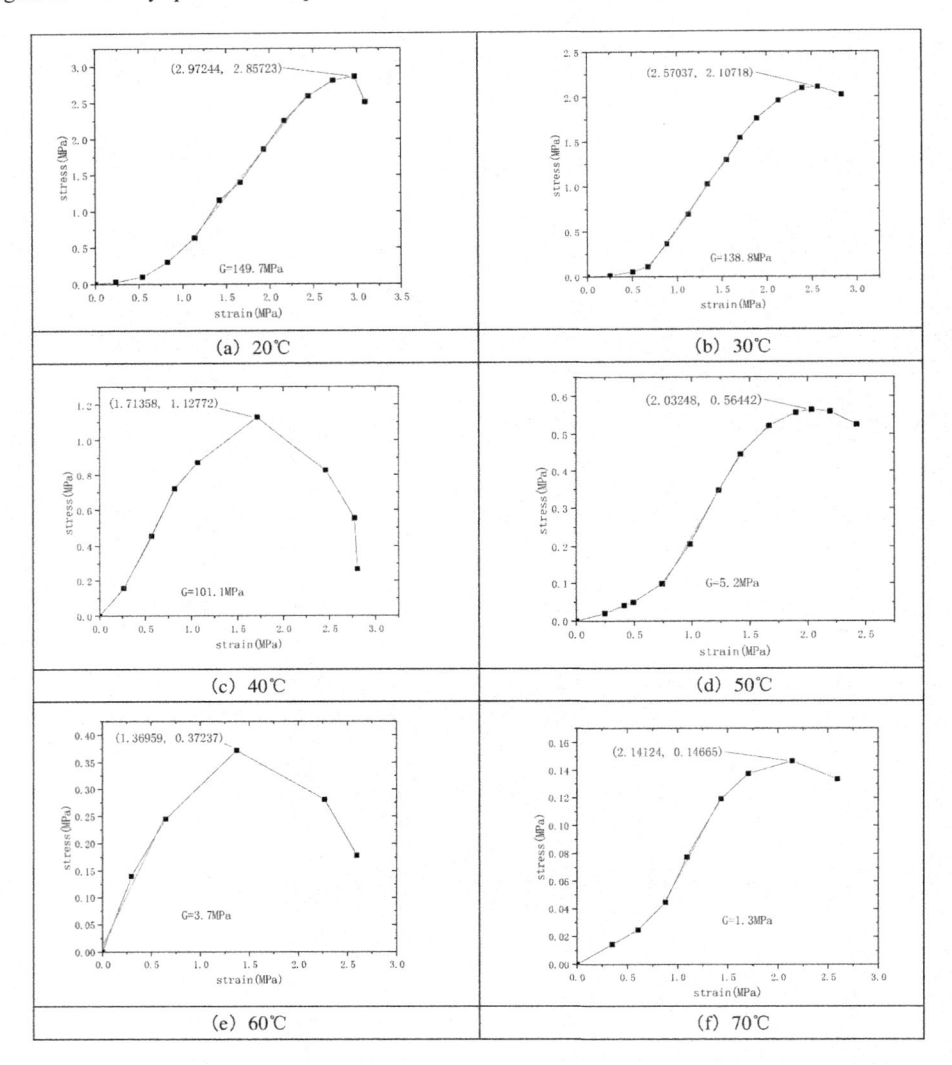

(a)  20℃

(b)  30℃

(c)  40℃

(d)  50℃

(e)  60℃

(f)  70℃

Figure 6.   Shear stress-strain curves of AC-13 asphalt mixture specimens at different temperatures.

It can be seen from Figure 6 that during the pure shear test, the internal shear stress change of the asphalt mixture specimen is divided into three stages. The first stage is the initial loading stage, and the shear stress increases slowly; the second stage is the shear. The stress rapidly increases stage and reaches the peak stress; after that, it enters the third stage, that is, the failure stage, where the shear stress drops rapidly, and the specimen is sheared and destroyed. In some cases, the first stage is short or does not appear, and on the whole, there will be a stage of rapid stress increase, peak stress, and rapid decline. It can be seen from Figure 6 that the shear stress peak is obvious in the pure shear test of the asphalt mixture.

Figure 7 shows that the shear strength of ARHM-13, AC-13, and AC-16 asphalt mixtures all show a downward trend with the increase or decrease of the test temperature. The order of the shear strength under each temperature condition is AC-16>ARHM -13>AC-13. The result shows: Comparing the shear strength of the three, it can be seen that when the nominal particle size of the asphalt mixture is large, the increase in shear strength is more obvious. Comparing ARHM-13 and AC-13, it can be found that rubber asphalt can improve the shear strength of the asphalt mixture. In addition, the differences in shear strength of ARHM-13, AC-13, and AC-16 asphalt mixtures obtained by the pure shear test are obvious, indicating that the developed pure shear test device and method can be used to evaluate the shear characteristics of asphalt mixtures, and it is simple and effective.

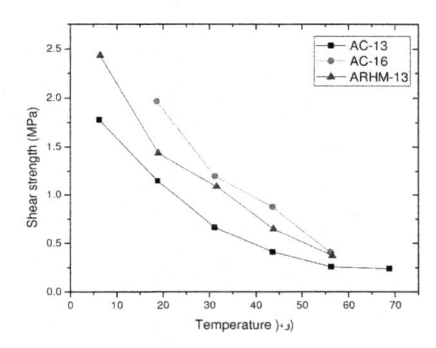

Figure 7.  Comparison of shear strength test results.

## 4  CONCLUSION

(1) In the process of the pure shear test, the change of shear stress in the shear plane of the asphalt mixture specimen is mainly divided into three stages, and the peak value of shear stress is obvious.
(2) The shear strength of ARHM-13, AC-13, and AC-16 asphalt mixtures decreased with the increase and decrease of test temperature, and the order of shear strength under different temperature conditions was AC-16 > ARHM-13 > AC-13.
(3) The developed pure shear test device and method can be used to evaluate the shear characteristics of asphalt mixture, and it is simple and effective.
(4) The shear strength and compressive modulus of asphalt mixtures with ARHM-13, AC-13, and AC-16 with different nominal particle sizes, different gradations, and different materials show a significant linear correlation.

The program is supported by Xingtai scientific program 2020ZC033 and program 2020ZC029.

REFERENCES

Bi Yufeng, Sun Lijun. Research on Test Method of Asphalt Mixture's Shearing Properties[J]. *Journal of Tongji University* (Natural Science), 2005, (08):1036–1040.
Han Ping, Duan Danjun. Analysis of the Shear Test Method For Asphalt Mixture[J]. *Highway Engineering* (Natural Science), 2014, 39(04):1–3+56.

Hao Peiwen, Wang Junbiao, Zeng Zhiwu, et al. Shear Performance of Different Types of High Modulus Asphalt Mixtures[J]. *Bulletin of the Chinese Ceramic Society*, 2020, 39(12):4054–4060+4067.

Li Jilu. *Analysis on Formation Mechanism and Prevention & Control Factors of Rutting Disease of Asphalt Pavement*[D]. Jilin University,2019.

Li Qiang, Hou Rui, Ma Xiang, et al . Testing Methods and Factors for Shear Properties of Asphalt Mixtures[J]. *Highway Engineering* (Natural Science), 2016, 41(04):50–54+66.

Wang Gang, Liu Liping, Sun Lijun. Research on Shearing Strength and Compressive Resilient Modulus Experiment of Asphalt Concrete[J]. *Journal of Building Materials* (Natural Science), 2012, 15(02):279–282.

Wang Ruilin, Yu Miao. Application of Inclined Plane Shear Test to Asphalt Mixture[J]. *Journal of Chongqing Jiaotong University* (Natural Science), 2009, 28(01):54–55+120.

Wang Shuiyin. Research on Test Method of Shear Strength Between Layers of Indoor Asphalt Concrete Pavement [J]. *Highway* (Natural Science), 2010, (02):144–147.

Wu Bangwei, Liu Liping, Sun Lijun. Influence of Different Parameters on Shear Performance of Asphalt Mixture[J]. *Journal of Highway and Transportation Research and Development*(Natural Science), 2019, 36(10):1-6+24.

Xie Jun, Huang Lin. Experimental Study on Shear Fatigue of Asphalt Mixture [J]. *Journal of China & Foreign Highway*(Natural Science), 2016, 36(01):229–234.

Civil Engineering and Urban Research – Mohamed & Hou (Eds)
© 2023 the Authors, ISBN 978-1-032-44487-1

# Analysis of bearing characteristics of rock-socketed pile based on laboratory model test

Haijun Yu*

*Chongqing Aerospace Polytechnic, Chongqing, China*

ABSTRACT: In order to analyze the distribution characteristics of negative friction of piles in deep backfill, the indoor scale model under similar conditions is used to test the soil around the pile under the condition of soil rock mixture. The test results show that the changes in the thickness of the backfill and the load on the top of the pile significantly affect the position of the neutral point; the axial force of the pile body first increases and then decreases along the depth direction of the pile body. The axial force will change abruptly. Reducing the pile top load and increasing the surface heap load will increase the pull-down load of the negative friction resistance of the pile, but the existence of the rock-socketed section of the pile body can effectively improve the vertical bearing performance of the pile.

## 1 INTRODUCTION

In western mountainous areas with complex topography, the construction method of "digging high and filling low" is often adopted in the construction of the project, forming a large number of deep backfill earthwork. Rock-socketed piles are widely used in this kind of site engineering because of their excellent performance in controlling foundation settlement. The self-consolidation settlement of the backfill will inevitably increase the negative friction resistance of the pile body(Ye 2019), which affects the safety of the pile foundation structure. At present, there is still a lack of systematic research on the distribution characteristics of the negative friction resistance of the pile under this condition. Although the on-site in-situ test method is the closest to the test working state of the pile, due to the large bearing capacity of the rock-socketed pile, the high test cost, and the difficulty in conducting destructive tests (Zhou 2019), and there are many uncontrollable factors in the field test, a certain method is required. Compared with the field test, the model test solves many problems that are difficult or impossible to realize in the field test under the premise of restoring the main geological environment characteristics and mechanical properties of the mountainous area (Wu 2021). In order to deeply analyze the distribution characteristics of negative friction resistance of rock-socketed piles in deep filling sites, an indoor scale test model was designed in which the soil around the piles was a soil-rock mixture. The proposed model has engineering value and theoretical significance.

## 2 INDOOR MODEL PILE TEST PROGRAM

### 2.1 *Pile fabrication*

PVC (inner diameter d = 70mm) was used as the pile casting mold, and four ribbed steel bars (d = 6mm) were used as the longitudinal reinforcement of the pile body. The length of the solid

---

*Corresponding Author: hijun_yu@qq.com

pile is 950mm, the soil layer thickness is 750mm, and the rock layer is 150mm. The loading test was carried out 28d after pouring.

## 2.2 Backfill material around piles

According to the relevant requirements of the "Highway Geotechnical Test Regulations" (JTG 3430-2020), the excavation and filling of soil-rock mixture in a construction site in Chongqing were selected, and the main components were silty clay, sandstone, and mudstone through experimental analysis, and the mechanical parameters were obtained. The cohesion is 34kPa, and the internal friction angle is 26.3°.

## 2.3 Making bedrock materials

Considering that the mix ratio of each material is different, the strength of the simulated rock mass will also be different. In order to determine the most suitable mix ratio of each material, several sets of mix ratio tests have been done before the test (Wang 2006). Finally, the mix ratio of the simulated bedrock material is determined as: Barite powder: quartz sand: rosin alcohol solution: gypsum = 10:3.3:2:1, and the mass fraction of rosin in the rosin alcohol solution is 10%. The prepared samples are shown in Figure 1, and the basic physical parameters of the prepared bedrock are shown in Table 1.

Table 1. Various mechanical parameters of weathered sandstone.

| Simulate bedrock | Unit weight $(kN/m^3)$ | Elastic Modulus (MPa) | Compressive strength (MPa) | Cohesion (kPa) | Internal friction angle(°) |
|---|---|---|---|---|---|
| Moderately weathered sandstone | 20.8 | 165 | 1.36 | 96.5 | 40.1 |

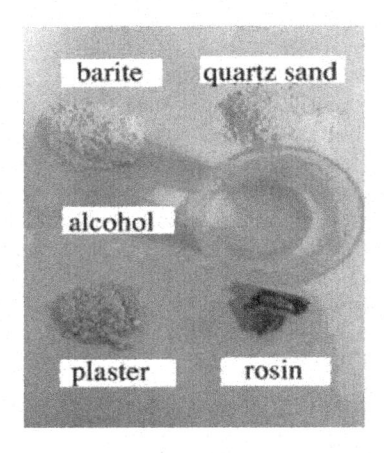

Figure 1. Bedrock preparation materials.

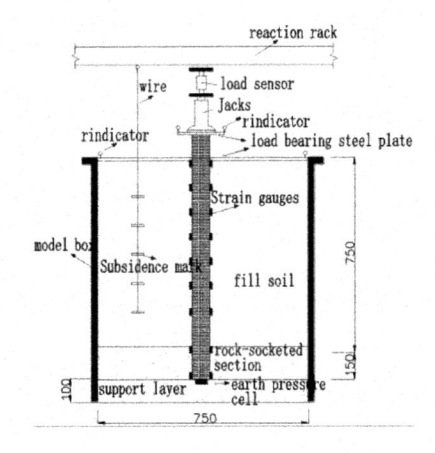

Figure 2. Schematic layout of model test piles.

## 2.4 Design scheme of test loading device

In order to study the compressive deformation characteristics of the pile body during the loading process(Gong 2014), it is necessary to measure the axial force and lateral friction resistance at different positions of the pile body. The specific measuring point layout is shown in Figure 2. To collect the pile body strain during the test, it needs to be connected with a wireless static strain tester. The strain gauges use 120-50AA concrete strain gauges to measure the strain of the model pile and use the DH3819 wireless static strain test system for data acquisition. Variation of pile tip resistance value under top load.

## 3 INDOOR MODEL FILLING

We pour into the model box according to the similar material of the rock layer at the pile end. Then we stop pouring when the pouring reaches 10cm, mark the placement position of the model pile, embed the earth pressure box to test the pile end's resistance, and place the model pile above the earth pressure box. In order to ensure the verticality of the pile body during the loading process, continue pouring to 25cm to stop. After the rock layer has a certain strength, the soil material is filled into layers, the rock content is controlled to 35%, the compaction degree is 0.85, the thickness of each layer is 10cm, and a settlement mark is buried according to the design position, as shown in Figure 2. During the filling process, compaction is carried out to achieve the purpose of simulating the self-weight consolidation effect of the filling site in the early stage. The filling process is shown in Figure 3–5.

Figure 3.  Bedrock filling.     Figure 4.  Soil filling.     Figure 5.  Model filling completed.

## 4 CALCULATION PRINCIPLES OF TEST DATA

The deformation of the concrete pile body satisfies Hooke's law. When the test is loaded, the strain gauges of each section of the pile body can be measured to obtain the strain data of the section, and the average value is obtained to ensure the accuracy of the measurement data. Then, the axial force variation of each section of the pile body can be obtained by using equations (1) and (2).

The lateral friction resistance of the pile body can be obtained by measuring the axial force of the pile body, and the average lateral friction resistance between each test section of the pile body can be obtained by using the formula (3).

$$\varepsilon_i = \frac{\varepsilon_{i-1} + \varepsilon_{i-2}}{2} \tag{1}$$

$$P_i = \varepsilon_i A_p E_p \tag{2}$$

$$\tau_i = \frac{P_i - P_{i+1}}{\pi d l_i} \tag{3}$$

Where $\varepsilon_i$ is the average strain value corresponding to the i-th section, $\varepsilon_{i-1}$ and $\varepsilon_{i-2}$ are the measured strain value of the pile section, $A_P$ is the net section area of the pile body; $P_i$ is the axial force of the i-th section pile body, $E_P$ is the elastic modulus of the pile concrete, d is the diameter of the pile, $\tau_i$ is the average side friction, and $l_i$ is the distance between the measuring points of the section.

## 5 INDOOR MODEL TEST DATA PROCESSING AND ANALYSIS

Under the condition of soil-rock mixture filling, the distribution characteristics of pile body settlement, axial force, and pile side friction resistance of rock-socketed piles were analyzed when only the heap load and the combined action of the heap load and the pile top load were carried out.

### 5.1 Displacement-settlement distribution under load

With the increase of the heap load level of the soil surface, the settlement of the soil increases with the increase of the heap load. As shown in Fig.6, the settlement of the soil surface is the largest and gradually decreases with the increase of the depth, the settlement rate is also gradually decreased, and the maximum value of the settlement is 11mm at 50kPa. Then we continue to stabilize it at 50kPa and apply pile top load and 0.5kN per stage until 2.5kN. The pile top load has little effect on the settlement of the soil layer around the pile and can be ignored. When the pile top load is 2.5kN, the pile body is not damaged and is in an elastic working state, and the influence of the pile top compression deformation on the pile top displacement can be ignored, so the top pile displacement is approximately equal to the settlement of the pile end.

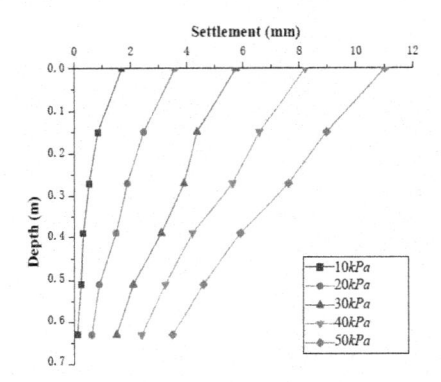

Figure 6.    Soil settlement curve under heap load.

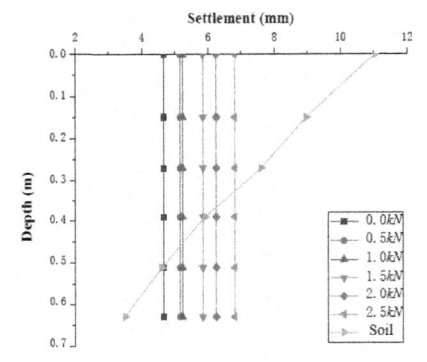

Figure 7.    Pile settlement and soil settlement relationship.

The relative displacement of the intersection point of the soil settlement curve and the pile settlement curve is zero, which is the position of the neutral point, as shown in Figure 7. It can be concluded that the settlement of the pile body increases with the increase of the pile top load, and the position of the neutral point also increases step by step, as shown in Figure 8.

### 5.2 Distribution law of axial force of pile body

The distribution trend of the axial force of the pile body is to increase first and then decrease, as shown in Figure 9. The axial force of the pile body increases with the depth in the range of the top half of the pile body, and the pile body is affected by negative friction within this range; the lower half of the pile body decreases with the depth, indicating that the pile side friction resistance is transformed into positive friction resistance. Since the axial force of a certain section of the pile body is measured, the maximum value of the obtained axial force is not necessarily the maximum value of the actual axial force. The average value method is used to infer that the maximum value of the axial force in the depth range of 0.43~0.63m is as follows: zero is the neutral position. Due to the large difference between the strength of the rock layer and the soil mass, a sudden change will occur at the junction. The two measuring points at the bottom of the pile body are in the rock-socketed section. It can be concluded that the axial force of the pile body gradually decreases after reaching the maximum value in the soil layer. When the soil layer transitions to

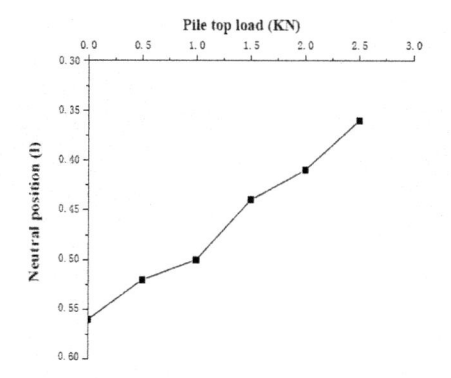
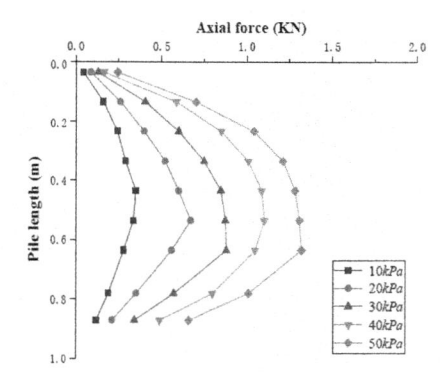

Figure 8. Neutral point position-pile top load relation.

Figure 9. Axial force of pile body under different pile loads.

the rock-socketed section, the axial force of the pile body decreases sharply, indicating that the existence of the rock-socketed section effectively improves the vertical bearing performance of the pile.

The pile load is continuously stable at 50kPa, and the pile top load is applied in stages, which shows that the axial force curve of the pile body first increases and then decreases along the pile body.

## 5.3 *Distribution law of friction resistance of pile body*

According to the measured value of the test, the axial force of the pile body is calculated according to formulas (1) and (2), and then the average lateral friction resistance between the two points of the pile body can be obtained by using formula (3).

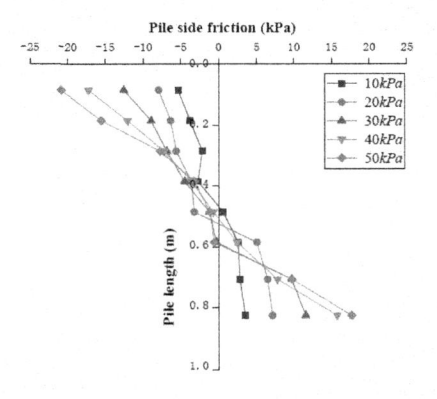
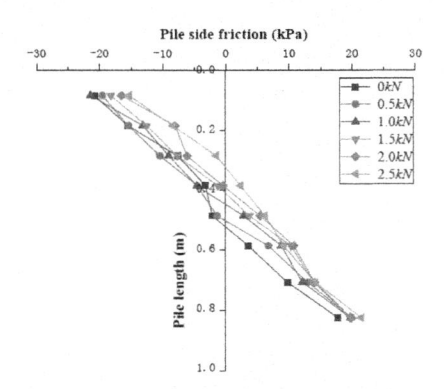

Figure 10. Pile side friction resistance under heap loading.

Figure 11. Pile side friction resistance under pile top load.

When the pile is loaded, the side friction resistance of the pile gradually transitions from the negative friction resistance to the positive friction resistance along the pile body, as shown in Figure 10. As the stacking load increases, both the negative frictional resistance and the positive frictional resistance tend to increase gradually, and the value of the negative frictional resistance is greater than that of the positive frictional resistance; at the same time, the neutral point begins to move downward from $0.54l$. The vicinity moves down to the vicinity of $0.64l$, resulting in a gradual increase in the range of negative frictional resistance of the pile body. The pile load is continuously stable at 50kPa, and the distribution curve of the pile side friction resistance under the combined action of the pile top load is shown in Figure 11.

## 6 CONCLUSIONS

(1) According to the indoor model test data, the law of soil settlement around the pile and the pile top displacement with the load is analyzed. The negative friction resistance in the backfill soil layer is larger than that of the rock layer. Reducing the pile top load and increasing the surface heap load will cause the pull-down load of the negative friction resistance of the pile to be increased, but the existence of the rock-socketed section of the pile body can effectively improve the vertical bearing performance of the pile.

(2) Under the condition of continuous and stable heap load, the axial force of the pile body will change abruptly when the soil layer transitions to the rock layer, and the axial force of the pile body will first increase and then decrease along the depth direction of the pile body; increase the pile top load or reduce the surface load, the pull-down load will be significantly reduced; the pile settlement increases with the increase of the pile top load, and the position of the neutral point begins to move up step by step.

## ACKNOWLEDGMENTS

This work was supported by the Scientific and Technological Research Program of the Chongqing Municipal Education Commission (No. KJQN202103004).

## REFERENCES

Gong C Z, *Theoretical analysis and experimental research on bearing characteristics of large-diameter rock-socketed pile foundations* [D]. Nanjing Southeast University, 2014.

Highway Science Research Institute, Ministry of Transport. JTG 3430-2020 *Highway Geotechnical Test Regulations* [S]. 2020

Wang H P, Li S C, Zhang Q Y, etc. Development of new geomechanical model test-like materials[J]. *Chinese Journal of Rock Mechanics and Engineering*, 2006(9):1842–1847.

Wu H J.*Research on the characteristics of negative friction resistance of rock-socketed piles in deep fill sites*[D].Chongqing Jiaotong University, 2021.

Ye G B, Zheng W Q, Zhang Z. Research on the distribution characteristics of negative friction resistance of friction piles in large-area filling sites[J].*Geotechnical Mechanics*, 2019,40(S1): 440–448.

Zhou B, Research on Bearing Calculation of Large Diameter Rock-Socketed Pile of a Bridge[J]. *Chinese and Foreign Architecture*, 2015(9):128–129.

Civil Engineering and Urban Research – Mohamed & Hou (Eds)
© 2023 the Authors, ISBN 978-1-032-44487-1

# Optimal design of transverse and longitudinal sections of U-shaped cross-section seepage control canals

Ling Liu*
*Tianjin Agricultural University, Computer and The Information Engineering Institute, Tianjin, China*

Shuhong Sun* & Mengjiao Zhang*
*Tianjin Agricultural University College of Hydraulic Engineering, Tianjin, China*

ABSTRACT: A U-shaped cross-section anti-seepage canal has good hydraulic conditions and can accommodate a large water-passing capacity. It is widely used in irrigation channel design. Under the requirement of design flow capacity, the optimized design of cross-section and vertical sections is of great significance in reducing construction investment. According to the Manning formula of hydraulics, a mathematical model for the optimization of the cross-section of the canal is established, and the optimal hydraulic section is optimized under the condition that the longitudinal slope is 1/1000, 1/2000, 1/3000, 1/4000, 1/5000, using MATLAB's SQP algorithm for planning and solving. It is concluded that under different longitudinal slope conditions of the U-shaped section, the angle between the center line of the tangent point and the horizontal is 90°, the design flow rate is $0.015 \text{m}^3/\text{s}$, the longitudinal slope i = 1/1000, and the bottom arc radius R = 2.18 m, the water depth h = 2.18 m, the minimum hydraulic wet circumference is 6.85 m, which is the optimal hydraulic section.

## 1 RESEARCH BACKGROUND

Water-saving irrigation mainly uses anti-seepage channels and low-pressure water transmission in the water transmission and distribution links. The water utilization coefficient of the anti-seepage channel can reach more than 0.75. It is convenient to construct, have low cost, and widely used, accounting for 70% of the water-saving irrigation area.

The commonly used cross-section forms of concrete anti-seepage canals are trapezoid and U-shaped. Among them, U-shaped cross-section anti-seepage canals have good hydraulic conditions, narrow openings, and occupy less land, which are the preferred forms in the design. Under the premise of ensuring the flow of water, the optimal cross-section and longitudinal slope of the anti-seepage channel can be determined through reasonable optimization design, which can reduce project investment, high water delivery efficiency, and reducing operation and management costs. The optimized design is paid more attention to, but most of the optimized designs are concentrated on trapezoidal cross-sections. Due to the complexity of the hydraulic calculation of U-shaped cross-section seepage control channels, there is little research on cross-section optimization design.

G.V.Loganathan (1993) proposed the optimal conditions for U-shaped channel design considering chord height, flow velocity, and size constraints, and the results showed that the slope parameter of the optimal hydraulic section was 0.514. M. Mohammad Rezapour Tabari et al. (2014) used MATLAB software to optimize the design of the channel with the goal of minimizing the total cost per unit length of the channel and finally gave the design parameters for the channel size under different flows and conditions.

---

*Corresponding Authors: liuling0709@126.com, hongss63@126.com and mengjiaoz@126.com

DOI 10.1201/9781003372417-23

Zhang Zhichang et al. (2001) calculated the relationship between water depth and discharge and the optimal hydraulic section of the U-shaped channel based on the theory of uniform flow in the open channel. Lv Hongxing, Zhou Weibo and Liu Haijun (2004) proposed the design method for the best hydraulic section of a U-shaped channel and derived the iterative formula for the hydraulic calculation of normal water depth and critical water depth of a U-shaped channel. Zhang Libing et al. (2005) used an immune genetic algorithm to optimize the bottom width and design depth of the trapezoidal and U-shaped irrigation channel sections. Zhang Xinyan, Lv Hongxing, and Zhu Delan (2013) used SAS software to program and pass the optimal fitting to establish the normal water depth hydraulic calculation formula of the U-shaped channel. Liu Zhipeng et al. (2017) used a two-stage artificial fish school algorithm that introduced coordinated behavior to optimize the cross-section design of the U-shaped channel. And Xue Zhenshan (2019), Using numerical simulation methods, the optimization design of the irrigation canal system was studied, and U-shaped section design parameters were obtained.

However, the joint optimization design of the longitudinal slope and cross-section of the anti-seepage canal has received little attention. This study has carried out joint optimization on the vertical and horizontal cross-section of the anti-seepage canal, which provides a better theoretical basis for the engineering design of the anti-seepage canal in actual work.

## 2 THEORETICAL BASIS AND DESIGN VARIABLES

### 2.1 *Hydraulic calculation*

A U-shaped channel is calculated according to the uniform flow of the open channel, and the flow formula is:

$$Q = Av \qquad (1)$$

In the formula:
Q—channel flow rate, $m^3/s$
A—channel cross section water area, m2

$$v = C\sqrt{RJ} \qquad (2)$$

In the formula:
v—channel flow velocity, m/s
C—Xie Cai coefficient
R—hydraulic radius, m
J—hydraulic slope
Since the hydraulic gradient J in the uniform flow of the open channel is equal to the longitudinal slope i, the formula (2) is incorporated into the formula (1), which can finally be written as:

$$Q = AC\sqrt{Ri} \qquad (3)$$

In the formula:
i—longitudinal slope
From Manning's formula:

$$C = \frac{1}{n}R^{1/6} \qquad (4)$$

In the formula:
n—open channel roughness
Putting Manning formula (4) into formula (3), we can get

$$Q = Av = AC\sqrt{Ri} = \frac{1}{n}AR^{2/3}i^{1/2} = \frac{1}{n}\frac{A^{5/3}i^{1/2}}{\chi} \qquad (5)$$

In the formula:

$\chi$ —hydraulic wet cycle, m

### 2.2 *U-shaped channel section design variables*

A U-shaped channel section with an arc at the bottom refers to a U-shaped section with a circular arc shape at the bottom and straight-line projections of the slope sections on both sides. The channel section is shown in Figure 1.

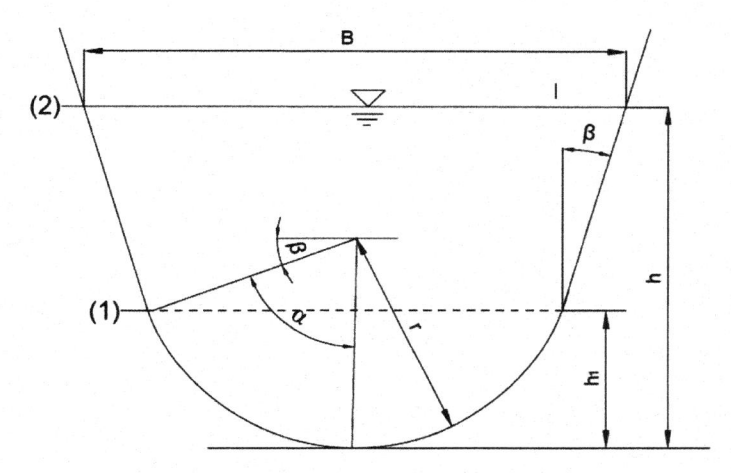

Figure 1.    U-shaped channel section of the circular arc.

The relationship between the hydraulic elements of the U-shaped channel section with the arc at the bottom is as follows:

1) When the water is deep, the water surface line is inside the arc, and the relevant hydraulic elements of the channel section at this time are as follows:
Cross-section area:

$$A = r^2 (\alpha - \cos\alpha \sin\alpha) \tag{6}$$

Wetted perimeter:

$$\chi = \min\{2r\alpha\} \tag{7}$$

In the water depth, the wet cycle $\chi$ of the U-shaped cross-section with the arc as the bottom is the constraint condition for the optimal section of the channel in the optimization design of the planning solution, and the optimal hydraulic section is the minimum wet cycle $\chi$.
Hydraulic radius:

$$R = \frac{A}{\chi} = \frac{r^2 (\alpha - \cos\alpha \sin\alpha)}{2r\alpha} = \frac{r (\alpha - \cos\alpha \sin\alpha)}{2\alpha} \tag{8}$$

2) When the water is deep and the water surface is on the upper part of the arc, the relevant hydraulic elements of the channel section at this time are as follows:
Cross-section area:

$$A = r^2 (\alpha - \cos\alpha \sin\alpha) + 2r (h - r + r\cos\alpha) \sin\alpha + (h - r + r\cos\alpha)^2 \cot\alpha$$
$$= r^2\alpha + (h^2 + 2r^2 - 2rh) \cot\alpha + \frac{2r (h - r)}{\sin\alpha} \tag{9}$$

Wetted perimeter:

$$\chi = \min\left\{2r\alpha + \frac{2\,(h - r + r\cos\alpha)}{\sin\alpha}\right\} \tag{10}$$

In the water depth, the wet cycle $\chi$ of the U-shaped cross-section with the arc as the bottom is the constraint condition for the optimal section of the channel in the optimization design of the planning solution, and the optimal hydraulic section is the minimum wet cycle $\chi$.

Hydraulic radius:

$$R = \frac{A}{\chi} = \frac{r^2\alpha\sin\alpha + \left(h^2 + 2r^2 - 2rh\right)\cos\alpha + 2r\,(h - r)}{2\,(r\alpha\sin\alpha + h - r + r\cos\alpha)} \tag{11}$$

### 2.3   *MATLAB solves non-linear programming*

#### 2.3.1   *Principles of non-linear programming*

Planning and solving are a part of a set of commands. Through the use of mathematical language, the actual problem is transformed into a mathematical model. Then, the optimal solution that satisfies the model is found according to the objective function expression and constraint conditions of the model. According to the difference in the objective function, the programming solution is divided into linear programming and nonlinear programming. It is very difficult to solve nonlinear programming problems using ordinary mathematical methods as they cannot give the same concise and clear results as linear programming problems.

The solution to the nonlinear programming problem is generally an iterative method, which is to find a feasible solution of the objective function according to certain rules under the condition of satisfying constraints and use the old value of the variable to repeatedly recurse the new value.

On MATLAB's mathematical platform, various instructions and functions can be used to perform corresponding mathematical calculations and drawing work, which is widely used in engineering. In general, nonlinear programming and optimization problems have one or more nonlinear functions in the objective or constraint function. The corresponding language of MATLAB is used to program the instruction, write the objective function and constraint function, and use function calls to solve the problem. The MATLAB algorithm used in this article is sequential quadratic programming or SQP algorithm for short.

The algorithm converts the problem into a series of quadratic programming problems, which it solves to obtain the optimal solution to the original problem. It then takes the quadratic value of the Lagrangian function to improve the approximation of the quadratic programming problem.

#### 2.3.2   *Application of solver tool*

MATLAB tools are used to compile the planning and solving process according to the cross-section characteristics of different channels. The creation process is mainly as follows:

(1) Compile the objective function expression;
(2) Compile the constraint expression;
(3) Compile the main program and obtain the optimal solution for the function.

This article uses MATLAB's SQP algorithm to provide the corresponding fmincon function to solve this problem. The basic call format is as follows:

[x,fval]=fmincon ('fun',$X_0$,A,b,Acq,bcq,lb,ub, 'nonlcon',options), x is the solution vector of the optimal solution, and fval is the optimal function value.

# 3 CALCULATION EXAMPLE

## 3.1 *Basic conditions*

The field area is 17.50 h/m$^2$, and the designed flow rate of the irrigation channel is Q = 0.015m$^3$/s. The first case of MATLAB programming and solving is: when the design longitudinal slope is i = 1/1000 and the roughness is n = 0.025, the solution arc is the optimal hydraulic section of the bottom cross-section channel.

## 3.2 *Solving process*

MATLAB uses the SQP algorithm to solve the problem as follows:

(1) The main program to establish the object for the objective function of wet cycle $\chi$:
```
function f=object(x)
r=x(1);h=x(2);J=x(3)*pi*(1/180);
f=2*r*J+2*(h-r+r*cos(J))*((sin(J))^-1);
end
```
(2) Establish the non-linear constraints of the subject:
```
function [g,ceq]= subject(x)
r=x(1);h=x(2);J=x(3)*pi*(1/180);
ceq=[sqrt(0.001)*(1/0.025)*((r*r*J+(h*h+2*r*r-2*r*h)*cot(J)+2*r*(h-r)*
((sin(J))^-1))^(5/3))*((2*r*J+2*(h-r+r*cos(J))*((sin(J))^-1))^(-2/3))-10; -h+r-r*cos(J)];
g=[];
end
```
(3) The main program to find the optimal solution is:
```
A=[];b=[];
Aeq=[];beq=[];
lb=[0;0;0];
ub=[];
[x,f,fval,exiflag]=fmincon('object',[1,1,45],A,b,Aeq,beq,lb,ub,'subject')
```

Similarly, when the longitudinal slope is 1/2000, 1/3000, 1/4000, or 1/5000, the SQP algorithm is used to nonlinearly program the optimal hydraulic section such that the bottom arc radius of the section, r (m), the water depth, h (m), and the angle, $\alpha$, between the center of the tangent point and the horizontal line are satisfied, as shown in Table 1.

Table 1. Arc is the optimal section parameter of the channel of the bottom u-shaped cross-section.

| Longitudinal slope i | Radius of bottom arc r (m) | Intersection distance h$_1$ (m) | Included angle $\alpha$ (°) | Wetted perimeter $\chi$ (m) |
|---|---|---|---|---|
| 1/1000 | 2.1799 | 2.1799 | 90 | 6.8484 |
| 1/2000 | 2.4825 | 2.4825 | 90 | 7.7989 |
| 1/3000 | 2.6786 | 2.6786 | 90 | 8.4149 |
| 1/4000 | 2.8270 | 2.8270 | 90 | 8.8813 |
| 1/5000 | 2.9478 | 2.9478 | 90 | 9.2608 |

# 4 CONCLUSION

(1) Establish a mathematical model for optimization of channel cross-section and longitudinal section through MATLAB's SQP algorithm (sequential quadratic programming) using the programming method.

(2) When the longitudinal slope of the U-shaped cross-section with the arc as the base is 1/1000, 1/2000, 1/3000, 1/4000, and 1/5000, the angle between the line of the tangent point and the horizontal is 90°, That is, the bottom of the optimal hydraulic section of the channel whose arc is the bottom cross-section is a semi-circular arc.

(3) Under the same flow condition, the vertical slope and the optimal section structure size are different. The vertical slope i=1/1000, the bottom arc radius R=2.18m, the water depth h=2.18m, and the tangent point center line connects with the horizontal included angle $\alpha$=90°, the minimum hydraulic wet circumference is 6.85m. When the longitudinal slope is i=1/5000, the bottom arc radius R=2.95m, the water depth h=2.95m, the tangent center line and the horizontal clamp angle $\alpha$=90°, the minimum hydraulic wet circumference is 9.26m.

In this study, the cross-section and the longitudinal slope are jointly optimized and designed, and the best hydraulic section and the best economic section are taken into consideration for later reference in practical applications.

## REFERENCES

G.V. Loganathan. Optimization design of U-shaped channel[J]. *Anti-seepage Technology*, 1993, 2:45–60.

Liu Zhipeng, Zhu Lili, Zhai Yaming, Chen Liang. Optimization design of U-shaped channel section based on two-stage artificial fish swarm algorithm [J]. *Water Conservancy and Hydropower Technology*, 2017, 48(01):149–153.

Lv Hongxing, Zhou Weibo, Liu Haijun. Hydraulic holding capacity and hydraulic calculation of U-shaped channel[J]. *Journal of Irrigation and Drainage*, 2004.

M. Mohammad Rezapour Tabari & Shiva Tavakoli &Mohsen Mazak Mari. Optimal Design of Concrete Canal Section for Minimizing Costs of Water Loss, Lining and Earthworks[J]. *Water Resources Management*, 2014, 28(10):3019–3034.

Tan Kai, Hu Qiushun. *Discussion on the Water Conveying Capacity of Small-scale Coagulation Main Channel and the Quantity of Lining Works in Hunan Water Conservancy*, 1996, 01:10-13, 28.

Xue Zhenshan. Research on optimal design of irrigation canal system based on genetic algorithm [J]. *Hebei Water Conservancy*,2019(06):38-39.4:50–52.

Zhang Libing, Cheng Jilin, Jin Juliang, Jiang Xiaohong. Application of Immune Genetic Algorithm in Channel Optimization Design [J]. *Journal of Yangzhou University*,2005,8(3):50–53.

Zhang Xinyan, Lu Hongxing, Zhu Delan. Direct hydraulic calculation formula of normal water depth of U-shaped channel[J].*Chinese Journal of Agricultural Engineering*,2013,14:115–119.

Zhang Zhichang, Li Yincai, Liu Yafei, et al. Calculation of optimal hydraulic section of U-shaped channel [J]. *Shaanxi Hydropower*, 2001, 02: 25–27.

*Civil Engineering and Urban Research – Mohamed & Hou (Eds)*
© 2023 the Authors, ISBN 978-1-032-44487-1

# Experimental study on the anti-horizontal force performance of flexural and non-flexural expanded stem piles with different strengths under horizontal loads

Qian Yongmei, Li Huaqiang, Wang Ruozhu* & Tian Wei
*Jilin Jianzhu University, Changchun, China*

ABSTRACT: With the rapid development of society, the demand for high-rise and super-high-rise structures, as well as the foundations they require, is increasing. Pile foundations have been notably developed as a favorite kind of foundation and have many years of history. Concrete expanded stem pile is a form of pile basis on multi-joint extruded and extended cast-in-place pile, which has been rising in latest years. When the pile is under load, the interplay between the pile and soil is now not only the friction pile side resistance but also the resistance of the aiding plate. This paper conducted a mannequin test to inspect the overall mechanical performance of concrete expanded stem piles in opposition to horizontal loads, especially the anti-horizontal force of flexural and non-flexural expanded stem piles. During the test, a modern test technique was used to recognize the changes in the pile-soil in the course of the loading technique by using a machine that can be located in actual time. In this paper, the bearing ability of the pile beneath horizontal load is analyzed in detail by using the horizontal bearing potential of exceptional strength flexural and non-flexural expanded stem piles and the state of destruction of the soil around the pile. This is done by enlarging the expanded plate and flexural and non-flexural elevated stem piles. The failure mechanism of flexural and non-flexural expanded stem piles beneath the motion of horizontal bearing potential is determined. The effect of the various parameters of flexural expanded stem piles is similarly studied. It presents reliable training for the extensive utility of concrete expanded piles in true initiatives and the research and development of pile foundations.

## 1 INTRODUCTION

A concrete extended pile (see Figure 1) is a new accelerated pile created with the aid of extruding, spinning, or drilling. It is an expanded pile with enhanced load-bearing and anti-horizontal capabilities. As a brilliant anti-settlement performance, it has various purposes, thanks to a growth plate as support (Zhao et al. 2009). The bearing-increasing plate can be flexibly modified under the field circumstance while bearing force, decreasing pile length, saving construction materials, and shortening development time simultaneously. In bearing capacity, the concrete expanded pile is higher than the concrete straight-hole pile foundation. The bearing potential of the straight gap pile is frequently composed of the pile end assisting force and the measured friction resistance of the pile. The measured friction resistance and the aiding force of the plate are composed of three parts, so the bearing potential of the concrete expanded plate pile is better than a straight hole pile foundation. At the same time, during the construction of the expanded concrete piles, there will be special drilling and expanding machines to clean up the soil produced by the drilling, so the construction environment is better. Scholars have performed preliminary theoretical and experimental studies on the horizontal load of concrete expansion piles, notably investigating the

---

*Corresponding Author: 306109357@qq.com

DOI 10.1201/9781003372417-24

consequences of a variety of load-bearing multiplied plate parameters such as the location of the plate, the gap of the plate, and the diameter of the plate on the destruction mode of rigid expanded piles, and the horizontal ultimate bearing capacity is determined on this basis (Kumar & Prakash 2004). There are few studies on the ultimate bearing capacity of flexural expanded piles and the damage of piles under horizontal loads. The research results obtained in this paper on the influence of flexural and non-flexural expanded stem piles on the bearing capacity of concrete extended stem piles and soil failure state will be used to supplement the design theory of extended stem piles and enrich the calculation methods to study the mechanical properties of concrete expansion piles under horizontal loads, particularly the anti-horizontal force of flexural and non-flexural expanded stem piles. In addition, the improvement and innovation of the test method will provide a reference for pile foundation research.

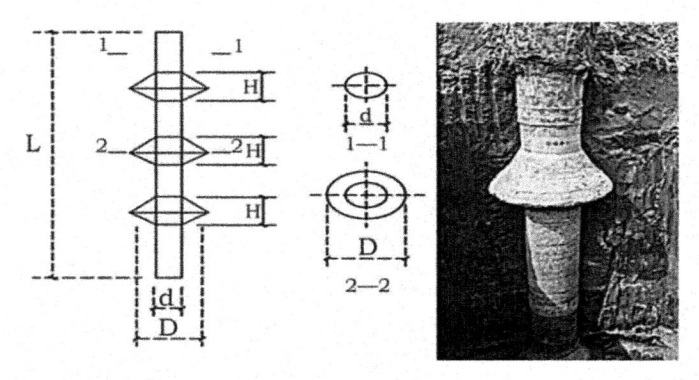

Figure 1.    Schematic diagram and actual pile diagram of concrete expansion pile.

This paper has the taking after advancement points:

(1) Based on the vertical loading test and loading equipment, the test equipment progressed to realize concurrent vertical and horizontal loading within the simulation test. At the same time, a small-scale model is utilized to test the piles. The investigation appears that the bearing execution of the small-scale model pile establishment is steady with the large-scale pile establishment (Qian et al. 2016).
(2) The whole test can be watched. This experiment innovatively proposes a visualization device that comprehensively considers the model size and boundary effects and designs a special dust collector. The half-faced pile was used instead of the whole pile for the test design, and the steel plate on the side of the dust collector was changed to a glass plate to observe the damage to the soil during the test process.
(3) The performance of flexural and non-flexural expanded stem piles against horizontal force is calculated according to their failure state.

## 2   EXPERIMENTAL STUDY

### 2.1   *Model pile specimens*

According to Figure 3, in this study, four samples were divided into two groups: (G3, R3) and (G4, R4). G3 and G4 were non-flexural, and R3 and R4 were flexural. The distance between the pile pinnacle and the load-bearing plate function of the mannequin pile was 40 mm and 160 mm. The diameter of the plate was D = 50(mm), the length of the pile was L = 227(mm), the top of the plate was H = 27(mm), the diameter of the pile was d = 16(mm), the overhanging length was R = 17 (mm), the slope angle of the plate was $\alpha$=39°, and the distance between the plate and the top of the pile was L1= 40 (mm). The genuine mannequin pile can be seen in Figure 2.

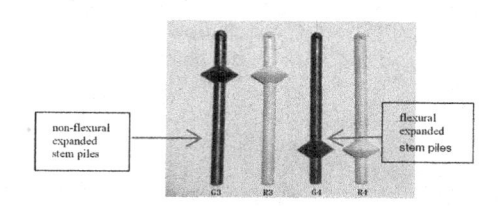
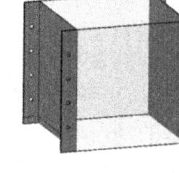

(a) Model for soil collector (b) Actual picture for soil collector

Figure 2.  Model of piles.

Figure 3.  Soil collector.

## 2.2  Design of soil collector

Q235 metal with excessive rigidity was chosen as the material of the soil collector, as shown in Figure 3. The size of the soil collector was 300mm*250mm*300mm, and the thickness of the plate was 5 mm. In order to prevent the deformation of the soil collector in the actual site, an obvious thick plate is usually used too increase the sidewall rigidity of the soil sampler. The lower edge of the plate was polished to a wedge shape. The four steel plates of the soil sampler were bolted and detachable, and both its front and back can be used as the test surface. A 300mm*300mm toughened glass can be mounted. To fix the tempered glass and prevent interference with the soil collector itself during the experiment, we used a restrictive clip and a constant clip.

## 2.3  Design of test rig

The loading machine in this test was a progressive multi-function loading rig (see Figure 4). They involved the design of vertical and horizontal loaders; the loading platform consisted of a working surface, columns, and reaction beams. Five rows of holes were reserved along with the columns. In the right column, we bolted a fixed pulley and set the pulley hole reserved for three rows in the test. The horizontal loading system consists of screws, jacks, displacement sensors, spacers, etc. The horizontal and vertical loading system adopts jack loading, the vertical load is transmitted through the bearing plate of the jack, and the horizontal load is transmitted through the screw rod and wire rope.

Figure 4.  Multi-function test rig.

Figure 5.  Undisturbed soil collection pictures.

## 2.4  Soil collection from undisturbed soil

Since this test is an in-situ soil model test, a soil extraction site that meets the test criteria needs to be selected. At the same time, the homemade soil extractor can be more convenient and quick to conduct soil extraction (see Figure 5).

## 2.5  Experimental process

To simulate the expanded disc pile in the actual project, when loading, if the vertical load on the expanded disc pile is lower than the limit value, the horizontal load will be applied in the horizontal direction of the expanded disc pile. The test device used in this test is self-developed, and the loading scheme has been applied for the national invention patent. The test loading process is divided into four parts (Orr 2016), i.e., installation of the soil extractor, installation of the vertical loading device, installation of the horizontal loading device, and starting of the loading device.

## 3 ANALYSIS OF TEST RESULTS

During the test, the soil ring shear samples were despatched to the laboratory to find out about soil-related parameters to forestall adjustments in soil-related parameters. Since the four groups of soil samples used in the test were in-situ soils, the land taken and the depth of the soil taken were the same; therefore, the parameters such as the angle of internal friction, water content, and density of the soil obtained from the experimental results were the same.

### 3.1 *Comparison of failure states of soil round piles*

The following comparison of the different damage states of the soil around G3, R3, and G4, R4 piles shows the different damage states of the expanded pile when the horizontal displacement of the top of the model pile reaches 14 mm under the horizontal load. During the test, both the displacement sensor and the camera recorded the pile displacement and the damage state of the soil around the pile in real-time. After the test, the experimental results were compiled, compared, and analyzed. Figure 7 shows the damage state of the soil around different model piles.

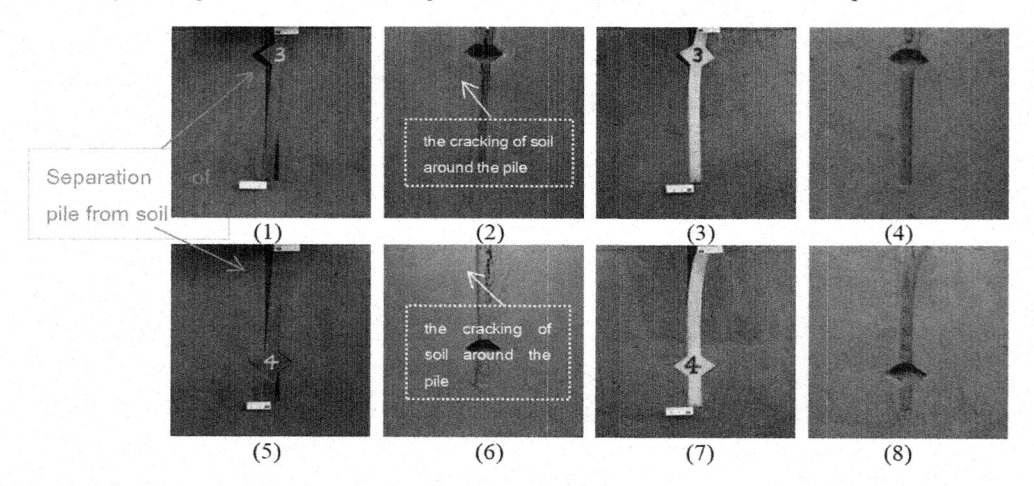

Figure 6.   Comparison of soil failure state around different piles.

As shown in Figure 6, the damage patterns of the soils around the four groups of model piles are the same. From the soil damage state, the soil in the upper part of the bearing plate and the right side of the pile were squeezed, while the soil in the upper part of the bearing plate on the left side was pulled. This is because the right side of the bearing plate plays a role. When the horizontal load is applied to the model pile, the pile has an extrusion effect on the upper part of the bearing plate on the right side and a pulling effect on the lower part of the bearing plate on the left side.

First, compare the damage state of soil around G3 and R3. When the horizontal displacement of the top of the model pile reached 14 mm, the degree of cracking of the soil around the pile and the deformation pattern of the pile were different. G3 mainly shows rigid body rotation. And the whole pile body rotates along a point on the bearing plate, and the upper soil on the right side is squeezed. The upper pile body of the bearing plate of the R3 pile had flexible local bending, while the lower part remained unchanged in most of the loading range. At the same time, as shown in Figure 6(3), no significant soil deformation around the pile's lower part was observed.

Second, the damage state of the soil around G4 and R4 is shown in Figures 6(5) and 6(7). It can be seen through the pictures that the soil cracks have become smaller. Especially for the flexural model pile R4, the upper part underwent pure bending deformation, while the soil on the right side was compressed. The role of the load-bearing plate in resisting the bending moment is very weak. G4 shows rigid body rotation.

Third, the comparison of the non-flexible model piles, G3 and G4, shows that both model piles underwent rigid body rotation at the end of the test, but the displacement of the top of each pile is different. Figures 6(2) (G3) and 6(6) (G4) show that the cracks formed on the soil around G3 are significantly larger than around G4. For non-flexible concrete expansion piles, the distance from the location of the bearing plate to the top of the pile affects its horizontal bearing capacity: the shorter the distance from the location of the bearing plate to the top of the pile, the higher the horizontal bearing capacity and the better the performance of the pile.

As shown in Figures 6(3) and 6(7), both the flexural model piles R3 and R4 exhibit bending deformation of the piles. The load-bearing plate of R3 rotates along a certain point, while the load-bearing plate of R4 is placed in the lower part and does not rotate at the end of the plate. The bearing plate of the bending model pile R4 played an insignificant role due to the influence of the horizontal load.

## 3.2 *Analysis of displacement-load data*

For a detailed analysis of the horizontal bearing capacity of the concrete expansion pile, the data of displacement and load were collected during the test and plotted in Figure 7.

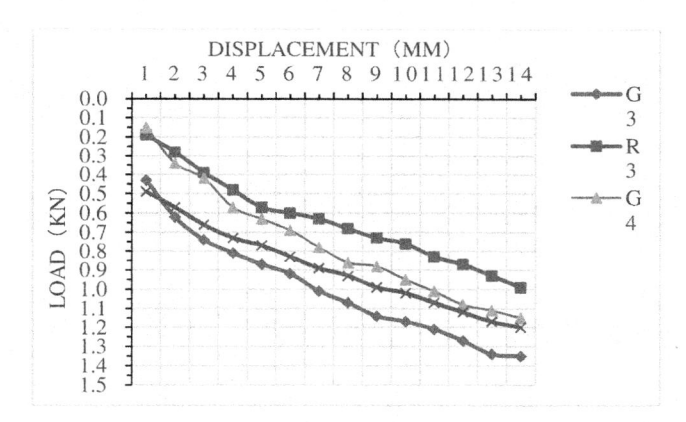

Figure 7. The displacement-load curves of the pile specimens.

From the load-displacement curves in Figure 7, it is clear that the four curves have an identical trend: as the horizontal displacement increases, the load on top of the pile also increases. At the early loading stage, the slope of the curve is extraordinarily steep because the load-bearing plate steadily takes effect as the horizontal load increases. In the later stages of loading, the slope of the curve becomes slower because the soil around the pile progressively fails as the horizontal load increases. The change in the slope of the curve is most reported in the R3 pile. This shows that the soil around the load-bearing plate pile has entered the failure state, ensuing in a gradual reduction in the force of the load-bearing plate. For the non-flexible growth pile, the distance between the bearing plate role and the pinnacle of the pile affects the horizontal bearing ability of the concrete expansion pile. In addition, the shorter the distance between the bearing plate region and the top of the pile, the higher the horizontal bearing potential and the higher the overall performance. For non-flexible expansion piles, the distance between the role of the bearing plate and the top of the pile affects the horizontal bearing ability of the concrete expansion pile. In addition, the shorter the distance between the load-bearing plate function and the pile top, the larger the horizontal bearing capacity that the growth pile can withstand, and the better the pile performance. When the pile top displacement is the same, the shorter the distance between the load-bearing plate and the pile top, the better the horizontal resistance of the non-flexible growth dry pile than the flexible growth dry pile. However, when the distance between the bearing plate function and the pile top is exceedingly large, the horizontal bearing potential of the non-flexible enlargement dry pile and the flexible expansion dry pile is not much different.

## 4 CALCULATION MODEL OF THE HORIZONTAL ULTIMATE BEARING CAPACITY FOR FLEXURAL AND NON-FLEXURAL EXPANDED STEM PILES

By studying the failure state of flexible and inflexible expansion piles under horizontal loads, the failure modes of soil around flexible and non-flexible expansion piles have been determined (Qian et al. 2015; Wei et al. 2020). Under the action of horizontal load, the stress patterns of the surrounding soil of rigid and flexible piles are shown in Figures 8 and 9.

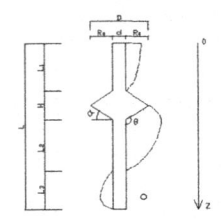

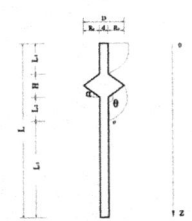

Figure 8. Stress mode of soil around rigid expanded pile.

Figure 9. Stress mode of soil around flexible expanded pile.

According to the damage mode of the soil around the non-flexible expanded pipe pile, combined with the slip line theory (Li et al. 2005; Yin et al. 2017), the calculation model of the ultimate bearing capacity of a single non-flexible expanded pipe pile against horizontal force is proposed as follows.

$$F = F_{ep} + F_{sp} \quad \text{and} \quad F \leq F_p \tag{1}$$

$$\text{where } F_{ep} = \pi \frac{D+d}{4} c \cot \phi R_0 (e^{2\theta \tan \phi} - 1), \ F_p = \frac{3EI\omega}{(L-L_3)^3} \tag{2}$$

$$\text{and } F_{sp} = F_1 + F_2 = \int_0^{L_1} mY_0 b_0 dz + \int_{L-L_3}^{L} mY_0 b_0 dz \tag{3}$$

According to the damage mode of the soil around the non-bending spread pile, combined with the slip line theory, the horizontal ultimate bearing capacity calculation model of a single-bending spread pile is proposed as follows.

$$F = F_{ep} + F_{sp} \quad \text{And} \quad F \leq F_p \tag{4}$$

$$\text{where } F_{ep} = \pi \frac{D+d}{4} c \cot \phi R_0 (e^{2\theta \tan \phi} - 1), \ F_p = \frac{3EI\omega}{(L-L_3)^3} \tag{5}$$

$$\text{and } F_{sp} = F_1 + F_2 = \int_0^{L_1} mY_0 b_0 dz + \int_{L_1}^{L-L_3} mY_0 b_0 dz \tag{6}$$

where $F_{ep}$—bearing capacity of the end of a plate;
$F_{sp}$—bearing capacity of the side of the aspect of the pile;
$F_p$—bearing capacity of pile body;
$F$—horizontal ultimate bearing capacity of a single pile;

c—cohesion of soil around piles;

d, D—diameter of pile body and bearing plate, respectively;

$\phi$—internal friction angle of soil around piles;

$\theta$—angle between the decreased surface of the plate and the vertical direction;

$F_1$—$L_1$ region's bearing capacity;

$F_2$—$L_2$ region's bearing capacity;

$Y_0$—horizontal displacement of piles;

$b_0$—calculated width of the pile;

L—full length of the expanded pile. Moreover;

m—proportional coefficient for resistance coefficient of soil varying with depth;

$L_1$—length of the vertical pile body on the plate;

$L_3$—length of the vertical pile body beneath the slipping line;

E—modulus of elasticity of materials;

I—moment of inertia of piles;

$\omega$—displacement of the pile top in the effect of the horizontal load.

The calculation models of horizontal force resistance for flexural and non-flexural expanded dry piles are summarized. In summary, the calculation models for flexible and non-flexible expanded dry piles are different in terms of lateral resistance while almost identical in terms of sub-slab resistance.

## 5  CONCLUSIONS

This paper studies the ultimate bearing capacity of flexural expanding piles and non-flexural expanding diameter piles under horizontal load through experimental research and numerical analysis. Through a small-scale half-section model pile test, the damage state and bearing mechanism of the soil around the flexible and non-flexible expansion piles were analyzed, and the following conclusions were drawn.

(1) When the distance between the location of the bearing plate and the top of the pile is relatively short, the non-flexible expanded pile shows rigid body rotation, while the flexibly expanded pile shows local bending in the upper part. The result remained constant over a large area under the bearing plate. At the same time, the soil around the lower part of the pile showed no significant deformation. For a specific location of the load-bearing slab, the greater the stiffness of the concrete expansion pile, the greater the counter-horizontal force of the expansion pile.

(2) Since the damage states of flexible and non-flexible expanded dry piles subjected to horizontal loads are significantly different, two independent calculation models are proposed for the inverse horizontal forces of flexible and non-flexible expanded dry piles, respectively. These models provide a theoretical basis for the design and application of expanded concrete piles.

This paper studies the effect of horizontal loads on the damage state of soil around the piles of flexible and non-flexible expanded piles of different strengths. Under the action of the horizontal load, the expansion plate plays a role, which increases the bearing capacity of the expansion pile. Future research on various parameters of flexible expansion piles should be further expanded, such as the influence of different shapes, positions, and quantities of load-bearing plates on the ultimate bearing capacity of expansion piles.

ACKNOWLEDGMENTS

This work was financially supported by the Jilin Provincial Department of Science and Technology (20210509042RQ).

REFERENCES

L. Z. Yin, X. G. Fan, and S. J. Wang, (2017) A study on the application of squeezed branch pile in clay soil foundation, *Earth and Environmental Science*, vol. 61, no.1, pp1–6.

Liping Zhao, Li Xiang, Chen Lu, Shilin Han. (2009) Analysis of stress characteristics of squeezed branch pile group under horizontal load [J]. *Industrial Building*, 39 (10): 52–56.

Qimin Li, Manchu He, Yueqing Tan, (2005) Experimental study on Extruded Branch Plate supporting pile [J]. *Geotechnical Mechanics*, (10): 146–149.

S. Kumar, S. Prakash. (2004) Estimation of Fundamental Period for Structures Supported on Pile Foundations[J]. *Geotechnical and Geological Engineering*, 22: 375–389.

T. Orr, (2016) Design examples: comparison of different national practices, in Proceedings of ISSMGE ETC 3 International Symposium on Design of Piles in Europe, Leuven, Belgium, vol. 1, pp.47–62.

Yingjie Wei, Duli Wang, Jiawang Li, et al., (2020) Evaluation of ultimate bearing capacity of pre-stressed high-strength concrete pipe pile embedded in saturated sandy soil based on in-situ test, *Applied Sciences*, vol. 10, no. 18, pp62–69.

Yongmei Qian, Jie Wang, Ruozhu Wang, Lian Zhai. (2016) Research status and working mechanism of concrete expanded pile under horizontal load [J]. *Shanxi Architecture*, 42 (13): 62–63.

Yongmei Qian, Rongzheng Zhai, Ruozhu Wang. (2015) Research on the Calculating Mode of the Uplift Bearing Capacity of the Concrete Expanded-Plates Pile, *Architectural Engineering and New Materials*, 2: pp240–247.

*Civil Engineering and Urban Research – Mohamed & Hou (Eds)*
*© 2023 the Authors, ISBN 978-1-032-44487-1*

# Study on seismic behavior of composite hollow high pier with functionally recoverable reinforced concrete columns and corrugated steel plates

Ziqi Li*, Yanyan Fan*, Li Wang* & Jihua Zeng*
*Lanzhou Jiaotong University, China National and Provincial Joint Engineering Laboratory of Road and Bridge Disaster Prevention and Control, Lanzhou, China*

ABSTRACT:    Due to the lack of research on seismic performance of reinforced concrete column corrugated steel plate composite hollow high pier under earthquake, this paper takes a railway reinforced concrete column corrugated steel plate composite hollow high pier bridge as the engineering background and carries out nonlinear analysis and research on the influence of corrugated steel plate wave height parameters on the seismic response of high pier. The results show that the reinforced concrete column corrugated steel plate composite hollow high pier is in an elastic state without damage and has excellent seismic performance under rare earthquakes. The analysis of the dynamic characteristics of the structure shows that the transverse stiffness of the structure is greater than the longitudinal stiffness, and the wave height of the corrugated steel plate has little effect on the dynamic characteristics. The results of nonlinear time history analysis show that under rare earthquakes, the lateral stress of the structure is relatively unfavorable. The optimal value of corrugated plate wave height is 25cm under lateral earthquake input and 20cm under longitudinal earthquake input.

## 1   INTRODUCTION

In recent years, the column slab high pier has been developed and applied in railway bridge engineering in China because of its large span capacity and novel structural form. China has a vast territory, many mountains and valleys, and special geological and hydrological conditions. Many large and high-level bridge construction projects rapidly extend to these areas with complex terrain. Due to the terrain constraints in these areas, building bridges with large spans and piers up to 100 meters is reasonable and economical.

A corrugated steel plate is a kind of structural steel plate with specific specifications, which compresses a certain size of structural steel plate into a corrugated shape. Because of the existence of corrugation, the bending moment of inertia of the steel plate is increased so that it has high bearing capacity and stability. A series of tests conducted by Chalmers Technical University in Sweden in early 1980 showed an interaction between local buckling and overall buckling of corrugated steel plates, and the shear capacity of corrugated steel plates was nearly twice that of flat steel plates. This point has been confirmed in the comparative stability test of corrugated steel plates and flat steel plates. There is no application of corrugated steel plate in this kind of pier. The selection of corrugated steel plate mainly considers its mechanical and economic advantages. In conclusion, corrugated steel plate is more suitable for high hollow piers than other materials mentioned above.

In terms of seismic test research of high pier bridges at home and abroad, shaking table tests and quasi-static tests are mostly used in the experimental research of the seismic performance of bridges.

---

*Corresponding Authors: liziqi@lzjtu.edu.cn, fyyanl@163.com, 823283485@qq.com and 858138601@qq.com

The shaking table test results of the scaled model of a rigid frame bridge with a pier height of 75m show that the effects of near-fault pulse earthquake and parameter changes should be considered (Yan 2020). According to the pseudo-static test results of the scale model of a hollow rectangular high pier (56.5m), it is shown that the high pier with energy dissipation continuous beam has better seismic performance (Wang 2020). Taking a 40m round high pier of a railway as an example, a scale model is designed for the shaking table test. The results show that the middle and upper area of the pier body is a potential plastic hinge area (Shao 2020). A scale-up rigid frame bridge model with a pier height of 12m is designed at the scale of 1/12. The seismic test of the bridge is carried out, and the results show that the bridge has good seismic performance (Wang 2016). The pseudo-static test is carried out based on the pier height, stirrup ratio, and other parameters of the box-type high pier. The results show that the failure mode of the box-type high pier is mainly the yielding or even fracture of the longitudinal reinforcement (Guan 2014). According to the pseudo-static test of PC Hollow High Pier and the comparison with the numerical analysis results, the results show that the two are in good agreement (Wang 2015). Considering the slenderness ratio and axial compression ratio, the seismic performance of hollow thin-walled rectangular high pier is studied. The results show that this type of high pier mainly has a bending failure, but the influence of shear cannot be ignored (Zong 2010).

From the seismic numerical simulation of the pier, the "response spectrum method" is applied to carry out seismic back analysis, design the damping scheme, and realize the balance of force and displacement (Liu 2021). Taking a high pier continuous rigid frame bridge as an example, the research results show that the greater the pier height, the worse the stability of the bridge under lateral earthquake, and the closer the pier height of the bridge is, the better the overall stiffness is (Ye 2020). The IDA method is used to study the influence of high-order mode shapes on the ductility of high piers. The calculation shows that the current pier displacement ductility calculation method does not apply to high piers (Li 2005). Fifty-three real seismic records are selected to study the seismic performance of RC high pier bridges using the IDA method. The results show that considering the high-order mode shapes can more effectively evaluate the seismic performance of high piers (Zhang 2017). The influence of the rocking isolation device on the seismic performance of the hollow high pier of the railway is analyzed. The results show that rocking isolation will change the vibration period of the high pier, and the isolation effect is good (Xia 2012). The numerical simulation method proves that the seismic isolation bearing has a certain seismic effect on high pier bridges (Chen 2008).

In the research of new type column slab pier, the seismic performance of new type column slab hollow high pier has been studied earlier abroad. The bending and shear failure mechanism of the hollow high pier column is tested and analyzed. The results show that the deformation and bending resistance of the new column slab hollow high pier under load has been significantly improved (Eric m 2006). The research on seismic behavior of Column Slab hollow pier in China is a new type of high pier, which is based on the research of foreign scholars and combined with rectangular high pier and frame shear wall structure. The domestic railway design institute proposed a new type of Column Slab Hollow High Pier Based on the foreign four-column hollow pier, in which the frame column and beam constitute the frame as the main load-bearing member of the pier, and the thin plate is used for connection, which not only meets the requirements of stiffness, but also increases the seismic performance of the pier (Zhang 2012). The seismic performance of the specimens is studied by a pseudo static test on the column slab of the new column slab hollow high pier and 8 1/10 scale model members of the frame (Li 2019).

Therefore, this paper takes a column slab high pier bridge as the background and replaces the concrete soil slab with a corrugated steel plate. The advantages and disadvantages of seismic response and seismic performance of this new composite structure are explored. It references the seismic design of high pier bridges with corrugated steel plate columns.

## 2 ENGINEERING EXAMPLE

The prototype bridge of the project background in this paper is the Zongmugou super large bridge on the Huang (Ling) - Han (Cheng) - Hou (MA) railway line. The Zongmugou bridge is located in the loess gully area, with deep valleys, deep beams, and narrow and deep main ditches. The geological structure within the bridge site is free of fault structure development. The main bridge adopts (78+2×136+78) m prestressed concrete continuous rigid frame scheme, as shown in Figure 1.

Superstructure: the box girder adopts the straight web single box single chamber variable height box section, with a total length of 429.8m and a calculated span of (78+2×136+78) m, the height of the fulcrum beam is 10.0m, the height of the mid-span beam is 5.0m, the lower edge of the beam body changes according to the parabola of 1.8 degrees, the equation is y=0.0030576 $x^{1.8}$, the top width of the box girder is 11.5m, and the bottom width is 7.2m. Two 2.0m thick end partitions are set at the top of three rigid frame piers of the box girder. The top plate of the box girder is 0.45m thick, the bottom plate is 0.5 ~ 1.1m thick, and the web plate is 0.5 ~ 1.1m thick.

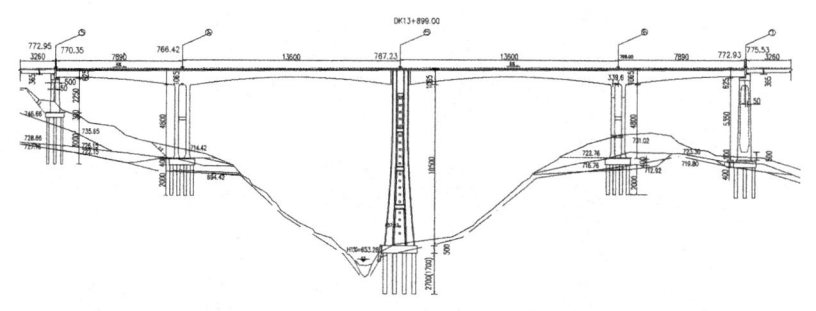

Figure 1.    General layout of the bridge.

Substructure: the new column corrugated steel plate hollow pier is adopted for the 5#main pier. Its structural feature is to set four columns with high stiffness, set transverse connecting beams to connect the columns within a certain distance, and set thin-walled slabs with openings between the columns and beams. The section of the whole pier body is shown in Figure 1. The section size of the pier top column is 300×300cm (vertical and horizontal). The section size of the pier bottom column is 550×550cm (Longitudinal × Horizontal). The longitudinal and transverse outer contour dimensions of the middle main pier shaft section shall be determined according to the change of the 1.6-degree parabola. The design concept is that in the normal use stage, the stiffness of the pier itself is large enough to bear the gravity, live load, and other forces of the superstructure. However, when the pier encounters a strong seismic load, the thin-walled plate between columns cracks first due to its small strength, dissipating most of the seismic input energy, so that the stiffness of the whole structure decreases rapidly, the period is prolonged, and the seismic input energy is greatly reduced. Finally, it can protect the main structure of the bridge from damage.

The secondary main piers 4# and 6# are rectangular double thin-walled piers with a width of 2.2m along the bridge direction and a center distance of 6.8m between the two walls. The piers are made of C55 concrete, and the pier body is changed by a quadratic parabola 45.5m above the transverse outer contour. Solid round end pier is adopted for side pier 3# pier, C35 concrete is adopted for pier shaft, and round end hollow pier is adopted for side 7# pier. C55 concrete is used for the main beam, C55 concrete is used for the top of 5# main pier, and C50 concrete is used for the following pier shafts. The pile foundation of the main bridge shall be designed as per the column pile, the 5# pier shall be excavated pile, and the other piers shall be a bored pile.

The section of the pier shaft of the corrugated column plate combined high pier is shown in Figure 2.

## 3   FINITE ELEMENT MODEL AND STRUCTURAL DYNAMIC CHARACTERISTICS ANALYSIS

The upper structure is (78+2×136+78) m prestressed concrete rigid frame bridge. The main beam adopts a single box and single chamber box section with a straight web. The beam height at the

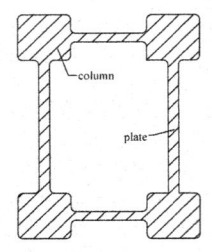

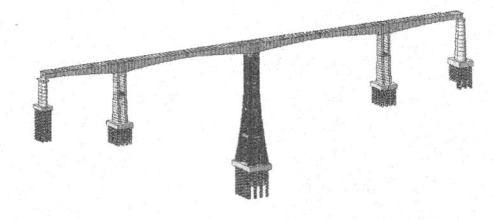

Figure 2.   Section diagram of the hollow pier.

Figure 3.   Schematic diagram of finite element model of the whole bridge.

fulcrum is 10m, the beam height at the beam end is 5m, the lower edge of the beam body changes according to the parabola of 1.8 degrees, the top width of the box girder is 11.5m, and the bottom width of the box girder is 7.2m. The thickness of the top plate of the main beam is 45cm. The bottom plate is 50 ~ 120cm thick. The web is 50cm ~ 110cm thick. The 5# main pier in the substructure adopts a new type of column corrugated steel plate hollow pier.

The spatial finite element model is used for seismic response analysis. According to the structural design scheme, the finite element model of the rigid frame bridge is established by using Midas software. The main beam, bearing platform, and pile adopt spatial beam elements, and the corrugated steel plate adopts plate elements. The restraint effect of soil on the pile is simulated by soil elastic modulus, and the elastic stiffness is calculated by the response spectrum method. The bridge is divided into 3911 nodes and 3150 elements, including 1968 beam elements and 1182 plate elements. The full bridge space finite element model is shown in Figure 3.

By changing the wave crest parameters of the corrugated plate, 15cm, 20cm, and 25cm are respectively taken for modeling.

The dynamic characteristics of the structure are analyzed. Only when the wave height is 20 cm the natural vibration period and vibration mode shape of the structure are shown in Table 1.

Table 1.   Calculation results of dynamic structural characteristics at a wave height of 20cm.

| Vibration mode | Period(s) | Vibration description |
| --- | --- | --- |
| 1 | 1.89 | Longitudinal bending vibration of main beam and pier |
| 2 | 1.87 | Bending vibration of structure along the bridge |
| 3 | 1.42 | Transverse bending vibration of main beam and pier |
| 4 | 1.02 | Transverse bending vibration of main beam and pier |
| 5 | 0.79 | Vertical bending vibration of the main beam |
| 6 | 0.77 | Transverse bending vibration of main beam and pier |
| 7 | 0.63 | Vertical bending vibration of the main beam |
| 8 | 0.60 | Transverse bending vibration of main beam and pier |
| 9 | 0.51 | Transverse bending vibration of 3# side pier |
| 10 | 0.50 | Transverse bending vibration of 7# side pier |

The calculation results show that: (1) the column corrugated steel plate composite high pier bridge belongs to a long period structure, and the structural flexibility is large. (2) The low-order vibration mode of the structure is forward vibration, the first-order vibration mode is the bending vibration of the main beam and pier along the bridge, and the second-order vibration mode is the bending vibration of the structure along the bridge. (3) The transverse stiffness of the structure is greater than the longitudinal stiffness. (4) The wave height of a corrugated steel plate does not affect the structure's dynamic characteristics.

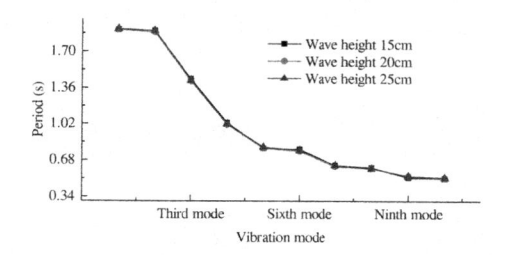

Figure 4. Comparison of natural vibration period of the structure under different wave heights.

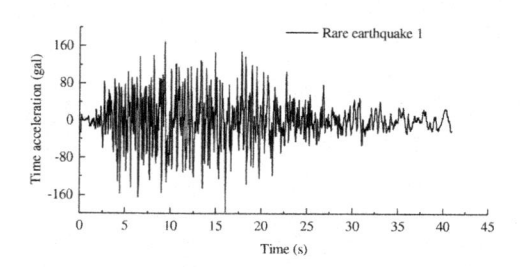

Figure 5. Rare earthquake 1.

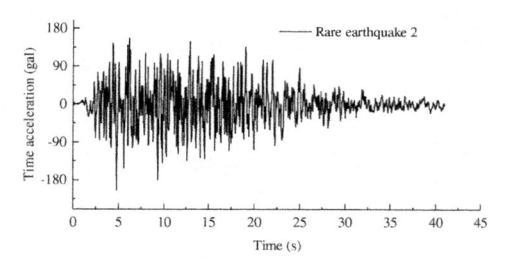

Figure 6. Rare earthquake 2.

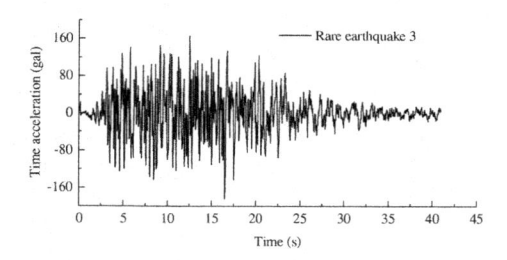

Figure 7. Rare earthquake 3.

## 4 SEISMIC INPUT

To study the nonlinear seismic response of the new column corrugated steel plate combined high pier under the rare earthquake, the time history response analysis of the structure is carried out by using the three seismic waves provided in "Report on seismic safety evaluation of key sites in Shaanxi section of Huangling Hancheng Houma Railway," as shown in Figure 5 to Figure 7. The seismic input directions are input horizontally and longitudinally, respectively.

## 5 NUMERICAL ANALYSIS OF NONLINEAR MODEL FOR A NEW TYPE OF COLUMN CORRUGATED STEEL PLATE COMPOSITE STRUCTURE

The column plate structure is similar to the shear wall in the building structure. When it encounters a strong earthquake, the thin-walled plate will crack first to consume seismic energy. In the process of continuous earthquake action on the structure, it is always necessary to ensure the safety and reliability of the column, so the thin-walled plate has become the key part of energy dissipation research.

For the simulation of a thin-walled plate, the solid model can accurately simulate the failure process, hysteretic characteristics, concrete spalling, and other failure characteristics of the plate, but the solid modeling process is complex, and the parameter calculation is not easy to master, so it is not suitable for practical engineering applications. Compared with the solid model, the bar model simplifies the modeling process, and the calculation principle is easy to master. Therefore, this chapter uses the mature bar model in the relevant shear wall structures for reference to conduct numerical analysis and Research on thin-walled plates.

For the seismic analysis of continuous steel structure bridges, the locations of pier bottom and pier top are potential plastic hinge areas, so the seismic response of column corrugated steel plate pier bottom and pier bottom is mainly analyzed. The nonlinear response internal force of the main pier under rare earthquakes is shown in Table 2 and Table 3.

Table 2. Internal force of each pier when the seismic wave is input along the bridge direction.

| Section position | | Column top | | Column bottom | |
|---|---|---|---|---|---|
| Internal force type | | Longitudinal bending moment (kN•m) | Longitudinal shear (kN) | Longitudinal bending moment (kN•m) | Longitudinal shear (kN) |
| Model operating conditions | Wave height 15cm | 55098 | 8178 | 47291 | 6017 |
| | Wave height 20cm | 44369 | 5862 | 46809 | 4749 |
| | Wave height 25cm | 60203 | 6333 | 58691 | 6081 |

Table 3. Internal force of each pier when the seismic wave is input across the bridge.

| Section position | | Column top | | Column bottom | |
|---|---|---|---|---|---|
| Internal force type | | Transverse bending moment (kN•m) | Transverse shear (kN) | Transverse bending moment (kN•m) | Transverse shear (kN) |
| Model operating conditions | Wave height 15cm | 66141 | 15800 | 53024 | 7349 |
| | Wave height 20cm | 64958 | 13718 | 67547 | 7481 |
| | Wave height 25cm | 59859 | 10289 | 53708 | 7085 |

It can be seen from Table 2 that the wave height of the corrugated steel plate is 20cm when the earthquake is input laterally. The wave height of the corrugated steel plate is 20cm, which is 15cm and 25cm higher than that of the corrugated steel plate. The column top bending moment is reduced by 24% and 36%, respectively, and the column bottom bending moment is reduced by 10% and 25%, respectively. The shear force at the top of the column is reduced by 39% and 8%, respectively, and the shear force at the bottom is reduced by 26% and 28%, respectively. Therefore, when the wave height is 20cm, the seismic response of the main pier is the smallest

It can be seen from Table 3 that when the earthquake is input longitudinally, the wave height of the corrugated steel plate is 25cm, which is the best. The wave height of the corrugated steel plate is 20cm, which is 15cm and 25cm higher than that of the corrugated steel plate. The column top bending moment decreases by 18% and increases by 36%, respectively, and the column bottom bending moment increases by 21% and 21%, respectively. The shear force at the top of the column decreases by 15% and increases by 8%, respectively, and the shear force at the bottom increases by 2% and 5%, respectively. Therefore, when the wave height is 25cm, the seismic response of the main pier is the smallest.

To sum up, for the high pier bridge with column corrugated steel plate combination, the transverse internal force response is greater than the longitudinal internal force response, so the transverse stress of the high pier is relatively unfavorable. When the wave height of the corrugated plate is taken as 25cm, the internal force response of the high pier is the smallest.

The waved steel plate is shifted at the top of the column at different wave heights, as shown in Table 4.

It can be seen from Table 4 that when the corrugated steel plate is taken as 15cm, the column top displacement is the smallest, but the influence of wave height on the longitudinal displacement of the column top is less than 5%, and there is no influence on the transverse displacement of the column top. Therefore, the wave height of the corrugated plate has little effect on the displacement of the high pier top.

It can be seen from Table 5 that the column corrugated steel plate composite high pier bridge has excellent seismic performance under rare earthquakes and keeps the elastic state without yielding.

Table 4. Maximum displacement of the pier (column) top.

| Displacement (mm) | Seismic wave input along the bridge | | | Seismic wave transverse bridge input | | |
| --- | --- | --- | --- | --- | --- | --- |
| | 1 | 2 | 3 | 1 | 2 | 3 |
| Wave height 15cm | 130 | 304 | 186 | 152 | 345 | 252 |
| | 133 | 316 | 184 | 153 | 358 | 246 |
| | 138 | 324 | 185 | 153 | 369 | 248 |

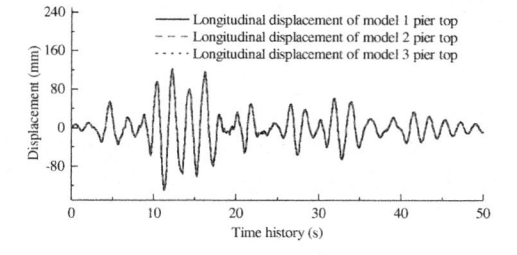

Figure 8. Longitudinal displacement of the column top.

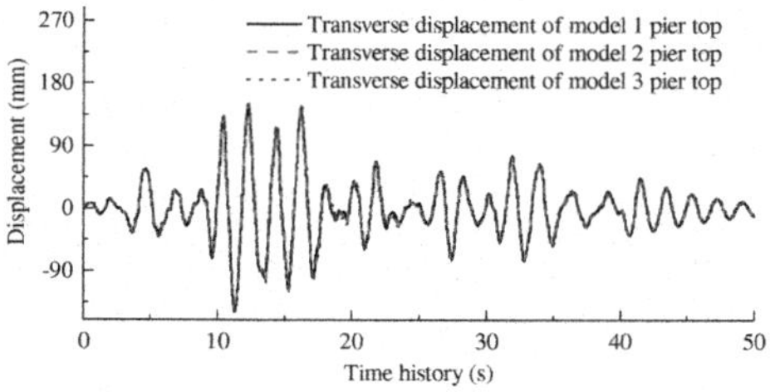

Figure 9. Transverse displacement of the column top.

Table 5. Checking calculation of columns under rare earthquake.

| Position | Along the bridge | | | Transverse bridge direction | | |
| --- | --- | --- | --- | --- | --- | --- |
| | Bending moment $M_{max}$ (kN•m) | $M_y$ (kN•m) | $M_{max} < M_y$ | Bending moment $M_{max}$ (kN•m) | $M_y$ (kN•m) | $M_{max} < M_y$ |
| Column top | 55098 | 71357 | YES | 66141 | 71357 | YES |
| Column bottom | 58690 | 292414 | YES | 67546 | 292414 | YES |

## 6 CONCLUSION

In this paper, a high pier bridge with column corrugated steel plate composite piers is taken as the research object, and the influence of corrugated plate parameters on the seismic performance of the high pier bridge with column corrugated steel plate composite piers is studied. The following conclusions are obtained:

(1) The high pier bridge with corrugated steel plate columns belongs to a long-period structure with large flexibility. The transverse stiffness of the structure is greater than the longitudinal stiffness. The wave height of the corrugated steel plate does not affect the dynamic characteristics of the structure.
(2) The optimum wave height of the corrugated plate is 25cm when the lateral earthquake is input. For longitudinal seismic input, the optimal wave height of the corrugated plate is 20cm.
(3) Under the rare earthquake, the lateral stress of the column corrugated steel plate composite high pier is unfavorable.
(4) The column corrugated steel plate composite high pier bridge has excellent seismic performance. It remains elastic without damage under rare earthquakes.

ACKNOWLEDGMENTS

This work was supported by Gansu Natural Science Foundation (Grant No. 20JR10RA237).

REFERENCES

Chen X. C. & Shang Y. Z. & Zhang Y. L. (2008). Seismic performance analysis of simply supported steel truss bridge with high piers and long span. *World Earthquake Engineering* (01): 6–11.

Guan Z. G. & Li X. B. & Li, J. Z. (2014). Experimental study on seismic behavior of reinforced concrete high pier structure *Earthquake Engineering and Engineering Vibration* 34 (S1) 663–668.

Hines, E. M. & Dazio, A., & Seible, F. (2006). Structural testing of new East Bay Skyway piers. *ACI Materials Journal*, 103(1), 103–112.

Li J. Z. & Song X. D. & Fan L. Z. (2005). Discussion on displacement ductility of high pier of bridge [J] *Earthquake Engineering and Engineering Vibration* (01), 43–48.

Li Z. Q. (2019). *Research on seismic performance test and numerical simulation method of new type column slab high pier of railway* Lanzhou Jiaotong University.

Liu, L & Li Z. Y. & Yuan G. G. (2021). Study on seismic measures of high pier and long-span continuous rigid frame *Highway* 66 (02), 174–177.

Shao C. J. & Qi, Q. M. & Wei W (2020). Study on shaking table model test of high hollow pier at round end of railway. *Journal of Civil Engineering*. 53 (02), 72–80.

Wang H. B. (2016). *Experimental study on seismic performance of high pier and long-span concrete continuous rigid frame bridge* Northeast Forestry University.

Wang, J. W. & Zhang W. G. & Li J. Z. (2015). Pseudo static test and numerical analysis of prestressed concrete hollow pier *Bridge Construction*, 45 (03), 63–69.

Wang Y & Wang T. Q. & Sun L.M. (2020). Experimental study on seismic behavior of rectangular hollow double column high pier with energy dissipation coupling beam *Engineering Mechanics*. 37 (07) 159–167.

Xia X. S. (2012). *Study on seismic design method of high railway pier* Lanzhou Jiaotong University.

Yan, W.M. & Luo Z. Y. & Xu W. B. (2020). Shaking table test of high pier continuous rigid frame bridge under near-fault impulse earthquake *Journal of Beijing University of technology*. 46 (08) 868–878.

Ye Y.L. (2020). Analysis of seismic characteristics of continuous rigid frame bridge with high and low piers in mountainous areas *Engineering Construction* 52 (07) 35–38+53.

Zhang C & Shen Y.L. (2017). Study on applicable earthquake intensity parameters for seismic performance evaluation of high piers. *Journal of Disaster Prevention and Mitigation Engineering* 37 (01) 9–16.

Zhang L (2012). *Research on seismic performance and thermal stress analysis of Column Slab hollow pier*. Lanzhou Jiaotong University.

Zong Z. H. & Chen S. H. & Xia Z. H. (2010). Study on two-way quasi-static test of reinforced concrete box type high pier. *Journal of Disaster Prevention and Reduction Engineering* 30 (04) 369–374.

*Civil Engineering and Urban Research – Mohamed & Hou (Eds)*
*© 2023 the Authors, ISBN 978-1-032-44487-1*

# Development and application of prefabricated building construction simulation training system

Jing Xiong*
*College of Science and Technology of Nanchang University, Nanchang, Jiangxi, China*

Kai Luo
*Jiangxi Port Group Co., Ltd, Nanchang, Jiangxi, China*

ABSTRACT: Prefabricated buildings refer to the production of prefabricated components, mainly in factories. A building is assembled, connected, and partly cast on site. Compared with traditional buildings, prefabricated buildings have the characteristics of saving energy and resources, and the construction process is more environmentally friendly, which further enhances the technological content of the construction industry. The prefabricated building is still in the initial stage of development. In order to make the prefabricated building continue to develop for a long time, the problem of construction safety management must be solved. Therefore, establishing a safety evaluation system for prefabricated buildings is crucial to safety management.

## 1 INTRODUCTION

A prefabricated building is a factory-produced prefabricated component hoisted, connected, and partially cast at the site. Compared with traditional buildings, prefabricated buildings have the characteristics of saving energy and resources, and the construction process is more environmentally friendly, further enhancing the construction industry's technological content. It fully conforms to the development strategy of green building design, construction and industrialization and can develop steadily for a long time. The industrialization of construction is also a global trend. To build a resource-saving and environment-friendly society, it is necessary to develop prefabricated buildings. The promotion and maturity of prefabricated buildings will greatly change building production. The construction speed of prefabricated buildings is fast, the construction period is shortened, the production efficiency is further improved, and the precision and quality of the components produced by the standardized operation of the factory are good. The advantage of prefabricated buildings is the reduction of manpower usage. Buildings 3 and 7 are prefabricated residences, the construction period is shortened by about 35%, the loss of concrete, steel, and construction water is reduced by 60%, construction waste is reduced by 80%, and on-site technicians are reduced by 60%. Realizing the industrialization, diversification, and industrialization of construction is the development goal of the construction industry in recent decades.

## 2 SYSTEMS THEORY

The point of view of the system theory is to identify the state of the system when an accident occurs and to distinguish the primary and secondary causes of the accident deeply. The representative

---

*Corresponding Author: O123456KYG@163.com

theories are trajectory intersection theory, human error model and its subordinate expansion theory, P theory, energy release theory, accident-causing catastrophe theory, etc. Figure 1:

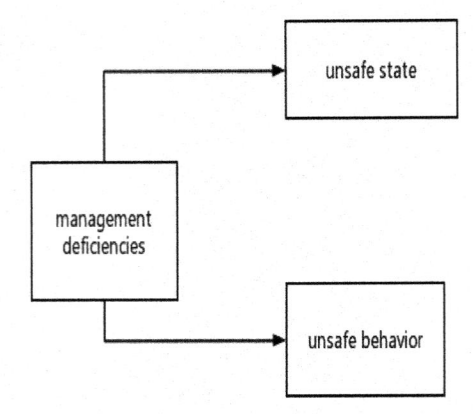

Figure 1.    Systems theory.

## 3   KNOWLEDGE EXPRESSION SYSTEM

In the system, x is the i-th prefabricated building construction project; ni, 2, 1, and n are the total number of target projects; rj is the j-th index; sj, 2, 1, and s are the number of indicators. The four-tuple fVRUS represents the above table, where U is the universe of discourse (object set), n xxxU 21 is the non-empty finite set of objects, and the prefabricated building construction safety evaluation system is the set of target items; R is the attribute set, s rrrR 21 represents the set of all index sets, attributes and indexes match each other one-to-one, and an equivalence relation is generated accordingly. The indexes and attributes in the evaluation system in this paper can be the same concept. Rr r VVVVr is the value range of attribute r; f represents VRU, as an information function set, it gives corresponding information value to each attribute, the expression is r VrxfUxRr upper approximation set and lower approximation set Let $R$ be an equivalence relation on $U$, for any $UX$, $XxUxXR$ R: or $XYRUYXR$ :/ is called the lower approximation set of $X$ with respect to; $XxUxXR$ R: or $XYRUYXR$ :/ is called the upper approximation set of X concerning R. Binary $XRXR$, forming a rough set. $XRXPOSR$ is the R · positive region of X; XRUXNEG R' is the R negative region of X; XRXRXBNR is the $R$· boundary region of $X$.

## 4   COMPREHENSIVE RISK ASSESSMENT

In the comprehensive risk assessment, the percentile numbers [0 40), [40 60), [60 80), [80 90), and [90 100) are five-level fuzzy membership intervals, and the corresponding evaluation levels are "Class I, Class II," III, IV, V. In order to simplify the calculation process, in the risk assessment of the prefabricated building construction stage, we take the median value of 20, 50, 70, 85, 95 for quantitative calculation. The evaluation vector calculation formula is as follows: $Yi$, $QiRi$, $i$, 1, 2, and $n$. A matrix represents the evaluation vectors generated by experts after evaluating the secondary indicators, and the primary indicator evaluation matrix is represented by the weight vector obtained in combination with the previous section. According to the formula:

$Y2$ $Q2$ $R2$ 0.215, 0.177, 0.214, 0.203, 0.191
0.336,0.219,0.219,0.142,0.100?
0.139,0.141,0.148,0.134

3 3 3 0.146,0.146,
*YQR*, 0.469, 0.229, 0.173, 0.072, 0.044.
*Yi · QiRi, ·i*1, 2, ·, *n.*
*R* 1 *R* 2 *R* 3 *R* 4 *R* 5 *R*

## 5 INTUITIVE FUZZY SETS AND BASIC PRINCIPLES

Based on fuzzy theory, intuitionistic fuzzy set theory has been further extended. In fuzzy set theory, the negative and positive are described, and the intuitionistic fuzzy set also includes the evaluator's hesitant state, which makes the decision-maker thinking more compatible. Therefore, intuitionistic fuzzy set theory has strong practicability and flexibility when describing uncertain and fuzzy information. To understand the theory of fuzzy sets, introduce the basic knowledge of intuitionistic fuzzy sets, algorithms, scoring functions, and weighted average operators. As shown below:

$$\mu_A : X \to [0,1] , \quad x \in X \to \mu_A(x) \in [0,1]$$

$$\nu_A : X \to [0,1] , \quad x \in X \to \nu_A(x) \in [0,1]$$

$$0 \le \mu_A(x) + \nu_A \le 1 , \quad x \in X$$

Figure 2. Intuitive fuzzy sets and basic principles.

## 6 INTUITIVE FUZZY HIERARCHICAL EVALUATION MODEL

When analyzing multi-objective problems, the quantification of each index is the difficulty of research. It must be carefully discussed whether it is decision-making or a risk analysis problem. In reality, when too much emphasis is placed on the process of quantification of decision-making or evaluation, many difficulties will be encountered in operation. At this time, the expert's score can be highlighted. Before the construction of prefabricated buildings, it was difficult to quantify the risk assessment factors in all aspects. However, by inviting relevant experts to investigate and analyze various risk factors, in addition to intuitive and simple scoring of the importance of each factor, combining scientific evaluation methods to calculate weights, and combining empirical and normative research, it is a more systematic, efficient, scientific and objective realization of risk assessment in the construction stage of prefabricated buildings.

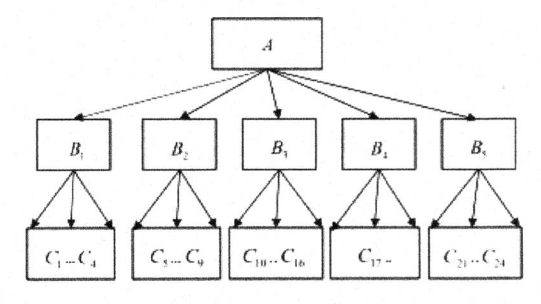

Figure 3. Hierarchical structure diagram.

# 7 CONCLUSION

The research of this paper adopts the intuitionistic fuzzy analytic hierarchy process to study the risk assessment and control of the prefabricated building construction stage. The research direction and focus of this paper were determined by reading a large number of domestic and foreign literature on the risk of prefabricated buildings in the construction stage and the intuition-fuzzy analytic hierarchy process. The relevant theoretical basis supporting the research of this paper is introduced in detail. We also identify the risks in the construction stage of prefabricated buildings with the help of fault tree analysis and study the causes of the risks in the construction stage of prefabricated buildings from five aspects: human risk, material risk, management risk, technical risk, and environmental risk. The risk assessment index system of the prefabricated building construction stage is established. This paper introduces the intuitionistic fuzzy analytic hierarchy process and adopts the method for the risk assessment model of the component prefabricated building construction stage.

## REFERENCES

Alberto Pavesc, DionysiosA. Bourns. Experimental assessment of the seismic performance of prefabricated concrete structural wall system[J]. *Engineering Structures*. 1998(41):2049–2062.

Cheng Kai. *Research on application condition evaluation of residential PCa technology based on fuzzy analytic hierarchy process* [D]. Hangzhou: Zhejiang University, 2012.

Faber M H. *Technology risk and safety in civil engineering*[R]. Switzerland: Swiss Federal Institute, 2001.

He Lingtong, Chen Yan. The present and future of construction industrialization [J]. *Engineering Quality*, 2013(02): 1–8.

He Zhengkai. *Development overview and development prospect of prefabricated buildings* [A]. April 2014 Academic Exchange of Construction Technology and Management.

Hyunsoo L, Hyunsoo K, Moonseo P, et al. Construction risk assessment using site influence factors [J]. *Journal of computing in civil engineering*, 2012(26): 319–330.

J James, LH Ikuma, I Nahmens, F Aghazadeh (2014). The impact of Kaizen on safety in modular home manufacturing[J]. *The International Journal of Advanced Manufacturing Technology* C70(1) F725–734.

LH Xian, Al Hussein Mohamed, Lei Zhen, Ajweh Ziad(2013): Risk identification and assessment of modular construction utilizing fuzzy analytic hierarchy process (AHP) and simulation [J] *Canadian Journal of Civil Engineering* C40(12): 1184–1195.

Li Yanyan. Research on the development status and countermeasures of housing industrialization [J]. *Architectural Engineering Technology and Construction*, 2013, 05(03):180–181.

Meng Yinzhong. *Construction engineering safety inspection, evaluation, and prevention* [D]. Beijing: Beijing University of Technology, 2010.

National Bureau of Statistics. *China Statistical Yearbook-2014*[R]. Beijing: Statistics Press, 2014.

Osama A J. Risks associated with trenching works in Saudi Arabia[J]: *Building and Environment*, 2008( 43): 776–781.

Proceedings of the conference [C]. 2014.

Wang Jiandong. *Research on Industrialized Housing Technology System-Based on "Vanke" Prefabricated Housing Design* [D]. Shanghai: Tongji University, 2009.

Zhao Jinhuai. *Research on sustainable development of the housing industry based on the circular economy* [D]. Xi'an: Xi'an University of Architecture and Technology, 2006.

Zhou Wenjun. *Research on the Influencing Factors of Housing Industrialization in the Promotion and Application of Public Rental Housing-Taking the Public Rental Housing in the Main Urban Area of Chongqing as an Example* [D]. Chongqing: Chongqing Jiaotong University, 2013.

*Civil Engineering and Urban Research – Mohamed & Hou (Eds)*
© 2023 the Authors, ISBN 978-1-032-44487-1

# Experimental study on the dynamic characteristics of pile foundation model for saturated soil and swelling soil

Jiaxing Li*, Qiang Li*, Xinyi Li* & Qian Wang*
*Department of Civil Engineering, Zhejiang Ocean University, Zhejiang, China*

ABSTRACT: Using the pile foundation vibration system, an indoor model test was carried out to investigate the difference between the dynamic response of pile foundation in expansive and saturated soil. The test shows that the resonant frequency of pile foundation in expansive soil is 54% higher and the amplitude is 59% lower than that in saturated soil. The test also shows that the acceleration increases with the increase of vibration frequency, and the acceleration decreases gradually when the resonant frequency is reached. The pile axial force decreases with the increase of pile depth, and the pile axial force in the swelling soil is less than that in the saturated soil. Besides, the decrease is larger when the depth is shallow, and the effect of swelling soil on axial force decreases with the increase of soil depth.

## 1 GENERAL INTRODUCTIONS

Swelling soil contains a large number of strong hydrophilic clay minerals, which can absorb a lot of water, and it is highly plastic, experiences violent expansion and contraction, and has other engineering characteristics. With water absorption, its swelling bearing capacity attenuation is obvious. And it also experiences water loss and dry shrinkage, among other characteristics. Due to this, the nature of the road and bridge built on it is extremely unstable, and the buildings situated nearby possess certain hidden dangers. Under the vibration load, it is easy to cause destabilization and damage to the soil foundation and even cause engineering accidents and other hazards.

Most of the research on swelling soil is still focused on the static aspects, and there is less research on the dynamic properties of swelling soil. Lei (2004) analyzed the static and dynamic properties of swelling soil and improved soil by comparing static and dynamic triaxial tests. Mao (2005) studied the permanent deformation and dynamic strength characteristics of swelling soil and modified swelling soil under dynamic loading under different circumferential pressure. Fattah (2020) studied that the adhesion factor decreases with the increase in the ultimate uplift stress (skin resistance) while it increases with the increase in the initial degree of saturation; Liu (2021) suggested that pile settlement increases due to water infiltration. Such a phenomenon can be attributed to pile shaft friction reduction in the active zone associated with the suction reduction due to water infiltration and volume expansion of the soil. Guillaume (2012) pointed out that lime modification can effectively prevent the swelling and shrinking of compacted soils by multi-dimensional analysis of lime-modified expansive soils. The author measured the dynamic response of pile foundation, pile shaft force, and pile bottom earth pressure in swelling soil and saturated soil, respectively. It is done to compare and analyze the dynamic characteristics of swelling soil and saturated soil under dynamic load and provide a reference basis for related projects.

---

*Corresponding Authors: 1010710485@qq.com, qiangli1972@163.com, lixinyi9510@163.com and wq2547557136@163.com

## 2 TEST EQUIPMENT

### 2.1 *Test setup*

The test device includes three parts: a vibration loading system, a signal acquisition system, and a model box system.

The vibration loading system consists of an electrodynamic exciter, DF1405 digital synthesis function signal generator, HEAS-50 power amplifier, and counterforce frame; the signal acquisition part mainly consists of the acceleration sensor, dynamic signal test analyzer, and server. The dynamic signal test analyzer is mainly used for the change signal of acceleration of the pile during the high-frequency vibration of the accelerometer.

The dimensions of the model box system were designed as a 1.5 m high and 1.4 m diameter drum, using filter material to wrap the inside and outside of the box, with a 10 cm aperture drainage hole reserved at the bottom of the box. The test setup is shown in Figure 1.

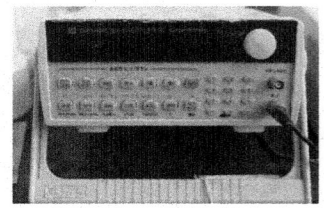 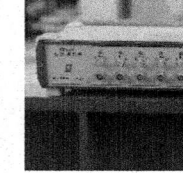 

(a) Digital signal generator      (b) Constant current adapter      (c) Model box

Figure 1. Test device system.

### 2.2 *Model piles*

The model pile material used in this test is a PVC pipe with a elastic modulus of 3000 MPa, a model pile length of 900 mm, and a pile diameter of 40 mm. The bottom of the pile is fixed with sealant on the bottom cover.

### 2.3 *Preparation and layering of expansive soils*

The model test uses sea sand, bentonite, and gypsum from the sea around Zhoushan. The model test uses the sand, bentonite, and gypsum mixture with a ratio of 7:10:3 as the similar model material for this test.

### 2.4 *Measurement program*

#### 2.4.1 *Pile end earth pressure measurement*
The test pressure sensor is buried at the bottom of the pile end, waterproofed with k-706 waterproof adhesive, covered with fine sand, and compacted.

#### 2.4.2 *Dynamic response of the pile*
The steady-state sinusoidal excitation force is applied to the top of the PVC pipe pile (top of the bearing), and the acceleration is measured by the accelerometer, which is fixed at the lower part of the bearing. The tests are conducted in the form of frequency sweep (150–400Hz and 380–450Hz sweep) for the monopile with bearing in expansive soil and saturated soil, respectively, to study the dynamic response law of PVC pipe pile when vibrating in sandy soil.

#### 2.4.3 *Pile shaft force measurement*
The model pile used for the test was made of a 40mm diameter PVC pipe, and the strain gauges were pasted along the pile body of the model pile. The strain gauges were 150mm from the end of

the pile and 150mm from the top of the pile, and the distance between the strain gauges varied for different pile lengths. The specific pasting positions are shown in Figure 2.

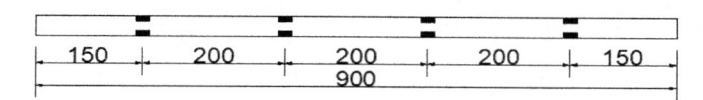

Figure 2. Model pile.

## 3 EXPERIMENTAL PROCESS

### 3.1 *Positioning of the pile*

An infrared laser is used to position the pile on the surface of the bearing layer, and the verticality of the pile is controlled using a crane.

### 3.2 *Embedding of the pile*

The fine sand transported by crane to the inside of the model box should first fall to the edge of the model box to avoid the fine sand falling around the pile and causing the pile itself to shift in position. Then, the fine sand is manually shoveled to form a cone around the bottom of the pile to better and faster add support to the pile itself. Every 25cm of sand fill is tamped, taking care not to touch the pile itself during tamping, and the pile bottom pressure sensor and the pile body strain gauge wires are connected to the signal receiver, respectively. The model box fill is roughly divided into three parts, with gravel of different particle sizes at the bottom as the back filter layer, fine sand of 80 cm thickness in the middle as the holding layer, and swelling soil of 30 cm at the top as the fill layer. First of all, a 15cm thickness of gravel with a 2–3cm diameter was laid at the bottom, then a 15cm thickness of gravel with 0.5-1cm diameter was laid as the back filter layer, and geotextile was laid at 30cm from the bottom of the model box to prevent the fine sand from blocking and speed up the drainage. Then, fill the middle bearing layer in layers, and tamp the soil layer with a rammer at each filling thickness of 25 cm. When the soil layer is filled to the corresponding thickness of the bearing layer, the surface of the soil layer is checked with a level to ensure its level.

### 3.3 *Loading of moving loads*

The motorized shaker is fixed vertically to the reaction frame, and the excitation load is in the form of sine wave loading. The load form is $Q(t) = Q + A sin(\omega t)$.

Where $Q(t)$ is the load acting on the top of the raft (kN), $Q$ is the constant load acting on the top of the raft (kN), $A$ is the amplitude of the dynamic load, $\omega$ is the angular velocity (rad/s), where $\omega = 2\pi f$, $f$ is the excitation frequency (Hz), and $t$ is the loading time (s).

The model test was conducted in two test conditions, in saturated soil and expansive soil, and the acquisition frequency of each condition was 3000 times per second, and the excitation frequencies were scanned from 150-400 Hz and 380-450 Hz, respectively.

## 4 ANALYSIS OF EXPERIMENTAL RESULTS

### 4.1 *Pile dynamic response*

Under sinusoidal cyclic load, the acceleration response time domain curve of pile foundation in expansive soil also shows obvious periodic acceleration and unloading process with the cyclic load addition and unloading process. With the increase of the excitation frequency, the acceleration is small to large and then gradually becomes smaller after reaching the resonance peak, showing a "spindle shape," and the time domain variation of the pile in the range of 300–450Hz is given in Figure 3.

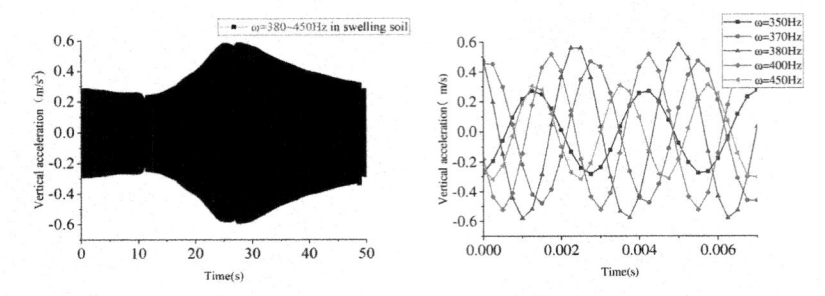

Figure 3.    Sweep and point frequency-time domain of pile foundations in Swelling Soil.

To observe the vibration of each frequency more specifically, the time domain response of the excitation frequency of 350Hz, 370Hz, 380Hz, 390Hz, 400Hz, and 450Hz is given in Figure 4, respectively, from which it can be seen that the vertical acceleration from 350Hz to 380Hz shows the law of gradual increase, and the vertical acceleration from 380Hz to 450Hz shows the law of gradual decrease.

In order to obtain the resonant frequency of the pile in the swelling soil, the time domain response was transformed into the frequency domain response by Fourier variation. As shown in Figure 5, it can be seen that the amplitude of the monopile in the swelling soil reaches a maximum value of 0.00597m at a frequency of 385.21 Hz. The time domain response of the monopile in the saturated soil and the amplitude-frequency curve. It can be concluded that the vertical amplitude reaches 0.00951m at a frequency of 250 Hz. The resonant frequency of monopile in saturated soil is lower, and the amplitude is increased by 59%, which is due to the increase in strength and volume expansion of the swelling soil compared with the saturated soil.

### 4.2    *Pile bottom soil pressure*

The time domain response curves of the soil pressure at the bottom of the pile in expanding and saturated soils in the frequency range of 150–400Hz are given in Figure 6. The pile bottom pressure will show a sinusoidal periodic pattern with the increase of external load, indicating that the excitation load causes the soil pressure at the bottom of the pile. No change in the values from the pile bottom earth pressure in expansive and saturated soils indicates that the soil medium does not change the pile bottom earth pressure.

### 4.3    *Pile shaft force*

The pile axial force distribution in the form of pile top load varies with the sinusoidal cycle and decreases with increasing depth in Figure 7. The axial force of a pile in swelling soil is less than that of a pile in saturated soil, the reduction down is larger when the depth is shallow, and the effect of swelling soil on axial force decreases with increasing depth.

## 5    CONCLUSION

In this paper, the dynamic response of pile foundations in saturated and expansive soils under sinusoidal excitation load and the soil pressure at the bottom of the pile is obtained, and the test results show the following points.

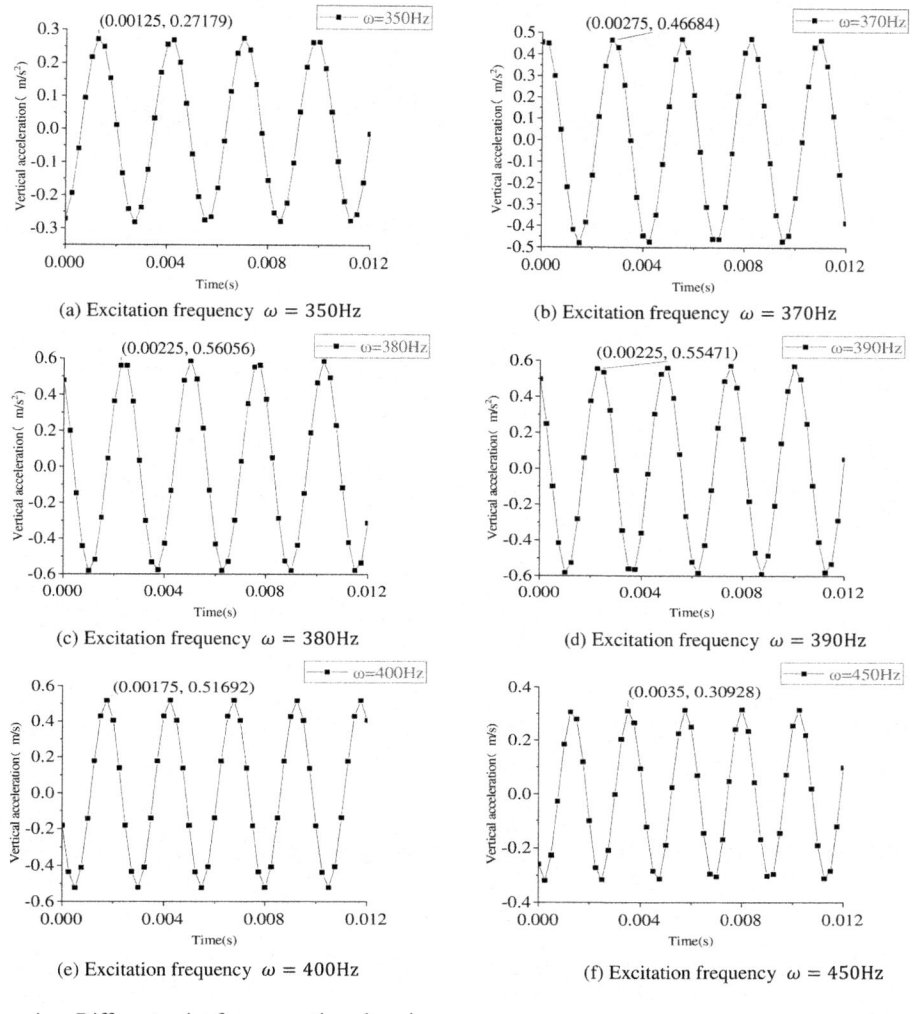

Figure 4. Different point frequency-time domains.

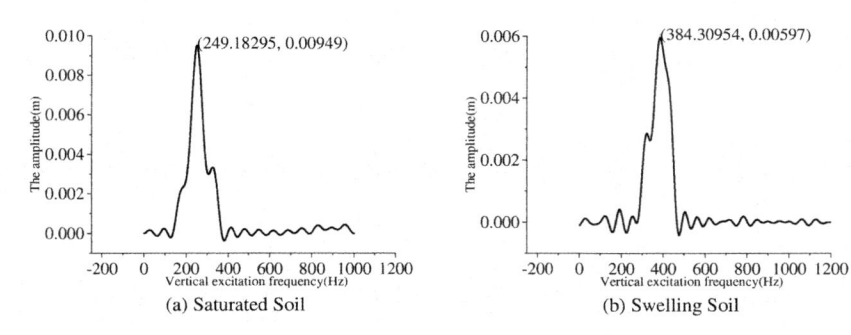

Figure 5. Amplitude-frequency curve.

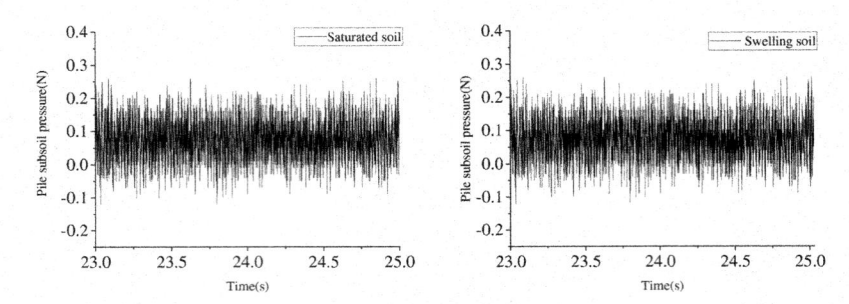

Figure 6. Variations of soil pressure under the pile tips.

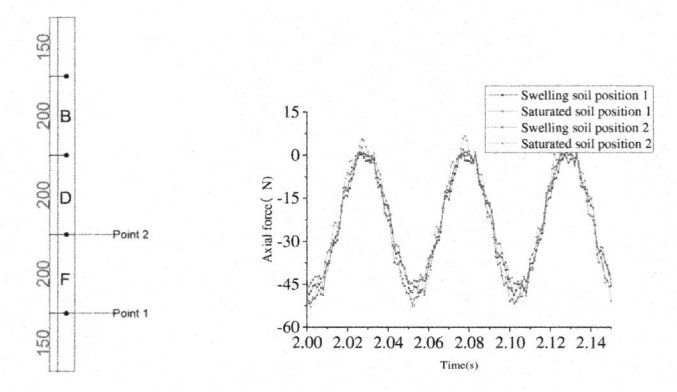

Figure 7. Change of axial force of pile.

(1) Regarding the resonant frequency, the frequency sweep of the monopile can better reflect the monopile vibration law. The resonant frequency of the expanded soil is significantly higher than that of the saturated soil, and the amplitude decreases by about 59%, increasing first with the excitation frequency and then decreasing rapidly when the resonant frequency is reached.

(2) The pile bottom pressure sensor data indicate that pile foundations have little effect in expansive soils compared to saturated soils.

(3) The axial force of piles decreases with increasing depth, and the axial force of piles in expansive soils is less than that of piles in saturated soils.

REFERENCES

Fattah M. Y. & R. R. Al-Omari(2020). Load sharing and behavior of single pile embedded in unsaturated swelling soil. *J. Eur. J. Environ. Civ. Eng.* 24(12), 1–26.

Guillaume S. & C. Olivier(2012). Multi-scale analysis of the swelling and shrinkage of a lime–treated expansive clayey soil. *J. Appl Clay Sci.* 61(61), 44–51.

Lei, S.Y. & H.Q. Hui (2004). Hdynamostatic analysis of properties of expansive soil and improved soil. *Rock Mech Rock Eng.* 17, 3003–3008.

Liu, Y. & S. K. Vanapalli (2021). Mechanical behavior of a floating model pile in expansive unsaturated soil associated with water infiltration: laboratory investigations and numerical simulations. *J. Soils Found.* 61(4), 929–943.

Mao, C. & Y.J. Qiu (2005). Experimental study on the dynamic properties of expansive soils and modified expansive soils. *J. Rock Mech Rock Eng.* 10, 1783–1788.

Civil Engineering and Urban Research – Mohamed & Hou (Eds)
© 2023 the Authors, ISBN 978-1-032-44487-1

# The application of high-density electrical method in karst roadbed investigation

Xiaolu Yan*, Yang Zhang* & Lu Yang*
*Geological Exploration Academy of China Metallurgical Geology Breau1'Baoding City, Hebei Province, China*

Xiaolei Sun*
*Aerospace Kaitian Environmental Technology Co.2, Ltd, Dongcheng District, Beijing, China*

ABSTRACT: In the carbonate development area, the landscape combination of topography, geomorphology, and hydrogeology formed by long-term dissolution and transformation of carbonate stratum by surface water and groundwater is generally called karst (also called karst). Karst development is easy to cause ground collapse, which is the main hidden danger of roads, bridges, and other engineering facilities, and seriously affects the safety of the project (Li 2019; Zhang 2019). In order to investigate the development and distribution of underground karst, the high-density electric method is applied for investigation, and reasonable working device and parameters are selected, good geological results can be obtained for exploring Quaternary soil cave, limestone karst, and fracture development (Ge 1999). The electrical characteristics of the abnormal body are not only related to the material composition of the medium itself, but also closely related to its structure, porosity, geometry, and natural water. Reflected in the resistivity parameter map, it is more likely that the abnormal body is at the lower part of the geophysical anomaly.

When the underground karst is strongly developed and the connectivity is good, a large area of low resistivity anomaly will be reflected on the electrical data, especially since the verification rate of vertical belt development anomaly is very high. That is to say, the abnormal reliability of this form is high. It is worth noting that not all the dissolution phenomena show low resistance reflection because limestone belongs to high resistance lithology, so the cavity with high resistance reflection is not easy to be found under DC survey conditions (Dong 2015).

## 1 INTRODUCTION

High-density electrical method is a new electrical exploration technology with dual electrical sounding and electrical profile characteristics. It is suitable for detecting karst, fault, fissure, cave, roadbed states, and occurrence of geological boundaries (Deng 2009). This method can be used as a coordination method in the preliminary investigation, detailed investigation, and construction of subgrade, bridge, and tunnel. The results can delineate the areas where underground karst and soil caves may develop (Li 2019; Zhang 2019). Through drilling verification, the accuracy rate of the results is about 70%, and the effect is obvious. However, some problems have been found in data understanding and interpretation, such as the scale of dissolution development, depth of interpretation, etc. These problems will be discussed below.

## 2 PRINCIPLES OF METHOD

The high-density electrical method is a new geophysical prospecting method developed based on the conventional direct current method, and its working principle is consistent with that of the

---

*Corresponding Authors: 303805547@qq.Com, 1595912377@qq.Com and 64963183@qq.Com

conventional direct current method (Ge 1999). It is based on the conductivity difference of the soil and rock media through the observation and study of the artificial underground stable current field distribution law to solve geological problems. Its working principle is to observe the potential difference between the measuring electrodes to calculate the apparent resistivity and then calculate, process and analyze the measured apparent resistivity value to obtain the distribution of the rock resistivity in the crust (He 2019), so as to investigate and delineate the karst. Compared with the conventional direct current method, the high-density current method has the characteristics of intuition, high efficiency, high resolution, and high accuracy.

## 3 BASIC PHYSICAL GEOGRAPHY AND GEOLOGY OF THE WORK AREA

The working area is located in Wuping County in southwest Fujian Province. Along the line, the mountains stretch, and the terrain changes greatly. Most of them are hills and low mountains. The highest elevation in the survey area is about 560 meters., and the maximum height difference is about 300 meters.

According to the existing open geological data, the work area is located in the Mingxi-Wuping depression belt and Hufang-Yongding uplift belt of the southwest depression belt, with complex regional structural conditions. NNE, NE, and NW structures are the most developed, followed by nearly EW and SN structures. The fault zone in the work area is generally characterized by a compressional fracture zone, altered fracture zone, schistosity zone, and fractured breccia zone.

The exposed strata mainly include Quaternary slope residual sandy clay, slope residual clay ($Q^{el-dl}$), and Holocene alluvial-proluvial soil ($Q^{al-pl}$). The pre-Quaternary strata are completely exposed and are distributed from Sinian (Z) to Baibanian (K) strata. The main stratum of this geophysical prospecting section is the Permian (P) stratum, and the lithology of bedrock is limestone, carbonaceous siltstone, gossamer, mudstone, etc. The intrusive rocks are mainly Yanshanian and Indosinian granite and granite porphyry ($\gamma_5^{1-2}$).

## 4 ANALYSES OF PHYSICAL PROPERTIES

According to the hydrogeological situation and geophysical data of the area, the electrical properties of each geological body are analyzed as follows (the overall resistivity value in the working area varies from tens to 10000 $\Omega \cdot m$):

(1) The resistivity values of quaternary alluvial diluvial subclay, sand, gravel, slope diluvial subclay, gravel-bearing subclay, and residual cohesive soil on the surface are tens to 5000 $\Omega \cdot m$, which varies greatly. Due to water content and good connectivity, the gully area shows low resistivity, generally in tens to 500 $\Omega \cdot m$. Due to poor water content, the structure is loose on the slope and shows high resistance, generally greater than 500 $\Omega \cdot m$.

(2) Limestone: in a limestone area, due to the influence of water content and filling materials, the resistivity is generally less than 300 $\Omega \cdot m$ or less than two times the background value, showing low or relatively low resistivity and different forms. Regions with strong dissolution development often form caverns and gaps of different sizes, which are characterized by large areas of low resistivity anomalies or obvious relatively low resistivity anomalies, generally less than 200$\Omega \cdot m$. The intact limestone shows high resistivity, with resistivity generally greater than 1000$\Omega \cdot m$.

(3) Structural fracture zone, intrusive contact zone, and lithologic contact zone: due to the large difference between rock mass fracture and rock physical properties, the dissolution zone is easy to form, generally showing the abnormal display of low and relatively low resistance extending to the deep. Due to contact metasomatism and differential weathering, the intrusive sites often show low resistance.

(4) The soft layer and soil hole in the overburden layer: due to the strong water content and good conductivity of the soft layer and the filled soil hole medium, there is a large electrical difference from the surrounding medium, showing a low resistivity anomaly, but the resistivity properties of the two are similar, it is not easy to distinguish; Empty soil holes will form high resistance characteristics, but due to the small size of empty soil holes and the poor sensitivity of the high-density electrical method to reflect high resistance body compared with low resistance body, the high resistance anomalies formed are not obvious and mixed with high resistance background, so it is very difficult to identify them.

(5) In basins or intermountain gullies, the overlying alluvial diluvial layer will reflect low resistivity due to water content. The filling karst caves developed near the bedrock interface will show a concave shape under the resistivity equivalence line, which can be used to identify the dissolution and occurrence of elephants exist. Still, the line is not easy to draw.

## 5 INTERPRETATION AND INFERENCE OF APPLICATION EXAMPLES AND WORK RESULTS

According to the morphology of resistivity anomaly and the characteristics of karst development, several examples are listed to illustrate the significance of this work to know.

Example 1

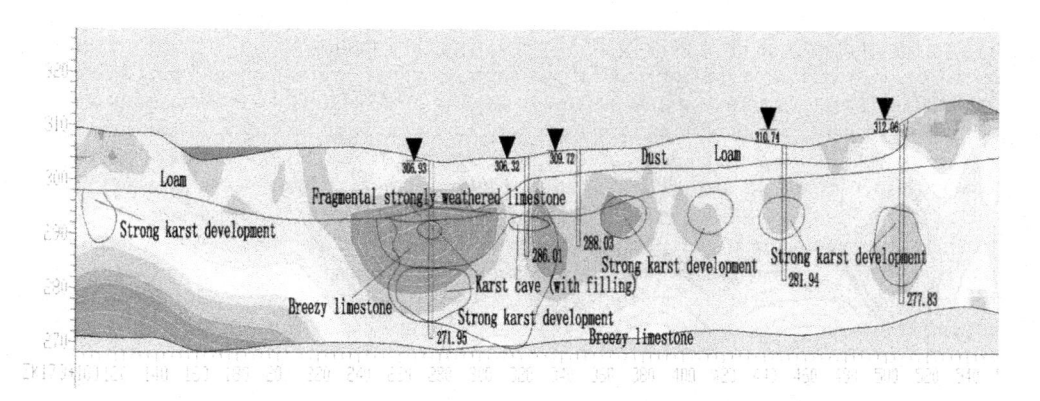

Figure 1.   K173+100-K173+540 high-density electrical profile interpretation map.

Figure 1 shows a resistivity and geological section at 8 m left of K173+100-K173+540. By the figure visible near the K173 + 270, abnormal low resistance isoline resistivity is the concave shape. After it has been verified by drilling, the elevation of 260-276 meters of drilling hole in the soil, consistent with resistivity overall form, but the position is located in the low resistance lower part of the core, in the work area several similar morphological abnormalities, abnormal body in low resistance lower part of the core, to analyze its reasons may be: the exception When the upper part of the body is covered with subclay and residual cohesive soil, the electrical method is not sensitive to the low resistance reflection of the lower part due to the distribution of water, and the volume effect is obvious. Most bodies are concave under the contour line, while the actual buried depth of abnormal bodies is larger.

Example 2

It can be seen from Figure 2 that a low resistivity anomaly occurs near K179 + 990-K180 + 010, mainly composed of loam clay and a large buried depth of bedrock. The low resistance anomaly near K180 + 110-K180 + 150 extends to the deep in a strip shape. Through drilling verification, both boreholes have encountered karst caves. The karst development in this part is strong, and the rate of karst caves drilled in this area is very high for the anomaly with a similar shape. It can be seen that this type of anomaly shows a strong development of dissolution. It can be used as a distinguishing mark in the work area.

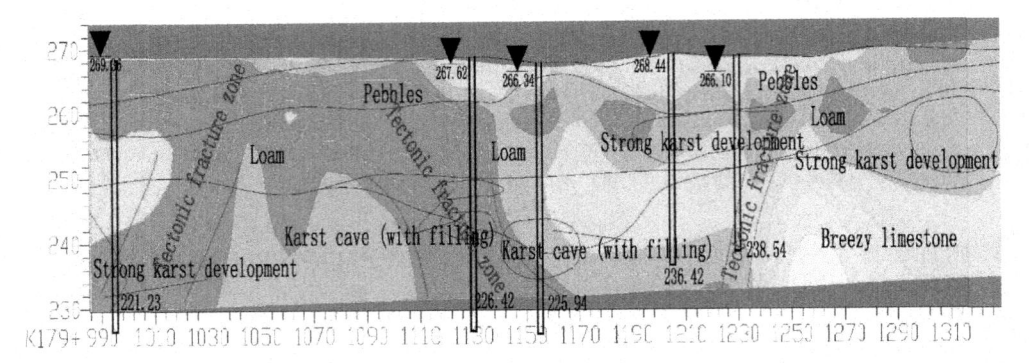

Figure 2.　K179+990-K180+310 high-density electrical profile interpretation map.

Example 3

As seen from Figure 3, in the K193+400-K193+460 range, the shallow layer is subclay and residual cohesive soil, which is in the uplift and shows high resistance, while the underlying bedrock is fragmentary granite porphyry and shows low resistance. K193+440 is the intrusion contact zone and influence zone, which is prone to dissolution due to the large lithology difference between the two sides. According to the drilling verification, there are several contact parts of granite and limestone intrusion in this area, and the granite body parts all show low resistance, which may be related to differential weathering. In the K193+590-K193+630 distance, the surface layer is low resistivity, and the resistivity isoline in the deep is concave. The soil hole is drilled at 268-274 meters elevation, consistent with the overall resistivity shape. Still, it is located at the lower part of the low resistivity core, which is inferred to be a karst development area. After drilling verification, this part is dissolved. The development was strong, which was consistent with the analysis results. In elevation 274-282 meters of drilling hole in the soil, consistent with resistivity overall form. The reasons may be analyzed as follows: when the distribution of water forms a low resistance reflection, the electrical method is not sensitive to the low resistance reflection at its lower part, the volume effect is obvious, and most of them are concave in contour. But the actual buried depth of the anomaly body is larger. There are many similar morphological anomalies in the work area, most of which are located in the lower part of the low-resistance core.

Example 4

It can be seen from Figure 4 that there is a large section of low resistance anomaly in mileage section K176 + 860-K176 + 960, and it is not closed downward. Through drilling verification, many karst caves have been drilled in this part, indicating that large areas of low resistivity anomalies are the reflection of strong development of dissolution. Based on this, it can be inferred that the corrosion development is strong in K176 + 780-K176 + 840 mileage section, and the karst densely developed area shows a large area and low resistivity anomaly under the investigation of the high-density electrical method.

Given the geological problems to be solved in this work area, that is, karst development and distribution characteristics in the limestone area, we have made a comprehensive and systematic analysis of geophysical data in this area. Through the comprehensive comparison of the electrical

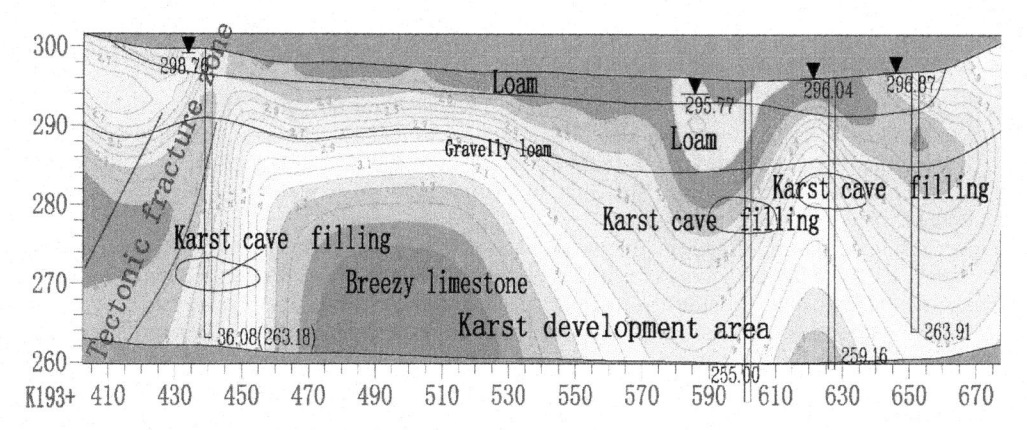

Figure 3.    K193+410-K193+670 high-density electrical profile interpretation map.

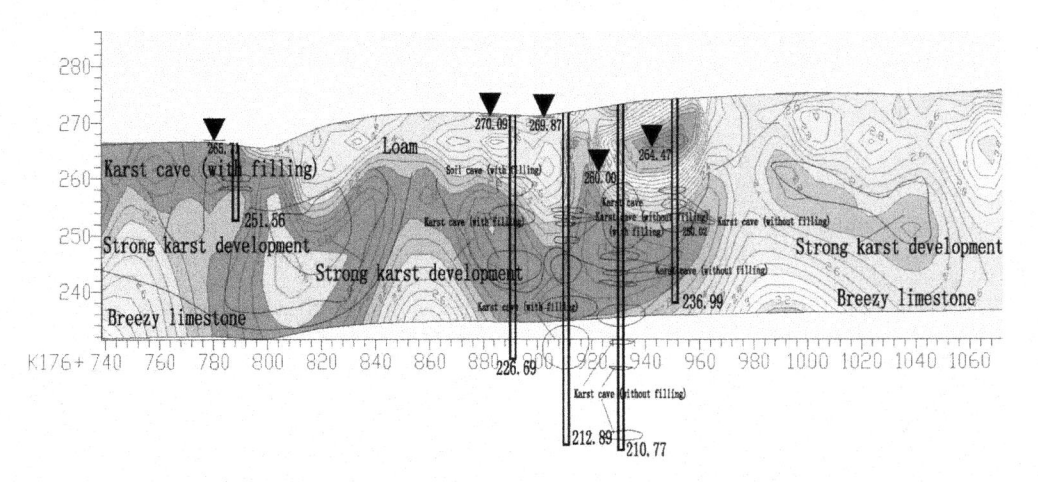

Figure 4.    K176+740-K175+060 high-density electrical profile interpretation map.

characteristics and abnormal forms of geophysical prospecting anomalies, the ratio of the vertical development depth to the horizontal development width of the anomalies (we call it the vertical and horizontal development ratio of the anomalies here temporarily); the vertical and horizontal development ratio is greater than 1, that is to say, the horizontal development width of the anomaly is greater than its vertical development depth; in this area, we think that this kind of anomaly is caused by the strong development of limestone karst at the corresponding depth. The second is vertical development, and the ratio of vertical to horizontal development is less than 1. That is to say, the abnormal horizontal development width is less than its vertical development depth. In this area, we think this kind of anomaly is caused by the strong vertical karst development caused by the corresponding depth of limestone structural fractures.

## 6   CONCLUSIONS

Through the objective analysis of the working area's topographic, geological, and physical conditions, the application of a high-density electrical method to solve the problem of roadbed karst can play a better role by selecting reasonable working devices and working parameters. Through this work, we get the following points of understanding:

1. The electrical characteristics of an underground medium are not only related to the material components of the medium, but also closely related to its structure, voidness, geometry, and water in nature. The position of the anomaly body and geophysical data anomaly is not necessarily coincident, and it is more likely that the anomaly body is under the geophysical anomaly in the resistivity parameter diagram.
2. According to the law found in this work, the verification rate of vertical belt development anomaly is very high. That is to say, the abnormal reliability of this form is relatively high; the anomaly is evident in the intrusive contact zone, and the drilling situation further proves that the dissolution phenomenon is more likely to occur in the contact zone of different lithology; affected by the phreatic water surface, the dissolution near the top interface of bedrock in some parts forms low resistance reflection together with the overlying rock and soil layer, which makes it very difficult to determine the true bedrock surface.
3. When the underground karst develops strongly, and the connectivity is good, a large area of low resistivity anomaly will be formed on the electrical data, so we should pay attention to the data understanding. Not all the dissolution phenomena show low resistance reflection because limestone belongs to high resistance lithology. So, the cavity with high resistance reflection is not easy to be found under DC survey conditions (Dong 2015).

## 7 ADVICE

1. In engineering investigation, any geophysical method is restricted by many factors, so does the high-density direct current method. Due to the influence of power supply current, cable length, and measurement sensitivity, the actual arrangement length, and survey depth are limited. When the exploration depth increases, the resolution decreases, and if the resolution is increased, the exploration depth is affected: the interpretation depth cannot accurately reflect the actual depth. Therefore, appropriate observation devices, power supply current, electrode distance, and grounding conditions should be selected according to the requirements of exploration targets and tasks.
2. In some areas with complex terrain and object conditions, multiple parallel survey lines and directions should be selected as far as possible, and comprehensive geophysical exploration should be carried out. At the same time, the depth should be corrected according to the borehole data. Therefore, in practical production work, various geophysical exploration methods should be used for exploration and comprehensive interpretation to learn from each other's strengths.

At the same time, we should combine geological data and drilling data to improve the authenticity and accuracy of data interpretation.

## REFERENCES

Den Ke, Jiang Chang-sen. *Calculation techniques for geophysical and geochemical exploration*, 2009, 31 (6):577~581.
Dong Maogan, Wu Shanshan, Li Jiawang, Application of high-density electrical method in karst development characteristics investigation [J]. *Chinese Journal of Engineering Geophysics*, 2015, 162(2): 194~199.
Ge Rubing, Huang Weiyi, Zhang Yuming, Application of high-density electrical method in limestone area [J]. *Geophysical and Geochemical Exploration*, 1999, 23[01]:28–37.
Guangzhou: Guangdong Science and Technology Press, 2019. 8.
He Qingli, Application of comprehensive geophysical prospecting method and technology in the treatment of highway subgrade settlement [J]. *Journal of Engineering Geophysics* (in Chinese), 2019.16(20): 230–236.
Zhang Xiujie, Li Hongzhong. *Key techniques and comprehensive investigation methods of highway engineering geological survey in scattered carbonate distribution area*.

*Civil Engineering and Urban Research – Mohamed & Hou (Eds)*
*© 2023 the Authors, ISBN 978-1-032-44487-1*

# Bearing performance analysis of CFG piles in Tidal Zone

Zhenxiang Shi, Wei Zhang & Whenzhen Li
*China Power Construction Municipal Construction Group Co., Ltd, Tianjin, China*

Haotian Luo* & Ke Wu
*School of Civil Engineering, Shandong University, Jinan, China*

ABSTRACT: Aiming to analyze the maintenance of cement fly-ash gravel (CFG) piles under the periodic tidal water level change after construction, the finite element simulation of soil seepage consolidation is used to study the changes and mechanisms of various parameters of CFG piles after construction and maintenance. The research results show that the pore pressure of the CFG pile after the immersed tube is poured under the condition of the water level difference will appear as the "slope umbrella surface" of the CFG pile, which is like the "umbrella" supporting the pore pressure boundary, and the lower part of the pile body is more obvious. The tidal change of the CFG pile in the shallower burial depth has a faster recovery rate of the effective stress in the early stage of maintenance, and the pressure difference becomes larger. Under periodic tides, the seepage direction in the soil changes from the outer sea to the inner sea to the two-way drainage of the CFG pile to the inner sea and the outer sea, which effectively improves the drainage effect in the middle of the CFG pile.

## 1 GENERAL INSTRUCTIONS

Cement fly-ash gravel (CFG) pile is made of gravel, stone chips, sand, and fly ash mixed with cement and water. It has many advantages, such as high bearing capacity, small settlement deformation, and low engineering cost (Poulos 1980; Zhao 2007). The principle is to use professional drilling rig equipment to fuse concrete with the original soil in the depths of the foundation, artificially make the top of the pile penetrate the cushion, and adjust the relative displacement between the pile and soil. Pile and soil can jointly bear vertical and lateral loads, and their sharing ratio can be adjusted, which plays an important role in reducing the stress concentration of the base (Filatov 1991; Lee 1968; Tong 2005). However, it is inevitable to encounter marine silty clay strata in civil engineering construction in coastal cities. The silty stratum not only has weak bearing performance but also easily flows with the seepage between soil bodies (Li 2020), which makes it difficult to effectively control the quality of CFG piles. The construction quality could not reach the expected target (Chen 2019; Wang 2020; Feng 2020). When CFG piles are used to deal with the inner beach's soft soil under tidal hydrodynamic pressure, the surrounding soft soil will be squeezed, disturbed, and destroyed. At the same time, the pore water pressure in the soft soil of the inner beach is dynamically changed due to the tidal action. It reduces the integrity and bearing capacity of CFG piles, increases the difficulty of construction of CFG piles, invisibly expands construction costs, and delays construction progress; if the roadbed is filled, CFG piles may be sheared, inclined, and bent. The damage phenomenon and the effect of dominoes cause the subgrade to quickly produce external extrusion, subsidence, and cracking and deformation damage, resulting in the foundation failing to meet the requirements of use (Liu 2019). Drilling will cause disturbance in

---

*Corresponding Author: luohaotian@mail.sdu.edu.cn

the surrounding soil, resulting in a new excess pore water pressure in the surrounding soil (Burland 1973). It prevents the design and construction personnel from starting with the layout, material selection, and construction method of CFG piles (Zhang 2009).

To explore the applicability of CFG piles in tidal areas, this paper takes the Fujian Funding Municipal Road Project as an example. Laboratory tests and numerical analysis methods were used to study the changes and mechanisms of various parameters of CFG piles after construction and during maintenance to comprehensively evaluate the periodicity. The pile-soil interaction under changing water level differences provides theoretical research results.

## 2 ANALYSIS METHOD

### 2.1 Numerical model parameters

Based on the relevant construction data and records of the first bid section of the second phase of Binhai Avenue in Fuding City, the maintenance of CFG piles under the periodic change of water level difference is analyzed, and the effective stress recovery problem is carried out in terms of time and cost. The construction section connects the two places across the sea, and the ocean is divided into the inner sea and the outer sea. The tide is a regular semi-diurnal tide, and the ebb tide speed is higher than the high tide. The highest tide level over the years is 4.25m, the lowest tide level over the years is -3.42m, the average high tide level over the years is 3.86m, the average low tide level over the years is -1.61m, the average tidal range over the years is 4.17m, and the average sea level is 0.53m. The ABAQUS finite element simulation of the simplified CFG pile and the surrounding soil is used to study the influence of the periodic change of the water level difference between the inner sea and the outer sea on the maintenance of the CFG pile under the action of the open sea tide. The foundation soil model, in which the foundation is 10m long, 10m wide, 15m deep, 0.6m in diameter of the CFG pile, and 10m deep. The CFG pile is divided into 1360 units, and the soil is divided into 292450 units. Vertical constraints are set at the bottom, and horizontal constraints are set on the sides to ensure the rationality of the simulation.

### 2.2 Model mechanical parameters

The basic mechanical parameters of the soil are obtained through the laboratory test of the field soil samples, and the modified Drucker-Prager cap model is used for simulation, where the elastic modulus is $E_s$=85MPa, the dry density is $\rho_d$ =1800kg/m$^3$, the viscosity Cohesion is $d$=60kPa, yield surface inclination is $\beta$=30° , cap eccentricity is 0.1, transition surface radius is 0.03, flow stress ratio $k$=1, and permeability coefficient is $k$=5.7×10$^{-10}$m/s, initial void ratio $e_0$=1.35, using the C3D20R Pore Fluid/Stress element type. The pile body adopts the elastic model, the elastic modulus E_"p" $E_p$=15GPa, the density $\rho$=2500kg/m$^3$, the C3D20R three-dimensional stress element type, and the friction coefficient between the pile and soil $\mu$=0.7.

During the construction of the CFG pile, the surrounding soil will be squeezed, and even lead to structural damage to the soil, and excess pore water pressure will be generated around the soil. The influence on the soil during the process and the maintenance of CFG piles with tidal changes (Wang 2004).

$$\Delta u \doteq \frac{1}{3lna}\left\{\left[2\left(2c_u - \tan^2(45° + \frac{\varphi_r}{2})r^{'}r_0tan\varphi_r\right)ln\frac{R_p}{r_0} + \frac{c_rz}{r_0}\right] + 2.7\partial_f c_r\right\}ln\frac{a}{\rho} \qquad (1)$$

$a$ is the multiple of the influence radius; $c_u$ is the undrained shear strength of the soil. The pore pressure changes outside nine meters are small, so the friction angle and cohesion are damaged. That is, the residual internal friction angle $\varphi_r$ and residual cohesion $c_r$, introducing softening coefficient $\gamma$, where $\varphi_r = (1 - \gamma)\varphi$, $c_r = (1 - \gamma)cr^{'}$ is the effective weight of soil; $r_0$ is the pile radius; $R_p = r_0\sqrt{E/2(1 + \mu)c_u}$ is the radius of the plastic zone; $E$ is the elastic modulus of the soil; $\mu$ is the Poisson's ratio of the soil; $z$ is the burial depth; $\partial_f = 0.707(3A_f - 1)$ is the Henkel pore water pressure parameter; $A_f$ is the Skempton pore water pressure parameter, $A_f = 0.85$; $\rho = \frac{r}{r_0}$ and $r$ is the calculation point. The distance to the center of the pile. The conversion relationship between the Mohr-Coulomb model and the Drucker-Prager model is:

$$\tan\beta = \frac{6\sin\varphi}{3 - \sin\varphi} \tag{2}$$

$$k = \frac{3 - \sin\varphi}{3 + \sin\varphi} \tag{3}$$

$$\sigma_c^0 = 2c\frac{\cos\varphi}{3 - \sin\varphi} \tag{4}$$

$\beta$ is the inclination angle of the yield surface, $\varphi$ is the friction angle, $k$ is the ratio of the triaxial tensile strength to the triaxial compressive strength, $\sigma_c^0$ is the initial uniaxial compressive strength, and $c$ is the cohesion of the Mohr-Coulomb model.

## 3 ANALYSIS OF RESULTS

### 3.1 Analysis of the influence of CFG pile construction on the foundation

When the CFG pile is immersed, the surrounding soil will be vibrated and compacted, and a short-term excess pore water pressure will appear, which is combined with the pore pressure of the soil at the tidal water level to form a new type of pore pressure distribution, as shown in Figure 1. The CFG pile is like the "umbrella" supporting the "inclined umbrella surface" of the pore pressure boundary. The pore pressure around the pile body increases greatly. The farther away from the CFG pile, the lower the increase in soil pore pressure. Under the influence of the water level difference, the distribution is low in the inner sea and higher in the outer sea. The increase of pore pressure at the top of the pile is smaller than that at the bottom, which indicates that the CFG pile has a larger effect on the bottom soil when the immersed pipe is poured, while the upper part has little effect, showing a larger local maximum value at the bottom of the CFG pile. With the increase of burial depth, the increase in pore pressure gradually slows, and the change of pore pressure caused by CFG pile construction at 15 m can be ignored.

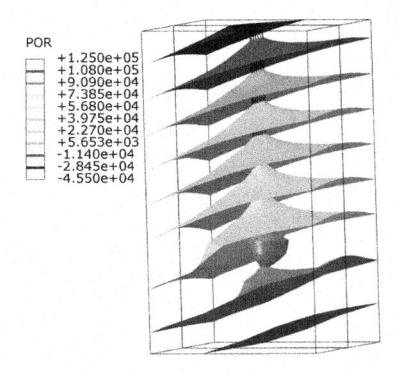

Figure 1.    Distribution of pore pressure after construction is completed.

### 3.2 Analysis of CFG pile maintenance

Through the "soil" analysis step in the ABAQUS finite element software, the maintenance situation of the CFG pile after seven days of construction is simulated, and the effective stress of the soil under the condition of seepage and drainage is analyzed by statistics of the soil pressure around the pile and the soil pressure inside the pile at the same level. Periodic tides cause the water level difference from the outer sea to be higher than the inner sea. The earth pressure around the CFG piles gradually recovers with time at a buried depth of 3.3 meters. The effective stress recovery speed is faster in the early stage of maintenance and the recovery speed after reaching the 7th day. It shows that the impact of CFG pile construction on the surrounding soil is recoverable, and the recovery effect is excellent in a short period. Under the tidal action, the pressure difference around the pile increased from 4.8% to 10.9%, and the CFG pile was subjected to uneven stress at the later stage, which greatly impacted the maintenance effect. At the buried depths of 6.6m and 10m, the internal soil pressure of the CFG pile is much greater than the lateral pressure of the surrounding soil. In the case of 7-day maintenance, the lateral pressure after the tidal water level at the buried depth of 6.6m is restored, only reaching 50% of the soil pressure inside the pile. 51.8%, indicating that the correlation between the change of the water level in the deeper burial depth and the maintenance effect of the CFG pile is reduced. The change in the water level mainly affects the upper part of the pile body, while the lower part mainly depends on the excess design amount of the cast-in-place pile body (1.2~1.5 times) to ensure the stability of the soil pressure at the bottom of the pile.

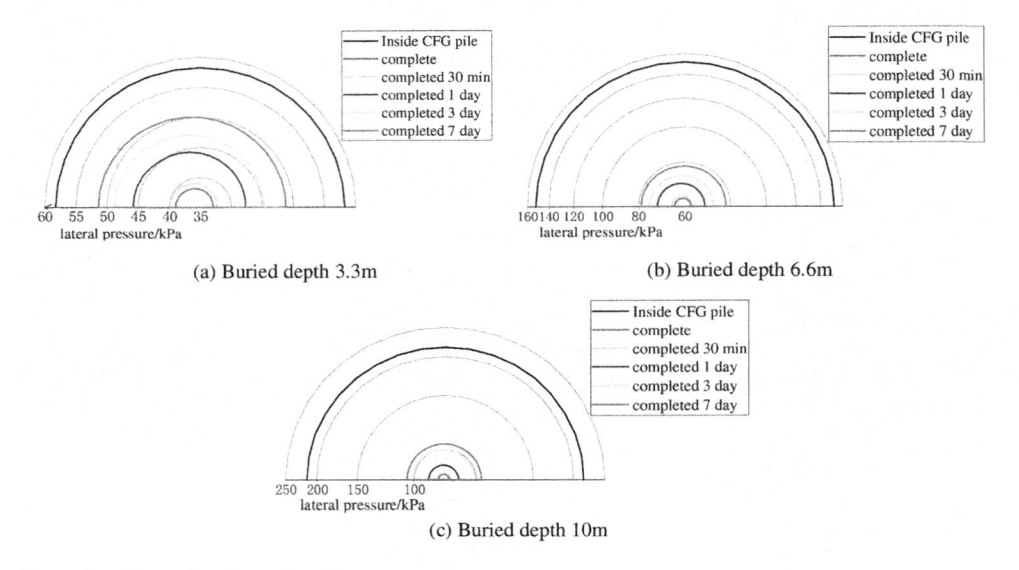

(a) Buried depth 3.3m          (b) Buried depth 6.6m

(c) Buried depth 10m

Figure 2.    Lateral pressure ring diagram.

### 3.3  *Analysis of CFG pile maintenance*

As shown in Figure 3, a burial depth of 5 m is selected to analyze the drainage effect of CFG piles under the tidal water level. The tidal water level fluctuates due to the periodic changes of the outer sea, and even the inner sea level is higher than the outer sea level. The foundation soil is subjected to cyclically changing water pressure, and the change of the drainage path directly affects the place far from the CFG pile, the pore pressure at the farthest dissipates rapidly, and the pore pressure changes significantly at the lowest cyclical water level. At the level of 2 m, the pore pressure changes to a smooth sliding, indicating that the change of pore water pressure has a hysteresis effect on the difference between high and low water levels. Under the influence of the change of pore pressure at the far end of the pile body, the direction of soil seepage is transferred from the outer sea to the inner sea to the two-way drainage of the CFG pile, thereby enhancing the drainage effect around the pile.

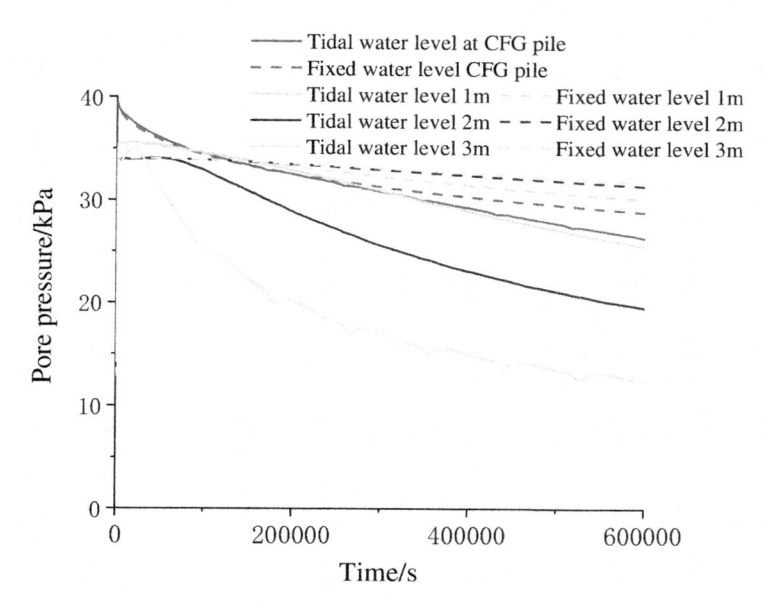

Figure 3. Variation of pore pressure at 5 m buried depth.

## 4 CONCLUSION

Aiming to analyze the maintenance of CFG piles under the periodic tidal water level change, the soil seepage consolidation is simulated by ABAQUS, and the changes and mechanisms of various parameters of CFG piles after construction and during maintenance are studied. The main research results are as follows :

1. Under the condition of water level difference, the pore pressure of the CFG pile will appear after the immersed pipe is poured. The CFG pile is like the "umbrella" supporting the "slope umbrella surface" of the pore pressure boundary, and the lower part of the pile body is more obvious. As the burial depth increases, the bottom of the pile body will gradually decrease from the area maximum.
2. The tidal change of the CFG pile in the shallower burial depth has a faster recovery rate of the effective stress in the early stage of maintenance, and the pressure difference becomes larger. The stress of the CFG pile in the later stage is uneven, which has a great impact on the maintenance effect of the pile. The lower part of the pile body mainly relies on the excess of the design amount of the cast-in-place pile body (1.2 to 1.5 times) to ensure the stability of the soil pressure at the lower part of the pile body.
3. Under the action of periodic tides, the direction of seepage in the soil changes from the outer sea to the inner sea to the two-way drainage of the CFG pile to the inner sea and the outer sea, which effectively improves the drainage effect in the middle of the CFG pile, and the short-term drainage effect is obvious.

## REFERENCES

Burland, J. B. Shaft friction of piles in clay – a simple fundamental approach. *Ground Engineering* 6: 30–42.
Chen, D. 2019. Common problems and optimization suggestions in the construction of CFG piles. *Doors & Windows* 16: 242–243.
Construction Technology 49: 71–75.

Feng, Z. & Zhang, Y.2020. Technology on quality defect treatment of CFG piles in a residential area. *Construction Technology* 49: 63–67.

Filatov A. V. 1991. Single-pile column foundations in industrial construction in Karaganda. *Soil Mechanics and Foundation Engineering* 28: 51–55.

Lee K. L. 1968. Buckling of partially embedded piles in sand. Journal of soil mechanics and foundation division, *ASCE*, 94:255–270.

Li, H. & Dong, Z. 2020. Key Technology of PTC Pile Net Composite Foundation Supporting Embankment in Deep Soft Silt Area.

Liu, Z.2019. Application research of CFG pile composite foundation in soft soil subgrade. *Journal of Highway and Transportation Research and Development* 11:84–87.

Poulos H. G.& Davis E. H. 1980. Pile foundation analysis and design. New York: John Wiley & Sons Inc.

Tong, J. & Hu, Z. (2005). Identification of the bearing capacity of CFG pile composite foundation. *China Civil Engineering Journal* 38: 87–91.

Wang, T. 2020. Construction method and matters needing attention of CFG piles for highway subgrade. *Intelligent City* 3: 179–180.

Wang, W. & Zai, J. 2004. 3D calculation of excess pore water pressure due to driving pile and its application. *Rock and Soil Mechanics* 25:774–777,779.

Zhang, L. & Li, Z. 2009. The application of dynamic-static drainage consolidation method on a silt ground disposal project in Guangzhou. *Rock and Soil Mechanics* 30:567–571.

Zhao, M. & Liu. Q. 2007. Consolidation analysis of composite ground with cement fly-ash gravel piles. *Journal of Hunan University (Natural Sciences)* 34: 1–5.

*Civil Engineering and Urban Research – Mohamed & Hou (Eds)*
© 2023 the Authors, ISBN 978-1-032-44487-1

# Study on mechanical properties of construction waste subgrade filler under long-term immersion condition

Siyu Cao, Jiaqi Li, Weihao Wang, Xinhang Ren & Congye Shen
*Chang'an-Dublin International College of Transportation, Chang'an University, Xi'an, China*

Xiaoyong Lai & Qian Wu*
*Highway School, Chang'an University, Xi'an, China*

ABSTRACT: To solve the settlement of construction waste subgrade filler in south rainy areas and provide a theoretical basis for recycling construction waste, compression tests of recycled construction waste were carried out after complete immersion treatment. The influence of adverse conditions on the long-term stability of construction waste subgrade filler was analyzed. The results show that the mechanical properties of construction waste subgrade fillers are greatly reduced after full immersion, the settlement is significantly increased, and the compression modulus is reduced compared with the reclaimed construction waste under normal conditions. The deformation of the specimen occurs at the moment of applying the load, and the cumulative deformation develops in a stepwise manner. Bricks and cement mortar in recycled materials crush more easily than stones, which reduces the overall mechanical properties of construction waste subgrade fillers. After 11 days of immersion, the sample can still meet the engineering requirements of the highway subgrade.

## 1 INTRODUCTION

With the continuous development of China's economy and the acceleration of infrastructure construction and urbanization, construction wastes generated by urban reconstruction and road reconstruction, such as discarded concrete, waste bricks, and dregs, are increasing yearly. According to statistics, the total amount of construction waste in China has reached 7 billion tons, and more than 400 million tons of new construction waste are produced yearly (Xiao et al. 2015). Open-air stacking and underground landfill, which serve as the main dealing methods of urban construction waste, not only occupy production land but also waste a lot of natural resources and aggravate ecological pollution (Fu et al. 2017). Therefore, effectively realizing the recycling and harmlessness of construction waste has become an urgent problem for government departments, experts, and scholars.

Construction waste is mainly composed of concrete, cement mortar, broken steel bar, dregs, and waste soil, and the proportion of each component varies with the development level of the region and the building structure. Highway construction has a great demand for natural sand and gravel materials. If the construction waste is recycled after classification, it can not only realize the reclamation treatment of construction waste but also meet the green and sustainable development of the strategy. To solve the contradiction between the sustainable development of the road engineering industry and the shortage of sand and gravel materials, scholars have conducted a lot of research. Based on the expressway project of Xi'an-Xianyang North Ring Road, Zhang et al. (2018) pointed out that the degree of compaction has the most significant impact on the compression performance

---

*Corresponding Author: 3027893185@qq.com

of recycled construction waste; when construction waste is used as subgrade filler, the content of brick slag should be strictly controlled not to exceed 50% to ensure the strength (Zhang et al. 2018). Although there are defects in the performance of construction waste soil, its strength still meets the needs of road construction projects; in the construction of high filling subgrade of urban roads, the mixing ratio of benign soil and construction waste shall be adjusted to ensure the overall strength of the subgrade (Song 2021).

Construction waste has the characteristics of low density and high water absorption. Impurities such as bricks, tiles, wood, and porous mortar in their composition are unfavorable to the mechanical properties of construction waste. The use of construction waste in roadbeds or road cushions with low index requirements can greatly improve economic and environmental benefits (Zou et al. 2017). Results of the performance tests of cement stabilized recycled aggregate, cement fly ash stabilized recycled aggregate, and lime fly ash stabilized recycled aggregate show that construction waste has good compressive strength, freezing stability, and scouring resistance, and this kind of recycled road material is conducive to the development of recycled green roads (Qian et al. 2021; Wu et al. 2016). The physical and mechanical tests show that construction waste has good strength, hardness, frost, and water resistance (Chen 2016; Han 2015; Li & Han 2014). Chang (2017) and Hong (2016) analyzed the settlement characteristics of construction waste using field monitoring data and numerical simulation methods and pointed out that the sum of error squares of the three-point Hoshino method and three-point modified hyperbolic method was the smallest, which means that its fitting effect was best (Chang 2017; Hong 2016). The compressibility and water stability of silt can be effectively improved by mixing construction waste, which is conducive to subgrade filling (Li 2005; Li & Wu 2015).

This paper collected the research object from the demolition site in Aodihu Village, Wenling City, Zhejiang Province. Firstly, the samples' basic physical and mechanical properties were determined, which was used as the basis for the study of the deterioration mechanism of the construction waste roadbed filler under the condition of complete immersion. To study the influence of complete immersion on the strength of construction waste subgrade filler, both microcosmic and macroscopic aspects are considered to discuss the influence mechanism of adverse conditions on construction waste subgrade filler.

## 2 TEST AND METHOD

### 2.1 *Test material*

The test materials were selected from the construction waste at the demolition site of Aodihu Village, Wenling City, Zhejiang Province, and most of the materials were brick-concrete structures. The construction waste is sorted out by manual work, crushed by the crusher after preliminary screening, and finally screened to prepare the roadbed filler according to the proportion of 11:4:5 ($0\sim10$ mm, $10\sim20$ mm, $20\sim40$ mm). The main physical parameters of the recycled material were as follows: the maximum dry density was 1.76 g/cm3, the optimum moisture content was 13.5%, and water absorption was 19.8%.

The construction waste comes from the demolished red brick house, mainly composed of concrete blocks, bricks, stones, cement mortar, and a small number of ceramic tiles and sawdust. Because the particle size of $0\sim10$ mm recycled construction waste is too small, it is difficult to analyze the material composition with naked eyes, so only the $10\sim20$ mm and $20\sim40$ mm filling materials are sorted manually. The obtained corresponding material composition is shown in Table 1.

According to the relevant provisions on the particle analysis test in the Standard for Soil Test Method (GB/T 50123-2019), the sieve analysis method is used to get related particle characteristics of the sample. The non-uniform coefficient Cu is 24.69, and the curvature coefficient Cc is 2.12. The grading is good, and the grading curve is shown in Figure 1.

Table 1. Proportion of each substance of construction waste subgrade filler in the initial state (%).

| Category / Grain file | | Coagulation Clod | Brick | Cement Mortar | Stone | Ceramic tiles | Wood chips | Other |
|---|---|---|---|---|---|---|---|---|
| Initial Status | 10~20mm | 31.22 | 26.38 | 15.53 | 24.33 | 2.09 | 0.15 | 0.3 |
| | 20~40mm | 36.12 | 17.23 | 13.75 | 30.26 | 2.12 | 0.17 | 0.35 |

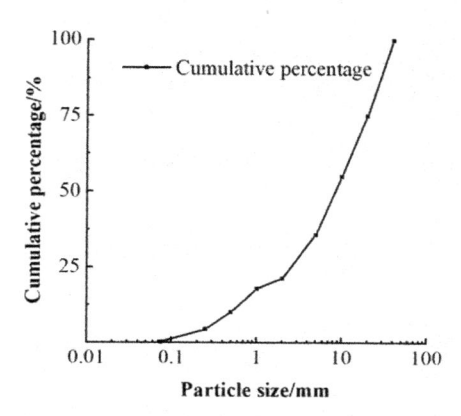

Figure 1. Grading curve of recycled construction waste in the initial state.

## 2.2 Test device

ZJ30-2 coarse-grained soil direct shear apparatus of Chengdu Donghua Zhuoyue Technology Co., Ltd. is used in the test, as shown in Figure 2. The size of the instrument is Φ300 mm × 300 mm, and the loading rate is 0.02~1 MPa/min. The instrument can be used to measure the compression characteristics of samples with a maximum particle size of 60 mm. The vertical and horizontal displacement division value of the measurement system is 0.01 mm. The maximum vertical and horizontal load that can be measured for horizontal thrust and axial load is 250 kN with an accuracy of 0.1 kN.

When the compression test is carried out, the sample is put into the shear box in three layers and tamped in each layer. The vertical displacement sensor is installed and pre-contacted. Then the vertical loading system is opened, and the load level is set. When the deformation of the sample is stable (the deformation is less than 0.01mm per hour), apply the next level of load. Repeat loading until the end of the test.

## 2.3 Test scheme and steps

To simulate the adverse humid environment, the construction waste was completely immersed in water. The compression curves of construction waste subgrade filler under different conditions were obtained to study the effect of complete immersion on the deterioration of samples. Two groups of compression tests were carried out: one group of test materials was under natural conditions, and one group of test materials was completely soaked in water for 11 days. Considering the influence of soil weight and additional stress, the load is increased step by step. The vertical load is 50 kPa, 100 kPa, 200 kPa, 300 kPa, 400 kPa, 500 kPa, 600 kPa, 700 kPa and 800 kPa in turn. When the deformation of the soil sample under each level of load does not change more than 0.01 mm per hour, it is deemed that the compression of the soil sample is stable, and the next level of load can be applied until the end of the preset loading process, and the deformation under each level of load is recorded.

Figure 2.    ZJ30-2 direct shear apparatus for coarse-grained soil.

## 3    TEST RESULTS AND ANALYSIS

### 3.1    *Change of settlement*

Figure 3 shows the compression deformation s-time t curve of samples under optimum moisture content state and immersion in water for 11 days under stepwise loading. It can be seen from Figure 3. that the cumulative deformation of the two samples increases with the gradual increase of the load (load levels are 50, 100, 200, 300, 400, 500, 600, 700, and 800kPa), and most of the deformation occurs at the moment of applying the load, forming a ladder-type, and the samples become denser. The amount of deformation produced by each level of load decreases progressively. When the load is less than 500 kPa, the deformation occurs at the moment of applying the load, and the cumulative deformation is between 0.5 and 3.5 mm. When the load is 500 kPa, the deformation is significantly reduced. This is because, under the action of the first five levels of load, the interstices of the particles are compressed, and the particles are compacted and filled with each other, occluded, dislocated, and broken to achieve a relatively dense state. When the load is greater than 500kPa, the equilibrium state between the particles is difficult to break. At this time, the particles are only fine-tuned based on the original flat state, showing a small macroscopic deformation, and the sample is pressed more densely.

By comparing the compression of the samples in the two states, it can be seen that the cumulative deformation of the samples after long-term immersion under each level of load is significantly greater than that of the samples without immersion under the same level of load. For this test result, the following explanation is concluded: long-term water immersion treatment will reduce the friction coefficient and particle strength between particles, making the particles more closely embedded. The particles are more likely to be destroyed and broken, resulting in an increase in cumulative deformation.

### 3.2    *Change in compression modulus*

Figure 4 shows the variation of the compressive modulus with load for the immersed and natural specimens. The analysis shows that under the same test conditions, the compression modulus of the sample increases with the growth of the vertical load and decreases significantly after soaking in water. The reason for this phenomenon may be that with the gradual increase of vertical load, the compressible pores in the soil sample are gradually reduced, the contact of the sample particles is more and more close, the sample is denser, and the compression becomes more difficult; therefore the compression modulus is gradually increased. In the process of water immersion loading, coarse particles have different degrees of volume deformation, among which bricks, cement mortar, and concrete blocks with 10~20 mm particles have been seriously damaged, and the component content

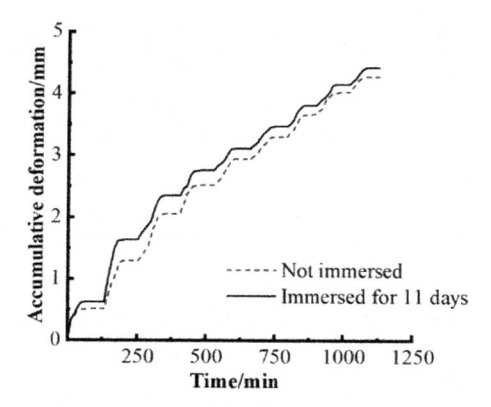

Figure 3. Settlement-time cumulative curve of graded loading of specimen.

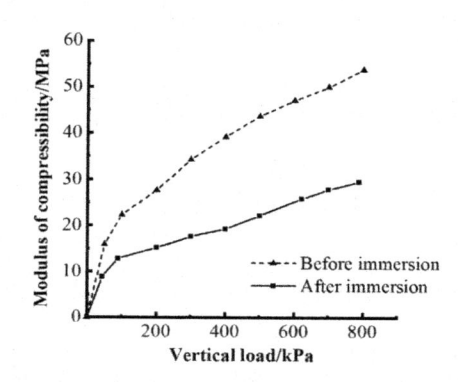

Figure 4. Es-P curves of specimens under various loads.

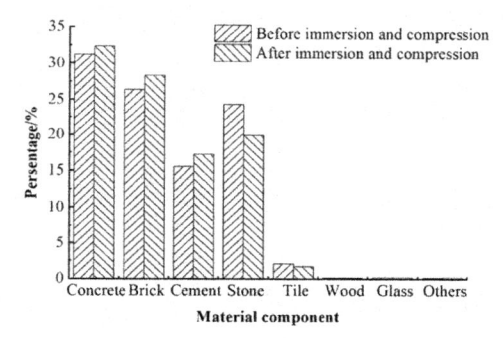

Figure 5. Composition changes of recycled materials of construction waste with grain size 10 20mm.

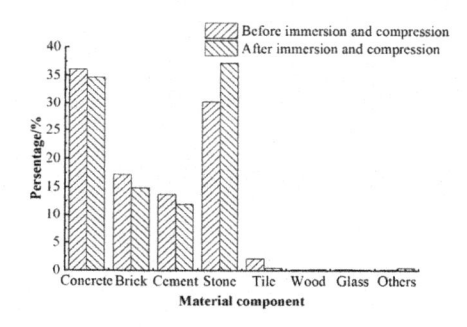

Figure 6. Composition changes of recycled materials of construction waste with grain size 20 40mm.

decreases by $2 \sim 3\%$. As a result, the skeleton structure of the sample is unstable, and the overall compression performance is greatly reduced.

### 3.3 *Composition and gradation evolution of filling materials before and after the test*

After the compression test, the recycled construction waste was sampled and dried for the screening test. The particle size content change and gradation change of $10\sim20$ mm and $20\sim40$ mm were analyzed. The change of material composition before and after the compression test for the immersion treated sample is shown in Figures 5 and 6.

It can be seen from Figure 5 and Figure 6 that among the bricks, cement mortar, and concrete blocks in the $20\sim40$ mm, blocks are broken after the immersion and compression. Among which bricks are broken most obviously, with a breakage rate of 1.96%, the damage of cement mortar is also obvious, with a breakage rate is 1.68%. Concrete took third place, with a breakage rate of 1.1%. Because of the high compressive strength of the stone, its damage is relatively small. Therefore, the relative percentage content of stone in $20\sim40$mm increases obviously. The relative percentage content of the concrete block, brick, and cement in the $10\sim20$ mm block increases slightly, which may be because the corresponding large-sized particles are broken into small-sized particles, resulting in the increase of the relative percentage content of the concrete block, brick and cement in the $10\sim20$ mm grain size group.

211

This phenomenon may be because the water absorption of bricks and cement mortar in recycled construction waste materials is large. The structure of bricks and cement mortar is softened, and the strength is reduced under long-term immersion conditions, the spatial position of different particles in the sample is rearranged under step-by-step loading conditions, and the more fragile particles (bricks, cement mortar) are damaged. The original position is transferred to the coarse particles with higher strength (stones, concrete blocks), and the small particles formed by the crushing of fragile particles are filled into the original voids.

It can be seen from Figure 7 and Table 2 that after the sample is immersed in water and compressed, the particles of the 20~40 mm group are significantly reduced compared with the initial state, while the percentage of particles of 0~10 mm is increased. Under the condition of immersion, the gradation of the sample changes little, the non-uniform coefficient is greater than 5, and the curvature coefficient is between 1 and 3, meaning the gradation after compression still meets the engineering needs. Although the large particles in the sample were significantly broken by water immersion, the construction waste subgrade filler could still maintain good stability and be used for urban road subgrade filling.

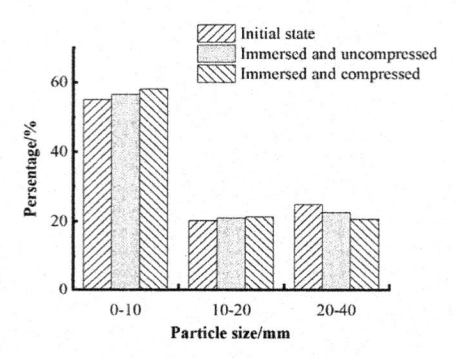

Figure 7.    Grading curve of specimen under different conditions.

Further analysis shows that when the construction waste recycled material sample is completely immersed in water, the soluble cement connecting the coarse material is easily dissolved, the cement's connection effect between particles is lost, and the external load-bearing capacity is lost is reduced. In the process of compression, the material with lower strength in the recycled material of construction waste is broken under the action of load, resulting in the relative increase of recycled material particles from 0 to 20 mm.

Table 2.    Non-uniformity coefficient and curvature coefficient of specimen.

|  | d10 | d30 | d60 | Cu | Cc |
|---|---|---|---|---|---|
| Initial state | 0.70 | 2.92 | 11.45 | 16.36 | 1.06 |
| Sample without immersion after compression | 0.59 | 2.7 | 10.87 | 18.42 | 1.14 |
| Sample being immersion after compression | 0.53 | 2.55 | 10.48 | 19.77 | 1.17 |

## 4   CONCLUSION

In this paper, the compression characteristics and deterioration mechanism of construction waste subgrade filler are analyzed by an indoor large-scale compression test, and the following conclusions are drawn:

1) In the compression test of construction waste subgrade filler, with the increase of load, the sample is compacted, the cumulative settlement increases, but the incremental settlement decreases

step by step. The deformation occurs at the moment of loading, and the cumulative deformation curve develops in a "step" manner. Compared with the samples without immersion, the settlement of the construction waste roadbed filler after long-term immersion is larger, and the compression modulus is significantly reduced. However, the gradation is still good after immersion compression, which meets the needs of highway subgrade engineering.

2) In the process of compression loading, the bricks and cement mortar are seriously damaged and broken into small particles, resulting in the increase of the relative content of particles in the range of $0 \sim 20$ mm. Immersion makes the coarse particle structure of construction waste roadbed filler soften, which is more likely to break under the same load, and the overall performance is reduced.

3) This paper provides some theoretical support for simulating the performance of construction waste roadbed filler under adverse conditions such as urban waterlogging, but it has limitations in reflecting more completed conditions like the dry-wet cycle. Further study can change the preprocessing procedures where construction waste is under the dry-wet cycle. Through these subsequent experiments, the impact of the dry-wet cycle on the mechanical properties of construction waste roadbed filler can be comprehensively considered to better reflect the reality.

## ACKNOWLEDGMENT

This paper is supported by the National Natural Science Foundation of China, No.5187081822; Research and Development Program of the Ministry of Housing and Urban-Rural Development, No.2020-K-078; Natural Science Basic Research Program of Shaanxi Province, No.2021JQ-244; Urban and Rural Planning Project of Zhejiang Provincial Department of Housing and Urban-Rural Development in 2020 (CTZB-2020050374 (3)); Study on the Long-term Properties of Construction Waste Subgrade Fillers (Projects in Key Supported Areas).

## REFERENCES

Chang Xiao. *Prediction and analysis of settlement law of roadbed filled with construction waste* [D]. Chang'an University, 2017.

Chen Dongmei. *Study on technical parameters and construction technology of construction waste recycled materials for treatment of wet soft loess foundation* [D]. Chang'an University, 2016.

Fu Yu, Liao Mingjuan, Chen Chuantao, Zhou Haijiang. Comparative analysis of construction waste treatment status at home and abroad and relevant suggestions for China [J]. *Resource Conservation and Environmental Protection*, 2017 (08): 79-82. DOI: 10. 16317/J. CN ki. 12-1377/x.2017.08.045.

Han Baogang. *Experimental study on mechanical properties and freeze-thaw microstructure of road construction waste under cyclic freeze-thaw dry and wet conditions* [D]. Chang'an University, 2015.

Hong Xuefeng. *Analysis of settlement characteristics of embankment filled with construction waste* [D]. Chang'an University, 2016.

Li Lihui, Wu Yongyan. Experimental study on modified silt from recycled construction waste [J]. *Highway*, 2015, 60 (08): 239–242.

Li Ning. Experimental study on modification of construction waste soil as roadbed filler. Shanxi Architecture, 2005 (05): 105–106.

Li Zhe, Han Baogang. Experimental analysis on compaction characteristics of road construction waste. *Transportation Energy Conservation and Environmental Protection*, 2014, 10 (06): 65–71.

Qian Biao, Fang Rui, Wang Biao, Ling Jiang, Wang Jianfeng, Zhang Bowen. Effect of different gradations on the optimum moisture content and maximum dry density of recycled aggregate [J]. *Jiangxi Building Materials*, 2021 (01): 26–28.

Song Gang. Discussion on resource reuse of construction waste as roadbed filler [J]. *Building Technology Development*, 2021, 48 (22): 107–108.

Standard for soil test method: GB/T 50123 — 2019 [S]. Beijing: China Planning Press, 2019.

Wu Yingbiao, Shi Jinjin, Liu Jinyan, Zhao Wen. Comprehensive application of construction waste in urban road engineering [J]. *Construction Technology*, 2016 (23): 33–36. DOI: 10. 16116/J. CNKI. Jskj. 2016.23.007.

Xiao Xuwen, Feng Dakuo, Tian Wei. Current situation and suggestions for construction waste recycling in China [J]. *Construction Technology*, 2015, 44 (10): 6–8.

Zhang Ximin, Zhang Mengke, Li Zhe. Large-scale consolidation test analysis of mixed construction waste recycled material [J]. *Transportation Energy Conservation and Environmental Protection*, 2018, 14 (05): 81–84.

Zou Guilian, Zhou Haohao, Peng Chaojie, Xu Luqiao. Investigation on production and technical index of recycled coarse aggregate from construction waste in Guangdong Province [J]. *Road Construction Machinery and Construction Mechanization*, 2017, 34 (01): 64–68+72.

*Civil Engineering and Urban Research – Mohamed & Hou (Eds)*
*© 2023 the Authors, ISBN 978-1-032-44487-1*

# Research on the light prestressed hollow flue slab structure of super-large diameter highway shield tunnel

Jian Li*, Lei Zhang*, Yulin Yin*, Bowen Tao* & Zhanhu Yao*
*CCCC Tunnel Engineering Company Limited, Beijing, China*

Yunli Li*
*Wuhan Institute of Technology, Wuhan, China*

ABSTRACT: A lightweight flue plate is studied based on Nanjing Heyan Road cross-river tunnel project, aiming at a series of problems caused by the large size and heavy structure of ordinary concrete flue plates. Based on the mechanical properties of new structures and new materials, the spatial layout, ultimate bearing capacity, and deflection of light flue slab are studied using mechanical theory. The results show that the optimized design adopts a SP18A8710 type flue slab, whose span is reduced to 8.7 m, and the weight per meter is only 2.71 t, which is 43.7% of the weight of the original ordinary concrete flue plate. The cost is reduced to 43.7 % of the original, and carbon emissions to 33 % of the original. Moreover, ten blocks can be stacked simultaneously, which greatly reduces the required area of the yard and does not require special hoisting equipment for transportation. The ordinary forklift in the tunnel can meet the installation requirements. The research results will provide a new application direction for the design of energy-saving and rapid construction of the internal structure of highway shield tunnels.

## 1 INSTRUCTIONS

With the rapid development of urbanization in my country, the river-crossing and sea-crossing tunnels have also been developed accordingly. In recent years, the ultra-large-diameter underwater and municipal tunnels constructed in China have almost encountered the problems of composite strata, fault fracture zone, shallow soil, high water pressure, and long-distance construction. In particular, the construction of ultra-large-diameter shield speed adjustment under such complex conditions poses many challenges to tunnel builders (Li 2021). By the end of 2021, there will be 63 tunnel projects with a diameter of more than 14 m, including 43 projects in China, accounting for about 68%. Moreover, there are more than 100 large-diameter shield tunnels with a diameter of more than 10 m in China (Chen 2020). With the large-scale construction of long shield tunnels in China, the rapid construction of internal structures has attracted more and more attention. Due to the current design scheme, the flue slab structure leads to the need for large area site stacking and special equipment for transportation and hoisting, seriously affecting construction efficiency. It is necessary to focus on its rapid construction technology.

The traditional form of flue slab is a prefabricated block assembled reinforced concrete arch plate. In order to ensure the sealing of smoke exhaust, the joints between flue slabs must be well treated with fire sealing (Liang 2021). Therefore, considering the working characteristics of flue slabs, the current research on the working performance of flue slabs in China and abroad is mostly based on their fire resistance (Sun 2021; Zhang 2014). For example, ZiyiAccording to Ziyi's (2021)

*Corresponding Authors: 278438671@qq.com, 350737300@qq.com, 519185903@qq.com, 1284584654@qq.com, 379358063@qq.com and 20605081@qq.com

DOI 10.1201/9781003372417-31

research suggestions, for the shield tunnel with centralized smoke exhaust at the top, the pavement segments are protected by fire retardant materials, and the segments at the vault of the flue plate are protected by concrete lining, and the flue slab does not need fire protection. Wu Dexing et al. (2011) carried out experimental research on the fire resistance of the key components of the top separator of the flue, including the fire resistance test of the top separator of the flue, the fire resistance test of the ham, and the pull-out test of the planting bar specimen under high temperature. The temperature field variation law and high-temperature mechanical response characteristics were obtained under high temperature and load. At present, there are few studies on the optimization of flue slab structure. Most studies start from the synchronous and rapid construction of the internal structure. For example, Li He (2011) studies the mutual interference between shield tunneling construction and internal structure construction from construction process management. To solve problems caused by the large size, heavy structure, and high arching height of the flue slab at present, such as great demand for stacking sites, special equipment for transportation and installation, narrow construction space of flue, high cost and low efficiency, a set of mature design and construction technology of lightweight flue plate for large-diameter highway shield tunnel is still needed.

As mentioned above, systematically studying the structural design optimization of flue slabs is necessary. In this paper, the lightweight theoretical research on the flue slab structure of highway shield tunnels is carried out. Through the investigation of new structures, new materials, and structural checking calculations, the flue slab structure that meets the bearing performance requirements is studied to solve the technical problems of high-efficiency synchronous construction of large-diameter shield tunnels and provide a theoretical basis for similar projects in the future.

## 2 PROJECT PROFILE

### 2.1 Based on project

The south section of the Nanjing Heyan Road shield tunnel passes through the Yangtze River by shield tunnel. According to the standard construction of an urban expressway and two-way six lanes, the design speed is 80 km/h. The project is about 5.723 km long and the tunnel section is about 4.215 km long. The outer diameter of the shield segment is 14.5 m, the excavation diameter of the shield machine is 15.03 m, and the designed span of the flue plate is 9.67 m. The tunnel passes through complex geological conditions, such as a strong permeable sand layer, full-face hard rock layer, soft and hard uneven strata, upper soft and lower hard strata, karst strata, and regional fracture. It is the most complex underwater shield tunnel project in China. The specific project and the situation of crossing strata are shown below.

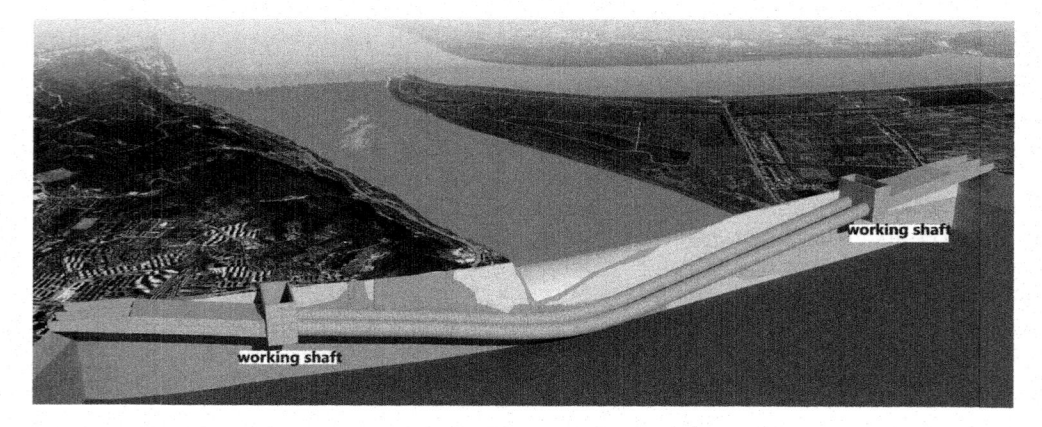

Figure 1.   Longitudinal section layout of shield tunnel.

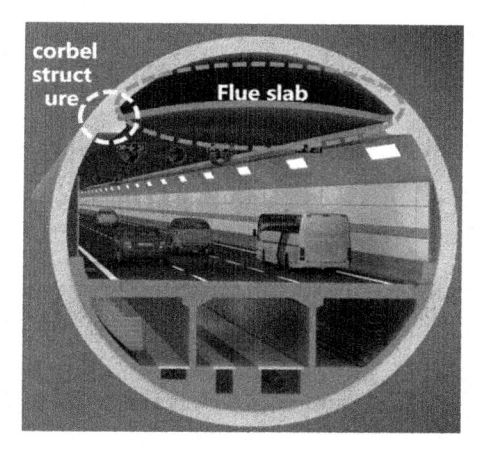

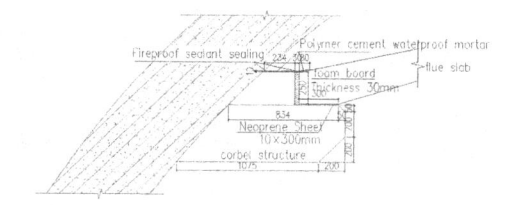

Figure 2. Traditional flue slab setting form of Shield tunnel.

Figure 3. Linking method of corbel and flue slab.

## 2.2 *Overview of flue slab design*

The initial design scheme of the tunnel flue slab was designed with a span length of 9.67 m, a width of each flue slab of 2 m, a thickness of the flue slab of 25 cm, and the arch height of 0.53 m. The C40 concrete was used for prefabricated reinforcement. The flue slab is divided into four types, A, B, C1, and C2, with a total of 1488 pieces, of which 1407 pieces of type A are used for general sections; 50 blocks of type B for exhaust valve section; 16 blocks of C1 and C2 type for fan section. In this design scheme, the corresponding weight of the flue slab is A 12.5 t, B 8.6 t, and C 12.8 t.

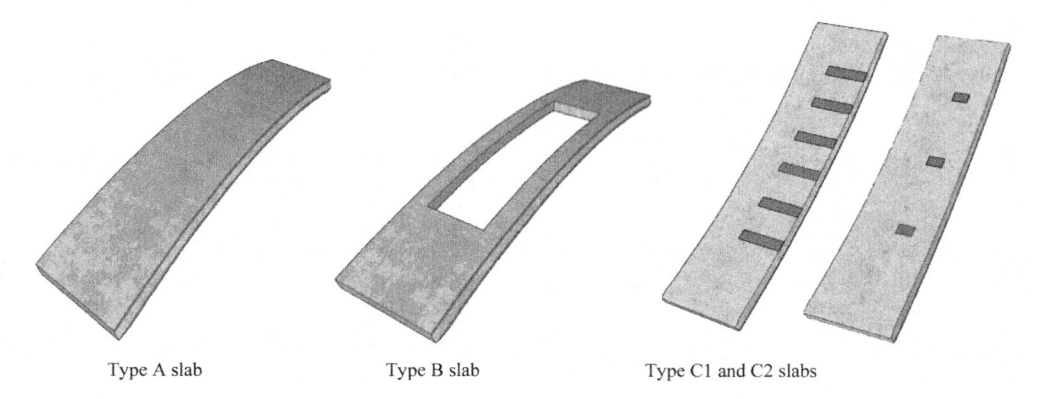

Type A slab                Type B slab                Type C1 and C2 slabs

Figure 4. Type A, B, C1, and C2 flue slab.

## 3 OPTIMIZED DESIGN

### 3.1 *The current problems*

The following deficiencies exist in the construction of flue slabs with a traditional design concept:

1. The structure of the flue slab is heavy, but its thickness is limited, resulting in limited strength. When stacking on-site layer by layer, it can use no more than five stacks, which requires sufficient

yard area for stacking. At the same time, because of the weight problem, we need to set up a special door crane for transportation.

2. Because of the heavy structure of the flue slab, it is necessary to cast-in-place 1 m wide and 0.45 m thick corbel structure at both ends of the tunnel segment to support. Before cast-in-situ construction of the corbel structure, large-scale chiseling of the tunnel segment's outer concrete is needed, and the planting reinforcement operation is needed too. These construction behaviors not only cause damage to the durability of the segment structure but also need a large amount of rebar planting and cast-in-situ concrete. They have low construction efficiency and are unfriendly to the environment.

3. Because the flue slab is heavy, the current installation method uses a door frame bracket, a type of special equipment, for positioning, installation, and splicing, with low construction efficiency.

4. When the flue slab splicing is completed, the construction personnel must also enter the exhaust duct for subsequent sealing procedures. But in the traditional flue slab design, its upper height is limited, the upper maximum is only 1.795 m, and the construction personnel's working space is very narrow. So, it's easy to cause safety accidents.

### 3.2 *Optimized design ideas*

Through the query specification, the design of tunnel flue slab mainly considers the ultimate state of structural bearing capacity and the deformation deflection of the plate under load. Because the flue slab is a subsidiary structure with 50-year durability and can be replaced, the structural performance design is mainly to optimize the size of the flue slab after meeting these two requirements.

There are three main ways to realize the lightweight optimization design of the structure:

1. Size: thinning, lightening;
2. Structure: dematerialization and hollowing out;
3. Material: high strength and lightweight.

Based on the above three criteria, the research focuses on optimizing the design of flue slab structures by adopting new structures, new materials, and light and thin dimensions.

## 4 OPTIMIZED SCHEME

By investigating UHPC, carbon fiber, PC slab, lightweight foam concrete, honeycomb beam, SP slab, and other structures and materials, the SP slab structure was finally selected for the optimal design of the flue board, considering the bearing capacity and economicalness. The design is based on the hollowing of the structure and does not use the traditional form of reinforcement. It adopts the prestressed structure of steel strands and replaces the original reinforced concrete structure by using large-span prestressed concrete hollow slabs.

### 4.1 *Space layout optimization*

In the layout of the tunnel space, the previously used arc-shaped reinforced concrete flue slabs with an arching height of more than 50 cm are discarded, as shown in the dotted line legend in Figure 5. After optimization, prestressed hollow flue slabs with a maximum arching height of no more than 5 cm are used. The structure has been rearranged, and the position of the flue slab is raised by 48cm. On the premise that the smoke exhaust area remains unchanged, the driving space under the slab remains unchanged, and the space for the jet fan remains unchanged. The length of the flue slab is effectively reduced to 8.7 m. The working space of the flue exhaust duct construction personnel between the slab and the segment has increased from 1.795 m to 1.915 m, significantly improving the construction safety index.

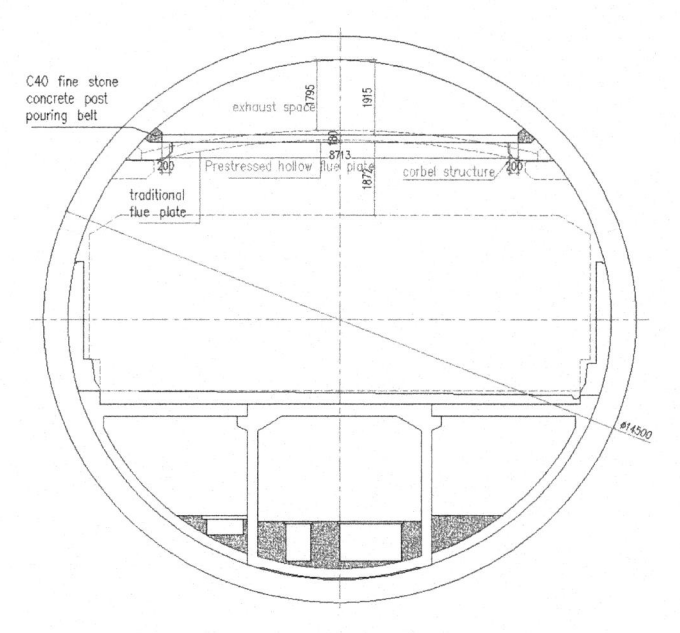

Figure 5.    Space layout optimization design drawing of flue plate.

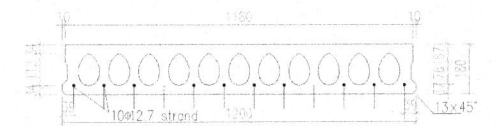

Figure 6.    Design schematic of prestressed hollow flue slab.

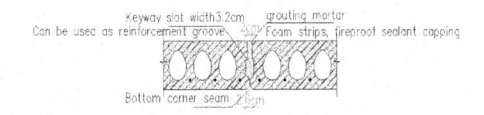

Figure 7.    Joint treatment of prestressed hollow flue slab.

### 4.2 *Flue slab structure optimization*

Under the premise of optimizing the design of the space layout of the flue slab inside the original highway shield tunnel, the model of the prestressed hollow slab that satisfies the load action and deflection control is calculated. After referring to the SP Prestressed Hollow Slab Technical Manual (99ZG408)(2005), the SP18A8710 type prestressed hollow slab is tentatively selected.

1. General segment flue slab (Type A slab)

SP18A8710 type prestressed hollow-core slab, the concrete grade used is C45 zero slump fine stone concrete, that is, dry hard concrete. The steel strands are ten low-relaxation prestressed steel strands with a diameter of 12.7mm formed after tensioning, as shown in Figure 6. The formed prestressed hollow panels should meet the requirements of surface roughness, normal bearing capacity, and fire protection.

For the treatment of the flue slab joint, it can be seen from the figure when the contact surfaces of the two flue slabs are aligned, a 2 cm wide slab joint is formed above the slab top, and a right-angle internal angle suture is formed below the slab joint. A key slot with an enlarged width of 3.2cm is formed between the bottoms of the slabs, which can be used for preembedding reinforcement grooves. C20 fine stone concrete or 1:3 cement mortar can be used to fill the joints between the slabs. The cement mortar must have certain workability to ensure the keyway between the slab joints is filled. It is easy to operate. After the grouting of the board joint is completed, it is the same as the traditional prefabricated slab design, supplemented by a fireproof sealant to seal the top.

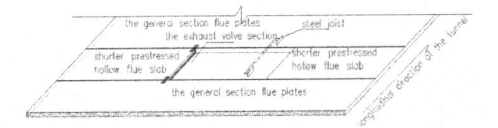

Figure 8. Treatment diagram of the lap joint of the flue slab at the exhaust valve.

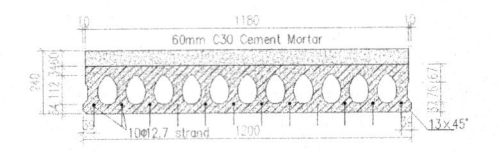

Figure 9. SPD slab for fan.

2. The flue slab for the smoke exhaust damper (type B slab)

According to the flexibility of the design of the prestressed hollow flue slab, two shorter prestressed hollow flue slabs can be designed at the flue slab of the exhaust valve section and then lapped to the general section flue plates on both sides through steel joists structurally, as shown in Figure 8.

3. The flue slab for the fan (type C slab)

The supporting project of this project needs to be equipped with three jet fans with a diameter of 1.1m. According to the SP Prestressed Hollow Slab Technical Manual (99ZG408), the concentrated load capacity of the hollow slab can be improved by adding a laminated layer to the prestressed hollow slab; that is, the prestressed laminated hollow slab structure is used to solve the problem. In this design optimization, this section adopts SPD18A8710 type prestressed hollow slab, and its single concentrated load bearing capacity is 3400kg.

## 5 THEORETICAL CALCULATION

### 5.1 *When the uniform load on the slab meets the following requirements*

According to the requirements of the specification for the limit state of the bearing capacity and the limit state of normal service, the bearing capacity, bending moment, shear force, and deflection of the prestressed hollow slab have the following requirements (1997):

When the uniform load on the slab meets the following requirements, $1kN/m^2 \leq$ variable load standard combination values $\leq 5kN/m^2$ and quasi-permanent combination design values $\leq 0.87$ standard combination design values. It can be calculated according to the standard combined design value of the load acting on the slab in the code and according to the allowable uniform load $[q_k]$ in the allowable load table, that is

$$G_k + \sum_{i-1}^{n} Q_{ki} \leq [q_k] \tag{1}$$

Where $G_k$ is the standard value of uniformly distributed permanent load applied to the surface of the slab, $Q_k$ is the standard value of the uniformly distributed variable load on the i-th slab, and $[q_k]$ is the allowable value of the standard combination of uniform load on the slab.

### 5.2 *When the uniform load on the slab doesn't meet the above conditions*

It can be selected according to the allowable bending moment values and allowable shear force values in the allowable load table, and the deflection check is performed:

$$M_u \leq [M_u] \tag{2}$$

$$M_k \leq [M_k] \tag{3}$$

$$M_q \leq [M_q] \tag{4}$$

$$V \leq [V] \tag{5}$$

Where $M_u$ is the design value of bending moment calculated according to the basic combination of load effects, $M_k$ is the design value of bending moment calculated according to the standard combination of load effects, $M_q$ is the design value of bending moment calculated by a quasi-permanent combination of load effects, $V$ is the design value of shear force based on basic combination of load effects, $[M_u]$ is the allowable bending moment values calculated for basic combinations of load effects, $[M_k]$ is the allowable bending moment values calculated by the standard combination of load effects, and $[M_q]$ is the allowable bending moment value calculated according to the quasi-permanent combination of load effects.

If deflection is required, the deflection calculation shall be carried out according to the formula in the SP Prestressed Hollow Slab Technical Manual (99ZG408).

The short-term back arch value of the prestressed hollow slab due to prestress is calculated as follows:

$$a_p = \frac{-N_{P0} e_{P0} l_0^2}{8 E_c I_0} \tag{6}$$

The long-term back arch value of the prestressed hollow slab due to prestress is calculated as follows:

$$a_{pl} = \frac{-2N_{p0} e_{P0} l_0^2}{8 E_c I_0} \tag{7}$$

The corresponding short-term deflection is:

$$a_s = a_p + \frac{5 M_k l_0^2}{48 B_s} \tag{8}$$

The corresponding long-term deflection is:

$$a_l = a_{pl} + \frac{5(M_q + M_k) l_0^2}{48 B_s M_k} \tag{9}$$

Where $B_S$ is the short-term stiffness of the prestressed hollow slab, $N_{p0}$ is the resultant force, $e_{p0}$ is the eccentric distance of the resultant force point, $l_0$ is the calculated length of the slab span, $E_c$ is the elastic modulus of concrete, and $I_0$ is the moment of inertia.

According to the calculation of the bearing capacity of the flue slab, when the live load of the slab surface is $2 kN/m^2$ personnel load, the quasi-permanent composite design value is $4.69$ $kN/m^2$ that is equal to 0.81 times the standard composite design value $\leq 0.87$ standard composite design value, which satisfies the first condition. According to the $[q_k]$ in the allowable load table in the specification, the SP18A8710 type prestressed hollow slab is selected, and its $[q_k] = 7.12 kN/m^2 \geq q_k = 2 kN/m^2$, which meets the bearing capacity requirements.

According to the flue plate deflection calculation, its short-term deflection $a_s = 18.24$ mm, long-term deflection $a_l = 28.89$ mm, which are all less than $l/250 = 34.8$ mm required by the specification and meet the deflection requirements.

## 6 APPLICABLE ANALYSIS

After obtaining the prestressed hollow slab design scheme through the optimized design, it is compared with the traditional structural flue slab and UHPC high-performance concrete flue slab in terms of technology, economy, and carbon neutrality, as shown in Table 1.

Through the above technical and economic comparison, it can be found that compared with the traditional concrete structure, the prestressed hollow flue slab not only greatly reduces the cost by 44% but also reduces the occupied area of the site and the investment in special construction equipment, and in terms of energy conservation and environmental protection, The carbon emission is greatly reduced, and the carbon emission of the prestressed hollow flue plate is only 33% of the ordinary concrete flue slab.

Table 1. Parameter comparison table.

| Type | Traditional structural flue slab | UHPC high-performance concrete flue slab | Prestressed hollow flue slab |
| --- | --- | --- | --- |
| Weight / per meter | 6.2t | 3.12t | 2.71t |
| Cost / per meter'− | 5500 | 12200 | 3100 |
| Yard area/m$^2$ | 750 | 750 | 325 |
| Convenience of construction | good | general | excellent |
| Carbon emission /T /CO$^2$ | 3.06 | 1.09 | 1.03 |

## 7  CONCLUSION

Using the optimization design of flue plate structure and theoretical research, the following main conclusions can be obtained:

(1) By lifting the design height of the flue slab and flattening the design of the flue slab, the span length of the slab is reduced, so that the flue slab can be designed with a thin prestressed hollow slab, which greatly reduces the weight of the flue slab. At the same time, due to the rationality of the structure design, ten pieces can be placed when storing the flue board, which can greatly reduce the yard demand area.

(2) After the design of the SP18A8710 prestressed hollow flue slab, the span is only 8.7 m, the weight per meter is only 2.71 t, which is 43.7% of the weight of the original concrete flue slab, and the cost is only 43.7% of the original ordinary concrete flue slab. After the optimization design, no special hoisting equipment is needed for transportation, and no special equipment is needed in the hole. The ordinary forklift can meet the needs of the installation.

(3) Due to the significant reduction of the weight of the flue slab, the structure size of the corbel was optimized. It can control the cast-in-place scale of corbels in a large range and effectively reduce the number of planting bars and pouring scale, the construction difficulty, the corresponding construction process, and the engineering cost of corbel parts. At the same time, it improves construction efficiency.

(4) The optimized design scheme improved the layout of the upper space in the tunnel, significantly increasing the working space of the flue from 1.795 m to 1.915 m and improving the construction environment.

(5) The design of a prestressed hollow flue slab can ensure the structure's safety, improve the construction progress, and save the project cost. At the same time, it can greatly reduce carbon emissions, and the emission is only 33% of the original ordinary concrete flue plate.

## REFERENCES

Chen Xiangsheng. & Xu Zhihao. (2020). Challenges and Technological Breakthroughs in Tunnel Construction in China *J. China Journal of Highway and Transport*, 33(12), 1–14.

Li He. (2018). Research on Synchronous Rapid Construction Technology for Internal Structure of Subway Shield Tunnel with Large Diameter Single-hole Double-line Composite Lining. *J. Railway Construction Technology*, 7, 56–59.

Li Jianbin. (2021). Current Status, Problems, and Prospects of Research, Design, and Manufacturing of Boring Machine in China. *J. Tunnel Construction*, 41(6), 877–896.

Liang Yang. & Wei Zhihua. (2021). Operation ventilation system design for Shantou Suai Tunnel. *J. Journal of HV&AC*, 51(3), 78–83.

Manual for Design of Hollow Core Slab (1997): PCI, USA.

Ministry of Construction of the People's Republic of China. (2005). *Prestressed hollow slab (05SG408)*. Bei Jing, China Planning Press.

Sun Ce. (2021). Fire Characteristics and Evacuation Rescue of Double-Deck Shield Tunnel with Super-Large Cross-Section: A Case Study on Shenzhen Bao'an International Airport Expressway Tunnel. *J. Tunnel Construction*, 41(8), 1297–1306.

Wu Dexing. & Li Weiping. (2011). Experimental study on fire resistance of exhaust flue structure of highway tunnel. *J. Highway*, 8, 282–286.

Zhang Xianfu. (2014) SUN Ce. (2021). *Study on Smoke extraction technology and Evacuation of Ma Wan cross-sea Tunnel*. D. Cheng Du, Southwest Jiaotong University.

Zi Yi. & Lu Zhipeng. (2021). Study on fire protection and key parameters of top centralized smoke exhaust shield tunnel structure. *J. Fire Science and Technology*, 40(6), 813–817.

*Civil Engineering and Urban Research – Mohamed & Hou (Eds)*
© *2023 the Authors, ISBN 978-1-032-44487-1*

# Research on pipeline deformation of foundation pit excavation

Hao Hu*
*Publishers Chinese China Railway 14th Bureau Group Mege Shield Construction Engineering Co., Ltd., Beijing, China*

Jin Pang*, LingChao Shou*, Ting Bao* & LiFeng Wang*
*Chinese Zhejiang Mingsui Technology Co., Ltd, Zhejiang, China*

ABSTRACT: Taking deep foundation pit as engineering background, the field measured data of settlement deformation of pipeline around foundation pit are sorted out and analyzed. And the deformation characteristics of the pipeline around the deep foundation pit are discussed. The research results indicate that the settlement of rigid and flexible pipelines increases with the increase of excavation depth. The settlement of rigid pipeline is basically between 0.3‰H~3.6‰H, the average value is 1.2‰H; the settlement of flexible pipeline is between 0.2‰H~3.2‰H and the average value is 1‰H, where H is the excavation depth of foundation pit. There is little correlation between the settlement of the flexible pipeline and the burial depth, but the settlement of the rigid pipeline increases with the increase of the buried depth. The distribution of the rigid and flexible pipelines is analyzed. The maximum settlement point is located at the 1.5H~2H around the foundation pit, and the mutation point is located at the 2.5H. Based on the ground settlement envelope obtained by Hashash, it is concluded that the proportion of sewage, water supply, gas, communication, and electric power pipelines is 43%, 38%, 34%, 33%, and 35.5% respectively, and the deeply buried sewage and water supply pipeline account for a large proportion in the surface settlement. The settlement of the pipelines is in accordance with the normal distribution.

## 1 INTRODUCTION

Recently, with the rapid economic development of China, urban traffic problems are getting worse. Vehicle roads only are unable to solve the urban congestion problems. It is necessary to develop underground traffic to alleviate the urban traffic congestion problems. As a city's artery, the subway can be more convenient for the city. It makes the design and construction of the foundation pit of the metro station develop towards more complex geological conditions when excavating foundation pits in areas where urban traffic is concentrated. Meanwhile, the accidents in the pits of metro stations should not be ignored (Yao 2008; Zhou 2009). Therefore, in the field of underground engineering, it is necessary to conduct statistical analysis on the measured data of a large number of foundation pit projects in the local area so that we can obtain more reliable information on guiding the design and construction of foundation pit excavation.

Currently, many scholars have done a lot of research on the characteristics of deep foundation excavation pits in subways through measured data. Liao Shaoming et al. (Liao 2015) and Li Lin et al. (Li 2007) did a comprehensive analysis of the deformation behavior of deep foundation pits with different retaining structures in Suzhou, Hangzhou, and Shanghai areas, respectively. Wei Shifeng et al. (Wei 2014) analyzed the differences in deformation behavior of deep foundation

*Corresponding Authors: 375110469@qq.com, pangj00@163.com, slczust@163.com, 1542528435@qq.com and wanglfzust@163.com

DOI 10.1201/9781003372417-32

pits with different construction schemes through measured data. Based on the monitoring data of station pipelines in Shanghai. Li Dapeng et al. (Li 2018) did research based on ground surface settlements of metro-station deep excavation predict building deformation. Liu Xuezhu et al. (Liu 2014) discussed the influence of deep foundation pit excavation on adjacent pipelines from the perspective of vertical displacement. Wu Fengbo et al. (Wu 2012) defined the control standards for rigid and flexible pipelines. However, the above studies are mainly focused on research on the inclination measurement, surface settlement, and axial force of the foundation pit retaining structure, and there is little analysis of the measured data of the pipeline.

We usually use numerical simulation methods to analyze pipelines. Du Jinlong et al. (Du 2009) use FLAC to analyze the influence of foundation pit excavation on different adjacent pipe diameters. For example, pipe-soil interaction should be considered when the pipe diameter is larger than 400mm. Jie Zhang et al. (Zhang 2018) established a three-dimensional model of pipeline and foundation pit and the effects of pipeline parameters, foundation pit parameters, soil parameters, and underground continuous wall on the stress, strain, and deformation of the pipeline were studied.

However, the current pipeline research work mostly predicts pipeline deformation based on the study of single foundation pits or based on complex numerical simulation calculations. Parameter values are more complicated, and there are still some problems in the practical application. This paper comprehensively considers the pipeline monitoring data of several subway foundation pits and studies the pipeline deformation from the aspects of foundation pit excavation depth, pipeline depth, and pipe diameter. It provides a meaningful reference for the protection of pipelines in excavation activities.

## 2 FOUNDATION PIT AND ENCLOSURE OVERVIEW

In this area, its formation unit is relatively centralized. The main formation of this site from top to bottom consists of layers of mixed fills, silty clay, clayey silt, sandy silt, and silt. The station floor is located in silty clay, silt layer, and clay layer, as shown in Figure 1.

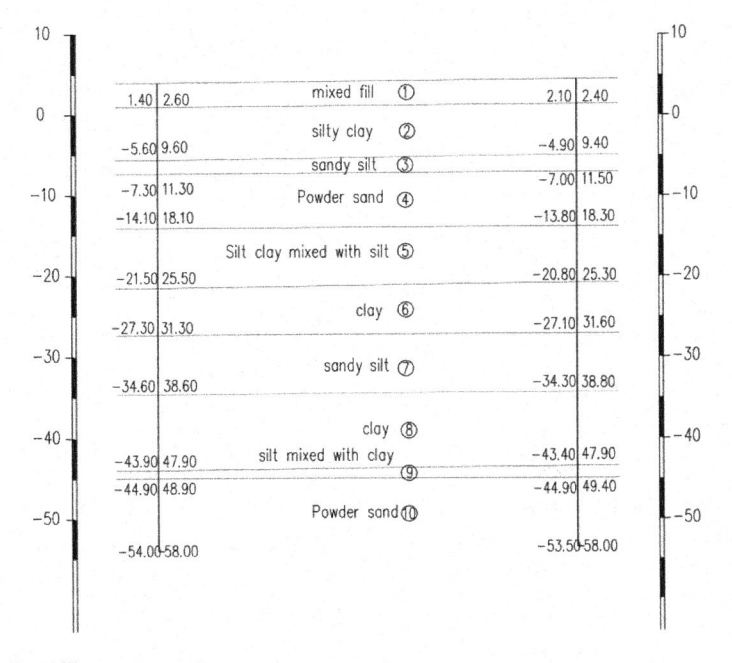

Figure 1.    Soil profile.

Foundation pits are constructed by the open excavation sequential operation method. The length of excavation is about 200m and the width is about 20m. The excavation adopts the diaphragm wall combined with the interior horizontal supports as Figure 2 shows. The excavation depth is 18.50~21.11m. The type of pipelines and their distance from the foundation pit are shown in Table 1. Table 1 shows that most of the pipelines are distributed within 30m from the foundation pit, rigid pipelines account for 76%, flexible pipelines account for approximately 86%, and 75.21% of flexible pipeline measuring points are distributed in the range of 10-30m.

## 3 ANALYSIS OF DEFORMATION CHARACTERISTICS OF UNDERGROUND PIPELINE

### 3.1 *Influence of excavation depth on pipeline settlement*

Urban underground pipelines usually include flexible pipelines and rigid pipelines, mainly depending on the relative stiffness of the underground pipeline and the surrounding soil. Generally speaking, reinforced concrete pipes, gas pipes and cast-iron pipes are all rigid pipelines, while power pipelines, and communication pipelines are flexible pipelines.

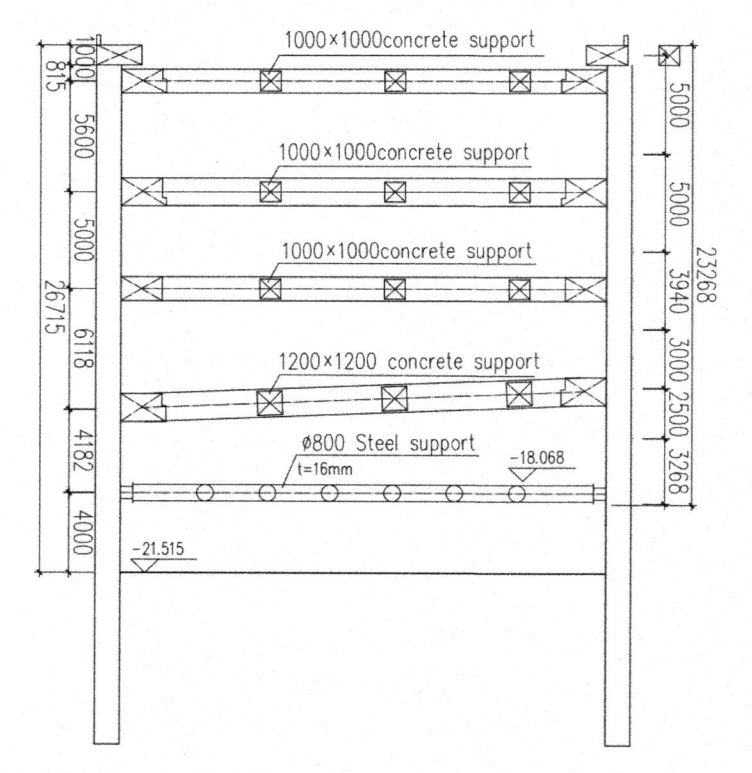

Figure 2.    Foundation pit enclosure profile.

Figure 3 is the relationship between the settlement S of the pipeline and the excavation depth H (the settlement is negative). As Figure 3(a) shows, the settlement of rigid pipeline is basically between 0.3‰H~3.6‰H (4.8mm~57.6mm), and the average value is 1.2‰H (19.2mm); figure 3(b) shows the settlement of flexible pipeline is between 0.2‰H~3.2‰H (3.2mm~51.2mm), and the average value is 1‰H(16mm). The data does not indicate significant differences in pipeline settlement due to differences in pipeline types. The foundation pits are constructed with the envelope structure of the diaphragm wall, which has strong integrity and a stronger ability to control the

Table 1. Statistical map of distance distribution of foundation pit with different pipeline distance.

| Pipeline type | Rigid pipeline measuring point (total 183) | | | | Flexible pipeline measuring point (total 117) | | | |
| --- | --- | --- | --- | --- | --- | --- | --- | --- |
| Distance from foundation pit | <10m | 10-20m | 20-30m | >30m | <10m | 10-20m | 20-30m | >30m |
| Quantity | 42 | 63 | 34 | 44 | 13 | 41 | 47 | 16 |
| percentage | 22.95% | 34.43% | 18.58% | 24.04% | 11.11% | 35.04% | 40.17% | 13.68% |

excavation deformation. Table 1 shows that 86% of the flexible pipelines in this paper exist within 30m from the foundation pit, restricted by the envelope structure. Therefore, the settlement of low stiffness flexible pipelines is less than the settlement of rigid pipelines.

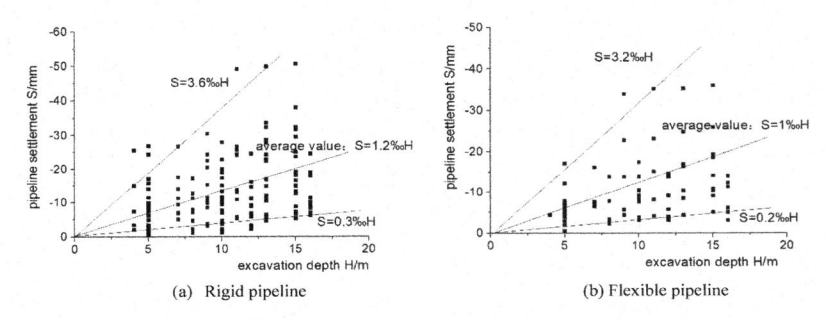

(a) Rigid pipeline  (b) Flexible pipeline

Figure 3. Relationship between the displacement of the pipeline and the excavation depth.

### 3.2 Influence of pipeline depth on pipeline settlement

Figure 4 is a diagram showing the relationship between relative settlement and pipeline depth, and the ordinate is the relative pipeline settlement (the ratio of settlement value to excavation depth). The flexible pipelines' buried depths are generally between 0.5 and 1.25m, and the relative pipeline settlement data are concentrated. The relationship between relative settlement and flexible pipeline buried depth is inconspicuous. The rigid pipelines' buried depths are generally between 1 and 4 m, and the relative settlement increases with the increase of buried depth.

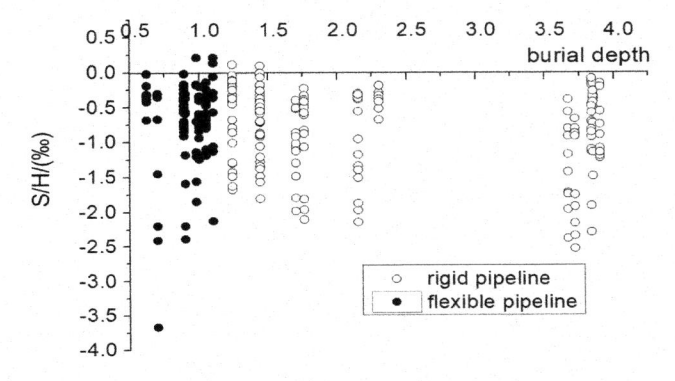

Figure 4. Relationship between relative settlement and burial depth of the pipeline.

The relative settlement of the rigid pipeline has a certain relationship with the buried depth, so taking the maximum relative settlement of the rigid pipeline and considering the common influence of buried depth and diameter. The relation chart between the ratio of the maximum relative displacement of the rigid pipeline to the outer diameter of the pipeline and the ratio of the buried depth h to the outer diameter D of the pipeline. Figure 5 shows the maximum relative settlement of rigid pipelines is roughly linear with h/D, and as the h/D increases, the maximum relative settlement of the pipeline increases linearly. The larger the diameter of the pipeline, the smaller the settlement; the larger the buried depth, the larger the settlement.

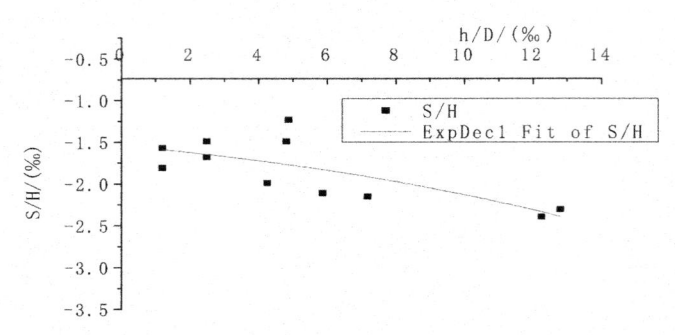

Figure 5.    Relationship between the maximum relative settlement of rigid pipeline and h/D.

### 3.3  Influences of distance on settlement

Figure 6 is the relationship between the pipeline maximum relative settlement (S/H) and the pipeline relative distance (L/H) (L is the distance between the pipeline and the foundation pit, H is the excavation depth, and the ordinate value is the thousands of score)

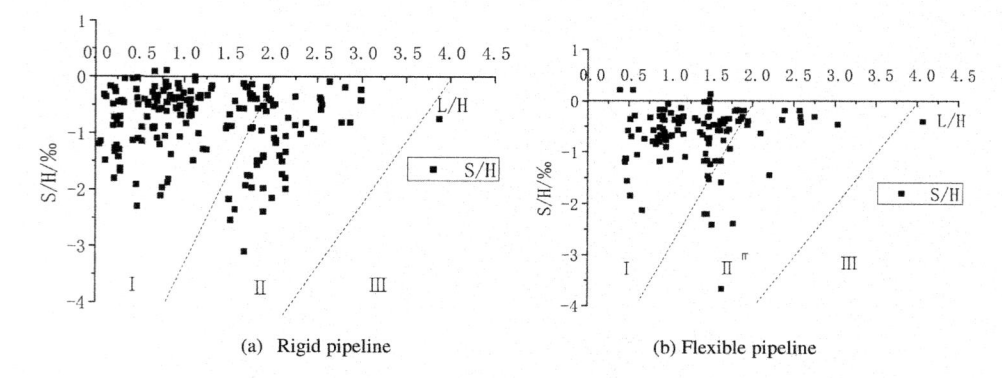

(a)  Rigid pipeline                          (b) Flexible pipeline

Figure 6.    L/H impact on the pipeline.

Annotation: Zone I in Figure is sand, hard clay, and soft clay; Zone II is soft clay and dead soft clay; Zone III is soft clay with deeper thickness and dead soft clay.

1) Figure 6(a) shows the relative settlement of rigid pipelines is dispersed, but the settlement is mostly located in the zone I proposed by Peck [PECK 1969], and there are still many settlements points located in Zone II; Figure 6 (b) shows the distribution of flexible pipeline relative settlement is more concentrated. The settlement points are mainly located in zone I, and only a small number of settlement points are located in zone II;

2) The settlement of the pipeline is a small groove type which is big in the middle, and small at the ends. Rigid and flexible pipelines are affected by excavation in the range of 3H;

3) The maximum settlement of the pipeline is approximately located at 1.5H outside the pit wall, and the settlement breakpoint is located near 2.5H.

The maximum settlements of rigid pipelines and flexible pipelines are different from the range affected by foundation pit excavation. But the distribution of its settlement is similar, the maximum settlement is close to the location of the settlement breakpoint, and the average value differs only by 0.2‰H. Therefore, all the pipeline settlements can be integrated to study the settlement distribution laws of pipelines and ground surface.

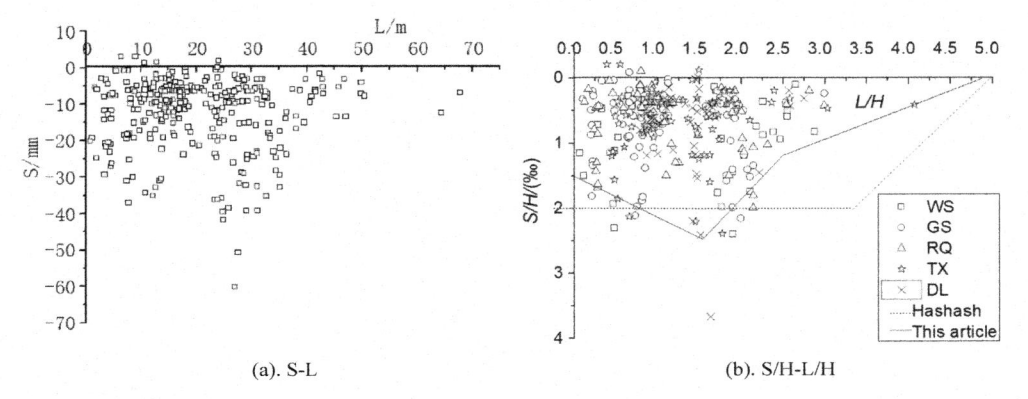

(a). S-L                                        (b). S/H-L/H

Figure 7.    Pipeline settlement pattern.

Figure 7 shows the settlement pattern of the post-wall. Figure 7 (a) shows the relationship between S and L. It shows that regardless of the influence of excavation depth or the pipeline itself, the influence range of the pipeline is generally within 50m from the pit wall to the foundation pit. The maximum settlement of the pipeline is about 40mm, and the maximum of the settlement is controlled within 20mm. It shows that the support system of the metro foundation pits in this paper can control the influence of foundation pit excavation on surrounding underground pipelines.

Figure 7(b) shows the relationship between pipeline S / H and L / H. Based on the measured data; this paper proposes the groove-shaped distribution pattern of the pipeline with a three-fold line. As the fig shows, in the range of excavation depth (0∼1.5) H, the pipeline settlement is large; in the range of excavation depth (1.5∼2.5) H, the pipeline settlement is reduced; when the excavation depth exceeds 2.5H, the settlement of the pipelines is small.

Figure 7(b) shows the envelope curve of the post-wall surface settlement proposed by Hashash et al. (Hashash 2008) based on the measured data of the foundation pit in the United States. Tan et al. (Tan 2012) proposed that the prediction curve of Hashash et al. can be applied to the surface settlements of the envelope behind the subway foundation pit wall in the Shanghai area (soft soil area). So, the envelope can also be used in the metro pits of this area. The figure shows the settlement pattern of the pipeline behind the wall is similar to the surface settlement. In the statistical analysis based on measured data, the average settlement values of sewage pipelines, water supply pipelines, gas pipelines, communication pipelines, and power pipelines are 0.86‰H, 0.76‰H, 0.67‰H, 0.66‰H, and 0.71‰H, respectively.

Since the surface settlement envelope proposed by Hashash et al. is a straight-line S/L=2‰ in the range of 3.5H, it can be seen that surface settlement is only related to the excavation depth H, and the settlements proportion of sewage, water supply, gas, communication, and power pipelines to ground surface settlement are 43%, 38%, 34%, 33%, and 35.5%, respectively. The settlement of sewage and water supply pipelines accounts for a large proportion of surface settlement because of the large depth of the sewage and water supply pipelines (buried depth is 3∼4m). The pipeline at a certain depth below the surface is more affected by the excavation, because the maximum settlement of the soil occurs at a certain depth below the surface (Liu 2014, OU 2000).

### 3.4  *Statistical analysis of pipeline settlement*

Figure 8 shows the normal distribution curve of pipelines settlement (the ordinate is the frequency and the corresponding ratio of frequencies)

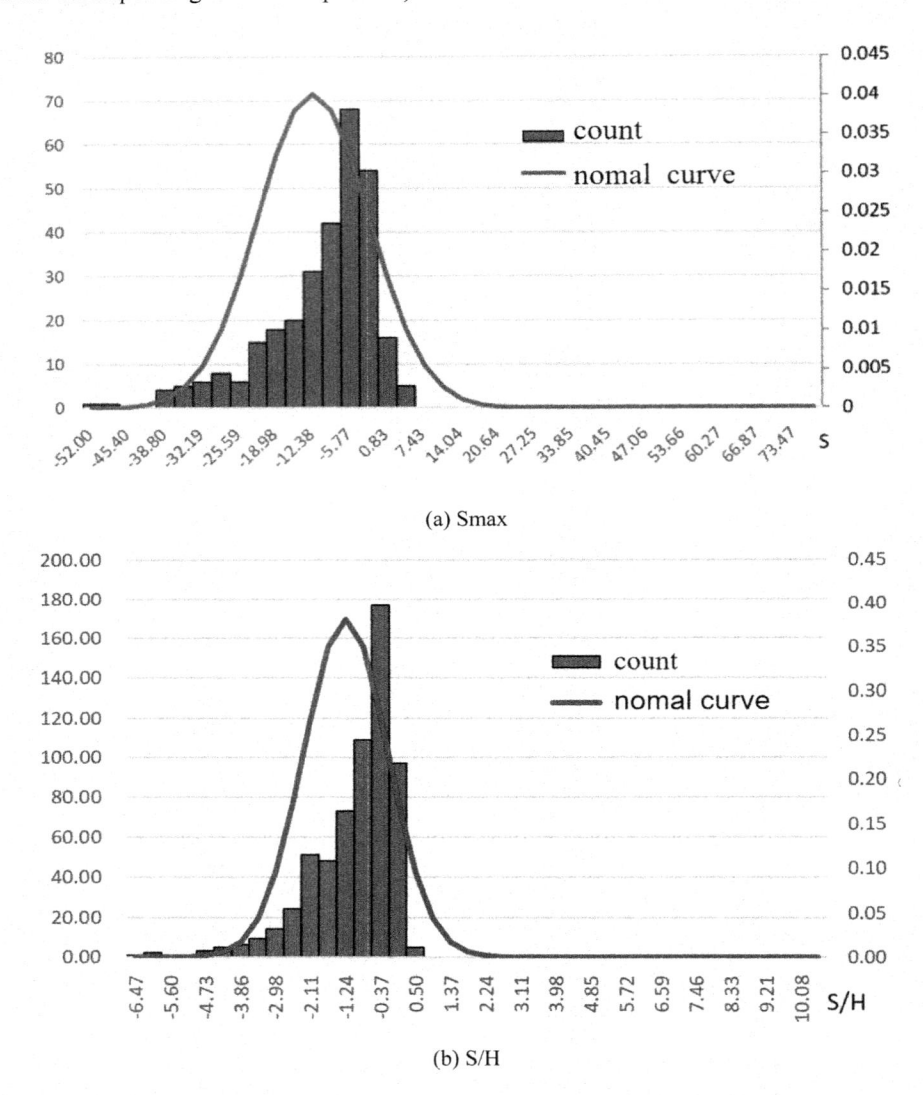

(a) Smax

(b) S/H

Figure 8.    Maximum settlement distribution map of pipelines.

Figure 8 (a) shows the distribution curve of pipeline settlement S when the foundation pit is excavated to the bottom. The abscissa is a settlement (in mm), and the ordinate is the frequency and ratio. Figure 8 (a) shows: 1) The pipeline settlement conforms to the normal distribution, and the number of measuring points with a settlement of about 6mm is the largest, and the frequency of occurrence reaches 22.67%. 2) The maximum value of pipeline settlement is about 42.1mm, and the pipeline settlement in the range of 15.68mm accounted for 62.93%, and the settlement in the range of 32.19mm accounted for 97.72%. Therefore, when the foundation pit is excavated to the bottom, most of the pipeline settlements are within the required range of control.

Figure 8 (b) shows the distribution curve of relative settlement S/H of pipelines at different excavation depths during foundation pit excavation, and the abscissa is thousandth. Figure 9 (b) shows: 1) The average value of pipeline settlement is 1.24‰H, which is at the center of the curve, but not the peak of the curve. The frequency of pipeline settlement of 0.37‰H was the highest, and the frequency of occurrence reaches 17.44%. 2) The maximum value of pipeline settlement is about 3‰H. The settlement which is less than the average value of 1.24‰H accounts for 73.76%, and the settlement exceeding 2‰H is only 18.56%. Therefore, during the excavation process, the pipeline settlements are within the required range of control.

## 4 CONCLUSION

The settlement of rigid and flexible pipelines increases with the increase of excavation depth. The settlement of rigid pipeline is basically between 0.3‰H~3.6‰H, the average value is 1.2‰H; the settlement of flexible pipeline is between 0.2‰H~3.2‰H and the average value is 1‰H, where H is the excavation depth of foundation pit. When the excavation depth of the foundation pit is similar, the settlement value of rigid pipeline and flexible pipeline are close.

The flexible pipeline's buried depth is generally 0.5~1.25m, which is not obvious with the buried depth. The rigid pipeline buried depth is generally $1 \sim 4$ m, and the settlement increases with the increase of buried depth. The rigid pipeline maximum settlement increases linearly with the increase of the ratio of the buried depth h to the pipeline outer diameter D.

The settlement of the pipeline is a small groove type which is big in the middle, and small at the ends. The settlement increases first, then decreases, and finally slowly decreases. The influence range of excavation is $0 \sim 3H$. and the maximum pipeline settlement is within the range of about 1.5H~2H outside the pit wall. The settlement abrupt point is located near 2.5H, and when L / H > 2.5, the pipeline settlement is small.

There is a certain correlation between pipeline settlement and soil settlement. The proportion of sewage, water supply, gas, communication, and electric power pipelines is 43%, 38%, 34%, 33%, and 35.5% respectively, and the deeply buried sewage and water supply pipelines account for a large proportion of the surface settlement.

The pipeline settlement is in accordance with the normal distribution, and the pipeline settlement in the range of 32.19mm due to excavation accounts for 97.72%. During the excavation the settlement was less than the average value of 1.24‰H, accounting for 73.76%.

## REFERENCES

Du, J. L. & Yang, M. (2009). Analysis of the influence of deep foundation pit excavation on adjacent buried pipe lines. *Journal of Geotechnical and Engineering*. 28(s1), 3015–3020.

Hashash Y, M. A., Osouli, A. & Marulanda, C. (2008). Central tunnel project excavation induced ground deformations. *Journal of Geotechnical and Civil Engineering*. 134(9), 399–1406.

Li, D. P. & Yan, C. H. (2018). Building deformation prediction based on ground surface settlements of metro-station deep excavation. *J. Advances in Civil Engineering*, Volume 2018.

Li, L., Yang, M. & Xiong, J. H. (2007). Analysis of deformation characteristics of deep foundation pit in soft soil area. *J. Journal of Civil Engineering*. 40(4), 66–72.

Liao, S. M., Wei, S. F. & Tan, Y. et al. (2015). Analysis of deformation behavior of large-scale deep foundation pit in Suzhou area. *Journal of Geotechnical and Engineering*. 37(3), 458–469.

Liu, N. W., Gong, X. N. & Lou, C. H. (2014). Influence of foundation pit excavation on deformation characteristics of surrounding facilities in soft soil area. *Journal of Zhejiang University* (Engineering Edition), 48(7), 1141–1147.

Liu, X. Z., Gu, M. N. & Wu, X. Z. et al. (2014). Monitoring analysis of the influence of subway deep foundation pit construction on adjacent underground pipeline. *J. Engineering Investigation*, (s1), 462–467.

Ou, C. Y., Liao, J. T. & Cheng, W. L. (2000). Building response and ground movements induced by a deep excavation. *J. Geotechnique*, 50(3), 209–220.

Peck, R. B. (1969). *Deep excavation and tunneling in the soft ground*, In Proceedings of the 7th International Conference on Soil Mechanics and Foundation Engineering. State-of-the-Art-Volume. Mexico City C, 225–290.

Tan, Y. & Wei, B. (2012). Observed behaviors of a long and deep excavation construction by cut-and-cover technique in Shanghai soft clay. *Journal of Geotechnical and Civil Engineering.* 138(1), 69–88.

Wei, S. F., Tan, Y. & Liao, S. M. et al. (2014). Analysis of deformation behavior of deep foundation pit by the method of combination of forward and inverse in QianJiang Tunnel. *Journal of Civil Engineering.* 47(8), 112–119.

Wu, F. B., Jinn, H. & Yang, H. T. et al. (2012). Monitoring control index of underground pipeline around urban rail transit project. *J. Construction Technique.* 41(379), 72–75.

Yao, G. W., Lv, G. F. & Yang, Y. P. et al. (2008). Construction safety problems caused by foundation pit collapse of a metro station. *J. Urban Rapid Rail.* (2), 71–74.

Zhang, J., Xie, R. & Zhang, H. (2018). Mechanical response analysis of the buried pipeline due to adjacent foundation pit excavation. *J. Tunnelling and Underground Space Technology.* 78, 135–145.

Zhou, H. B., Cai, L. B. & Gao, W. J. (2009). Statistical analysis of foundation pit accidents in the metro station. *J. Hydrogeological Engineering Geology.* 36(2), 7–71.

*Civil Engineering and Urban Research – Mohamed & Hou (Eds)*
© *2023 the Authors, ISBN 978-1-032-44487-1*

# Study on properties of recycled aggregate and recycled concrete modified by chemical modification reagent and biotechnology

Chun-hui Lan*
*School of Electric Power, Civil Engineering and Architecture, Shanxi University, Taiyuan, Shanxi, China*
*School of Human Settlements and Civil Engineering, Xi'an Jiaotong University, Xi'an, Shaanxi, China*

Nan-xing Wang
*School of Electric Power, Civil Engineering and Architecture, Shanxi University, Taiyuan, Shanxi, China*
*Beijing Construction Engineering Group, Beijing, China*

ABSTRACT: Polyaluminium sulfate (PA) and water glass (WG) as the chemical modification reagents and DSM8715 strains as mineralization bacteria preparing for the MICP mineralization solutions were used to modify the recycled aggregates. The effects of solution concentration of polyaluminium sulfate, modulus of water glass, and concentration of bacterial in solutions on the crushing index and water absorption of recycled aggregate were researched, and at the same time, the influence of the chemical enhancing agents and biomineralization solutions on the mechanical strength and durability of recycled concrete was studied as well. The results show that the crushing index and water absorption of recycled aggregate are reduced effectively when the modulus of water glass is 2.8 (the solution concentration is kept at 10.0%), and the mechanical strength (the 3d and 28d compressive strength are enhanced by 58.44% and 52.36%, respectively) and durability of recycled concrete are also enhanced distinctly; the enhancing effect of polyaluminium sulfate is not as good as water glass, but the properties of recycled aggregate and recycled concrete (the 3d and 28d compressive strength are enhanced by 33.77% and 27.47%, respectively) are enhanced in varying degrees. The modification effects of biotechnology are obviously better than that of both PA and WG, with the apparently improved properties of recycled concrete (the 3d and 28d compressive strength are enhanced by 61.50% and 57.20%, respectively). Therefore, the enhancing effect of MICP mineralization solutions on the recycled aggregate is more prominent than the use of chemical enhancing agents, but water glass is also the ideal choice for recycled aggregates modification.

## 1 INTRODUCTION

Recently, with the development of the construction industry, tremendous amounts of concrete were utilized, which cause the increased consumption of nature sandstone aggregates. Because the nature sandstone aggregates resources are extensive and low cost, it's widely used in the construction industry. But the over-exploitation of nature sandstone aggregates has aroused many problems such as environmental pollution and resource exhaustion. Meanwhile, the construction and demolition (C&D) waste are mostly disposed of by piling up or land-filling, which takes high cost and occupied vast land resources and further result in resource- waste (Aïtcin2000; Hideo Ogawa et al. 2010; Salomon & Paulo 2004). In recent years, the demands of high strength, high-performance, as well as energy-saving and environment-friendly concrete in the construction industry, are getting more and more attention, and the concept of green concrete is gradually coming into our mind.

---

*Corresponding Author: lanchunhui@163.com

DOI 10.1201/9781003372417-33

The utilization of C&D waste as recycled aggregates could not only reduce the exploitation of natural sandstone but also mitigate the environment and air pollution, which served as the main measures for the development of green concrete. Compared with nature sandstone aggregates, the high porosity and water absorption, as well as the low density and strength of recycled aggregates, have hindered the utilization of C&D waste in the construction industry (Achal et al. 2010; Bru et al. 2014; Lotfi et al. 2015; Zhang et al. 2015).

Many studies are conducted to modify the properties of recycled aggregates and recycled concrete. Except for traditional chemical modification methods, based on microbial-induced carbonate precipitation (MICP) technology, some scholars have tried to use microbial modification by microorganisms to improve the properties of recycled aggregates (Feng et al. 2020; Julia García-González et al. 2017; Sun & Huang 2019; Zeng 2019). When the microorganisms that produced urease exist in an environment which was full of urea and calcium ions ($Ca^{2+}$), they could produce urease by its metabolism, and the urease could catalyze the urea decompose into ammonia gas ($NH_3$) and carbon dioxide ($CO_2$), further increased the local concentration of $CO_3^{2-}$. At the same time, the $Ca^{2+}$ can be chelated by the cell wall which was full of a negative charge, and finally, the mineralization precipitation calcite ($CaCO_3$) was formed. Calcite is not only insoluble in water but also has positive compatibility with cement matrix. The chemical equations can be concluded as below:

$$CO(NH_2)_2 + 2H_2O \rightarrow 2NH_4^+ + CO_3^{2-} \tag{1}$$

$$Cell + Ca^{2+} \rightarrow Cell - Ca^{2+} \tag{2}$$

$$Cell - Ca^{2+} + CO_3^{2-} \rightarrow Cell - CaCO_3 \tag{3}$$

Because the surface of recycled aggregates was covered by old mortar and exists lots of micro-cracks and even pores, the water absorption is obviously higher than that of natural aggregates. When recycled aggregates absorbed urease bacterial was put into calcium salt and urea mixed solutions, the micro-organisms could catalyze urea and form the $CO_3^{2-}$, then the $CO_3^{2-}$ will combine $Ca^{2+}$ and finally form $CaCO_3$, which could remedy the micro-cracks and pores within recycled aggregates, whereby reduce its water absorption and improve its mechanical properties. In our study, the mechanism of biotechnology modifying recycled aggregates was investigated, and the comparison of modification effects between biotechnology and chemical methods was also studied, the theory of biotechnology modifying RA was shown in Figure 1.

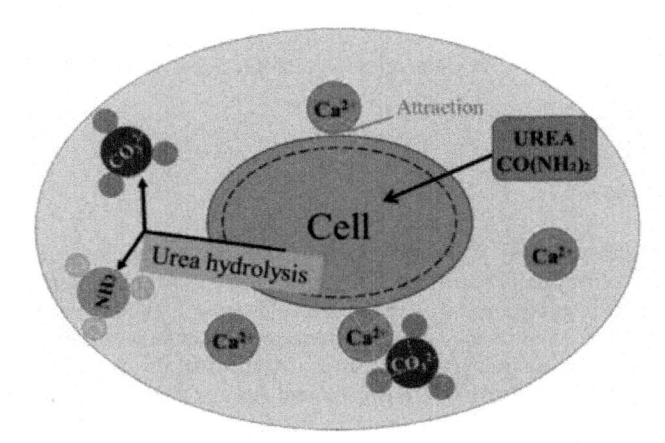

Figure 1.    Schematic of bio-methods: bacterially induced $CaCO_3$ precipitation by urea hydrolysis (Hao et al. 2019).

In this paper, the differences between the chemical modification method and the bio-method were studied. In order to improve the mechanical properties of recycled aggregates and recycled concrete,

Polya-aluminum sulfate (PA) and water glass (WG) as the chemical modification reagents were used, and the improvement effects of these two chemical modification reagents on the mechanical properties and durability of recycled aggregates and recycled concrete were studied. Meanwhile, the modification effect of the micro-interface of recycled aggregates was also studied.

## 2 METHODS AND MATERIALS

### 2.1 *Raw materials*

The natural coarse aggregates (NA) in this experiment are gravel limestone and the particles are continuous grading, the size is 5~20mm. Recycled coarse aggregates (RA) are from crushing C&D waste and the size of continuous grading particles is also 5~20mm. The properties of NA and RA were shown in Figure 2 and Table 1. Compared with nature aggregate, recycled aggregate has a higher porosity and water absorption as well as lower bulk density and strength. The size of river sand particle is 0.14~ 5.00 mm, fineness modulus is 2.49 and apparent density is 2.70g/cm$^3$. The basic properties of experimental cement were shown in Table 2, and the solid content and water-reducing rate of superplasticizer are 40.0% and 35.0% respectively. The PH and relative density of white powdery chemical enhancing agents PA are 3.5~5.5 and 2.1 respectively and the solid content and modulus of WG are 35.1% and 2.0 respectively. The WG in different modulus was prepared by sodium hydroxide (NaOH). For MICP methods, the *sporosarcina pasteurii* (ATCC11859) was selected as the experimental bacteria, the original bacteria were cultivated in the culture solution, the culture solution was composed of A solution and B solution (CAPS), and the concentration of bacteria solution was kept up $10^8$cells/mL. Then the bacteria solution ($1 \times 10^8$ cells/mL, marked as 1#) was double concentrated to get concentrated bacteria solution ($2 \times 10^8$ cells/mL, marked as 2#). Meanwhile, the 1# bacteria solution was also diluted into $5 \times 10^7$ cells /mL to get diluted bacteria solution (marked as 3#). The mineralization solution was also prepared at the same time. The components of culture solution and mineralization solution were shown in Table 3 and Table 4 respectively.

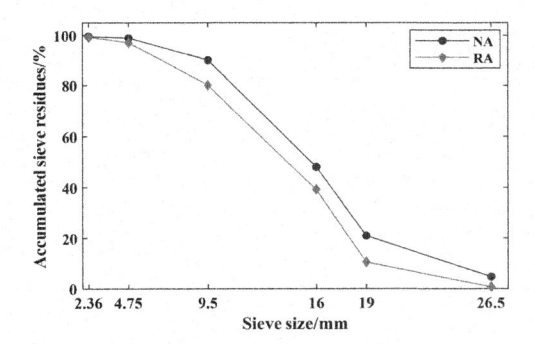

Figure 2.    Gradation curves of coarse aggregate.

Table 1.    Properties of natural coarse aggregate and recycled coarse aggregate.

| Aggregates | Bulk density/ ($kg/m^3$) | Apparent density/ ($kg/m^3$) | Moisture content/ % | Porosity/ % | Water absorption/ % | Crush index/ % |
|---|---|---|---|---|---|---|
| NA | 1545 | 2589 | 1.07 | 37.98 | 0.46 | 4.42 |
| RA | 1402 | 2254 | 3.29 | 39.19 | 2.91 | 14.74 |

Table 2. Physical and mechanical properties of cement.

| Density/ $(kg/m^3)$ | Specific surface area/ $(m^2/kg)$ | Standard water consumption % | Compressive strength/MPa | | Flexural strength/MPa | |
|---|---|---|---|---|---|---|
| | | | 3d | 28d | 3d | 28d |
| 3250 | 361.0 | 30.0 | 23.1 | 47.2 | 5.6 | 8.9 |

Table 3. The components of culture solution.

| Solution | Components | Contents |
|---|---|---|
| A solution | Peptone | 3g |
| | Beef extract | 10g |
| | Agar powder | 15g |
| | Distilled water | 850mL |
| B solution (CAPS) | CAPS | 22.13g |
| | Distilled water | 150mL |

Table 4. The components of mineralization solution.

| Solution | Components | Contents |
|---|---|---|
| A solution | Calcium nitrate ($Ca(NO_3)_2$) | 0.5M |
| | Distilled water | 850mL |
| B solution (CAPS) | CAPS | 22.13g |
| | Distilled water | 150mL |

## 2.2 Experiments methods

### Test of physical properties of aggregate

The bulk density, apparent density, water content, porosity, water absorption and crush index of both natural coarse aggregate and recycled coarse aggregate were tested following the GB/T 14685-2011 *specification of pebble and crushed stone for builds*

### The modification of recycled aggregate

Considering the physical modification theory, the recycled coarse aggregate was firstly stirred in the concrete mixer for about 20mins, and then the aggregate with particles size larger than 5 mm was collected, in order to remove the residual slurry that adhered to the surface of the aggregate particles. Finally, the collected recycled aggregate was immersed in WG solution (10% concentrations) in different modulus (1.0, 1.2, 1.5, 2.8) and at the same time, the same amount of recycled aggregate was immersed in PA solution with 5%, 10%, 15%, and 20% concentration respectively. Every 30mins, the mixed solution should be thoroughly stirred and the process lasts for 2h. After 2h, the recycled coarse aggregate was taken out and aired for further experiments. For the MICP method, the aggregate was immersed in culture solution for 24h, then taken out and divided into three parts with equal amounts. The three parts of aggregate were immersed into mineralization solution respectively for another 24h, and finally, taken out and aired for further tests. All the aggregate should be totally immersed in the solution.

### Preparation of recycled concrete specimens

The proportion of natural and recycled concrete can be seen in Table 5. After the concrete specimens were made, the slump was measured, and then molded and compacted. After curing for 1d at room temperature, the specimens were demold, and then cured the specimens for 3d and 28d respectively at standard conditions (20±2°C, relative humidity 95%). The size of specimens is 150mm×150mm×150mm.

Table 5. Mix properties of natural and recycled concrete.

| Components | Cement | Sand | Coarse aggregate | Water | Water reducer |
| --- | --- | --- | --- | --- | --- |
| Amounts/(kg/m$^3$) | 415 | 625 | 1210 | 170 | 2.08 |

**Workability and mechanical property of concrete**

The test of workability of concrete was followed by the Chinese specification GB/T 50080-2016 *Standard for test method of performance on ordinary fresh concrete,* which is similar to BS EN 12350- 2:2009, and the test of mechanical property of concrete was followed by the Chinese specification GB/T 50081-2019 *Standard for test method of mechanical properties on ordinary concrete.* The side surface of *the* specimen as a pressure-bearing surface was put on the backing plate of the hydraulic press. Then starting the machine, both upper and lower surfaces of the specimen should have uniform contact with the upper and lower pressure-bearing plates. The loading speed was 0.3MPa/s~0.8MPa/s. When the specimen was destroyed, it is needed to stop loading and record the loading value. The compressive strength should be calculated as below:

$$f_{cc} = \frac{F}{A}$$

where:
$f_{cc}$ = compressive strength of concrete (MPa);
$F$ = failure load;
$A$ = area of the pressure-bearing surface.

**Properties of anti-carbonation and anti-freezing of recycled concrete**

The test of concrete-accelerated carbonation was followed by the Chinese specification GB/T 50082-2009 *standard test methods for long-term performance and durability of ordinary concrete.* The three pieces of specimens were put into a carbonization box, and the intervals of specimens were not less than 50mm. The carbonization box was sealed after putting into the specimens, and then $CO_2$ was aerated into the box. The concentration of $CO_2$ should be kept up to 20±3%, the temperature was kept up to 20±2°C and relative moisture was around 70±5%. The carbonation depth of specimens was measured every 3d, 28d, and 60d. The mean carbonation depth of specimens should be calculated as below:

$$\overline{d_t} = \frac{1}{n}\sum_{i=1}^{n} d_i$$

where:
$\overline{d_t}$ = the mean carbonation depth of specimens after t days;
$d_i$ = the carbonation depth of each test point;
$n$ = the number of test points.

The freeze-thaw test followed the same Chinese specification as the test of concrete-access-elevated carbonation. The size of the specimen was 150mm×150mm×150mm and all the specimens were divided into three groups (the recycled concrete modified by WG, PA, and MICP methods respectively), each group contains three pieces of specimens. It is needed to measure the mass loss of specimens after different amounts of cycles of freezing and thawing and the compressive strength of recycled concrete after 150 times of cycles of freezing and thawing.

# 3 RESULTS AND DISCUSSION

## 3.1 *Influence of WG, PA, and MICP methods on water absorption and crush index of RA*

The recycled aggregate without both chemical and MICP modification was seen as the control group.

The first group of recycled aggregate was equally divided into four parts and each part was immersed in water glass solution with a modulus of 1.0, 1.2, 1.5, and 2.8 respectively (marked as WG 1, WG2, WG3, and WG4), and the second group of recycled aggregate was also equally divided into four parts and each part was immersed into Polyaluminium sulfate solution with a concentration of 5.0%, 10%, 15.0%, and 20.0% respectively (marked as PA1, PA2, PA3, and PA4). The third group of recycled aggregate was equally divided into three parts and each part was immersed in culture solution with different concentrations of bacteria (mentioned in 2.1, 1#, 2#, and 3#) respectively and then put into mineralization solution for 24h, marked as M1, M2, and M3 respectively. The enhancing effects of the WG, PA, and MICP method on water absorption and crush index of RA were shown in Tables 6, 7, and 8. It can be seen that with the increase of modulus of WG and concentration of PA, both the water absorption and crush index of RA were decreased. But the RA immersed in 1# culture solution has a lower water absorption and crush index than that of RA immersed in 2# and 3# culture solution. Both chemical and MICP methods could modify the recycled aggregate.

Table 6. Enhancing effect of water glass on the crushing index and water absorption.

| Modulus of WG | Control | 1.0 | 1.2 | 1.5 | 2.8 |
|---|---|---|---|---|---|
| Crushing index/% | 14.65 | 13.32 | 13.08 | 12.65 | 12.33 |
| Water absorption/% | 2.94 | 2.49 | 2.37 | 2.08 | 1.98 |

Table 7. Enhancing effect of Polyaluminium sulfate on the crushing index and water absorption

| Concentration of PA | Control | 5.0 | 10.0 | 15.0 | 20.0 |
|---|---|---|---|---|---|
| Crushing index/% | 14.65 | 14.39 | 14.15 | 13.87 | 13.59 |
| Water absorption/% | 2.94 | 2.73 | 2.47 | 2.42 | 2.20 |

Table 8. Enhancing effect of MICP methods on sulfate on the crushing index and water absorption.

| Concentration of bacteria | Control | $5 \times 10^7$ cell/mL | $1 \times 10^8$ cell/mL | $2 \times 10^8$ cell/mL |
|---|---|---|---|---|
| Crushing index/% | 14.65 | 12.80 | 11.70 | 12.93 |
| Water absorption/% | 2.94 | 2.23 | 2.03 | 2.56 |

## 3.2 *Influence of WG, PA, and MICP method on mechanical property of recycled concrete*

The recycled concrete made in RA was modified by the chemical method and MICP method, the relationship between water absorption of RA and slump of recycled concrete can be seen in Figures 3 and 4 respectively. It can be seen that with the decreasing of water absorption of RA, the slump of recycled concrete was increased linearly, but the recycled concrete made in MICP modified RA has a higher linear correlation and lower dispersion. Both chemical and MICP methods could improve the workability of recycled concrete.

The enhancing effects of chemical agents WG & PA and MICP methods on mechanical properties of recycled concrete were shown in Figures 5, 6, and 7 respectively. It can be seen that the mechanical properties of recycled concrete were apparently improved by both chemical and MICP modification methods. The 3d and 28d compressive strengths of recycled concrete were increased by 58.44% and 52.36% respectively modified by WG with modulus 2.8, and the counterparts of recycled concrete were increased by 33.77% and 27.47% respectively modified by PA with 20.0% concentration. While the recycled aggregate modified by culture solution with $1 \times 10^8$ cell/mL has the 3d and 28d compressive strength increased by 61.50% and 54.1%. It's obvious that both chemical and MICP methods could effectively improve the compressive strength of both RC and NRC, but the MICP-modified aggregate has higher compressive strength than that of aggregate modified by WG and PA. This was because the $CaCO_3$ precipitation produced by bacteria metabolism covered the surface of aggregate and filled the micro-cracks within aggregate, the $CaCO_3$ precipitation also has hardness and viscosity which could remedy the micro-cracks of aggregate and further improve the compressive strength of recycled aggregate. Meanwhile, the hard $CaCO_3$ could also strengthen the aggregate particles and combine with aggregate which commonly promotes the compressive strength of recycled concrete.

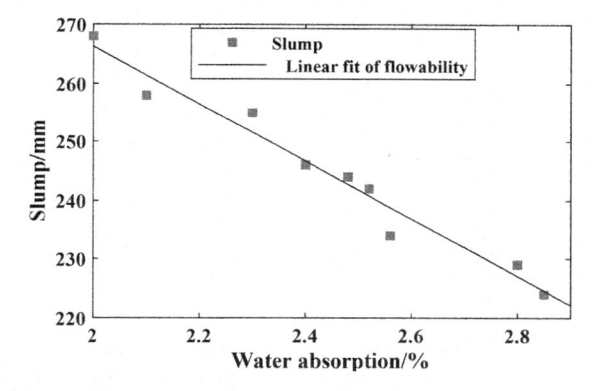

Figure 3.    Relation between the water absorption of chemical method enhanced recycled aggregate and the slump of concrete.

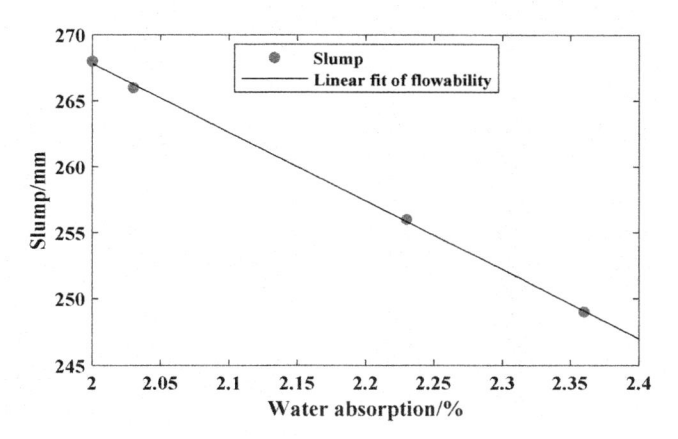

Figure 4.    Relation between the water absorption of MICP method enhanced recycled aggregate and the slump of concrete.

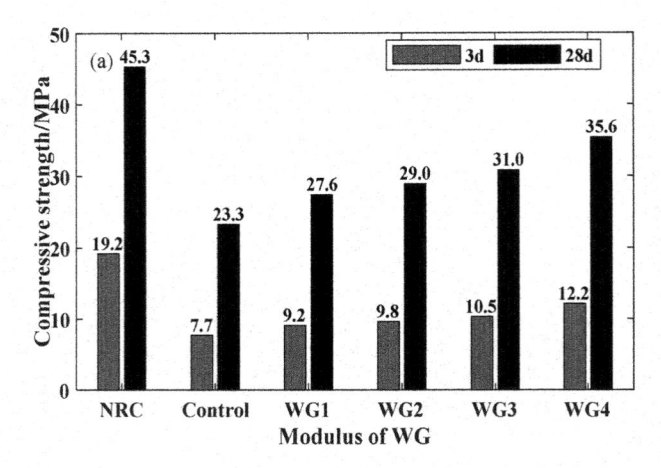

Figure 5.    Effect of WG on compressive strength of RC.

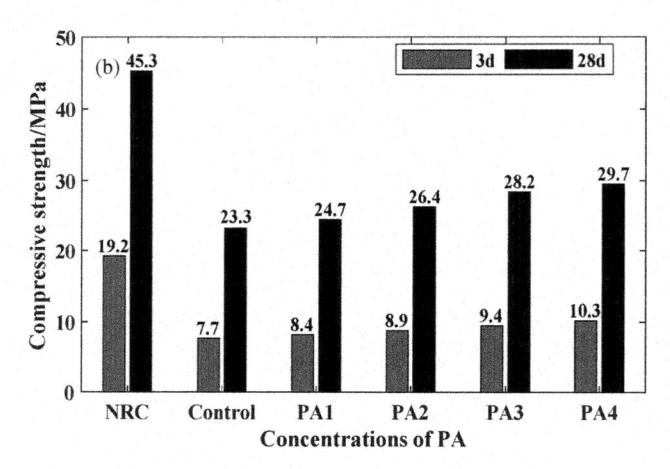

Figure 6.    Effect of PA on compressive strength of RC.

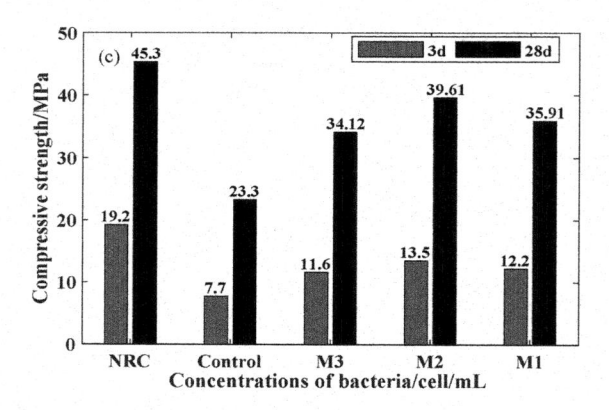

Figure 7.    Effect of MICP on compressive strength of RC.

In order to study the influence of chemical enhancing agents and the MICP method on the durability of recycled concrete, the properties of anti-carbonation and anti-freezing of recycled concrete were studied.

The depth of carbonation of recycled concrete modified by WG, PA, and MICP methods in different ages were shown in Figure 8(a), (b) and (c) respectively. It can be seen that the ability of anti-carbonation of natural aggregate concrete was better than that of recycled concrete. After modification by chemical and MICP methods, the depth of carbonation of recycled concrete was decreased, accordingly, the ability of anti-carbonation was strengthened. The optimum effect of anti-carbonation of recycled concrete was modified by WG in modulus 2.8 and PA in 20% concentration, meanwhile, the recycled concrete modified by culture solution with $1 \times 10^8$ cell/mL has the best effect of anti-carbonation in all modification methods, and its ability of anti-carbonation was even better than that of natural aggregate concrete. This was that compared with chemical enhancing agents modified recycled concrete, the RA modified by MICP method could produce $CaCO_3$ precipitation, and the products of concrete carbonation were also the $CaCO_3$, the MICP method produced $CaCO_3$ could distribute on the surface and in the micro-cracks of aggregate particles and further prohibit the process of carbonation.

The relationship between cycles of freezing and thawing and mass loss of recycled concrete modified by chemical enhancing agents and MICP methods were shown in Figure 9(a), (b), (c) respectively.

It can be seen that the mass loss of natural concrete aggregate was lower than that of recycled concrete. But with the increase of WG modulus and concentrations of PA, the mass loss of modified RC was decreased. The RC modified by WG in modulus 2.8 and PA with 20% concentration has the lowest mass loss, only 1.86% and 1.90% mass loss at the 150th cycle of freezing and thawing respectively. While the RC modified by culture solution with $1 \times 10^8$ cell/mL has only 1.80% mass loss. This was that the $CaCO_3$ precipitation produced by urease bacteria adheres to the surface within the micro-cracks of RA, and the $CaCO_3$ precipitation possessing high strength and viscosity could strengthen the structure stability of RA and further improve the compactness of RC. Meanwhile, the $CaCO_3$ precipitation has good compatibility with the concrete matrix, which could weaken the carbonation of concrete and regulate and control the process of concrete hydration. In summary, the RC could keep up the high strength and integrity during the process of cycles of freezing and thawing and prohibit the expansion shrinkage failure and deformation of concrete structure, reducing the mass loss.

According to the effect of WG, PA, and MICP method on mass loss of RC under cycles of freezing and thawing, the compressive strength of modified RC under the 150th cycle of freezing and thawing was measured. From Figure 10(a) and (b) we can see that both NRC and RA have lower compressive strength after the 150th cycle of freezing and thawing, but with an increase in modulus of WG and concentration of PA, the strength loss rate of modified RC were continuous decreased. The RC modified by WG in 2.8 moduli and PA in 20% concentration has only 26.9% and 28.3% strength loss respectively, compared with the control specimen has nearly 45.7% strength loss. From Figure 10(c) we can also see that the RC modified by culture solution with $1 \times 10^8$ cell /mL has only 19.1% strength loss after the 150th cycle of freezing and thawing. It can be seen that both chemical enhancing agents and the MICP method could reduce the strength loss after a cycle of freezing and thawing, but the MICP method has a better effect than that of chemical enhancing agents.

The interfacial microstructure of recycled aggregate modified by chemical enhancing agents and the MICP method was characterized by a scanning electron microscope (SEM). The images of interfacial micro-structure of recycled aggregate modified by WG in different modulus were shown in Figure 11, and the counterparts of recycled aggregate modified by PA and MICP method were shown

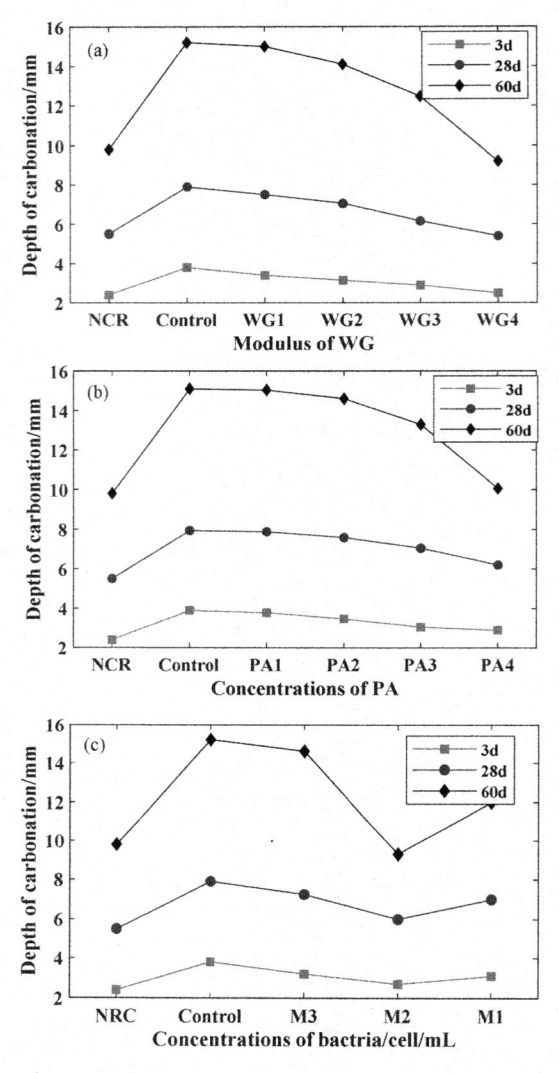

Figure 8.  Effect of chemical and MICP methods on the depth of carbonation (a) WG, (b) PA, (c) MICP.

in Figure12 and Figure13 respectively. It can be seen from Figure 11 that the aggregate particles without modification have a rough and uneven surface, and it was covered by lots of hardened cement paste and hydrate crystals. The porous structure with a large number of holes linking together within the recycled aggregate particles, and exist obvious micro- cracks ((Hao et al. 2019; Zhu & Wu 2017). The density of RA was improved after modification by WG and with the increasing modulus of WG, the hydration product gradually covered the surface of aggregate particles, then more and more hydration products remedied the micro-cracks and deficiency of aggregate and intensively covered the particle surface. The WG could promote the hydration reaction of concrete and benefit the remedy of the concrete matrix (Grabiec et al. 2012). The improvement of the density of RA was apparent when modified by WG in modulus 2.8, at the same time, the micro-cracks were nearly fully filled by new hydration product and the needle-like substance and other hydration products adhered to the surface of RA were increased the roughness of the surface of RA together,

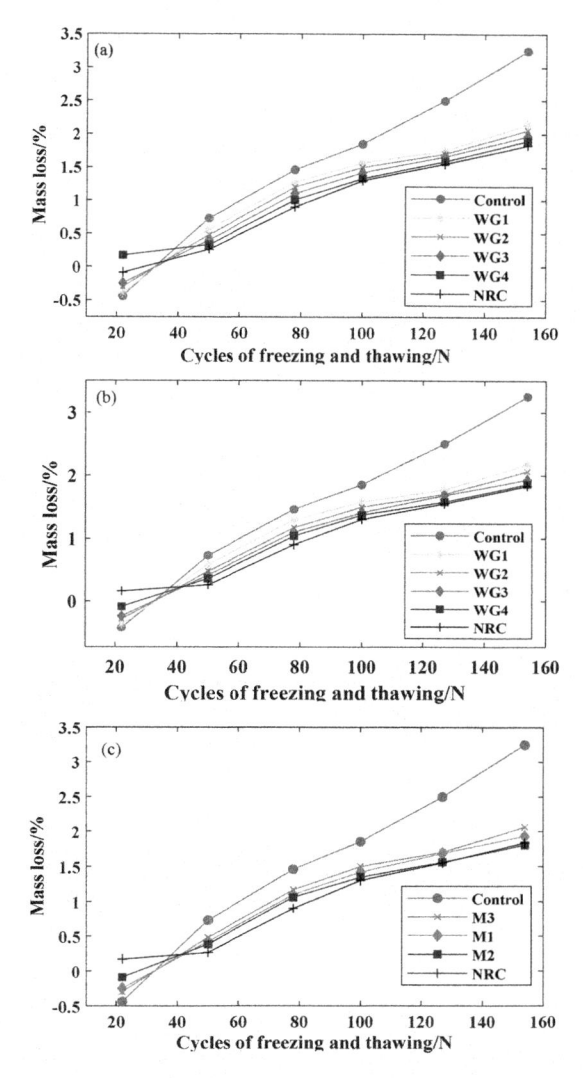

Figure 9. Mass loss of freeze-thaw cycles with enhancing agents and MICP method (a) WG, (b) PA, (c) MICP.

which benefit to the adhesion of mortar to the surface of RA and further improve the strength of RA. From figure 12 we can see that the density of interfacial RA modified by PA was increased compared with that of RA without modification. With the increase of concentration of PA, the Gypsum dihydrate and small amounts of Ettringite was formed and covered by the surface of RA as well as filling the defects of RA particles, and the increasing of hydration pro- ducts further promote the Ettringite separate out and fill the micro-crack distributed on the surface of RA. The RA modified by PA with 20% concentration has the highest density.

The RA modified by bacterial solution with different concentrations also has different surface morphology. It can be seen that the $CaCO_3$ precipitation induced by culture solution in low bacterial concentration ($5 \times 10^7$ cell /mL) was in a regular diamond or cube, while the $CaCO_3$ precipitation induced by high bacterial concentration culture solution ($2 \times 10^8$ cell /mL) was shaped in a sphere and overlap each other. Meanwhile, the moderate concentration culture solution ($1 \times 10^8$ cell /mL)

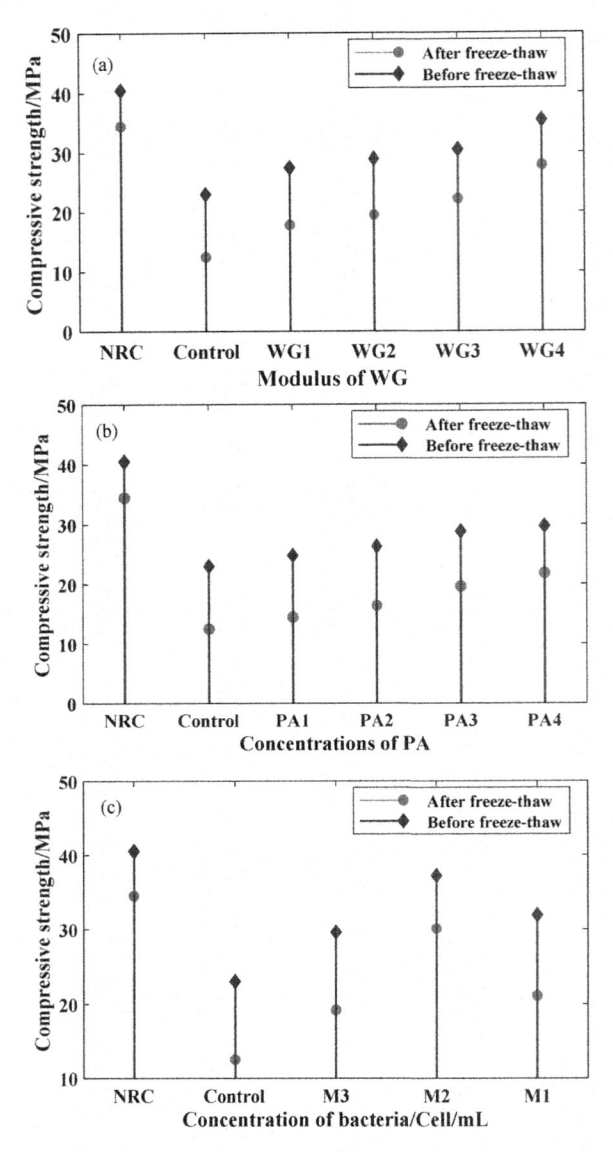

Figure 10.  Compressive strength with chemical and MICP methods after 150 times of freeze-thaw cycles.

generally induced the $CaCO_3$ precipitation has not only a regular cube but a sphere. All these precipitations were closely adhered to the surface of RA and wrapped in the RA particles. Low bacterial concentration solution mainly induced aragonite, while culture solution in high bacterial concentration always induced calcite. This was because, in the low bacteria concentration solution, the distance between the nucleation site of different organic matters was far away, and the growth of a crystal has sufficient space. The shapes of $CaCO_3$ crystals were mainly like cubes and diamonds. In the high bacteria concentration solution, the bacteria were agglomerated together (Yi et al. 2019). The polar group of the surface of bacteria was intertwined with organic macro- molecules secreted by bacteria and in solution, and finally formed the curved surface structure, so the spherical $CaCO_3$ crystal was easily formed.

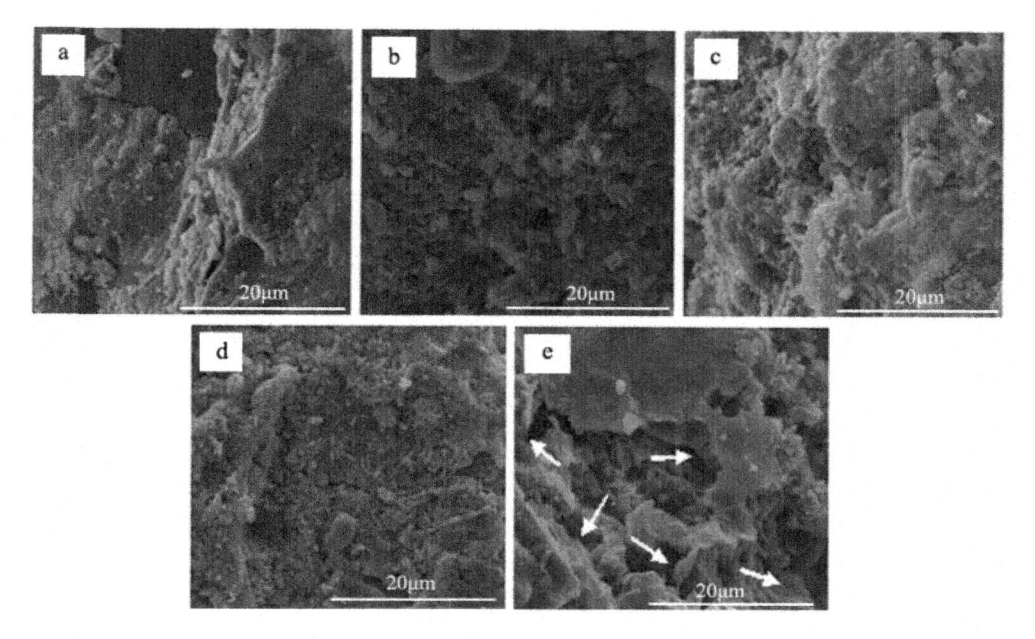

Figure 11. Interface of recycled aggregate by the enhancement of water glass (WG) (a) modulus 1.0; (b) modulus 1.2; (c) modulus 1.5; (d) modulus 2.8; (e) control group.

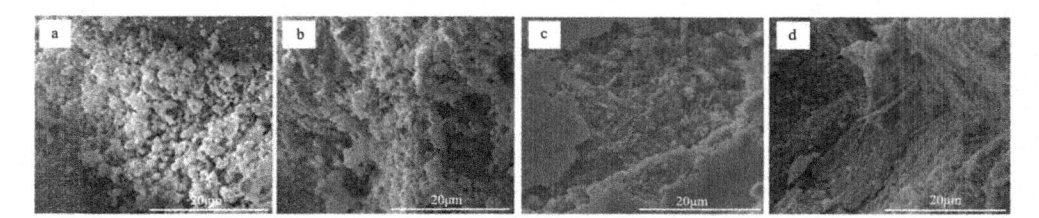

Figure 12. Interface of recycled aggregate by the enhancement of polyaluminium sulfate (PA) (a) concentration 5.0%; (b) concentration 10.0%; (c) concentration 15.0%; (d) concentration 20.0%.

Figure 13. Interface of recycled aggregate by the enhancement of MICP method bacteria concentration (a) $5 \times 10^7$ cell/mL; (b) $1 \times 10^8$ cell/mL; (c) $2 \times 10^8$ cell/mL.

## 3.5 *Analysis of the mechanism of chemical and MICP modification on RA*

(1) The WG with a modulus less than 1.5 could stimulate the activity of inert substance on mortar adhered to the surface of RA, promote the de- polymerization of aluminosilicate (Al-Si) glass phase of surface mortar and form the hydro-colloid precipitation phase, and further fill the pores of RA,

improve the density of the surface of RA. But the compressive strength of RA was not prominently promoted. With the increasing modulus of WG, the contents of $SiO_2$ and viscosity of the solution were also increased, when the solution of WG hardening, the silica gel was separated out and the micro-pores of RA were blocked by silica gel, further improving the micropore structure of RA. The hydration pro- duct $Ca(OH)_2$ formed in the old mortar of original concrete will react with WG and form the high-strength hydraulic calcium silicate colloid which could fill the pores of RA and further improve the density of RA (Cheng et al. 2008, Chen & Zhang 2020). The WG could also promote the hydration of unhydrated cement particles and activate the hydration of mineral admixtures in the original RA. And with the increasing modulus of WG, the amount of silica gel separated out was increased, more and more micro-pores within RA were blocked, and the density of RA was thus improved. All of these could refine the distribution of pores within concrete thus increasing the crushing index and strength of RA as well as reducing the water absorption of RA. And the compressive strength and durability of RC could also be improved.

(2) When immersing into the PA solution, the bubble appeared on the surface of RA and with the increasing concentration of RA, the reaction was more intense and lots of bubbles were produced, so it could assert that the surface of RA was partially carbonated, the original $Ca(OH)_2$ on the surface was formed into $CaCO_3$, the formed $CaCO_3$ were firstly react with slightly acidic liquid of PA and produce $CO_2$. After the $CaCO_3$ on the surface were exhausted, the PA will react with $Ca(OH)_2$ and form ettringite or gypsum. When the concentration of PA was increasing, the large amounts of directionally growing $Ca(OH)_2$ on the surface and $Al_2(SO4)_3$ in the PA will react with $Ca(OH)_2$ in the interface of RA and finally form the ettringite. At the same time, the gypsum dihydrate and $Al(OH)_3$ will also be formed (Song et al. 2015). All these reactions consume the interfacial weak phase $Ca(OH)_2$ and form the high-strength ettringite constantly filling the micro-cracks of RA, thereby improving the properties of both RA and RC.

(3) The produced $CaCO_3$ precipitation were in different morphology, size, and amounts under the different concentration of bacteria solution. In this study, the moderate bacteria concentration $(1 \times 10^8$ cell/mL) could form the calcite with moderate size and high strength as well as high hardness which could closely attach to the surface of RA and fill the micro-cracks within RA. The strength of RA and RC was greatly improved. While RA modified by culture solution with lower $(5 \times 10^7$ cell /mL) or higher $(2 \times 10^8$ cell /mL) bacteria concentration has a worse modification effect. When the RA is modified by a high concentration of bacteria solution, the formed $CaCO_3$ precipitation has a bigger size, and it's difficult to attach to the micro-cracks within RA, and the competitive effect will happen in a high concentration of bacteria, the bacteria would compete for oxygen and nutrition thus it can inhibit the growth of bacteria and further influence the analysis of bacteria, the decomposition of urea and formation of $CaCO_3$ precipitation will also be inhibited. The high concentration of bacteria will promote the amount and precipitation rate of $CaCO_3$ but weaken the mineralizing ability of single bacteria and the formed $CaCO_3$ precipitation was in powder and has low viscosity and hardness. In summary, the RA and RC modified by culture solution with a lower and higher concentration of bacteria will have a worse modification effect than that of RA and RC modified by culture solution with a moderate concentration of bacteria.

## 4 CONCLUSIONS

In this paper, both the chemical modification method and bio-method were conducted to modify the properties of recycled aggregates and recycled concrete, meanwhile, the difference between these two methods was also studied. The main conclusions can be summarized as follows:

(1) The WG could greatly improve the mechanical properties and compressive strength of RA and RC, the RC modified by WG solution with modulus 2.8 has the optimum improvement in compressive strength. The 3d and 28d compressive strength were increased by 58.44% and 52.36% respectively. The properties of anti-carbonation and anti-freezing also have a prominent promotion.

(2) The WG could greatly improve the mechanical properties and compressive strength of RA and RC, the RC modified by WG solution with modulus 2.8 has the optimum improvement in compressive strength. The 3d and 28d compressive strength were increased by 58.44% and 52.36% respectively. The properties of anti-carbonation and anti-freezing also have a prominent promotion.

(3) The RA and RC modified by the MICP method have the most prominent improvement in mechanical properties and compressive strength than that of RA and RC modified by chemical methods. The 3d and 28d compressive strength were increased by 61.50% and 54.1% respectively. The properties of anti- carbonation and anti-freezing could be greatly improved.

## REFERENCES

Achal V, Mukherjee A, Reddy M S. Microbial Concrete (2010): Way to Enhance the Durability of Building Structures[J]. *Journal of Materials in Civil Engineering*, 23 (6):730–734.

Aïtcin P C (2000). The cement of yesterday, concrete of tomorrow[J]. *Cement Concrete Research* 30: 1349–1359.

Bin Sun, Tintin Huang (2019). Effects of recycled aggregate by bacillus cohnii's mineralization on compressive strength of concrete[J]. *Journal of Yangtze River Scientific Research Institute*, 12(9):50–57.

Bru K, Touze S, Bourgeois F, et al. (2014). Assessment of a Microwave-assisted Recycling Process for the Recovery of High-quality Aggregates from Concrete Waste[J]. *International Journal of Mineral Processing*, 126:90–98.

Grabiec, Justyna Klama, Daniel Zawal, et al. (2012). Modification of recycled concrete aggregate by calcium carbonate biodeposition[J]. *Construction and Building Materials*, 34:145–150.

Haili Cheng, Di Zhang (2008). Discussions on frost resistance and resistance to sulfate attack of water glass reinforced reclaimed aggregate concrete[J]. *New Building Materials*, 35(08); 5–7.

Hideo Ogawa, Toyoharu Nawa, Kazu Ohya (2010). Research on Characterization of Recycled Fine Aggregate[J]. *Doboku Gakkai Ronbunshuu E*, 66(1):107–118.

Julia García-González, Desirée Rodríguez- Robles, et al (2017). Quality improvement of mixed and ceramic recycled aggregates by bio-deposition of calcium carbonate[J]. *Construction and Building Materials*, (154):1015–1023.

Liyang Yi, Chaosheng Tang, Yuehan Xie, et al. (2019). Factors affecting improvement in engineering properties of geomaterials by microbial-induced calcite precipitation[J]. *Rock and Soil Mechanics*, 7(40):2525–2546.

Lotfi S, Eggimann MW, Eckhard (2015). Performance of recycled aggregate concrete based on a new concrete recycling technology [J]. *Construction and Building Materials*, (95):243–256.

Pin Chen, Jingxu Zhang (2020), Experimental study on the behavior of recycled aggregates strengthened by microbial induced carbonate precipitation[J]. *Journal of Zhejiang Sci-Tech University* (Natural Sciences Edition), 43(01):122–129.

Salomon ML, Paulo H (2004). The durability of recycled aggregates concrete: a safe way to sustainable development[J]. *Cement and Concrete Research*, 34(11): 1975–1980.

Song X F, Qiao P Z, Wen H F (2015). Recycled aggregate concrete enhanced with polymer aluminum sulfate[J]. *Magazine of Concrete Research* 1:1–7.

Weilai Zeng (2019). Using microbial carbonate precipitation to improve the properties of recycled aggregate[J]. *Construction and Building Materials*, 26(8):1–9.

Xiaohu Hao, Jiaguang Zhang et al. (2019). Study on the effect of calcium ion on the performance of microbial mineralized modified recycled aggregate[J]. *New Building Materials*, 46 (09):84–87.

Yaguang Zhu, Yankai Wu (2017). Effect of microbial mineralization on properties of recycled fine aggregate[J]. *China Concrete and Cement Products*, (12):93–96.

Zhang J, Shi C, Li Y, et al. (2015). Performance of Enhancement of Recycled Concrete Aggregates Through Carbonation [J]. *Journal of Materials in Civil Engineering*, 27(11):1–6.

Zhangyao Feng, Yuxi Zhao, Weilai Zeng (2020). Using microbial carbonate precipitation to improve the properties of recycled fine aggregate and mortar [J]. *Construction and Building Materials*, 230: 1–8.

*Civil Engineering and Urban Research – Mohamed & Hou (Eds)*
*© 2023 the Authors, ISBN 978-1-032-44487-1*

# Experimental investigation of mechanical properties of sleeve grouting materials

Yuliang Qi & Keke Huang
*Guangzhou Research Institute of Construction Industry Co., Ltd., Guangzhou, P.R. China*
*Guangzhou Construction Engineering Co., Ltd., Guangzhou, P.R. China*

Zhanzhong Li*
*School of Civil Engineering, Guangzhou University, Guangzhou, P.R. China*

ABSTRACT:   In order to study the mechanical properties of grouting materials for rebar sleeve splicing in assembled members, the strength test of grouting materials was carried out, and 12 specimens were designed and made for compressive strength test and splitting test. The failure modes of grouting materials were observed. The compressive test showed a large amount of grouting materials peeling off, and the splitting test showed cracking in the middle. At the same time, the corresponding strength was measured. The results show that the displacement and the strain increase with the increase of load, but when the load reaches the ultimate load, the crack develops rapidly, leading to the failure of the specimen. The model of grout sleeve splicing of rebars was established and the three-dimensional finite element analysis of the tensile loading test was carried out. The simulation results are basically consistent with the experimental results.

## 1   INTRODUCTION

Prefabricated building is a new type of building that is different from traditional cast-in-place concrete construction. Building materials are pre-produced in factories into components and then transported to the construction site. These lightweight prefabricated components are combined by using some light steel structures (Jiang et al. 2017; Qi & Zhang 2015). Due to its unique working mechanism, grout sleeve splicing of rebars is one of the most commonly used reinforcement connection methods in the construction of prefabricated components. In recent years, many scholars have studied the structural properties of grout sleeve splicing of rebars (Zheng et al. 2018, Zhao et al. 2019). Li et al. (2021) pointed out that the mechanism of the grout sleeve splicing of rebars is that when the grouting material in the sleeve is restrained by the sleeve, a large normal stress is generated. Then the steel bar generates a frictional force on its surface through the normal stress and transmits the axial stress of the steel bar.

Sun et al. (2017) pointed out that the humidity of the curing environment has a great influence on the strength of the grouting material. In practical applications, the grouting material is embedded in the sleeve of the concrete member, and its water loss is less. Therefore, the actual strength is closer to the hydroponic condition at room temperature. Xiong et al. (2019) pointed out that the shape of the grouting material has a great influence on its compressive strength, while the size has no significant effect.

According to the existing research, the strength and mechanical properties of grouting materials for rebar sleeve splicing under axial compression have not been fully studied. In this paper, on the basis of literature (Liu et al. 2020), the compressive strength and tensile strength of grouting

---

*Corresponding Author: 2112116188@e.gzhu.edu.cn

DOI 10.1201/9781003372417-34

materials were measured. Through the analysis of test data, accurate data is provided for the subsequent numerical simulation of material parameters, and the accuracy of the rebar grouting sleeve model is verified. The strong test data and theoretical support are provided for the engineering application of reinforced sleeve grouting connectors.

## 2 EXPERIMENTAL PROGRAM

### 2.1 *Test design*

By analogy to the concrete test principle, the uniaxial compression test and splitting test are used to measure the compressive and tensile mechanical behavior of the grouting materials. Two batches of tests are designed in this paper. The first batch of experiments mainly studies the uniaxial compressive strength, and the second batch mainly studies the tensile strength.

### 2.2 *Test specimens*

In this strength test of grouting material, specimens were made according to the specification and maintained in the Structural Laboratory of South China Agricultural University. Standard cylindrical specimens with dimensions of 150 mm×300 mm are used for the uniaxial compression test, and cubic specimens with dimensions of 150 mm×150 mm×150 mm are used for the splitting test. The grouting material is injected into the cylindrical and cube-shaped specimen mold. At that time, the mold should not be vibrated until the grouting material and the upper edge of the mold are at the same level. The standard strength is achieved after curing for 28 days in the standard curing chamber. Two groups of 12 specimens are completed, with 6 specimens in each group. Uniaxial compressive strength test and splitting test are conducted respectively, (as illustrated in Figure 1. Grouting materials specimen.). Two groups of parallel experiments are used in this experiment.

Figure 1. Grouting materials specimen.

### 2.3 *Test method*

#### 2.3.1 *Compressive strength test*
The compression test was carried out on the WHY-5000 automatic pressure testing machine in the Structural Laboratory of the South China University of Technology. Before the test, three vertical

strain gauges and one transverse strain gauge were installed on the side of the specimen to measure the longitudinal and transverse deformation of the sample, as shown in Figure 2. During the test, a cylindrical specimen of 150 mm×300 mm was used, which was loaded with the forming surface. The loading surface was ground flat before loading. In the test, the specimens were loaded at the center above it and loaded under a constant displacement loading rate of 1 MPa/s. During the test, the specimen should be kept horizontal, and the force plane should be parallel to the horizontal plane. When the specimen is close to failure and begins to deform rapidly, the test throttle is stopped.

Figure 2.   Configuration of the uniaxial compression test.

### 2.3.2 *Splitting test*

The tensile strength of concrete has always been the focus of many researchers. Since the tensile strength of concrete is not easy to obtain, the splitting strength is usually considered the tensile strength of concrete. The tensile strength of the specimens was measured by a splitting test. As with the compressive test, there are a total of 5 specimens. Before the experiment, two strain gauges were installed on two sides of the sample. Among them, one side is installed at the upper and middle positions, while the other side is installed at the middle and lower positions, as shown in Figure 3. It is used to observe the crack development law of grouting materials specimen in splitting test. In the test, the sample should be fixed at the center of the plate. The specimens were loaded under a constant displacement loading rate of 0.1 MPa/s until the sample is destroyed.

## 3   TEST RESULTS AND ANALYSIS

The tested mechanical strength of the specimens is shown in Table 1, and the failure modes are presented in Figure 4.

### 3.1 *Compressive strength test*

Among the 6 specimens, CY1 is used as the preloaded specimen, whose data is not reliable. So, no response data of CY1 is collected. The remaining five test specimens were used for analysis. As illustrated in Figure 5, the load-displacement curves of grouting materials in the uniaxial compression test are similar to that of standard concrete. The load increases with the displacement.

Figure 3.  Configuration of the splitting test.

Table 1.  Test results of the mechanical properties.

| Test | Compression test | | Splitting test | |
|---|---|---|---|---|
| Specimen | Failure load javascript:void(0); (kN) | Axial compressive strength (MPa) | Failure load javascript:void(0); (kN) | Tensile strength (MPa) |
| 2 | 833.31 | 47.16 | 114.02 | 3.23 |
| 3 | 1198.96 | 67.85 | 140.02 | 3.96 |
| 4 | 1213.92 | 68.69 | 127.734 | 3.64 |
| 5 | 1173.17 | 66.39 | 93.18 | 2.64 |
| 6 | 1313.58 | 74.33 | 116.872 | 3.30 |
| Average | 1146.59 | 64.88 | 118.37 | 3.35 |

When the maximum load is reached, the cracks of the sample appear and increase sharply. Then the cracks run through the two loading surfaces quickly and form more vertical cracks. As a result, the bearing capacity of grout test blocks begins to decrease. Subsequently, a large amount of grouting materials began to flake, which means that specimens are completely destroyed. It can be observed that the ultimate load of CY2 is lower than that of other specimens. This phenomenon may be due to the excessive addition of water in the preparation of the specimen, leading to the reduction of load-bearing capacity. There is not much difference between the other groups of data.

During the experiment, a strain tester was used to collect the vertical and transverse strain data of the grouting materials specimens during the compression process. A total of 5 sets of strain data were taken, and the specimens CY4 and CY5 were selected for discussion. It can be seen from Figure 6 that the vertical strain and transverse strain increase with load in the initial stage, with the vertical strain being larger than the transverse strain. However, when the load reaches the ultimate load, the cracks of the specimen increase, the strain appears abruptly, the transverse strain increases sharply, and soon exceeds the vertical strain. The specimen begins to fail in a large area, and the axial strain gauge falls off with the failure part. Among 5 specimens, CY5 has an early mutation of

(a)  Cylindrical specimen.  (b) Cube specimen.

Figure 4.    Illustration of failure modes.

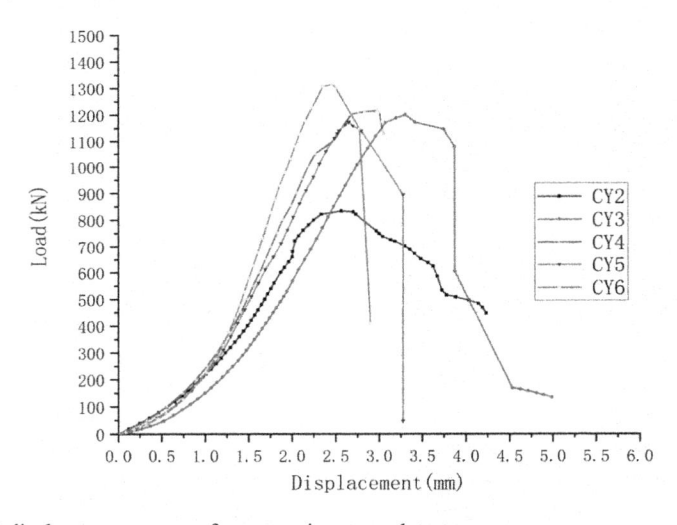

Figure 5.    Load-displacement curves of compressive strength test.

the transverse strain. This phenomenon may be due to the early peeling off a little surface grouting materials at the pasting strain, resulting in a large deformation of the strain gauge earlier.

### 3.2  *Splitting test*

Figure 7 shows the load-displacement curves of the splitting test. Before the load reaches the ultimate load, the load is positively correlated with the displacement, and the cube specimen has no visible cracks at the initial stage of loading. When the ultimate load is reached, the cracks in the middle of the specimen develop rapidly, and the bearing capacity of the specimen dropped sharply. Then the specimens cracked and failed. Among 5 specimens, the ultimate load of CY5 is relatively low, which may be caused by the operation error during the test or it is not placed in the center. There is not much difference between the other groups of data.

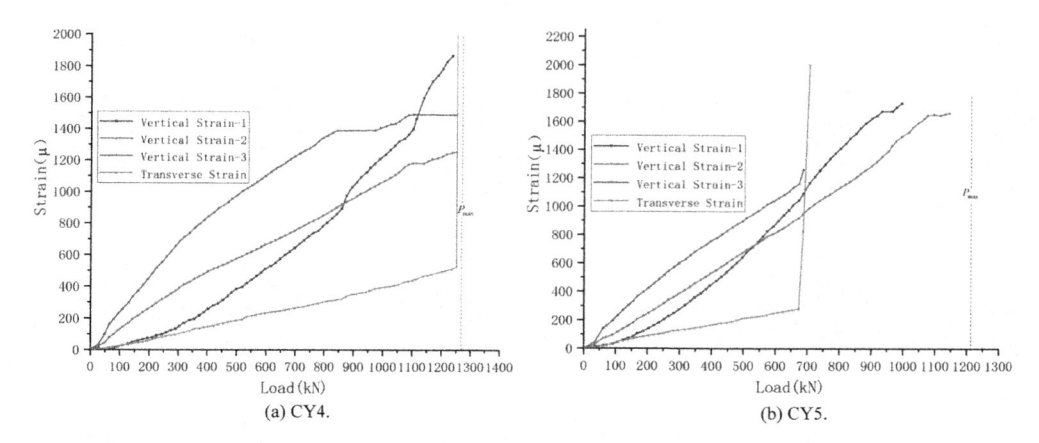

(a) CY4.                    (b) CY5.

Figure 6.    Deformation of cylindrical specimens in a uniaxial compression test.

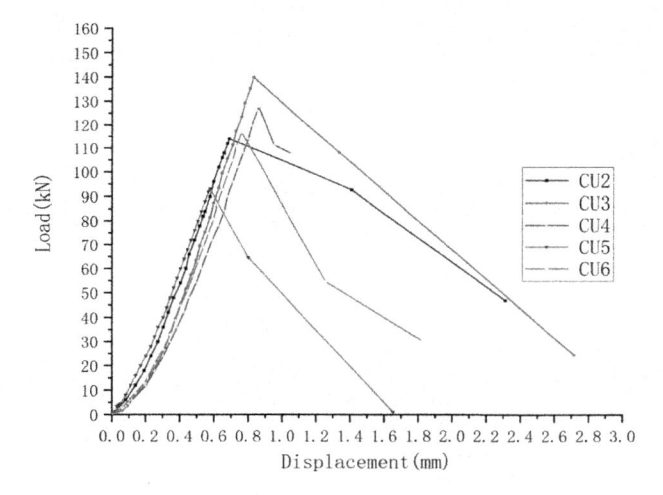

Figure 7.    Load-displacement curves of splitting test.

A total of 5 sets of strain data were taken, and the specimens CY4 and CY5 were selected for discussion. It can be seen from (a) CY4. (b) CY5.

Figure 8 that in the initial stage, the four strain gauges all increase with the increase of load. At that time, the strain at the middle position on both sides increase faster, and the tensile stress concentration appeared in the test specimen. When the load reaches the ultimate load, the strain gauges in the middle position increase sharply, showing that the cracks in the middle of the test block increase. Then the cracks develop on both sides of the sample. When the cracks develop into through cracks, some strain gauges are pulled off. As a result, the grouting material cracked and failed.

## 4    NUMERICAL ANALYSIS

### 4.1    *Model introduction*

The general-purpose finite element program ABAQUS was employed for the FE analysis in this study. In order to achieve the actual effect as much as possible, the finite element model is established

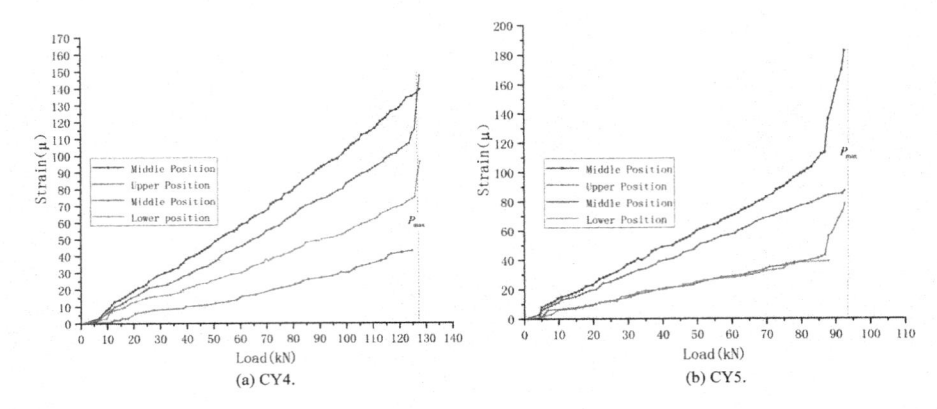

(a) CY4.

(b) CY5.

Figure 8.  Deformation development of cube specimens in splitting test.

based on the test specimen. HRB400 rebar with a diameter of 18 mm was selected as the simulation object. The length of the sleeve in the FE models is 340 mm, the outer diameter is 50 mm with 38mm of inner diameter. Also, there are 8 shear grooves (as illustrated in Figure 9). The tension of the connector is carried out by applying displacement load to the reinforcing bars at both ends in the model.

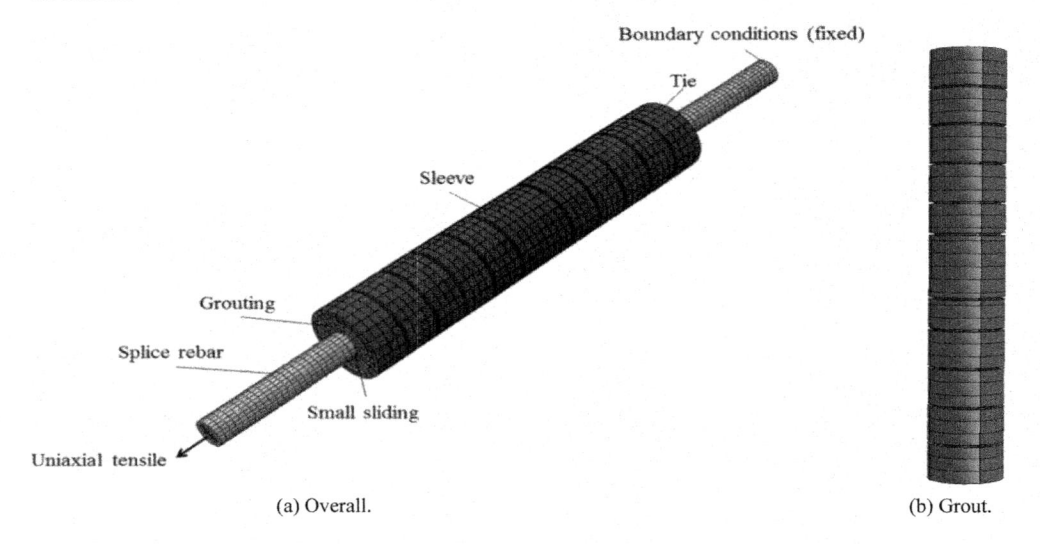

(a) Overall.

(b) Grout.

Figure 9.  Finite element model.

## 4.2  *The constitutive model of material*

In the simulation, the isotropic hardening trilinear model is adopted for rebar and sleeve with yield strength and ultimate strength.

Based on the above uniaxial compression and splitting test studies, the properties of grouting materials are similar to those of concrete. Therefore, the concrete damage plasticity (CDP) model is used as the constitutive model of the grouting materials. CDP model that in the finite element considers the plasticity and damage of the material at the same time. It is assumed that the concrete material in the CDP model has two main failure modes: tensile cracking and compression crushing.

The yield surface evolution is controlled by two hardening variables, $\varepsilon_t^{\sim ck}$ and $\varepsilon_c^{\sim in}$, which correspond to tensile and compressive failures, respectively. In the CDP model, two damage factors, namely compressive damage factor and tensile damage factor, are used to describe the damage behavior of concrete materials in the inelastic stage, as shown in Figure 10.

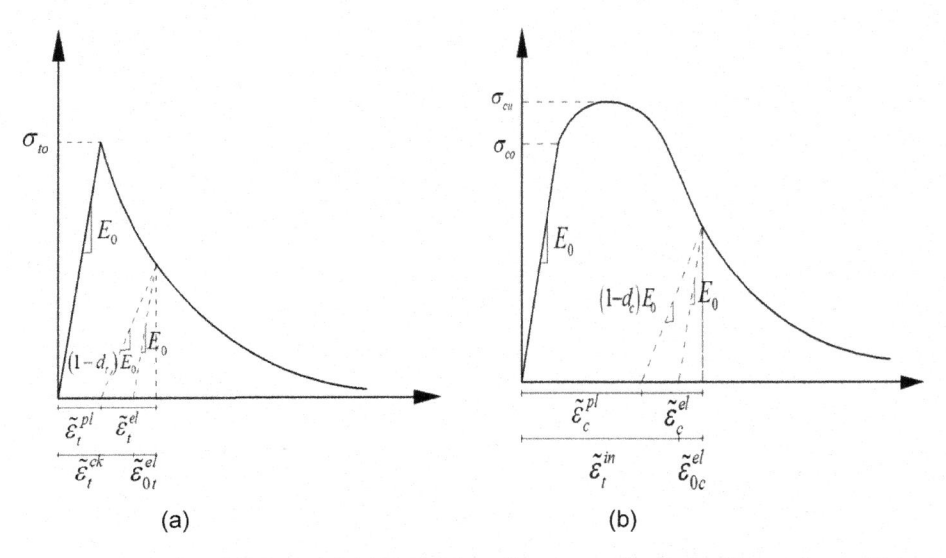

(a)　　　　　　　　　　(b)

Figure 10.　Description of CDP model: (a) Uniaxial tensile stress-strain relationship and the cracking strain $\varepsilon_t^{\sim ck}$; (b) Stress-strain relationship under uniaxial compression and compression inelastic strain diagram $\varepsilon_c^{\sim in}$.

### 4.3　*Numerical results and discussion*

Figure 11 illustrates the comparison of experimental and numerical results. It can be observed that the simulation results are basically consistent with the laboratory test results. It shows that rebars and sleeves simulated by linear elasticity and grouting materials simulated by the CDP model for analysis are feasible. It lays the foundation for further research such as parameter analysis.

## 5　CONCLUSIONS

In this paper, a series of tests were conducted to reveal the mechanical properties and failure mechanism of the cementitious grout commonly employed in prefabricated building engineering applications. Based on the test results, a three-dimensional FE model was established and validated. The following conclusions can be drawn:

(1) In the compressive strength test, the load increases with the increase of the displacement. When the load reaches the maximum load, the cylinder sample is damaged, showing a large amount of grouting material peeling off.
(2) In the splitting test, the tensile stress concentration appears in the test block when loading, and when the load reaches the ultimate load, the cube sample is damaged, manifesting as intermediate cracking.
(3) The numerical simulation results are basically consistent with the laboratory test results.

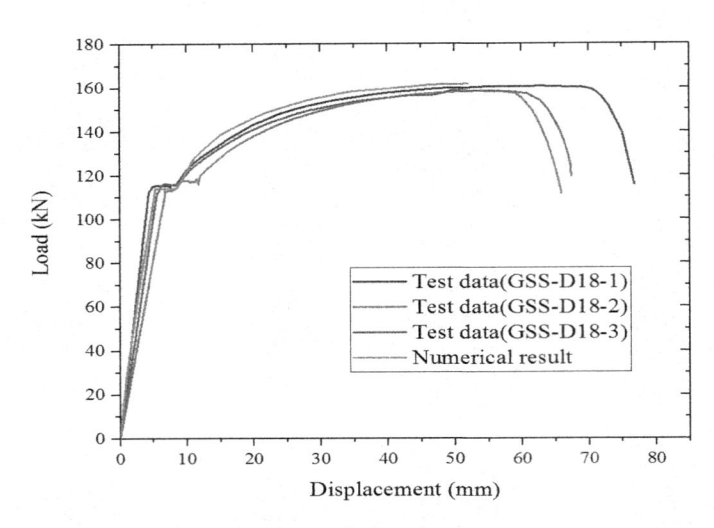

Figure 11.    Comparison of experimental and numerical results.

## ACKNOWLEDGMENTS

This work was supported by special funding for Assembly Buildings of Guangzhou Municipal Construction Group Co., Ltd (2019-KJ019; 2019-KJ033).

## REFERENCES

Jiang, W.L., Li, L., Hou, J., Zhang, Z.L., Guo, Z.Y. (2017) Research on the integration of design and construction of precast residence. *J. Construction Technology*, 46(22): 72–74.

Li, S.Q., Wang, Y.J., Feng, C. (2021) Summarization of research on mixing ratio and properties of sleeve grouting. *J. China Concrete* (Prefabricated Building), 39–43.

Liu, C., Pan, L.F., Liu, H., Tong, H.W., Yang, Y.B., Chen, W. (2020) Experimental and numerical investigation on mechanical properties of grouted-sleeve splices. *J. Sci. Construction and Building Materials*, 260.

Qi, B.K., Zhang, Y. (2015) Prefabricated construction development bottleneck and countermeasures research. *J. Journal of Shenyang Jianzhu University* (Social Science), 17(2): 156–159.

Sun, B., Mao, S.Y., Zhang, J.F., Yang, B. (2017) Experimental investigation of impacts on compression strength of grout for sleeve splicing of grouting rebars. *J. Engineering Quality*, 35(6): 25–28.

Xiong, Y., Li, J.H., Sun, B., Mao, S.Y. (2019) Strength and influence factors of sleeve grouting materials in prefabricated building. *J. Journal of Building Materials*, 22(2): 272–277.

Zhao, C.F., Zhang, Z.D., Wang, J.F., Wang, B. (2019) Numerical and theoretical analysis on the mechanical properties of improved CP-GFRP splice sleeve. *J. Sci. Thin Wall Struct*, 137: 487–501.

Zheng, Y.F., Guo, Z.X., Guan D.Z., Zhang, X. (2018) Parametric study on a novel grouted rolling pipe splice for precast concrete construction. *J. Sci. Construction and Building Materials*, 166: 452–463.

*Civil Engineering and Urban Research – Mohamed & Hou (Eds)*
*© 2023 the Authors, ISBN 978-1-032-44487-1*

# Research on influence of grouting defects on mechanical properties of sleeve joints

Yuliang Qi & Keke Huang
*Guangzhou Research Institute of Construction Industry Co., Ltd., Guangzhou, P.R. China*
*Guangzhou Construction Engineering Co., Ltd., Guangzhou, P.R. China*

Yequan Zhan*
*School of Civil Engineering, Guangzhou University, Guangzhou, P.R. China*

ABSTRACT: In this paper, a series of tensile loading tests were conducted on a group of grouted sleeve splices. Three-dimensional finite element analysis of the uniaxial tensile test was carried out considering grouting plasticity and damage, and the simulation results were in good agreement with the experimental data. At the same time, grouting defects were set at the end of the grouting sleeve, and the influence of different types of grouting defects on the mechanical properties of the grouting connection joint of the steel sleeve was studied. The numerical results reveal that the bearing capacity of the grouting sleeve will decrease exponentially with the increase of the volume ratio of grouting defects, and the failure mode will change from the tensile fracture of rebar to the pulling out of rebar. When the grouting defect sleeve is slipped and damaged, the anchorage length between the vertical direction and the upper steel bar should be greater than 6.5d, and the transverse defect thickness should be controlled within 0.13D.

## 1 INTRODUCTION

It is difficult to control the grouting quality in the sleeve grouting connection. Gao et al. (2019) investigated the effect of central defects on the uniaxial tensile properties of fully grouted sleeve joints, and it was concluded that the maximum length of central defects is about 50%~66.7% of the maximum length of end defects. Zheng et al. (2020) found that with the decrease in the anchorage length of the steel bar, the failure mode of the defective specimen may change from the tensile fracture of the steel bar to the interfacial bond-slip failure of the steel bar. Liu et al. (2020) investigated the failure process of the grouting sleeve joint under uniaxial tensile load and concluded that the anchoring length of greater than 7d is sufficient to prevent slip failure between the grout and steel bar. In this paper, based on literature (Liu et al. 2020), the influence of different sizes of defects on the uniaxial tensile properties of fully grouted sleeve joints with a diameter of $d = 18\,cm$ and an anchorage length of 8d is studied.

## 2 EXPERIMENTAL PROGRAM

### 2.1 Specimen

HRB400 deformed rebars with a diameter of 18 mm were used in tests. The sleeves were grouted manually and maintained for 28 days. Details of the specimen are shown in Figure 1 and Table 1.

---

*Corresponding Author: 2112016280@e.gzhu.edu.cn

DOI 10.1201/9781003372417-35
257

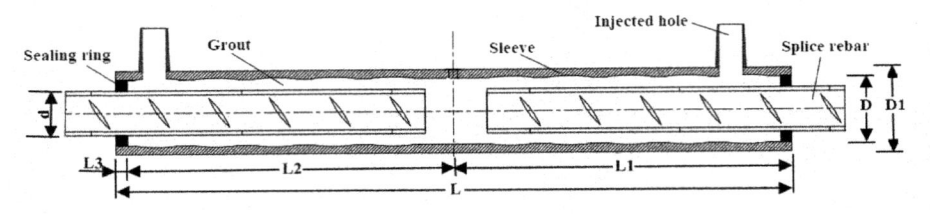

Figure 1.  Specimen detail.

Table 1.  Specimen dimensions.

| Steel bar diameter (mm) | Sleeve length L(mm) | Spliced bar L1 (mm) | Sealing ring size L3 (mm) | Sleeve outer diameter D1(mm) | Sleeve inner diameter D(mm) |
| --- | --- | --- | --- | --- | --- |
| 18 | 340 | 170 | 10 | 50 | 38 |

## 2.2  *Test results*

In order to measure the yield strength and ultimate strength of the specimen, a uniaxial tensile test was carried out on the specimen using an electro-hydraulic servo universal testing machine. In this test, the displacement-controlled loading method was adopted, and the loading speed was 5 mm/min. The loading was performed until the steel bar was broken or the steel bar was slipped and damaged, and the relevant data was recorded with a static strain tester.

During the stretching process, there will be a little grouting material falling off at both ends of the sleeve. The failure modes of the specimens are all steel bar breaking, and there is no phenomenon of steel slip failure. The ultimate bearing capacity of the specimen depends on the strength of the connecting reinforcement. The uniaxial tensile mechanical parameters of the specimens are shown in Table 2. Figure 2. shows the load-displacement curve generated during the tensile test, which

Table 2.  Test results of the mechanical properties.

| Specimen | Diameter d(mm) | Yield capacity (kN) | Ultimate capacity (kN) | Maximum elongation ratio (%) | Failure modes |
| --- | --- | --- | --- | --- | --- |
| D18-1 | 18 | 112.1 | 160.64 | 14.3 | Rebar fracture |
| D18-2 | 18 | 112.57 | 158.42 | 14.6 | Rebar fracture |
| D18-3 | 18 | 111.78 | 158.68 | 12.2 | Rebar fracture |

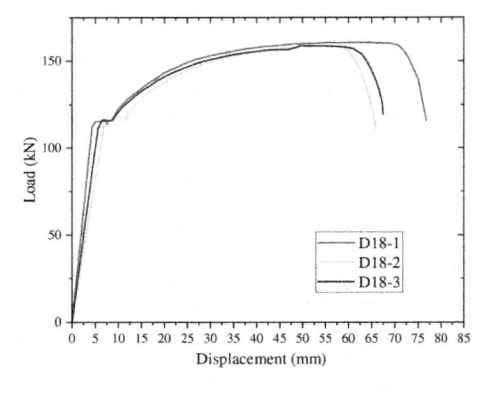

Figure 2.  Load-displacement curves of specimens.

is similar to the trend of the load-displacement curve of the steel bar. First is the elastic stage, then the yield stage, and finally reaches the ultimate strength and fails, which also means that no obvious slip failure occurs inside the sleeve. When the tensile strength of the specimen reaches the peak load, the stress of the steel bar reaches the ultimate strength and obvious necking deformation occurs, which can be judged as the fracture failure of the rebar.

## 3 FINITE ELEMENT MODEL

### 3.1 Model geometry, and boundary conditions

The grouting sleeve models were generated based on the finite element program ABAQUS, and the geometric parameters are shown in Table 1. Since the FE analysis required high calculation accuracy, a total of 26,784 elements of type C3D8I were used in the FE models to remove shear locking and reduce volumetric locking during the simulation. Reference points were established at both ends of the reinforcement to simulate the uniaxial tensile test of the specimen. Incremental displacement load is applied on the top of one bar with the bottom of the other bar fixed. The schematic diagram of the model is shown in Figure 3.

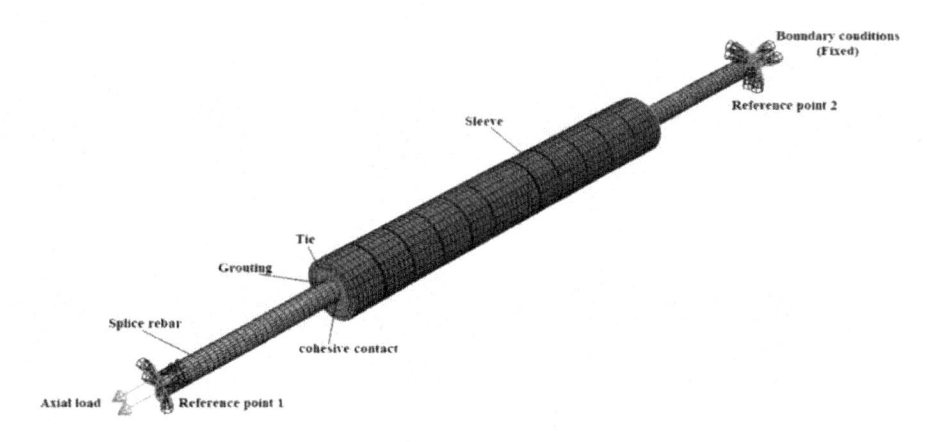

Figure 3. Finite element model.

### 3.2 Material and interaction properties

The uniaxial compressive test and splitting test of concrete show that the properties of the grouting material are similar to that of concrete, so the concrete damage model (Concrete Damage Plasticity) is used as the constitutive model of the grout. The isotropic hardening trilinear model is adopted for rebar and sleeve with yield strength and ultimate strength. The material parameters are shown in Table 3 and 4.

Where $\psi$ is the expansion angle, $\epsilon$ is the flow potential offset, $f_{b0}/f_{c0}$ is the ratio of the ultimate compressive strength of two axes to the ultimate compressive strength of one axle, $\kappa$ is the ratio of the second stress invariant on the tension meridian plane to the compression meridian plane, and $\mu$ is the viscosity coefficient.

According to the failure mode of the specimen in the test, there is no obvious slip failure between the sleeve and the grout, so the sleeve and the grouting material are tied together. The shear failure or the splitting failure of the grouting material between the rebar and the grout is essentially due to insufficient bond strength, so the cohesive contact between the rebar and the grout is used.

Table 3. Basic mechanical parameter.

| Materials | Elastic modulus E (MPa) | Poisson's ratio $\nu$ | Density $\rho$ (kg/m$^3$) | Yield strength $f_v$ (MPa) | Ultimate strength $f_u$ (MPa) | Axial compressive strength $f_c$ (MPa) | Tensile strength $f_t$ (MPa) |
|---|---|---|---|---|---|---|---|
| Rebar | 206 000 | 0.3 | 7850 | 445 | 602 | | |
| Sleeve | 206 000 | 0.3 | 7850 | 400 | 600 | | |
| Grout | 34 500 | 0.2 | 2539 | | | 69.3 | 3.5 |

Table 4. CDP model parameters for grout material.

| $\psi$ | $\epsilon$ | $f_{b0}/f_{c0}$ | $\kappa$ | $\mu$ |
|---|---|---|---|---|
| 30° | 0.1 | 1.16 | 0.667 | 0.0005 |

### 3.3 *Validation of FE model*

Figure 4. shows the finite element results of the load-displacement curve of the D18 specimen. It can be seen from the figure that the simulation results of the tensile load-displacement curve shape of each specimen are basically consistent with the experimental results. As can be seen from Table 5, the maximum difference between the yield load and ultimate load of the numerical simulation and test results is 2.3% and 2.1%, respectively. The failure modes of the samples are all steel bar tensile failure.

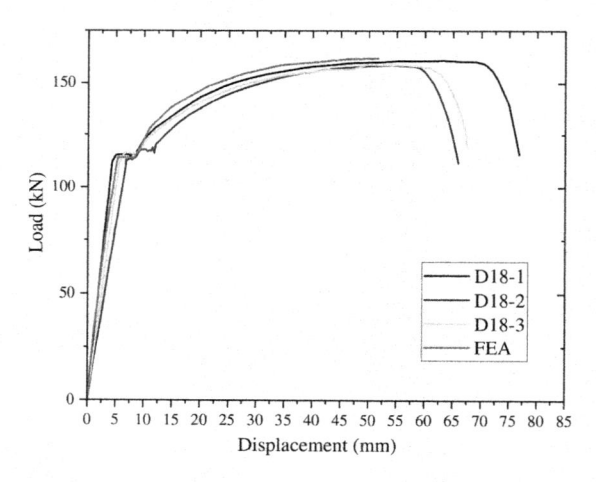

Figure 4. Comparison of experimental and numerical results.

Table 5. Comparison of experimental and numerical results.

| Specimen | Yield capacity (kN) | | Ultimate capacity (kN) | | | | Failure modes | |
|---|---|---|---|---|---|---|---|---|
| | FE/$P_y'$ | Test/$P_y$ | FE/$P_u'$ | Test/$P_u$ | $P_y'/P_y$ | $P_u'/P_u$ | FE | Test |
| D18-1 | | 112.1 | | 160.64 | 1.018 | 1.007 | | Rebar fracture |
| D18-2 | 114.13 | 112.57 | 161.73 | 158.42 | 1.014 | 1.021 | Rebar fracture | Rebar fracture |
| D18-3 | | 111.78 | | 158.68 | 1.023 | 1.019 | | Rebar fracture |

# 4 GROUTING DEFECT MODEL

## 4.1 *Model design*

Based on the above-verified model, different types of grouting defects were set at the end of the model. Use the defect-free model as the reference group (Q-8d). There are two types of grouting defects: vertical and horizontal, as shown in Figure 5. For the vertical direction group, the anchoring length of the upper steel bar is 36mm (2d), 72mm (4d), and 108mm (6d). The horizontal direction group is based on the vertical direction group, and the upper surface of each sample is covered with a thickness of 3.8mm (0.1D) to 19mm (0.5D). A total of 19 sets of samples were produced, as shown in Table 6.

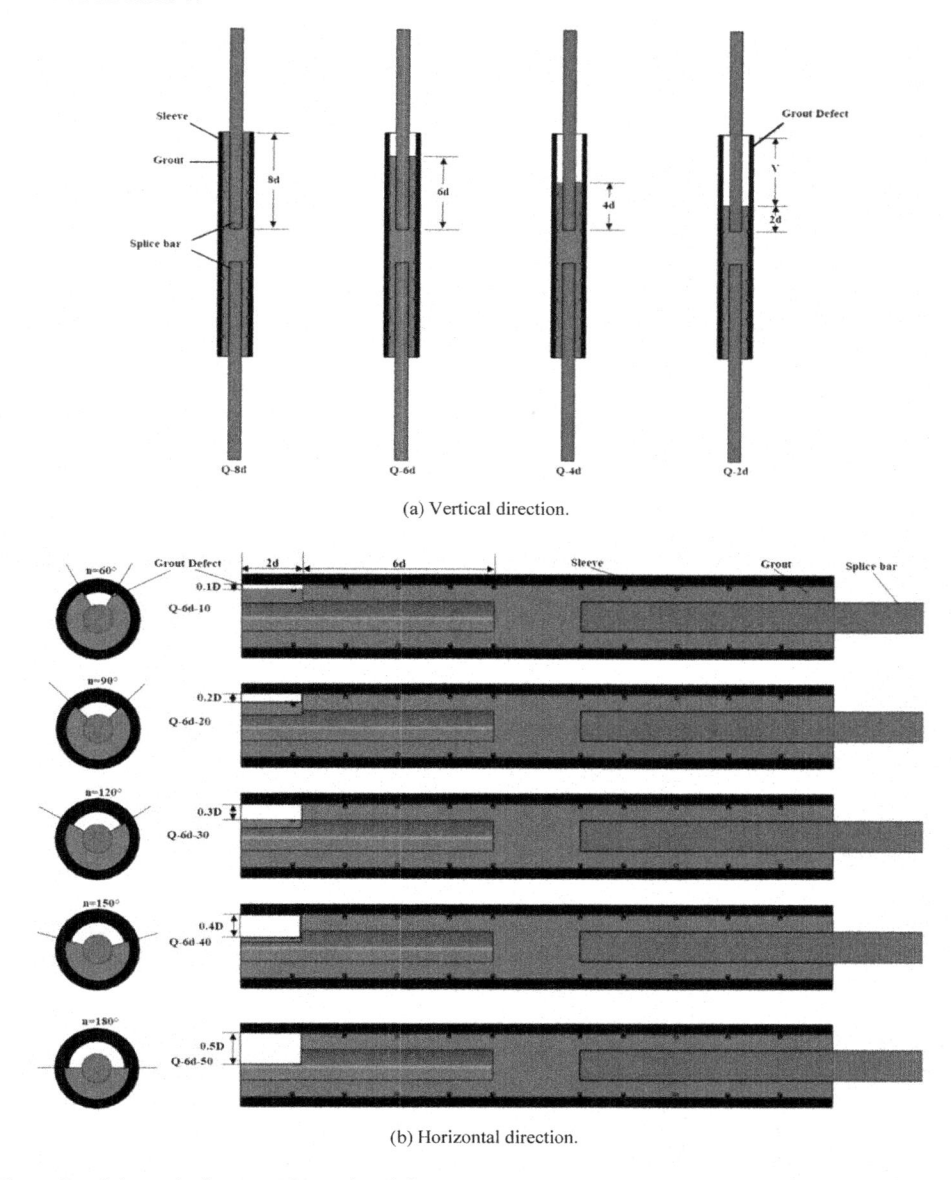

(a) Vertical direction.

(b) Horizontal direction.

Figure 5.   Schematic diagram of grouting defects.

Table 6. Specimen details.

| Specimens | | Vertical defect length V (mm) | Horizontal defect thickness H (mm) | Defect volume ratio (%) |
|---|---|---|---|---|
| (a) | Q-2d | 108(6d) | 38(D) | 30.19 |
| | Q-4d | 72(4d) | 38(D) | 20.61 |
| | Q-6d | 36(2d) | 38(D) | 9.87 |
| | Q-8d | 0 | 0 | 0 |
| (b) | Q-2d-10 | 108 (6d) | 3.8(0.1D) | 5.03 |
| | Q-2d-20 | 108(6d) | 7.6(0.2D) | 7.55 |
| | Q-2d-30 | 108(6d) | 11.4(0.3D) | 10.06 |
| | Q-2d-40 | 108(6d) | 15.2(0.4D) | 12.58 |
| | Q-2d-50 | 108(6d) | 19(0.5D) | 15.09 |
| (c) | Q-4d-10 | 72(4d) | 3.8(0.1D) | 3.43 |
| | Q-4d-20 | 72(4d) | 7.6(0.2D) | 5.15 |
| | Q-4d-30 | 72(4d) | 11.4(0.3D) | 6.87 |
| | Q-4d-40 | 72(4d) | 15.2(0.4D) | 8.59 |
| | Q-4d-50 | 72(4d) | 19(0.5D) | 10.30 |
| (d) | Q-6d-10 | 36(2d) | 3.8(0.1D) | 1.65 |
| | Q-6d-20 | 36(2d) | 7.6(0.2D) | 2.47 |
| | Q-6d-30 | 36(2d) | 11.4(0.3D) | 3.29 |
| | Q-6d-40 | 36(2d) | 15.2(0.4D) | 4.11 |
| | Q-6d-50 | 36(2d) | 19(0.5D) | 4.94 |

## 4.2 *Failure mode*

There are only two modes of failure of the specimen simulation results: pull-out after yielding and pull-out without yielding. As shown in Figure 6, the defective specimen was pulled out of the sleeve prematurely due to insufficient bonding ability with the rebar, and the ultimate tensile capacity of the rebar was not reached. As shown in Table 7, most of the specimens with a volume ratio of grouting defects greater than 5% failed to reach the yield of the rebar and failed. Specimen Q-6d failed because the anchorage length with the rebar was more than 5.5d and reached yield (Ji & Liu 2020).

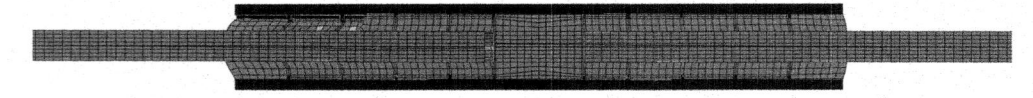

Figure 6. Bond-slip failure of specimens.

Table 7. Simulation results.

| Specimens | | Yield capacity $P_y$ (kN) | Yield displacement $S_y$ (mm) | Ultimate capacity $P_u$ (kN) | Ultimate displacement $S_u$ (mm) | Failure modes |
|---|---|---|---|---|---|---|
| (a) | Q-2d | / | / | 35.39 | 2.75 | P[a] |
| | Q-4d | / | / | 87.35 | 5.24 | P[a] |
| | Q-6d | 114.15 | 6.04 | 135.82 | 15.17 | PY[b] |
| | Q-8d | 114.13 | 5.35 | 161.73 | 51.56 | TF[c] |

*(continued)*

262

Table 7. Continued.

| Specimens | | Yield capacity $P_y$ (kN) | Yield displacement $S_y$(mm) | Ultimate capacity $P_u$ (kN) | Ultimate displacement $S_u$(mm) | Failure modes |
|---|---|---|---|---|---|---|
| (b) | Q-2d-10 | 114.16 | 6.04 | 127.47 | 11.79 | PY[b] |
|  | Q-2d-20 | / | / | 81.77 | 3.86 | P[a] |
|  | Q-2d-30 | / | / | 65.14 | 3.06 | P[a] |
|  | Q-2d-40 | / | / | 55.04 | 2.61 | P[a] |
|  | Q-2d-50 | / | / | 46.08 | 2.07 | P[a] |
| (c) | Q-4d-10 | 114.07 | 5.81 | 143.85 | 19.21 | PY[b] |
|  | Q-4d-20 | 114.07 | 5.90 | 132.16 | 13.47 | PY[b] |
|  | Q-4d-30 | / | / | 114.16 | 6.33 | P[a] |
|  | Q-4d-40 | / | / | 70.77 | 3.29 | P[a] |
|  | Q-4d-50 | / | / | 52.04 | 2.31 | P[a] |
| (d) | Q-6d-10 | 114.04 | 5.46 | 119.66 | 9.68 | PY[b] |
|  | Q-6d-20 | 113.8 | 5.49 | 114.16 | 5.66 | PY[b] |
|  | Q-6d-30 | 113.02 | 5.50 | 114.16 | 5.74 | PY[b] |
|  | Q-6d-40 | / | / | 86.29 | 3.91 | P[a] |
|  | Q-6d-50 | / | / | 66.16 | 3.03 | P[a] |

[a] P denotes the pullout failure without yielding rebar
[b] PY denotes the pullout of rebars after yielding
[c] TF denotes the tensile fracture of rebars

### 4.3  Load-displacement curve

Figure 7(a) shows the load-displacement curve of each specimen in the vertical direction. As can be seen from the figure, with the increase of the defect volume ratio, the failure mode of the specimen ranges from the tensile fracture failure of the rebar to the yield pull-out failure of the rebar, and finally to the unyielding pull-out failure of the rebar. When the volume ratio increases from 0% to 30.19%, the ultimate load of the specimen decrease from 161.73kN to 35.39kN, and the decrease is 78.12%. When the volume ratio is 0% and 9.87%, the curves of the specimens in the elastic and yield stages are very close. With the increase in the defect volume ratio, the extreme point decreases by 16.2%. When the volume ratio increases from 20.61% to 30.19%, the extreme point decreases significantly, and there is no yield point.

Figure 7(b)–(d) are the load-displacement curves of each specimen in the horizontal direction. As can be seen from the figure, the vertical defect length V=6d (Figure 7b), and when the defect volume ratio increases from 5.03% to 15.09%, the ultimate load of the specimen decrease from 127.47kN to 46.08kN, with decreasing amplitude of 63.86%. The specimen reaches the yield point only when the volume ratio is 5.03%. The vertical defect length V=4d (Figure 7c), and when the volume ratio increases from 3.43% to 10.3%, the ultimate load of the specimen decreases from 143.85kN to 52.04kN, with a decrease of 63.82%. The specimen reaches the yield point when the volume ratio is 3.43% to 5.15%, and the curve begins to decline at the yield point when the volume ratio is 6.87%. The vertical defect length V=2d (Figure 7. d), and when the volume ratio increases from 1.65% to 4.94%, the ultimate load of the specimen decreases from 119.66kN to 66.16kN, with a decrease of 44.71%. The specimen reaches the yield point when the defect volume ratio is 1.65% to 3.29%.

### 4.4  Discussions

Xu ( 2018) et al. proposed that the failure mode of the grouting defect sleeve is greatly affected by the volume ratio of the grouting defect, and is less affected by the grouting defect position. Figure 8

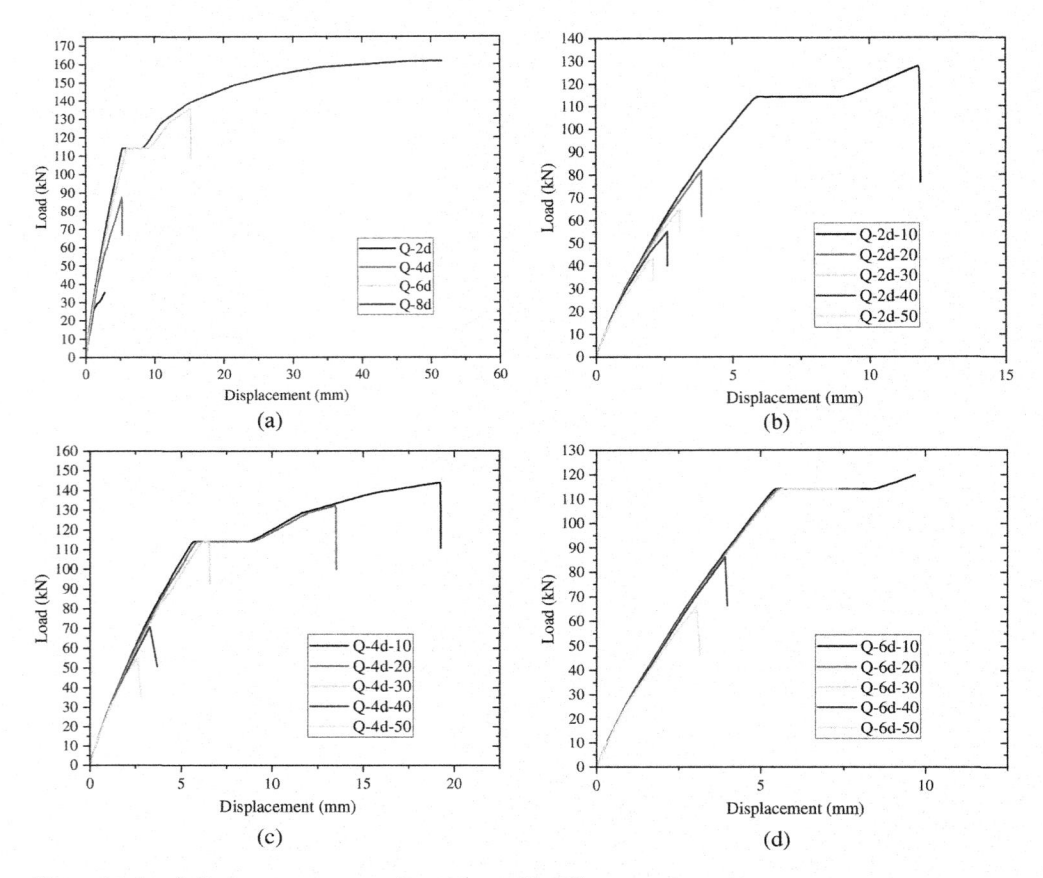

Figure 7. Load-displacement curves of specimens with different numbers.

shows the relationship between the defect volume ratio and the ultimate capacity corresponding to different types of specimens. Regardless of the type of defect, the ultimate capacity decreases as the volume ratio of the defect increases. The drop of the specimen in the vertical direction is gentler than that in the horizontal direction. The fitting equation is given as (1)–(4):

$$P_a = -0.06864x^2 - 2.14717x + 162.2414 \tag{1}$$

$$P_b = -0.90317x^2 - 25.70902x + 230.91771 \tag{2}$$

$$P_c = -0.96536x^2 - 1.00645x + 160.74216 \tag{3}$$

$$P_d = -5.98585x^2 + 23.02907x + 97.2336 \tag{4}$$

where $P_i$ is the ultimate load, and $x$ is the defect volume ratio.

Zheng (2017) et al. defined the failure of grouting sleeve specimens as when the steel bar was broken or the load dropped to less than 85% of the ultimate bearing capacity after slippage occurred. The ultimate capacity of the specimen used in this paper is $P_u = 161.73$kN and the ultimate defect volume ratio can be obtained by the formula as $x_a = 8.97\%$, $x_b = 4.28\%$, $x_c = 4.42\%$, $x_d = 1.01\%$, respectively. The volume ratio in the vertical direction is approximately regarded as the ratio of length, which is converted into the limit defect length $V_a = 1.5$d, that is, the anchorage length with the rebar should be greater than 6.5d. The limit defect thickness converted in the horizontal direction is $H_b = 0.12$D, $H_c = 0.18$D, $H_d = 0.1$D. To sum up, when the defective sleeve in the

264

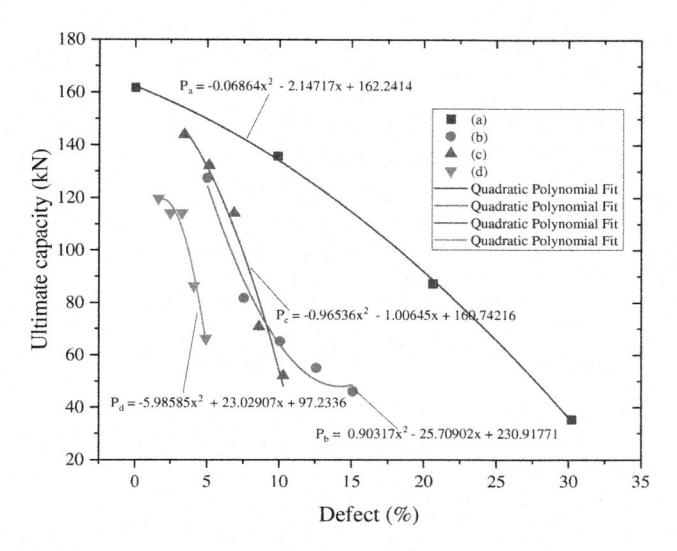

Figure 8. Relationship between defect volume ratio and ultimate capacity.

vertical direction (a) is slipped and damaged, the anchorage length with the upper steel bar should be greater than 6.5d to ensure the performance of the specimen. When the defect sleeve in the horizontal direction (b)-(d) is slipped and damaged, the thickness of the horizontal defect should be controlled at 0.1D.

## 5 CONCLUSION

1) Using the cohesive contact of ABAQUS to simulate the bond-slip effect of the rebar and the grout, the bearing capacity and deformation of the model are basically consistent with the actual situation, and the failure mode and ultimate capacity can be accurately judged;
2) The numerical simulation results are roughly consistent with the actual uniaxial tensile test results, which can be used for further defect analysis and research;
3) As the defect volume ratio increases, the failure mode of the specimen changes from the tensile failure of the rebar to the yield pull-out failure of the rebar, and finally to the unyielding pull-out failure of the rebar, and the ultimate capacity decreases exponentially;
4) To prevent the slippage damage of the defective grouting sleeve, the anchorage length between the vertical direction and the upper rebar should be greater than 6.5d, and the defect thickness in the horizontal direction should be controlled at 0.1D.

ACKNOWLEDGMENTS

This work was supported by special funding for Assembly Buildings of Guangzhou Municipal Construction Group Co., Ltd (2019-KJ019; 2019-KJ033).

REFERENCES

Gao, R.D., Li, X.M., Zhang, F.W. (2019) Study on the influence of grouting defects in different positions on the strength of steel sleeve grouting joints. *J. Construction Technology.*, 48(18): 116–119+124.

Zheng, G.Y., Kuang, Z.P., Xiao, J.Z., Pan, Z.F. (2020) Mechanical performance for defective and repaired grouted sleeve connections under uniaxial and cyclic loadings. *J. Sci. Construction and Building Materials*, 233(C): 117233–117233.

Liu, C., Pan, L.F., Liu, H., Tong, H.W., Yang, Y.B., Chen, W. (2020) Experimental and numerical investigation on mechanical properties of grouted-sleeve splices. *J. Sci. Construction and Building Materials*, 260.

Ji, Y., Liu, W. (2020) Numerical simulation analysis of the influence of grouting defects on the mechanical properties of connecting sleeves. *J. Jiangsu Construction*, (06): 64–66.

Xu, F., Wang, K., Wang, S.G., Li, W.W., Liu, W.Q., Du, D.S. (2018) Experimental bond behavior of deformed rebars in half-grouted sleeve connections with insufficient grouting defect. *J. Sci. Construction and Building Materials*, 185: 264–274.

Zheng, Q.L., Wang, N., Tao, L., Xu, W.J. (2017) Experimental study on the effect of grouting defects on the properties of steel sleeve grouted connection specimens (in Chinese) *J. CSCD. Building Science*, 33(05): 61–68.

*Civil Engineering and Urban Research – Mohamed & Hou (Eds)*
© 2023 the Authors, ISBN 978-1-032-44487-1

# Research on influencing factors on the mechanism of multi-ribbed composite wall

Junbin Gu*

*School of Civil Engineering, Inner Mongolia University of Science and Technology, Baotou, China*

ABSTRACT: Two different modes of Multi-ribbed composite walls are tested under vertical and horizontal load, so as to study the cooperative working mechanism of the multi-ribbed composite wall and the main factors affecting the cooperative working of the structure, and then reveal the mechanical properties of the multi-ribbed composite wall. The experiments show that different multi-ribbed composite wall stress, strain, and deformation are very different in the same experimental conditions. Under the exterior loads, the box of IFH-11 mainly undertakes the external load, block bears almost zero, box and block of IFH-12 as a whole share the external load, so IFH-12 of stress diagram is consistent with multi-ribbed composite wall-Frame tie model for oblique; IFH-11 deformation under external force, the box and block was isolated, while IFH-12 of box and block bound together as a good structure to take force and deformation properties, which offers a theoretical basis for analysis and design of the Multi-ribbed composite wall.

## 1 INTRODUCTION

Multi-ribbed wall slab lightweight frame structure study began in 1990. The new technology subject was hosted and completed by the new technology research institute about the construction of Xi'an University of Architecture and technology. After 20 years of hard research, the research in theoretical research and applied research has a large number of detailed and fruitful work, and we achieved the breakthrough so that the research results are systematic, theoretical, and practical (Chen et al. 2021; Gu & Han 2020; Liang & Wang 2021; Zhang et al. 2020).

The multi-ribbed composite wall is a basic bearing component of the multi-ribbed wall slab lightweight frame structure system, and it includes the reinforced concrete frame beam, column, and embedded with a light block (Ding & Jiang 2019; Gu 2018; Liang 2019; Li 2019; Shun et al. 2016; Zhao et al. 2019). After joining the filling block, the bearing capacity of the multi-ribbed composite wall improved obviously because the filling block and the frame are interacting. On the one hand, the filling block is restricted from the frame, and the bearing capacity of the filling block was increased under the condition of bi-directional compression. On the other hand, the frame is also restricted from the filling block, and the stiffness of frame structure was increased. The filling block and frame give mutual support to each other to full play to their respective characteristics (Guo et al. 2016; Jia et al. 2016; Wang et al. 2019). In short, multi-ribbed composite wall is a new force-bearing component. It is effectively converted the frame and the filling block by special tectonic, and the mechanical properties of the frame and the filling block are quite different, and is mainly reflected in the mechanical properties of the frame and the filling block and the mutual coordination work of the frame and the filling block. We mainly discuss the main influence factors for the coordination work mechanism of the frame and the filling block, such as the differences in material performance of the frame and filler block, the material section characteristics of the frame, and other factors.

---

*Corresponding Author: gu613@163.com

DOI 10.1201/9781003372417-36

## 2 THE MODELING PROCESS OF MULTI-RIBBED COMPOSITE WALL

The multi-ribbed composite wall is a new type of composite wall, and choosing a reasonable calculation model to simulate is the first. The multi-ribbed composite wall is more complex, and it consists of a reinforced concrete frame and filled block, so using a simplified calculation model to solve is feasible.

### 2.1 The entity model and its material properties

The multi-ribbed composite wall is composed of two parts: frame unit and block unit, so the test specimen of the multi-ribbed composite wall is named the IFH specimen, and the physical model is shown in Figure 1.

Table 1. Mechanical properties of concrete and building block.

| Material | Performance | | | |
|---|---|---|---|---|
| | Compressive strength (MPa) | Tensile strength (MPa) | Bulk density $(kN/m^3)$ | Elasticity modulus (MPa) |
| Concrete | 30 | 1.43 | 24 | $3.0\times10^4$ |
| Aerated concrete | 7.5 | 0.65 | 7 | $3.0\times10^3$ |

Where the thickness of the filled block is 100mm, the ratio of height to thickness is 4, b is 50mm, h is 40 mm, B*H is 400mm*300mm of IFH-1 and B*H is 400mm*400mm of IFH-2, and the load is selected monotonically loaded. The materials' mechanical properties of the frame and filled block are shown in Table 1.

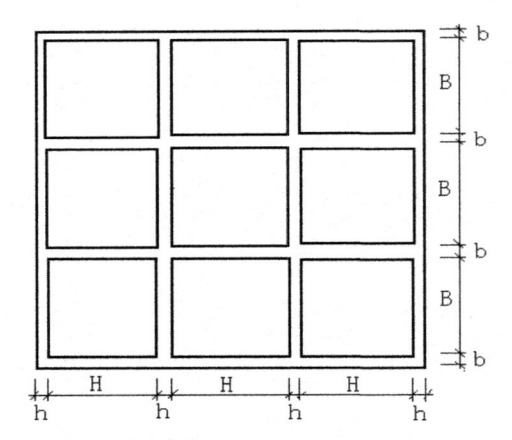

Figure 1. Schematic map of the sample series IFH.

### 2.2 The model of the contact element

The concrete style of the frame and the filling block adapt to setting the contact element in a multi-ribbed composite wall because the reinforced concrete frame in conjunction with the filling block ignores the tensile stress of the contact elements. It is supposed that the force of the frame and the filling block is the contact stress in multi-ribbed composite wallboard and assumed that the contact style between the frame and the filling block is the rigid contact of surface-surface. For example the calculation model of the contact element, Specimen IFH, is shown in Figure 2. The calculation model loaded the level load of 20 kN and the vertical load of 15 kN, and the loading model is shown in Figure 3.

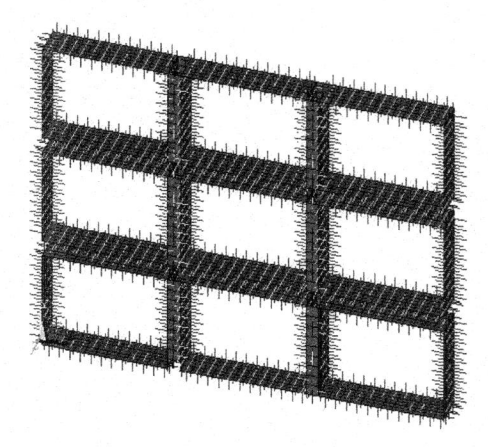 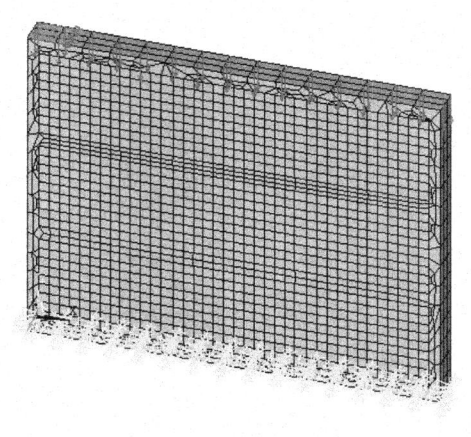

Figure 2. The calculation model of contact element.   Figure 3. The loading model of the contact element.

## 3   THE CALCULATION RESULTS AND DISCUSSION

### 3.1   *The effect of the filling block on mechanical properties of structural coordination*

The coordination work of the frame and the filling block fully embodies the superiority of a multi-ribbed composite wall. The filling blocks were restrained by the frame under external loads, On the other hand, the frame was also restrained by the filling block, and the interaction of the filling blocks and the frame together to resist the external force for full play to their respective performance. The filling block is a mechanical component and its influence cannot be ignored.

Table 2.   Comparison with the principal stress of IFH.

| The specimen number | The first principal stress (MPa) | The third principal stress (MPa) |
| --- | --- | --- |
| IFH-1 | 160 | 1.04 |
| IFH-2 | 69.6 | 0.3 |

The specimen mode of IFH-1 and IFH-2 were studied by ANSYS, and they can reflect the effects of the filling block on the coordination of structure, so we mainly study the influence of the section size of the filling block on the performance of the structure. The size of the specimen IFH-1 for 400mm *300mm, the size of the specimen IFH-2 for 400mm*400mm, and the thickness of the IFH-1 and IFH-2 is 100mm. The principal stress value of IFH-1 and IFH-2 is shown in Table 2.

Due to specimen IFH-2 better simulated equivalent oblique compression bar than specimen IFH-1, so the contact effect of the filling block and the frame is ideal under external loads. Compared with the specimens of IFH -1, the first principal stress peak value of IFH-2 is only 44% and the third principal stress peak value is only 29%. Figures 4 and 5 show respectively the stress nephogram of IFH-1 and IFH-2.

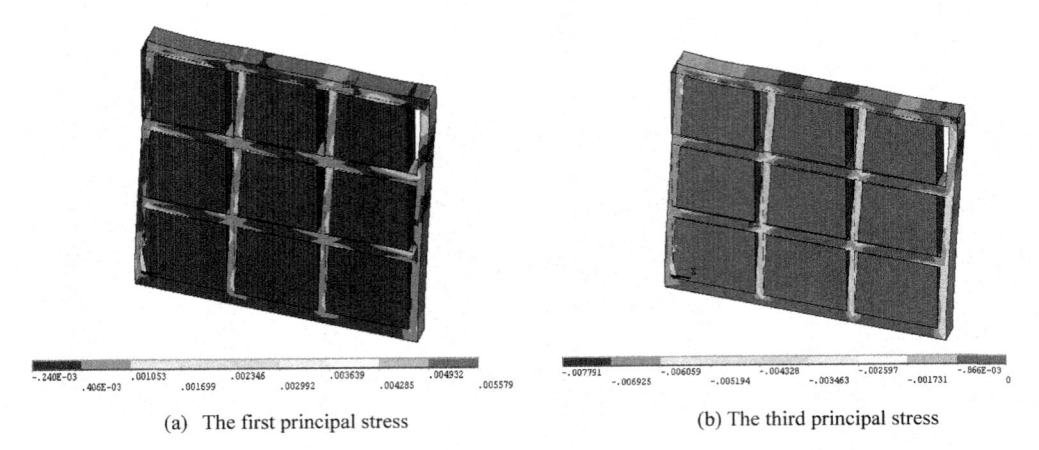

(a) The first principal stress

(b) The third principal stress

Figure 4. The first and third principal stress of IFH-1.

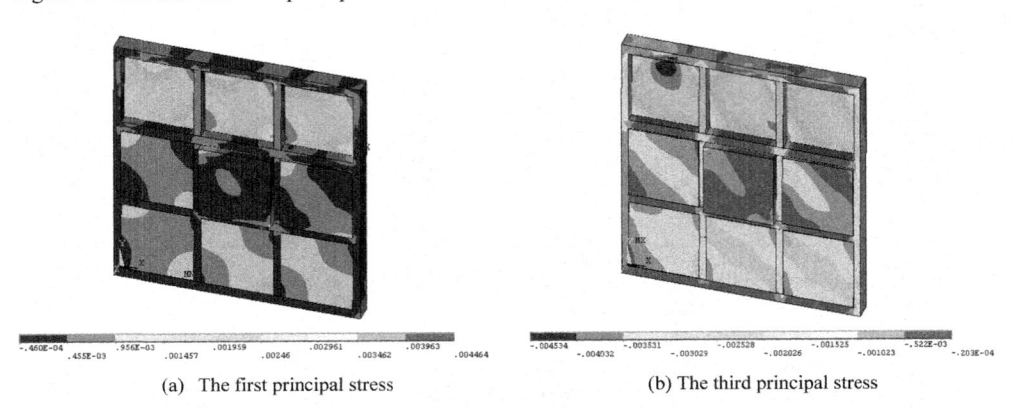

(a) The first principal stress

(b) The third principal stress

Figure 5. The first and third principal stress of IFH-2.

## 3.2 The effect of the filling block on structural deformation of structural coordination

The filling block of the multi-ribbed composite wall adopts an aerated concrete block, and we choose different sizes of the filling block to study the structure deformation of coordination work under the elastic modulus, Poisson ratio and density stay the same. Here we study the coordination work of the frame and the filling block through displacement deformation of IFH. Table 3 shows the displacement value of IFH-1 and IFH-2.

Table 3. Comparison with the displacement of IFH.

| The specimen number | X-direction displacement (mm) | Y-direction displacement (mm) |
| --- | --- | --- |
| IFH-1 | 0.016 | $0.1 \times 10^{-3}$ |
| IFH-2 | 0.004 | $0.6 \times 10^{-6}$ |

The stress nephogram of IFH shows that the IFH-2 specimens of the frame and the filling block have close contact together and they work under the same horizontal and vertical load, which is conducive to a wide range of cracking of specimens. The displacement diagram of IFH shows that the specimen deformation is mainly horizontal deformation because vertical deformation is small and we can ignore it. At the same time, the contact effect of the IFH-2 is good, and the frame and the filling block without separation and work together under the loads to effectively resist the horizontal deformation of the specimens, which further illustrates that IFH-2 is more aligned with the equivalent oblique compression bar model and the rigid-oblique compression bar model.

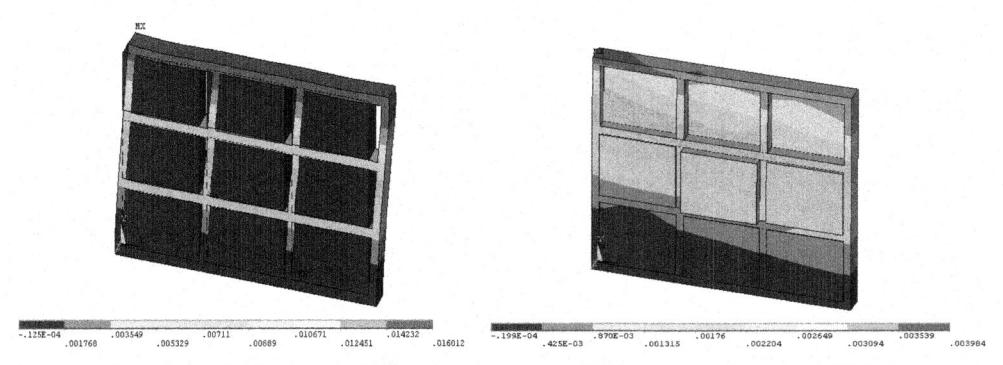

Figure 6. X-direction displacement of IFH-1.  Figure 7. X-direction displacement of IFH-2.

## 4  CONCLUSION

The size selection of the filling block is different for the coordination work of the multi-ribbed composite wall has a significant influence. (1) From the point of force, the frame and the filling block of IFH-2 together bear the external force, but the frame of IFH-1 only bears the external force and the stress of the filling block is almost zero, IFH-2 is more aligned with the equivalent oblique compression bar model than IFH-1. (2) From the deformation, the frame and the filling block of IFH-2 are closely in contact together, but the frame and the filling block of IFH-1 are separate, so it is verified that IFH-2 has a good performance of structural deformation. At the same time, when IFH-2 was destroyed, the filling block have two main cracks on both sides of the diagonal, and the characteristics of the oblique compression bar are obvious. In short, test and theoretical analysis have shown that the filling block can be equivalent to the oblique compression bar hinged on the concrete frame structure. The closer the size of the filled blocks B and H is, the more the ribbed composite wall panel conforms to the rigid frame-diagonal model.

## ACKNOWLEDGMENTS

This work was supported by the scientific research project of higher education institutions in the Inner Mongolia autonomous region (NJZY21378).

## REFERENCES

Chen, Y.T., Lu, J.W., Sun, W., Qiang, Z.H., Qian, W.Y. (2021). Design and Consideration of Prefabricated Structure of Guangzhou International Campus of South China University of Technology phase I Project, *SichuanBuilding Materials*, 47(11):43–44.

Ding, Y.G., JIANG, S. (2019). Effect of the filled block on the mechanical behavior of ribbed composite wall under grain load. *Journal of Henan University of Technology,* 40:100–114.

Gu, J.B. (2018). Influence of the filling blocks on multi-ribbed composite walls. *Quarterly Journal of Indian Pulp and Paper Technical Association*, 30:507–511.

Gu, J.B., Han, X.L.(2020). Effect of the number of ribbed columns on the mechanical properties of multi-ribbed composite walls. *Proceedings of the 26th Annual Conference of Beijing Force Society*, pp.1089–1091

Guo, M., Liu Z.Y., Huang, W., et al. (2016). Experiment on Seismic Performance of Earthquake Damage Frame Strengthened with Aerated Concrete Block Composite Wall. *Journal of the South China University of Technology*, 44(10), pp:117–124.

Jia, S. Z.*, Cao W.l., Yuan, Q., et al. (2016). Experimental study on frame-supported multi-ribbed composite walls under low-reversed cyclic loading. *European Journal of Environmental and Civil Engineering*, 20(3), pp: 314–331.

Li, L. (2019). Experimental study on bearing capacity of precast ultra-high performance concrete light and dense ribbed floor slab. *Industrial building*, 49:92–96

Liang, X.J., Wang, Y. (2021). Progress in Semi-rigid Connection of Assembled Steel Structures [J]. *Progress in Building Steel Structures*, (11):1–15.

Liang, Y. (2019). Research on Mechanical Behavior and Computer-Aided Design of Multi-ribbed Composite Wall (Master's Thesis, Beijing Jiaotong University).

Shun, X., Jin, T., Ma, Y.H., et al. (2016). Torque analysis on the bionic model of bamboo weevil rostrum based on ANSYS. *Transactions of the Chinese Society of Agricultural Engineering*, 32(12), pp: 11–16.

Wang, F., Wei, H., Luo, Z.F., Li, X.S. (2019). Design and Consideration of Prefabricated Structure of Guangzhou International Campus of South China University of Technology phase I Project, *Building structure*, 49(11):56–61.

Zhang, Z.Y., Deng, K.L., Xu, T.F. (2020). Research progress of prefabricated concrete bridge structure in 2019. *Journal of Civil and Environmental Engineering*, 42(05):183–191.

Zhao, Y.Y., Yuan, Q., Zhu, H.L. (2019). Study on the Seismic Behavior of Monolithic Precast Shape Steel Oblique Multi-Ribbed Composite Wall [J]. *Science of Advanced Materials*, 11(11): 1632–1646.

*Civil Engineering and Urban Research – Mohamed & Hou (Eds)*
*© 2023 the Authors, ISBN 978-1-032-44487-1*

# Numerical simulation of seismic performance for a light-weight steel frame with steel tube column filled with aluminum foam under different structural parameters

Qinglan Liu* & Jianhua Shao*
*Department of Civil Engineering, Jiangsu University of Science and Technology, Zhenjiang, China*

Zhanguang Wang*
*Department of Civil Engineering, Kaili University, Kaili, China*

Jiangcheng Man
*Guizhou Zhongjian Weiye Construction (Group) Co., Ltd, Kaili, China*

ABSTRACT:   Three parameters affecting the light-weight steel frame structure with square steel pipe columns filled with aluminum foam had been investigated and analyzed by using Abaqus finite element software. The results indicated that the overall bearing capacity of the finite element model for a lightweight steel frame with a square steel tube column filled with aluminum foam was gradually decreased with the increase in axial compression ratio. The hysteresis performance of light-weight steel frames was not significantly affected by the beam-column linear stiffness ratio, and changing only the frame beam span length did not produce an effective increase in the restraint effect of aluminum foam on frame columns. The seismic performance of light-weight steel frames decreased with the increase in porosity of aluminum foam, whereas the difference was not significant. An important reference for the study of aluminum foam-filled light-weight steel frames and such filled light-weight steel frames, meanwhile some prospects and fields for subsequent research were proposed.

## 1   INTRODUCTION

The aluminum foam was already utilized in the construction field as a filling material and decorative material. Its light mass and low relative density, low strength, and large deformation under external load can effectively enhance the bearing capacity of the overall frame structure, meanwhile, the aluminum foam also had the functions of damping vibration attenuation, noise reduction, and heat insulation, electromagnetic shielding, etc. It had better application prospects in aerospace science and technology as well as industrial buildings. The research on foamed metals conducted by various scholars and experts in recent years had been mainly carried out from two aspects. On the one hand, it involves the study of the structural parameters for the foamed metal itself (Banhart & Baumeister 1998; Liu et al. 2016), such as porosity, relative density, etc. On the other hand, it includes research on the performance applications of the foamed metal (Wang et al. 2018; 2021; Zhang 2019), such as seismic performance, energy dissipation, and vibration reduction, etc. However, the research on metal foam was still incomplete and needed further study, especially for the application of aluminum foam in building structures in the field of seismic performance. Therefore, it was particularly important to investigate the seismic performance of lightweight steel frames with aluminum foam-filled square steel columns.

---

*Corresponding Authors: andmoon0@outlook.com, shaojianhua97@163.com and wzg3262396@163.com

DOI 10.1201/9781003372417-37

There were many factors influencing the seismic performance of light-weight steel frame structure with aluminum foam-filled square steel tube column, such as axial compression ratio, beam-column linear stiffness ratio, the porosity of aluminum foam, etc. Parametric analysis of several significant factors affecting the steel frame system with aluminum foam-filled square steel tube column was conducted to investigate the seismic performance of light-weight steel frame under different parameters. An important reference for the study of aluminum foam-filled light-weight steel frames was provided.

## 2 PARAMETER SELECTION

The seismic performance of the frame under the influence of different structural parameters was analyzed by Abaqus finite element software. The axial compression ratio, beam-column linear stiffness ratio, and porosity of aluminum foam were selected as the control parameters, which were analyzed and compared with the calculated results.

In order to facilitate the comparison between the effects of each control parameter on the seismic performance of light-weight steel frames with aluminum foam-filled square steel columns, a basic structural type was established in each parameter group, i.e., all the parameter models involved based on this basic structural type. The specific geometry and material parameters of the basic structural type were shown in table 1, and the finite element model was shown in Figure 1 (a). The horizontal load loading method was shown in Figure 1(b). The mechanical parameters of aluminum foam were obtained from the literature (Wang & Tu 2008). Because the aluminum foam was prepared by the foaming method, then cut and filled into the structure, therefore the aluminum foam has integrity. The connection with the square steel column using ordinary connectors could not achieve the effect of restraining the buckling of the frame column, so the connection with the frame column was made by polyurethane bonding. The aluminum foam was connected to the frame column by binding constraints in the finite element software.

Table 1. Geometry and material parameters of basic structural type.

| | Section size | $E$ (MPa) | $E_s$ (MPa) | $f_y$ (MPa) | $f_y^*$ (MPa) | $f_u$ (MPa) |
|---|---|---|---|---|---|---|
| Frame beam | 350×200×9×10 | 206000 | 2060 | 235 | – | 390 |
| Frame column | 250×250×2 | 206000 | 2060 | 235 | – | 390 |
| Aluminum foam with 80% porosity | 246×246 | 253 | – | – | 6.29 | – |
| Steel cover plate | 300×300×30 | 206000 | 2060 | 235 | – | 390 |

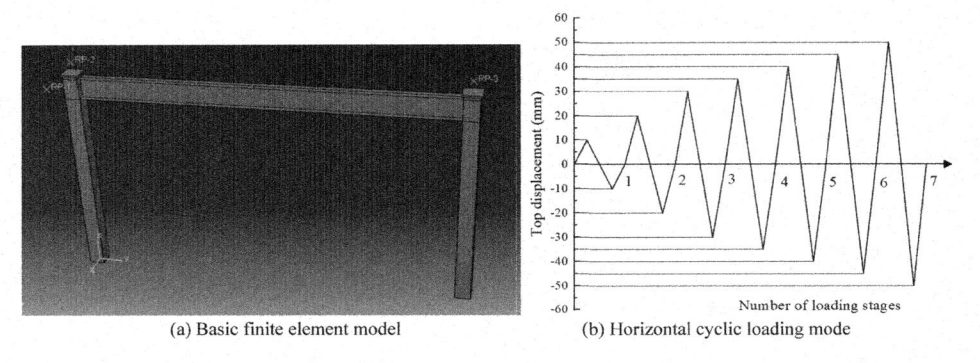

(a) Basic finite element model          (b) Horizontal cyclic loading mode

Figure 1.   Model and loading method.

# 3 EFFECTS OF DIFFERENT PARAMETERS ON LIGHT-WEIGHT STEEL FRAME

## 3.1 *The effect of axial compression ratio*

The first group was modeled with different axial compression ratios of 0.24, 0.35, 0.41, and 0.47, respectively. The axial compression ratio was expressed by $n$.

The hysteresis curves under different axial compression ratios were shown in Figure 2. It can be seen from the figure that the shapes of the hysteresis curves derived from the finite element models under the action of four different axial compression ratios were all extremely full, reflecting the strong plastic deformation capacity of the whole structure or member, which had excellent seismic performance and energy dissipation capacity. In addition, with the increase in axial compression ratio, the overall bearing capacity of the finite element model of a light-weight steel frame with a square steel tube column filled with aluminum foam was gradually decreased, which was due to the restraining effect of aluminum foam on the frame column, as with the increase of axial compression, thus leading to the weakening of the restraining capacity of aluminum foam and the decrease of the overall bearing capacity of the light-weight steel frame.

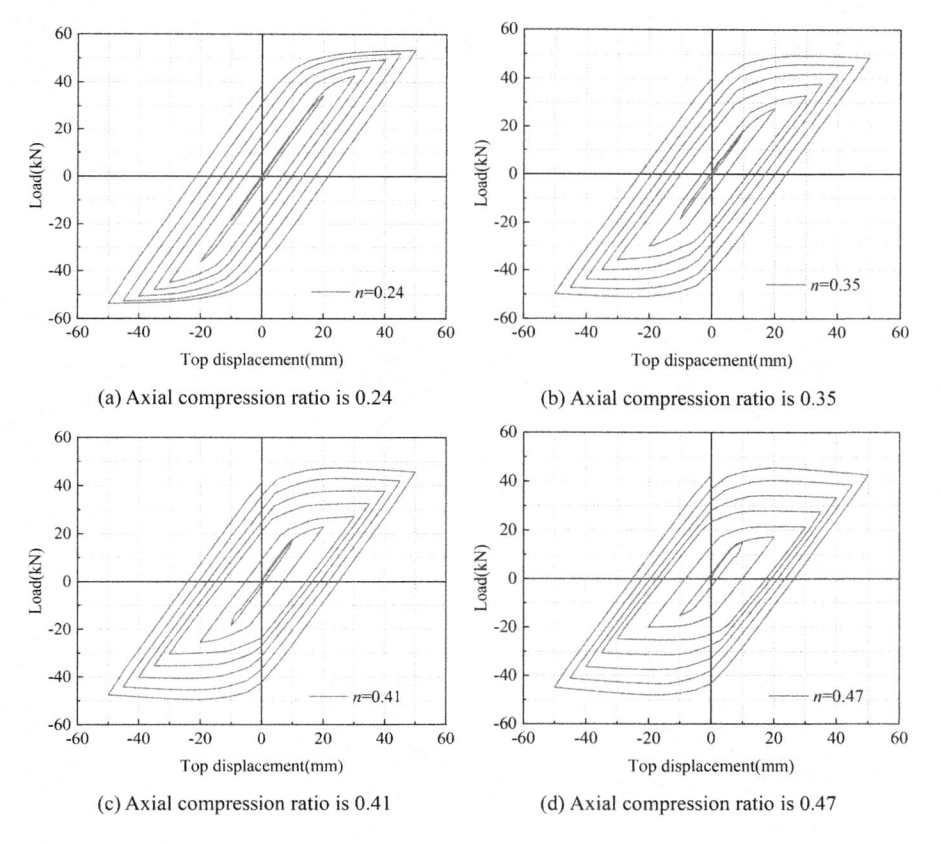

(a) Axial compression ratio is 0.24

(b) Axial compression ratio is 0.35

(c) Axial compression ratio is 0.41

(d) Axial compression ratio is 0.47

Figure 2.    Hysteresis loop.

## 3.2 *The effect of beam-column linear stiffness ratio*

The second group was modeled with different beam-column linear stiffness ratios of 4.97, 4.52, 4.15, and 3.83, respectively. The beam-column linear stiffness ratio was expressed by $\alpha$.

The hysteresis curves under different beam-column linear stiffness ratios were shown in Figure 3. It can be seen from the figure that the shapes of the hysteresis curves originated from the finite element models under the action of four different beam-column linear stiffness ratios were all extremely full, reflecting the strong plastic deformation capacity of the whole structure or member, which had excellent seismic performance and energy dissipation capacity. Additionally, with the decrease of beam-column linear stiffness ratio, the overall load-carrying capacity of the finite element model of light-weight steel frame with aluminum foam-filled square steel tube column gradually decreased, but it was not obvious, because the restraining effect of aluminum foam on frame column did not produce large changes due to the change of span length for frame beam. Overall, the hysteresis performance of the whole light-weight steel frame structure was not significantly affected by the beam-column linear stiffness ratio.

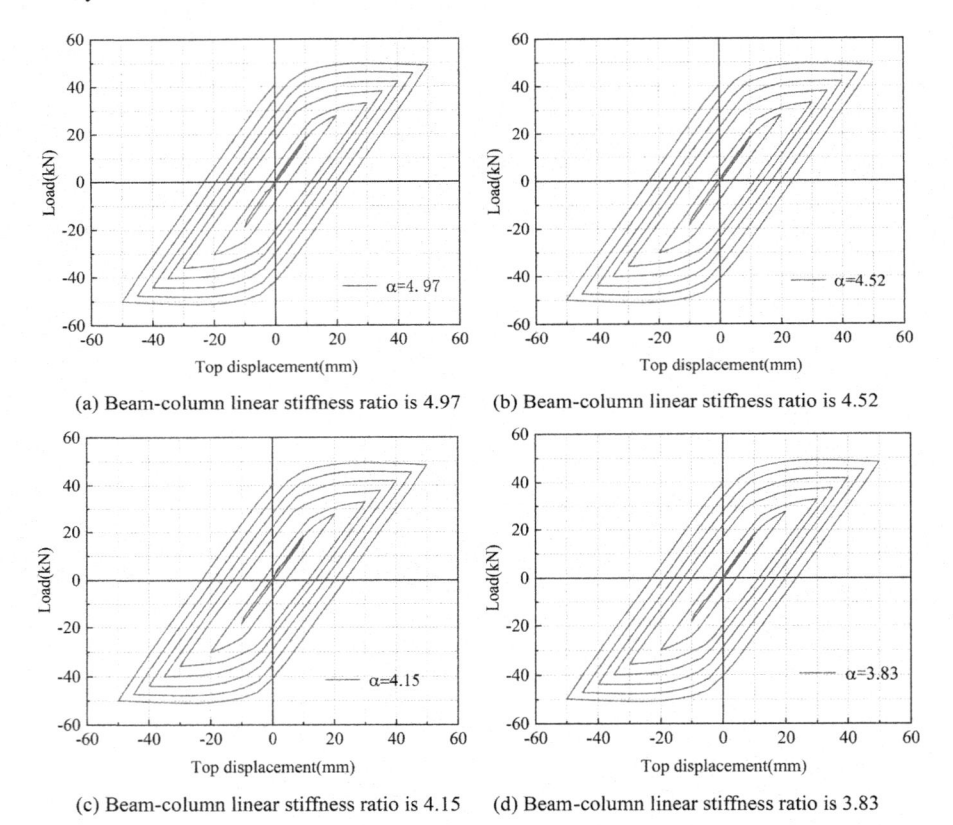

(a) Beam-column linear stiffness ratio is 4.97    (b) Beam-column linear stiffness ratio is 4.52

(c) Beam-column linear stiffness ratio is 4.15    (d) Beam-column linear stiffness ratio is 3.83

Figure 3.   Hysteresis loop.

### 3.3   *The effect of porosity of the aluminum foam*

The third group was modeled with different porosity of aluminum foam with 60%, 70%, 80%, and 90% porosities of aluminum foam, respectively.

The hysteresis curves under different porosity were shown in Figure 4. It can be seen from the figure that the shapes of the hysteresis curves for the finite element models of light-weight steel frames filled with four different porosities of aluminum foam were all extremely full, reflecting the strong plastic deformation capacity of the whole structure or member, which had excellent seismic performance and energy dissipation capacity. In addition, with the increase of porosity of aluminum foam, that was, the increase of aluminum foam cell pores, the overall bearing capacity of the finite element model for the light-weight steel frame with aluminum foam-filled square steel tube column was progressively reduced due to the restraining effect of aluminum foam on the frame column, which led to the weakening of the restraining capacity for aluminum foam and the reduction of the overall bearing capacity of the light-weight steel frame.

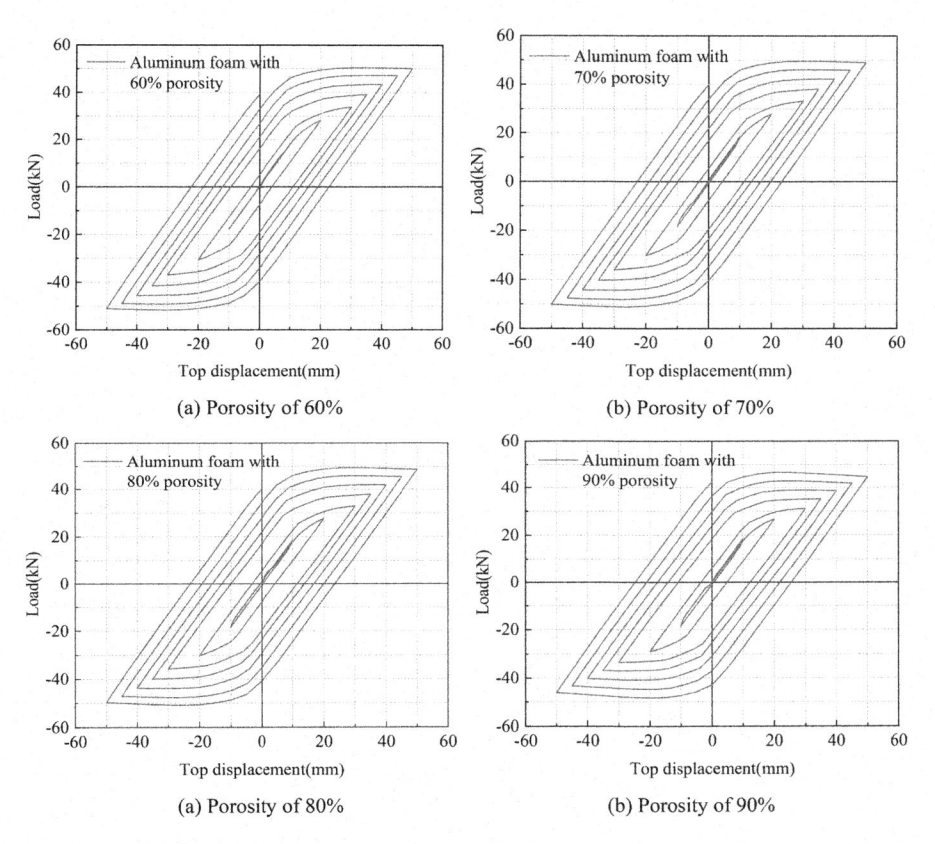

(a) Porosity of 60%

(b) Porosity of 70%

(a) Porosity of 80%

(b) Porosity of 90%

Figure 4.    Hysteresis loop.

## 4   CONCLUSION

Based on the results and discussions presented above, the conclusions were obtained as below:

(1)  With the increase in axial compression ratio, the overall load capacity of the finite element model for a lightweight steel frame with a square steel tube column filled with aluminum foam was gradually decreased.

(2)  The hysteresis performance of light-weight steel frames was not significantly affected by the beam-column linear stiffness ratio, and changing only the frame beam span length did not produce an effective increase in the restraint effect of aluminum foam on frame columns.

(3)  The seismic performance of light-weight steel frames decreases with the increase in porosity of aluminum foam, however, the difference was not substantial.

(4)  Aluminum foam and frame column in the finite element software is adopted by the binding connection, however, the actual construction process of aluminum foam and frame column will produce slip phenomenon and contact issues between the two components. The subsequent research on aluminum foam can explore the two contacts and force.

(5) The beam span and column height were taken for the parametric analysis of single-story single-span light-weight steel frame in this paper were limited. The influence law of beam-column linear stiffness ratio on the seismic performance of light-weight steel frame structure with square steel tube column filled with aluminum foam was not obtained, which lacked guidance for practical engineering. However, the wall thickness of the frame column and the cross-sectional dimensions of the foamed aluminum will also change when the cross-sectional dimensions of the beam and column were modified, therefore, these factors should be taken into account in the subsequent study.

## ACKNOWLEDGMENTS

This work was financially supported by the Science and Technology Plan Program of Qiandong-nan State (project number: Qiandongnan Science and Technology Cooperation J [2021]034), Postgraduate Research and Practice Innovation Program of Jiangsu Province (project number: KYCX21_3495), Science and Technology Program of Guizhou Province (Grant No. [2019]1288).

## REFERENCES

Banhart J and Baumeister J. (1998) Deformation characteristics of metal foams. *Journal of Materials ence*, 33(6): 1431–1440.

Liu C J, Zhang Y X and Yang C H. (2016) Numerical Investigation on Mechanical Behaviour of Closed-cell Aluminium Foams Using a Representative Volume Element Method. *MATEC Web of Conferences*, 65: 3003.

Wang R B and Tu Y Q. (2008) Experimental study on hysteresis behaviors of foamed-aluminum filled square steel tubular columns. *World Earthquake Engineering*, 24(3): 68–73.

Wang T T, Shao J H, Xu T and Wang Z G. (2021) Study on axial compression properties of aluminum foam-filled steel tube members after high temperature. *Iranian Journal of Science and Technology - Transactions of Civil Engineering*, 7: 1–18.

Wang Z G, Wang Y, Pan C R and Li J. (2018) Research on the torsional properties of aluminum foam-filled steel tube after a fire. *Journal of Asian Architecture and Building Engineering*, 17(3), 525–531.

Zhang J Y. (2019) Study on mechanical properties and energy absorption characteristics of foam metal filled steel tube components. Jiangsu: Jiangsu University of Science and Technology.

Civil Engineering and Urban Research – Mohamed & Hou (Eds)

# Mechanical properties and numerical simulation analysis of v-shaped steel joints

Wenting Liu, Xiaomeng Zhang* & Qingying Ren
*China Architecture Design & Research Group, Beijing, China*

Shengze Li
*Beijing Jiaotong University, Beijing, China*

ABSTRACT: A new type of intersecting joint is presented, and its finite element analysis and mechanical properties are analyzed. Through experiments and finite element analysis, the mechanical behavior and failure mode of the reactor are studied, and the correctness of finite element analysis is verified by comparison with experimental results.

## 1 INTRODUCTION

### 1.1 Structure introduction

In this paper, a new type of intersecting joint is proposed. The structure form and force transmission mode of intersecting joint are very simple and direct, and the bearing capacity of the structure is stronger than other similar joints. The intersecting node has a simple appearance and no obvious concave and convex. Intersecting nodes can save space and make full use of the node interior. Typical intersecting nodes are shown in Figure 1A, and scientists in many countries have carried out various studies on this. Japanese scholar Paul et al. (1994) conducted experimental studies on 58 spatial KK and TT nodes, analyzed the test results, and obtained the calculation formula of the ultimate bearing capacity of nodes through multiple linear regression analysis, and summarized the influence on engineering practice. Lee et al. (Lee & Wilmshurst 1996; Lee & Wilmshurst 1997) used finite element analysis software to analyze emphatically the wall thickness ratio of the branch head of the KK node and obtained the bearing capacity formula of this type of node according to the numerical simulation results. In this paper, ABAQUS (Wang et al. 2021) is used to carry out numerical simulation and parametric analysis of the bearing capacity and seismic performance of the new V-pillar joints.

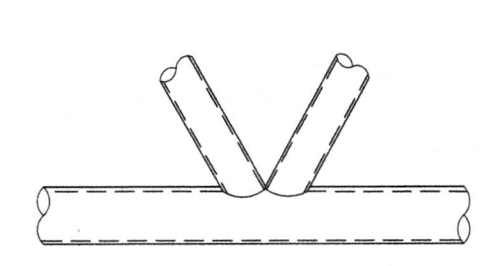

Figure 1. New intersecting node.

---

*Corresponding Author: 155203255@qq.com

## 2  FINITE ELEMENT ANALYSIS OF KEY NODES

Q355B steel is used for v-shaped steel joints. In this paper, ABAQUS/Standard is used to establish the finite element model analysis of assembled steel joints. The nonlinear finite element analysis of prefabricated steel joints was carried out by simulating the boundary conditions and loading conditions similar to the actual working conditions, and the working mechanism of prefabricated steel joints (beam V-pillar) was analyzed by using ABAQUS software.

In order to verify the rationality of finite elements, the stress states of uniaxial tensile specimens LV-1 and LV-2 and low-cyclic reciprocating specimens V-1 and V-2, load-displacement curves, skeleton curves, and bearing capacity obtained after treatment were compared with the finite element results.

### 2.1  *Comparison of uniaxial tension results of nodes*

As can be seen from the finite element analysis results in Figure 2, the overall stress of the branch pipe is significantly greater than that of the square steel pipe, and the stress in the area near the side wall of the branch pipe and the upper and lower areas of the square steel pipe is larger, while the stress in other areas is smaller, which is consistent with the stress state between the branch pipe and the beam of the square steel pipe in the test process.

Figure 2.    Finite element calculation results of LV-1.

Figure 3 shows the comparison between the uniaxial tensile test and finite element simulation curves of beam V-pillar joints. The yield load simulated by finite element is different from the yield load in the test. This is because bolts are used to connect the fixed end of the branch pipe string with the beam in the test, while the influence of bolt slip is ignored in the finite element and the fixed support is directly used for processing. As a result, the stiffness in the elastic stage of finite element simulation is greater than that in the test process.

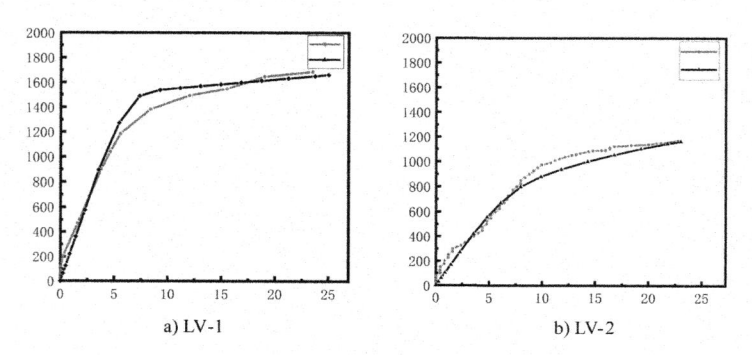

a) LV-1                                b) LV-2

Figure 3.    Load-displacement curve comparison of LV-1 and LV-2.

Figure 3 shows that the finite element simulation analysis results of lV-1 and LV-2 specimens are basically consistent with the test curve. However, the errors of lV-1 and LV-2 of ultimate bearing capacity obtained from finite element simulation and test in Table 3-1 are only 1.5% and 0.5%. Based on the above comparison, it is considered that the results obtained by finite element are reliable, and the modeling and loading methods of this node can be used for parametric analysis of the node in subsequent analysis.

Table 1. Test and finite element bearing capacity comparison of LV-1 and LV-2.

| Specimen number | Bearing capacity test value (kN) | Finite element analysis value of bearing capacity (kN) | Difference value |
|---|---|---|---|
| LV-1 | 1685.90 | 1659.89 | 1.5% |
| LV-2 | 1168.39 | 1161.81 | 0.5% |

## 2.2 Comparison of results of nodes subjected to low cyclic reciprocating load

Figure 4 shows the finite element simulation results of beam V-pillar joints, and the stress states of V-2 joints under tension and compression in finite element simulation respectively. It can be seen that the stress on the upper and lower sides of the square steel tube beam at the joint of the square steel tube beam and the branch pipe string is obviously greater than that on the other parts, and the stress near the head of the branch pipe string is also greater than that at the fixed end. Because the finite element ignores the influence of residual stress of crack, the stress state of V-2 in the finite element is consistent with the stress state of the upper and lower sides of the square steel tube beam at the junction of the branch head in the test process.

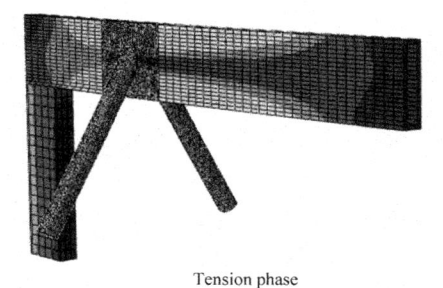 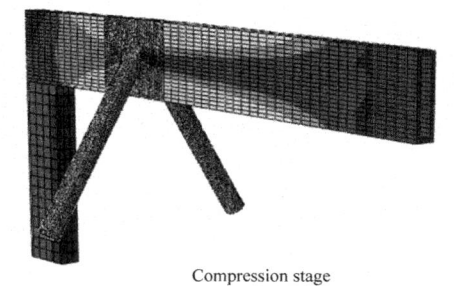

Tension phase                                              Compression stage

Figure 4. Finite element calculation results of V-2.

Figure 5a) shows the hysteretic curves obtained by the v-1 node test and finite element analysis. It can be seen that the hysteretic curves of the two are relatively consistent under forwarding compression, while obvious errors occur under negative tension. The reason is that the hysteretic curves obtained in the finite element analysis are symmetric in positive and negative directions. However, in the test process, due to the gap between the steel plate and the I-beam, the displacement of the loading end increases faster in the negative tension than in the positive tension, leading to the asymmetry of the positive and negative hysteretic curves.

Figure 5b) shows the hysteretic curves obtained from the V-2 node test and finite element analysis. It can be seen that there is a declining section in both the finite element simulation and test load-displacement curves under forwarding compression, while there is no declining section in the finite element simulation load-displacement curve under tension. The phenomenon of "pinching" of test curves of V-1 and V-2 joints is obvious, because the influence of cracks, bolt slip of fixed end of joints and Bauschinger effect of steel are ignored in the finite element simulation analysis.

Bauschinger effect generally does not appear in single crystal materials, while steel is polycrystalline material, and the residual stress generated by the interaction between crystals generally needs to consider the Bauschinger effect. Polycrystal metal material after enters the yield stage, reverse loading again after unloading the yield limit has obvious drop, unloading and reverse loading again yield limit will fall further, therefore in the process of test of the load-displacement curve has an obvious phenomenon of "pinch approach", while in the finite element assuming steel for isotropic materials, Therefore, the load-displacement curve obtained by finite element analysis is relatively full.

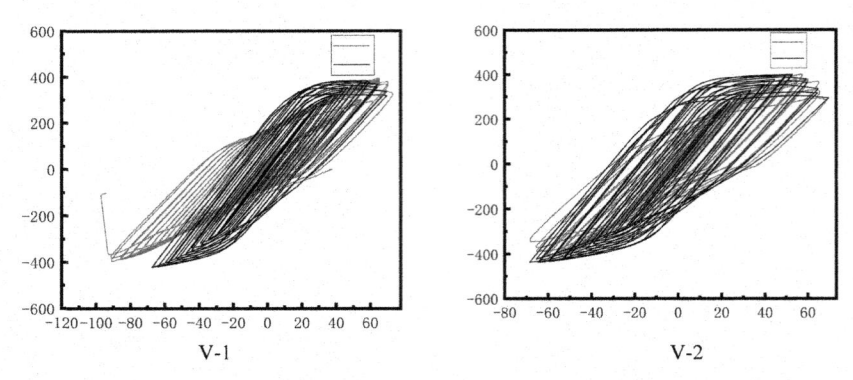

Figure 5.    Load-displacement curve comparison of V-1 and V-2.

Figure 6(a) shows the comparison of skeleton curves obtained by the V-1 node test and finite element simulation. The two curves fit well in the initial stage. The ultimate bearing capacity obtained by the test under forwarding compression is +400.78kN, and the ultimate bearing capacity obtained by finite element analysis is +383.07kN, with an error of 4.4%. The ultimate bearing capacity of the negative tension test is -395.88kN, and the ultimate bearing capacity of finite element analysis is -412.96kN, and the error between them is 4.1%. Figure 6(b) shows the comparison between the skeleton curves of the V-2 node obtained from the test and the finite element simulation. It can be seen that the two are in good agreement. The ultimate bearing capacity obtained from the test under forwarding compression is +403.36kN, and the ultimate bearing capacity obtained from the finite element analysis is +395.81kN, with an error of 2.0%. The ultimate bearing capacity of the negative tension test is -410.40kN, and the ultimate bearing capacity of finite element analysis is -431.42kN, and the error between them is 4.9%.

Table 7 shows the comparison between the ultimate bearing capacity obtained from V-1 and V-2 finite element analysis and the test. It can be seen that the stiffness obtained from the test and the finite element analysis is similar in the loading process of the node, and the ultimate bearing capacity error is within 5% when the node is subjected to tension and compression. Therefore, the reliability and rationality of the finite element analysis can be preliminarily verified.

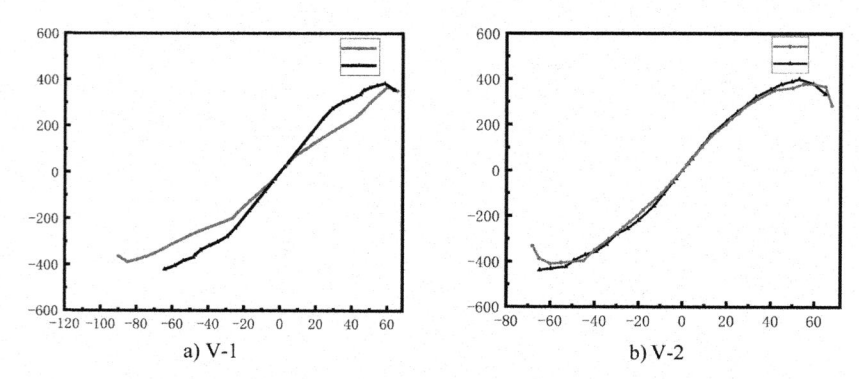

Figure 6.    Skeleton curve comparison of V-1 and V-2.

Table 2. Test and finite element bearing capacity comparison of V-1 and V-2.

| Specimen number | Loading direction | Test ultimate bearing capacity (kN) | Finite element ultimate bearing capacity (kN) | Difference value |
|---|---|---|---|---|
| V-1 | + | +400.78 | +383.07 | 4.4% |
| V-1 | − | −395.88 | −412.96 | 4.1% |
| V-2 | + | +403.36 | +395.81 | 2.0% |
| V-2 | − | −410.40 | −431.42 | 4.9% |

## 3 CONCLUSION

This chapter introduces the basic principles and characteristics of ABAQUS, provides a theoretical basis for the simulation calculation of beam V-pillar joints in the software ABAQUS, and gives a detailed description of the node modeling process, basic assumptions in the finite element, the determination of material parameters according to the material test, the determination of boundary conditions, and the introduction and division of grid types. Then the stress state of LV-1 and LV-2 nodes and load-displacement curves, hysteretic curves, skeleton curves, the stress state of V-1 and V-2 nodes, and test results were compared. The conclusions are as follows:

(1) The difference between the ultimate bearing capacity obtained by LV-1 and LV-2 finite element analysis and the test result is only 1.5% and 0.5%, the load-displacement curve is basically consistent with the test result, and the stress state of the node obtained by finite element analysis is consistent with that obtained by strain analysis of the test node.
(2) The errors between the bearing capacity of v-1 and V-2 finite element analysis and the test results at the stage of tension and compression are all within 5%. The hysteresis curves calculated in the finite element are not pinched due to the neglect of bolt slip and Bauschinger action, but the skeleton curves are consistent with the test results.

The above comparison results can prove the rationality and reliability of finite element analysis and provide a theoretical basis for subsequent parameter analysis.

ACKNOWLEDGEMENTS

This paper is supported by the Advanced Advanced Innovation Center for Future Urban Design at the Beijing University of Civil Engineering and Architecture (UDCGJ002).

REFERENCES

Lee M M K, Wilmshurst S R. Parametric study of strength of tubular multiplanar KK-joints[J]. *Journal of Structural Engineering*, 1996, 122(08): 893–904.
Lee M M K, Wilmshurst S R. Strength of multiplanar tubular KK-joints under antisymmetrical axial loading[J]. *Journal of Structural Engineering*, 1997, 123(06): 755–764.
Paul J C, Makino Y, Kurobane Y. Ultimate resistance of unstiffened multiplanar tubular TT-and KK-joints[J]. *Journal of Structural Engineering*, 1994, 120(10): 2853–2870.
Wang J, Zhang K, Shu G, et al. Investigations on stainless steel T-and Y-joints in cold-rolled circular hollow sections[J]. *Journal of Constructional Steel Research*, 2021, 177(05): 106462.

*Civil Engineering and Urban Research – Mohamed & Hou (Eds)*
© 2023 the Authors, ISBN 978-1-032-44487-1

# Research on strength and stability of underwater ring-ribbed cylindrical shell structure considering corrosion effect

Chuang You* & Feiyu Chen

*713th Research Institute of China State Shipbuilding Corporation Limited, Zhengzhou, China*
*Henan Key Laboratory of Underwater Intelligent Equipment, Zhengzhou, China*

ABSTRACT: Underwater vehicles are playing an increasingly important role in the exploitation of deep-sea resources and energy utilization. In order to ensure the reliability and multiple reuse requirements under the coupling of complex loads of pressure structure of the underwater vehicle, strength, and stability of underwater ring-ribbed cylindrical shell structure considering corrosion effect were studied in this paper. The finite element model was established by ABAQUS at first, and then the failure criteria of strength and stability are given. The geometric model of pitting was established according to the aperture and depth of pitting on this basis and the surface corrosion model was established by selecting typical positions in the structure to analyze the influence of different corrosion depths and corrosion radiuses on the strength and stability of the structure. The results show that the stress concentration occurs around the axial bus of pitting and the value decreases rapidly along the axis. Besides, the critical buckling load of the structure decreases with the increase of the corrosion depth when the corrosion radius reaches a certain value.

## 1 INTRODUCTION

With the exploitation of marine resources and the development of marine science, the role of underwater vehicles is increasingly prominent. However, owing to the extreme working conditions, underwater vehicles are often subjected to complex alternating loads in addition to an underwater high-pressure environment. The underwater pressure structure of the vehicle is threatened by corrosion, fatigue crack, and other damages, which may lead to local buckling of the structure and seriously affect the safety and reliability of the equipment. Among them, corrosion damage is the most common in marine structures under extreme loads (Yu et al. 2018). Therefore, it is necessary to study the influence of corrosion effect on underwater pressure structure to ensure the safe navigation of the underwater vehicle.

Pitting corrosion and stress corrosion are the main corrosion forms of aluminum alloy in the marine environment, and the sensitivity and the extent of corrosion intensify with the increase in water depth (Hou et al. 2015). Duo et al calculated the ultimate strength of the plate under the pitting effect and obtained the formula for ultimate strength reduction under the multi-parameter comprehensive influence (Ok et al. 2007). Yao et al calculated the ultimate strength of longitudinal stiffeners due to pitting based on data from prototype testing and proposed an empirical formula for the ultimate strength of stiffened plates with two parameters (Yao et al. 2018). Wang et al studied the buckling of 2d and 3D pipeline models considering external random pitting by numerical simulation and experiment (Wang et al. 2018, 2018). Yu et al simulated and analyzed the collapse pressure of the 2D ring pipe model caused by internal random pitting under external pressure. Their results show that corrosion area ratio, diameter-thickness ratio, and ellipticity are the main factors affecting collapse pressure, and empirical formulas are established to describe the effects of three parameters on collapse pressure reduction (Yu et al. 2019).

---

*Corresponding Author: chuangyou@hrbeu.edu.cn

DOI 10.1201/9781003372417-39

In conclusion, scholars have done a lot of research on the structural strength and buckling considering the corrosion effect, but the research on the ring-ribbed cylindrical shell structure of the underwater vehicle is relatively few. In this paper, the underwater ring-ribbed cylindrical shell structure is investigated. A simulation model considering the effect of corrosion is established by ABAQUS. The influence of the corrosion caused by geometric defects on the strength and stability of the underwater ring-ribbed cylindrical shell structure is studied and a method considering the effect of the corrosion model for strength and stability assessment of underwater pressure-resistant structures is put forward.

## 2 NUMERICAL MODEL

The ring-ribbed cylindrical shell structure is the main structure of the underwater vehicle. In this paper, a defect-free ring-ribbed cylindrical shell structure is firstly established according to the ratio of diameter to thickness, then the pitting corrosion model is established according to the aperture and depth of pitting corrosion, and the surface corrosion model is established by selecting typical positions in the structure. Finally, the solution is carried out by dividing mesh, defining the material properties of the model, and setting initial conditions and boundary conditions.

The material property of the ring-ribbed cylindrical shell is aluminum alloy ZL 114A. The material parameters are shown in Table 1, and Figure 1 gives the semi-sectional view of the ring-ribbed cylindrical shell structure.

Table 1. Material property.

| Aluminum | Elasticity modulus (GPa) | Poisson's ratio | Yield stress $\sigma_y$ (MPa) |
|---|---|---|---|
| ZL 114a | 71 | 0.33 | 280 |

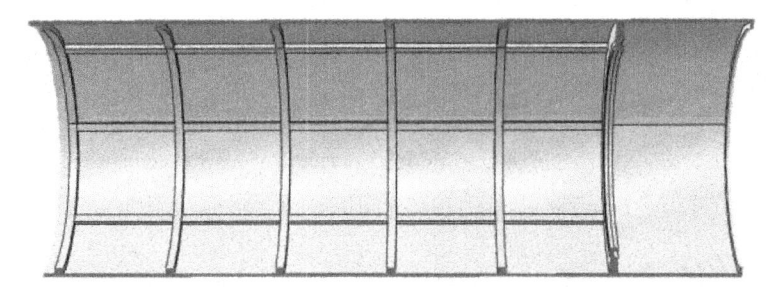

Figure 1. Semi-sectional view of the ring-ribbed cylindrical shell structure.

The environment load of an underwater vehicle is very complicated in practical conditions, among which the most important one is static pressure. In this paper, it is set as 1.5 Mpa and distributed uniformly in the radial direction along the outside surface of the component.

Due to the complexity of the model, the partition tool is used to segment the model, and tetrahedral mesh is used to improve the computational accuracy, efficiency, and convergence of simulation. The approximate global size is set to 2 mm to mesh the parts.

### 2.1 Failure criterion of strength

For corrosion defects, the local maximum stress of corrosion is calculated as follows:

$$\sigma_{\text{cor}} = K_{\text{cor}} K_0 K_1 \frac{P_j R}{t} \tag{1}$$

Where $K_{cor}$ is the coefficient of corrosion correction, $K_0$ denotes the reinforcement coefficient of longitudinal ribs, is set to 1.07, $K_1$ is the coefficient of intensity correction, and $P_j$ and $R$ are calculated pressure and mean radius of cylindrical shell respectively.

$$K_{cor} = 0.04r + 0.07d + 1 \tag{2}$$

Where $r$ is the corrosion radius and $d$ denotes the corrosion depth.

For the corrosion defect model, the strength failure occurs when the maximum local stress of corrosion meets the following equation:

$$\sigma_{cor} = K_{cor}K_0K_1\frac{P_jR}{t} \geq 0.85\sigma_y = 238\text{MPa} \tag{3}$$

### 2.2 Failure criterion of stability

When there is no ellipticity in the model, the elastic critical pressure of the pressure structure is calculated as follows:

$$P_e = E(\frac{t}{R})^2\frac{0.6}{u - 0.37} \tag{4}$$

Where $E$ is the elastic modulus, $t$ denotes the shell thickness, and $u = 0.643l\,(Rt)^{1/2}$.
When $P_j \geq P_e = 16.39\text{Mpa}$, the stability failure occurs.

## 3 STRUCTURAL STRENGTH AND STABILITY ANALYSIS CONSIDERING CORROSION EFFECT

### 3.1 Structural strength analysis considering corrosion effect

The pitting corrosion and surface corrosion are considered in this paper in view of the conditions of corrosion of underwater pressure structures. According to engineering experience, generally, the aperture of pitting corrosion is less than 1mm, and the depth is greater than the aperture. Therefore, this paper carries on the simulation analysis of the pitting corrosion where the aperture is 1mm and depth is 4mm. On account that the range of influence of stress distribution is limited due to the small pitting radius, a ring model with an outer diameter of 324 mm, a height of 10mm, and a thickness of 5mm is selected for simulation, and the mechanical environment of the whole structure is simulated by setting the fixed boundary conditions.

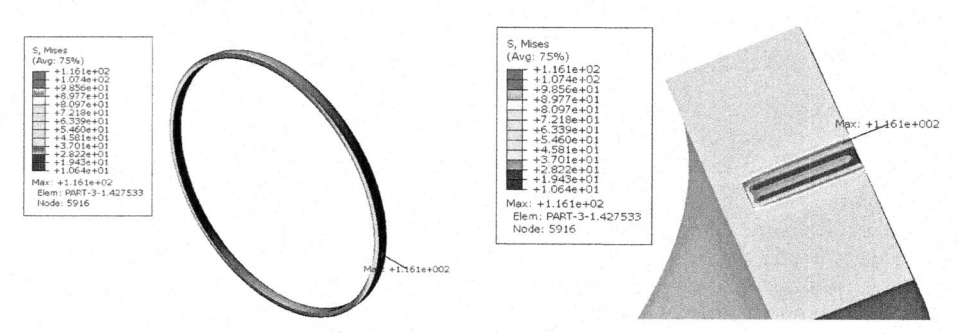

Figure 2.  Stress nephogram of the model.             Figure 3.  Stress nephogram of axial section of pitting.

It can be seen from Figures 2 and 3 that the influence range of pitting corrosion on the stress distribution of the model is very limited. Because of pitting pits, the stress is concentrated around the axial bus of pitting pits, and the stress decreases rapidly along the axial direction. The maximum stress is 116.1Mpa and lower than the yield strength of the material, so in the condition of pitting corrosion of 1mm aperture and 4mm depth, the model will not be damaged due to the stress concentration of pitting corrosion.

For surface corrosion, different degrees of surface corrosion may occur in any position of the model in actual working conditions. As shown in Figure 4, in order to facilitate the analysis, two typical positions in the ring-ribbed cylindrical shell are selected to add a corrosion model respectively for corrosion analysis.

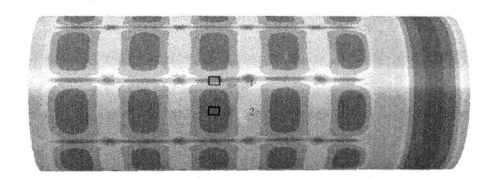

Figure 4.   Location of surface corrosion.

Corrosion depth and corrosion radius are two parameters to characterize the degree of corrosion in simulation. The simulation model is numbered A-B-C, Where A represents the corrosion position, B represents the corrosion radius, and C represents the corrosion depth. Two groups of 12 corrosion models are set as follows: 1-3-2.5, 1-5-2.5, 1-7.5-2.5, 1-10-2.5, 1-5-4, 1-10-4, 2-3-2.5, 2-5-2.5, 2-7.5-2.5, 2-10-2.5, 2-5-4, and 2-10-4.

The overall nephogram is obtained by the simulation analysis of ring-ribbed cylindrical shell structure with a corrosion model. The stress nephogram of the corroded part can be obtained by magnifying the corroded part. Simulation results of cases 1-5-2.5 and 1-10-2.5 are shown in Figure 5 and Figure 6 respectively.

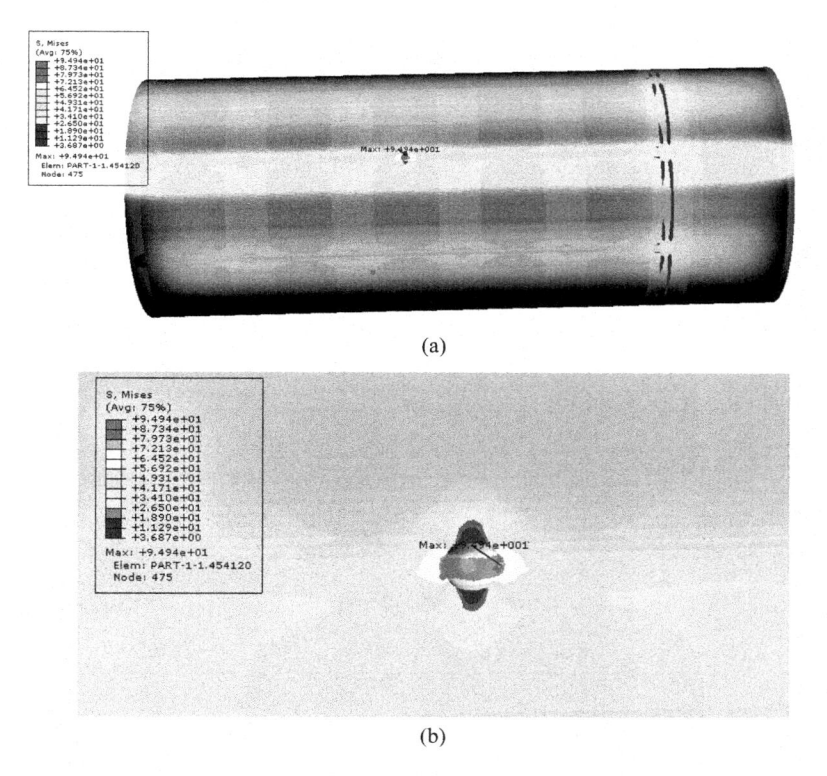

(a)

(b)

Figure 5.   Global and local stress of the case of 1-5-2.5.

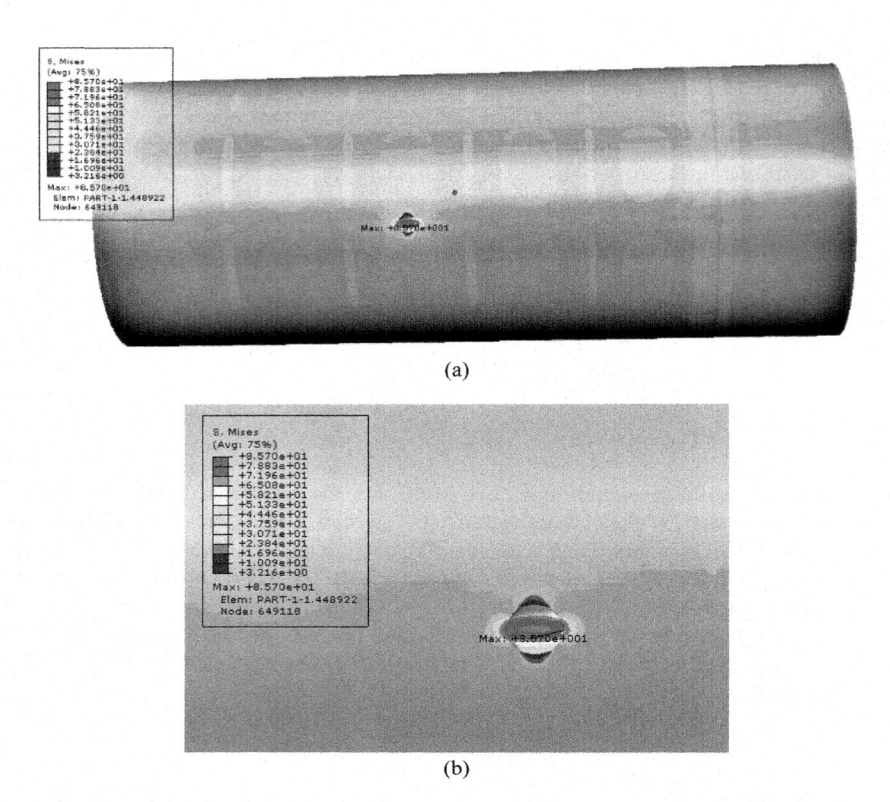

(a)

(b)

Figure 6.    Global and local stress of the case of 1-10-2.5.

The numerical results of all corrosion models of corrosion depth of 2.5 mm and 4 mm are statistically summarized, as shown in Tables 2 and 3, and Figure 7.

Table 2.    The maximum stress in case of corrosion depth of 2.5 mm (MPa).

| Corrosion radius (mm) | Position 1 | Position 2 |
|---|---|---|
| 3 | 61.43 | 67.69 |
| 5 | 94.94 | 76.52 |
| 7.5 | 87.74 | 80.53 |
| 10 | 85.70 | 85.80 |

Table 3.    The maximum stress in case of corrosion depth of 4 mm (MPa).

| Corrosion radius (mm) | Position 1 | Position 2 |
|---|---|---|
| 3 | – | – |
| 5 | 104.1 | 118.0 |
| 7.5 | – | – |
| 10 | 104.6 | 145.7 |

As can be seen from Figures 5–7 and Tables 2–3, when corrosion occurs on the surface of a ring-ribbed cylindrical shell structure, the stress is concentrated near the corrosion point. From

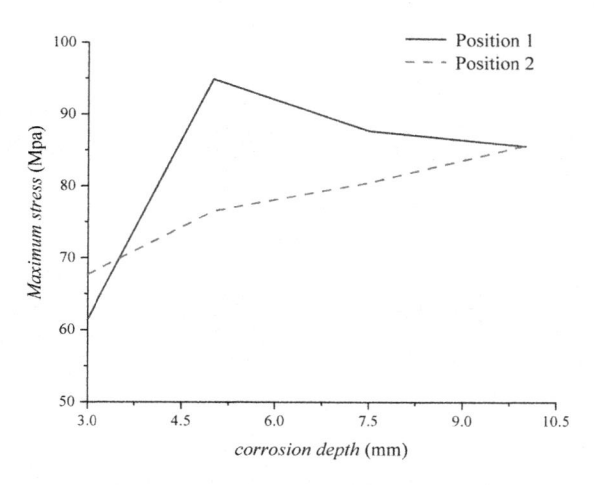

Figure 7.   The maximum stress of all corrosion models.

the corrosion pit outward, the axial stress slowly decreases to the distant normal area, and the circumferential stress first plummets and then slowly rises to the distant normal area. The internal support rib plate has a significant effect on the stress distribution near the corrosion point. Under the same corrosion radius, the stress concentration intensifies with the growth of the corrosion depth ($\leq$5mm), and the maximum stress increases. At the same corrosion depth, the maximum stress magnifies as the corrosion radius increases. When the corrosion depth reaches 4 mm (the thickness of the structure is 5 mm) and the corrosion radius reaches 10mm, the maximum stress in the structure is 145.7 Mpa, which is less than the yield strength of 280 Mpa. Therefore, in the above corrosion conditions, no structural failure occurs in the underwater ring-ribbed cylindrical shell structure.

## 3.2   *Structural stability analysis considering corrosion effect*

Corrosion-buckling analysis models are established by adding corrosion damage of radius of 5 and 10 mm and depth of 2.5, 3 and 4 mm respectively to the outer surfaces of the intercostal models.

Here, a case of corrosion damage with a radius of 10 mm and a depth of 3 mm is selected for corrosion-buckling analysis. When the initial arc length increment is set as 0.1 and the increment step is 50, the buckling stress nephogram and pressure-displacement curve of the ring-ribbed cylindrical shell model are shown in Figure 8.

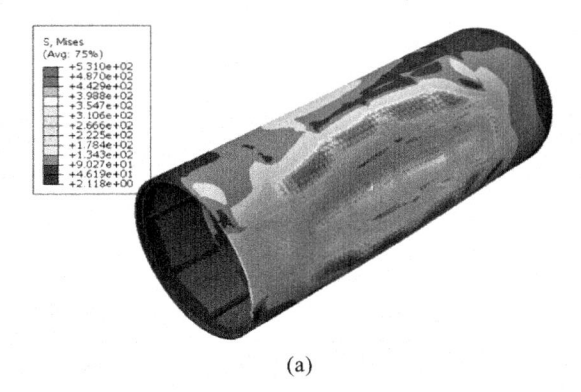

(a)

Figure 8.   The buckling stress and pressure-displacement curve of the ring-ribbed cylindrical shell model.

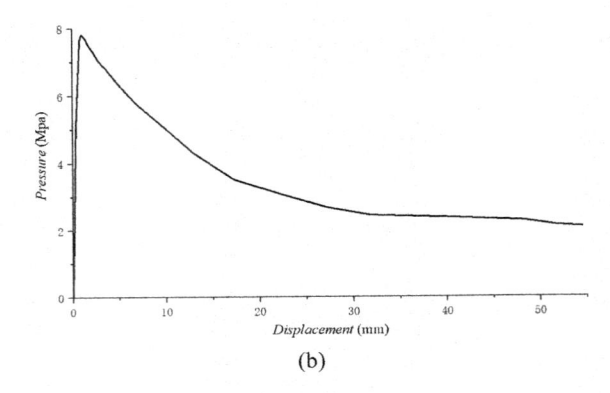

(b)

Figure 8. Continued.

Results of the corrosion analysis are summarized as shown in Table 4, and the results of the corrosion-buckling analysis are shown in Figure 9.

Table 4. Buckling analysis of corrosion model (unit: MPa).

| Corrosion radius(mm) | Corrosion depth (mm) | | |
|---|---|---|---|
| | 2.5 | 3 | 4 |
| 5 | 7.87 | 7.86 | 8.10 |
| 10 | 7.96 | 7.82 | 7.71 |

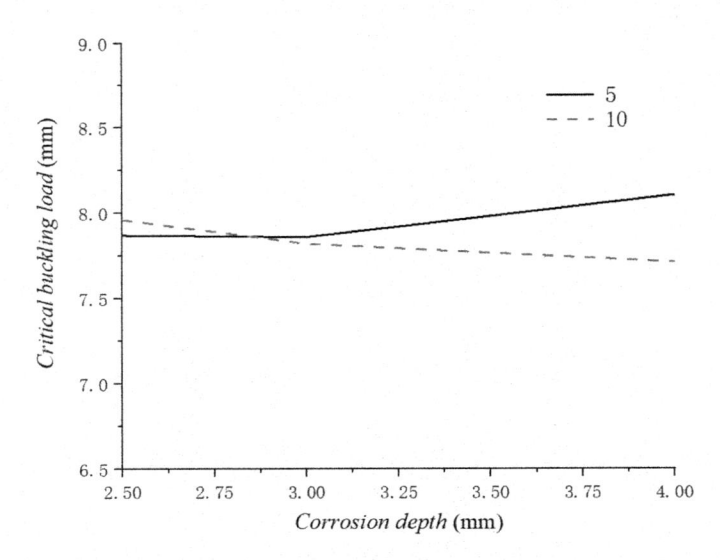

Figure 9. Critical buckling load of corrosion-buckling analysis.

As can be seen from Table 4 and Figure 9, when the corrosion radius is larger, the critical buckling load of the ring-ribbed cylindrical shell model decreases as the corrosion depth increases, which is consistent with common sense. When the corrosion radius is small, the relationship between the critical buckling load and the corrosion depth is not obvious. In general, corrosion has no obvious influence on the critical buckling load of the ring-ribbed cylindrical shell structure in the above conditions.

## 4 CONCLUSION

In this paper, a finite element model is established for the underwater ring-ribbed cylindrical shell structure considering the corrosion effect, and the strength and stability of the structure under the action of pitting and surface corrosion are studied. The conclusions are as follows:

(1) the stress concentration appears around the axial bus of the pitting pit, and the value decreases rapidly along the axial direction under the influence of pitting. When the pitting corrosion depth is less than 5 mm, under the same corrosion radius, the extent of stress concentration intensifies and the maximum stress increases as the corrosion depth increases; and at the same corrosion depth, the maximum stress increases as the corrosion radius increases.
(2) When the corrosion radius is small, the influence of corrosion effect on the critical buckling load of the underwater ring-ribbed cylindrical shell is not obvious. When the corrosion radius reaches a certain value, the critical buckling load of the structure decreases with the increase of the corrosion depth.
(3) The simulation analysis method established in this paper can provide technical support and program strategies for the structural strength and stability evaluation of underwater vehicles and underwater intelligent equipment. The strength and stability of underwater pressure structure under the combined action of multiple factors can be considered for further study.

## REFERENCES

Hou J, Zhang P H and Guo W M 2015 Study on corrosion of aluminum alloys for ship applications in marine environment *Equipment Environmental Engineering* 12(2) 59–63.
Ok D, Pu Y Y and Incecik A 2007 Computation of ultimate strength of locally corroded unstiffened plates under uniaxial compression *Mar. Struct.* 20 100–14.
Wang H K, Yu Y, Yu J X, Duan J H, Zhang Y, Li Z M and Wang C M 2018 Effect of 3D random pitting defects on the collapse pressure of pipe (Part I): Experiment *Thin Wall Struct.* 127 512–26.
Wang H K, Yu Y, Yu J X, Jin C H, Zhao Y, Fan Z Y and Zhang Y 2018 Effect of 3D random pitting defects on the collapse pressure of Pipe (Part II): Numerical analysis *Thin Wall Struct.* 129 527–41.
Yao Z G, Lyu Y S, Hua L and Zhang Y C 2018 Ultimate strength calculation of stiffened plates under longitudinal compression with pitting damage *Ship Engineering* 40(supplement1) 238–42.
Yu J X, Jin C H, Yu Y, Wang H K and Tan Y N 2019 Plastic collapse capacity of the 2D ring with internal random pitting corrosion defects *Journal of Tianjin University Science and Technology* 12(52) 1220–26.
Yu J X, Wang H K, Yu Y, Luo Z, Liu W D and Wang C M 2018 Corrosion behavior of X65 pipeline steel: Comparison of the wet-dry cycle and full immersion *Corros. Sci.* 133 276–87.

*Civil Engineering and Urban Research – Mohamed & Hou (Eds)*
*© 2023 the Authors, ISBN 978-1-032-44487-1*

# Deformation and failure mechanism of surrounding rock considering the influence of tensile yield on mechanical properties of rock mass

Hai-feng Li, Biao Wang, Lei Chen, Yang Wang & Shuai Tao*
*POWERCHINA Huadong Engineering Corporation Limited, Hangzhou, Zhejiang, China*
*Zhejiang Huadong Engineering Consulting Corporation Limited, Hangzhou, Zhejiang, China*

ABSTRACT: Based on the Mohr-Coulomb constitutive model, using the nonlinear CWFS criterion of friction strengthening and cohesion weakening, according to the degradation behavior of tensile strength and stiffness modulus of intact rock mass after tensile yield, considering the dilatation effect of intact surrounding rock, the failure and deformation of tunnel surrounding rock during excavation unloading are simulated. The results show that the nonlinear CWFS model is used to simulate the formation of a 'V' shape plastic zone in the surrounding rock of tunnel top and bottom of the tunnel, which is consistent with the failure zone observed in the field. Controlling the attenuation of Young's modulus by tensile plastic strain threshold, a single crack with pure tensile yield in the horizontal direction of the tunnel surrounding rock can be simulated. The decrease of Poisson's ratio causes the sharp angle of the 'V' shape plastic zone formed by shear yield to expand and the surrounding rock changes from brittleness to ductility. In schemes 1 - 6, the tensile plastic strain increases slightly with the decrease of residual elastic modulus, and the maximum values are $1.8132 \times 10^{-2}$, $1.8154 \times 10^{-2}$, and $1.8204 \times 10^{-2}$, respectively. Under the same residual elastic modulus, the smaller the Poisson's ratio, the smaller the extreme value of the maximum principal stress. Compared with Poisson's ratio of 0.25, when the Poisson's ratio is 0.2, the extreme value of the maximum principal stress decreases by about 0.7%. When Poisson's ratio is 0.2, the influence of residual elastic modulus on maximum principal stress decreases.

## 1 INTRODUCTION

With the increasing improvement and development of China's economic level, the demand for energy is becoming increasingly urgent. With the implementation of the dual carbon policy, green and clean energy projects have ushered in the construction climax, such as water conservancy power generation, pumped storage, water resources allocation, and other projects. As the basic structure of water conservancy projects, hydraulic tunnels will inevitably be widely used. The stability of tunnel surrounding rock is one of the key factors affecting the design and construction of hydraulic tunnels. The failure and deformation mechanism of surrounding rock and the stress and strain evolution of surrounding rock caused by excavation unloading are crucial to the stability analysis of surrounding rock, especially the complete and hard surrounding rock (Jiang et al. 2008; Tao et al. 2011).

In recent years, many scholars have carried out a lot of research on the yield failure mechanism of hard rock, but they mainly focus on the shear yield failure of rock mass, such as strain softening yield criterion based on the Mohr-Coulomb model, CWFS model of cohesive softening and friction strengthening; nonlinear yield criterion, softening model and CWFS model based on Hoek-Brown constitutive model; the composite nonlinear yield criterion based on Mohr-Coulomb and Hoek-Brown constitutive models (Hajiabdolmajid et al. 2002; Hajiabdolmajid & Kaiser 2003; Hoek & Martin 2014; Martin 1997; Tao et al. 2011). A large number of studies have been carried out on the

---

*Corresponding Author: tao_s@hdec.com

 DOI 10.1201/9781003372417-40

linear and nonlinear deterioration and attenuation mechanisms of cohesion and friction angle with equivalent plastic strain or minimum principal stress in the shear yield process of the surrounding rock, which are applied in the numerical calculation (Patel & Martin 2018; Pate & Martin 2018; Patel & Martin 2020; Renani & Martin 2018).

Although the understanding of the progressive failure mechanism of hard surrounding rock is gradually deepened, the deterioration mechanism of stiffness modulus and tensile strength caused by tensile yield during crack initiation and development has not been fully revealed. The constitutive model used in numerical calculation directly affects the accuracy of calculation results. The previous numerical simulation studies can not fully reveal the brittle failure mechanism of rock. Although the V-type failure zone formed by a brittle failure of hard brittle surrounding rock with high geostress can be simulated, the influence of stiffness modulus and tensile strength attenuation after the tensile yield on the deformation and failure of surrounding rock is not considered.

In this paper, based on the Mohr-Coulomb strain softening constitutive model, the nonlinear model of cohesion and friction angle with the degradation and enhancement of equivalent plastic strain is adopted. The influence of plastic tensile strain caused by the tensile yield of surrounding rock on the stiffness modulus and tensile strength of rock mass is considered. The plastic zone, stress, and strain evolution of surrounding rock under different tensile softening modulus and Poisson's ratio are analyzed. The failure mechanism of surrounding rock with tensile strength and stiffness attenuation with tensile yield is studied, which provides a reference for the tunnel design and construction.

## 2 CALCULATION MODEL, CONSTITUTIVE MODEL AND CALCULATION SCHEME

The plane strain model with length ($x$ direction) and height ($y$ direction) of 20 m is divided into 40000 rectangular grids. The original rock stress is obtained by no excavation and first loading. When the calculation reaches equilibrium, the whole section excavation is carried out. The circular tunnel with a diameter of 3 m is excavated. The geometric model before and after excavation is shown in Figure 1. The calculation was carried out under small deformation conditions, and the model was subjected to the horizontal compressive stress $P_h$ and the vertical compressive stress $P_v$ around it.

The yield condition in the constitutive model is the same as that in the M-C softening model. Considering the dilatation characteristics of rock mass, the non-associated flow rule is adopted. The deterioration law of cohesion and friction angle with plastic strain adopts CWFS nonlinear relationship (Renani, Martin, 2018), where,

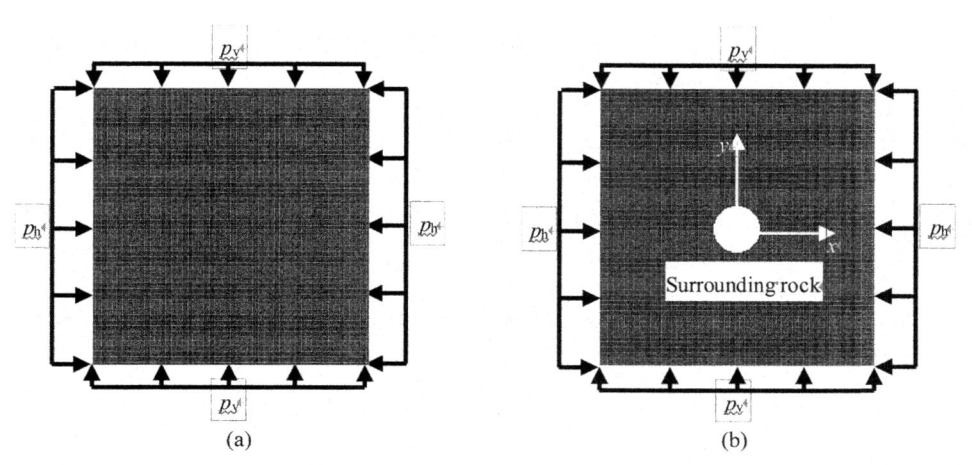

Figure 1. Geometric model and boundary conditions: (a) Not excavated; (b) After excavation.

$$c = c_r + (c_i - c_r) \times (2 - 2/(1 + e^{-5 \times \varepsilon p/\varepsilon r})) \tag{1}$$

$$\varphi = \varphi_i + (\varphi_r - \varphi_i) \times (2/(1 + e^{-5 \times \varepsilon p/\varepsilon r}) - 1) \tag{2}$$

Other parameters are shown in Table 1.

Horizontal and vertical loads $P_h$ and $P_v$ are 60 MPa and 11 MPa, respectively (Renani & Martin 2018). The initial values of cohesion $c$ and internal friction angle $\varphi$ are 55 MPa and 0, respectively. The residual cohesion $c_r$ and friction angle $\varphi_r$ are 5.5 MPa and 42, respectively.

Table 1. Mechanical parameters of surrounding rock.

| Scheme | $E_0$ (GPa) | $E_r$ (GPa) | $c$ (MPa) | $c_r$ (MPa) | $\sigma_t$ (MPa) | $\sigma_{tr}$ (MPa) | $\sigma_{ttr}$ (MPa) | $\varphi_0^\circ$ | $\varphi_r^\circ$ | $\psi(\circ)$ | $\varepsilon_{sr}$ | $\varepsilon_{tr}$ | $\upsilon$ |
|---|---|---|---|---|---|---|---|---|---|---|---|---|---|
| Scheme 1 | 60 | 55 | 55 | 5.5 | 11.6 | 1.6 | 0 | 0 | 42 | 30 | 0.005 | 0.00028 | 0.25 |
| Scheme 2 | 60 | 50 | 55 | 5.5 | 11.6 | 1.6 | 0 | 0 | 42 | 30 | 0.005 | 0.00028 | 0.25 |
| Scheme 3 | 60 | 42.4 | 55 | 5.5 | 11.6 | 1.6 | 0 | 0 | 42 | 30 | 0.005 | 0.00028 | 0.25 |
| Scheme 4 | 60 | 55 | 55 | 5.5 | 11.6 | 1.6 | 0 | 0 | 42 | 30 | 0.005 | 0.00028 | 0.20 |
| Scheme 5 | 60 | 50 | 55 | 5.5 | 11.6 | 1.6 | 0 | 0 | 42 | 30 | 0.005 | 0.00028 | 0.20 |
| Scheme 6 | 60 | 42.4 | 55 | 5.5 | 11.6 | 1.6 | 0 | 0 | 42 | 30 | 0.005 | 0.00028 | 0.20 |

The attenuation of cohesion is consistent with the corresponding plastic strain $\varepsilon_{sr}$ when the friction angle after friction starts reaches a stable value, respectively, which is $5 \times 10^{-3}$ (Renani & Martin 2018). Considering the effect of tensile plastic strain on the modulus and tensile strength of surrounding rock, according to the relevant experimental results reported in the literature, it is assumed that when the plastic tensile strain is not zero, the elastic modulus of surrounding rock decreases from the initial value $E_0$ to the residual value $E_r$, and the tensile strength $\sigma_t$ decreases to $\sigma_{tr}$ when the tensile plastic strain reaches the residual value $2.8 \times 10^{-4}$ (Renani & Martin 2018), that is 1.6 MPa. When the tensile plastic strain exceeds the residual value, the crack propagates, and the tensile strength $\sigma_t$ decreases to $\sigma_{ttr}$, which is assumed to be 0 MPa in this paper. Figure 2 shows the evolution of cohesion and internal friction angle with plastic strain according to Formulas (1) and (2).

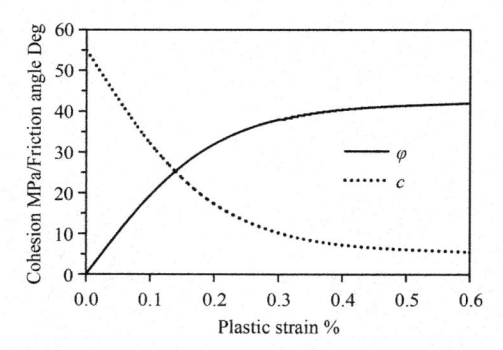

Figure 2. Cohesion degradation and friction mobilization.

In this paper, six calculation schemes are used to analyze and compare the elastic modulus attenuation after tensile plastic strain reaches the threshold and the deformation and failure of tunnel surrounding rock under different Poisson's ratios. The initial elastic modulus of the six schemes is all 60 GPa. The elastic modulus of schemes 1–3 is 55 GPa, 50 GPa, and 42.4 GPa respectively when Poisson's ratio is 0.25 and tensile plastic strain reaches the threshold. Schemes 4–6 are the calculation results when the Poisson's ratio is 0.2 and the elastic modulus decreases with the tensile plastic strain.

# 3 RESULT ANALYSIS AND DISCUSSION

## 3.1 *Distribution of plastic zone of the surrounding rock*

Limited to space, this paper only gives the results when the calculation reaches the static equilibrium. The distribution of the plastic zone and the evolution of the maximum unbalanced force with the time step calculated by schemes 1–6 are given in Figure 3. In the figure, the dark grey area is the shear plastic zone, s - n means that the surrounding rock is shear yielding, the black area is the tensile plastic zone, and t - n means that the surrounding rock is tensile yielding. It can be found from Figure 3 (a–c) that when the tensile plastic strain of surrounding rock reaches the threshold in Schemes 1–3, the attenuation of elastic modulus does not have an obvious influence on the distribution of the plastic zone of surrounding rock. However, due to the influence of tensile yield on the stiffness modulus of the surrounding rock, a single crack along the horizontal direction appears in the surrounding rock on the horizontal symmetry axis of the model. In the past, the calculation results of other models cannot simulate such phenomena. In addition, the comparison schemes 4–6 are shown in Figure 3 (d-f). It can be found that the decrease of Poisson's ratio leads to a slight change in the form of the shear yield zone. Due to the decrease of Poisson's ratio, the tip of the 'V' shape plastic zone in the surrounding rock tends to be flat, and the surrounding rock changes from brittle failure to ductile failure.

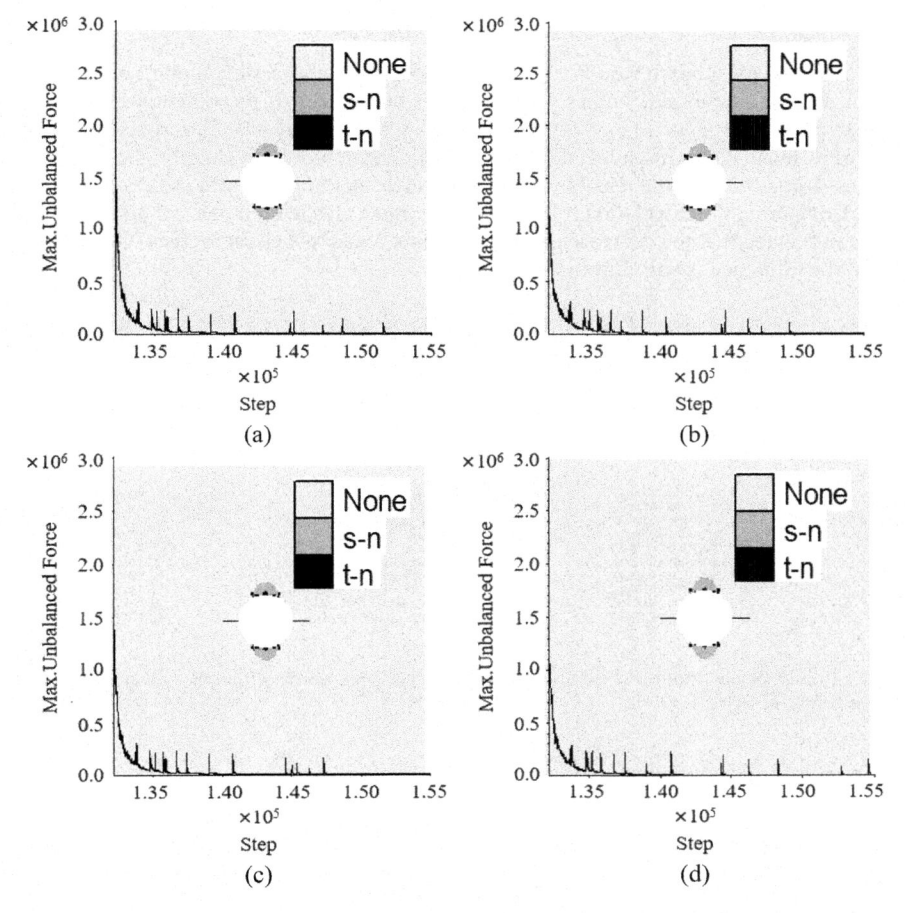

Figure 3.   Plastic zone distribution of surrounding rock under different schemes: (a)–(f) is the calculation results of Schemes 1–6.

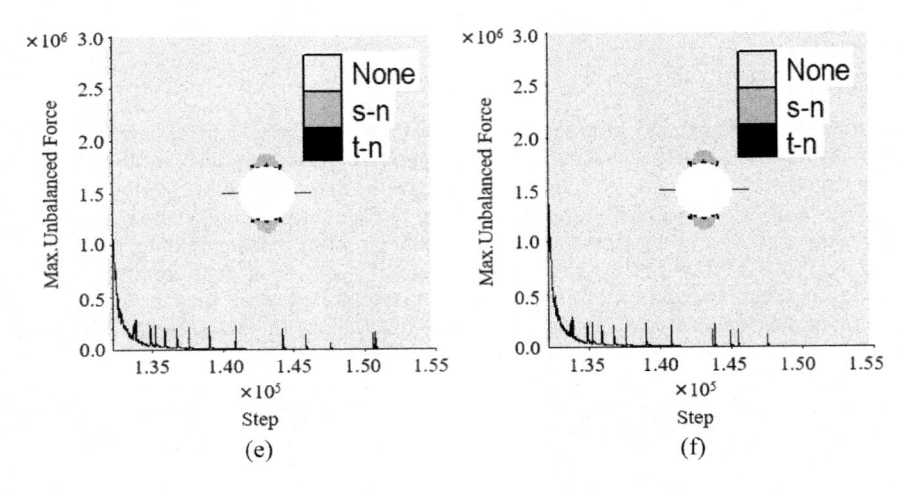

(e) (f)

Figure 3. Continued.

## 3.2 *Distribution of tensile plastic strain of surrounding rock*

Figure 4 gives the calculation results of the tensile plastic strain of surrounding rock in schemes 1–6. From Figure 4 (a-c) and Figure 4 (d-f), it can be found that the maximum tensile plastic strain in schemes 1–3 is $1.8132 \times 10^{-2}$, $1.8154 \times 10^{-2}$ and $1.8204 \times 10^{-2}$, respectively. With the decrease of residual elastic modulus, the tensile plastic strain increases slightly; the Poisson's ratio in schemes 4–6 is 0.2, and the tensile plastic strain increases slightly with the decrease of elastic modulus. Comparing the calculation results under the same elastic modulus and different Poisson's ratios, it can be seen that the decrease of Poisson's ratio leads to a slight increase in tensile plastic strain, but the influence is very limited.

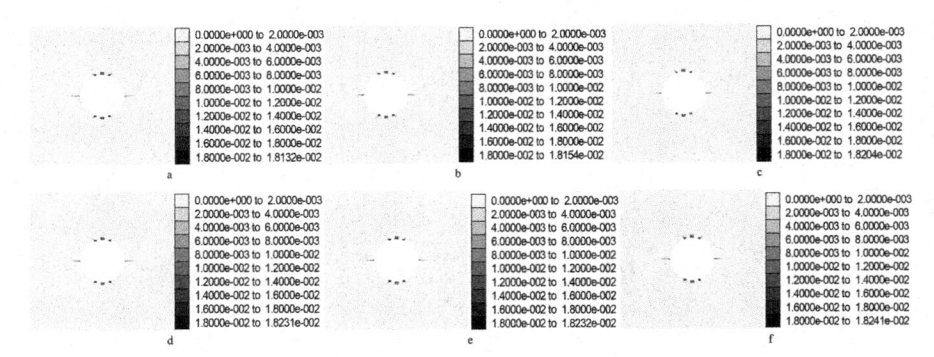

Figure 4. Tensile plastic strain distribution of surrounding rock under different schemes: (a)–(f) is the calculation results of Schemes 1–6.

## 3.3 *Incremental distribution of shear strain of surrounding rock*

Figure 5 shows the distribution of shear strain increment in surrounding rock calculated by schemes 1–6. It can be found that the high-value area of shear strain increment penetrates into the surrounding rock and forms the 'V' type area similar to the shear plastic yield zone. Under the same Poisson's ratio, with the decrease of residual elastic modulus, the high shear strain area increases gradually, and the maximum shear strain increment in schemes 1–3 increases from about $2.33 \times 10^{-2}$ to $2.35 \times 10^{-2}$. Under the same residual elastic modulus, the smaller the Poisson's ratio is, the larger the shear strain increment is. Under the same conditions, the Poisson's ratio decreases from 0.25 to 0.2, and the corresponding shear strain increment increases by about 0.1%.

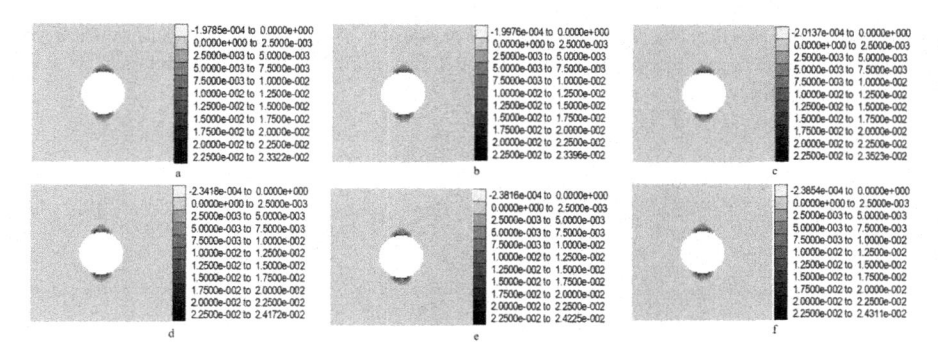

Figure 5.    Shear strain increment of surrounding rock under different schemes.

## 3.4    *Distribution of maximum principal stress of surrounding rock*

Figure 6 shows the distribution of the maximum principal stress in the surrounding rock calculated by schemes 1–6. It can be found that the maximum principal stress reaches the extreme value in the surrounding rock. Under the same Poisson's ratio, with the decrease of residual elastic modulus, the maximum principal stress increases gradually. The maximum principal stress in schemes 1–3 increases from about 42.06 MPa to 42.09 MPa. Under the same residual elastic modulus, the smaller the Poisson's ratio is, the smaller the maximum principal stress is. Under the same conditions, the Poisson's ratio decreases from 0.25 to 0.2, and the corresponding maximum principal stress decreases by about 0.7%. However, when Poisson's ratio is 0.2, the influence of residual elastic modulus on the maximum principal stress decreases.

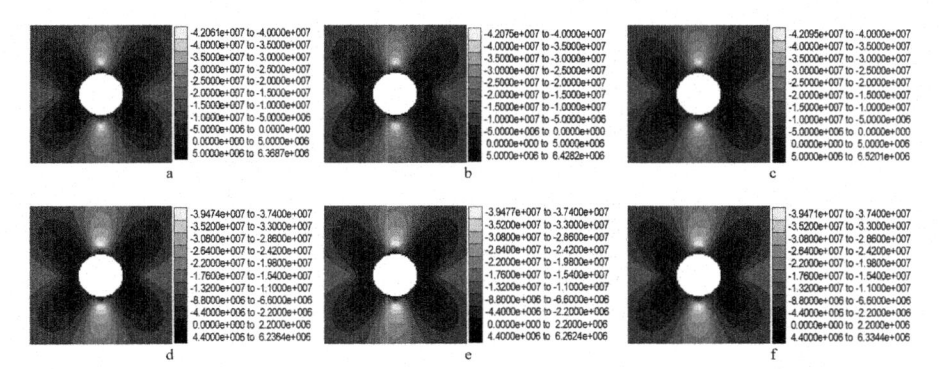

Figure 6.    Maximum principal stress of surrounding rock under different schemes.

## 3.5    *Distribution law of maximum principal stress of surrounding rock*

Figure 7 shows the measured results of rock burst morphology of field test tunnel in References (Hajiabdolmajid et al. 2002; Hajiabdolmajid & Kaiser 2003; Martin 1997). It can be found that the rock burst zone of the surrounding rock of the tunnel is similar to the calculation results of the high-value area of the shear strain increment in Figure 5, but the simulation results considering the influence of the tensile plastic strain on the strength and stiffness of the surrounding rock are not observed in the test tunnel. The reason is that the stress applied is different from the actual stress direction, but the influence of the tensile yield on the surrounding rock cannot be ignored. In the calculation results of this paper, the horizontal tensile plastic zone inside the surrounding rocks on both sides is simulated. The simulation results are similar to the cracking phenomenon in the Brazilian splitting test (Patel & Martin 2018; Pate & Martin 2018), and the simulation method can be further applied to the numerical simulation of the Brazilian splitting test(Patel & Martin 2018; Pate & Martin 2018).

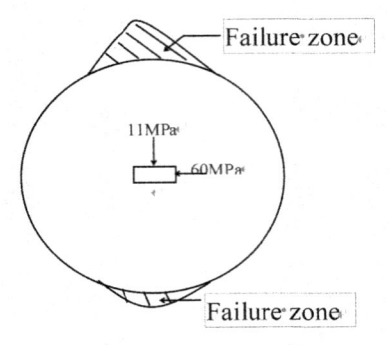

Figure 7.    The failure zone of the circular test tunnel.

## 4    CONCLUSIONS

Based on the nonlinear CWFS model, considering the influence of tensile plastic strain on the tensile strength and Young's modulus of surrounding rock, by controlling the shear and tensile plastic strain threshold, the degradation and attenuation of surrounding rock strength parameters in the tensile and shear yield process are realized in the calculation software. Six calculation schemes were carried out. Based on the nonlinear attenuation and enhancement of cohesion and friction angle with shear plastic strain in the CWFS model, the attenuation of elastic modulus was controlled by the tensile plastic strain threshold, and the influence of different Poisson's ratios on the stability of surrounding rock was compared. The calculation results show that when the tensile plastic strain of surrounding rock reaches the threshold, the attenuation of elastic modulus has no obvious influence on the distribution of plastic zone of surrounding rock, but because the tensile yield affects the stiffness modulus of surrounding rock, there is a single crack in the horizontal direction of surrounding rock on the horizontal symmetry axis of the model, and the calculation results of other models in the past cannot simulate such phenomena; Due to the decrease of Poisson's ratio, the tip of 'V' plastic zone formed by shear yield tends to flat, and the surrounding rock changes from brittle failure to ductile failure; In schemes 1–6, the tensile plastic strain increases slightly with the decrease of residual elastic modulus, and the maximum values are $1.8132 \times 10^{-2}$, $1.8154 \times 10^{-2}$ and $1.8204 \times 10^{-2}$, respectively; Under the same residual elastic modulus, the smaller the Poisson's ratio is, the smaller the maximum principal stress is. Under the same conditions, the Poisson's ratio decreases from 0.25 to 0.2, and the corresponding maximum principal stress decreases by about 0.7%. However, when Poisson's ratio is 0.2, the influence of residual elastic modulus on the maximum principal stress decreases.

## REFERENCES

Hajiabdolmajid V, Kaiser P K, 2003. Brittleness of rock and stability assessment in hard rock tunneling[J]. *Tunneling and Underground Space Technology*, **18(1)**: 35–48.
Hajiabdolmajid V, Kaiser P K, Martin C D, 2002. Modeling brittle failure of rock[J]. *International Journal of Rock Mechanics and Mining Sciences*, **39(6)**: 731–741.

Hoek E, Martin C D, 2014. Fracture initiation and propagation in intact rock-a review[J]. *Journal of Rock Mechanics and Geotechnical Engineering*, **6(4)**: 287–300.

Jiang Q, Feng X T, Chen G Q, 2008. Study on the constitutive model of hard rock considering surrounding rock deterioration under high geostresses[J]. *Chinese Journal of Rock Mechanics and Engineering*, **27(1)**: 144–152.

Martin C D, 1997. Seventeenth Canadian geotechnical Colloquium: the effect of cohesion loss and stress path on brittle rock strength[J]. *Canadian Geotechnical Journal*, **34(5)**: 698–725.

Patel S, Martin C D, 2018. Application of flattened Brazilian test to investigate rocks under confined extension[J]. *Rock Mechanics and Rock Engineering*, **51(12)**: 3719–3736.

Patel S, Martin C D, 2018. Evaluation of tensile Young's modulus and Poisson's ratio of a bi-modular rock from the displacement measurements in a Brazilian test[J]. *Rock Mechanics and Rock Engineering*, **51(2)**: 61–373.

Patel S, Martin C D, 2020. Impact of the initial crack volume on the intact behavior of a bonded particle model[J]. *Computers and Geotechnics*, **127**: 1–10.

Renani H R, Martin C D, 2018. Cohesion degradation and friction mobilization in brittle failure of rocks[J]. *International Journal of Rock Mechanics and Mining Sciences*, **106**: 1–13.

Tao S, Wang X B, Pan Y S, Wang W, 2011. Effects of critical stress corresponding to the tangential point between linear and nonlinear yield functions on the mechanical behavior of tunnel surrounding rock[J]. *Journal of Water Resources and Water Engineering*, **22(3)**: 31–36.

Civil Engineering and Urban Research – Mohamed & Hou (Eds)
© 2023 the Authors, ISBN 978-1-032-44487-1

# Design and verification of towing system of resistance model test for high-speed crafts

Yongshun Wu
*Marine Design & Research Institute of China, Shanghai, China*
*Shanghai Key Laboratory of Ship Engineering, Shanghai, China*

Sujun Yang*
*Marine Design & Research Institute of China, Shanghai, China*
*Shanghai Key Laboratory of Ship Engineering, Shanghai, China*
*Laboratory of Science and Technology on Water Jet, Beijing, China*

Panhao Shi & Pan Yan
*Marine Design & Research Institute of China, Shanghai, China*

Guozhao He
*Shanghai Changzhang Manager Human Co., LTD, Shanghai, China*

ABSTRACT: As a high-performance ship with a simple structure, the planning craft is in the planning stage at a high speed, and the total resistance is correspondingly reduced, which is easy to achieve immense speed. In order to study the performance index of ship speed performance, it is necessary to accurately determine its resistance for the planning craft. At present, the towing modes of planning crafts and the amphibious vehicle were described in detail by ITTC and ship signs. However, some requirements were provided. Based on the two common towing modes of the tank model test, a new towing mode with an active control towbar was proposed. Combined with the latest test equipment and control technologies, the towing device design, resistance and motion attitude measurement, and control principle were studied, a towing system suitable for high-speed boat resistance test was developed, and the advantages of the system were analyzed. The comparison test of three towing modes was further conducted on a planning craft model, the suggestions for the towing mode of the planning craft resistance model test were proposed by the test results.

## 1 INTRODUCTION

As a relatively simple form of a high-performance ship, the planning crafts are widely used in every walk of life for their miniaturization, high speed, and flexibility characteristics, including antismuggling patrol boats, service boats, rescue boats, investigation boats, military missile boats, torpedo boats, civil entertainment boats, sports competitive boats, and others. During high-speed navigation, the planning surface could be formed so that the bottom of the planning craft contacts the water surface with a certain attack angle, the chord part of the ship detaches from the water surface with the lifting hull, and the drainage volume decreases greatly with the decreased resistance. While the weight of the planning craft is mainly supported by the hydraulic lift, and the total weight ratio of the hydrostatic buoyancy occupying craft is very small. Therefore, the high-speed, motorized, and intelligent planning crafts can be used in civil and military fields such as law enforcement,

---

*Corresponding Author: sujuny@163.com

 DOI 10.1201/9781003372417-41

rescue, detection, and combat tasks, with vast application prospects. Amphibious vehicles are one of the main equipment for amphibious warfare.

Speed performance is an important factor affecting the viability of high-speed planning crafts and amphibious vehicles, there are many factors affecting the rapidity of high-speed crafts and amphibious vehicles, such as navigation resistance, propeller efficiency, and engine output power (Hu 2008; Yuan et al. 2006; Zheng et al. 2015). Thus, resistance performance is an important indicator to be considered during the early design of high-speed planning crafts and amphibious vehicles. The resistance of high-speed craft and amphibious vehicles is greatly affected by the navigation posture, the designers usually improve the navigation posture and optimize the resistance performance of the line shape of crafts and amphibious vehicles by adding appendages (Fu 2015; Wang et al. 2012; 2020). As the speed increases, the dolphin motion occurs when the speed of high-speed planning crafts and amphibious vehicles reaches the critical speed. Dolphin motion is one of the most well-known dynamic unstable movements of planning crafts, it could occur in static water without excitation. Dolphin motion is a periodic and bounded movement in the vertical plane. The severe longitudinal bump will induce discomfort to the passenger. Also, dolphin motion will lead to serious water drilling off the bow so that the structure of the planning crafts and the amphibious vehicle would be damaged, while the crew will be seasick and affecting the personnel's control of the planning crafts and amphibious vehicles, so the designer will optimize the navigation posture and inhibit or delay the occurrence of dolphin motion by optimizing the type of craft and other means at the early stage of design (Jiang et al. 2015; Ling & Wang 2014; Zhu et al. 2013).

## 2 TOWING SYSTEM DESIGN PRINCIPLE

### 2.1 *System measurement principle*

The active control towing device can be applied in all tests of ship model resistance. Moreover, it can also be applied to the operating conditions of wave attack and following waves in common ship model wave resistance tests. However, in the consideration of testing efficiency, the measurement equipment should be simplified as far as possible. It is still recommended that the mature products should be used for the rapidity test of common transport ships, while the active control towing mode is mainly applied to the model tests of planning crafts and amphibious vehicles that have large changes in a posture with the forward speed, which is shown in Figure 1.

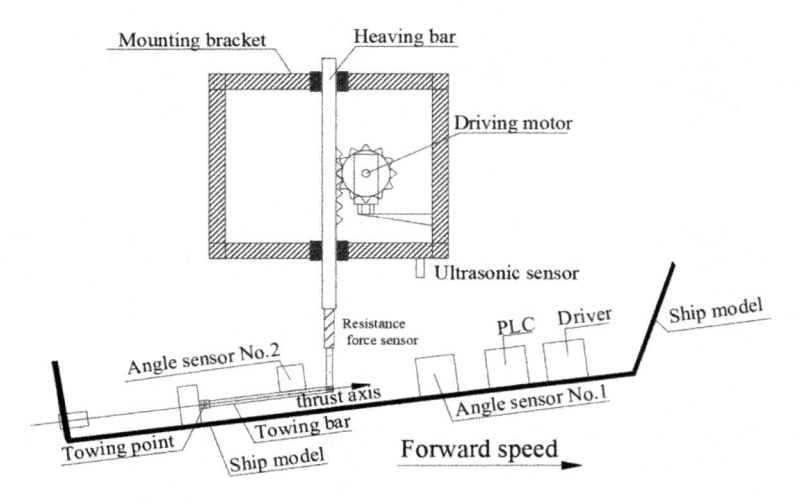

Figure 1. Measurement principle of towing system.

The frame structure is fixed on the measuring bridge of the trailer, and the No.1 angle sensor is installed on the ship model to measure the trim angle of the ship model. The ultrasonic sensor is installed on the frame structure to irradiate the ship model deck surface, which can measure the heave at the center of gravity of the ship model. The No. 2 angle sensor is installed on towing bar to measure the angle of the trailing rod, and the angle and heave signals can be outputted to the PLC. The driving motor is controlled by PLC, and the motor drives the vertical reciprocating movement induced by the heaving bar, which drives the lift rod movement. The towing bar is connected with a force sensor in serial connection, which is connected with the towbar through a universal joint. The heaving and towing bar is connected with the towing points. The rotation shafts of the towing points are located at the intersection point between the longitudinal center of gravity and the extension line of the thrust axis. According to the corresponding rule of the ship model test, the direction of resistance is consistent with the forward speed, which has nothing to do with the attitude of the model. The towbar conducts the physical strength of the ship to the force sensor, and the force sensor measures the ship's model resistance.

## 2.2 *Principle of active control towing system*

When the ship model is static, the force, displacement, and angle zero point acquisition are performed. With the change in flight speed, the posture, and angle of the ship model, the displacement sensor will collect the posture change of the model in real-time. The height change of the lift rod is actively controlled so that the towbar always coincides with the direction of the thrust line. The height variation of the lift rod is controlled by the inclination of the ship model and the heave, and the calculation formula is specified below:

$$H_1 = L * (K_1 * V_1 + B_1)/57.289; \qquad (1)$$

$$H_2 = K_2 * V_2 + B_2; \qquad (2)$$

$$H = H_1 + H_2; \qquad (3)$$

In the above formula,
$H_1$—the heaving height caused by the dip angle of the ship model;
$H_2$—the heaving height caused by lifting and sinking of ship model;
H—the total vertical sway value of the ship model, that is, the displacement of the lift rod;
$K_1$—the sensitivity coefficient of the angle sensor;
$K_2$—the sensitivity factor of the displacement transducer;
$V_1$—the voltage value of the angle sensor;
$V_2$—the voltage value of the displacement sensor;
$B_1$—zero compensation value of angle sensor;
$B_2$—zero compensation value of displacement sensor;
L—length of towbar
When the ship model sails with the trailer at high speed, the vertical oscillation of the ship model can be obtained through the above formula. After the signals are obtained, the corresponding action is performed by the motor-driven lift rod. Afterward, the full freedom of the ship model posture could be realized. It is used to simulate the posture of the live ship navigation.

## 3  SYSTEM REALIZATION AND FEATURE

### 3.1  *System measurement principle*

In view of the shortcomings of the common towing modes for the high ship model resistance test, an active control tow bars towing system is proposed. The active control towing device is mainly composed of a fixed bracket, servo motor & driver, PLC, transmission gear, tow bars, towing point, ultrasonic sensor, angle sensor, and force sensor.

When the ship model is dragged by the trailer at a fixed forward speed, the attitude of the ship model will change. The displacement and lift at the center of gravity of the ship model are measured by the ultrasonic sensor, and the trim angle can be measured by the angle sensor. The heave signal can be gained by the data acquisition module and sent to the PLC controller. When the servo motor receives the signal, the heaving bar will move up and down based on the heave motion of the model. The resistance force can be gained by the sensor, which is installed on the heaving bar, and its direction is consistent with the forward speed.

## 3.2 *Main performance parameters*

According to the load capacity of the trailer measuring bridge, the self-weight of the active control trailing device is determined by combining it with essential equipment, such as a navigation device; according to the height of the trailer measuring bridge from the water surface of the tank, its own vertical lifting capacity and the deck height of the trailer, the self-height and vertical lifting range of active control device is determined by combining with the travel of crane; the Z-direction and X-direction load capacity of the active control device is determined in accordance with the test capacity of the tank. For example, the trailer speed and trailer traction, the following control speed, and acceleration are determined in accordance with the positioning of equipment.

If only the equipment is located for the ship model resistance test, because the posture of the model tends to be stable, when the model reaches the predetermined speed, except for the dolphin motion phenomenon. While the control speed and acceleration can be set smaller. If the parameters are blindly increased, the servo system with greater power is inevitably required, which will increase the self-weight and own volume of the equipment. The main performance parameters of the active control towing device introduced in the article are shown in Table 1.

Table 1. Main performance parameters.

| No. | Parameters | Unit | Value | No. | Parameters | Unit | Value |
|---|---|---|---|---|---|---|---|
| 1 | Height | mm | 1100 | 2 | The vertical displacement | mm | 600 |
| 3 | Weight | kg | 280 | 4 | Speed of the vertical drive | mm/s | 600 |
| 5 | Rated Max Z force | N | 500 | 6 | Acceleration of the vertical drive | m/s2 | 0.8 |
| 7 | Rated Max X force | N | 2000 | 8 | Control precision | mm | ±0.1 |

## 3.3 *System features*

Active control of tug towing mode will not produce additional resistance for the amidship towing points. Meanwhile, the active control of the height change of the lift rod could ensure that the towbar could coincide with the direction of the thrust line so that the additional torque or other factors which bind the movement of the model will not generate. The physical quantity shall be inputted for the active control towbar of the towing mode, for which the angle and lift of the ship model are physical quantities to be measured in the test. Compared with the conventional towing mode, the active control towing mode is relatively simple in the process of test preparation and test, the test efficiency is higher, and it can better simulate the high-speed navigation state of planning crafts. Therefore, the results of model tests are more credible.

## 4  MODEL TEST VERIFICATION OF TOWING SYSTEM

### 4.1  *Test scheme and operating conditions of the model*

A specific planning craft was selected as the research object, and the model tests of three towing modes were conducted. In view of the correlation between the resistance of the planning craft and the navigation posture, and the phenomenon of dolphin motion could occur, thus, the inclination angle of the ship model and the heaving of the center of gravity were measured in the test, except for the resistance of the ship model. On the consideration that there is a larger change of posture of planning crafts and amphibious vehicles along with the change of speed, the test speed in low speed, test speed near resistance peak, the planning speed, and critical speed of dolphin motion was selected as 2.5, 3.5, 4.0, 4.5, 7.0, 8.0, 8.4 and 8.8 m/s, respectively.

### 4.2  *Test results and analysis*

The dolphin motion occurred at the model speed of 8.4 m/s in the outboard towing line and the towing mode of active control towbar. However, no dolphin motion occurred at the model speed of 8.8 m/s in the in-boat towing mode. Towing test results of high-speed crafts under different towing modes are shown in Figure 2.

For the model resistance test data, the model resistance of the outboard towing mode is significantly greater than that of the other two modes in the low-speed segment, combined with the heaving oscillation data, the main reason is that the tugging point of the outboard towing mode and the bayonet of the wire rope contact with the water at a low speed so that additional resistance could be obtained; in the peak-crossing stage, the model resistance of the amidships towing line mode is significantly greater than that of the other two models, mainly due to the small lift of the model under the friction resistance of the lift rod; as for the sliding speed, the three towing modes are almost the same, which indicates that there is no problem in the three towing modes in the sliding state. The obvious dolphin motion phenomenon occurs in the outboard towing and the active control towing mode at the re-acceleration time, which is protective equipment and cannot be used for data collection.

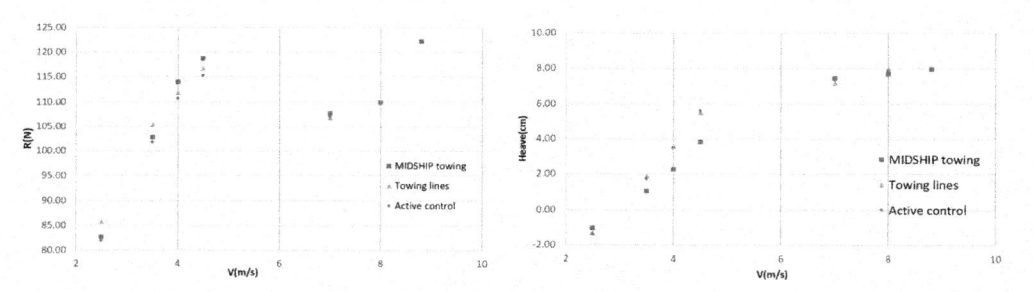

Figure 2.   Resistance and heave of model test results under different towing modes.

## 5  CONCLUSION

The navigation posture of the planning crafts is greatly affected by the speed, while the navigation posture has a greater impact on the model resistance. Therefore, it should be avoided as far as possible to restrain the craft model when the test device is being tested, and the additional resistance of the test device is generated during the test. Aiming at the advantages and disadvantages of two towing modes, it is designed that the towing mode test by active control will not restrain the model posture. Moreover, it will not produce additional resistance, with the high test effect. The towing device is verified by model test. From the test results, it can be found that the test results are highly consistent when the test model reaches the sliding speed stage. If the test results are required, all three towing modes could comply with the requirements. In the consideration of constantly striving for perfection, it is recommended to apply the towing mode of active control towbar. If there is no towbar drag device of active control, the corresponding towing mode can be selected according to the above conclusions in combination with the test demands.

# REFERENCES

Fu Liang. Resistance test and attachment design of new planning crafts model [J]. *China Water Transport* 2015 (12): 5–6.

Hu Jian. Survival Ability Analysis of Group Solider Scout Equipment [J]. *Journal of Harbin Engineering University*, 2008, 29 (1): 11–15.

Jiang Zhengqiao, Zhu Qidan, Liu Zhilin, et al, Simulation and Verification for Stability Analysis of Porpoising of Planning Vessels [J]. *Ship Engineering*, 2015 (4): 4–8.

Ling Hongjie, Wang Zhidong. Real-Time Numerical Prediction Method of "Dolphin Motion" High-Speed Planning Craft [J]. *Journal of Shanghai Jiaotong University*, 2014 (1): 106–110.

Wang Wenjiang, Zong Zhi, Ni Shaoling, et al, Model Tests of Effects of Interceptor on Resistance of a Semi-Planning Ship [J]. *Chinese Journal of Ship Research*, 2012 (2): 18–22.

Wang Zhipeng, Ni Yang, Qiu Gengyao, et al, *Analysis of the influence of sailing posture on the resistance performance of amphibious vehicles* [C]. Proceedings of the 31st National Hydrodynamics Symposium, 2020,1163–1169.

Yuan Tao, Liu Faming, Liu Yunbi. Survival Ability Analysis of Group Solider Scout Equipment [J]. *Ordnance Industry Automation*, 2006, 25 (7): 19–20.

Zheng Xiangyu, Fang Linghui, Wang Chen, et al. Research on Rapidity Design of Amphibious Vehicles [J]. *Journal of Sichuan Ordnance*, 2015 (11): 34–37.

Zhu Qidan, Jiang Zhengqiao, Liu Zhilin. Simulation and analysis for proposing of planning vessels by periodic force control [J]. *Journal of Harbin Engineering University*, 2013 (9): 1147–1164.

*Civil Engineering and Urban Research – Mohamed & Hou (Eds)*
*© 2023 the Authors, ISBN 978-1-032-44487-1*

# Research and treatment measures on track deformation of bridge crane in main powerhouse of LD hydropower station

Zheng Si* & Yanbing Wu
*Institute of Water Resources and Hydro-electric Engineering, Xi'an University of Technology, Xi'an, China*

Bin Duan & Xingwei Hu
*Northwest Engineering Corporation Limited, Xi'an, China*

Rongping Tang & Huiyuan Zhang
*Huaneng Lancang River Hydropower Inc, Kunming, China*

ABSTRACT: LD Hydropower Station adopts a riverbed type power station powerhouse. After the impoundment of the reservoir, the dam section of the powerhouse has sustained deformation. As of June 2021, the maximum cumulative horizontal deformation of the dam crest of the powerhouse dam section has reached 25.00mm. During the operation, the bridge crane of the main powerhouse once had a slight "rail gnawing" phenomenon. In this paper, the finite element method is used to calculate the powerhouse deformation when the reservoir water level rises to the normal pool level. Based on the calculation results, the causes of rail gnawing of bridge cranes are analyzed, and the countermeasures are put forward. The calculation results show that the main reason for the rail gnawing of the bridge crane is the uneven deformation along the river of the bent column upstream and downstream of the powerhouse, which leads to the narrowing of the track gauge of the bridge crane. The influence on the operation of the bridge crane can be eliminated by adjusting the gauge.

## 1 INTRODUCTION

As common mechanical lifting equipment, a bridge crane has been widely used in hydropower plants to undertake the lifting and maintenance of generators, turbines, and other equipment (Jiang & Lin 2020). The rail gnawing problem is often encountered during the operation of the bridge cranes. If it is not found and taken measures in time, the service life of wheels and rail will be reduced rapidly (Tang 2021; Zhong 2021). There are two main causes of rail gnawing. One is the wheel defect, and the other is the original defect or deformation of the track (Gao 2020; Lin 2021; Lan 2020; Pan et al. 2021). The track deformation of the bridge crane is often affected by the deformation of the concrete structure at its location. With the help of finite element calculation, this paper reflects the track deformation of the bridge crane from the concrete structure deformation of the powerhouse dam section, so as to analyze the causes of bridge crane rail gnawing. The research content has practical engineering significance for solving the rail gnawing problem of bridge cranes.

## 2 ENGINEERING OVERVIEW

LD Hydropower Station is a riverbed type power station. The powerhouse dam section is arranged in the middle of the river. The maximum dam height is 76.6m, the dam crest elevation is 1820.50m,

---

*Corresponding Author: sz123hlz@163.com

 DOI 10.1201/9781003372417-42

and the foundation surface elevation is 1749.00m ~ 1743.90m. Three ZZD706B-LH-740 axial flow paddle turbine generator units are installed in the plant, with a single unit capacity of 140MW and a total installed capacity of 420mw. The net span of the main powerhouse is 23.50m and the total length is 85.80m. The elevation of the foundation surface of the main powerhouse is 1743.90m, and the total height of the powerhouse is 78.3m. The main powerhouse adopts the layout of one machine and one joint, and the unit spacing is 28.6m. The installation elevation of the unit is 1770.40m (guide vane centerline). A bridge crane with a capacity of QE300/50t+300t-23.5m is installed in the plant. Each unit section is equipped with four span crane beam, double columns are used at the joint, and the bent column section is 1.2m × 2.0m. The bent column bracket elevation of the main powerhouse is 1810.70m, and the crane beam is a T-shaped beam, with a height of 1.5m. The bridge crane track is arranged on the top surface of the crane beam, and the top elevation of the bridge crane track in the plant is 1812.20m. See Figure 1 for the schematic diagram of the powerhouse dam section.

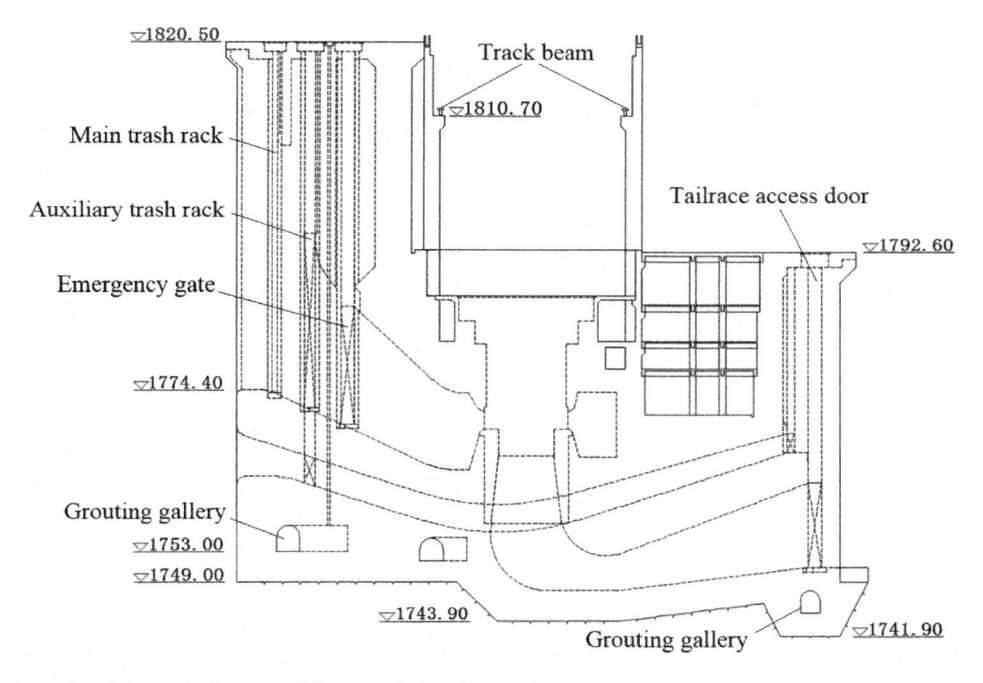

Figure 1.    Schematic diagram of the powerhouse dam section.

## 3    CALCULATION AND ANALYSIS OF TRACK DEFORMATION OF BRIDGE CRANE

### 3.1    *Calculation model and parameters*

The calculation model of the powerhouse dam section is shown in Figure 2. The origin of the coordinate system is located at the 1770.40m elevation (turbine installation elevation) of the central axis of the generator and volute. The cross river direction is the x-axis direction, and it is positive when pointing to the right bank; The direction along the river is the y-axis direction, and the direction downstream is positive; The vertical direction is the z-axis direction, and the upward direction is positive.

According to the concrete test of LD Hydropower Station, the calculation parameters are as follows: the elastic modulus E(28) of C25 concrete is 28.72 GPa and Poisson ratio is 0.167. The deformation modulus of the foundation rock mass of the powerhouse is 4~6 GPa, and the Poisson ratio is 0.25. The specific heat of C25 concrete in the powerhouse dam section is 0.971 kJ/(kg·°C), the temperature conductivity coefficient is 0.0037314 m²/h, the thermal conductivity coefficient is 8.696 kJ/(m·h·°C), and the linear expansion coefficient is $5.7 \times 10\text{-}6/°C$.

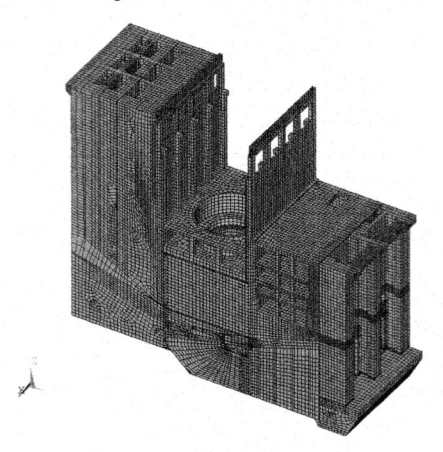

Figure 2.    Finite element calculation of dam powerhouse model.

## 3.2    *Calculation and analysis of track deformation of bridge crane*

During the operation of LD Hydropower Station, the bridge crane of the main powerhouse has a slight "rail gnawing" phenomenon. The bridge crane of the main powerhouse has a slight "rail gnawing" phenomenon. The track installation and wheel quality of the bridge crane meet the specification requirements without installation defects. The reason for rail gnawing is preliminarily analyzed, which may be related to the deformation and extrusion of the dam crest. In this paper, considering the influencing factors such as water pressure, temperature, and aging, the finite element simulation calculation, and analysis of the powerhouse dam section are carried out. The purpose is to find out the cause of the rail gnawing of bridge crane tracks by analyzing the deformation of the concrete structure.

The width of the powerhouse in the dam section of a single unit is 28.6m, and five bent columns are arranged at equal intervals on the upstream and downstream sides of the powerhouse. Corbels are set at the elevation of 1810.7m of the bent column to erect the cross beam. The overhead crane track is laid on the top surface of the beam. The elevation of the overhead crane track top is 1812.20m. When the reservoir water level reaches 1818.00m, the calculation results of the bent column displacement at the elevation of the overhead crane track top are shown in Figures 3~4.

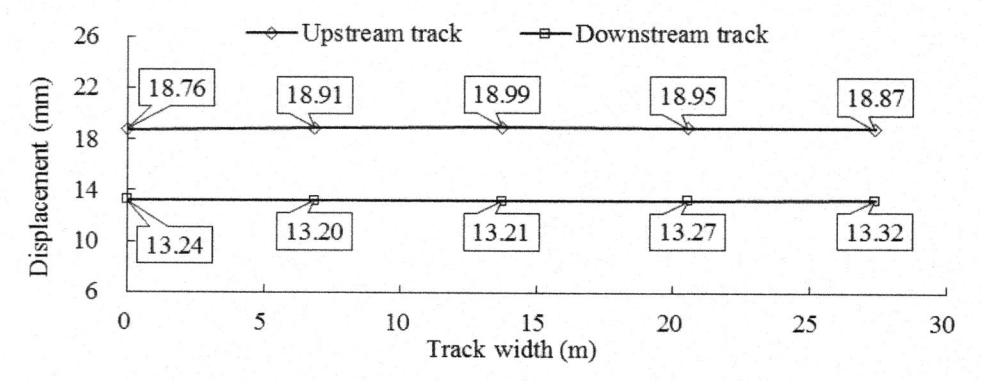

Figure 3.    Displacement of bridge crane track along the river.

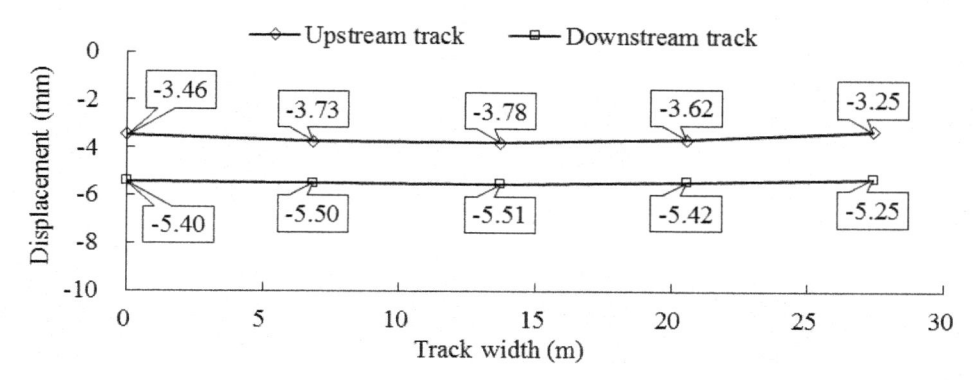

Figure 4. Vertical displacement of bridge crane track.

According to the design drawings, the allowable values of track deformation of the bridge crane are as follows:

(1) The deviation of the actual center line of the track from the design center line shall not be greater than ± 3mm, and the allowable deviation of the gauge on both sides shall be ± 3mm.
(2) The longitudinal unevenness of the track shall not be greater than 1 / 1500, the total length shall not be greater than 10mm, and the transverse inclination of the track shall not be greater than 0.7mm;
(3) On the same section of the plant, the relative difference between the top elevation of the two rails shall not exceed 5mm.

According to the calculation results:

(1) There is unequal downward deformation of the upstream and downstream tracks in the vertical direction. The deformation of the upstream track is 3.57mm (average value), and the deformation of the downstream track is 5.42mm (average value). The elevation of the top surface of the upstream and downstream tracks produces a relative difference of about 2mm, which is less than the allowable value of 5mm. The track unevenness on the upstream side is 1/22344 and that on the downstream side is 1/27500, which is far less than the allowable value of 1/1500.
(2) With the overall inclination of the powerhouse dam section to the downstream, the track displacement along the river on both sides is large. As shown in Figure 1, the top of the bent column on the upstream side of the main powerhouse is in contact with the dam crest concrete. Under the influence of dam crest deformation, the bent column on the upstream side is squeezed by the dam crest concrete, resulting in the deformation of the upstream track being greater than that of the downstream track. Therefore, the upstream and downstream tracks in the horizontal direction have unequal downstream displacement. The upstream track moves 18.90mm downstream (average value) and the downstream track moves 13.25mm downstream (average value), narrowing the bridge crane gauge by 5.65mm (average value) and the maximum value is 5.78mm. If the gauge change exceeds the allowable value of 3mm, the bridge crane will gnaw the rail.

As shown in Figure 5, the steel rail of the bridge crane is pressed on the embedded foundation plate by the pressing plates on both sides. The opening of the pressing plate is flat and round, and 15mm adjustment space is reserved at the upstream and downstream to ensure that the center line of the rail has a certain adjustable range. The first purpose of this adjustment range is to correct the deviation of the foundation embedded parts during embedding; Second, the deformation of the plant in use can be adjusted. The maximum narrowing of the bridge crane gauge of the main powerhouse of LD Hydropower Station is 5.78mm, which can be corrected by adjusting the tracks on one or both sides.

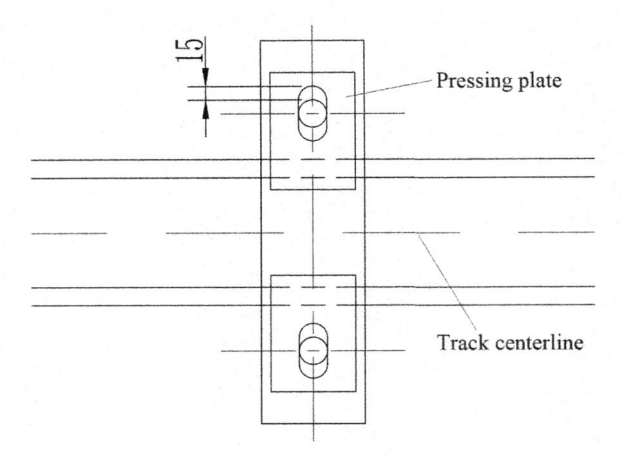

Figure 5. Track pressing plate of bridge crane.

## 4 CONCLUSIONS AND SUGGESTIONS

According to the finite element calculation results, the main reason for the rail gnawing of the bridge crane track is that the bent columns upstream and downstream of the main powerhouse produce uneven deformation along the river, resulting in the narrowing of the bridge crane track gauge. The maximum value of gauge narrowing is 5.78mm, and the bridge crane will have a slight "rail gnawing" phenomenon. The adverse impact on the operation of the bridge crane can be eliminated through the adjustment of the gauge, but it will not have a substantial impact on the operation of the plant bridge crane.

It is suggested to strengthen observation and adjust in time to avoid the bridge crane running in the rail gnawing state.

## ACKNOWLEDGMENTS

This research was supported by the science and technology project of Huaneng Group headquarters (HNKJ21-HF239).

## REFERENCES

Gao L. Reasons, detection, prevention and countermeasures of bridge gantry cranes gnawing on rails [J]. *Chemical Management*, 2020(12):155–156.

Jiang Z Z, Lin C Z. Cause analysis and treatment of rail gnawing of 250t bridge crane[J]. *China Plant Engineering*, 2020(16):144–145.

Lan Y L. Research on Causes and solutions of rail gnawing of bridge crane [J]. *Modern Manufacturing Technology and Equipment*, 2020, 56(09):83–97.

Lin Y B. Discussion on the cause analysis and treatment method of bridge crane's "track gnawing" phenomenon [J]. *China Plant Engineering*, 2021(02):149–151.

Pan C W, Dai J X, Mai X J. Discussion on Cause Analysis and Treatment Countermeasures of Gantry Crane's Rail Gnawing[J]. *China Plant Engineering*, 2021(13):178–179.

Tang H H. Analysis and Control of Rail Gnawing with Large Wheels of Gantry Cranes[J]. *Intelligent City*, 2021, 7(21):84–85.

Zhong W R. Inspection, analysis, and discussion on rail gnawing phenomenon of crane [J]. *Mechanical & Electrical Engineering Technology*, 2021, 50(12):317–319.

*Civil Engineering and Urban Research – Mohamed & Hou (Eds)*
© *2023 the Authors, ISBN 978-1-032-44487-1*

# Experimental study on the load-bearing performance of extra-long piles in saturated sandy soils

Wang Yifei*
*Wang Yifei Publishers Zhejiang Ocean University, Zhoushan, China*

Li Qiang*, Hu Lin*, Qiu Yiqin* & Zhang Yongyuan*
*Zhejiang Ocean University, Zhoushan, China*

ABSTRACT: On the basis of a similar theory, this paper conducts model tests on static-pressurized ultra-long piles of different lengths in saturated sand soil, and summarizes the objective laws of the bearing performance of single piles of different piles' lengths through the load and settlement curves obtained by static load tests. Through the strain data of the pile body, the changes in axial force and friction resistance of the piles in different sections of different pile lengths under different loads were studied. Based on the data obtained by the pile bottom pressure sensor, the law of pile end resistance of different pile lengths under different loads is obtained. It provides effective data test support for the engineering application of ultra-long piles in pile foundation engineering.

## 1 INTRODUCTION

With the development of ultra-long cross-sea bridges and ultra-high-rise buildings, extra-long piles are also applied more and more widely. In the research of pile foundation engineering, more research results have been achieved in different pile types such as pipe piles and perfusion piles. (Komurka V E 2003; Linehan P W; Wong I H 1999) But in the research of super-long piles, the research is often limited to some extent due to the difficulty of field tests and the experimental conditions are not easy to control. Zhu(2003) et al. studied the bearing characteristics of large-diameter extra-long grouted piles on the basis of field tests. Li et al (Li 2019; Tang 2020) studied the bearing properties and load transfer rule of extra-long piles in pulverized sand stratum in an indoor model test. The model test method can study the intrinsic laws of the model with a lower cost and experimental period. In this paper, the model test of the extra-long pile was designed and the model test of the hydrostatic extra-long pile in saturated sandy soil was carried out to study the load-bearing properties and load transfer law of the model pile with different pile lengths based on the similarity theory.

## 2 THE PILE DYNAMIC MODEL TEST

### 2.1 *The derivation of similarity ratios for model tests*

When the permanent load working on the pile-soil structure, the general form of the relationship between the physical parameters is

$$f\left(L, E_p, \rho_p, P, f, F, s, \gamma_d, E_s, \rho_s\right) = 0 \tag{1}$$

*Corresponding Authors: wangyifeilovestudy@163.com, qiangli@zjou.edu.cn, linghu821@163.com, qiuyiqin8845@163.com, 836135880@qq.com

DOI 10.1201/9781003372417-43

Where the physical parameters of the pile structure are geometry (L), pile modulus of elasticity ($E_p$) pile density ($\rho_p$), pile bottom pressure ($F$), and upper bearing platform self-weight ($P$); physical parameters of the soil structure are the dry weight of soil ($\gamma_d$), the density of soil ($\rho_s$) and compression modulus of soil ($E_s$); physical parameters of the pile-soil structure under the action of constant load are pile side resistance ($f$) and displacement (s).

By using mechanical dimension analysis, F, L, and T as the basic dimension, the similarity criterion can be obtained in the following form according to the Buckingham $\pi$ theorem:

$$[\pi] = [L]^{\alpha 1} \left[E_p\right]^{\alpha 2} \left[\rho_p\right]^{\alpha 3} [P]^{\alpha 4} [f]^{\alpha 5} [F]^{\alpha 6} [s]^{\alpha 7} [\gamma_d]^{\alpha 8} [E_s]^{\alpha 9} [\rho_s]^{\alpha 10} \tag{2}$$

The $\pi$ matrix can be obtained from the dimensional matrix, so the similarity criterion is:

$$\pi_1 = \frac{\gamma_d L}{E_p}, \pi_2 = \frac{E_s}{E_p}, \pi_3 = \frac{\rho_s}{\rho_p}, \pi_4 = \frac{s}{L}, \quad \pi_5 = \frac{P}{E_p}, \pi_6 = \frac{F}{E_p}, \pi_7 = \frac{f}{E_p} \tag{3}$$

So the relationship between the similarity constants can then be obtained as:

$$\frac{C_{\gamma_d} C_L}{C_{E_p}} = 1, \frac{C_{E_s}}{C_{E_p}} = 1, \frac{C_{\rho_s}}{C_{\rho_p}} = 1, \frac{C_s}{C_L} = 1, \frac{C_P}{C_{E_p}} = 1, \frac{C_F}{C_{E_p}} = 1, \frac{C_f}{C_{E_p}} = 1 \tag{4}$$

The ratio of the prototype pile to the model pile in the test is 30:1, so $C_L = 30$; the modulus of elasticity of the field pile is 32,500 MPa, and the modulus of elasticity of the model pile is chosen to be 3,753 MPa, $C_{E_p} = 8.66$; the density of the field pile is 2500kg / m$^3$, and the material density of the model pile is 1390kg / m$^3$, $C_{\rho_p} = 1.8$.

By taking the geometry, elastic modulus, and density of the site and model piles as the basic physical quantities, the geometric similarity ratio: $C_L = 30$, the elastic modulus ratio: $C_{E_p} = 8.66$, and the density ratio: $C_{\rho_p} = 1.88$, which can be found as follows:

$$C_{\gamma_d} = 0.289, C_{E_s} = 8.66, C_{\rho_s} = 1.8, C_P = 8.66, C_F = 8.66, C_f = 8.66, C_s = 30 \tag{5}$$

### 2.2 Experimental materials

This experiment uses marine sand from the sea areas surrounding Zhoushan, and the model box (L*W*H=2000*1500*3000 mm) is constructed by steel with a drainage device at the bottom, and the experimental setup is schematically shown in Figure 1(a). The soil filling of the model box is divided into three parts, which are laid from the bottom of the model box to the top in turn. At the bottom, gravel of different particle sizes was used as the inverted layer, and geotextiles were placed on top of the inverted layer to prevent the drainage channels from being blocked by fine sand; in the middle, fine sand of 50–150 cm thickness was considered as the supporting course, and at the top, fine sand of varying thickness with pile length was used as the fill layer. The soil layer is compacted once for each filling thickness of 0.25 m. The foundation soil filling in the model box is shown in Figure 1(b).

### 2.3 Steps of loading experiments

Referring to the relevant regulation, the slow maintenance load method was adopted for loading and unloading, and the whole process was divided into five steps: graded loading $\rightarrow$ observation of settlement $\rightarrow$ termination of loading $\rightarrow$ graded unloading $\rightarrow$ observation of settlement. Firstly, the load-bearing of each group test is estimated, and the load is loaded by 1/10 of the total load; when the settlement s<0.1 mm/h, it is considered to have reached stability, and the test of the next level of load is begun, while the time of each level of constant load in the early stage should not be less than 30 min.

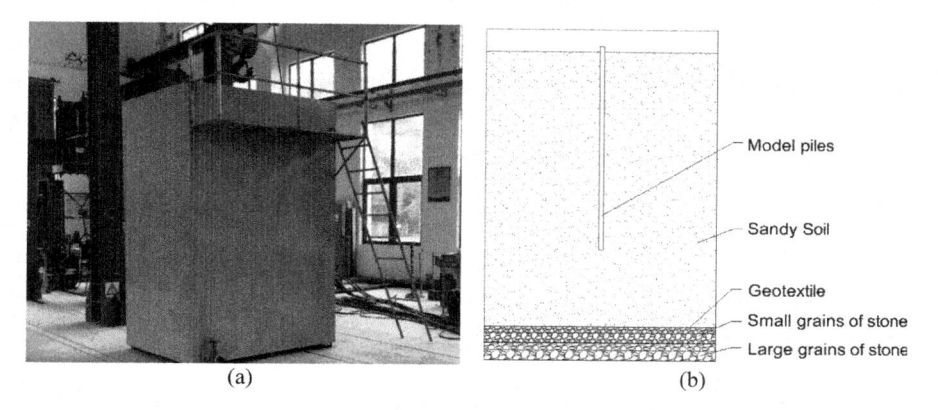

(a) (b)

Figure 1. Model experimental setup and model piles.

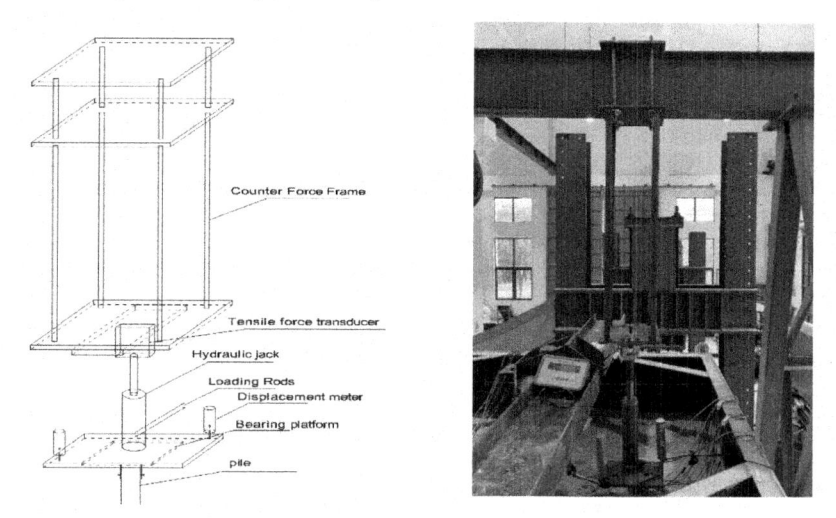

Figure 2. Hydraulic jack.

The pile top is loaded manually by hydraulic jack, as shown in Figure 2, which is divided into a total of 10 levels of loading, 2–1.5 m and 1.25–1 m each with a load of 250 N and 100 N. After each level of loading, the strain gauges are measured and the displacement meter data are observed to check whether there is any change. If the settlement at the top of the pile does not exceed 0.1 mm, it means that the settlement is stable, then the next step of loading can be performed, and when the pile settles at a later stage, the loading time will be extended (60 min can be used as the first level of loading time, and the measurement and reading will be performed at 5 min, 10 min, 15 min, and 30 min), and then the next step of loading will be performed after the settlement is stable. After the settlement is stabilized, then proceed to the next step of loading.

## 3 ANALYSIS OF RESULTS

### 3.1 *Loading and settlement curves*

The pile static load test was loaded and unloaded by the slow maintenance loading method, and the loads were 250 N and 100 N per stage for 2–1.5 m and 1.25–1 m. The corresponding amount of settlement changes were obtained under each load, and the Q-S curves were plotted in Figure 4 below.

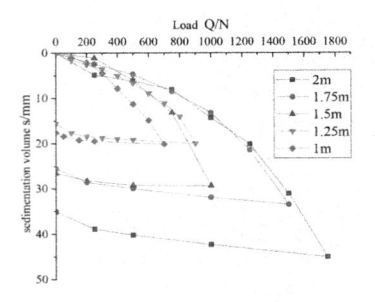

Figure 3.    Load and settlement curves for different pile lengths.

As can be seen from Figure 3, it is found that the bearing capacity of a single pile of 2-1m is about 1750 N,1500 N,1000 N,900 N and700 N respectively. The bearing capacity of a single pile increases with the increase of pile length, while the cumulative settlement decreases continuously when reaching the ultimate bearing capacity, and the settlement is basically between 20–45 mm.

During the loading process, we found that the static load curves of the piles with different pile lengths have stage changes, 2-1 m all go through the elastic stage in the early stage, which is basically linear; as the load increases, the accumulated sedimentation volume also increases at the same time, and the slope of the static load curve will become larger, which is the non-linear stage; when the pile reaches the ultimate limit states, the sedimentation volume of the single stage is often twice as large as the settlement of the previous stage . the settlement does not reach relative stability even after 1 hour and the pile is in the stage of piercing damage. The sedimentation volume of the pile increases sharply, and the soil on the side of the pile and at the end of the pile reaches the yielding state, which is especially obvious for the 1.5 m pile compared with other groups of piles.

During the unloading process, there is a certain amount of rebound for both 2–1 m, with the largest rebound for 2 m and the smallest rebound for 1 m.

### 3.2    *The pile axial force, pile side frictional resistance, and bottom pressure curve*

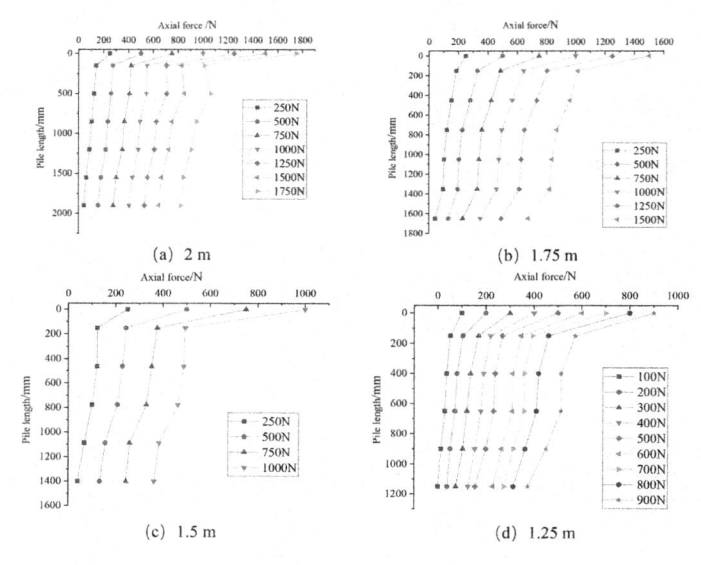

Figure 4.    Single pile body axial force at different pile lengths.

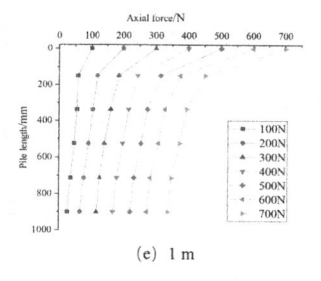

(e) 1 m

Figure 4.    Continued.

As can be observed from Figure 4, the distribution pattern of 2-1 m axial force is basically the same, decreasing down along the pile length in order. In the preliminary stage of loading, the change of axial force in the upper part of the pile body is rather obvious, which indicates that the lateral frictional resistance at that point of the pile was fully exploited. In the process of the static pile test, with the increase of load, the axial force at different measurement points of the pile increased simultaneously, the axial force at the top of the pile was the maximum, and the axial force of the pile decreased with the increase of the pile length of the same pile. This is due to the fact that in the process of static load test, the pile is displaced relative to the soil, and the presence of the soil will produce a certain amount of frictional resistance to the pile, and when the load is transmitted down

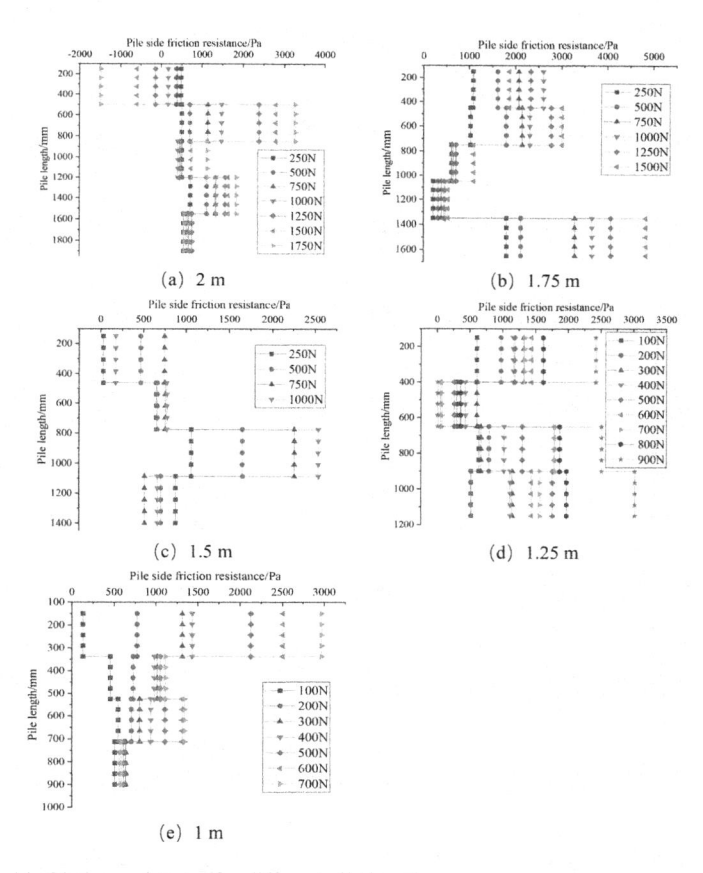

Figure 5.    Pile side friction resistance for different pile lengths.

the pile gradually need to overcome more frictional resistance, so the axial force falls gradually down the pile body.

From Figure 5, it can be shown that 2-1 m in the pile top area, with the increase of the pile top load, the pile side friction resistance also rises, the side friction resistance of the upper part of the pile side is more effective compared with other pile sections, and the maximum pile side friction resistance of 2 m and 1.5 m appears in the middle. With the increase of the pile top load, the maximum pile side friction resistance also becomes bigger gradually. the pile side resistance of 2 m, 1.75 m, and 1.25 m at a certain depth below the top of the pile is close to zero, and there is no relative displacement between the pile and the soil in this section. After that, the pile shows a tendency to move downward relative to the inter-pile soil due to the relatively large sinking rate of the pile, which makes the pile lateral friction resistance increase and prevents the pile from sinking, so that the stress in the pile diffuses into the inter-pile soil.

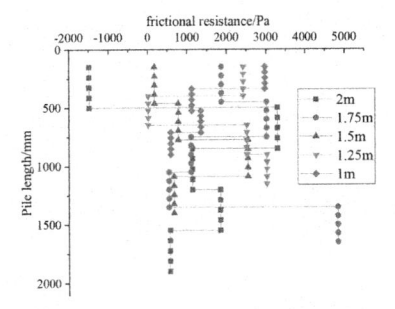 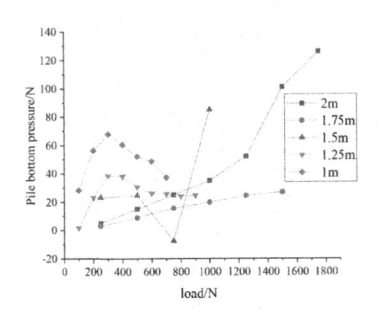

Figure 6. Pile side frictional resistance for different pile lengths.

Figure 7. Pile bottom reaction force for different pile lengths.

Figure 6 shows the pile side friction resistance of the pile under different pile length limit loads. In the section near the top of the pile, the resistance at 1 m is the highest, while the resistance at 2 m is the lowest. Due to the presence of negative resistance in the upper part of the 2 m pile and positive resistance in the lower part, there are obvious "double peaks" in the 500–750 mm and 1250–1500 mm sections respectively. The peaks in the 1.75-1 m range from 150–750 mm, 776–1088 mm, 900–1150 mm, and 150–327.5 mm.

The curve of the reaction force at the bottom of a single pile with different pile lengths varies with the increase of pile top load, which is shown in Figure 7. The pressure at the bottom of 2 m and 1.75 m piles increases with the increase of pile top load. With the increase of pile top load, the pressure at the bottom of the 1.5m pile also increases in general. With the increase of load, the pressure at the bottom of 1.25 m and 1m piles increases continuously at the beginning, and then decreases and tends to slow down.

## 4  CONCLUSION

In this paper, by setting up model pipe piles with different pile lengths for static load tests, the following conclusions were obtained.

(1) The bearing capacity of the pile of 2-1 m is about 1750 N, 1500 N, 1000 N, 900 N, and 700 N respectively. the bearing capacity of the pile keeps increasing with the growth of pile length, while the accumulated settlement when reaching the ultimate bearing capacity keeps decreasing, and the amount of settlement is basically between 20-45 mm.

(2) In the process of the single pile static load test, with the increase of load, the axial force at different measurement points of the pile body also increases simultaneously, the axial force

at the top position of the pile is the maximum, the axial force of the pile decreases with the increase of the same pile length.

(3) 2-1 m in the pile top section, with the increase of the pile top load, the pile side frictional resistance also increases in sequence.

(4) In the graph of the pile side friction resistance of a pile under different pile length limit loads, it is known that the friction resistance of 1 m is the largest in the section near the top of the pile, while 2 m is the smallest.

(5) In the variation curve of pile bottom reaction force of a single pile with increasing pile top load at different pile lengths, it can be seen that the pressure at the bottom of the pile has a tendency to fall back locally and increase generally with the increase of pile top load.

## REFERENCES

Komurka V. E. & Wagner A. B. (2003). *Estimating soil/pile set-up*. R. Madison, WI: Wisconsin Highway Department of Transportation.

Li X. K., Li Y. F. & Chen L. J. (2019). An indoor modeling study on the bearing characteristics of extra-long piles in silt strata. *J. Cem Concr Res*, (4):5.

Li Z. G. (1982). *Similarity and Modelling* (Theory and Applications).M. National Defense Industry Press.

Linehan P. W. & Longinow A. (1992). Pipeline response to pile driving and adjacent excavation. *J. J Geotech Geoenviron*. 118(2): 300–316.

Tang M. X., Hu H. S., Liu C. L., Yue Y. P. & Hou Z. K. (2020). Comparative experimental study on the bearing performance of different pipe pile types in sandy soils. J/OL. *Geotechnics*, (S2): 1–9[2021-12-11].

Wong I H & Chua T S (1999). Ground movements due to pile driving in an excavation in soft Soil. *J. Can Geotech J*. 36(1): 152–160.

Zhu Xiang Rong & Fang Peng Fei (2003). Research on the super-long pile in soft clay. *J.CJGE*. 25(1): 76–79.

*Civil Engineering and Urban Research – Mohamed & Hou (Eds)*
© 2023 the Authors, ISBN 978-1-032-44487-1

# Study on size effect of concrete porosity measured by water saturation method

Xiaozhong Zhang* & Guomin Sun
*School of Civil Engineering and Architecture, Guizhou University of Engineering Science, Bijie, Guizhou, China*

ABSTRACT: The application of the water saturation method to determine the porosity of concrete test blocks with different specifications and sizes is a subject of great research significance in the research of concrete materials. In order to accurately determine the porosity of concrete and study the influence of concrete test blocks with different sizes on its porosity, the water saturation method is used to determine the porosity of concrete test blocks with different sizes. First, the test block with a fixed mix proportion is made. In order to study its size effect, three prism test blocks of dimensions 50mm × 50mm × 75mm, 75mm × 75mm × 100mm, and 100mm × 100mm × 150mm are tested with three different specifications and sizes. Next, three different sizes of test blocks are tested and measured by the water saturation method, so as to obtain the law of the effect of different sizes on the porosity of concrete. Finally, based on the comprehensive mechanical experimental research, the relationship between the size of porosity and concrete strain energy is studied, and the side effect of porosity is explained by the energy method. Finally, the energy method is used to determine the best water saturation method to determine the size of concrete.

## 1 INTRODUCTION

The traditional methods of studying concrete porosity include; Small angle X-ray diffraction (saxrd), gas adsorption (BET), electron microscope observation (SEM), mercury injection (MIP), bubble method, methanol method, transmission method, nuclear magnetic resonance method (Cui et al. 2009), etc. Among them, the determination of concrete porosity by the water saturation method is favored by concrete technicians because of its simple operation, high measurement accuracy, and stability. It also attracts more and more technicians at home and abroad to carry out experimental research on concrete porosity based on the water saturation method.

First, scholars have done a lot of research on the determination of porosity by water saturation method at home and abroad (Denget al. 2013; Han et al. 2014; Liu et al. 2017; Sheng et al. 2014). Its basic principle is to vacuumize and saturate the pores in the rock through full pressure and fill the pores in the rock to a saturated state. The vacuum water saturation test can save the test time and determine whether the aggregate can reach the effective void absorption saturation state within 24 hours. The saturation degree of water in rock pores has a great influence on rock strength and elastic modulus. A large number of experiments show that the saturation effect of measuring rock porosity by vacuum water saturation method is better. Due to the simple operation and good water saturation effect, some concrete material researchers began to apply the water saturation method to the determination of concrete porosity.

In foreign studies, Guo Liping et al. (2009) used high-resolution X-ray microfocus CT (x-ray micro - CT) images to study four series of high-performance concrete with different cementitious materials and compared with the test results of normal temperature water saturation method.

---

*Corresponding Author: zhangxiaozhong@126.com

 DOI 10.1201/9781003372417-44

The influence law of fly ash and ground slag on the initial defects of concrete microstructure is revealed. It is preliminarily found that the water saturation method has great operability and practicability.

In China, Zhang Qiang et al. (2017) used the water saturation method to determine the concrete porosity, obtained the influence of different test block sizes on the concrete porosity measurement results, and suggested that the optimal test block size for determining the concrete porosity based on the water saturation method is 100mm × 100mm × 100mm cube. Ma Xinli (2019) conducted vacuum water saturation and conventional immersion tests on two different sizes of cylindrical concrete test blocks. It was found that compared with the conventional immersion method, the vacuum water saturation method has a greater degree of water saturation and a shorter time. The larger the size of the test piece, the more obvious the advantages of the vacuum water saturation method.

Generally speaking, scholars at home and abroad have carried out a series of studies on the determination method of concrete porosity based on the water saturation method. However, at present, the sizes of test blocks used to determine concrete porosity by the water saturation method are different in different regions, and the measurement accuracy is not constrained by corresponding specifications, which greatly limits the application of the water saturation method in the determination of concrete porosity. In order to conveniently determine the best side effect of the water saturation method, based on the research of other scholars, this paper applies the basic principle of energy dissipation to explore the best side effect of the water saturation method in determining concrete porosity. This study is a preliminary exploration of the comprehensive application of the energy method and water saturation method to the determination of concrete porosity.

## 2 RESEARCH METHODS

The theoretical calculation formula for the determination of concrete porosity by the water saturation method refers to the percentage of the pore volume of the test block VK, in the total volume of the test block V, which is composed of pore volume and solid volume. The pore volume of the test block is calculated according to the weighing method. First, the test block is saturated with water by the water saturation method, and then the test block is dried in a drying oven. Through the change of mass difference before and after, the pore volume, VK of the test block is obtained, the calculation formula is as follows:

$$V_k = \frac{M_{kw}}{\rho_w} = \frac{M_2 - M_3}{\rho_w} \quad (1)$$

Where $M_{kw}$ is the mass of filling water in the pores of the test block, $M_2$ is the mass when the saturated surface of the test block is dry, $M_3$ is the mass of the test block after drying, and $\rho w$ is the density of water.

$$M_1 g + M_2 V_s g = M_2 g \quad (2)$$

$$V_s = \frac{M_2 - M_1}{\rho_W} \quad (3)$$

Where M is the mass of the test block when suspended in water.

In the process of water saturation and drainage, concrete is a process of energy absorption, dissipation, and release. The energy dissipation reflects the continuous closure and evolution of the pores in the concrete material. It needs to absorb energy to fill the pores with water, and it also needs energy to remove the water in the pores. The energy conversion relationship of concrete in the determination process of the water saturation method is shown in Formula (4):

$$W = W_e + W_d \quad (4)$$

Where $W$ is the work done by external force on concrete, i.e. energy input from outside, J;
$W_e$ is the energy applied externally during pore water saturation, J;
$W_d$ is the energy dissipated in the process of pore water dissipation, J.

## 3 MIX PROPORTION CALCULATION

According to the current specifications, the detailed proportions of E, F, and G are calculated in Tables 1 and 2:

Table 1. Calculation mix proportion table.

| grade | Cement (kg/m$^3$) | Mineral admixture (kg/m$^3$) | Water (kg/m$^3$) | Coarse aggregate (kg/m$^3$) | Fine aggregate (kg/m$^3$) | Water reducing agent (kg/m$^3$) | Admixture content |
|---|---|---|---|---|---|---|---|
| E | 282.28 | 31.37 | 153.75 | 1294.84 | 637.76 | 6.273 | 10% |
| F | 261.99 | 65.50 | 153.75 | 1285.57 | 633.19 | 6.550 | 20% |
| G | 244.31 | 104.70 | 153.75 | 1271.15 | 626.09 | 6.980 | 30% |

Table 2. The mix proportion design of concrete (mass radio).

| Strength grade | | Water binder ratio mix proportion (Cementitious material: Sand: Stone: water) | Stone particle size (mm) | slump control (mm) |
|---|---|---|---|---|
| C40 | 0.49 | 1.00: 2.03: 4.13: 0.49 | 5~20 | 55 |
| C40 | 0.47 | 1.00: 1.93: 3.93: 0.47 | 5~20 | 55 |
| C40 | 0.44 | 1.00: 1.79: 3.64: 0.44 | 5~20 | 55 |

## 4 EXPERIMENTAL STUDY

In this test, the concrete cured under standard conditions for 28 days is 150 mm × 150mm × 150mm original test block. For the innovation and diversity of experiments, 150 mm × 150mm × 150mm cut to 50mm × 50mm × 75mm, 75mm × 75mm × 100mm, 100mm × 100mm × 150mm prisms of three sizes are shown in Figure 1.

According to the test operation steps required by the specification, the test of 50mm was carried out × 50mm × 75mm, 75mm × 75mm × 100mm, 100mm × 100mm × The porosity test of three kinds of prismatic test blocks with different sizes of 150mm under the mix proportion of different mineral admixtures of 0%, 10%, 20%, and 30% respectively. There are 3 test blocks of each size under one admixture mix proportion, 12 test blocks in four groups, and 36 prismatic test blocks of three different sizes. The data results of the final test are shown in Table 3. The variation trend of porosity of three different sizes of concrete prism test blocks under different admixtures is shown in Figure 1. The energy consumption of three different sizes of test blocks is shown in Figure 2.

Table 3. Summary of measured values of porosity of concrete prismatic blocks.

| Dimension | Mineral admixture | | | |
|---|---|---|---|---|
| mm | 0% | 10% | 20% | 30% |
| 50 × 50 × 75 | 0.134 | 0.112 | 0.087 | 0.084 |
| 75 × 75 × 100 | 0.121 | 0.115 | 0.100 | 0.095 |
| 100 × 100 × 150 | 0.105 | 0.104 | 0.091 | 0.088 |

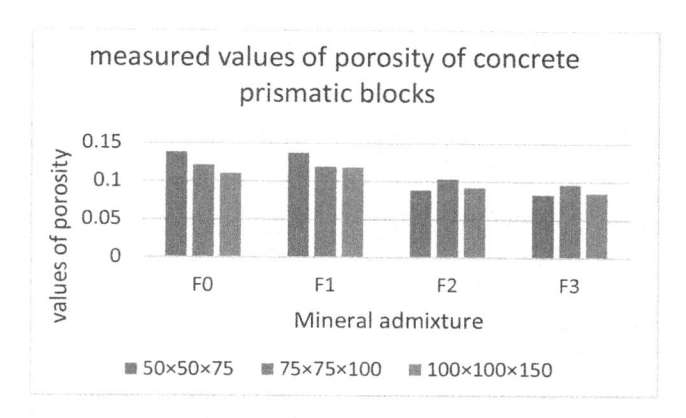

Figure 1. Porosity variation trend chart of different size test blocks.

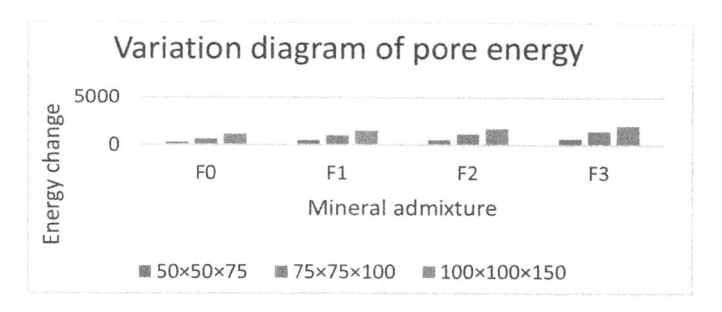

Figure 2. Measured value of energy change of concrete prism test block.

## 5 CONCLUSION

According to the test results of the measured value of porosity and the broken line diagram of the changing trend of the measured value, the L50 test block is 50mm × 50mm ×75mm. The porosity of the test block with the size of 75mm is the largest, and the changing trend is also the largest. From 0 to 30% of the mineral admixture, the porosity decreases rapidly, which is finally higher than that of L75, that is, 100mm × 100mm ×150mm. The porosity of the 150mm test block is still low. In contrast, 75mm × 75mm × 100mm and 100mm × 100mm ×150mm. The porosity of the 150mm test block is low, and the downward trend is slow and stable.

The test results show that the measured value of porosity of a small-size prismatic concrete test block is slightly larger than that of a large-size test block, and the decrease in porosity is obvious. This result may be due to many internal pores of small-size test blocks and higher efficiency of water saturation and drying due to small size. However, the internal pore distribution of large-size test blocks is uneven, with many places and few places. Due to the large size, the efficiency of water saturation and drying is not high. Due to various factors, the final large-size porosity is the lowest.

According to the energy change value, with the increase of mineral admixture, the energy released or absorbed during water saturation of the test block tends to increase, 75mm × 75mm × 100mm. The energy change of the 100mm test block is the smallest, in terms of energy change 75mm × 75mm × 100mm test block is most suitable for measuring porosity by the water saturation method.

## ACKNOWLEDGMENT

This work was financially supported by a New engineering construction project of the Ministry of Education: Construction of innovation and entrepreneurship education curriculum system for civil engineering majors in Western Local Universities under the background of new engineering construction, Project No.: ECXCYYR202000945. The work was financially supported by the Guizhou Natural Science Foundation (QKHJC-[2018]-1056).

## REFERENCES

Cui Jingjie, HE Wen, Liao Shijun, et al. Measuring and analyzing techniques of structural features for porous materials research. *Material Review*, vol. 23(13), pp:82–86. 2009

Deng Huafeng, Yuan Xianfan, Li Jianlin, He Ming, Luo Qian, Zhu Min. Experimental study on the effect of water saturation on longitudinal wave velocity and strength of Sandstone. *Chinese Journal of Rock Mechanics and Engineering*, vol. 32(08), pp:1625–1631, 2013

GB/T50082-2009, *Standard method of test for long term and durability of ordinary concrete*[S].

Guo Liping, Andrea Carpinteri, Sun Wei, Qin Wenchao. Measurement and analysis of defects in high-performance concrete with three-dimensional microcomputer tomography. *Journal of Southeast University*, vol. 25(01), pp:83–88, 2009

Han Xuehui, Li Fengbi, Dai Shihua, Zhang Juanjuan, Tang Jun, Wang Xueliang, Wang Hongliang. Study on core saturation method of low permeability reservoir based on CO2 replacement. *Experimental Petroleum Geology*, vol. 36(06), pp: 787–791, 2014.

Liu Xiumin. Jiang Xuanwei, Chen Congxin, Xia Kaizong, Zhou Yichao, Song Xugen. Experimental study on creep of gypsum rock under natural and saturated conditions. *Rock and Soil Mechanics*, vol. 38(S1), pp: 277–283, 2017

Ma Xinli. Applicability study of concrete vacuum saturation method based on CT technology [J]. *Shanxi Building*, vol. 45(04), pp:126–128, 2019

Sheng Yanping, Yin Huiqing, Wang Guanghui, Li Haibin. The density of coarse aggregate was determined by the vacuum saturation method. *The Chinese and Foreign Road*, vol. 34(06), pp:206–209, 2014.

Zhang Qiang, Liu Hongbiao. Study on optimization of test block size for determination of concrete porosity by saturated water method. *Channel Port*, vol. 38 (06), pp:604–609, 2017

*Civil Engineering and Urban Research – Mohamed & Hou (Eds)*
© *2023 the Authors, ISBN 978-1-032-44487-1*

# Influence of slenderness ratio on eccentric compression performance for GFRP confined reinforced concrete columns

Jingshan Jiang

*School of Civil Engineering and Architecture, Nanjing Institute of Technology, Nanjing, China*
*Nanjing Geo Underground Space Technology Co., Ltd, Nanjing, China*

Xin Huang*

*Key Laboratory of Ministry of Education for Geomechanics and Embankment Engineering,*
*Hohai University, Nanjing, China*

Zhihua Wang

*Nanjing Geo Underground Space Technology Co., Ltd, Nanjing, China*

Chao Zhang & Youxin Wei

*School of Civil Engineering and Architecture, Nanjing Institute of Technology, Nanjing, China*

ABSTRACT: Six eccentric compression numerical models of GFRP confined reinforced concrete columns with different slenderness ratios were established using finite element analysis software ABAQUS. The influence of slenderness ratio on eccentric compression performance of GFRP confined reinforced concrete long columns was investigated based on finite element simulation results. The results show that the increase in the slenderness ratio significantly reduces the bending stiffness of the column. With the increase in slenderness ratio, the ultimate bearing capacity of long columns decreases linearly. The slenderness ratio has an obvious influence on the lateral deflection for the long column under eccentric compression.

## 1 INTRODUCTION

In recent years, glass fiber reinforced polymer (GFRP) confined concrete structures have received a lot of attention from scholars due to their high pressure-bearing capacity, high ductility, and good corrosion resistance, which are widely used in high-rise structures and offshore buildings (Zeng et al. (2014). Based on the previous research, many new types of GFRP confined structures have been proposed and studied by scholars (Bazli et al. 2020; Jin et al. 2020; Zhang et al. 2019). It should be noted that the influence of eccentric compression was not considered in most of the above studies. In fact, almost all of the building components are under an eccentric compression state. Therefore, it is necessary to study the eccentric compression performance of columns.

In this paper, numerical analysis models for eccentric stressing of GFRP confined reinforced concrete long columns were established using the finite element software ABAQUS considering the slenderness ratio as the main variable, and the influence of slenderness ratio on eccentric compression performance of columns was studied.

## 2 MODEL DESIGN

In order to investigate the influence of the slenderness ratio on the eccentric performance of GFRP confined reinforced concrete columns, 6 sets of cylindrical column eccentricity simulation specimens were designed with the slenderness ratio $\lambda$ (12~52, 8 increase at a time) as the variable.

---

*Corresponding Author: wzhnjut@163.com

DOI 10.1201/9781003372417-45

The columns all have core concrete diameter $D$ of 200 mm, GFRP tube wall thickness $t$ of 5 mm, and laying angle $\theta$ of $\pm 57.5°$. The height $L$ of the columns ranges from 600mm to 2600mm, and the slenderness ratio is $\lambda = 4L/D$. All columns were loaded at an eccentric distance of 60mm. The thickness of the reinforced protective layer was 25mm, 6 longitudinal reinforcements were arranged circumferentially, and 10 stirrups at a spacing of 180 mm. The diameter of longitudinal reinforcements was 12mm and the yield strength $f_{y1}$ was 365MPa. The diameter of the stirrups was 6.5 mm, and the yield strength $f_{y2}$ was 260 MPa. The specific layout of longitudinal reinforcements and stirrups is shown in Figure 1.

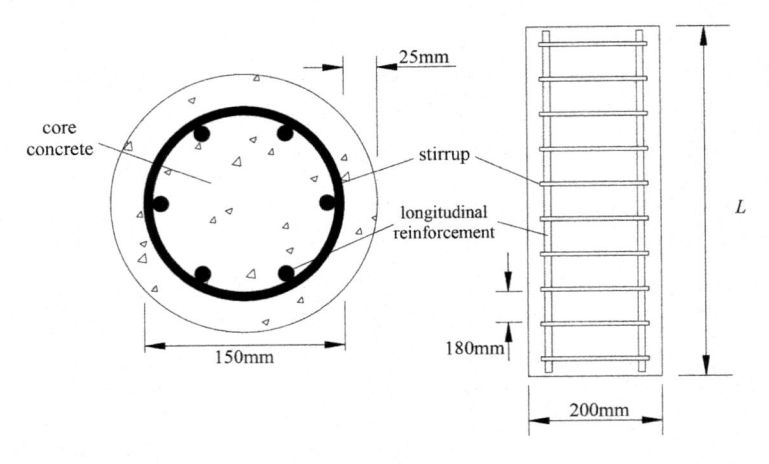

Figure 1. Reinforcement of columns.

## 3 ESTABLISHMENT OF FINITE ELEMENT MODEL

### 3.1 Constitutive model

#### 3.1.1 Concrete
The concrete damaged plasticity (CDP)model embedded in ABAQUS was used to describe the compressive and tensile behavior of concrete. The model defines the evolution of the concrete damage surface as controlled by the compressive equivalent plastic strain $\varepsilon_c^{pl}$ and the tensile equivalent plastic strain $\varepsilon_t^{pl}$ (hang et al. 2017). When GFRP-confined reinforced concrete columns are under compression, the inner core concrete is in a three-dimensional compression state, and the ordinary concrete constitutive model is not applicable.

The core concrete was defined using the compression and tensile constitutive models for restrained concrete proposed by Han Linhai (Han 2007), where the compression constitutive model is:

$$y_1 = \begin{cases} 2x_1 - x_1^2 & (x_1 \leq 1) \\ \frac{x_1}{\beta(x_1-1)^2 + x_1} & (x_1 > 1) \end{cases} \tag{1}$$

Where $y_1$ = normalized value of compressive stress, $y_1 = \sigma_c/\sigma_0$; $x_1$ = normalized value of compressive strain, $x_1 = \varepsilon_c/\varepsilon_0$; $\sigma_c$ = compressive stress; $\sigma_0$ = compression peak stress, $\sigma_0 = [1 + (-0.054 \times \xi^2 + 0.4 \times \xi) \times (24/f_c)] \times f_c$; $\varepsilon_c$ = compressive strain; $\varepsilon_0$ = compression peak strain, $\varepsilon_0 = 1300 \times 10^{-6} + 12.5f_c \times 10^{-6} + [1400 + 800 \times (f_c/24 - 1)] \times \xi^{0.2} \times 10^{-6}$; $\beta = (2.36 \times 10^{-5})^{[0.25+(\xi-0.5)^7]}f_c^2 \times 5.51 \times 10^{-4}$; $f_c$ = axial compression strength of concrete, $f_c = 51.1$MPa; $\xi$ = constraint effect coefficient, $\xi = A_{gfrp}f_{gfrp}/(A_cf_c)$; $A_{gfrp}$ = surface area of GFRP tube, mm$^2$; $f_{gfrp}$ = axial tensile strength of GFRP tube, $f_{gfrp} = 680$MPa, $A_c$ = core concrete area, mm$^2$.

The tensile constitutive model is:

$$y_2 = \begin{cases} 1.2x_2 - 0.2x_2^2 & (x_2 \leq 1) \\ \frac{x_2}{\alpha(x_2-1)^{1.7}+x_2} & (x_2 > 1) \end{cases} \tag{2}$$

Where $y_2$ = normalized value of tensile stress, $y_2 = \sigma_t/\sigma_1$; $x_2$ = normalized value of tensile strain, $x_2 = \varepsilon_t/\varepsilon_1$; $\sigma_t$ = tensile stress; $\sigma_1$ = axial tensile peak stress, $\sigma_1 = 3.1$ MPa; $\varepsilon_t$ = tensile strain; $\varepsilon_1$ = axial tensile peak strain, $\varepsilon_1 = 43.1 \times \sigma_1 \times 10^{-6}$; $\alpha = = 0.31 \times \sigma_1^2$.

Compression and tensile damage factors are used to describe the stiffness degradation of concrete after damage. The expression of compression damage factor dc is:

$$d_c = \frac{(1-\beta_c)\,\varepsilon_c^{in}E_0}{\sigma_c + (1-\beta_c)\,\varepsilon_c^{in}E_0} \tag{3}$$

Where $\varepsilon_c^{in}$ = compressive inelastic strain, $\varepsilon_c^{in} = \varepsilon_c - \sigma_c/E_c$; $\varepsilon_c$ = compressive strain; $\sigma_c$ = compressive stress; $E_c$ = compressive elastic modulus of concrete; $\beta_c$ = ratio of compression plastic strain to inelastic strain, $\beta_c = 0.70$ (Liu et al. 2014).

The expression of tension damage factor dt is:

$$d_t = \frac{(1-\beta_t)\,\varepsilon_t^{ck}E_0}{\sigma_t + (1-\beta_t)\,\varepsilon_t^{ck}E_0} \tag{4}$$

Where $\varepsilon_t^{ck}$ = tensile cracking strain, $\varepsilon_t^{ck} = \varepsilon_t - \sigma_t/E_t$; $\varepsilon_t$ = tensile strain; $\sigma_t$ = tensile stress; $E_t$ = tensile elastic modulus of concrete; $\beta_t$ = ratio of tensile plastic strain to cracking strain, $\beta_t = 0.95$ (Liu et al. 2014).

A number of other parameters need to be defined when using the CDP model: the expansion angle $\varphi$ was 30°, the plastic potential offset m was 0.1, the ratio of biaxial to uniaxial ultimate compressive strength fb0/fc0 was 1.16, the ratio K of the second stress invariant on the meridian plane of tension and compression was 0.6667, and the viscosity coefficient $\mu$ was 0.0005.

### 3.1.2 *GFRP*

GFRP is a special anisotropic elastic brittle material, and its constitutive model is defined by Lamina linear elastic model. The fiber direction is defined as 1 axis, the direction perpendicular to the fiber direction on the GFRP surface is defined as 2 axes, and the direction perpendicular to 1 axis and 2 axes is defined as 3 axes. After trial calculation, the elastic modulus $E_1$, $E_2$ of 1 axis and 2 axes, the main Poisson's ratio $v_{12}$, the shear modulus $G_{12}$ in the face of 1 axis and 2 axes, the shear modulus $G_{13}$ in the face of 1 axis and 3 axis and the shear modulus $G_{23}$ in the face of 2 axes and 3 axes were defined. The specific values are shown in Table 1.

The lay-up design of the GFRP tube was carried out using composite layup units. 10 layers of GFRP tube were set up with a thickness of 0.5 mm per layer and a lay-up angle of ±57.5°.

Table 1. Engineering elastic constants of GFRP tube.

| Index | $E_1$/MPa | $E_2$/MPa | $v_{12}$ | $G_{12}$/MPa | $G_{13}$/MPa | $G_{23}$/MPa |
|-------|-----------|-----------|----------|--------------|--------------|--------------|
| Value | 22000 | 12100 | 0.3 | 2080 | 2080 | 1480 |

### 3.1.3 *Steel bars*

The ideal elastic-plastic bilinear model was used to define the constitutive model of steel bars, with the slope of the reinforcement section taken as one percent of the elastic modulus.

### 3.2  Grid division

High-quality grid division is the key to 3-D model research. After trial calculations, a grid size with better accuracy and calculation costs was determined: 30mm for the core concrete, 35mm for the GFRP tube, and 10mm for the reinforcement. The reinforcement, core concrete, and GFRP tube were set as Truss units, Solid units C3D8R, and Shell units S4R respectively.

### 3.3  Interaction and boundary conditions

The contact between the GFRP tube and the core concrete was described using a finite slip equation and a Coulomb friction model with a friction coefficient of 0.6 (Zhou & Li 2013), and the normal behavior was set to 'hard contact'. The reinforcement was embedded in the concrete. The load and boundary conditions for GFRP confined reinforced concrete long columns are defined at reference points *Load* and *Fix*. Coupling constraints are applied to the upper and lower surfaces of the long column using a kinematic coupling, and the control points are *Load* and *Fix*, respectively. As a result, the load can be applied at different eccentric distances by changing the coordinates of *Load* and *Fix*.

In order to simulate the eccentric compression condition of the long column hinged at both ends, the U1 (x-axis direction), U2 (y-axis direction), UR2 (rotation around y-axis), and UR3 (rotation around z axis) of the Load were constrained, and the displacement load of 60 mm was applied in the U3 (z-axis direction), and the U1, U2, U3, UR2 and UR3 of the Fix were constrained, the specific load and boundary conditions are shown in Figure 2.

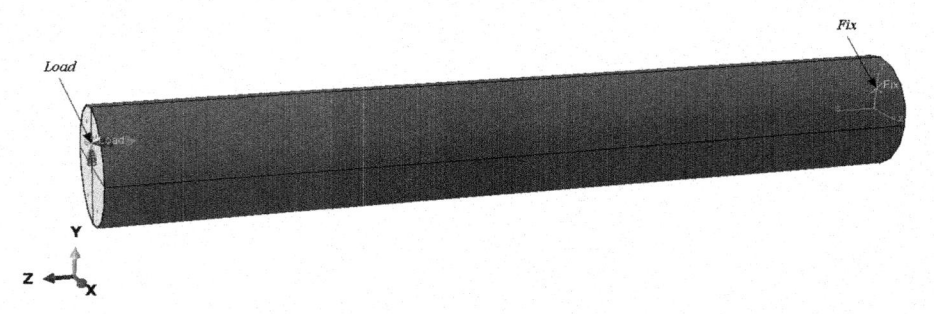

Figure 2.  Load and boundary conditions for GFRP confined reinforced concrete columns.

## 4  FINITE ELEMENT ANALYSIS RESULTS

The load-midspan deflection curves for long columns of different slenderness ratios λ are plotted in Figure 3. It can be seen that when the GFRP confined concrete column is a medium-long column (λ = 8 ∼ 30), the eccentric compression performance of the column in the elastic stage is not significantly different. With the increase in slenderness ratio, the slope of the decline section of the load-midspan deflection curves is basically the same, which is relatively flat. When the GFRP confined basalt fiber reinforced concrete column is a long column (λ > 30), the increase in the slenderness ratio will lead to the decrease of stiffness of the composite column in the eccentric compression process. With the increase in slenderness ratio, the composite column will enter the elastoplastic stage later.

Figure 4 shows the curve of ultimate bearing capacity for long columns under eccentric compression versus slenderness ratio λ. It can be seen that the ultimate bearing capacity of GFRP confined concrete columns under eccentric compression decreases with the increase of λ, and the decreasing trend is close to a linear relationship. The maximum decrease of the ultimate bearing capacity is 38.5%.

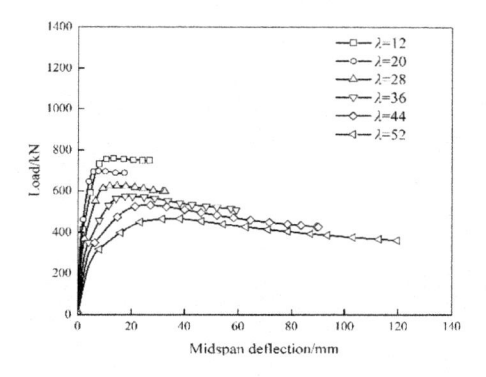
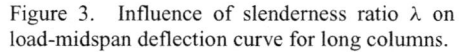
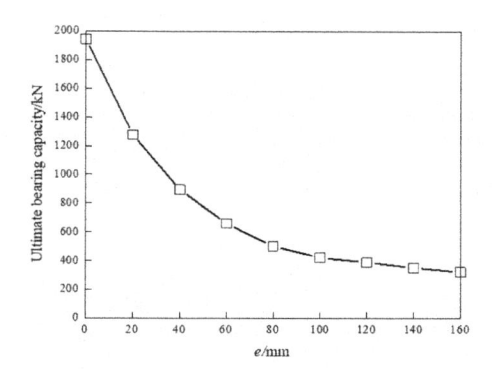

Figure 3. Influence of slenderness ratio λ on load-midspan deflection curve for long columns.

Figure 4. Influence of slenderness ratio λ on the ultimate bearing capacity for long columns.

Figure 5 shows the lateral deflection curves of columns for long columns under eccentric compression versus slenderness ratio λ. In order to facilitate comparison, the column height L of each column is taken as 1, the ordinate is set to La / L, and La is the section height of the current measuring point. It can be seen that the lateral deflection curve of the column gradually transits from a gentle curve to a sinusoidal half-wave curve with the increase of λ. When λ is less than 44, the ultimate mid-span deflection increases slowly with the increase of λ; While when λ increases from 44 to 52, the ultimate mid-span deflection increases sharply, and the amplification is 71.4%. This phenomenon shows that the slenderness ratio has a significant effect on the lateral deformation of the column under eccentric compression.

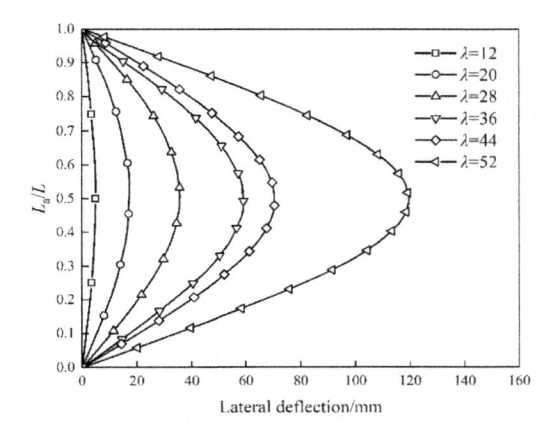

Figure 5. Influence of slenderness ratio λ on a lateral deflection for long columns.

## 5 CONCLUSION

In this paper, we use finite element analysis software ABAQUS to simulate the whole process of GFRP confined reinforced concrete column under eccentric compression. With the increase in slenderness ratio, The stiffness of GFRP confined concrete columns decreases continuously. The ultimate bearing capacity of composite columns decreases with the increase in slenderness ratio, and the decreasing trend is basically linear. The slenderness ratio can significantly affect the lateral deflection of the composite column. The larger the slenderness ratio, the greater the lateral ultimate deflection.

# ACKNOWLEDGMENTS

This work was supported by Jiangsu Industry-University-Research Cooperation Project (BY2021044).

# REFERENCES

Bazli, M., Zhao, X. L., Raman, R. K. S., et al (2020) Bond performance between FRP tubes and seawater seas and concrete after exposure to seawater condition. *Constr Build Mater*, 265:120342.

Cao, S. T., Li, Z. S. (2017) An elastoplastic damage constitutive model for confined concrete under uniaxial load. *Eng Mech*, 34: 116–125.

Han, L. H. (2007) *Concrete filled steel tubular structures from theory to practice*. Science Press, Beijing.

Jin, L., Chen, H., Wang, Z., et al (2020) Size Effect on the axial compressive failure of CFRP-wrapped square concrete columns: tests and simulations. *Compos Struct*, 254:112843.

Liu, W., Xu, M., Chen, Z. F. (2014) Parameters calibration and verification of concrete damage plasticity model of ABAQUS. *Ind. Constr*, 44:167–171, 213'D

Zeng, L., Li, L.J., Chen, G.M., et al (2014) Experimental study on the mechanical behavior of GFRP-recycled concrete-steel tubular columns under axial compression. *Chin. Civ. Engrg. J*, 47: 21–27.

Zhang, B., Wei, W., Feng, G. S., et al (2019) Influences of fiber angles on axial compressive behavior of GFRP-confined concrete stub column. *J. Bldg Struct*, 40: 192–199.

Zhang, J., Wang, Q. Y., Hu, S. Y., et al (2008) Parameters verification of concrete damaged plastic model of ABAQUS. *Bldg Struct*, 38: 127–130.

Zhou, C. D., Li, C. L. (2013) Strain evolution of concrete square columns confined by GFRP. *J. Bas Scie Engg*, 21: 137–146.

*Civil Engineering and Urban Research – Mohamed & Hou (Eds)*
*© 2023 the Authors, ISBN 978-1-032-44487-1*

# Experiment on mechanical properties of sisal fiber-basalt fiber reinforced concrete

He Xiang
*School of Civil Engineering and Architecture, Nanjing Institute of Technology, Nanjing, China*

Jingshan Jiang*
*School of Civil Engineering and Architecture, Nanjing Institute of Technology, Nanjing, China*
*Nanjing Geo Underground Space Technology Co., Ltd, Nanjing, China*

Xin Huang
*Key Laboratory of Ministry of Education for Geomechanics and Embankment Engineering, Hohai University, Nanjing, China*

Zhaoyue Zhu, Yuan Zhang & Yuelai Qiao
*School of Civil Engineering and Architecture, Nanjing Institute of Technology, Nanjing, China*

ABSTRACT: Different volume fractions of sisal fiber (0%, 1%, 2%, 3%) and basalt fiber (0%, 0.1%, 0.2%, 0.3%, 0.4%) were added into concrete to prepare 12 groups of sisal-basalt fiber (SF-BF) concrete and 1 group of C30 reference concrete. The cube compressive strength and splitting tensile strength of SF-BF concrete were tested to research the effect of SF and BF volume fractions on the mechanical properties of SF-BF concrete. The test results show that appropriate SF and BF can enhance the cube compressive strength and splitting tensile strength of concrete, but excessive fiber leads to low strength. The maximum increase of cube compressive strength and splitting tensile strength is 22.2% and 11.8%, respectively. SF has little effect on cube compressive strength. The splitting tensile strength decreases first and then decreases with the increase of SF volume fraction. BF can have a good positive mixing effect with SF. For cube compressive strength and splitting tensile strength, the optimal BF volume fraction is 0.2%. The test results experiment were regressed, and the prediction models of the cube compressive strength and splitting tensile strength of the SF-BF concrete were obtained, and the accuracy of the model is high.

## 1 INTRODUCTION

Concrete as a composite material has poor crack resistance and is prone to brittle damage. Fiber reinforced concrete is prepared by adding a single fiber into concrete. Fiber can better improve the characteristics of brittle failure of concrete (Huang & Deng 2010). The addition of steel fibers in concrete can play a role in toughening and crack resistance of concrete. Basalt fibers(BF) are inexpensive, have a high modulus of elasticity and low environmental pollution, and can be added to concrete to improve the tensile properties and durability of concrete (Wang et al. 2019). Studies have shown that a single fiber has great limitations for the improvement of concrete performance, two fiber blends can complement each other to better play the 'positive hybrid effect' (Quan et al. 2019).

Sisal fiber(SF) is a natural fiber, added to concrete that has the effect of anti-cracking and toughening, heat preservation, and energy saving, but there are problems of fiber performance degradation and weakened interfacial bond. After weak acid treatment of BF added to concrete can effectively enhance the interfacial bond strength of fiber and concrete. In addition, most of

*Corresponding Author: jingshanjiang@njit.edu.cn

DOI 10.1201/9781003372417-46

the current research and applications on hybrid fiber concrete are steel-polypropylene hybrid fiber reinforced concrete (Gao et al. 2018), and there are fewer studies on SF and BF hybrid blended into concrete. Therefore, in this paper, SF and BF are compounded into concrete to prepare SF-BF concrete, in order to play a 'positive mixing effect' of the two fibers. At the same time, the SF-BF concrete prepared in this paper is inexpensive and meets the concept of green building. In order to study the effect of SF and BF on the mechanical properties of concrete, 13 sets of SF-BF concrete collapse, cubic compressive strength, and splitting tensile strength tests were carried out.

## 2 MATERIALS AND METHODS

### 2.1 *Materials*

Cement is P.O42.5 ordinary Portland cement. The SF is the finished product of sisal fiber produced by Guangxi Jianma Group. It was cut into short-cut SF with a length of 15 mm in the laboratory, and the length of BF is 12 mm. Table 1 shows the main performance parameters of SF and BF. The sand used in the test is standard sand, and the fineness modulus is 2.61. The coarse aggregate is $5\sim25$mm continuously graded gravel, the apparent density is 2635 kg/m$^3$, the bulk density is 1575 kg/m$^3$, and the mud content is 1.1 %. The water reduction rate is 37 %.

Table 1. Main performance parameters of sisal fiber and basalt fiber.

| Fiber | Density/ (g·cm$^{-3}$) | Tensile strength/MPa | Modulus of elasticity/GPa | Monofilament diameter/$\mu$m | Elongation at break/% |
|---|---|---|---|---|---|
| Sisal fiber | 1.4 | 4700 | 25 | 400 | 3.5 |
| Basalt fiber | 2.8 | 4000 | 100 | 12 | 3.1 |

### 2.2 *Test design*

The strength grade of the reference concrete is C30, and the fiber volume fraction is defined as the ratio of the fiber volume to the total volume of the concrete, with SF volume fractions ($V_s$) of 0.1%, 0.2%, and 0.3%, and BF volume fractions ($V_b$) of 0.1%, 0.2%, 0.3%, and 0.4%. Table 2 shows the mix proportion of reference concrete.

Table 2. The proportion of C30 reference concrete kg·m$^{-3}$.

| Type | Cement | Sand | Stone | Water | Water reducer |
|---|---|---|---|---|---|
| H1 | 400 | 660 | 1240 | 225 | 7 |

### 2.3 *Sample preparation and test method*

BF was modified with 10% glacial acetic acid solution before use. The fiber was mixed by the dry mixing method, and it was found that the fiber volume fraction was too large to cause fiber agglomeration during the trial mixing process, so sodium carboxymethyl cellulose was added in the preparation process at 0.7% of the total mass of the cementitious material. The main role of sodium carboxymethyl cellulose is to make the fiber in a monofilament state in the concrete fully dispersed, to improve the fiber reinforcement, the impact on the performance of the concrete itself can be ignored (Wang et al. 2007). Preparation of the cementitious materials, fibers, and sodium carboxymethyl cellulose mixed dry for 1.5 min, then add sand and stone to continue dry mixing for 2 min, and finally add water reducer and water mixed wet for 2.5 min, and finally made SF-BF concrete. For each proportion, three cube specimens of 100 mm × 100 mm × 100 mm were prepared for the cube compressive strength test and splitting tensile strength test, respectively. The specimens were tested after 28 d of standard indoor curing under natural conditions, with the curing temperature varying from 16 to 24° C and humidity greater than 95%.

## 2.4 Test method

According to GB/T 50081-2019, the YAW-4206 pressure testing machine of MTS Industrial System (China) Co., Ltd. was used for the compressive strength and splitting tensile strength test of concrete cube. For each proportion, take the average strength of three samples as the final strength of the proportion. Since the specimens were 100 mm × 100 mm × 100 mm nonstandard specimens, the cube compressive strength and splitting tensile strength were converted to standard test block results by multiplying the size conversion factor. The size factor of cube compressive strength was 0.95 and the size factor of splitting tensile strength was 0.85 (Zhou et al. 2013).

## 3 RESULTS AND DISCUSSION

Table 3 is the cube compressive strength and splitting tensile strength of reference concrete and SF-BF concrete. It can be seen that with the increase of fiber volume fraction, the cube compressive strength and splitting tensile strength increase first and then decrease. The cube compressive strength and splitting tensile strength of the H2 specimen are the highest, which are increased by 22.2% and 11.8% respectively compared with the reference concrete.

Table 3. Test results of the strength of C30 reference concrete and sisal fiber(SF)-basalt fiber(BF) concrete.

| Type | Volume fraction of SF/% | Volume fraction of BF/% | Cube compressive strength/MPa | Splitting tensile strength/MPa |
|------|------|------|------|------|
| J1 | 0 | 0 | 27.99 | 2.63 |
| H1 | 1 | 0.1 | 34.2 | 2.72 |
| H2 | 1 | 0.2 | 34.6 | 2.94 |
| H3 | 1 | 0.3 | 31.2 | 2.83 |
| H4 | 1 | 0.4 | 20.1 | 2.01 |
| H5 | 2 | 0.1 | 18.9 | 1.87 |
| H6 | 2 | 0.2 | 31.6 | 2.91 |
| H7 | 2 | 0.3 | 28.2 | 2.44 |
| H8 | 2 | 0.4 | 19.1 | 1.74 |
| H9 | 3 | 0.1 | 26.4 | 2.62 |
| H10 | 3 | 0.2 | 31.3 | 2.86 |
| H11 | 3 | 0.3 | 30.5 | 2.69 |
| H12 | 3 | 0.4 | 19.2 | 1.89 |

## 3.1 Cube compressive strength

Figure 1 shows the curves of the effects of $V_s$ and $V_b$ on the compressive strength of SF-BF concrete. Figure 1 (a) shows that with the increase of $V_s$, the overall change of cube compressive strength is not significant. Only at the $V_b$ of 0.1%, the cube compressive strength decreases and then increases with the increase of $V_s$, with a decrease of 42.8%. The reason may be a test error. The effect of SF on the cube compressive strength of SF-BF concrete is small. It can be seen from Figure 1 (b) that the cube compressive strength increases slightly at first, and then decreases approximately linearly as the increase of $V_b$. The main reason is that too many fibers will cause them to gather into clusters inside the concrete, forming stress concentration points and leading to a decrease in the cube compressive strength. $V_b$ has a relatively obvious effect on the cube compressive strength of SF-BF concrete. For cube compressive strength, the optimal volume fraction of BF is 0.2%.

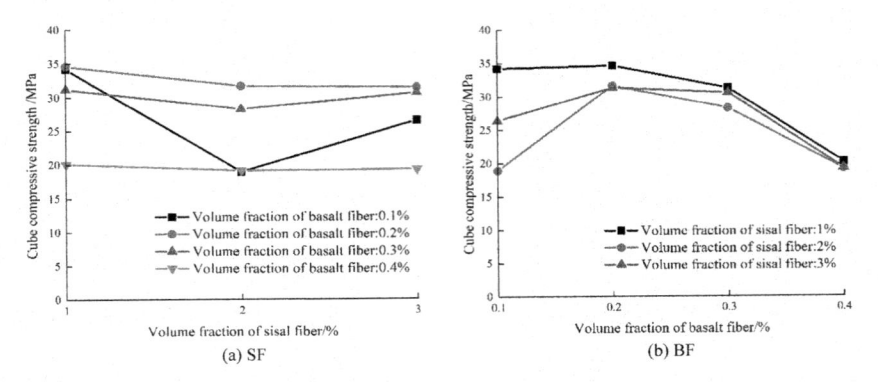

(a) SF  (b) BF

Figure 1.    Effects of volume fraction of SF and BF on cube compressive strength of SF-BF concrete.

### 3.2  *Splitting tensile strength*

Figure 2 shows the curves of the effect of $V_s$ and $V_b$ on the splitting tensile strength of SF-BF concrete. Figure 1 (a) shows that with the increase of $V_s$, the splitting tensile strength decreases first and then increases. It shows that SF and BF have a positive hybrid effect and increase the splitting tensile strength of concrete. The addition of BF makes up for the defect that SF will lead to the decrease in the splitting tensile strength of concrete. Figure 2 (b) shows that with the increase of $V_b$, the splitting tensile strength of SF-BF concrete increases first and then decreases, with a maximum increase of 55.6 %. The reason for the decrease in splitting tensile strength with increasing fiber volume fraction is that too many fibers will lead to more harmful pores inside the concrete, making the splitting tensile strength decrease. The effect of BF on the splitting tensile strength of concrete is more obvious than that of SF. For splitting tensile strength, the optimal volume fraction of BF is 0.2%.

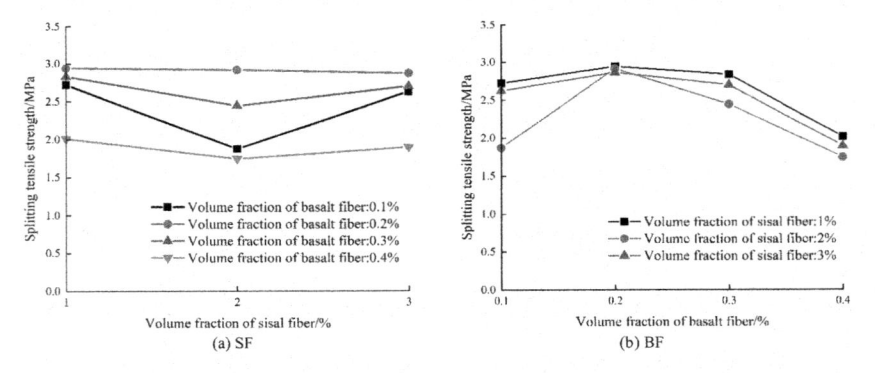

(a) SF  (b) BF

Figure 2.    Effects of volume fraction of SF and BF on splitting tensile strength of SF-BF concrete.

## 4   STRENGTH PREDICTED MODEL OF SF-BF CONCRETE

According to the mechanical theory of composite materials, it can be assumed that the strength of SF-BF concrete consists of three parts: matrix strength, SF reinforcement, and BF reinforcement (Dong & Li 2001). The least square method can be used to estimate the strength model. Firstly, assuming that the strength predicted model of SF-BF concrete is linearly composed of matrix strength, $V_s$ and $V_b$:

$$f^1 = \alpha_0 + \alpha_1 x_1 + \alpha_2 x_2 + \varphi \tag{1}$$

Where $f^1$ is the cube compressive strength or splitting tensile strength of concrete, MPa; $\alpha_i$ is regression coefficient, $i=0, 1, 2$, and 3; $\varphi$ is test parameter; $x_1$ is the volume fraction of SF, vol%; $x_2$ is the volume fraction of BF, vol%.

Substituting the data in Table 3 into Eq (1) and conducting the least squares estimation of $\alpha_i$. The linear strength predicted model of cube compressive strength and splitting tensile strength of SF-BF concrete can be obtained as follows:

$$f_{cu}^1 = 36.192 - 158.75x_1 - 2363.3x_2 \ (R^2 = 0.691) \tag{2}$$

$$f_{ts}^1 = 3.025 - 5.5x_1 - 182x_2 \ (R^2 = 0.713) \tag{3}$$

Where $f_1^{cu}$ is the cube compressive strength of concrete, MPa; $f_1^{ts}$ is the splitting tensile strength of concrete, MPa; $R^2$ is the decision coefficient.

The decision coefficient of Eq (2) and Eq (3) is low, indicating that the linear strength predicted model is discrete. Figure 3 is the comparison between the predicted values and measured values of SF-BF concrete mechanical properties obtained by strength prediction according to Eq (2) and Eq (3). It can be seen from Figure 3 that for the cube compressive strength, the relative errors of H2, H3, H4, H8, H10, and H11 are larger than ±8 %. For the splitting tensile strength, the relative errors of H2, H3, H6, H8, H10, and H11 are larger than ±10 %. The relative error between the predicted values and measured values is large, indicating that the accuracy of the linear strength predicted model is low. The reason may be that the hybrid effect between different fibers will affect the strength.

In order to accurately predict the strength of concrete, based on the least square method, considering the hybrid effect between fibers, the strength prediction model is defined as a quadratic polynomial:

$$f^2 = \beta_0 + \beta_1 x_1 + \beta_2 x_2 + \beta_3 x_1 x_2 + \beta_4 x_1^2 + \beta_5 x_2^2 + \varphi \tag{4}$$

Where $f^2$ is the cube compressive strength or splitting tensile strength of concrete, MPa; $\beta_j$ is regression coefficient, $j = 0, 1, 2, 3, 4$, and 5.

Substituting the data in Table 3 into Eq (4) and conducting the least squares estimation of $\beta_j$, the secondary strength predicted model of cube compressive strength and splitting tensile strength of SF-BF concrete can be obtained as follows:

$$f_{cu}^2 = 28.613 - 16.403x_1 + 179.553x_2 + 8.537x_1x_2 + 3.209x_1^2 - 438.443x_2^2 \ (R^2 = 0.891) \tag{5}$$

$$f_{ts}^2 = 2.618 - 1.437x_1 + 13.834x_2 - 0.001x_1x_2 + 0.345x_1^2 - 31.342x_2^2 \ (R^2 = 0.889) \tag{6}$$

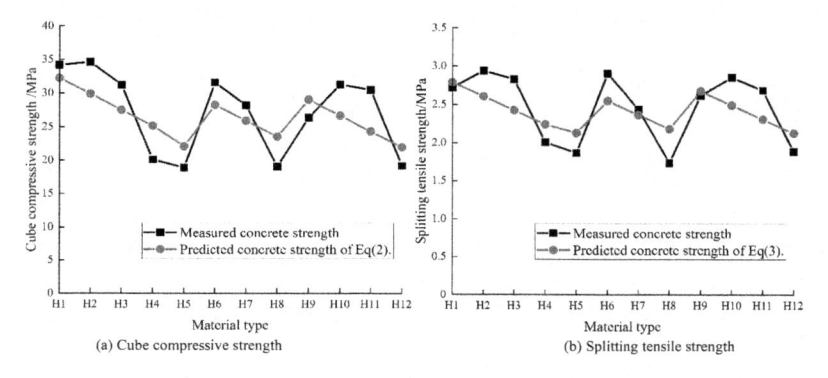

(a) Cube compressive strength      (b) Splitting tensile strength

Figure 3.    Comparison of predicted values with measured values of linear strength predicted model.

Figure 4 is the comparison between the predicted values and measured values of SF-BF concrete mechanical properties obtained by strength prediction according to Eq (5) and Eq (6). It can be seen from Figure 4 that the relative error between the predicted value and the measured value of the

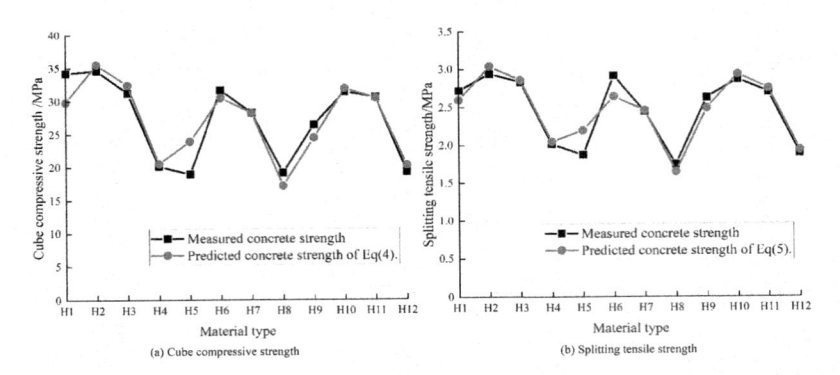

Figure 4. Comparison of predicted values with measured values of secondary strength predicted model.

secondary strength model is less than ±5%, and only a large relative error occurs in H5, which can be regarded as caused by the test error. The secondary strength model can better predict the cube compressive strength and splitting tensile strength of SF-BF concrete, and the predicted values can meet the engineering needs.

## 5 CONCLUSION

We show that a reasonable fiber volume fraction can improve the cube compressive strength and splitting tensile strength of concrete. The effect of sisal fiber (SF) on the strength of concrete is not obvious. Basalt fiber (BF) is an obvious factor affecting the strength of concrete. For cube compressive strength and splitting tensile strength, the optimum volume rate of basalt fiber is 0.2 %. Based on the test results, we established the SF-BF concrete cube compressive strength and splitting tensile strength secondary predicted models with SF volume fraction and BF volume fraction as variables. The accuracy of the model meets the engineering requirements.

## ACKNOWLEDGMENTS

This work was supported by Jiangsu Industry-University-Research Cooperation Project (BY2021044) and Innovation and Entrepreneurship Training Program for College Students in Jiangsu Province (202111276022Z, 202111276116H, 202011276001Z).

## REFERENCES

Dong, Z. Y., Li, Q. B. (2001) Fiber reinforced brittle composite several advances on mesoscopic mechanics. *Adv Mech.*, 31(4): 555–582.

Gao, D. Y., Jing, J. H., Zhou, X. (2018) Reinforcing mechanism and calculation method of compressive behavior of hybrid fiber reinforced recycled brick aggregates concrete. *Acta Materiae Compositae Sinica*, 35(12): 3441–3449.

Huang, K. J., Deng, M. (2010) Basalt fiber alkali resistance and its influence on the mechanical properties of concrete. *Acta Materiae Compositae Sinica*, 27(1): 150–154.

Quan, C. Q., Jiao, C. J., Yang, Y. Y., et al (2019) Orthogonal experimental study on mechanical properties of hybrid fiber reinforced concrete. *J. Bldg Mater*, 22(3): 363–370.

Wang, C., Wang, A. L., Zhang, X. S. (2007) Effect of dispersion of short carbon fibers on the mechanical properties of CFRC composites. *MTLS Rev*, 21(5): 125–128.

Wang, D., Ju, Y. Z., Shen, H., et al (2019) Mechanical properties of high-performance concrete reinforced with basalt fiber and polypropylene fiber. *Constr Build Mater*, 197: 464–473.

Zhou, L., Wang, X. C., Liu, H. T. (2013) Experimental study of mechanical behavior and failure mode of carbon fiber reinforced concrete. *Eng Mech*, 30(s1): 226–231.

*Civil Engineering and Urban Research – Mohamed & Hou (Eds)*
*© 2023 the Authors, ISBN 978-1-032-44487-1*

# Development and performance evaluation of a new permeable pavement material

Liu Yugui
*Chongqing Zhixiang Paving Technology Engineering Co., Ltd, Chongqing, China*
*Chongqing Jiaotong University, Chongqing, China*

Xu Jianhui
*Chongqing Jiaotong University, Chongqing, China*
*Chongqing Expressway Group Co., Ltd, China*

Yue Xiaowen, Zhao Yun & Wang Jie
*Chongqing Zhixiang Paving Technology Engineering Co., Ltd, Chongqing, China*

ABSTRACT:   In order to improve the durability of pervious pavement materials for slow-moving systems, a new pervious pavement material composed of stain-resistant binder, sand-based aggregate, functional additives, and other raw materials was developed, and the pervious and mechanical properties of the material were evaluated. The test results show that the water permeability of the material is 1.65cm/min, the water retention rate is $0.31g/cm^3$, the water permeability aging is greater than 10 times, and the compressive strength is 13.78MPa, and the water permeability still meets the specification requirements after 6 months of natural treatment, which proves that the material has excellent water permeability and meets the mechanical properties of the pavement application of the slow-moving system.

## 1   INTRODUCTION

Pervious pavement in urban built-up areas plays an important role in infiltrating rainwater and alleviating urban waterlogging. At the same time, it can alleviate the heat island effect and rain island effect (Brattebo & Booth 2003). It is known as "breathing" ground pavement and has good ecological and environmental benefits. It is an important technology in the construction of sponge cities and is widely used in urban roads, squares, parking lots, and other slow-moving systems.

Pervious paving materials mainly include pervious brick, cement pervious concrete, pervious asphalt concrete, etc., among which the pervious brick pavement is the most widely used in the slow-moving system. At present, the permeable pavement of the slow traffic system has been well applied in some foreign cities. The domestic slow traffic system using permeable pavement mainly includes municipal sidewalks, business district footpaths, etc., but the application proportion is still low, and there are application problems in varying degrees, such as insufficient durability of permeable performance, poor blocking resistance, insufficient durability. (Andrea et al 2013; Alalea et al. 2017; Montes & Haselbach 2006; Su et al. 2014).

According to the problems existing in the existing pervious pavement technology of the slow-moving system, this paper develops a new pervious sand-based pavement material to ensure the service performance of the pavement material and improve the water permeability, water retention performance, and service durability at the same time.

DOI 10.1201/9781003372417-47

## 2 RAW MATERIALS AND PREPARATION

### 2.1 *Raw materials*

(1) Binder: the binder used for permeable sand-based pavement material is a stain-resistant modified waterborne polymer material, which has the characteristics of low surface energy, stain resistance, low viscosity, and easy mixing construction.

(2) Aggregate: the aggregate used for permeable sand-based pavement material is sand-based aggregate. In this paper, 80~100 mesh quartz sand is selected as sand-based aggregate.

(3) Additives: the additives used are mainly anti-aging agents, mildew inhibitors, algae inhibitors, colorants, etc.

### 2.2 *Preparation of permeable sand-based materials*

Prepare stain-resistant binder, add various auxiliaries and mix evenly for reserve. Add aggregates to the mixing pan, then add metered binder, and discharged after mixing uniformly. Different types of permeable sand-based mixtures were molded and demoulded after 24 h of normal temperature curing reaction, and a performance evaluation test was carried out.

## 3 PERFORMANCE EVALUATION

### 3.1 *Determination of the oil-stone ratio*

The oil-stone ratio has a very important influence on the performance of the mixture. Choosing the proper oil-stone ratio can not only ensure the better comprehensive performance of the mixture, but also control the amount of binder and save cost. In this paper, the Kentuburg flying test is selected to study the stability of permeable sand-based materials to determine a better oil-rock ratio. Cylindrical specimens with a diameter of 101.6mm and height of 63.5mm were formed by Marshall test mold, and the wet density of the mixture was controlled to 1.7g/cm3. The test was carried out after curing and forming strength. Kentuburg flying test was carried out with a Los Angeles abrasion tester. Stop the test when rotating 300 revolutions at a speed of 30-33 rpm, weigh the quality of the specimens before and after the test, and determine the best oil-rock ratio by calculating the dispersion loss.

Pictures of test specimens for the Kentuburg flying test are shown in Figure 1 and the test results are shown in Table 1. According to the experimental results, with the increase of the oil-rock ratio, the damage degree and flying loss of the specimens gradually decrease, and the stability of the permeable sand-based mixture gradually increases. When the oil-stone ratio is more than 10%, the dispersion loss tends to be flat. Considering that an excessive oil-stone ratio can block the void, the recommended oil-rock ratio for this test is 10%.

Figure 1.    Photos of specimens before (A) and after (B) flying loss test.

Table 1. Test results of dispersion loss.

| Oil-stone ratio/% | 8 | 9 | 10 | 11 |
| --- | --- | --- | --- | --- |
| Flying loss/% | 23.7 | 12.9 | 11.7 | 11.5 |

## 3.2 *Evaluation of water permeability*

Water permeability is a key assessment index of permeable pavement materials. Especially for the permeable sand-based pavement materials developed in this project, the excellent water permeability can ensure that there is no ponding in the slow-moving system and enhance the pedestrian experience. In this paper, a 2cm thick permeable sand-based specimen is selected as the research object to evaluate the comprehensive permeability of the specimen, mainly including three indexes: permeability, water retention, and permeability aging. Figure 2 shows the process of the water permeability test, Table 2 shows the test results of permeability. Considering that the specification of sand-based permeable brick requires (JG/T 376-2012) that the water permeability shall not be less than 1.5cm/min, the water retention rate shall not be less than $0.06g/cm^3$, and the water permeability aging shall not be less than 10 times. The performance evaluation results show that the comprehensive water permeability of sand-based permeable material is excellent and can meet the needs of practical applications.

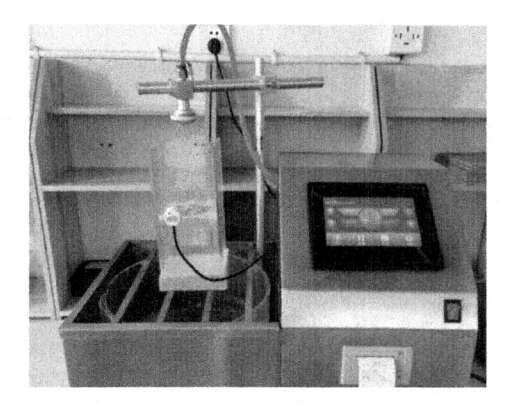

Figure 2. Photo of water permeability test process.

Table 2. Test results of water permeability.

| Permeability/(cm/min) | Water retention rate/(g/cm$^3$) | Permeable aging / time |
| --- | --- | --- |
| 1.65 | 0.31 | >10 |

## 3.3 *Evaluation of compressive performance*

The mechanical properties of road materials can reflect the durability of materials to a certain extent. Therefore, the compressive strength performance evaluation of sand-based permeable materials has been carried out in this project. In this paper, 100*100*100mm specimens are formed. The oil control stone ratio is 10% and the compacted wet density is $1.7g/cm^3$. The compressive strength performance evaluation is carried out after the strength is formed by curing. The compressive strength test was conducted at the pressure rate of 0.5MPa/s. During the test, it was found that the failure process of the specimen was central cracking, and the average compressive strength of the specimen was 13.78MPa. The photo of the test process is shown in Figure 3. Although the compressive strength of the specimen only reaches Cc10, it can be used in the slow-moving system for people to walk, which is enough to meet the application requirements.

Figure 3.   Photos of the specimen (A) before and (B) after the compressive strength test.

### 3.4  *Water permeability durability evaluation*

The permeable sand-based specimen is placed in the outdoor walking area. After 6 months of natural condition treatment, the permeability rate and permeability aging difference of the specimen before and after natural condition treatment are compared to reflect the influence of the blocking behavior of the specimen in the actual use on the permeability. Photos before and after natural disposal are shown in Figure 4.

It is observed that the erosion of rain and soil makes the surface of the test piece absorb fine particles of impurities, and the color of the test piece is deepened, but the surface of the test piece is free of mold, moss, and other blocking gaps. Table 3 shows the comparison of water permeability before and after treatment. It can be seen from the data in the table that the water permeability of the test piece decreased by 7.3%, but the water retention rate increased slightly by 9.7%, and the water permeability aging is still greater than 10 times. The above performances meet the specification requirements.

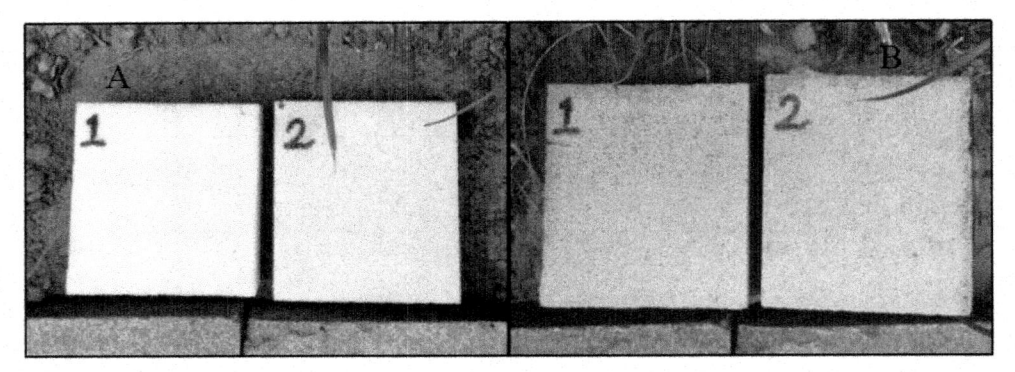

Figure 4.   Photos of test pieces (A) and (B) 6 months before and after natural treatment.

## 4  CONCLUSION

(1) By studying the influence of different asphalt stone ratios on the stability of permeable sand-based mixture, it is determined that the best asphalt stone ratio for mixture forming is 10%.

Table 3. Comparison of water permeability of specimens before and after natural treatment.

| Performance index | Before disposal | After disposal |
|---|---|---|
| Permeability/(cm/min) | 1.65 | 1.53 |
| Water retention rate (g/cm$^3$) | 0.31 | 0.34 |
| Permeable aging/time | $\geq 10$ | $\geq 10$ |

Through tests, it is verified that the mixture has mechanical properties that meet the use of a slow-moving system, and the compressive strength can reach 13.78MPa;

(2) The permeability, water retention, and permeability aging of permeable sand-based specimens are evaluated. The results are 1.65cm/min, 0.31g/cm$^3$, and more than 10 times respectively, indicating that the specimens have excellent comprehensive permeability;

(3) The permeability durability of permeable sand-based materials in the application process is studied, and the changes in permeability, water retention, and permeability aging before and after natural treatment for 6 months are evaluated. The test results show that the performance indexes after treatment meet the specification requirements, which proves that permeable sand-based materials have good permeability durability.

## REFERENCES

Alalea Kia and Hong S. Wong and Christopher R. Cheeseman. Clogging in permeable concrete: A review[J]. *Journal of Environmental Management*, 2017, 193: 221–233.

Andrea L W, Jennifer K G, Leslie M, et al. Examination of the Material Found in the Pore Spaces of Two Permeable Pavements[J]. *Journal of Irrigation and Drainage Engineering*, 2013, 139(4): 278–284.

Brattebo B O, Booth D B. Long-term stormwater quantity and quality performance of permeable pavement systems[J]. *Water Research*, 2003, 37(18):4369–4376.

Felipe Montes and Liv Haselbach. Measuring Hydraulic Conductivity in Pervious Concrete[J]. *Environmental Engineering Science.*, 2006, 23(6): 960–969.

Su Y M, Hsu C Y, Lin J D. Clogging evaluation of porous asphalt concrete cores in conjunction with medical x-ray computed tomography[J]. *Proceedings of SPIE – The International Society for Optical Engineering*, 2014, 9063.

*Civil Engineering and Urban Research – Mohamed & Hou (Eds)*
*© 2023 the Authors, ISBN 978-1-032-44487-1*

# Preparation and performance evaluation of stain-resistant coating materials

Liu Yugui, Xu Jianhui, Chen Cheng, Dai Jianfeng & Wang Jie
*Chongqing Zhixiang Paving Technology Engineering Co., Ltd, Chongqing, China*

ABSTRACT: In order to protect buildings from pollution, a contamination-resistant coating material is developed in this paper, which can be widely used to protect road traffic and buildings. Through the comprehensive evaluation of the dilution, water demand, emulsifying effect, mechanical properties, and other performance parameters of the three self-emulsifying curing agents, the influence of the stirring speed on the particle size distribution and size of the coating material during the preparation was studied by the particle size analyzer, and the interfacial tension of the material before and after the addition of stain-resistant additives was evaluated by the gas-liquid interfacial tension instrument. The results show that curing agent 3 has better suitability. High-speed stirring is conducive to better emulsification of lotion, making the particle size distribution more uniform and smaller. The use of stain-resistant additives can significantly reduce the interfacial tension of coating materials and improve stain resistance.

## 1 INTRODUCTION

Coating materials are widely used in road traffic, housing construction, and other fields. They can cover subtle cracks, and decorate and protect buildings. However, when the coating materials are exposed to the outdoor environment for a long time, they are prone to aging and color change, or ash staining due to insufficient stain resistance. Especially in highly polluted areas, the pollution problems caused by serious haze, dust, and industrial emissions further pose a great challenge to the durability of the coating materials (Ji 2021).

Stain-resistant coating materials have the ability to resist the fading caused by the deposition of pollutants in the environment. At present, there are usually two ways to improve the stain resistance of coating materials (Steffen 2013; Zheng 2013). One is to make the coating surface highly hydrophobic, that is, the hydrophobic self-cleaning coating with low surface energy. Water droplets form a round bead on the coating surface, which is easy to roll down and take away pollutants. This way is usually to use silicone modified resin or silicone additives in the coating formulation. Another way is to make the coating surface hydrophilic. Water droplets will spread on the coating surface, and the surface is easy to be cleaned by rainwater. Fluorine-containing surfactants improve the stain resistance of the coating in this way (Renner 2006; Schellenberger Reiner et al 2013).

In this paper, the stain-resistant coating is prepared through the above two stain-resistant ways, and the properties of the coating are compared and evaluated, including adhesion, stain resistance, and aging resistance. Stain-resistant coating material with excellent performance is obtained through research.

## 2 RAW MATERIALS AND PREPARATION

### 2.1 *Raw materials*

(1) Main binder: as the dispersion medium of coating materials, the main binder needs to choose materials with good compatibility with other constituent materials and good mechanical properties

DOI 10.1201/9781003372417-48

as the dispersion medium has. In this study, the widely used waterborne epoxy resin is selected as the dispersion medium of coating materials.

(2) Stain-resistant additives: different types of stain-resistant coating materials were prepared by using hydrophobic and hydrophilic stain-resistant additives respectively. In this study, silicone additives and fluorine-containing surfactants were selected as stain-resistant additives.

(3) Anti-aging agent: the anti-aging agent selected in this study includes an ultraviolet absorber and hindered amine light stabilizer as anti-aging additives for coating materials.

## 2.2 *Preparation of coating materials*

First, the metered epoxy resin and emulsified curing agent are uniformly dispersed at a high speed for 3min to obtain the waterborne epoxy lotion, then different types of stain-resistant additives and aging agents are added to continue dispersing for 1min, that is, the lotion material for stain-resistant coating is obtained, and the coating material is immediately used for specimen fabrication or performance evaluation.

## 3 PERFORMANCE EVALUATION

### 3.1 *Performance evaluation and screening of curing agent*

Three commonly used self-emulsifying waterborne epoxy curing agents are selected. Through the evaluation of their properties, the one with the best performance is selected as the coating material. This paper mainly evaluates the emulsifying effect and bonding strength of the curing agent. The three curing agents are named curing agent 1, curing agent 2, and curing agent 3 respectively. Curing agent 1 has a large viscosity, which needs to be diluted with water before use, and then added with epoxy resin for emulsification. In order to ensure that the aqueous epoxy lotion has a suitable viscosity for construction, a large amount of water needs to be added to dilute the curing agent, but with the increase of water volume, the time required for curing the lotion gradually increases, And the bonding strength with cement concrete substrate is low, so curing agent 1 is not suitable for coating materials.

Curing agent 2 has low viscosity and low water consumption for dilution, but the stability of the waterborne epoxy lotion obtained after emulsification is poor, and water emulsion separation occurs (Figure 1), indicating that curing agent 2 is also not suitable for coating materials. The viscosity of curing agent 3 is between curing agent 1 and curing agent 2. It has a good emulsifying effect on epoxy resin and good bonding strength with cement concrete substrate. It can be used to prepare coating materials. Table 1 shows the comprehensive performance comparison of the three curing agents.

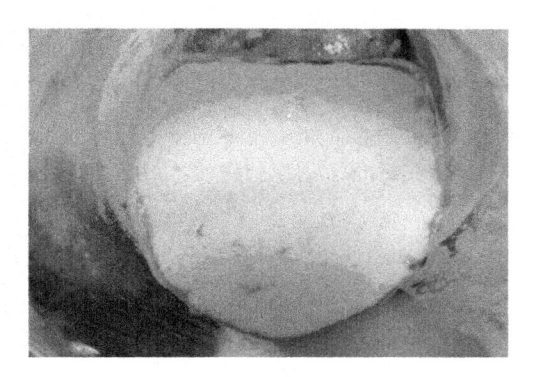

Figure 1.    Water emulsion separation phenomenon of aqueous epoxy lotion prepared by curing agent 2.

Table 1. Comparison of comprehensive properties of three curing agents.

| | Dilubility | Water demand | Emulsifying effect | Mechanical property |
|---|---|---|---|---|
| Curing agent 1 | Bad | Large | Commonly | Bad |
| Curing agent 2 | Good | Low | Bad | Bad |
| Curing agent 3 | Commonly | Moderate | Good | Good |

### 3.2 Influence of mixing process on the particle size of lotion

The size and size distribution of dispersed phase particles not only affect the stability of waterborne epoxy lotion itself, but is also the most important factor affecting the performance of the waterborne epoxy resin. The results show that the smaller the particle size of the dispersed phase, the narrower the distribution range, the more complete the curing, and the better the physicochemical properties of the film.

The mixing process is the key control factor of epoxy resin emulsification, including the selection of mixing rotor and mixing speed, which has an important impact on the particle size distribution of lotion. In this paper, the dispersion disk with gear is selected as the stirring rotor, and curing agent 3 is selected as the curing agent used in the research. Different lotion material samples are dispersed by high-speed stirring (about 1000rpm) and low-speed stirring (about 3000rpm), respectively. The particle size distribution of the two lotion samples is detected to study the effect of stirring speed on the particle size.

Figure 2 (a) and Figure 2 (b) show the test results of the particle size distribution of lotion obtained by low-speed and high-speed stirring respectively. According to the relevant data and Figure 2, it can be analyzed that there are two peaks in the particle size distribution diagram of lotion in Figure 2 (a), and D10 and D90 are 4.089 $\mu$m and 44.022 $\mu$m respectively. The volume average particle size D (4, 3) is 28.792 $\mu$m. The particle size distribution width span is 2.752. Figure 2 (b) D10 and D90 of lotion are 0.562 respectively $\mu$m and 1.978 $\mu$m. The volume average particle size D (4, 3) is 1.243 $\mu$m. The particle size distribution width span is 1.173. The comparison between the two shows that the particle size of lotion prepared by high-speed stirring is significantly smaller. The calculation shows that the volume average particle size of lotion prepared by low-speed stirring is basically more than 20 times that prepared by high-speed stirring. In addition, the overall particle size distribution of lotion prepared by high-speed stirring is more uniform.

### 3.3 Evaluation of stain resistance

Interfacial tension is the force of liquid surface shrinkage. The smaller the surface tension is, the smaller the shrinkage force is. Therefore, the easier it is to spread on the surface and the easier it is for the solution to wet the solid. Therefore, it is generally considered that the material with lower gas-liquid interfacial tension is more resistant to contamination than the material with higher interfacial tension. Therefore, this paper chooses to add stain-resistant additives to improve the stain resistance of coating materials, and selects the appropriate stain-resistant reagents by evaluating the gas-liquid interfacial tension.

Three kinds of stain-resistant additives were selected for this study, namely, additive 1 (organosilicon additive), additive 2 (fluorosurfactant), and additive 3 (fluorosurfactant). The stain-resistant coating materials were added to the lotion according to the usage measurement during the preparation of the lotion, and the interfacial tension of the materials was measured by the interface tension meter. The sample numbers of the above three additives were a, b and c respectively, The blank sample without stain-resistant additives is NO d and the water sample is NO e. Table 2 shows the test results of interfacial tension. The results show that the interfacial tension of No a, b, and c is significantly lower than that of No d, indicating that the addition of stain-resistant additives

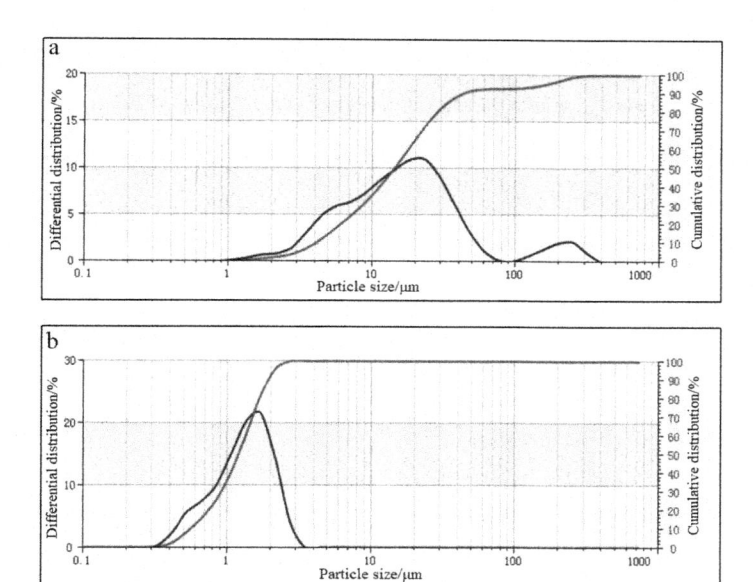

Figure 2.    (a) Particle size distribution of lotion stirred at low speed (b) particle size distribution of lotion stirred at high speed.

can effectively reduce the interfacial tension of coating materials to achieve the effect of stain resistance.

Table 2.    Test results of interfacial tension of coating materials.

| Sample number | NO a | NO b | NO c | NO d | NO e |
| --- | --- | --- | --- | --- | --- |
| Interfacial tension/ (mN/m) | 27.71 | 25.52 | 28.04 | 41.62 | 72.81 |

## 4   CONCLUSION

(1) Three kinds of self-emulsifying curing agents were optimized through performance evaluation. Finally, curing agent 3 was selected as the emulsifier of epoxy resin for the preparation of coating materials.

(2) The particle size of lotion can be effectively reduced by increasing the stirring speed in the preparation process of lotion, and the volume average particle size is 1.243 $\mu$m. It is beneficial to improve the comprehensive properties of a lotion.

(3) By adding stain-resistant additives, the interfacial tension of the coating material can be significantly reduced, which is conducive to improving the stain resistance of the coating material.

## REFERENCES

Ji X.H. (2021)Influence of fluorosurfactants on dirt pick-up resistance of exterior paints. *coating and protection*, 42(4): 25–28.

Renner R. The long and the short of perfluorinated replacements[J]. *Environmental Science & Technology*, 2006, 40(1):12–3.

Schellenberger S, Pahnke J, Reiner R, et al. A New Generation of High-Speed Fluorosurfactants[J]. *Paint & Coatings Industry*, 2013, 29(5) : 22–26.

Steffen S. (2013)A new generation of high-speed fluorosurfactants. *Journal of Paints &Coatings Industry*, 29(5): 22.

Zheng X. (2013)Preparation and properties of hydrophilic self-cleaning fluorocarbon coatings. *Paint&coating industry*, 43(1): 59–62.

# Effects of grinding aids on properties of cement clinker

Lang Du
*Chengdu Institute of Product Quality Inspection and Research Co., Ltd., Chengdu, Sichuan, China*
*National Center for Building Material Quality Supervision and Inspection (Sichuan Division), Chengdu, Sichuan, China*

Jing Peng*
*Chengdu Institute of Product Quality Inspection and Research Co., Ltd., Chengdu, Sichuan, China*
*National Center for Building Material Quality Supervision and Inspection (Sichuan Division), Chengdu, Sichuan, China*
*Sichuan Product Quality Supervision, Inspection and Testing Institute, Chengdu, Sichuan, China*

Xiang Zhou
*Chengdu Institute of Product Quality Inspection and Research Co., Ltd., Chengdu, Sichuan, China*
*National Center for Building Material Quality Supervision and Inspection (Sichuan Division), Chengdu, Sichuan, China*

Liang Li
*Chengdu Institute of Product Quality Inspection and Research Co., Ltd., Chengdu, Sichuan, China*
*National Center for Building Material Quality Supervision and Inspection (Sichuan Division), Chengdu, Sichuan, China*
*Sichuan Product Quality Supervision, Inspection and Testing Institute, Chengdu, Sichuan, China*

Xiaomin Zhang, Shuangfu Zhou, Zhao Shao, Honggen Chen, Shuang Li & Xiao Xiao
*Chengdu Institute of Product Quality Inspection and Research Co., Ltd., Chengdu, Sichuan, China*
*National Center for Building Material Quality Supervision and Inspection (Sichuan Division), Chengdu, Sichuan, China*

ABSTRACT: The purpose of this paper is to study the effect of adding cement grinding aids on the properties, physical properties, and strength of Portland cement powder. The particle size distribution, standard consistency water requirement, setting time, and flexural and compressive strength of the cement clinker powder after grinding were tested by adding grinding aids during the grinding process of cement clinker. The test results show that after using the two grinding aids, the percentages of <5um and 5~30um in each group of experiments have increased compared with the control group (without grinding aids). The repose angle, collapse angle, and compression degree of cement clinker changed significantly after adding grinding aid. After adding the No. 1 grinding aid, the strength of cement mortar increases. When the dosage is 0.035%, its strength increases significantly. Compared with the control group, it increases by about 10Mpa. With the increase of content, its strength no longer increases but tends to decrease. The conclusion shows that the use of grinding aids can improve the particle size distribution of cement clinker and improve the mortar strength of cement to a certain extent.

## 1 INTRODUCTION

Energy conservation and emission reduction are the focus of various industries. Adding a small amount of grinding aids in the cement production process can increase the cement grinding efficiency, reduce the production cost of enterprises, save electric energy and protect the environment

*Corresponding Author: 1181708197@qq.com

(Celik 2009; Li 2020; Qiu et al. 2011). China is a big country in cement production, so the use of grinding aids is of great significance (Sottilil 2000). The grinding aid can be attached to the surface of the particles, so as to reduce the electrostatic adsorption of the powder and the adhesion of the powder to the wall panel; The grinding aid also changes the granulometric distribution of the powder, making the distribution more uniform. On the other hand, grinding aids can change the shape of powder particles and indirectly lead the changes in the hydration rate and strength of cement (zhao & gao 2011).

## 2 EXPERIMENT

### 2.1 *Raw material*

Cement clinker: the clinker produced by Jiangyou Hongshi Company is adopted. The Chemical and mineral composition of clinker are shown in Table 1. Fly ash: purchased from Fengyuan Fly Ash Trading Co., Ltd., Wutongqiao District, Leshan City. Gypsum: from the building materials market in Youxian District, Mianyang City. Grinding aids: from Xintongling Building Materials Group Co., Ltd. In this experiment, there are No. 1 and No. 2 grinding aids.

Table 1. Chemical and Mineral Composition of Clinker %.

| LOI | $SiO_2$ | CaO | $Al_2O_3$ | $Fe_2O_3$ | MgO | $SO_3$ | f-CaO | Σ | $C_3S$ | $C_2S$ | $C_3A$ | $C_4AF$ |
|---|---|---|---|---|---|---|---|---|---|---|---|---|
| 0.17 | 21.96 | 63.41 | 5.22 | 4.06 | 3.15 | 0.49 | 0.66 | 98.46 | 46.4 | 27.95 | 6.95 | 12.34 |

### 2.2 *Methods*

Those without grinding aid are the control group. The grinding aids are No. 1 and No. 2. The two grinding aids grind the materials according to the dosage of 0.020%, 0.025%, 0.035%, 0.040%, and 0.045% respectively. The weight of each grinding is 5 kg. The SM-500 ball mill of Wuxi Jianyi Instrument Machinery Co., Ltd. is used to grind materials, and then the operation is stopped after 30min.

## 3 EXPERIMENTAL RESULTS AND ANALYSIS

### 3.1 *Effect of grinding aids on the granulometric distribution of cement clinker*

The granulometric distribution of cement determines its strength, fluidity, and other physical properties of cement (Sun et al. 2015). In this experiment, the laser particle size analyzer MS-2000 produced by Malvin, UK is used for the granulometric test. The experimental results are shown in Tables 2 and 3.

Table 2. Percentage content of each particle size range of cement clinker added with No. 1 grinding aid.

| No. 1 grinding aid dosage | less than 5um | 5~30um | 30~60um | More than 60um |
|---|---|---|---|---|
| 0.020% | 29.94 | 48.89 | 19.82 | 1.34 |
| 0.025% | 25.6 | 54.48 | 19.12 | 0.80 |
| 0.035% | 24.79 | 53.88 | 20.05 | 1.29 |
| 0.040% | 24.47 | 52.79 | 19.65 | 3.09 |
| 0.045% | 27.34 | 50.77 | 19.95 | 1.95 |
| Control group | 23.19 | 48.89 | 23.89 | 2.18 |

It can be seen from Tables 2 and 3 that after the use of two grinding aids, the percentage content of < 5um and 5 ~ 30um in each experimenting group is higher than that in the control group (no grinding aids), while the range of 5 ~ 30um particle size plays a major role in cement strength (Qian et al. 2014).

Table 3. Percentage content of each particle size range of cement clinker added with No. 2 grinding aid.

| No. 2 grinding aid dosage | less than 5um | 5~30um | 30~60um | More than 60um |
|---|---|---|---|---|
| 0.020% | 27.52 | 51.50 | 20.98 | 0.01 |
| 0.025% | 27.76 | 49.79 | 20.95 | 1.71 |
| 0.035% | 23.69 | 51.85 | 22.29 | 2.18 |
| 0.040% | 22.17 | 50.86 | 24.52 | 2.48 |
| 0.045% | 23.76 | 52.06 | 22.22 | 1.96 |
| Control group | 23.19 | 48.89 | 23.89 | 2.18 |

It can be seen from Table 2 that the mass number of cement with particle size in the range of 5 ~ 30um after grinding with No. 1 grinding aid increases first and then decreases, and the volume of particles with 0.045% content is greater than that with 0.020% content.

### 3.2 Effect of grinding aids on the properties of cement clinker powder

In this experiment, the BT-1000 powder comprehensive characteristic tester jointly developed by Dandong Baite Instrument Co., Ltd. and the split Technology Development Department of Tsinghua University is used to test the angle of repose, angle of fall, and compressibility of the ground cement clinker.

The angle of repose is the angle between the surface layer of the cone formed by the free accumulation of powder and the plane. The angle of repose has a great influence on the fluidity of powder. The smaller the angle of repose, the better the fluidity of powder. The angle of fall is the angle between the surface of the collapsed cone and the plane after the free accumulated powder is impacted to a certain extent. The difference between the angle of repose and the angle of fall is called the phase difference angle The larger the difference angle is, the better the fluidity of the powder. As shown in Table 4 and Table 5, the difference angle of cement clinker added with No. 1 and No. 2 grinding aids also increases with the increase of grinding aids. This shows that the fluidity of cement clinker becomes better with the increase of grinding aid content (Doncaster 2004).

Table 4. Powder properties of cement clinker after grinding with No. 1 grinding aid.

| No. 1 grinding aid dosage | Angle of repose/° | Angle of fall/° | Phase difference angle/° | Bulk Density/ $g/cm^3$ | Tap density/ $g/cm^3$ | Compression |
|---|---|---|---|---|---|---|
| 0.020% | 49 | 14 | 35 | 0.86 | 1.62 | 47.20% |
| 0.035% | 51 | 14 | 37 | 0.97 | 1.81 | 46.30% |
| 0.045% | 52 | 15 | 37 | 1.03 | 1.85 | 44.40% |

Loose density is the density of the powder after it falls naturally and fills the container. The vibrating density is the density of filling the gap between the powders through vibration so that the powder can tightly fill the container. The degree of compressibility can be obtained by subtracting the loose density from the compacted density and dividing it by the compacted density. Compressibility can also explain the fluidity of powder. The smaller the compressibility, the better the fluidity. As shown in Tables 4 and 5, the compressibility of cement clinker added with No. 1 and

Table 5. Powder properties of cement clinker after grinding with No. 2 grinding aid.

| No. 2 grinding aid dosage | Angle of repose/ ° | Angle of fall/ ° | Phase difference angle/° | Bulk Density/ g/cm³ | Tap density/ g/cm³ | Compression |
|---|---|---|---|---|---|---|
| 0.020% | 50 | 19 | 31 | 0.99 | 1.83 | 45.60% |
| 0.035% | 49 | 17 | 32 | 1.07 | 1.86 | 42.60% |
| 0.045% | 47 | 15 | 32 | 1.09 | 1.83 | 40.40% |

No. 2 grinding aids also decreases with the increase of grinding aids. This shows that the fluidity of cement clinker becomes better with the increase of grinding aid content (Zhang et al. 2010).

### 3.3 Effect of grinding aids on standard consistency and water demand of cement clinker

After grinding, the particle size and shape of the powder change, which will affect the water consumption of cement. Water consumption will also affect the hydration, hardening, setting time, and stability of the slurry. The determination of water consumption of standard consistency is of great significance to practical application and construction. The effects of grinding aids on the standard consistency and water demand of cement clinker are shown in Figure 1.

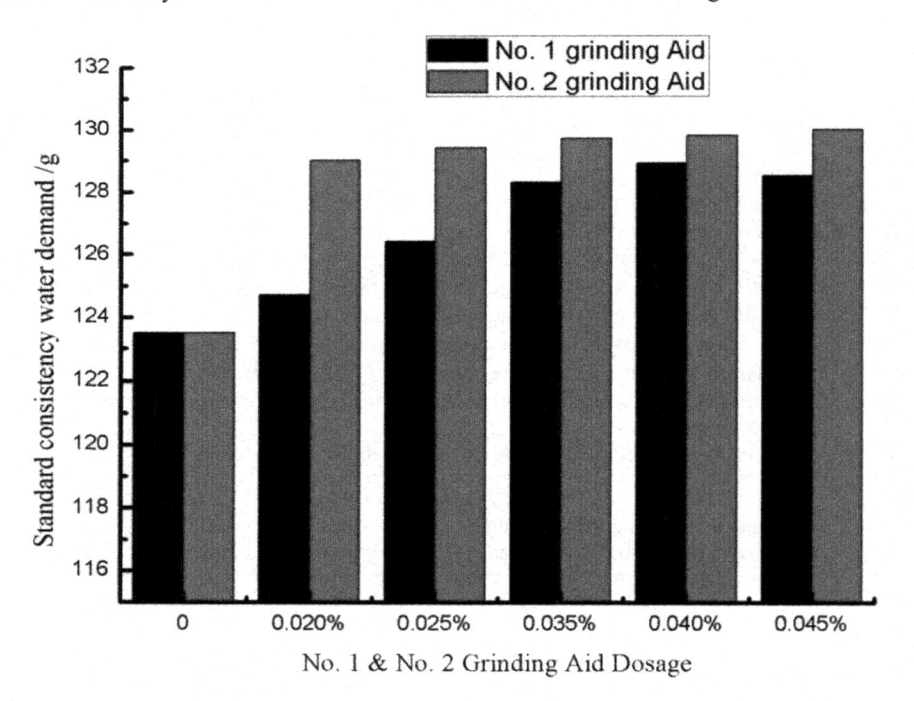

Figure 1. Standard consistency water demand of cement clinker with No. 1 & 2 grinding aid.

It can be seen from Figure 1 that after adding grinding aids, the water demand of standard consistency of each group increases to a certain extent, and the water demand of standard consistency of the experimental group with No. 2 grinding aid increases more than that of the experimental group with No. 1 grinding aid. In addition, with the increase of grinding aid content, the standard consistency water demand of the two experimental groups also showed an increasing trend. From the

348

analysis of the chemical composition of grinding aid, this may be because there are polyol amines in the grinding aid, which leads to the increase in water demand; On the other hand, the water demand for cement is related to the fineness of particles. When the fineness of cement particles is finer and the surface area is larger, the water demand is more[12].

## 3.4 *Effect of grinding aids on the strength of cement clinker*

The effects of grinding aids on the flexural and compressive strength of silicate cement clinker 3d and 28d colloidal mortar are shown in Tables 6 and 7.

It can be seen from Table 6 that the strength of cement colloidal mortar increases after adding the No. 1 grinding aid. When the content of grinding aid is 0.035%, its strength increases obviously, but with the increase of content, its strength does not increase, but decreases. Therefore, in terms of strength, adding 0.035% of the No. 1 grinding aid can improve the strength to the best. It has the best effect in improving particle size distribution and promoting material hydration, and increasing the content of 5 ~ 30um particle size plays a major role in cement activity (Lv & Sheng 2010). It also increases the overall surface area of cement particles, optimizes the gradation of grinding particles, and maximizes the strength of cement (BCelik 2008).

Table 6.    3d and 28d strengths of cement clinker with No. 1 grinding aid added.

| No. 1 grinding aid dosage | Compressive strength | | Flexural strength | |
| --- | --- | --- | --- | --- |
| | 3d | 28d | 3d | 28d |
| 0.02% | 5.58 | 7.42 | 26.81 | 42.72 |
| 0.025% | 5.11 | 7.75 | 26.90 | 46.62 |
| 0.035% | 6.17 | 7.99 | 33.12 | 53.65 |
| 0.04% | 5.32 | 7.13 | 29.01 | 52.88 |
| 0.045% | 5.17 | 6.05 | 25.60 | 47.82 |
| Control group | 4.77 | 6.26 | 25.62 | 43.58 |

Table 7.    3d and 28d strengths of cement clinker with No. 2 grinding aid added.

| No. 1 grinding aid dosage | Compressive strength | | Flexural strength | |
| --- | --- | --- | --- | --- |
| | 3d | 28d | 3d | 28d |
| 0.02% | 5.02 | 6.67 | 25.41 | 44.52 |
| 0.025% | 5.09 | 6.69 | 28.12 | 43.15 |
| 0.035% | 5.13 | 6.63 | 27.40 | 47.67 |
| 0.040% | 5.05 | 6.70 | 27.81 | 48.27 |
| 0.045% | 5.40 | 7.13 | 28.52 | 48.93 |
| Control group | 4.77 | 6.26 | 25.60 | 43.58 |

It can be seen from Table 7 that the strength of cement colloidal mortar increases after adding the No. 2 grinding aid. When the content of the No. 2 grinding aid is 0.045%, the strength of colloidal mortar increases obviously, and its strength decreases with the decrease of content. Therefore, in terms of strength, adding 0.045% No. 2 grinding aid has the best effect on improving the strength of cement.

# 4 CONCLUSION

Based on the results and discussions presented above, the conclusions are obtained as below:

(1) Different grinding aids have different effects on the granulometric distribution and powder properties of cement clinker. The colloidal mortar strength of cement clinker ground with No. 1 grinding aid is stronger than that of cement clinker ground with No. 2 grinding aid. The appropriate content of each grinding aid is also different. From the granulometric distribution and the colloidal mortar strength of cement clinker, the appropriate content of the No. 1 grinding aid is 0.035%, and that of the No. 2 grinding aid is 0.045%.

(2) The use of grinding aids will not only affect the size of powder particles but also affect their shape and voids, resulting in changes in fluidity, density, and compressibility. With the increase of grinding aid content, the fluidity of powder also tends to be improved.

## REFERENCES

BCelik. The effects of particle size distribution and surface area upon cement strength development[M]. *Powder Technology*, 2008.

Celik I B. The effects of particle size distribution and surface area upon cement strength development [J]. *Powder Technology*, 2009, 188(3):272–276.

Doncaster J. Research on the mechanism of strength enhancement of cement paste [J]. *Cement and Concrete Research*, 2004, 34(6):973–976

Li Xianbiao. The characteristics of different types of cement grinding aids and their influence on cement properties [J].*Sichuan cement*, 2020(2):7.

Lv Wanheng, Sheng Jingsheng. Test and Application of Cement Grinding Aid [J] *Building Materials Technology and Application*, 2010, 9:12.

Qian Hui, Fang Ying, Li Zhen, et al. Influence of monomer grinding aids on properties of Portland cement [J]. *China Powder Technology*, 2014, 20(3):48–51.

Qiu X Q, Peng X Y, Yi C H, et al. Effect of side chains and sulfonic groups on the performance of polycarboxylate-type superplasticizers in concentrated cement suspensions [J].*Journal of Dispersion Science and Technology*, 2011, 32(2):203–212.

Sottilil, Padovanid. Effect of grinding admixtures in the cement industry [J].*ZKG International*, 2000, 53(10):568–575.

Sun Xiaowei, Chen Yanwen, Pan Wenhao et al. Influence of grinding aids on properties of slag Portland cement [J]. *Bulletin of the Chinese Ceramic Society*, 2015, 34(8):2083–2088.

Zhang Guangyu, Lin Dong, Jiang Guihua Experimental Research on Cement Grinding Aids [J] *Guangdong Building Materials*, 2010, 9:910.

Zhao Xiaodong, Gao Zhuofang Optimum Selection and Practical Application of Cement Grinding Aids [J]. *Cement Guide for New Epoch*, 2011, 5:52–53.

# Study on polyurethane thermal insulation materials for prefabricated buildings in cold areas*

Wei Sisheng

*Shandong Huayu Institute of Technology, Shandong University Clean Air Conditioning Engineering Technology R&D Center, Henan, China*

ABSTRACT: With the help of experiments, all water polyurethane thermal insulation materials in fabricated thermal insulation building materials are analyzed, their foaming characteristics and problems are studied, and the application conclusion of all water polyurethane as thermal insulation building materials is obtained.

## 1 INTRODUCTION

Restricting and forbidding the use of nonenvironmental protection building insulation materials has long since become a policy measure in China. It should take all kinds of measures to ensure comprehensive elimination. This will promote the development of China's construction industry and polyurethane industry, and can achieve foreign advanced level. The development of new technology for the foaming system with polyurethane foam has become the main trend of technical innovation in the world-building insulation industry.

## 2 PRODUCT DEVELOPMENT TREND

The hard foam elimination technologies widely used in China include cyclopentane foaming technology, water foaming technology, and HCFC-141b foaming technology.

2.1 Cyclopentane technology has been commercialized and widely used in Europe. It belongs to zero ODP and zero GWP substances, but it is flammable. The safe operation of flammable foaming agents should be carefully considered before use (Chen 2008).

2.2 Water foaming technology is also a zero ODP technology and has been commercialized in China. It is mainly used to produce foam with low requirements for thermal insulation performance. The disadvantage of this method is that it requires the transformation of the foaming machine (material ratio) and the change of raw material formula.

2.3 As a transitional technology (ODP is not zero), HCFC-141b technology will still be phased out in the end, and once this technology is selected, the multilateral fund will not provide funding for the second phase-out.

In view of the analysis of the above alternative technologies, we chose to apply all water-foaming technologies to the development and research of prefabricated building thermal insulation materials in cold areas.

---

*Based on the project research of "exploration and localization of passive ultra-low energy consumption buildings with integrated assembly in cold areas – Taking Shandong as an example" (No. 2020-r2-3)

2.4 HFC245fa technology is an alternative technology with zero ODP value, a high price, and low requirements for equipment transformation. At present, HFC245fa has not been commercially produced in China. However, its foam has excellent thermal insulation performance and is widely recognized and used in Europe and the United States.

2.5 ODS elimination in the foam industry is a compulsory policy in China

In view of the analysis of the above alternative technologies, we chose the whole water foaming technology to be applied to the development and research of prefabricated building insulation materials for cold areas (Liu 2018).

## 3 EXPERIMENTAL PART

### 3.1 *Main raw materials*

Table 1. Composite polyether index.

| Brand | SDG/P-05-401 |
| --- | --- |
| Appearance | brown |
| Hydroxyl value/mg KOH.g$^{-1}$ | 399±30 |
| Acid value/mg KOH.g$^{-1}$ | ≤ 0.51 |
| Water content/% | 4.5±0.5 |
| Viscosity (25°/mPa.S) | 1110±310 |
| pH value | 10~12 |

### 3.2 *Main experimental equipment and instruments*

The main experimental equipment and instruments were micro moisture analyzer, constant temperature water bath, automatic mixer, 60 cannon high-pressure foaming gun, etc.

### 3.3 *Experimental methods and steps*

It is needed to adjust the temperature of materials A, B, and mold, weigh materials a and B according to the experimental requirements, mix them for 5 ~ 10s, immediately pour them into the mold for foaming, successively measure the foaming parameters, measure the foam height index (foam height/foam mass), unit cm/g, then divide the foam into 20 parts according to the same height, measure their density respectively, and then calculate the root mean square deviation of density, that is, the density distribution coefficient (Polyurethane industry 2019).

### 3.3.1 *Manual foaming (Experimental process parameters)*

Table 2. Foaming parameters.

| Serial number | Proportion | | Parameter Milky white/ gel / non stick(s) | Free bubble density (kg/m$^3$) | The product has no black-and-white crazing, uniform, and dense bubbles, which are white to light yellow. |
| --- | --- | --- | --- | --- | --- |
| | Black material (g) | White material (g) | | | |
| 1 | 193.5 | 145 | 21/55/90 | 36.8 | |
| 2 | 196 | 145 | 32/65/101 | 37.05 | |
| 3 | 200 | 145 | 39/68/142 | 39.02 | |
| 4 | 215 | 145 | 20/63/177 | 40.1 | |
| 5 | 210.3 | 145 | 18/65/109 | 40.5 | |
| 6 | 223.5 | 145 | 20/63/125 | 42.7 | |

According to the parameters in the above table, the reasonable proportion of black and white materials is 1.45:1.

### 3.3.2 *Liquidity test*

Table 3. Test data of flow index and density distribution coefficient are as follows.

| System | Density kg/m$^3$ | Liquidity index | | Density distribution coefficient | |
|---|---|---|---|---|---|
| | | Sampling value | Average value | Sampling value | Average value |
| System 1 | 39 | 0.96 | 0.99 | 2.37 | 1.66 |
| | | 1.02 | | 1.69 | |
| | | 0.98 | | 0.88 | |
| System 2 | 36 | 0.97 | 1.02 | 2.32 | 1.15 |
| | | 1.017 | | 0.55 | |
| | | 1.016 | | 0.63 | |
| System 3 | 31 | 0.08 | 1.06 | 0.76 | 0.57 |
| | | 1.09 | | 0.43 | |
| | | 1.06 | | 0.53 | |
| CFC-F$_{11}$ System | 31 | 0.92 | 0.93 | 4.44 | 4.64 |
| | | 0.95 | | 4.79 | |
| | | 0.97 | | 4.77 | |

It can be concluded that the larger the flow index, the better the fluidity, and the smaller the density distribution coefficient, the more uniform the density distribution of the system. It can be seen from the above table that the comprehensive performance of raw materials of all water system 3 is good. We chose system 3 as the test material.

### 3.3.3 *High pressure foaming machine test*

Table 4. High-pressure polyurethane foaming test.

| Type of foaming agent | All water foaming | HCFC-141b | CFC-F$_{11}$ |
|---|---|---|---|
| Milky white / gel / non stick(s) | 17/59/84 | 17/56/71 | 18/50/75 |
| Free bubble density (kg/m$^3$) | 36 | 34 | 31 |
| Box density (kg/m$^3$) | 54 | 46 | 45 |

### 3.3.4 *Physical properties of foam*

Table 5. Comparison of properties of polyurethane rigid foam.

| Performance / PU type | | All water foaming | HCFC-141b | CFC-F11 | Basis standard |
|---|---|---|---|---|---|
| Dimensional stability (%) | $-25°C \times 48h$ | 0.45 | 0.45 | 0.45 | GB/T8811-1988 |
| | $70°C \times 48h$ | 0.95 | 0.95 | 0.95 | |
| Average daily efficiency (%) | | 54.5 | 54.5 | 54.5 | GB/T12915-1991 |
| Average heat loss coefficient W/m$^2 \cdot °C$ | | 0.61 | 0.61 | 0.64 | |
| Thermal conductivity (25°C)W/m$^2 \cdot °C$ | | 0.026 | 0.023 | 0.023 | GB/T10295-1988 |
| Compressive strength (MPa) | | 0.211 | 0.141 | 0.15-0.21 | GB/T8813-1988 |
| Density (kg/m$^3$) | | 36~41 | 31~36 | 31~36 | Internal control standard |
| Low-temperature test of water tank: $-25°C \times 72h\%$ | | | Crack size$\leq$0.45 | | Internal control standard |

The test results of all test items in the above table are within the product design and belong to the design requirements.

## 4 CHARACTERISTICS OF ALL WATER FOAMING

### 4.1 *Advantages*

Water is used as a foaming agent to replace chlorofluorocarbon foaming, which has no pollution to the atmosphere and truly realizes the green product of solar energy. It is suitable for high fluorine (high-pressure foaming machine) foaming lines without increasing equipment investment. By adjusting the formula, it can meet various process requirements.

### 4.2 *Disadvantages*

The viscosity of the composite material is relatively high. Other disadvantages are poor fluidity during molding, poor adhesion to the substrate, and poor dimensional stability. Carbon dioxide foam has slightly higher thermal conductivity and requires higher density and compressive strength to maintain its performance (Polyurethane industry 2019).

## 5 PROBLEMS SOLVED

Water and carbon dioxide reflect the formation of more urea bonds, which reduces the bonding performance of the foam. At the same time, the diffusion rate of carbon dioxide from the inside to the outside of the cell is greater than that of air to the inside of the cell, which reduces the pressure in the cell, resulting in the shrinkage of foam and the reduction of dimensional stability. Therefore, the key to all water-foaming rigid polyurethane foam technology is to develop new low-viscosity and high-performance foam polyether polyols (Zheng 2001).

### 5.1 *Effect of isocyanate index on properties of rigid polyurethane foam*

As there are a large number of rigid polyurea chain segments in the rigid polyurethane foam of the water foaming system, the brittleness of the foam increases with the increase of the isocyanate index, which leads to the decrease in the toughness of the foam, the decrease of the volume change rate at high and low temperatures, and the increase of the dimensional stability.

### 5.2 *Effect of water content and density on properties of rigid polyurethane foam*

When the amount of water in the combined polyether increases, it reacts with isocyanate to generate more $CO_2$ and release more heat, which reduces the density of the foam. The change in density has a great impact on the properties of foam plastics.

### 5.3 *Effect of water content and density on thermal conductivity of the foam*

The pore diameter of closed-cell rigid polyurethane foam is very small, and its thermal conductivity mainly depends on the thermal conductivity of gas, resin solid, and radiation in the cell, while the thermal conductivity of gas accounts for 60% of the thermal conductivity of foam. Therefore, the content of insulating gas in the foam will be the key to affecting the overall thermal conductivity of the foam. When the density of the foam is greater than $30kg/m^3$, the thermal conductivity of the resin solid is basically fixed, while the effect of radiation heat transfer thermal conductivity is small, the increase of water will reduce the density of the foam, which is conducive to improving the initial thermal insulation performance of foam (Zhang et al. 2013).

### 5.4 Selection of surfactants

In the water foaming system, the polarity of the foaming system is increased by using water with strong polarity as a chemical foaming agent, and the polarity of the traditional methyl siloxane oxidized olefin copolymer surfactant is strong. Therefore, its emulsification and nucleation effect on the all-water rigid polyurethane foam foaming system with the same strong polarity is relatively weak, and it is difficult to form a fine honey and uniform closed cell structure foam.

### 5.5 Catalyst selection

In the aqueous foaming system, the traditional tertiary amine catalysts, such as n,n-dimethyl cyclo-hexylamine, triethylenediamine, dimethylaminoethanol, etc., mainly promote the foaming reaction of polyurea and $CO_2$ between isohydric acid ester and water but do not promote the gel reaction. In order to adjust the balance between foaming and gel, triazine catalysts, alkali metal salts, and other gel catalysts can be used together to improve the brittleness of all water foam and enhance its adhesion to the substrate (Fu 2013).

## 6 CONCLUSION

6.1 To sum up, for the hard foam made with water as a blowing agent, the polyether polyol, isocyanate, surfactant, and catalyst must be improved accordingly according to the different application objects, and the formula and process conditions must be adjusted to produce qualified all water PU foam products. Therefore, there is no need to worry about the impact of water foaming instead of the CFC-fl1 system on the performance of the foam.

6.2 Considering policy and environmental protection, water foaming technology is a necessary technology.

6.3 The popularization and application of all water foam can completely eliminate ozone-depleting substances, make the polyurethane industry a truly clean, safe, and effective industry, accelerate the replacement process of CFCs in China, and promote China's polyurethane industry to catch up with the international advanced level.

## REFERENCES

Chen Dingnan. *The production process of polyurethane products* [M] Beijing: Chemical Industry Press. 2008.
Fu Dongsheng. *Fly ash modified polyoxynitride technology and its application, plastic technology*, 2013.
Influence of lianglongqiang polyurethane foam layer on damping performance of isolation composite damping material [J] *Polyurethane industry*, 2019, issue 5.
Liu Bin. research progress of polyurethane foam [J] *China chemical trade*. 2018 issue 29.
Zhang Yan, Wu Xi, Wang Huan, et al. *Research on UV curable waterborne polyurethane based on different diols, coating industry*. 2013.
Zheng Ruicheng. *Engineering technical manual of the solar hot water system in civil buildings*, chemical industry press, 2001.

*Civil Engineering and Urban Research – Mohamed & Hou (Eds)*
*© 2023 the Authors, ISBN 978-1-032-44487-1*

# Study on the formation and transmission characteristics of soot particles in Chinese cooking

Yang Yuan*
*China Academy of Building Research, Chaoyang, Beijing, China*

Neng Zhu*
*Tianjin University, Nankai, Tianjin, China*

Jing Liu*
*University of Science and Technology Beijing, Haidian, Beijing, China*

Zhiqiang Li*, Chunlong Li* & Zhengzheng Zhang*
*China Academy of Building Research, Chaoyang, Beijing, China*

ABSTRACT: In this paper, the physical experimental platform was built to test the generation and propagation characteristics of soot particles in the cooking process of different dishes. The results show that the food materials have an obvious influence on the particle size distribution and time evolution of particulate matter emitted during cooking; when cooking different dishes, there are not only differences in the characteristics of particulate matter emitted, but also differences in the evolution law of particulate matter transmission; under the test conditions in this paper, increasing the exhaust speed of the range hood can significantly reduce the particle concentration above the frying pan, but the effect on the particle concentration at the mouth and nose of the operator and the center of the kitchen is not obvious.

## 1 GENERAL INSTRUCTIONS

As an important indoor pollution source, cooking fume has a significant negative effect on indoor air quality and residents' health (Cao 2016; EPA 1998). Studies show that the cooking process of the cooking oil fuel gas contains a variety of toxic and harmful substances, including $CO_2$, $SO_2$, CO, NOx, inhalable particulate matter, polycyclic aromatic hydrocarbons, and a variety of strong carcinogens (Faber 2015; Muleski 2005). Therefore, effective control of cooking fume is an important way to improve indoor air quality.

Particulate matter is an important pollutant produced in the cooking process and also the focus of cooking fume pollution research (Shi 2014; Wang 2015, 2017; Xiao 2017; Zhang 2017; Zhou 2015). Compared with western cooking, the amount of edible oil used in the Chinese cooking and frying process is large, the oil temperature is higher, and the amount of soot particles produced in the cooking process is significantly higher than in the steaming process. In this study, an experimental bench was built in accordance with the spatial scale of a typical residential kitchen. Through the experimental research, the pollution characteristics of soot particles in real Chinese cuisine cooking materials were analyzed from the aspects of soot particle emission quantity, particle size distribution, and particle propagation.

---

*Corresponding Authors: 13810969897@163.com, nzhu@tju.edu.cn, 13718127285@163.com. 147938949@qq.com. 353356558@qq.com and 984311284@qq.com

DOI 10.1201/9781003372417-51

## 2 FIELD MEASUREMENT METHOD

### 2.1 *Test instruments and measuring point layout*

Typical kitchens in Beijing were selected for testing in this paper (see Figure 1). Graywolf PC3016-IAQ particulate matter detector was used to detect indoor particulate matter concentration. The temperature of the oil pot during cooking was measured by infrared thermography. Three measuring points were arranged for particulate matter detection, namely A1, A2, and A3 (see Figure 2). Point A-1 is set right above the pot, 58cm away from the desktop, the sampling port of the dust particle counter is connected with the rubber hose, and the air inlet is placed horizontally and perpendicular to the direction of flue gas flow. Sampling point A-2 is set 50cm away from the north side of the hearth, 70cm away from the desktop, and the air inlet section is vertically downward to simulate the concentration of cooking smoke at the operator's mouth and nose. Sampling point A-3 is set in the center of the kitchen, 220cm above the ground, and the air inlet is vertical upward.

Figure 1.    Schematic diagram of the kitchen.

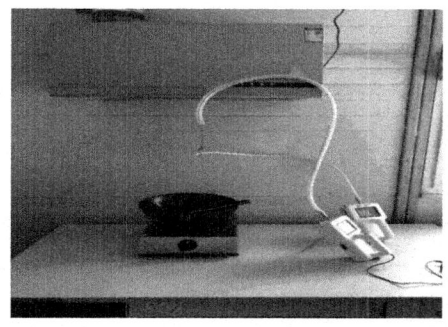 

(1) A-1 and A-2 sampling points                 (2) A-3 sampling point

Figure 2.    Sampling points layout.

### 2.2 *Test Method*

In the test, green pepper and pork were selected as food materials to analyze the effects of the single vegetable, single meat, and vegetable plus meat on the emission characteristics of particulate matter in cooking. Table 1 shows the number of ingredients, oil, and spices for different types of food ingredients.

Table 1. Amount of food ingredients under different food ingredient types.

| Dish number | Ingredients and dosage | Edible oil consumption (g) | Soy sauce dosage (g) |
|---|---|---|---|
| 1 | 100g green peppers | 26.0 | 3.6 |
| 2 | 50g pork | 26.0 | 3.6 |
| 3 | 50g pork+100g green peppers | 26.0 | 3.6 |

In the measurement, it is needed to first open the range hood, and heat the pot to dry, with the infrared thermometer to observe the temperature of the iron pot, when the pot temperature reached 220°C, add cooking oil and open 3 dust particle counters at the same time. For working condition 1, it needs to heat it on medium heat for 20 seconds, then add sliced green pepper and soy sauce, stir fry it for 120 seconds, and then remove it from the pot. For working condition 2, heat it on medium heat for 20 seconds, then add sliced meat and soy sauce, stir fry it for 120s, and then remove it from the pot. For working condition 3, it is required to heat it on medium heat for 20 seconds, then add the shredded meat and soy sauce, stir fry it for 60 seconds, then add green pepper, stir fry it for 60 seconds, and then remove it from the pot.

## 2.3 Test conditions

The results of soot emission and soot particle size distribution in different cooking processes can be obtained through the above experimental process. The cooking of the three dishes combined with the working modes of the range hood can form six scenarios (see Table 2).

Table 2. Test conditions.

| Scenario | Dishes | Wind speed of range hood |
|---|---|---|
| 1 | Dish 1 | Low speed |
| 2 | Dish 1 | High speed |
| 3 | Dish 2 | Low speed |
| 4 | Dish 2 | High speed |
| 5 | Dish 3 | Low speed |
| 6 | Dish 3 | High speed |

## 3 ANALYSIS OF MEASURED RESULTS

### 3.1 Influence of cuisine types on particulate-forming characteristics

Figures 3 to 5 show the concentration evolution of particles with different particle sizes at the A-1 sampling point under working conditions 1, 3, and 5. The range hood is running at low wind speed in the above three working conditions.

As can be seen from Figure 3, under the condition of working condition 1, the count concentration of soot particles with particle size less than $0.3\mu m$ does not rise but drop at the beginning of cooking, and then drops slightly. After cooking, the concentration value returns to the initial level. At point A-1, the counting concentration and weight concentration of soot particle size above $0.3\mu m$ showed a process of rising firstly, then decreasing, and finally decreasing to the background value, but the numerical change process of soot particle concentration was different for a different particle size range. However, the counting concentration of soot particle size less than $0.3\mu m$ was "decreased firstly and then increased". The reason is that a large amount of water vapor is generated and promotes the coagulation of particles with particle size less than $0.3\mu m$ during cooking, and the amounts of particles which particle size less than $0.3\mu m$ decrease.

Under working condition 3, the count concentration of soot particles less than $0.3\mu m$ fluctuated repeatedly, decreased overall, and then gradually increased. The counting concentration and weight concentration of lampblack particles above $0.3\mu m$ showed a process of vibration rising, then falling, and finally falling to the background value (see Figure 4).

Under condition 3, the peak time of soot particles with different size range at the A-1 point generally lags behind the peak time of soot particles in condition 1. There is no significant difference between the soot particle concentration in the size range of $0.5$-$1.0\mu m$ and $1.0$-$2.5\mu m$ and that in the working condition 1. However, the peak concentration of soot particles smaller than the size range is significantly higher than that of working condition 1, while the peak concentration of soot particles larger than the size range is significantly smaller than that of working condition 1. Compared with operating condition 1 and preliminary analysis, other conditions remain the same, the variety of dishes, pot temperature, the cooking ingredients, etc. are the main factors leading to the difference in the process of soot particle concentration change.

Under working condition 5, the count concentration of $0.3$-$0.5\mu m$ particles presents a characteristic of "turbulently decrease-increase", while the count concentration of other soot particles in the particle size range is basically a process of "turbulently increase-decrease" among which the count concentration of $0.5$-$1.0\mu m$ particles increases most dramatically. The concentration fluctuation of counted particles in the size range of $1.0$-$2.5\mu m$ was next (see Figure 5).

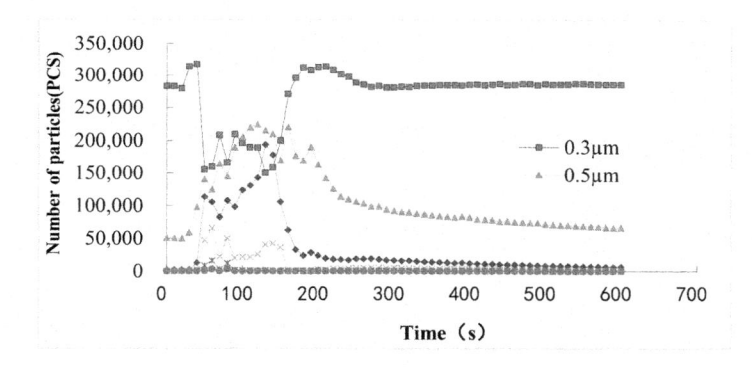

Figure 3. Counting concentration of soot particles at A-1 of measuring point in working condition 1.

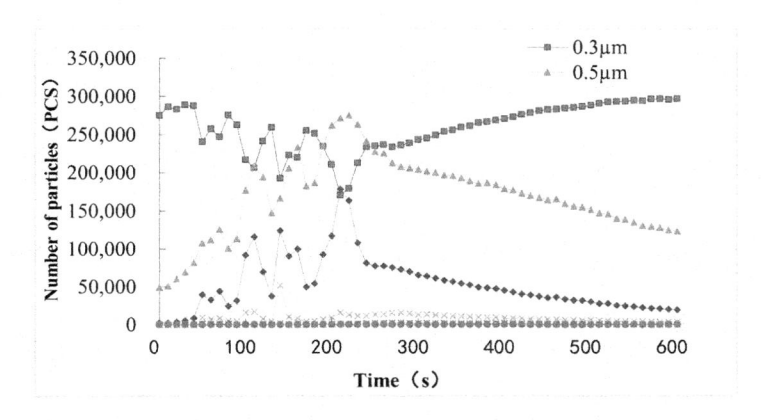

Figure 4. Counting concentration of soot particles at A-1 of measuring point in working condition 3.

Further analysis of the test data shows that the peak concentration of soot particles is significantly different under different working conditions. In the particle size range of $2.5\mu m$ to $5.0\mu m$ and

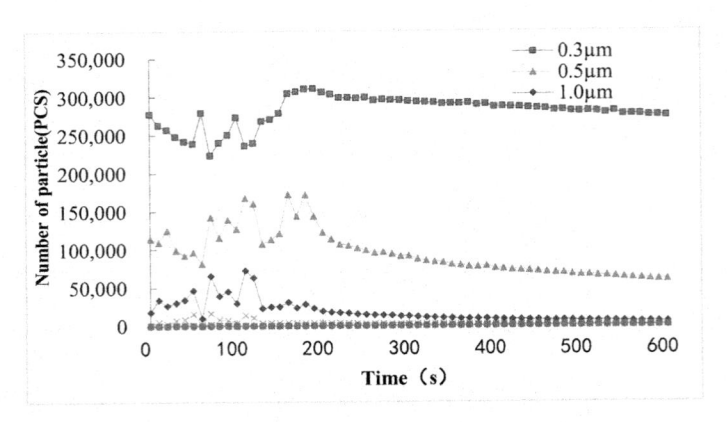

Figure 5. Counting concentration of soot particles at A-1 of measuring point 5 in working condition.

5.0μm to 10μm, the concentration peak of operating condition 1 was much higher than that of operating condition 3 and condition 5. This may be because the cooking materials in working condition 1 are vegetables, and the amount of water vapor generated is larger than that in working condition 3 and condition 5. Higher water vapor content promotes coagulation of small particle size.

Table 3. Soot particles under different working conditions.

| Working condition | Particle diameter | Peak concentration (PCS/L) |
|---|---|---|
| 1 | 0.3μm~0.5μm | 224260 |
| | 0.5μm~1.0μm | 193874 |
| | 1.0μm~2.5μm | 65798 |
| | 2.5μm~5.0μm | 16183 |
| | 5.0μm~10μm | 3254 |
| 3 | 0.3μm~0.5μm | 275178 |
| | 0.5μm~1.0μm | 178051 |
| | 1.0μm~2.5μm | 52123 |
| | 2.5μm~5.0μm | 2315 |
| | 5.0μm~10μm | 111 |
| 5 | 0.3μm~0.5μm | 172284 |
| | 0.5μm1.0μm | 72479 |
| | 1.0μm~2.5μm | 16441 |
| | 2.5μm~5.0μm | 1561 |
| | 5.0μm~10μm | 114 |

### 3.2 Difference in concentration distribution of soot particles in the cooking process of different dishes

Figure 6 shows the concentration peak distribution of particles with different particle sizes above the frying pot (A-1 point), at the mouth and nose of the operator (A-2 point), and at the center of the kitchen (A-3 point) in the cooking process of different types of dishes under the low-grade exhaust of the range hood. As can be seen from the figure, there are significant differences in the characteristics of the peak concentration of particulate matter at the above locations when different dishes are cooked. Above the frying pot, for example, when the cooking dish is fried with green pepper (working condition 1), the mass concentration of particles with particle sizes between $1.0\mu m$ and $2.5\mu m$ is the largest. However, when the cooking dishes were fried with meat (working condition

3) and fried with green pepper (working condition 5), the maximum particle mass concentration ranged from $1.0\mu$m to $2.5\mu$m (see Figure 6 (1)). At the same time, the mass concentration of $5.0\mu$m to $10\mu$m when frying green pepper is much higher than in the other two conditions. That shows that the stir-frying process of pure vegetables will generate more particles with large particle sizes. This may be because vegetables contain more water than meat and are more likely to evaporate during cooking. However, the larger moisture content will increase the coagulation of particles (including the coagulation of droplets) and generate more particles with a large particle size.

Compared with the air above the frying pot, the particle size distribution of the operator's mouth and nose changed significantly. When cooking dishes with fried green pepper (working condition 1), the highest mass concentration of particulate matter at the operator's mouth and nose is $0.5\mu$m to $1.0\mu$m. When frying meat (working condition 3), the highest mass concentration of particulate matter at the operator's mouth and nose was $2.5\mu$m to $5.0\mu$m. Under the condition of meat frying green pepper (condition 5), compared with the air above the frying pot, the concentration ratio of $0.5\mu$m to $1.0\mu$m and $2.5\mu$m to $5.0\mu$m at the operator's mouth and nose increased (see Figure 6 (2)).

In the center of the kitchen, the particle size ranges with the largest proportion of mass concentration in fried green pepper (working condition 1) and fried meat (working condition 3) are $5.0\mu$m to $10\mu$m. However, in the condition of fried green pepper with meat (condition 5), the concentrations of particles with different particle sizes from $1.0\mu$m to $2.5\mu$m, $2.5\mu$m to $5.0\mu$m, $5.0\mu$m to $10\mu$m were similar.

There are not only different characteristics of particles emitted by different dishes, but also different evolution rules of particle propagation. It should be noted that under the three working conditions, the mass concentration ratio of large particle size ($5.0\mu$m to $10\mu$m) in the center of the kitchen was higher than that above the frying pot and at the mouth and nose of the operator. This indicates that there is obvious coagulation in the process of kitchen particulate matter propagation, which should be paid attention to in the follow-up study.

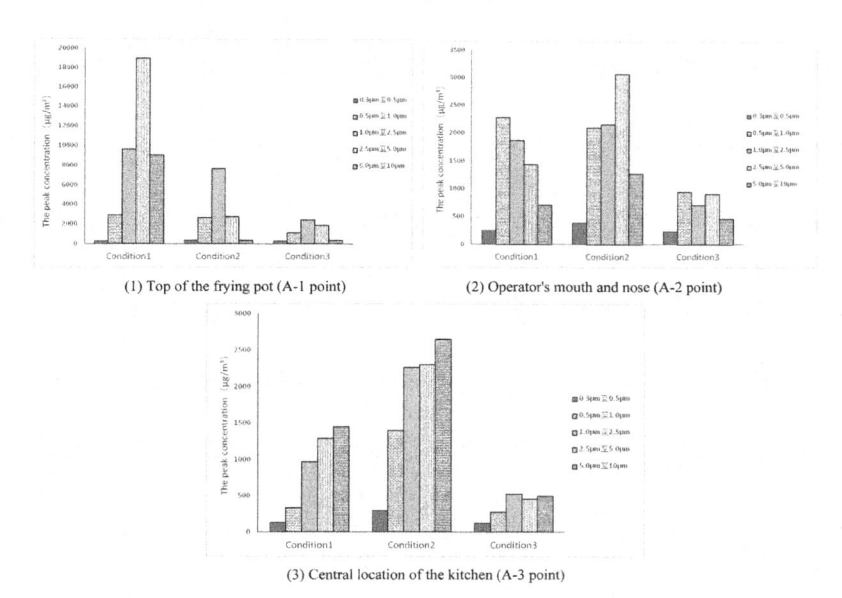

(1) Top of the frying pot (A-1 point)  (2) Operator's mouth and nose (A-2 point)

(3) Central location of the kitchen (A-3 point)

Figure 6.  Particle size distribution at different locations under different working conditions.

Based on the particle concentration in the air above the frying pot (A-1 point), Figure 7 shows the changes in particle concentration at the mouth and nose of the operator (A-2 point) and the center of the kitchen (A-3 point) relative to the air above the frying pot, taking working condition 1 as an example.

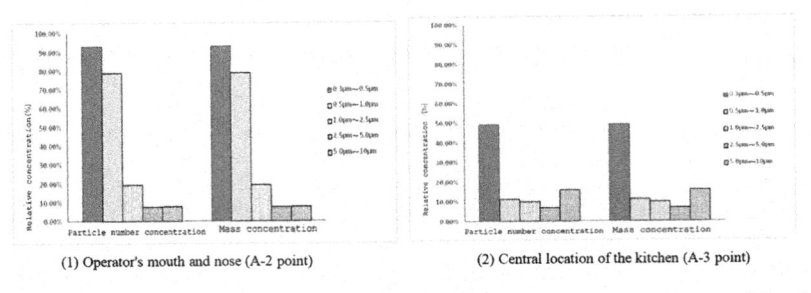

(1) Operator's mouth and nose (A-2 point)　　　　(2) Central location of the kitchen (A-3 point)

Figure 7. Relative changes of concentrations at different measuring points in working condition 1.

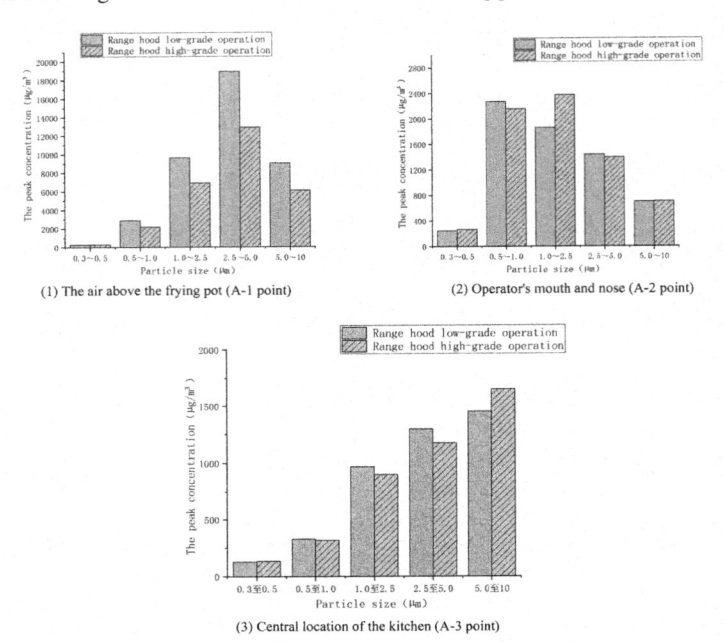

Figure 8. Particle size distribution at different positions of stir-fried green pepper range hood at different working conditions (working conditions 1 and 2).

As can be seen from Figure 7, the variation trend of mass concentration and count concentration of particles with different particle sizes between different locations is exactly the same. For particles with particle sizes ranging from $0.34\mu$m to $5.0\mu$m, the relative concentration at different measuring points showed the same pattern, that is, above the frying pot (A-1 point) > at the operator's mouth and nose (A-2 point) > at the center of the kitchen (A-3 point). For particles with particle sizes ranging from $5.0\mu$m to $10\mu$m, the concentration at the center of the kitchen (A-3 point) is higher than that at the operator's mouth and nose (A-2 point), but much smaller than the top of the frying pot (A-1 point).

For cooking operators, the concentrations of $0.34\mu$m to $0.54\mu$m and $0.54\mu$m to $1.0\mu$m particles at the nose and mouth (A-2 points) were 93.0% and 78.8% higher than those above the frying pot, respectively. This indicates that the range hood cannot effectively reduce the impact of small particle size on human health under low operating conditions. At the same time, the concentration of particulate matter with particle size between $0.3\mu$m and $0.5\mu$m in the center of the kitchen reached 49.4% of that above the frying pot, indicating that a large number of particulate matter with small particle size were not discharged outdoors by the range hood. These particles are likely to enter other indoor spaces, and then cause comprehensive pollution of the indoor environment of the family.

### 3.3 Influence of exhaust velocity on concentration distribution of soot particles

Figure 8 shows the distribution of particulate matter concentration at each sampling point under different range hood operating gears when the food material is green pepper.

As shown in Figure 8, when the operating gear of the range hood is changed from low to high to increase the emission speed, the concentration of particles in other particle size ranges above the frying pot (A-1 point) decreases significantly except for $0.3\mu$m to $0.5\mu$m particles. However, the concentration of particulate matter at the operator's mouth and nose (A-2) and the center of the kitchen (A-3) did not decrease significantly, and the concentration of particulate matter in some particle size range even increased (see Figure 8 (2) and (3)). The reason for the above phenomenon is that the exhaust direction is the same as the flow direction of the flue gas" hot plume". After increasing the exhaust speed, the rising rate of soot particles generated in the frying pot is accelerated, resulting in the reduction of local concentration. However, the increase in exhaust speed of the range hood is accompanied by the increase in fluid turbulence intensity, which enhances the flow pulsation, thus causing the escape of particles and reducing the control effect of the range hood. At the same time, artificial disturbance in cooking operation will also cause the diffusion of particulate matter to the outside of the exhaust hood. Figure 9 shows the variation range of particle concentration at different locations after increasing exhaust velocity.

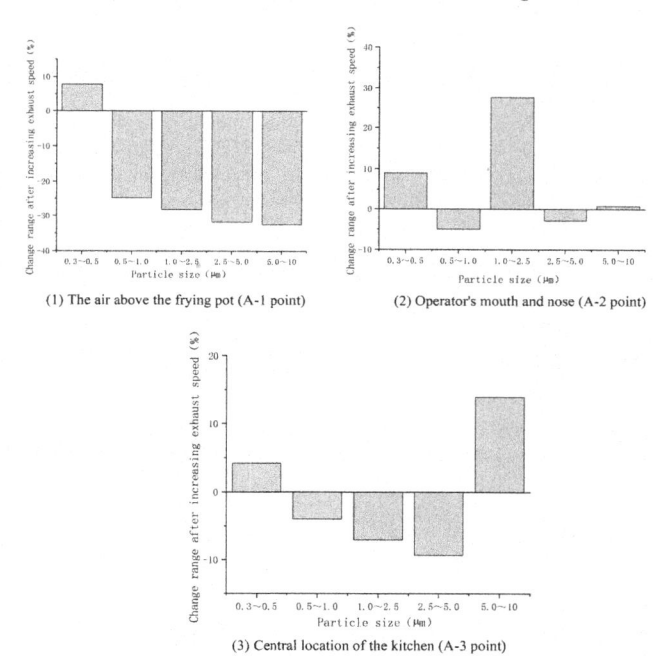

(1) The air above the frying pot (A-1 point)   (2) Operator's mouth and nose (A-2 point)

(3) Central location of the kitchen (A-3 point)

Figure 9.    Influence of increasing exhaust velocity on particle concentration at different locations.

## 4   CONCLUSION

In this paper, the particulates emission characteristics of vegetable, meat, and vegetable plus meat were studied by field measurement method when stir-frying was used as cooking. Through the research of this paper, the following conclusions can be obtained:

(1) The food material had a significant effect on the particle size distribution and time evolution of particles emitted during cooking.
(2) The characteristics of particulate matter emitted by different dishes are different, and the evolution of particulate matter propagation is also different.

(3) Under the test conditions in this paper, improving the exhaust speed of the range hood can significantly reduce the concentration of particulate matter above the frying pot, but the reduction effect on the concentration of particulate matter at the operator's mouth and nose and the center of the kitchen is not obvious.

## ACKNOWLEDGMENTS

This work was financially supported by the Applied Research Project on the distribution regularity of particulate matter in the kitchen and its control mechanism Department of science and technology, Ministry of housing and urban-rural development (2018-R1-015).

## REFERENCES

Cao T, Durbin T D, Russell R L, et al. (2016). Evaluations of in-use emission factors from off-road construction equipment [J]. *Atmospheric Environment*, 147: 234–245.

Faber P, Drewnick F, Borrmann S. (2015). Aerosol particle and trace gas emissions from earthworks, road construction, and asphalt paving in Germany: Emission factors and influence on local air quality[J]. *Atmospheric Environment*, 122:662–671.

Muleski G E, Cowherd C, Kinsey J S. (2005). Particulate emissions from construction activities. [J]. *Journal of th Air & Waste Management Association*, 55(6): 772–783.

Shi H, Wang Y, Huisingh D, et al. (2014). On moving towards an ecologically sound society: with a special focus on preventing future smog crises in China and globally [J]. *Journal of Cleaner Production*, 64:9–12.

U.S. EPA. (1998). *National air pollutant emission trends, procedures document*, EPA-454/R-98-008[R]. Research Triangle Park, NC: U.S. Environmental Protection Agency.

Wang L, Bi X, Liu B, et al. (2017). Source Directional Apportionment of PM2.5 in Heze City[J]. *Research of Environmental Sciences*, 30:1849–1858.

Wang X H, Zhao Q B, Cui H X. (2015). Source analysis of PM2.5 in winter in Suburban Shanghai based on online monitoring. *Journal of Nanjing University* (Natural Sciences), 51(03):517–523.

Xiao M A, Li T W, Simeng M A, et al. (2017). Spatial and temporal distribution and source analysis of components in PM2.5, Handan[J]. *Environmental Chemistry*, 36, 1932–1940.

Zhang S, Wang Y, Li Y, et al. (2017). Spatial distribution of haze pollution and its influencing factors [J]. *China Population Resources & Environment*, 27:15–22.

Zhou Y, Zhang H, Wang Q, et al. (2015). Pollution characteristics and source apportionment of PM2.5 from Qinshan District in Wuhan during the winter[J]. *Environmental Science & Technology*, 38:159–164.

# Effect of different admixtures on early static strength of geopolymer mortar

Yipeng Ning*, Ao Yao*, Erlei Bai*, Zhihang Wang* & Biao Ren*
*Air Force Engineering University, Xi'an, China*

ABSTRACT: Taking geopolymer mortar as the research object, this paper tests its early static strength under the action of different admixtures and analyzes its changes. It is found that different amounts of water-reducing agents, lithium carbonate, triethanolamine, and triisopropanolamine will have an impact on the compressive and flexural strength of geopolymer mortar. Among them, water reducing agent has the greatest impact on geopolymer mortar. Excessive use of water reducing agent will reduce the 3D compressive and flexural strength of mortar; when the dosage of lithium carbonate is 0.3% and 0.5%, the effect of strengthening the early strength of mortar is the best; with the increase of the amount of triethanolamine and triisopropanolamine, the changing trend of the strength curve is first increased and then decreased, and the strength is the highest when the amount of triethanolamine and triisopropanolamine is 0.08%.

## 1 GENERAL INSTRUCTIONS

Cement mortar is a kind of material widely used in construction, transportation, and other fields. Because of its low price and easy access to raw materials, it is mainly used in foundation construction, masonry pouring, and other projects in the process of civil construction. As the most widely used hydraulic cementitious material in the market, cement mortar has a low cost and has a significant impact on the whole society and the national industrial system. However, the manufacturing of cement mortar will consume a lot of energy, and a lot of non-technical minerals such as clay and limestone will be consumed in the production process of cement mortar, Cement mortar is being used by the industrial system with a rapidly increasing consumption. At the same time, the manufacturing of cement mortar will also consume a lot of energy, including power and manpower required in the production process of cement mortar, high pollution emissions in the production process, and a large number of pollutants, including dust and PM2 5. PM10 and other pollutants, which is inconsistent with the relevant concepts of green, low-carbon, energy conservation, and emission reduction we currently advocate, and also runs counter to the current policy of carbon neutralization. In order to solve the corresponding problems of cement mortar, researchers can prepare a cementitious material instead of cement through alkali-activated cementitious material, with the support of certain technical means and using slag and fly ash as the main raw materials. Using geopolymer (Hardjito et al. 2008) as a kind of cementitious material can effectively reduce the pollution of cement mortar in the production process. It can not only achieve the cheap effect economically but also solve the development of green, low-carbon, and environmental protection from the perspective of ecological protection, so as to make the cementitious material go better on the road to sustainable development.

The main materials of geopolymers (Zhang et al. 2018) include silica fume and volcanic ash, among which the products of Al and Si are the main components excited by alkali. Comparing

---
*Corresponding Authors: ningyipeng1998@163.com, 1124038494@qq.com, 1119771625@qq.com, 1305352550@qq.com and 1484531731@qq.com

DOI 10.1201/9781003372417-52

geopolymer mortar with cement mortar, it can be found that the unit weight of raw materials compared with cement mortar is less than that of ordinary portland cement. As a widely used Geopolymer Material (Görhan & Kürklü 2014), the basic unit weight of fly ash is only half of that of 425 cement. Secondly, the volume stability of geopolymer mortar is superior. For conventional cement mortar, during the hydration process after mixing, the temperature change during hydration will lead to thermal expansion and cold contraction caused by a mismatch with the external temperature. Therefore, geopolymer has good volume stability, and its shrinkage capacity is less than that of ordinary portland cement in 1D, 7d, and 28d.

The early strength of mortar is of great significance to the construction performance and structural formwork removal time in the construction field. Improving the early strength of geopolymer mortar can make the surface of geopolymer mortar not dehydrate and sand after pouring because of its low strength. When pouring building foundation and pile foundation, the early strength of geopolymer mortar is too low, resulting in irregular settlement on the upper surface of the foundation, which has a great impact on the appearance of the whole foundation and the bearing capacity of the upper part. The low strength after sand well pouring may also prevent the construction personnel from carrying out relevant work too early on the construction surface, which will affect the overall construction. In view of this, the in-depth study of the early strength of mortar is of great significance for the field of civil engineering and construction and has an indispensable impact on construction safety, construction progress, and quality problems in the construction process.

For geopolymer mortar, hardjito D et al. (Sathonsaowaphak et al. 2009) studied the related properties of geopolymer mortar mixed with low calcium powder in different states; Zhang P et al. (Al-Majidi et al. 2016; Chithambaram et al. 2019; Pan et al. 2009; Panda et al. 2018) studied the relevant properties of geopolymer mortar under the action of fly ash at different temperatures; Panda B et al. (Shadnia et al. 2015; Adak et al. 2014, 2014) proposed and established a strength prediction model for geopolymer mortar at ambient temperature; Zhang H Y et al. (Budh & Warhade 2014; Elyamany et al. 2018; Kantarcı et al. 2019; Sata et al. 2012; Zhang et al. 2016) studied the relevant mechanical properties of different types of geopolymer mortar, including compressive strength, flexural strength, etc., and determined the optimal content through experiments; SATA V et al. (Gouny et al. 2012; Kantarcı et al. 2019; Mermerdaş et al. 2017) carried out a large number of tests to solve the related properties of geopolymer mortar under other salt corrosion. The results show that the mechanical properties of geopolymer mortar for civil structures will be reduced to a certain extent under the action of sulfate corrosion.

However, there is not much research on the early strength of geopolymer mortar under different admixtures. Based on this, this paper studies the early performance strength of geopolymer mortar under the action of different admixtures, in order to obtain the corresponding laws and the change laws of geopolymer mortar with different admixtures.

## 2 TEST MATERIAL

The specific parameters of silica fume used in this test are shown in Table 1.

Table 1. Chemical composition of silica fume.

| Chemical composition | Calcium oxide | Silicon dioxide | Alumina | Ferric oxide | Sulfur trioxide | Magnesium oxide | Potassium oxide | Sodium oxide |
|---|---|---|---|---|---|---|---|---|
| Content (%) | 1.85 | 93.9 | 0.01 | 0.59 | 0.86 | 0.27 | 0.01 | 0.17 |

The slag used in this test is the slag produced by Jinan rouang new building materials Co., Ltd. This type of slag is formed by refining the slag in the blast furnace, finally drying at low temperature, putting it into the ball mill for rough grinding, and putting it into the grinder for fine grinding after the rough grinding of the ball mill. Its main chemical components include silicon dioxide, aluminum oxide, iron oxide, calcium oxide, and other substances. The table shows the specific parameters.

Table 2. Chemical composition of slag.

| Chemical composition | Calcium oxide | Silicon dioxide | Alumina | Ferric oxide | Sulfur trioxide | Magnesium oxide | Potassium oxide | Sodium oxide |
|---|---|---|---|---|---|---|---|---|
| Content (%) | 38.6 | 19.4 | 5.8 | 5.8 | 2.6 | 2.8 | 0.1 | 0.2 |

The sand used in this test is river sand. Table 3 describes the results after screening the residue screen.

Table 3. Sieve residue of sand and gravel.

| Mesh size | Percentage of sieve residue | Cumulative sieve residue percentage |
|---|---|---|
| 1.18 | 11.6 | 11.6 |
| 0.63 | 18.0 | 29.6 |
| 0.315 | 36.4 | 66.0 |
| 0.16 | 32.4 | 98.5 |
| <0.16 | 1.5 | 100 |

## 3 TEST METHOD

In order to explore the effect of additives on the performance of geopolymer mortar, the orthogonal test is set in this test, in which the cementitious material is 10% silica fume and 90% slag, and the alkali activator is sodium hydroxide solution with a concentration of 30%, and the sol ratio is 0.55 (0.58). In the process of preparing geopolymer, the selected cementitious sand ratio is 1:2, that is, 500g cementitious material + 1000g sand, and the mass ratio of water glass and sodium hydroxide solution is 1:1. The water reducer selected in this test is naphthalene water reducer. The size of the test piece is 160 * 40 * 40MM. Four influencing factors are set in this test. The specific parameters of different influencing factors are shown in Table 4, and the grouping of each test is shown in Table 5.

Table 4. Test influencing factors.

| Factor | Water reducing agent (A) | Lithium carbonate (B) | Triethanolamine (C) | Triisopropanolamine (D) |
|---|---|---|---|---|
| 1 | 0.6% | 0.3% | 0.06% | 0.04% |
| 2 | 0.8% | 0.5% | 0.08% | 0.06% |
| 3 | 1% | 0.7% | 0.1% | 0.08% |

## 4 TEST RESULT

The test results are shown in Table 6.

As shown in Figure 1 and Figure 2, in the early compressive strength, 1D strength is mainly affected by water reducing agent, and the strength will change greatly with the change of water

Table 5. Test grouping.

| Number | Water reducing agent (A) | Lithium carbonate (B) | Triethanolamine (C) | Triisopropanolamine (D) |
|---|---|---|---|---|
| Z-1 | A1 (0.6%) | B1 (0.3%) | C1 (0.06%) | D1 (0.04%) |
| Z-2 | A1 (0.6%) | B2 (0.5%) | C2 (0.08%) | D2 (0.06%) |
| Z-3 | A1 (0.6%) | B3 (0.7%) | C3 (0.1%) | D3 (0.08%) |
| Z-4 | A2 (0.8%) | B1 (0.3%) | C2 (0.08%) | D3 (0.08%) |
| Z-5 | A2 (0.8%) | B2 (0.5%) | C3 (0.1%) | D1 (0.04%) |
| Z-6 | A2 (0.8%) | B3 (0.7%) | C1 (0.06%) | D2 (0.06%) |
| Z-7 | A3 (1%) | B1 (0.3%) | C3 (0.1%) | D2 (0.06%) |
| Z-8 | A3 (1%) | B2 (0.5%) | C1 (0.06%) | D3 (0.08%) |
| Z-9 | A3 (1%) | B3 (0.7%) | C2 (0.08%) | D1 (0.04%) |

Table 6. Compressive strength, flexural strength, and fluidity of test piece.

| Number | Mobility (mm) | 1d | | 3d | | 7d | |
|---|---|---|---|---|---|---|---|
| | | Compression | Bending resistance | Compression | Bending resistance | Compression | Bending resistance |
| Z-1 | 126 | 13 | 3.6 | 28.5 | 5.8 | 28.6 | 4 |
| Z-2 | 123 | 12.8 | 3.0 | 27.1 | 5.6 | 30.1 | 3.8 |
| Z-3 | 124 | 9.1 | 2.3 | 24.5 | 5.8 | 25.7 | 4.6 |
| Z-4 | 125 | 10.3 | 2.9 | 29 | 4.9 | 22.5 | 4 |
| Z-5 | 124 | 10.2 | 2.4 | 13.8 | 2.4 | 24.7 | 3.8 |
| Z-6 | 124 | 9.1 | 2.2 | 8.3 | 3 | 27.4 | 4.6 |
| Z-7 | 141 | 12.4 | 2.9 | 7.6 | 2.8 | 18.6 | 6.2 |
| Z-8 | 135 | 12.5 | 2.3 | 8.7 | 3.2 | 26.3 | 6.8 |
| Z-9 | 132 | 10.5 | 2.2 | 12.8 | 2.8 | 31.2 | 5.5 |

reducing agent; The 3D strength is mainly affected by the use of a low amount of water reducing agent, which will improve its 3D compressive strength. When the use amount of water reducing agent is 1%, its compressive strength decreases to 10MPa, reaching the minimum value of compressive strength among many test groups. It can be found that lithium carbonate and triethanolamine have little influence on it by analyzing other influencing factors; For the 7d compressive strength analysis, the water reducing agent can improve the 7d performance when it is used in a large amount. Under the influence of different factors, the response trend of early flexural strength and compressive strength remains the same. Through the analysis of the fluidity of the test piece, it can be found that the early strength has been greatly improved with the increase of the use of a water-reducing agent. The specific parameters can be seen in the three groups of test pieces z-7, Z-8, and Z-9. The average fluidity has increased by 15mm, while the fluidity of the other six test groups has basically remained at about 120mm. When the dosage of lithium carbonate is 0.3% and 0.5%, the effect of enhancing the early strength of mortar is the best.

5 CONCLUSION

The use and amount of water-reducing agents, lithium carbonate, triethanolamine, and triisopropanolamine will have an impact on the compressive and flexural strength of geopolymer mortar. Among them, water reducing agent has the greatest impact on geopolymer mortar. Excessive use of water reducing agent will reduce the 3D compressive and flexural strength of mortar; The early

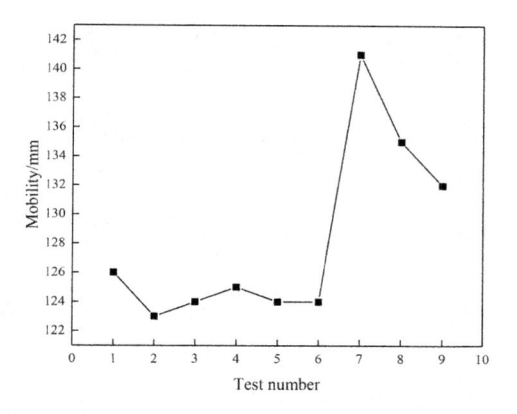

Figure 1.   Bending strength of test piece.

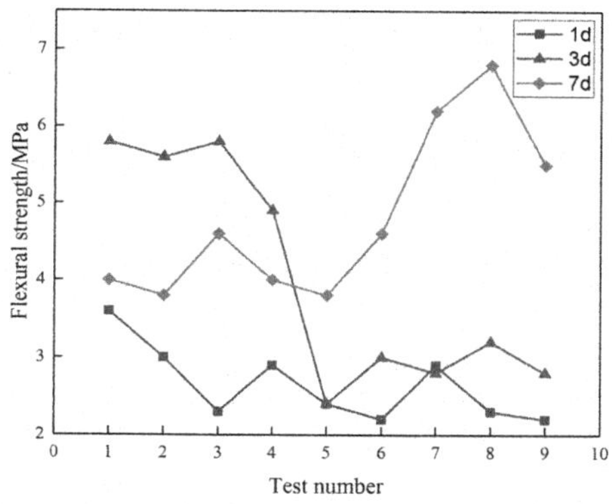

Figure 2.   Compressive strength of test piece.

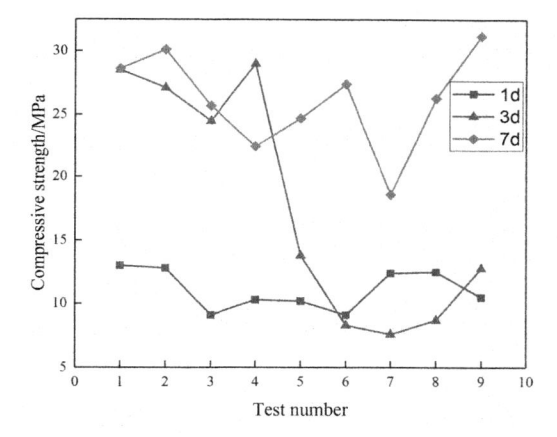

Figure 3.   Mobility of each group.

strength effect of lithium carbonate is better when the dosage is 0.3% and 0.5%; With the increase of the amount of triethanolamine and triisopropanolamine, the changing trend of the strength curve is first increased and then decreased, and the strength is the highest when the amount of triethanolamine and triisopropanolamine is 0.08%. Therefore, in practical work, it is necessary to mainly control the dosage of water-reducing agents and add other admixtures appropriately to ensure the best strength of geopolymer mortar.

## FUNDING STATEMENT

This research was funded by Project approved by the State Civil Air Defense Office, grant number RF20SC01J—S0.

## REFERENCES

Adak D, Sarkar M, Mandal S. (2014). Effect of nano-silica on strength and durability of fly ash-based geopolymer mortar[J]. *Construction and Building Materials*. 70: 453–459.

Adam A Défense, Horianto X Haryanto X. (2014). The effect of temperature and duration of curing on the strength of fly ash-based geopolymer mortar[J]. *Procedia Engineering*. 95: 410–414.

Al-Majidi M H, Lampropoulos A, Cundy A, et al. (2016). Development of geopolymer mortar under ambient temperature for in situ applications[J]. *Construction and Building Materials*. 120: 198–211.

Budh C D, Warhade N R. (2014). Effect of molarity on compressive strength of geopolymer mortar[J]. *International Journal of Civil Engineering Research*. 5(1): 83–86.

Chithambaram S J, Kumar S, Prasad M Chidambaram. (2019). Thermo-mechanical characteristics of geopolymer mortar[J]. *Construction and Building Materials*. 213: 100–108.

Elyamany H E, Abd Elmoaty M, Elshaboury A M. (2018). Magnesium sulfate resistance of geopolymer mortar[J]. *Construction and Building Materials*. 184: 111–127.

Görhan G, Kürklü G. (2014). The influence of the NaOH solution on the properties of the fly ash-based geopolymer mortar cured at different temperatures[J]. *Composites part b: engineering*. 58: 371–377.

Gouny F, Fouchal F, Maillard P, et al. (2012). A geopolymer mortar for wood and earth structures[J]. *Construction and Building Materials*. 36: 188–195.

Hardjito D, Cheak C Cheek, Ing C H L. (2008). Strength and setting times of low calcium fly ash-based geopolymer mortar[J]. *Modern applied science*. 2(4): 3–11.

Jindal B Cheek. (2019). Investigations on the properties of geopolymer mortar and concrete with mineral admixtures: A review[J]. *Construction and Building Materials*. 227: 116644.

Kantarcı F, Türkmen İ, Ekinci E. (2019). Optimization of production parameters of geopolymer mortar and concrete: A comprehensive experimental study[J]. *Construction and Building Materials*. 228: 116770.

Mermerdaş K, Manguri S, Nassani D E, et al. (2017). Effect of aggregate properties on the mechanical and absorption characteristics of geopolymer mortar[J]. *Engineering Science and Technology, an International Journal*. 20(6): 1642–1652.

Pan Z, Sanjayan J G, Rangan B V. (2009). An investigation of the mechanisms for strength gain or loss of geopolymer mortar after exposure to elevated temperature[J]. *Journal of Materials Science*. 44(7): 1873–1880.

Panda B, Paul S C, Mohamed N A N, et al. (2018). Measurement of tensile bond strength of 3D printed geopolymer mortar[J]. *Measurement*. 113: 108–116.

Sata V, Sathonsaowaphak A, Chindaprasirt P. (2012). Resistance of lignite bottom ash geopolymer mortar to sulfate and sulfuric acid attack[J]. *Cement and Concrete Composites*. 34(5): 700–708.

Sathonsaowaphak A, Chindaprasirt P, Pimraksa K. (2009). Workability and strength of lignite bottom ash geopolymer mortar[J]. *Journal of Hazardous Materials*. 168(1): 44–50.

Shadnia R, Zhang L, Li P. (2015). Experimental study of geopolymer mortar with incorporated PCM[J]. *Construction and building materials*. 84: 95–102.

Zhang H Y, Kodur V, Wu B, et al. (2016). Thermal behavior and mechanical properties of geopolymer mortar after exposure to elevated temperatures[J]. *Construction and Building Materials*. 109: 17–24.

Zhang P, Zheng Y, Wang K, et al. (2018). A review on properties of fresh and hardened geopolymer mortar[J]. *Composites Part B: Engineering*. 152: 79–95.

*Civil Engineering and Urban Research – Mohamed & Hou (Eds)*
*© 2023 the Authors, ISBN 978-1-032-44487-1*

# Example analysis of the virtual displacement principle of the rigid body system in structural mechanics

Ting Kang*
*The Aerospace Engineering Institute, Air Force Engineering University, Xi'an, Shaanxi, China*

Qiqi Sun
*Mechanical and Electrical Engineering Institute, Xi'an University of Engineering University, Xi'an, Shaanxi, China*

Erlei Bai & Huixiang Sun
*The Aerospace Engineering Institute, Air Force Engineering University, Xi'an, Shaanxi, China*

ABSTRACT: The application of the principle of virtual displacement is an important content in structural mechanics teaching. The concept of virtual displacement is very abstract, and there is little specific analysis of the concept in general textbooks, which is often difficult for students to master. In this paper, through a simple and easy-to-understand example, the virtual work equation of the rigid body system is derived from the static balance equation. According to the conditions of the virtual work equation, the specific concept of virtual displacement is clarified. The virtual displacement is an infinitesimal quantity of a rigid body system. And it must meet the constraints. Each key problem of the virtual displacement is explained in detail, which simplifies the principle of virtual displacement. Finally, two key problems in the application of the virtual displacement principle, that is, how to remove constraints to obtain an equilibrium force system and how to suppose the virtual displacement, are discussed, and examples are given. This paper can provide references for the majority of teaching workers.

## 1 INTRODUCTION

Using the virtual displacement principle to solve the internal force of statically determinate structure is an important content in the teaching of structural mechanics. The virtual displacement principle has been strictly proved in the textbook of theoretical mechanics. Many scholars have analyzed the application of the virtual displacement principle (He et al 2020; Ou et al 2021; Xue & Su 2017; Wang 2011; Zhou et al 2020), Shu kai'ou and Yanxia Xue et al. (ou et al 2021; Xue & Su 2017) analyzed the virtual displacement calculation method to solve the internal force of statically indeterminate structure, Huanding Wang, Yanqing He and Xinzhu Zhou et al. (He et al 2020; Wang 2011; Zhou et al 2020) analyzed the application of virtual displacement principle in structural internal force calculation through examples.

However, few works of literature introduce the concept of virtual displacement combined with examples and how to suppose the virtual displacement. Therefore, students often feel that the virtual displacement principle is very abstract and difficult to understand. In the teaching of structural mechanics, it is often difficult for students to solve the internal force of the statically determined structures with the virtual displacement principle. This paper will prove the virtual displacement principle of the rigid body system through a simple and easy-to-understand example. The specific

---

*Corresponding Author: bysapple@126.com

DOI 10.1201/9781003372417-53

concept of virtual displacement is clearly put forward according to the virtual work equation. Some key problems in the application of the virtual work principle to solve the internal force of structures are discussed. It is helpful for scholars to deeply understand and apply this principle.

## 2 EXAMPLE ANALYSIS OF THE VIRTUAL DISPLACEMENT PRINCIPLE OF A RIGID BODY SYSTEM

The virtual displacement principle of a rigid body system was derived from the static equilibrium condition through a simple example. Figure 1 (a) shows a simply supported beam AB, on which a vertical concentrated force $F_{p1}$ acts at point C. Now, the question is to calculate the support reaction force $F_{RB}$. According to the static balance principle, the moment balance equation $\sum M_A = 0$ is taken at point B, and it is to obtain

$$F_{RB}l + F_{p1}a = 0 \qquad (1)$$

The support reaction force of the simply supported beam AB can be obtained by solving Equation (1). This is obviously a static calculation problem.

In order to prove the virtual displacement principle, both sides of Equation (1) are divided by l, and the equation $F_{RB} + F_{p1}\frac{a}{l} = 0$ can be obtained. Both sides of the equation are multiplied by any non-zero real number $\delta$ at the same time, and the equation is still valid. Equation (2) is available.

$$F_{RB}\delta + F_{p1}\frac{a}{l}\delta = 0 \qquad (2)$$

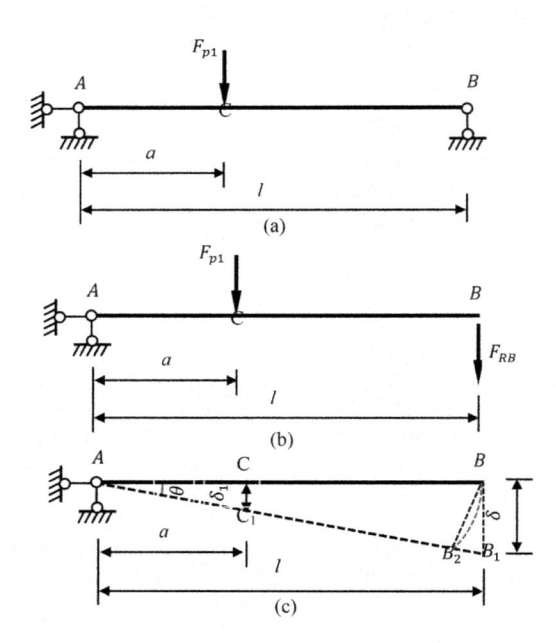

Figure 1. The simply supported beam AB.

Since $\delta$ can be any non-zero real number, it can be assumed that $\delta$ is the displacement in the same direction as the support reaction force $F_{RB}$ after removing the corresponding support shown in Figure 1(c). The simply supported beam AB becomes an unstable system after removing the constraints corresponding to the support reaction force $F_{RB}$ shown in Figure 1(b). Its degree of freedom is 1. The possible displacement of the system is that the beam AB can only rotate around point A. It can be assumed that the downward displacement of point B is $\delta$, then the displacement of point C is $\delta_1$. Thus, the virtual work W of the equilibrium force system in Figure 1 (b) corresponding to the displacement in Figure 1 (c) is:

$$W = F_{RB}\delta + F_{p1}\delta_1 \tag{3}$$

Is virtual work w equal to "0"? What are the conditions? Compared with Equation (2), if $\delta_1 = \frac{a}{l}\delta$, It can be derived that w = 0. What are the conditions for this conclusion $\delta_1 = \frac{a}{l}\delta$?

In Figure 1 (c), when point B rotates around point A to B2, the displacement of point B is the oblique line segment $\overline{BB_2}$ but not the vertical line segment $\overline{BB_1}$. The relationship between them is $\overline{BB_2} = \overline{BB_1}\cos\theta$. When $\theta \to 0$, $\overline{BB_2} = \overline{BB_1}$ can be obtained. This is the first condition. That is, when the displacement is an infinitesimal quantity, the displacement $\delta$ of point B is vertically downward, and the displacement $\delta_1$ of point C is also vertically downward. In addition, the rod AB is a rigid body and there is no strain during rotation around A, and $\triangle ACC_1 \cong \triangle ABB_1$ can be derived, so the relation $\delta_1 = \frac{a}{l}\delta$ can be obtained. By substituting it into equation (3) and comparing it with equation (2), the virtual work $W = 0$ is obtained.

It can be seen that virtual displacement has three characteristics. It must meet the constraints of the unstable system and may occur, and it is an infinitesimal displacement, and it is a displacement of a rigid body system. Firstly, assume the virtual displacement $\delta$. Secondly, the displacement $\delta_1$ corresponding to the external load $F_{p1}$ can be calculated according to the geometric conditions. Finally, the support reaction force can be calculated according to the virtual work equation. This is the virtual work principle of a rigid body system, which transforms the original mechanical calculation problem into a geometric calculation problem.

From the above analysis, it can be seen that the virtual work equation and the static equilibrium equation are equivalent. The concept of virtual displacement is specific and clear, which is much simpler than the abstract text proof in the textbook of theoretical mechanics.

## 3   KEY PROBLEMS IN THE APPLICATION OF THE VIRTUAL DISPLACEMENT PRINCIPLE

Two key problems in the application of the virtual displacement principle are removing constraints and assuming the virtual displacement. The constraint corresponding to the force to be calculated will be removed. After removing the corresponding constraint, the statically determinate structure becomes a geometrically unstable system. The force to be calculated must be added to the system. And a balance force system is formed together with the external load. If the sectional shear force is to be calculated, the constraint against shear deformation will be removed. If the sectional bending moment is to be calculated, the constraint against bending deformation will be removed.

How to assume the virtual displacement? After removing the constraint corresponding to the force to be calculated, the original statically determinate structure becomes a geometrically unstable system with a degree of freedom. The correspondence here refers that the dot product of the force and the displacement is work. In order to simplify the calculation, its direction is often assumed to be consistent with the direction of the force to be calculated.

Taking the multi-span beam as an example shown in Figure 2(a), the bending moment M of section D is to be calculated. There are two key steps to solve with the virtual displacement principle. The first step is removing the constraint against bending deformation. The rigid joint at section D becomes the hinge joint shown in Figure 2(b). The second step is to analyze the possible movement of the unstable system. The beam AC cannot move. The rod CD can rotate around Point C, and the angle is $\theta_A$. The rod BD can rotate around Point B, and the angle is $\theta_B$. Point D can move slightly in the vertical direction. The virtual displacement $\delta$ corresponding to $M_D$ is the relative rotation angle of sections on both sides at point D. The virtual work equation $q \cdot \frac{l_3\Delta}{2} + M_D\delta = 0$ is established. $BB_1 = l_3\delta$ can be calculated due to the micro-displacement $\delta$.

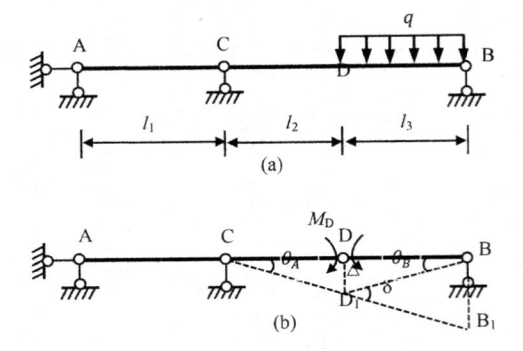

Figure 2.  The multi span beam.

The vertical displacement of point D is $\Delta = \frac{l_2 l_3}{l_2 + l_3}\delta$. It can be calculated that the bending moment $M_D$ of section D is $M_D = -\frac{q l_2 l_3^2}{2(l_2 + l_3)}$, where the sign '-' indicates that the real direction is opposite to that assumed in Figure 2(b). In fact, the underside of section D is tensioned.

## 4   CONCLUSIONS

The principle of virtual displacement is analyzed through the simple and understandable examples in this paper. Two difficulties: removing constraints and assuming the virtual displacement are discussed. And the specific meaning of virtual displacement is clarified. This paper is hoped to provide a reference for the majority of structural mechanics teaching workers.

REFERENCES

Huanding Wang. (2011) Another discussion on the virtual work principle of the deformable body. *Mechanics in Engineering*, 39(2):93–95.
Shu Kai ou, Zitao Guo, Zhong Yang, et al. (2021) Analysis of virtual displacement principle of statically determinate structure equilibrium problem based on kinematics. *Mechanics in Engineering*, 43(6): 976–980.
Xinzhu Zhou, LinJiahui, Zhewen Hu, et al. (2020)*Application of the principle of virtual work to the computation of internal forces of structural bars*, 48(2):212–216.
Yanqing He, Jiajun Wu, Xiangyue Li, et al. (2020) Application of virtual displacement principle in solving internal force equation, *Journal of Taizhou University*, 42(6):58–61.
Yanxia Xue, Zhenchao Su. (2017) Solutions to internal forces of bars of a planar truss by the principle of virtual displacement. *Mechanics in Engineering*, 39(6): 627–628.

*Urban planning and construction and environmental engineering analysis and management*

*Civil Engineering and Urban Research – Mohamed & Hou (Eds)*
© 2023 the Authors, ISBN 978-1-032-44487-1

# Remediation of different concentrations of cadmium and lead-polluted water by *Spathiphyllum kochii*

Mengjie Lou, Kaiweng Li, Shukang Liu, Changzhe Luo & Junying Zhao*
*School of Environmental Science and Safety Engineering, Tianjin University of Technology, Tianjin, China*

ABSTRACT: In view of the current difficult situation of heavy metal polluted water treatment, the remediation effect of *Spathiphyllum kochii* in heavy metal polluted water was analyzed, and the removal efficiency and accumulation characteristics of *Spathiphyllum kochii* were further explored by setting different gradient concentrations of Cd and Pb. The results showed that the removal efficiency of Cd and Pb both decreased by increasing of $Cd^{2+}$ and $Pb^{2+}$ concentration in the solutions and *Spathiphyllum kochii* exposed to 5 mg/L $Cd^{2+}$ exhibited the greatest Cd removal rate with 97.5%. The potential of *Spathiphyllum kochii* to accumulate of Pb and Cd in its root was greater than in shoots. The amount of BCF for Cd and Pb was more than 1, but the amount of TF of all samples were less than 1. It was concluded that *Spathiphyllum kochii* had the potential to remove Cd and Pb from contaminated water. This study has certain reference significance for further promoting and developing plant intervention in the treatment of heavy metal-polluted water environment. With the deepening of practice, it will have a profound impact on environmental protection and governance.

## 1 INTRODUCTION

Heavy metal contamination in water environment is a worldwide concern. Heavy metals come from industrial activities such as manufacturing, mining, metallurgy, chemical industry, and agricultural activities such as the application of a large number of chemical fertilizers and pesticides, which causes heavy metal pollution in water environment in China. The pollution rate of sediments in rivers, lakes, and reservoirs is as high as 0.1% (Chengfang et al. 2021). Moreover, these heavy metals are not easily degraded in the environment, and produce a high risk to human health and safety through the food chain. Therefore, how to remove heavy metals from aqueous system is one of the hot issues. Traditional repair methods include precipitation, ion exchange, adsorption, and so on (He et al. 2008; Zhou et al. 2021). These ways have high costs and sometimes lead to secondary pollution. Phytoremediation is an environment-friendly biotechnology that can use plants to absorb, transfer, or transform pollutants to reduce the harm of pollutants to the ecological environment and human health (Lin et al. 2021). The key to phytoremediation is to screen plants that are high tolerance to pollutants and can remove pollutants from aqueous system or soil system significantly. Hyperaccumulators are the most appropriate plants for phytoremediation technology. However, hyperaccumulators have strong selectivity, such as As hyperaccumulator *Pteris vittata* L. (Lena et al. 2001), Cr hyperaccumulator *Leersia hexandra* Swartz (Wen et al. 2018), Zn hyperaccumulator *Sedum alfredii* Hance (Yang et al. 2002), Mn hyperaccumulator *Phytolacca americana* L., Cr hyperaccumulator *Solanum nigrum* L. and so on (Tie et al. 2005; Wei et al. 2004). The growth of these plants is affected by geographical conditions. Screening hyperaccumulators from ornamental plants to remediate heavy metal-contaminated water is low-cost and conducive. In the

---

*Corresponding Author: zhao_jy2@163.com

studies reported, *Chlorophytum comosum* can tolerate Pb under hydroponic conditions. With the extension of exposure time, the bioconcentration factors of Pb in roots decreased and those in leaves increased, which can be candidate forremediation of Pb-contaminated wastewater (Tang et al. 2018). *Scindapsus aureum* has the capacity of accumulating Mn from water and its transfer coefficient of Mn increases with the increase of exposure concentration. It can be used for Mn pollution remediation, which also has a certain removal effect on N and P from it (Chen et al. 2011; Ye et al. 2020).

*Spathiphyllum kochii* is a perennial ornamental plant. Nevertheless, there is limited information reported about the removal and accumulation characteristics of Pb and Cd. Therefore, this work aimed to evaluate the Pb and Cd removal rate with different heavy metal concentrations, and the accumulation content in the shoot and root of *Spathiphyllum kochii*.

## 2 MATERIALS AND METHODS

### 2.1 *Experimental materials*

*Spathiphyllum kochii* used in this research were purchased from a flower market in Tianjin. *Spathiphyllum kochii* were rinsed with water and then washed with deionized water. After about a week of cultivating in deionized water, the *Spathiphyllum kochii* seedlings with roughly the same height were selected for the experiment.

In addition, both Pb nitrate and Cd nitrate were analytically pure which was used in the experiment.

### 2.2 *Experimental design*

Dissolve Pb $(NO_3)_2$ and Cd $(NO_3)_2$ with deionized water, and the volume of the solution was 500 ml. The Pb concentration gradient in the cultured solution was 0 mg/L, 4 mg/l, 7 mg/l, 10 mg/l (calculated as $Pb^{2+}$), and the Cd concentration gradient was 0 mg/L, 5 mg/l, 10 mg/l, and 25 mg/l (calculated as $Cr^{2+}$). Deionized water without Pb and Cd was used as the blank control (CK). Each concentration gradient was repeated in 3 groups, and each treatment was equipped with two *Spathiphyllum kochii* seedlings to ensure that the biomass of the three replicates of each gradient was similar. The experiment lasted 10 days.

### 2.3 *Processing and determination of plant samples*

After 10 days, parts of the roots and shoots of each group of *Spathiphyllum kochii* were cut off. They were cleaned with deionized water. Subsequently, the samples were dried to a constant weight by the electric oven at the temperature of 70°C for three days. Finally, they were ground and passed through a 150-mesh sieve for use. A total of 2 mg of each sample was weighed, and 8 ml of $HNO_3$–$HClO_4$ solution with a ratio of 5: 1 was added. The prepared mixture was placed on an electrothermal digestion instrument for digestion and acid removal. At last, the mixture was diluted to 10 ml with distilled water.

ICP-OES was used to determine the content of total Pb and Cd in different parts of *Spathiphyllum kochii*. In order to ensure the quality of the test, national standard plant samples were used for controlling quality during the analysis process.

$$C_p = \frac{C_w \times V}{W_p} \times 1000 \tag{1}$$

where:

$C_p$: the concentration of total lead or cadmium in plants (mg/kg);

$C_w$: the concentration of total lead or cadmium in the water sample after digestion (mg/L);

V: the volume of constant volume after cooking (ml);
$W_p$: the mass of the digested plant sample (g);

## 2.4 *Determination of Pb and Cd in water*

The content of Pb and Cd in the solution was determined by an inductively coupled plasma atomic emission spectrometer.

## 2.5 *Evaluation of plant enrichment capacity*

The ability of plants to bioaccumulate heavy metals was evaluated by Bioconcentration Factor (BCF) and Transfer Factor (TF).

BCF is the ratio of the content of heavy metals in the plant to that in the cultured solution. The BCF is expressed by the formula as $BF = C_p/C_s$, where $C_p$ is the concentration of $Pb^{2+}$ or $Cd^{2+}$ in *Spathiphyllum kochii* and $C_s$ is the concentration in cultured solution.

TF refers to the ratio of heavy metal content in the aerial part of plants to that in the roots of plants. TF is expressed by the formula as $TF = C_1/C_2$. In the formula, $C_1$ and $C_2$ are $Pb^{2+}$ or $Cd^{2+}$ concentrations (mg/kg) in the aerial part and roots of *Spathiphyllum kochii*. If the transfer coefficient is greater than 1, it indicates that the plant has the characteristics of hyperaccumulation of this heavy metal (Chancy 1997).

$$R = \frac{C_1 - C_2}{C_1} \times 100\% \qquad (2)$$

where:

R: the removal rate for the heavy metal;
$C_1$: the initial concentration of Pb or Cd in each treatment (mg/L);
$C_2$: the concentration of Pb or Cd in each treatment after the experiment (mg/L);

## 3 RESULTS AND DISCUSSION

### 3.1 *Removal rate of Pb and Cd by Spathiphyllum kochii*

The content of Cd and Pb in the solutions after ten days were measured and the removal efficiency of Cd and Pb from water was calculated (Table 1).

Table 1. Removal rate of Cd and Pb by *Spathiphyllum kochii*.

| Cd treatment (mg/L) | Cd concentration after 10d (mg/L) | Cd removal rate (%) | Pb treatment (mg/L) | Pb concentration after 10d (mg/L) | Pb removal rate (%) |
|---|---|---|---|---|---|
| $Cd^{2+}$-5 mg/L | $0.1224 \pm 0.0105$ | 97.5 | $Pb^{2+}$-4 mg/L | $0.0453 \pm 0.0095$ | 88.7 |
| $Cd^{2+}$-10 mg/L | $1.8042 \pm 0.5131$ | 81.9 | $Pb^{2+}$-7 mg/L | $1.2982 \pm 0.0803$ | 81.5 |
| $Cd^{2+}$-25 mg/L | $9.7015 \pm 2.1401$ | 61.2 | $Pb^{2+}$-10 mg/L | $3.9288 \pm 0.9115$ | 60.7 |

As can be seen in Table 1, the removal rates of Pb from water by *Spathiphyllum kochii* are from 60.7% to 88.7% and those of Cd are from 61.2% to 97.5%. They decrease with the increase of heavy metal concentrations. The efficiency of heavy metal remediation by aquatic plants is affected by heavy metal concentration, exposure time, the form of heavy metals and the interaction between heavy metals, the amount of plant and plant age, pH, Eh, temperature, light, and other factors (Jiang et al. 2016). In this study, when the content of $Cd^{2+}$ in solutions was the same as $Pb^{2+}$, the removal efficiency of Cd by *Spathiphyllum kochii* was higher than that of Pb, which indicates that the uptake amount of Cd by *Spathiphyllum kochii* is higher. The results show that *Spathiphyllum kochii* is a good candidate for remediating water polluted by Cd and Pb.

### 3.2 *Bioaccumulation characteristics of Pb and Cd in water by Spathiphyllum kochii*

The content BCFs and TFs for Pb and Cd in different parts of *Spathiphyllum kochii* are shown in Figures 1 and 2.

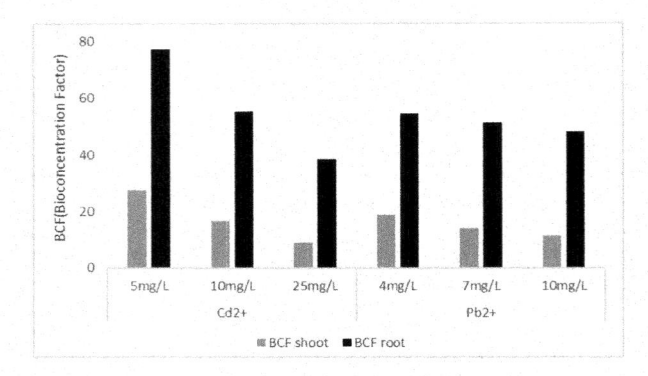

Figure 1.   Bioaccumulation Factor of *Spathiphyllum kochii* under different $Cd^{2+}$ and $Pb^{2+}$ concentrations.

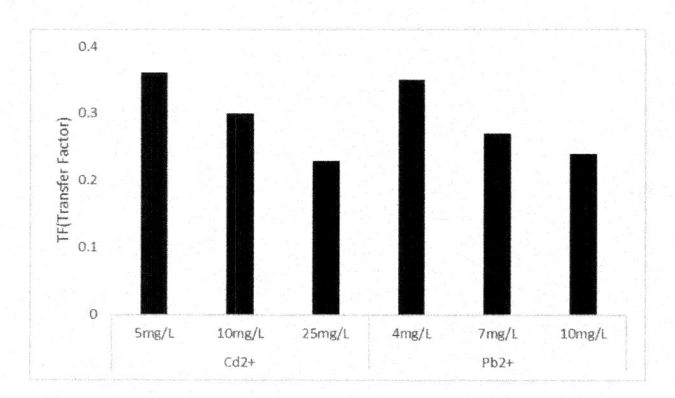

Figure 2.   Transfer Factor of *Spathiphyllum kochii* under different $Cd^{2+}$ and $Pb^{2+}$ concentrations.

### 3.2.1 *Bioaccumulation of Cd and Pb in Spathiphyllum kochii*
The amount of accumulation of Cd and Pb in the aerial parts and roots of *Spathiphyllum kochii* increased with the increase of $Pb^{2+}$ and $Cd^{2+}$ concentration and the accumulation amount in the roots is much greater than that in the aerial part. The amount of Cd accumulation in plants is involved in the concentration of Cd and its biological effectiveness, pH, soil redox potential, temperature, and other element concentrations (Li et al. 2006), and the quantities of Pb uptake by aquatic plants are related to plant species, pH, temperature, concentration and other factors (Chu & Zhao 2014). The results in this experiment testify again that aquatic plants accumulated more with higher heavy metal concentrations (below the lethal threshold) ability.

BCF is the ratio of the heavy metal content in the plant to the heavy metal concentration in the solution, which is used to express the enrichment ability of different parts of *Spathiphyllum kochii*. BCF of *Spathiphyllum kochii* decreased with the increase of heavy metal concentration, indicating that the increased times of Cd and Pb accumulation in roots is smaller than the increased times of Cd and Pb concentration in the cultured solutions. Hyperaccumulators should have a high tolerance to heavy metals and their BCFs are dozens or even tens of thousands of times higher than that of ordinary plants (Xie et al. 2020). BCFs of Cd and Pb are all greater than 1.0, which reaches the standard that the BCF of Cd and Pb for hyperaccumulators should be greater than 1.0.

The removal efficiency of Cd by *Spathiphyllum kochii* is higher than that of Pb when *Spathiphyllum kochii* is in the solution with the same concentration of 10 mg/L $Pb^{2+}$ and $Cd^{2+}$. BCF for Cd by *Spathiphyllum kochii* is also higher than that of Pb and the accumulation amount of Cd in plant body also bigger than that of Pb. Plants can show certain avoidance and tolerance under heavy metal stress (Huang et al. 2006). Avoidance means that plants do not absorb harmful substances in the environment so that plants maintain healthy growth. Tolerance means that plants can absorb harmful substances and survive by detoxification mechanism. *Spathiphyllum kochii* can grow healthily in a higher concentration of Cd solution, indicating that it survives in high Pb stress, and it is not carried out in a high avoidance way, but through some detoxification method to reduce the toxicity of Cd to the plants.

### 3.2.2 *The ability of to transfer Cd and Pb*

As shown in Table 2 and Table 3, *Spathiphyllum kochii* concentrates a large amount of the metals in its root. The content of Pb in the root of *Spathiphyllum kochii* accounts for 74.3%–90.8% of the total accumulation and the content of Cd accounts for 73.8%–81.1%. TF reflects the plant's ability to transfer heavy metals from the roots to the aerial part. The higher the value is, the stronger the plant's ability to transfer heavy metals from the roots to the aerial part (Zhu et al. 2014). TFs for Cd and Pb of *Spathiphyllum kochii* are between 0.23 and 0.36 which are all less than 1. This indicates that the plant has not been successful in transporting Pb and Cd to the aerial part.

The maximum translocation factor of Pb by *Spathiphyllum kochii* is 0.35 when it is in 500 ml solution with 4 mg/L $Pb^{2+}$. The minimum translocation factor of Cd by *Spathiphyllum kochii* is 0.23 when it is in 500 ml solution with 25 mg/L $Cd^{2+}$. The translocation factors are less than 1 and they decreased with the increase of the heavy metal concentration. This may be because when the stress concentration of heavy metals increases, in order to reduce its toxic effect, *Spathiphyllum kochii* combines with excess nonprotein substances in cells after the heavy metals entered the roots to form complexes, most of which are bound to the roots, and rarely to other organs (Stolts and Gregor 2004).

## 4 CONCLUSIONS

In the present study, it was concluded that the removal rate decreases with the increase of $Pb^{2+}$ and $Cd^{2+}$ concentration. *Spathiphyllum kochii* has the highest removal rate at 97.5% in the solution with 5 mg/L $Cd^{2+}$ and the lowest removal rate at 60.7% in the solution with 10 mg/L $Pb^{2+}$. *Spathiphyllum kochii* showed good potential to accumulate Cd and Pb in its root and the number of its BCFs for the elements is more than 1. When $Cd^{2+}$ concentration in solution is the same as $Pb^{2+}$, *Spathiphyllum kochii* has a strong ability to absorb Cd. The translocation factors of all treatments are less than 1, indicating that although *Spathiphyllum kochii* is not a hyperaccumulator of Cd and Pb, it still has the potential to treat Cd or Pb-contaminated water. The effect of the removal of Cd and Pb from soil by *Spathiphyllum kochii* requires further investigation.

## ACKNOWLEDGEMENTS

The work was funded by the College Students Innovation and Entrepreneurship Training Program Project, Tianjin University of Technology College (No. 202110060027) and by the Open Fund of Key Laboratory for Environmental Factors Control of Agro-Product Quality Safety, Ministry of Agriculture and Rural Affairs (2020-hjyzfp-70201801).

# REFERENCES

Chancy R. L. & Malik M. & Y. M. Li (1997). Phytoremediation of soil metals. *J. Current Opinions in Biotechnology*, (8): 279.

Chen Y. Y. & X. Cui & B. Dong, et al. (2011). Study on Purification Effect of Experimental Wastewater by Three Aqua cultured Ornamental Plants. *J. Journal of Soil and Water Conservation*, 25 (2): 253–257.

Chengfang B. B. & T. T. Gan & N. J. Zhao (2021). Adsorption Characteristics of Four Freshwater Microalgae to Heavy Metal Lead in Water. *J. Environmental Science and Technology*, 34 (06): 1–6.

Chu Y. D. & S. M. Zhao (2014). The Effects of Lead on Aquatic Plants. *J. Guangdong Chemical Industry*, 41 (16): 98–99.

He G. & C. G. Geng & R. Luo (2008). The Effect of Heavy Metals on Water Plants and Its Control. *J. Guizhou Agricultural Sciences*, 36 (3): 147–150+153.

Huang Y. J. & D. Y. Liu & Y. B. Wang, et al. (2006). Heavy metals accumulation by hydrophytes. *J. Chinese Journal of Ecology*, (05): 541, 545.

Jiang X. & C. Wen & S. S. Cao, et al. (2016). Research progress on the phytoremediation of water bodies contaminated by heavy metals. *J. Applied Chemical Industry*, 45 (10): 1982–1985.

Lena Q. & Ma Kenneth & M. Komar, T. Cong, et al. (2001). A fern that hyperaccumulates arsenic. *Nature*, 409 (6820): 579–579.

Li H. F. & Y. S. Bai & W. H. Fan, et al. (2006). Influence of Selenium on Germination Percentage of Soybeans Seeds. *J. Journal of Shanxi Agricultural University: natural Science Edition*, 26 (3): 256–258.

Lin H. & C. J. Liu & B. Li, et al. (2021). *Trifolium repens* L. regulated phytoremediation of heavy metal contaminated soil by promoting soil enzyme activities and beneficial rhizosphere associated microorganisms. *Journal Hazard Materials*. 402, 123829, ISSN 0304-3894

Stolts E. & Gregor M. (2004). Accumulation properties of As, Cd, Pb and Zn by four wetland species growing in submerged mine tailings. *J. Environmental and Experimental Botany*, 47 (3): 27128.

Tang Y. Y. & Y. N. Gui & Y. B. Wang, et al. (2018). Tolerance and uptake ability of *Chlorophytum comosum* in Pb-polluted water. *J. Journal of Shanghai Jiao Tong University* (Agricultural Science), 36 (4): 89–94.

Tie B. Q. & M. Yuan & M. Z. Tang (2005). *Phytolacca americana* L.: A new manganese accumulator plant. *Journal of Agro-Environment Science*, 24 (2): 340–343.

Wei S. H. & Q. X. Zhou & X. Wang, et al. (2004). A newly discovered Cd hyperaccumulation plant *Solanum nigrum* L. *J. Chinese Science Bulletin*, 49 (24): 2568–2573.

Wen X. B. & X. H. Zhang & J. Liu (2018). A Comparative Study on the Disposal of Harvested Products of Cr Hyper – accumulator *Leersia hexandra Swartz* by Incineration and Pyrolysis. *J. Industrial Safety and Environmental Protection*, 44 (03): 73–77.

Xie Y. H. & X. H. Ji & J. M. Wu, et al. (2020). The "Three Highs" Hyperaccumulators Screening and Repair Cost Analysis of Cadmium and Arsenic Contaminate Soil. *J. Environmental science & technology*, (S1): 116,121.

Yang X. & X. X. Long & W. Z. Ni, et al. (2002). *Sedum alfredii* H: A new Zn hyperaccumulating plant first found in China. *Chinese Science Bulletin*, 47 (19): 1634–1637.

Ye M. J. & Y. Y. Liu & Y. M. Yang, et al. (2020). Effects of Three Hydroponic Plants on the Growth, Absorption and Enrichment Under Heavy Metal Mn Stress. *J. Journal of Chengdu Normal University*, 36 (11): 103–113.

Zhou X. Y. & X. Y. Han & H. Luo (2021). Research Situation and Treatment Technology on Heavy Metal Pollution in Water Environment. *J. Guangdong Chemical Industry*, 48 (19): 128–128141.

Zhu X. Q & C. Y. Wang & H. Chen, et al. (2014). Effects of arbuscular mycorrhizal fungi on photosynthesis, carbon content. and calorific value of black locust seedlings. *Photosynthetica*, 52 (2): 247–25.

*Civil Engineering and Urban Research – Mohamed & Hou (Eds)*
© *2023 the Authors, ISBN 978-1-032-44487-1*

# Park green space from the perspective of an aging society: Taking Dongguan Botanical Garden Park as an example

Rongbing Mu*
*Guangdong University of Science and Technology, Dongguan, China*

Yuanlong Tan*
*Guangdong University of Science and Technology, City University of Macau, Dongguan, China*

ABSTRACT: The phenomenon of aging is one of the characteristics of modern social phenomena. As one of the important places for outdoor activities and leisure for the elderly, park green space assumes an important role in satisfying the spiritual and material needs of the elderly. The purpose of this paper is to investigate the accessibility analysis of park green spaces and accessibility strategies based on the social perspective of the aging population, taking the Dongguan Botanical Garden Park as a research model, and based on the needs of the elderly in terms of their behavioral patterns and landscape preferences. This paper aims to provide ideas for parkland to better serve the development needs of the elderly. It is also important to improve the use of parkland and the layout of urban parks to build a harmonious and comfortable aging society.

## 1 INTRODUCTION

Data from the 7th National Population Census shows that the proportion of elderly people aged 65 and above in China will be about 13.5% in 2020, and China will officially enter an aging society soon. According to the World Population Ageing 2019 (United Nations 2019) published by the United Nations, China's population enters a phase of accelerated aging at the beginning of the 21st century. As a globally populous country, China's total population growth has slowed down in recent years while population aging has continued to deepen. It is urgent to optimize the living and activity space for the elderly. Optimizing the accessibility of green space in response to the behavior of the elderly can help them improve their quality of life.

## 2 ANALYSIS OF THE CURRENT SITUATION OF DONGGUAN'S AGEING SOCIETY

### 2.1 Current research

Dongguan, located in the Pearl River Delta region of Guangdong Province, is an industrial city centered on manufacturing development and is one of the largest manufacturing bases in the world. Guangdong Province is a major labor-importing province in China, and the influx of foreign labor has resulted in an aging population that is significantly lower than the national average, while Dongguan has one of the youngest population age structures in Guangdong Province. With the massive importation of labor, the aging population in Dongguan has not been given sufficient attention. Dongguan's household population is small and has long maintained a low growth rate, and the foreign population has been the main source of population float data for the city.

---

*Corresponding Authors: 985613213@qq.com and U21092120241@cityu.mo

Further collation of the data shows that the internal structure of Dongguan's working-age population is quietly changing, with the internal structure of the working-age population aging significantly. In recent years, the proportion of Dongguan's resident population aged 15 to 24 has declined significantly, the proportion of people aged 25 to 29 has decreased, while the proportion of people aged 40 to 54 has been increasing (Figure 1), reflecting the accelerated rate of aging and the seriousness of the aging problem in the city of Dongguan (Michael J. Annear, Grant Cushman 2014).

| Year | Age Structure Chart (%) | | | | | | | | | |
|------|-------|-------|-------|-------|-------|-------|-------|-------|-------|-------|
| | 15~19 | 20~24 | 25~29 | 30~34 | 35~39 | 40~44 | 45~49 | 50~54 | 55~59 | 60~64 |
| 1987 | 16.29 | 16.80 | 11.50 | 13.44 | 11.53 | 8.02 | 6.27 | 5.90 | 5.80 | 4.44 |
| 1995 | 21.64 | 24.27 | 13.29 | 9.14 | 8.94 | 6.59 | 4.43 | 4.00 | 4.06 | 3.61 |
| 2005 | 16.61 | 27.05 | 18.38 | 15.40 | 9.93 | 5.74 | 2.47 | 2.23 | 1.34 | 0.86 |
| 2010 | 10.36 | 21.95 | 17.84 | 14.55 | 13.19 | 9.96 | 5.87 | 2.68 | 2.20 | 1.38 |
| 2015 | 7.61 | 12.60 | 16.76 | 15.11 | 12.81 | 12.94 | 10.22 | 6.14 | 3.14 | 2.66 |

Source: 1987 data from "China 1987 1% Population Sample Survey (Guangdong Province Sub-book)", 1995 data from "1995 National 1% Population Sample Survey (Guangdong Province Sub-book)", 2005 data from "2005 Guangdong Province National 1% Population Sample Survey", 2010 data from "Dongguan City 2010 Population Census", 2015 data from "2015 Guangdong Province 1% Population Sample Survey".

Figure 1. The internal age structure of the resident population aged 15−64 in Dongguan over time.

## 2.2 *Behavioral patterns of older people*

With the serious problem of aging in society, the proportion of older people has increased significantly. The creation of behavioral patterns and habits of older people is inextricably linked to their psychological and physiological behavior. Changes in the social and family roles of older people are the main external and internal factors affecting the physiological and psychological situation of older people. Firstly, older people have the following physiological characteristics: deterioration of the five senses, deterioration of muscular abilities, reduced mobility, and reduced sensitivity to external perceptions. Secondly, old people have a sense of insecurity, loss, loneliness, and place attachment. Psychological factors and physiology should be considered and researched in the relevant design. The landscape design should consider the relevant factors for the elderly. For example, the acoustic environment, the light environment, and accessibility need to be fully considered (Barbosa O et al. 1997).

## 2.3 *Green space preferences for older people in parks*

Based on the psychological and physiological characteristics of the elderly, research on the behavioral preferences of the elderly was conducted as a reference factor for green space preference selection. The behavior of elderly people who move around in the green spaces of Dongguan Botanical Garden Park was categorized and studied. Among them, walking and sightseeing were the most frequent but only accounted for 20% of the total. Besides, there were relatively more behaviors such as exercising, chatting, singing, and dancing. The statistics show that the behavioral activities of older people are predominantly spontaneous, with over 70%, while social behavior is relatively low, at less than 30% as shown in (Figure 3). Based on the data and the analysis of the physiological and psychological characteristics of older people, older people arrive at parks and green spaces with a high frequency of activity and at a more regular time. Due to the deterioration of their physiological muscles, most older people prefer bright and open public green space facilities, such as open woodland cover and large open water areas. These landscape preferences can be used to inform the design of parkland accessibility routes (Youssoufi S et al. 2013).

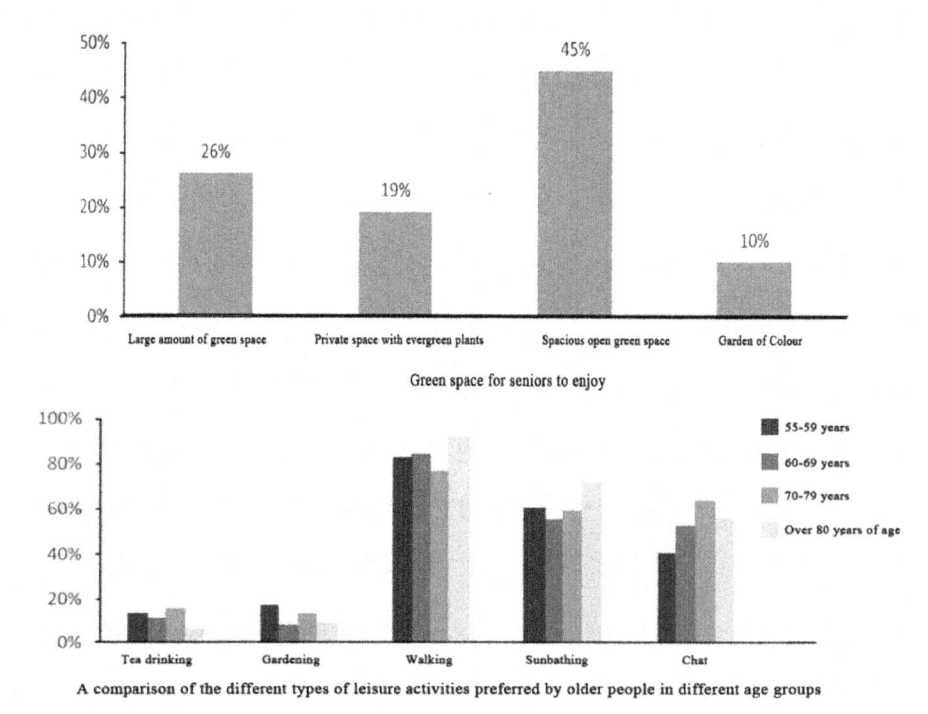

A comparison of the different types of leisure activities preferred by older people in different age groups

Figure 2. Survey on the behavioral preferences of older people for park activities.

## 3 PARK GREEN SPACE BASED ON THE PERSPECTIVE OF AGEING SOCIETY-DONGGUAN BOTANICAL GARDEN AS AN EXAMPLE

### 3.1 *Analysis of the current situation of green space in Dongguan Botanical Garden Park*

In addition to the development of industries, Dongguan also focuses on the construction of the environment. The city is positioned as a "national manufacturing city and a modern ecological city". The city is committed to building an ecological and economical urban gardening system to protect and improve the ecological environment. Dongguan Botanical Garden, also known as Green World, is located in the southern city of Dongguan. Located in the southern city of Dongguan, the botanical garden has a planned area of 203 hm². At present, the main tourist area is about 43 hm², while the rest of the area is ecological public welfare forest and scientific research land. The park's recreational lawn spaces are distributed throughout the park, with different types of recreational lawn spaces staggered and the overall arrangement well-proportioned.

### 3.2 *Frequency and mode of travel to parks and green spaces by older people*

Older people of all ages are habitually active in the park's green spaces. Younger seniors under the age of 70 make up most of the population. There are also fewer intermediated older people in park green spaces, and no intermediated older people are present. Among the public green space travel choices of older people, pleasure gardens, integrated parks, and community parks are the most common types of park green space chosen by older people. Older people mainly walk to park green spaces. The length of time spent in a park varies by age group. Walking time of 0-15 minutes is the most comfortable length of trip for older people of all ages. In addition, the maximum acceptable length of time for older people to reach different types of public green space varies. The longest acceptable walking time for older people to reach different types of public green spaces is shown in (Figure 3), which will be used as a measure to calculate the accessibility of public green spaces for older people (Barbosa O 2007).

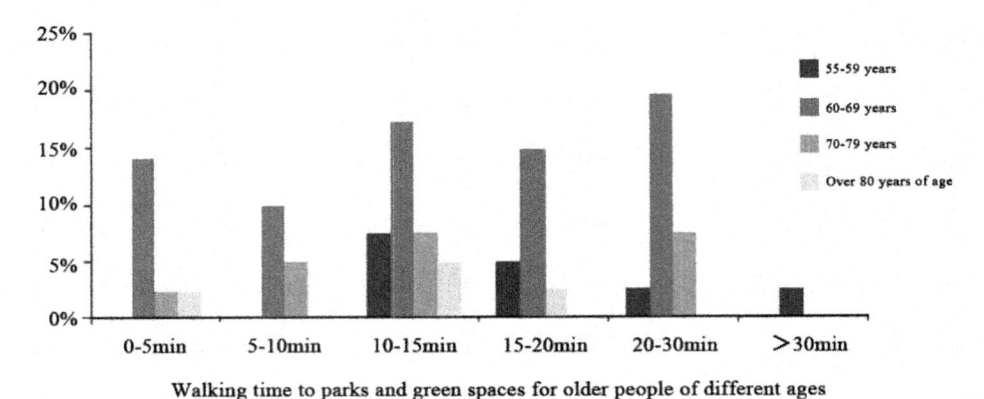

Walking time to parks and green spaces for older people of different ages

Figure 3.   Survey on the behavioral preferences of older people for park activities.

### 3.3 *Analysis of the accessibility needs of older people to parks and green spaces*

The overall demand for public green space by the elderly in the vicinity of the Dongguan Botanical Garden was divided into seven classes according to different walking accessibility times. According to the research (see Figure 4), the distribution of park green space accessibility is analyzed in comparison with the spatial distribution of the elderly population. According to the results of the analysis, Dongguan Avenue has a high density of the elderly population and is densely populated by nearby residential areas with a low rate of public green space. For example, the walking distance from Dingfeng Yuanzhu to Dongguan Ecological Park is only 1 km, which can be reached in ten minutes. However, the main attractions at the northern entrance are the Rock Garden, the Jukebed Plant Area, and the Lychee Garden, which are slightly underdeveloped concerning the landscape

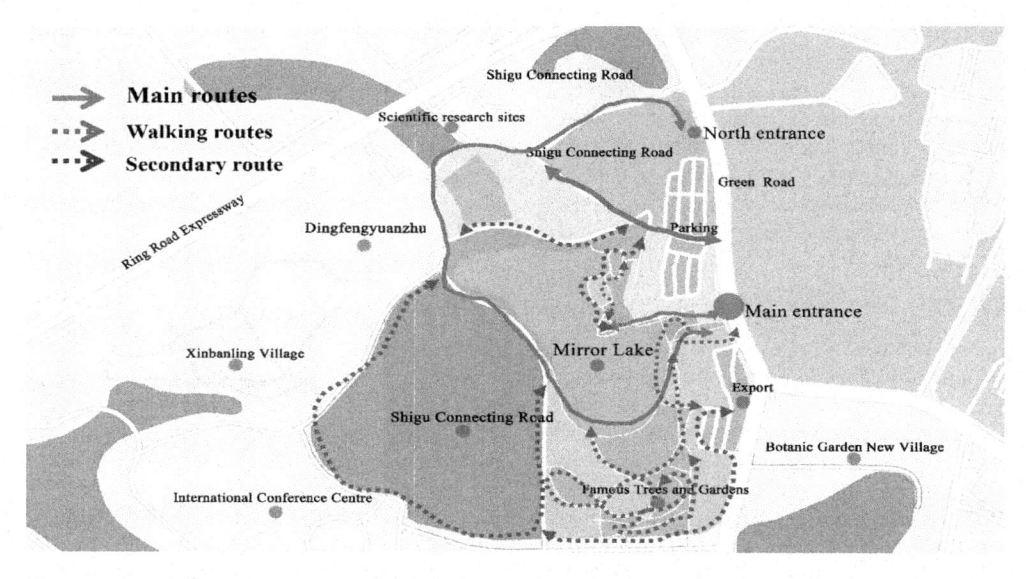

Figure 4.   Landscape route analysis map of Dongguan Botanical Garden Park.

preferences and accessibility of the elderly. The disadvantage is that these parks do not have a high level of parkland coverage and a single variety of plants. Although the rock gardens are beautiful, the landscaping in these areas is not suitable for elderly people and poses a safety risk. Some parks, such as the Rhododendron Slope, the Aromatic Garden, the Cherry Garden, and the Guan Xiang Garden, are relatively far away and less accessible based on the preferences and behavioral accessibility of the elderly (Safeeq M 2016).

## 4 PARKLAND ACCESSIBILITY STRATEGIES IN THE PERSPECTIVE OF AN AGEING SOCIETY

The planning and design of accessible routes to park space should be in line with an aging perspective. Targeted allocation of green space resources and improved park layout routes. To ensure the efficiency and effectiveness of the use of green space. Based on the above analysis, the author believes that the following aspects should be emphasized in the accessibility planning of park green space from an aging perspective (Cetin & Mehmet 2015).

### 4.1 *The overall layout of the space needs to consider the characteristics of the surrounding elderly population density distribution*

The spatial distribution of the elderly population shows a high density of concentration in urban centers and dispersion at the periphery. However, there are individual cases of concentrated density at a fixed point, as shown in the comparison in Figure 5, where the accessibility of green spaces

**Targeted allocation of green space resources and improved park layout routes.**

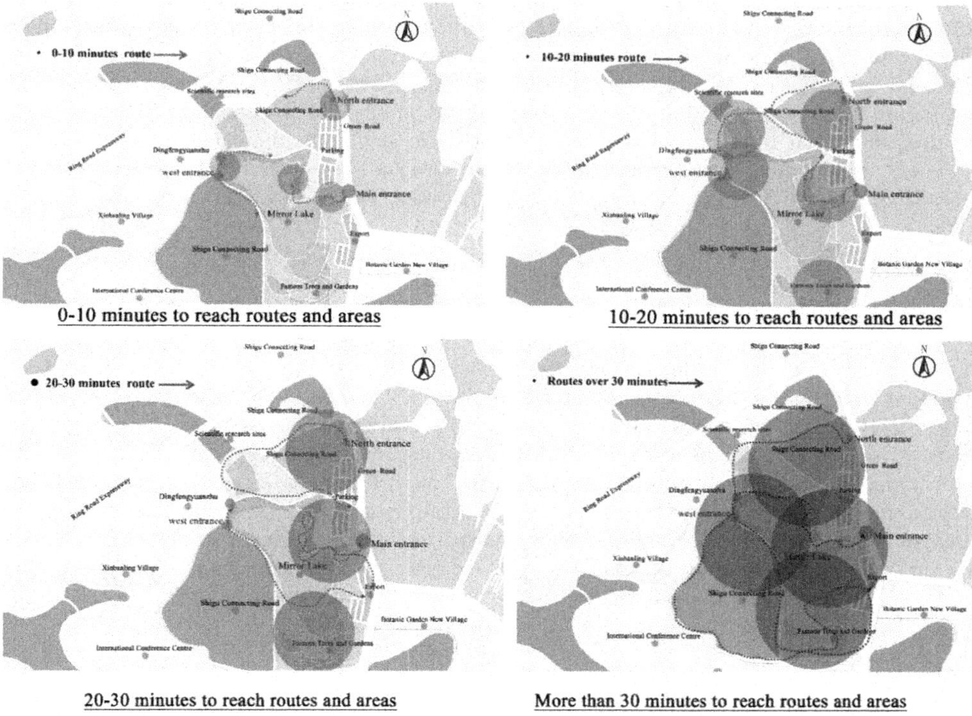

0-10 minutes route
North entrance
Main entrance
Mirror Lake
0-10 minutes to reach routes and areas
10-20 minutes route
10-20 minutes to reach routes and areas
20-30 minutes route
20-30 minutes to reach routes and areas
Routes over 30 minutes
More than 30 minutes to reach routes and areas


Figure 5. Comparison of accessibility zones at different times.

varies at different times. Therefore, the spatial layout of the park should first consider the characteristics of the dense distribution of the surrounding elderly population. The main scenic areas of green space will be located close to residential areas with a high density of elderly people. The elderly and children are the people who use the most parkland, and they have a more urgent need for green space. Some existing public green spaces have entrances and exits that are not located in line with residential areas. Although the park is close by, elderly people must take a detour to reach it, which results in additional time and labor costs. In this case, it is possible to adjust or add entrances in conjunction with the location of the residential area to make it easier for residents to reach the park. Increasing the accessibility of green spaces can meet the physiological and psychological needs of the elderly, increase the frequency of their outings, and enhance the development of their personal physical and mental health.

## 4.2 *Enhancing the service functions of green spaces attached to residential areas*

Field research revealed that some street areas have public green areas but their internal facilities are dilapidated and the greenery is poor. It is difficult for older people to engage in activities inside. As the elderly prefer bright and open parkland facilities, they also need spaces containing rich planting, water features, vignettes in a variety of materials, and walking gardens. Therefore, it is possible to renovate poorly maintained public green spaces and residential green spaces in some areas where the elderly population is concentrated. A certain amount of activity facilities for the elderly can be added to make up for the lack of services that cannot be provided by parkland.

## 4.3 *Spatial planning and design of parks and green spaces need to meet the real needs of the aging population*

As the physiological and psychological changes of the elderly in different age groups lead to different behavioral activities, the accessibility of the elderly in different age groups should be fully considered in the planning and design of green spaces in parks. For example, when planning accessibility facilities for the elderly, the service radius should be reduced and the range of activities and leisure areas should be reduced by increasing the number of fragmented green spaces. In addition, the visual and tactile design should emphasize physical and mental healing functions. In the case of the younger age groups, additional leisure and fitness facilities can be provided to enhance the interactivity of the elderly and to meet their diverse needs (Van Herzele A & Wiedemann T 2003).

## 5 CONCLUSION

In the context of an increasingly aging society, the optimization of the living environment is an important element in the future development of new cities in China. The spatial environment of living for an accelerating aging population and the differences in the needs of the aging population has been widely discussed in academic circles. From the perspective of humanistic care, building a new mode of planning and construction that is "people-oriented and symbiotic" with the needs of the aging population in mind is a necessary path for the construction and development of the future human living environment. Parks and green spaces are an important part of the urban public service system. Improving the accessibility of green spaces and parks plays a significant role in improving the living environment and enhancing the quality of life of the elderly.

## REFERENCES

Barbosa O, Tratalos J A & Armsworth P R. Who benefits from access to green space? A case study from Sheffield, UK. *Landscape and Urban Planning*, 2007, 83 (2-3): 0–195.

Bohannon R W. Comfortable and maximum walking speed of adults aged 20-79 years: Reference values and determinants [J]. *Age & Ageing*, 1997,26(1): 15–19.

Cetin, & Mehmet. Using GIS analysis to assess urban green space in terms of accessibility: case study in Kutahya. *International Journal of Sustainable Development & World Ecology*, 2015, 22(5): 1–5.

Michael J. Annear & GrantCushman. A place for visual research methods in the field of leisure studies Evidence from two studies of older adults' active leisure. *Leisure Studies*, 2014,33(6).

Safeeq M, Grant G E & Lewis S L. Predicting landscape sensitiv- ity to present and future floods in the Pacific Northwest, USA. *Hydrological Processes*, 2016,29(26): 5337–5353

Van Herzele A, & Wiedemann T. A monitoring tool for the provision of accessible and attractive urban green spaces. *Landscape and Urban Planning*, 2003,63(2): 109–126.

Youssoufi S, Foltête & Jean-Christophe. Determining appropriate neighborhood shapes and sizes for modeling landscape satisfaction. *Landscape and Urban Planning*, 2013,110: 12–24.

*Civil Engineering and Urban Research – Mohamed & Hou (Eds)*
*© 2023 the Authors, ISBN 978-1-032-44487-1*

# Developing a recycling logistics network for the disposal of urban construction waste

Yunjin Yang* & Yanlin Zhao*
*Panzhihua University, Panzhihua, China*

ABSTRACT: In China, with the progress of urbanization, the amount of construction waste is growing rapidly. At present, the comprehensive utilization rate of construction waste in Chinese cities is low, which has caused certain damage to the environment. Therefore, we should reduce the adverse impact of construction waste on the city as much as possible, and recycle it to take a sustainable development path. This paper summarizes the problems existing in China's current construction waste management by consulting literature and research materials. This paper proposes a 3R model and builds a construction waste recycling network model based on the 3R model. Then the construction waste recycling logistics network model is constructed to minimize the comprehensive cost, and the purpose is to analyze the location of the recycling network nodes. Finally, the paper takes S City as an example, collects data to simulate 7 urban areas in S City, solves the optimal total cost of 2,984.999 million yuan, and obtains the optimal layout of the construction waste recycling logistics network in S City. This paper integrates theoretical analysis and case simulation, and the results can provide a reference for construction waste recycling strategies.

## 1 INTRODUCTION

In recent years, with the continuous increase of construction projects and sustainable economic development, the output of construction waste in China has increased, but the reuse rate of construction waste is very small. These construction wastes occupy urban land and can also cause serious damage to the environment (Zhang 2021). Most housing construction and infrastructure projects consume a lot of natural resources, such as sand and cement, and make some natural resources depleted (Gao 2021). At present, the treatment method of construction waste in China is relatively backward, and a large amount of construction waste is disposed of on-site, which not only hinders urban construction, but also directly affects the utilization rate of construction waste, and at the same time cannot guarantee the quality of the obtained products. Faced with the problem of urban construction waste, it should be treated from the source, and it is urgent to reduce the amount of construction waste at the source.

In the special treatment plan for construction waste, many countries such as Japan, France, and Singapore have mentioned different recycling and treatment technology measures for centralized management of construction waste, but there are still many imperfections in China (Huang 2006). Pan (2016) proposed the working principle of classifying the sources and production of various large-scale construction waste landfills. Zhu (2007) pointed out that to improve the recycling rate of urban construction waste, we must first form a differentiated construction waste market. Chen (2004) suggested that the active response mode of source control and whole-process control should be adopted. In large developed cities abroad, the forms of construction waste recycling and terminal recycling are very common. Urban construction waste is treated, recycled, and reused at rates of

---

*Corresponding Authors: duandanlu1995@163.com and 914347668@163.com

DOI 10.1201/9781003372417-56

80% and 90%, respectively, on a daily basis (Rong 2022). Therefore, it is necessary to study the construction waste disposal strategy and logistics network planning in a certain city in this paper and to establish a reference for construction waste disposal methods in other cities in China.

## 2 CONSTRUCTION WASTE TREATMENT STRATEGY

### 2.1 *Problems with construction waste disposal*

At present, China's construction waste disposal methods are simplified and disordered, coupled with low utilization efficiency. Compared with foreign countries, the development level of the domestic building materials industry is still very low, and the development situation of S City is not optimistic. The relevant laws and regulations formulated by the relevant departments of S City do not have mandatory provisions for the recycling of construction waste. There are no relevant specifications in the special standards of S City for recycling building materials in the links of classification, production, and monitoring. In terms of policy support, the policies of S City are relatively simple, and there is no appropriate adjustment and guidance on prices. In addition to the above reasons, the construction waste recycling enterprises in S City are still in the stage of blind development, and China's resource chemical industry has not been systematically and scientifically planned.

In the market environment, it is difficult to form a complete industrial chain for the recycling and reuse of construction waste, and construction waste recycling enterprises with processing capacity are in trouble due to the lack of building materials. In China, construction waste has not been mandated to be recycled, and construction waste manufacturers often bury and dump it. Therefore, many construction waste companies have been forced to suspend production and incur losses because of a shortage of raw materials.

The traditional construction waste disposal can only be carried out superficially, and in the new era, the disposal of construction waste must be transformed from the traditional palliative treatment to the fundamental solution to the problem of construction waste. In this paper, a method based on hierarchical analysis is proposed to address some of the current problems in S City.

### 2.2 *Construction waste treatment and recycling levels*

The utilization level of construction waste refers to the recycling of construction waste using different utilization methods, as well as evaluating the advantages and disadvantages of each method under specific environmental conditions (Zhao 2016). The 3R model can reduce energy consumption, extend life, reuse, remanufacture and recycle. Reuse is the simple recycling of the original use of raw materials and products. Remanufacturing is to send the material or product to the processing workshop, and then divide it into usable parts as required, and then use it for production. Recycling refers to the recycling of materials or products and pulverizing them to turn them into basic production materials, thereby replacing the original raw materials to manufacture new products. In terms of cost savings and resources saved, recycling is better than recycling when considering the energy requirements for recovery, transportation, and processing into usable products. The 3R model is shown in Figure 1.

## 3 CONSTRUCTION OF RECYCLING LOGISTICS NETWORK

### 3.1 *Descriptions for symbols*

$A$: the set of collection point; $A= \{1, 2, 3, ..., i\}$.
$B$: the set of the transshipment plant; $B= \{1, 2, 3, ..., j\}$.
$C$: a collection of classified processing plants; $C= \{1, 2, 3, ..., k\}$.
$X$: the set of resale markets; $X= \{1, 2, 3, ..., a\}$.

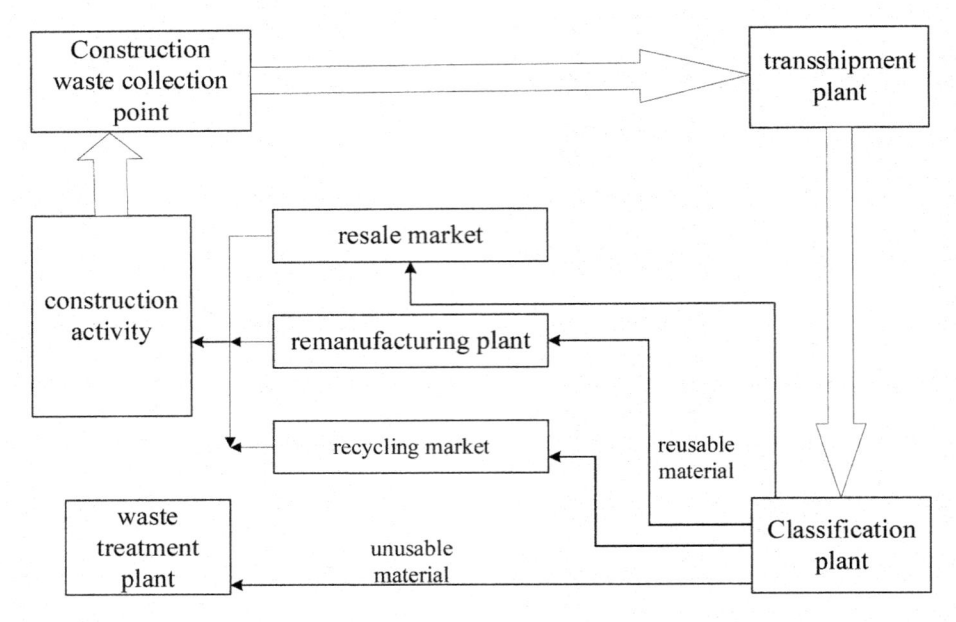

Figure 1.　3R model.

$Y$: the set of remanufacturing factories; $Y = \{1, 2, 3, \ldots, b\}$.

$Z$: the set of recycling plants; $Z = \{1, 2, 3, \ldots, c\}$.

$W$: the collection of abandoned markets; $W = \{1, 2, 3, \ldots, d\}$.

$R$: the collection of construction waste types; $R = \{1, 2, \ldots, r\}$.

$U$: the scale collection of construction waste transfer plants; $U = \{1, 2, \ldots, t\}$.

$V$: the scale collection of construction waste classification and treatment plants; $V = \{1, 2, \ldots, v\}$.

$S_{rjk}^{BC}$: the unit transportation cost of different types of construction waste from $B_j$ to $C_k$. By analogy, the meaning of symbols of the same category.

$O_{rij}^{B}$: the unit operating cost of different types of construction waste for transshipment plant $B_j$. By analogy, the meaning of symbols of the same category.

$D_{jt}^{B}$: construction costs of $B_j$. By analogy, the meaning of symbols of the same category.

$P_{jt}^{S}$: for the 0-1 variable of the transshipment plant $B_j$, take 1 when building a transshipment plant of scale $t$ at location $j$; otherwise, take 0. By analogy, the meaning of symbols of the same category.

$T_{ri}^{A}$: the recycling number of different types of construction waste at collection point $A_i$. By analogy, the meaning of symbols of the same category.

## 3.2　Mathematical model

$$
\begin{aligned}
\min = &\sum_{R=1}^{r} \sum_{A=1}^{i} \sum_{B=1}^{j} S_{rij}^{AB} T_{rij}^{AB} + \sum_{R=1}^{r} \sum_{B=1}^{j} \sum_{C=1}^{k} S_{rjk}^{BC} T_{rjk}^{BC} + \sum_{R=1}^{r} \sum_{C=1}^{k} \sum_{X=1}^{a} S_{rka}^{CX} T_{rka}^{CX} \\
&+ \sum_{R=1}^{r} \sum_{C=1}^{k} \sum_{Y=1}^{b} S_{rkb}^{CY} T_{rkb}^{CY} + \sum_{R=1}^{r} \sum_{C=1}^{k} \sum_{Z=1}^{c} S_{rkc}^{CZ} T_{rkc}^{CZ} + \sum_{R=1}^{r} \sum_{C=1}^{k} \sum_{W=1}^{d} S_{rkd}^{CW} T_{rkd}^{CW} \\
&+ \sum_{R=1}^{r} \sum_{A=1}^{i} \sum_{B=1}^{j} \sum_{U=1}^{t} T_{rij}^{AB} O_{rij}^{B} P_{jt}^{B} + \sum_{R=1}^{r} \sum_{B=1}^{j} \sum_{C=1}^{k} \sum_{V=1}^{v} T_{rjk}^{BC} O_{rjk}^{V} P_{kv}^{C} \\
&+ \sum_{B=1}^{j} \sum_{U=1}^{t} D_{jt}^{B} P_{jt}^{B} + \sum_{C=1}^{k} \sum_{V=1}^{v} D_{kv}^{C} P_{kv}^{C}
\end{aligned}
\tag{1}
$$

$$\begin{cases} \sum_{t=1}^{l} T_{rij}^{AB} = \sum_{k=1}^{n} T_{rjk}^{BC} \\ \sum_{j=1}^{m} T_{rjk}^{BC} = \sum_{d=1}^{h} T_{rkd}^{CW} + \sum_{a=1}^{e} T_{rka}^{CX} + \sum_{b=1}^{f} T_{rkb}^{CY} + \sum_{c=1}^{g} T_{rkc}^{CZ} \\ \sum_{j=1}^{m} T_{ij}^{AB} \le T_{ri}^{A} \\ \sum_{k=1}^{n} T_{rka}^{CX} \le T_{ra}^{X} \\ \sum_{k=1}^{n} T_{rkc}^{CZ} \le T_{rc}^{Z} \\ \sum_{k=1}^{n} T_{rkd}^{CW} \le T_{rd}^{W} \end{cases} \qquad (2)$$

The following relevant factors in construction waste recycling and disposal activities need to be considered in this model.

For example, the flow constraint is that the construction waste flow in the entire network is in equilibrium, which means that the total inflow at a node is equal to the total outflow at that node. The processing capacity constraint is that the processing equipment of the transshipment plant and the classification processing plant has the limitation of the minimum and maximum processing capacity. Under a certain scale, it can match a certain flow of construction waste to ensure the recycling of construction waste. Meanwhile, it will not be used. It will cause a waste of resources due to the large scale. In addition, some other constraints are added (including recycling factor constraints, operational constraints, cost constraints, waste mass balance constraints, design capacity constraints of treatment sites, and location constraints) to make the model effective and complete.

## 4   SITE SELECTION OF CONSTRUCTION WASTE LOGISTICS NETWORK NODES

S City is divided into 20 districts, of which S City has seven main urban areas, A, B, C, D, E, F, and G. This paper uses the seven main urban areas as the scope of the recycling logistics network to construct the construction waste recycling logistics network system in S City. It is assumed that these urban areas need two resale markets, remanufacturing plants, recycling markets, and waste treatment plants for building renewable resource materials.

In view of the increasing trend of the amount of construction waste and the service life of the equipment and other factors, this paper is based on the construction waste data of S City in 2020. The relevant data are shown in Table 1 and Table 2. Due to space limitations, other data tables are not shown in this article.

Table 1.   The amount of construction waste in different areas of S City (unit: ton).

| Area | Demolition waste | New waste | Area | Demolition waste | New waste |
| --- | --- | --- | --- | --- | --- |
| A | 290 | 12.4 | E | 680 | 6.6 |
| B | 600 | 9.8 | F | 270 | 7.6 |
| C | 280 | 6.9 | G | 550 | 12.5 |
| D | 560 | 11.1 | | | |

Table 2. Shipping price between collection point and transshipment plant (unit: yuan/ton).

|   | A | B | C | D | E | F | G |
|---|---|---|---|---|---|---|---|
| 1 | 14 | 17 | 17 | 22 | 21 | 22 | 24 |
| 2 | 21 | 18 | 18 | 14 | 21 | 24 | 25 |
| 3 | 22 | 19 | 23 | 24 | 17 | 14 | 14 |
| 4 | 20 | 16 | 20 | 20 | 14 | 16 | 14 |

## 5 OPERATION AND RESULT ANALYSIS

Solve the case model, convert the constraints required by the model into Lingo's mathematical language, and input it into the computer Lingo system. The result of the construction waste recycling logistics network in S City is shown in Figure 2.

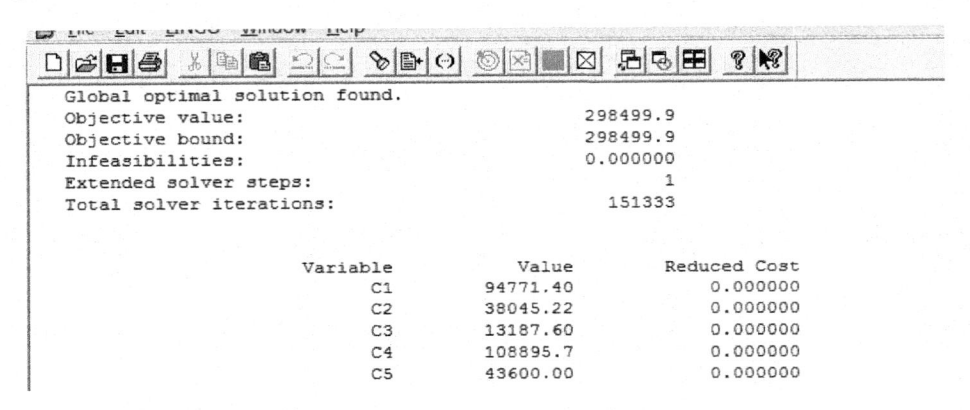

Figure 2. Lingo simulation results.

Under the goal of minimum recycling cost of urban construction waste in S City, various cost prices are given first. When all constraints are met, the model obtains the optimal solution of the optimal total cost of 2,984.999 million yuan that satisfies the goal. At the same time, the simulation gives the layout of the construction waste recycling logistics network in S City, that is, the location and scale of the transshipment plant and the sorting plant. The calculation results show that transshipment plants should be established in four locations: 1, 2, 3, and 4. Among them, a medium transshipment plant with a scale of location 2 should be established in location 3, and a large transshipment plant with a scale of location 3 should be established in the remaining locations. Similarly, classification plants are built in three locations: 1, 2, and 3. A medium-sized classification plant with a scale of location 3 is built in location 1, and a large-scale classification plant with a scale of location 2 is built in location 2.

## 6 CONCLUSIONS

In this paper, the idea of the 3R model is used to construct a collection transport sorting processing redistribution construction waste recycling network system. The mathematical model is constructed according to the recycling logistics network. First, the cost generated by each node of the construction waste recycling network is analyzed, and then the function construction of the construction waste recycling logistics network model and the setting of constraints are carried out to minimize the comprehensive costs. Lingo is used for the numerical simulation analysis of the construction waste recycling logistics network in S City, and the optimal solution is obtained for the site selection of construction waste nodes in S City. This paper integrates theoretical analysis and case simulation, and the results can provide a reference for construction waste recycling strategies. Although this paper has made some explorations on construction waste treatment strategies and construction waste recycling logistics planning, due to the constraints of time, epidemic environment, research scope, and other factors, the research in this paper still has many shortcomings. This paper ignores the losses caused by the cyclic network, and there is currently no unified operating specification in China, which makes it difficult to quantify and generalize some cost and expense items. Therefore, future research should refer to relevant foreign operating specifications.

## ACKNOWLEDGMENTS

This work was supported by the Sichuan Province Key Laboratory of Higher Education Institutions for Comprehensive Development and Utilization of Industrial Solid Waste in Civil Engineering (Grant No. SC_FQWLY-2019-Y-01).

## REFERENCES

Chen, L. & W. Chen (2004). Management of construction waste in Hong Kong SAR. *J. Journal of Kunming University of Science and Technology* (Science and Technology Edition). 02, 07–11.

Gao, Q. & H. H. Li (2021). Research, application and development of construction waste. *J. Sichuan Building Materials*. 47, 32–34.

Huang, S. C. & M. W. Peng (2006). Application of asphalt pavement regeneration technology at home and abroad. *J. Highway Traffic Science and Technology*. 11, 5–8.

Pan, F. Q. (2016). Current status of urban solid waste treatment and utilization. *J. Prospects of Science and Technology*. 26, 268–272.

Rong, Y. F. & X. Y. Zhang (2022). Research on source reduction planning of construction waste based on the concept of green development. *J. Journal of Beijing University of Architecture*. 38, 9–17.

Zhang, H. Chen (2021). Research on the status of urban construction waste treatment and resource utilization. *J. Green Building Materials*. 12, 31–32.

Zhao, H. & Y. X. Guo (2016). Research on reuse of reverse logistics network site selection. *J. Railway Procurement and Logistics*. 02, 55–57.

Zhu, L. B. & H. Ren (2007). Research on building material saving based on circular economy. *J. Ecological Economy*. 01, 185–187.

*Civil Engineering and Urban Research – Mohamed & Hou (Eds)*

# Design and effect analysis of Southward External Shading of college classroom in Guangzhou

Yang Wang* & Rui Hou*
*Architectural Design & Research Institute of SCUT Co., Ltd., Guangzhou, China*
*School of Architecture, South China University of Technology, Guangzhou, China*

Xijia Sun*
*Department of Architecture, Southeast University, Nanjing, China*

ABSTRACT: Guangzhou belongs to the south subtropical monsoon climate. The demand for building shading is obvious due to the hot climate. As a learning place for undergraduates, classroom in colleges and universities has higher requirements for an interior luminous environment. Rational external shading design can solve the problem of shading while optimizing the interior luminous environment. This paper takes the Southward External Shading design of the Shensi Building of the South China University of Technology as a case study and carries out an analysis based on Autodesk Ecotect Analysis to quantitatively evaluate its performance from the perspective of building technology. Through the calculation of the external shading coefficient, the shading device can withstand more than 74% of the total solar radiation in an average year and can meet the different needs of solar radiation in cold and hot months. Through the analysis of interior natural lighting, it is found that after using the shading device, the daylight factor and natural lighting illuminance are increased by more than 0.036 to balance the overall luminous environment of interior space and reduce the glare problem. Furthermore, it seeks a kind of design of a Southward External Shading device that balances the natural lighting and shading effect to provide a certain reference for the Southward External Shading design of colleges and universities in Guangzhou.

## 1 INTRODUCTION

Guangzhou has unique climatic characteristics, and the climate adaptability design is an important factor in the building design here. College classrooms are one of the important places for college students to study. A comfortable interior climate is a necessary condition for long-term study and scientific research. Shading is one of the passive strategies in building climate adaptability design. Effective shading design can significantly reduce the impact of solar radiation on interior climate, and it is also an effective way to actively explore the reduction of building energy consumption. College students spend more time in the classroom, so lighting requirements in the classroom are higher than in other rooms. Building shading can also affect interior lighting while shading solar radiation, but a reasonable design of shading devices can form a positive response, which will help to improve the interior luminous environment. Besides, a proper design of shading devices can form a positive response, which will help to improve the interior luminous environment. The same design also reduces glare while guaranteeing interiors' natural illumination and connection to the outside which increases occupants' physiological and psychological well-being (Francesco 2022).

---

*Corresponding Authors: yangwang804@126.com, hour1996@163.com and 337621262@qq.com

    DOI 10.1201/9781003372417-57

## 2 ANALYSIS OF CLIMATE CONDITIONS IN GUANGZHOU

Guangzhou belongs to the south subtropical monsoon climate. It is characterized by abundant rainwater resources, abundant light and heat, and long summer months. High humidity throughout the year, with an annual average relative humidity of 78%, and the flood season from April to September, with abundant precipitation. Combined with the data of the Guangzhou Meteorological Observatory, the annual temperature in Guangzhou is analyzed. The annual average temperature in Guangzhou ranges from 21.5°C to 22.2°C. The temperature from May to October is relatively high. The hottest month is July, with a monthly average temperature of 28.7°C; the temperature from December to February is a low-temperature month, and the coldest month is January, with a monthly average temperature of 13°C. Therefore, reasonable shading measures can be used to reduce the sunlight in hot months, and reduce the interior solar radiation to lower temperature and improve interior comfort. In winter, more solar radiation can be obtained interior, thus increasing interior temperature.

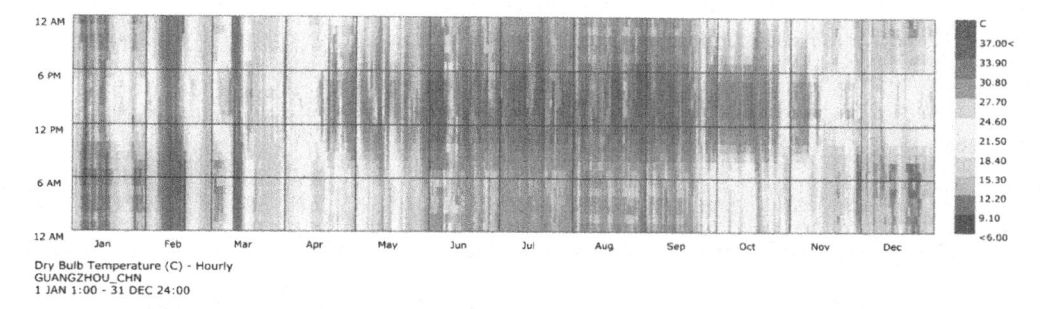

Dry Bulb Temperature (C) - Hourly
GUANGZHOU_CHN
1 JAN 1:00 - 31 DEC 24:00

Figure 1.    Annual temperature distribution map of Guangzhou.

## 3 SOUTHWARD EXTERNAL SHADING DESIGN OF SHENSI BUILDING OF THE SOUTH CHINA UNIVERSITY OF TECHNOLOGY

South China University of Technology (SCUT) is located in Guangzhou, Guangdong Province, China. Shensi Building is located next to West Lake on the Wushan Campus of SCUT. It is a group of teaching buildings composed of four buildings in series. The external Shading design of its southward windows forms a unique facade effect.

### 3.1  Design features of Southward External Shading design of Shensi Building

The main form of southward shading design is combined shading. Two windows are a shading unit, which is combined with the building structure. The concrete frame is used for shading, and a vertical component is designed in the middle of the frame for flank shading. The overhang shading is set above the window opening fan. And it is designed as an inner grille and an outer aluminum plate. The lower part is the air-conditioning outdoor unit, which is separated by a horizontal concrete slab, and the outer cover is a vertical louvered aluminum plate.

The southward shading design is based on the frame shading, and the shading device is comprehensively and flexibly arranged. The building uses the cast-in-place reinforced concrete frame as its external facade. While shading, it cleverly hides or combines the beams, columns, slabs, and other building structures, and reserves the slots of shading components on the frame. The frame makes the building facade design logical and delineates the shading units. In this way, the shading component inside the unit can be quickly produced for each unit, and dynamically adjusted according to the size of different rooms. In the internal shading component of the unit, a prefabricated aluminum plate and a grille plate are used for shading. The gap between the grille plates

397

can improve the interior ventilation and the structure of the storage, and cooling problems, using aluminum alloy materials with faster heat transfer speed so that the accumulated heat is easy to release. The lower air conditioner outlet is separated by a concrete slab from the window, and the vertical blinds aluminum slab of the cover is tilted downwards, exporting the hot air from the outside of the air conditioner while sheltering the rain, which makes the heat not easy to accumulate inside the shading unit.

Figure 2. Analysis of southward shading device of Shensi Building.

## 4 ANALYSIS AND SIMULATION OF SOUTHWARD EXTERNAL SHADING DESIGN OF SHENSI BUILDING

### 4.1 *Ideal model establishment*

To facilitate the simulation calculation and reduce unnecessary calculations, the model of the shading device is simplified, and the standard room on the middle floor of the No. 33 Building in the Shensi Building is selected as the analysis model. The size of the single room is 11.8 m × 16.2 m, the net height of the floor is 3.9 m, the depth of the southward shading system frame is 800 mm, and the size of the single window is 2.2 m × 3.4 m. The internal grille is 160 mm away from the window, 1,500 mm away from the lower window sill, and each grille plate spacing is 80 mm, a total of 6 grids, 100 mm high. The external horizontal aluminum plate is picked out 500 mm and 2,900 mm wide. The outer corridor is 2.1 m wide, and the two windows on the north side are 0.9 m high from the ground, 2,200 mm high, and 3,500 mm wide.

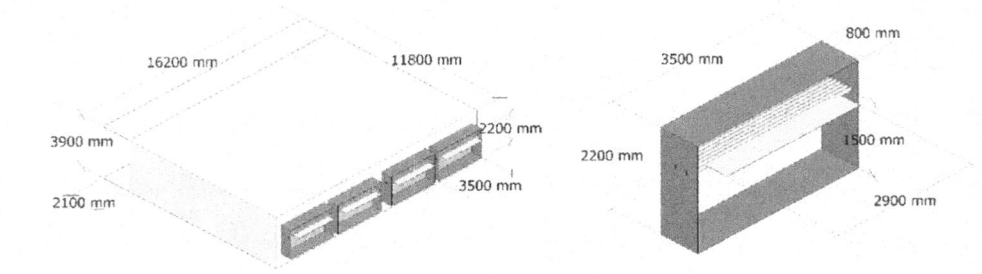

Figure 3. The monthly solar radiation and external shading system of shading devices.

### 4.2 *Analysis of solar radiation in southward window*

To better illustrate the shading situation, Autodesk Ecotect Analysis is used to calculate the external shading coefficients of the windows without and with shading respectively.

Because the South shading unit contains a complex combination of overhang, flank, and blade shading, it is not possible to calculate the accurate outer external shading coefficient using the current Chinese standards. According to the research by Professor Zhao Lihua of SCUT, we can calculate the solar radiation heat for a certain direction through a shaded outer window and an unshaded outer window based on the local solar radiation data. The ratio of the two is approximately the external shading coefficient (Zhao 2010).

External shading coefficient (SC) = cumulative solar radiation of external windows with shading (Wh)/cumulative solar radiation of external windows without shading (Wh).

From the calculation formula, the smaller the coefficient is, the smaller the cumulative solar radiation of the outer window is, and therefore, the better the shading capability of the shading device is. Through the software, the external shading coefficients of the shading device of the Shensi Building are calculated quarterly and monthly to judge the effectiveness of the shading component, then calculate the direct solar radiation and scattered radiation, and further explore the shading performance.

### 4.2.1 Quarter-by-quarter analysis of solar radiation in southward windows

Table 1. Calculation of external shading coefficient of shading device of Shensi House by quarter (Radiation unit: Wh).

| Time* | | Winter | Spring | Summer | Autumn | Annual |
|---|---|---|---|---|---|---|
| TSR* | No visor | 95838716 | 90466548 | 107074737 | 142460559 | 435840561 |
| | Have visor | 30318716 | 24058639 | 26741647 | 33262330 | 114381331 |
| | SC*** | 0.32 | 0.27 | 0.25 | 0.23 | 0.26 |
| DSR* | No visor | 42814408 | 26763644 | 35791597 | 71724303 | 177093953 |
| | Have visor | 10426671 | 160739 | 0 | 6726061 | 17313471 |
| | SC*** | 0.24 | 0.01 | 0.00 | 0.09 | 0.10 |

*Winter runs from December to February, Spring from March to May, Summer from June to August, and Autumn from September to November.
**TSR is total solar radiation, and DSR is direct solar radiation.
**SC is the external shading coefficient.

The calculation results by season are shown in Table 1. Combined with the solar radiation characteristics of the southward windows in Guangzhou, it can be found that the shading device of the Shensi Building has an obvious effect of blocking solar radiation, and can block more than 74% of the radiation on average throughout the year. During the winter and spring, when the temperature is low, the shading coefficient is relatively large, at 0.32 and 0.27, which allows more radiation to enter the room during these periods; because Guangzhou is located on the Tropic of Cancer, the sun's altitude angle is higher in summer, so this season mainly effectively shields scattered radiation, and direct radiation relies more on building self-shading. From September to November, when the temperature is higher and the solar radiation is stronger, the external shading coefficient is the smallest among the four seasons, and the shading effect is more obvious, which can block more than 77% of the radiation on average. It is calculated that the shading member will block more than 65% of the scattered radiation energy on average throughout the year.

### 4.2.2 Month-by-month analysis of solar radiation in southward windows

Monthly calculation results are shown in the chart. From the calculation of total solar radiation, after adding a shading device, the monthly solar radiation received by the window is much lower than that without shading. Especially from August to October, when the temperature is relatively high and the sun radiates the most, the shading coefficient maintains a relatively low level. From December to February when the temperature is low, the shading coefficient is above 0.3, which allows more solar radiation to enter the room and raises the interior temperature. From the calculation of direct solar radiation, the shading device is the most efficient for shielding direct solar radiation. From January to July in Guangzhou, the weather is cloudy and rainy, and the direct solar radiation is relatively small. The weather from August to December is usually sunny, so there is higher Solar

Table 2. Calculation of external shading coefficient of shading device of Shensi House by month (Radiation unit: Wh).

| | Month | Jan. | Feb. | Mar. | Apr. | May | Jun. |
|---|---|---|---|---|---|---|---|
| TSR* | No visor | 31057994 | 24438764 | 24630675 | 29853463 | 35982409 | 21495381 |
| | Have visor | 9925868 | 7360341 | 6393116 | 8170709 | 9494812 | 8063930 |
| | SC** | 0.32 | 0.30 | 0.26 | 0.27 | 0.26 | 0.38 |
| DSR* | No visor | 12498468.5 | 7455968.64 | 8017557.12 | 8073216 | 10672871 | 0 |
| | Have visor | 2963243.52 | 989262.72 | 160738.56 | 0 | 0 | 0 |
| | SC** | 0.24 | 0.13 | 0.02 | 0.00 | 0.00 | - |
| | Month | Jul. | Aug. | Sep. | Oct. | Nov. | Dec. |
| TSR* | No visor | 38011271 | 47568084 | 47166526 | 51679843 | 43614190 | 40341957 |
| | Have visor | 9958705 | 8719009 | 9775975 | 11737082 | 11749271 | 13032506 |
| | SC** | 0.26 | 0.18 | 0.21 | 0.23 | 0.27 | 0.32 |
| DSR* | No visor | 11465112 | 24326484 | 21983644 | 25591605 | 24149053 | 22859971 |
| | Have visor | 0 | 0 | 328809 | 1950192 | 4447059 | 6474165 |
| | SC** | 0.00 | 0.00 | 0.01 | 0.08 | 0.18 | 0.28 |

*TSR is total solar radiation, and DSR is direct solar radiation.
**SC is the external shading coefficient.

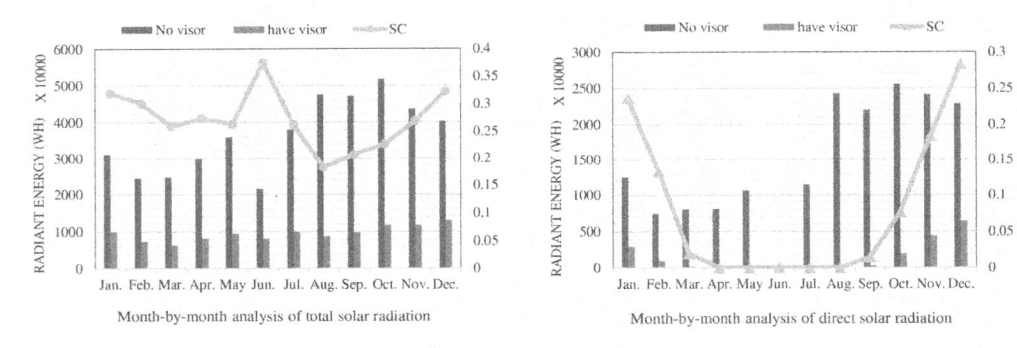

Month-by-month analysis of total solar radiation    Month-by-month analysis of direct solar radiation

Figure 4. The monthly solar radiation and external shading system of shading devices.

direct radiation. Because Guangzhou is located on the Tropic of Cancer, the sun turns to the north around the summer solstice, so there is no direct radiation from the south in June (Qi 2008). This is also the reason for the high shading coefficient before and after June in the total radiation chart. The shading member is efficient for direct radiation blocking, but the monthly average value for blocking scattered radiation is about 65%. From April to August, the shading device can block all the direct radiation out of the window, including the reason for the large sun altitude angle. The direct shading coefficient from September to October is also relatively small, which satisfies the shielding of direct light from May to October, which is hot in Guangzhou and has a good cooling effect on the interior climate. In January and December when the temperature is low, the relative maximum external shading coefficients are 0.24 and 0.28, and the direct radiation shielding is the least so that more sunlight can be injected into the room to meet the need for more solar radiation in winter.

### 4.3 *Analysis of interior natural lighting*

Building shading not only blocks solar radiation but also blocks sunlight, which will have a certain impact on interior lighting. To comprehensively evaluate the shading device, it is necessary to simulate and analyze the interior lighting conditions. And mainly from the selected 9 test points, analyze and compare the daylight factor and natural lighting illuminance on the working face plane with the height of 0.9 m.

According to the lighting and climate division of China, the Guangzhou area belongs to the IV category, the daylight climate coefficient K is taken as 1.1, and the design illuminance of external daylight is 13,500 lx. In the experiment, the outdoor environment was set to the CIE full cloudy mode. Chinese specifications require that the lighting illuminance of ordinary classrooms is not lower than Class III, which means that the standard value of daylight factor for side lighting is 3%, and the standard value of interior daylight illuminance is 450 lx. In this experiment, the uniformity of illuminance (minimum illuminance value/average illuminance value) and the uniformity of daylight factor (minimum daylight factor/average daylight factor) will be used as the criteria for judging the quality of lighting effects. The higher the uniformity, the more balanced the overall luminous environment of the space (Yang 2017). The result of the experiment is that the one with the best uniformity of illuminance and daylight factor is the best.

Table 3. Comparison of daylight factor and natural lighting illuminance at 0.9 m in the room.

| | Natural lighting illuminance (lx) | | | Daylight factor (%) | | |
|---|---|---|---|---|---|---|
| | Minimum | Standard | Uniformity | Minimum | Standard | Uniformity |
| No visor | 178.43 | 822.08 | 0.2170 | 1.32% | 6.09% | 0.2167 |
| Have visor | 171.82 | 952.06 | 0.1805 | 1.27% | 7.05% | 0.1801 |
| Difference | 6.61 | −129.98 | 0.0365 | 0.05% | −0.96% | 0.0366 |

From the calculation results, after the shading device is applied, the standard daylight factor is 6.09%, and the standard value of interior daylight illuminance is 822.08 lx, both of which meet the requirements of 3% and 450 lx in the standard. Compared with the case of no shading, the standard values of the two indicators have a slight decrease, but the lowest values have increased, and the uniformity of the two indicators has been greatly improved compared with the case of no shading. By using this shading device, the extreme value area near the window position is significantly reduced, thereby improving the overall luminous environment of the interior space.

The lighting conditions of all interior test points are good, and large differences occur near the window, such as test points A, B, and C. After applying shading devices, the illumination is reduced by 200 lx, and the daylight factor is reduced by about 1%. With the increase of the distance from the window, the influence of shading on daylighting decreases gradually, which is similar to that without shading devices. From this analysis, it can be concluded that the interior luminous environment is better after adding shading devices. The illumination value and daylight factor can

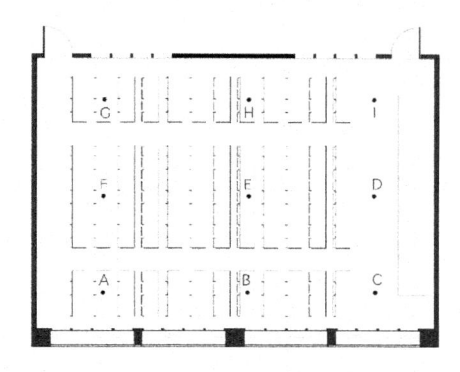

Figure 5. Test point distribution.

Table 4. Comparison of daylight factor and natural lighting illuminance at test point.

| | | A | B | C | D | E | F | G | H | I |
|---|---|---|---|---|---|---|---|---|---|---|
| $E_n$ * | No visor | 1021 | 867 | 1083 | 486 | 454 | 476 | 720 | 385 | 666 |
| | Have visor | 1227 | 1080 | 1353 | 535 | 430 | 540 | 627 | 353 | 650 |
| | Difference | −206 | −212 | −270 | −48 | 24 | −63 | 92 | 32 | 16 |
| C * | No visor | 7.57 | 6.43 | 8.02 | 3.60 | 3.37 | 3.53 | 5.33 | 2.86 | 4.94 |
| | Have visor | 8.55 | 8.14 | 9.48 | 4.06 | 3.36 | 4.15 | 5.43 | 2.65 | 4.28 |
| | Difference | −0.98 | −1.71 | −1.46 | −0.46 | 0.01 | −0.62 | −0.1 | 0.21 | 0.66 |

*$E_n$ is natural lighting illuminance (lx); C is daylight factor (%).

be appropriately reduced through the shading devices at the position with too strong light to reduce the impact of glare.

## 5 CONCLUSION

In this paper, the Southward shading device of the Shensi building at the South China University of technology is selected as the research object. Through experimental simulation, its solar radiation shielding ability and its impact on the interior luminous environment are quantitatively evaluated. It is concluded that this kind of shading device is economical and efficient, can block more than 74% of the radiation on average throughout the year, and can dynamically adapt to different interior solar radiation needs in different seasons and months, which reflects the dynamic adaptability of climate. The application of the shading device can make the overall luminous environment of the interior space more balanced. Therefore, the reasonable design of building external shading can solve the problem of shading and optimize the interior luminous environment, which provides a certain reference for the design of southward external shading of college classrooms in Guangzhou and Lingnan areas.

## REFERENCES

A. A. Freewan, L. Shao & S. Riffat. (2009). Interactions between louvers and ceiling geometry for maximum daylighting performance. *Renew. Energy* (34), 223–232.

Chen Qianrong. (2011). *Preliminary study on the development process of climate-adaptive design of building skins in hot regions* (1930–1980) (Master's thesis, South China University of Technology)

Francesco De Luca, Abel Sepúlveda & Toivo Varjas. (2022). Multi-performance optimization of static shading devices for glare, daylight, view and energy consideration. *Building and Environment* (217), 109–110.

He Xiaoqian & Wang Yang. (2021). Optimal Design of Fixed Shading for Classroom Window in Primary and Secondary Schools in Guangzhou. *Building Energy Efficiency* (Chinese and English) (05), 21–27+36.

Liu Lina. (2011). *Analysis and comprehensive evaluation of the influence of external shading on lighting and ventilation of residential buildings in Guangzhou area* (Master's thesis, South China University of Technology).

M. David, M. Donn, F. Garde & A. Lenoir. (2011). Assessment of the thermal and visual efficiency of solar shades. *Build. Environ* (46), 1489–1496.

Qi Baihui. (2008). *Analysis of Shading Technology in Xia Changshi's Works in Early Modern Architecture in Lingnan* (Master's Thesis, South China University of Technology).

Xia Changshi. (1958). The cooling problem of subtropical buildings—shading, heat insulation and ventilation. *Chinese Journal of Architecture* (10), 36–39+42.

Yang Jinchun. (2017). *Research on shading optimization design of south-facing windows of office buildings in Xuzhou* (Master's thesis, China University of Mining and Technology).

Zhang Shuyang & Wang Yang. (2019). Dynamic external shading design and shading effect analysis in southeast and southwest of Guangzhou. *Building Energy Conservation* (08), 91–95+100.

Zhao Lihua, Qi Baihui & Xiao Yiqiang. (2010). Analysis of Shading Effect of "Xia's Shading". *Green Building* (01), 23–26.

*Civil Engineering and Urban Research – Mohamed & Hou (Eds)*
*© 2023 the Authors, ISBN 978-1-032-44487-1*

# Design of shallow foundation applied to large-area recent filling areas

Yi Zhou*
*China Nonferrous Metal (Guilin) Geology and Mining Co. Ltd., Guilin, Guangxi, China*

Feng Liang
*Guilin Institute of Architectural Design and Research, Guilin, Guangxi, China*

ABSTRACT: Based on the principle of foundation depth correction, combined with relevant documents and the Code for Design of Building Foundation, this paper interprets the depth of foundation in a large-area fill area and illustrates two important links in the design of shallow foundation in large-area new fill area through examples. The purpose of this article is to discuss how to adopt a reasonable depth of foundation design based on engineering feasibility. Considering the additional stress created by the newly added large-area fill, we develop a deformation control design; this ensures that the corresponding structural measures are considered reasonable.

## 1 INTRODUCTION

According to the *Code for Design of Building Foundation* (abbreviated as the Foundation Code), the embedded depth $d$ of the foundation is summarized as follows. It may be calculated from the ground level of the fill in the leveled areas, but it is preferable to calculate it from the natural ground once the superstructure is constructed. There are three scenarios for the time of backfilling in the actual projects: (1) before excavation of the foundation trench; (2) after the construction of the superstructure; (3) after the construction of the foundation but before the construction of the superstructure. The third scenario is not specifically stipulated in the Foundation Code. The calculation of the embedded depth of the foundation relates to the correction value of foundation bearing capacity, determines the area of the foundation, the size of the reinforcement, and even the form of the foundation, and closely relates to the foundation cost and building safety of the project. Reasonable application of the Foundation Code is of great significance. In this paper, the relevant literature and examples are considered for further discussion, and the deformation control that is often overlooked in the foundation design of large-area recent filling areas is also elaborated.

## 2 PRINCIPLE OF CORRECTION OF EMBEDDED DEPTH OF FOUNDATION

It is pointed out in the Foundation Code that when the foundation width is greater than 3 m or the embedded depth of the foundation is greater than 0.5 m, the characteristic value of foundation bearing capacity determined from the load test or other in-situ tests, empirical values, or other methods should be corrected as follows: $f_a = f_{ak} + \eta_b \gamma (b - 3) + \eta_d \gamma_m (d - 0.5)$

As per the Terzaghi bearing capacity theory, the theoretical plastic flow boundary when the foundation fails is shown in Figure 1.

---

*Corresponding Authors: wymfy126vip@126.com and 2776687132@qq.com

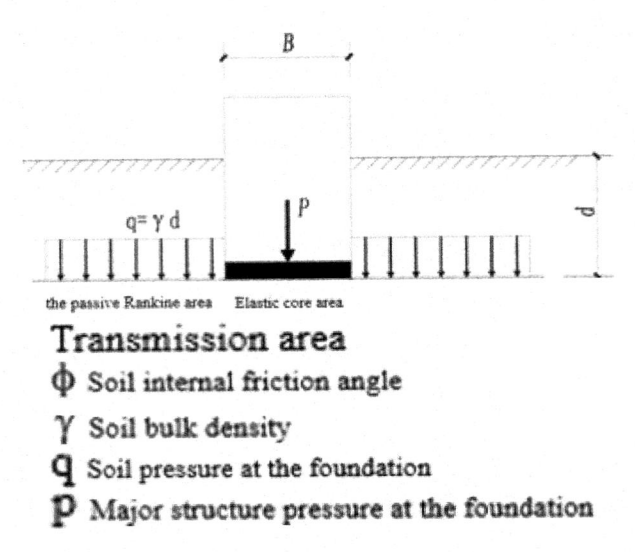

Transmission area

$\phi$ Soil internal friction angle

$\gamma$ Soil bulk density

$q$ Soil pressure at the foundation

$p$ Major structure pressure at the foundation

Figure 1. The overload and foundation failure formed by the embedded depth of the foundation.

The gravity formed by the soil around the foundation is the overload ($q = \gamma d$), which can resist the upward sliding of the soil mass, thereby providing the stability of the soil mass. The third item in the above formula represents the effect of the overload ($q = \gamma d$) on both sides of the foundation. Only the overload located on the passive Rankine zone in Figure 1 can improve the foundation bearing capacity. The size of the passive Rankine zone may be subject to the description in the Foundation Code that "at present, there are a large number of structures with the main buildings integrated with the annex buildings in the construction project". For the correction of the foundation bearing capacity of the main structure, the load within the range above the bottom of the foundation should be considered as the overload on both sides of the foundation. When the overload width is greater than twice the foundation width, the overload can be converted into the thickness of the soil mass as the embedded depth of the foundation, and when the overload on both sides of the foundation is not equal, the lower value should prevail." Thus, it is believed that the foundation bearing capacity can be improved only when the overload width is greater than twice the foundation width. As per the failure mode of the foundation, the foundation will fail in the most unfavorable direction, so the embedded depth $d$ of the foundation should be the minimum embedded depth around the foundation.

## 3 INTERPRETATION AND ENGINEERING APPLICATION OF EMBEDDED DEPTH OF FOUNDATION IN LARGE-AREA FILLING AREAS IN FOUNDATION CODE

It is recommended that the embedded depth of foundation (m) should be calculated from the outdoor elevation. It can be calculated from the ground level of the fill in the leveled areas, but when the fill is completed after the construction of the superstructure, it should be calculated from the natural ground level. The Foundation Code only addresses the following two cases.

### 3.1 When the recent fill is filled before the excavation of the foundation trench, it can be calculated from the ground level of the fill

It is believed among some insiders that only the soil that has been consolidated can be included in the calculation of the embedded depth $d$ of foundation, which confuses the concepts of correction of the embedded depth of foundation and consolidation of the soil. It is stated in the study of

Shi Chunle (2012) that the correction of the embedded depth of the foundation is specific to the contribution of the surrounding overload to the improvement of the stability and bearing capacity of the foundation soil under the base, which is only related to the weight of the soil above the base, regardless of whether this part of the soil has been consolidated. The consolidation reinforcement of soil refers to the consolidation and compaction of the soil under long-term upper pressure, which increases the bearing capacity of the soil under the base. These are two concepts with essential differences. Based on the recognition of this point, most designers express no objection to the provisions in the Foundation Code.

The Baohu Education Park Project is a good example of this kind. Since most of the project site is low-lying land with a depth of 2-3 m, the Owner and the Construction Contractor have completed the site leveling before excavation considering the convenience of mobilization of the construction machinery. The project is a 5-6-story teaching building, which adopts an independent column foundation and uses shallow hard plastic clay as the bearing stratum, and the embedded depth of the foundation is calculated from the filled ground.

### 3.2 When the fill is filled after the construction of the superstructure, it should be calculated from the natural ground level

This is because the pressure of the base under the action of the superstructure is close to the maximum at this moment, but the recent fill has not formed an overload, which does not contribute to the stability or bearing capacity of the foundation soil. As a result, the embedded depth of the foundation can only be calculated from the natural ground level, which is also recognized by the insiders. It results in a lot of inconveniences throughout the construction process, and the author has never experienced such a situation in any of the previous projects.

### 3.3 When the recent fill is filled after the construction of the foundation and before the construction of the upper main structure, the embedded depth of the foundation should be corrected

This situation happens to most projects because it facilitates the construction activities, avoids the excavation after filling the recent fill, and reduces the construction cost. However, if a shallow foundation is used in this situation, how should the embedded depth of the foundation be calculated is not specifically stipulated in the Foundation Code and the insiders have different points of view on it. Lou et al. (2013) found that only when the recent fill is completed before the foundation construction can the embedded depth $d$ of foundation for correction be calculated from the ground level of the recent fill, and in this case, it can only be calculated from the natural ground level. The author considers it from the perspective of design and emphasizes the significance of safety. However, it is discussed in the research by Zhang (2014) that the recent fill is backfilled immediately after the foundation construction, and the embedded depth $d$ of the foundation for correction may be calculated from the ground level of the recent fill. The authors agree with this point of view overall because when the recent fill is filled immediately after the construction of the foundation and the upper load has not acted, the overload around the foundation has been formed, which contributes to the bearing capacity and stability of the foundation soil. However, the embedded depth of the foundation may be different in actual engineering design.

The Fuli Town Junior Middle School in Fu Yangshuo County is designed as a teaching building, which is a 5-6-story frame structure and mainly uses an independent column foundation, with shallow hard plastic clay as the bearing stratum $f_{ak} = 180kPa$. Since the site is below the flood level, the entire site is filled with a height of 2 m for convenience of use. The natural ground is excavated during construction, and the interior of a single building is backfilled to $\pm 0.000$ m after the construction of the foundation. To facilitate the erection of the outside scaffold, the outdoor ground is also filled with a 2 m thick recent fill, but it is only 1.5 m-2 m wider than the outer wall. The site beyond 2 m is not filled with recent fill, and the original ground is maintained. After the roofing of the main structure is completed, the outside scaffold is removed, the outdoor pipe

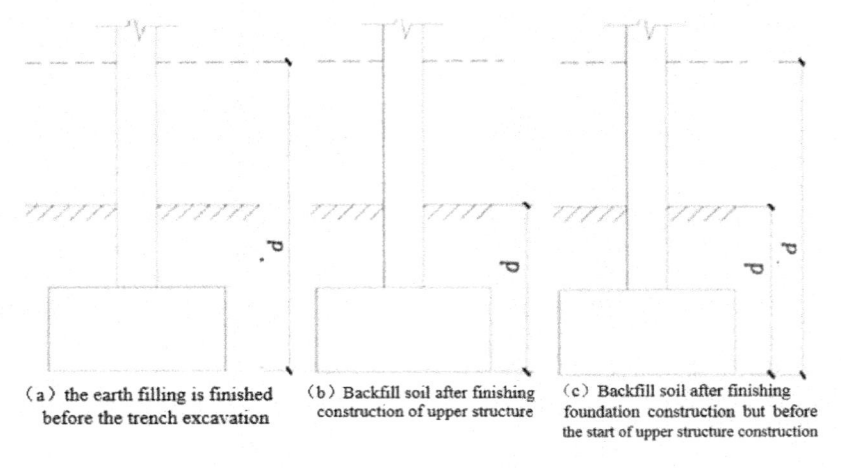

(a) the earth filling is finished before the trench excavation

(b) Backfill soil after finishing construction of upper structure

(c) Backfill soil after finishing foundation construction but before the start of upper structure construction

Figure 2. The recent fill of the foundation trench.

network (2 m to 3 m from the outer wall) is laid, and then 2 m thick recent fill is backfilled, as shown in Figure 3 below.

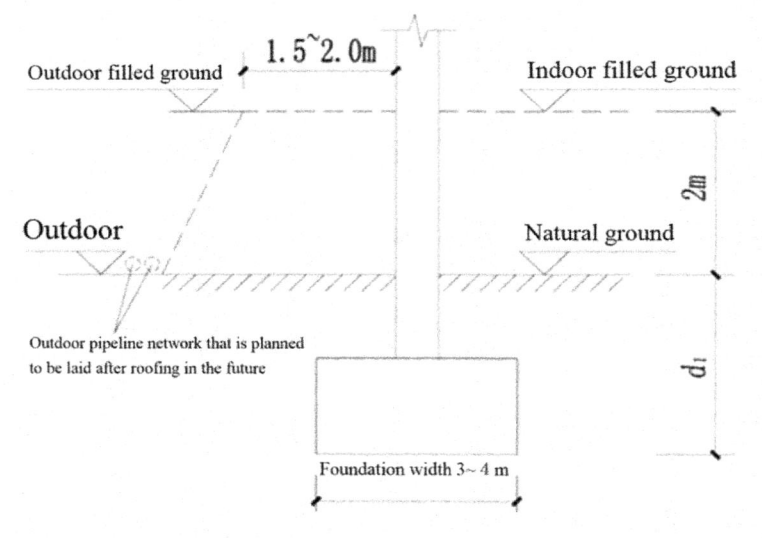

Figure 3. Main structure of the teaching building.

In this case, it is not safe that the calculated embedded depth of the foundation is $d = d_1 + 2\text{m}$ (Zhang 2014) because the foundation bearing capacity can be improved only when the overload width is greater than 2 times the foundation width according to Section 1. This requirement is not satisfied apparently in the above case. However, when $d = d_1$, it is too conservative and it leads to too low correction value $f_a$ of the foundation bearing capacity and too large foundation area, which is not economical. The designers refer to Page 169 of *Unified Technical Measures for Structural Design 2018 Edition* (China Design Institute Limited 2018) prepared by *China Architecture Design & Research Group* and it is $d = 0.5 (d_1 + d_2)$ for the exterior wall foundation in Figure 4. Therefore, the embedded depth of foundation used in this project is $d = 0.5 (d_1 + d_1 + 2\text{m}) = d_1 + 1\text{m}$. Furthermore, as shown in Figure 5, the additional compressive stress of the base is caused by the fill of the back subgrade of an abutment according to *Soil Mechanics* (3rd ed.) (2010):

$$P_{01} = \alpha_1 \gamma_1 H_1$$

where $H_1$ is the height of fill at the rear edge of the base; $\gamma_1$ is the weight density of filled earth; $\alpha_1$ is the vertical additional pressure coefficient.

When the fill height is smaller than or is equal to 5 m, $\alpha_1 = 0.44$, which means that the pressure caused by the self-weight of the fill should be reduced, being the same as the principle of the above measures. At present, Fuli Town Junior Middle School has been completed, and every building is in good condition. Both theoretically and practically, the basic design of the project has been proved safe and rational.

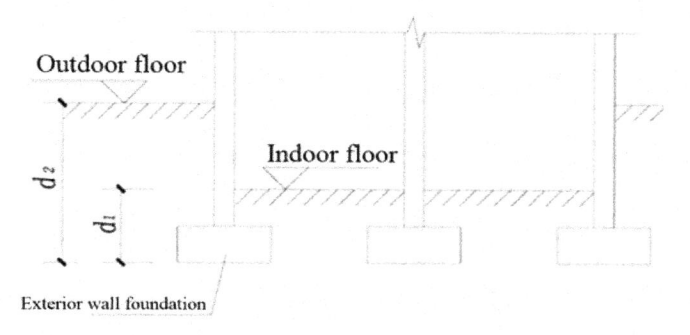

Figure 4.   The exterior wall foundation.

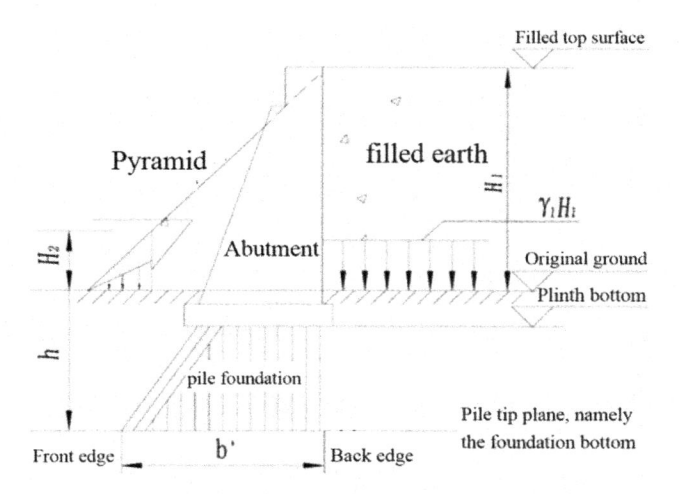

Figure 5.   The fill of the back subgrade of abutment.

Therefore, we believe that when the recent fill is filled after the construction of the foundation and the superstructure (similar to the situation of Fuli Town Junior Middle School), the embedded depth of the foundation can be calculated using the following expression: $d = d_1 + 0.5 \times$ the thickness of the recent fill, where $d_1$ refers to the natural ground.

This implies that the embedded depth of shallow foundations in large-area recent filling areas should depend on the construction requirements and filling stage, as appropriate. The designers should make judgments according to different situations.

# 4 DESIGN OF FOUNDATION IN THE CASE OF RECENT FILL THICKNESS (LARGER THAN FOUR METERS)

As an example, in a mountainous area where the fill thickness is greater than or equal to 4 m, the site leveling should be carried out first to facilitate the mobilization of construction equipment. When designing a foundation, the key consideration is not the depth of the embedded foundation, but rather how well the natural foundation fits. In most cases, the treatment of recent fill and the use of composite foundation are considered such as Jiangdong Resettlement Community of Guanyang County Relocation of Impoverished Residents. It is not the main topic of this paper.

# 5 DEFORMATION CONTROL DESIGN

In addition to the rational use of calculated embedded depth of foundation for the foundation design on the natural foundation covered by large-area recent fill, there is a very important process: deformation control design. The bearing capacity checking control determines the size of the foundation, which is the first step in the foundation design and may meet the requirements of bearing capacity of a large-area recent filling site, but the deformation may not meet the requirements. Under normal circumstances, the foundation that meets the deformation requirements will not have strength damage, while the foundation that does not consider deformation control often produces excessive settlement, resulting in tilting and cracking of the building, and such engineering accidents occur a lot. In the actual engineering design, deformation control is a process that many designers tend to ignore, but it is very important and will directly affect the size and form of the foundation, and even the safety of the building.

## 5.1 Calculation method

If the natural soil layer is the soil that has been consolidated, the self-weight stress does not cause the settlement of the building foundation and only the additional stress of the base can cause the settlement of the foundation. The large-area recent fill in the soil has not been consolidated, and the recent fill is the additional load for the lower natural soil layer, which produces additional stress on the natural foundation to increase the settlement of the foundation. As a result, it is pointed out in the Foundation Code that if there is thick or uneven fill in the foundation, and the self-weight consolidation has not been completed, the deformation checking should be done. As described in the study by Zhang (2004), "the additional settlement $S_g'$ caused by the ground load can be simplified as follows: it can be calculated according to the Foundation Code, but not multiplied by the empirical coefficient. The recent fill load is considered as the uniformly distributed load, and its acting surface is $P_0 = \gamma'H'$ at foundation ground." In addition, for the recent fill of the entire site, we can assume that it has an infinite uniformly distributed load, which is transmitted downward without diffusion. According to the Foundation Code, the formula for calculating the final deformation of a foundation is as follows:

$$S = \psi \sum_{i=1}^{n} \frac{P_0}{E_{si}} (z_i \overline{\alpha_i} - z_{i-1} \overline{\alpha_{i-1}}) \tag{1}$$

Equation (1) is used for large-area recent fill, and the final deformation of the foundation is calculated as follows:

$$S_g' = \sum_{i=1}^{n} \frac{P_0}{E_{si}} (z_i - z_{i-1}) \tag{2}$$

where $P_0 = \gamma'H'$; $\gamma'$ is the weight of recent fill; $h'$ is the thickness of recent fill.

## 5.2 Calculation example

For the new filling site of Yangshuo County Fuli Town Junior Middle School above, the average thickness of the recent fill is 2 m, the unit density after compaction is $\gamma = 18$ kN / m³, and the groundwater is usually below the natural ground. A typical soil layer profile is shown in Figure 6 below. Since there are two layers of soft substrata, the depth of foundation deformation is calculated until it reaches the rock surface, as described in Article (1): $P_0 = \gamma'h' = 18 \times 2 = 36$ kPa.

$$S = \sum_{i=1}^{n} \frac{P_0}{E_{si}} (z_i - z_{i-1}) = \frac{36}{7.0} \times 5 + \frac{36}{5.0} \times 1.2 + \frac{36}{3.0} \times 1 = 46 \text{ mm}$$

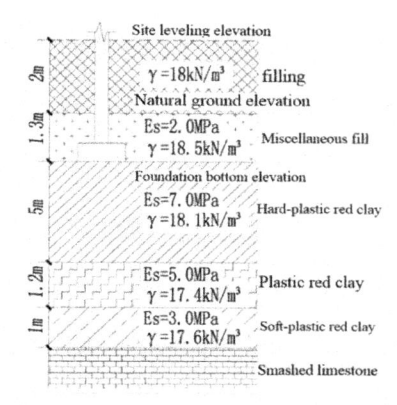

Figure 6.   A typical soil layer profile.

For the single foundation under the column mentioned above, the settlement of the column foundation calculated by JCCAD in PKPM is about 40 to 60 mm, and the total settlement of the column foundation of the building is 86 to 106 mm. The foundation settlement caused by the large-area recent fill cannot be ignored, especially when there is a soft substratum. The settlement caused by the recent fill plus the settlement of the column foundation of the building itself will greatly exceed that of the non-filled site. If it is ignored in the design, the excessive deformation of the foundation may occur, and accidents such as foundation inclination may happen. Since there is the excessive settlement of some buildings of Yangshuo County Fuli Town Junior Middle School, and the independent column foundation is changed to the strip foundation under the column, the deformation control design of the large-area recent filling site decides the size and form of the foundation.

## 6   CONCLUSIONS

In conclusion, in the design of foundation in large-area recent filling areas, if the fill is not too thick and the natural foundation is considered, the filling plan for the recent fill should be determined as required by the owner and according to the feasibility of construction, and the corresponding calculated embedded depth of foundation should be rationally used. The size and form of the foundation should be determined through the bearing capacity calculation and deformation control design (He 2021; Zhu & Zhang 2017). In addition, the following structural measures are recommended to avoid excessive deformation of the building foundation of the large-area recent filling site (Cui 1994; Shi et al. 2012).

(1) It is mentioned in the Foundation Code that large-area filling should be completed three months before foundation construction. Where conditions permit, it is recommended to complete backfilling three months before excavation, and use surcharge preloading to accelerate foundation consolidation and reduce deformation during use;
(2) Increase the elevation of the top of the column in advance;
(3) Select a lower base pressure, that is, increase the base area to reduce deformation;
(4) Increase the overall stiffness of the building, add a layer of straining beams on the ground floor, and if conditions permit, add a layer of boards. Avoid ground cracking caused by the subsidence of the backfill due to excessive thickness;
(5) Carry out settlement observation because the theoretical basis and assumed conditions of settlement calculation differ much from the actual situation of soils. The consolidation settlement of recent fill is also an unknown factor, and there are many uncertainties with weather changes. Settlement observation is an auxiliary means for safety monitoring of building usage.

## REFERENCES

China Design Institute Limited. (2018) *Unified technical measures for structural design* [M]. Beijing: China Construction Industry Press.

Cui Kai Calculation of additional settlement of foundation caused by large-area new fill and corresponding engineering measures [J] *Northwest Geology*, 1994, 15(3): 4.

He Mingyue Design of local thick miscellaneous fill complex foundation in thick self weight collapsible loess site [J] *Building structure*, 2021, 51(S01): 5.

Lou Yu, Yang Qi, Zhu Bingyin. (2013) *Design method and example analysis of building foundation 2nd Edition* [M]. Beijing: China Construction Industry Press.

Shi Chunle, Wang Pengfei, Wang Xiaojun, et al Measured analysis of large deformation and damage of adjacent buildings caused by deep foundation pit excavation [J] *Journal of Geotechnical Engineering*, 2012 (S1): 7.

Southeast University, Zhejiang University, Hunan University. (2010) *Soil mechanics. 3rd Edition* [M]. Beijing: China Construction Industry Press.

Weibin Zhang. (2004) *Questions and Answers on Concrete Structure Design* [M]. Beijing: China Construction Industry Press.

Zhang Hongming, Zhu Jiejing Foundation treatment and pile foundation design of large area filling [J] *Building structure*, 2017(S2): 4.

*Civil Engineering and Urban Research – Mohamed & Hou (Eds)*
© *2023 the Authors, ISBN 978-1-032-44487-1*

# The configuration characteristics and spatial relationships of indoor plants in large airports of China

Yang Keran*
*School of Journalism and Communication, Peking University, Peking, China*

Wuyun Bagen*
*School of Architecture, Huaqiao University, Xiamen, China*

Kato Shoko*
*Department of Design for Contemporary Life, Gifu City Women's College, Gifu, Japan*

Li Lu*
*School of Architectural Science and Civil Engineering, Xiamen Institute of Technology, Xiamen, China*

ABSTRACT: With the development of China's society and economy, people also have placed higher quality requirements for the indoor environment of airports. The introduction of plant landscapes into the airport space can improve indoor air quality and enrich the interior space design. Through the visual, tactile, and taste experience of plants, passengers can relieve their physical and mental pressures. In this study, the current situation of plant configuration was investigated and analyzed for representative airport terminals in North China and East China, which provided a relevant reference for the configuration and design of indoor plants in airports in the future. The results show that there are 40 species of indoor plants (21 families and 34 genera) in an airport in North China, and 31 species of indoor plants (13 families and 26 genera) in an airport in East China. We have been familiar with the configuration characteristics and spatial relationship of indoor plants in major airports in China.

## 1 INTRODUCTION

For a city, the airport is an important transportation hub and a window for displaying images and even an exchange place (Zhang 2021) for foreign cultures in a country. Although the airport interior space is the most frequently contacted by passengers, its greening design is often neglected. With the higher requirements for the quality of the living environment, introducing plants in the natural world into the airport indoor space becomes an indispensable and important design element to create a high-quality and comfortable airport indoor environment.

Changi Airport in Singapore has been honored as the "Best Airport in the World" by Skytrax for seven consecutive years. It created a three-dimensional indoor garden called "Green Tapestry", forming a "Forest Valley" (Lee 2006) of plants. Incheon International Airport in South Korea has applied a large area of vertical greening in the interior, and combined the plant landscape with artistic paintings and iron art, demonstrating Korean culture and characters with the decorative. Nakamura and other scholars have monitored the environmental conditions and physiological characteristics

*Corresponding Authors: 934388813@qq.com, wuyun_bagen@hotmail.com, h7632013@gmail.com and h7632013@gmail.com

of plants in the passenger terminals of Kansai International Airport (1998). From the above foreign cases, we can see that the plant landscape in airport design is normal and valued.

The research on airport landscape design in China mainly focuses on the planning and design of the surrounding space nodes such as the airport airline overlook area (Tao 2015), traffic roads (Wang 2012; Zhu 2018), and the ecological environment (Yu & Zhang 2019). In the research on the indoor plant landscape of the airport, Tang (2012) pointed out that there is solitary planting, coupled planting, and group planting that can be selected for airport indoor plants and plants can be arranged in point, line, and plane. Zhao (2017) also pointed out the basic forms and layout types of indoor plant landscape applications in airports, and the application methods of indoor plants. While analyzing the application of indoor plant landscape in a terminal building in East China, Wu (2019) summarized the daily maintenance and design forms of plants. Yu (2010) proposed that when designing the plant landscape for the airport indoor space, the selection of plants should consider the location of the airport, the climate, the spatial structure, and overall style of the indoor functional divisions, and harmonize and unify with the architectural style. The above research mainly focuses on the types of airport indoor plants, plant maintenance management, greening forms, and design methods. However, a detailed investigation of the current situation of plants in the airport has not been carried out, and no specific data.

To recap, it is the basis for plant landscape research in the airport to understand and grasp the actual status quo of application in the airport. In the research, the plant landscape inside the airport is taken as the object, a case study of indoor plant configuration status quo in the airports of 2 cities in North China and East China was used to investigate and analyze, which is aimed to understand plant configuration species, design forms and other features in the indoor space of China's airport to explore the status quo and rules of indoor plant landscape and provide reference and basis for the future design of indoor plant landscape in the airports of China.

## 2 METHOD

### 2.1 *Survey object*

Considering the climate and plant growth characteristics in different regions, this survey selected representative airports in North China and East China as the survey objects. At the same time, to avoid using the specific names of the airports, the airport in North China is abbreviated as airport A and the airport in East China as airport B. Airport A is located in North China, where the habitat is a typical warm temperate semi-humid continental monsoon climate (Pan 2018). The airport owns 214 parking spaces and a terminal area of about 3.14 million square meters. As an important main airport in North China, airport A flies 100.013 million persons annually. Airport B is located in East China with a subtropical monsoon climate, where there are four distinct seasons (Liu & Zhang 2019) and abundant precipitation. There are a total of 218 parking spaces in airport B, with a terminal area of about 622,000 square meters, and flies 76.1534 million persons annually.

### 2.2 *Survey methods*

From June 16, 2020, to June 28, 2021, field research was conducted on the indoor plant landscape design that appeared in airports A and B. To correctly identify the species, the survey conducted preliminary identification and recording of plants species using on-site observation and photography and confirmed plant information such as species and classification by referring to the illustrated book (2016) and Drude's life-type system (Wang 1987) in the future. Due to the huge size of the airport terminals and their ancillary facilities, the research is limited to the public area of terminal 3 of airport A and terminal 2 of airport B, including the departure floor, arrival floor, waiting area, security check area, dining, and leisure area, shopping mall and connecting channels.

Table 1. Statistics of indoor plant species in airport A.

| Latin name | Family |
| --- | --- |
| *Cordyline fruticosa* (L.) A. Cheval. | Liliaceae |
| *Chlorophytum comosum* var. Variegatum | Liliaceae |
| *Ravenala madagascariensis* Adans. | Musaceae |
| *Wisteria villosa* Rehd. | Leguminosae |
| *Euphorbia pulcherrima* Willd. et Kl. | Euphorbiaceae |
| *Billbergia pyramidalis* (Sims) Lindl. | Bromeliaceae |
| *Strelitzia reginae* Aiton | Strelitziaceae |
| *Fittonia verschaffeltii* (Lemaire) van Houtte | Acanthaceae |
| *Cymbidium faberi* × *hybridum* | Orchidaceae |
| *Oncidium flexuosum* Lodd. | Orchidaceae |
| *Phalaenopsis aphrodite* Rchb. F. | Orchidaceae |
| *Aglaia odorata* Lour. | Meliaceae |
| *Fagraea ceilanica* Thunb. | Gentianaceae |
| *Sansevieria trifasciata* var. *laurentii* N. E. Brown | Agavaceae |
| *Pachira glabra* Pasq. | Bombacaceae |
| *Gardenia jasminoides* Ellis | Rubiaceae |
| *Rosa chinensis* Jacq. | Rosaceae |
| *Nephrolepis auriculata* (L.) Trimen | Nephrolepidaceae |
| *Beaucarnea recurvata* Lem. | Asparagaceae |
| *Dracaena angustifolia* Roxb. | Asparagaceae |
| *Dracaena deremensis* Engl. | Asparagaceae |
| *Dracaena fragrans* var. *massangeana* Hort. | Asparagaceae |
| *Aglaonema modestum* cv. Red Valentine | Araceae |
| *Aglaonema modestum* cv. Silver Queen | Araceae |
| *Aglaonema modestum* Schott ex Engl. | Araceae |
| *Anthurium andraeanum* Linden | Araceae |
| *Dieffenbachia picta* (Lodd.) Schott Bunting | Araceae |
| *Epipremnum aureus* (Linden et André) Bunting | Araceae |
| *Philodendron gloriosum* | Araceae |
| *Spathiphyllum kochii* Engl. et Krause | Araceae |
| *Syngonium podophyllum* Schott | Araceae |
| *Hedera helix* var. Discolor | Araliaceae |
| *Heteropanax fragrans* (Roxb.) Seem. | Araliaceae |
| *Schefflera octophylla* (Lour.) Harms | Araliaceae |

## 3 RESULTS AND DISCUSSION

### 3.1 *Status quo of indoor plant configuration in airports*

A field survey was conducted in airport A, and the types of plants recorded in the survey are shown in Table 1. There are 40 species of commonly used green plants in airport A (including subspecies, and varieties), in a total of 21 families and 34 genera. Among them, there are 10 families, 12 genera, 12, species of dicotyledons, and 10 families, 21 genera, and 27 species of monocotyledons. Among them, foliage plants accounted for 67.50%, while flowering plants accounted for only 32.50%. Plants with higher frequency include *Cymbidium*, mosaic ivy, *Dieffenbachia*, and *Scindapsus aureus*.

According to Drude's life-type system, plants can be divided into five types: arbors, shrubs, annual herbs, perennial herbs, and vines. It can be seen from Table 2 that the indoor green plants in airport A are dominated by herbs, accounting for about 52.50%; followed by shrubs, accounting for about 25.00%, and arbors and vines accounting for 17.50% and 5.00%, respectively.

Table 2. The main families and life forms of greening plants in airport A.

| The main families | Number of general | Proportion in total general (%) | Species number | Proportion in total general (%) |
|---|---|---|---|---|
| Orchidaceae | 3 | 8.82 | 3 | 7.5 |
| Asparagaceae | 2 | 5.88 | 4 | 10 |
| Araceae | 7 | 20.59 | 9 | 22.5 |
| Araliaceae | 3 | 8.82 | 3 | 7.5 |
| Marantaceae | 1 | 2.94 | 3 | 7.5 |
| Total | 16 | 47.05 | 22 | 55 |

| Life-type | Species number | Proportion (%) |
|---|---|---|
| Arbors | 7 | 17.5 |
| Shrubs | 10 | 25 |
| Annual herbs | 0 | 0 |
| Perennial herbs | 21 | 52.5 |
| Vines | 2 | 5 |
| Total | 40 | 100 |

A field investigation was conducted in airport B, and the types of plants recorded in the investigation are shown in Table 3. There are 31 species of commonly used green plants in airport B (including subspecies and varieties), in a total of 13 families and 26 genera. Among them, there are 6 families, 8 genera, 8, species of dicotyledons, and 7 families, 18 genera, and 23 species of monocotyledons. Among them, foliage plants account for 75.00%, while flowering plants only

Table 3. Statistics of indoor plant species in airport B.

| Latin name | Family |
|---|---|
| *Salvia japonica* Thunb. | Labiatae |
| *Codiaeum variegatum* (L.) A. Juss. | Euphorbiaceae |
| *Bambusa multiplex* (Lour.) Raeusch. ex Schult. | Gramineae |
| *'Fernleaf'* R. A. Young | |
| *Billbergia pyramidalis* (Sims) Lindl. | Bromeliaceae |
| *Pleioblastus amarus* (Keng) keng | Gramineae |
| *Strelitzia reginae* Aiton | Strelitziaceae |
| *Pachira glabra* Pasq. | Malvaceae |
| *Kalanchoe blossfeldiana* L. | Crassulaceae |
| *Dendrobium nobile* Lindl. | Orchidaceae |
| *Phalaenopsis aphrodite* Rchb. F. | Orchidaceae |
| *Dracaena angustifolia* Roxb. | Asparagaceae |
| *Dracaena deremensis* Engl. | Asparagaceae |
| *Dracaena deremensis* 'Lemon Lime' | Asparagaceae |
| *Aglaonema modestum* cv. Pattaya Beauty | Araceae |
| *Aglaonema commutatum* cv. Pseudo-bracteatum | Araceae |
| *White Rajah* | |
| *Anthurium andraeanum* Linden | Araceae |
| *Dieffenbachia amoena* cv. Camilla | Araceae |
| *Dieffenbachia amoena* cv. Tropic Snow | Araceae |
| *Dieffenbachia picta* (Lodd.) Schott Bunting | Araceae |
| *Dieffenbachia sequina* (L.) Schott | Araceae |
| *Epipremnum aureus* (Linden et André) Bunting | Araceae |
| *Philodendron imbe* Schott ex Engl. | Araceae |
| *Spathiphyllum kochii* Engl. et Krause | Araceae |

*(continued)*

Table 3.    Continued.

| Latin name | Family |
|---|---|
| *Zamioculcas zamiifolia* Engl. | Araceae |
| *Hedera nepalensis* var. *sinensis* (Tobl.) Rehd. | Araliaceae |
| *Schefflera octophylla* (Lour.) Harms | Araliaceae |
| *Ctenanthe oppenheimiana* (E. Morren) K. Schum. | Marantaceae |
| *Maranta bicolor* Ker | Marantaceae |
| *Caryota ochlandra* Hance | Palmaceae |
| *Collinia elegans* (Mart.) Liebm. | Palmaceae |
| *Phoenix hanceana* Naud. | Palmaceae |

account for 25.00%, which is 7.50% less than that of airport A. Plants with a higher frequency include *Chamaedorea elegans*, *Billbergia pyramidalis*, *Bambusa multiplex*, *Codiaeum variegatum*, and *Scindapsus aureus*.

As shown in Table 4, the indoor green plants in airport B are mainly herbs, accounting for 64.51%, of which 61.29% are perennial herbs, and annual herbs only account for 3.22%. Followed by shrubs, accounting for 16.13%, and arbors and vines accounted for 9.68% of the total.

Table 4.    The main families and life forms of plants in airport B.

| The main families | Number of general | Proportion in total general (%) | Species number | Proportion in total general (%) |
|---|---|---|---|---|
| Gramineae | 2 | 7.69 | 2 | 6.45 |
| Orchidaceae | 2 | 7.69 | 2 | 6.45 |
| Asparagaceae | 2 | 7.69 | 3 | 9.68 |
| Araceae | 7 | 26.92 | 11 | 35.48 |
| Araliaceae | 2 | 7.69 | 2 | 6.45 |
| Marantaceae | 2 | 7.69 | 2 | 6.45 |
| Total | 19 | 73.06 | 24 | 77.41 |

| Life-type | Species number | Proportion (%) |
|---|---|---|
| Arbors | 3 | 9.68 |
| Shrubs | 5 | 16.13 |
| Annual herbs | 1 | 3.22 |
| Perennial herbs | 19 | 61.29 |
| Vines | 3 | 9.68 |
| Total | 31 | 100 |

Judging from the configuration characteristics of indoor plants in airports A and B, there are 31 to 40 plant species, mainly herbs, followed by shrubs. The main families of indoor plants in the two airports are highly similar. The appearance of palm plants in airport B reflects the characteristics of plants in different regions, but there is no major difference viewing from other plants. In terms of the number of plant species, the number of indoor plants is less than that in commercial spaces in Hangzhou (80 species) (Fan et al. 2015). and Xiamen (120 species in total) (Wuyn et al. 2015). In addition, the two airports do not make good use of local plants, almost all of them are exotic species and highly similar. Taking airport B as an example, local plants with high ornamental value including Ginkgo biloba, silk tree, wild chrysanthemum, and radix ophiopogonin (Tian 2003) are not used. Chung (2004) and Zhu (2018) advocated that greening should be dominated by local plants, emphasizing regional cultural characteristics, supplemented by imported species so that the diversity of plants can be fully reflected, and the best ecological benefits can be realized. Although there are some gardens in the airports, most of them are simply piling plants and local characteristics cannot be reflected. Nanjing Lukou International Airport is prominent in terms of reflecting regional characteristics. The waterscape design focuses on the atrium garden connecting the arrival hall and the departure hall. To combine with traditional culture, sculpture pieces with "Konghou" elements are arranged, surrounded by tropical foliage plants such as monstera and star anise (Sun 2012). The combination of waterscape, sculptures, and plants reflects the characteristics of a typical water town and reminds the historical sense of the ancient capital of the six dynasties. The regional cultural design of the airport indoor plant landscape is very important for improving the cultural atmosphere of the airport space and shaping the city image.

### 3.2 Configuration characteristics and spatial relationship of indoor plants in airports

The plant landscape configuration of airport A and airport B are distinct in different indoor functional spaces. The halls on the departure floor and arrival floor are the spaces firstly impressing the passengers, which, usually, are high and large with the design of large-scale landscape combinations. Large solitary arbors are set up at the flow intersection of the halls and used as a sign for identification and diversion. Taking the landscape design of the welcome hall of Airport B as an example, the plant group is set in the central area. A *Phoenix loureiro* with a height of more than 3 m is planted in the upper area, and a small amount of strelitzia, as well as *Malabar chestnut* is in the medium, combined with the arrangement of a small number of scenery stones. Plants in the lower area are the foliage plants including *Codiaeum variegatum, Calaihea leopardina, Aglaonema Snow White, Fragrant dracaena*, as well as ornamental plants such as jonquil. Surrounded by wooden benches, the whole plant design is naturally arranged.

Most of the plant bonsai in the waiting area and luggage area are arranged as foliage plants with a height of more than 1 m, which are generally distributed as solitary potted plants to reduce the floor space. Taking airport A as an example, in the same waiting area, *Ginkgo biloba* is distributed in a solitary form. Plants are not commonly used in the whole area for the main reason of saving space and placing more seat chairs for passengers when they are waiting.

Due to the high flow rate of passengers in the security check area, there are few plant landscapes for safety reasons. Generally, potted foliage plants are placed on both sides of the team. The plant landscape in the dining and leisure area appears in groups, whose overall size and height are moderate. Generally, perennial herbs or small shrubs and chairs are used to seal the periphery of the landscape. Taking Airport B as an example, the landscape is dominated by flowering plants in rich colors. Equipped with vines such as *Scindapsus aureus* and ivy, combined with medium- and low-sized plants such as *anthurium andraeanum*, it is the brightest area with indoor plants in the airport. At the same time, with seasonal flowering plants, the booths are enclosed to form medium-sized pieces, which are ornamental and practical, providing a resting place for travelers.

Most of the plants in the channel connection area are distributed in lines, and the potted herbs or small shrubs are placed equidistantly on both sides of the channels, mainly for guiding and leading. *Chamaedorea elegans, Codiaeum variegatum*, and arrowroot are used in airport A, which are coupled and planned in pots on both sides of the passage to guide the flow, or placed in a single line in the center of the passage to divide and guide the flow.

Viewing the characteristics of different plant landscape configurations in each functional area of Airport A and B, we can comprehend the characteristics and spatial relationship of indoor plant configuration in airports, thus providing a reference for the optimization and design of airport indoor green landscapes.

## 4 CONCLUSIONS

With people's increasing demand for environmental quality, based on meeting the main functions, people's demand for returning to nature should be considered in airport interior design. This research is focused on the indoor plant landscape of airports, in which the current situation of airport indoor plant configuration is investigated and analyzed in two cities of North China and East China to comprehend the types and design characteristics of airport indoor configuration nowadays.

The survey results of airport A (North China) and airport B (East China) show that there are 40 indoor plant species in airport A, of which *Orchidaceae, Asparagaceae, Araceae, Araliaceae,* and *Marantaceae* account for the major part. There are 31 indoor plant species in airport B, of which *Asparagaceae, Araceae,* and *Palmae* account for the major part. The plants in the two airports are mainly herbs, followed by shrubs, but almost all of them are exotic species, and native plants are not effectively used. Meanwhile, by viewing the characteristics of different plant landscape configurations in each functional area of airport A and airport B, we can comprehend the characteristics and spatial relationship of indoor plant configuration in airports.

From the above results, we can comprehend the current situation of the configuration and design of airport indoor plant landscapes, meanwhile clearly see the lack of cultural connotations in domestic airport indoor plant landscapes, especially the lack of diversity and locality of plant species, meanwhile the styles of species and landscape gardens are relatively simple. Therefore, for the design of airport indoor plant landscapes in China, natural elements should be actively introduced, and airports should strive to improve the quality of indoor plant landscapes as well as the design of cultural connotations to promote passengers' physical and emotional health by considering the color richness, intimacy, and interaction of plant landscapes.

## ACKNOWLEDGEMENTS

This work was supported by the education research project for young and middle-aged teachers of Fujian Province (Grant No. JAT170043)) and the high-level talent introduction and research start-up fund project launched by Huaqiao University (Grant No. Z14Y0010).

## REFERENCES

Chung, Sung-Hye (2004). Interior Landscape Design for Samsung Semiconductor Gymnasium. *Journal of Korean Society for Plants, People and Environment*, 64–69.

Edited by the editorial board of color atlas (2016). *higher plants in China Color atlas of higher plants in China.* Beijing: Science Press (11), 1.

Fan, L. P. & Yu, L. X., etc (2015). Analysis of application of plant landscape in indoor commercial space – Taking the main urban area of Hangzhou as an example. Chinese society of landscape architecture, *Proceedings of 2015 annual meeting*, 486–489.

Lee (2006). *Changi Airport – connecting Singapore with the world*, Civil aviation of China, (12): 38–40.

Liu, M. & Zhang, D. S. (2019). Impact of climate warming on urban garden trees. *Science*. 71 (03): 25–28.

Nakamura, A., Abe, M., Okumura, S., Nakagawa, I., & Morimoto, Y. (1998). Characteristics of transpiration and photosynthesis of 11 planted evergreen tree species in the passenger terminal building of kansai international airport and their environmental factors. *Journal of the Japanese Society of Revegetation Technology*, 24 (2), 71–79.

Pan, L. G. (2018). *Application Research of Household Low Ring Temperature Air Source Heat Pump Heating in Beijing.* Shenyang: Shenyang Architecture University.

Sun, X. J. (2012) *Research and analysis of interior plant landscape design of the public transportation space in Nanjing.* Nanjing University of technology.

Tang, S. L. (2012). *Application of indoor plant configuration in improving the environment of airport waiting hall, agriculture and, technology*, (05): 185.

Tao, J. L. (2015). Preliminary study on landscape style planning of Fuzhou Changle airport route overlooking area. *Southern agriculture*, 9 (21), 105,107.

Tian, D. K. (2003). Protection and utilization of local plants in Shanghai, *Garden*, (08), 63.

Wang, B. S. (1987). *Plant Community Science Higher Education*

Wang, Z. H. (2012). Slide slope greening technology of Airport Expressway. *Journal of Green Science and Technology*. (02), 67–68.

Wu, J. F. (2019) *Research on the application of plant landscape in interior design, urban housing*, (03), 164–166.

Wuyun, B. G., Kato, S. K., Zhao, Y., etc (2018). Xiamen City commercial facilities indoor plants utilization real state, *Journal of Japan Greening Engineering Association*, (01), 237–240.

Yu, F. S. & Zhang, Y. (2019) The Planning and Construction of the Wetland Ecological Belt of the Airport? ChannelsofWuhan. *Garden*, (04), 1216.

Yu, L. P. (2010). Landscape design of Shanghai Hongqiao International Airport expansion project. *Shanghai construction technology*, 5, 47–49.

Zhang, F. (2021). *Improve airport service quality and promote the construction of civil aviation power*, Transportation enterprise management, 27 (12): 53–54.

Zhao, X. (2017). *Application of plants in indoor environment of terminal, modern horticulture*, 11, 111–113.

Zhu, G. F. (2018). Discussion on landscape greening design of Shang ri La airport. *Forestry construction*, (03), 73–77.

*Civil Engineering and Urban Research – Mohamed & Hou (Eds)*
*© 2023 the Authors, ISBN 978-1-032-44487-1*

# Research on the design of urban road systems under the concept of sponge cities

Hongli Huang*
*Jiangxi University of Engineering, Xinyu, Jiangxi, China*

Liangsong Li*
*Sponge City Research Institute, Pingxiang University, Pingxiang, Jiangxi, China*

ABSTRACT: Because traditional cities have ecological security problems such as rainwater resource loss and pressure to flood discharge, it is easy to cause the life and property safety of urban residents to be guaranteed. Therefore, the design and application of an urban road system based on the concept of a sponge city, a traffic trunk road as the engineering case, cross-section, and interchange as the angle of system optimization use the SWMM model to simulate and test the urban road system. The final test results also show that the urban road system based on sponge city has been optimized to the greatest extent, the rainwater resources have been recycled, and the urban flood discharge capacity has been greatly improved. This provides a certain research basis for the subsequent optimization research of the urban road system.

## 1 INTRODUCTION

There are many problems with traditional urban roads. Whether it is waterlogging or the loss of rainwater resources, this is a certain threat to the ecological security of the city. In serious cases, it may also cause the loss of life and property of urban residents. Therefore, some scholars put forward the concept of sponge city and try to find the key points of urban road system optimization based on the concept of sponge city. Some scholars also put forward corresponding suggestions on the development and optimization of sponge city from the micro perspective of urban blocks and constructed a block urban design system oriented by sponge city. However, there are few engineering cases for sponge city road design optimization, and the evaluation standards are not standard enough, so the sponge city road system optimization studied in this paper is of great engineering significance.

## 2 EXAMPLE ANALYSIS OF SPONGE CITY ROAD SYSTEM OPTIMIZATION PROJECT

### 2.1 Spirits of the design

The design idea of the sponge city road is to realize the road ecological drainage of sponge city through the permeable pavement and rainwater wetland based on ensuring urban ecological safety, as shown in Figure 1 (Wang 2022).

---

*Corresponding Authors: 2449645938@qq.com and 2780874560@qq.com

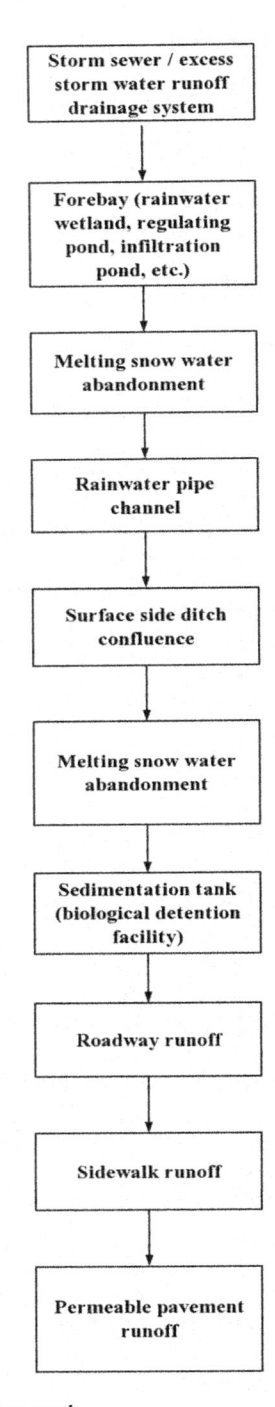

Figure 1.　The drainage system in urban road.

## 2.2 Relying on the project overview

Relying on a certain traffic trunk road, the concept of a sponge city is analyzed (Chen 2022). The total length of the road is 8.1 km, mainly using nonpermeable asphalt concrete pavement. The cross-section is shown in Figure 2.

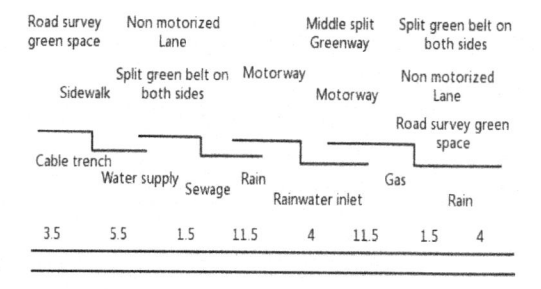

Figure 2. Original cross-section design of a road (unit: m).

It can be seen that too many rainwater discharge facilities on the road will not only occupy the road surface but also lead to damage to the ecological environment to a certain extent. Therefore, the design of the sponge city road engineering system should be considered from both cross-section and interchange (Jiang 2022).

## 2.3 Cross-sectional optimization design based on LID

Because of the high rainfall, large rainfall areas should use permeable asphalt concrete pavement for motor vehicles and nonmotor roads. A 4-cm thick OGFC mixture can reduce the proportion of pavement seepage, and a drainage pipe buried in the permeable layer can guide infiltrating rainwater into the pipe, and it can be connected to both sides of the green belt (Zhang 2022). Therefore, it is necessary to set up rainwater outlets in the green belt on both sides, as shown in the figure below.

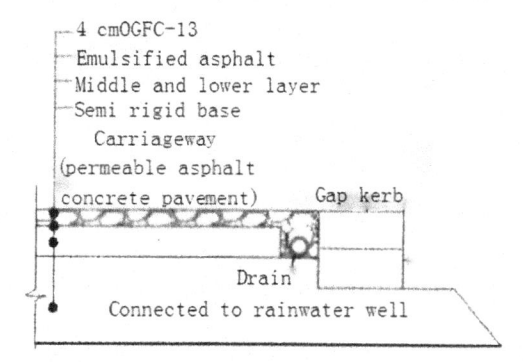

Figure 3. Motor vehicle lane.

## 2.4 Sidewalk optimization

The pavement is paved with permeable brick. The permeable brick through the surface layer is 6 cm, the horizontal layer is 3cm and the permeable base is 25 cm (Xiang 2022). Meanwhile, a drainage hose is also set at the permeable base of the sidewalk to connect the green belt on the roadside, as shown in Figure 4.

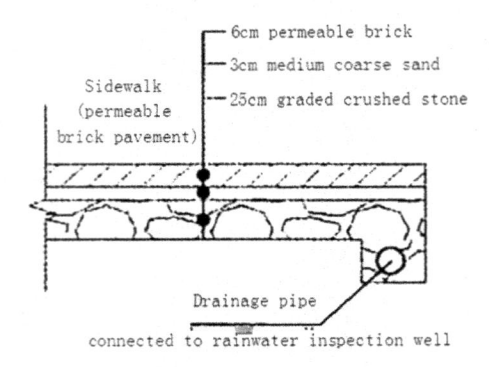

Figure 4. Pavement.

## 3 OPTIMIZATION EFFECT TEST OF SPONGE CITY ROAD SYSTEM

### 3.1 *SWMM model*

The SWMM model mainly determines the study area by dividing the sink area and calculates the runoff process and the drainage process to realize the simulation of the ground runoff process (Liao 2022). The first step in applying SWMM models in sponge urban road optimization is to collect relevant research data from the sink area and to determine the outlet, hinge, and LID facilities, as well as cohesion. It is also necessary to determine the physical and mechanical parameters of each element's location and conduct simulation calculations and analysis of the results (Dong 2022).

### 3.2 *Simulation test of the optimization effect of the sponge city road system*

The simulated section selected for the test is 686.5 m long, the cross-section is 43 m, the east and west green areas of the red line are 2.8 hm$^2$, and the impermeability is 5% before development. After inputting these attributes and calculating, the surface runoff error, which is within the permitted range, is $-0.28\%$, and in addition, the flow calculation error does not exceed 0.025%. However, the surface runoff error and flow calculation error of the optimized sponge city road system were calculated by the SWMM model, and the results were 0.21% and $-0.08\%$, respectively, which were within the permitted range (Mo 2021).

Table 1. Flow simulation results before and after road optimization under the SWMM model.

| Working condition | Time/s | Flow LPS |
| --- | --- | --- |
| Site developer | 0.15 | 0 |
| | 0.45 | 0 |
| | 1 | 0 |
| | 1.15 | 1.58 |
| | 1.45 | 8.01 |
| | 2 | 35.68 |
| | 2.15 | 140.14 |
| | 2.45 | 265.4 |
| | 3 | 357.23 |
| | 3.15 | 357.23 |
| | 3.45 | 357.23 |
| | 4 | 357.23 |
| | 4.45 | 357.23 |

(*continued*)

Table 1. Continued.

| Working condition | Time/s | Flow LPS |
|---|---|---|
| Sponge City Road | 0.15 | 0 |
| | 0.45 | 0 |
| | 1 | 0 |
| | 1.15 | 8.73 |
| | 1.45 | 46.74 |
| | 2 | 66.85 |
| | 2.15 | 129.68 |
| | 2.45 | 146.78 |
| | 3 | 157.89 |
| | 3.15 | 269.65 |
| | 3.45 | 304.78 |
| | 4 | 389.64 |
| | 4.45 | 389.64 |

In the first hour, the flow was small, rising after the hour. At three hours, the flow reached 357.23 LPS and almost no longer floated thereafter, with a total runoff of 2,237 LPS (Zhou 2022).

Sponge city road system optimization should be occured like this: small flow within an hour, but the arrival of the peak flow was delayed an hour. In four hours, it peaks at 389.64 LPS, and the total runoff was reduced by 17.1% before development. It can be said that the optimized road system will delay peak time, reduce the total runoff, and, to a certain extent, reduce the risk of urban waterlogging.

## 4 CONCLUSION

Sponge city road design needs to follow the principle of ecological priority, ensure the implementation of urban waterlogging prevention measures to a certain extent, and ensure the recycling of urban rainwater while protecting the ecological environment. Among them, the combination and choice of LID facilities are the most critical. Sponge urban road design should realize ecological drainage on the premise of meeting the needs of safety, beauty, and practicality, and the cross-section can be optimized by using LID facilities. The final system test also showed that optimizing the sponge city roads can delay the peak time and reduce the rainwater pipe network pressure. The next step needs to continue to deepen the optimization of the urban road system.

ACKNOWLEDGMENTS

This work is part of the Study on Landscape Evaluation of Pingxiang Expressway Based on Digital Mode (Grant No. 2020H0115) and is supported by the Jiangxi Provincial Department of Science and Technology (Grant Nos. 2021B0501 and 2021B0502).

REFERENCES

Baiguo Jiang. Application of Sponge City in Municipal Road Water Supply and Drainage Design [J]. *Career*, 2022 (03): 118–120.

Cheng Yao. Research on sponge City Planning System and implementation in the background of Space Governance [J]. *Chinese Building Decoration*, 2022 (04): 36–37.

Cong Wang, Weiqiang Han, Yang Wang, Shuhui Cheng, Tong Sha. — takes Beijing A New City as an example [J]. *Sichuan Environment*, 2022, 41(01):18–23. DOI:10. 14034/j. cnki. schj. 2022. 01. 004.

Feiyong Chen, Jinyu Li, Jin Wang, Jian Liu, Lingyi Wu, Bo Liu, Sisi Xu, Xue Shen, Yang Song. Enlightenment of rain and flood disaster response experience in Ishikawa Basin of Japan on the development of sponge city in China [J / OL]. *News of Yangtze River Academy of Sciences*: 1–11 [2022-05-09]. http://kns. cnki. net/kcms/detail/42. 1171. TV. 20220307. 0954. 020. html

Heng Zhang, Tianci Lin, Bingyun Zheng, Genli Tang. — takes Tianjin S of VFM in Sponge City PPP Project as an example [J]. *Journal of Tianjin University* (Social Science edition), 2022, 24 (03): 212–222.

Hongwu Xu. Road plant selection and application based on sponge city theory — takes Qishan Avenue in Chizhou city as an example [J]. *Agriculture and Technology*, 2022, 42(05):122–126. DOI:10. 19754/j. nyyjs. 20220315030.

Hui Wen. Based on the selection and design of Plants in the construction of "Sponge City" in Guangzhou, — takes Pocket Park as an example [J]. *Modern gardening*, 2022, 45(02):57–59. DOI:10. 14051/j. cnki. xdyy. 2022. 02. 021.

Jian Zhou. Research on the road reconstruction and upgrading methods of old urban residential areas under the sponge city concept [J]. *Building Safety*, 2022, 37 (01): 31–35.

Juhu Xiang, Tao Yuan. Chinese Sponge City Construction under the Vision of Ecological Civilization: International Experience and Path Selection [J]. *Innovation*, 2022, 16 (01): 109–116.

Notice of Zhumadian Municipal People's Government Office on Printing and Issuing of Zhumadian Sponge City Construction Management Regulations (Interim) [J]. *Zhumadian Municipal People's Government Bulletin*, 2022 (01): 32–37.

Puna Liao, Hang Li, Ruilai Yi. Construction and application of the implementation effect evaluation system of sponge city construction [J]. *Water Conservancy Planning and Design*, 2022 (04): 29–32 + 73.

Wujuan Dong. Research on sponge City Planning System and implementation in the background of Space Governance [J]. *Scientific consulting* (Science and technology management), 2022 (03): 38–40.

Xiong Xiong, Yiming Liu, Yang He, Dingfang Xu, Tie Pang, Chuanghua Cao. Study on the suitability evaluation of geological conditions of sponge city construction in Changde City [J]. *Groundwater*, in the 2022, 44(02):16–20. DOI:10. 19807/j. cnki. DXS. 2022-02-005.

Xue Bai, Hui Yu. Analysis of Application Strategy of Sponge City Concept in Urban Waterfront Landscape Design [J]. *Modern gardening*, 2022, 45(07):142–144. DOI:10. 14051/j. cnki. xdyy. 2022. 07. 075.

Yan Yang, Wenbin Deng. Research on Risk Management of PPP Model in Sponge City from the View of Secondary Financing [J]. *Commercial Economy*, 2022(02):131–133. DOI:10. 19995/j. cnki. CN10-1617/F7. 2022. 02. 131.

Zulan Mo, Pan Wang, Wuxing Zhu, Songlei Han, Jieran Li, Xinyi Xiang, Chen Zhang, Yongpeng Lv. Research on the volume conversion method of sponge city facilities and rainwater pipeline upgrading and storage facilities [J]. *Water Supply and Drainage*, 2022, 58(03):29–34. DOI:10. 13789/j. cnki. wwe1964. 2021. 11. 19. 0004.

*Civil Engineering and Urban Research – Mohamed & Hou (Eds)*
*© 2023 the Authors, ISBN 978-1-032-44487-1*

# Development of urban public management platform based on geographic information technology

Jing Wu*

*Nanchang Institute of Technology, Nanchang, Jiangxi, China*

ABSTRACT: Geographic information technology plays an important role in public management and services. Geographic information technology can integrate multi-source data, and use powerful geographic information processing capabilities to analyze data to obtain hidden information in the data. This paper constructs a public management platform of geographic information serving the people. On this platform, a lot of information during the operation of the city is recorded, such as urban traffic information, e-government information, and so on. This information can help governments manage public affairs more efficiently and provide better public services to people. This public management platform adopts open standard protocols such as Web Service so that other government affairs systems can connect with this platform. This platform can also share distributed geographic information and provide users with an application development environment, making the platform more widely used.

## 1 INTRODUCTION

With the development of science and technology, computer technology and its extended science and technology have been widely used in many fields. In the process of managing and serving the public, the national government also uses computer technology to improve the efficiency and quality of service management. Nowadays, when people use computer technology, geographic information has become an important source of information in daily life. There are already many companies in the market that provide users with channels to use geographic information. For example, on takeaway platforms, users can check the geographic location information of takeaways and restaurants at any time. In social software, users are provided with the function of making friends based on geographic location information. In the process of government management, geographic location information is an important basis for government decision-making. At present, government departments lack a unified data source and the geographic information collection systems constructed by each department are not the same, and the data cannot be integrated and interacted with, which seriously affects the effect of geographic information technology application. To provide the public and government departments with a high-efficiency and high-quality public service and management platform, this research builds a geographic information public service platform that can integrate various types of geographic information resources. The geographic information public service and management platform constructed in this study can provide the public, enterprises, and government departments with various types of services such as querying geographic information, analyzing geographic information, and spatial positioning. The platform has an open development interface and is extensible the space provides a basic guarantee for future updates and expansions.

---

*Corresponding Author: jxnc2222@163.com

DOI 10.1201/9781003372417-61

Geospatial refers to a coordinate positioning datum and the geographic entities to which it is attached. Geospatial data is the expression of geospatial data, including all data with geographic coordinates in the fields of resources, environment, economy, and society. Geospatial data makes the core of geographic information technology (Li 2014). The object of geographic information technology operation is geospatial data. To design a system using geographic information technology, the first step is to create a geospatial database (Liu 2000).

The geospatial data is classified according to the topology type, which can be divided into point data, line data, polygon data, etc. Remote sensing and mapping techniques can be used to obtain geospatial data. Today's modern devices such as GPS, RTK, and CORS base stations can also acquire accurate geospatial data. Geospatial data has the characteristics of multi-source and heterogeneity (Xue 2022).

The multi-source of geospatial data is reflected in the multi-source of data acquisition means, multi-source of storage format, rich semantics, multi-temporal and multi-scale, and so on. There are many ways to obtain geospatial data, such as satellite remote sensing, digital aerial photography, GPS measurement, stereo image mapping, and so on (Jia 2000). Geospatial data is obtained in various ways, which leads to the diversity of storage formats. Different data acquisition methods will lead to great differences in geospatial data processing, processing, organization, storage, and management. People have different understandings of geospatial elements, which leads to different geospatial data models. Multi-temporality refers to the fact that geospatial data has practical and spatial properties. Geospatial data describes the state characteristics of geographic entities at a specific point in time and a specific location. Multiscale refers to the diversification of geospatial data hierarchies. Different uses of geospatial data result in different scales of geospatial data. Geospatial data is also divided into one-year data and two-year data (Zhu 2022).

The heterogeneity of geospatial data refers to the heterogeneity caused by differences in data standard specifications and data understanding. Different geospatial data have different precisions, units, types, classification methods, spellings, etc. These differences can lead to heterogeneity of geospatial data (Huang 2018).

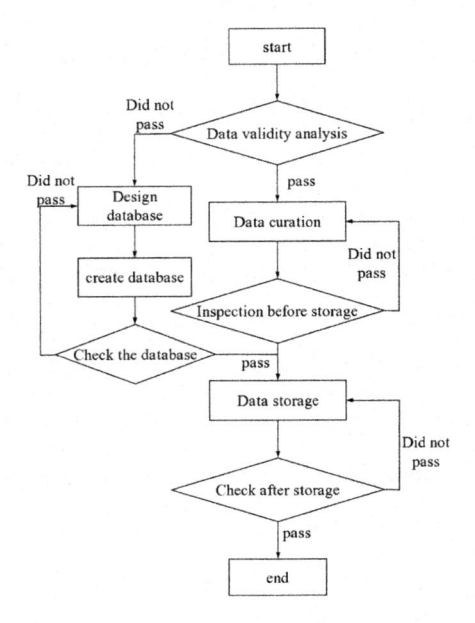

Figure 1.   Geospatial data integration and storage process.

Multi-source heterogeneous geospatial data integration is a key technology for the efficient utilization of geospatial data. Multi-source heterogeneous geospatial data integration refers to the integration of geospatial data of different types, sources, properties, and formats to form a unified whole. Integrating heterogeneous geospatial data from multiple sources enables software to use geospatial data in a transparent and coordinated manner.

## 3 OVERALL ARCHITECTURE OF THE GEOGRAPHIC INFORMATION PUBLIC MANAGEMENT PLATFORM

To efficiently utilize geographic information in public management and services, the public management platform based on geographic information technology should integrate and process geospatial data according to the specialized needs of the public and government staff, and be logically standardized. The platform should have standardized interfaces for docking with relevant professional information, use the Internet for interconnection, and design an integrated geographic information service platform (Geng 2014). This public management platform needs to consider three roles of platform, namely geographic information service provider, applicable party, and management party. To facilitate the management of the platform, the three roles should be assembled during the development process, so that the three roles have a unified registration entrance, and three roles should be authorized by levels (Song 1998).

According to the needs and goals of the development of the geographic information public management platform, this research has carried out a three-tier architecture design for this platform. The specific architecture of the platform is shown in Figure 2.

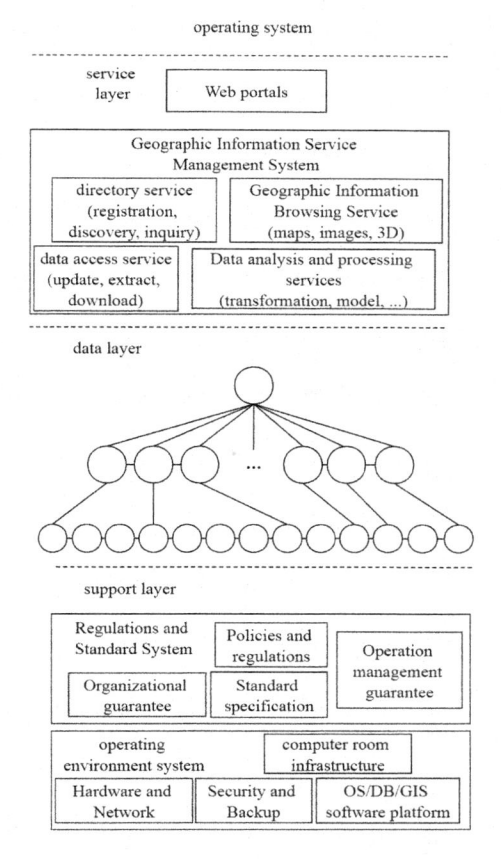

Figure 2. The architecture of geographic information public service platform.

## 4 DATA LAYER DESIGN OF GEOGRAPHIC INFORMATION PUBLIC MANAGEMENT PLATFORM

There are three main ways to integrate multi-source data using geographic information technology today, which are the integration method based on data format conversion, the integration method based on direct data access, and the integration method based on data interoperability (Chen 2003). The integration method based on data format conversion requires many conversion tools, and the data conversion process is complex and inefficient. The integration method based on data format conversion cannot provide users with real-time updated data, and different conversion tools express different levels of geographic data information, which leads to the loss or error of geographic data information during the conversion process. The integration process based on data format conversion is shown in Figure 3.

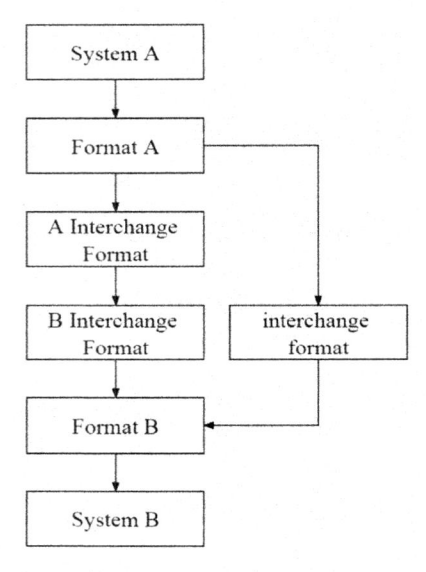

Figure 3. Integration method based on data format conversion.

The integration method based on direct data access refers to the direct access to geographic information data in other formats in a geographic information system (Zhang 2022). This method is simpler and faster than the integration method based on data format conversion, but it has certain requirements for the data format and cannot fully meet the data of all formats. If a GIS application's data version is upgraded or its format changed, another system may not be able to access the data. The integration process based on direct data access is shown in Figure 4.

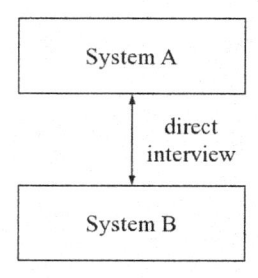

Figure 4. Integration based on direct data access.

The integration method based on data interoperability is to establish a standard framework so that geographic information data can be applied to various Web, wireless, LBS, and other mainstream information technology applications. This data integration method enables GIS to organize and integrate multi-source geographic data through the Internet. This method is based on Web services, avoids errors between data format transfer in and out, and preserves the content of geographic information data. As long as the Web services invoked by the geographic information system meet specific interoperability standards, data organization and integration can be achieved between different systems at the Web service level. The integration process based on data interoperability is shown in Figure 5.

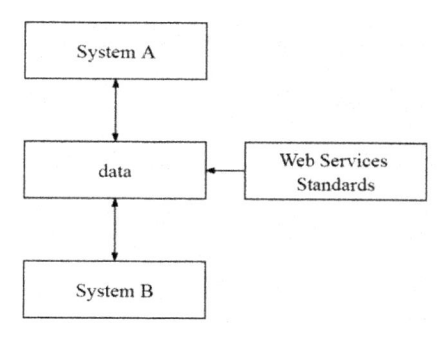

Figure 5.    Integration based on data interoperability.

This system is a large-scale geographic information system based on urban geographic information data, so users maintain and update business attribute data very frequently. The system stores and maintains business attribute data and spatial data separately to avoid mutual interference between the two types of data and reduce the cost of data maintenance and update. The integration method based on data interoperability is used in this system. The system divides the data layer into the integrated spatial database engine and the attribute database engine and uses the data association engine to combine the two. An integrated approach using data interoperability has the potential to confuse spatial data with business attribute data. This problem can be effectively solved by storing the two kinds of data separately (Qiu 2022).

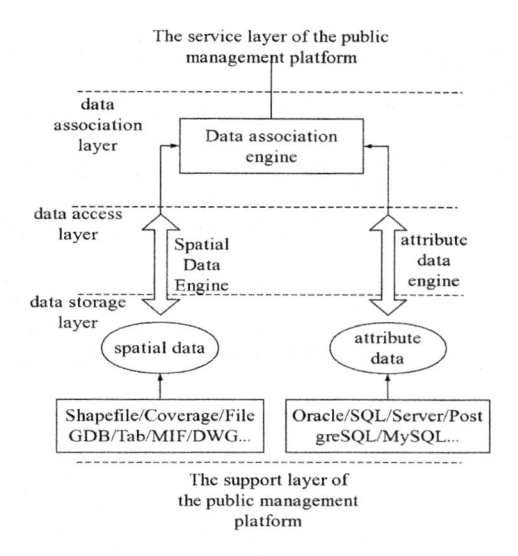

Figure 6.    The data layer of public management system based on geographic information technology.

## 5 CONCLUSIONS

The public management platform is an important way to stabilize the country and society. Public management is not only an important way to maintain the stability of the city, but also an important channel to provide services to residents. This research proposes to apply geographic information technology in public management and builds a public management platform based on geographic information technology. This platform can give full play to the maximum benefit of geographic information resources, provide urban residents with detailed geographic information data, and improve the management and service level of the city. In this study, the integration of multi-source geographic information is also set up, and the integration method based on data interoperability is improved. This research uses a variety of data classification storage and builds a data association engine between databases to make up for the defects in the data interoperability process so that the platform can be more perfect and provide users with better services. Next, this research will pilot the platform in some regions, and make detailed improvements to the platform through real feedback from users. The application prospect of this research is very broad, it has strong use value, and it can play an important role in public service and management.

## ACKNOWLEDGMENT

This work was supported by the 14th Five-Year Plan of Education Science in Jiangxi Province– Research on the Construction of Teachers' Professional Well-Being System in Private Colleges and Universities under the Background of the New Era (Grant No. 21YB233).

## REFERENCES

Chen Baiming, Liu Xinwei, Yang Hong. Review of the latest progress in LUCC research [J]. *Advances in Geographical Sciences*, 2003(01):22–29.

Geng Zhi. Opportunities and challenges faced by statistics in the era of big data [J]. *Statistical Research*, 2014, 31(01): 5–9. DOI: 10.19343/j.cnki.11-1302/c.2014.01.002.

Huang Liwei, Jiang Bitao, Lv Shouye, Liu Yanbo, Li Deyi. A Review of Recommendation System Research Based on Deep Learning [J]. *Journal of Computer*, 2018, 41(07): 1619–1647.

Jia Yonghong, Li Deren, Sun Jiapin. Multi-source remote sensing image data fusion [J]. *Remote Sensing Technology and Application*, 2000(01): 41–44.

Li Deren, Zhang Liangpei, Xia Guisong. Automatic analysis and data mining of remote sensing big data [J]. *Journal of Surveying and Mapping*, 2014, 43(12): 1211–1216. DOI: 10.13485/j.cnki.11-2089.2014.0187.

Liu Shenghe, Wu Chuanjun, Shen Hongquan. GIS-based urban land use expansion model in Beijing [J]. *Acta Geographica Sinica*, 2000(04):407–416.

Qiu Weiwei, Chen Congxi, Xiang Jiayu, Gao Yu, Qi Shuhua, Zhu Xianyun. A review of the application of 3D GIS in land spatial planning [J/OL]. *Land and Resources Information*: 1–6 [2022-05-07]. http://kns.cnki.net/kcms/detail/11.4481.N.20220427.1615.026.html

Song Guanfu, Zhong Ershun. Research and Development of Component Geographic Information System [J]. *Chinese Journal of Image and Graphics*, 1998(04):53–57.

Wu Xincai. The basic technology and development trend of geographic information system [J]. *Earth Science*, 1998 (04): 5–9.

Xue Tianhan, Jin Zhefei, Yao Haiyuan, Qi Yue, Wang Dachuan, Su Mengchao. Port space planning and design technology method based on BIM+GIS~[J/OL]. *Water Transport Engineering*: 1–5 [2022-05-07]. DOI: 10.16233/j.cnki.issn1002-4972.20220418.032.

Zhang Minggang, Wei Changshou. Analysis of temporal and spatial variation of groundwater storage in North China Plain based on multi-source data [J]. *Geodesy and Geodynamics*, 2022, 42(05): 505–509. DOI: 10.14075/j.jgg.2022.05.012.

Zhu Haoliang, Wang Yousong. Intelligent assessment of structural safety performance based on multi-source data [J/OL]. *Building Structure*: 1–10 [2022-05-07]. DOI: 10.19701/j.jzjg.20220432.

*Civil Engineering and Urban Research – Mohamed & Hou (Eds)*
© *2023 the Authors, ISBN 978-1-032-44487-1*

# Research on the renewal strategy of urban typical vitality space based on typology theory

Jialei Li & Hongmei Li*
*The School of Fine Arts, University of Jinan, Jinan, China*

ABSTRACT: To protect and balance settlements in the city, the author adopts the form of dynamic space in the strategy to form a model, which promotes the dynamic spread of organisms. achieve the urgent need for multiple contradictions in urban space, and explore suitable updates. The model is aimed at the renewal of the typical urban vitality space, and under the guidance of the typology theory model, the typical urban vitality space should pay attention to the restoration of the production vitality value and the translation of the endogenous power of the space, and finally expounds the systematic update and design strategy.

## 1 INTRODUCTION

In the process of modern urban development, it has changed from "incremental space" construction mode to "stock space" development mode. Each city has a typical dynamic area in the process of time and space evolution, that is, a dynamic energy space that supports urban development and assumes forward factors. The performance of energy aggregation includes the improvement of production kinetic energy, the improvement of cultural added value, and the strengthening of social cohesion.

The research on typical urban vitality space is to activate the potential energy of urban development, provide visual data utilizing sampling surveys, and analyze the typical spatial texture. The research goal is to solve the problem of the slow spatial development of kinetic energy from a micro perspective, including the exploration of regional culture, architectural expression, environmental behavior, and other fields.

## 2 THE APPLICATION OF TYPOLOGY IN THIS DISCIPLINE

Architectural space typology is a discipline that understands the essence of architecture by classifying buildings. Architectural typology theorist De Quincy believes that types are different from models and cannot be imitated and copied. Types are implicit, unchanging principles held by changing objects (Wang 2003).

## 3 THE MAIN PROBLEMS FACED BY TYPICAL URBAN DYNAMIC SPACES

### 3.1 *Lack of continuity of traffic flow lines in space*

As the main crowd space in the city, the typical urban vitality space has publicity and entertainment, and the two together constitute the basic attributes of the space. In the expression of spatial semantics, both attributes cover the universality for the public, that is, a wide range of spatial use functions and spatial distribution functions. The use of functions enables the public to achieve

---

*Corresponding Author: sa_lihm@ujn.edu.cn

DOI 10.1201/9781003372417-62

specific spatial cognition and spatial behavior driving conditions in space. The specific cognition is the type of recognition of space and the behavior-driven condition is the premise required for the occurrence of individual actions. To sum up, the main premise is to sort out the spatial order in advance, that is, to identify the area through the traffic flow lines of the space. In medium and large cities, urban planning managers have defects in the continuity design of traffic flow lines in the typical dynamic space of the city, so that the experience cannot achieve a clear judgment of the space environment in the space, and the use of each building in the space is also lacking. Understanding, the brief identification of individual small spaces produces understanding biases. Taking the old commercial port of Jinan City as an example, the survey found that there is a break in thematic space in the site, which cannot connect individuals to achieve the overall effect perception, which requires the design manager to reorganize the traffic flow in the space. with redesign.

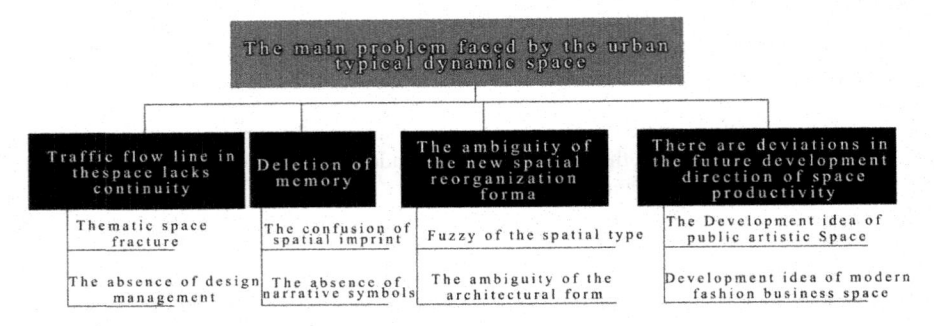

Figure 1.   A logical framework of typology in urban typical vitality space renewal.

Figure 2.   The calorific value of Jinan vitality space.

### 3.2   *Memory loss*

Historical blocks are an intuitive material mapping of a region's traditional industries, living situations, and regional culture, and this mapping is a dynamic and cumulative system evolution process. During the evolution process, the original spatial imprints should be preserved, including architectural imprints, lifestyle imprints, cultural and ecological imprints, etc. First, the traditional part should be inherited and developed in a contextualized way under the condition of following historical traces (Ao 2021); secondly, the material space mixed with old and new should be integrated into modern urban functions in a symbiotic mode and radiate new vitality.

In the research, it is necessary to systematically excavate and extract information such as historical events and story elements using historical research and field research interviews, construct representative narrative symbols, and express them using spatial formal language so that residents can enjoy leisure activities in the community space. People can feel the historical and cultural characteristics of the community, arouse the collective memory of residents, and play a positive role in cultural heritage.

### 3.3 *The ambiguity of new spatial reorganization forms*

The use functions and types of urban spaces have changed over time. In the current space reorganization, due to the different functions used, the departmental building spaces have begun to be redesigned and externally improved according to the requirements of users. The lack of integrity blurs the boundaries of space types.

### 3.4 *There are deviations in the future development direction of space productivity*

The development idea of typical vitality space is to pursue the combination of modern commercial space and memory space, and it is based on the humanization of the object of care. It means that the creation of urban vitality space starts from the relationship between audience and memory and analyzes the scope of the audience based on the type, fully considering the audience's emotional and psychological reactions (Li et al. 2022). Form a consumption space dominated by psychological perception, emotional counseling, and physical experience.

## 4 THE INTERVENTION OF TYPOLOGY THEORY AND COPING STRATEGIES FOR SPATIAL EVOLUTION

### 4.1 *Extraction of architectural space types*

The three most typical buildings in the space are selected, the space is re-divided according to the architectural texture, the language of one level is used to describe the language of another level, and the buildings are restored to the basic elements one by one. The vocabulary is composed of a complete architectural space through a certain grammar.

The extraction of architectural space types should be based on factors such as the area and height of the building, and consider whether the public's vision can transmit the information of the building into the human brain and whether it can map the spatial atmosphere to the heart in a short time.

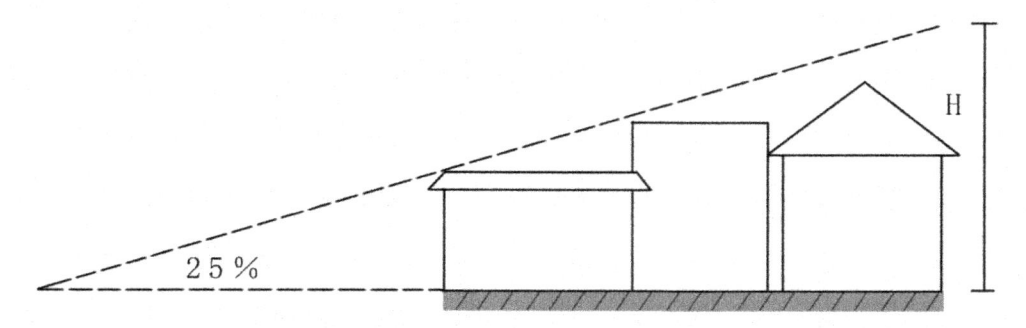

Figure 3.   Schematic diagram of building height control.

### 4.2 Extraction of texture types

Extract the architectural space elements with retained value in the space, and be good at giving them the external form and content in line with economic development. At the same time, the small-scale and gradual micro-renewal method is more conducive to the preservation of the extracted architectural space texture (Han & Shen 2022).

### 4.3 Coping strategies for spatial transmutation

It can be seen from the figure that when the space changes, the spatial arrangement of the original individual buildings cannot meet the changed conditions. Therefore, the different representative buildings will be rearranged to create a sustainable development space.

(1) Focus on the opening of nodes at the junction of internal and external spaces. Based on the establishment of the internal opening system of the space, re-plan the junction of the boundaries of each space atmosphere, select appropriate spaces to shape the nodes, and gradually blur the separated boundaries (Zhang & Yang 2021).

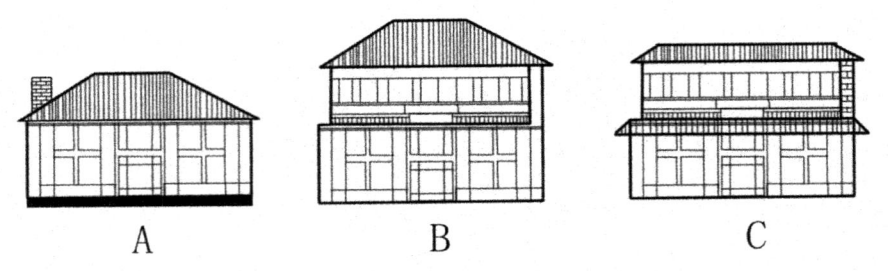

Figure 4. Representative building model.

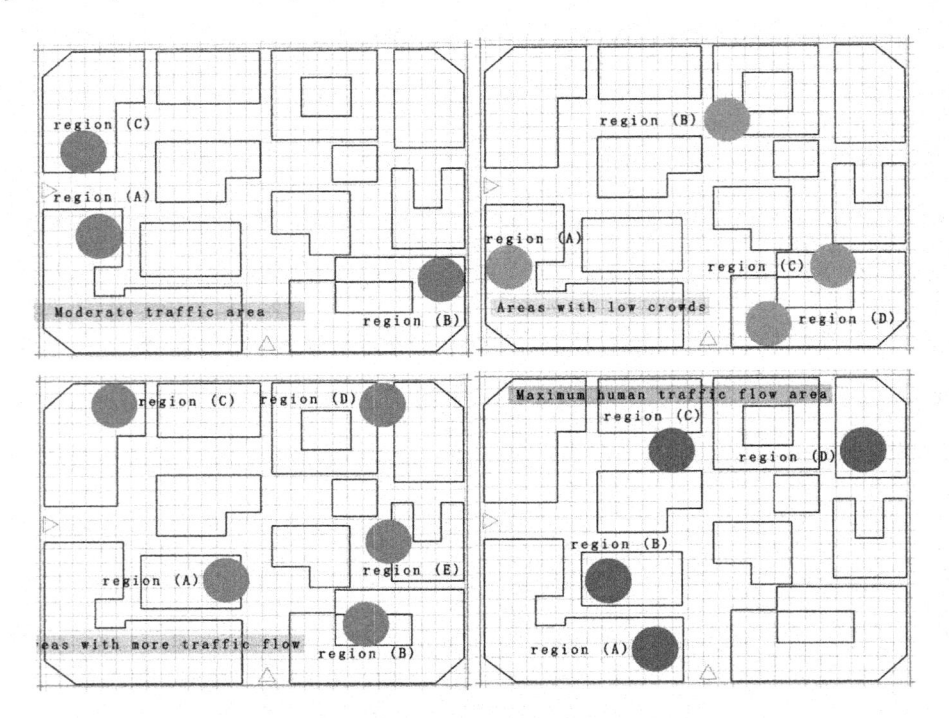

Figure 5. Color analysis map of thermal value of Jinan old commercial port space.

(2) Weakening its closedness, creating public spaces such as public spaces inside, and making organic connections between them, so that the open space becomes a complete system, creating more penetration points for the integration of the inner and outer spaces of the old commercial port.

## 5 CONCLUSION

Through the classification and integration of the architectural texture in the space, the integration and analysis of the vitality space are re-analyzed, the characteristics of the scale and perspective of the vitality space are extracted, and the narrative memory type of the traditional public space is preserved to ensure that the vitality space can be transformed in the future. It is important to ensure that the original vitality method can be maintained when the vitality space changes in the future, and to avoid the generation of the "easy to get sick" space and the "oblivion" space.

## REFERENCES

Architecture and Culture. (2021). Reconstruction of Tracing, Coexistence of Old and New: Integrated Design of Architecture and Landscape in the Renewal of Historic District. J. AoLei. (07):123–124.

Architecture and Culture. (2022). The exploration of Suzhou traditional street space from the perspective of typology. J. Han Zixuan, Shen Shaojie. (05):110–112.

Journal of Beijing University of Architecture Li Chunqing. (2022). Research on the renovation and design ofpublic space in old residential areas in Beijing from the perspective ofnarratology. J. Li Ying, Kong Jie, Zhang Man. 38(03):10–20.

Journal of Suzhou University of Science and Technology (Engineering Technology Edition). (2021). Research on the micro-renewal method of urban villages under the guidance of typological thinking—Taking Guangzhou as an example. J. Zhang Yuan, Yang Zhongwei. 34(01): 63–73.

Tianjin University. (2003). *Research on generalized architectural typology*. D. Wang Lijun. 5–7.

*Civil Engineering and Urban Research – Mohamed & Hou (Eds)*
*© 2023 the Authors, ISBN 978-1-032-44487-1*

# Research on integrated scheduling strategy of inter-terminal container transportation

Sun Jiuzeng* & Ding Yi*
*Shanghai Maritime University, Shanghai, China*

ABSTRACT:    The container transportation in the port is divided into inter-terminal transportation (ITT) and internal transportation. The former is completed by the truck company's external truck entrusted by the terminal, and the latter is completed by each terminal's internal truck. Due to the uneven distribution of workload in the terminal, the terminal's free internal trucks can participate in ITT scheduling. This paper proposes a strategy that truck companies and terminals provide trucks together for ITT scheduling, and an integer programming ITT scheduling model based on a space-time graph is created. The model aims at minimizing the total cost, considers the uneven busy degree in each terminal, completes vehicle dispatching in periods, and maximizes resource utilization on the premise of ensuring the normal work in each terminal. This paper makes a comparative analysis based on the separate scheduling strategy of external trucks. The results show that the integrated scheduling model with the participation of internal trucks can reduce costs and improve resource utilization.

## 1    GENERAL INSTRUCTIONS

With the further development of globalization, the total amount of global container transportation is still rising year by year, and the increasing number of containers has brought great challenges to every port. As large ports grow and the number of terminals increases, ports need to improve efficiency not only within internal transportation but also inter-terminal transportation, which Duinkerken (2006) et al. called inter-terminal Transportation ITT (Inter-terminal Transportation). ITT has become a key issue in the competition between ports, with limited potential for efficient development due to scarce land resources and expensive equipment such as cranes.

Container transport in port is divided into inter-dock transport and dock transport. Inter terminal transportation is completed by an external container truck dispatched by the truck company entrusted by the dock. Due to the long distance between the truck company and the port and the continuous increase of ITT orders, ITT scheduling costs are increasing year by year. Reducing ITT costs has become a key issue for the port to improve operational efficiency. internal transportation is completed by intra-dock container trucks. To ensure the smooth progress of internal work, the total amount of trucks in each dock can meet its peak demand. However, due to the differences in ship arrival time and transport volume, the workload of each period in the dock is not uniform, resulting in the waste of truck resources in some periods. Multi-party resource sharing is an important research topic to improve ITT efficiency.

Scholars at home and abroad have carried out many studies on resource sharing among terminals. Heilig (2017) et al. conducted a comprehensive literature review on ITT issues, and for the first time identified resource sharing and collaboration as important research directions in ITT. Jin (2018) et al. proposed the strategy of resource integration based on the shared order of card collecting companies and constructed the integer programming model based on the space-time network

---

*Corresponding Authors: 15553697179@163.com and yiding@shmtu.edu.cn

 DOI 10.1201/9781003372417-63

graph. This study only analyzed the cooperation between different card collecting companies and did not consider the available resources for card collecting on the wharf. The problem of sharing internal trucks among multiple container terminals (SIMT) was proposed for the first time by He (2013) et al. and a linear programming model was created to solve SIMT. The optimal solution was obtained by combining genetic algorithm and simulation optimization. This study only considers inter-terminal set card sharing and does not apply the set card to ITT scheduling. The multi-manager system proposed by Nabais (2013) et al. solves the problem of multi-terminal cooperation. The control manager is responsible for the dispatching of controlled vehicles at the terminal, and the coordination manager is responsible for coordinating the cooperation among control managers. The paper did not consider the possibility of external collection card companies joining the overall cooperation. Herbert (2016) et al. proposed three scenarios for ITT resource sharing and cooperation strategy. After data experimental evaluation and analysis, relevant suggestions were given according to terminal operators' preferences. This study does not analyze the cooperation between the company and the wharf. ITT model construction is based on the network graph. Ding (2021) et al. established an integer programming ITT model based on the spatiotemporal extended graph. This study considered the price comparison problem of various transportation equipment and the layout problem of the road of the new port and proposed a set of effective inequalities to accelerate the convergence of the calculation. The study did not consider factors such as resource sharing and cooperation.

The work in the dock mainly includes unloading containers from the ship, loading containers to the ship and temporary storage of containers, etc. The collection card in the dock is responsible for transporting containers for loading and unloading operations. The docking time of ships at wharves is different, and the overall transport volume is also different due to the size of ships, so the loading and unloading operations at various wharves are not even. At the same time, to ensure that the terminal can always operate, the number of terminal cartridges can meet peak workload demands, which makes it possible to use idle terminal trucks to participate in ITT scheduling during off-peak periods. To ensure both the dispatch and the normal work in the wharf, the planning time is equally divided into N periods, and two terms are put forward: wharf demand workload and wharf available workload. The former includes loading and unloading operations, while the latter refers to the sum of the product of all internal trucks and their stay time in the dock at this period. Each period should ensure that the amount of work available at the wharf is greater than the amount of work required at the wharf.

## 2 MATHEMATICAL MODEL

In this paper, based on the time-space network diagram to create a set card company together with the terminal to provide vehicles to complete the ITT integer programming model of scheduling, the planning period can be divided into $n$ periods, at each period of guarantee available workload of each terminal workload under the premise of providing vehicles is greater than demand. This ensures the terminal internal work can run normally. Considering congestion caused by busy ports, vehicle travel times are different at different times. The model also considers a variety of cost problems, including the delay cost due to the time window of the order, the external truck occupancy cost, truck wage cost, and driving cost, and makes the following assumptions:

(1) Intra-dock trucks can be used for inter-dock container transportation.
(2) The trucks in the dock cannot perform the internal operations of other terminals.
(3) The workload required at each period of the wharf shall not be greater than the maximum available workload.
(4) The number of available external trucks is theoretically unlimited.

The parameters, sets, and variables used in the mathematical model are shown in Tables 1 to 3.

Table 1. Model-related parameters.

| Symbol | Meaning | Symbol | Meaning |
|---|---|---|---|
| $t, t_1$ | Time | $u$ | Period |
| $a, b$ | Location, including docking area and dock | $i$ | Unit, including truck company and dockside |
| $n$ | Total time length | $m$ | Period length |
| $e$ | The order | $r_e$ | Number of containers ordered $e$ |
| $s_e$ | The earliest pickup time of an order $e$ | $f_e$ | The latest delivery time of an order $e$ |
| $q_1^i$ | $i$ refers to the cost per vehicle per hour | $q_2^i$ | $i$ refers to the wage cost per vehicle per hour |
| $q_3$ | Cost per unit time delay | $q_4$ | Out-of-collection card occupancy cost per unit time |
| $q_5$ | Driving cost per unit outside collection card to the parking area | $s_{abt}$ | The total time from a to b, including time to load at $a$, time to travel at $b$, and time to unload from a to b |
| $k_1^i$ | $i$ refers to the average loading time for unit internal operations | $k_2^i$ | $i$ refers to the average uninstallation time of internal jobs |
| $d_u^i$ | $i$ refers to the load per period $u$ | $g_u^i$ | $i$ refers to the uninstallation effort per period $u$ |
| $v_i$ | The total number of vehicles for unit $i$ | | |

Table 2. Model-related set.

| Symbol | Meaning | Symbol | Meaning |
|---|---|---|---|
| $C_1$ | Set of docks; $C_1 = \{1, 2, 3, \ldots, C_1^{\max}\}$ | $U$ | Set of periods; $U = \{1, 2, \ldots, n/m\}$ |
| $C_0$ | Truck companies; $C_0 = \{0\}$, and when $a = 0$ or $b = 0$ represents the stopping point of truck companies | $C$ | Set of units; $C = C_1 \cup C_0$ |
| $T$ | Set of time; $T = \{1, 2, \ldots, t_{\max}\}$ | $T_u$ | The set of the period $u$; $T_u = \{m(u-1) + 1, m(u-1) + 2, \ldots, mu\}$ |
| $P_u$ | The time set of the previous period $u$; $P_u = T_1 \cup T_2 \cup T_3 \cdots \cup T_u$ | $E$ | Set of orders; $E = \{1, 2, \ldots, e_{\max}\}$ |
| $A(a, b)$ | Collection of orders from $a$ to $b$ | | |

Table 3. Model-related variables.

| Symbol | Meaning | Symbol | Meaning |
|---|---|---|---|
| $W_{abt}^i$ | The number of vehicles unit $i$ from $a$ to $b$ at the time $t$ | $M_{et}^i$ | The number of containers dispatched to order $e$ for the unit $i$ at the time $t$ |
| $Y$ | Number of trucks dispatched by vehicle companies | | |

The mathematical model is shown below.

$$\min Z = \left[ (q_2^0 + q_4) n + 2q_5 \right] Y + \sum_{t \in T} \sum_{a \in C} \sum_{b(\neq a) \in C} \left( \sum_{i \in C} q_1^i W_{abt}^i s_{abt} + \sum_{i \in C_1} q_2^i W_{abt}^i s_{abt} \right)$$
$$+ \sum_{t \in T} \sum_{a(\neq i) \in C} \sum_{i \in C_1} q_2^i W_{aat}^i + \sum_{i \in C} \sum_{a \in C} \sum_{b(\neq a) \in C} \sum_{e \in A(a,b)} \sum_{t \geq f_e - s_{abt}} (t + s_{abt} - f_e) M_{et}^i q_3 \qquad (1)$$

$$\sum_{b \in C_1, t=1} W_{ibt}^i = Y, i \in C_0 \tag{2}$$

$$\sum_{a \in C_1, t_1 = n+1-s_{ait_1}} W_{ait_1}^i = Y, i \in C_0 \tag{3}$$

$$\sum_{b \in C_1, t=1} W_{ibt}^i = v_i, i \in C_1 \tag{4}$$

$$\sum_{a \in C_1, t_1 = n+1-s_{ait_1}} W_{ait_1}^i = v_i, i \in C_1 \tag{5}$$

$$W_{ii(t-1)}^i + \sum_{a \in C, t_1 = t - s_{abt_1}} W_{ait_1}^i = \sum_{b \in C} W_{ibt}^i, t \in T, i \in C_1 \tag{6}$$

$$\sum_{i(\neq a) \in C} W_{aa(t-1)}^i + \sum_{i(\neq a) \in C} \sum_{b(\neq a) \in C, t_1 = t - s_{abt_1}} W_{bat_1}^i = \sum_{i(\neq a) \in C} \sum_{b(\neq a) \in C} W_{abt}^i, t \in T, a \in C_1 \tag{7}$$

$$\left( v_i + \sum_{t \in P_{u-1}} \sum_{a(\neq i) \in C, t_1 = t - s_{ait_1}} W_{ait_1}^i - \sum_{t \in P_{u-1}} \sum_{b(\neq i) \in C} W_{ibt}^i \right) m + \sum_{t \in T_u} \sum_{a(\neq i) \in C, t_1 = t - s_{ait_1}} W_{ait_1}^i (mu - t)$$

$$- \sum_{t \in T_u} \sum_{b(\neq i) \in C} W_{ibt}^i [t - (m-1)u] \geq k_1^i d_u^i + k_2^i g_u^i, u \in U, i \in C_1 \tag{8}$$

$$\sum_{e \in A(a,b)} M_{et}^i \leq W_{abt}^i, t \in T, i \in C, a \in C, b \in C \tag{9}$$

$$\sum_{i \in C} \sum_{t \geq s_e} M_{et}^i = r_e, e \in A(a,b), a \in C, b \in C \tag{10}$$

$$W_{abt}^i \leq v_i, t \in T, i \in C_1, a \in C, b \in C \tag{11}$$

$$W_{abt}^i \leq Y, t \in T, i \in C_0, a \in C, b \in C \tag{12}$$

$$\sum_{t < s_e} M_{et}^i = 0, e \in A(a,b), a \in C, b \in C \tag{13}$$

$$W_{abt}^i, M_{et}^i, Y \geq 0, \text{integer} \tag{14}$$

Equation (1) minimizes the total cost including various costs, including the total cost of external trucks, the total cost of internal trucks and the total cost of order delay. The total cost of the external trucks includes the round-trip expenses between the external truck warehouse and the docking area, the personnel salary expenses within the whole planning time and the driving expenses during the transportation period. The total cost of the internal trucks includes the personnel salary expenses during the assignment period and the driving expenses during the transportation period. The total delay cost refers to the delay cost exceeding the specified latest arrival time. Equations (2) and (3) ensure that the external trucks at the beginning and end of the planning time are uniformly concentrated at the docking area, and the total number of vehicles is equal to the number of external trucks dispatched by the truck company. Equations (4) and (5) ensure that the trucks of terminals at the beginning and end of the planning time are at their respective terminals, and the number is equal to the number of free vehicles at each terminal. Equation (6) ensures that the flow of owned vehicles in and out of each wharf is balanced, that is, the sum of owned vehicles entering and staying at the wharf in the previous period is equal to the sum of owned vehicles leaving and staying at the wharf in the next period. We stipulate that all owned vehicles parked at the wharf are allowed to work inside the wharf. Equation (7) ensure a balanced flow of non-owned vehicles in and out of all terminals.

Equation (8) ensure that the available workload at each period of the wharf is greater than the required workload at that period. The required workload is the sum of loading workload and unloading workload, and the available workload refers to the sum of the product of all internal trucks and their residence time in the wharf at that period. Equation (9) ensure that the number of vehicles in each transport is not less than the number of containers undertaken for the transport. Equation (10) ensure that all containers for each order can be delivered after the earliest release time of the order. Equations (11) and (12) ensure that the number of vehicles per unit run is not greater than the maximum number of vehicles per unit. Equation (13) ensure that the vehicle cannot pick up the container until the earliest release date of the order.

## 3 EXAMPLE ANALYSIS

Due to the competition among ports, ITT orders are entrusted by different truck companies to dispatch vehicles to complete scheduling. Gharehgozli et al. (2016) proposed a way for truck companies to share orders to realize resource integration, the essence of which is to realize unified scheduling by truck companies (hereinafter referred to as unified scheduling). On this basis, the paper considers the available resources of idle internal container trucks, proposes the comprehensive scheduling strategy (hereinafter referred to as integrated scheduling) of vehicles jointly provided by container trucks and terminals, and establishes an integer programming model. In this paper, the advantages and disadvantages of unified scheduling and comprehensive scheduling are compared and analyzed from three aspects: total cost, number of external trucks, and resource utilization rate, and a total of 18 scenarios are comprehensively compared based on different numbers of terminals.

Input the set created by the demand generator and the relevant parameters of transportation equipment to solve and analyze 18 scenarios. PYTHON programming was adopted and a GUROBI solver was used to solve the problem. Intel Core I5, 1.80 GHz CPU, and 8G memory were used for computing. The results are summarized in Table 4.

Table 4. Model-related results.

| | | | | Unified scheduling | | | Integrated scheduling | | |
|---|---|---|---|---|---|---|---|---|---|
| Number | Terminal | Demand | Time | Total cost per yuan | Number of the external collection vehicle | Resource utilization | Total cost per yuan | Number of the external collection vehicle | Resource utilization |
| 1 | 3 | Low | 6 | 13928 | 40 | 71.6% | 8694 | 29 | 81.1% |
| 2 | 3 | Normal | 6 | 22626 | 62 | 71.6% | 17514 | 45 | 86.5% |
| 3 | 3 | Peak | 6 | 35724 | 96 | 71.6% | 30726 | 72 | 92.8% |
| 4 | 3 | Low | 12 | 26161 | 42 | 72.4% | 17789 | 32 | 82.6% |
| 5 | 3 | Normal | 12 | 44319 | 68 | 72.4% | 36341 | 52 | 86.3% |
| 6 | 3 | Peak | 12 | 69758 | 109 | 72.4% | 62085 | 87 | 92.4% |
| 7 | 4 | Low | 6 | 15057 | 44 | 71.7% | 9185 | 32 | 79.6% |
| 8 | 4 | Normal | 6 | 32370 | 72 | 71.7% | 24601 | 49 | 86.5% |
| 9 | 4 | Peak | 6 | 41364 | 111 | 71.7% | 34332 | 81 | 91.3% |
| 10 | 4 | Low | 12 | 29500 | 48 | 72.9% | 20355 | 37 | 84.4% |
| 11 | 4 | Normal | 12 | 46785 | 78 | 72.9% | 37428 | 58 | 90.5% |
| 12 | 4 | Peak | 12 | 83936 | 126 | 72.9% | 73024 | 99 | 84.3% |
| 13 | 5 | Low | 6 | 16046 | 47 | 71.3% | 9467 | 32 | 82.1% |
| 14 | 5 | Normal | 6 | 28274 | 78 | 71.3% | 20357 | 51 | 85.2% |
| 15 | 5 | Peak | 6 | 45609 | 121 | 71.3% | 35118 | 86 | 89.8% |
| 16 | 5 | Low | 12 | 31398 | 53 | 72.1% | 20722 | 40 | 79.3% |
| 17 | 5 | Normal | 12 | 41797 | 86 | 72.1% | 31347 | 63 | 84.5% |
| 18 | 5 | Peak | 12 | 90957 | 138 | 72.1% | 73675 | 107 | 88.6% |

## 4 CONCLUSION

In this paper, the problem of multi-terminal container transportation in ports is studied, and a strategy of ITT scheduling based on integer programming based on a space-time network graph is proposed. The model takes the minimum total cost as the goal, considers the uneven busy degree of each dock, completes vehicle dispatch in periods, and achieves the maximum resource utilization on the premise of ensuring the normal work of each dock. This paper illustrates the scheduling model with a case and compares the advantages and disadvantages of unified scheduling and integrated scheduling from three aspects: total cost, number of calling external trucks, and resource utilization, and makes a comprehensive comparison based on a total of 18 scenarios, such as different number of terminals, different number of ITT orders and different planning time lengths. According to the results, the integrated scheduling model with integration trucks can greatly reduce cost and improve resource utilization. This study also has shortcomings. For example, the ITT model does not consider the problem of information opacity between terminals, nor does it combine intelligent optimization algorithms to conduct a further study on the ITT mathematical model, which is a director of the research on the ITT scheduling strategy of ports.

## REFERENCES

Ding Yi, He Lemei, Sha Mei. Optimization of multi-terminal container transportation based on spatio-temporal extended graph model. *China Navigation*, Vol. 44, No. 1, March 2021.

Duinkerken MB, Dekker R, Kurstjens STGL, Ottjes JA, Dellaert NP (2006) Comparing transportation systems for inter-terminal transport at the Maasvlakte container Terminals. *The OR Spectr* 16(2):469–493.

Gharehgozli, A. H., de Koster, R., and Jansen, R. (2016), Collaborative solutions for inter terminal transport, *International Journal of Production Research*, 1–20, Online publication, Doi: 10.1080/00207543.2016.1262564.

Heilig, L, Lalla-Ruiz E, STEFAN VOβ. Port-IO: An Integrative Mobile Cloud Platform for Real-time Inter-Terminal Truck Routing Optimization [J]. *Flexible Services and Optimization Manufacturing Journal*, 2017, 29(3/4):504–534.

Heilig, L, Stefan Voß. A Cloud-Based SOA for Information Exchange and Decision Support in ITT Operations [C] *International Conference On Computational Logistics*, 2014:112–131.

Heilig, L. and Voß, S. (2017), Inter-Terminal Transportation: An annotated bibliography and research agenda, *Flexible Services and Manufacturing Journal*, 29(1), 35–63.

Herbert Kopfer, Dong-Won Jang (2016), *Scenarios for Collaborative Planning of Inter-Terminal Transportation* Springer International Publishing Switzerland 2016 A. Paias et al. (Eds.): ICCL 2016, LNCS 9855, pp. 116–130.

João Lemos Nabais, Rudy R. Negenborn (2013), *Setting Cooperative Relations Among Terminals at Seaports Using a Multi-agent System,Proceedings of the 16th International IEEE Annual Conference on Intelligent Transportation Systems* (ITSC 2013), The Hague, The Netherlands, October 6–9, 2013.

Junliang He, Weimin Zhang (2013), A simulation optimization method for internal trucks sharing assignment among multiple container Terminals, *Advanced Engineering Informatics* 27(2013), 598–614.

Leedh, Jinig, Chengh. Terminal and Yard Allocation Problem for a Container Transshipment Hub with Multiple [J]. *Transportation Research Part E: Transportation Review*, 2012, 48(2):516–528.

Xuefeng Jin, Kap Hwan Kim (2018), Collaborative Inter-Terminal Transportation of Containers, *Industrial Engineering & Management Systems* Vol 17, No 3, September 2018, pp.407–416.

*Civil Engineering and Urban Research – Mohamed & Hou (Eds)*
*© 2023 the Authors, ISBN 978-1-032-44487-1*

# Study on the control of contaminated gas in the negative pressure ward combined with local exhaust air

Chenxu Zhou, Xiaoyong Peng* & Hao Zhang
*School of Civil Engineering, University of South China, Hengyang Hunan, China*

ABSTRACT: During the outbreak of the Corona Virus Disease 2019, negative pressure wards played a major role in treating patients with infectious diseases. However, when medical staff enter and leave the negative pressure ward, it may cause the leakage of contaminated gas and increase the risk of cross-infection in the hospital. Therefore, this paper takes the corridor-buffer room-negative pressure ward as the research object, proposes a form of air organization combined with local exhaust air, and uses CFD numerical simulation method to study the pressure gradient between different areas and the distribution of contaminated gas concentration fields in the ward when the buffer room door is closed or opened. The simulation results show that when the buffer room door is fully closed, the pressure gradients between the three areas of the corridor-buffer room-negative pressure ward are greater than 5 Pa. When the buffer room door is opened, the pressure difference between the connected areas gradually decreases to 0 Pa. When the buffer room door is closed, the contaminated gas released from the source is distributed on both sides of the beds and near the head of the beds; when the buffer room door is opened, the concentration of contaminated gas in the ward is significantly lower compared with the condition when the buffer room door is closed. When the buffer room door is opened, the concentration of contaminated gas in the ward is significantly reduced compared with the working condition when the buffer room door is closed, and it is only distributed in the area near the bed head of the hospital bed, which effectively controls the diffusion of contaminated gas to the outside and overcomes the problem of contaminated gas spreading to the buffer room when the room door is opened.

## 1 INTRODUCTION

A negative pressure ward is a significant place for treating patients with respiratory infectious diseases. It plays a crucial role in controlling airflow between the ward and the surrounding area to avoid cross-infection. At the same time, it speeds up the discharge of contaminated gas to protect health care workers from the infection of germs (Dong 2020; Ling 2014). Zhao Y used the numerical simulation method to study the effect of different forms of air organization on the diffusion distribution of contaminated gas in an isolation ward and analyze the mechanism of contaminated gas diffusion between the ward and its adjacent rooms under different pressure differences when the door of isolation ward is open or closed (Zhao 2011). Ji Y Z used numerical simulation combined with experimental methods to study the airflow trajectory of polluted gas through the door cross-section when the ward door is opened and to propose effective control methods for reducing the diffusion of polluted gas to the outside. It is concluded that the contaminated gas inevitably flows out from the ward during the door opening process. And the methods of increasing the ward exhaust air volume, increasing the front room air supply volume, or decreasing the ward air supply volume can effectively inhibit the outward diffusion of contaminated gas (Ji 2005, 2006 & 2009). Deepthi

---

*Corresponding Author: pengxiaoyong@126.com

DOI 10.1201/9781003372417-64

Sharan Thatiparti studied the air organization in the infectious isolation room and the behavior of aerosol transport during coughing in patients after infection with influenza (Deepthi 2016). Suvanjan Bhattacharyya used CFD numerical simulation method to simulate the diffusion distribution of air delivered by air conditioning and contaminated aerosol gas after mixing the contaminated gas in the isolation ward. And the analysis obtained that disinfection devices arranged in the area with a large distribution of contaminated gas concentration in the isolation ward can effectively kill the COVID-19 virus (Suvanjan 2020). Scholars at home and abroad have never stopped researching negative pressure wards. However, there has been less research on the opening of negative pressure ward doors and the effective control of contaminated gas diffusion to the outside.

According to current research, pollutant gas diffusion would be detected in the ward whether the buffer room door is closed or opened. This paper proposes the idea of an integrated corridor-buffer room-negative pressure ward. Combined with the method of local ventilation, the exhaust air outlet is set near the source of contamination to control the diffusing contaminated gas in the negative pressure ward. In this way, the inhibiting effect on the diffusion of contaminated gas in the negative pressure ward is studied under different working conditions when the buffer room door is closed or opened. It is expected to effectively solve the problem of pollutant gas diffusion to the buffer room when the door is opened.

## 2 CONTROL EQUATIONS AND NUMERICAL METHODS

### 2.1 Control equations

The airflow in the room is a three-dimensional constant incompressible turbulent flow with the control equations as a system of N-S equations. Numerical simulations are performed using the standard turbulence model with the equations shown below. Representation of concentration changes during diffusion of contaminated gas in the ward using component transporters.

### 2.2 Numerical methods

Simulation calculations use the SIMPLE algorithm, and the discretization method is the finite volume method, which is pressure based on the standard discrete format. The continuity equation, the energy equation, and the turbulence equation are in first-order windward format, and the momentum equation is in second-order windward format. The structured mesh is used to divide the whole mesh, and the local mesh is encrypted at the door seams and air openings, which can effectively improve calculation efficiency. The velocity inlet boundary condition is used for the air openings in each room and the patient's mouth in the ward; the floor, floor, exterior wall, and human body are set as solid wall boundary conditions with heat flux.

## 3 PHYSICAL MODEL AND DESIGN CONDITIONS

### 3.1 Physical model

Figure 1 shows the physical model of the corridor-buffer room-negative pressure ward. The corridor size is $4.5m \times 1.8m \times 2.5m$, with two top air supply outlets, which are $0.2m \times 0.2m$. The buffer room is between the corridor and the negative pressure ward, and its size is $2m \times 1.5m \times 2.6m$. The size of the ward is $4.5m \times 3m \times 2.8m$, and the ward adopts the top air supply/combined with the local exhaust airway. The ward size is $4.5m \times 3m \times 2.8m$, and the ward adopts the air organization form of top air supply/combined with local exhaust, i.e., the air vent is set near the breathing area of the patient's bed to form local exhaust, and the top air supply is arranged on the opposite wall, where the size of the air supply is $1.2m \times 0.16m$, and the size of the air vent is $0.63m \times 0.25m$. The size of the bathroom is $2.3m \times 1.65m \times 2.4m$, and the design of the ceiling exhaust is $0.2m \times 0.2m$.

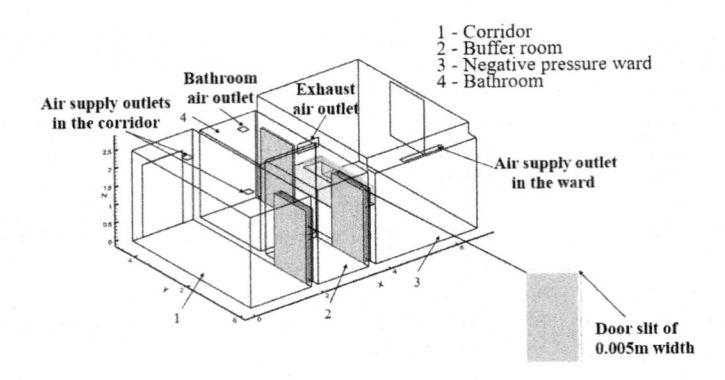

Figure 1. Physical model of the negative pressure ward.
(P.S.: The window size is 1.5m×1.3m, the height from the ground is 1.2m, the single bed size in the ward is 2m×0.9m×0.5m, and the human model is simplified to a rectangular body of 1.65m×0.5m×0.2m, room door size is 2m×1.2m, door slit width is 0.005m.)

### 3.2 *Design working conditions*

When there is an outbreak of respiratory infectious disease, infectious disease hospitals must use all fresh air conditioning systems. Take Hengyang City as an example. According to the "Practical Heating and Air Conditioning Design Manual" (Lu 2008), the summer interior design temperature of the negative pressure ward and corridor are 24°C and 27°C respectively. And the room cooling load can be obtained, the air supply temperature is chosen to be 20°C, and the air supply volume of the ward and corridor are calculated to be 450m³/h and 120m³/h, respectively. The "Requirements of environmental control for hospital negative pressure isolation ward" stipulate that the pressure gradient between zones should be no less than 5Pa (Jiang 2017). The minimum infiltration air volume through the doorway when the door is closed is 120m³/h. The exhaust air volume of the bathroom is set to 140m³/h, and the number of air changes in the ward is guaranteed to be 12 times per hour, so the exhaust air volume is 480m³/h. The difference between the air supply and exhaust air volume of the ward is 170m³/h, which is greater than the minimum infiltration air volume through the doorway. The designed exhaust air volume is verified to meet the specification of forming a pressure gradient of no less than 5Pa in the room. The contaminated gas exhaled from the patient's mouth is simulated with $SF_6$ tracer gas for its diffusion distribution (Hang 2014). The average breathing volume of an adult is 8L/min (Duan 2018). The mass fraction of exhaled contaminated gas is 0.04 (Jinkyun 2019). The design working conditions are shown in Table 1.

Table 1. Design working conditions.

| Name | Size (m×m) | Velocity (m/s) | Name | Size (m×m) | Heat flux (W/m²) |
|---|---|---|---|---|---|
| Air supply outlet in the ward | 1.2×0.16 | 0.65 | Roof | 4.5m×2.75m | 12.6 |
| Air supply outlets in the corridor | 0.2×0.2 | 0.42 | Ground | 4.5m×3.15m | 13.0 |
| Pollution source | 0.02×0.02 | 0.33 | Window | 1.5m×1.3m | 60.1 |
| Exhaust air outlet | 0.4×0.2 | 0.85 | Window wall | 4.5m×2.8m | 7.8 |
| Bathroom air outlet | 0.2×0.2 | 0.97 | Human body | 1.65m×0.5m×0.2m | 5.2 |

# 4 SIMULATION RESULTS AND ANALYSIS OF BUFFER ROOM DOOR CLOSING CONDITIONS

## 4.1 *Numerical simulation results and analysis of the flow field*

Figure 2 shows the flow diagram in the corridor-buffer room-negative pressure ward. The airflow from the corridor air supply outlet circulates in the room, part of which penetrates the buffer room through the doorway of the room, and then the airflow in the buffer room penetrates the negative pressure ward through the doorway, meeting the design requirement of directional flow from the clean area to the contaminated area; after the airflow from the ward, air supply outlet reaches the ground, a small part of the airflow flows forward along the ground and is obstructed at the bed to form a vortex. Most of the airflow is directed towards the walls on both sides of the ward and flows against the wall to the exhaust air outlet above the beds, which can form a better air organization.

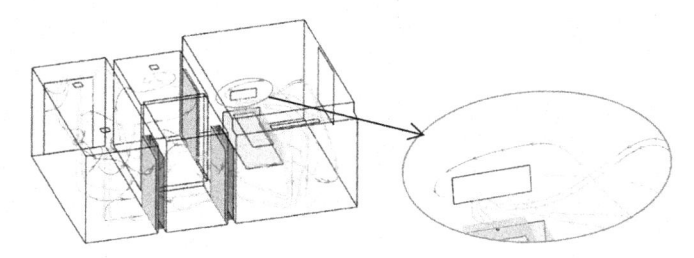

Figure 2.    Flowline distribution in the corridor-buffer room-negative pressure ward.

## 4.2 *Simulation results and analysis of pressure gradient*

When the door of the buffer room is fully closed, the infiltration air volume mainly enters the room through the door slit, intercepting the pressure cloud map at Z=0.003m at the door slit and observing the pressure change at the door slit as the transition zone of the two rooms, as shown in Figure 3(a). Z=0.7m in the ward is the average breathing height of the patient when lying down, so the air pressure distribution cloud map of this height plane is intercepted for analysis, as shown in Figure 3(b). The negative pressure value of the corridor-buffer room-negative pressure ward has obvious gradient changes, and the average value of the calculated plane pressure is −0.017Pa for the static pressure value of the corridor, −5.78Pa for the static pressure value of the buffer room, -11.21Pa for the isolation ward, and −16.26Pa for the bathroom. The pressure difference between each area is greater than 5pa, which meets the design requirements that the pressure gradient between corridor, buffer room, and negative pressure ward shall not be less than 5pa as specified in the Requirements of environmental control for hospital negative pressure isolation ward (Jiang 2017).

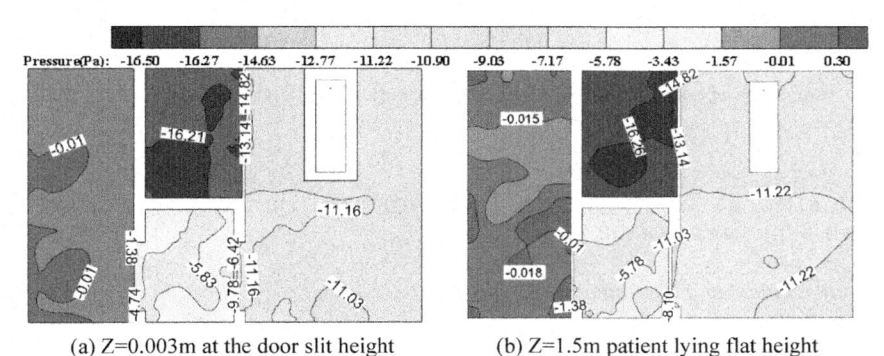

(a) Z=0.003m at the door slit height          (b) Z=1.5m patient lying flat height

Figure 3.    Pressure section cloud in the Z direction.

### 4.3 *Simulation results and analysis of pollution gas concentration field*

Figure 9 shows a cloud plot of the percentage mass of the contaminated gas equivalent surface. At the mass fraction of $1 \times 10^{-7}$, the contaminated gas is distributed between the hospital bed and the bathroom room door. At the selected mass fraction of $1 \times 10\text{-}9$, the diffusion of contaminated gas is reduced and distributed between the hospital bed and the air vent. There is no widespread diffusion of contaminated gas in the ward. Figure 5 shows the cloud diagram of pollution gas concentration distribution in different cross-sections. At the patient's lying height Z=0.7m, the pollution gas released from the source is distributed on both sides of the bed and near the head of the bed, and at the cross-section of the pollution source X=5.15m, the pollution gas diffuses vertically upward to the exhaust vent. Combined with the ventilation scheme of the local exhaust mode, the contaminated gas released from the pollution source is directly suppressed from the exhaust outlet, which can effectively inhibit the diffusion of contaminated gas.

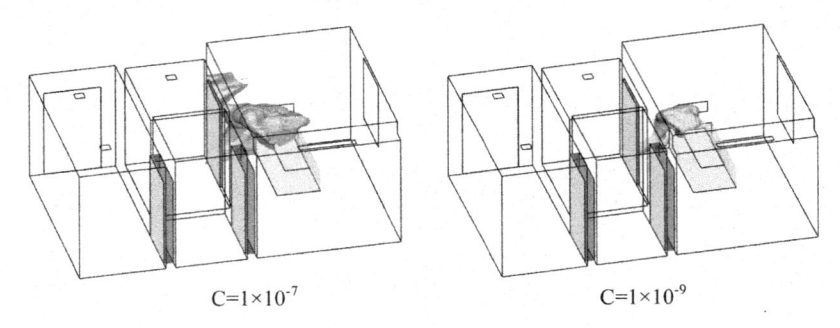

$C=1 \times 10^{-7}$ $C=1 \times 10^{-9}$

Figure 4.    Contaminated gas equivalent surface mass percent cloud map.

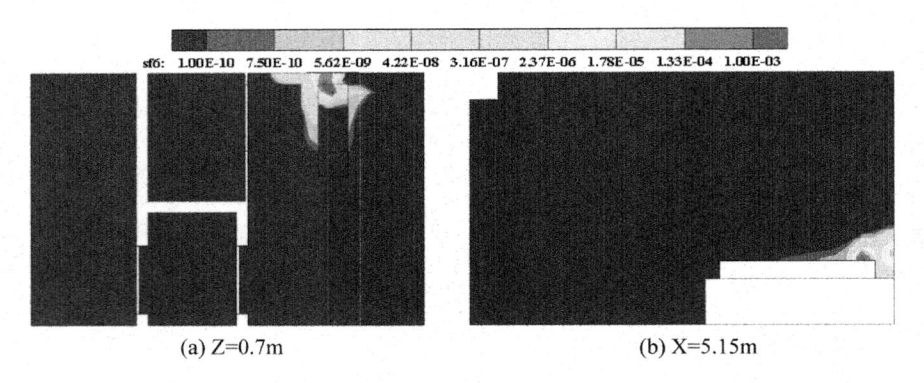

sf6:  1.00E-10  7.50E-10  5.62E-09  4.22E-08  3.16E-07  2.37E-06  1.78E-05  1.33E-04  1.00E-03

(a) Z=0.7m (b) X=5.15m

Figure 5.    Cloud map of pollution gas concentration distribution in different sections.

## 5    SIMULATION RESULTS AND ANALYSIS UNDER THE OPENING CONDITION OF THE BUFFER ROOM DOOR

### 5.1 *Physical model and operating conditions*

When the door of the buffer room is fully closed, an orderly pressure gradient is formed between the corridor- buffer room - negative pressure ward, which prevents the contaminated gas from escaping. However, when the door of the buffer room is opened, the contaminated gas in the ward will diffuse to other areas with indoor airflow. Therefore, it is necessary to study the effect of room door opening on the diffusion of contaminated gas. The CFD technique is used to simulate the effects of two conditions: the door between the buffer room and the ward is opened, and the door of the buffer room is fully opened on the pressure gradient and the diffusion of contaminated gas in the room. The model size and simulation boundary conditions are not changed, and the room door boundary conditions are changed from the solid wall boundary to the interior boundary in the simulation calculation, and the physical model is shown in Figure 6.

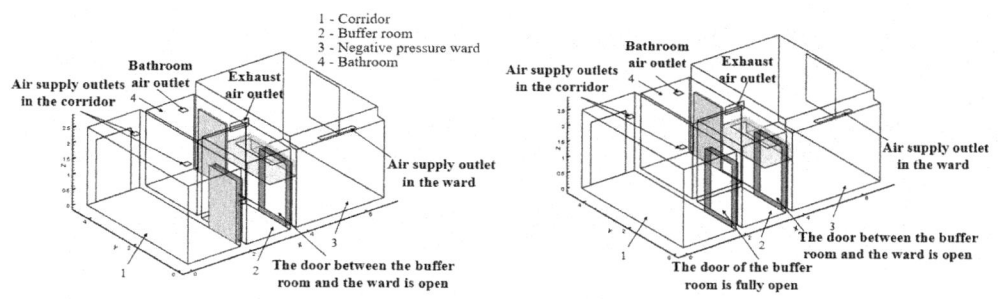

(a) The door between the buffer room and the ward is open (b) The door of the buffer room is fully open

Figure 6.    Physical model.

## 5.2    *Numerical simulation results and analysis of the flow field*

As shown in Figure 7(a), when the door between the buffer room and the ward is opened, the airflow from the corridor air supply port penetrates the buffer room through the doorway, and the airflow in the buffer room enters the negative pressure ward through the opened the door to participate in the air circulation in the ward; after the airflow from the ward air supply port reaches the ground, part of the airflow enters the buffer room from the lower side of the opened door. As shown in Figure 7(b), when the door of the buffer room is fully opened, part of the airflow from the air supply outlet of the corridor enters the buffer room and the negative pressure ward through the opened door. A small part of the airflow from the air supply outlet of the ward enters the buffer room and the corridor through the lower side of the door, and most of it enters the air exhaust outlet, forming a directional air organization. When the door of the buffer room is opened, the airflow entering the

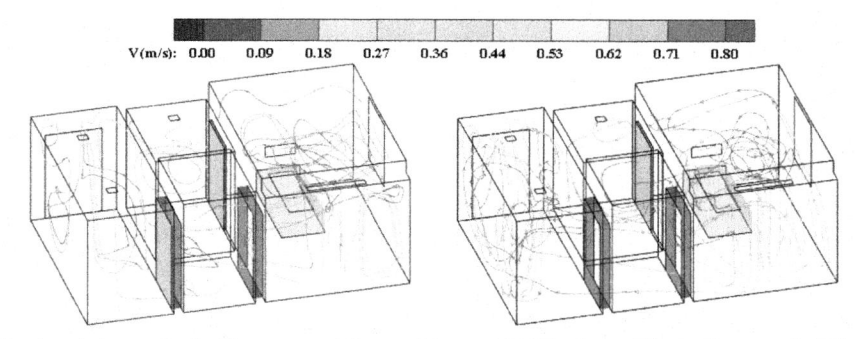

(a) The door between the buffer room and the ward is open (b) The door of the buffer room is fully open

Figure 7.    Flow chart after opening the buffer room door.

ward through the door increases, and the directional air organization formed in the ward makes the airflow entering the local exhaust port larger than the working condition when the door of the buffer room is all closed.

### 5.3 Simulation results and analysis of pressure gradient

As shown in Figure 8(a), when the door between the buffer room and the ward is opened, the airflow between the buffer room enters the negative pressure ward through the room door. The pressure difference between the two rooms gradually decreases and reaches a steady state; the pressure distribution between the negative pressure ward and the buffer room is generally similar. The pressure difference is maintained at 0Pa, and a pressure difference of 5.7Pa is formed with the corridor. As shown in Figure 8(b), when the doors of the buffer room are all opened, the airflow between the corridor-buffer room-negative pressure ward circulates with each other. The total air supply and exhaust air volumes are equal to reach the air volume balance, and the pressure distribution between the regions does not change much. The pressure difference is maintained at about 0Pa.

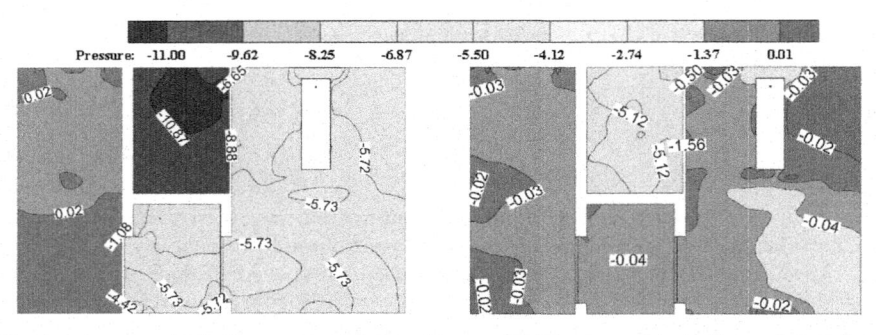

(a) The door between the buffer room and the ward is open (b) The door of the buffer room is fully open

Figure 8.    Pressure distribution cloud at Z=0.7m section.

### 5.4 Simulation results and analysis of pollution gas concentration field

Figure 9 shows a cloud plot of the percentage mass of the contaminated gas equivalent surface when the buffer room door is open. As the mass fraction of the contaminated gas decreases from $1 \times 10^{-7}$ to $1 \times 10^{-9}$, its diffusion range also decreases. When the door between the buffer room and the ward is opened, the concentration of contaminated gas in the ward is significantly higher than when the buffer room door is fully open. Figure 10 shows the concentration of contaminated gas at the cross-section of the patient lying down at the height of Z=0.7m. When the door between the buffer room and the ward is opened, the contaminated gas is distributed in the area around the bed, and when the door of the buffer room is fully opened, the contaminated gas is only distributed in the area near the head of the bed, and the diffusion range at the height of the patient's breathing plane is reduced. Section 5.3 shows that the pressure gradient between the regions to ensure airflow tends to 0 Pa, compared with the room door all closed. When the buffer room door is opened, there is no diffusion of contaminated gas from the ward to other clean areas. Therefore, arranging the exhaust air outlet near the contamination source can effectively inhibit the contamination gas diffusion.

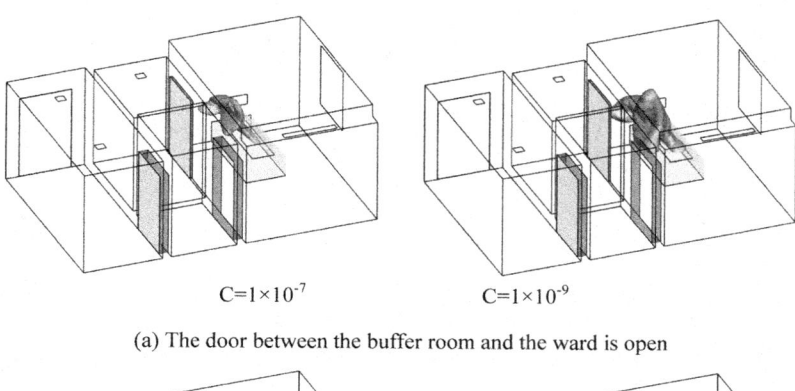

$C=1\times10^{-7}$  $C=1\times10^{-9}$

(a) The door between the buffer room and the ward is open

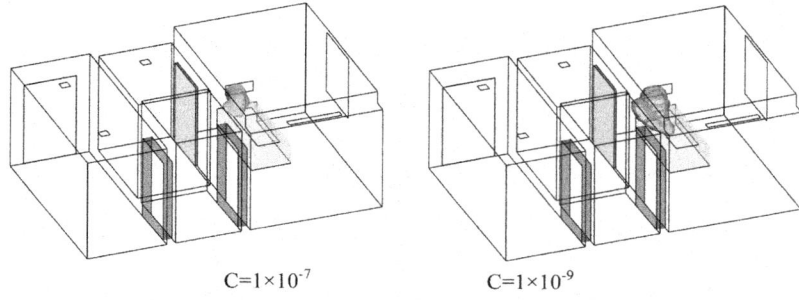

$C=1\times10^{-7}$  $C=1\times10^{-9}$

(b) The door of the buffer room is fully open

Figure 9.   Contaminated gas equivalent surface mass percent cloud map.

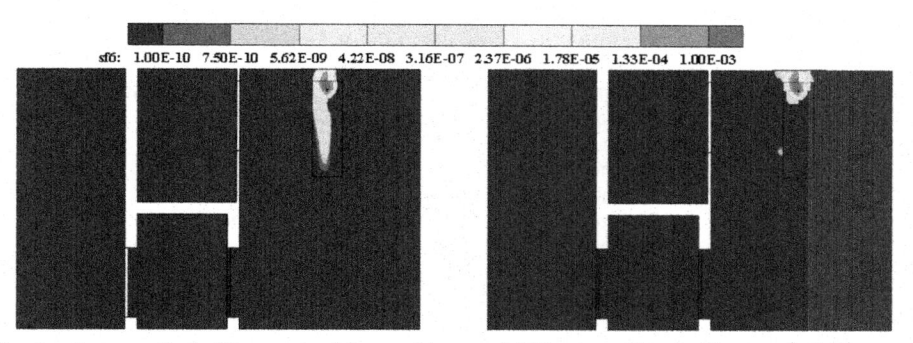

sf6:   1.00E-10  7.50E-10  5.62E-09  4.22E-08  3.16E-07  2.37E-06  1.78E-05  1.33E-04  1.00E-03

(a) The door between the buffer room and the ward is open (b) The door of the buffer room is fully open

Figure 10.   Cloud map of pollution gas concentration distribution at Z=0.7m cross-section.

## 6   CONCLUSIONS

(1) When the door of the buffer room is fully closed, the air organization of the top air supply/local exhaust can form a unidirectional airflow to the local exhaust in the ward; when the door of the buffer room is opened, the airflow entering the ward through the door increases, and the directional air organization formed in the ward makes the airflow entering the local exhaust larger than the working condition when the door of the buffer room is fully closed.

(2) When the door of the buffer room is fully closed, the minimum pressure difference between the corridor-buffer room-negative pressure ward is 5.43Pa, which meets the requirement of the pressure gradient of negative pressure ward of more than 5Pa. When the door of the buffer

room connected with the ward is opened, the pressure in the two rooms gradually converges and finally forms a pressure difference of 5.71Pa with the corridor. When the door of the buffer room is fully opened, the pressure difference between the corridor-buffer room-negative pressure ward gradually decreases to ensure that the pressure gradient of airflow between the regions disappears, and there is no contaminated gas diffusion from the ward to other clean areas.

(3) Combined with the local exhaust air organization, the exhaust air outlet is arranged near the contamination source, which can ensure the distribution of contaminated gas near the patient's breathing area, no matter the working condition when the buffer room door is closed or opened. When the buffer room door is opened, the air organization has better control over the diffusion of contaminated gas than the working condition when the buffer room door is fully closed.

(4) This paper provides suggestions for the future design of negative pressure wards. On the one hand, a reasonably designed air conditioning system of the corridor-buffer room-negative pressure ward could form an orderly pressure gradient, eliminating the possibility of backflow of contaminated gas into the clean area. On the other hand, it is considered to combine local exhaust air to effectively control the diffusion of contaminated gas when designing the air organization of the negative pressure ward.

## ACKNOWLEDGMENTS

The authors wish to acknowledge assistance from Professor Peng X Y, special work by financial support from the Postgraduate Scientific Research Innovation Project of Hunan Province (CX20210940).

## REFERENCES

Deepthi S. T. & Urmila G. (2016). Assessing the Effectiveness of Ceiling-Ventilated Mock Airborne Infection Isolation Room in Preventing Hospital-Acquired Influenza Transmission to Health Care Workers. *ASHRAE Trans*, 122(2): 35–46.

Dong L D, Zhou Y and Yang J X (2020). The numerical investigation of pressure field and air distribution in the negative pressure isolation ward. *Clean and Air Conditioning Technology*, 04, 102–107.

Duan D D. (2018). *Mechanism respiration intensity on the lower respiratory tract exposure of particulates*. Xi'an University of Architecture and Technology, 29–36.

Hang J. & Li Y (2014). The influence of human walking on the flow and airborne transmission in a six-bed isolation room: tracer gas simulation. *Build. Environ.* 77: 119–134.

JI Y Z. & Tu G B (2006). *Numerical study of contaminated air leakage in an infectious isolation ward with fully open doors*. The 18th International Conference on Pollution Control, 861–869.

JI Y Z. & Tu G B et al. (2009). Numerical study of air distribution in door opening process of isolation wards *Journal of HV&AC*, 39(04):90–93.

JI Y Z. (2005). *Numerical study of air distribution and negative pressure monitor in an infectious isolation room*. Tianjin University, 45–51.

Jiang W. K. & Bing S. T et al (2017). *Requirements of environmental control for hospital negative pressure isolation ward*. China Architecture & Building Press, Beijing, 4–9.

Jinkyun C. (2019). Investigation on the contaminant distribution with the improved ventilation system in hospital isolation rooms: Effect of supply and exhaust air diffuser configurations. *Applied Thermal Engineering*, 148: 208–218.

Ling J H. & Yu H Y et al (2014). Effects of airflow distribution on the contaminant removal efficiency in an infectious isolation room. *Journal of Tianjin University* (Natural Science and Engineering Technology Edition), 47(02):174–179.

Lu Y Q. (2008). *Practical Heating and Air Conditioning Design Manual* (2nd ed.). China Architecture & Building Press, Beijing, 216–255.

Suvanjan B. & Kunal D. et al (2020). A novel CFD analysis to minimize the spread of the COVID-19 virus in a hospital isolation room. *Chaos, Solitons & Fractals*, Volume, 139, 110294.

Zhao Y. (2011). *Research on the influence of air distribution on pollution diffusion in a negative pressure isolation room*. Tianjin University, 15–24.

Civil Engineering and Urban Research – Mohamed & Hou (Eds)
© 2023 the Authors, ISBN 978-1-032-44487-1

# Transformation of the old industrial zone in Dongguan City from the perspective of urban renewal

Le Li*

*Guangdong Institute of Science and Technology, Guangdong, China*

ABSTRACT: Dongguan, once the "factory of the world," is one of the most important cities in the Greater Bay Area for exporting foreign trade goods. Dongguan's industrial structure has transformed processing foreign trade into high technology. Currently, it is moving towards an open economic system through the upgrading and transformation of high-tech industrialization. Although the city is developing through the upgrading and transformation of industrial structure and urban renewal and development, the following problems still exist: the idleness of the original land and old industrial areas in the city, the waste of space, and the inability to match the development of the city. In this paper, we analyze and design the old industrial area renovation project in Dongguan City that combines the regional characteristics, cultural contents, and creative design of the city and make suggestions on the strategy for using the old space. The purpose is to explore the impact that the old industrial area renovation approach can bring to the new urban development. The significance is to maximize the utilization rate of urban space.

## 1 INTRODUCTION

In the context of consumption upgrading and industrial transformation, urban renewal has become an important proposition in urbanization. To alleviate the conflict between land space and urban renewal development, we analyze the transformation of old industrial areas in Dongguan City and study the impact of urban old industrial areas on the renewal design of the urban environment through upgrading and transformation, spatial reshaping, and facility configuration. The renovation of the old industrial area of Flounder Island combines urban memory, urban culture, functional panels, and recreational facilities and will be based on the preservation of the original architectural structure for the renovation of the whole area of the old industrial area.

In 2021, the Office of the Dongguan Municipal People's Government issued the "Notice of the Office of the Dongguan Municipal People's Government on the Issuance of the Operational Guidelines for Land Preparation and Development for Municipal Enterprises in Dongguan City Urban Renewal (for Trial Implementation)," which aims to encourage municipal enterprises to participate in urban renewal. Focusing on the promotion of continuous "work to work," it clarifies the implementation path and operation rules for land preparation and development of municipal enterprises. Through the government's main leadership and industry priority, the municipal government will be deeply involved in urban renewal, expand and optimize urban development space, promote the continuous "work to work," and the integration of industry and the city as the main direction to help Dongguan City to build a new industrial dynamic. The reasonable use of the old industrial areas in the city and the reshaping of its space have become a necessity for urban renewal.

---

*Corresponding Author: 405704804@qq.com

## 2 DONGGUAN OLD INDUSTRIAL ZONE DEVELOPMENT STATUS AND NEW FEATURES OF INDUSTRIAL TRANSFORMATION

### 2.1 *Status of development of old industrial areas*

The State Council issued the "Trial Measures for Conducting Foreign Assembly Business" on July 15, 1978. Dongguan City has policy support, coupled with a superior geographical location. Guangdong Province developed an industry of foreign materials and local processing. The successful establishment of Dongguan Taiping Handbag Factory also brought the management model of Hong Kong to Dongguan City. Foreign materials and local processing also became synonymous with Dongguan City. With the development of reform and opening up over the past forty years, there has been a significant industrial upgrading in Dongguan due to low-end manufacturing, foreign materials, local processing, Dongguan City's environmental requirements for enterprises, the global economic downturn, and the increase in labor costs. More and more processing-oriented enterprises are leaving Dongguan City, and the old industrial areas and plants are abandoned and idle. These factories and plants, which once brought economic benefits to the city, can now only become a historical symbol of the city for a generation. One is the old industrial zone in Boxia Community, Guancheng Street, Dongguan City.

The original industrial zone factory buildings had the following problems: closed space, old infrastructure facilities, and lack of capacity. The old industrial area buildings were interspersed with parts of the city's development and appeared to be mismatched.

### 2.2 *Industrial transformation methods*

With the development of China's overall urbanization, land in cities becomes more and more precious, and it becomes necessary to improve land utilization in urban spaces. As the area where the reform and opening up began the most, the urbanization process of Dongguan City has also transformed with the development of policies. The industrial zone of Flounder Island, located next to Dongjiang Avenue, has been preserved intact after years of abandonment. It has become a remnant of the old industrial area that witnessed the reform and opening-up process of Dongguan City. From the perspective of urban renewal, spatial upgrading and industrial transformation become the key issues in the spatial remodeling of old industrial areas.

There are three main ways of industrial upgrading and transformation of the old industrial area as follows: overturning the original industrial factory buildings and rebuilding the planned spatial functions; replacing the spatial functions by protecting the architectural structure of the old industrial area and establishing museums, exhibition halls, and other cultural display spaces with educational significance; building in the form of cultural and creative tertiary industries, combining the cultural and creative industries based on retaining the architectural structure of the original industrial area and establishing multi-functional space complex. The functions of these old industrial areas are adjusted and transformed for reuse. The industrial functions carried out by the project are transformed and upgraded to meet the new development needs of the city.

## 3 DIFFICULTIES IN URBAN RENEWAL PRACTICE

With the development of the global economy, the industrial structure of cities has been upgraded. Some successful cases have also emerged in transforming old industrial areas in cities, such as Beijing Scene Experience Hotel—Capital Pick Up Hotel, Wuhan Liang You Hong Fang Cultural Art Community Landscape Transformation, Shanghai Happy Park Joint Office, and Guangzhou City Panyu Shiqiao Street Xifang Compound.

Urban renewal will rise to a national strategy through land space design. Since the renovation of old industrial areas is a systematic project, there are many problems: large capital requirements, many major members with interests involved, complex planning procedures, long development, and operation cycles, return on revenue, and the need to consider the comprehensive strength of enterprises and local governments. Although the old urban industrial area has a lot of space for transformation potential, it also has problems: aging physical form, backward functional structure, insufficient supporting facilities, and fire safety (Xie Zhanghui et al. 2021). There are still some difficulties in trying to carry out urban renewal and transformation from an overall perspective.

## 3.1 *The danger of homogeneous replication*

As a result of the industry upgrade, under the role of urban renewal relationship, real estate developers and local governments use the trend of the Netflix effect for marketing. Urban space renewal becomes commercially homogenized in project positioning, planning and design, and mainstream aesthetic boosting (Ji Ronghua 2021). By shaping the concept of consumer punch cards, online marketing is allowed to cater to the aesthetic sense of contemporary youth. What is needed from the perspective of urban renewal is design, creativity, the cultural connotation of the city, humanistic values, and not homogenized Netflix businesses. The city of Xiamen, as the first generation of Netflix city, has serious homogenization and replication, and in recent years there is also the problem of tourism decline.

## 3.2 *Difficulties of industry introduction*

The early urban renewal methods were to demolish old buildings and rebuild new ones, following a crude model of old reform. As real urban renewal is not only the shaping of space form but also the upgrading of industrial content. The tertiary industry is the best way to upgrade the industry. In terms of content and meaning, the government and the original space users pay more attention to the industrial transformation and upgrading in the urban renewal process (Gao Ang et al. 2017). The government and real estate enterprises, as the actual investment and operation subjects, need to pay more attention to the industrial occupation.

## 4 EMPOWERING OLD SPACES WITH NEW FUNCTIONS THROUGH URBAN RENEWAL AND CREATIVITY

### 4.1 *Urban space renewal*

Flatfish Island carries the past industrial civilization of Dongguan City and the memories of many Dongguaners. Renovating and revitalizing the use of the industrial heritage of Flounder Island has become a breakthrough point in developing the waterfront area along the six banks of the three rivers and a demonstration project of the three-year urban quality improvement plan of Dongguan City. Resting for nearly ten years, the main body of the development of Flounder Island, the industrial park, was created by the East Solid Group. Creative design is used to give new vitality to this old industrial area.

Flatfish Island connects the past and modern regional culture by keeping the old industrial area architecture. It resurrects the old industrial area through local shaping and creates a new cultural landmark in Dongguan. From the industrial function planning using functional plates combined with supporting pleasure gardens, it will be divided into cultural tourism space, industrial space, commercial space, and riverfront pleasure gardens to create. Guancheng culture will be inherited. By setting up the intangible cultural heritage display square, ground paving, landscape sketches, iconic spiritual architecture landscape, and characteristic landscape, a cultural and artistic atmosphere of the renewal of the industrial park is created. The comprehensive design uses the arc that echoes the circular shape of the vertical cylinder bank to divide the form of paving. In terms of material, red bricks are used as the main material of the area to echo the façade of the chimney. The rusty steel plate historical and cultural scenic wall is used as a cultural carrier to reflect the strong atmosphere of the times. Starting from the multi-functional exhibition area, the commercial

Figure 1.    Original topography (Photo credit: Self-drawn).

Figure 2.    Functional space division (Photo source: Self-drawn).

space is used as the starting point of the cultural display of the area, which is used to introduce the development history of Flatfish island (Miao Chunsheng 2014). The design of the riverfront ecological and cultural tour park creates a water-watching and water-friendly atmosphere and integrates the history and culture of Flounder Island.

Using the original state of Flatfish Island, the government and enterprises invest and operate to renew it through urban renewal. Part of the area was the former site of the Customs office in Flatfish Island before and after the transformation, i.e., the Tmall Dongguan Service Center. Part of the area was the workshop of Guangdong Concentrated Premix Feed Factory before and after the transformation, i.e., the M.ICITY Misty Electric Theatre. Part of the area was the dormitory of Guangdong East Rice and Noodle Products Factory before and after the transformation, i.e., a

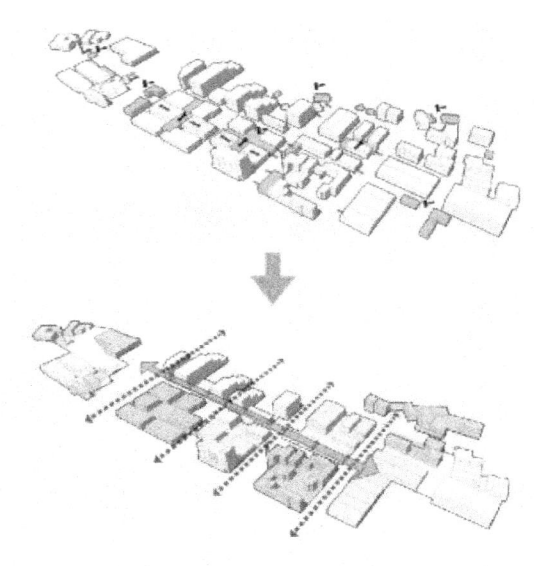

Figure 3.   Spatial combing of the remains (Source: Urbanity Reconstruction and Local Shaping of Waterfront Industrial Remains – A Summary of Dongguan's Approach to the Regeneration of Floundering Fish Island).

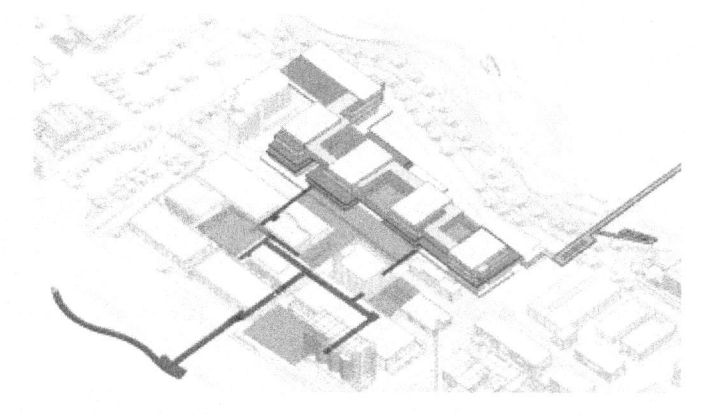

Figure 4.   New buildings and pedestrian bridges (Source: Urbanity Reconstruction and Local Shaping of Waterfront Industrial Remains – Summary of Dongguan's Approach to Regenerating the Flatfish Island).

hotel. Part of the area was the office of Dongtai Feed Factory before and after the transformation, i.e., a commercial building. China's outstanding basketball Yi Jianlian's salary fire camp (Flatfish Island) basketball training center successfully moved in, and the new design and corporate IP give the Dongguan Flatfish Island industrial zone fresh blood.

## 4.2   *Dongguan city transformation plan*

From the Dongguan Municipal Government on March 22, 2021, published data and work reports can be seen to promote the urban renewal process. Based on land spatial planning in Guangdong Province, more use of micro-renovation approach was used to promote urban renewal. To prevent the emergence of quick and large demolition of destructive "construction" problem, urban space was optimized. The model for organic renewal was from the new urban construction land and the

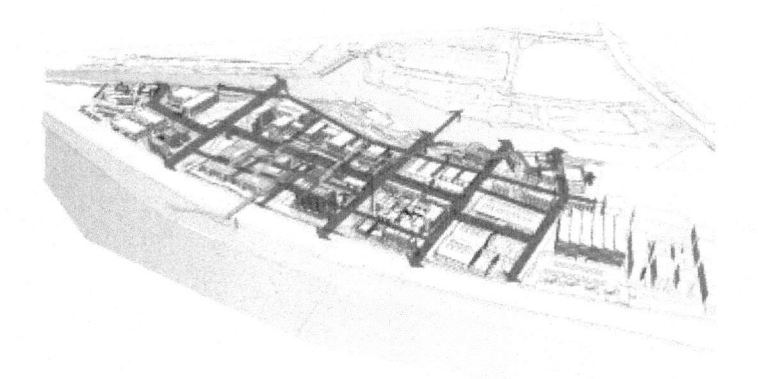

Figure 5. Three-dimensional roaming system (Image source: Urbanity Reconstruction and Local Shaping of Waterfront Industrial Remnants – Summary of Dongguan's Fish Island Renewal Approach).

urban space under the influence of inertia aging. (Wang Peng & Shan Liang 2018) Only by using a combination of old and new space, reasonable use of historical buildings combined with innovative design can the advantages of urban renewal design be reflected.

### 4.3 *The significance of the new functional space*

Standing from the perspective of the Greater Bay Area and looking at Flatfish Island, this is a historic opportunity. As the epitome of the spirit of Dongguan, cultural reset and commercial revival are the new paths of Flatfish Island. Continuing the overall characteristics of the old industrial area, the organic integration of old and new creates both ecological and humanistic complex business. It will turn the old industrial area into a cultural and creative park with business and leisure. From the perspective of the meaning of public space, this will also become a good place for Dongguan citizens to relax and entertain. Cultural and creative activities, cultural exhibitions, and creative bazaars in the park let the participating Dongguan citizens feel the collision between the new trend and industry and a new culture.

The highly mixed use of land is the main trend of urban transformation and development.(Zhuang Shaopang et al. 2021). With the upgrading and transformation of industries, the rise of R & D, logistics, and creative industries also makes the mixed use of land in industrial areas inevitable. In order to achieve the purpose of mixed development in response to the transformation and upgrading of industrial areas and to adapt to the current advocated concept of green development, as far as possible, no demolition and reconstruction. With flexible use of comprehensive remediation and functional change, composite renewal mode will gradually become the direction of renewal development of old industrial areas. The industrial development of many large and medium-sized cities in China is undergoing or will undergo the development process of transformation and upgrading of old industrial areas at home and abroad. This process is difficult to replicate due to differences in urban development level, resources, location factors, and social demographic structure (Wang Wei 2020). However, there are still some domestic and foreign experiences and lessons to be learned in the transformation and upgrading of old industrial areas.

## 5 CONCLUSION

The upgrading of urban areas has become an inevitable way of urban renewal and development. There is a good side to optimizing the quality of the urban environment through the transformation and upgrading of old industrial areas, but there is also a certain danger of homogeneous replication of the design and transformation methods. With the change of time, new requirements are put forward for urban space, and the inheritance adjusts the way of shaping urban space and the precise positioning of the city's own development and construction in order to regain a new life.

## REFERENCES

Gao Ang & Zou Bing&Liu Chengming.(2017). Exploring the renewal mode of Shenzhen old industrial area from "single" to "composite," *Planner*, 33(5):114–119.

Ji Ronghua. (2021).Analysis of renovation strategies of existing industrial buildings. *Engineering Seismic and Reinforcement Transformation*, 43(1): 169.

Miao Chunsheng. (2014). *A review of Shenzhen's urban renewal for more than thirty years and its next stage of reflection*. Urban and Rural Governance and Planning Reform – 2014 China Urban Planning Annual Conference.

Wang Peng&Shan Liang. (2018). Regeneration of old industrial areas under stock planning-an example of urban renewal in old industrial areas in Shenzhen. *Urban Architecture*,(1):62–65.

Wang Wei. (2020) Research on renovation and protection of existing industrial buildings in urban centers. *Green Building*, 12(6): 48–50+57.

Xie Zhanghui & Zhou Shutong & Yu Yong & Yang Fengping & Zhang Tongwei & Fu Shuman. (2021).*A study on the renewal strategy of existing industrial building park – A case study of Dongguan Flatfish Island Cultural and Creative Industrial Park*.10.005.

Zhuang Shaopang & Gao Kun & Wang Jing & Zhong Guanqiu.(2021). Urbanity Reconstruction and Local Shaping of Waterfront Industrial Remains – A Summary of the Regeneration Approach of Dongguan Fengyuzhou, *Southern Architecture*, (5):30–37.

*Civil Engineering and Urban Research – Mohamed & Hou (Eds)*
*© 2023 the Authors, ISBN 978-1-032-44487-1*

# The impact of urban green space on health at different spatial scales

Tan Yuan Long*

*Guangdong University of Science and Technology, City University of Macau, Zhuhai, China*

ABSTRACT: With the global outbreak of the new crown pneumonia, the issue of symbiosis between human and green space environments is gaining attention. As an important part of urban development, urban green space research has an intervention effect on health. In this paper, we use the information visualization software CiteSpace to analyze the health-related research of urban green space and study the impact of urban green space on health through the health effects of urban green space at different spatial scales, starting from the availability and accessibility of urban green space.

## 1 RESEARCH BACKGROUND

Since 2020, the novel coronavirus-infected pneumonia (Covid-19) epidemic has had a wide range of impacts worldwide and has become a global public health safety event with a high impact in recent years. In the context of the rapid development of healthy cities, urban green spaces are exploring how to strive for a healthy and sustainable habitat and asking key questions related to human health impacts.

Urban green spaces, as an important part of the built environment, affect human health. Some studies have shown that green space in the natural environment positively affects health (HAQ 2011). In order to have a clear understanding of the development trend of the research field of urban green space on health, this study uses spatial analysis as an entry point to sort out the impact of urban green space on health and provide theoretical guidance on the development of green space in healthy cities.

## 2 RESEARCH METHODOLOGY

Bibliometrics is an interdisciplinary discipline that uses mathematical and statistical methods to quantitatively analyze all carriers of knowledge. Through the use of bibliometrics, it is possible to describe, evaluate and predict the research hotspots, evolution, and development trend of a research field. With the urgent need for data acquisition and the development of information technology background, we use the data processing nodes such as "author," "keyword," and "institution" in CiteSpace software. The evolution of the research hotspots in the discipline, the internal connection between different research directions, and the cooperation network among author teams can be presented visually through the combination of nodes and lines, and the research frontiers and hotspots of urban green space can be analyzed.

---

*Corresponding Author: U21092120241@cityu.mo

 DOI 10.1201/9781003372417-66

# 3 BASIC ANALYSIS OF RESEARCH IN THE FIELD OF URBAN GREEN SPACE

## 3.1 *Analysis of the number of articles published*

In the 1990s, the concept of "healthy cities" gained attention in China. According to the literature statistics, the annual number of articles on health in urban green spaces at different spatial scales increased from three in 2000 to 78 in 2022, which is a significant growth rate (Figure 1). Based on the overall trend of annual publication volume, urban green space in China can be divided into three stages in the health field at different spatial scales.

(1) From 2000 to 2003, the average annual number of articles was three, and the development trend was slow. Given the widespread phenomenon of urban air pollution, which seriously affects the quality of the global ecological environment, domestic scholars put forward suggestions and countermeasures of biological control based on the ecological and environmental effects of the urban green space system (JERRETT 2003), which started the research of domestic scholars on the health problems of urban green space, but the research has not become systematic, and the number of relevant publications is still very small.

(2) The rapid development phase from 2004 to 2016, with an annual average of more than 50 papers, including the gradual growth trend of master's and doctoral dissertations from 2006 to 2010, indicates that the research on the issue of urban green space on health has attracted the attention of many scholars. The health effects of urban green space on residents vary according to the characteristics of urban green space, demographic characteristics, socioeconomic characteristics, and built environment characteristics, and therefore produce different health promotion effects (Urban green spaces and health: a review of evidence 2016), and academic research related to both urban green space and human health has received attention from many disciplines involving urban planning, landscape architecture, and medical health.

(3) 2016–2019 is a stable development phase, with an average annual volume of 125 articles, and the research heat from 2017-2020 is an active period for research in this field, with literature mostly on urban parks, protected green spaces, waterfront green spaces, urban greenways, and pedestrian green trails (Ye Lin 2018).

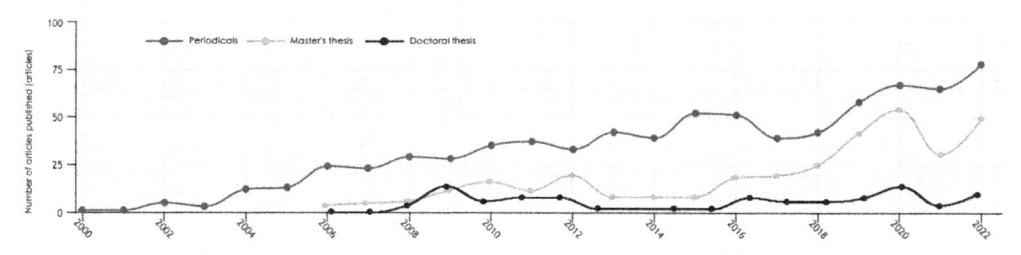

Figure 1. Statistical chart of annual publication distribution of research papers on urban green space on health from 2000 to 2021.

## 3.2 *Analysis of representative people and cooperation network*

The author's co-occurrence graph in CiteSpace can identify key people in the research field and their degree of association. The network knowledge map (Figure 2) was obtained with 215 nodes and 258 links. On the whole, the distribution of authors shows a "small concentration and large dispersion," with 253 authors in the research area and only 706 publications, indicating the small size of the core author group; in terms of author cooperation, Yuanrun Zheng and Jihua Zhou work closely together, followed by the team led by Zhiqiang Liu, Jundi Wang, and Huanwei Hong, while other teams are not closely connected. The other teams are not closely connected.

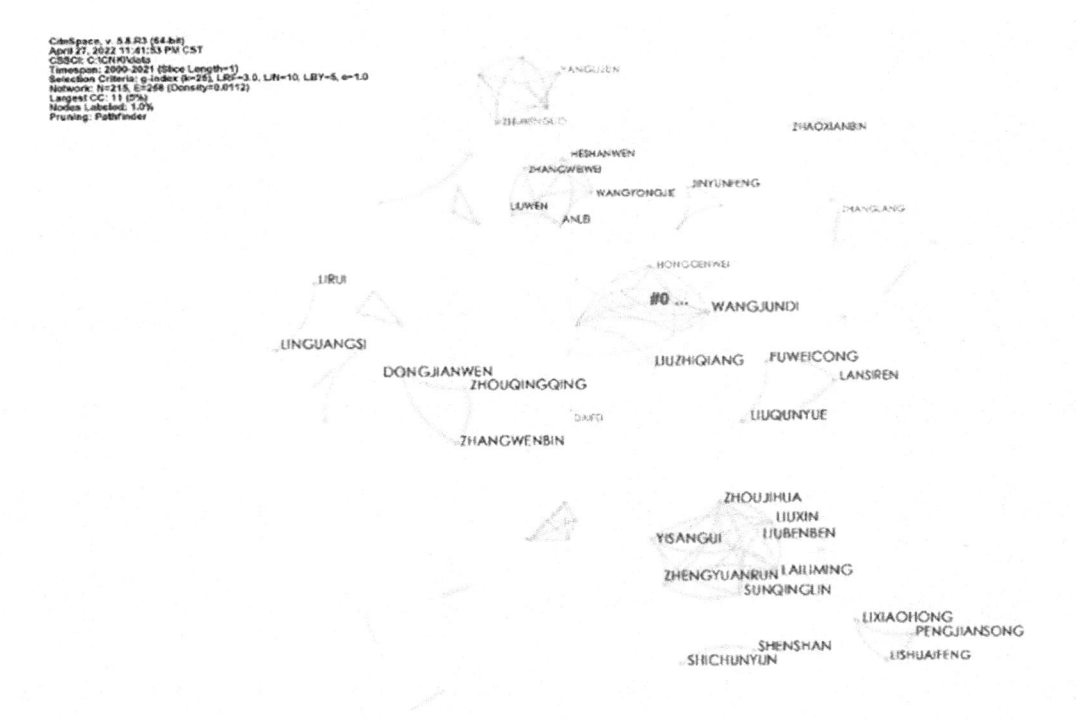

Figure 2.   Knowledge map of the collaborative network of urban green space research authors.

## 4   ANALYSIS OF THE IMPACT MECHANISM OF URBAN GREEN SPACE ON HUMAN HEALTH

### 4.1   *Urban green space and human health*

Urban green space is used in Chinese cities to protect or improve the natural ecology of Chinese cities, to protect the environment, to provide a variety of recreational and leisure activities for the daily life of Chinese urban residents, and protect and beautify the city. Currently, urban buildings and green spaces are becoming more and more densely populated, the living green space is gradually reduced in large quantities, the natural environmental resources are seriously damaged, etc. These problems have brought about certain adverse effects on the development of urban human health, and from now on, scholars in urban construction have carried out research on urban green spaces for health needs.

Human health is defined as an individual in good physical, psychological and social condition. The existing research questions on the effect of urban green space on human health are mainly three categories of urban green space, physical health, psychological health, and social health. It is also found that researchers have analyzed the relationship between these green space-related elements and human health to investigate whether there is a direct impact of urban green space elements on human health, based on various characteristics of green space, such as green space area, quantity and quality of green space size, and accessibility.

### 4.2   *Influence mechanism*

The impact of urban green space on human health can be expressed at both macro and micro levels.

(1) From the macro level, the layout of urban green space has an important impact on human health. Combined with China's urban green space classification standards, it is mainly expressed on three scales in terms of health (Figure 3): First, it can form a green belt around the city with agricultural land, forests, and other urban green space to guide and control the healthy development of urban green space; second, it can provide residents with green space for recreation and healthy activities and improve the health of urban residents; third, urban green space as a natural resource, can regulate the urban climate and improve the heat island effect, to maintain a healthy urban environment.

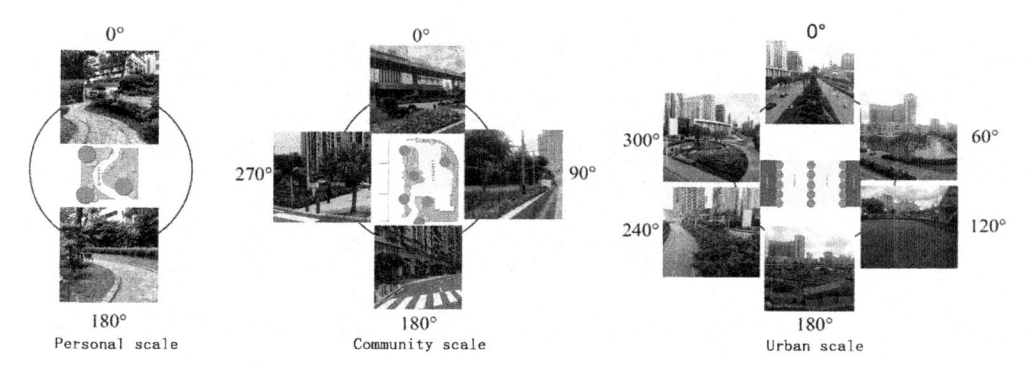

Figure 3. Three-dimensional spatial analysis of three scales of healthy urban green space.

(2) At the micro level, the impact on residents' health mainly manifests in three aspects. First, it promotes residents' health activities, and green space is more attractive than indoor space for residents to carry out health activities and reduce the risk of chronic diseases. Second, as a positive environmental element, green space can improve mental health, relieve residents' stress and reduce the risk of mental illness. Third, as a space for residents to interact with each other, green space can help increase social ties and residents' communication and enhance neighborhood relations.

## 5 RESEARCH ON THE IMPACT OF URBAN GREEN SPACE ON HUMAN HEALTH BASED ON DIFFERENT SPATIAL SCALES

At present, more and more experts and scholars are paying attention to research on urban green space and human health in China. The study of the effects of urban green space on human health using spatial analysis methods is mainly based on the spatial interaction between human and urban green space. This study takes Macau as an example and analyzes the impact of urban green space on human health from two aspects: the impact of urban green space on human health and the common spatial data and analysis methods in spatial analysis methods, and the impact of urban green space on human health at different spatial scales.

### 5.1 *Common data and methods in spatial analysis of urban green space for human health*

One of the basic methods for analyzing data related to different green space types in studies of the impact of urban green space on human health is GIS technology, a form that enables data processing of urban green space and is used to examine the interrelationships between urban space and other geoenvironmental variables. GIS can capture, process, collect and analyze information on green space at different scales with different green space types and environments in real-time. Labib summarized the GIS tools and techniques commonly used in green space and health research (Labib 2020). Using spatial regression, Wu found a significant correlation between urban green spaces and human health (Wu 2017). Zhai and Baran clarified the relationship between accessibility and the function and configuration of paths in the park on residents' walking behavior using a spatial syntax approach using the depth map software (Zhai 2016). By making the green space visible and accessible and focusing on more details about the green space.

## 5.2 *Impact of urban green space on human health at different spatial scales*

Spatial scale is a concept of geospatial volume. The space where human behavioral activities in the green space interact with the scale of urban green space can be roughly divided into the following three aspects from a micro to a large macro: personal scale, community scale, and urban scale (see Table 1)

(1) From the aspect of personal scale, personal scale refers to the space of people's living environment, such as the landscape environment of residential areas. The distance between people is 5~50m. In this reachability range, people are easily affected by the environment.
(2) From the aspect of community scale, community scale refers to the area where people can carry out small activities within a certain range; generally, such scale is divided into 500, 1000 m, and above 1000, such as living settlements, park areas, neighborhoods, etc.
(3) On an urban scale, most scholars have focused their research on human health in urban spaces. Yao Yanan investigated the information on urban green space use within 300, 1000, and 3000 m around the workplace location of Internet information technology workers in Beijing and

Table 1. Health effects of urban green space at different spatial scales.

| | Site plan | 3D space picture | Green space health effects |
|---|---|---|---|
| Personal scale | | | Small-scale green spaces can promote residents' healthy activities, attract residents' healthy exercise, and reduce the risk of residents' chronic diseases. |
| Community scale | | | Community-scale green space helps to increase the connection and communication between residents and enhance the relationship between neighbors. Its natural elements can relieve the pressure and fatigue of residents and provide leisure and entertainment space for fast-paced urban life. |
| Urban scale | | | At the urban scale level, green space guides urban development and protects the ecological environment. Through green belts combined with the direction of urban development, it can reasonably guide cities to form a development axis in green space, taking into account the control of urban scale and urban development patterns. |

found that urban green space in the work environment is positively associated with human physical and mental health (Yao 2018). Zandieh applied GPS technology to investigate in-depth residents' perceptions of their perceptions of green space in the neighborhood (Zandieh 2017), from which it can be learned that green space within the neighborhood space has a positive activity influence on the health status of people in the neighborhood. In addition, Sun Peijin and Lu Wei conducted an in-depth study on the direct correlation between urban green space and residents' physical exercise activities and residents' body weight health index in Dalian (Sun 2019), which showed that the size of the area and the number of green spaces in urban green space for residents' fitness and recreation facilities and activities and their positive influence on the human body were significant.

## 6 CONCLUSION AND DISCUSSION

The article takes urban green space and human health as the entry point and studies the data and methods in spatial analysis and different spatial scales of the impact of urban green space on human health from different spatial analysis methods. In general, the factors of two dimensions of availability and accessibility of green space can produce beneficial effects of promoting physical health, psychological health, and social health at different spatial scales, which provide references to improve urban planning and green space system planning and enhance the health benefits of urban green space.

### 6.1 Study on the health of the human body by focusing on healthy city-oriented urban green space planning

People are gradually exploring and thinking about building healthy cities based on the global outbreak of new coronavirus pneumonia. The rapid development of urbanization has led to many social health problems. Human health depends on factors such as the living environment and the economic level of the city. Healthy city planning should address human health issues from all aspects such as environment, social system, culture, etc. A healthy environment better promotes human health.

### 6.2 Integrated spatial analysis from different scales

The construction position of urban green space is crucial in the whole process of building a healthy city. Through the research analysis, at the level of spatial scale, overseas analysis of the impact of green space on human health research study is usually achieved at the scale of the community. China's research scope is broader, mostly based on the scale of large cities; the difference in the scale of analysis can jeopardize the link between green space and health index middle of the compressive strength. At the level of analysis methods, several studies based on GIS profiling have turned out to be one of the most common ways used by the natural environment clinical epidemiology and related industries, and spatial analysis methods are widely used for the extraction of characteristics of green spaces in large cities. In summary, the impact of green spaces on health hazards has long been confirmed in various aspects. Future research should consider the assessment of green spaces on personal scale and should also pay attention to the correlation between green spaces and more optimal health indicator values.

### 6.3 Improve the availability and accessibility of urban green space

The availability of urban green space is one of the basic index values of the urban green space system in overall planning under a healthy city orientation. Planners should clarify the index values of green space availability when carrying out overall urban planning and urban green space system planning to enhance the availability of urban green space to meet the specific requirements for residents' use. The improvement of urban green space accessibility can promote residents' active participation in physical and cultural activities and have good opportunities for their health.

## REFERENCES

HAQ S M A. (2011) Urban Green Spaces and an Integrative Approach to Sustainable Environment. *Journal of Environmental Protection*. 2(5): 601–608.

JERRETT M & BURNETT R& GOLDBERG M. (2003). Spatial Analysis for Environmental Health Research: Concepts, Methods, and Examples. *Journal of Toxicology and Environmental Health Part A*, 2003, 66(19): 1783–1810.

LABIB S M & LINDLEY S & HUCK JJ. (2020). Spatial Dimensions of the Influence of Urban Green-Blue Spaces on Human Health: A Systematic Review. *Environmental Research*. 180: 108869.

Sun Peijin & Lu Wei (2019). Study on the correlation between urban green space and residents' physical activity and body mass index: a case study of Dalian City. *Southern Architecture*. 34–39.

Urban green spaces and health: a review of evidence. ( 2016). *Copenhagen: WHO Regional Office for Europe.*

WU & JACKSON. (2017). Inverse Relationship Between Urban Green Space and Childhood Autism in California Elementary School Districts.*Environment International*. 107: 140–146.

Yao YAN & Huang QIUYUN & Li SHUHUA.(2018). Research on the relationship between green space in work environment and physical and mental health: an example of Beijing IT industry population. *China Garden*. 34 (9): 15–21.

Ye Lin & Xing Zhong & Yan Wentao. (2018). Discussion on Urban Green Space Planning Approach to Justice. *Urban Planning Journal*. 57–64.

ZANDIEH R & FLACKE J & MARTINEZ J. (2017). Do Inequalities in Neighborhood Walkability Drive Disparities in Older Adults'Outdoor Walking? *International Journal Environment Research and Public Health*. 14(7): 740.

ZHAI & BARAN.(2016). Do Configurational Attributes Matter in the Context of Urban Parks? *Park Pathway Configurational Attributes and Senior Walking. Landscape and Urban Planning*. 148: 188–202.

*Civil Engineering and Urban Research – Mohamed & Hou (Eds)*
*© 2023 the Authors, ISBN 978-1-032-44487-1*

# Urban small house storage space design under the concept of multi-functional design

Wei Dai* & Xinru Mu*
*Shenyang Jianzhu University, Shenyang, Liaoning Province, China*

ABSTRACT:   With the increasing urbanization, many people are moving into cities. The small house with limited living area and many limitations for storage has become the first choice of most people. In this paper, we take the design of small house interior space in old urban residential buildings as the research object and the multi-functional storable home as the research carrier and conduct in-depth research and analysis on the design method of urban small house storage interior space under the concept of multi-functional design. Through this study, we aim to form a complete and guiding theoretical system of urban small house interior space design under the concept of multi-functional design and conclude that the storage space under the concept of multi-functional design has a certain ability of reconciliation and balance, which can improve the narrow and depressing living environment of small houses. Finally, the analysis and personal understanding of the paper will be combined with the design practice, hoping to play a role in the design of small house interior spaces and the study of the relationship between furniture and space in many new first-tier and second-tier cities, such as Tianjin, Chengdu, Xining, Shenyang, and Chongqing, where the economy is developing rapidly.

## 1   INTRODUCTION

Under the social environment of increasing pressure on young people to purchase houses, many young home buyers in first-tier, new first-tier, and second-tier cities prefer to purchase small homes with relatively low prices (Hua 2021). Through the study of the concept of livable home and space design, combined with the psychology of young home buyers pursuing fashion, novelty, and diversity, the creative design of small home storage space under the concept of multi-functional design is a topic worthy of study (Leng 2019; Zhou 2019).

## 2   RELEVANT OVERVIEW OF THE MULTI-FUNCTIONAL DESIGN CONCEPT

### 2.1   *The concept of multi-functional design*

The multi-functional design concept combines function and design in the face of different preferences of people, atmosphere, and time. Functional integration is one of the starting points of the multi-functional design concept.

### 2.2   *The merit of multi-functional design*

The multi-functional design concept needs to focus on several key factors such as design, economy, and environmental protection to meet the growing needs and trust of consumers (Peng 2022; Xu 2022). In order to achieve a high level, the functions and shapes of the designed products need to be changed accordingly.

---

*Corresponding Authors: 14106361@qq.com and mxr386007166@126.com

DOI 10.1201/9781003372417-67

### 2.2.1 *Design aspect*

The performance of design excellence is in the function and functional value to be able to do with consumers to carry out a reasonable and effective communication embodiment. In the design, we need to always be in the leading position so that consumers can identify with product performance and process design. The accurate and complete research and prediction of consumer needs is an important process that must be taken before the design to be full of unique personality. We try to meet the needs of consumers in all aspects and encourage them to love the products we design.

### 2.2.2 *Economic aspect*

The economic advantages of products designed under the multi-functional design concept are mainly reflected in the consumers' willingness to consume and satisfaction. They can buy multiple single-function products to meet their needs or buy products designed with the multi-function design concept to meet their needs. In contrast, the price of a multi-functional product is more affordable than the total price of multiple single-function products. Buying multi-functional products can not only meet a variety of needs, but also save costs.

### 2.2.3 *Environment-friendly aspect*

In today's society, excessive consumption and waste of resources and the environment is a major problem that plagues people's lives. Multi-functional design is a significant way of addressing the above problems, for it is highly compatible with the concept of green design. Through functional integration, multi-functional design can make the utmost of the resources in the product manufacturing process (Shen 2020; Wang 2020) and reduce the waste of materials. Compared with other single-function products, multi-functional design alleviates environmental damage and waste of resources in product manufacturing, use, and waste.

## 3 THE SIGNIFICANCE OF STORAGE AND STORAGE SPACE

Storage means organizing and putting away. As a way to receive or collect things, proper storage makes it easier to store and use things in daily life.

Storage space refers to the parts of a space that are scattered in an independent space, which has both storage and decoration functions. By studying the structure, function, and use of space as an entry point, we can create a fully functional, comfortable, and convenient storage system and improve the quality of living with a unified overall layout and refined design. The design of storage space is inseparable from the overall concept of interior space, individual space, style, and color, and is a very important part of residential interior design.

## 4 DESIGN PRINCIPLES

The design of interior storage space under the multi-functional design concept requires not only a certain amount of space to be cut out for storage, but also a certain amount of innovation and ability to control the overall situation in the interior space design.

### 4.1 *Principles of space integration*

To ensure the integrity and unity of the overall interior space, it shall achieve a reasonable division of the storage space while maintaining the harmonious development of function and aesthetics, which is inseparable from the deep excavation and research of the overall space and layout of the residence to ensure the coordination and unity between different functional areas. It's necessary to implement humanized planning and design, thus enhancing the utilization rate and effect of the overall storage.

## 4.2 Follow the principles of praxiology and psychology

The major functional systems of storage space are closely related to human behavior and psychological patterns, and human physical and mental patterns will change due to the change of various complex conditions of users, but there are still similarities between the two, and the commonality is the most basic applicability and comfort. The behavior, daily habits, and physiological conditions of people need to be combined with various important external factors to create a multi-functional design concept of a small house storage space that meets the behavioral habits and physical and mental rules of people.

## 4.3 Environmental safety principles

The safety design of the storage space has always been an ethical and moral factor that needs to be focused on, so it is necessary to choose safe and reasonable environmentally friendly green materials, strictly evaluate and monitor whether there are a series of problems such as substances that are harmful to people's health and serious safety hazards, and resist them firmly (Dai 2020; Zhu 2020). In the design, we prioritize the needs of the elderly and children in all aspects so that the storage space has human characteristics and is in line with the multi-functional design concept. Designers must design storage items for younger, active children without sharp edges, durable and clean, safe and non-toxic. At the same time, designers need to design spaces for the elderly over 65 years old who are bothered by a series of problems such as slowing down and declining intelligence, especially those who live alone (Chen 2021; Li 2021), to improve their quality of life following their special ergonomic scale.

## 5 METHODS AND STRATEGIES

Storage space in small houses can improve people's living quality, and "reasonable space" is a kind of space closely related to living quality, and the characteristics of "reasonable space" are consistent with "functionality" and "complementary" to the overall comfort of the space. This characteristic is especially prominent in the storage space, which fully reflects the importance of exploring the layout strategy and method of urban small house space design under the concept of multi-functional design.

## 5.1 Precise overall floor plan

The space of a small house containing a large number of irregular residential structures can easily affect the overall space utilization, "reduce space," that is, reduce the redundant and complicated level of space (Song 2021; Zang 2021), and let the overall visual effect of the space is neat and unified.

The proportion of space that can be stored should be maximized in the space to create a neat and unified layout of functional space. The raised part of the space can be transformed into storage space, such as the protruding wall into cabinets, niches, etc. This multi-functional integrated furniture can not only separate the storage space, but also reasonably group products. For example, the "niche" design in the projection of the wall in Figure 1 has storage, display, lighting, decoration, and other functions. While saving space, the recessed design produces a visual illusion of forward extension and reduces the oppressive feeling of a small indoor space.

Another strategy for planning space is to "add space." The point of "adding" is to make the space richer at all levels and to maximize the use of the extra space, so that the storage space and the functional space in a small house complement each other. The visual effect of "big in small" in a successful small house space is the proper use of "framing." The use of a large volume of partitions to divide the major space areas with different functions makes the connection of each functional space more intertwined and interoperable. To achieve the effect of "constant separation," "holes"

(1) Niches with lighting and extended visual functions. (2) "Cabinet-type" niches embedded in the wall.

Figure 1.　"Niche" design.

*https://www.zgsjlm.cn/wap_sn991.html.

can be made in the non-supporting walls. Figures 2 and 3 mainly use the "doorway" frame view and "window" frame view to combine the inside and outside of the two spaces to connect the different functional spaces. The small indoor space environment of the oppressive feels disappeared, both beautiful and can be used for display, storage, etc.

Figure 2.　"Door hole" frame view.　　　　Figure 3.　"Window frame" frame view.

*https://www.puxiang.com/

In this article, the author designed and renovated a small residential space of 59 square meters for a newlywed couple with limited funds living in Jinniu, an old neighborhood on Beiguan Street in Xining's Chengdong District. As shown in Figure 4, the rectangular solid wood bar can be used to divide the overall living space. The existing living room, which accounts for less than 50 percent of the space can be divided and reorganized for projection. The rest space, the large wooden table, and the wall are connected into a complete unified body as a place to sleep, rest, work, play, and study. Multiple composite kitchen and bathroom spaces, storage cabinets, and other furnishings are arranged against the wall to reduce these redundant and protruding parts, so the overall layout of the space is more standardized and unified. "Increasing space" is widely used, such as the "frame table" placed at the entrance, so people can have a broader overall view. In order to increase the sense of air permeability of the room, the space area near the window is designed into different forms of "window pavilion" chosen in the form of a round arch. The eco-friendly material home is designed as a "frame" in the space, as shown in Figure 5.

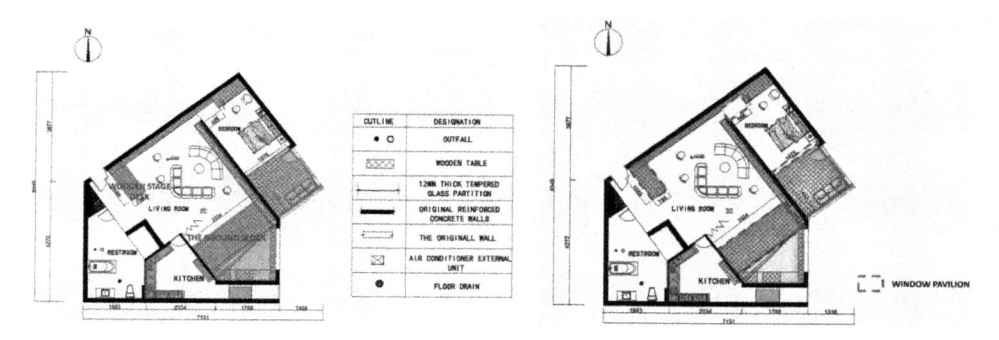

Figure 4.  Design of apartment storage space.          Figure 5.  Apartment view space design.
*Self-drawn by the author

However, at this stage, most of the homes for rent and sale in some first-tier cities are two-bedroom apartments of limited size, and the overall floor area is often no more than 30-40 square meters. Most of the spaces are relatively flexible and versatile in terms of functional conversion, and many functional spaces will generally overlap. The functional layout of storage space and other living spaces can be intermingled and converted to create a multi-functional design concept of a rich form of storage space for small homes.

The picture below shows a small top-floor house of only 36 square meters in Tiexi District, Shenyang. Before redecoration, the interior space was the most common traditional layout design, very crowded and difficult to transform the construction by themselves. The wall structure is difficult to demolish, move, and transform. It only has a floor height of 2.6 m–2.8 m, the width is also about 3 m, and the narrow layout is very inconvenient for this family with many children.

This paper designed the interior space layout of this old residential building, reinforcing the diagonal of the house, dividing the core area into three main functional spaces, and connecting their corners to design a diagonal folding corridor. The overall space looks much larger due to the repeated appearance of the same sequence and less standard layout, and the overall sight distance is expanded. At the same time, the vertical three-dimensional room is not affected by the diagonal space layout, and many compound parts of the flat space can be remodeled, expanding the overall area of the storage space and increasing its proportion in the overall space so that the storage space occupies about 3/5 of the broad residential area, as shown in Figure 6.

### 5.2  *Ingenious three-dimensional layout with vertical settings*

Compared with flat planning, the three-dimensional space of a small house is variable in terms of planning. The storage space in the two-dimensional flat space of a small house is very limited, so to further make full use of the rational use of the overall space and create a wide variety of storage space, you can make the three-dimensional space of a small house play a greater role. This requires designers to maximize the use of various spaces.

#### 5.2.1  *Wall space utilization*

Walls can usually be designed for vertical storage, and in small homes with a limited overall area, there are also relatively large, continuous, and well-formed walls. The effective use of the wall makes the space look unified and expands the storage volume. As shown in Figure 7, the "hole board" wall storage is easy to install, will not damage the wall, and can be adjusted according to individual needs. The "cork board" wall storage is mainly used for picture walls and workbench

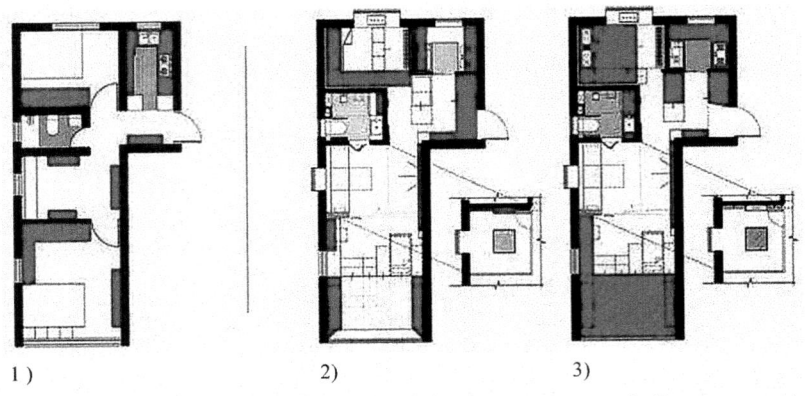

1) 6.8 m² storage space before renovation. 2) 6.8 m² storage space in the upper part after renovation. 3) 14 m² storage space in the lower part after remodeling.

Figure 6.   Storage space of the old house in Tiexi District, Shenyang, after renovation.
*Self-drawn by the author

decoration and storage and is fast and easy to replace. The "iron mesh" wall storage can store various small objects and is cost-effective. In the corner of the wall, we design "shelves" wall storage, whose layered design makes the shelves rich in form and decorative.

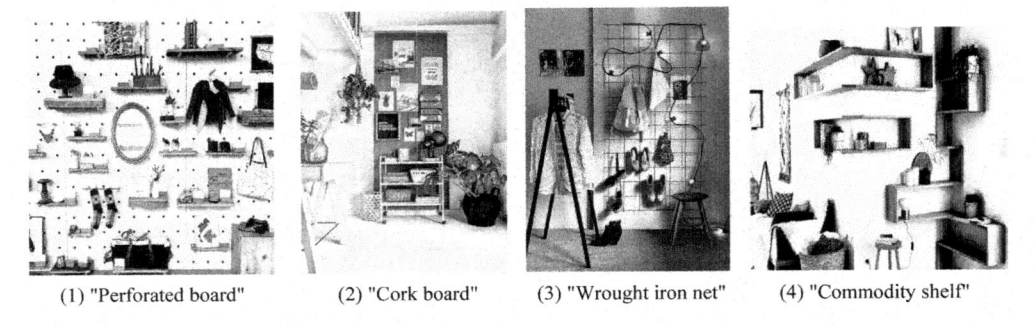

(1) "Perforated board"        (2) "Cork board"        (3) "Wrought iron net"        (4) "Commodity shelf"

Figure 7.   A variety of wall storage forms.
*https://www.puxiang.com/

### 5.2.2   *Wall body utilization*

(1) Wall body expansion

The further expansion of the wall can create new storage space in the interior, which can make full use of the thickness of the building wall, mainly indoor non-load-bearing walls. The expansion of the wall mainly plays the role of dividing space, storage, and display. As shown in Figure 8 and Figure 9, the expansion of the wall not only divides and isolates the dining room, living room, bathing area, and toilet to a certain extent in the small house space, respectively (Wang 2021; Zheng 2021), but also plays the role of the display, lighting, storage, and decoration.

(2) Wall body dislocation

The staggered design of the wall is to optimize the space further so that two adjacent spaces can share a set of floor plans, flexibly divide and regulate the layout of different functional rooms inside the house, achieve fuller utilization and saving of space, and make the functional storage space

Figure 8. Division of living room and restaurant.

Figure 9. Separation of bathing area and toilet.

*https://www.zgsjlm.cn/wap_sn991.html

easy to use. As shown in Figure 10, there is a small house located in Chongqing Nanan District Changjiahui SOHO transformation program. Before the renovation, the aisle width of the entrance is narrow. In order to solve this problem, the aisle wall is pushed 50-60 cm to the adjacent space, so there is enough space to install the embedded shoe cabinet, and the dislocation of the adjacent space can be embedded into the large-volume washing machine or refrigerator.

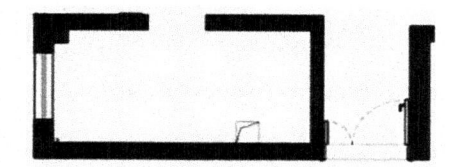

(1) Before wall renovation

(2) After wall renovation

Figure 10. Wall dislocation design plan of Changjiahui SOHO space in Nanan District, Chongqing.
*Self-drawn by the author

### 5.2.3 *Ground surface utilization*

The main purpose of the floor is to make full use of the space on the ground for storage and storage design, and furniture such as beds, storage cabinets, and sitting on the floor can replace furniture to meet the concept of multi-functional design. The space can be designed as a whole or as a partial floor, which is economically very practical and allows the space to be used effectively and reasonably, mainly for small, micro-family houses in cities.

Pull-out floors, lift-up floors, and side-opening floors are the three main forms of flooring design.

1) Pull-out floor: The floor has drawers on the side.
2) Lift-up floor: The cover is set up as a door that can be lifted, and the storage compartment is made inside the box, following the design form of Japanese tatami.
3) Side opening floor: A floor that can be opened with a side door.

As shown in Figure 11, the height difference between the tatami and the floor is 85 cm, and the upper floor is full of storage space. Three steps (four drawers) can be removed, six drawers can be pulled out, and the floor of the tatami can be lifted to accommodate luggage.

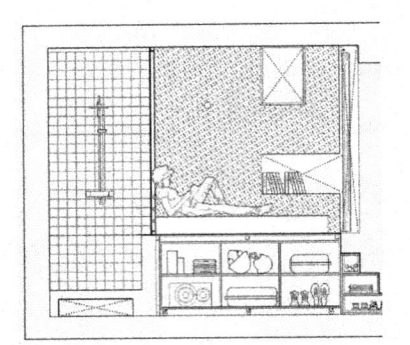

(1) Tatami Profile Elevation Drawing.

(2) Flip-up Platform.

(3) Side-open Platform, Pulling Type Platform.

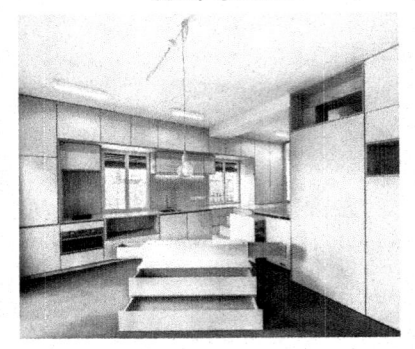

(4) Side-open Platform, Pulling Type Platform.

Figure 11.   Ground storage of small unit space.
*https://www.puxiang.com/

### 5.2.4   *Top surface utilization*

The indoor roof space of the small house storage space is generally above 2.2 meters in height, so the available volume is not only large but will not interfere with the life of the occupants. It is not easy to implement, but it can be displayed in small amounts to store some infrequently used objects, both in line with the multi-functional design concept and make the design of the living room full of interest. As shown in Figure 12, the partial return ceiling can also be used for storage space, does not occupy the floor height, has a display effect, and can be used to store infrequently used items. As the floor height of the kitchen is relatively low, the ceiling can be used to store some common kitchen utensils, which is very convenient (see Figure 13).

Further expanding the design of the top space of a small-sized house will help us fully utilize the interior space and increase the storage area. Especially in a small-sized house, it's essential to utilize the roof space. Interior ceiling spaces can be covered with clapboards extending to the top of the room, which will not block views with many decorations.

### 5.3   *Scalable and flexible layout*

The flexible detail space is used throughout the small house design, which can improve the use of the small house space and expand the space again. Some of the staircases in small homes can be used to create additional storage space, and doors and storage hooks can be combined with the staircase to create storage drawers according to their sizes (Shi 2022; Xu 2022), which is practical, beautiful and convenient. As shown in Figure 14, the triangular space at the bottom of the stairs

Figure 12. Partial back-shaped ceiling storage design.

Figure 13. Kitchen ceiling storage design.

*https://www.zhihu.com/

can be used not only for storage and display but also transformed into a small study with storage functions.

(1) Cabinet and drawer combination staircase. (2) Reading area and storage cabinet combination staircase.
(3) Wall and staircase bottom combination storage.

Figure 14. Small-sized stairs storage.
*https://www.zgsjlm.cn/wap_sn991.html.

### 5.4 *Utilization of spare space*

To further explore the interior of the fragmented corner space, you can solve the dilemma of the lack of storage in small homes and expand the limited space, personalized design, and decorative role of beauty. As shown in Figure 15, you can set aside a width of 15–20 cm behind the sink, which can be used as a temporary storage place for washing dishes. The space behind the door can be installed as a storage cabinet for keys, backpacks, documents, and other personal items, which can be seen once you go out, as shown in Figure 16.

## 6   CASE ANALYSIS OF STORAGE SPACE DESIGN IN SMALL-SIZED HOUSE

Single people with limited capital in big cities are the main force living in small and old neighborhoods. To adapt to the high-pressure urban life rhythm and environment, they mostly choose

Figure 15.　Storage design of the space behind the Sink.

Figure 16.　Storage design of the space by the door.

*https://www.zgsjlm.cn/wap_sn991.html.

small homes of 20 m²–50 m² with more convenient locations and transportation. The interior of a small rectangular 50m² house is designed in the 13th section of Dingzigu Street, Hongqiao District, Tianjin, for a single woman who had just graduated and could only purchase a 30-year-old second-hand house. Since most young single white-collar workers have high requirements for the aesthetics and storage functions of the living environment, the living necessities of the occupants will accumulate more and more, and the gender, age, and personality differences of the occupants will also have different storage needs. Therefore, we should strengthen the target in the design and consider more for the customers to meet the various needs of different people. As shown in the figure, the space in the house before the renovation is small and confined. The kitchen area, which is not commonly used by young people, takes up a lot of space, and there is not enough space for storing clothes, placing washing machines, and refrigerators, as shown in Figure 17.

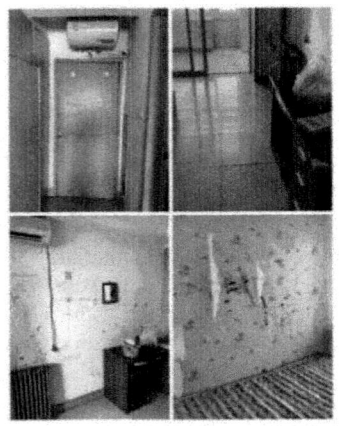
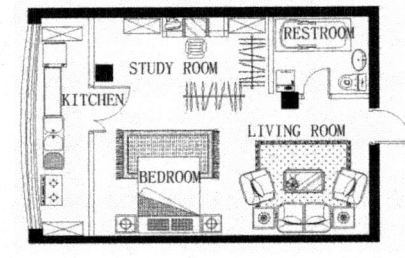
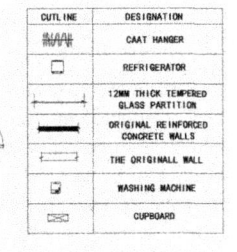

(1) Real-life photography before renovation.　　(2) Floor plan before renovation.

Figure 17.　Before the transformation of small house space in the 13th section of Dingzigu street neighborhood in Hongqiao District, Tianjin.

*Self-drawn by the author.

While maintaining the overall layout, the author made some flexible changes to the space with a minimum budget and reduced the number of parts that needed demolishing. The work and reception areas were designed to be the main functional spaces in the small house, which are the most important areas for single white-collar workers with high work pressure. Therefore, we need to make the living room occupy nearly 45% of the total area and set up a special sunken area to make the original single space more diversified (Shen 2021; Wan 2021). At the same time, a large amount of storage space will be left in the overall space, such as the storage floor, to solve the situation of white-collar workers with too many documents and objects. Since single white-collar workers are busy and seldom use the stove, this plan reduces the overall area of the kitchen and dining room, and the kitchen and dining area can be transformed into storage space for the washing machine and clothes drying. The bathroom and washroom area are also designed as multi-functional storage spaces using the wall and façade. The bedroom resting area is isolated and divided as a separate functional area full of privacy ( Meng 2020; Zhu 2020). Because the area of the residence is so small, some space areas may all be similar in function, which is to reserve more space and possibilities for further storage design and multi-functional design transformation, as shown in Figures 18 (1), (2), (3).

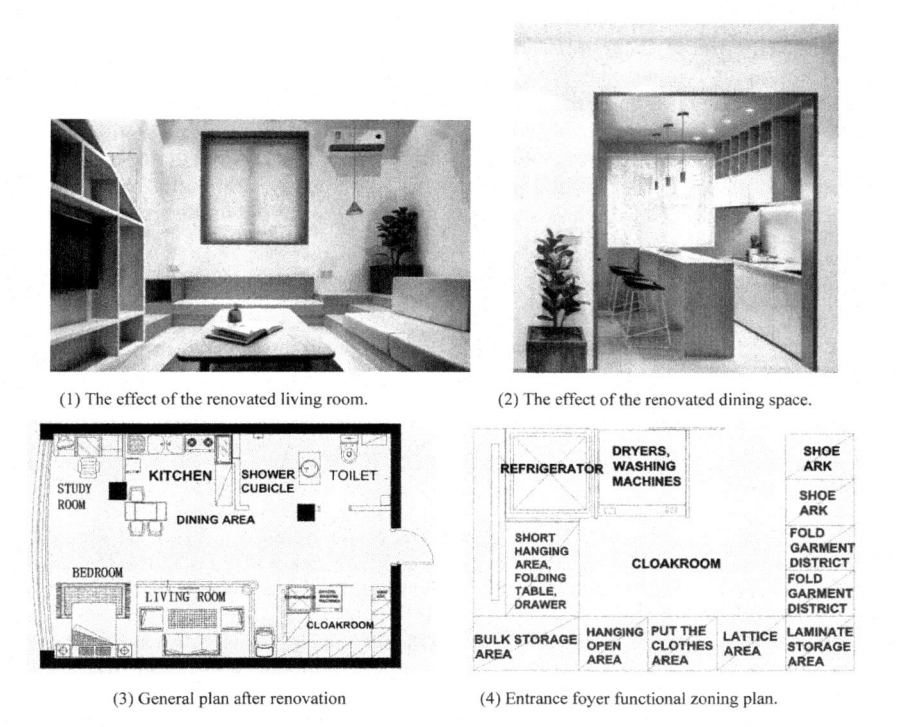

(1) The effect of the renovated living room.

(2) The effect of the renovated dining space.

(3) General plan after renovation

(4) Entrance foyer functional zoning plan.

Figure 18.    Effect diagram after renovation.

*Self-drawn.

The entrance foyer is set up with a square box type checkroom, integrating the dryer with shoe storage and clothes storage, and the washed clothes can be quickly done for storage and finishing. The stacking of washing machines can save and expand the space of the balcony and solve the problem of difficult drying clothes in small households. The countertop of the hanging area can be used as a stacking table simultaneously, in line with the concept of multi-functional design. A row of cabinets is placed near the door, overhanging the middle and bottom to place daily odds and ends such as keys, cell phones, and frequently worn shoes, as shown in Figure 18(4).

## 7 CONCLUSIONS

This paper concludes for the first time that the limitation of spatial form and scale is the main spatial layout element limiting the storage space design in small urban houses. The limitation of the difference between the age and gender of residents and some special groups is the main demographic difference element limiting the storage space design in small urban houses. The main technology element limiting the storage space design in small urban houses is the difference between the development conditions of materials and various processing technologies and means. The differences between materials and the development conditions of various processing technologies and methods are the main technological factors limiting the design of storage space in small urban houses.

This paper presents for the first time the many principles and norms of flexibility, layout planning, green design, care for humanity, etc., that need to be followed in the design of small residential storage spaces in old residential buildings under the multi-functional design concept applicable to new first-tier and second-tier cities with rapid economic development, such as Shenyang, Xining, Tianjin, Chongqing, etc.

This paper presents for the first time the combination of humanized design, multi-functional design concept, space, product, energy, sustainable development concept principles, and different complex and diverse process technology and other elements of many new first-tier, second-tier cities in Tianjin, Chongqing, Xining, Shenyang City as an example of the old residential building small residential storage space design in the overall space layout, changeable and adjustable, growable, humanized universal design and green safety, low-carbon design on the design strategy.

In order to live in the small area of the old residential building and ensure a good living experience, it is necessary to ensure the economy, applicability, practicality, simplicity, convenience, and efficiency of the small house, which is the key to the quality of living in the small space.

## FUND PROJECT

Liaoning Provincial Department of Education School-Enterprise Alliance Project – Research on the linkage construction of experimental base and professional disciplines in the elderly industry in colleges and universities (xqlm2019005).

## REFERENCES

Dai, X. D. Zhu, Z. H. Zeng, X. et al. (2020). Research on environmental protection and fire safety in the construction of furniture industry cluster. *Furniture and Interior Decoration* 11, 42–44.

Hua, J. X. (2021). Horseshoe Method. Furniture design based on small house. *Furniture and Interior Decoration* 1, 35–37.

Leng, C. X. Zhou, C. M. Zhang, Y. H. (2019). Furniture and Interior Decoration 8, 30–31.

Li, Z. R. & Chen, B. L. (2021). Research on the health needs of empty nesters in the smart home scenario. *Furniture and Interior Decoration* 8, 1–6.

Meng, F. Y. & Zhu, X. (2020).Research on the influence of young women's home living habits on furniture design. *Furniture and Interior Decoration* 12, 82–85.

Peng, R. & Xu, W. (2022). *Furniture* 43(01), 54–57.

Shen, L. & Wan, H. (2021). Design of social space in youth apartment. *Furniture and Interior Decoration* 04, 97–99.

Shi, Z. B. & Xu, B. M. (2022). Research on Elastic Response Design of Furniture in children's Room. *Furniture and Interior Decoration* 29(03), 32–35.

Wang, S. X. Shen, L. M. Tan, L. S. Wang, Y. (2020). Research progress of functional sofa design. *Furniture* 41(04), 6–10.

Zang, L. & Song, R. Q. (2020). The application strategy of minimalism in brand construction. Art Design Research 5, 85–90.

Zheng, Y. W. & Wang, X. H. (2021). Furniture 42(01), 63–66+75.

Civil Engineering and Urban Research – Mohamed & Hou (Eds)

# Research on integration evaluation and development model of traffic and tourism in non-core urban areas: A case study of Pingyao Ancient City

Shali Zhou*
*Design and Research Institute of CCCC Second Navigation Engineering Bureau Co., Ltd.,
Deputy Chief Planner of the Planning Consulting Institute
Research Direction: Urban Planning and Design, Urban Renewal, Rural Environment Research,
Wuhan, China*

Weiyang Luo*
*Design and Research Institute of CCCC Second Navigation Engineering Bureau Co., Ltd. and
The Planning Consulting Institute
Research Direction: Urban Planning and Design, Urban Renewal, and Rural Environment Research,
Wuhan, China*

ABSTRACT: The integrated development of tourism is an important path to achieving the harmonious development of tourism, transportation, economy, and society. Taking the ancient city of Pingyao as the empirical research object, the combined qualitative and quantitative methods are used to measure the integration and development level of traffic and tourism in non-core areas of the city based on the coupling coordination degree model. The result is that from 2016 to 2020, the ancient city of Pingyao transitioned from serious imbalance coupling to primary coupling coordination, and finally stabilized in the stage of excellent coupling coordination, which was first developed by transportation, then developed by special tourism, and finally transformed into a mode of transportation-led development, in which the local per capita economic production and the surrounding economic production had a positive promotion effect. Finally, a new idea of integrated development of communications and tourism is proposed.

## 1 INTRODUCTION

Since the reform and opening up, China's national economy has continued to thrive, the consumption of residents has also increased, and tourism has gradually changed from the activities of some groups to the daily activities of the mass. At the same time, with the advent of the era of mass tourism and the implementation of the global tourism strategy, the demand for tourism travel has begun to grow rapidly, which has stimulated the economic growth of the cities where the scenic spots are located, and the status and role of tourism in the national economy have become more and more important. Tourism, like other industries, requires support from the transport sector.

In order to better realize the integration of transportation and tourism, this study follows the development trend of the times, takes the non-core area of the city as the main research scope, and studies the integrated development mode of transportation and tourism. The integrated development of tourism and transportation is realized at the micro level, and the harmonious development of tourism and economy, and society is realized at the macro level. Through a large number of investigations and literature investigation and combing, and refining and summarizing, based on the definition of relevant concepts and the ancient city of Pingyao as an empirical research

---

*Corresponding Authors: zhoushali1@ccccltd.cn and 2021009606@ccccltd.cn

object, customized evaluation mechanism for comprehensive evaluation and analysis, to explore the innovative model and ideas of the integrated development of communication and tourism.

## 2 DEFINITION OF RELEVANT CONCEPTS

### 2.1 *Conceptual definition of urban non-core areas*

At present, there is no relevant theoretical definition of the non-core area of the city in the academic community, and this paper takes the boundary of the core area of the city as the starting point to derive the concept of "non-core area."

The core area of the city is the core of the city, which can reflect the image of a city, and it undertakes important functions in government, commerce, religion, scenery, and so on. In general, the core area of the city is the main public facility of the city, the place where people and the foreign population carry out various activities and exchanges, and the center of the city's social activities.

Non-core areas are often understood narrowly as follows: the difference between non-core areas and core areas is that non-core areas are "land use transformation zones that change land use, social and demographic characteristics, between built-up areas and suburbs, and between non-agricultural, non-agricultural and non-agricultural land."

Since the core content of this paper is the integration of urban non-core areas and cultural tourism, the research scope is increased to the marginal urban areas and rural areas outside the urban core areas, collectively referred to as non-core areas of the city.

### 2.2 *Conceptual definition and classification of transport resources*

Zhang Ning pointed out that "transportation resources are the entities that make up multiple modes of transportation and all the external factors associated with them, and their combination, determine the technical-economic links and constraints of different modes of transportation." The definition of a thing is not whether it is perfect or not, but its essence. Fundamentally, the way and means of distribution of transportation resources cannot be incorporated into the transportation resources themselves, and they cannot be used as a form of production factor and the role they play in transportation. In terms of space and time, a certain amount of transportation resources provides a certain degree of traffic capacity for people's travel and economic activities, and the size of traffic flow refers to the total amount of transportation resources. Through the analysis of urban road capacity, the basic principle of urban road capacity allocation is derived; that is, the market demand in urban road planning is the main factor affecting the allocation of urban road traffic resources. From the perspective of market law, transportation resources are the transportation capacity provided for people's travel and economic activities and are transportation infrastructure formed in a specific time and space.

In summary, "transport resources" refers to the various entities of road transport and the various external factors associated with them, which constitute the technical-economic relations and constraints of road transport models.

## 3 CONSTRUCTION OF A COMMUNICATION AND TOURISM INTEGRATION EVALUATION SYSTEM IN NON-CORE AREAS OF CITIES

### 3.1 *Evaluation principles*

Based on Maslow's hierarchy of needs, when people meet lower-level needs, they turn to higher-level needs.

After low-level needs such as "accessibility" are met, more advanced services such as "security" and "convenience" will emerge. Especially in the context of the integration of transportation and tourism, people's demand for transportation will extend to other aspects, and it is hoped that the transportation function can be combined with other functions to provide better services. From the perspective of the integrated development of transportation and tourism, the traffic demand level of urban roads has developed from a low level to a high level.

Traditional tourism is dominated by providing tourism resources and services to ensure the basic needs of "software and hardware." With the development of tourism resources, people stay longer and longer in tourism resource areas, and their behaviors are becoming more and more diversified to seek higher tourism quality. This requires a better experience, better information management, and more humane services. In the context of the integration of transportation and tourism, people expect that the benefits of tourism development can be extended to more places, promoting the development of transportation, people's livelihood, environment, and other aspects, rather than simply improving the level of regional tourism. From the perspective of transportation integration, transportation demand includes accessibility, safety, convenience, comfort, and integration; tourism demand includes tourism resources, service supply, slow travel experience, information management, and sharing.

Technical principles are generally followed when designing and carrying out comprehensive assessment questions: purposefulness, operability, and independence. (1) The principle of purpose means that the assessment indicators can truly reflect the comprehensive assessment goals, accurately characterize the characteristics of the target system, and contain the necessary elements required to assess the goals. (2) The principle of operability means that each evaluation index, whether qualitative or quantitative, needs to have indicators that can be observed and measured; that is, the evaluation data of the evaluation indicators can be obtained. It should be as open and objective as possible. In addition, data on assessment indicators should usually be easy to collect and not too expensive to observe. (3) The principle of significance, that is, in the evaluation, the main important indicators should be maintained, and the secondary and unimportant indicators should be excluded. The key factor of the indicator is its contribution to the overall assessment, and the greater its contribution, the higher its importance and can be used as a key indicator; conversely, it is a non-key indicator.

## 3.2 *Evaluation methods*

### 3.2.1 *Comprehensive evaluation based on coupling coordination*

The so-called "coupling" is the interaction between two or more systems, a dynamic relationship that affects each other, is interconnected, interdependent, coordinate with each other, and promotes each other. When the coupling effect makes two or more systems with certain connections form a new overall system, it is called system coupling. Under the integration mechanism of transportation and tourism, the two major systems of transportation and tourism have formed a process of interaction and mutual influence. It is defined as a road traffic-tourism integration development coupling.

On this basis, the traditional comprehensive assessment method is used to obtain a comprehensive assessment result of a complete traffic-tourism integration system, which only reflects the overall development level of the transportation and tourism integration system. Still, it does not reflect the degree of coupling and coordination of the road-tourism system. The coupling degree and coordination degree of each subsystem better grasp the development trend and problems of the system. Compared with a single comprehensive assessment, coupled collaborative scheduling has stronger relevance, coordination, and integration, which is an improvement of a single comprehensive assessment, which can supplement its inherent defects, thus providing a strong reference for better formulation of more effective optimization strategies.

### 3.2.2 *Subsystem comprehensive development level evaluation index*

In general, the transportation-tourism system comprises two subsystems: transportation and tourism. The subsystem obtains the standard value of the indicator and the combined weight and then substitutes the comprehensive evaluation results of the system into the appropriate comprehensive model, so as to obtain the comprehensive evaluation results of the subsystem. This method can not only evaluate the coupling and collaborative scheduling of transportation tourism, but also evaluate the overall development degree of the two subsystems and their development degree and trend, so that different coupling synergy models and hierarchical relationships can be distinguished in a certain period of time.

The Ui calculation method of the single-system comprehensive development evaluation index is as follows:

$$\text{Ui} = \sum_{i=1}^{k} \alpha i x i \tag{1}$$

where $\alpha i$—the combined weight of the ith indicator
$xi'$—the standard value of the ith indicator.

### 3.2.3 *Coupled coordination degree model*

The degree of coupling stems from the concept of physics. Under this coupling effect, there are complex factors between the flow of matter, energy, and information, and the study of the degree of coupling is an important way to study the coupling development law and coupling effect of the system. The coupling coordination model is:

$$C = \left\{ \frac{U_1 \times U_2}{\left((U_1 + U_2)/2\right)^2} \right\}^k \tag{2}$$

$$D = \sqrt{(C \times T)}, T = \alpha \times U_1 + \beta \times U_2 \tag{3}$$

Where:

D – Coupling coordination;

C – Subsystem coupling;

T – Subsystem comprehensive coordination;

U1 – Evaluation index of development level of highway traffic subsystems;

U2 – Tourism subsystem development level evaluation index;

Coordination refers to the harmonious, coordinated, and virtuous circle of development of the two subsystems, including the coordination of structure, functionality, time and space, etc., with the inherent attributes of high comprehensive development. By introducing the concept of coordination in the coupling degree model, a more scientific solution method is derived, which can better reflect the degree of coupling and coordination between systems.

### 3.2.4 *Multiple linear regression model*

In order to further investigate the influence of transportation construction and tourism economy on the integration of transportation and tourism, a multivariate linear regression equation can be constructed, and the general form of a multivariate linear regression model can be constructed:

$$y_i = \alpha_0 + \alpha_1 x_{1i} + \alpha_2 x_{2i} + \cdots + \alpha_k x_{ki} + \varepsilon_i \, i = 1, 2, \cdots, n \tag{4}$$

In the model, y is the dependent variable, while x_1i, x_2i, $\cdots$, and x_ki are the independent variables. When k=1, it is a univariate linear regression model. When k>2, it is called a multiple linear regression model. The dependent variable y_i is a linear function of $\varepsilon$_i and k independent variables by the error term random variable

$$\alpha_0 + \alpha_1 x_{1i} + \alpha_2 x_{2i} + \cdots + \alpha_k x_{ki} \tag{5}$$

## 4 EMPIRICAL RESEARCH: EVALUATION OF THE INTEGRATION OF TRAFFIC AND TOURISM IN PINGYAO ANCIENT CITY

### 4.1 *Tourism resources of Pingyao Ancient City*

Pingyao Ancient City, Jinzhong City, Shanxi Province, with a history of more than 2,700 years, together with Langzhong in Sichuan, Lijiang in Yunnan, and Yixian in Anhui, is known as the "Four

Ancient Cities" and is the only ancient city in the country that has successfully declared itself a world cultural heritage.

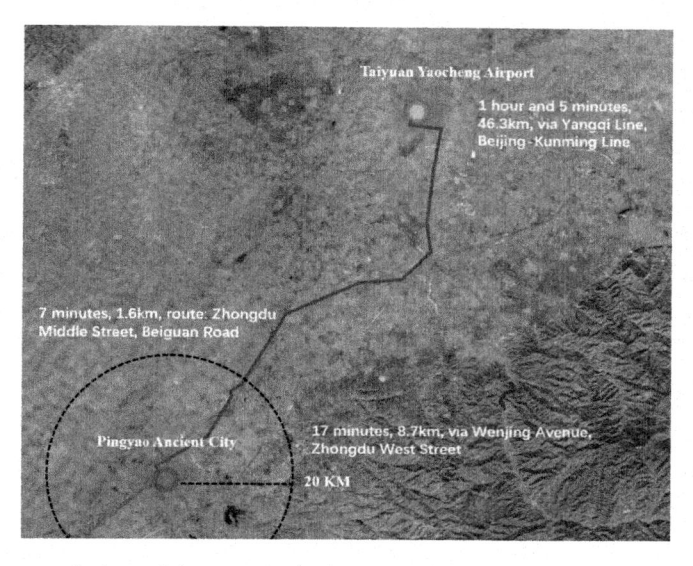

Figure 1. Distance analysis graph between the destination and the transportation hub.
Source: Self-drawn.

Table 1. Flow of people and economic income of Pingyao Ancient City. Source: Self-drawn.

| | Traffic during National Day (10,000 people) | Income during National Day (10,000 yuan) | Year-round traffic (10,000 people) | Annual revenue (10,000 people) |
|---|---|---|---|---|
| 2017 | 57.24 | 1547.2 | 228.21 | 11604.7 |
| 2018 | 57.88 | 1932.4 | 254.67 | 13309.62 |
| 2019 | 59.03 | 2097.4 | 267.95 | 14381.2 |

## 4.2 Evaluation of the integration and development level of Pingyao Ancient City traffic and tourism

Through multiple regression analysis (Tables 4–5), the local per capita income (10,000 yuan) will significantly impact the evaluation index of the comprehensive development level of tourism.

Table 2. Results of the evaluation of the integration of traffic and tourism.

| Evaluation results | 2016 | 2017 | 2018 | 2019 | 2020 |
|---|---|---|---|---|---|
| Evaluation Index of Comprehensive Development Level of Transportation Construction | 0.0858 | 0.2239 | 0.4520 | 0.8122 | 0.9987 |
| Tourism comprehensive development level evaluation index | 0.0279 | 0.2327 | 0.5175 | 0.6032 | 0.9360 |
| The level of communication and tourism integration (Coupling coordination) | 0.1766 | 0.4776 | 0.6931 | 0.8229 | 0.9828 |

However, the GDP of the local region (100 million yuan) will not impact the evaluation index of the comprehensive development level of tourism. The per capita income of the surrounding area (10,000 yuan) does not affect the coupling coordination relationship. However, the gross domestic product of the surrounding area (100 million yuan) will have a significant positive impact on the degree of coupling coordination.

Table 3. The impact of the local economy on the integration of transportation and tourism.

| | Linear regression analysis results (n=5) | | | | | | | | |
| | Non-standardized coefficients | | Normalized coefficients | | | | | | |
| | B | Standard error | Beta | t | p | VIF | $R^2$ | Adjust $R^2$ | F |
| Constant | −2.122 | 0.172 | – | −12.371 | 0.006** | – | 0.992 | 0.985 | F (2,2)=130.842, p=0.008 |
| Local Income per Capita ($10,000) | 0.734 | 0.163 | 0.612 | 4.508 | 0.046* | 4.861 | | | |
| Gross Local Product (RMB100 million) | 0.001 | 0.000 | 0.411 | 3.029 | 0.094 | 4.861 | | | |

Dependent variable: coupling coordination
D-W value: 2.584
*p<0.05 **p<0.01

### 4.3 *Analysis of evaluation results*

The calculation results show that from 2016 to 2020, the ancient city of Pingyao transitioned from severely offset coupling to primary coupling coordination and finally stabilized in the excellent coupling coordination stage. The first is traffic-led development, then tourism-led development, and then transportation-led development. Through the calculation of various sub-indicators of Pingyao Ancient City, it can be seen that a set of effective hierarchical tourism transportation planning systems has not yet been formed, and there is a lack of hierarchical distribution and intermodal transportation hubs. In terms of transportation, the construction of transportation facilities, the accessibility of scenic spots, and the highway landscape need to be optimized urgently, and in terms of tourism, it is necessary to strengthen the service capabilities such as resource richness, parking satisfaction rate, and slow travel system. Only by starting from the actual situation of the local economy and the surrounding economy and the two levels of highway transportation and tourism, can the "transportation-tourism" coupling and coordinated development level develop to a higher stage.

## 5 CONCLUSION

Based on the conclusions obtained, this paper believes that reflecting the development and evolution of tourism transportation accessibility and tourism regional economic links from the perspective of space has important theoretical significance and practical value for promoting the flow of tourism elements between regional cities, optimizing the allocation of resources between tourism regions, and realizing the integration and development of tourism, transportation and tourism industries, and demonstrating the role of tourism transportation development in promoting tourism development in non-core urban areas. In addition, due to the certain limitations of the collection of some transportation and tourism index data, this paper should conduct more in-depth research in the future, improve the evaluation system of the integrated development of transportation and tourism to measure the level of integration of transportation and tourism more objectively, and put forward the idea of integrated development of transportation and tourism in non-core areas.

## REFERENCES

Jiang Yuan, Gao Jinjin, Wang Guosheng, Zeng Jiao, Lu Chunfang (2021). Research on the Integration and Development of Tourism and Transportation: A Case Study of Fujian Province[J]. *Integrated Transport*, 43(03): 111–118.

Liu Hongfang, Zhou Xiaoqin, Ming Qingzhong, Li Huayong (2019). Review and trend of tourism and transportation integration development in Yunnan[J]. *Resources Development and Marketing*, 35(04): 578–584.

Wang Lan, Liu Jie (2021). Research on the Planning Target System of Transportation and Tourism Integration Development[J]. *Highway*, 66(03): 187–192.

Zhang Xu, Zhang Shizhi, Tu Jingyu (2017). Integrated Development and Planning Response of Transportation and Tourism[J]. *Integrated Transport*, 39(06): 28–32.

*Civil Engineering and Urban Research – Mohamed & Hou (Eds)*
*© 2023 the Authors, ISBN 978-1-032-44487-1*

# An investigation and quantitative analysis of color landscape of countryside residences in Northeastern China

Zhihui Wang*, Xu Lu* & Shan Guan*
*Department of Architecture and Urban Planning, Shenyang Jianzhu University, Shenyang, Liaoning, China*

ABSTRACT: Using a combination of qualitative and quantitative analysis methods, this paper studies the color landscape of rural architecture in northeastern China based on The China Building Color Card and discusses the influence of building materials on the rural color environment. The results show that color differences and strong contrasts produced by different building materials can lead to a confusing rural architectural color environment, which is a common problem in the rural color environment and provides a basis for color control and material selection in rural architectural landscape planning.

## 1 INTRODUCTION

For buildings, the expression of architectural color characteristics is determined by multiple dimensions, including different materials and surrounding environments (Boyd et al. 1987). In Japan and some western countries, more attention is paid to the shaping of the color landscape in the countryside. A habitable environment was created by planning and designing the color landscape of buildings, vegetation, roads, etc., in the countryside (Prieto 1995). Take Japan as an example. After the "Japan Landscape Law" had been officially implemented in 2004, it gradually worked to offer more comprehensive suggestions to different regions of Japan and emphasized more on investigating and protecting the color landscape of the "traditional construction group conservation area" located in the countryside (Caivano 2006). China's color research is late and mainly focuses on urban areas, with less research on the color landscape of the countryside. This has led to the loss of the original characteristics and conventionalities of the local color landscape of the country (Pelli 1996), resulting in the phenomenon of "thousand villages with the same appearance, and in lack of culture" during the country's transformation and renovation (Zhan et al. 2006). Therefore, in order to accurately control the rural color landscape planning, detailed investigation and analysis can help to better summarize the laws of rural color, which is one of the reasons why China's rural color landscape research is weaker than foreign countries. This study selected some ordinary countryside residences in some villages of northeastern China as the research objects. The study explicitly reflected the overall picture of their color landscape through concise investigation and analysis. In addition, the reasons behind it were proposed to pave the way for the future planning and regulation of color landscape in the countryside (Haifacree 2006).

## 2 COLOR DATA INVESTIGATION AND PROCESSING METHODS

### 2.1 Research object

This color investigation was mainly conducted in randomly selected 15 villages from four cities located in northeastern China, including four villages in Tonghua City, five villages in Chaoyang

---

*Corresponding Author: 1264221167@qq.com

DOI 10.1201/9781003372417-69

City, three villages in Dandong City, and three villages in Panjin City. The typical country residences in these areas were studied as the major research objects, and their color performance was further analyzed.

## 2.2 *Color research methods and data processing methods*

This color investigation followed the GB / T18922—2008 "China Building Color Card" and conducted color comparisons directly on the site. We recorded the color attribute number closest to the color card while taking pictures with the camera for further analysis. In order to avoid uncertainties such as light and shadows, which might generate errors in performing color acquisition, the investigation time is limited between 10: 00-15: 00. Moreover, the color distance between those investigated buildings was controlled in a reasonable range as well.

The initial color classification will be carried out when sampling colors on the spot first. First, we divided the building into both facade colors and roof colors. Second, the facade color is further categorized into two kinds of major colors and auxiliary colors. Then, according to different classification schemes, the corresponding color attribute number, the number of colors, and the materials corresponding to the colors were collected for subsequent quantitative analysis of the color data.

After the on-site investigation, we started to quantify various types of data, and the data processing software includes Excel, Origin, and Photoshop. The data processing was mainly to classify the recorded color data in Excel, import it into the Origin software, and generate the Value-Chroma and Hue-Chroma distribution charts based on the Mushell data so that further explicit analysis of color data could be conducted.

## 3 ANALYSIS OF RESEARCH RESULTS

### 3.1 *Overall investigation of architectural color landscape in the countryside*

Through the investigation of these buildings from the 15 villages, from the perspective of the manifestation of the roof and the façade, the matching between the two is observed to be more random, and there are no obvious connections. In terms of architectural expressions, traditional facades match not only the color of steel roofs but also the high-brightness, high-chroma facades in combination with flat roofs, etc. Based on architectural color expression, many buildings are presenting contrastive chroma between the facades and roofs. The color landscape generated by its matching makes the architectural style of the entire countryside messy (Table 1).

### 3.2 *Analysis of the overall situation of architectural color*

By observing the architectural Chroma-Hue and Value-Chroma data distribution charts drawn by Origin software (Figures 1 and 2), we can find the following features: (1) The major colors tend to be presented in high-hue, low-chroma and low-hue, low-chroma, while the auxiliary colors are mainly tended to high-hue and low-chroma. The hue values of the major colors were between the ranges of 2.5–3.5 and 6.5–9, and the chroma dispersed throughout the range of 0–4. When it comes to the auxiliary colors, the hue values ranged between 6.5 and 9, and the chroma values were mainly distributed between 1 and 3.5. (2) The major colors are more inclined to be either warm or cold, while the auxiliary colors tend to be warm tones. The hue of the major color is mainly distributed between 10YR-2.5R and 10B-2.5BG.

The above situations generally reflected the warm hue with high-value, and low-chroma of the building is more suitable for the architectural color environment of countryside residences in northeast China, but there are still some high-value and high-chroma cold colors mixed in the current color environment. Combined with previous conclusions on the architectural color status, it can be found that these cold colors are used on a large area on a limited number of building

Table 1. The color landscape of country residences in northeastern China.

| City | Building status quo | Façade main colors Munsell value | Materials | Auxiliary colors Munsell value | Materials | Roof colors Munsell value | Materials |
|---|---|---|---|---|---|---|---|
| Tong hua | | 0.6RP 7/1 | Outer walls - stone | 4R 7.5/1 | Doors and windows - glass | 6.9YR 8/1 | Grey tiles |
| | | 1.3GY 9/1 | Doors and windows - glass | 4R 7.5/1 | Doors and windows - glass | 10R 7/7.2 | Red tiles |
| Chao yang | | 5.6R 7/6.8 | Doors and windows - glass | _____ | _____ | 4.4PB 7/7.2 | Color steel tiles |
| | | 2.75 | Outer walls - paint | 7.5B 7.5/1 | Outer walls - cement | 4.4PB 7/7.2 | Color steel roof |
| Dan dong | | 6.9GY 8.5/1 | Doors and windows - glass | 3.1B 7.5/1.4 | Doors and windows - glass | 3.8YR 6.5/6.8 | Red tiles |
| | | 5.6GY 9/1 | Outer walls - tile | 3.1B 7.5/1.4 | Doors and windows - glass | 10R 7/7.2 | Red tiles |
| Pan jin | | 8.8Y 9/3.6 | Doors and windows - glass | _____ | _____ | 4.4PB 7/7.2 | Color steel tiles |
| | | 7.5BG 8.5/1.8 | Outer walls - red Brick | 4R 7.5/1 | Doors and windows - glass | _____ | Flat roof |

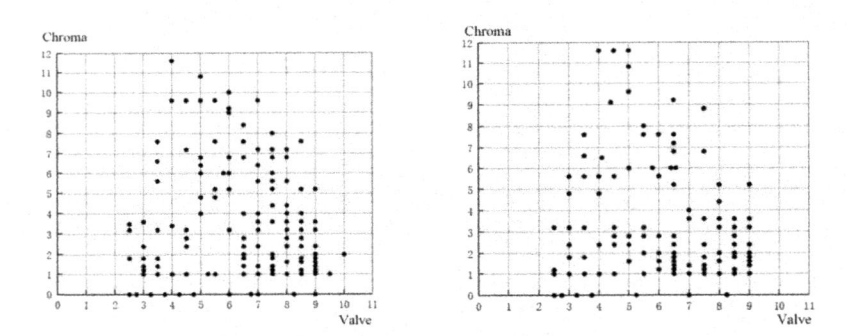

Figure 1. Architectural Value-Chroma distribution figure (Left: major; Right: auxiliary).

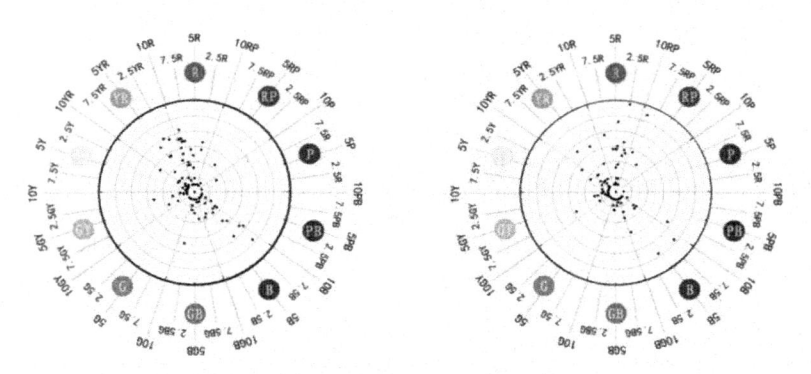

Figure 2. Architectural Chroma-Hue distribution figure (Left: major; Right: auxiliary).

facades, and they cannot mediate the high-chroma warm colors by forming contrasts. Therefore, to a certain extent, it greatly destroys the overall rural color landscape.

### 3.3 *Analysis of the colors of different materials*

To better understand the cause of this color phenomenon, the author analyzed the color tendencies of various materials through the Chroma-Hue distribution diagram. There are six major categories of building facade materials in rural areas in various regions. According to the number of quantities, the ranking from higher to lower is the coating, tiles, glass, cement, red bricks, and steel. Their usages and color expressions are different. The use of coatings is extensive, with a wide spectrum of colors. It is distributed in red, orange, and blue ranges (10YR-0RP, 5B-10B), and red-yellow and red tones outperformed the blue tones in chroma and types. The color characteristics of tiles and cement are more similar, mostly distributed between yellow and red-yellow tunes (5Y-5YR). Though glasses accounted for a large portion in the form of windows and large-scale lobbies in the facades, the color difference is not very large for various types of residential buildings. Therefore, the hue of the colorful glass is mainly distributed in blue-green and blue-purple ranges (5PB-7.5B, 2.5BG-10G), while steel and red bricks are mainly based on the attributes of their original materials. Therefore, the red bricks mainly presented red and yellow-red tones, while steel mainly showed blue-purple and blue tones (Figure 3).

Generally speaking, in the currently investigated villages, the coating is the material having the richest spectrum of color. The major color tone is not clear and difficult to control, but tiles and cement tend to be low-chroma with gray and white tones. The rest of the materials were based on their characteristics. For example, steel and glass are mostly cold-colored, while red bricks tend to present in warm colors. Although coatings and tiles constitute the predominant tendency of color landscapes in the villages, some overgrown conspicuous color coatings will strongly interfere with surrounding buildings, and large-area low-chroma grey-white tiles will lead to the lack of distinctive features due to the monotonous color.

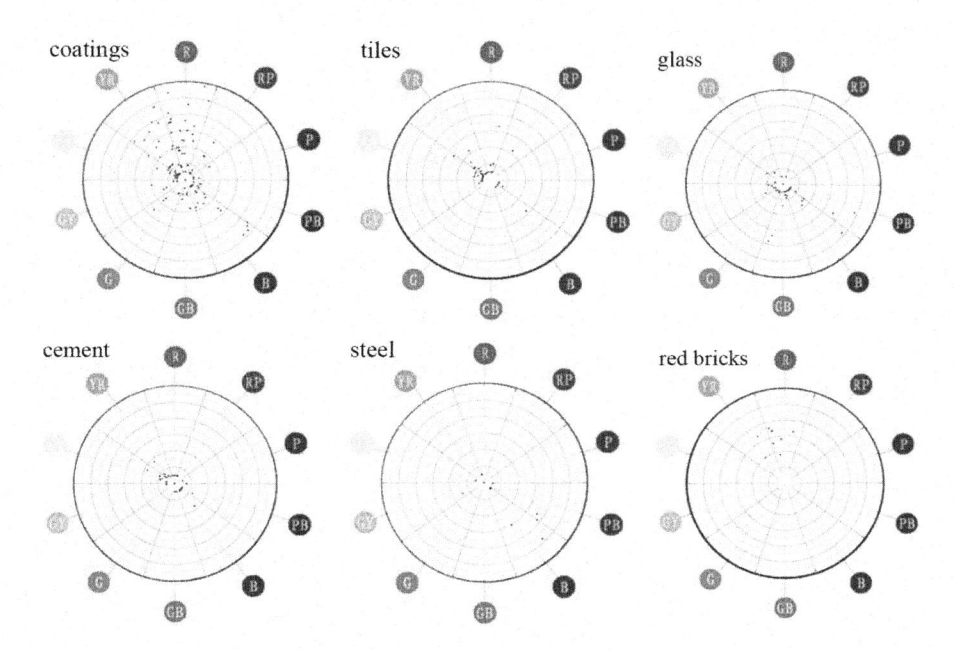

Figure 3. Architectural Chroma-Hue distribution for different architectural materials.

## 4 CONCLUSIONS

The results of the above survey and analysis of the current state of rural architectural color can also be referred to by the general rural areas in northeast China. (1) The results of the survey on the color of rural buildings reveal the tendency and limitations of different building materials in terms of color rendering. For example, most tiles and glass tend to have cooler tones and have very limited ability to express warm colors, while the widespread use of white tiles and glass in the cold northeastern region exhibits a color tone that leans toward cool gray. (2) The color difference between traditional and modern materials is that traditional materials such as brick and tile, and cement belong to low color and warm tones, while glass and paint are easily used in large areas to produce obvious cold and warm color differences and strong contrast of tones with traditional building materials.

In summary, this study uses the China Building Color Card as a survey tool for rural architecture in Northeast China and also uses software such as Origin and Excel to conduct quantitative and qualitative analysis through color scatter diagrams and other forms. This method of overall color environment evaluation helps to grasp the basic color situation of the study area. The current color environment influence factors are mainly based on building materials for analysis, which has certain limitations. Color planning can take this result as a reference and then combine it with building function, location, and other factors for further analysis, which helps to better coordinate the color relationship of each component. This kind of color survey and research method is very common in cities at home and abroad, but it is less used in rural color landscape planning. We can summarize the basic consensus of various methods, formulate technical guidelines for the overall working procedure and key technical links, make the color survey and quantitative analysis a standardized and easy-to-promote technique, and apply it to all levels of rural color landscape planning.

## ACKNOWLEDGMENTS

The authors acknowledge financial support from the National Natural Science Foundation of China(No.51778375)

## REFERENCES

Boyd, A. & Xie, M.C. & Song, S.Y. (1987). *Chinese architecture and town planning*. Taibei: Nantian Publishing House.
Caivano, J.L. (2006). *Research on color in architecture and environmental design: A brief history, current developments, and possible future*. 31(4), 350–363.
Haifacree, K. (2006). *Rural space: Constructing a three-ford architecture*. London: Sage Publications.
Pelli, C. (1996). Designing with color. Arch Design. 120, 26–29.
Prieto, S. (1995). The color consultant: a new professional serving architecture today in France. *Color Res Appl.* 20, 4–17.
Zhan, X.Y. & Fu, S.Z. & Liang, J.Q. (2006). *Chinese brick and tile history*. Beijing: Chinese Building Material Publishing House.

Civil Engineering and Urban Research – Mohamed & Hou (Eds)

# Research on urban color characteristics of Hailar District under the influence of regional culture

Xuan Tang*, Xu Lu* & Jiayi Dai*

*Department of Architecture and Urban Planning, Shenyang Jianzhu University, Shenyang, Liaoning, China*

ABSTRACT: The urban development of Hailar District is accompanied by the integration of diverse cultures, and its urban color has rich connotations and uniqueness. Qualitative (color spectrum induction) and quantitative (color quantitative analysis) methods are used to research the relationship between urban color and regional culture. The results show that the current overall color characteristics of Hailar District lack signs, and the color is not coordinated among various functional areas. Considering all factors, the keynote of Hailar's future color development tends to be high value, low chroma, and warm tones, and the corresponding improvement suggestions are put forward. The research results can lay a foundation for investigation, analysis, and scientific planning of urban color.

## 1 INTRODUCTION

"Sketching gives form to an object, and color gives it life." Urban color generally refers to the sum of various constituent elements of a city in the public space. Urban color research helps bridge the rational design of contemporary cities with the humanistic spirit of traditional ones (Minah 2008). Urban color contains many complex and variable elements, and it should be scientifically investigated and analyzed to plan and guide its development effectively.

The objective of the work is to clarify the overall tone of urban color, which is significant for the plan to reflect different humanistic characteristics in Hailar District, and to create a clear color landscape for the region.

## 2 RESEARCH METHODS AND MATERIALS

The core of the urban color survey is visually comparing the building color card with the survey object and recording the color sample code closest to the survey object. According to the needs of this study, a digital camera and survey cards are used to record the pattern, location, functionality, and materials of urban space elements. Based on the existing color survey cases in China, this paper follows the binary classification of "main colors - embellishment colors," considering climate, natural environment, urban historical contexts, and function partitions to analyze (Caivano 2006). The main chromatographic concept of "basic color – auxiliary color" is formed by fusion, which provides a direct reference for research. Combine the information collected from research, organize them into corresponding color codes (H V/C method) and color block samples, and further analyze them through Origin, a technology mapping software.

---

*Corresponding Authors: 1847562651@qq.com, luxupku@163.com and 728615083@qq.com

DOI 10.1201/9781003372417-70

# 3 CURRENT URBAN COLOR IN HAILAR DISTRICT

## 3.1 *Natural environment colors*

Natural color is the sum of natural elements in the city, such as soil, rocks, lawn, trees, water, and sky. The Hailar soil is distributed in the junction zone of chernozem and chestnut soil, and the geological environment is mostly warm dark chestnut. Hailar district is rich in water resources. There are two large rivers, Hailar River and Yimin River, flowing through this district. Along the river and high plain are lakes of different sizes left over from the old course. Typical grassland vegetation and soil characteristics are mainly dry steppes, forming a composite landscape with different landscape patches and mosaic distribution. The landform of soil and water blending also makes its natural environment color unique.

Table 1. Overview of natural environment colors in Hailar District.

| Basic forms | Meadow soil, dark chestnut soil, sand, rock, bank line, grass seed, bush, arbor |
| Typical places | the Isle of Yimin River, Argun River, the Hulun Buir Prairie |

| Basic colors | Auxiliary colors |
| --- | --- |
| | |

## 3.2 *Local humanistic architecture*

During the Ming and Qing dynasties, Hailar was dominated by religious buildings and commercial buildings, with warm colors such as red and yellow; during the Republican period and after the liberation, the architectural colors were mainly gray and red colors of masonry materials; after the reform and opening up, the diversification of building materials and the change of aesthetic concepts showed a variety of color characteristics. The color perception and interpretation of the representative traditional humanistic buildings and structures in Hailar District are summarized in the following three local humanistic color tones.

First, the warm gray tones of Hulunbuir Old Town. The ancient city covers an area of more than 100,000 square meters, with all kinds of traditional buildings with grey walls and red eaves. Its landmark decorated archway at the entrance is inlaid with warm gold carved vignettes, recreating the historical appearance of the Qing Dynasty.

Second, the red and yellow tones of ancient Buddhist architecture. Hulunbuir's "One Tower and Two Temples" are representative buildings. Located at the confluence of the Hailar River and the Yimin River, the Ciji Vajra Tower is a landmark lama tower in Hailar, with the reputation of "the heart of Hailar." You can see this sacred tower from afar, which is white all over and reflects the light of the Buddha. The main architecture of Darjarin Temple and Ten Thousand Buddhas Temple is the Han-Tibet mixture style, with golden glazed tiles, white walls, red columns, and white marble and granite steps.

Third, the silver-gray color of the Hasar Bridge. Located on the Yimin River, it is a landmark bridge and gateway landscape building in Hulunbuir. It has an exquisite design, integrating architectural art and national culture. The silver-gray style of the bridge has an important influence on the color tone of the city and adds a modern touch to the urban landscape of Hailar.

## 3.3 *Urban center landscape*

The color landscape of the city center shows different color gradients according to its different functions and distribution locations. The color of buildings located in the core area is a relatively uniform bright gray, while the colors of buildings outside the core area differ because of the different

Table 2. Color characteristics of humanistic architecture.

| Representative buildings | Hulunbuir Ancient City, Ciji Vajra Tower, Darjarin Temple, the Temple of Ten Thousand Buddhas, Ciji Vajra Stupa, Hasar cable-stayed bridge |
|---|---|
| The overall hue | Red, yellow, grey and white |

| Basic colors | Auxiliary colors |
|---|---|
| | |

building functions. For example, cultural and educational buildings and residential buildings are mainly in warm colors such as red and yellow. Commercial buildings, transportation architecture, and other public buildings are mainly white, gray, blue, etc.

Table 3. Color Landscape of Hailar City Center. (Source: Taken by the author).

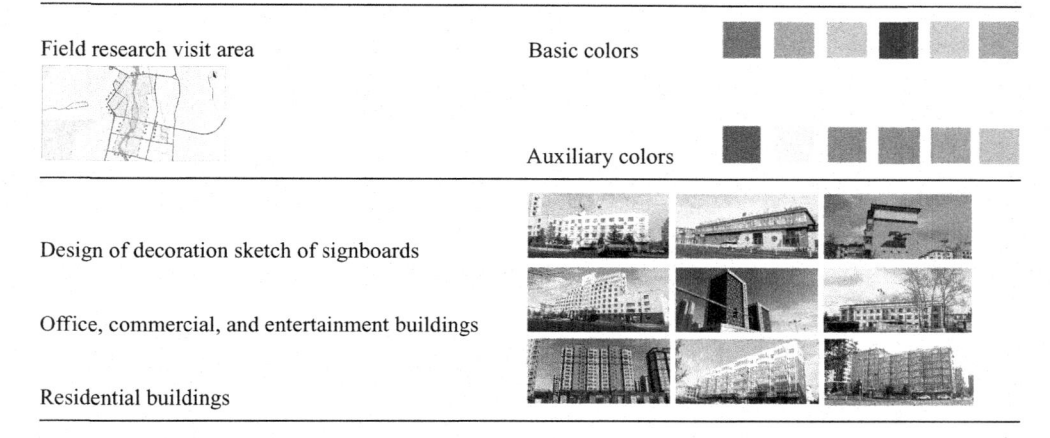

| Field research visit area | Basic colors |
|---|---|
| | Auxiliary colors |
| Design of decoration sketch of signboards | |
| Office, commercial, and entertainment buildings | |
| Residential buildings | |

Artificial environment color, as a more stable color composition element in the overall color of the city, mainly contains the color landscape of the material entities such as city streets, buildings, and structures, and pavement for squares. Among them, architectural color is its main component. Although landscape sketches and signboards are smaller than buildings, many signboards often use bright colors to break the dullness in China's cold cities. Such a unique landscape in Hailar also highlights the overall rhythm and soul of the city color.

## 3.4 *Tourism landscape features*

As a tourist city, Hulunbuir has rich and diversified tourism resources in its Hailar District. The study analyzes the color characteristics of the tourism landscape in representative public places such as Two Rivers Sacred Mountain Scenic Area and the Genghis Khan Square in the Hailar District. The tourism landscape promotes the combination of color art and urban design culture. Integrate grassland nationalities' plastic arts in clothing, painting, carving, totem, utensils, etc. As well as consciousness and thought reflected in religious belief and living customs in landscape environment and urban color. Reconciling urban color landscape according to tourism landscape features is significant to enrich the grassland folk culture and promote further development of the local tourism economy.

The Hailar region is close to the eastern part of Russia and is home to three major ethnic groups: the Mongolians, the Ewenki, and the Oroqens. During its development, the city has developed a unique colorful style, incorporating its profound and distinctive regional and ethnic cultural traits (Lenclos 2005). Many Russian-style buildings located in the center of the city are diverse and unique, with a combination of yellow, red, blue, and white colors of lower brilliance and brightness, and the white relief figures or floral forms set into the facades of the buildings are classical. This combination gives people a feeling of elegance, cleanliness, brightness, and grandness every time they look from afar or close, and the style is distinctive and impressive.

Mongolian architectural forms tend to be spacious, reflecting the Mongolian people's ruggedness and sincerity. The theme color of Mongolian-style buildings is embellished with white. Besides, blue, yellow-based Mongolian pattern laces are mainly used in the eaves, which seem like the headdress of Mongolian girls. The unique city color fully reflects the Mongolian culture and provides a rich source for shaping city characteristics.

The traditional transportation of Ewenkis mainly includes reindeer and birch bark boats. They have lived in harmony with nature for a long time and transform the symbolic patterns that best reflect the national culture into architectural components for application through architectural art creation, highlighting cultural characteristics and traditional style. Hunting culture has dominated the Oroqen people for a long time, "cuo luo zi" is the main housing at that time. The overall shape of the building is cone-shaped, mainly made of several round logs. The exterior of the building is

Table 4.   Color Interpretation from the Perspective of National Characteristics and Regional Culture. (Source: Taken by the author).

| Color Interpretation | Characteristics | The color style of Hailar under the influence of humanistic characteristics |
| --- | --- | --- |
| Russian-style | elegant and grand | |
| Matching yellow, red, blue, etc., whose chroma and value are relatively low, with white. | | |
| Mongolian | Broad and stretching spatial form and scale | |
| Using Mongolian pattern laces mainly in blue and yellow in Mongolian buildings. | | |
| Ewenki | Reindeer herders in the northern forest | <br>Chromatograms in buildings and costumes |
| 1)Transforming floral and grass patterns and animal patterns, etc. into architectural components through the creation of architectural art. 2) Distinctive and colorful traditional costumes. | | |
| Oroqen | Hunting culture | |
| Presenting a natural color that blends with the natural environment. | | |

covered with different materials, and the color of the building is seasonal with different materials, presenting the personalities of the Oroqen people.

## 5   OVERALL COLOR ANALYSIS AND DISCUSSION

Based on the color research and data collation, the distribution of buildings' primary color samples on the color ring is uneven. The samples with high brightness ($V > 5$) account for 82%, those with a low color ($C < 7$) account for 72%, and the percentage of the samples with both attributes is 70%. This concentrated distribution is related to the characteristics of the building color itself, and the light hue is easily harmonious with the outdoor color environment.

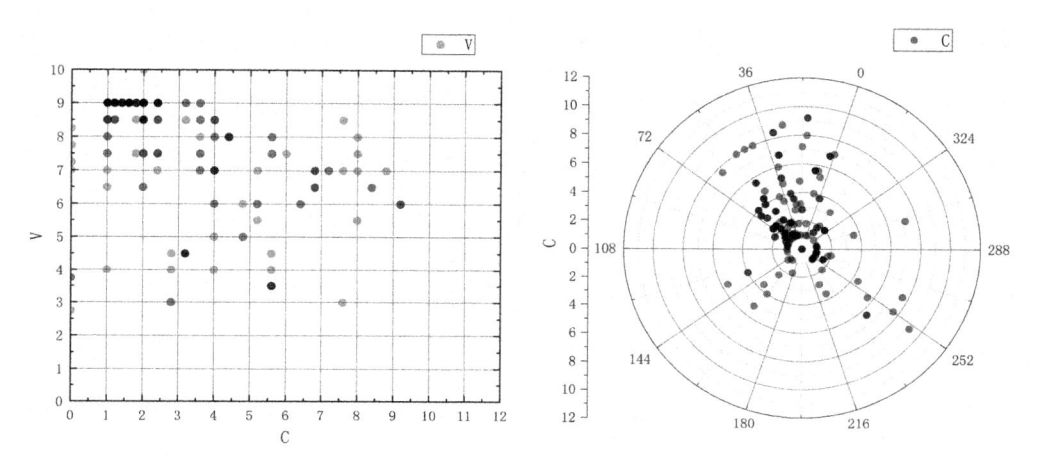

Figure 1.   Buildings' main color chroma-value and color hue-chroma distribution map. (Source: Drawn by the author)

In addition to the concentration of high brightness and low chroma, the color samples also show a tendency to concentrate in both warm and cold directions. Overall, warm tones dominate with about twice the number of samples as cool tones, but the difference in number is still insignificant. From the perspective of planning, the characteristics of urban color in Hailar District are relatively unambiguous, and the matching between the cool and warm tones forms a comparative harmony in some parts. It should be determined that the tone of the future color development in Hailar tends to be high value, low chroma, and warm tones, which is in harmony with the natural environment of the northern inland cities and reflects the historical heritage of Hailar. However, there are also many color landscape problems in the study area:

To begin with, color samples with high chroma and warm tones and the contrast and mixed distribution among cold, warm, and grey colors have resulted in a chaotic feeling in the street. Besides, the results show that glass and paint are the most widely used materials for building facades, while the rich and varied paint colors are most likely to lead to uncontrolled color effects in the urban center. Thirdly, obvious differences exist between commercial and tourism color tendencies and that of office buildings. The color of certain sections is quite different from the overall situation.

## 6   CONCLUSIONS

Through the above investigation and analysis, the following conclusions are drawn: 1) The overall color characteristics of Hailar District are lack of signs, and the color of each functional area is not coordinated; 2) The strong color difference generated by building materials leads to the confusion of urban color environment; 3) By synthesizing the influence of regional cultural characteristics on urban color and clarify the color orientation in Hailar's humanistic tradition, that should tend to be high brightness, low color, and warm colors.

The research under the influence of regional culture also has a certain reference significance for highlighting the color characteristics of other cities. In this new era, higher requirements have been put forward for the shaping of urban features and urban environmental quality construction. The characteristics of a city should be a perfect combination of its regional culture and urban construction, so that they are integrated and benefit by associating together (Boeri 2017). Since areas with different backgrounds and different functional positioning are bound to have different color tendencies.

In a further study, the concept of "basic color – auxiliary color" can be introduced in color planning. The functional positioning of the area to which the building belongs in the whole city should be comprehensively considered, so that the color of the urban space not only reflects the natural and cultural characteristics of the area, but also presents a unique color character. To sum up, urban color planning should combine natural conditions and cultural characteristics and highlight its regional characteristics to achieve a harmonious, unified and pleasing urban environment.

## ACKNOWLEDGMENT

The authors would like to acknowledge financial support from the National Natural Science Foundation of China. (No.51778375)

## REFERENCES

Boeri, C. (2017). Color Loci Placemaking: Urban Color Between Needs of Continuity and Renewal. *Color Research & Application.* 42,641–649.

Caivano, J. L. (2006). *The research on environmental design: Brief history, current development, and possible future.* Proceedings of 10th congress of the international color association. 705–713.

Lenclos, J. P. (2005). *The geography of color. Proceedings of 10th Congress of International Color Association.* Granada, Spain. 307–315.

Minah, G. (2008). Color as an idea: The conceptual basis for using color in architecture and urban design. *Journal of Color Design & Creativity.* 3(2),1–9.

Civil Engineering and Urban Research – Mohamed & Hou (Eds)
© 2023 the Authors, ISBN 978-1-032-44487-1

# Inheritance of summer natural ventilation in traditional dwellings of Miao nationality in Qiandongnan, China

Heyu Hao, Yuan Li* & Yawei Yang
*School of Civil Engineering, Guizhou University, Guiyang, China*

ABSTRACT: With the rapid development of China's village economy, the income level of residents is gradually increasing, and people's demand for residential comfort is also rising, leading to an increase in energy consumption in village houses. At the same time, the "double carbon" target has put forward new requirements for the low carbon transformation of construction and other industries, and how to reduce energy consumption in buildings is of great significance to achieving the task of energy saving and emission reduction. The Miao dwellings in Qiandongnan of Guizhou province are among the best representatives of ethnic minority architecture in China, with their unique external form, internal structure, building materials, and spatial layout. The structural design of the stilted building takes full advantage of the local climate, especially the natural ventilation, to achieve energy savings and improve the indoor microclimate environment. It is an important reference for the inheritance and development of China's excellent national architectural culture, the stilted building of the Miao Nationality dwelling in Guizhou.

## 1 INTRODUCTION

The energy crisis has intensified in recent years, and the construction industry accounts for a huge share of energy consumption. Guizhou's 14th Five-Year Plan proposes "promoting the energy-saving renovation of existing buildings, adhering to the principle of adapting to local conditions, exploring the technical routes of energy-saving renovation of existing residential buildings that meet the climatic characteristics and living habits of residents in each region, improving the energy efficiency and indoor comfort of buildings, and promoting the energy-saving renovation of existing buildings" (Guizhou Provincial Department of Housing and Urban-Rural Development 2021). People are gradually recognizing the importance of energy-efficient building design and are keen to learn from the experience of traditional residential buildings. As a representative of green architecture with a long history and many advantages, the stilted building has always been an important inspiration for contemporary architecture.

This paper takes the Miao hanging foot building in the Qiandongnan region as the main research object and uses Ansys CFX numerical simulation techniques to explore the heritage value of the stilted building of Miao Nationality in Qiandongnan in terms of natural ventilation and to provide guidance for the design and construction of new farmhouses.

## 2 BASIC INFORMATION AND MIAO VILLAGES OF QIANDONGNAN REGION

### 2.1 Geographical area and natural environment of Qiandongnan region

Located in the southeastern part of China's Yunnan-Guizhou Plateau, the Qiannan region covers a total area of about 30,000 square kilometers and is situated in the Miao mountainous region, which

---

*Corresponding Author: 402597063@qq.com

DOI 10.1201/9781003372417-71

has long stretches of mountains and rivers, a pleasant climate, a wide variety of scenery, a wide variety of resources and a rich and unique ethnic flavor. It is one of the largest Miao settlements in China and has the best preserved Miao traditional villages. Miao settlements are mainly located in Leishan, Congjiang, and Taijiang counties, and most of them gather in the Miao hinterland of Kaili, Majiang, Leishan, and Danzhai along the Leigongshan Mountains, forming a kind of Miao living culture circle centered on Leishan (Ai 2020). Figure 1 shows the geographical location of Qiandongnan Prefecture.

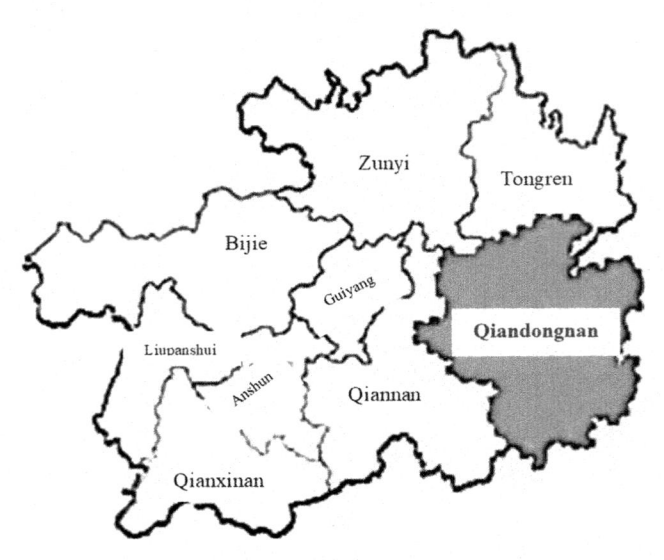

Figure 1.   Geographical location of Qiandongnan prefecture.

## 2.2   Climatic characteristics of the Qiannan region

The climate of the Qiandongnan region is a humid central subtropical monsoon climate, with no severe cold in winter, no scorching heat in summer, rain, and heat at the same time, less rain in winter and more in summer, and a comfortable and pleasant climate (Xiao 2016). The climate of the region is very pleasant in winter and summer.

In general, the average annual temperature in the Qiannan area ranges from 14°C to 18°C (Pang 2019). The general trend is that the temperature in the east and south is higher than in the west and north. The annual sunshine hours are between 1068h and 1296h, the frost-free period from 270d to 330d, the rainfall from 1000mm to 1500mm, and the relative humidity from 78% to 84%.

## 2.3   Optimising the characteristics of traditional Hmong dwellings

The Miao, Dong, Buyi, and Shui ethnic groups, of which the Miao and Dong make up the majority, are the main minority groups in Qiandongnan, where traditional dwellings stand near the water and are built on the mountains.

In order to adapt to the harsh geographical environment, the Miao compatriots in Guizhou have successfully and cleverly combined the southern dry-rail architecture with the local geographical environment, creating a unique kind of dry-rail dwelling with an elevated bottom floor (Shi 2013). The structure is ingenious, the appearance is beautiful, and the layout can be freely arranged regardless of the environment. It protects the land effectively and helps to ventilate and protect against dampness, animals, and insects, making it one of the best representatives of ethnic dwellings in China. Figure 2 shows a picture of the Xijiang MiaoVillage.

Figure 2. Houses of the Xijiang Miao Village.

## 3 OVERVIEW OF THE TRADITIONAL STILTED BUILDINGS OF QIANDONGNAN

### 3.1 *The structure of a traditional stilted building*

The Hmong dwellings are mainly tall dangling houses with a wooden structure of a pierced frame. The main house is built on the ground, with one side of the compartment connected to the main house on the ground and the other three sides overhanging, mainly supported by pillars, as shown in Figure 3. The hammock is generally divided into two or three storeys. If it is two storeys: the lower storey is used for keeping livestock or storing miscellaneous items and is generally unoccupied, while the upper storey is mainly used for living accommodation. If it is a three-storey building, the ground floor is also used for keeping livestock and poultry, stacking firewood and agricultural tools, etc.; the first floor is the main place for eating and living, with the central area of the house being the hall, mainly used for relaxing and entertaining guests, and the small rooms on the left and right sides being used as bedrooms, kitchens, toilets or places for recreation, as shown in Figure 4. The second floor usually has an external corridor with elegant and chic curved fence seats on the periphery, as shown in Figure 5. The third floor is mainly used for storage of food or sundries and can also be used as a bedroom. If the rooms are not large enough for the initial construction or additional rooms for other functions are desired, a 'wing' or annex can be attached to the side of the main building.

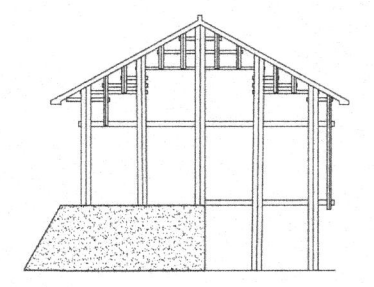 

Figure 3. Section of a Miao stilted buildings.

Figure 4. Plan of the Miao stilted buildings.

Figure 5. External curved rail seating.

### 3.2 *Energy-efficient structural design of a suspended footbridge*

(1) Flexible construction in response to the terrain

The stilted buildings emphasize dependence on and conformity to nature, focusing on the balance between architecture and the spatial environment of the mountains and therefore making full use of the natural space of the mountains (Tian 2020). Most of the areas in Qiandongnan, China, are characterized by high mountains and dense forests, with ravines and gullies. As there is little arable land, stilted buildings are often built on steep slopes or wet waterfronts to conserve limited land resources. At the same time, the dwellings are built partly on flat land and partly suspended by pillars, thus extending the living space. The structure and function of the houses are adapted to the situation, and the layout of the houses is not confined to the environment.

(2) Exquisite techniques and materials taken from nearby locations

The floor structure is mostly built with mortise and tenon work for piercing and fastening, except for the citron and purlins, which need a few locust nails for fixing, but hardly a nail or iron is used. The forest coverage rate in Guizhou province has always been at the forefront of the country, from 50% in 2015 to 60% in 2020 (Zhou et al. 2021), and is planned to reach 64% in 2025, which shows that its natural resources are still very considerable, so most of the construction materials for the stilted building are made from existing local surplus resources, and trees, rattan, rocks, and straw can be used as raw materials for the construction of houses. In areas where forest resources are abundant, almost all of the houses are made of cedar wood, and the roofs are mostly covered with green tiles or cedar bark, while the floors are mostly covered with mineral rock.

(3) Spatial stratification adapted to the climate

As mentioned above, the most basic feature of the stilted building is that the main house is built on the ground, with one side of the wing connected to the main house and the other three sides overhanging and supported by pillars. This structure allows the interior to be ventilated and dry, highly adaptable to the environment, protected from poisonous snakes and wild animals, etc. The ground floor can also be used as a pig and cattle pen or for storing miscellaneous items.

## 4 ANALYSIS OF THERMAL INSULATION PROPERTIES OF VENTILATION ATTIC OF STILTED BUILDING

### 4.1 *Energy saving principle of ventilation attic roof*

The attic floor between the slope roof and the upper floor of the living room can be regarded as an air layer. At the same time, this roof often opens holes in the eaves, roof ridges, or gable, which can make the air layer achieve ventilation and heat dissipation. The air in the attic space forms an insulation layer (Sun et al. 2016). Part of the heat transferred downward from the roof is blocked in the attic space, so that the living environment of the living space below is cooler and more comfortable than that of the attic.

### 4.2 *Establish the mathematical model*

When the southern buildings are completely exposed to sunlight in summer, the heat load from the roof to the room accounts for as high as 36.7%. In practice, however, the walls are not exposed directly due to the self-shielding and greening of buildings, so the actual heat from the roof into the room will be greater (Hu 2016). The roof of the house must become the most important part of heat insulation and energy saving in summer. This paper studies the thermal insulation performance of flat and sloping roofs with the attic. In the simulation, the material layer above the air layer is set as plate A, and the material layer below the air layer is set as plate B. The basic roof structure studied in this paper is shown in Figure 6.

Compared with solar radiation, the radiation heat transfer of the roof to the sky can be ignored. Only the solar radiation absorbed by the roof is considered, and the comprehensive temperature can be simplified as follows:

$$T_r = T_o + a \cdot r_{ou} \cdot I \tag{1}$$

where, $T_o$ is the temperature of outdoor shade, °C; a is absorptance of solar radiation on the roof surface; $I$ is the solar radiation intensity, W/m$^2$; flat roofs receive more solar radiation than attic tiles in a day (Wang 2013).

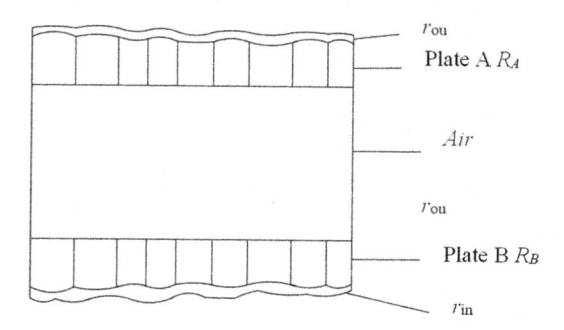

Figure 6.  Ventilation slope roof of the attic.

The total thermal resistance $R_{t1}$ and $R_{t2}$ refer to the total thermal resistance of flat roofs and ventilation slope roofs with attics, respectively. The calculation is as follows:

$$R_{t1} = r_{ou} + R_A + R_B + r_{in} \tag{2}$$

Calculation of $R_{t2}$ by Heat Balance Formula:

$$Q_1 = (T_{in} - T_r) \cdot F_1 / r_{in} + R_B \tag{3}$$

$$Q_2 = (T_r - T_e) \cdot [F_2 / (R_A + r_{ou}) + nvp \cdot C_p] \tag{4}$$

$$Q_3 = (T_{in} - T_e) \cdot F_1 / R_{t2} \tag{5}$$

From $Q_1 = Q_2 = Q_3$, $R_{t2} = (R_B + r_{in}) + F_1 \cdot (R_A + r_{ou})/F_2 + nvp \cdot C_p \cdot (R_A + r_{ou})$

In the formula, $R_A$ and $R_B$ are the thermal resistance of plate A and plate B, m$^2$·K/W; $R_{air}$ is the thermal resistance of the atmosphere, m$^2$·K/W; $r_{in}$ and $r_{ou}$ are the surface thermal resistance of the upper surface of the roof and the lower surface of the roof base (ceiling surface), m$^2$·K/W; $Q_1$ is the heat flowing indoors to the roof, W; $Q_2$ is the heat outflow from the attic to the outdoor, W; $Q_3$ is the heat from indoor to outdoor, W.

Therefore, through two different roofs, the indoor heat is:

(1) Flat roof

$$Q_a = \frac{T_r - T_{in}}{R_{t1}} \tag{6}$$

(2) Ventilation roof with attic

The outlet temperature of the air layer is calculated as follows (Ciampi et al. 2003):

$$T_{ol} = T_m + (T_0 - T_m) \times e^{-v} \tag{7}$$

In the formula, $T_{ol}$ is the outlet temperature of the air inter-layer, $T_m = z \cdot T_{in} + (1 - z) \cdot T$ and $\mu = \frac{1}{C \cdot R_{t1}[H + z(1-z)]}$, $z = \frac{R_{Al}}{R_{tl}}$ $C = \frac{M \times C_p}{A}$, $z$ is the dimensionless parameter, $M$ is the airflow, $C_P$ is air heat capacity, $A$ is roof area, $H = \frac{r_A \times r_B}{\Phi \times R_{t2}}$ is a relative correction factor and a radiation factor because it is not enough to calculate the heat transfer of the air inter-layer surface with radiation heat transfer only by using the surface heat transfer resistance (rather than convective resistance) Therefore, such a radiation factor is introduced (Ciampi et al. 2003).

The $r_1$ and $r_2$ are the surface thermal resistance of the upper and lower surfaces of the air layer. $r_1$, $r_2$, and $\Phi$ are calculated as follows (Chen et al. 2004):

$$r_1 = \frac{r_A \times \Gamma}{\Phi}, r_2 = \frac{r_B \times \Gamma}{\Phi}, \Phi = r_A + r_B + \Gamma \tag{8}$$

$\Gamma$ is the radiative thermal resistance of the air layer, which can be calculated as follows:

$$\Gamma = \frac{\frac{1}{\varepsilon_1} + \frac{1}{\varepsilon_2} - 1}{4\sigma \overline{T}^3} \tag{9}$$

In the formula, $\varepsilon_1$ and $\varepsilon_2$ are the emissivities of the upper and lower surfaces of the air layer; $\sigma$ is the Stefan-Boltzman constant; $\overline{T} = \frac{T_1 + T_2}{2}$ is the average of $T_1$ and $T_2$ which are the surface temperatures of the upper and lower surfaces of the air layer, i.e. plates A and B; $r_A$ and $r_B$ are convective heat transfer resistance of the air layer.

Therefore, the average heat flux of ventilated attic roof into the room is as follows:

$$Q_b = \frac{T_r - T_{in}}{R_{t_2}} - z \cdot C \cdot (T_{ol} - T_o) \tag{10}$$

Compared with $Q_a$ and $Q_b$, $Q_b$ is obviously much smaller than $Q_1$ because $Q_b$ subtracts one more $z \cdot C \cdot (T_{ol} - T_o)$, which represents the heat taken by the air layer. It can be seen that the ventilated attic roof forms an air layer, and the ventilation takes away wind heat, which is greatly reduced compared with the flat roof. Traditional stilted buildings have ventilation attic settings. Its impact on heat transfer below will be simulated by CFD software.

## 5 THERMAL ENVIRONMENT SIMULATION ANALYSIS OF THE INDIVIDUAL TYPICAL RESIDENTIAL STILTED BUILDING

### 5.1 Analysis of climate-appropriate constitution of stilted buildings

In order to adapt to the local hot and humid climate and achieve the purpose of ventilation and cooling, and dehumidification in summer, inhabitants of the Miao nationality have created numerous semi-outdoor spaces with great regional characteristics, such as halls, back halls, and attics. In these spaces, air can flow freely without obstruction, creating a cool and comfortable living environment for the occupants and having a sense of architectural beauty and regional characteristics. The wind environment condition of the attic and hall space is very important to the cooling and heat dissipation of the whole building (Sun et al. 2016). This section focuses on how these building sectors can achieve the purpose of cooling and ventilation and aids the analysis by computer CFD software and how it compares to a typical modern farmhouse.

### 5.2 Extraction of the analytical model

The model retains the outer envelope of the stilted buildings, the doors and windows are set to be fully open, the sloping roof is separated from the using rooms by floor slabs to form the attic, and the eaves are retained. The building faces south, and the bottom is directly on the concrete floor by comparing and analyzing the temperature distribution and wind environment of these two buildings when they are exposed to solar radiation simultaneously. Thus, we know how the hall and attic contribute to the cooling and heat insulation.

Selection of analysis parameters: The initial air temperature was set to 29°C according to the calculated outdoor temperature of summer ventilation in Qiandongnan prefecture. The wind direction was set as positive south wind. The incoming wind velocity is calculated according to the number of room air exchanges 0.45 times per hour. The radiation intensity of the roof is set to 1.25W/m$^2$.

### 5.3 Analysis of natural ventilation in the hall and attic

From Figure 7, part of the summer wind enters the room through the window of the fire pit on the left side of the parsonage and then blows into the parsonage through the door of the fire pit; part of it enters the bedroom along the middle of the parsonage through the door of the west side bedroom,

Table 1. Extraction of traditional dwelling and its analytical model control group.

| Item | Traditional dwelling | Modern farmhouse |
| --- | --- | --- |
| Model comparison | | |

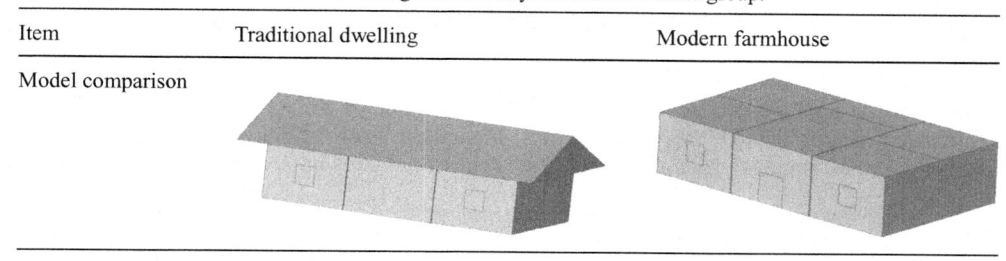

then blows along the south side wall of this bedroom with the west side wall, and finally blows out by the window on the north side wall. A part of the summer wind enters the room through the window of the kitchen on the right side of the parsonage and then blows into the parsonage through the door of the kitchen; another part enters the bedroom along the middle of the parsonage through the door of the east side bedroom, then blows along the south side wall of this bedroom with the east side wall, and finally blows out by the window on the north side wall; the monsoon wind that does not blow into the bedroom goes along the north side wall of the parsonage, through the door of the storage room, enters the storage room, and finally blows out by the window on the north side wall blows out of the outdoor area by the window.

The open hall has a guiding effect on the wind and brings the wind into the room, which satisfies the ventilation deep inside the hall on the one hand. On the other hand, for stilted buildings with a complicated environment, setting up a hall makes it more convenient for the wind to blow to the rooms on both sides in summer, which is especially important for houses with front and back hall patterns.

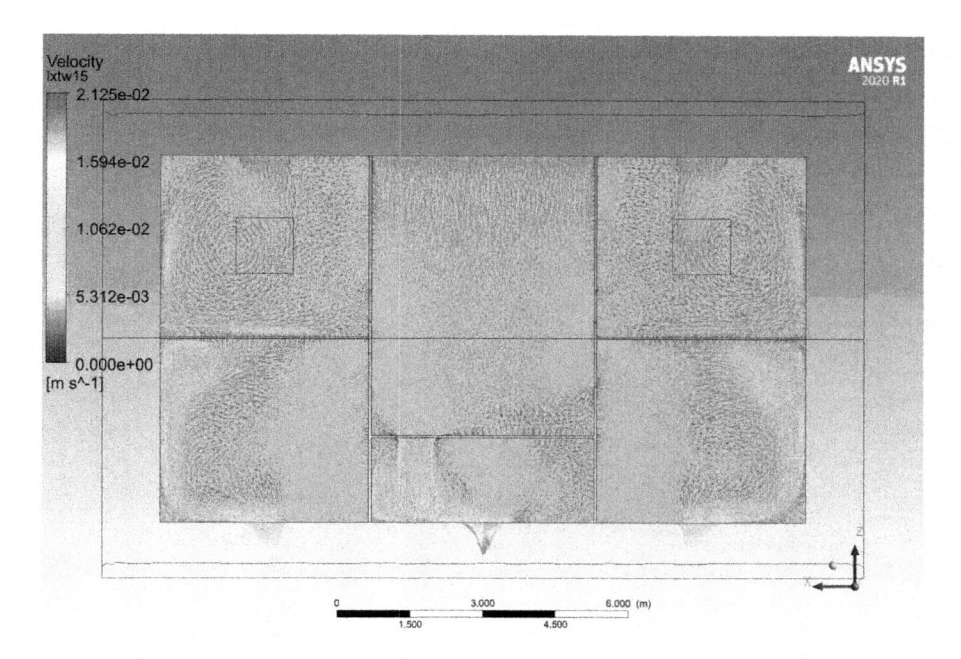

Figure 7. Vector diagram of wind speed at 1.5 m inside a traditional residential house.

Most new farmhouses are designed with flat roofs, while traditional stilted buildings have attics and open holes in the roof. As seen from the comparison of Figure 8 and Figure 9, the attic can serve as a shade and take away the heat from the roof through ventilation, so the indoor temperature can be reduced. The holes opened in the roof. And it is mainly opened over the fire pit, which is convenient for smoke dissipation. At the same time, the area of the static wind area is reduced compared with that without the hole, and the indoor ventilation of the fire pit will be greatly improved, and subsequently, the whole indoor wind environment will be improved.

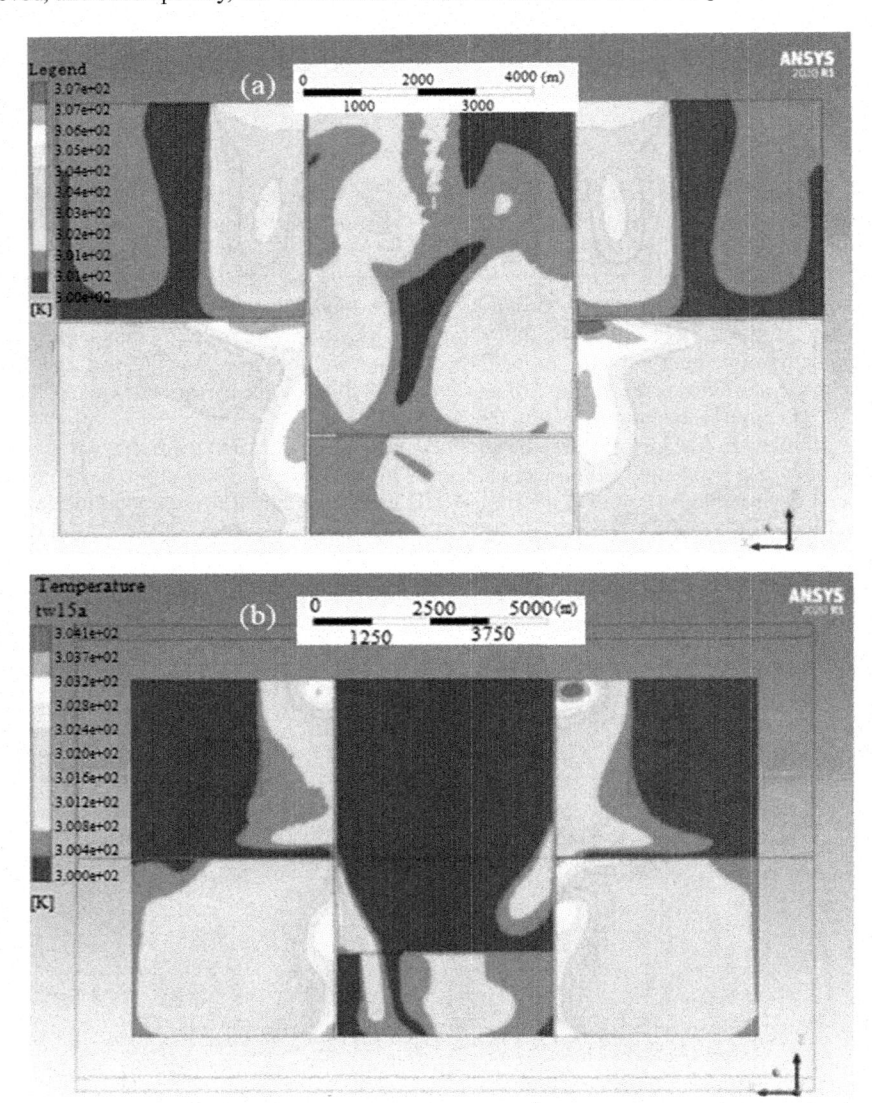

Figure 8.  Temperature cloud at 1.5 m in the room with and without attic and hole.

Nowadays, new farmhouses with sloping roofs can be designed with ventilated attic space, which plays the role of a thermal buffer layer in summer. When the room does not have conditions to open side windows or open window area does not have conditions to open large, can consider opening skylights on the roof to promote vertical ventilation (Dong 2020). The design of modern

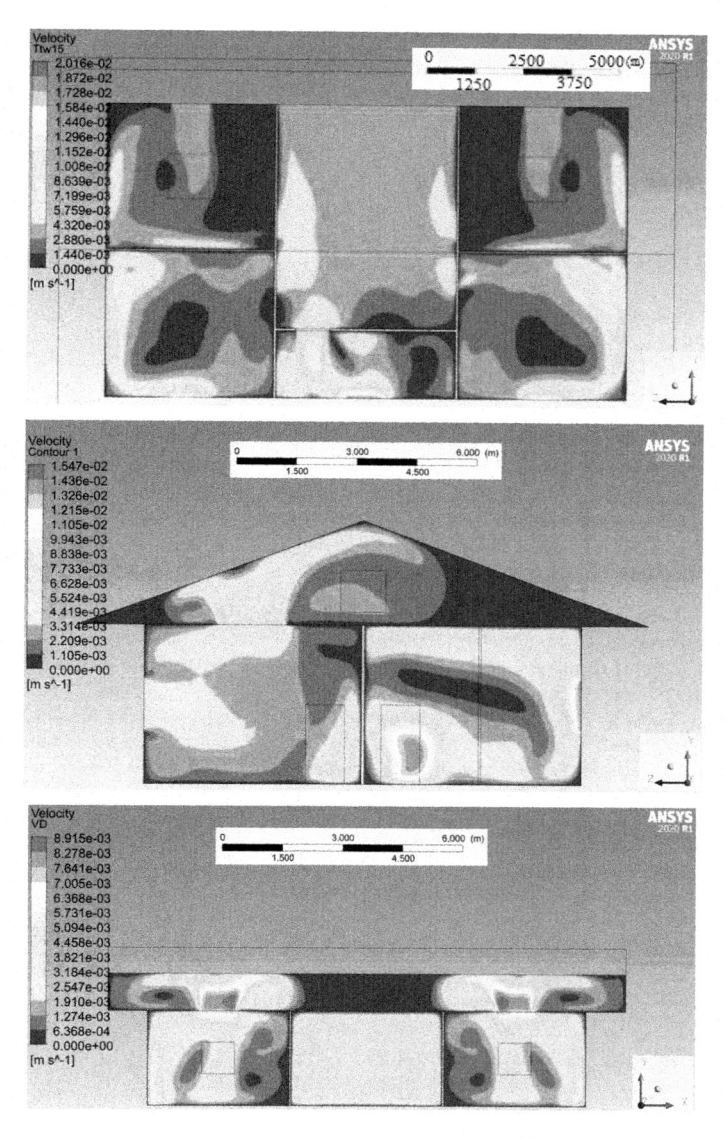

Figure 9.    Wind speed cloud at 1.5 m inside with and without attic and hole.

farmhouses can combine the high stairwell with the ventilation attic to achieve vertical ventilation through the stairwell. The stairwell is set up with an air inlet to the outside world, and a hole is set up at the height of the stairwell attic so that the stairwell and the attic can be connected (Li 2016). Air is heated up by entering the stairwell, reaching the attic, and being discharged by the air outlet.

## 6    CONCLUSIONS

This paper uses numerical simulation as the main technical means to conduct a comparative study on the summer indoor ventilation condition of Miao nationality traditional dwellings and modern farmhouses to summarize the ecological wisdom of Miao nationality traditional dwellings, so that the construction of new farmhouses can be informed in practice and achieve the purpose of energy saving through passive technical means.

The low-level ecological techniques in traditional dwellings are plain and simple, and the clever space organization for ventilation guidance brings good environmental effects. New technologies should be valued and widely applied, and these traditional technical means should be inherited and innovated to bring out the glory in modern architecture.

## REFERENCES

Ai, L. (2020). *Study on the Miao Traditional Villages in Guizhou* [D]. Guizhou University.

Chen, Y. M. & Wang, S. W. (2004). *A new method for unsteady heat transfer analysis of building envelope.* Beijing: Science Press:150.

Ciampi, M., Leccese F. & Tuoni G. (2003). Ventilated facades energy performance in summer cooling of buildings. *Solar Energy*, 75(6):491–502.

Dong, R. J. (2020). *Research on natural ventilation inheritance of traditional villages and hanging foot buildings in western Hunan based on wind environment simulation*[D]. Hunan University, 001480.

Guizhou Provincial Department of Housing and Urban-Rural Development. (2021). *Provincial Department of Housing and Urban-Rural Development on the issuance of the "Fourteenth Five-Year Plan" of Guizhou Province Construction Technology and Green Building Development Plan"* [Z].

Hu, Q. M. (2016). *Research on heat insulation and cooling of rural residential roofs in Chongqing*[D]. Chongqing University.

Li, X. W. (2016). *Research on the natural ventilation technology of traditional dwellings of Tujia family in southeast Chongqing*[D]. Chongqing University.

Pang, M. (2019). *Spatial texture identification of traditional villages in karst areas of Guizhou and its differentiation*[D]. Guizhou University.

Shi, H. F. (2013). *Study on the morphology and structure of hanging foot tower in Xijiang Miao Village, Qiandongnan* [J]. Shenzhou, (04):197.

Sun, Y. & Li, X. W. (2016). Simulation of summer wind environment and summer thermal environment of a typical southeast Chongqing Tujia settlement[J]. *Western Journal of Habitat Environment*, 31(02):96–101.

Tian, M. J. (2020). Tujiajiajiajiajiajiaolou: standing in a dangerous place, melting in the landscape[J]. *Xinxiang Review*, (15):46–47.

Wang, J. (2013). *Research on real-time roaming and infrared simulation of virtual battlefield based on VR technology*[D]. Nanjing University of Science and Technology.

Xiao, Y. J. (2016). The current situation of ethnic music culture inheritance and protection in Qiandongnan[J]. *Contemporary Music*, (24):43+45.

Zhou, Q. P. (2021). Zhang Pengbin, et al. Ecological and environmental changes in the Yangtze River Economic Zone in the past 20 years[J]. *China Geology*, 48(04):1127–1141.

Civil Engineering and Urban Research – Mohamed & Hou (Eds)
© 2023 the Authors, ISBN 978-1-032-44487-1

# Research on the forecasting settlement method for high-filled highway subgrade

Dahai Zhang*
*Bureau of Transport of Jinzhong, Jinzhong, Shanxi, China*

ABSTRACT: Based on the development principal for highway subgrade settlement, the Pearl curve model and the Gompertz curve model are established. Combined with samples on site, they can reflect the developing process of high embankment settlement, and the prediction results of the model are very reliable and accurate. So the new method will provide meaningful reference value and guiding significance.

## 1 INSTRUCTION

Compared with the general subgrade, the high-filled subgrade has the following characteristics: the filling width is relatively large. According to the relevant theory of soil mechanics, the higher the embankment filling height, the higher the corresponding requirements for the overall stability of the embankment and subgrade stiffness, and the larger the filling section area (Fang Lei 2019). The high-filled embankment has a higher filling height, usually more than 20 meters. Combined with the cross-section of the highway, its filling area is often larger because the self-weight of the embankment is also larger. The self-weight stress is also large; The cumulative settlement of the embankment itself is large. The embankment filler has a large amount of filling, many components, and is miscellaneous, which will inevitably lead to a large amount of post-construction settlement. There is a difference in the stiffness between the high-filled embankment and the general subgrade. After it is put into use, the subgrade soil will produce a larger primary consolidation and secondary consolidation settlement for the high-filled embankment area under the action of the pavement structure and additional loads, and a smaller primary consolidation and secondary consolidation settlement for the general subgrade area, resulting in uneven settlement of the overall structure of the subgrade. In practical engineering, the settlement of high-filled embankment often has the following hazards: the overall settlement of the subgrade, resulting in a differential settlement with other sections, which seriously destroys the road performance. The local settlement of the subgrade causes the ups and downs of the pavement, such as vehicles jumping at the bridgehead and other engineering problems, longitudinal and transverse cracks of subgrade, subgrade sliding and subsidence, and even slope instability.

## 2 MECHANISM OF HIGH-FILLED SUBGRADE SETTLEMENT

The settlement of subgrade is caused by the effect of gravity itself and additional external stress. Excessive stress leads to the compression deformation of the subgrade (Gao Dazhao 1999). In practical engineering, when the settlement deformation of subgrade soil has been completed, and the state is relatively stable, the maximum settlement represents the settlement of subgrade, and the

---

*Corresponding Author: 463322321@qq.com

maximum settlement reached by the consolidation of subgrade under the joint action of external force and its load (Li Guangxin 2004). Generally, the settlement deformation of subgrade under the action of load can be divided into three main settlement stages.

Instantaneous settlement (Sa) is the settlement deformation that occurs instantly when the external load is just loaded. At this time, the pore water in the pores of the soil cannot be discharged in an instant. The settlement deformation that occurs in the soil is shear deformation, and the way and rate of load application affect the initial settlement. Consolidation settlement (Sb), the main consolidation settlement, is the settlement deformation caused by the dissipation of excess pore water pressure as the water in the soil pores begins to drain, which is usually the main component of subgrade soil settlement deformation. In the deformation of consolidation settlement, the main deformation is the volume deformation of soil, but a small part of the deformation is the shear deformation shape. Secondary consolidation settlement (Sc) is mainly caused by the settlement deformation caused by the deformation characteristics of the soil skeleton. Generally, the amount of deformation is small. At this time, the pore water in the soil pores is completely discharged, the excess pore water induced pressure is completely eliminated, and the effective stress has no longer changed. The secondary consolidation settlement is generally slow and lasts for a long time, which will change slowly with time (Pi Daoying 1996). The final settlement is the sum of instantaneous settlement, consolidation settlement, and secondary consolidation settlement (Bates J M 1969).

The Pearl curve model is expressed below.

$$y(t) = \frac{L}{1 + ae^{-bt}}$$

Where: L, a, and b are three parameters of the model, in which a>0, b>0. Figure 1 is the schematic diagram of a typical Pearl curve.

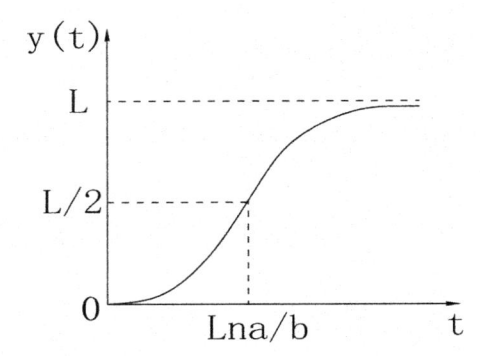

Figure 1. Pearl curve.

Let S1, S2, and S3 represent three section numerical sums

$$S_1 = \sum_{t=1}^{r} \frac{1}{y(t)}, S_2 = \sum_{t=r+1}^{2r} \frac{1}{y(t)}, S_3 = \sum_{t=2r+1}^{3r} \frac{1}{y(t)}$$

$$\frac{1}{y(t)} = \frac{1}{L} + \frac{ae^{-bt}}{L}$$

$$b = \frac{\ln \frac{(S_1 - S_2)}{(S_2 - S_3)}}{r}$$

$$L = \dfrac{r}{S_1 - \dfrac{(S_1 - S_2)^2}{(S_1 - S_2) - (S_2 - S_3)}}$$

$$a = \dfrac{(S_1 - S_2)^2 \left(1 - e^{-b}\right) L}{((S_1 - S_2) - (S_2 - S_3)) e^{-b} \left(1 - e^{-rb}\right)}$$

So far, three parameters are calculated by the formula, which can be obtained from the Pearl prediction model.

Gompertz method is proposed by the statistician and mathematician B. Gompertz. The three parameters in the model can be obtained using the three estimations.

$$y_t = e^{(k + ab^t)}$$
$$y' = k + ab^t$$
$$y_1 = k + ab^1$$
$$y_2 = k + ab^2$$
$$y_3 = k + ab^3$$
$$y_t = k + ab^T$$

$$\sum\nolimits_1 y_t = \sum_{t=1}^{n} y_t = nk + ab(b^0 + b^1 + b^2 + \cdots + b^{n-1})$$

$$\sum\nolimits_2 y_t = \sum_{t=n+1}^{2n} y_t = nk + ab^{n+1}(b^0 + b^1 + b^2 + \cdots + b^{n-1})$$

$$\sum\nolimits_3 y_t = \sum_{t=2n+1}^{3n} y_t = nk + ab^{2n+1}(b^0 + b^1 + b^2 + \cdots + b^{n-1})$$

$$(b^0 + b^1 + b^2 + \cdots + b^{n-1}) = \frac{b^n - 1}{b - 1}$$

$$\sum\nolimits_1 y_t = nk + ab\frac{b^n - 1}{b - 1}$$

$$\sum\nolimits_2 y_t = nk + ab^{n+1}\frac{b^n - 1}{b - 1}$$

$$\sum\nolimits_3 y_t = nK + ab^{2n+1}\frac{b^n - 1}{b - 1}$$

$$b = \sqrt[n]{\frac{\sum_3 y_t - \sum_2 y_t}{\sum_2 y_t - \sum_1 y_t}}$$

$$a = \frac{b - 1}{(b^n - 1)^2 b}\left(\sum\nolimits_2 y_t - \sum\nolimits_1 y_t\right)$$

$$k = \frac{1}{n}\left(\sum\nolimits_1 y_t - ab\frac{b^n - 1}{b - 1}\right)$$

$$k = \frac{1}{n}\left[\frac{\sum_1 y_t \sum_3 y_t - (\sum_2 y_t)^2}{\sum_1 y_t + \sum_3 y_t - 2\sum_2 y_t}\right]$$

$$a = \frac{b - 1}{(b^n - 1)^2 b}\left(\sum\nolimits_2 \ln y_t - \sum\nolimits_1 \ln y_t\right)$$

$$b = \sqrt[n]{\frac{\sum_3 \ln y_t - \sum_2 \ln y_t}{\sum_2 \ln y_t - \sum_1 \ln y_t}}$$

$$k = \frac{1}{n}\left[\frac{\sum_1 \ln y_t \sum_3 \ln y_t - (\sum_2 \ln y_t)^2}{\sum_1 \ln y_t + \sum_3 \ln y_t - 2\sum_2 \ln y_t}\right]$$

## 3  SAMPLES IN ENGINEERING

Combined with the subgrade settlement observation data in the expressway, the measured point is chosen as the case study.

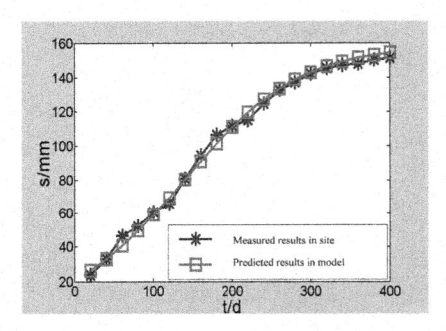

Figure 2. Comparison between measured points and predicted points.

Table 1. Parameters.

| Model | Parameter | $L$ | $a$ | $b$ |
|---|---|---|---|---|
| Predicted Model | | 158.7895 | 6.5942 | 0.2730 |

Table 2.  Measured points and predicted points.

| Time (d) | Predicted results(mm) | Measured results(mm) | Difference (mm) | Time (d) | Predicted results(mm) | Measured results(mm) | Difference (mm) |
|---|---|---|---|---|---|---|---|
| 20 | 26.3828 | 23.9 | 2.5 | 220 | 119.6399 | 114.9 | 4.7 |
| 40 | 32.9468 | 32.9 | 0.0 | 240 | 127.1288 | 124.6 | 2.5 |
| 60 | 40.6426 | 46.3 | 5.7 | 260 | 133.4881 | 132.7 | 0.8 |
| 80 | 49.4299 | 52.1 | 2.6 | 280 | 138.7712 | 137.0 | 1.8 |
| 100 | 59.1656 | 60.6 | 1.5 | 300 | 143.0809 | 142.3 | 0.8 |
| 120 | 69.5986 | 65.7 | 3.9 | 320 | 146.5447 | 145.2 | 1.3 |
| 140 | 80.3868 | 80.9 | 0.5 | 340 | 149.2954 | 147.1 | 2.2 |
| 160 | 91.1385 | 94.4 | 3.3 | 360 | 151.4591 | 148.0 | 3.5 |
| 180 | 101.4672 | 106.3 | 4.9 | 380 | 153.1483 | 150.2 | 2.9 |
| 200 | 111.0450 | 111.6 | 0.6 | 400 | 154.4594 | 151.4 | 3.1 |

According to the analysis of the above chart, the Pearl curve is a good reflection of S shape relationship between settlement and time of high embankment settlement process, and more importantly, it can be accurately applied to predict the settlement of subgrade.

Table 3. Results of Gompertz and forecasting and measuring.

| Time (d) | Predicted results(mm) | Measured results(mm) | Difference (mm) | Time (d) | Predicted results(mm) | Measured results(mm) | Difference (mm) |
|---|---|---|---|---|---|---|---|
| 20 | 23.6187 | 23.9 | 0.3 | 220 | 118.1724 | 114.9 | 3.2 |
| 40 | 32.1777 | 32.9 | 0.7 | 240 | 124.9498 | 124.6 | 0.3 |
| 60 | 41.7554 | 46.3 | 4.6 | 260 | 130.9612 | 132.7 | 1.7 |
| 80 | 52.0064 | 52.1 | 0.1 | 280 | 136.2502 | 137.0 | 0.7 |
| 100 | 62.5736 | 60.6 | 2.0 | 300 | 140.8721 | 142.3 | 1.4 |
| 120 | 73.1273 | 65.7 | 7.5 | 320 | 144.8878 | 145.2 | 0.3 |
| 140 | 83.3896 | 80.9 | 2.5 | 340 | 148.3601 | 147.1 | 1.3 |
| 160 | 93.1464 | 94.4 | 1.3 | 360 | 151.3503 | 148.0 | 3.3 |
| 180 | 102.2482 | 106.3 | 4.1 | 380 | 153.9166 | 150.2 | 3.7 |
| 200 | 110.6040 | 111.6 | 1.0 | 400 | 156.1125 | 151.4 | 4.8 |

## 4 CONCLUSION

Among the prediction methods, both the Pearl curve and the Gompertz curve can show an S-shape relationship between settlement and time during the process. They can not only reflect settlement developing principal, but also can be used in prediction for high embankments.

In order to obtain a good prediction effect, it is necessary to use the Pearl model and Gompertz model to forecast the roadbed settlement. The measured settlement value should be checked in advance, and obvious errors should be deleted before adoption.

The measured samples on site should be renewed and updated during the settlement process so that the prediction accuracy will be improved and refined.

## REFERENCES

Bates J M, Granger C W J. The combination of forecasts [J]. *Operations Research Quarterly*, 1969 (12):319–325.

Fang Lei, Huang Xiaoming. *Influence of uneven subgrade settlement on Pavement Engineering* [M]. Beijing: People's Communications Press, 2019

Gao Dazhao, Yuan Juyun. *Soil Science in soil mechanics* [M]. Beijing: People's Communications Press, 1999

Li Guangxin. *Advanced soil mechanics* [M]. Beijing: Tsinghua University Press, 2004

Pi Daoying, Sun Youxian. An algorithm of [J]. *Control and Decision of Multi-Model Adaptive Control*, 1996,11 (1): 77–80.

*Civil Engineering and Urban Research – Mohamed & Hou (Eds)*
© *2023 the Authors, ISBN 978-1-032-44487-1*

# Integrated construction of a prefabricated residential building project from the perspective of high-quality

Shengping Tang*
*School of Civil Engineering, Hunan University, Changsha, China*

Xiaoyi Hu*
*Chongqing Open University, Chongqing, China*

ABSTRACT: To realize construction 4.0 and solve the dissociation of design, production, transportation, and assembly from the perspective of high-quality development, the integrated construction technology of prefabricated concrete buildings was studied. Taking a residential building project as an example, efficient design standards were established, and the design was standardized and performed in advance. An informationalized management platform was built, and the building information model (BIM) and Internet of Things (IoT) were used in combination. By applying the BIM and radio frequency identification (RFID) to links, including the design, production, transportation, and assembly of the prefabricated concrete building, the quality of the prefabricated building was improved. The promotion and application of the integrated construction technology of prefabricated buildings have prospected, and reasonable development suggestions were proposed.

## 1 INTRODUCTION

Currently, the construction industry in China has entered the stage of supply chain reform. To realize Construction 4.0, change the extensive development mode, and realize high-quality development, the mode of construction has to be changed. The integrated construction of prefabricated buildings refers to the use of systematic thinking and method, and informationalized management technologies and means to optimize all factors in links, including the design, production, transportation, and assembly, to maximize benefits in the construction process (Ye 2017). The key to prefabricated buildings is the standardized design, which enables the subsequent intelligent and mass production of components and parts. The basic unit of prefabricated buildings is prefabricated components. All links from the production to transportation and stacking in the middle and finally to the assembly of prefabricated components on site are keys to the management of prefabricated components. However, the lack of traceability for prefabricated components and the high cost of application of technologies pertaining to prefabricated components intensify people's repellency to new technologies (Ye 2018). Taking a prefabricated residential building project as the research object, the research explored the integrated construction of prefabricated buildings, which is of practical significance for promoting and applying the integrated construction of prefabricated buildings.

## 2 INTEGRATED CONSTRUCTION APPROACH OF PREFABRICATED BUILDINGS

Prefabricated components are generally in a large quantity and a wide variety, so they are difficult to manage from production to assembly. Moreover, many participants are involved in prefabricated buildings, and their coordination also faces certain difficulties (Wang 2015). The following cases

---

*Corresponding Authors: tshp66@163.com and 312788068@qq.com

DOI 10.1201/9781003372417-73

may occur: produced components cannot be assembled on-site, and the assemblability of components cannot be completely guaranteed, so it is unable to ensure the overall construction quality of prefabricated buildings.

## 2.1 *Establishment of an informationalized management platform*

Information technology (IT) is an important tool in the integrated construction mode from the design, component production, and transportation to assembly (Sun 2020). To realize the integration of the design, production, transportation, and assembly of prefabricated buildings, the building information model (BIM) and Internet of Things (IoT) can be used to build an informationalized management platform. Based on the platform and modern IT, the information can be collected and real-timely exchanged, and information sharing and collaboration of participants can be realized conveniently and effectively, which solves the dissociation of the design, production, transportation, and assembly (Figure 1).

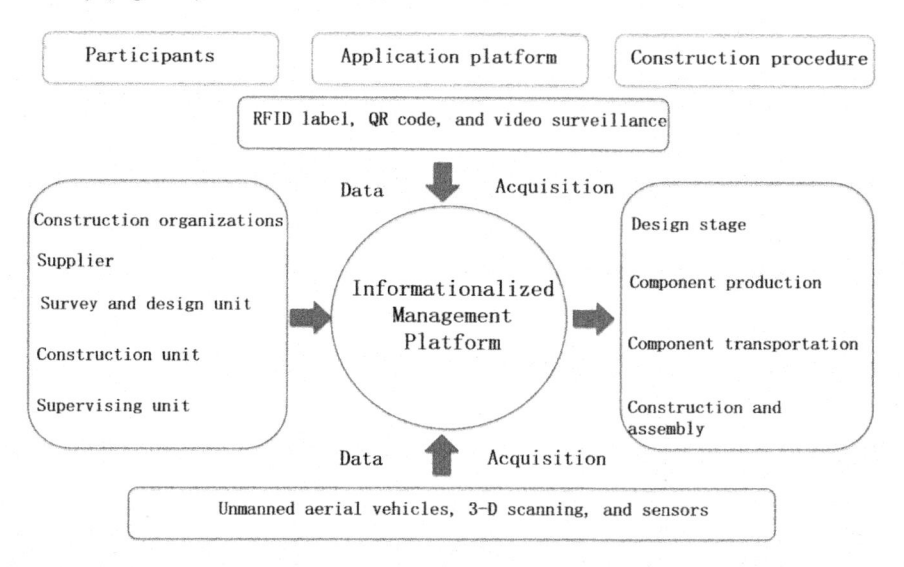

Figure 1.   An intelligent construction system of prefabricated buildings based on the building information model and the internet of things.

## 2.2 *Design of prefabricated components*

The standardized design of planes, facades, spaces, components, and parts is a key to the integrated construction of prefabricated buildings. However, the design, component production, and construction are often interrupted due to the time sequence of labor division and bidding and purchasing (Yu 2015). To meet these requirements and perform design in advance, requirements for detailed design, production, fabrication, and transportation of components and parts, and even construction are considered in advance in the design, so as to improve the construction efficiency and save cost using efficient technologies conducive to production and construction.

## 2.3 *Production of prefabricated components*

The accurate design information is timeously transmitted to factories using digital technology to realize intelligent production. This further achieves the intelligent production of construction products by industrial means. The use of radio frequency identification (RFID) and BIM technologies in the production link of prefabricated buildings will substantially improve the information acquisition capacity and render attribute parameters, including the material, size, and quality of prefabricated components to be traceable.

## 2.4 Transportation of prefabricated components

The transportation cost is mainly related to the transportation distance, so the selection of the factory for prefabrication should consider whether the transportation distance is reasonable and economically efficient or not. It would be better if the maximum distance between the factory of prefabrication and the construction site is no longer than 80 km. Before transportation, the prefabricated components should be identified using quick response (QR) codes and RFID (Zhang 2019), and the instant status information needs to be written into the prefabricated components. In addition, the management information in the transportation process should also be collected.

## 2.5 Installation of prefabricated components

To optimize the construction scheme, the construction process is simulated from multiple dimensions using the BIM+VR/AR technology to make the construction organization and design more reasonable, take a shorter period of time, and save cost. Using the BIM5D platform, the information related to the project, including the material, quality, progress, and cost, is shared. Participants of the project timeously deal with the corresponding work according to their responsibility, which can substantially improve the working efficiency and reduce labor intensity simultaneously.

## 3 ENGINEERING APPLICATION CASE

The project covers a $2.8 \times 10^5 \mathrm{m}^2$ multistory residential building with a prefabrication rate of 60% and an assembly rate higher than 90%. Prefabricated panels and slabs are used, and the facade of buildings is constructed with architectural concrete, which calls for high construction precision. A BIM informationalized management platform is established, and RFID is used to track the whole process from production to transportation, and finally to the assembly of the prefabricated components. Various participants can obtain information in the whole process of corresponding components, including the design, production, transportation, and construction, by establishing the informationalized management platform.

### 3.1 Design quality and efficiency of the construction drawing

The component and part families of prefabricated buildings are constructed. The standardized component family is embedded in the design software to provide options for architects. Requirements for indexes, including the plot ratio and proportion of house types of the whole project, are satisfied by combining rooms of different bays and depths (Figure 2). The concept of building blocks is used in the framework of standardized design to create variable house types in the full life cycle and design large indoor space and variable space (Zeng 2019). The designed architectural form is simple and smooth, which is easily industrially prefabricated.

The universalization and standardization are highlighted in the design, and standardized prefabricated components are directly used. Types of vertical and horizontal components are reduced through optimization, which is conducive to industrial mass production. On the premise of achieving the scale effect, the design scheme is optimized, and the types of components are decreased (Table 1), reducing the amortization expense of components, lowering the construction cost, and saving funds.

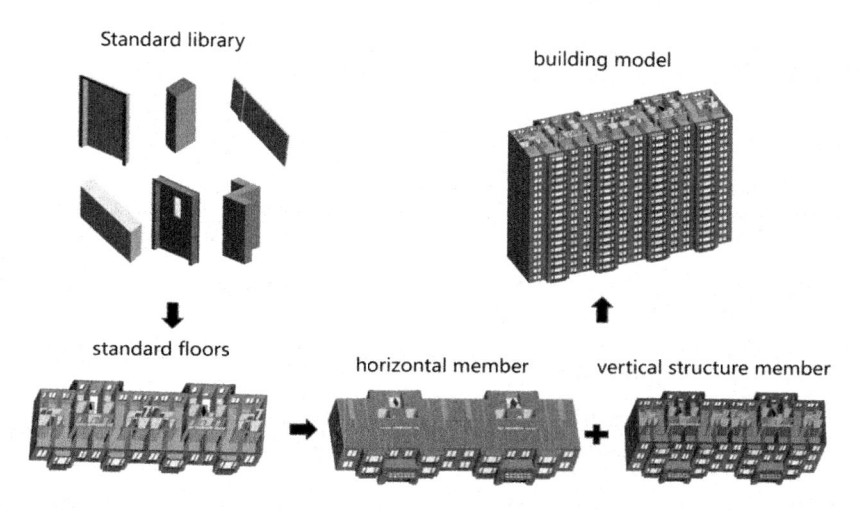

Figure 2. Selection and design of standardized components.

Table 1. Comparison of types of components before and after optimization of the design scheme.

| Types of prefabricated components | | Prefabricated facade panels | Prefabricated internal panels | Composite floor slabs | Prefabricated balcony slabs | Prefabricated air conditioning boards | Prefabricated stairs | PCF slabs | Total |
|---|---|---|---|---|---|---|---|---|---|
| Number of standard layers | Before optimization | 20 | 9 | 10 | 1 | 1 | 2 | 2 | 45 |
| | After optimization | 14 | 7 | 8 | 1 | 1 | 2 | 1 | 35 |
| Decrease amplitude | | 30% | 22% | 20% | – | – | – | 50% | 22% |
| Precast concrete volume (m³) | | 890 | 680 | 430 | 90 | 22 | 33 | 11 | 2156 |

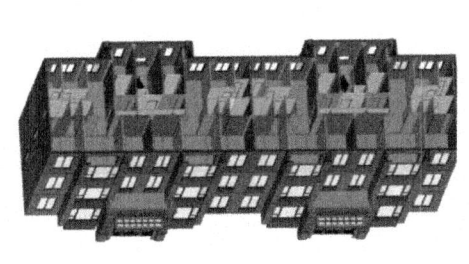

Material Statistics Sheet

| number | Material Name | Material specification | Quantity (t) | shape of bars |
|---|---|---|---|---|
| 1 | | 6 | 1.21 | |
| 2 | | 8 | 0.86 | |
| 3 | HRB400 | 10 | 1.51 | |
| 4 | | 12 | 1.86 | |
| 5 | | 14 | 3.64 | |
| 6 | | 16 | 4.12 | |
| 7 | | 18 | 12.22 | |

| number | Material Name | Material specification | Quantity (m³) | Note |
|---|---|---|---|---|
| 1 | concrete | C30 | 680 | 3F-8F |
| 2 | | C40 | 890 | 1F-2F |

Figure 3. Extracting production data and information from the model.

## 3.2 Production of prefabricated components

The data of the design model are real-timely interacted with the production equipment to realize high-precision control of the component quality and reasonable arrangement of production progress, as well as automatic and efficient production. Industrial robots are used for repetitive assembly-line operations such as automatic blanking and melting of reinforcements to realize intelligent production of construction products using industrial methods (Figure 3). Meanwhile, the IoT is used to assign the unique code to each component, which allows tracking and management of the whole-process information of each component while realizing efficient and automatic production.

### 3.3 *Transportation of prefabricated components*

Before the transportation of prefabricated components, the transportation schemes are compared to determine the optimal transportation route and formulate an economical, safe, and reasonable transportation scheme. Prefabricated components are identified using QR codes and RFID before transportation, and the status information of components—including production and quality inspection, information about transportation vehicles, and information about transportation managers—are transmitted to the IoT collaborative management platform. Global position system (GPS) vehicle locators are installed on vehicles to track the locations of vehicles and acquire dynamic information on prefabricated components, so as to realize the transportation and tracking management of components.

### 3.4 *Installation of prefabricated components*

After transporting components to the construction site, the appearance quality, component dimensions, reserved holes, and reserved length of reinforcements are inspected to record quality inspection information and upload such information to the system server. When installing the prefabricated components, the information, including the design, production, and transportation, is shared using the system management system based on technologies including the BIM, RFID, and mobile terminals. This realizes the digital assembly, reasonably formulates the construction scheme and schedule of prefabricated buildings, achieves informationalized management of labor, materials, equipment, and methods. It also improves the efficiency of the integrated construction of prefabricated buildings.

## 4 OUTLOOK OF INTEGRATED CONSTRUCTION OF PREFABRICATED BUILDINGS

To promote the development of prefabricated buildings and meet requirements for high-quality development of the construction industry in China, the single-step research, improvement, and promotion of integrated construction technologies of prefabricated buildings can be conducted from the following aspects.

### 4.1 *Multi-specialty collaboration and detailed design*

The construction process of prefabricated buildings generally can be divided into stages: design, production, transportation, and installation. Design is the key link. The collaboration of various specialties, including architecture, structures, water supply and drainage, and heating and ventilation, should be completed in the design and factory production links. However, the design standards of prefabricated buildings have many differences from those of existing buildings, which sets higher requirements for architects and brings more difficulty to the subsequent construction precision and seamless connection of specialties. Therefore, multiple specialties need to be collaborated to achieve a detailed design.

### 4.2 *Standardized design and mass production*

The design process of prefabricated buildings should strictly follow the modular standard, which is conducive to the intelligent and mass production of structural components and component parts. However, the standards in China are still not unified and incomplete, and the production tools are less versatile, so prefabricated components cannot be produced at scale. Therefore, it is urgent to formulate relatively complete standards and norms to promote mass production, reduce costs, and improve efficiency.

### 4.3 *Policy support and ordered popularization*

The development of prefabricated building systems needs a giant input. The cost of prefabricated buildings is 10%~30% higher than traditional buildings, so enterprises are not active in developing prefabricated buildings, which calls for policy support to orderly popularize such buildings. Developing the construction industry of prefabricated buildings is not only a matter of enterprise but also needs vigorous support from the government. Social cohesion has to be developed to form a development mode of the entire industrial chain involving the planning, design, production, construction, and management and a unified, integrated platform to enable the coordinated development of all participants involved in the design, production, and construction. With the wide application of prefabricated buildings and the vigorous support from the government, the unique advantages of prefabricated buildings will gradually be highlighted.

## 5 CONCLUSIONS

To overcome the drawback of the less close connection between the design and construction of prefabricated buildings, BIM and IoT have gradually been applied in the construction industry. The research proposed to carry out the design in advance, collected and analyzed restrictions and technical demands of each participant, and then established the informationalized management system based on BIM. This solves the problem of the unreasonable design of prefabricated components when using the traditional design mode. Information including the design, production, and transportation was shared via the informationalized management platform to share and track information on-site during the whole assembly process. Through practice in the integrated construction of a prefabricated residential building project, the experience of integrated construction of prefabricated buildings, including standardized design, industrialized production, construction, and informationalized management, is accumulated. This is of certain theoretical and practical significance for promoting the high-quality development of the construction industry in China.

## ACKNOWLEDGMENT

This work was funded by the Research Foundation of Chongqing Technology and Business Institute (Grant No. NDYB2020-10).

## REFERENCES

Sun Hui, Feng Weidong, Chen Wei. (2020). Practice of integrated construction of prefabricated buildings on campuses. *Construction Technology*.10,111–114

Wang Lanzhi, Qie Ze. (2015).BIM numerical simulation of the construction of prefabricated buildings. *China Housing Facilities*.3, 81–83.

Ye Haowen, Wang Bing, Tian Zixuan. (2018). Study and Expectation of prefabricated concrete building integration key technologies. *Construction Technology*.6,66–69.

Ye Haowen, Zhou Chong, Fan Zesen, Liu Chengwei. (2017). Thinking and application of integrated digital construction to prefabricated buildings. *Journal of Engineering Management*. 5,85–89.

Yu Longfei, Zhang Jiachun. (2015). Computer integrated construction system based on BIM. *Journal of Civil Engineering and Management*. 32,73–77

Zeng Qi, Li Xinwei, Zheng Shengbo. (2019). Research on the construction scheme of prefabricated building component library based on BIM.*Construction Technology*.48,19–22.

Zhang Yingying. (2019). *Research on tracking and positioning technology of structural components in the whole life cycle of prefabricated buildings, Doctoral dissertation*, Southeast University.

*Civil Engineering and Urban Research – Mohamed & Hou (Eds)*
© 2023 the Authors, ISBN 978-1-032-44487-1

# A research of urban blue corridor planning from the perspective of public gealth — Case study of Stalin Park in Harbin

Guanyan Xiao
*School of Architecture, Harbin Institute of Technology, Harbin, China*

Binxia Xue* & Tongyu Li*
*School of Architecture, Harbin Institute of Technology/Key Laboratory of Cold Region Urban and Rural Human Settlement Environment Science and Technology, Ministry of Industry and Information Technology, Harbin, China*

Taorong Liu
*School of Architecture, Harbin Institute of Technology, Harbin, China*

Siyuan Guo
*School of Architecture, Harbin Institute of Technology/Key Laboratory of Cold Region Urban and Rural Human Settlement Environment Science and Technology, Ministry of Industry and Information Technology, Harbin, China*

ABSTRACT: The blue corridor is a kind of urban public space with multiple functions, which integrates natural ecology, cultural connotation, and social environment, as well as a crucial carrier to realize public health. In addition to introducing the blue corridor and its value for public health, this paper summarizes the operational mechanism of the blue corridor on public health based on the research achievements. Combining with the field research of Stalin Park in Harbin, the author explores the design strategies of the blue corridor based on public health from the view of recovery environment theory. This paper intends to provide references and illustrations for the blue landscape design in China and facilitates the implementation of the healthy China strategy.

## 1 INTRODUCTION

Health has always been an important proposition accompanying the progress of human civilization and an important goal of urban evolution and development. The World Health Organization (WTO) defines the term health as "health is not only the absence of disease and physical defects, but also the complete physical and mental state and good social adaptation." (Berg et al. 2010) With the rapid progress of urbanization and the change in world population structure (e.g., aging), the number of people living in densely populated urban areas is increasing yearly (United Nations 2015). Cities provide people with sufficient material resources and public services, but urban life with high-stress, fast-paced, and far-from-nature poses a threat to public health (Patel 2018). Urban dwellers experience increased stress and loneliness, and sedentary lifestyles reduce their opportunities to engage with nature, triggering and exacerbating public health problems (World Health Organization 2017) and affecting social and interpersonal interactions (Ottoni et al. 2016).

The blue corridor is a common way for cities to use river resources. As an important component of the urban built environment, it has empirical health benefits. Since the early 19th century, western countries have tried the practice of urban blue-green space with health as the premise (CORBURN 2004) in reconnecting urban planning and public health, and its availability and composition play

---

*Corresponding Authors: binxia68@126.com and li_tonghe@126.com

DOI 10.1201/9781003372417-74

an important role in determining the quality of public space (Roman et al. 2021). Although the city is constantly reclaiming the sea, as the population of the waterfront city continues to increase, the residents' lives are increasingly in contact with the Landau, and the blue corridor will become a more important place for life and entertainment (Simon et al. 2019).

As an important part of the ecosystem, the blue corridor has brought regulation services to human beings, which is reflected in the purification of water quality and the regulation of climate (He 2022). Chase, Sen A, et al. demonstrates that the blue corridor plays a vital role in purifying water quality, both as a recipient of untreated sewage and as a provider of biological sewage treatment (Chase 2015; Sen 2020). Anna W demonstrates the important role of the blue corridor in ecological regulation and cultural services (Anna et al. 2021). Lin Y explores the cooling effect and efficiency of the blue corridor in the Pearl River Delta metropolitan area (PRD), proving that the blue corridor is beneficial for mitigating the urban heat island effect and improving urban livability (Lin et al. 2020). In addition, the combination of the blue corridor and green space brings co-benefits to the urban ecosystem (Bockarjova et al. 2020; Kershaw et al. 2017).

The blue corridor is important in promoting health (Völker et al. 2015). In terms of physical health, Volker S found that the blue corridor is a possible factor in reducing the temperature (Volker et al. 2013), which is beneficial in reducing the morbidity and mortality of temperature-related diseases (World Health Organization 2013). Braubach M et al. demonstrate that Landau provides citizens with opportunities for physical exercise and recreational activities (Braubach et al. 2017), motivating physical activity, especially participating in land-based outdoor activities such as walking (Pasanen et al. 2019). Vert C confirms the alleviating effect of the blue corridor on chronic diseases such as cardiovascular (Vert et al. 2020). Regarding mental health, Lee A C confirms that Landau can improve emotional state and relieve stress, anxiety, and depression (Lee et al. 2015). Nieuwenhuijsen MJ uses remote sensing and smartphone technology to explore the potential mechanisms between blue roads and health (including stress reduction, recovery, physical activity, and social interaction) (Nieuwenhuijsen et al. 2014). Gascon M is the first to systematically review quantitative evidence on the benefits of outdoor blue spaces on human health and wellbeing (Gascon et al. 2017). Dzhambov AM studies the mediating mechanism and variables between home blue-green space and anxiety and depressive symptoms. The results show that blue road is good for mental health (Dzhambov et al. 2018). In terms of social adaptability, Rostami R confirms that Randall promotes social interaction and the creation of collective memories of coexistence with nature (Rostami et al. 2015). ASHBULLBY K J finds that a blue corridor is an important place for active socializing with friends and family (ASHBULLBY et al. 2013). Siân B confirms that the blue corridor is more conducive to social interaction in older people (Siân et al. 2017). In addition, long-term evidence shows that residents near or appreciatively the blue corridor have higher life satisfaction than those living inland (Wheeler et al. 2012).

Although some attention has been paid to the ecological, social, and economic impacts of waterfront regeneration (Jones 1998; Sairinen 2006), the research on its potential impact on public health and wellbeing only began almost ten years ago (Völker et al. 2011), and there are not many studies on the construction of specific site environmental quality. In early 2020, the ravages of the new crown epidemic highlighted the need to study healthy environments. Therefore, this research aims to strengthen the connection between the urban built environment and public health and improve the planning and design of the blue corridor. Taking the Stalin Park in Harbin as an example, it proposes strategies for organizing the spatial nodes of the blue corridor through research to further call for the concept of a healthy city and guide the active lifestyle of the public.

## 2 CONNOTATION OF THE BLUE CORRIDOR

### 2.1 *Definition and application*

The concept of the blue corridor originated from the practice of Simmonds' revival of the central riverfront design project, combining the design of the water system and green space, so that they can be connected in different dimensions and open spaces to obtain a global water development system that connects points, lines, and planes. Since then, the urban blue corridor has gradually developed into a linear landscape extending from the edges of urban surface water bodies and shorelines such as rivers, rivers, floodplains, wetlands, and lakes. Based on the ecology of the water system, it integrates surrounding natural resources, human landscapes, and other spatial elements to form an ecological restoration. The waterfront open space with recreational and ornamental functions has the potential to promote human health and wellbeing.

According to the spatial scale, the blue corridor space can be divided into large water bodies such as seas, bay shorelines, etc.; inland watersheds such as rivers, rivers, canals, streams, etc.; inland static water bodies such as lakes, reservoirs, wetlands, etc.; small water bodies and water systems such as pools, swimming pools, fountains, etc.

According to the function orientation, the blue corridor space can be divided into four types: natural resource conservation, leisure and viewing, historical and cultural landscape, shipping, transportation, and logistics.

1. Natural resource conservation type: the ecological elements of the water system are relied on for ecological restoration and conservation of water resources and the surrounding natural resources through river bank safety protection, water source purification, flood regulation, storage, etc.
2. Recreation and viewing type featured pleasant scenery, rich water landscape, and entertainment and leisure facilities. People's behavior characteristics are mainly leisure and recreational activities, such as blue corridor spaces, urban riverbank parks, lakes, creeks and scenic spots, seaside resorts, etc.
3. Historical and cultural landscape type: the blue corridor space passes through the historical and cultural protection area. In cities with a long history and profound cultural heritage, rivers and waters are often used together with historically protected buildings and cultural relics as important protection resources to witness the development of the city and the lives of the residents of the past dynasties, such as moats, water systems in Suzhou gardens, etc.
4. Shipping and transportation logistics: mainly for shipping and transportation functions, such as the Beijing-Hangzhou Grand Canal, the Yangtze River, the Huaihe River, and the navigable sea areas.

## 2.2 Environmental characteristics

Arriaza, Wang, Zhao, et al. proposed that the blue corridor is composed of a natural ecological environment, social and cultural environment, and artificial material environment (Arriaza et al. 2004; Wang et al. 2016; Zhao et al. 2013), and the natural ecological environment includes water bodies, river banks, waterfront greening, and river breeze; the social and cultural environment includes cultural taste, leisure and entertainment, and management mechanisms; the artificial material environment includes waterfront land, road traffic, Building space and environmental facilities.

Combining the literature to refine the composition of the blue corridor space, interview feedback, and on-site observation, four types of environmental elements are summarized, as shown in Table 1.

## 3 THE ACTION MECHANISM OF URBAN BLUE CORRIDOR ON AFFECTING PUBLIC HEALTH

### 3.1 Blue corridor's influence on physiological health

The blue corridor is rich in negative oxygen ions. According to the standards formulated by the World Health Organization, fresh air is the air with the content of negative oxygen ions higher than 1000 to 1 500/cm$^3$. In blue corridors with water, negative oxygen ions can reach 100,000 to 1

Table 1. Elements of the blue corridor.

| Elements | Content | Space elements |
| --- | --- | --- |
| Landscape environment | Natural landscape | River channel scale, river channel shape, river water quality Green space scale, green layout, plant configuration, landscape seasonality |
| | Urban built environment | Riverside walk, riverside plaza, riverside park, riverside pavilion Leisure cafes, outdoor sports venues |
| Regional humanities | Historical context | Memorial square, cultural corridor, landscape sketches |
| | Neighborhood identity | Celebration organization, creative space |
| Water-city interaction | Riverfront land use | Social activity venues, physical activity venues |
| | Transportation network | Pedestrian path environment |
| | Public space system | Natural landscape recreation space |
| Activity groups | Activity venues needs | Sporty, cultural |
| | Public facilities needs | Rest, aesthetic |

million/cm$^3$, indicating rich content (Garrett et al. 2019). Negative oxygen ions directly act on the human central and blood circulation systems, soothing fatigue and promoting metabolism.

The blue corridor is easy to form a comfortable and pleasant microclimate. A comfortable and pleasant microclimate benefits human physiological health, mainly reflected in temperature and humidity. The ratio of water is higher, and the climate is cooling and warming at night due to the heat circulation. Due to the evaporation of water, a blue corridor with large humidity and small temperature difference helps improve the human body's heat dissipation and metabolism, physical fitness, and disease resistance.

### 3.2 *The blue corridor's influence on mental health*

Reduce psychological pressure. Standing nearby the water body can maintain a healthy spirit; the higher the water surface, the less psychological pressure (Mustafić et al. 2012). After walking in the blue corridor, happiness and emotions improve immediately, and the activity of the sympathetic nervous system increases; that is, a suitable walk in the blue corridor benefits psychological health (WHO 1997).

Soothing mental fatigue. The blue corridor presents the clear water body and green vegetation, which can make people feel peaceful. In addition, its environmental characteristics and constituent elements also affect its psychology through various sensory organs of the human body. Natural sounds can eliminate tension and regulate psychological states by creating blue road health landscapes, soothing mental stress and changing emotions, and promoting public health.

Bring a sense of natural regression. Human beings are hydrophilic, and 65% to 70% of the human body consists of water, and approaching water can give people psychological satisfaction. The shallow-level hydrophilic behavior is to perceive water, such as walking and leisure on the blue road; deep-level hydrophilic behavior is exposed to water, such as water, water swimming, and rafting, thereby meeting people's emotional needs.

### 3.3 *The impact of the blue corridor on social adaptability*

Bring a sense of social belonging. The blue corridor provides places and social opportunities for individuals and groups to formally or informal leisure activities to enhance the interaction of community neighborhoods. It is an important way to establish a sense of belonging and local identity in the community. The blue-green space has attracted the crowd, promoted people's familiarity with the same work or living environment, and guided them to participate in more outdoor gathering activities.

## 4 FIELD RESEARCH RESULTS AND ANALYSIS OF THE BLUE CORRIDOR IN STALIN PARK

Stalin Park is located on the south bank of the Songhua River, with a total length of 1750 meters, and covers an area of about $105000m^2$. It is across the river from the well-known Sun Island. It is a blue corridor space representing the Songhua River for the city of Harbin with both tourism and service functions. It comprises lawns, flower beds, trees, sculptures, classical toy buildings in Russia, Roman Corridor, seats, garden lights, guards, and Harbin people's flood prevention commemorative towers. The number of shrubs is large, the lawn is large with rich color, and the shape of the forest crown line is ups and downs, which is the desirable space of an urban blue corridor (see Figure 1). This article considers Stalin Park as the research scope. The author mainly investigates the satisfaction of tourists' satisfactory environmental elements and proposes venue problems in combination with on-site surveys in the form of questionnaires.

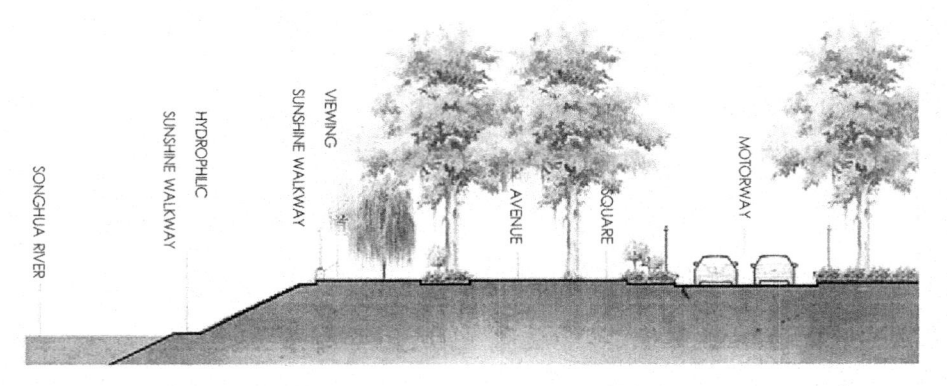

Figure 1. Stalin Park's section.

### 4.1 Survey results of satisfaction for environmental elements of the blue corridor

A total of 165 valid questionnaires are obtained from the survey, of which 78.18% of the respondents are 18–25 years old, 12.12% are 26–45 years old, and 9.7% are from other age groups. Among the respondents, college students in Harbin account for 68.48%, citizens (non-college students) accounted for 18.18%, and tourists accounted for 13.33%.

Combined with the specific site characteristics of Stalin Park, the elements in Table 1 are screened and optimized, and the respondents' satisfaction with each indicator of Stalin Park's landscape environment, regional culture, water-city interaction, and activity groups is evaluated in a neutral way of evaluation. The indexing measurement adopts the Likert 5-level scale. The larger the value, the higher the satisfaction. The satisfaction survey results are shown in Figure 2. Regarding the elements of the landscape environment, the respondents are satisfied with the water quality of the river and the scale of greening but are less satisfied with the layout of greening and plant configuration. The respondents are satisfied with the landscape structure and festival organizations regarding the regional and cultural elements. Regarding the interactive elements of the water city, the respondents are relatively satisfied with the riverside trails, riverside plazas, and social places but are less satisfied with the surrounding chronic road network and are the least satisfied with the sports venues. Regarding the activity group element, respondents are more satisfied with lighting and safety facilities and less satisfied with public seating.

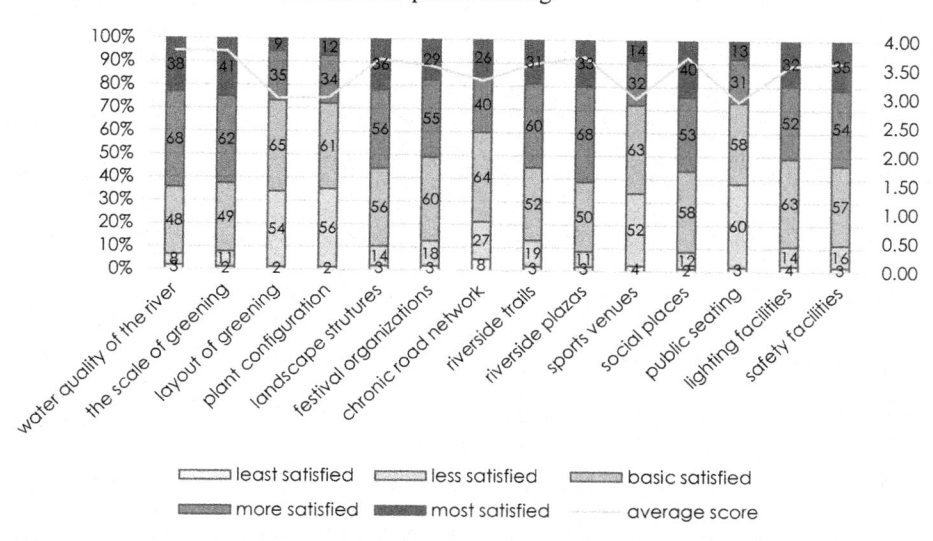

Figure 2.  Survey satisfaction results for environmental elements of the blue corridor in Stalin Park.

## 4.2  *On-site investigation*

Statistically sort out the facilities of Stalin Park through on-site and previous investigation.

Seats: about 600 seats; the benches in the park are mainly wooden backrest seats (for four people), a total of 120, that is, 480 seats; 20 double stone benches, 40 seats; 7 sets of square tables and chairs with 28 seats; 3 sets of tree pond seats, 18 seats; 20 sets of umbrellas, tables, and stools for cold drink booth, 80 seats. These facilities can meet people's use in normal times, but they cannot be satisfied during the peak crowd in the evening or during holidays.

Trash cans: There are 60 trash cans evenly distributed throughout the park. The quantity distribution is reasonable, and the park is in good sanitary condition, but the trash cans are of ordinary style and have no special features.

Lighting: The park is well lit at night. There is a high-pole street light every 30m. low-pole lighting and lawn lights are evenly scattered, and there are special spotlights for flowers, trees, and sculptures. The lights in the park are bright enough, and the layers are slightly monotonous, which can meet various needs in the park.

Fitness equipment: There are 100 sets of fitness equipment and four sets of table tennis tables in the whole park; the stone pavement, running through the east and west, provides a good venue for roller skating enthusiasts; several small squares under the forest provide venues for dancing and boxing.

Children's entertainment venues and facilities: There is one dedicated venue with eight sets of children's entertainment equipment such as slides. The venue is small, the facilities are not complete, it cannot meet the needs of children, and it is extremely crowded in the evening or on holidays.

Handwashing pools: There are eight handwashing pools in the park that are clean and convenient and add some fun.

## 4.3  *Site problems*

Based on the survey results of the respondents' environmental satisfaction with Stalin Park and the on-site investigation, the following site problems are obtained.

The landscape layout is not comprehensive. Lack of private space, the environment of the entire park is relatively homogeneous, and there is a lack of interesting spaces such as centralized sunshine lawns; there are insufficient structures with roofs, the sun is severely exposed in summer, and the places for tourists to rest and play in winter lack thermal facilities.

The plant configuration is not perfect. Plants are mainly composed of trees and shrubs, with few grass and flowers. The seasonal landscape design is insufficient, and the scenery in autumn and winter is bleak. The scale is unreasonable, and the landscape on the side of the waterfront trail is mostly small flower beds, which have a low recovery effect.

Insufficient sports grounds. The space for children's activities is insufficient, and the programs and facilities for young people's activities are extremely lacking.

Insufficient public seats. There are fewer centralized backrest seats, which cannot meet the needs of citizens during the peak crowd in the evening or during holidays.

## 5  STRATEGIES

This study proposes strategies for specific site design and spatial node organization based on the survey results to create a healing environment and improve the quality of the environment. For the "healing environment," Kaplan S believes that four conditions need to be met: being away, attraction, extent, and compatibility (Kaplan 1995). The urban blue corridor has good unity in these four levels; "In theory, people like places that provide shelter and ensure safety, so urban blue corridor also needs to consider safety.

### 5.1  *Healing environment design strategy of the blue corridor*

#### 5.1.1  *Design of being away*
Watercolor and water quality. Natural and harmonious colors make people secrete substances that are beneficial to health, making people full of energy and happy. The water color changes of the blue corridor are mainly affected by the external environment color, light intensity, water depth, and underwater substances. The design should mainly consider the color matching and coordination of the plants around the water body to create a comfortable living environment and space far away from the noisy city. In addition, the water quality also directly reflects the overall quality of the blue corridor. The later maintenance and management of the water body should be strengthened, and the ecological purification effect of plants should be considered simultaneously.

Landscape signage facilities. Text, graphics, symbols, and other identification facilities should be combined with functions and images, and the scale, shape, color, and material selection should be unified. While providing clear guidance for tourists, it should also offer services related to material and spiritual needs, reduce artificial traces, and set off Natural wildness.

#### 5.1.2  *Design of fascination*
Water sound. The most attractive sound among all kinds of natural sounds is the sound of water, so the blue corridor can add a waterscape design to the auxiliary space other than natural water and consider the setting of living water, such as the fountain, spring, falling water, pipe flow, water curtain, etc. We can also set up still water according to the atmosphere of the venue so that people have a sense of relaxation in nature.

Selection of plant species. Aesthetic and health needs should be met by choosing plants with rich color changes and health benefits. The flowers, fruits, leaves, and other organs of many plants have high ornamental value. The natural and artistic beauty of the design and collocation can please the body and mind and relieve fatigue. Different colors of plants have different health care effects: red invigorates the spirit; white calms the mood; yellow makes the mood happy; blue relieves nervous tension and regulates body temperature; green is full of hope and slows down the pulse rate.

### 5.1.3 Extensible Design

Layers of plants. The space level creation is coordinated with the style of the waterscape, and the natural and beautiful forest edge and canopy lines are formed through the combination of trees, shrubs, and grass, creating spatial changes and visual extension. When many plants are planted on and around the edge of a large water area, the forest edge line and the canopy line are particularly important. The proper matching will make the overall space of the water area clear and full of rhythm. In addition, the rationality of the opposite scenery should be considered, a broad perspective should be provided, and the scenery viewed should be organized mainly by adding ecological floating islands, arranging water activities, setting dynamic and still waterscapes, and contrasting the density of plants.

Venue openness. Given the lack of existing group activities, the connection between the blue corridor and urban open space and community open space should be strengthened, as shown in Figure 3, avoiding too tortuous paths, shortening the topological distance between the small-area blue corridor, avoiding too much space at the end, and improving the openness of the space. Taking Stalin Park as an example, the direct connection should be strengthened between the Flood Control Memorial Tower and other waterfront spaces, the public transportation convenience with Zhaolin Park and Zhongyang Street, realizing the direct walking distance with the community within 1000m, and the indirect walking distance with the 1500m community.

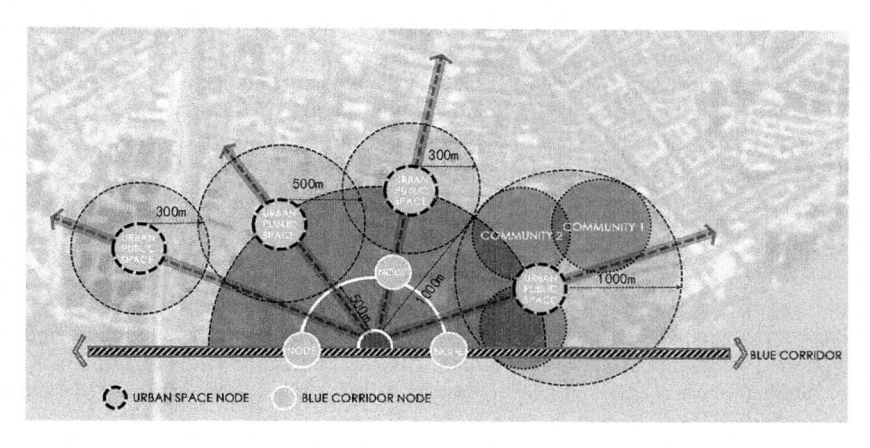

Figure 3.    Strengthening the openness.

### 5.1.4 Compatibility design

Tour facilities. The tourist facilities mainly meet people's needs for movement and sports. The hydrophilic platform, waterscape plank road, cruise ships, etc., not only improve the visual aesthetics, but also promote movement and increase the interaction between people and the water-scape. While satisfying safety, rationality, and humanization, it should coordinate with the water environment in form and play a role in enriching spatial forms and increasing levels.

Rest facilities. Rest facilities mainly meet the needs of people to watch and rest. The blue corridor needs to provide various forms of rest facilities: the main seats are the more intuitive seats in the blue corridor, and enough seats can create more humanized rest conditions for users; auxiliary seats are mostly landscape pavilions, waterfronts steps, and the outer edge of the tree pond can not only provide people with rest, but also provide people with opportunities to get close to nature as landscape sketches. In addition, horticultural therapy planters, faucets, leaf collectors, pergolas, etc., can be added to rest facilities to create places that support people's interests.

Various event spaces. In addition to chatting, playing, exercising, sitting on the lawn, sunbathing, and whispering in the activity square, people should also design some isolated and private spaces to provide visitors with a place to relax and think to relieve stress.

## 5.2 *The blue corridor's environment design strategy of safety*

Safety facilities are the "lifeline" of the blue corridor and a necessary condition for providing pressure relief functions. Their perfection is directly related to the safety of people's lives and property. The blue corridor safety facilities mainly include warning signs, warning banners, life buoys, lifeboats, safety guardrails, etc., which not only ensure functionality, but also pay attention to the era background of different classes, highlight humanistic care, and consider differences in gender, age, height, etc. In addition, more seats with backrests should be provided so that people can enjoy the view in complete relaxation.

## 6 CONCLUSION

The blue corridor is a multi-functional complex urban public space integrating natural ecology, cultural heritage, and social environment. It is not only a place for people's daily leisure and exercise, but also promotes physical and mental health in more aspects. The close relationship between the blue corridor and public health has made developed countries such as the United States, the United Kingdom, Canada, and Copenhagen begin to re-examine the development and utilization of the local blue corridor. From the perspective of rehabilitative landscape design, based on the relationship and influence mechanism of the blue corridor and public health, combined with the research on the blue corridor space in Harbin Stalin Park, this paper proposes an environmental construction strategy, which provides a reference for the blue corridor's landscape design in China and promotes the implementation of the Healthy China Strategy. However, since the research is only carried out in Harbin, there are site limitations. The subsequent research should be conducted from other sites, such as southern cities, to make the conclusions more regional.

## ACKNOWLEDGEMENT

The study was supported by the Key Entrusted Project of Teaching Reform Research Project of Heilongjiang Provincial Higher Education (No. SJGZ20200196); Heilongjiang Provincial Natural Science Foundation of China (No. LH2021E068, LH2020E052); the Opening Fund of Key Laboratory of Interactive Media Design and Equipment Service Innovation, Ministry of Culture and Tourism (No. 20201,20206).

## REFERENCES

Anna Wilczyńska, Izabela Myszka, Simon Bell, Małgorzata Slapińska, Nasime Janatian, Axel Schwerk(2021). *Exploring the spatial potential of neglected or unmanaged blue spaces in the city of Warsaw*, Poland, UrbanForestry & Urban Greening, Volume 64. 127252,ISSN 1618–8667.

Arriaza M, Cañas-Ortega J F, Cañas-Madueño J A, et al. (2004). Assessing the visual quality of rural landscapes [J]. *Landscape and Urban Planning*. 69(1): 115–25.

ASHBULLBY K J, PAHL S, WEBLEY P, et al. (2013). The Beach as a Setting for Families Health Promotion: A Qualitative Study with Parents and Children Living in Coastal Regions in Southwest England[J]. *Health Place*. 23:138–147.

Berg A E V D, Maas J, Verheij R A, et al. (2010). Green space as a buffer between stressful life events and health[J]. *Social Science & Medicine*. 70(8):1203~1210.

Bockarjova M, Botzen W, Koetse M J(2020). Economic valuation of green and blue nature in cities: A meta-analysis[J]. *Ecological Economics*. 169.

Braubach, M., Egorov, A., Mudu, P., Wolf, T., Thompson, C.W., Martuzzi, M(2017). *Effects of urban green space on environmental health, equity, and resilience*. In Nature-Based Solutions to Climate Change Adaptation in Urban Areas; Springer: Berlin, Germany. pp. 187–205.

Chase, Dagmar(2015). Reflections about blue ecosystem services in cities[J]. *Sustainability of Water Quality & Ecology*. 5:77–83.

Corburn J(2004). Confronting the challenges in reconnecting urban planning and public health[J]. *American journal of public health*. 94(4): 541–546.

Dzhambov AM(2018). *Residential Green and Blue Space Associated with Better Mental Health: A Pilot Follow-up Study in University Students*[J].Arhiv Za Higijenu Rada I Toksikologiju-Archives of Industrial Hygiene and Toxicology. 69(4):340–349.

Garrett Joanne K, White Mathew P, Huang Junjie, Ng Simpson, Hui Zero, Leung Colette, Tse Lap Ah, Fung Franklin, Elliott Lewis R, Depledge Michael H, Wong Marti C S(2019). Urban blue space and health and wellbeing in Hong Kong: Results from a survey of older adults.[J]. *Health & place*. 55.

Gascon M, Zijlema W, Vert C, et al (2017). Outdoor Blue Space, Human Health and Well-being: A Systematic Review of Quantitative Studies [J].*International Journal of Hygiene and Environmental Health*. 220(8):1207–1221.

He Tianjiao, He Qixiao, Tan Shaohua (2022). Research on Planning Paths for Urban Blue Spaces to Promote Elderly Health[J]. *South Architecture*. (5):5-4-63.

Jones A (1998). Issues in Waterfront Regeneration: More Sobering Thoughts: A UK Perspective[J]. *Plan Prac Res*. 13: 433–442.

Kaplan S (1995). The restorative benefits of nature: Toward an integrative framework[J]. *Journal of environmental psychology*.15(3): 169–182.

Kershaw, T, Wells, et al. (2017). Utilizing green and blue space to mitigate urban heat island intensity[J]. *Science of the Total Environment*.

Lee, A.C.; Jordan, H.C.; Horsley, J (2015). Value of urban green spaces in promoting healthy living and wellbeing: Prospects for planning. Risk Manag. *Healthc. Policy*. 8, 131–137.

Lin Y, Wang Z, Chi Y J, et al (2020). Water as an urban heat sink: Blue infrastructure alleviates urban heat island effect in mega-city agglomeration[J]. *Journal of Cleaner Production*. 121411.

Mustafiæ, H., Jabre, P., Caussin, C., Murad, M.H., Escolano, S., Tafflet, M., Perier, M.C., Marijon, E., Vernerey, D., Empana, J.P., Jouven, X (2012). Main air pollutants and myocardial infarction a systematic review and meta-analysis. *J. Am. Med. Assoc*. 307, 713–721.

Nieuwenhuijsen MJ, Kruize H, Gidlow C, et al. (2014). Positive Health Effects of the Natural Outdoor Environment in Typical Populations in Different Regions in Europe(PHENOTYPE): A Study Programme Protocol[J]. *Bmj Open*. 4(4).

Ottoni C A, Sims-Gould J, Winters M, et al. (2016). "Benches become like porches": Built and social environment influences older adults' experiences of mobility and wellbeing [J]. *Social Science & Medicine*. 169:33–41.

Pasanen T P, White M P, Wheeler B W, et al. (2019). Neighbourhood blue space, health, and wellbeing: The mediating role of different types of physical activity[J]. *Environment international*. 1313: 105016.

Patel V, Saxena S, Lund C, et al. (2018). The Lancet Commission on global mental health and sustainable development[J]. *The Lancet*. 392(10157):1553–1598.

Roman Fedorov (2021). Zooming in on Arctic urban nature: green and blue space in Nadym, Siberia[J]. *Environmental Research Letters*. 16(7):075009 (10pp).

Rostami, R., Lamit, H., Khoshnava, S.M., Rostami, R., Rosley, M.S. (2015). Sustainable cities and the contribution of historical urban green spaces: A case study of historical Persian gardens. *Sustainability*. 7, 13290–13316.

Sairinen R, Kumpulainen S (2006). Assessing Social Impacts in Urban Waterfront Regeneration [J]. *Environmental Impact Assessment Review*. 26(1): 120–135.

Sen A, H Nagendra (2020). Local community engagement, environmental placemaking, and stewardship by migrants: A case study of lake conservation in Bengaluru, India [J]. *Landscape and Urban Planning*. 204:103933.

Siân de Bella, Grahamb H, Jarvisb S, et al. (2017). The importance of nature in mediating social and psychological benefits associated with visits to freshwater blue space[J]. *Landscape & Urban Planning*. 167: 118–127.

Simon Bell, Chen Yiyan, Chen Zheng (2019). Health and Well-being Aspects of Urban Blue Space: The New Urban Landscape Research Field[J]. *Landscape Architecture*. 26(9): 119–131 (In Chinese).

United Nations (2015). World Urbanization Prospects: The 2014 Revision. New York. p. 493.

Vert C, Gascon M, Ranzani O, et al (2020). Physical and Mental Health Effects of Repeated Short Walks in a Blue Space Environment: A Randomised Crossover Study[J]. *Environmental Research*. 188.

Völker S, Kistemann T (2011). The Impact of Blue Space on Human Health and Wellbeing-salutogenic Health Effects of Inland Surface Waters: A Review[J]. *Int J Hyg Environ Health*. 214: 449–460.

Volker S, Baumeister H, Classen T, et al (2013). Evidence for the Temperaturemitigating Capacity of Urban Blue Space-A Health Geographic Perspective[J]. *Erdkunde*. 67(4):355–371.

Völker, S.; Kistemann, T (2015). Developing the urban blue: Comparative health responses to blue and green urban open spaces in Germany. *Health Place*. 35, 196–205.

Wang R, Zhao J, Liu Z (2016). Consensus in visual preferences: The effects of aesthetic quality and landscape types [J]. *Urban Forestry & Urban Greening*. 20(2):10–7.

Wheeler B W, White M, Stahl-Timmins W, et al (2012). Does Living by the Coast Improve Health and Wellbeing?[J]. *Health Place*. 18: 1198–201.

WHO Region Office for Europe (1997). *Twenty Steps for Developing a Healthy Cities Project*. Spon Press. 15~20 3~6.

World Health Organization (2013). Health 2020. *A European Policy Framework and Strategy for the 21st Century*[R]. Copenhagen: WHO.

World Health Organization (2017). *Depression and other common mental health disorders: global health estimates*[R]. Geneva: World Health Organization.

Zhao J, Luo P, Wang R, et al. (2013). Correlations between aesthetic preferences of river and landscape characters [J]. *Journal of Environmental Engineering and Landscape Management*. 21(2): 123–32.

*Civil Engineering and Urban Research – Mohamed & Hou (Eds)*
*© 2023 the Authors, ISBN 978-1-032-44487-1*

# Design of fire detection and early warning system for minority nationality buildings in Southwest China

Mingxuan Li* & Xiujuan Mei*
*Sichuan Fire Research Institute, Ministry of Emergency Management, Sichuan, China*

ABSTRACT: Given the frequent occurrence of fires in minority nationality villages in recent years, the current situation of fire prevention in minority nationality villages in southwest China is clarified through field investigations. According to the current situation, fire detection and early warning system design is proposed, which can detect early fires as soon as possible and issue a rapid early warning. Hopefully, this work can improve fire detection and early warning in minority nationality villages.

## 1 INTRODUCTION

Because most of the minority nationality villages in southwest China have high cultural relic value and commercial tourism value and are non-renewable, local governments and fire departments attach great importance to their fire safety (Tian 2016). In those areas, the fire separation distance is small, and many houses are close. Once a fire breaks out in one building, it will likely spread to others (Xu 2019). The dire situation of fire prevention poses a challenge to the fire department. Therefore, for this type of civil building, prevention should be the priority, and early detection and early warning should be given by technical measures, which can put out fires in the early phase and dramatically reduce property loss and mortality (Cao 2014). To make this happen, a field visit and investigation of minority nationality villages in southwest China were conducted to accurately figure out the current situation of fire prevention in minority nationality villages, as well as the housing layout and how the living habits affect the fire detection (Xiong 2021). Early fire detection and warnings are supposed to improve fire prevention and protection in minority nationality villages.

## 2 FIRE SITUATION IN MINORITY NATIONALITY VILLAGES

### 2.1 Fire resistance rating

Affected and restricted by natural conditions, most people in those minority nationality villages were relatively poor. On top of that, the wood was easy to process and obtain. Therefore, most of the walls and beams use wood as the major building material, as shown in Figure 1. So the building fire-resistance rating is low, and the fire protection capability is poor.

### 2.2 Fire load

Usually, minority nationality villages are located in remote rural areas, and their buildings are supposed to meet the needs of daily life and working, not only have the functions of living, leisure,

---

*Corresponding Authors: limingxuan@scfri.cn and 2993350228@qq.com

Figure 1. Dong and Miao nationality buildings.

and workplace, but also serve as a place storing agricultural tools, food, and raising livestock (Yuan 2021), as shown in Figure 2.

Figure 2. Farming tools and food storage.

## 2.3 *House spacing and density*

Historically, to gather strength and defend against the enemy, most of the residents of minority nationality villages live in groups, and the distance between a large number of houses was very narrow, and some different buildings may have no distance at all, as shown in Figure 3. Concentrated residential areas, such as the "Thousand Household Miao Village" in Guizhou province, have a large population, and if a fire breaks out, it will endanger the whole village.

## 2.4 *Buildings on slopes*

Because of the topography and geomorphology of the southwest region of China, especially the minority nationality areas are mostly located in mountainous areas, the buildings in those villages are built on hills, so the buildings are at different heights, as shown in Figure 4. In case of fire, it is easier to spread to other buildings than the building on flat ground.

Figure 3.  Fire separation distance.

Figure 4.  Buildings on mountains.

### 2.5  *Misuse of fire and electricity*

Due to the economic situation and the natural environment, most people in minority nationality villages cook food with firewood, which very likely causes fires, as shown in Figure 5. At the same time, electrical wires are installed without any requirements and can cause short circuits very easily after those wires are exposed to air, which are the factors that cause electrical fires. According to statistics, more than 56% of the fires were caused by electricity in minority nationality villages in Guizhou.

To sum up, minority nationality villages have problems such as low fire resistance level, large fire load, narrow spacing between houses, high density, sloping buildings mostly built in remote mountainous areas, and misuse of fire and electricity. In case of fire, the fire department will take a long time to reach the building, and the fire is not easy to control. It must be dealt with quickly before the fire grows out of control.

Figure 5.    Electricity wires and cooking.

## 3   THE DESIGN OF FIRE DETECTION AND EARLY WARNING SYSTEM

To design fire detection and early warning system suitable for minority nationality villages, the technology roadmap is taken as follows:

The first step is to conduct field research on fire protection situations in minority nationality villages in southwest China and analyze the causes of fire in those areas, fire characteristics, and the impact of geographical environment characteristics on fires. The second step is to research IoT technology, artificial intelligence technology, fire detection, and early warning technology. Based on the fire and geographical environment characteristics of minority nationality villages in China, choose the fire detection and early warning technologies suitable for minority nationality villages. The third step is to research how to make detection technology, IoT, and AI technology work together and verify the solution. Based on the fire detection and early warning technology that integrates the Internet of Things and artificial intelligence, the fire detection, and early warning equipment are built to realize real-time dynamic collection and processing of fire information, remote fire control, and early warning.

### 3.1   *Choosing fire detectors*

The fire detector is the most crucial component of the fire detection and early warning system. It is related to whether it can quickly and accurately detect the fire and give an alarm. It is necessary to fully consider the actual situation of the detection target scene and make a wise selection. Because the residents of minority nationality need to cook food by burning wood in the building, there is usually a certain amount of smoke in the kitchen. Therefore, it is not suitable to use smoke detectors in the kitchen; otherwise, it is easy to trigger false alarms, and it is more suitable to use heat detectors. In addition, the heat detector cannot be installed directly above the stove because the temperature directly above the stove is very high when people are cooking, and it is easy to give the heat alarm. During the field investigation, it was found that many fire detectors were installed close to the stove. After frequent false alarms, the resident removed the detector without any other choice and completely lost the fire detection and alarm function. Except for the kitchen, other rooms can use smoke detectors. At present, the relevant institutes have developed a type of flame and heat detector, which can not only sense the far-infrared light generated by the flame but also alarm by sensing the temperature, mainly by coating the transparent plastic shell of the detector with a layer of Photosensitive coating, this kind of photosensitive material will change its light transmittance at a higher temperature. There is an optical receiving module inside the detector,

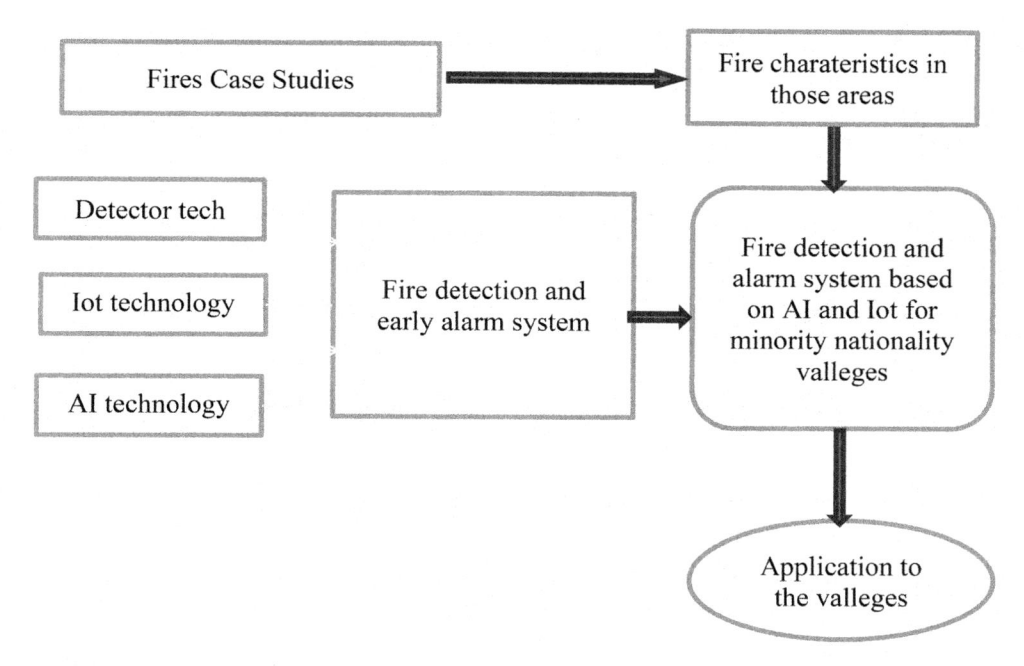

Figure 6.    Technology roadmap.

which can sense optical changes and can quickly convert temperature signals into optical signals, and small temperature changes can also cause its optical Signal response, its alarm response time is faster than the ordinary detector, and can be used as a kitchen fire detector in ethnic villages. At the same time, because the wires are connected carelessly, and aging is serious, a short-circuit extinguishing device should be installed within a certain range to prevent electrical fires. Due to the level of economic development, there is no mandatory requirement for minority nationality villages to install fire detection systems according to the codes and regulations. At the same time, because residents do not know the importance of the fire alarms system if fire detection alarms are wired, which is not convenient, residents will not agree to install them. During the field investigation, it was found that most fire detectors are independent and not connected to the Internet, so there is no fire alarm function. The fire alarm is limited to single-family residents, and it cannot give early warning to other residents. Their limitations are obvious. After a fire occurs, the fire warning system sends an alarm signal to the fire department and the nearby residents through the Internet to evacuate as soon as possible.

### 3.2   Network of the system

Because the independent fire detector is not connected to the Internet, in case of fire, it will only inform the residents through the sound and light signals, but other people outside of the building cannot receive the alarm signals and cannot have early warning. Therefore, The detector needs to be connected to the Internet. There are currently two options connecting with the Internet for fire detectors. One is the NB-IoT solution. However, the signal is weak in countryside areas. Besides that, this method requires regular payment, so residents do not prefer this solution. After field investigation, it was found that the existing NB method of installing fire detectors, after the first 1 or 2 years, no one would pay for maintenance later on, and the detectors don't work any longer, which means this solution is not fit in these places. The other solution is the adoption of Lora and 4G/5G communication service. The advantage of this solution is that the sensing layer equipment

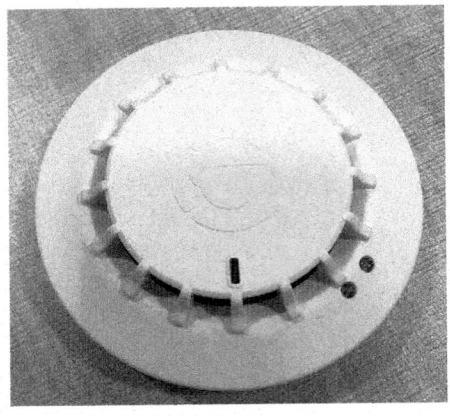

(A)Smoke detector           (B)Heat detector

Figure 7.   Wireless fire detector.

does not need to pay communication service fees. Once installed, it can be used for a long time. The signal is connected to the wireless gateway through Lora, and the smart gateway is connected to the Internet through 4G/5G wireless communication service, then connected to the cloud platform, and then linked to the control terminal. The fire department can get an early warning through SMS, app, WeChat, telephone, etc., and can send control instructions to the fire fighting equipment through the cloud platform and wireless gateway, as shown in Figure 8. This Lora wireless transmission distance is generally not more than 15km. It is necessary to use an intelligent wireless gateway to connect the fire detectors. At the same time, safety redundancy should be considered. The fire detector should be able to connect with multiple wireless gateways. Suppose there is a problem with a wireless gateway. In that case, the fire detector can be linked with other nearby wireless gateways to solve the problem that the detector cannot be connected to the Internet and improve the system's reliability.

### 3.3   *System structure*

The fire detection and early warning platform were based on IoT and artificial intelligence, which are divided into four layers according to the network structure: sensing layer, communication layer, platform layer, and application layer, as shown in Table 1.

Table 1.   System structure.

| Layers | Functions |
| --- | --- |
| Sensing layer | Wireless smoke detector, heat detector, and short-circuit detector |
| Communication layer | Comunication with Lora and 4G |
| Server layer | Computing and data storage |
| Application layer: | Fire information displayed and sent by Apps, Wechat, SMS |

1) Sensing layer: It is the bottom layer of the fire detection and early warning platform. It collects information through various sensors, including smoke detectors, heat detectors, etc.
2) Communication layer: Through the communication protocol, the data collected by the sensing layer is transmitted to the platform, and the data sent by the platform is forwarded. Lora transmitted information from sensing to the gateway, and information transmission from the

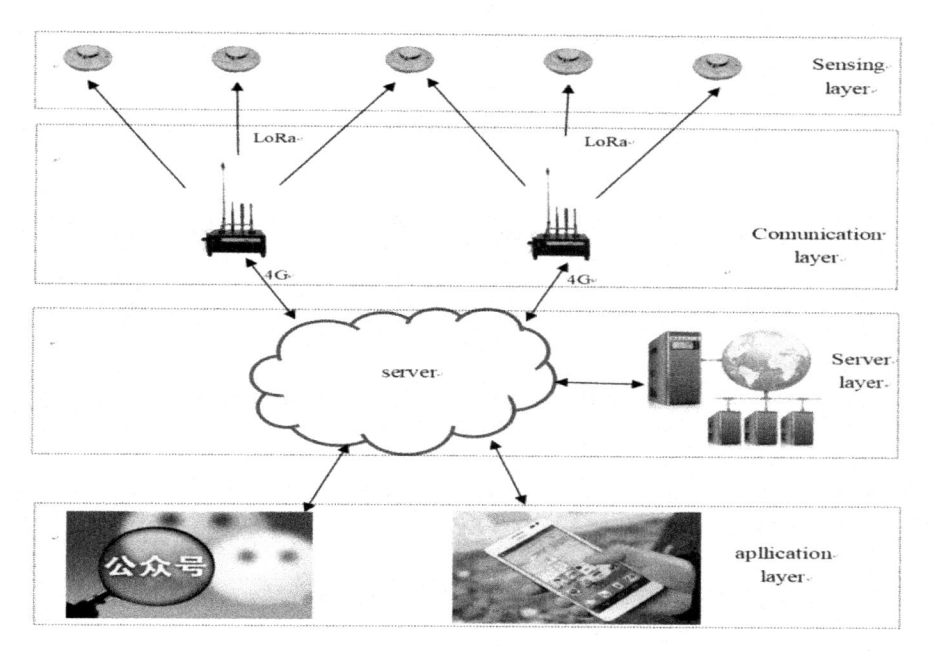

Figure 8. Layers of the system.

gateway to the server adopts 4G. The gateway at the network layer has the function of computing, which can make a preliminary judgment on the alarm information.

3) Server layer: with cloud computing and storage as the core, the data collected by the sensor is stored and processed. The platform layer is the core of the system.

4) Application layer: It is the top layer of the system, directly facing the application and needs of users. Its functions include user information management, fire protection equipment management, alarm reception and distribution, and communication with third-party platforms.

### 3.4 *Functions of the system*

The fire detection and early warning system need to detect the fire quickly and accurately and distribute the fire incident information to the residents in time through SMS, phone calls, apps, WeChat, and so on. Therefore, residents can evacuate quickly and simultaneously send the alarm information to the fire department. Then, the fire department receives the fire alarm information and dispatches firefighters to the building after confirming whether it is a false alarm through remote video or in person. To prevent the fire from further spreading, and at the same time, the proper fire devices and facilities can be turned on through a remote control system, such as activating sound and light alarms and other fire fighting equipment.

## 4 CONCLUSIONS

Field investigation on the minority nationality villages in southwest China was conducted, and it is clear that the villages in these areas have low fire resistance levels, large fire loads, narrow fire separation distance, likely misuse of firewood and electricity, and buildings on slopes. Combined with the living and working habits of residents, the requirements of fire detectors and the design of fire alarm and early warning systems were proposed, which can be an alternative design guide for the design of fire detection, alarm, and early warning systems in minority nationality villages in China, which can issue a rapid-fire alarm and early warning, and reduce mortality and property losses.

## ACKNOWLEDGMENTS

The authors gratefully acknowledge the financial support of the National Key Research and Development program under Grant No. 2020YFD1100702-2.

## REFERENCES

Cao Shuangyou. (2014) Construction of spring fire barrier in southwestern Guizhou. *J. China Fire Protection.* 6, 9–11.

Tian Cong, Zhang Weihua, Wang Wenqing. (2016). Analysis of fire in contiguous wooden structure villages and discussion on prevention and control measures. *J. Fire Science and Technology.* 4, 576–578.

Tian Cong, Zhang Weihua, Wang Wenqing. (2016). Research on early warning, prevention, and control system of fire in contiguous wooden structure villages. *J. Science and Technology Bulletin.* 2, 209–212.

Xiong Feng, Wu Xiao, Liu Jinfeng, et al. (2021) Research concept and achievement prospect of comprehensive disaster prevention technology for southwest ethnic villages. *J. Engineering Science and Technology.* 54, 13–22.

Xu Mengyi, Zhao Jiayi. (2019). Fire Prevention Countermeasures for Minority Diaojiao Buildings. *J. New Materials and New Decoration.* 1, 125–126.

Yuan Shasha, Han Tengben, Yan Feng, et al. (2021) Investigation and Research on Fire Loads in Southwest Ethnic Villages. *J. Engineering Quality.* 30, 23–27.

*Civil Engineering and Urban Research – Mohamed & Hou (Eds)*
*© 2023 the Authors, ISBN 978-1-032-44487-1*

# Research on green architecture in China based on sustainable development theory

Yan Yu*

*University of Glasgow, Glasgow, Scotland, UK*

ABSTRACT: In this study, the existing theoretical framework of sustainable construction development is combined with the actual situation of China's construction industry to establish a new energy system for construction. Based on the existing research, qualitative and quantitative methods are used to study the current situation of urban green building projects and the possibility of sustainable development of future buildings. It is found that China's construction industry consumes a lot of energy. Although drip and sprinkler irrigation have good applications in agricultural irrigation, the shortage of water resources limits the development of water energy in the application of new energy. Solar energy and wind energy can be converted into electric energy or thermal energy through corresponding conversion equipment, which can be well applied in buildings. This research suggests that with the rapid development of China's construction industry, the renewable clean energy system in construction must be adopted to ensure the development of the construction industry based on not damaging the environment.

## 1 INTRODUCTION

China has a large population, relatively insufficient resources, and weak environmental carrying capacity. With the rapid development of the economy, population growth, and industrialization, the contradiction between the supply and demand of resources and environmental pressure is growing. At present, our country is in the stage of rapid development, no matter the level of economy or science and technology, there is a significant improvement, especially when the rise of the construction industry is obvious, which also marks the development of the construction industry in our country is constantly moving forward. However, in the development of the construction industry, there are many problems with energy consumption, many of which are constantly wasted. According to the data obtained by Lai X et al. (2019), from 2008 to 2014, the annual building area is increased from 2.24 billion square meters to 42.3 billion square meters, and the energy consumption increased from 3.825 million tons to 72.601 million tons. According to the National Bureau of Statistics (2011), carbon dioxide emissions have also increased yearly. How to realize the sustainable development of architecture has become one of the hot issues in modern architectural design. The main purpose of this research is to establish a renewable energy system in line with the sustainable development of China's construction industry. Under the existing theoretical framework of new energy application and sustainable development, this research aims to answer the following three research questions:

1. What are the current problems facing China's construction industry?
2. How to improve the efficiency of urban water use?
3. How to build a convenient renewable energy system that integrates wind, solar, and water energy?

*Corresponding Author: ericyu1996@163.com

This research will combine the actual situation of our country, discuss the sustainable development of our country's construction industry, and discuss the hope to increase the feasibility of resource utilization. It is hoped that this research can help the government to develop renewable energy in areas with insufficient resources, reduce water use pressure, and improve resource utilization rates.

## 2 LITERATURE REVIEW

### 2.1 *Development status of China's construction industry*

Sustainable development is often discussed as a hot topic, while the construction industry is often raised as an industry that relies heavily on limited resources. China is a big country in the construction industry. According to Lai, Lu, and Liu (2019), from 1990 to 2004, the development of the construction industry reflects a high dependence on fossil fuel energy. From 2008 to 2014, the annual construction area increased by 40.06 billion square meters. According to the National Bureau of Statistics (2015), energy consumption is undoubtedly high. With the acceleration of China's urbanization process, the energy consumption and carbon dioxide emissions of the construction industry have increased dramatically (Wang & Feng 2018). A large amount of energy consumption will undoubtedly cause great environmental pollution. Wang and Feng (2018) believed that saving energy and reducing pollutant emissions in the Chinese construction industry is significant to China's sustainable development. How to save energy and reduce environmental pollution and promote the sustainable development of China's construction industry will be part of this research.

### 2.2 *Sustainable development in construction engineering*

With the growth of urbanization in various countries, energy consumption and environmental problems are becoming more and more prominent. There is much discussion around how to achieve sustainable development of cities. Transformation of regions and energy is one of the most commonly used methods in the conversion process of sustainable urban development. Cattaneo et al. (2016) noted that sustainable urban development could be achieved through regional transformation. Paszkowski and Golebiewski (2017) indicated that environmental quality could be improved by reshaping cities and creating green spaces that can be integrated into urban structures. On the contrary, according to Pollice (2016), the transformation of degraded green space into an urban park will periodically expose the architectural beauty of the urban center in the process of urban transformation, and the areas involved in urban reconstruction will be accompanied by modernization, and the contrast between the old and new areas will lead to inequality in all aspects of society. In view of the comparison between the old and new areas proposed by Pollice (2016), Cattaneo et al. (2016) pointed out that regional transformation should expand the communication boundary, promote regional cooperation, and establish corresponding work, business, and living environments. In addition to the regional transformation, Paszkowski and Golebiewski (2017) also referred to the noise generated by the existing infrastructure, and pollution can be transformed into a component of the green urban foundation. Although the transformation of the old city and the transformation of existing energy sources have reduced energy consumption and environmental pollution to a certain extent, the impact on the sustainable development of the city takes a long time and sometimes even destroys the original resource. One of the focuses of this research is how to influence the sustainable development of cities fundamentally.

### 2.3 *The application of water, wind, and solar energy*

Urban planning is the prerequisite for sustainable urban development, and energy planning in urban planning is even more important. According to Cattaneo et al. (2016), urban planning is the premise of urban sustainable development, and urban planning must become a tool that can reduce or limit land and energy consumption and promote the common development of all social resources. Renewable resources can fundamentally solve the problem of energy consumption. Paszkowski and Golebiewski (2017) pointed out that energy consumption can be reduced by developing and applying renewable energy systems, including solar energy, wind energy, water energy, etc. According to Mathew (2007), wind energy is the fastest-growing energy source and one of the energy sources with commercial feasibility and economic competitiveness. Tian and Qin (2007) pointed out that wind energy is often used to generate electricity and improve indoor air quality in China, and its disadvantages are too large equipment and greatly affected by air volume. In addition to wind energy, solar energy is also often used. Tian and Qin (2007) and Mackay (2015) pointed out that solar energy is used for heating and power generation, and Mackay (2015) also pointed out that solar energy is divided into high-quality energy and low-quality energy. Although the application of wind energy and solar energy has matured in today's society, it is mostly used in rural areas because of its large size and inconvenient equipment.

## 3 METHODOLOGY

This research is based on past theory and uses qualitative and quantitative methods to study the current situation and future development possibilities of urban green building projects. The data used in this study comes from journals, books, websites, and academic papers, which are integrated and analyzed to provide new conclusions for the current new energy-building market. The resources currently used are mainly related to the problems faced by the Chinese construction market in recent years, the analysis of new energy, and the exploration of the application of new energy in buildings. This method is useful for this report because it provides a lot of data on new energy applications and their advantages and disadvantages.

## 4 RESEARCH FINDINGS AND DISCUSSION

### 4.1 *The current problems of China's construction industry*

The construction industry is one of China's pillar industries (Xue et al., 2015). In 2014, China's construction industry realized a total output value of 447.9 billion yuan, an increase of 9.76% over the previous year (Lai et al. 2019). The original data related to the construction industry of this research is selected from the relevant data of the China Statistical Yearbook for the past nine years (2011–2018). Total energy consumption, total coal consumption, coke consumption, gasoline consumption, etc., are directly derived from statistical yearbooks. The carbon emissions can be calculated by multiplying each energy consumption (standard coal) and the corresponding carbon emission coefficient, as shown in Eq. 1.

$$C = \sum Ei * \lambda i * \alpha i \tag{1}$$

Where
C denotes carbon emissions
Ei denotes energy consumption
$\lambda i$ denotes the carbon emission coefficient of i energy
$\alpha i$ denotes the standard coal Conversion coefficient of i energy
Data and calculation results are shown in Table 1.

According to Wang and Feng (2018), with the acceleration of China's urbanization process, the energy consumption and carbon dioxide emissions of the construction industry have increased dramatically. The scatter diagram of carbon emissions from 2009–2017 is shown in Figure 1. It can be seen that the carbon emissions gradually increase with the year, and the increase in carbon emissions also indicates that energy consumption is increasing year by year, and the total construction area is also expanding.

Table 1. Energy consumption, carbon emission from 2009–2017 (10K tons of standard coal)

| Year | 2009 | 2010 | 2011 | 2012 | 2013 | 2014 | 2015 | 2016 | 2017 |
|---|---|---|---|---|---|---|---|---|---|
| Total energy consumption | 4562 | 6226.3 | 5872.2 | 6167.4 | 7017 | 7519.6 | 7696.4 | 7990.9 | 8554.5 |
| Coal consumption | 635.59 | 718.91 | 781.81 | 753.41 | 811.39 | 913.6 | 878.07 | 805.29 | 732.82 |
| Coke consumption | 5.68 | 5.81 | 4.81 | 6.31 | 7.69 | 9.69 | 6.68 | 7.05 | 12.57 |
| Carbon emission | 3965.5 | 4563.2 | 4817.2 | 4768.6 | 5361.2 | 5430.4 | 5773.9 | 5805.2 | 5988.7 |

Remark: above data are collected from the China Statistical Yearbook (2011–2018), translated by the author.

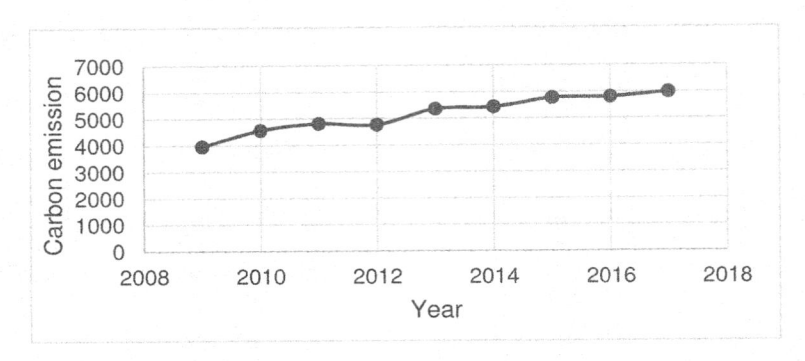

Figure 1. Carbon emission trend from 2009–2017.

As the literature review and data show, the problem China's construction industry currently faces is excessive energy consumption. Houses and residences are people's great needs. Although the increase in the total construction area marks the rapid development of China's construction industry, the current development of the construction industry relies heavily on nonrenewable and high-carbon emissions resources such as coal. So finding clean and renewable resources is imminent for China's construction industry.

### 4.2 The sustainable development of construction engineering

The city is made up of various buildings. To seek the sustainable development of architecture is to seek the sustainable development of the city. As Paszkowskiand and Golebiewski (2017) and Cattaneo et al. (2016) said, There are five ways to achieve sustainable urban development:

(1) Create a new "ecological space" in the city center.

Reconstructing urban facilities that have been contaminated or have divided the urban space to form new green buildings that are conducive to sustainable urban development.

(2) Reduce energy consumption in urban space.

Minimize the energy consumption of buildings through construction and maintenance, develop and apply systems based on renewable energy, and optimize urban development by introducing natural resources.

(3) Optimize urban development.

In the existing cities, use the city's resource advantages to build new cities to maximize the use of urban resources.

(4) Rational urban planning

Rational urban planning is the premise of urban sustainable development.

(5) Renewable energy system

The local renewable energy shall be used according to the local situation to realize the sustainable development of the region.

Paszkowskiand and Golebiewski (2017), Pollice (2016), and Cattaneo et al. (2016) pointed out that renewable energy is important for the construction industry, and construction projects should adopt renewable resources according to the actual situation.

According to the actual situation in China, this research takes hydropower, solar energy, and wind energy as examples to explore the possibility of renewable resources in China.

(1) Water energy

According to Jiang (2009), China has been facing a serious water shortage, especially in North China. Take the water supply in North China in 2006 as an example, as shown in Table 2. In 2006, 63.3% of the water supply in North China came from surface water and 35.7% from underground water, while renewable water resources only accounted for 36.9% of the surface water and occupied 36.3% of the underground water.

Table 2.   China's water supply in 2006 and renewable water resources, adopted from Jiang (2009).

| | Water supply (%), $10^9$ $m^3$ | | | Water resource use rate,% | | |
|---|---|---|---|---|---|---|
| Region | Surface | Aquifer | Total | Surface | Aquifer | Total |
| North | 166.4(63.3) | 92.6(35.7) | 259.0(100) | 36.9 | 36.3 | 48.3 |
| Northwest | 52.7(84.9) | 9.4(15.1) | 62.1(100) | 45.3 | 10.8 | 47.6 |
| South | 304.3(95.6) | 14.0(4.4) | 318.3(100) | 13.5 | 2.4 | 14 |
| Southeast | 31.5(96.6) | 1.1(3.4) | 32.6(100) | 12.3 | 1.8 | 12.6 |
| Southwest | 9.9(96.5) | 0.4(3.5) | 10.2(100) | 1.7 | 0.2 | 1.7 |

According to the current situation of water resources in China, it is very important to use water efficiently. Agriculture is a sector that is often criticized for excessive water use (Serra-Wittling et al. 2019), so in agricultural irrigation, water-saving measures are already being used.

Umair et al. (2019) pointed out that the production and irrigation efficiency of agricultural water resources can be improved by an effective irrigation system such as drip irrigation and sprinkler irrigation. In addition, according to Salvador et al. (2011) and Serra-Wittling et al. (2019), several irrigation methods (surface sprinkler irrigation and drip irrigation) were compared and found that drip irrigation was more efficient than surface irrigation, and also pointed out that drip irrigation can reduce water loss and increase grain yield.

On the contrary, Zhang et al. (2018) believe that sprinkler irrigation is more suitable for the agricultural field, with strong adaptability and simple technology, which can effectively solve the problem of blockage of drip irrigation equipment.

In general, drip irrigation is undoubtedly the best way to save water in areas where water resources are scarce. For urban buildings, a method similar to drip irrigation can be used. For example, drip irrigation can be connected to the domestic water system in terms of landscaping or three-dimensional greening of the home. The used water can be filtered again and used as landscaping water through the drip irrigation system.

(2) Solar energy

Solar energy is the most extensive resource in people's life. According to Mackay (2015), the sun produces amazing energy daily. One hour is enough to meet the energy needs of everyone in the world. He also pointed out that solar technology is still a newly developed technology, and the biggest challenge of solar technology is how to use efficient and low-cost equipment to obtain energy.

According to Mackay (2015), the principle of solar energy is to convert the high heat of solar energy into low heat through the conversion equipment (Carnot cycle system), then store the power for solar cells, and finally convert it into electric energy for home use, as shown in Figure 2.

According to Tian and Qin (2007), solar energy utilization can be divided into passive solar energy technology, active solar energy technology, and solar photovoltaic technology. No matter

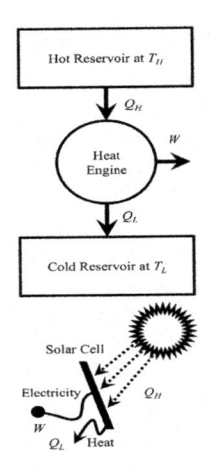

Figure 2.  Adopted from Mackay (2015)

what system is adopted, there must be five essential equipment, as shown in Figure 3, including collector, absorber, thermal mass, distribution, and control system.

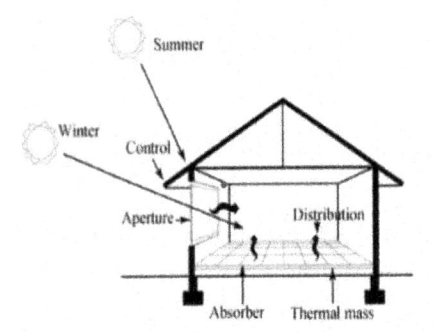

Figure 3.  Adopted from Tian and Qin (2007).

To sum up, the current solar energy technology is still in the early stage of development. Its basic working principle is that the collector will collect solar energy. Then, the absorber will convert the solar energy into heat energy and store it by thermal mass. Finally, the energy will be transmitted to the power distribution and control system.

(3) Wind energy

According to Tian and Qin (2007), wind energy began as early as 5000 BC, where wind power was used to fetch water, grind corn, etc., and many devices are still in use. However, this equipment is mostly used in villages without electricity. It is rare in cities because wind power equipment is too large, has large noise, and large batteries are needed. After solving the above problems, wind power can be widely used in urban buildings.

According to Mathew (2007), the basis of wind energy conversion is the energy available in the wind, which is the kinetic energy of the wind. The basic principle of wind power generation is that the blades of a wind turbine receive kinetic energy, which is then converted into mechanical or electrical energy depending on the end use. The conversion efficiency of wind energy depends mainly on the efficiency of the turbine rotor interacting with the wind current, that is, the speed of the wind turbine.

In summary, wind energy has been used for a long time, whether it is simple water intake measures or complex power supply facilities. Most wind power is used in rural areas because of its large facilities, noise, and the need for large batteries. If wind energy is used for power generation in urban construction, the problems that need to be overcome most are complex and oversized equipment.

## 5  CONCLUSION

In general, China's urban development is facing huge energy consumption. In order to achieve sustainable urban development, we can start from three aspects: old city renovation, energy efficiency improvement, and new energy applications. The corresponding planning or renovation methods should be adopted according to the actual situation of the planning area. In order to be more beneficial to the development of China's construction industry, this research will propose a new type of green energy building system that integrates wind, water, and solar energy. In the early stage of architectural design, renewable resources or local resources from the origin are considered as building materials; in the process of construction, waste gas generated by energy consumption is strictly controlled; in the later stage of construction, solar energy is considered as power supply facilities and heating facilities, and drip irrigation equipment is connected with domestic water system to realize wastewater reuse to save water resources, and micro wind turbines assist the solar system in providing power. Renewable resource systems can be used not only in buildings but also in many areas, such as agriculture and landscaping.

## REFERENCES

Cattaneo, T., Giorgi, E., Ni, M., and *Manzoni*, G.D. (2016). Sustainable development of rural areas in the EU and China: a common strategy for architectural design, research practice, and decision-making. *Buildings, 6*(4), 42.

Jiang, Y. (2009). China's water scarcity. *Environmental Management, 90,* 3185–3196.

Lai, X., Lu, C., and Liu, J. (2019). A synthesized factor analysis on energy consumption, economic growth, and carbon emission of the construction industry in china. *Environmental Science and Pollution Research, 26,* 13896–13905.

Mackay, M. (2015). *Solar Energy: An Introduction.* London: Oxford University Press. Available at: https://www-oxfordscholarship-com.ezproxy.lib.gla.ac.uk/view/10.1093/acprof:oso/9780199652105.001.0001/acprof-9780199652105 [Accessed June 5, 2022].

Mathew, S. (2007). *Wind energy: fundamentals, resource analysis, and economics.* Heidelberg: Springer-Verlag.

National Bureau of Statistics (2011). *China Statistical Yearbook.* Beijing: China Statistics Press. Available at: http://www.stats.gov.cn/tjsj/ndsj/2011/i ndexch.htm[Accessed 01 June 2022].

National Bureau of Statistics (2012). *China Statistical Yearbook.* Beijing: China Statistics Press. Available at: http://www.stats.gov.cn/tjsj/ndsj/2012/i ndexch.htm [Accessed 01 June 2022].

National Bureau of Statistics (2013). *China Statistical Yearbook.* Beijing: China Statistics Press. Available at: http://www.stats.gov.cn/tjsj/ndsj/2013/i ndexch.htm [Accessed 01 June 2022].

National Bureau of Statistics (2014). *China Statistical Yearbook.* Beijing: China Statistics Press. Available at: http://www.stats.gov.cn/tjsj/ndsj/2014/i ndexch.htm [Accessed 01 June 2022].

National Bureau of Statistics (2015). *China Statistical Yearbook.* Beijing: China Statistics Press. Available at: http://www.stats.gov.cn/tjsj/ndsj/2015/i ndexch.htm [Accessed 01 June 2022].

National Bureau of Statistics (2016). *China Statistical Yearbook.* Beijing: China Statistics Press. Available at: http://www.stats.gov.cn/tjsj/ndsj/2015/i ndexch.htm [Accessed 01 June 2022].

National Bureau of Statistics (2017). *China Statistical Yearbook.* Beijing: China Statistics Press. Available at: http://www.stats.gov.cn/tjsj/ndsj/2015/i ndexch.htm [Accessed 01 June 2022].

National Bureau of Statistics (2018). *China Statistical Yearbook.* Beijing: China Statistics Press. Available at: http://www.stats.gov.cn/tjsj/ndsj/2015/i ndexch.htm [Accessed 01 June 2022].

Paszkowski, Z. W., and Golebiewski, J. I. (2017). The renewable energy city within the city. the climate change-oriented urban design – Szczecin green island. *Energy Procedia,* 115, 423–430.

Pollice, F. (2016). Urban planning and architectural design for sustainable development. *Procedia – Social and Behavioural Sciences, 216*, 6–8.

Salvador, R., Martínez-Cob, A., Cavero, J., Playán, E. (2011). Seasonal on-farm irrigation performance in the EBRO basin (Spain): Crops and irrigation systems. Agric. *Water Manag*, 98, 577–587.

Serra-Wittling, C., Molle, B., and Cheviron, B. (2019). Plot level assessment of irrigation water savings due to the shift from sprinkler to localized irrigation systems or the use of soil hydric status probes. Application in the French context. *Agricultural Water Management, 223*, 105682–105694.

Sun, C., Zhao, L., Zou, W., and Zheng, D. (2014). Water resource utilization efficiency and spatial spillover effects in china. *Journal of Geographical Sciences*, 24(5), 771–788.

Tian, L. and Qin, Y. (2007). Utilization of renewable energy in architectural design, *Frontiers of Architecture and Civil Engineering in China*, 1(1), pp. 114–122.

Umair, M., Hussain, T., Jiang, H., Ahmad, A., Yao, J., Qi, J., Zhang, Y., Min, L, and Shen, Y. (2019). Water-Saving Potential of Subsurface Drip Irrigation For Winter Wheat. *Sustainability*, 11(10), 2978 Available at: https://doi.org/10.3390/su11102978 [Accessed 13 December 2019].

Wang, M., and Feng, C. (2018). Exploring the driving forces of energy-related $CO_2$ emissions in China's construction industry by utilizing production-theoretical decomposition analysis. *Cleaner Production, 202*, 710–719.

Xun, X., Wu, H., Zhang, X., Dai, J., and Su, C. (2015). Measuring energy consumption efficiency of the construction industry: the case of China. *Journal of Cleaner Production*, 107(2015), 509–515.

Zhang, G., Xie, C., Lai, H., and Li, X. (2018). Buried Lifting Sprinkling Irrigation Device. *Irrigation and Drainage Engineering*, 144(1).

Civil Engineering and Urban Research – Mohamed & Hou (Eds)
© 2023 the Authors, ISBN 978-1-032-44487-1

# Research on influencing factors and strategies of historical and cultural districts renewal—Take Yangzhou Nanhexia historic district for an example

Xuyuan Zhang*
*Anhui Communications Vocational and Technical College, Anhui, China*

Jianming Su*
*School of Architecture and Art, Hefei University of Technology, Hefei, China*

ABSTRACT: The protection and renewal of historical and cultural blocks, serving as critical resources for urban construction, is a professional and complex social system project. It is an insurmountable step to realizing cities' modernization and sustainable development. Given the present condition of historical and cultural blocks, this paper analyzed the influencing factors during the update through professional investigation and evaluation. Besides the evaluation index system of influencing factors being constructed as per the setting principle and causal analysis method, the analytic hierarchy process (AHP) software is adopted to calculate the weight of the evaluation index. The renewal strategies thus formulated will provide a reference for developing comprehensive and systematic overall protection and organic renewal planning of the historic district.

## 1 INTRODUCTION

### 1.1 Research background

The preservation of historical and cultural heritage is critical in the development of civilization, and the cultural heritage in the scope of historic districts is a non-renewable multi-dimensional cultural resource. As a concentrated unit of the traditional urban landscape and an organic component of the region, the historic heritage contained in the district has a key root of memory and long-term value in urban development. In the conservation practice, due to some lack of understanding, inadequate methods, and imperfect regulations, historic and cultural districts often turn into the areas with the most prominent development conflicts and relatively concentrated influencing factors.

As a national-level historical and cultural city, Yangzhou City flourished several times during the Han Dynasty, Sui and Tang Dynasties, and Kangxi-Qianlong Period of the Qing Dynasty. There are four distinctive blocks in the ancient city, among which Nanhexia Historical and cultural District, as a cultural protection unit in the old city, is a block with the most concentrated historic buildings, relatively complete block style, and well-preserved spatial pattern, which is included in the list of the first batch of China historical and cultural blocks. The total planned land area of the conservation area is 42.03ha, and the planning scope is shown in Figure 1. Nanhexia Historical and cultural District is designated as the key protection area within the district (Figure 1), covering an area of 22.35ha. The main social function of this district is residence and tourism, well retaining the original traditional street style.

In 2006, Yangzhou City won the United Nations Best Habitat Award, but the development process of urban construction in the old city has also highlighted various urban system problems, including

---

*Corresponding Authors: xuyuansundae@126.com and jianmingsu0923@163.com

but not limited to unsatisfactory traffic conditions with imperfect road and parking system, failure of infrastructure and landscape to meet the requirements of residents' living circle, the contradiction between modern city construction and traditional street style preservation, and symbiosis and coexistence of culture and commerce. All these problems hinder the development of Yangzhou's living environment and create many unfavorable factors for the protection of Yangzhou's historic and cultural districts and the continuation of the traditional context.

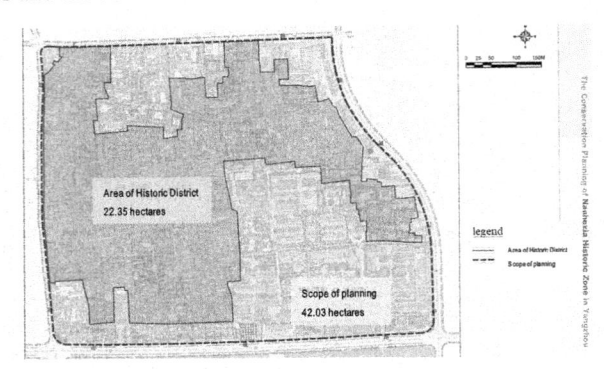

Figure 1.    Scope of the Nanhexia historic district.

## 1.2  *Previous studies*

This paper combines field research on the historical and cultural district of Nanhexia, synthesizes the factors affecting the renewal of the historical and cultural district from previous studies, categorizes each factor according to its relevance using hierarchical analysis based on in-depth visits and data collection combined with the actual use of guidelines, and uses YAAHP software to analyze the data correlation between the obtained program factors and the influencing factors of the renewal of the historical and cultural district.

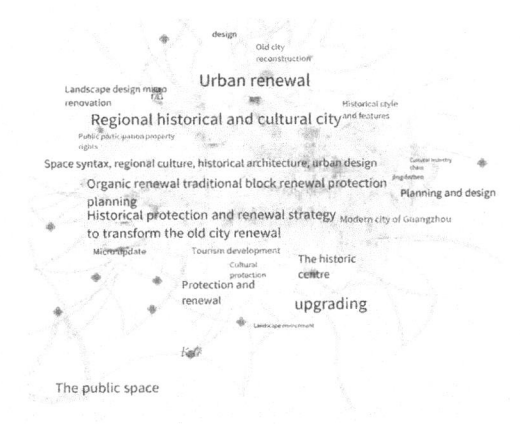

Figure 2.    Keyword mapping for documentary analysis of historical and cultural districts.

## 2   ANALYSIS OF FACTORS INFLUENCING THE RENEWAL OF HISTORICAL AND CULTURAL DISTRICTS

### 2.1  *Determination of influencing factors*

After conducting previous studies, this paper outlines five first-criterion-level factors and 20 second-criterion-level factors from the factors that may affect the renewal of historical and cultural districts. Although these are not enough to cover all the problems in each stage of the renewal of historical and cultural districts, they are the problems that appear more frequently and play a more important role in the previous studies of experts and scholars and are somewhat representative. By classifying the criterion layers, the hierarchical division of the analysis model of influencing factors for the renewal of historical and cultural districts is shown in Table 1.

Table 1. Hierarchical comparison of factors influencing the renewal of historical and cultural districts.

| Objectives | Criterion Layer (I) | Criterion Layer (II) |
| --- | --- | --- |
| Influencing Factors of Renewal of Historical and cultural Districts | Living Conditions | Residential Buildings |
| | | Open Space |
| | | Infrastructure |
| | | Landscape Environment |
| | Commercial Business | Local Characteristics |
| | | Commercialized Blocks |
| | | Tourism Development |
| | | Spatial Productivity |
| | District Style | District Skyline |
| | | Color Tone |
| | | Material Texture |
| | | Opposite Scenery |
| | Location Context | Traditional Architecture |
| | | Location Spirit |
| | | History Context |
| | | Heritage Authenticity |
| | Traffic Status | Traffic Location |
| | | Fire Safety |
| | | Traffic Construction |
| | | Transportation Needs |

## 2.2 Public satisfaction survey

Based on extensive literature analysis, the author visited the Nanhexia Historical and cultural District to conduct research and communicate with some of the residents of the district to understand the current development status of the district in terms of housing, commerce, transportation and culture, and to have a basic understanding of the district's appearance through field visits, which was analyzed and summarized based on interviews. The current development status of Nanhexia's historical and cultural district has these basic characteristics.

*Living Conditions*: The basic structure of the residences in the Nanhexia District is old, and it is not easy to improve the living conditions while preserving the original style and appearance; the construction of infrastructure and public service facilities still needs further implementation of planning; there are practical living problems for residents such as drying and garbage disposal; the residents are mainly elderly-led family units.

*District Style*: Many traditional buildings remain in the Nanhexia District, including many precious historical and cultural heritage, such as the residence of the salt merchant surnamed Liao, the residence of the salt merchant surnamed Zhou, the residence of the salt merchant surnamed Lu, the gatehouse of Lingnan Hall, the gatehouse of Hunan Hall, Erfenmingyue Pavilion, Xiaopangu Garden, Si'an Gongsuo, Ho Family Garden, etc. Some significant buildings have been destroyed in the years, and the remains are now mostly for tourism or private property. The district style is mainly composed of traditional Ming and Qing architecture, and the authenticity of the heritage in the historic district is high.

*Location Context*: The residents are mainly elderly-led family units with harmonious neighborhood relations, and the long-time inheritance of the residential environment has formed a place quality for the residents to integrate with the neighborhood culture, and the architectural function here is more of a centripetal spiritual value. However, the combination of traditional residential houses and salt merchant culture is not reflected more intuitively and deeply, and the cultural basis of district renewal needs to be put into further practice.

## 3 EVALUATION OF FACTORS INFLUENCING THE RENEWAL OF HISTORICAL AND CULTURAL DISTRICTS

### 3.1 *The application of AHP hierarchical analysis to calculate index weights*

In this paper, the subjective evaluation of the importance of influencing factors includes the subjective evaluation of non-measurable and observable experiences, and the importance of influencing factors is assigned by distributing expert questionnaires to relevant students and researchers. A total of 30 questionnaires were distributed (of which 26 were returned and 25 were valid), and the ranking weights of the influencing factors were derived as shown in Tables 2 and 3.

Table 2. Ranking weights of elements in the 1st criterion layer for objectives.

| Elements of the Criterion Layer (I) | Weights |
| --- | --- |
| **Location Context** | **0.4093** |
| **District Style** | **0.2652** |
| **Living Conditions** | **0.1693** |
| Commercial Business | 0.1007 |
| Traffic Status | 0.0555 |

Table 3. Ranking weights of elements in the 2nd criterion layer for objectives.

| Elements of the Criterion Layer (II) | Weights |
| --- | --- |
| **Location Spirit** | **0.1976** |
| **Material Texture** | **0.1300** |
| **Historic Context** | **0.1054** |
| **Residential Buildings** | **0.0911** |
| Opposite Scenery | 0.0613 |
| Heritage Authenticity | 0.0602 |
| Traditional Architecture | 0.0462 |
| Infrastructure | 0.0434 |
| District Skyline | 0.0433 |
| Local Characteristics | 0.0397 |
| Spatial Productivity | 0.0379 |
| Color Tone | 0.0306 |
| Traffic Construction | 0.0247 |
| Landscape Environment | 0.0209 |
| Tourism Development | 0.0157 |
| Fire Safety | 0.0143 |
| Open Space | 0.0139 |
| Transportation Needs | 0.0105 |
| Commercial Blocks | 0.0074 |
| Traffic Location | 0.0060 |

# 4 RESEARCH ON PRACTICAL RENEWAL STRATEGIES

After combining the case studies and conducting a comprehensive analysis of the data, this paper comes up with program-level indicators in five areas: planning and design, street vibrancy, spatial scale, regulations, and capital investment. The evaluation yields the weighting relationships of the five indicators, as shown in Table 4, and the synergy is carried out so that the practical strategies for the renewal of historical and cultural districts can be analyzed.

Table 4. Ranking weights of elements in the program level for objectives.

| Program-level Elements (Renewal Strategies) | Weights |
| --- | --- |
| **Planning and Design** | **0.3256** |
| **Street Vibrancy** | **0.2224** |
| Spatial Scale | 0.1743 |
| Regulations | 0.1712 |
| Capital Investment | 0.1066 |

## 4.1 Planning and design

The design of regional regeneration should focus on the principle of being people-oriented. Firstly, it is important to consider the cultural practices of groups and individuals of cultural producers, as well as various historical practices, including political, economic, and social practices; secondly, it is needed to fully understand all the expressions and skills of culture itself, including intangible cultural heritage; thirdly, there is a necessity to dig deeper into the historical origins behind the historical and cultural remains, such as the salt merchants' culture of Nanhexia, and incorporate them into the urban design. More importantly, it is necessary to create a physical environment with certain spatial forms for people's various cultural and historical practices, including various buildings, municipal facilities, landscaping, and other aspects, so that social, economic, functional, and aesthetic requirements can be comprehensively reflected. The purpose of the design is to drive the overall renewal through local renewal, improve the overall image of the city and environmental aesthetics, boost people's quality of life, and facilitate the historical and cultural city to protect and inherit and enhance its vibrancy (Lu 2019).

## 4.2 Street vibrancy

To create the source of vibrancy, we need to be based on each node of the historical and cultural district and explore the basic elements of street vibrancy from its architecture and spatial form, such as the architecture of the district, historical relics, human awareness, and participation, etc. Social vitality is the core of vitality enhancement mainly expressed in stimulating people's passion for activities and building good interactive relationships, such as creating an open communication platform, creating diverse social activities, and forming a good community layout structure by refining the public space in Nanhexia district. Economic vitality is expressed in enhancing spatial productivity and promoting the agglomeration and transformation of economic space, including the agglomeration of material and energy flows and the agglomeration of activity behaviors, which constitute the basis and driving force of vibrancy enhancement. Cultural vibrancy can stimulate people's thinking, memory, and resonance, demonstrating the readability and uniqueness of historical and cultural districts.

## 4.3 Spatial scale

When the connectivity and functionality of the street space are guaranteed, the existing transitional spaces of Nanhexia street are to be renovated (Figure 3). These spaces have a high degree of freedom and provide places for people to communicate, pass and stop. The space of Nanhexia street is small, and there is no space to build a leisure platform; while these transitional spaces with practical functions can simultaneously show the regional culture and individual culture and can be transformed into an outdoor space suitable for modern living needs, such as the outdoor ancient well platform in the street and alley, the space between the building and the street, and the bridge space of the building (Zhang 2013).

Figure 3.    Existing transitional spaces in the Nanhexia street.

## 5   CONCLUSIONS

The overall research on the comprehensive evaluation system, methods, and strategies for the influencing factors of renewal of historical and cultural districts is carried out precisely to respond to the requirements of heritage protection and development regarding the adaptation to urbanization development, to guide the districts to cope with the unavoidable change process, and to alleviate to a certain extent the challenges brought by the urbanization process to the conservation and utilization development of historic districts. Specifically, through the comprehensive evaluation of residence, culture, commerce, style, and transportation, an effective decision-making mechanism for the renewal and development of historical and cultural districts is established, and the institutionalized, scientific and rational decision-making mechanism is realized to promote the organic renewal of historical and cultural districts, thus leading to the renewal of urban development order.

## REFERENCES

Cao C. Z. (2012). Theoretical Issues on the Protection of Historical and Cultural Cities in China. R. *Proceedings of the Conference on Urban Planning Development and Planning*, 1–16.
K. Lynch. *Urban Morphology*. (2002). M. Trans. By Lin Q. Y., et al. Beijing: Huaxia Publishing House.
Li L. Xu X. (2008). Analysis of Urban Organic Renewal Theory and Its Practical Significance. *J. Modern Garden*, 07, 25–27.
Lu J. W. et al. (2019). Historical and cultural City Preservation and Modern Urban Design. *J. China Ancient City*. 01, 4–8.
Outline of Detailed Control Plan for 11# Neighborhoods in the Old City of Yangzhou. (2004 version).
Steven Thiesdell, et al. (2006). *The Revival of Urban Historic Districts*. M. Trans. by Zhang M. Y. et al. Beijing: China Architecture & Building Press.
Zhang J. (2013). *Conservation and Renewal of the Nanhexia Street in Yangzhou*. D.05.
Zhao H. X. (2019). *Research on Historic Districts in the Context of "Authenticity" Theory: The Case of Shichahai and Nanluoguxiang in Beijing*. J. Human Geography, 02, 47–54.

# Research on intelligent operation and platform of urban large bridge and tunnel cluster project based on BIM technology

Ying Gu*
*Chongqing Zengjiayan Bridge Construction Management Co., Ltd., Chongqing, China*

Youhui Yang*, Renfu Li*, Peng Zhang*, Dayang Liu* & Xinhua Si*
*China Merchants Chongqing Highway Engineering Testing Center Co., Ltd., Chongqing, China*

ABSTRACT: In view of the problems of large quantities, high operation safety risk, and difficult maintenance in the later stage of urban large-scale bridge and tunnel cluster project, including multiple disciplines such as bridge, tunnel, road, electromechanical and slope, as well as the difficulties such as the collaborative integration of multi-business data such as daily inspection, inspection and detection, structural monitoring and maintenance, this paper expounds the design of intelligent operation and maintenance platform of urban large-scale bridge and tunnel cluster project based on BIM, including asset twin intelligent maintenance, monitoring and early warning, equipment integration and control, emergency management and control and other functions as one of the application research, so as to realize the integration of project information, digitization of project delivery and accuracy of operation and maintenance.

## 1 INTRODUCTION

In recent years, with the sustainable development of China's economy, the national investment in the field of transportation infrastructure has been further strengthened, making the economic development and infrastructure construction of western mountain cities with complex geological conditions promote each other and achieve sound development. In view of the characteristics of the geographical environment of mountainous cities, bridges and tunnels in urban infrastructure construction are not separate individuals. Bridges and tunnels are interconnected to form a complex of large bridge and tunnel clusters. At the same time, due to the different years of each bridge and tunnel, the different management and maintenance units, detection and monitoring units, and the different standards and specifications of relevant information, the bridge and tunnel during the operation period are in the problem of "one bridge, one tunnel, and one system", resulting in the long management chain, heavy tasks and repeated waste of relevant system development resources of the operation and maintenance unit; manual inspection and real-time detection operate independently. There are limitations in the state perception of bridges and tunnels, and the state evaluation is relatively one-sided; the low level of digitalization is not conducive to the intelligent, information-based, and efficient management of bridges and tunnels (Wu 2018; Yue 2015).

With the strategic goal of big data and intelligence of infrastructure put forward by the state and the mature application of new generation information technology, artificial intelligence, big data, and other high-tech in some industries, it brings opportunities for intelligent operation and maintenance of infrastructures such as bridges and tunnels. At the same time, the intelligent transportation development action plan of the Ministry of Transport (2017-2020) points out that BIM Technology

*Corresponding Authors: 10344061@cmhk.com, yangyouhui@cmhk.com, lirenfu90@cmhk.com, 10330182@cmhk.com, liudayang@cmhk.com and sixinhua@cmhk.com

is encouraged to be used in the design, construction, operation, and maintenance stages of major infrastructure such as expressways, super large bridges, and long tunnels. Among them, Zhangguizhong and others (Zhang 2019) proposed the design scheme of digital bridge management and maintenance platform based on BIM to meet the modern operation, maintenance, and management needs of long-span bridges, and defined the basic functions and physical architecture of the platform. Zou et al. (2019) integrated the real-time risk decomposition processing system of bridge structure into BIM and proposed the risk visualization and information management method based on BIM, which effectively improved the quality and efficiency of bridge management and maintenance. Gou Hongye et al. (2018) built a high-speed railway disaster big data analysis platform and developed an engineering practical intelligent evaluation system for high-speed railway bridge traffic safety. However, there is little research on intelligent operation and maintenance platforms based on BIM Technology in urban large-scale bridge and tunnel cluster projects. Therefore, the construction of an intelligent operation and maintenance platform for urban large-scale bridge and tunnel cluster projects based on BIM Technology is of great scientific and practical significance for the comprehensive acquisition of bridge and tunnel-related structural information, timely and effective early warning, ensuring the safe operation of bridge and tunnel clusters, and realizing the clustering (Zhang 2018), integration (Han 2019), and intelligent (Luo 2017) maintenance and management of bridges and tunnels.

## 2 DESIGN OF INTELLIGENT OPERATION AND MAINTENANCE PLATFORM

In this paper, the research on the intelligent operation and maintenance platform of the urban large-scale bridge and tunnel cluster project based on BIM technology mainly depends on the Chongqing Zengjiayan Jialing River Bridge Project, which is located in the center of the main urban area of Chongqing. It is composed of the first rigid suspension bridge for both public and rail in China, an overpass, and five connecting tunnels, which is a typical large-scale bridge and tunnel cluster project in mountainous cities. The engineering route corridor is narrow, the terrain is undulating obviously, the five tracks are vertical and horizontal, the highway track is built together, the bridge station (track station) is built together, and the bridge and tunnel are integrated. After it is completed and opened to traffic, the traffic flow is large, the operation safety risk is high in the later stage, and the maintenance is difficult. Therefore, combined with the actual needs of the design stage, construction stage, and operation stage of the Zengjiayan Jialing River Bridge project, this paper deeply studies the application of BIM technology in the Zengjiayan Jialing River Bridge project. Aiming at the research on the intelligent operation and maintenance platform of the urban large-scale bridge and tunnel cluster project, it focuses on the design of the platform around the goals of the breakthrough application, energy and risk reduction, durable cost reduction, and intelligent efficiency increase.

Among them, the breakthrough application focus is to break through the full life cycle BIM Technology Application and demonstration, and carry out multi-stage BIM Technology digital application, intelligent management and control, intelligent operation and maintenance and other application research; Increasing energy and reducing risk is to improve the emergency response capability of project management and reduce the operation safety risk; The purpose of durability and cost reduction is to enhance the durability and reliability of infrastructure and reduce the life-cycle cost by improving the preventive maintenance technology level of tunnel cluster projects; The intelligent efficiency increase rate refers to the free flow of engineering information in the project design, construction and operation and maintenance stages, solves the problem of "information island effect" in the management and maintenance of engineering projects, and improves the efficiency of business collaboration and treatment.

In view of the above design objectives, combined with BIM, GIS, Internet of things, and other technologies, the system architecture of intelligent operation and maintenance platform for urban large-scale bridge and tunnel cluster project based on BIM Technology proposed in this paper is shown in Figure 1.

Based on the digital and intelligent standard system, the intelligent operation and maintenance platform adopts the c/s and b/s architecture to separate the front and rear ends for development. Through the integrated application of BIM, Internet of things, GIS, big data analysis, and other technologies, and based on the multi-end collaboration of app, web, sensors, cloud services, etc., it proposes the bridge and tunnel structure disease diagnosis and intelligent decision-making technology, based on intelligent identification, GIS information The rapid positioning method of components in the 3D virtual scene based on the combination of route stake number and geographic information, the intelligent monitoring and management technology of electromechanical equipment of municipal engineering based on BIM Technology, and the life-cycle monitoring and dynamic evaluation method of bridge and tunnel structures based on BIM Technology have comprehensively improved the intelligent level of project operation and maintenance. The functions of each component of the platform are as follows:

1) Data sources mainly include the maintenance management system, the structural monitoring of bridge and tunnel system, traffic monitoring system, overload control system, human management system, and other systems.
2) Data exchange mainly includes shared exchange services and data conversion services.
3) The data center is composed of the basic database, subject database, business database, shared database, unstructured meta database, rule database, model database, etc.
4) Application support includes ad hoc query, multi-dimensional analysis, interactive charts, semantic mapping, GIS Engine, and structural early warning.
5) The function application module is composed of asset twinning, intelligent maintenance, monitoring and early warning, equipment integration, and monitoring, emergency management, etc.
6) The client includes a BIM command center, PC workbench (PC and web), and mobile platform (IOS and Android).

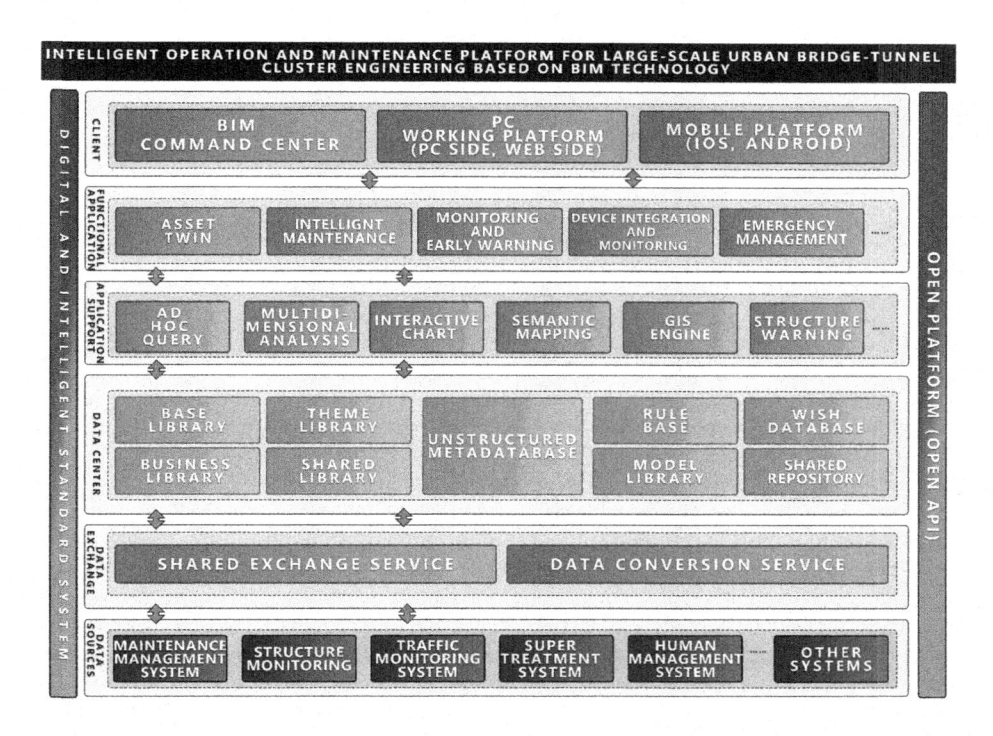

Figure 1.   The system architecture of the intelligent operation and maintenance platform based on BIM.

Based on the technical specifications for highway operation and maintenance, the platform takes the BIM model as the carrier and integrates all professional and all business data islands. It has carried out application research integrating the functions of asset twinning, intelligent maintenance, monitoring and early warning, equipment integration and control, emergency management and control, and realized the integration of project information, digitalization of project delivery, and accuracy of operation and maintenance. The function realization of the intelligent operation and maintenance platform is shown in Figure 2.

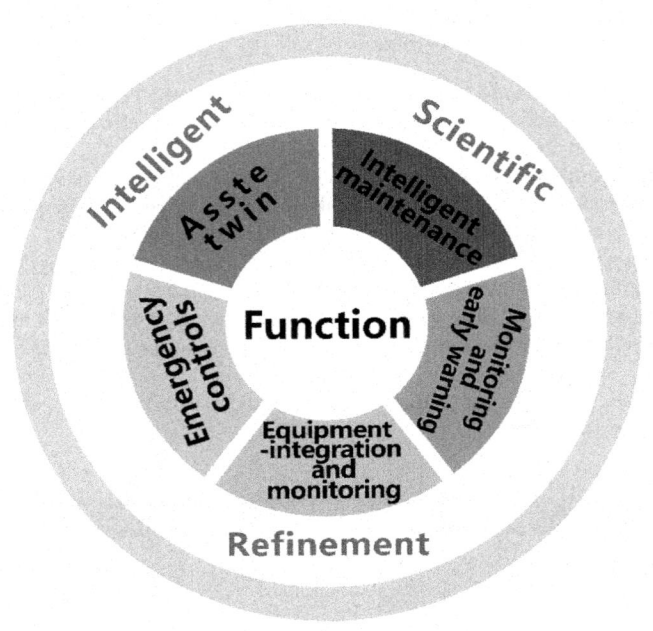

Figure 2.    Intelligent operation and maintenance platform function realization.

## 3.1 *Asset twin*

Based on the industry maintenance standards and specifications, the application research of BIM-based facility operation and maintenance is further refined. According to the coding standard for highway infrastructure maintenance, the platform establishes a unique identity (smart information tag) for each component based on two-dimensional code and radio frequency identification technology (RFID). The smart information tag is shown in Figure 3.

The intelligent identity tags of all the single components in the bridge and tunnel cluster project can be identified and exchanged through the mobile APP, which establishes the engineering twin digital assets at the component level, realizes the digitization of the project infrastructure, and lays a model foundation for the digital management of the infrastructure in the whole life cycle.

The intelligent operation and maintenance platform realizes the unified component level management of all bridges, tunnels, roads, electromechanical, and other ancillary facilities, and integrates the relevant information of the whole life cycle of structural design, construction, operation, and maintenance (including drawings, patrol inspection reports, patrol inspection records, structural monitoring data, etc.). The model of bridge and tunnel cluster also has spatial geographic information and realizes the mutual transformation between geodetic coordinates, longitude and latitude,

Figure 3.    Intelligent information label.

world coordinates, and route station, which is convenient for the rapid positioning of the structure. The digital model of project assets is shown in Figure 4.

Figure 4.    Asset digitization model diagram.

When digitizing the project infrastructure, it involves the corresponding relationship between the component or unit and the station and the map coordinates. The positioning research is carried out through the GPS joint station. This paper is intended to study the local benchmark coordinates of the project construction, obtain the corresponding relationship between the construction coordinates and the pile number (the precision of the pile number is the corresponding construction coordinates every 20m), obtain the longitude and latitude of the construction control point, establish the corresponding relationship between the construction coordinates and longitude and latitude of the whole project, and finally combine the pile number and GPS to locate specific components or units to solve the problem of correspondence between the tunnel pile number (no signal) and the map coordinates.

In order to meet the operation and maintenance needs of the Zengjiayan bridge and tunnel cluster and the requirements of BIM digitization and informatization, the principle and depth of component analysis of infrastructures such as roads, bridges and tunnels are studied, and the corresponding coding structure and system are set to realize the integration of multi-source data. The coding design rules are shown in Figure 5.

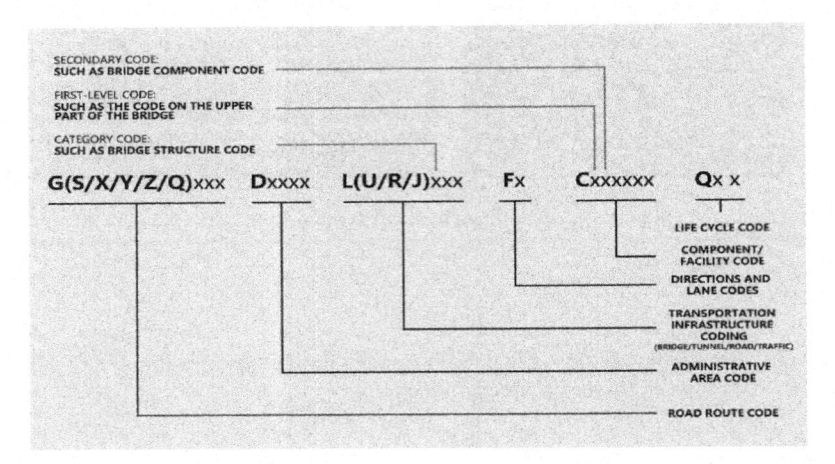

Figure 5. Coding rules.

## 3.2 *Intelligent maintenance*

The intelligent management and maintenance platform of the Zengjiayan Jialing River Bridge Project Based on BIM Technology is a data island integrating all disciplines and businesses by using the 3D spatial display ability of BIM Technology, using GIS, Internet of things, artificial intelligence, big data, and other technologies, and taking BIM model as the carrier. Based on BIM Technology, the platform realizes intelligent distribution and guidance of maintenance tasks, automatic component identification, multi-terminal real-time collaboration, intelligent diagnosis and decision analysis, component degradation law prediction, and realizes closed-loop management of maintenance operations.

1) Task intelligent distribution guidance: The maintenance task is automatically distributed according to the maintenance plan platform, and the app obtains the maintenance task and plans the path (the BIM on the app side shows the planned path).
2) Component intelligent identification: Inspectors arrive at the site, sense the intelligent information tag through the mobile app, and obtain the basic information, design and construction information, historical diseases, etc. of components; If a new disease is found, the disease type and its parameters will be automatically read through the image recognition technology, which will be checked and uploaded by the patrol inspectors to realize the visual collection of the disease (the map and disease data of the new disease can be displayed in the BIM of the mobile terminal).
3) Multi-terminal real-time collaboration: It includes the real-time collaboration and update of data at the mobile terminal, Web terminal, and BIM operation and maintenance platform. Amplification of the operation and maintenance platform can realize maintenance collection message prompt, rapid disease location, disease details view, and other functions.
4) Intelligent disease diagnosis and maintenance decision-making: You can request the disease disposal scheme with one click and automatically send it to the maintenance personnel for

disease disposal. After the disposal is completed, the mobile app can collect the disease disposal results on site and upload them.

The platform can not only visually display the whole process of the identified specific disease treatment from the micro perspective, but also visually manage the maintenance management from the macro perspective, including the full data source dynamic evaluation, roaming, and patrol inspection, report downloading, and other functions. The full data source dynamic assessment is based on the results of routine inspection, regular inspection, regular inspection, cleaning and maintenance, and structural monitoring data to comprehensively assess the service status of the structure, and display the holographic cloud picture of the technical status of the structure in the BIM model; roaming inspection is to visually distinguish the inspection type and severity of diseases by means of the fixed path and free roaming. At the same time, it is available to click the disease tag to view the information on the whole process of disease disposal (discovery, decision-making, disposal, and acceptance); the downloading of relevant reports and statements is sorted out according to the maintenance specifications, "one bridge and tunnel case", so as to realize the downloading of reports and statements such as daily patrol, regular inspection, regular inspection, and structural monitoring.

### 3.3 *Monitoring and early warning*

Based on the integrated application of "bim+ Internet of things" technology, the platform realizes the visualization of the bridge, tunnel, and slope monitoring and structural early warning, and visually views the sensor layout, monitoring data, and real-time structural early warning in the facilities. At the same time, the minute-level real-time data collection is synchronized to the BIM model, and the real state of facility operation is controlled through the combination of dynamic real-time monitoring + patrol maintenance.

The data used for monitoring and early warning come from sensors widely distributed in bridge and tunnel cluster projects. The data collected by the sensor is transmitted to the cloud in real-time. The cloud server preprocesses the massive data and synchronously associates the data with the BIM model. Users can master the operation status of the bridge and tunnel cluster online; At the same time, the security status of the structure operation is automatically pre-warning and evaluated, and the visual display is carried out on the BIM model in real time; When an early warning occurs, it can quickly locate the early warning component, view the early warning parameters, early warning values and historical diseases of the component, and automatically push the early warning information to relevant management personnel to timely confirm and handle structural abnormalities.

### 3.4 *Equipment integration and monitoring*

Through the private network, the data integration and control of more than 3,000 electromechanical equipment in the project are realized, and the visualization of traffic control is realized by modifying the information board display content on the BIM model, viewing the closed-circuit television monitoring, changing the lane indicator status, etc.

### 3.5 *Emergency controls*

The BIM information model is used to carry out emergency response strategy drills, formulate emergency plans and implant them into the platform, so as to realize the visualization of emergency management and control and the intelligentization of emergency disposal. After the monitoring system gives an alarm, it can accurately and quickly locate the accident location in the platform 3D model according to the site stake number obtained by the sensor, view the upstream and downstream videos, and confirm the accident type. After the platform automatically pushes the emergency plan according to the obtained site information, and after confirmation by the workers, it can be executed with one click until the site disposal is completed and the traffic is restored, which greatly improves the efficiency of on-site emergency management and control.

## 4 CONCLUSIONS

The intelligent operation and maintenance platform of a large-scale urban bridge and tunnel cluster project based on BIM was applied in the Chongqing Zengjiayan Jialing River Bridge project, involving the BIM design of construction drawings of Zengjiayan Jialing River Bridge, the pilot application of BIM Technology in project construction, and the transportation and nutrition protection system of major municipal projects based on BIM technology. The successful demonstration and application of the research results have improved the intelligent operation and management level of the operating company, improved the operation and service capacity, responded to the national "Internet +" strategy, promoted the in-depth integration of the Internet industry and the operation and maintenance of transportation infrastructure, and comprehensively improved the development level of intelligent transportation, with huge social benefits.

In addition, the BIM Technology Model of the research results of this project has the ability of parametric updating, the BIM Technology Application standard system meets the industry requirements, the supporting platform is expandable, and the research methods can be used for reference. Therefore, the results of the project have good applicability, practicality, and reference, and have wide application value. At the same time, the team has been focused on intelligent transportation infrastructure built custody platform and system equipment research and application, bound to undertake the project research and transformation of scientific and technological achievements, and gradually formed a "traffic infrastructure equipment detection, monitoring technology and intelligent system" and "based on 'GIS +BIM technology' of high-speed road network intelligent maintenance key technology and common platform", such as research, it not only can be used in the field of highway maintenance, the future can also be applied to intelligent transportation, auxiliary public travel, its huge social and economic benefits.

## ACKNOWLEDGMENTS

Fund project: Chongqing talent plan, Project No.: cstc2021ycjh-bgzxm0264.

## REFERENCES

Gou Hongye, Pu Qianhui, Yan Meng, et al. *Intelligent evaluation and early warning system for traffic safety on high-speed railway bridges*: China, 2019v1.0sr100.

Han Yixuan, Ju Yongfeng, Du Kai, et al. The Bridge Health Monitoring System based on BIM [J]. *Electronic Design Engineering*, 2019, 27 (7): 48–52.

Luo Fuyuan, Yu Tingjia. The development of Bridge Intelligence in China [J]. *Prestressed technology*, 2017 (5): 12–15.

Wu Zhixia, Han Peng, Zheng Shuquan, et al. Design and implementation of bridge health monitoring and maintenance platform [J]. *Computer applications and Software*, 2018 Magi 35 (5): 115–120.

Yue Qing, Wu Laiyi, Zhu Liming, et al. Research on Long-span Bridge Operation Monitoring System of High-speed Railway [J]. *Railway Construction*, 2015 (12): 1–6.

Zhang Guanhua. Clustering of bridge monitoring [J]. *China Highway*, 2018, 520 (12): 71–73.

Zhang Guizhong, Zhao Weigang, Zhang Hao. Design and development of digital operation and maintenance system of Shanghai-Tong Yangtze River Bridge [J]. *Railway Journal*, 2019, 41 (5): 16–26.

Zou Y, Kiviniemi A, Jones S W, et al. Risk information management for bridges by integrating risk breakdown structure into 3D/4D BIM[J]. *KSCE Journal of civil engineering*, 2019, 23(2): 467–480.

*Civil Engineering and Urban Research – Mohamed & Hou (Eds)*
*© 2023 the Authors, ISBN 978-1-032-44487-1*

# Application of implicative analysis principle in the design of artistic painting retaining wall

Yangyang Tan*
*School of Fine Art and Design, Kunming University, Kunming, China*

Weiwei Zhu
*School of Architecture and Civil Engineering, Kunming University, Kunming, China*

ABSTRACT: In order to study the ways to achieve the function of retaining walls as the carrier of painting art in road engineering, the implicative analysis principle in Extenics is introduced. Through a real project analysis, this paper expounds on the application method of this principle in the design of artistic painting retaining walls. The main steps are as follows: list the goal basic elements of slope engineering or retaining wall engineering to be analyzed; according to the known information and implicative analysis principle, combined with engineering knowledge, the implicative system of the goal is established; by achieving the lower basic-elements, the upper basic-elements is achieved, so as to solve the problem.

## 1 INTRODUCTION

In modern road engineering, the design of retaining walls is a multi-objective design. Artistic painting has been very common in urban road retaining walls. It is of great significance to advocate civilization, publicize public welfare and show urban culture (Ostendorf 2021). The conventional retaining wall in the project is composed of a foundation and wall body. This structure only considers its use in the project itself, that is, supporting and retaining rock and soil mass to maintain the stability of the subgrade. When it is also used as the carrier of artistic painting, the firmness, and aesthetics of wall painting patterns will be greatly reduced due to the influence of rain and sunshine. In order to solve the problem, this paper introduces the basic theory of Extenics, combined with an invention example of a real retaining wall project, and puts forward the specific application method of implicative analysis principle in the design of artistic painting retaining wall, which can be used as a reference for engineers and technicians.

## 2 PRINCIPLE USED IN THIS STUDY: IMPLICATIVE ANALYSIS PRINCIPLE

Extenics is an original interdisciplinary subject proposed by Chinese researcher Cai Wen and other scholars in 1983 (Cai 1983). With a formal model, it discusses the extension possibility of things and the laws and methods of development and innovation, and is used to solve the contradictions in reality. Extenics has the characteristics of formalization, logicalization and mathematicization, where there are contradictions, and Extenics can be used. After 40 years of research, Extenics has developed from a person's academic thought and a paper to a new discipline with a more mature theoretical framework (Cai 2013). At present, Extenics has entered the stage of combining

---

*Corresponding Author: tanyang2006@126.com

DOI 10.1201/9781003372417-79

applied research with theoretical research (Yang 2013). The theoretical system of Extenics includes basic-element theory, extension set theory, and extension logic (Yang 2014).

The implicative analysis principle is a part of the extension analysis principle in the basic-element theory, which can be expressed as:

(1) If basic-element $B_1$ is achieved, it must have basic-element $B_2$ achieved, written as $B_1$ @ must have $B_2$ @, or $B_1 \Rightarrow B_2$, which means $B_1$ implies $B_2$.

(2) If $B_1$ @ must have $B_2$ @ under condition $l$, it is called $B_1$ implies $B_2$ under condition $l$, written as $B_1 \Rightarrow B_2$.

(3) If both $B_1$ and $B_2$ are achieved, there must have B achieved, it is called $B_1$ AND $B_2$ implies B, written as $B_1 \wedge B_2 \Rightarrow B$.

(4) If $B_1$ or $B_2$ is achieved, there must have B achieved, it is called $B_1$ OR $B_2$ implies B, written as $B_1 \vee B_2 \Rightarrow B$.

(5) If $B_1 \Rightarrow B_2$, $B_2 \Rightarrow B_3$, then $B_1 \Rightarrow B_3$, or written as $B_1 \Rightarrow B_2 \Rightarrow B_3$.

In the above principle of implication analysis, the basic element on the left of the symbol "$\Rightarrow$" is called the lower basic element, and the basic element on the right is called the upper basic element.

## 3 APPLICATION OF IMPLICATIVE ANALYSIS PRINCIPLE: A CASE STUDY OF ARTISTIC PAINTING RETAINING WALL DESIGN

According to the implicative analysis principle in Extenics, when the upper goal is not easy to achieve, we can find its lower goal. If the lower goal is easy to achieve, we can confirm that we have found a way to solve the contradiction.

For example, the soil excavation slope of a road has a stability safety factor of 1.15 and a high foundation bearing capacity. It is proposed to adopt a gravity retaining wall for reinforcement. Considering that there are too many masonry protection works along the road, which are too rigid, it is planned to create some wall painting art landscapes on the retaining wall. In this way, the contradiction comes out, because the retaining wall itself is only an engineering reinforcement structure, only considering its engineering use, that is, supporting the retaining soil and maintaining the stability of the slope, which itself is not suitable as the carrier of painting. When used as the carrier of painting, how to ensure the paintability, firmness, cleanliness, and appreciability of the pattern still need to be solved.

According to the implicative analysis method, the design idea of retaining walls for artistic painting is analyzed here.

The goal matter element (abbreviated as $M_g$) of this problem is

$$M_g = \begin{bmatrix} \text{wall painting pattern,} & \text{carrier,} & \text{wall surface} \\ & \text{artistic landscape effect,} & \text{good} \end{bmatrix},$$

Through implicative analysis of the goal matter element, the goal implicative system of the problem can be obtained, as shown in Figure 1:
In Figure 1,

$$M_1 = \begin{bmatrix} \text{wall painting pattern,} & \text{carrier,} & \text{wall surface} \\ & \text{paintability,} & \text{good} \\ & \text{firmness of color,} & \text{high} \\ & \text{cleanliness of pattern,} & \text{high} \\ & \text{appreciability,} & \text{good} \end{bmatrix},$$

$$M_{11} = (\text{retaining wall, characteristic, flat wall surface}),$$

$$M_{12} = (\text{retaining wall, shading property, good}),$$

$$M_g \Leftarrow M_1 \Leftarrow AND \begin{cases} M_{11} \Leftarrow M_{111} \\ M_{12} \Leftarrow M_{121} \\ M_{13} \Leftarrow (M_{131} \wedge M_{132}) \\ M_{14} \Leftarrow M_{141} \\ M_{15} \Leftarrow M_{151} \end{cases}$$

Figure 1.   Implicative system of the goal (M is the abbreviation of matter-element).

$$M_{13} = (\text{retaining wall, rainproof property, good}),$$

$$M_{14} = (\text{retaining wall, characteristic, wall surface is almost vertical}),$$

$$M_{15} = (\text{retaining wall, function, providing lighting}),$$

$$M_{111} = (\text{retaining wall, material, neat block stone} \vee \text{concrete}),$$

$$M_{121} = (\text{shading cantilever plate, location, top of retaining wall}),$$

$$M_{131} = (\text{canopy cantilever plate, location, top of retaining wall}),$$

$$M_{132} = (\text{drain hole, location, bottom of retaining wall}),$$

$$M_{141} = (\text{wall surface, inclination, } 70° - 90°),$$

$$M_{151} = (\text{lamp holder, location, wall body}).$$

Considering matter-element

$$M_{16} = \begin{bmatrix} \text{contilever plate,} & \text{location,} & \text{top of retaining wall} \\ & \text{role,} & \text{shading} \wedge \text{rain protection} \wedge \text{lights installing} \end{bmatrix},$$

So there is $M_{16} \Rightarrow M_{121} \wedge M_{131} \wedge M_{151}$.

It can be seen from the above analysis that as long as the lowest basic-elements $M_{111}$, $M_{132}$, $M_{141}$ and $M_{16}$, and $M_g$ are achieved, the goal basic element (the highest basic element) can be guaranteed achievement, namely:

$$\left. \begin{array}{l} M_{16} \Rightarrow M_{121} \wedge M_{131} \wedge M_{151} \\ M_{132} \end{array} \right\} \Rightarrow M_{12} \wedge M_{13} \wedge M_{15} \left. \begin{array}{l} \\ M_{111} \\ M_{141} \end{array} \right\} \Rightarrow M_1 \Rightarrow M_g.$$

Thus, a concrete retaining wall for artistic painting can be designed (Zhu 2018), as shown in Figure 2.

Based on the same idea, when the retaining wall is constructed with block stone, a material with an uneven surface, it can be seen from the analysis of the implicative system of the goal matter-element $M_{g2}$ that the goal $M_{g2}$ can be achieved by adding a painting board on the block stone wall. At this time, in order to strengthen drainage, drain holes can be set at the back of walls at all levels, but the water flow should not affect the wall painting pattern, as shown in Figure 3. The reasoning process is shown in Figure 4.

In Figure 4,

$$M_{g2} = \begin{bmatrix} \text{painting pattern,} & \text{carrier,} & \text{painting board} \\ & \text{artistic landscape effect,} & \text{good} \end{bmatrix},$$

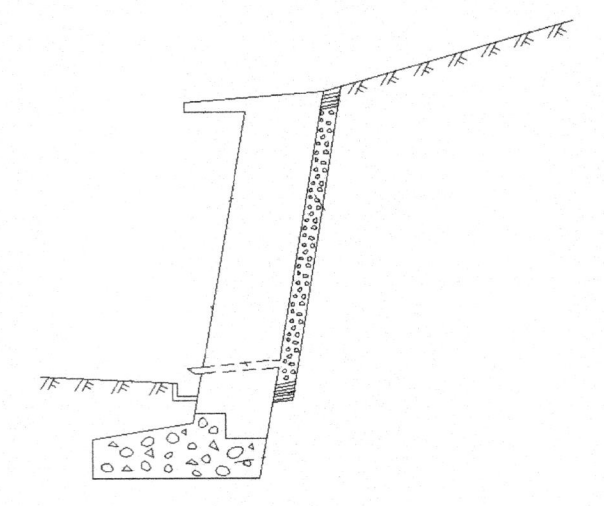

Figure 2.    Concrete retaining wall for artistic painting.

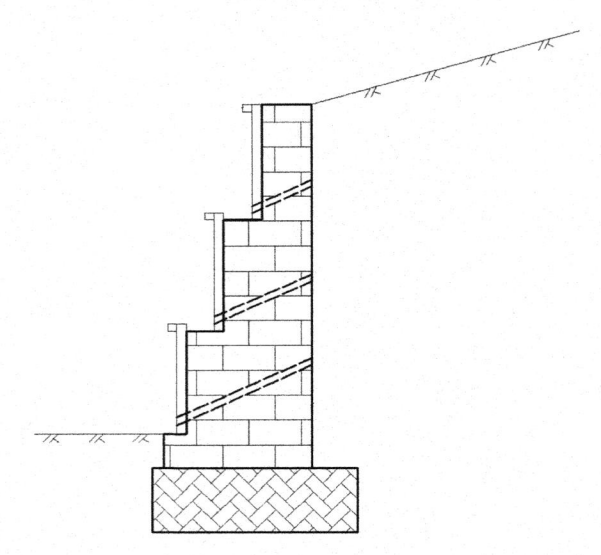

Figure 3.    Block stone retaining wall for artistic painting.

$$\mathbf{M}_{g2} \Leftarrow \mathbf{M}_2 \Leftarrow \mathbf{AND} \begin{cases} \mathbf{M}_{21} \Leftarrow \mathbf{M}_{211} \\ \mathbf{M}_{22} \Leftarrow \mathbf{M}_{221} \\ \mathbf{M}_{23} \Leftarrow \mathbf{M}_{231} \\ \mathbf{M}_{24} \Leftarrow \mathbf{M}_{241} \\ \mathbf{M}_{25} \Leftarrow \mathbf{M}_{251} \end{cases}$$

Figure 4.    Implicative system of the goal.

$$M_2 = \begin{bmatrix} \text{painting pattern,} & \text{carrier,} & \text{painting board} \\ & \text{paintability,} & \text{good} \\ & \text{firmness of color,} & \text{high} \\ & \text{cleanliness of pattern,} & \text{high} \\ & \text{appreciability,} & \text{good} \end{bmatrix},$$

$$M_{21} = (\text{retaining wall, characteristic, has even painting board}),$$

$$M_{22} = M_{12}, M_{23} = M_{13},$$

$$M_{14} = (\text{retaining wall, characteristic, painting board is vertical}),$$

$$M_{25} = M_{15},$$

$$M_{211} = (\text{retaining wall, material, block stone}),$$

$$M_{221} = M_{121}, M_{231} = M_{131},$$

$$M_{241} = (\text{painting board, inclination, } 90°),$$

$$M_{251} = (\text{lamp holder, location, painting board}).$$

## 4 CONCLUSION

To sum up, the implicative analysis method is the application of the implicative analysis principle. The implicative analysis method is a way to analyze the goal or condition of a contradictory problem according to the implicative analysis principle in Extenics theory, so as to find the path to solve the problem. To sum up, the main steps of the application of this method in retaining wall design are as follows:

(1) The goal basic elements of slope engineering or retaining wall engineering to be analyzed (usually a slope engineering goal or its sub-goal expected to be achieved) are listed;
(2) According to the known information (design requirements, engineering geological conditions, technical strength of relevant units, etc.) and implicative analysis principle, combined with engineering knowledge, the goal implicative system is established;
(3) By achieving the lower basic element, the upper basic element is achieved, so as to solve the problem.

## ACKNOWLEDGMENTS

This paper was financially supported by the Basic Research Project of Yunnan Province (202101AT070144) and the Scientific research fund project of the Yunnan Provincial Department of Education (2021J0717).

## REFERENCES

Cai W. 1983 Extension set and incompatible problem *Journal of Scientific Exploration* vol 1 pp 83–97.
Cai W and Yang C Y. 2013 Basic theory and methodology on Extenics *Chinese Science Bulletin* vol 58 pp 1190–1199.
Ostendorf M, Morgan S, Celik S, and Retzlaff W. 2021 Evaluating the potential stormwater retention of a living retaining wall system *Journal of Living Architecture* vol 8 pp 1–18.
Yang C Y and Cai W. 2013 *Extenics: Theory, Method and Application* chapter 2 pp 23–26.
Yang C Y and Cai W. 2014 Extenics and intelligent processing of contradictory problems *Science and Technology Review* vol 32 pp 15–20.
Zhu W W, Tang Y Y, Yang P, and et al. 2018 Concrete retaining wall for artistic painting China Patent ZL2018204622055, 2018-04-03.

*Civil Engineering and Urban Research – Mohamed & Hou (Eds)*
*© 2023 the Authors, ISBN 978-1-032-44487-1*

# Analysis of the impact of urbanization on precipitation characteristics and trends in Shanghai

Sihui Dong*
*School of Mechanical and Mining Engineering, The University of Queensland, Brisbane QLD, Australia*

Tianya Xu*
*Business school, Central University of Finance and Economics, Beijing, China*

ABSTRACT: Based on the daily precipitation observation data of Xujiahui, Baoshan, and Chongming meteorological stations in Shanghai from 1960 to 2020, the variation of precipitation in Shanghai during the past 61 years was analyzed. It is found that the variation trend of annual total precipitation is not obvious with climate change, but the number of downpour days increases. Since 1997, the annual precipitation in Xujiahui was significantly different from the other two places, and the increase rate was 9.33mm/a, indicating that there was an obvious urban rain island effect. From the analysis of precipitation level days and rainfall load, it was found that the annual precipitation days observed in Xujiahui gradually increased from 2005, and the annual growth rate of heavy downpour rainfall load reached 6.8%/10a, while the annual increase rate of total precipitation was 20.27mm/a, with a significant seasonal difference of precipitation. Urbanization resulted in an increase in the number of heavy rainfall days, such as moderate rain, heavy rain, and heavy downpour, compared with 1960-1979, by 16%, 33%, and 75%, respectively. In contrast to the period before rapid urbanization, from 1984 to 2007, the urbanization process led to a marked trend of increasing precipitation in Shanghai, and the contribution of urbanization to the increase of urban precipitation in Shanghai was 40.89%.

## 1 INTRODUCTION

With economic development, global urbanization is also accelerating, with 4.46 billion people living in urban areas worldwide by 2021 (UN Department of Economic and Social Affairs 2021). Urban diseases and climate change caused by urbanization and human activities have received widespread attention (Chapman 2017; Satterthwaite 2009). According to the statistical analysis of urban land area expansion by the National Bureau of Statistics of China, urban land expansion in southern China and coastal cities is exceptionally rapid, with the maximum rate reaching 30% per year. From 1989 to 1997, China's urban area expanded by 1.2 million hectares, of which 867,000 hectares of arable land area were used for urban construction (Jin & Shepherd 2005). Since 1978, China's Yangtze River Delta region (including Shanghai, Zhejiang, Jiangsu, and Anhui provinces) has witnessed rapid economic development and urban land expansion, which has formed the distribution characteristics of urban agglomeration in the Yangtze River Delta. According to the Shanghai Statistical Yearbook, during the 12 years from 1996 to 2007, the built-up area of the city rose from 412 km² to 885.67 km², and the arable land area rose from 30.06 km² to 20.6 km², and the built-up area was 549.58 km² in 2003 and 885.67 km² in 2007, with an increased rate of 84.02 km²/year, with many suburban areas gradually becoming urban land. The Yangtze River Delta region has accelerated the transformation of rural areas into towns, with the number of

---

*Corresponding Authors: sihui.dong@uq.net.au and gregxu0105@gmail.com

DOI 10.1201/9781003372417-80

established towns in Shanghai rapidly increasing from 26 in 1985 to 208 in 1995, 1,085 in Jiangsu Province, a surge to 1,006 in Zhejiang Province in 1998, and 970 in Anhui Province from 118 in 1982 to 2003 (national bureau of statistics). By 2015, the population of Shanghai, Nanjing, and Hangzhou will reach 24.153 million, 6.534 million, and 5.329 million, respectively, making them centers and growth poles for national and regional development. According to the 2010 census data, the urbanization rates of the three central cities of Shanghai, Nanjing, and Hangzhou were 89.30%, 77.94%, and 73.2%, respectively, while the overall urbanization rate of the Yangtze River Delta was 47.81% (Luo 2018).

While people improve the quality of life, they also bring many unexpected consequences to the urban environment and climate. In urban areas, both atmospheric composition and underlying surface have been considerably changed by human activities, and the impact of human activities on climate is particularly prominent in urban areas. In the context of regional climate represented by suburban areas, urbanization of the underlying surface has led to the formation of particular local meteorological environments in cities, and the climatic and environmental changes brought about by urban development have been confirmed by numerous observations and simulations. In the highly urbanized Yangtze River Delta region, the urban meteorological phenomena and the changes caused by urban development are particularly prominent. For example, Ren Chunyan (2006) noted significant heat island, rain, dry, and dark island effects in large cities in northwest China. When discussing the impact of urbanization on short-term precipitation in Jeddah, Saudi Arabia, Luong, et al. (2020) revealed that the urban effect caused an increase in rainfall during heavy rainfall, which was most significant in urban areas. Some studies statistically showed that urbanization caused a significant decrease in the relative humidity in Cairo, with a decreasing trend of 0.55%/decade (Mahmoud 2018). Relevant studies (Argüeso 2014; Oleson 2015) found that the relationship between urban development and the urban meteorological environment is complex. In the process of urban development and changing the present situation of the urban meteorological environment, the status quo will, in turn, affect the sustainable development of the city, urbanization on the city's air temperature, wind speed, and precipitation have influence, and bring a series of city meteorological problems, such as "urban heat island", "urban rain island", "urban pollution" and other city-specific phenomena (Yu 2017), and the climate and environmental problems manifested by urbanization in different regions are different. The influence of urbanization on rainfall is a controversial topic in urban climatology. It has been studied by many experts worldwide (Hejazi 2009; Shastri 2015), who have concluded that cities have the effect of increasing precipitation in urban areas and their downwind directions. The Yangtze River Delta region experienced the third sea-level rise in the early 20th century, with a total of 14 cm of sea-level rise in China, whereas more than 20 cm of coastline rise in the Yangtze River Delta region (Gu 2011).

Shanghai is one of mainland China's most economically developed and densely populated cities. Torrential rain is the primary meteorological disaster in Shanghai during the summer half-year, which causes flood disasters such as flooding of roads, houses, and farmland. For example, in 1999, Shanghai was hit by severe plum rain that had never happened in a century, frequent heavy rainfall flooded more than 85,000 hectares of farmland and 60,600 homes for more than 7 days. There have been some analyses and studies on the rainstorm in Shanghai (Liang 2017; Qi & Zhao 2004), such as the analysis of the process of a hefty rainstorm in Shanghai from August 5 to 6, 2001, and the study of the characteristics and causes of the rainstorm caused by typhoons in Shanghai. However, most studies focus on individual case analysis and research, and few articles specifically analyze and study rainfall characteristics caused by urbanization in Shanghai from the perspective of climate change. A deeper understanding of the impact of urbanization on precipitation in a modern metropolis is required to be explored. The UHI-influenced area in Shanghai increased from 1709.81 $km^2$ to 1585.84 $km^2$ from 1984 to 2014, an increase of 81.03%, and the annual average relative humidity generally showed a decreasing trend year by year (Zhao 2016). Therefore, under the global climate change, studying the impact of urbanization on urban precipitation in the Yangtze River Delta, such as Shanghai, has some guiding significance for urban development and disaster weather forecasting.

## 2 VARIATION CHARACTERISTICS OF PRECIPITATION

This paper analyzes the precipitation differences between urban and suburban Shanghai based on the precipitation data of three meteorological stations in Xujiahui, representing the urban area, Baoshan and Chongming, representing the suburban area, from 1960 to 2020 (data from http://data.cma.cn/, National Meteorological Science and Data Center of China), and investigates the impact of urbanization on precipitation in Shanghai over 61a years. Xujiahui station (31.24°N, 121.43°E), Baoshan station (31.25°N, 121.28°E) and Chongming station (31.67°N, 121.49°E) have similar latitude and longitude positions, and their location distribution and related information are shown in Figure 1 and Table 1. The three places belong to the same meteorological division with similar background climates. Since both temperature and precipitation data exceed 50 sets, the Kolmogorov-Smirnov test found that neither of the two had a normality trait. After the Spearman correlation analysis of the observed precipitation and temperature data of the three stations from 1960 to 2020 with two comparisons, the results in Table 2 show that the correlation coefficients of Xujiahui and Chongming stations, Xuhui and Baoshan stations, and Chongming and Baoshan stations are all greater than 0.8, and all reach the significance level of 0. 01. As shown in Figure 2, by comparing the climate change trends of the three stations, it is observed that the temperature change trends of the three stations are consistent, and the warm and cold periods are also the same. The trend characteristics of characteristic air temperature at Baoshan station and Chongming Station well represent the change characteristics of regional background air temperature: In the regional background, the annual mean minimum temperature and annual mean maximum temperature increased the most rapidly in spring and winter, followed by the annual mean minimum temperature and the annual mean maximum temperature. This is consistent with the conclusion of global and regional climate change under the 1.5°C path described by IPCC (Masson-Delmott 2018). Therefore, Baoshan station and Chongming station can be used as the regional background of urban comparison, i.e., urban climate, to analyze the characteristics of Shanghai urban climate represented by Xujiahui Station.

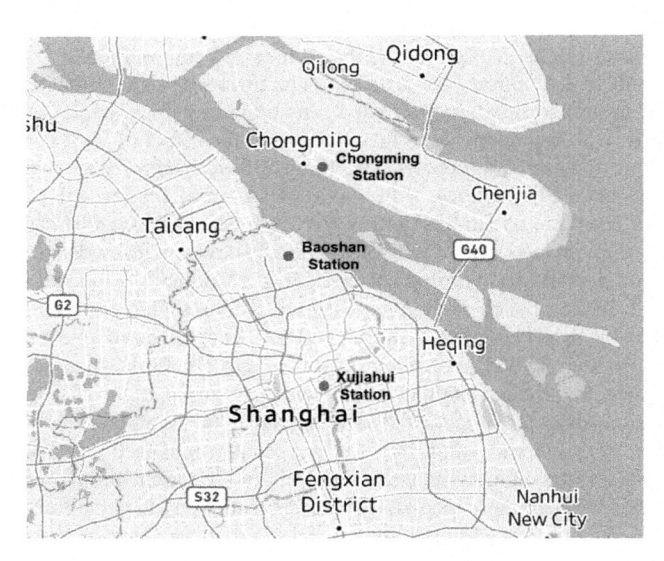

Figure 1.    The geographical location of Xujiahui, Baoshan, and Chongming meteorological station.

The daily precipitation data of Xujia station, Baoshan station, and Chongming station from 1960 to 2019 were analyzed for annual precipitation statistics, respectively, and the results are shown in Table 3 and Figure 3. Table 3 shows that the precipitation of each station fluctuates and roughly increases with the years, among which the rainfall of Xujiahui was always more prominent than the other two stations; the annual precipitation of the three stations in the 10a period of 1980-1989 and 2010-2019 increased significantly relative to the previous years, while Baoshan considerably increased, with an increase of 190.37 mm relative to 1960s, and nearly 141 mm relative to 2000s. The annual rainfall in Baoshan has increased by 190.37 mm relative to the 1960s and by nearly 141.28 mm relative to the 2000s. Therefore, from the analysis of the total annual precipitation, the precipitation in Shanghai from 1960 to 2019 was stable except for the peak in the 1980S and 2010S. However, it can be seen that the precipitation in the urban area was more than that in the surrounding areas, indicating that the urban rain island effect exists in Shanghai and has decreased in recent years. It means that there is some improvement in urban air quality.

Table 1. The specific information of Xujiahui, Baoshan, and Chongming meteorological station.

| Station | Latitude | Longitude | Elevation(m) | Duration | Landscape |
|---|---|---|---|---|---|
| Xujiahui | 31.24°N | 121.43°E | 4.6 | 1960-2020 | Urban |
| Baoshan | 31.25°N | 121.28°E | 5.5 | 1960-2020 | Peri-urban |
| Chongming | 31.67°N | 121.49°E | 4.5 | 1960-2020 | Rural |

Table 2. Spearman correlation analysis of temperature and precipitation in Xujiahui, Baoshan, and Chongming.

| Coefficient correlation | Xuhui-Chongming | Xuhui-Baoshan | Chongming-Baoshan |
|---|---|---|---|
| Spearman (temperature) | 0.995** | 0.998** | 0.997** |
| Spearman (precipitation) | 0.834** | 0.855** | 0.861** |

Table 3. Annual average precipitation and temperature of Xujiahui, Baoshan, and Chongming stations (1960-2019).

| Period | Xujiahui | | Baoshan | | Chongming | |
|---|---|---|---|---|---|---|
| | Temperature | Precipitation | Temperature | Precipitation | Temperature | Precipitation |
| 1960-1969 | 15.75 | 1013.44 | 15.77 | 985.79 | 15.30 | 952.58 |
| 1970-1979 | 15.67 | 1060.34 | 15.70 | 1007.46 | 15.22 | 960.33 |
| 1980-1989 | 15.83 | 1199.53 | 15.73 | 1197.83 | 15.06 | 1130.60 |
| 1990-1999 | 16.88 | 1266.69 | 16.66 | 1146.29 | 15.69 | 1141.92 |
| 2000-2009 | 17.82 | 1320.23 | 17.49 | 1174.45 | 16.49 | 1146.18 |
| 2010-2019 | 17.70 | 1375.99 | 18.24 | 1315.73 | 16.25 | 1235.70 |

The total annual average precipitation in Xujiahui, Baoshan, and Chongming is 1206.04mm, 1137.93mm, and 1094.55mm, respectively, with Xujiahui getting more precipitation than the remaining two areas, with an annual average difference of 89.80mm. From the 10a moving average, precipitation in all places fluctuated with time, but after 1997, Xujiahui was significantly different from the other two places, showing a slight increase in precipitation. After the linear fitting of precipitation in Xujiahui, it is found that the annual increase rate of precipitation is 9.33mm/a, which may be under the direct effect of urban heat island and other factors. The temperature field distribution and circulation conditions in urban areas are changed, making air stratification unstable and conducive to thermal convection. The convective weather is enhanced, and the spatial and temporal distribution of urban rainfall changes, which makes precipitation change. Precipitation in the Yangtze River Delta region is mainly associated with large-scale circulation, subtropical monsoon, and heat island effect. In contrast, large-scale weather systems are unlikely to change significantly within 60 a. The increase in urban land use in Shanghai and the influence of the surrounding urban subsurface, coupled with more condensation nuclei in the urban air, are favorable to the formation of clouds (especially low clouds), in which some condensation nuclei with larger particle sizes (such as nitrates) can promote the warm cloud precipitation, resulting in a slight increase in annual precipitation in Shanghai.

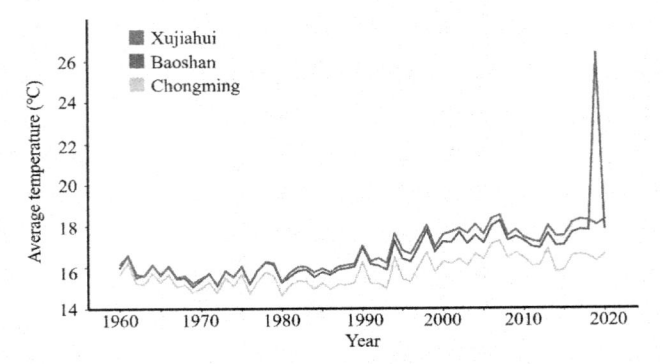

Figure 2.    Annual average temperature trend of Xujiahui, Baoshan, and Chongming.

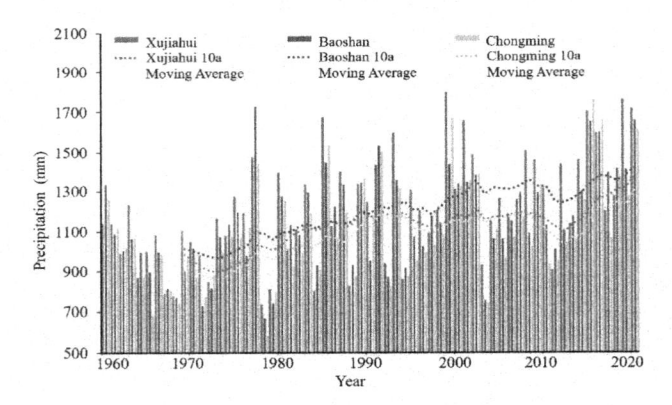

Figure 3.    The annual total precipitation performance in Xujiahui, Baoshan and Chongming.

## 3    COMPARISON OF CHANGES IN PRECIPITATION LEVELS

From meteorology, it can be roughly divided into 7 levels according to the classification of precipitation grade of the China Meteorological Administration, as shown in Table 4. Urbanization is a gradual process. Based on the daily precipitation data from 1960 to 2020, the statistics of the number of days with different precipitation levels were made for Xujiahui, which represents the city, and Baoshan and Chongming, which represent the suburbs, respectively, according to different precipitation levels in different periods, as shown in Table 5-7.

As seen from Table 5, in the urban area, the number of days of moderate, heavy rain and downpour in Xujiahui from the 1970s onwards is significantly greater than in the 1960s. However, in the 1990s, the number of days of moderate and heavy rainfall shows a decreasing trend, while

Table 4.   Precipitation classification.

| 24h Precipitation | Level |
|---|---|
| L<10.0 mm | Light rain |
| $10.0 \leq L \leq 24.9$ mm | moderate rain |
| $25.0 \leq L \leq 49.9$ mm | heavy rain |
| $50.0 \leq L \leq 99.9$ mm | downpour |
| $100.0 \leq L \leq 199.9$ mm | heavy downpour |
| $\geq 200.0$mm | heavy rainfall |

the number of days of downpour reaches a peak. Entering the 21st century, the number of days of moderate, heavy rain and heavy downpour increased significantly during 10a and reached the highest value in history, among which the number of days of heavy rain was 117 d, 34.25 d higher than the average number of days from the 1960s to 1999, while the number of days of the occurrence of downpour reached the lowest compared with other years, and the trend of decrease was apparent. In the 2010s phase, except for the significant increase in light rain days, other rainfall days show a gradually decreasing trend with the chronological period. From the 10a average, the number of days of moderate rain, heavy rain, and heavy downpour occurrence in Xujiahui is all increasing with the chronological period, which increased by 16%, 33%, and 75%, respectively, compared with that within 1960-1979.

Table 5.   Statistics of precipitation days in different periods and different orders of magnitude in Xujiahui (unit: d).

| Period | Drizzle | Moderate rain | Heavy rain | Downpour | Heavy downpour | Heavy rainfall |
|---|---|---|---|---|---|---|
| 1960-1969 | 891 | 203 | 76 | 24 | 2 | 1 |
| 1970-1979 | 940 | 228 | 81 | 26 | 2 | 0 |
| 1980-1989 | 920 | 241 | 94 | 34 | 3 | 0 |
| 1990-1999 | 906 | 232 | 80 | 45 | 4 | 0 |
| 2000-2009 | 875 | 259 | 117 | 19 | 8 | 1 |
| 2010-2019 | 982 | 242 | 100 | 39 | 7 | 0 |
| (1990-2019)/(1960-1989) | 1.00 | 1.09 | 1.18 | 1.23 | 2.71 | 1.00 |

According to Table 6-7, the number of days of heavy downpour in Baoshan and Chongming also increases most significantly with age from 2000 to 2009, and the increase is higher than that of Xujiahui in the urban area. Besides, the number of days of heavy rain also increases with age, but the increase is minimal, while the number of days of light rain is similar to that of Xujiahui, showing a gradual decrease. Figures 4-5 show the number of heavy rain days in Xujiahui and Baoshan and heavy downpours in Baoshan and Chongming has increased significantly since the 1990s, especially since 2003, the growth rate of heavy rain or above in Xujiahui is 18.2 times per 10a. Moreover, the variation pattern of heavy rain days in Baoshan and Chongming is basically the same as that in Xujiahui, which remained at 2.8 times /a without significant increase or decrease. Compared with the three stations, the rate of increase in heavy downpour days in Xujiahui is more noticeable than in Baoshan and Chongming.

From the perspective of annual precipitation days, Xujiahui observed a gradual increase in the number of annual precipitation days from 2005, while the increase rate of its total annual precipitation was 20.27 mm/a. In addition, the seasonal differences in precipitation were significant, mainly concentrated in summer, indicating a trend of concentrated precipitation in Shanghai, which inevitably led to increased precipitation intensity. From the above results, it can be concluded that urbanization has a more significant impact on the extreme precipitation weather in cities, which makes the precipitation concentrated precipitation trend and increases the probability of large

Table 6. Statistics of precipitation days in different periods and different orders of magnitude in Baoshan (unit: d).

| Period | Drizzle | Moderate rain | Heavy rain | Downpour | Heavy downpour | Heavy rainfall |
|---|---|---|---|---|---|---|
| 1960-1969 | 1308 | 211 | 63 | 22 | 2 | 0 |
| 1970-1979 | 1091 | 185 | 71 | 23 | 1 | 1 |
| 1980-1989 | 926 | 225 | 93 | 38 | 3 | 0 |
| 1990-1999 | 879 | 231 | 79 | 28 | 5 | 0 |
| 2000-2009 | 850 | 210 | 109 | 23 | 7 | 0 |
| 2010-2019 | 931 | 246 | 106 | 28 | 8 | 0 |
| (1990-2019)/(1960-1989) | 0.80 | 1.11 | 1.30 | 0.95 | 3.33 | 0.00 |

Table 7. Statistics of precipitation days in different periods and different orders of magnitude in Chongming (unit: d).

| Period | Drizzle | Moderate rain | Heavy rain | Downpour | Heavy downpour | Heavy rainfall |
|---|---|---|---|---|---|---|
| 1960-1969 | 912 | 210 | 71 | 19 | 1 | 1 |
| 1970-1979 | 974 | 200 | 73 | 15 | 2 | 1 |
| 1980-1989 | 964 | 235 | 79 | 28 | 3 | 0 |
| 1990-1999 | 945 | 223 | 73 | 34 | 3 | 0 |
| 2000-2009 | 912 | 209 | 85 | 27 | 7 | 0 |
| 2010-2019 | 931 | 236 | 85 | 28 | 10 | 0 |
| (1990-2019)/(1960-1989) | 0.98 | 1.04 | 1.09 | 1.44 | 3.33 | 0.00 |

precipitation processes such as heavy rain and heavy downpour in cities. To conclude, with the development of urbanization, annual precipitation in urban areas has a slight increase trend, and the number of heavy rainfall days such as heavy rain and heavy rainstorm increases significantly, which means the occurrence probability of heavy downpours increases. From the 10a moving average trend line, the annual trend of rainfall load of heavy downpour level in Baoshan and Chongming suburban stations is not apparent, with no noticeable increasing or decreasing trend; the rainfall load of downpour level in two suburban stations fluctuates more, with a significant increase from the 1980s onwards, but with a gradually decreasing trend from 2000s onwards. When comparing the annual trends of rainfall loads in Xujiahui, Baoshan, and Chongming, it can be seen that the rainfall loads of moderate and light rain in the suburbs of Shanghai show a slight decreasing trend while the rainfall loads of moderate and light rain in the urban area (Xujiahui) are relatively stable without prominent chronological trends; the downpour rainfall loads in the urban area change significantly with chronology, especially from 1984 to 2007 when Shanghai was in the rapid growth period of urbanization. The increasing trend of its downpour rainfall load also reaches a rate of 6.8%/10a, and the further away from the urban area, the lower the fluctuation of rainstorm precipitation load.

Therefore, the precipitation changes in Shanghai, the comparison of precipitation changes at the three stations, and the precipitation levels show a marked trend of increasing extreme precipitation weather in Shanghai after accelerated urbanization. It is likely caused by the uneven height of urban buildings, whose roughness is larger than that of the suburbs, which not only causes turbulence but also obstructs the stable and slow precipitation system. The time in urban tributaries is longer, leading to increased precipitation intensity in urban areas and the extension of precipitation time.

## 4 VARIATION IN RAINFALL LOAD AT DIFFERENT LEVELS

From the above analysis, it can be seen that urbanization increases the probability of occurrence of heavy precipitation in cities. Daily precipitation data from 1959 to 2009 are used to make statistics

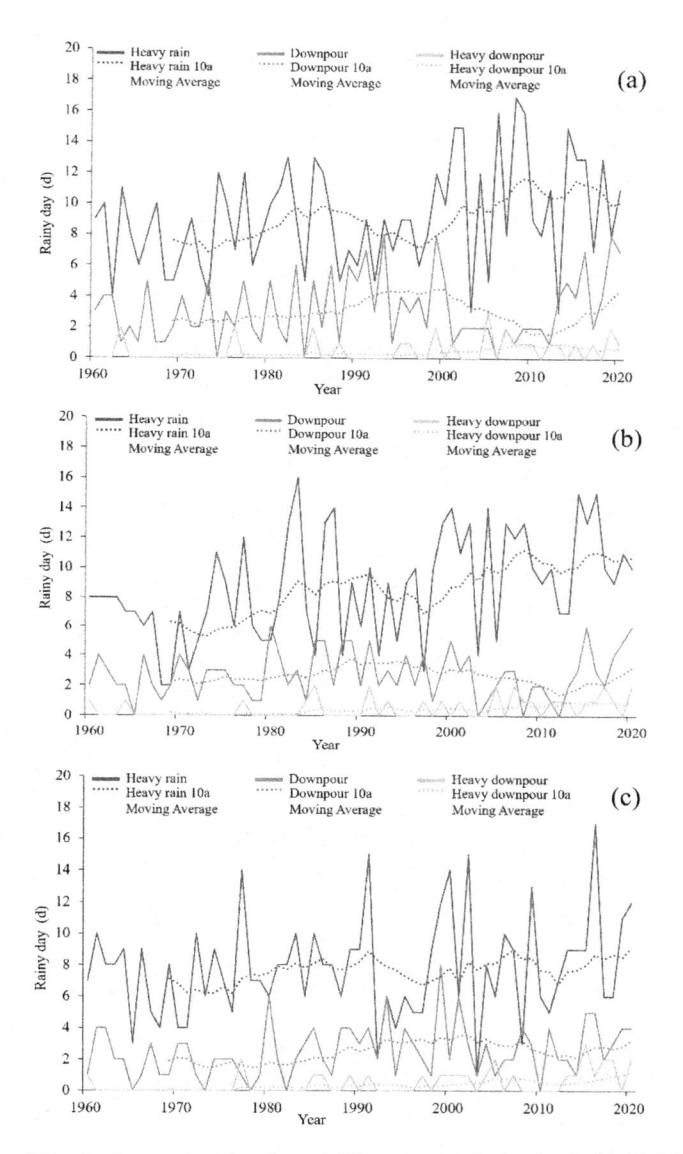

Figure 4. The variation tendency of raining days at different precipitation levels (a) Xujiahui, (b) Baoshan, (c) Chongming.

on Xujiahui, Baoshan, and Chongming according to different precipitation levels in different periods. By comparing 1 urban station and 2 suburban stations, the annual rainfall load of each precipitation level was analyzed. This paper defines the percentage of annual precipitation and total annual precipitation of a certain precipitation level as the rainfall load of the precipitation level. Thus, the rainfall load can be expressed as Equation 1:

$$h = R/R_t \times 100\%$$ (1)

where h = rainfall load; R = total annual precipitation for a certain precipitation level; $R_t$ = total annual precipitation. Following the above results, the precipitation load statistics for each level

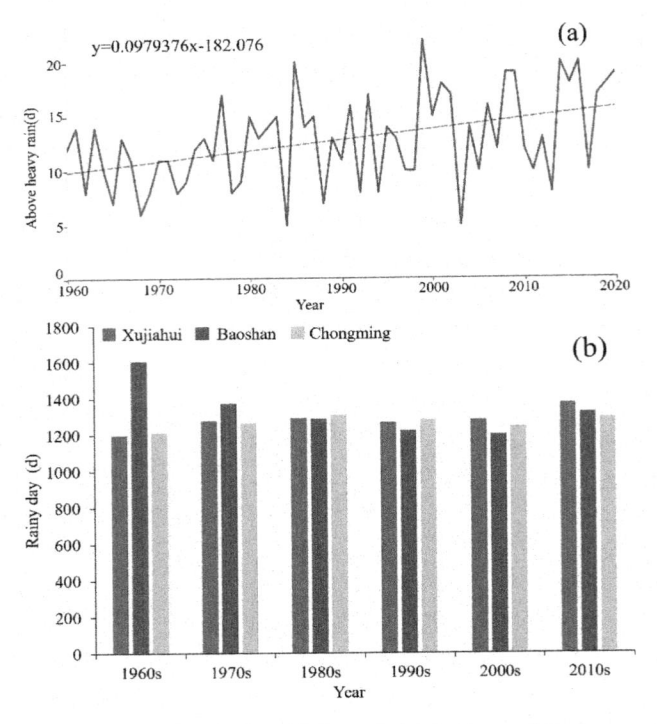

Figure 5. The variation tendency of t rain days (a) the variation tendency of the heavy rain days since 1997 in Xujiahui (b) the total precipitation days.

of precipitation are shown in Figure 6, from which it can be seen that the annual average values of rainfall load for heavy rain in Xujiahui, Baoshan, and Chongming are 26.62%, 26.63%, and 25.34% respectively. In comparison, the annual average values of rainfall load for downpour are 17.23%, 16.17%, and 15.58%, respectively.

## 5 CONTRIBUTION OF URBANIZATION TO PRECIPITATION

Regarding the contribution of urbanization to precipitation, this paper uses the calculation of He Yun-ling & Lu Zhihai (2012) of the contribution of urbanization to climate change to explore the impact of urbanization on precipitation in Shanghai. Taking Xujiahui's 1960 annual average precipitation as the benchmark, the change of precipitation in other years relative to the 1960 precipitation is $\Delta R_{xj}$, and the change of regional background climate Baoshan and Chongming annual average precipitation relative to the 1960 precipitation is $\Delta R_{bs}$ and $\Delta R_{cm}$, respectively. Then the urbanization influence on Shanghai precipitation is $\Delta R_{u1}$ and $\Delta R_{u2}$. Since the change of precipitation in Shanghai is influenced by regional background climate and urbanization together, the influence of urbanization on precipitation can be expressed as Equations 2 and 3:

$$\Delta R_{u1} = \Delta R_{xj} \Delta R_{bs} \tag{2}$$

$$\Delta R_{u2} = \Delta R_{xj} \Delta R_{cm} \tag{3}$$

Where $\Delta R_{xj}$ series is calculated from the annual mean precipitation series in Xujiahui, $\Delta R_{bs}$ and $\Delta R_{cm}$ indicate the influence of regional background, and their time series over the years are calculated from the annual mean precipitation series of Baoshan and Chongming, respectively,

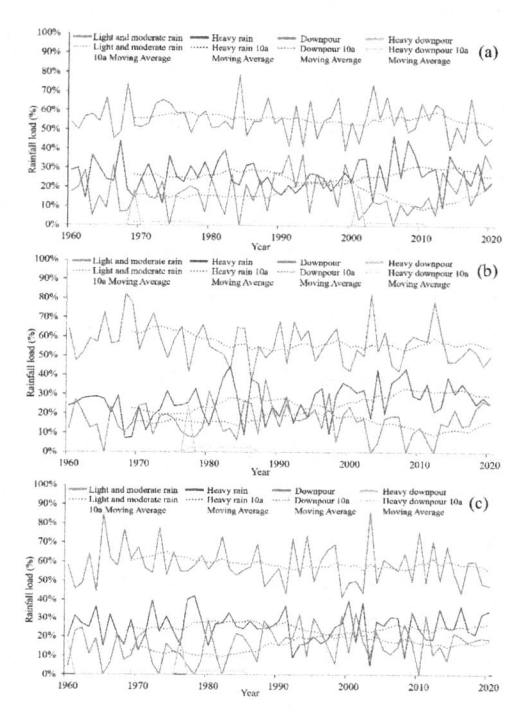

Figure 6.    The annual variation of rainfall load (a) Xujiahui, (b) Baoshan, (c) Chongming.

as shown in Figure 7. From Figure 7, it can be seen that urbanization-influenced precipitation increased between 1990 and 2014, except for 1977, when a decrease in urbanization-influenced precipitation occurred (called a negative effect, and vice versa, a positive effect), and the amount of precipitation was enormous. Also, the average annual precipitation change in Xujiahui is no longer the same as the regional background precipitation change trend, which indicates that precipitation in urban areas is increasingly influenced by urbanization, and the influence of urbanization on rainfall determines the trend of precipitation change. The annual rainfall increase rate in Shanghai can be expressed as Equations 4 and 5:

$$\Delta Q_1 = \Delta Q_{xj} \Delta Q_{bs} \tag{4}$$

$$\Delta Q_2 = \Delta Q_{xj} \Delta Q_{cm} \tag{5}$$

where $\Delta Q_{xj}$ = annual precipitation variation rate of Xujiahui, $\Delta Q_{bs}$ = annual precipitation variation rate of Baoshan, $\Delta Q_{cm}$ = annual precipitation variation rate of Chongming. Since precipitation is an increasing variable for both urban and regional backgrounds, the contribution rate of urbanization to precipitation change is shown in Equation 6:

$$P = |\Delta Q / \Delta Q_{xj}| \times 100\% \tag{6}$$

Based on the calculation, the annual average precipitation increase rate in Xujiahui from 1960 to 2020 is 9.49mm/a, while the average precipitation increase rate in Baoshan and Chongming is 5.61mm/a in the regional background. Therefore, the urbanization precipitation increase rate in Shanghai is 3.88mm/a, while the contribution rate of urbanization precipitation increase in Shanghai is 40.89%. As there are many possible mechanisms affecting urban precipitation, for example, urban heat island makes the air stratification unstable, and low-level water vapor rises to form convective precipitation; the blocking effect of different heights of urban buildings leads to prolonged precipitation time; aerosols emitted from urban life and production may affect precipitation from microphysical processes, atmospheric dynamics processes, and cloud precipitation, leading to lightning and convective heavy precipitation. Therefore, the influence of cities on precipitation is the product of many factors, and the mechanism of urban development on urban precipitation in Shanghai needs to be further discussed in future work.

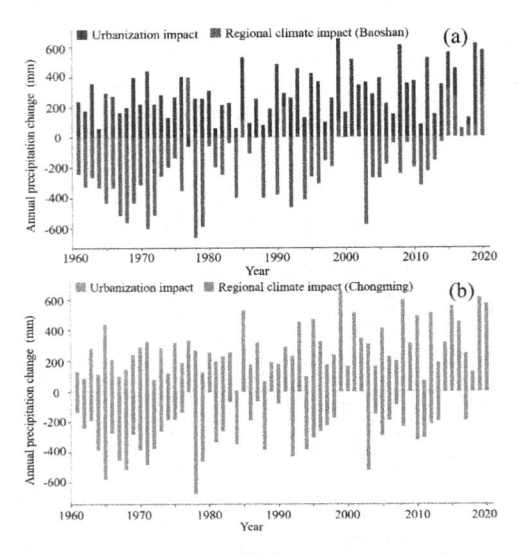

Figure 7. Impacts of both urbanization and regional climate on urban precipitation in Shanghai (a) Xujiahui-Baoshan, (b) Xujiahui-Chongming.

## 6 CONCLUSION

Based on the precipitation observation data from 1960 to 2020, the annual variation characteristics of precipitation in Shanghai and the influence of urbanization on precipitation change are analyzed. After analysis and discussion, the following conclusions are drawn:

(1) The annual variation of total precipitation in the urban area of Shanghai represented by Xujiahui and the suburban areas of Shanghai represented by Baoshan and Chongming are fluctuating, with Xujiahui receiving slightly more precipitation than the remaining two areas, with an average difference of 89.80 mm. The annual precipitation growth rate in the city (Xujiahui) is 0.37 times higher than that in the suburbs (Chongming), with an increased rate of 9.33 mm/a. The more distant the suburbs, the lower the annual precipitation growth rate. In terms of total annual precipitation, urbanization has a significant impact on urban precipitation, and the rain island effect exists in Shanghai.

(2) In terms of precipitation levels, the frequency of heavy rain and heavy downpour in Shanghai city showed a more apparent increasing trend, with an increase of 16%, 33%, and 75% compared with 1960-1979, respectively. In particular, the growth rate of heavy rain and above in Xujiahui has been 18.2 times/10a since 2003, while the rainfall in suburban areas is relatively stable.

(3) From the perspective of rainfall load, the annual heavy rain and heavy downpour load of Xujiahui in Shanghai is 26.62% and 17.23%, respectively, and the annual growth rate of rainfall load for heavy rain in Xujiahui reached 6.8%/10a during the rapid urbanization phase of Shanghai from 1984 to 2007, indicating that urbanization has an enormous impact on the temporal distribution of urban precipitation, which makes urban precipitation concentrated. The trend of precipitation, with the total number of annual rainfall days remaining basically the same, and the decreasing trend

of moderate and light rainfall loads in suburban areas indicate that the frequency of catastrophic weather in Shanghai has increased.

(4) Overall, the contribution of increased rainfall from 1960 to 2000 urbanization to the increase of urban precipitation in Shanghai is 40.89%. From the 1980s, urbanization-influenced urban precipitation began to increase, and the amount of precipitation was also larger. The change of precipitation affects the supply and demand of regional water resources and impacts regional flood control in addition. The temporal and spatial variability of precipitation trends in Shanghai should be taken seriously. Further research is needed to develop water resources management and flood response strategies adapted to climate change for the causes and effects of the changes in precipitation trends and spatial variability.

## REFERENCES

Argüeso, D., Evans, J. P., Fita, L., & Bormann, K. J. (2014). The temperature response to future urbanization and climate change. *Climate Dynamics*, 42(7), 2183–2199.

Chapman, S., Watson, J. E., Salazar, A., Thatcher, M., & McAlpine, C. A. (2017). The impact of urbanization and climate change on urban temperatures: a systematic review. *Landscape Ecology*, 32(10), 1921–1935.

Gu, C., Hu, L., Zhang, X., Wang, X., & Guo, J. (2011). Climate change and urbanization in the Yangtze River Delta. *Habitat International*, 35(4), 544–552.

Hejazi, M. I., & Markus, M. (2009). Impacts of urbanization and climate variability on floods in Northeastern Illinois. *Journal of Hydrologic Engineering*, 14(6), 606–616.

Jin, M., & Shepherd, J. M. (2005). Inclusion of urban landscape in a climate model: How can satellite data help?. *Bulletin of the American Meteorological Society*, 86(5), 681–689.

Liang, P., & Ding, Y. (2017). The long-term variation of extremely heavy precipitation and its link to urbanization effects in Shanghai during 1916–2014. *Advances in Atmospheric Sciences*, 34(3), 321–334.

Luo, J., Xing, X., Wu, Y., Zhang, W., & Chen, R. S. (2018). Spatio-temporal analysis on built-up land expansion and population growth in the Yangtze River Delta Region, China: From a coordination perspective. *Applied Geography*, 96, 98–108.

Luong, T. M., Dasari, H. P., & Hoteit, I. (2020). Impact of urbanization on the simulation of extreme rainfall in the city of Jeddah, Saudi Arabia. *Journal of Applied Meteorology and Climatology*, 59(5), 953–971.

Mahmoud, S. H., & Gan, T. Y. (2018). The long-term impact of rapid urbanization on urban climate and human thermal comfort in a hot-arid environment. *Building and Environment*, 142, 83–100.

Masson-Delmotte, V., Zhai, P., Pörtner, H. O., Roberts, D., Skea, J., Shukla, P. R., ... & Waterfield, T. (2018). *Global warming of 1.5 C*. An IPCC Special Report on the impacts of global warming of, 1(5).

National Bureau of Statistics, https://data.stats.gov.cn

Oleson, K. W., Monaghan, A., Wilhelmi, O., Barlage, M., Brunsell, N., Feddema, J., ... & Steinhoff, D. F. (2015). Interactions between urbanization, heat stress, and climate change. *Climatic Change*, 129(3), 525–541.

Qi Linlin, Zhao Sixiong. An analysis of mesoscale features of heavy rainfall in Shanghai on 526 August 2001 [J]. *Chinese Journal of Atmospheric Sciences*, 2004, 28 (2): 2542268. (in Chinese)

REN Chunyan, WU Dianting, DONG Suocheng. The influence of urbanization on the urban climate environment in Northwest China. *Geographical Research*( in Chinese) (2006)25(2): 233–241.

Satterthwaite, D. (2009). The implications of population growth and urbanization for climate change. *Environment and urbanization*, 21(2), 545–567.

Shanghai statistical yearbook,http://tjj.sh.gov.cn/tjnj/index.html

Shastri, H., Paul, S., Ghosh, S., & Karmakar, S. (2015). Impacts of urbanization on Indian summer monsoon rainfall extremes. *Journal of Geophysical Research: Atmospheres*, 120(2), 496–516.

UN Department of Economic and Social Affairs. (2018). *World Urbanization Prospects: The 2018 Revision*.

Yu, Y., Liu, J., Yan, S., & Yang, Z. (2017, August). *The analysis of the "urban rain island effect" in Jingjinji District of China*. In Proceedings of the 2017 International Conference on Material Science, Energy and Environmental Engineering (MSEEE 2017).

Yun-Ling, L. H., & Zhi-hai. (2012). Air temperature changes and urbanization contribution ratio in Kunming city in recent 50 years. *Meteorological and Environmental Research*, 3(1), 41–44.

Zhao, M., Cai, H., Qiao, Z., & Xu, X. (2016). Influence of urban expansion on the urban heat island effect in Shanghai. *International Journal of Geographical Information Science*, 30(12), 2421–2441.

*Civil Engineering and Urban Research – Mohamed & Hou (Eds)*
*© 2023 the Authors, ISBN 978-1-032-44487-1*

# Design of highway maintenance decision system based on Nifi

Juming Hao
*Gansu Highway Traffic Construction Group Co., Ltd, Lanzhou, China*

Xinxiu Zhang*
*Gansu Hengshi Highway Inspection Technology Co., Ltd, China*
*Gansu Key Laboratory of Highway Network Monitoring, Lanzhou, China*

ABSTRACT: Highway intelligent maintenance decision-making has been perplexing to experts and scholars, mainly how to integrate detection data, highway construction data, historical maintenance data, traffic volume data, meteorological data, and geological data, and develop an intelligent maintenance platform. The Nifi visual data flow is applied to integrate the collected external and internal maintenance data into the PostgreSQL database. Based on the relevant theories of big data analysis, the key factors of pavement disease causes and the internal relationship between various factors are studied. Nifi has rich data processing functions, including numerical value, mathematical calculation, string, and other processing functions. It supports user-defined scripts, not only its own script language but also the current mainstream Python script language.

## 1 INTRODUCTION

On September 14, 2019, the CPC Central Committee and the State Council issued the outline of building a transportation power. Building a transportation power is the leading field of building a modern economic system, important support for building a socialist modern power in an all-round way, and the general starting point for doing a good job in transportation work in the new era. The construction and development of roads not only provide safe, convenient, and fast road conditions for people to travel, but also provide transportation guarantees for accelerating the allocation of regional market resources and promoting the efficient development of the local economy. The state has put forward higher requirements for highway construction and maintenance (https://baike.so.com/doc/28833835-30298915.html, 2019).

In the research of pavement maintenance decision-making, in the 1960s, Canada first proposed the concept of "pavement management" and established the OPAC system in Ontario (Abaza & Khaled 2021; Aboah et al. 2020). Subsequently, the United States carried out a 15-year SHRP project research on this. Through the analysis and comparison of the gradual decline of pavement performance and the gradual increase of maintenance and repair costs, it was concluded that the service life of highway asphalt pavement can be extended by 15 years after more than four times of preventive maintenance and repair (Elkhawaga et al. 2020; Fani et al. 2020).

Many domestic experts and scholars seek big data technology to solve the problem of maintaining multi-source heterogeneous data warehousing. Zhu ye and others researched and designed the big data acquisition and storage system of injection molding equipment based on Kubernetes, build a Nifi data stream processing cluster based on the Nifi big data integration tool, and deploy a variety of data stream processors to complete the access to multiple data sources of injection molding production data and visual data stream processing (Zhu 2019). Ren Nuer and others store

---

*Corresponding Author: 1822543914@qq.com

 DOI 10.1201/9781003372417-81

and analyze the data generated by the automobile industry based on Gambari's data processing platform, and use Nifi to import the data configuration (Ren et al. 2019). Ma Zhonghao's master's thesis focuses on the theory of "real-time data acquisition and computing system optimization of industrial Internet", and puts forward a real-time plant equipment acquisition and computing system based on Nifi, miNifi, spark streaming, HBase, and hive (Ma 2020).

In view of the above research, the realization of intelligent maintenance decision-making must integrate multi-source heterogeneous data and can add data sources and data types in real-time. In the face of such multi-source heterogeneous data, how to integrate and store the construction, maintenance, geography, meteorology, and other data, and use Nifi technology to solve the maintenance problem. The problem of data warehousing can be solved. Based on the powerful function of Nifi, a pavement maintenance data management system is built. Nifi has the following advantages:

- In the process of highway intelligent maintenance decision data warehousing evaluation, the real-time monitoring of the decision data warehousing process is realized. In case of emergency, it is not necessary to stop the system service, but only need to control the Nifi node to interrupt or stop the data flow, so as to improve the high reliability of the system;
- Nifi adjusts the data processing workflow in real-time, and no data will be lost in the process of adjusting the workflow;
- Nifi has rich data processing functions, including numerical value, mathematical calculation, string, and other processing functions. It supports user-defined scripts, not only its own script language but also the current mainstream Python script language.

## 2 DATA ANALYSIS OF HIGHWAY INTELLIGENT MAINTENANCE DECISION

### 2.1 *Specification requirements*

According to the technical specification for highway asphalt pavement maintenance JTG 5421-2018, after the pavement performance is evaluated and analyzed by using the pavement technical condition data, the corresponding sections can be subject to preventive maintenance, functional repair, or structural repair according to the evaluation results (JTG 5142-2019). In order to achieve accurate and scientific highway intelligent maintenance decision-making, the internal and external factors affecting pavement diseases are combined to conduct systematic analysis, modeling, calculation, establish an automatic regression model, and issue maintenance decision-making reports. The external factors affecting pavement performance mainly include natural conditions, traffic conditions, maintenance information, economic parameters, and other uncertain factors.

### 2.2 *Maintenance decision data flow*

The internal factors affecting pavement performance mainly include construction data, design indicators, geological data, and other factors. Therefore, it is necessary to integrate the external cause data, internal cause data, and road disease data over the years into a database for mining and analysis. The data flow based on Nifi visualization can realize the automatic warehousing processing of multi-source heterogeneous data for highway intelligent maintenance decision-making. Nifi works based on the Web and is scheduled on the server in the background. It is a real-time data acquisition and integration tool for data flow design. Users can define data processing as a process and then process it. The background has a data processing engine, task scheduling, and other components. It supports data routing, transformation, and system mediation logic of a highly configurable indicator diagram, and supports dynamic pulling of data from multiple data sources. The data source of intelligent maintenance decisions is shown in Figure 1.

# 3 DATA PROCESSING OF HIGHWAY INTELLIGENT MAINTENANCE DECISION

## 3.1 *Nifi data processing flow*

Nifi supports diversified data acquisition. It can collect unstructured data including video, audio, and pictures, semi-structured data including text and XML / JSON documents, and structured data represented by a relational database. At the same time, it can monitor data quality and acquisition exceptions in the acquisition process to ensure the high stability and reliability of the acquisition process. In the process of data warehousing, the system platform also supports data fusion, compression, splitting, and verification, and supports the embedding of processed data into nonrelational databases or data warehouses, such as HBase, hive, Kafka, and other big data storage components. The process of Nifi processing highway intelligent maintenance decision data is shown in Figure 2:

The specific data processing steps are as follows:

Users upload the road condition detection data, meteorological data, traffic volume data, construction data, and geological data of manual survey and automatic detection equipment in Excel format, e.g., .csv, .txt, and .dat;

The server of the evaluation system judges the manual investigation and automatic detection equipment according to the uploaded data. In the case of automatic detection, the system automatically discriminates the detection type according to the data content, and finally creates a data processing task according to the judgment results and triggers the Nifi processing task;

Nifi starts the data processing task and downloads the original data files.

Data format conversion is to convert excel and txt data into csv. If it is csv format, go to the next step directly.

The data in csv, txt, and data formats are divided into multiple pieces of data according to the separator (the manual questionnaire calls Python script for segmentation), and the data is transferred to the next step one by one.

Data processing mainly carries out data verification and null value check, and then converts the data text into a data object.

The transformed data object is transferred to the warehousing interface to realize data warehousing. After data warehousing, indicator calculation and data summary are triggered.

Apply Nifi to process management of highway intelligent maintenance decision-making data; Monitor the inflow and output dynamics of multi-source heterogeneous data in real-time; Flexibly configure data parameters, table attribute information, etc. The Nifi table is developed for the investigation of complex pavement diseases, and the manual processing of the total value of 100 meters cannot be realized by using the Nifi table. This kind of manual processing technology is developed for the total value of 100 meters of pavement diseases, which cannot be processed manually.

## 3.2 *Nifi warehousing data instance*

In 2021, the 567.084 km ordinary trunk line managed and maintained by the Linxia highway development center was tested and evaluated (JTG 5210-2018). The sections include: G248 Lanzhou-Maguan K104+944-K122+175, K156+171-K161+800, K166+ 600-K212+757 (69.017km); G309 Qingdao-Lanzhou K2103+ 050-K2168+367, K2169+565-K2170+355 (66.107km); G310 Lianyungang-Gonghe K1783+893-K1789+493, K1794+645-K1877+851, K1890+600-K1964+327 (162.533km); G568 Lanzhou-Luqu K92+169-K149+961, K179+408-K210+650 (89.034km); S105 Lanzhou-Jishishan K0-K15+619, K27+910-K47+659 (35.368 km); S106 Lanzhou-Linxia K2+703-K34+ 275, K43+411-K79+549, K114+450-K135+945 (89.205km); S230 Honggu-Mingxian K157+547-K180+329 (22.782km); S232 Daban-Hezuo K94+774-K97+979 (3.205km); S309 Gancaodian-Jishishan K80+779-K81+306,K81+627-K86+800 (5.700km); S311 Dingxi-Hezheng K107+376-K113+136, K147+634-K166+007 (24.133km).

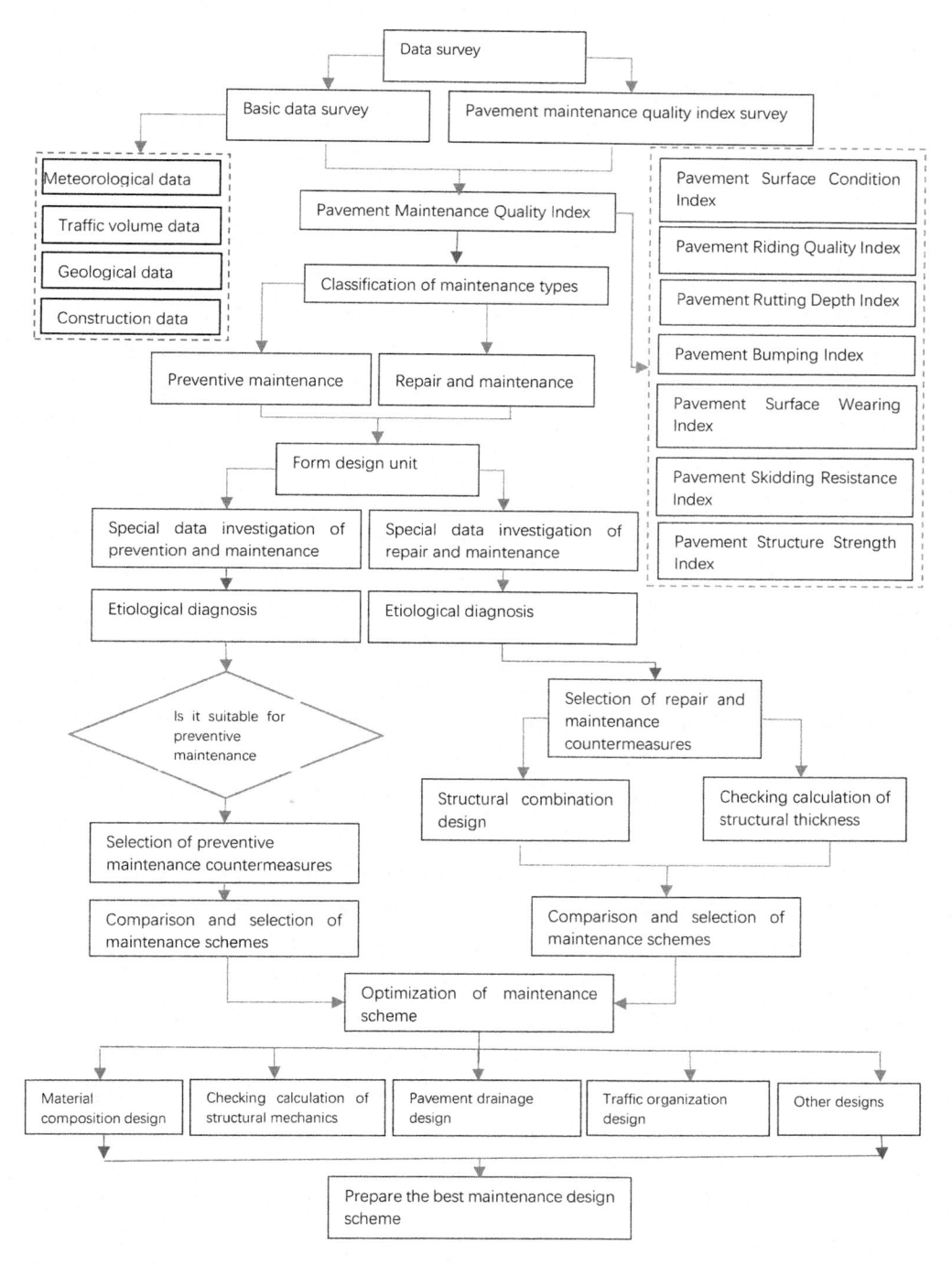

Figure 1. Intelligent maintenance decision flow chart.

The pavement damage index PCI of G248 highway is 91.72, which is evaluated as excellent. There are strip repair diseases in some sections and a small number of transverse and longitudinal cracks

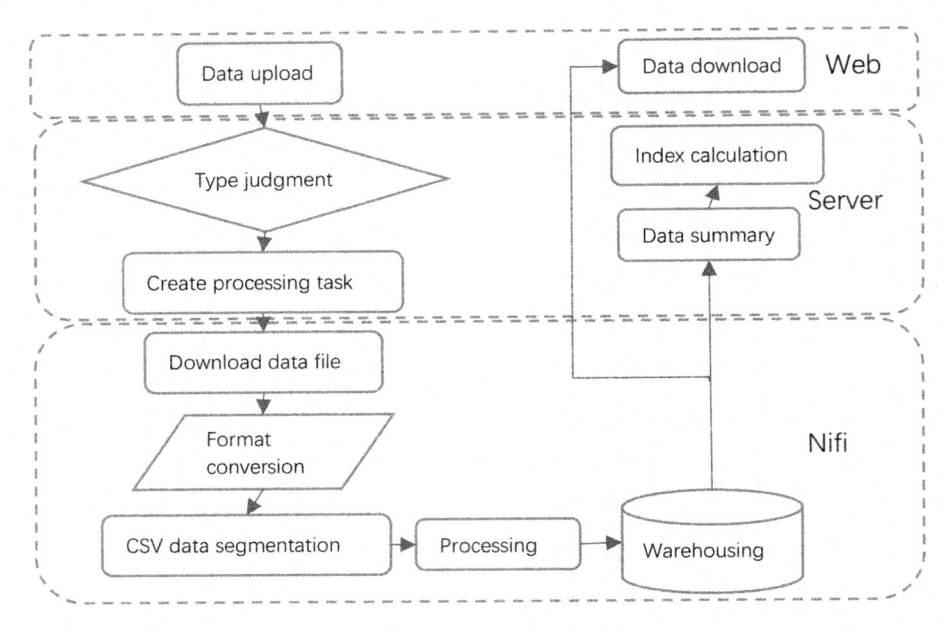

Figure 2.  Nifi data processing flow.

Table 1.  Linxia trunk highway.

| Number | Number | Name | Start station | End station | Mileage (km) |
|---|---|---|---|---|---|
| 1 | G248 | Lanzhou-Maguan | K104+944 | K122+175 | 69.017 |
| | | | K156+171 | K161+800 | |
| | | | K166+600 | K212+757 | |
| 2 | G309 | Qingdao-Lanzhou | K2103+050 | K2168+367 | 66.107 |
| | | | K2169+565 | K2170+355 | |
| 3 | G310 | Lianyungang-Gonghe | K1783+893 | K1789+493 | 162.533 |
| | | | K1794+645 | K1877+851 | |
| | | | K1890+600 | K1964+327 | |
| 4 | G568 | Lanzhou-Luqu | K92+169 | K149+961 | 89.034 |
| | | | K179+408 | K210+650 | |
| 5 | S105 | Lanzhou-Jishishan | K0+000 | K15+619 | 35.368 |
| | | | K27+910 | K47+659 | |
| 6 | S106 | Lanzhou-Linxia | K2+703 | K34+275 | 89.205 |
| | | | K43+411 | K79+549 | |
| | | | K114+450 | K135+945 | |
| 7 | S230 | Honggu-Mingxian | K157+547 | K180+329 | 22.782 |
| 8 | S232 | Daban-Hezuo | K94+774 | K97+979 | 3.205 |
| 9 | S309 | Gancaodian-Jishishan | K80+779 | K81+306 | 5.700 |
| | | | K81+627 | K86+800 | |
| 10 | S311 | Dingxi-Hezheng | K107+376 | K113+136 | 24.133 |
| | | | K147+634 | K166+007 | |
| | | summation (km) | | | 567.084 |

and crazing diseases. The strip repair diseases of K104+944-K121, K156+171-K161, and K1+99-K203 sections are relatively concentrated, and the cracking, transverse and longitudinal cracks of K105-K106 and K208-K212 sections are relatively concentrated. The disease characteristics of the G248 trunk highway are shown in Figures 3-5.

Through the statistical analysis of the number and types of damage diseases in asphalt pavement, it can be seen that the main disease of asphalt pavement with the G248 line is cracking, of which the proportion of cracking diseases is 63.57%. Figure 3 is the statistics of the proportion of various diseases in the pavement damage rate DR. The main disease of G248 trunk highway asphalt pavement is shown in Figure 6.

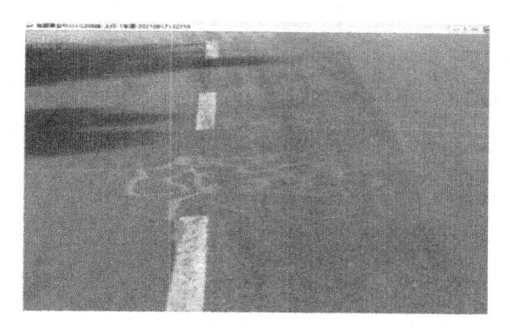

Figure 3.    Spaghetti repair of Asphalt Pavement SK105+364.

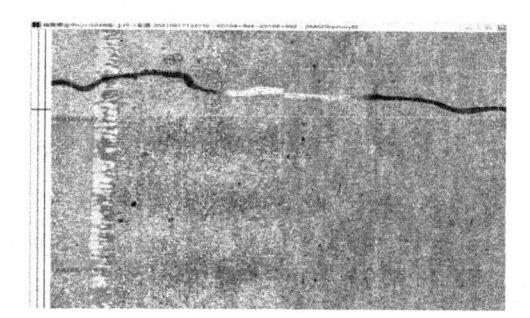

Figure 4.    Spaghetti repair of Asphalt Pavement SK105+516.

Through the statistical analysis of the number and types of damage diseases of cement pavement, it can be seen that other diseases of the G248 line mainly refer to cracks and repair diseases with cement concrete pavement K22+2-K35+7, K61+5-K63. See Figure 4 for the statistics of the proportion of various diseases of cement pavement in the pavement damage rate DR. The proportion of G248 trunk highway cement pavement diseases is shown in Figure 7.

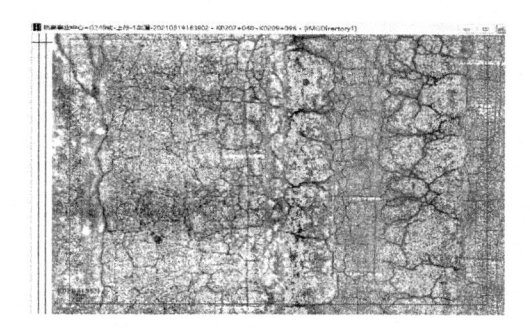

Figure 5.    Cracking of cement pavement SK105+111.

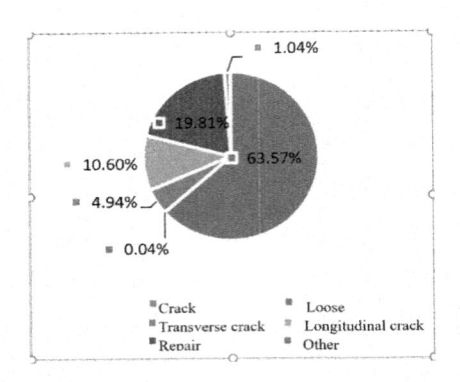

Figure 6.  G248 asphalt pavement.

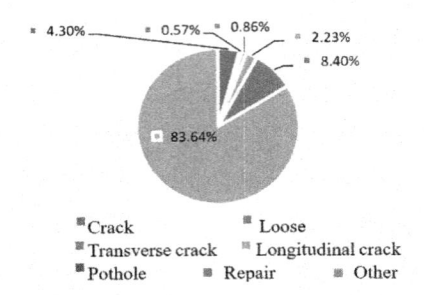

Figure 7.  G248 cement pavement.

Nifi converts excel into csv format, and the python script eliminates empty rows and non-data rows in the table. Nifi divides the results into multiple strings again and stores them in the warehouse. Nifi will script excel columns and add, delete and edit the information of each column according to the standard change content. Nifi is an easy-to-use, powerful and reliable data pulling, processing, and distribution system, which can match and adjust the changed content, or customize the changed content. There are various types of Nifi warehousing data, including picture format, excel table, text, and csv data of road comprehensive inspection vehicles, which provides rich data sources for intelligent maintenance decision-making for analysis, processing, and statistics.

## 4  COMPREHENSIVE ANALYSIS AND APPLICATION

Based on Nifi, the automatic processing of highway intelligent maintenance decision data ware-housing is realized, and the multi-source heterogeneous data affecting intelligent maintenance decisions are integrated into the PostgreSQL database. The advantage of PostgreSQL is that it can build extension components in many ways to realize big data management. It can support the query of Jsons data, while the meteorological data, engineering construction data, and geological data collected and downloaded online are Jsons data, which is convenient to use SQL and has the highest query efficiency.

We develop Gansu intelligent maintenance decision-making data platform, which is developed and deployed by using big data processing technology. Gansu intelligent maintenance decision-making data platform has designed and developed three modes: web terminal, mobile terminal, and data cockpit. Meanwhile, the cloud online "one picture" browsing and application mode centered on a Web map. Combined with the road condition diseases, influence factors, and evaluation results, the technical conditions of the road network data are rendered according to different technical condition indicators, which realizes the display of the road condition index performance thematic map of the road network, and supports the display and viewing of the excellent, good, medium, secondary and poor road conditions of each route in the map.

## FUND PROJECT

Project supported by Gansu Provincial Department of science and technology (21YF11GA008).

## REFERENCES

Abaza, Khaled A. "Optimal Novel Approach for Estimating the Pavement Transition Probabilities Used in Markovian Prediction Models." *International Journal of Pavement Engineering*, 2021, pp. 1–12.

Aboah, Armstrong, and Yaw Adu-Gyamfi. "Smartphone-Based Pavement Roughness Estimation Using Deep Learning with Entity Embedding." *Advances in Data Science and Adaptive Analysis* 12, 2020, no. 03, pp. 2050007.

Elkhawaga, M., S. Elbadawy, and A. Gabr. "Comparison of Master Sigmoidal Curve and Markov Chain Techniques for Pavement Performance Prediction." Arabian Journal for Science and Engineering. Section A, *Sciences*, 2020, vol. 45, no. 5 pp. 3973–3982.

Fani, Amirhossein, Amir Golroo, S. A. Mirhassani, and Amir Gandomi. "Pavement Maintenance and Rehabilitation Planning Optimisation under Budget and Pavement Deterioration Uncertainty." *International Journal of Pavement Engineering*, 2020, pp. 1–11.

https://baike.so.com/doc/28833835-30298915.html, 2019.

JTG 5142-2019, *technical specification for highway asphalt pavement maintenance* [S]

JTG 5210-2018, *highway technical condition evaluation standard* [S]

Ma Zhonghao *Optimization of real-time data acquisition and computing system for the industrial Internet.* Beijing University of technology, 2020.

Ren Nuer, Wei Jinjin, Cai Jianjun Data processing platform based on ambari. *Computer knowledge and technology*, 2019, vol. 15, no. 31, pp. 1–5.

Zhu Ye *Research and design of big data acquisition and storage system for injection molding equipment based on Kubernetes South China University of technology*, 2019.

*Civil Engineering and Urban Research – Mohamed & Hou (Eds)*
*© 2023 the Authors, ISBN 978-1-032-44487-1*

# Influence on water environment improvement of diversion water from Nanjiang sluice based on two-dimensional numerical simulation analysis

Zhihao Fang

*School of Naval Architecture and Maritime, Zhejiang Ocean University, Zhoushan, China*
*The Key Laboratory for Technology in Rural Water Management of Zhejiang Province, Zhejiang University of Water Resources and Electric Power, Hangzhou, China*

Dongfeng Li*

*The Key Laboratory for Technology in Rural Water Management of Zhejiang Province, Zhejiang University of Water Resources and Electric Power, Hangzhou, China*

ABSTRACT: Yubei plain is located in the water network plain in the north of Shangyu District, Shaoxing City. It is adjacent to Cao'e River in the west, Fenghui Town in the south, Yuyao City in the east, and Hangzhou Bay in the north. The city is located in the north-central part of Zhejiang Province and on the south bank of the estuary of the Qiantang River. As it is located near the estuary of the Yangtze River, Yubei plain is not only threatened by the flood tide of the outer river, but also has the hidden danger of waterlogging. In this paper, the two-dimensional mathematical model of the river and lake hydrodynamic water environment is established, and the Sanhui District is taken as the research object. Through analysis, it is concluded that the water diversion scheme can improve the water environment of the river network, and the water diversion scheme is feasible.

## 1 INTRODUCTION

The study area is the Lihai District, Sanhui District, and the northern coastal district, among which the Sanhui District is planned for the riverside district of Binhai New Town. The main sluices and selected lines of the Sanhui river network are shown in Figure 1. The existing main rivers in Sanhui are Qiliuqiu-Beitang River, Central River, Sanhui Central River, Huantang West River, etc. During water diversion, the maximum diversion flow can reach 47.52m by controlling the opening height of the sluice.

Due to the insufficient measured data on river networks, a large number of hydrodynamic factors such as water level and velocity can be obtained by using the calculation of the river network mathematical model (Chen et al. 2014; Han et al. 2020; Zhu et al. 2009). Using these spatio-temporal data, the impact of sluice opening and closing on the water environment of the river network can be analyzed. The mathematical model is an important research tool and has been widely used. However, in the study of the river network, there are more one-dimensional mathematical models and fewer two-dimensional mathematical models, but the two-dimensional mathematical model has many advantages (Cheng 2005; Tong et al. 2016; Yang et al. 2015). It is of great significance to use the two-dimensional mathematical model to study the improvement of the pump sluice water environment of the river network in the Yubei plain (Yang et al. 2021). In order to improve the water environment quality in the plain river network, the Sanhui District of Binhai New Town is selected to study the diversion scheme. The impact of the scheme on the line

---

*Corresponding Author: lidf@zjweu.edu.cn

DOI 10.1201/9781003372417-82

is analyzed by analyzing the changes in water level elevation, flow velocity, BOD concentration, and DO concentration at each point on the line.

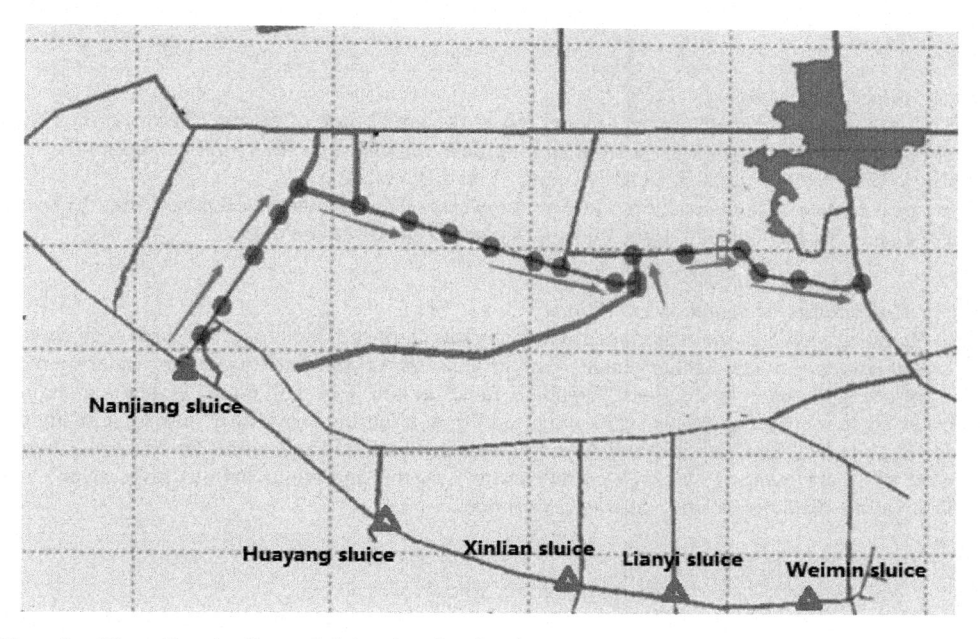

Figure 1.   Water diversion line and sluices location drawing.

## 2   TWO-DIMENSIONAL HYDRODYNAMIC PLANE MATHEMATICAL MODEL OF PLAIN RIVER NETWORK

### 2.1   *The basic equation of the two-dimensional model of water flow in the river network*

#### 2.1.1   *Basic equation*
Flow continuity equation:

$$\frac{\partial z}{\partial t} + \frac{\partial (hu)}{\partial x} + \frac{\partial (hv)}{\partial y} = 0 \tag{1}$$

X. Y-direction flow equation:

$$\frac{\partial u}{\partial t} + u\frac{\partial u}{\partial x} + v\frac{\partial u}{\partial y} - fv + g\frac{\partial z}{\partial x} + g\frac{u\sqrt{u^2+v^2}}{c^2 h} = \lambda \Delta u \tag{2}$$

$$\frac{\partial u}{\partial t} + u\frac{\partial v}{\partial x} + v\frac{\partial \beta v}{\partial y} + fu + g\frac{\partial z}{\partial y} + g\frac{v\sqrt{u^2+v^2}}{c^2 h} = \lambda \Delta v \tag{3}$$

Where: x and y are spatial coordinates; z is the water level; t is the time; u, v is the velocity component in the X direction and Y direction respectively; h is the water depth; $f$ is Coriolis force, $f = 2W \sin \Phi$; $W$ is the rotation velocity of the earth; $\Phi$ is the latitude of the earth; c is Xie Cai coefficient; g is the gravitational acceleration.

Water quality equation:

$$\frac{\partial hC}{\partial t} + \frac{\partial uhC}{\partial x} + \frac{\partial vhC}{\partial y} = \frac{\partial}{\partial y}\left(E_x h\frac{\partial C}{\partial x}\right) + \frac{\partial}{\partial y}\left(E_x h\frac{\partial C}{\partial y}\right) + S + F(C) \tag{4}$$

Where C is the pollutant concentration (mg /L); u and v is the velocity component in X and Y directions respectively; $E_x$ and $E_y$ are diffusion coefficients in X and Y directions respectively $(m/s^2)$; S is the source and sink item $(g / m^2/ s)$; F (C) is biochemical reaction term.

### 2.1.2 *Initial conditions*

The initial value of the mathematical model, given the initial time of the river network, the water level is the constant water level of 2.8m and the flow velocity is 0; the initial condition of water quality is class V water, that is, BOD is 10 mg / L and DO is 2 mg / L.

In this calculation, the introduced clean water is class III water, that is, BOD is 4 mg / L, DO is 5 mg / L, and the planned and designed flow is given at the diversion outlet.

### 2.1.3 *Computational boundary conditions*

Take Nanjiang sluice as the main water diversion source of the Sanhui section, and four culvert sluices of Huayang sluice, Xinlian sluice, Lianyi sluice and Weimin sluice as standby water sources. By controlling the opening height of the sluice, the diversion flow shall not be higher than 30 m³/ s. Water diversion for three days (9:00 a.m. – 5:00 p.m.). During water diversion, Qianjin sluice, sluice 1, sluice 2, Sifeng sluice, and Anding sluice are closed. The Central River and Liujiuqiu Central River are pumped into high-quality water flow through pump stations respectively. See Table 1 for the diversion scheme of Nanjiang sluice.

Table 1. Diversion scheme of Nanjiang sluice

| Sluice | Flow (m³/s) |
| --- | --- |
| Nanjiang sluice | 27 |
| Huayang sluice | 5.13 |
| Xinlian sluice | 5.13 |
| Lianyi sluice | 5.13 |
| Weimin sluice | 5.13 |

## 3 ANALYSIS OF CALCULATION RESULTS

It can be seen from the variation diagram of water level and elevation along the way in Figure 2 that the water level and elevation rise gradually within 2 days and 18 hours after water diversion, and the water level and elevation of each distance are almost the same, indicating that the water level and elevation of the line is relatively stable.

It can be seen from Figure 3 that the flow velocity increases 0-2km away from the diversion outlet and close to the Nanjiang sluice. The flow velocity changes gently at 2km-5km and increases at 5km-6km, which may be due to confluence and flow increase. After 6.8km, the flow is close to 0, because the Yuwei sluice is not opened and there is no live water flow.

It can be seen from Figure 4 that the overall BOD concentration has decreased. 0-6km away from the water intake, the BOD concentration has decreased to about 3.5 mg/L, which has reached the standard of class III water, and the water quality has improved better. However, after 6.8 km, the BOD concentration has decreased to about 5.5 mg/L, which is the standard of class IV water, and the water quality has not improved significantly. It can be seen from Figure 5 that after 2 days and 6 hours, the DO concentration in most waters is more than 6.5 mg/L, reaching class III water. After 6.8 km, the DO concentration reaches 4.5 mg/L. Although the water quality has not improved much, it has also reached the standard of class IV water.

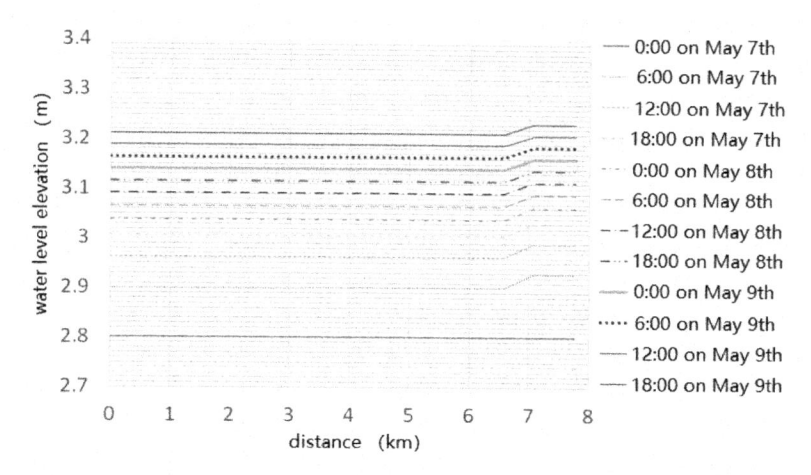

Figure 2. Variation diagram of water level elevation with distance.

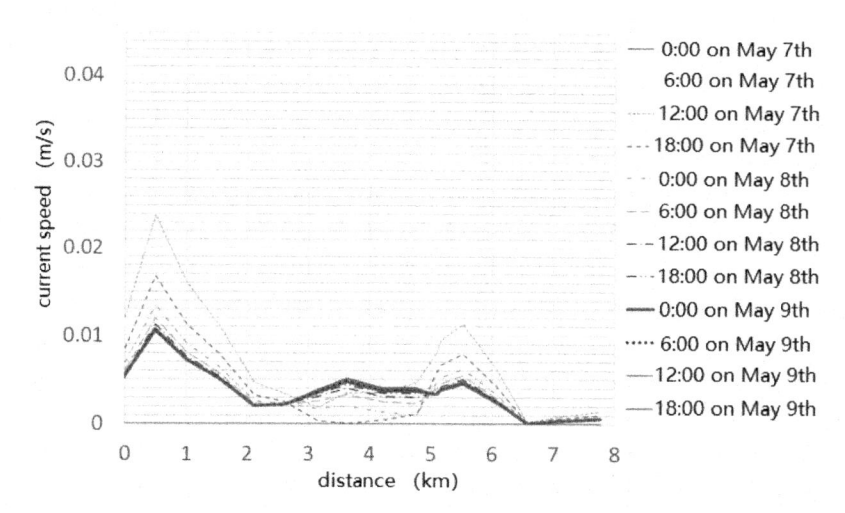

Figure 3. Variation of velocity with distance.

## 4 CONCLUSION

In this paper, a diversion scheme is proposed, and a route is selected for two-dimensional numerical simulation. Through calculation, it is found that after water diversion, the water level of the route gradually rises and is relatively stable, and the flow velocity generally increases, but the increasing degree also changes with the change of distance. The overall BOD concentration decreased, the DO concentration increased, and the water quality was improved. Through the analysis, it is concluded that the scheme can improve water quality, and the scheme is feasible. Due to the limited space, this paper only puts forward one diversion scheme. In the future, a variety of diversion schemes will be selected and compared, and the scheme that can optimize the improvement of the water environment can be selected.

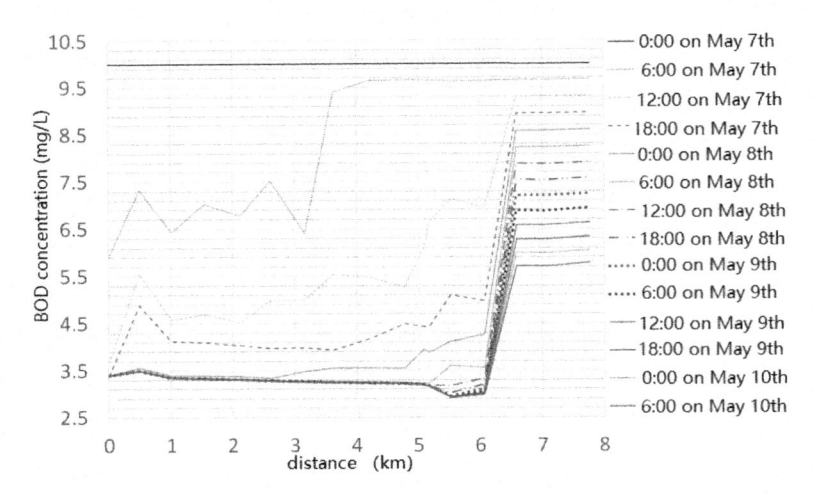

Figure 4. Variation diagram of BOD concentration with distance.

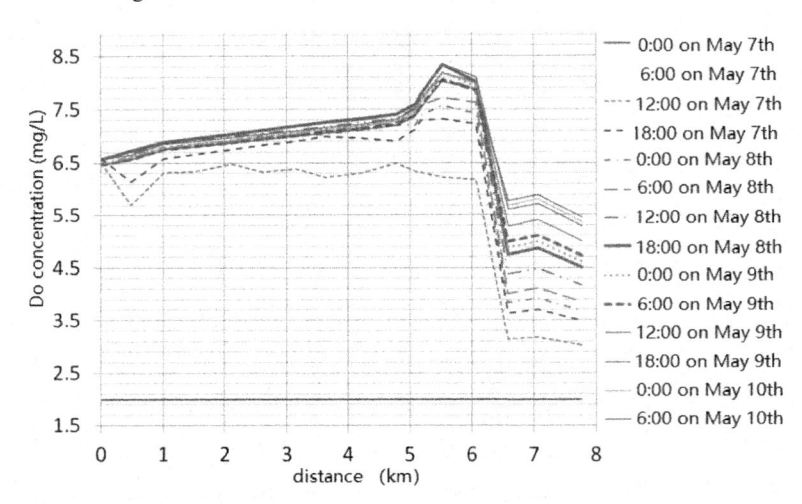

Figure 5. Variation of DO concentration with distance.

## ACKNOWLEDGMENTS

This research was supported by the Funds Key Laboratory for Technology in Rural Water Management of Zhejiang Province(ZJWEU-RWM-202101), the Joint Funds of the Zhejiang Provincial Natural Science Foundation of China (No. LZJWZ22C030001, No. LZJWZ22E090004), the Funds of Water Resources of Science and Technology of Zhejiang Provincial Water Resources Department, China (No.RB2115, No.RC2040), the National Key Research and Development Program of China (No.2016YFC0402502), the National Natural Science Foundation of China (51979249).

## REFERENCES

Chen W.J, Li D.F, Zhang H.W. (2014) Analysis of control conditions of a hydrodynamic model for flood control and drainage of the river network in Shaoxing Plain.*Journal of Zhejiang University of water resources and hydropower*, 26 (03): 38–41.

Cheng K.Y. (2005) Discussion on a mathematical model of river network flow simulation. *Northeast water resources and hydropower*, (03): 3–5 + 55.

Han F, Zhao G. R, Li L. (2020) Study on water resources regulation effect in the plain river network area. *People's Yellow River*, 42 (S2): 77–79.

Tong Y.Y, Li D.F, Nie H. (2016) Study on a hydrodynamic and plane two-dimensional mathematical model of the river network in Datian Plain. *Journal of Zhejiang University of water resources and hydropower*, 28 (01): 14–17.

Yang C.J, Ying Z.L, Chen F. (2021) Study on precise diversion scheme of Shaoxing plain river network diversion project. *Journal of Zhejiang University of water resources and hydropower*, 33 (04): 20–26.

Yang F.X, Zhang P.J, Chen W.J, Li D.F, Zhang H.W. (2015) Analysis of a hydrodynamic two-dimensional numerical model of the river network and lakes of Qingshui project in Keqiao main urban area. *Journal of Zhejiang University of water resources and hydropower*, 27 (03): 18–21.

Zhu C.X, Huang C.Y, Yin T, Zhang W.J. (2009) Research progress of river network water quality model. *Sichuan environment*, 28 (02): 66–69.

*Civil Engineering and Urban Research – Mohamed & Hou (Eds)*
*© 2023 the Authors, ISBN 978-1-032-44487-1*

# Research on three-station integration construction scheme

Qifeng Zou*, He Chen & Youchao Wu
*Beijing Smartchip Microelectronics Technology Company Limited, Changping District, Beijing, China*

ABSTRACT: The future Energy Internet is a smart grid integrating energy flow, business flow, and data flow. In order to cooperate with the construction of the energy Internet, this paper proposes a three-station integration construction scheme that can realize resource reuse of substations, energy storage stations, and data center stations. First of all, the scheme puts forward the construction targets that meet the actual requirements based on the number, distribution location, and covered area of substation stations. Secondly, according to the voltage level of substations, the specific schemes of three-station integration in 10kV, 110kV, and 220kV substations are described in detail. The feasibility of practical engineering construction is verified through the study of a three-station integration scheme.

## 1 INTRODUCTION

At present, the decentralized construction of the substation, energy storage station, and data center station is not conducive to the intensive and efficient utilization of resources. This also makes it difficult to develop a highly integrated energy system CPSS (Yang et al. 2016).

In order to change this situation, the development of a highly integrated power grid physical information system and the promotion of the new mode of "three-station integration" are the future development trend (Huang et al. 2020).

The research on the three-station integration construction scheme provides a feasible reference for practical engineering construction. First of all, the actual area and geographical location of the substation are defined by means of data and field investigation. The scheme calculates the available land and power resources of the substation according to the actual conditions. Secondly, the function orientation of the new substation is defined on the basis of geographical location. The scheme selectively integrates the energy storage station or data center station according to the station function. Finally, the construction plan and construction scale are determined comprehensively considering the reconstruction difficulty, construction cost, and other factors. This paper puts forward three different construction schemes of three-station integration.

## 2 THREE-STATION INTEGRATION FRAMEWORK

This solution is based on the existing substation and integrates energy storage stations and data centers to achieve the goal of three stations in one. The integration of three stations can complement the advantages of each station:

The substation provides ready-made land, power, manpower, and perfect network conditions for energy storage stations and data center stations, and can also reduce power transmission consumption. Energy storage stations can solve the problems of peak and frequency regulation in substations and the consumption of new energy. It can also be used as a backup power source for data center

---

*Corresponding Author: zouqifeng@sgitg.sgcc.com.cn

 DOI 10.1201/9781003372417-83

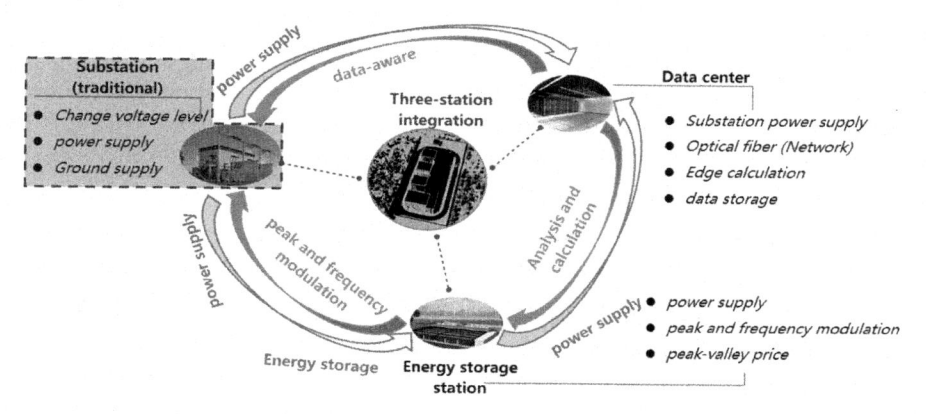

Figure 1. Three-station integration framework.

stations. The data center station can collect and send various data from substations and energy storage stations, and provide edge computing services for the power grid. In addition, it serves as a node of a smart city to provide data storage and edge computing services to the outside world, increasing profitability. Figure 1 is the overall framework of three-station integration.

## 3 CONSTRUCTION CONTENT OF THREE-STATION INTEGRATION

### 3.1 Substation construction

A substation is a place used to change the voltage during the transmission of electrical energy. Substations selected by three-station integration are step-down substations close to the user side. According to the construction goal of three-station integration, the focus is on the construction plan of 10kv, 110kv, and 220kv substations.

The location of the substation is mainly related to the load. An important principle when selecting a site is to be close to the load center (Li 2016). When the load capacity is 3MW-5MW, a 10kv substation will be built within 5-15km of the load center; when the load capacity is 10MW-50MW, a 110kv substation will be built within 50-150km of the load centers; when the load capacity is 50MW-200MW, a 220kv substation will be built within 150-300km of the load center. In terms of area, 10kv substation is generally about 200 square meters;110kv outdoor substation is generally about 4000 square meters, indoor substation is generally about 2000 square meters;220kv substation is generally about 20,000 square meters.

Therefore, the 10kv substation is closest to the electrical load, with the largest number and the smallest area; the 110kv substation is located in the urban area, with a small number and a moderate area; the 220kv substation is located on the edge of the city, with the smallest number and the largest area.

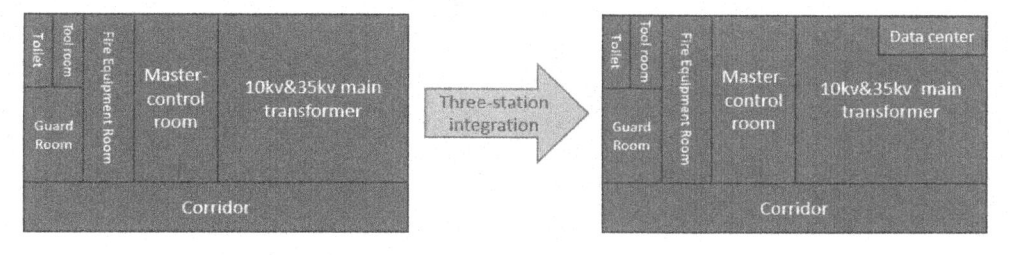

Figure 2. 10kv substation transformation plan.

The 10kv substation is a natural edge computing node because of its large number and location in the center of the city, which is conducive to the construction of data centers. At the same time, a 10kv substation cannot meet the requirements of energy storage station construction because its space is small and cannot be expanded. In addition, there is no demand for energy storage in the 10kv substation, so only a data center station needs to be added for its transformation. The transformation scheme of the 10kv substation is shown in Figure 2. The existing 10kv substation has greater redundancy in the area of the main transformer room. It is possible to free up 20-30 square meters of space from the main transformer room to build a mini data center composed of 5 cabinets. The transformed 10kv substation can provide data services in addition to the original functions. Internally, it can meet the needs of supporting integrated energy service terminal monitoring and power intelligent inspection; externally, it can provide data transfer services for Internet businesses such as smart healthcare, smart transportation, and smart security.

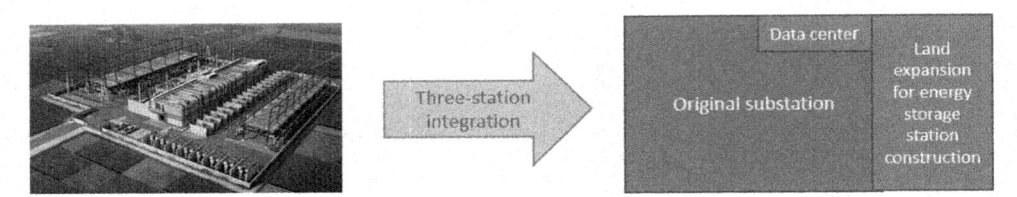

Figure 3.　110kv substation transformation plan.

Important financial districts, industrial parks, and science and technology parks are generally powered by special 110kv substations. The integration of data centers within the stations is conducive to expanding substation external business. At the same time, 110kv substations really need an energy storage station to improve the power supply quality, so a data center station and energy storage station are added at the same time for the transformation of 110kv substations. The transformation scheme of the 110kv substation is shown in Figure 3. The indoor 110kv substations are mostly intelligent substations without too much available space. Usually, 200 square meters can be set aside for the construction of a small-sized or medium-sized data center composed of 30-50 cabinets. Because the land price of the 110kv substation is relatively cheap, the original substation can be expanded to provide space for the construction of energy storage stations. The outdoor 110kv substations are mostly traditional substations, which occupy a large area. After the intelligent transformation of these stations, the volume of equipment in the station is greatly reduced (Yang 2018) and more space can be freed.

In addition, the intelligent substation does not need personnel on duty. The former staff dormitory and office area can be used to build a data center station and energy storage station without additional expansion. After the three-station integration transformation, the 110kv substation has more abundant functions. Internally, it can realize peak and frequency modulation, new energy consumption, and regional power Internet of Things data processing support. At the same time, it can also work as a spare node for power dispatching and a spare sink node for power communication. Externally, it can provide infrastructure services for governments and enterprises such as smart cities, industrial Internet, Internet business, IDC rental, and cloud services.

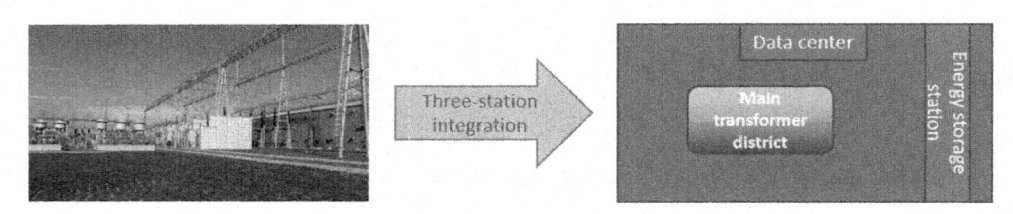

Figure 4.　220kv substation transformation plan.

kv substations choose to add energy storage stations and data center stations at the same time as 110kv substations. The area of the 220kv substation is already large, and there is enough space to build large energy storage stations and data center stations without additional expansion. Due to the location of the 220KV substation, it mainly provides internal service after the transformation of the three-station integration. Large energy storage stations are used to access new energy, reduce peak load and make up for the shortage of power; large data center stations are used to collect data from data centers in 10kV and 110kV substations. Carrying out data analysis and data mining on the basis of these data to effectively promote the confluence of energy flow, business flow, and data flow.

## 3.2 *Energy storage station*

Energy storage stations can store the excess electric energy in the low time and release it during peak time to relieve the pressure of electricity consumption. In addition, due to the instability of wind and solar energy, energy storage stations can make efficient use of these new energies and promote new energy consumption (Li et al. 2020). For the three-station integration scheme of increasing energy storage stations, the first problem that needs to be solved is the construction scale of energy storage stations. Taking the 110kv substation with a load capacity of 10-50MW as an example. Assuming that the load is overloaded by 10% for two hours at the peak of power consumption, then the energy gap of the substation during peak hours is 2-10MWh.A 40-inch lithium battery storage container has a capacity of 1MWh and covers an area of 40 square meters. Therefore, it is necessary to build an energy storage station with a capacity of 10MWh and an area of 400 square meters for the 110kv substation.

## 3.3 *Data Center Station*

Based on the construction of substations and energy storage stations, data center stations will be built according to the coverage of the Internet of Things terminals and the scale of bearer services.

Table 1. Datacenter station construction.

| Voltage level | Substation characteristics | Construction scheme | Construction scale | Support services |
|---|---|---|---|---|
| 10kv | Low-voltage substation, great quantity, close to the user, and small free space | Mini data center station | 5–20 cabinets | **Internal service:** power terminal monitoring, power intelligent inspection, power data collection, and simple data processing; **External service:** data transfer services for Internet businesses; |
| 110kv | High voltage substation, medium quantity, urban areas, and enough free space | Small-sized data center station | 30–50 cabinets | **Internal service:** power data summary, comprehensive processing of power data, power dispatch standby node, and power communication backup aggregation node; |
| 220kv | High voltage substation, small quantity, urban fringe, and enough free space | Medium-sized data center station | 5–100 cabinets | **External service:** infrastructure services such as smart cities, industrial Internet, Internet business, IDC rental, and cloud services. |

The data center station provides integrated data output services by deploying servers, storage equipment, network equipment, security equipment, and basic environmental facilities. It changes the multi-system collection and storage mode of various business data by fully accessing various types of Internet of Things data in the covered area, and realizes multi-source data collection at one time and sharing of collected data. At the same time, the data center station provides comprehensive support services for the power grid business, energy Internet business, and new Internet business. By providing localized lightweight computing support, it responds to local service needs quickly. The specific construction plan is shown in Table 1.

## 4 CONCLUSION

This paper gives the overall framework of the three-station integration and explains the specific construction model of the three-station integration. Scheme includes the selection of substations, the determination of the capacity of supporting energy storage stations, and the construction of data center stations. The main conclusions can be summarized as follows:

(1) Due to the floor area, the 10kV substation is only equipped with a micro data center station;
(2) 110kV substation integrates 10MWh energy storage station and small data center station;
(3) 220kV substation integrates 100MWh energy storage station and middle-large size data center station;

In terms of future work, the above-mentioned construction scheme should be carried out to enhance the integration of the Internet of Things and smart grid, and promote the construction process of energy Internet.

## ACKNOWLEDGMENT

Thank my colleagues in the project team for their help in writing my paper.

## REFERENCES

Huang Yi, Zhang Chuanyu, LIU Tianming. Research on "three stations in one" mode of ubiquitous Power Internet of Things[J]. *Journal of Internet technology*, 2020, 10 (01): 44–47. DOI: 10.16667 / j.i SSN. 2095-1302.2020.01.012.

Li Fa Wen. Discussion on City Substation Location and Environment impact[J]. *Telecom Power Technology*, 2016, 33(03):130–131.

Li Jian Lin, Li Ya Xin, Zhou Xi Chao. Summary of Research on Grid-Side Energy Storage Technology[J]. *Electric Power Construction*, 2020, 41(06):77–84.

Yang DESheng, Sun Fei, Wu Hongxia. Research on parallel grid System Framework for global Energy Internet [J]. *Microcomputer applications*, 2016, 32 (10): 54–58.

Yang Li Xing. *Research on the intelligent transformation of conventional substations* [D]. Nan chang University, 2018.

*Civil Engineering and Urban Research – Mohamed & Hou (Eds)*
© 2023 the Authors, ISBN 978-1-032-44487-1

# Study on treatment of printing and dyeing wastewater by different adsorption combined processes

Hongcui Li*, Yue Wu, Chenjia Zang & Boran Xie
*Qilu Institute of Technology, Jinan City, Shandong Province, China*

ABSTRACT: Taking methyl orange simulated printing and dyeing wastewater as the research object, activated carbon, bentonite, $MnO_2$, and $Mg(OH)_2$ were used to treat printing and dyeing wastewater, and the optimal experimental conditions for each adsorption material were determined. Orthogonal experiments show that the optimal reaction times of activated carbon, bentonite, $MnO_2$, and $Mg(OH)_2$ adsorption process are 45, 30, 30, and 40 min, respectively; the dosage is 0.06, 0.6, 0.03, and 0.12 g; , 20, 20, 20°C; pH values are 4.00, 6.00, 3.00, 12.00. Six different adsorption combined processes were designed to treat printing and dyeing wastewater, the COD of printing and dyeing wastewater before and after adsorption was measured, and the COD removal rates of methyl orange printing and dyeing wastewater by different adsorption combined processes were compared. The COD removal rate of the simulated printing and dyeing wastewater was the highest, which was 88.6%; the COD removal rate of the methyl orange simulated printing and dyeing wastewater by the combined process of bentonite and $Mg(OH)_2$ adsorption was the lowest, which was 74.3%.

## 1 INTRODUCTION

According to statistics, the annual discharge of printing and dyeing wastewater in my country can reach one billion tons, accounting for about one-third of the entire industrial wastewater discharge (Zhai 2021). If the wastewater is not properly treated, it will cause serious water pollution and destroy the water ecosystem. Therefore, it is of great significance to strengthen the treatment of printing and dyeing wastewater (Xue 2021). Methyl orange is widely used in papermaking, printing, and other industries. Methyl orange is not easy to degrade. The azo and quinoid structures under acidic and alkaline conditions are the main structures of dye compounds, which are representative (Li et al. 2020). There are many methods for treating printing and dyeing wastewater at home and abroad, and the commonly used methods are chemical coagulation, biological methods, membrane filtration, advanced oxidation, and adsorption (Sun et al. 2018). Among them, the adsorption method is a physical and chemical separation technology for dissolved pollutants, which has the advantages of low cost, simple operation, and regenerable adsorbent (Ma 2018).

In order to further improve the treatment effect and increase the efficiency of removing pollutants, the combined adsorption process has been applied and developed in water treatment (Maharani et al. 2020; Wei et al. 2020; Wan 2020). In the experiment, the adsorption process of activated carbon, bentonite, $MnO_2$, and $Mg(OH)_2$ was selected to treat printing and dyeing wastewater, and the optimal treatment conditions were determined by an orthogonal experiment. Under the optimal treatment conditions, the treatment effects of different adsorption combined processes in the treatment of printing and dyeing wastewater were compared and analyzed.

---

*Corresponding Author: lhc0827@163.com

DOI 10.1201/9781003372417-84

## 2 MATERIALS AND METHODS

### 2.1 Experimental materials and instruments

pH meter (Shanghai Sanxin Instrument Factory), COD digester (Shanghai INESA Scientific Instrument Co., Ltd.), constant temperature oscillator (Guohua Enterprise), COD analyzer (Shanghai INESA Scientific Instrument Co., Ltd.).

Methyl orange simulated printing and dyeing wastewater: COD is 500 mg/L.

Activated carbon, Bentonite, $MnO_2$, $Mg(OH)_2$

### 2.2 Experimental method

#### 2.2.1 Single factor adsorption experiment

Single-factor adsorption experiments were carried out to explore the effects of reaction time, adsorbent dosage, pH, and temperature on the COD removal rate of printing and dyeing wastewater. The experimental process is as follows: Pipette 100 mL of simulated printing and dyeing wastewater, add it to 250 mL conical flasks, add different qualities of activated carbon, bentonite, $MnO_2$, $Mg(OH)_2$, adjust the pH, time, and temperature to be the same, measure The COD removal rate before and after the reaction was used to determine the optimal dosage. In the same way, the optimal reaction time, pH value, and temperature were determined.

Table 1. The best single factor for different adsorption materials to adsorb printing and dyeing wastewater.

| Adsorbent material | Time (min) | Dosage (g) | pH | Temperature (°C) |
|---|---|---|---|---|
| Activated carbon | 50 | 0.060 | 4 | 30 |
| Bentonite | 30 | 0.060 | 5 | 25 |
| $MnO_2$ | 25 | 0.050 | 2.5 | 20 |
| $Mg(OH)_2$ | 40 | 0.120 | 11 | 25 |

#### 2.2.2 Orthogonal experiment

In the actual sewage treatment process, not all factors work alone, but a variety of factors work together. Therefore, taking the reaction time (A), dosage (B), pH value (C), and temperature (D) as the factors to be investigated, the $L_9(3^4)$ orthogonal experiment was designed on the basis of the single factor experiment. The factor level table is shown in Table 1, and the determination method is the same as that of the adsorption experiment.

#### 2.2.3 Treatment of printing and dyeing wastewater by different adsorption combined processes

The four adsorbent materials are combined in pairs to obtain activated carbon, bentonite adsorption combination, activated carbon, $MnO_2$ adsorption combination, activated carbon, $Mg(OH)_2$ adsorption combination, bentonite, $MnO_2$ adsorption combination, bentonite, $Mg(OH)_2$ adsorption combination, $MnO_2$, six different combinations of $Mg(OH)_2$ adsorption combinations were used, and the adsorption combination experiments were carried out respectively. Experimental process: take the adsorption combination of activated carbon and bentonite as an example, take three 250 mL conical flasks, pipette 50 mL of simulated printing and dyeing wastewater respectively, and carry out the adsorption combination according to the optimal reaction conditions obtained from the orthogonal experiment of activated carbon and bentonite in 2.2.2 In the experiment, the COD before and after adsorption was measured, and the COD removal rate was calculated.

Table 2. Orthogonal experimental factor level table for different adsorption materials to adsorb printing and dyeing wastewater.

| Adsorbent material | Factor<br>Level | Time (min) A | Dosage (g) B | pH C | Temperature (°C) D |
|---|---|---|---|---|---|
| Activated carbon | 1 | 45 | 0.040 | 3 | 25 |
| | 2 | 50 | 0.060 | 4 | 30 |
| | 3 | 55 | 0.080 | 5 | 35 |
| Bentonite | 1 | 25 | 0.400 | 4 | 20 |
| | 2 | 30 | 0.600 | 5 | 25 |
| | 3 | 35 | 0.800 | 6 | 30 |
| $MnO_2$ | 1 | 35 | 0.100 | 10 | 20 |
| | 2 | 40 | 0.120 | 11 | 25 |
| | 3 | 45 | 0.140 | 12 | 30 |
| $Mg(OH)_2$ | 1 | 35 | 0.100 | 10 | 20 |
| | 2 | 40 | 0.120 | 11 | 25 |
| | 3 | 45 | 0.140 | 12 | 30 |

## 3 RESULTS AND DISCUSSION

### 3.1 *Determination of optimal reaction conditions for different adsorption processes*

The experimental results show that:

(1) In the activated carbon experiment, through the range analysis of each factor, it can be concluded that the primary and secondary order of every single factor is: activated carbon dosage &gt; solution pH &gt; temperature &gt; reaction time. The best orthogonal experimental combination of activated carbon adsorption of printing and dyeing wastewater is $A_1B_2C_2D_2$, that is, the reaction time is 45min, the dosage is 0.060 g, and the pH value of the solution is 4 and the reaction temperature is 30°C.

(2) In the bentonite experiment, through the range analysis of various factors, it can be concluded that the primary and secondary order of every single factor is: bentonite dosage &gt; pH &gt; reaction time &gt; temperature. The best orthogonal experimental combination of bentonite adsorption of printing and dyeing wastewater is $A_2B_2C_3D_1$, that is, the reaction time is 30 min, the dosage is 0.600 g, and the pH value of the solution is 6, and the reaction temperature is 20°C.

(3) In the $MnO_2$ experiment, through the analysis of the range of each factor, it can be concluded that the primary and secondary order of every single factor is: solution pH &gt; $MnO_2$ dosage &gt; reaction temperature &gt; reaction time. The best orthogonal experimental combination of $MnO_2$ adsorption of printing and dyeing wastewater is $A_3B_1C_3D_2$, that is, the reaction time is 30 min, the dosage is 0.030 g, and the pH value of the solution is 3 and the reaction temperature is 20°C.

(4) In the $Mg(OH)_2$ experiment, through the analysis of the range of each factor, it can be concluded that the primary and secondary order of every single factor is: reaction temperature &gt; $Mg(OH)_2$ dosage &gt; solution pH &gt; reaction time. The best orthogonal experimental combination of $Mg(OH)_2$ adsorption of printing and dyeing wastewater is $A_2B_2C_3D_1$, that is, the reaction time is 40min, the dosage is 0.120 g, the pH value of the solution is 12 and the reaction temperature is 20°C.

### 3.2 *Comparison of treatment effects of different adsorption combination processes on printing and dyeing wastewater*

It can be seen from Figure 1 that the treatment effect of the combined adsorption process on printing and dyeing wastewater is better than that of the single adsorption process. The combined process of activated carbon and $MnO_2$ adsorption has the highest COD removal rate of 88.6%

Table 3. Orthogonal experiment results of four kinds of adsorption materials adsorbing printing and dyeing wastewater.

| Group number | Factor | Time (min) A | Dosage (g) B | Temperature pH C | (°C) D | COD removal rate % Activated carbon | Bentonite | MnO$_2$ | Mg(OH)$_2$ |
|---|---|---|---|---|---|---|---|---|---|
| Activated carbon | 1 | 1 | 1 | 1 | 1 | 48.6 | 27.7 | 54.4 | 53.1 |
| | 2 | 1 | 2 | 2 | 2 | 67.2 | 30.8 | 50.2 | 46.2 |
| | 3 | 1 | 3 | 3 | 3 | 53.2 | 31.5 | 61.2 | 52.3 |
| | 4 | 2 | 1 | 2 | 3 | 60.4 | 27.2 | 58.3 | 47.5 |
| | 5 | 2 | 2 | 3 | 1 | 58.6 | 35.9 | 53.4 | 58.7 |
| | 6 | 2 | 3 | 1 | 2 | 51.4 | 22.6 | 50.0 | 53.2 |
| | 7 | 3 | 1 | 3 | 2 | 53.3 | 30.1 | 66.5 | 50.1 |
| | 8 | 3 | 2 | 1 | 3 | 57.2 | 32.3 | 55.2 | 53.3 |
| | 9 | 3 | 3 | 2 | 1 | 50.4 | 29.6 | 58.6 | 54.7 |
| | Mean 1 | 0.5633 | 0.5410 | 0.5240 | 0.5253 | | | | |
| | Mean 2 | 0.5680 | 0.6100 | 0.5933 | 0.5730 | | | | |
| | Mean 3 | 0.5363 | 0.5167 | 0.5503 | 0.5693 | | | | |
| | Range | 0.0317 | 0.0933 | 0.0693 | 0.0477 | | | | |
| Bentonite | Mean 1 | 0.3000 | 0.2907 | 0.2753 | 0.3107 | | | | |
| | Mean 2 | 0.2857 | 0.3300 | 0.2920 | 0.2783 | | | | |
| | Mean 3 | 0.3067 | 0.2790 | 0.3250 | 0.3033 | | | | |
| | Range | 0.0210 | 0.0510 | 0.0497 | 0.0324 | | | | |
| MnO$_2$ | Mean 1 | 0.5527 | 0.5973 | 0.5320 | 0.5547 | | | | |
| | Mean 2 | 0.5390 | 0.5293 | 0.5570 | 0.5557 | | | | |
| | Mean 3 | 0.6010 | 0.5660 | 0.6037 | 0.5823 | | | | |
| | Range | 0.0620 | 0.0680 | 0.0717 | 0.0276 | | | | |
| Mg(OH)$_2$ | Mean 1 | 0.5053 | 0.5023 | 0.5320 | 0.5550 | | | | |
| | Mean 2 | 0.5313 | 0.5273 | 0.4947 | 0.5320 | | | | |
| | Mean 3 | 0.5270 | 0.5340 | 0.5370 | 0.5103 | | | | |
| | Range | 0.0217 | 0.0317 | 0.0423 | 0.0447 | | | | |

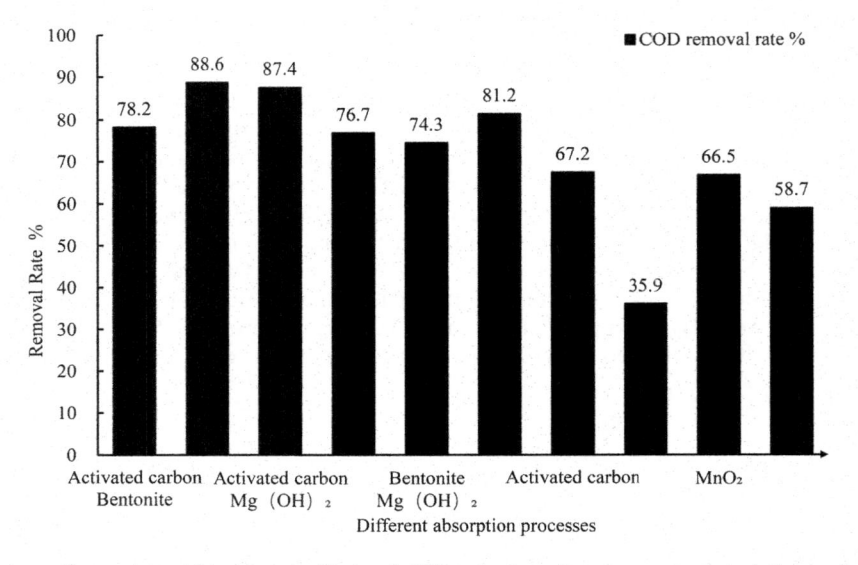

Figure 1. Comparison of treatment effects of different adsorption processes on printing and dyeing wastewater.

for methyl orange simulated printing and dyeing wastewater; The combined process of bentonite and $Mg(OH)_2$ adsorption has the lowest COD removal rate of 74.3% for methyl orange simulated printing and dyeing wastewater. The removal of COD from printing and dyeing wastewater by adsorbent is mainly due to the synergistic effect of adsorption and photocatalysis. The combination of activated carbon with high specific surface area and adsorption performance and $MnO_2$ with good photocatalytic performance can play a synergistic optimization and complementary effect. Therefore, the combined process of activated carbon $MnO_2$ adsorption can significantly improve the COD removal rate of printing and dyeing wastewater (Zheng et al. 2019). The unique performance of natural bentonite determines that it is especially suitable for the removal of heavy metals and other pollutants in wastewater. However, due to the strong hydrophilicity of its surface silica structure and the hydrolysis of interlayer cations, its surface is usually covered with a thin water film. Therefore, the unmodified bentonite has a poor ability to adsorb organic matter, but it is used to treat dye wastewater after modification.

## 4 CONCLUSION

(1) The order of COD removal rate of methyl orange simulated printing and dyeing wastewater adsorbed by a single adsorption material is: activated carbon &gt; $MnO_2$ &gt; $Mg(OH)_2$ &gt; bentonite.
(2) The order of COD removal rate of methyl orange simulated printing and dyeing wastewater by each adsorption combination process is: activated carbon and $MnO_2$ combination process &gt; activated carbon and $Mg(OH)_2$ combination process &gt; $MnO_2$ and $Mg(OH)_2$ combination process &gt; activated carbon and bentonite combination process &gt; bentonite, $MnO_2$ combination process &gt; bentonite and $Mg(OH)_2$ combination process.
(3) Compared with the single adsorption process, the combined adsorption process has significantly improved the treatment effect of printing and dyeing wastewater.

## REFERENCES

Li X L, Deng L X, Huang J M, Xin L. (2020) Study on decolorization of Methyl Orange Wastewater by photocatalytic hydrogen peroxide [J]. *Environmental Science and Technology*, 33 (05): 35–39 +43.
Ma A. (2018) Application and research of adsorption method in sewage treatment [J]. *Aging and application of synthetic materials*, 47 (02): 119–123.
Maharani S, Nailufhar L, Sugiarti Y. (2020) Adsorption effectivity of combined adsorbent zeolite, activated charcoal, and sand in liquid waste processing of agroindustrial laboratory [J]. *IOP Conference Series: Earth and Environmental Science*, 443.
Sun X X, Li X Y, Xu J. (2018) Experimental study on treatment of printing and dyeing wastewater by different activated carbon adsorption combined processes [J]. *Industrial water treatment*, 38 (07): 78–80.
Wan Y. (2020) *Experimental study on potassium permanganate powder activated carbon combined treatment of source water of South-to-North Water Transfer* [D]. Hebei University of engineering.
Wei H Y, Li X, Li R Y, Liu H. (2020) Research progress of activated carbon adsorption combined process in the treatment of printing and dyeing wastewater [J]. *Shandong chemical industry*, 49 (06): 83–84 + 92.
Xue G. (2021) Progress in treatment technology of printing and dyeing wastewater [J]. *Industrial water treatment*, 41(09): 10–17.
Zhai J. (2021) Study on treatment process of reactive dye printing and dyeing wastewater [J]. *Green building materials*, 12: 33–34.
Zheng C, Fang S G, Li W, Zhan L W. (2019) Preparation and properties of activated carbon/manganese dioxide composites [J]. *Journal of Fujian Institute of engineering*, 17 (01): 1–6.

*Architectural model research and environmental numerical monitoring and analysis*

*Civil Engineering and Urban Research – Mohamed & Hou (Eds)*
*© 2023 the Authors, ISBN 978-1-032-44487-1*

# Power load forecasting based on canonical-correlation analysis and LSTM networks

Duanxu Liu, Bin Wu* & Li Sun
*School of Business, Shanghai Dianji University, Shanghai, China*

ABSTRACT:   Under the premise of limited data, to improve the accuracy of short-term power load forecasting, a data-enhanced load forecasting method using a long-short-term memory network combined with canonical-correlation analysis is proposed. Firstly, canonical-correlation analysis is performed on the data to determine the correlation of each influencing factor in the data set to load prediction. The data is then augmented based on the correlations to reconstruct the dataset. Finally, the prediction model of the long short-term memory network is created, and the reconstructed data is added for power load prediction. The model load prediction methods such as RNN, LSTM, and GRU are compared. The results show that LSTM has a better stability. After adjusting and training with the canonical-correlation analysis method, the prediction accuracy of the model is further improved.

## 1   INTRODUCTION

Power load forecasting plays a very important role in the security, stability, and economic operation of the power grid. Load forecast can be divided according to the length of forecast time, short-term load forecast can be from one hour to one week (Lahouar & Slama 2015). The accuracy of short-term load forecasting will directly or indirectly affect the effectiveness of economic dispatch (Quan et al. 2014).

With the rapid development of the contemporary economy, the factors affecting the load are increasing, and some traditional forecasting methods cannot meet the environmental requirements of the current period (Wu et al. 2016). Therefore, more and more comprehensive algorithms are applied to load forecasting. Feng et al. compared load forecasting methods such as random forest and linear regression and used a random forest model with improved local forecasting to improve the accuracy and stability of short-term power load forecasting (Feng e tal. 2021). Although the single-step prediction accuracy has been improved, the prediction accuracy of multi-step prediction is lower, and the influence of different feature factors on the prediction results is not considered. In recent years, as the research interest in deep learning continues to rise, there has also been a research upsurge in the field of power load forecasting. Wang et al. used the improved BP neural network for prediction (Wang et al. 2021). Although the model is simple and the training convergence speed is fast, they did not consider the time series information between the load data. Wang Ting used the Pearson correlation coefficient to analyze the correlation between different influencing factors and load, and then the features with low correlation were discarded, and the features with high correlation were selected as the data for predictive analysis (Wang 2020). The background of this study is the fact that there is a variety of characteristic data, and data enhancement is not considered for the high-correlation data. Wu et al. used the method of principal component analysis combined with LSTM for load forecasting (Wu et al. 2021). The main information of the data was extracted,

---

*Corresponding Author: wub@sdju.edu.cn

so the data features were reduced and the prediction speed was improved. However, this study did not consider the impact of data augmentation based on the correlation of data features on the accuracy of load forecasting.

In addition, for deep learning algorithms, a large amount of data training is often used to improve their prediction accuracy. However, the data for power load prediction is usually limited in practical use. Based on the above research and practical problems, this paper proposes to use canonical-correlation analysis to identify the data information of data samples in the case of limited forecast data. Then data enhancement is carried out, and finally, the LSTM algorithm is combined to improve the accuracy of short-term power load forecasting.

## 2 METHOD AND THEORY BASIS

### 2.1 Long short-term memory networks (LSTMs)

LSTM is a temporal recurrent neural network whose design structure is more complex and unique than RNN. There are three gates in its single neuron, namely forget gate $f$, input gate $i$, and output gate $o$. After the information $C_t$ is input to this neuron, it will be processed in turn through these three gates to generate the output $C_{t-1}$. The state update equation of the basic unit of LSTM is shown in Equations (1) to (5):

$$i_t = sigmoid(W_i \cdot [h_{t1}, x_t] + b_i) \tag{1}$$

$$a_t = tanh(W_c \cdot [h_{t1}, x_t] + b_c) \tag{2}$$

$$C_t = f * C_{t-1} + i_t * a_t \tag{3}$$

$$a_t = sigmoid(W_o \cdot [h_{t1}, x_t] + b_o) \tag{4}$$

$$h_t = o_t * +tanhC_t \tag{5}$$

where $W$ is the weight matrix; $b$ is the deviation vector; $x_t$ is the input vector; $h_{t-1}$ is the output of the previous moment; $h_t$ is the output of the hidden layer unit at the current moment.

### 2.2 Canonical-correlation analysis

Canonical-correlation analysis is a multivariate statistical method to study the correlation between two groups of variables. It measures the strength of the connection between two groups of variables through the canonical-correlation coefficient. The correlation between the two groups of variables can be reflected by the strength of this connection between the two groups of variables.

Therefore, we can make the following assumptions: $X^{(1)} = \left(X_1^{(1)}, X_2^{(1)}, \ldots, X_P^{(1)}\right), X^{(2)} = \left(X_1^{(2)}, X_2^{(2)}, \ldots, X_q^{(2)}\right)$. They are interrelated random variables; each of them is selected as representative comprehensive variables $U_i$ and $V_i$ from two random variables. Let each comprehensive variable be a linear combination of the original variables, then we can obtain Equations (6) and (7)

$$U_i = a_1^{(i)} X_1^{(1)} + \ldots + a_p^{(i)} X_p^{(1)} \stackrel{\text{def}}{=} a^{(i)'} X^{(1)} \tag{6}$$

$$V_i = b_1^{(i)} X_1^{(2)} + \ldots + a_q^{(i)} X_q^{(2)} \stackrel{\text{def}}{=} b^{(i)'} X^{(2)} \tag{7}$$

If there are constant vectors $a^{(1)}$ and $b^{(1)}$, in $D(a^{(1)'} X^{(1)} = D(b^{(1)'} X^{(2)}) = 1$ so that $\rho(a^{(1)'} X^{(1)}, b^{(1)'} X^{(2)})$ reaches the maximum, then it can be called the first pair of canonical correlation variables. After that, the second and third pairs of canonical-correlation variables that are not correlated with each other are similarly obtained, reflecting the linear correlation between $X^{(1)}$ and $X^{(2)}$.

# 3 CCA-BASED LSTM MODEL

## 3.1 *Model structure*

The model of CCA-LSTM is divided into an input layer, hidden layer, and output layer. Here, we use early stopping to observe the loss curve and add random deactivation layers to prevent overfitting. The rectified linear activation function or ReLU for short is used to prevent gradient disappearance and gradient explosion, and the Adam algorithm is used for the optimization algorithm. A grid search method is used to traverse all permutations of hyperparameters. A cross-validation approach is applied to find the optimal hyperparameter combination for the model. The trained model is added to the test set data to obtain the prediction result.

## 3.2 *CCA-LSTM model prediction process*

The process of the CCA-LSTM load prediction model proposed in this paper is as follows:

Step 1: Firstly, the acquired existing load data is analyzed by the method of CCA so that the typical correlation coefficient between the power load data and other various characteristic data is obtained. Then, the features with a strong correlation to the coefficient results are selected for data enhancement processing.

Step 2: To eliminate the adverse effects caused by singular sample data and speed up the solution of gradient descent, normalize the enhanced reconstructed data set.

Step 3: Divide the normalized reconstruction data set into a training set and test set. Input the data from the training set into the model for training and optimize the model.

Step 4: Add test set data to the tuned model to obtain prediction results.

The process of CCA-LSTM prediction is detailed in Figure 1.

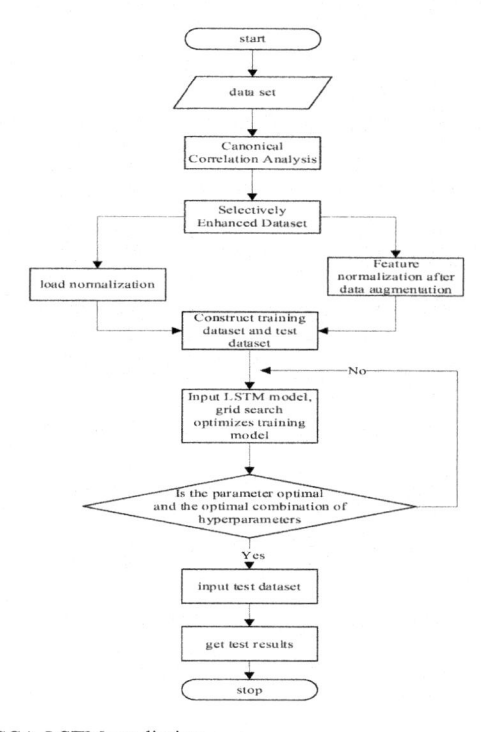

Figure 1.   Flow chart of CCA-LSTM prediction.

### 3.3 Experimental evaluation indicators

(1) The mean absolute percentage error (MAPE) is expressed as:

$$\text{MAPE} = \sqrt{\frac{1}{n}\sum_{i=1}^{n}|y^{\wedge} - y_i|^2} \tag{8}$$

(2) The mean squared error (RMSE) is expressed as:

$$RMSE = \frac{1}{n}\sum_{i=1}^{n}\frac{|y^{\wedge} - y_i|}{y_i} \tag{9}$$

(3) The mean absolute error (MAE) is expressed as:

$$MAE = \frac{1}{n}\sum_{i=1}^{n}|y^{\wedge} - y_i|^2 \tag{10}$$

(4) The coefficient of determination ($R^2$) is expressed as:

$$R^2 = 1 - \frac{\sum_{i=1}^{n}(y_i - y_i^{\wedge})^2}{\sum_{i=1}^{n}(y_i - y_i^{-})^2} \tag{11}$$

where $y^{\wedge}$ represents the predicted value of the power load; $y_i$ represents the actual value of the power load; $y_i^{-}$ represents the average value of the time series; $n$ represents the number of sample data in the test set.

## 4 EXPERIMENTAL RESULTS AND ANALYSIS

### 4.1 Data sources

The proposed method is tested and validated using the electricity load dataset published by the Australian Energy Market Operator (AEMO). The public data set includes the actual load data from 2006 to 2011. The data features are composed of dry bulb temperature, dew point temperature, wet bulb temperature, humidity, electricity price, and electricity load. The data sampling frequency is 30 min, indicating that 48 data are collected every day, and there are 87,648 data in the entire data set. Single-step prediction takes the first 87,647 data samples as the training set and the last data as the test sample. The multi-step prediction uses the first 87,600 data samples as the training set and the last 48 data samples as the test samples.

### 4.2 Perform a canonical-correlation analysis of the data

The results of the canonical-correlation analysis show that different features have different effects on model prediction. SPSS is used to carry out canonical-correlation analysis on data characteristics, in which dry bulb temperature and electricity load and electricity price and electricity load are positively correlated, and the other three characteristics are negatively correlated with electricity load, and the results are shown in Table 1.

### 4.3 Comparison of forecast results

Firstly, according to the results, we select the two features with the largest correlation coefficient with power load, namely dry bulb temperature and electricity price, as Group A, and select the two

Table 1. Results of the canonical-correlation analysis.

| Dual feature name | Pearson correlation coefficient |
|---|---|
| Dry bulb temperature and electrical load | 0.098** |
| Dew point temperature and electrical load | −0.111** |
| Wet bulb temperature and electrical load | −0.025** |
| Humidity and electrical load | −0.271** |
| Electricity price and electricity load | 0.156** |

features with the smallest correlation coefficient with power load, namely dew point temperature and humidity, as Group B. The LSTM model is joined for single-step prediction after other features have been removed. The prediction results and model evaluation results are shown in Tables 2 and 3.

Table 2. Prediction results for dual characteristic loads.

| Actual load | Group A | Group B |
|---|---|---|
| 8063.36 | 8114.404 | 7826.899 |

Table 3. Evaluation results for the dual-feature prediction model.

| Evaluation indicators | Group A | Group B |
|---|---|---|
| MAPE | 0.00633 | 0.02933 |
| RMSE | 51.04381 | 236.46107 |

Then, the original dataset with five features is put into LSTM, RNN, and GRU models for optimization training and single-step prediction. To verify the relative advantage of the LSTM prediction model in single-step prediction, we compare the obtained single-step prediction results. Then, based on the experimental results in Tables 2 and 3, features from Group A are added to the original data set again for data enhancement, and the data set with seven features are reconstructed. Finally, the reconstructed data set is used to train the CCA-LSTM, CCA-RNN, and CCA-GRU models and the single-step prediction results obtained are compared with the five-feature model. The single-step prediction results of the six models and the single-step prediction evaluation results of the six models are shown in Tables 4 to 7.

Table 4. Single-step prediction results for five-feature models.

| Actual load value | LSTM predictions | RNN predictions | GRU predictions |
|---|---|---|---|
| 8063.36 | 8052.6787 | 8006.9136 | 8075.971 |

Similar to single-step prediction, multi-step prediction is also performed using data sets for the two characteristics. The results and the multi-step prediction evaluation results of the six models are shown in Table 8, Table 9, and Figure 2.

The smaller the RMSE and MAPE values, the better the prediction effect of the model. Analysis of Table 2 and Table 3 shows that the prediction accuracy of the LSTM network model with positive correlation and dual features is higher, indicating that different factors have a great influence on the prediction results of power load. Therefore, considering the prediction speed and accuracy of the model, the features of Group A are used for data enhancement.

Table 5. Single-step prediction and evaluation results for five-feature models.

| Evaluation indicators | LSTM | RNN | GRU |
|---|---|---|---|
| MAPE | 0.00132 | 0.00700 | 0.00156 |
| RMSE | 10.68129 | 56.44642 | 12.61119 |

Table 6. Single-step prediction results for seven-feature models.

| Actual load value | CCA-LSTM predictions | CCA-RNN predictions | CCA-GRU predictions |
|---|---|---|---|
| 8063.36 | 8061.387 | 7580.7617 | 8077.8823 |

Table 7. Single-step prediction and evaluation results for seven-feature models.

| Evaluation indicators | CCA-LSTM | CCA-RNN | CCA-GRU |
|---|---|---|---|
| MAPE | 0.00024 | 0.05985 | 0.00180 |
| RMSE | 1.97279 | 482.59828 | 14.52232 |

Table 8. Multi-step prediction and evaluation results of seven-feature species models.

| Evaluation indicators | LSTM | GRU | RNN |
|---|---|---|---|
| $R^2$ | 0.8812 | 0.8852 | 0.8752 |
| MAE | 444.0989 | 422.9293 | 409.9098 |
| MAPE | 0.0527 | 0.0475 | 0.0470 |
| RMSE | 507.7674 | 499.0002 | 520.2676 |

Table 9. Multi-step prediction evaluation results of seven-feature models.

| Evaluation indicators | CCA-LSTM | CCA-GRU | CCA-RNN |
|---|---|---|---|
| $R^2$ | 0.9181 | 0.8338 | 0.8990 |
| MAE | 358.7575 | 521.6878 | 383.2558 |
| MAPE | 0.0426 | 0.0589 | 0.0450 |
| RMSE | 421.5813 | 600.4647 | 468.0810 |

By analyzing Table 4, Table 5, Table 6, and Table 7, we found that in the single-step prediction, among the five feature prediction models LSTM, RNN, and GRU, each evaluation index of LSTM is optimal. However, compared with the prediction results of CCA-LSTM, CCA-RNN, and CCA-GRU of seven features, the indicators of CCA-LSTM have improved, but the indicators of CCA-RNN and CCA-GRU have not been significantly improved. CCA-LSTM has the best performance in single-step prediction. Compared with LSTM, its RMSE is reduced by 8.7085, and MAPE is reduced by 0.108%.

By analyzing Table 8, Table 9, and Figure 2, it can be concluded that in the prediction model LSTM, RNN, and GRU of 5 features, each evaluation index of LSTM is not optimal in multi-step prediction. However, compared with CCA-LSTM, CCA-RNN, and CCA-GRU with seven features, we found that CCA-LSTM is still the best among the six models. Compared with the LSTM with five features, the RMSE of CCA-LSTM decreased by 86.1861, the MAPE decreased by 1.01%, the value of $R^2$ increased by 0.0369, and the MAE decreased by 85.3414. From this, it can be

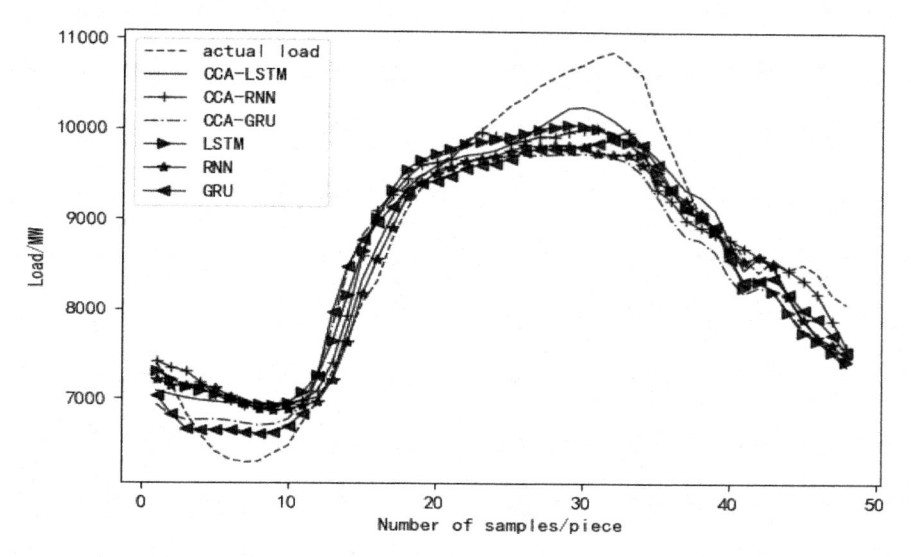

Figure 2.    Multi-step prediction results of the six models.

concluded that the validity of the CCA-LSTM model prediction method can improve the accuracy of power load prediction.

## 5    CONCLUSIONS

Considering the limited data, this paper proposes a load forecasting model based on the CCA-LSTM network. The CCA-LSTM proposed in this paper is verified by the simulation modeling of the data and the following conclusions are obtained.

(1)  By comparing the experimental results, it is found that the positive correlation characteristic factors have a greater impact on the load forecasting results than the negative correlation.
(2)  LSTM predicts multi-feature data better than LSTM, RNN, GRU, CCA-RNN, and CCA-GRU when compared with single-step prediction and multi-step prediction of the CCA-LSTM network. The proposed data-enhanced model can provide higher prediction accuracy.

Therefore, the load forecasting under the deep learning model in the future can be further studied in combination with canonical-correlation analysis.

## ACKNOWLEDGMENT

This work was supported by the Humanity and Social Science Youth Foundation of the Ministry of Education, China (Grant No. 20YJCZH027).

## REFERENCES

Feng, Z., Wang, Y., Yuan, B., Feng, X., Yao, Z., Wu, Z. (2021) Short-term power load forecasting based on random forest and improved local forecasting. *J. Water Resources and Hydropower Technology* (Chinese and English), 52 (S2): 300–305. DOI: 10.13928/j.cnki.wrahe. 2021. S2.067.

Jiang, M., Xu, L., Zhang, K., Ma, Y. (2022) Recurrent Neural Network Wind Speed Time Series Prediction Based on Seasonal Index Adjustment. *J. Chinese Journal of Solar Energy*, 43 (02): 444–450. DOI: 10.19912/j. 0254-0096. tynxb. 2020-0389.

Lahouar, A., Slama, J. H. (2015) Day-ahead load forecast using random forest and expert input selection. *J. Energy Conversion & Management*, 103: 1040–1051.

Liu, X. (2021) *Short-term load forecasting of power users based on deep learning*. D. North China Electric Power University, DOI: 10.27139/d.cnki.ghbdu. 2021.000520.

Ma, T., Wu, Y. (2021) Load forecasting of distribution network based on grey relational analysis and BP neural network. *J. Rural Electrification*, 2021 (12): 17–20. DOI: 10.13882/j.cnki.ncdqh. 2021.12. 004.

Quan, H., Srinivasan, D., Khosravi, A. (2014) Short-term load and wind power forecasting using neural network-based prediction intervals. *J. IEEE transactions on neural networks and learning systems*, 25 (2): 303–15.

Wang, L., Yang, H., Li, J. (2021) Research on power load forecasting based on improved BP neural network algorithm. *J. New Industrialization*, 11 (01): 112–113. DOI: 10.19335/j.cnki. 2095-6649.2021.1.044.

Wang, T. (2020) *Power Load Forecasting Based on LSTM Deep Network*. D. Shanxi University, DOI: 10.27284/d.cnki.gsxiu. 2020.000860.

Wu, D., Zhong, J., Wang, X., Xiang, J., Zeng, F., Hu, K., Chen, C. (2021) Principal Component Analysis and Power Load Prediction for Long Short-Term Memory Networks. *J. Internet of Things Technology*, 11 (08): 47–51. DOI: 10.16667/j.issn.2095-1302.2021.08.015.

Wu, Q., Gao, J., Hou, G., et al. (2016) Short-term load forecasting support vector machine algorithm for multi-source heterogeneous fusion of influencing factors. *J. Automation of Electric Power Systems*, 40 (15): 67–72.

Yang, D., Yang, J., Hu, C., Cui, D., Chen, Z. (2021) Short-term power load forecasting based on improved LSSVM. *J. Electronic Measurement Technology*, 44 (18): 47–53. DOI: 10.19651/j.cnki.emt. 2107628.

Civil Engineering and Urban Research – Mohamed & Hou (Eds)

# Mechanism of traffic safety culture on traffic accidents based on a structural equation model

Zhenqing Hao*

*Bureau of Transport of Jinzhong, Jinzhong, Shanxi, China*

ABSTRACT: To explore the influence of traffic safety culture on traffic accidents, this paper constructed a theoretical model of traffic safety culture-dangerous driving behavior-traffic accident, and used the structural equation model (SEM) to quantify the relationship between traffic safety culture, driving behavior and traffic accidents based on the data from 288 questionnaires of drivers.

## 1 INTRODUCTION

Among the most important public health issues in the world is road traffic safety. According to the report of the World Health Organization, 1.24 million people die in traffic accidents every year, and about 20 million to 50 million people suffer from non-fatal injuries, resulting in economic losses of more than a 150billion US dollars. In recent years, with the acceleration of urbanization in China, the number of motor vehicles and drivers has increased dramatically, and the traffic safety problem has become increasingly prominent. China is one of the countries that suffer the most serious damage from road traffic safety in the world. In 2021, the death toll of traffic safety accidents in China ranked first in the world.

Traffic accidents seriously threaten the safety of people's lives and property and hinder the sustainable development of the economy and society. Therefore, the Chinese government has issued a series of traffic regulations and strengthened law enforcement to reduce the incidence of traffic accidents. At the same time, the continuous development of automobile safety technologies, such as seat belts, airbags, driving images, and electronic stability control systems, has made the automobile safety factor higher and higher. However, the casualty ratio of traffic accidents remains high. Although vehicle conditions such as driving speed, acute technical failure, and tire burst, as well as external environments such as traffic density, traffic flow, slippery road surface, and adverse weather conditions, may cause traffic safety accidents, human factors are often the most important (Noland R B 1995). Almost 90% of accidents are caused by human factors such as violation of rules, distraction, and emotional and aggressive driving. According to the risk compensation theory, the increased sense of security will make individuals bear more risks, which means that vehicles with higher safety performance and a more suitable driving environment will make drivers underestimate the risk of accidents, leading to more radical and risky driving behavior. Dangerous driving is one of the main causes of traffic accidents (Ward N J 2010).

## 2 THEORETICAL ASSUMPTIONS

### 2.1 *Traffic accidents and traffic safety culture*

The research on Safety Science in the context of road traffic has gone through four stages. The initial research focus is on technical safety measures. In the second stage, individual and behavioral factors become the main research objects. In the third stage, the ergonomics and social technology

---

*Corresponding Author: haozhenqing1972@163.com

system attracted the attention of scholars. In the fourth stage, the impact of traffic safety culture and atmosphere will become the focus (Eagly A H 1993; Rakauskas M E 2009; Özkan T 2012). Traffic safety culture refers to the traffic safety concept recognized by the whole people. It is an explicit culture composed of consciously followed traffic safety rules, regulations, and behavior norms, including cognition, emotion, and behavior attitude. It can be described as the thoughts and feelings of road users on the surrounding traffic and their possible behavior intentions. It is an important guarantee for the whole people to participate in traffic safety practices. Traffic safety culture is a powerful comprehensive concept, which helps to explain the dangerous driving behavior of drivers. To investigate the traffic safety culture of the Chinese public, this paper discusses the four main aspects of traffic safety culture, including road safety attitude, safety risk awareness, dangerous driving behavior norms, and traffic law recognition.

## 2.2 *Traffic safety culture and dangerous driving behavior*

The human factors that cause traffic accidents mainly include wrong driving and dangerous driving, which are two distinct behavior types (Deery H A 1999). This kind of behavior is caused by non-subjective operation errors, which are difficult to avoid, and is mostly seen by novice drivers. And dangerous driving refers to "deviating from the behaviors necessary to maintain the safe operation of vehicles", such as dangerous violations, over-speed driving, and drunk driving (Kircher K 2013). More than 80% of the stories are caused by dangerous driving. Only by reducing and eliminating dangerous driving behaviors can we fundamentally reduce traffic safety accidents.

Iversen uses seven dimensions to measure traffic risk behaviors, including violating traffic rules/speeding, reckless driving, not using seat belts, cautious/vigilant driving, drinking driving, distracted driving, and driving below the speed limit (Iversen H 2004). Lawton described three types of dangerous driving behaviors: aggressive driving, distracted driving, and violating traffic rules while driving. Based on Weina's research results, this paper defines dangerous driving behaviors as four types, namely aggressive driving (behaviors that intentionally harm others physically or mentally, such as shouting or swearing at other drivers, flashing high beam lights, honking, blocking, etc.), risky driving (behaviors that are unacceptable to the society and may have negative consequences, such as running red lights, crossing the road and speeding) Negative cognitive/emotional driving (driving with anger, frustration, provocation and anger, such as losing temper while driving), and drinking driving.

## 3 DATA AND METHODS OF RESEARCH

### 3.1 *Study samples and data sources*

This sample selects the drivers in the Jinzhong Area of Shanxi Province as the research object. During the period from July to August 2019, the respondents must have a driving license and have driving experience in the last year. The questionnaire mainly includes four aspects: demographic characteristics, traffic safety culture survey, dangerous driving behavior survey, and traffic accident survey. A total of 400 questionnaires were distributed and 320 were recovered. After excluding the questionnaires with outliers, 288 valid questionnaires were obtained. The average time to complete the questionnaires was 5.5 minutes. The demographic characteristics of the respondents are shown in Table 1.

## 4 RESULT ANALYSIS

### 4.1 *Exploratory factor analysis results*

The samples formed by the effective questionnaire were randomly divided into half and half. First, half of them (n = 144) were used for exploratory factor analysis (EFA) to explore the internal

Table 1. Demographic characteristics of respondents.

| Variable | Index | Sample size | Proportion (%) |
|---|---|---|---|
| Gender | Male | 223 | 77.43 |
| | Female | 65 | 22.57 |
| Age | 18-35 | 92 | 31.94 |
| | 35-50 | 122 | 42.36 |
| | 50 | 74 | 25.7 |
| Age | 1-2 | 60 | 20.83 |
| | 3-5 | 109 | 37.84 |
| | 6-10 | 79 | 27.43 |
| | 10 | 40 | 13.9 |
| Education | Primary school | 52 | 18.06 |
| | Secondary & high school | 109 | 37.85 |
| | University | 127 | 44.09 |
| Marriage | No | 120 | 41.67 |
| | Married without children | 103 | 35.76 |
| | Married with children | 65 | 22.57 |

structure of traffic safety culture in China. The 23-traffic safety culture test items were evaluated by varimax-rotated principal component analysis, and the results are shown in Table 2. The KMO test value of the questionnaire is 0.927, greater than 0.7, and Bartlett's test value is 3409.916, which is significant at the 1% level, indicating that the questionnaire is suitable for factor analysis. According to the Kaiser principle, four factors with characteristic root greater than 1 were extracted, which explained 64.117% of the total variance in total, meeting the requirements of the social survey scale. It is found that the factor load of some items is less than 0.4 or multiple heavy loads are related

Table 2. Structure matrix of principal component analysis with Promax rotation (N = 144).

| Item | Factored load F1 | Item | Factored load F2 | Item | Factored load F3 | Item | Factored load F4 |
|---|---|---|---|---|---|---|---|
| TSC1 | 0.751 | TSC6 | 0.640 | TSC9 | 0.563 | TSC16 | 0.705 |
| TSC2 | 0.762 | TSC7 | 0.621 | TSC13 | 0.768 | TSC17 | 0.682 |
| TSC3 | 0.697 | TSC8 | 0.519 | TSC14 | 0.757 | TSC18 | 0.765 |
| TSC4 | 0.676 | TSC10 | 0.822 | TSC15 | 0.678 | TSC19 | 0.766 |
| TSC5 | 0.731 | TSC11 | 0.736 | | | TSC20 | 0.659 |
| | | TSC12 | 0.684 | | | | |
| Factor Naming | Traffic risk perception | Road safety attitude | | Driving behavior specification | | Traffic law recognition | |
| Characteristic Value | 9.129 | 1.757 | | 1.534 | | 1.045 | |
| Variance Contribution rate (%) | 43.469 | 8.367 | | 7.030 | | 4.978 | |
| Cumulative Variance contribution rate (%) | 43.469 | 51.836 | | 59.139 | | 64.117 | |
| KMO | | | 0.927 | | | | |
| Bartlett | | | 3409.916; DF = 210; significance = 0.0001 | | | | |

to different factors. After these items are gradually deleted and factor analysis is repeated, the final remaining 20 items cover four aspects: traffic risk perception, road safety attitude, driving code of conduct, and traffic law recognition. This is consistent with the research results of Timmermans et al., indicating that the traffic safety culture is also applicable to traffic safety research in China.

### 4.2 *Confirmatory factor analysis results*

Before analyzing the structural equation model, the structural validity of confirmatory factor analysis (CFA) should be tested, including convergent validity and discriminant validity. In this study, the convergent validity was first studied. The results showed that the factor load was between 0.5 and 0.9, the combined reliability (CR) was between 0.55 and 0.9, and the mean-variance extraction (AVE) values were between 0.45–0.7, all of which exceeded the standard values. The discriminant validity test shows that the diagonal value of the average variance of each potential variable is greater than the non-diagonal value representing the square of the correlation coefficient of each potential variable, indicating that the discriminant validity of the scale data is good (see the appendix for the detailed results due to space constraints).

## 5 CONCLUSION

The purpose of this study is to verify the psychometric characteristics of the traffic safety culture scale (TSCs) in the Chinese context and to deeply understand the causal relationship between traffic safety culture, dangerous driving behavior, and traffic accidents. The results confirm the reliability and validity of the traffic safety culture scale in the Chinese context. Traffic safety culture plays a key role in safe driving behavior, and dangerous driving behavior mediates the relationship between traffic safety culture and traffic accidents.

## REFERENCES

Deery H A, Fildes B N. Young novice driver subtypes: Relationship to high-risk behavior, traffic accident record, and simulator driving performance [J]. *Human factors*, 1999, 41 (4): 628–643.
Eagly A H, Chaiken S. *The psychology of attitudes* [M]. Harcourt brace Jovanovich college publishers, 1993.
Iversen H, Rundmo T. Attitudes towards traffic safety, driving behaviour and accident involvement among the Norwegian public [J]. *Ergonomics*, 2004, 47 (5): 555–572.
Kircher K, Andersson J. Truck drivers' opinion on road safety in tanzania—a questionnaire study [J]. *Traffic injury prevention*, 2013, 14 (1): 103–111.
Noland R B. Perceived risk and modal choice: risk compensation in transportation systems [J]. *Accident Analysis & Prevention*, 1995, 27 (4): 503–521.
Özkan T, Lajunen T. *Person and environment: Traffic culture* [M]//Handbook of traffic psychology. Academic Press, 2011: 179–192.
Rakauskas M E, Ward N J, Gerberich S G. Identification of differences between rural and urban safety cultures [J]. *Accident Analysis & Prevention*, 2009, 41 (5): 931–937.
Ward N J, Linkenbach J, Keller S N, et al. White paper on traffic safety culture [J]. *White Paper*, 2010, 2.

Civil Engineering and Urban Research – Mohamed & Hou (Eds)

# Numerical simulation and analysis of propylene pipeline leakage accident based on CFD

Shilin Chen
*Nanjing University of Technology, Nanjing, China*

Wei Ma
*Jiangsu Corechem Co., Ltd, Shaoxing, China*

Gang Tao* & Lijing Zhang
*Nanjing University of Technology, Nanjing, China*

ABSTRACT: To study the diffusion law of propylene pipeline rupture leakage, a numerical simulation study was carried out for an accident of propylene pipeline leakage by using CFD software Fluent. The study shows that the propylene leak diffusion process is a typical flow field jet process. In the early stage of propylene leak diffusion, $C_3H_6$ mainly gathered in a very small area around the leak, and in the static wind state, the flow field of propylene leak diffusion showed an obvious symmetric distribution, and the velocity direction dissipated in all directions, suggesting the diffusion direction of propylene. The concentration of propylene is concentrated and has a large gradient at the near-source release, and the concentration gradient becomes smaller as the distance increases. When the wind speed is 3 m/s, from the calculation results, the propylene diffusion in the windy condition is influenced by the wind field on the left side, which will diffuse to the right front and form a near-ground diffusion under the influence of gravity, while a vortex-like curve will be formed, which is because of a long-range wind absorption and wind field. The greater wind speed has a great influence on the diffusion of leaking gas, and its danger area is quite related to the wind speed. In general, the corresponding hazard area is becoming larger compared to the static wind state. The results of the study can be helpful for the analysis of hazardous gas leakage accidents.

## 1 INTRODUCTION

In recent years, due to the frequent occurrence of various pipeline accidents, which have caused serious casualties and property damage consequences, scholars at home and abroad have conducted a lot of research on this issue to study the diffusion law after gas leakage. On June 9, 1980, nine LNG diffusion tests were conducted in China Lake, California, to study the diffusion law of gas leakage under gravity. In 1991, Gkonig-Langlo and Schatzman used a wind tunnel (Luketa 2005) (Koopman 2007) test to simulate the combustible distance of combustible gas under different meteorological conditions, for both cases of transient source as well as the continuous source to verify that the density ratio is secondary. Winston L. Sweatman et al. used a wind tunnel test to verify the diffusion phenomenon of the transient source of heavy gas (Zhuang 2008), and mainly studied the variation of concentration with time. Qin Song et al. (2007) analyzed the study of heavy gas leakage by brine simulation method, derived the criterion number of the simulation experiment, and confirmed the phenomena of gravitational sedimentation, density stratification, and fork splitting in the near-source area in heavy gas diffusion. Huang Qin et al. (2008) used the

---

*Corresponding Author: taogangs@163.com

DOI 10.1201/9781003372417-87

turbulence model in Fluent software to simulate the diffusion of heavy gas transient and continuous leakage and predicted the changes of parameters in the process of heavy gas diffusion. Wei Tan et al. used CFD full-size simulated the diffusion of ammonia leakage in a plant area (Tan 2020), and studied the effect of pressure on the diffusion of ammonia, which increased with the concentration at the near source, and considered the effect of obstacles on the diffusion, showing that the gas cloud would gather above and in front of the obstacles at the initial stage, and the diffusion range became larger after crossing the obstacles.

At present, there is a relative lack of research on propylene pipeline accidents. In this paper, a pipeline leakage accident that occurred on July 28, 2010, in Wanshou Village, Qixia District, Nanjing City, is used as an example to study the gas diffusion law after pipeline rupture by Fluent software. Compared with the traditional empirical equations as well as theoretical model analysis methods, numerical simulation can solve more complex fluid problems.

## 2 NUMERICAL SIMULATION CALCULATIONS

### 2.1 *Initial conditions*

According to the accident information, the diameter of the propylene pipeline with orifice leakage is 159 mm, the operating pressure of the pipeline is 2.2 MPa, and the temperature of the transported gas is 298.15 K. Without considering the presence of other gases such as hydrogen sulfide, carbon dioxide and ethane in the pipeline, the remaining gases are all $C_3H_6$. The mass flow rate of gas leakage in the pipeline is related to the flow state, and for the propylene pipeline, the calculation for the leakage rate is described in Equation (1) (Pan 2011).

$$Q = C_d AP \sqrt{\left[ \frac{HM}{RT} \left( \frac{2}{K+1} \right)^{\frac{K+1}{K-1}} \right]} \tag{1}$$

where Q is the mass flow rate of gas (kg/S); $C_d$ is the gas leakage coefficient, and the round orifice is taken as 1; A is the area of the leakage port ($m^2$); P is the pressure in the pipe (Pa); H is the adiabatic index of gas, and the propylene is taken as 1.3; M is the molecular weight; R is the gas constant, usually taken as 8.314 kJ/(kmol•K); T is the adiabatic temperature of the gas (K). The calculated mass flow rate is 20.7 kg/s.

### 2.2 *Diffusion modeling*

In this paper, numerical simulation was carried out using CFD technology, and GAMBIT software was used for modeling. The two-dimensional model is used to study the diffusion law of propylene after leakage, with a 50 m × 50 m diffusion area with a leakage aperture of 0.06 m, the left side is the inlet side, the wind speed is 3 m/s, the upper side, as well as the right side, is the pressure outlet, the bottom is the leakage port, and the parts are named as wind-in, out-top 1, out-top 2, out-top 3, and right-out. The physical model is shown in Figure 1.

To facilitate analytical calculations and ensure convergence of calculation results, it is necessary to make the following assumptions based on site conditions (Gavelli 2008).

(1) There are no obstacles in the vicinity of the leak source and the area is open.
(2) The natural gas is leaking continuously, and the physical parameters of the leak are unchanged, the wind speed is horizontal and constant.
(3) The gas leak is turbulent, ignoring its compressibility, and no chemical reaction between the gas leak and air occurs.

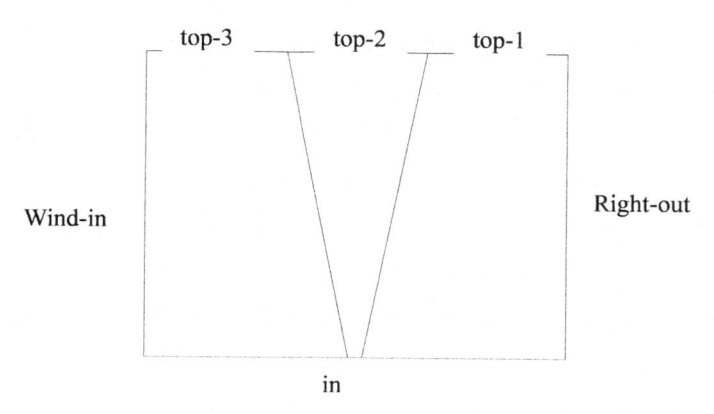

Figure 1.    Physical model.

## 2.3  *Grid division*

Since the flow state at the leakage orifice is more complicated compared with other zone places, it is necessary to encrypt this part locally. In this paper, the simulation condition with a leak orifice diameter of 0.06 m is illustrated as follows (Liu 2020).

(1) The line grid is divided as follows. In the two-dimensional plane model of the leak at the equal segment division, select the interval count, with the value of 12; on the left side of the wind speed inlet and the right side of the pressure outlet at the equal segment division, select the interval count, with the value 65; on the upper side of the pressure outlet (top-1, top-2, and top-3), select equal segment division and select the interval count with the values of 30, 24, and 30; for the middle part of the consideration of the diffusion area, select the interval count of 65 and 70. Scale factors for the line grid are set to 1.

(2) From the analysis of the results of grid division, we can see that the average value of cell mass is close to 1; the multilayered ratio is 1.1066, which is much smaller than 100; the tilt is 0.13513; the orthogonal mass is 0.9367. The results of model grid division are shown in Figure 2.

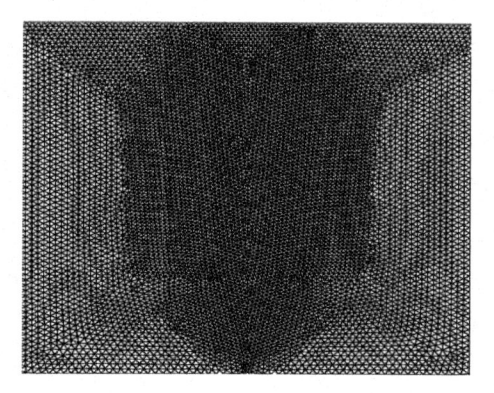

Figure 2.    Grid division model.

## 3   SIMULATION RESULTS AND ANALYSIS

After the model parameters are set, the residual curve of the numerical calculation is shown in Figure 3, indicating that the results of the numerical simulation are convergent. Therefore, there is no problem in the selection of the calculation model and parameter setting, and the obtained results meet the calculation requirements.

(1) Concentration cloud analysis. Figure 3 shows the flow field distribution of propylene diffusion from the leak in the static wind state when the wind speed is 3 m/s, the concentration field distribution is shown in Figures 4–6 below. In the static wind state, propylene gas is in the process of diffusion movement after the leak, due to the influence of the pressure in the pipe, and the airflow from the leak at the mouth of the high-speed jet out. At this time, the flow field diffusion presents a highly left and right symmetric state (Gavelli 2008). Based on the cloud diagram, propylene gas diffuses from the central part of the diffusion outward, and the concentration decreases with increasing distance until the height of the cloud reaches 20 meters. The concentration of propylene is about 0.2 when the height of the gas cloud reaches 20 m. As the distance increases, the diffusion range of the flow field increases and the concentration of propylene decreases, and the concentration of propylene is about 0.1 at the height of 50 m.

Due to the influence of wind speed, the first stage of propylene leakage diffusion is affected by gravity more than wind speed. The leaking propylene in the near ground movement during a period and air dilution mixed with neutral gas continue to diffuse movement. Based on simulation results, Figure 5 shows that in the early stages of propylene leakage diffusion, the influence of gravity is greater. According to Figure 6, the airflow upward jet forms a mushroom shape in the outer region of the diffusion because of the high-speed flow of gas along the coil suction effect. In the first period, pressure velocity is impacted, and then after a period of upward diffusion (Yang 2016), propylene gas sinks due to its greater density than air, which leads to a greater surface concentration of propylene gas. The cloud chart shows that when propylene concentration is about 0.2, the height of the central flow field is about 15 meters, which is lower than in the static wind state when propylene gas is at the same concentration. In addition, the tilt angle of the ground is greater. From the two cloud charts, the concentration of propylene is high near the leak source regardless of whether it is windy or still, and that its diffusion distance decreases near the leak, along with its risk area. In Figure 6, the density of the propylene gas cloud is close to air in the late stages of the leak and switches from heavy gas diffusion to neutral gas diffusion.

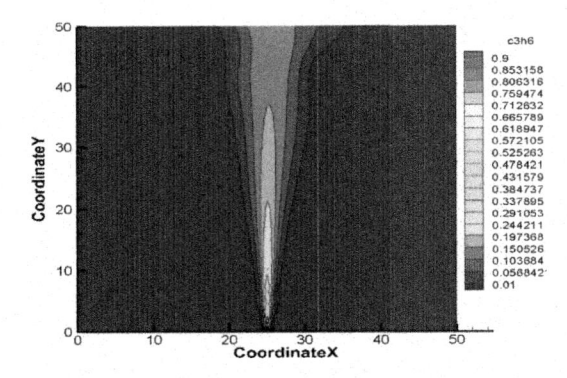

Figure 3.    Cloud map of propylene concentration distribution under static wind conditions.

(2) Contour analysis. As can be seen from the contour map of propylene, the diffusion cloud map reflects the danger region of propylene diffusion. Propylene has an explosion limit of 2% to 11%; therefore, the contour data reflects the danger region of propylene diffusion. According to Figure 7, the hazardous area of propylene in the static wind state exhibits an umbrella shape, which is caused by high pressure inside the pipe (Liu 2011) and reflects the trajectory of the propylene gas movement. The contour cloud in Figure 8 shows that the hazardous area has a small width at the distance from the leak, and its width increases as the distance from the leak increases and it has a symmetrical shape from left to right. Propylene gas concentration gradients are much higher near the source than in the upper area, and the high concentration of propylene gas gathers

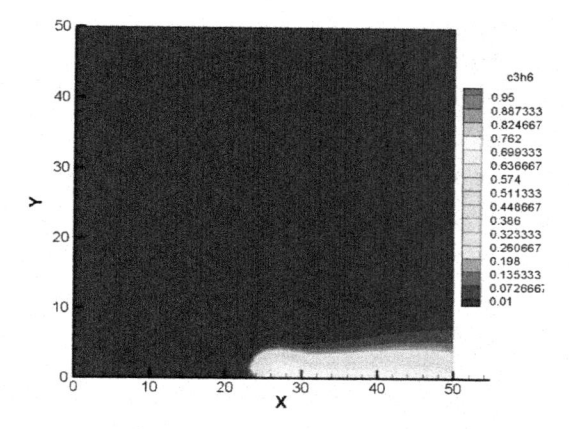

Figure 4.    Cloud map of propylene concentrations under windy conditions.

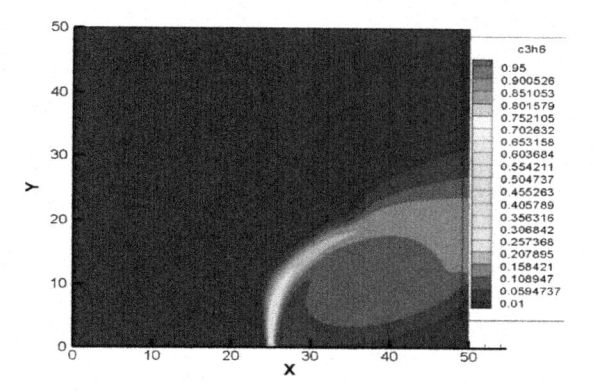

Figure 5.    Cloud map of propylene concentrations under windy conditions.

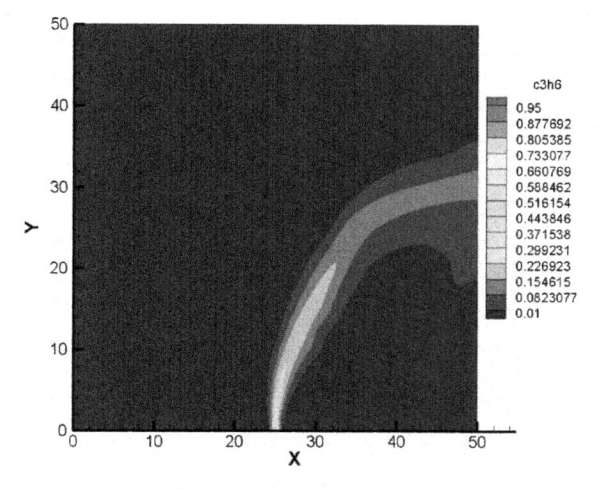

Figure 6.    Cloud map of propylene concentration distribution under windy conditions.

around the leak at the beginning. Figure 9 illustrates the danger area of propylene diffusion under windy conditions, from which the danger area will shift obviously to the lower right front under the influencing influence of wind speed. Propylene gas, being a heavy gas, will be influenced by gravity when it reaches a height of about 15 meters. Furthermore, because of the wind turbulence intensity, a wind suction back will occur in the right front 10 m, thus increasing the danger area of the wind suction. It can be seen in Figure 10 that propylene will be gathered in near-leakage, concentration gradient concentration, and at the place where suction reflux has the greatest effect, the pressure will be significantly higher than at other places, resulting in the development of localized high-pressure points. This phenomenon accords with the theory of fluid jets.

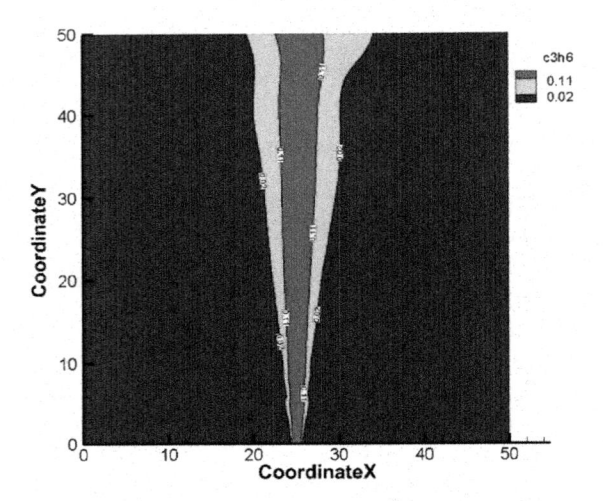

Figure 7.    Distribution of acrylic hazard areas under static wind conditions.

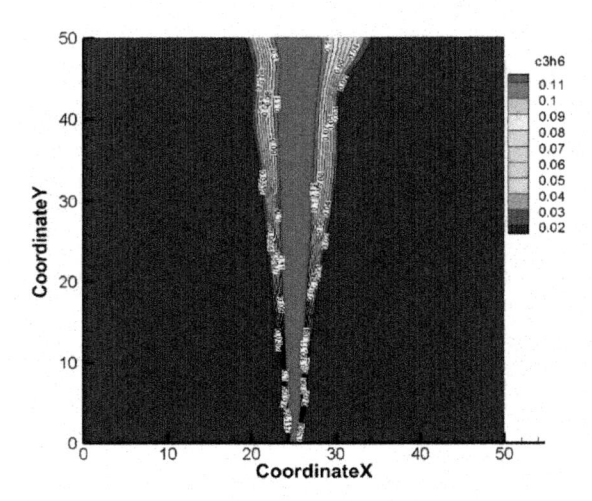

Figure 8.    Contour distribution of propylene concentration under static wind conditions.

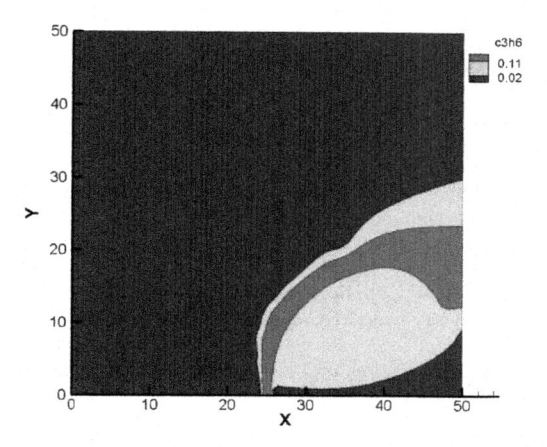

Figure 9. Distribution map of the danger areas under windy conditions.

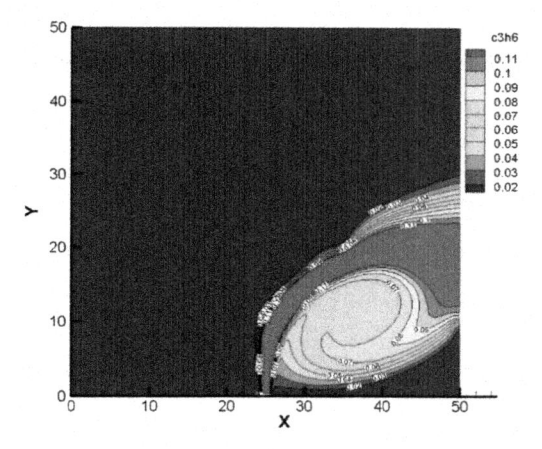

Figure 10. Contour distribution of propylene concentration under windy conditions.

## 4 CONCLUSIONS

(1) The diffusion of propylene gas in static wind shows a high degree of left-right symmetry, and its concentration value decreases as the diffusion range increases, while the danger area increases.

(2) The propylene gas diffuses from right to left near the ground while being influenced by the wind field. As the wind speed increases, the propylene expands more rapidly downwind and exhibits a greater angle of inclination towards the ground.

(3) When there is a wind, the danger area shifts to the right, and during diffusion movement, it causes the danger area to increase at the recoil, and the pressure there is greater than in other places. This creates a local area of high pressure.

ACKNOWLEDGMENT

This study was supported by the Postgraduate Research and Practice Innovation Program of Jiangsu Province (Grant No. SJCX21_0511).

# REFERENCES

Filippo Gavelli 2008. Application of CFD (Fluent) to LNG spills into geometrically complex environments. *Journal of Hazardous Materials J,.* 159 (2008) 158–168.

Huang Q, Jiang JC 2008. Validation of computational fluid dynamics (CFD) simulation for heavy gas leakage dispersion experiment. *Chinese Journal of Safety Science*, 01.013.

Hui-Ning Yang 2016. Confined vapor explosion in Kaohsiung City e A detailed analysis of the tragedy in the harbor city. *Journal of Loss Prevention in the Process Industries J* 41 (2016) 107e120.

Koopman R P, Ermak D L. 2007. Lessons learned from LNG safety research. *Journal of Hazardous Materials J*, 140 (3), 412–428.

Liu bing 2020.*fluent 2020 fluid simulation from beginner to master D.* Beijing: Tsinghua University.

Luketa-Hanlin A 2005.review of large-scale LNG spills: experiments and modeling. *Journal of Hazardous Materials*, 132 (2–3): 119.

Qinggong Li, 2013. Numerical Simulation of Liquid Chlorine Tank Leak Based on FLUENT. *Third International Conference on Information Science and Technology J* March 23–25, 2013.

Tatiana Flechas 2020. A 2-D CFD model for the decompression of carbon dioxide pipelinesusing the Peng-Robinson and the Span-Wagner equation of stateTatiana. *Process Safety and Environmental Protection J* 140 (2020) 299–313.

Wei Tan, Dong Lv 2020. Accident consequence calculation of ammonia dispersion in factory area. *Journal of Loss Prevention in the Process Industries J* 67 (2020) 104271.

Xiao Liu 2011. Modeling the two-phase cloud evolution from instantaneousflashing release using CFD. *Journal of Loss Prevention in the Process Industries J* 24 (2011) 420e425.

Zhuang X. Q., Gao H. H., Sun D, 2008. A review of accidental spill dispersion process of LNG ships. *China Navigation* (03): 280–283.

*Civil Engineering and Urban Research – Mohamed & Hou (Eds)*
*© 2023 the Authors, ISBN 978-1-032-44487-1*

# Empowering the dual-carbon policy: A study on the development of new information modeling of smart city in China

Yunlong Li*
*Jiangxi Science and Technology Normal University, Nanchang, China*

Luge Xing*
*Chengdu University of Technology, Chengdu, China*

Tianxiang Zhang*
*West Anhui University, Liuan, China*

ABSTRACT: China ushered in a new era of national green development during the 14th Five-Year Plan. The integration of low carbon, digitalization, and intelligence has become the key development direction in the future, and the construction of new smart cities will provide an important starting point for the realization of the double carbon targets. First, the article elaborates on the research background and research objectives. The release of the double carbon policy puts forward new requirements for the upgrading of sustainable urban development. Then, through the analysis of the new information model of the double carbon strategy and the construction of a smart city, a new form of combining the double carbon strategy and the construction of the smart city is obtained. This includes technical support for smart city construction systems and the development of potential application scenarios. Finally, we conclude the research findings from the three perspectives of government, enterprises, and individuals. Firstly, the development and application of the future smart city information model can help promote the development of cities to a higher level of management. Second, a smart city can facilitate the development of high-end architecture in enterprises. The third benefit is that a smart city can foster a low-carbon, green city construction consciousness to make the implementation of the double carbon policy more effective and efficient.

## 1 INTRODUCTION

At the 75th session of the United Nations General Assembly, China actively contributed to the implementation of the United Nations 2030 Agenda for Sustainable Development and committed to the people of the world to "strive to achieve the peak of world carbon emissions by 2030 and reach carbon neutrality in 2060". Therefore, this means that low carbon and digitalization are the two driving forces for China to achieve the "double carbon" strategic goal. While bringing urban management efficiency and convenient citizens' lives, the construction of new smart cities has also completed low-carbon energy conservation and emission reduction in different living scenarios with the help of the digital transformation of future cities (Chu, Z., Cheng, M., & Yu, N. N. 2021).

The purpose of this research is to confirm that the development of new smart cities will need to be integrated with new generation network technologies, such as new technologies such as the Internet of Things, big data, and virtual reality, based on the "double carbon" environmental background, by proposing a new model analysis of smart city development models. This study will

---

*Corresponding Authors: 924052505@qq.com, 1281246965@qq.com and 1799044840@qq.com

DOI 10.1201/9781003372417-88

also comprehensively promote the implementation of the national dual-carbon strategy, and promote energy conservation and emission reduction in multiple scenarios such as excellent governance, industrial development, and benefiting the people. Vigorously explore new paths and models for the development of the dual-carbon strategy driven by technological innovation. During the 14th Five-Year Plan, China entered a new stage of high-quality development, and this study once again proves that adhering to the development of new smart cities and implementing the concept of innovation, integration, green development, openness, and resource sharing will promote the sustained and healthy development of China's economic development, which is a major task for China's economy to enter a new stage of development.

## 2 ANALYSIS OF NEW INFORMATION MODELING OF DOUBLE CARBON STRATEGY AND SMART CITY CONSTRUCTION

The new smart city provides strong support in empowering the double carbon strategy, promoting the modernization of the urban governance system, improving the quality of life of citizens, and developing a green economy. It has become a new model of urban operation, a new way of urban management, and a new mechanism of urban construction in the current era. Continuing to deepen the construction of new smart cities (Camero, A., & Alba, E. 2019), while conscientiously implementing the new development concept to build a livable, innovative, smart, and low-carbon city to help achieve the goal of double carbon, has also promoted the green transformation of urban development methods, and achieved efficient utilization and recycling of resources.

After years of exploration and practice, China's new smart city construction has entered the stage of large-scale landing and deepening. The new smart city model constructed by the institute will place more emphasis on building a "cloud pipe end" infrastructure system 5G Internet of Things. The integration of new generation information technology such as artificial intelligence, big data, and blockchain, the support of common enabling platforms such as data intelligence and urban information model, and the operation and management mode of digital twin collaboration and intelligence are developing in the direction of "three integrations and five spans". Under the background of "double carbon", the construction of new smart cities should not only strengthen urban governance capabilities but also facilitate residents' lives and enhance the level of the industrial economy through advanced technologies. It is also necessary to use digital means to promote energy conservation and emission reduction in life and production, and to transform the driving force of economic development to guide citizens to change the concept of life consumption (Kim, M. J., & Jun, H. J. 2021), and explore a win-win path of low-carbon energy conservation and emission reduction and economic growth (Figure 1).

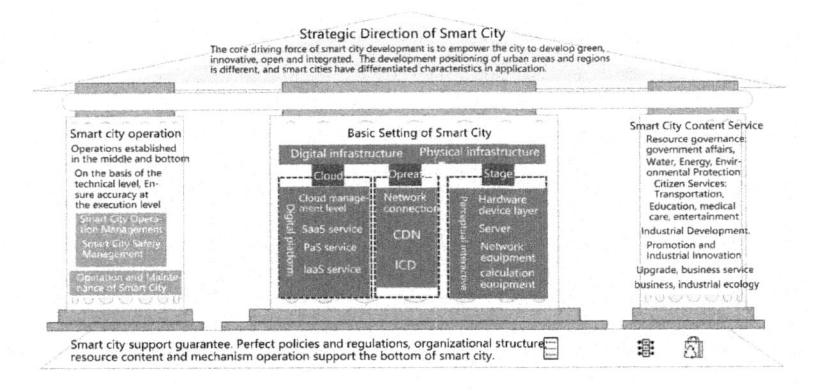

Figure 1.    Architecture analysis of new information modeling of a smart city (self-drawn).

## 3.1 *Technical support analysis of smart city construction*

From the perspective of planning and design, in the early stage of smart urban planning and design, we can actively use social networks, big data, satellite positioning, geographic information systems, sensors, virtual reality, and other technologies to carry out highly integrated data and information collection on regional and urban border conditions and citizen needs; then analyze and evaluate landscape processes and behavior habits of different groups; finally, promote public participation in design. This might include, for example, the use of geographic information systems (GIS) to analyze the ecological landscape of the site or the construction of digital elevation models. To create an effective green space in an urban setting, it is imperative that information is collected, processed, managed, analyzed, expressed, and applied so that the preliminary work is more solid, and can facilitate more evidence-based planning and design around meeting the diverse needs of citizens and site needs.

Planning and design include the development of spatial models and databases for spatial deliberation and environmental simulation, including developing urban spatial models, deliberating the integration effect of landscape and architecture, and simulating the impact of different landscape structures on the surrounding urban climate, thus improving the scientific nature of design, and paving the way for future urban management, repair, and update. In addition, the parametric design improves the accuracy and efficiency of designs, and it is also possible to design tense shapes. BIM technology combines the Internet of Things, sensors, satellite positioning (Anthopoulos, L. G. 2015), remote sensing, and other technical means to control the entire construction process (Figure 2).

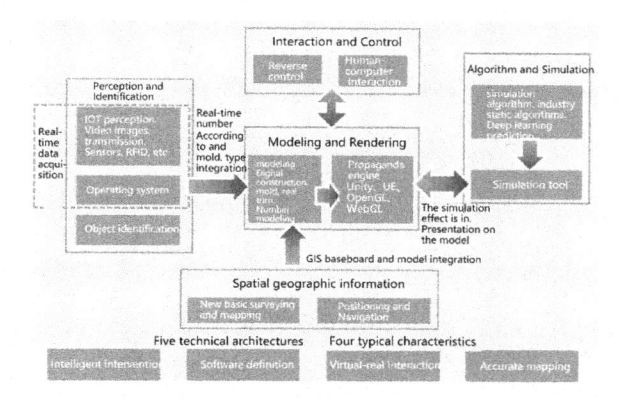

Figure 2. Technical architecture analysis of a smart city (self-drawn).

Intelligent operation management focuses on city managers, not only based on the subjective experience of managers for management and operation but also using big data, artificial intelligence, cloud computing, etc. Data mining and processing are performed dynamically in the city, providing technical support for city management and decision-making, such as the development of intelligent comprehensive information platforms, big data visualization (Sarker, I. H. 2022), intelligent ecological monitoring, and intelligent update management.

In terms of artificial intelligence, it is the driving force for the development of smart cities. This is due to the accelerated deployment of artificial intelligence infrastructure, such as data centers, the expansion of the development and application of various algorithm models, and the improvement of the energy consumption of electric utilities. Through the extensive use of algorithms for deep learning and artificial intelligence network open platforms, energy can be maximized, and low-carbon operation of data center and model development can be realized. At the same time, artificial intelligence in cities can use detection platforms such as smart sensors and satellite remote sensing to track, learn, and model carbon footprints to help smart cities to provide environmental warnings and optimize carbon emission activities more efficiently. In the field of "double carbon" of a smart city, "AI+" is the future development trend, optimizing its path in areas dealing with high carbon emissions such as electricity, heat, construction, and manufacturing, enhancing operational efficiency in transportation, gardening, environmental protection, water affairs, realizing resource recycling (Angelidou, M. 2014), and empowering the development of various industries (Figure 3).

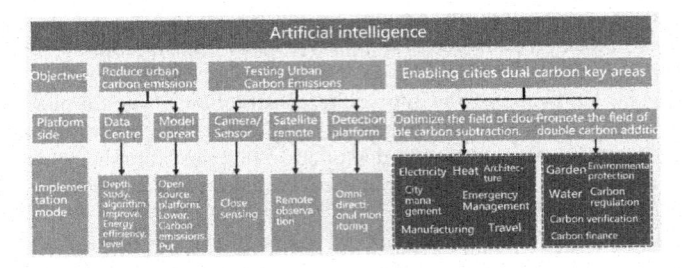

Figure 3.    Analysis of artificial intelligence framework for the construction of a smart city (self-drawn).

From the perspective of IoT technology, it constitutes the nervous system of smart cities and collects detailed data in real time using IoT technology, cameras, and specialized sensors, which are then transmitted to government agencies, enabling them to quickly make resource allocation decisions, and IoT actuators automatically initiate response measures in emergencies to ensure the healthy operation of the city. With the help of IoT, the government can quickly adjust road congestion, and water pollution, and detect data such as energy consumption, helping to achieve the "double carbon" goals from the source (Sarker, I. H. 2022). At the same time, its application potential in vertical industries such as entertainment, social public services, finance, and social networking has yet to be developed, and the needs of different scenarios for IoT and blockchain technology are different (Figure 4).

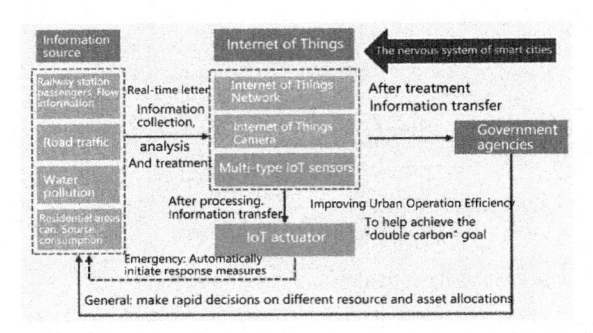

Figure 4.    Technical analysis of IoT in the construction of a smart city (self-drawn).

### 3.2   *Analysis of potential scenarios for smart city construction*

A smart city is a complex system of digitization and informatization, which is closely linked to urban geography, life, resources, and environment. Digitalization and intelligence are the practical basis for the management of smart city infrastructure and life services.

From the perspective of data fusion application scenarios, to solve the problems of lack of docking and application of government data and social data docking mechanisms in the existing data platform, smart cities strengthen the concept of data fusion, and the integration and sharing of government and enterprise information data have become the core of promoting digital management.

Data resource sharing has become the core driving force for the development of smart cities in the new era, providing a massive, standardized, and highly intelligent information resource platform for the development of new smart city applications (Figure 5).

From the perspective of ecological creation application scenarios, ecological openness is an important foundation for the development of smart cities, platform enterprises are "points", digital economy industry parts are "lines", digital economy industrial clusters are "surfaces", and points, lines, and surfaces are combined to achieve the optimal allocation of information resources to build an efficient development of smart cities (Yang, S., & Chong, Z. 2021). The composition of the smart city system is more complex, covering several links such as pre-design, technical consultation, investment and financing management, project transactions, project landing operation, and healthy development of industrial ecology, platform enterprises give full play to the radiation driving ability of "points", and the elements of the digital economy industrial chain are pulled by the "line". Finally, the cluster of digital economy industrial firms forms a "surface" and unfolds into a smart city life circle that is built "from points and lines" (Camero, A., & Alba, E. 2019).

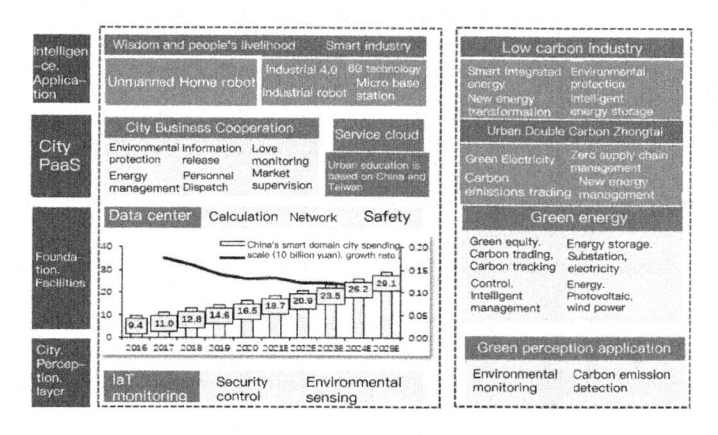

Figure 5.   Analysis of potential application scenarios of a smart city (self-drawn).

## 4   A NEW PATH FOR THE IMPLEMENTATION OF THE INFORMATION MODEL OF SMART CITIES IN THE FUTURE

### 4.1   *In the future, the smart government will empower the "double carbon" strategy and actively promote the development of cities through refined management*

In the future, the smart government will formulate a relevant refined standard system for the construction of a smart city empowerment "double carbon" strategy, and build a smart ecological community, a smart ecological industrial zone, and a smart digital transportation system based on the original old community, abandoned industrial zone, and old transportation system in the form of demonstration areas or demonstration clusters to provide suitable policy and infrastructure conditions for the promotion of smart cities. At the same time, the government should actively break down the barriers to information and exchange between different departments and institutions, establish a super smart circulation database and platform (Li, S. 2021), and help smart cities achieve sustainable development based on greatly improving the efficiency of office practice.

### 4.2   *In the future, smart enterprises will empower the double carbon policy and actively promote the development of cities with high-end architecture*

High-end architecture and smart city construction rely on the coordination of the whole industry chain, and future industrial development will flock to the trend of full scenario, all elements, and full cycle, adding bricks and tiles to the high-end urban architecture and stimulating more domestic demand potential. The smart city is a high-end architecture composed of four modules: intelligent interaction, intelligent connection, intelligent hub, and intelligent application, which has great advantages in data, ecology, technology, and security. The high-end architecture composition empowers the panoramic ecological construction of smart cities, from simply isolated scenario construction iteration to urban global aggregation innovation, maintaining the "full-cycle management awareness" of smart cities to create the innovation potential of sustainable urban competition (Li, S. 2021).

### 4.3 *In the future, smart cities will empower the double carbon policy and actively establish the awareness of individual active participation in the construction*

In the process of empowering the "double carbon" strategy of smart cities in the future, individual citizens in society must also consciously practice low-carbon survival, starting from the small things around them, starting from the drops, and striving to be the pioneers of low-carbon survival. People are encouraged, for example, to try to use public transportation while traveling, make plans for driving or going out shopping, try to buy enough at one time, walk more, ride bicycles, take the light rail subway, and carry out smart low-carbon life science activities and competitions (Dulschi, O. 2021).

## 5 CONCLUSIONS

In summary, the 14th Five-Year Plan opened a new chapter in China's green development, and the integration of low-carbon, digitalization, and intelligence has become the key development direction in the future. Through analysis, we can obtain the following research findings. Development and application of the future smart city information model are beneficial for several reasons, including firstly, improving the city's management, secondly, improving enterprise architecture, and thirdly, promoting the awareness of low-carbon, green city construction.

For the future research plan, this article will first, based on the development of new technologies, further study the interrelationship between information space, physical space, and social space, and deepen the intelligent management, space design, and social life content of future cities. Secondly, the plan also includes improving the new smart city information model, applying it to real-life cases, collecting feedback data, and further upgrading and optimizing it.

## REFERENCES

Angelidou, M. (2014). Smart city policies: A spatial approach. *Cities*, *41*, S3–S11.
Anthopoulos, L.G. (2015). Understanding the smart city domain: A literature review. *Transforming City Governments for Successful Smart Cities*, 9–21.
Camero, A., & Alba, E. (2019). Smart City and information technology: A review. *Cities*, *93*, 84–94.
Chu, Z., Cheng, M., & Yu, N. N. (2021). A smart city is a less polluted city. *Technological Forecasting and Social Change*, *172*, 121037.
Dulschi, O. (2021). Smart City-the city of the future. *In Teoria şi practica administrării publice* (pp. 152–155).
Kim, M. J., & Jun, H. J. (2021). Towards a sustainable life: Smart and green design in buildings and community. *Sustainability*, 13(3), 1022.
Li, S. (2021). Research on the development trend of smart community. *International Journal of Social Science and Education Research*, 4(7), 13–17.
Sarker, I. H. (2022). Smart city data science: Towards data-driven smart cities with open research issues. *Internet of Things*, *19*, 100528.
Yang, S, & Chong, Z. (2021). Smart city projects against COVID-19: Quantitative evidence from China. *Sustainable Cities and Society*, 70, 102897.

*Civil Engineering and Urban Research – Mohamed & Hou (Eds)*
*© 2023 the Authors, ISBN 978-1-032-44487-1*

# Numerical simulation of aerosol generation and distribution in sewage pipe networks

Zixin Liu*

*An De College, Xi'an University of Architecture and Technology, China*

ABSTRACT: This study aims to evaluate the environmental and health hazards caused by the urban sewage networks, which are a potential source of virus-causing pollution. A method incorporating CFD 3D numerical simulation, VOF, and DPM models is used to analyze virus aerosol under different sewage flow rates. The simulation analysis of the generation rules and the characteristics of the impact on virus transmission finally generated a model of virus aerosol's generation and diffusion mechanism in the sewage pipe network. The results show that the changes in water flow velocity and turbulence kinetic energy affect the production and diffusion characteristics of viral aerosols. With the increase of the water flow velocity, a higher mass concentration of the viral aerosol particles was produced and released in the sewage pipeline, and the diffusion range of the aerosol particles was greater. Affected by the gas exchange inside and outside the pipeline, the turbulent kinetic energy in the converging area of the inspection wells increased significantly, and the virus aerosol particles spread out of the outside atmosphere in the area. This suggests that sewage inspection wells could be a route for viral aerosols to spread out of the atmosphere, which may pose a risk of virus exposure to the inspection well workers. Thus, this article provides a theoretical basis for the study of the risk of viral infection in sewage networks.

## 1 INTRODUCTION

Fecal contaminants generated in daily life need to be transported through municipal sewer networks to sewage treatment plants for centralized and uniform treatment (Cahill & Morris 2020), making the municipal sewer drainage system a potential reservoir for many bacterial and viral pathogens (Liu & He 2021). According to the data survey, 80% of human diseases worldwide are water-related, and typical *enteroviruses, noroviruses, hepatitis A* viruses, and *adenoviruses* have been detected in urban domestic sewage (Kaas et al. 2018), and the mode of water transmission of the virus mainly includes direct contact with a respiratory intake of virus-containing aerosol particles (Foladori et al. 2020; Lahrich et al. 2021). The turbulence of sewage and gas exchange inside and outside the pipeline provide favorable conditions to produce viral aerosols, the inspection well interchange pipe section is the main place of turbulence, but also an important channel for gas exchange in the pipeline, the release of aerosol particles containing potentially pathogenic microorganisms will be transported into the external environment through air convection, which greatly threatens human health and safety.

The results of the study show that viral loads in sewer network systems are mainly derived from the flushing process in hospitals and residential areas, the volume of viral aerosols is usually small, and the bioaerosols containing viruses can travel through the air for several hours in drainage systems and vents (Usman et al. 2021). Sewage aeration conditions, turbulence kinetic energy, and environmental factors (wind speed, temperature, humidity, etc.) have an important impact on

---

*Corresponding Author: liuzixin@xauat.edu.cn

DOI 10.1201/9781003372417-89

the production of aerosols (Michał et al. 2018), while gas-liquid two-phase confluence conditions affect the pressure and turbulence intensity of sewage, especially gas convection inside and outside the inspection well, which increases the concentration of viral aerosols in the area. However, studies have focused only on the analysis of the health risks of people exposed to sewage treatment plants and their surrounding areas, ignoring the possible health risks associated with sewage transport. Therefore, this study simulates and analyzes the influence of water flow in the intersection area of inspection wells on the generation and distribution of virus aerosols in the sewage pipe network, which has important guiding significance for the formulation of protection policies.

In this paper, we examine the hydrodynamic laws in manholes, as well as the characteristics of aerosol particle generation and diffusion. By establishing a model of aerosol generation and diffusion in sewage inspection wells, the effects of flow velocity and turbulent kinetic energy changes on aerosol generation and diffusion were studied. It aims to reveal the health risks of sewage inspection well workers and nearby contacts that may be caused by viruses present in sewage pipelines, and to provide scientific basis and theoretical support for the study of virus transmission in sewage pipelines.

## 2 MATERIALS AND METHODS

### 2.1 *Numerical model building and solution methods*

This paper uses the Navier-Stokes equation as the control equation, and the SST k-omega turbulence model establishes a closed system of equations, which shows more powerful advantages for turbulent simulation of non-round tube flow, with higher computational accuracy and a wider range of applications. The method of combining the VOF and DPM model was used to simulate the generation of viral aerosols at different sewage flow rates and the characteristics of their influence on virus transmission. VOF models are suitable for transient and steady-state tracking of stratified flows, gravitational flows, the movement of large bubbles in fluids, and flow at the liquid-gas interface (Beg et al. 2020). Flows with a dispersed phase volume fraction of less than or equal to 10% can be modeled using DPM (Li et al. 2015). Since the municipal pipe network sewage conforms to the gravity non-pressure circular pipe flow, and the volume fraction of the aerosol dispersed phase generated in the sewage pipe network is much less than 10%, the method of combining the VOF and DPM model is used to simulate the generation and diffusion exposure of viral aerosol particulate matter in the combined area of pipeline and inspection well.

Using the finite volume method of discrete calculation domain, a three-dimensional sewage pipeline model with a pipe diameter of 800 mm was established, the inspection well diameter was 1200 mm, the total height of the inspection well was 3 m, and the inspection well was connected to the sewage pipeline at a 90°C right angle. The fullness of the pipe is 0.7, the upper layer is the gas phase, and the lower layer is the liquid phase. The simulation model calculation area consists of two parts: The aerosol generation area of the upstream and downstream 5 m pipeline range connected to the manhole catchment area. The outer ($5 \times 5 \times 3m^3$) rectangular space aerosol diffusion area leaves the manhole outlet. The polyhedral mesh is divided by Fluent meshing, the number of meshes is 387707, and the maximum distortion rate is 0.394, which meets the requirements of Fluent simulation calculation. The model calculation area and meshing results are shown in Figure 1.

To simplify the calculation, aerosol particles are assumed to be incompressible spherically rigid particles and the interactions between particles are ignored. The model solution is divided into two stages, the first one is steady-state continuous-phase fluid calculation. The turbulent flow of a continuous-phase fluid is simulated using the two-phase VOF model and the SST k-omega turbulence model. The implicit VOF formula performs steady-state calculation of the fluid, opens the two-way coupling of gas and liquid, considers the numerical diffusion caused by the turbulent effect, and discretizes the volume fraction equation, and the surface tension value is 0.072. The pressure-speed coupling calculation is carried out by the coupled method, the sewage density is 1028.58 kg/m$^3$, and the operating pressure is 101,325 Pa. In addition to the wall boundary

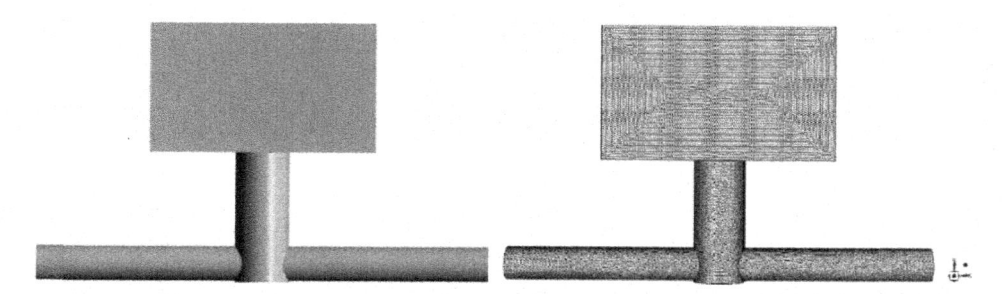

Figure 1.    Computational model and meshing.

conditions, six open boundary conditions are also set in the model, namely the water and air inlet boundary in the pipeline, the water and air outlet boundary in the pipeline, the air inlet in the diffusion area, and the outlet boundary in the diffusion area. All wall surfaces are provided with 3 boundary layers, wall roughness n equal to 0.014, and the hydraulic slope is 0.003. The second one is discrete phase particle tracking. Aerosol particles carry out the rigid spherical hypothesis. After the flow field calculation is stable, the DPM model is opened, considering Saffman lift and virtual mass forces. Particles inject from the inlet-water boundary for non-point source incidence, particle density of 1,100 kg/m$^3$, the average particle size range of 1-100 μm, particle mass flow of 1.51E-5 kg/s, continue 500 steps of iterative calculation to complete all particle tracking.

## 2.2    Experimental conditions and analytical methods

The working conditions studied in this paper are as follows: pipe roughness 0.005, pipe slope 5‰, fullness 0.7. Under this working condition, the water flow status and aerosol generation and distribution of the well intersection area were studied under the conditions of 0.6 m/s, 0.8 m/s, 1.0 m/s, and 5.0 m/s inlet water flow rate, respectively. The simulation adopts a three-dimensional coordinate system, the X-axis is the direction of the sewage pipeline water flow, the Y-axis is the direction of the inspection well height, the Z-axis is the direction of the diameter of the inspection well, and the coordinate origin is in the intersection center of the inspection well and the sewage pipeline. By collecting the water flow velocity of different sections selected from the X-axis of the sewage pipeline and the water depth direction of the Y-axis, the computer simulation data is verified and analyzed, in which the water depth is measured by the measuring pin, and the flow rate is measured by ultrasonic doppler flowmeter (ADV).

## 2.3    Numerical simulation verification

Based on the idealization of computer simulation, this study starts from reality and verifies the computer simulation results by establishing a set of sewage pipeline busting experimental simulation devices. The model device body is a plexiglass pipe; to simulate accurately the running state of the actual sewage inspection well pipe section, the bus simulation experimental device is set with pipe diameter, flow rate, slope, and pipe wall roughness similar to the numerical simulation of the manhole bus pipeline. In each test, the experimental analysis and verification are carried out by changing the flow rate conditions of the inlet water without changing the remaining parameter settings. The scheme of the facility is shown in Figure 2.

Numerical simulation mainly uses the inlet and outlet flow difference and residual diagram as the main judgment basis for convergence. Each numerical simulation first performs a 500-step iterative calculation to achieve steady-state conditions and then injects discrete particles into the calculation area to continue the 500-step iterative calculation of trace particles. Through comparative analysis, the error value of the calculation result of the experimental simulated inlet and outlet mass flow under each working condition is within 0.4%, and the parameters in the residual diagram have

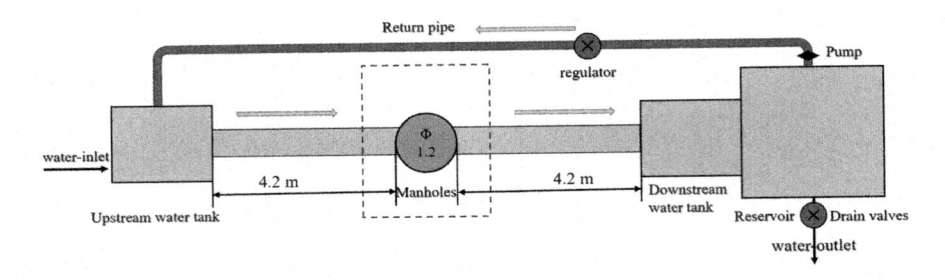

Figure 2. Schematic diagram of the interchange pipe section of the sewage inspection well.

reached less than $10^{-3}$, so the judgment model has converged, and the results obtained by the numerical calculation can be used for research and analysis. Figure 3 below uses 0.8 m/s as an example to verify the numerical simulation results.

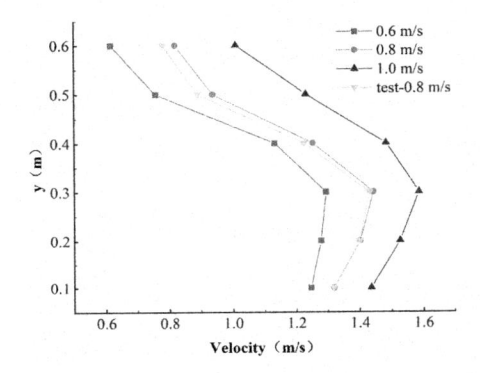

Figure 3. Contrast of velocity at flow depth direction under different working conditions.

As can be seen from Figure 3, on the whole, the law of water velocity change under different flow rate conditions is consistent, the bottom speed of the pipeline is larger than that of the water surface, and the law of first increasing and then decreasing from the bottom surface of the pipeline to the surface of the water flow is presented, and the maximum values are 1.29, 1.44, and 1.59 m/s are reached at a water depth of 0.3 m. Comparing the laboratory measurement data and computer simulation results under the condition of 0.8 m/s, it can be seen that the experimental and numerical simulation results have a high degree of fitting, which indicates that the flow state simulation of the intersection area of the sewage pipe network inspection well is reliable by applying this simulation method.

## 3 RESULTS AND ANALYSIS

### 3.1 *Pressure and turbulence kinetic energy in sewage pipe cross-section*

Figure 4 simulates the vector distribution of the resulting velocity of the center section and the manhole outlet section using 0.8 m/s as an example. The simulated region exhibits a distinct velocity partition, which is consistent with the study results of other researchers (Beg et al. 2020). Affected by the intersection area of the inspection well, the gas exchange in this area is strengthened, the external gas enters the sewage pipe network through the inspection well, and the gas in the corresponding part of the pipeline diffuses out of the inspection well.

Figure 5 shows the simulation results of the change in pressure and turbulence kinetic energy in the sewage pipe under different flow rate conditions. For the flow of non-pressure circular pipes,

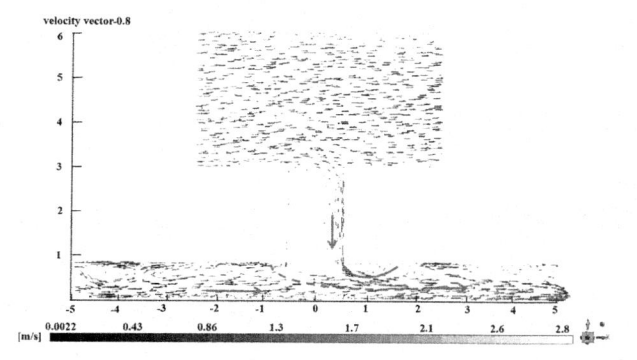

Figure 4. The velocity vector distribution of the center section.

overall, the total pressure value is positively correlated with the speed, the overall pressure change in the sewage pipeline is gentle, and the pressure fluctuation of the inspection well intersection pipe section changes sharply. The size of turbulence kinetic energy changes gently before the intersection, the intersection area increases sharply, violent turbulence occurs, and the turbulence kinetic energy leaving the intersection area decreases rapidly and eventually stabilizes and restores laminar flow.

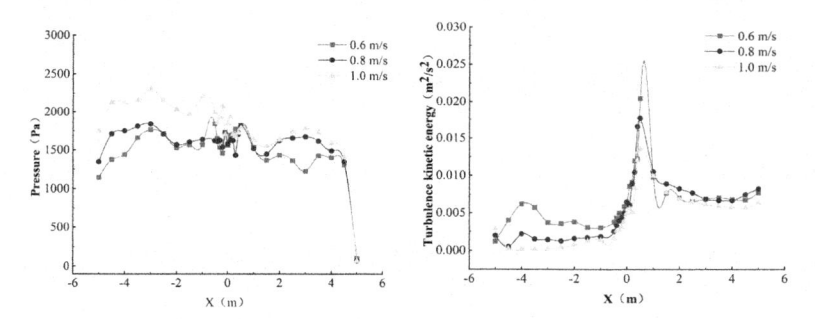

Figure 5. The change of pressure and turbulent kinetic energy in the X-axis.

### 3.2 Analysis of particle traits of sewage pipe virus aerosols

Aerosol production is related to sewage velocity, turbulence kinetic energy, temperature, humidity, and other factors, which together affect the production and diffusion exposure of aerosol particles (Li et al. 2021). Through the simulation of the physical parameters of aerosol particles generated under three flow rate conditions, the relationship between velocity, turbulence kinetic energy, and aerosol particles was explored.

Figure 6 below takes the velocity of 0.8 m/s as an example to analyze the trajectory streamline diagram of the aerosol particles, and the color represents the size of the turbulent kinetic energy, and it can be seen that the turbulent kinetic energy of the inspection well convergence area and the upper gas phase area of the sewage pipeline is significantly greater than that of the liquid phase area of the lower layer of the pipeline, which is consistent with the results of the turbulence kinetic energy change law analyzed above.

### 3.3 Sewage pipe virus aerosol motion diffusion characteristics

The DPM model can be used to simulate and analyze the concentration distribution of aerosol particulate matter. The simulation results of the aerosol particle distribution generated at different

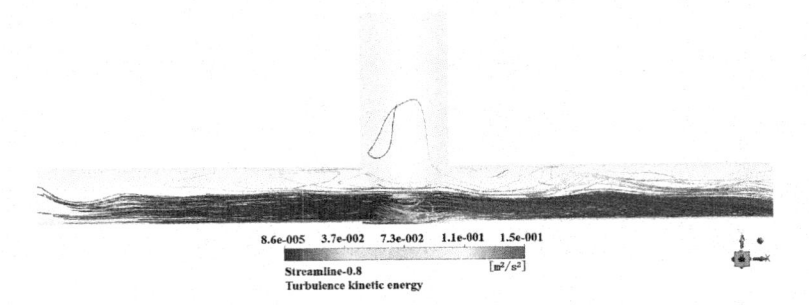

Figure 6.    The streamline of aerosol particle trajectory.

speed conditions are summarized in Figure 7 below. More than 99% of the aerosol particles are present in the sewage pipeline, and as the flow rate increases, the concentration of aerosol particles in the upper layer of the pipeline and the inspection well pipe increases accordingly. In addition, when the water flow velocity reaches 5 m/s, a small number of particles spread upward through the manhole into the outside atmosphere.

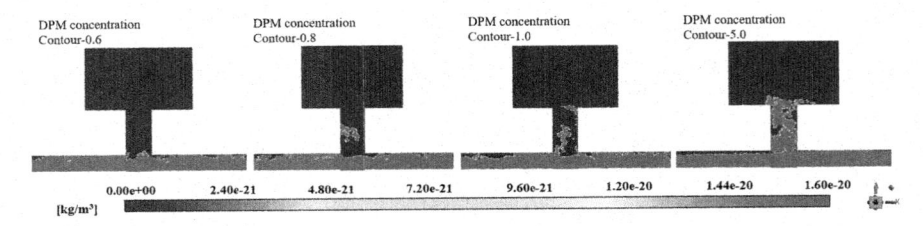

Figure 7.    The aerosol distribution under different working conditions.

Figure 8 summarizes the results of the average aerosol particle size distribution at different outlet interfaces. As can be seen from the figure, the aerosol content in the sewage is the highest, followed by the gas in the upper layer of the pipeline, and the lowest distribution in the inspection well pipeline. Under the three-speed conditions of 0.6, 0.8, and 1.0 m/s, no aerosol diffusion of the cross-section of the inspection well outlet was found, and the escape of viral aerosol particles was detected at the inspection outlet at 5 m/s. This suggests that the velocity of water flow has an important influence on the spread of viral aerosols in the manhole junction area. The specific magnitude of the impact deserves more in-depth study.

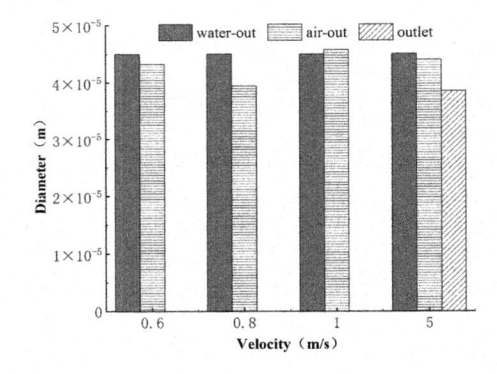

Figure 8.    The aerosol particle size distribution at various outlet interfaces under different velocity.

Sewage discharge sources, air diffusion mechanisms, and environmental conditions all affect the composition, size, and concentration of viral aerosol particles. In this study, a simplified sewage inspection well busting model was proposed to simulate and analyze the generation and diffusion of virus aerosols in the intersection area of a sewage pipe and inspection well under different water flow speed conditions, taking the virus concentration load in the average sewage as the input value. Figure 9 summarizes the flow regime and viral aerosol particle distribution characteristics under four water velocity conditions based on the DPM analysis results.

The transport of materials in the manhole confluence area is a process of horizontal flow and diffusion of longitudinal manhole pipelines, and the viral aerosol particles released by the sewage pipe network are transported to the downstream pipe section through the water flow on the one hand, and enter the manhole pipe through the gas-liquid interaction in the intersection area on the other hand, until they spread out of the outside atmosphere. The simulation results of the turbulence kinetic energy in Figure 9 show the complex gas-liquid two-phase fluid interaction in this area, the flow structure of the simulated area is consistent under the four working conditions, the flow partition is obvious, and the turbulence kinetic energy in the manhole intersection area is significantly increased.

From the analysis of viral aerosol particle traits, the particle size of viral aerosol is distributed in the range of 1 to 100 $\mu$m, and the overall distribution of particle size gradually decreases from bottom to top. Compared with different water flow speed conditions, at a lower water flow rate of 0.6 m/s, the mass concentration of viral aerosols produced is small and only distributed in the sewage pipeline. When the flow rate gradually increases, as shown in the simulation results of 0.8 m/s and 1.0 m/s, the pressure and turbulence kinetic energy also increase accordingly, and the aerosol particles gradually enter the inspection well pipeline with the exchange of airflow in the intersection area. At a maximum permissible sewage flow rate of 5.0 m/s, more and more particles diffuse into the manhole pipe and then diffuse out of the outside atmosphere as the longitudinal velocity increases.

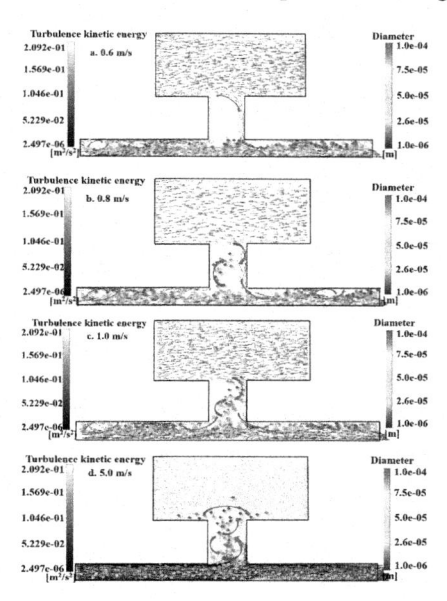

Figure 9.    Generation and diffusion mechanism of viral aerosol under different working conditions.

In this study, the influence characteristics of the flow structure of the combined area of sewage inspection wells formed under different speed conditions on the production and distribution of viral aerosols were simulated and analyzed to explore the health risks caused by the spread of human viral aerosols in the sewage pipe network. The influencing factors of the generation and diffusion of viral aerosols in the sewage pipe network include but are not limited to changes in the speed of water flow, and the influence of sewage temperature and pipeline connection form requires a more comprehensive analysis model to quantify this risk.

## 4  CONCLUSIONS

1. Numerical simulation results show that the method of combining the VOF and DPM model is adopted, the simulation and experimental fit degree are high, and the flow state of the intersection area of sewage inspection wells can be studied and analyzed by applying this method. The regional flow state simulation results show that for gravity non-pressure circular tube flow, the flow structure of the near-surface layer and near the bottom layer of the intersection section shows a regular distribution, and the overall flow rate shows a gradually decreasing trend from bottom to top.
2. The production and particle size distribution of viral aerosols in sewage pipelines under different speed conditions are different. The overall distribution law is manifested as the highest concentration in the sewage, followed by the gas phase area in the upper layer of the pipeline, and the least in the inspection well pipeline. The simulation results showed that the virus aerosol spread at the check well outlet interface did not occur at 0.6, 0.8, and 1.0 m/s, and the virus aerosol spread occurred at a high flow rate of 5.0 m/s.
3. The generation and distribution of aerosol particles in sewage pipelines have a significant correlation with the amount of turbulence kinetic energy. The influx of the external atmosphere in the manhole intersection area strengthens the gas-liquid interaction, and correspondingly, the turbulence effect between the gas phase area in the upper layer of the pipeline and the intersection area of the manhole increases, and the mass concentration and y-direction velocity of the aerosol particles also increase significantly.

## REFERENCES

Beg M N A et al. (2020). CFD modelling of the transport of soluble pollutants from sewer networks to surface flows during urban flood events [J]. *Water* (Basel). 12 (9): 2514.
Cajill N & Morris D. (2020). Recreational waters – A potential transmission route for SARS-CoV-2 to humans? [J]. *Science of Total Environment*. 740: 140122.
Chuang GAN et al. (2021). Numerical analysis of unstable seepage field of CSG dam based on VOF method [J]. *Journal of Changchun Institute of Technology: Natural Science Edition*. 22 (2): 96–102.
Foladori, P et al. (2020). SARS-CoV-2 from faeces to wastewater treatment: What do we know? A review. *Science of Total Environment*. 743: 140444.
Kaas, L et al. (2018). Detection of Human Enteric Viruses in French Polynesian Wastewaters, Environmental Waters and Giant Clams. *Food and Environmental Virology*. 11: 52–64.
Lahrich, S et al. (2021). Review on the contamination of wastewater by COVID-19 virus: Impact and treatment. *Science of Total Environment*. 751: 142325.
Lanzhou LIU & Ning HE. (2021). Transmission and prevention and control of viruses in the process of urban domestic sewage treatment and reuse [J]. *Water treatment technology*. 47 (11): 31–35.
Li, L et al. (2015). Large Eddy Simulation of Bubbly Flow and Slag Layer Behavior in Ladle with Discrete Phase Model (DPM)–Volume of Fluid (VOF) Coupled Model. *JOM* (1989). 67 (7): 1459–67.
Michał, M et al. (2018). The Variability of the Concentration of Bioaerosols Above the Chambers of Biological Wastewater Treatment. *Ecological chemistry and engineering*. 25 (2): 267.
Usman, M et al. (2021). Exposure to SARS-CoV-2 in aerosolized wastewater: Toilet flushing, wastewater treatment, and sprinkler irrigation. *Water* (Basel). 13 (4): 436.
Xue LI et al. (2021). Coronavirus aerosol transmission and environmental impact factors [J]. *Environmental science*. 42 (7): 3091–3098.

*Civil Engineering and Urban Research – Mohamed & Hou (Eds)*
*© 2023 the Authors, ISBN 978-1-032-44487-1*

# Research on discrimination model of seismic failure modes of RC columns based on K-nearest neighbor algorithm

Ao Lu & Chunhua Zhang
*CCCC Second Highway Consultants Co., Ltd., Wuhan, Hubei, China*

Fang Huang*
*Naval University of Engineering, Wuhan, Hubei, China*

Shuang Wang
*Wuhan Longfang Engineering Technology Co., Ltd., Wuhan, Hubei, China*

Ying Liu & Zi Wu
*CCCC Second Highway Consultants Co., Ltd., Wuhan, Hubei, China*

ABSTRACT: To accurately distinguish the seismic failure mode of reinforced concrete (RC) columns, a discriminant model of seismic failure modes of RC columns based on the K-nearest Neighbor algorithm (KNN) is proposed. Firstly, the optimal characteristic parameters for distinguishing flexure failure from non-flexure failure and flexure-shear failure from shear failure are selected; Then combining the optimal characteristic parameters and KNN algorithm, a two-stage discriminant model of seismic failure modes for RC columns is established; Finally, the factors affecting the precision of the discrimination model are analyzed. The results show that the relative amount of longitudinal stirrup, shearing resistance demand to shear capacity ratio, and shear span ratio are the optimal characteristic parameters to distinguish flexure failure and non-flexure failure; and the optimum characteristic parameters for identifying flexure-shear failure and shear failure are shear span ratio, the relative amount of longitudinal stirrup and stirrup spacing/effective height of the section. With the increase of the K value, the discriminant accuracy of the proposed model gradually increases and tends to be stable. The proposed model has a high precision of 90%, 83%, and 87% in identifying flexure failure, flexure-shear failure, and shear failure of RC columns.

## 1 INTRODUCTION

Reinforced concrete (RC) columns have different failure modes under the action of the earthquake, such as flexure failure, flexure-shear failure, and shear failure. A brittle failure such as flexure-shear failure and shear failure seriously influence the safety of the structure and should be paid special attention to in the design stage. Therefore, it is of great significance to establish the discriminant model of seismic failure mode for RC columns and to connect the design parameters with the failure mode construction for the protection of existing and newly built columns.

According to the number of characteristic parameters considered, the existing seismic failure mode identification methods for reinforced concrete columns can be divided into single parameter methods and multi-parameter methods. The single parameter method is to distinguish the failure mode by establishing the relationship between a single characteristic parameter and the failure mode of RC columns. For example, ductility coefficient (Ghee 1989), shear-span ratio (Wan 2012), and shear demand to shear capacity ratio (ASCE/SEI 2017) were used as the characteristic parameters

---

*Corresponding Author: 150149184@qq.com

DOI 10.1201/9781003372417-90

of seismic failure mode of RC column. However, many factors affect the seismic failure mode of RC columns, and different combinations of these uncertain factors lead to different failure modes of columns (Ma 2012), so the single-parameter discrimination method is prone to produce errors.

Therefore, the multi-parameter method considering various characteristic parameters is proposed, which can be divided into the empirical discriminant method and machine learning discriminant method according to different analysis principles. The empirical discriminant method is based on engineering experiences or experimental data to construct the relationship between characteristic parameters and failure modes (Ma 2018; Qi 2013). Although this method is simple and practical, the discrimination accuracy is not high (Yu 2021). The machine learning discriminant method is to build the relationship between characteristic parameters and seismic failure mode by combining experimental data and various machine learning algorithms (Li 2022; Mangalathu 2019; Xie 2021). The existing machine learning discriminant method also has some deficiencies. For example, the different influences of different characteristic parameters on failure modes are ignored, resulting in low computational efficiency and a complicated calculation process.

Therefore, the sensitivity of parameters is firstly analyzed in this paper, and the sensitive parameters of bending failure and non-bending failure and bending shear failure and shear failure are selected respectively. Then, based on the K-nearest neighbor algorithm, the discrimination model of the failure mode of RC columns is proposed, and the optimal K value is trained to make the model accuracy reach the optimal.

## 2 K-NEAREST NEIGHBOR ALGORITHM FOR SEISMIC FAILURE MODES DISCRIMINATION

K-nearest Neighbor (KNN) (Cover and Heart 1967) is one of the optimal machine learning algorithms for text classification processing, and its theory is mature and easy to implement, achieving good results in face recognition, word recognition, and medical image processing.

The principle of the KNN algorithm is to determine the classification of the object to be classified according to the classification of most samples in the nearest neighbor samples set in training data. The accuracy of the KNN algorithm is mainly affected by the following three points. The first point is the value range of the nearest neighbor sample set (also known as the value of K). Different value ranges will change the number of nearest neighbor samples, and the category of objects to be classified may change accordingly. Secondly, the method of calculating the distance between the object to be classified and all samples in the training data will also affect the calculation accuracy. Thirdly, decision methods are the key to determining the classification of objects to be classified. Using different decision methods may have different classification results.

Aiming at the above problems, this paper studies the influence of different K values on the accuracy of the discriminant model; the distance between the object to be classified and another sample in the training data is calculated by the Euclidean distance method; the nearest neighbor samples were weighted by weight method, and the failure mode with the highest total weight was selected as the failure mode of the samples to be classified.

Considering that the K-nearest neighbor algorithm is a binary classification algorithm, this paper divides the discrimination into two stages. Firstly, the flexure failure and non-flexure failure are discriminated against, and then the flexure shear failure and shear failure are discriminated against. Therefore, the failure mode identification process presented in this paper is as follows.

Step 1: Establish the flexure failure non-flexure failure training group (as the first training group) and the flexure-shear failure and shear failure training group (as the second training group).

Step 2: Calculate the distance between the samples to be classified and the samples in the first training group (arranged according to the distance from small to large), and take the first K samples to form the nearest sample set.

Step 3: According to the total weight of each failure mode (the weight is assigned according to the distance, the larger the distance, the smaller the weight), the category of the object to be discriminated against is determined.

Step 4: If the object is to be classified as a non-flexure failure, the distance between the sample to be classified and each sample in the second training group is calculated, and Step 3 is performed.

## 3 OPTIMAL CHARACTERISTIC PARAMETERS ANALYSIS BASED ON K-NEAREST NEIGHBOR ALGORITHM

There are many characteristic parameters affecting the seismic failure modes of RC columns, including shearing resistance demand to shear capacity ratio ($V_p/V_n$), hoop spacing to depth ratio ($s/h_0$), the relative amount of longitudinal stirrup ($\omega$), longitudinal Bars parameters ($\rho_l$), stirrup parameters ($\rho_v$), shear-span ratio ($\lambda$), and axial compression ratio ($n$).

Different characteristic parameters have different effects on the seismic failure modes of RC columns. Since the KNN algorithm is to search for major categories in the nearest adjacent sample set, the sensitivity of characteristic parameters directly affects the effectiveness of the algorithm. Therefore, the sensitivity analysis of characteristic parameters is carried out in this paper, which is used as the basis for selecting characteristic parameters in this paper.

### 3.1 Characteristic parameters for distinguishing seismic failure modes

The characteristic parametric analysis of seismic failure modes of RC columns has been extensively studied by scholars at home and abroad. However, there is no consensus on the important parameters affecting RC columns.

In this paper, 173 groups of experimental data of RC columns with circular cross-sections were collected from *PEER-Structural Performance Database* (Berry, 2004), and the influence law of the above seven characteristic parameters on seismic failure modes of RC columns was analyzed statistically. The basic parameters of specimens were as follows.

$$16MPa \leq f_c' \leq 119MPa; \qquad 249MPa \leq f_{yv} \leq 1424MPa;$$
$$318MPa \leq f_y \leq 587MPa; \qquad 1.01 \leq \rho_l \leq 6.03;$$
$$0.07 \leq \rho_v \leq 2.24; \qquad 0.036 \leq V_p/V_n \leq 1.765;$$
$$1 \leq \lambda = a/h \leq 10; \qquad 0.035 \leq s/h_0 \leq 0.667;$$
$$0.068 \leq \alpha_l = \rho_l f_y/f_c \leq 0.739; \ 0.274 \leq \alpha_v = \rho_v f_{yv}/f_t \leq 9.986;$$
$$0 \leq n = P/f_c' A_g \leq 0.8; \qquad 0.059 \leq \omega = A_{sl} f_y/A_{sv} f_{yv} \leq 3.738$$

where $f_c'$ is the compressive strength of a concrete cylinder; $f_y$ is the yield strength of longitudinal reinforcement; $f_{yv}$ is the yield strength of stirrup; $P$ is the axial pressure; $A_g$ is the cross-sectional area of the columns; $A_{sl}$ is longitudinal bars' area; $A_{sv}$ is longitudinal bars' area of the stirrup.

Half violin plot with various characteristic parameters in different seismic failure modes as shown in Figure 1, the rate of change of each characteristic parameter when different seismic failure modes change as shown in Figure 2. The left side of the half violin plot is the frequency distribution diagram, and the right side is the violin diagram. The external shape of the violin is the kernel density estimation diagram. The white dot in the middle represents the mean and the black line represents the range of standard deviations. F stands for flexure failure; FS stands for flexure-shear failure; S stands for shear failure.

### 3.2 Optimal characteristic parameters of different seismic failure modes

It can be seen from Figures 1 and 2 that $\omega$, $V_p/V_n$, and $\lambda$ are suitable for the discrimination of flexure failure and non-flexure failure. Here, $\omega$, $\lambda$ and $s/h_0$ are selected as characteristic parameters to distinguish flexure-shear failure and shear failure.

It can be seen from Figure 2 that the change rate of $n$ and $\rho_v$ is greater than $\lambda$ during the transition from flexure-shear failure to shear failure, but there are more overlaps between $n$ and $\rho_v$ relative to $\lambda$. Therefore, $s/h_0$ was selected instead of $V_p/V_n$ in this study.

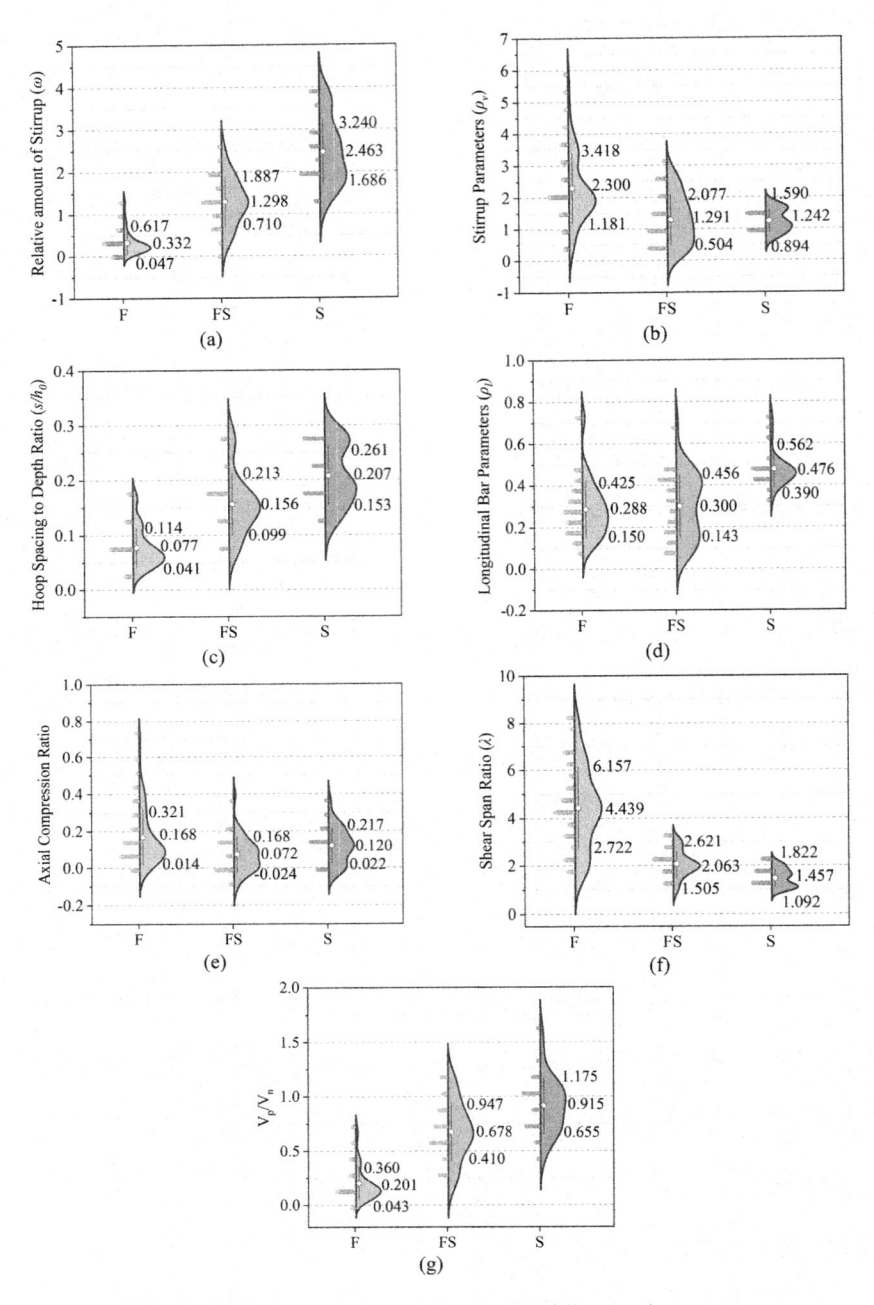

Figure 1.  Half violin diagram of parameters in different seismic failure modes.

## 4  DISCRIMINATION MODEL OF SEISMIC FAILURE MODES BASED ON K-NEAREST NEIGHBOR ALGORITHM

In this research, a total of 114 groups of data (57 groups of bending failure and 57 groups of non-bending failure) were randomly divided into three groups, two of which were used as training

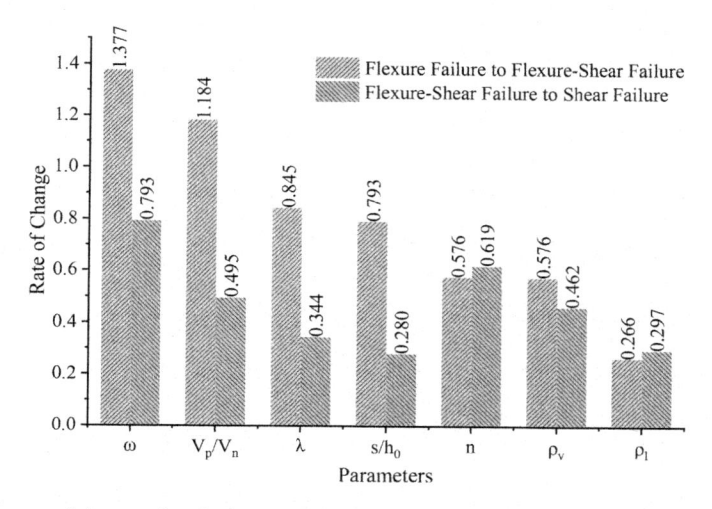

Figure 2. The rate of change of each characteristic parameter when different seismic failure modes change.

groups and one as a test group in turn. The average error rate of the training results for the three groups is taken as the result. The test results vary with the K value as shown in Table 1 and Figure 3.

Table 1. Flexure failure mode and non-flexure failure mode training results.

| K | Number of errors (Group 1) | Number of errors (Group 2) | Number of errors (Group 3) | The average number of errors | Average error rate |
|---|---|---|---|---|---|
| 1 | 6 | 6 | 3 | 5 | 0.17 |
| 3 | 6 | 7 | 2 | 5 | 0.17 |
| 5 | 6 | 8 | 2 | 5.33 | 0.18 |
| 7 | 6 | 8 | 2 | 5.33 | 0.18 |
| 9 | 5 | 8 | 2 | 5 | 0.17 |
| 11 | 5 | 4 | 2 | 3.66 | 0.12 |
| 13 | 4 | 3 | 2 | 3 | 0.10 |
| 15 | 4 | 3 | 2 | 3 | 0.10 |
| 17 | 3 | 3 | 2 | 2.66 | 0.10 |
| 19 | 4 | 3 | 2 | 3 | 0.09 |
| 21 | 3 | 3 | 2 | 2.66 | 0.10 |
| 23 | 3 | 4 | 3 | 3.33 | 0.09 |
| 25 | 2 | 4 | 3 | 3 | 0.11 |
| 27 | 2 | 3 | 3 | 2.66 | 0.10 |
| 29 | 2 | 3 | 3 | 2.66 | 0.09 |
| 31 | 2 | 3 | 3 | 2.66 | 0.09 |
| 33 | 2 | 3 | 3 | 2.66 | 0.09 |

When discriminating flexure failure and non-flexure failure, the error rate of the discrimination model shows a decreasing trend with the increase of the K value. Finally, when the K value reaches a certain value, the error rate of the discrimination model tends to be stable, and the lowest error rate can be as low as 9% (the accuracy is about 90%).

In our study, a total of 57 groups of non-flexure failure data were randomly divided into three groups, two of which were used as a training group and one as a test group in turn. The average error rate of the training results for the three groups is also taken as the result. The test results vary with the K value as shown in Table 2, Figure 4 and Figure 5.

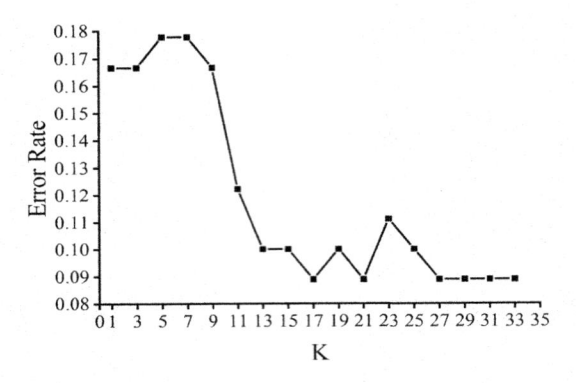

Figure 3. Relation between error rate and K value.

Table 2. Flexure-shear failure and shear failure model training results.

| K | Group 1 | | Group 2 | | Group 3 | | The average number of errors of FS | The average error rate of FS | The average number of errors of S | The average error rate of S |
|---|---|---|---|---|---|---|---|---|---|---|
| | Number of errors of FS | Number of errors of S | Number of errors of FS | Number of errors of S | Number of errors of FS | Number of errors of S | | | | |
| 1 | 2 | 4 | 4 | 0 | 0 | 5 | 2 | 0.25 | 3 | 0.38 |
| 3 | 1 | 0 | 2 | 0 | 0 | 6 | 1 | 0.13 | 2 | 0.25 |
| 5 | 1 | 0 | 2 | 0 | 0 | 5 | 1 | 0.13 | 1.67 | 0.21 |
| 7 | 1 | 0 | 4 | 0 | 0 | 4 | 1.67 | 0.21 | 1.33 | 0.17 |
| 9 | 1 | 0 | 5 | 0 | 0 | 3 | 2 | 0.25 | 1 | 0.13 |
| 11 | 1 | 0 | 5 | 0 | 0 | 3 | 2 | 0.25 | 1 | 0.13 |
| 13 | 1 | 0 | 3 | 0 | 0 | 3 | 1.33 | 0.17 | 1 | 0.13 |
| 15 | 1 | 0 | 3 | 0 | 0 | 3 | 1.33 | 0.17 | 1 | 0.13 |
| 17 | 1 | 0 | 3 | 0 | 0 | 3 | 1.33 | 0.17 | 1 | 0.13 |
| 19 | 1 | 0 | 3 | 0 | 0 | 3 | 1.33 | 0.17 | 1 | 0.13 |
| 21 | 1 | 0 | 3 | 0 | 0 | 3 | 1.33 | 0.17 | 1 | 0.13 |

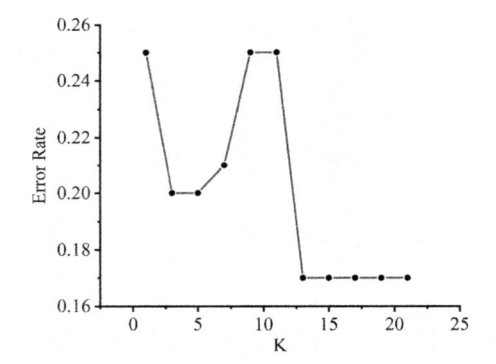

Figure 4. Training results for flexure-shear failure mode.

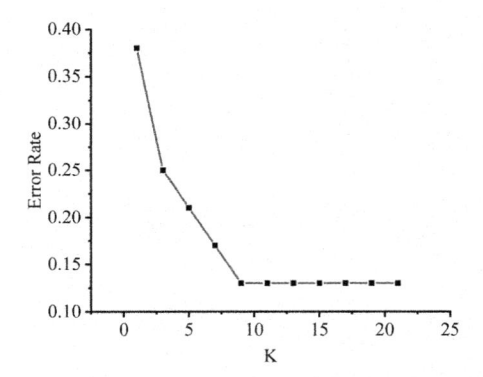

Figure 5. Training results for shear failure mode.

It can be seen that similar to the training results of flexure failure and non-flexure failure discrimination model, the error rate of discrimination model of flexure-shear failure and shear failure mode shows a decreasing trend with the increase of K value, and finally tends to be stable, the discriminant error rates of flexure-shear failure and shear failure are 17% (accuracy 83%) and 13% (accuracy 87%), respectively.

## 5 CONCLUSIONS

Based on the KNN algorithm, a discriminant method for RC seismic failure mode was proposed, and the discriminant accuracy was analyzed based on the existing RC column test data. Here are the conclusions we obtained.

(1) The optimal characteristic parameters for distinguishing flexural failure and non-flexural failure are the relative amount of longitudinal stirrup, shear capacity/shear demand, and shear span ratio. The optimal characteristic parameters for identifying flexural-shear failure and shear failure are the relative amount of longitudinal stirrup, shear span ratio, and hoop spacing to depth ratio.
(2) The accuracy of the discriminant model increases with the increase of the K value and eventually tends to be stable.
(3) The discriminant model presented in this paper has good discriminant accuracy, and the discriminant accuracy for three seismic failure modes of RC columns is 90%, 83%, and 87%, respectively.

## REFERENCES

ASCE/SEI 41-17. (2017). *Seimic evaluation and retrofit of existing buildings. Virginia: The American Society of Civil Engineerings.*

Berry M., Parrish M. & Eberhard M. (2004). *PEER structural performance database user's manual.* University of California, Berkeley.

Cover, T., & Hart, P. (1967). Nearest neighbor pattern classification. *IEEE transactions on information theory*, 13 (1), 21–27.

Ghee, A. B., Priestley, M. N., & Paulay, T. (1989). Seismic shear strength of circular reinforced concrete columns. *Structural Journal*, 86 (1), 45–59.

Li Q. M., Yu Z. C. & Ning C. L. (2022). Two-Stage Support Vector Machine Method For Failure Mode Classification of Reinforced Concrete Columns. *Engineering Mechanics*, 39 (02): 148–158.

Ma Y. & Gong J. X. (2018). Probability Identification of Seismic Failure Modes of Reinforced Concrete Columns based on Experimental Observations. *Journal of Earthquake Engineering*, 22 (8–10): 1881–1899.

Ma Y. (2012). *Study on Failure Modes and Seismic Behavior of Reinforced Concrete Columns.* Dalian: Dalian University of Technology.

Mangalathu, S., & Jeon, J. S. (2019). Machine learning–based failure mode recognition of circular reinforced concrete bridge columns: Comparative study. *Journal of Structural Engineering*, 145 (10): 04019104.

Qi Y. L., Han X. L. & Ji J. (2013). Failure mode classification of reinforced concrete column using Fisher method. *Journal of Central South University* 20 (10): 2863–2869.

Wan H. T., Han X. X. & Ji J. (2010). Analyses of reinforced concrete columns by performance-based design method. *Journal of Central South University* (Science and Technology) 41 (04): 1584–1589.

Xie L. L., Yu Z. C. &, Yu B. (2022). Classification method for failure modes of reinforced concrete columns based on imbalanced datasets. *Journal of Building Structures*.

Yu Z. C, Li Q. M., Xie L. L., & Yu B. (2021). Discrimination Model of Seismic Failure Mode of RC Columns Based on Exhaustive Search Strategy and Logistic Regression Algorithm. *Engineering Mechanics*, 39, 1–12.

*Civil Engineering and Urban Research – Mohamed & Hou (Eds)*
*© 2023 the Authors, ISBN 978-1-032-44487-1*

# Identification and cracking strategies of the constraints on the promotion of prefabricated buildings under "double carbon" policy: An empirical study of the DEMATEL-ISM method

Jingsheng Yang*
*CET-College of Engineering and Technology, Southwest University, Chongqing, China*

Qiyuan Xu & Jing Gao
*College of Economics and Management, Southwest University, Chongqing, China*

ABSTRACT: Given the promotion of prefabricated buildings after the release of the "dual-carbon" policy, a total of 22 constraints in five dimensions of policy, technology, society, economy, and management were identified and refined through literature analysis, questionnaire survey, and expert interviews, and then DEMATEL was used. The purpose of this study is to apply the ISM model, conduct systematic and hierarchical research on the relationship between blocking factors, develop a hierarchical structure model, and explore the blocking path for the promotion of prefabricated buildings. The study found that insufficient financial subsidies, lack of building codes and standards, backward prefabricated technology, insufficient technology research and development of prefabricated building products, and lack of designers in the design of prefabricated buildings are the key obstacles to the promotion of prefabricated buildings. The key obstacle elements form a dominant obstacle path by influencing the direct layer elements through the transmission of the intermediate layer elements, which should be given priority when formulating the promotion strategy of prefabricated buildings.

## 1 INTRODUCTION

Achieving carbon peaking and carbon neutrality is an important part of promoting ecological civilization construction and sustainable development (Li). As a basic industry related to residents' lives and economic operations, the construction industry has long been high in energy consumption and carbon emissions. In 2022, the "Opinions of the Central Committee of the Communist Party of China and the State Council on Completely and Accurately Implementing the New Development Concept and Doing a Good Job in Carbon Peaking and Carbon Neutralization" pointed out that we should vigorously develop energy-saving and low-carbon buildings, continuously improve energy-saving standards for new buildings, and accelerate the large-scale development of ultra-low energy, near-zero energy, and low-carbon buildings. Therefore, the way out of the "tough battle" of energy conservation and emission reduction in the construction industry is to take green and low carbon as a guide, reduce the amount of control from the source, strengthen process control and regulation, and speed up the construction of low-energy consumption and low-emission integration to meet the "double carbon". "Double-low" building system (Song 2021), and prefabricated buildings play a crucial role in it (Cao 2021; Lv 2021).

Existing research on the development of prefabricated buildings mainly focuses on the following two aspects. The first one is the identification of obstacles to the development of prefabricated buildings, mainly including high supply costs (Lu 2021), technical limitations, and imperfect

---

*Corresponding Author: yangjingsheng2002@163.com

 DOI 10.1201/9781003372417-91

industrial chains (Sn et al. 2022). There are biases in public perception in terms of demand (Chen 2021), and there are shortcomings in policy objectives, systems, and implementation (An 2020). The second one is the control and optimization of the prefabricated building supply chain, including improving the collaborative efficiency of supply chain entities (An 2020), and steadily resolving supply chain risks (Chen 2021; Huang 2020; Sun 2020), and comprehensively reduce supply chain costs (Xu 2021; Zhao 2020). The existing literature is still biased toward theoretical elaboration and normative analysis of the existing problems in the promotion of prefabricated buildings, lacking in-depth discussions on the relative relationship of obstacle factors and empirical measurement of important orders, and mainly proposes coping and optimization strategies from a single perspective.

Hence, the purpose of this paper is to (1) analyze the subjective and objective factors contained in each dimension that restrict the promotion of prefabricated buildings and (2) build a comprehensive evaluation system. DEMATEL is combined with the ISM method to provide empirical support for the identification of outstanding problems and the direction of intervention by analyzing the logical relationships and effect strength among factors at all levels. Steady promotion and high-quality development can provide useful management inspiration.

## 2 FACTOR IDENTIFICATION

This paper mainly uses the literature research method, questionnaire survey method, and expert interview method to identify the constraints. Based on the literature research method, this paper firstly identifies the constraints on the promotion of prefabricated buildings through databases such as CNKI, and WOS and then revises and simplifies the identified constraints through expert interviews and questionnaires, and finally extracts 22 constraints under the five dimensions of policy, technology, society, economy, and management were developed, and the evaluation system for the factors restricting the promotion of prefabricated buildings was constructed as shown in Table 1.

## 3 MODEL BUILDING

Combining the two methods of DEMATEL and ISM cannot only achieve complementarity, but also identify the key constraints and causal relationships between the promotion of prefabricated buildings, and clarify the hierarchical structure and action paths between the influencing factors, improving the scientificity of complex system analysis and decision-making. This paper uses the DEMATEI-ISM model to analyze the constraints on the promotion of prefabricated buildings. The specific modeling steps are as follows.

(1) Construct the direct influence matrix $A$. Using the expert scoring method, compare the strength of the association between $x_i$ and $x_j$, and quantify it numerically.

$$A = \begin{bmatrix} 0 & x_{12} & \cdots & x_{1n} \\ x_{21} & 0 & \cdots & x_{22} \\ \vdots & \vdots & \ddots & \vdots \\ x_{n1} & x_{n2} & \cdots & 0 \end{bmatrix} \tag{1}$$

(2) Calculate the direct influence matrix A to get the normalized direct influence matrix $B$.

$$B = A/s \tag{2}$$

(3) Calculate the normalized direct influence matrix $B$ to obtain the comprehensive influence matrix $T$, which represents the comprehensive effect of the mutual influence among the factors.

$$T = (B + B^2 + \cdots B^n) = \sum_{n=1}^{\infty} B^k = B(E - B)^{-1} \tag{3}$$

Table 1. Restrictive factor system of promoting the prefabricated building.

| Dimension | Factor | Factor explanation |
|---|---|---|
| Policy | Financial subsidy (B1) | Incentive support and subsidies for the prefabricated construction industry are not perfect |
| | Evaluation standard (B2) | Lack of building codes and standards |
| | Implementing regulations (B3) | Lack of relevant policy implementation details |
| Technology | Professional (B4) | Lack of professionals for the construction |
| | Component connection (B5) | The connection of prefabricated buildings is complex |
| | Technology R & D (B6) | Insufficient research and development of prefabricated building products |
| | Structural properties (B7) | The integrity of the prefabricated building structure is not high enough |
| | Safety performance (B8) | The safety performance design of prefabricated buildings is complex (seismic and fire resistance) |
| | Design level (B9) | Designers lack the design of prefabricated buildings |
| | Appropriate Technology (B10) | The technology of prefabricated type is backward |
| | BIM application (B11) | The promotion of BIM is not strong enough, and the drawings of design, construction, and assembly production are not universal enough |
| Society | Cognitive level (B12) | Consumers still have doubts about the quality and safety assurance system of prefabricated buildings |
| | Business transformation (B13) | Construction enterprises are not very enthusiastic about transformation and reform |
| | Project tendering (B14) | Traditional project bidding fails to take care of the needs of prefabricated buildings |
| Economy | Construction cost (B15) | Higher construction costs (high construction costs and additional transport costs) |
| | Cost of capital (B16) | Higher capital cost (equipment input cost) |
| | Economic benefit (B17) | Prefabricated buildings fail to reduce costs and increase efficiency |
| Manage | Supply integration (B18) | Poor integration of prefabricated building supply chain |
| | The idea is backward (B19) | The management concept is seriously out of touch with the actual work |
| | Mismanagement (B20) | The management system of prefabricated construction projects is not perfect |
| | Post-maintenance (B21) | The post-use and maintenance system of prefabricated buildings is not perfect enough |
| | Practitioners (B22) | Extremely lack of employees |

(4) Calculate the fourth degree.

$$D_i = \sum_{i=1}^{n} t_{ij} \tag{4}$$

$$C_i = \sum_{i=1}^{n} t_{ij} \tag{5}$$

$$M_i = D_i + C_i \tag{6}$$

$$R_i = D_i + C_i \tag{7}$$

(5) Draw a cause-effect diagram combining centrality and causality to identify the causal attributes of factors.

Table 2. The comprehensive influence matrix of the constraints on the promotion of prefabricated buildings.

| FACTORS | B1 | B2 | B3 | B4 | B5 | B6 | B7 | B8 | B9 | B10 | B11 | B12 | B13 | B14 | B15 | B16 | B17 | B18 | B19 | B20 | B21 | B22 |
|---|---|---|---|---|---|---|---|---|---|---|---|---|---|---|---|---|---|---|---|---|---|---|
| B1 | 0.028 | 0.065 | 0.056 | 0.120 | 0.070 | 0.136 | 0.095 | 0.076 | 0.103 | 0.126 | 0.144 | 0.069 | 0.169 | 0.107 | 0.132 | 0.127 | 0.132 | 0.127 | 0.081 | 0.097 | 0.096 | 0.120 |
| B2 | 0.057 | 0.044 | 0.080 | 0.123 | 0.111 | 0.142 | 0.138 | 0.122 | 0.138 | 0.140 | 0.148 | 0.110 | 0.138 | 0.102 | 0.121 | 0.120 | 0.131 | 0.119 | 0.114 | 0.125 | 0.124 | 0.109 |
| B3 | 0.044 | 0.052 | 0.024 | 0.077 | 0.066 | 0.092 | 0.079 | 0.068 | 0.075 | 0.086 | 0.099 | 0.073 | 0.113 | 0.081 | 0.082 | 0.073 | 0.081 | 0.081 | 0.076 | 0.095 | 0.082 | 0.096 |
| B4 | 0.028 | 0.045 | 0.036 | 0.062 | 0.068 | 0.086 | 0.101 | 0.071 | 0.074 | 0.084 | 0.099 | 0.074 | 0.119 | 0.072 | 0.103 | 0.081 | 0.094 | 0.085 | 0.090 | 0.094 | 0.093 | 0.115 |
| B5 | 0.033 | 0.070 | 0.046 | 0.143 | 0.076 | 0.136 | 0.156 | 0.134 | 0.131 | 0.154 | 0.158 | 0.099 | 0.163 | 0.114 | 0.146 | 0.136 | 0.150 | 0.130 | 0.108 | 0.109 | 0.115 | 0.125 |
| B6 | 0.049 | 0.090 | 0.070 | 0.142 | 0.144 | 0.103 | 0.167 | 0.152 | 0.157 | 0.170 | 0.185 | 0.114 | 0.181 | 0.118 | 0.167 | 0.151 | 0.171 | 0.154 | 0.124 | 0.125 | 0.130 | 0.118 |
| B7 | 0.030 | 0.047 | 0.037 | 0.098 | 0.107 | 0.096 | 0.066 | 0.106 | 0.092 | 0.101 | 0.106 | 0.077 | 0.119 | 0.086 | 0.096 | 0.092 | 0.111 | 0.093 | 0.075 | 0.082 | 0.085 | 0.079 |
| B8 | 0.031 | 0.063 | 0.041 | 0.116 | 0.112 | 0.096 | 0.120 | 0.063 | 0.098 | 0.118 | 0.123 | 0.089 | 0.137 | 0.088 | 0.125 | 0.120 | 0.124 | 0.108 | 0.093 | 0.094 | 0.096 | 0.087 |
| B9 | 0.042 | 0.065 | 0.051 | 0.117 | 0.117 | 0.136 | 0.126 | 0.076 | 0.134 | 0.146 | 0.091 | 0.145 | 0.095 | 0.128 | 0.113 | 0.137 | 0.122 | 0.101 | 0.097 | 0.106 | 0.091 |
| B10 | 0.038 | 0.063 | 0.043 | 0.117 | 0.113 | 0.119 | 0.129 | 0.111 | 0.111 | 0.082 | 0.141 | 0.083 | 0.141 | 0.097 | 0.122 | 0.109 | 0.129 | 0.117 | 0.103 | 0.097 | 0.103 | 0.096 |
| B11 | 0.044 | 0.064 | 0.046 | 0.115 | 0.121 | 0.116 | 0.135 | 0.112 | 0.118 | 0.120 | 0.095 | 0.103 | 0.155 | 0.104 | 0.143 | 0.134 | 0.150 | 0.128 | 0.112 | 0.107 | 0.120 | 0.116 |
| B12 | 0.026 | 0.026 | 0.025 | 0.056 | 0.029 | 0.055 | 0.042 | 0.036 | 0.047 | 0.043 | 0.049 | 0.025 | 0.094 | 0.053 | 0.051 | 0.037 | 0.042 | 0.042 | 0.033 | 0.042 | 0.049 | 0.068 |
| B13 | 0.040 | 0.046 | 0.044 | 0.125 | 0.075 | 0.127 | 0.103 | 0.083 | 0.098 | 0.120 | 0.132 | 0.075 | 0.096 | 0.095 | 0.111 | 0.100 | 0.115 | 0.118 | 0.109 | 0.118 | 0.106 | 0.127 |
| B14 | 0.031 | 0.042 | 0.035 | 0.086 | 0.054 | 0.073 | 0.066 | 0.059 | 0.079 | 0.098 | 0.083 | 0.057 | 0.117 | 0.043 | 0.069 | 0.063 | 0.078 | 0.078 | 0.064 | 0.071 | 0.070 | 0.081 |
| B15 | 0.031 | 0.035 | 0.037 | 0.082 | 0.064 | 0.100 | 0.084 | 0.067 | 0.084 | 0.093 | 0.115 | 0.058 | 0.147 | 0.090 | 0.068 | 0.103 | 0.122 | 0.112 | 0.082 | 0.086 | 0.094 | 0.090 |
| B16 | 0.035 | 0.034 | 0.035 | 0.074 | 0.054 | 0.099 | 0.076 | 0.062 | 0.079 | 0.099 | 0.103 | 0.049 | 0.134 | 0.076 | 0.111 | 0.058 | 0.119 | 0.102 | 0.074 | 0.082 | 0.086 | 0.075 |
| B17 | 0.033 | 0.036 | 0.033 | 0.078 | 0.057 | 0.100 | 0.078 | 0.066 | 0.084 | 0.095 | 0.102 | 0.057 | 0.145 | 0.081 | 0.121 | 0.104 | 0.070 | 0.104 | 0.077 | 0.090 | 0.099 | 0.088 |
| B18 | 0.038 | 0.046 | 0.036 | 0.094 | 0.080 | 0.095 | 0.103 | 0.078 | 0.079 | 0.095 | 0.110 | 0.071 | 0.117 | 0.077 | 0.107 | 0.100 | 0.112 | 0.064 | 0.075 | 0.079 | 0.093 | 0.075 |
| B19 | 0.028 | 0.037 | 0.035 | 0.099 | 0.060 | 0.067 | 0.079 | 0.064 | 0.066 | 0.091 | 0.093 | 0.056 | 0.101 | 0.061 | 0.085 | 0.075 | 0.092 | 0.081 | 0.047 | 0.099 | 0.082 | 0.067 |
| B20 | 0.033 | 0.044 | 0.039 | 0.094 | 0.075 | 0.078 | 0.095 | 0.078 | 0.078 | 0.098 | 0.110 | 0.076 | 0.114 | 0.083 | 0.099 | 0.089 | 0.105 | 0.101 | 0.098 | 0.058 | 0.098 | 0.089 |
| B21 | 0.031 | 0.035 | 0.031 | 0.070 | 0.049 | 0.065 | 0.066 | 0.059 | 0.061 | 0.068 | 0.073 | 0.088 | 0.108 | 0.076 | 0.079 | 0.063 | 0.086 | 0.078 | 0.065 | 0.081 | 0.047 | 0.075 |
| B22 | 0.032 | 0.041 | 0.037 | 0.123 | 0.072 | 0.104 | 0.088 | 0.071 | 0.083 | 0.085 | 0.097 | 0.068 | 0.122 | 0.077 | 0.090 | 0.081 | 0.092 | 0.078 | 0.079 | 0.095 | 0.095 | 0.057 |

(6) Calculate the overall influence matrix $H$.

$$H = T + E \tag{8}$$

(7) Determine the reachability matrix $F$.

$$f_{ij} = \begin{cases} 1, & e_{ij} \geq \lambda(i,j,=1,2,\ldots,n) \\ 0, & e_{ij} < \lambda(i,j,=1,2,\ldots,n) \end{cases} \tag{9}$$

(8) Establish antecedent set and reachable set.

$$S(x_i) = \{x_j \in X | f_{ij} = 1\} \tag{10}$$

$$R(x_i) = \{x_j \in X | f_{ij} = 1\} \tag{11}$$

(9) According to the above calculation results, draw the multi-layer hierarchical structure model diagram of the constraints of the development of prefabricated buildings.

## 4 MODEL CALCULATION AND ANALYSIS

### 4.1 *Importance of measurement of constraints and causality test based on DEMATEL method*

#### 4.1.1 *Calculation of comprehensive correlation matrix*

According to the above-established system of restrictive factors for the promotion of prefabricated buildings, 15 experts were invited to evaluate the relationship between the obstacles. The rules were no impact "0", low impact "1", and general impact "2", a high degree of influence "3", and an extremely high degree of influence "4". We summarized the scoring table of all experts, and arithmetically averaged the scores of all experts on the same factor to obtain a scoring summary table, and then calculated the comprehensive correlation matrix. The results are shown in Table 2.

#### 4.1.2 *Centrality calculation and analysis*

Centrality is the combined strength of a factor's influence and influence, reflecting the strength of its association with other factors. According to the comprehensive correlation matrix, combined with Equations (4) to (7), the influence degree, influence degree, centrality degree and cause a degree of each obstacle factor are calculated, as shown in Table 3.

Table 3. DEMATEL calculation results.

| Obstructing factor | Influence R | Influenced D | Centrality f | Degree of cause r | Factor attribute |
|---|---|---|---|---|---|
| B1 | 0.783 | 2.278 | 3.061 | −1.495 | result |
| B2 | 1.091 | 2.557 | 3.647 | −1.466 | result |
| B3 | 0.918 | 1.695 | 2.614 | −0.777 | result |
| B4 | 2.211 | 1.777 | 3.988 | 0.434 | reason |
| B5 | 1.776 | 2.633 | 4.409 | −0.857 | result |
| B6 | 2.223 | 2.984 | 5.207 | −0.761 | result |
| B7 | 2.204 | 1.881 | 4.085 | 0.323 | reason |
| B8 | 1.863 | 2.141 | 4.004 | −0.278 | result |
| B9 | 2.012 | 2.375 | 4.387 | −0.364 | result |
| B10 | 2.301 | 2.264 | 4.565 | 0.037 | reason |
| B11 | 2.512 | 2.457 | 4.969 | 0.054 | reason |
| B12 | 1.662 | 0.969 | 2.631 | 0.693 | reason |
| B13 | 2.876 | 2.162 | 5.039 | 0.714 | reason |
| B14 | 1.876 | 1.496 | 3.372 | 0.380 | reason |
| B15 | 2.355 | 1.844 | 4.199 | 0.511 | reason |
| B16 | 2.129 | 1.715 | 3.844 | 0.414 | reason |
| B17 | 2.443 | 1.799 | 4.243 | 0.644 | reason |
| B18 | 2.220 | 1.826 | 4.047 | 0.394 | reason |
| B19 | 1.881 | 1.563 | 3.444 | 0.318 | reason |
| B20 | 2.023 | 1.830 | 3.853 | 0.193 | reason |
| B21 | 2.069 | 1.456 | 3.524 | 0.613 | reason |
| B22 | 2.044 | 1.768 | 3.812 | 0.277 | reason |

It can be seen from Table 3 that the factors with higher centrality are B6, B13, B11, B10, B5, B9, B17, B15, B7, B18, and B8. These barriers greatly affect or are affected by other barriers and should be addressed in the short term.

### 4.1.3 *Calculation and analysis of cause degree*

The degree of cause is an index to divide the attributes of factors. The positive value of the degree of cause is the causal factor, and the opposite is the resulting factor. There are 15 cause factors and 7 result factors. Referring to Zhan Yi (Zhan 2017), Pan Yuhong (Pan 2017), and other scholars, draw a cause-result diagram (Figure 1) of the constraints on the promotion of prefabricated buildings and identify key obstacles.

It can be seen from Figure 1 that the average centrality is 3.952, and 12 factors are greater than the average centrality. These factors not only play an obvious hindering role in the promotion of prefabricated buildings but also actively affect other factors, which are key hindering factors. The remaining 10 hindering factors, although the centrality is relatively high, are greatly influenced by other factors and are not regarded as key hindering factors.

### 4.2 *Hierarchical structure analysis of constraints based on ISM method*

### 4.2.1 *Calculation of reachability matrix*

According to the comprehensive influence matrix, the accessibility matrix of the restrictive factors for the promotion of prefabricated buildings is calculated. First, use Equation (4) to calculate the overall influence matrix $H$; secondly, calculate $\lambda = \alpha + \beta$ ($\alpha$ and $\beta$ refer to the mean and standard deviation of the comprehensive influence matrix $T$, respectively); finally, determine the threshold $\lambda = 0.1227$, and calculate with Equation (5). A reachable matrix is obtained, as shown in Table 4.

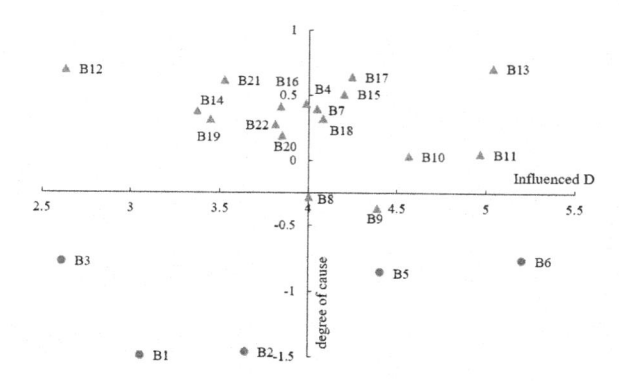

Figure 1. Reason-result diagram of the restrictive factors for the promotion of prefabricated buildings.

Table 4. Restriction factors reachable matrix of prefabricated building promotion.

| Factors | B1 | B2 | B3 | B4 | B5 | B6 | B7 | B8 | B9 | B10 | B11 | B12 | B13 | B14 | B15 | B16 | B17 | B18 | B19 | B20 | B21 | B22 |
|---|---|---|---|---|---|---|---|---|---|---|---|---|---|---|---|---|---|---|---|---|---|---|
| B1 | 1 | 0 | 0 | 0 | 0 | 1 | 0 | 0 | 0 | 1 | 1 | 0 | 1 | 0 | 1 | 1 | 1 | 1 | 0 | 0 | 0 | 0 |
| B2 | 0 | 1 | 0 | 0 | 0 | 1 | 1 | 0 | 1 | 1 | 1 | 0 | 1 | 0 | 0 | 0 | 1 | 0 | 0 | 1 | 1 | 0 |
| B3 | 0 | 0 | 1 | 0 | 0 | 0 | 0 | 0 | 0 | 0 | 0 | 0 | 0 | 0 | 0 | 0 | 0 | 0 | 0 | 0 | 0 | 0 |
| B4 | 0 | 0 | 0 | 1 | 0 | 0 | 0 | 0 | 0 | 0 | 0 | 0 | 0 | 0 | 0 | 0 | 0 | 0 | 0 | 0 | 0 | 0 |
| B5 | 0 | 0 | 0 | 1 | 1 | 1 | 1 | 1 | 1 | 1 | 1 | 0 | 1 | 0 | 1 | 1 | 1 | 1 | 0 | 0 | 0 | 1 |
| B6 | 0 | 0 | 0 | 1 | 1 | 1 | 1 | 1 | 1 | 1 | 1 | 0 | 1 | 0 | 1 | 1 | 1 | 1 | 1 | 1 | 1 | 0 |
| B7 | 0 | 0 | 0 | 0 | 0 | 0 | 1 | 0 | 0 | 0 | 0 | 0 | 0 | 0 | 0 | 0 | 0 | 0 | 0 | 0 | 0 | 0 |
| B8 | 0 | 0 | 0 | 0 | 0 | 0 | 0 | 1 | 0 | 0 | 0 | 0 | 1 | 0 | 1 | 0 | 1 | 0 | 0 | 0 | 0 | 0 |
| B9 | 0 | 0 | 0 | 0 | 0 | 1 | 1 | 1 | 1 | 1 | 1 | 0 | 1 | 0 | 1 | 0 | 1 | 0 | 0 | 0 | 0 | 0 |
| B10 | 0 | 0 | 0 | 0 | 0 | 0 | 1 | 0 | 0 | 1 | 1 | 0 | 1 | 0 | 0 | 0 | 1 | 0 | 0 | 0 | 0 | 0 |
| B11 | 0 | 0 | 0 | 0 | 0 | 0 | 1 | 0 | 0 | 0 | 1 | 0 | 1 | 0 | 1 | 1 | 1 | 1 | 0 | 0 | 0 | 0 |
| B12 | 0 | 0 | 0 | 0 | 0 | 0 | 0 | 0 | 0 | 0 | 0 | 1 | 0 | 0 | 0 | 0 | 0 | 0 | 0 | 0 | 0 | 0 |
| B13 | 0 | 0 | 0 | 1 | 0 | 1 | 0 | 0 | 0 | 0 | 1 | 0 | 1 | 0 | 0 | 0 | 0 | 0 | 0 | 0 | 0 | 1 |
| B14 | 0 | 0 | 0 | 0 | 0 | 0 | 0 | 0 | 0 | 0 | 0 | 0 | 0 | 1 | 0 | 0 | 0 | 0 | 0 | 0 | 0 | 0 |
| B15 | 0 | 0 | 0 | 0 | 0 | 0 | 0 | 0 | 0 | 0 | 0 | 0 | 1 | 0 | 1 | 0 | 0 | 0 | 0 | 0 | 0 | 0 |
| B16 | 0 | 0 | 0 | 0 | 0 | 0 | 0 | 0 | 0 | 0 | 0 | 0 | 1 | 0 | 0 | 1 | 0 | 0 | 0 | 0 | 0 | 0 |
| B17 | 0 | 0 | 0 | 0 | 0 | 0 | 0 | 0 | 0 | 0 | 0 | 0 | 1 | 0 | 0 | 0 | 1 | 0 | 0 | 0 | 0 | 0 |
| B18 | 0 | 0 | 0 | 0 | 0 | 0 | 0 | 0 | 0 | 0 | 0 | 0 | 0 | 0 | 0 | 0 | 0 | 1 | 0 | 0 | 0 | 0 |
| B19 | 0 | 0 | 0 | 0 | 0 | 0 | 0 | 0 | 0 | 0 | 0 | 0 | 0 | 0 | 0 | 0 | 0 | 0 | 1 | 0 | 0 | 0 |
| B20 | 0 | 0 | 0 | 0 | 0 | 0 | 0 | 0 | 0 | 0 | 0 | 0 | 0 | 0 | 0 | 0 | 0 | 0 | 0 | 1 | 0 | 0 |
| B21 | 0 | 0 | 0 | 0 | 0 | 0 | 0 | 0 | 0 | 0 | 0 | 0 | 0 | 0 | 0 | 0 | 0 | 0 | 0 | 0 | 1 | 0 |
| B22 | 0 | 0 | 0 | 0 | 0 | 0 | 0 | 0 | 0 | 0 | 0 | 0 | 0 | 0 | 0 | 0 | 0 | 0 | 0 | 0 | 0 | 1 |

### 4.2.2 Determination of the hierarchical structure

Combined with the reachability matrix (Table 4), the antecedent set $S(x_i)$ and the reachability set $R(x_i)$ of the constraints on the promotion of prefabricated buildings are obtained. To make the explanatory structural model more reasonable, a threshold $\lambda$ is introduced at this time, and $\lambda$ is used to identify the factors with less influence and make a choice. The larger the value of $\lambda$, the simpler the level of the structural system, but the relationship between factors is also blurred; the smaller the value of $\lambda$, the more detailed the level of the structural system, but the lack of integrity. After inspection, when $S(x_i) \cap R(x_i) = R(x_i)$ is satisfied, the hindering factors can be divided into 7 levels (Figure 2).

It can be seen from Figure 2 that the constraints on the promotion of prefabricated buildings are divided into 7 levels. The surface layer (i.e., the root layer) of the first level is the direct factor, and the intermediate factors pass through the second to sixth layers (i.e., the indirection layer) to the final Deep (i.e., the direct layer) root factors. The promotion of BIM at the fundamental level is not strong enough, the policy incentives and subsidies are not perfect, and the lack of building codes and standards are the key obstacles. It can be seen from the explanatory structural model that

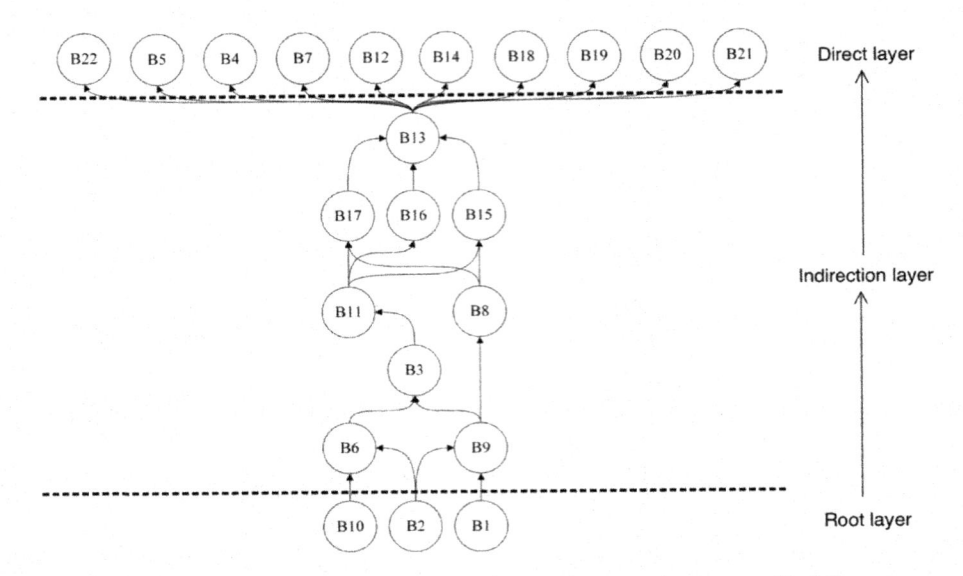

Figure 2.   Multi-level hierarchy diagram of constraints on the spread of prefabricated buildings.

there is a complex relationship between the constraints on the promotion of prefabricated buildings, which are embodied in the following three paths.

(1) Obstruction path 1: B10→B6→B3→B11→ (B17+B16+B15) →B13→ (B22, B5, B4, B7, B12, B14, B18, B19, B20, B21). The lack of BIM promotion has led to insufficient research and development of prefabricated building products and the lack of corresponding policy implementation rules, which impose a negative impact on different stakeholders and hinders the promotion of prefabricated buildings. Therefore, vigorously promoting BIM is one of the important means to promote prefabricated buildings.

(2) Obstruction path 2: B2→ (B6+B9) →B3→B11→ (B17+B16+B15) →B13→ (B22, B5, B4, B7, B12, B14, B18, B19, B20, B21). Insufficient government attention and lack of corresponding policies and regulations have resulted in insufficient research and development of prefabricated building products, and a lack of design of prefabricated buildings by designers, resulting in the lack of relevant policy implementation rules and the willingness of the construction industry.

(3) Obstruction path 3: B1→B9→ ((B3→B11→ (B17+B16+B15) + (B8→ (B17+B15))) →B13→ (B22, B5, B4, B7, B12, B14, B18, B19, B20, B21). The imperfect government incentives, support, and subsidies have led to the lack of design of prefabricated buildings by designers, resulting in the lack of detailed rules for the implementation of relevant policies, and at the same time, it is not conducive to enterprises to reduce costs and increase efficiency and reduce the number of people engaged in the willingness of the prefabricated construction industry.

## 5   CONCLUSIONS AND COUNTERMEASURES

### 5.1   *Research conclusions*

This paper empirically analyzes the internal relationship and logical correlation of the factors hindering the promotion of prefabricated buildings through the combined model constructed by the Decision Experiment and Evaluation Laboratory (DEMATEL) and the Interpretive Structural Model (ISM). The research results are as follows.

(1) Eleven factors such as insufficient technical research and development of prefabricated building products have a high centrality in the evaluation system, and their interaction with other factors is stronger, which should be solved in a timely and effective manner; construction enterprises are not highly motivated to transform and reform eight the centrality of the factor is greater than the average value and the cause degree is positive, which has a strong restrictive effect on the promotion of prefabricated buildings and has a high effect on other factors, which is a key obstacle.

(2) Based on the division of the hierarchical structure, the promotion of BIM is not strong enough, the generality of the drawings for design, construction, and assembly production is not enough, and the incentive support and subsidies for the prefabricated building industry are not perfect, and the lack of building codes and standards are three factors as the fundamental factors. Layer factors play a decisive role in the entire system. From these three factors, three main paths that hinder the promotion of prefabricated buildings are extended.

## 5.2 *Management implications*

(1) Standard first to achieve "global coverage". Based on the basic building codes, benchmark the building development goals that are compatible with the "dual-carbon" era, improve the whole process and type standard system of prefabricated buildings, and strengthen the promotion and integration of standards within the industry; focus on safe production, green and low-cost carbon, economical and practical mainline, standardize and refine the source of standard parameters, strengthen quality monitoring and control, adjust and optimize parameter structure promptly, and build a solid foundation for the healthy development of prefabricated buildings.

(2) Policy blessing to activate "nerve endings". Further, optimize the fiscal and tax subsidy policies in the field of prefabricated buildings, and exert efforts at the same time in intensity and level to relieve the rigid problem of the high cost of prefabricated buildings; strengthen the "dual carbon" goal orientation, and stimulate the internal power of the transformation and reform of construction enterprises and other micro-subjects, guide the expansion and extension of the building structure to "double low" and increase the proportion of prefabricated buildings; according to the actual needs, improve the professional training system for prefabricated buildings, create a team of professionals with rich reserves and strong skills, and provide basic manpower for the steady promotion of prefabricated buildings capital support.

(3) Leading by technology and strengthening "horizontal linkage". Make every effort to promote BIM technology, effectively embed it in the sub-sections of prefabricated building construction, twist the core element of information, strengthen the integration and unified and efficient management of the entire industry chain, and improve the applicability of components and drawings in various scenarios, environments, and subjects nature, and the pertinence of technical disclosure; starting from customer needs, increase the research and development of prefabricated building technology products, and strengthen the green and low-carbon orientation, insist that quantity follows quality, progress follows actual effect, from the perspective of depth, breadth and fit Improve the energy level of the technical system and promote the large-scale and intensive development of prefabricated buildings.

## REFERENCES

Guilin Huang, C. Z. (2020) Green supply chain risk of prefabricated buildings based on SNA. *Journal of civil engineering and management* 37 (02): 41–49.

Hui An, Y.-X. K., Wen-Jing Yang, Ling Song (2020) Analysis of supply chain integration motivation of prefabricated building based on SEM. *Journal of civil engineering and management* 37 (01): 50–56.

Li, G. "Dual carbon" targets guide new development *China Environmental Management.*

Rongxiu Lu, K. Q., Yuang Tan, Songlin Xu (2021) Main problems and Countermeasures of prefabricated building development from the perspective of developers. *Building Structure* 51 (S2): 1134–1138.

Sn, A., As, B., Gz, A., Kn, A., Sv, A. & Kp, C. (2022) The challenges confronting the growth of sustainable prefabricated building construction in Australia: Construction industry views. *Journal of Building Engineering* 48: 103935.

Song, G. (2021) The main actors and measures of China's implementation of carbon peak and carbon neutral goals. *Urban and Environmental Research* 04 (47–60).

Wan-Ching Chen, Z. -B. W. (2021) The practical dilemma and Improvement path of China's prefabricated building Policy Implementation: Based on the Perspective of Mitt-Horn Policy Implementation Model. *Building Economy* 42 (08): 77–80.

Xi Cao, C. M., Haitao Pan (2021) Comparative Analysis and research on carbon emission of prefabricated concrete and cast-in-place buildings based on carbon Emission Model. *Building structure* 51 (S2): 1233–1237.

Xu, Z. (2021) Research on The Cost Management of the Whole industrial chain of assembly Building. *Building Economy* (42 (02)): 81–85.

Yanli Zhao, C. Z., Yifan Liu (2020) Research on cost chain structure and Optimization direction of prefabricated building: Based on the perspective of real estate enterprises *Building Economy* 41 (11): 57–62.

Yaqi Sun, Y. T. (2020) Research on Key Risk of Assembly Building Supply Chain Based on Complex Network Theory. *Building Economy* 41 (11): 79–83.

Yuanfang Lv, S. Z., Kangjie Zhang, Haiyang Wang, Lei Fan, Tian-Feng Yang (2021) Research on Energy Conservation and Environmental Protection of prefabricated Buildings. *Building Economy* 42 (S1): 186–188.

Yuhong Pan, Y. Z., Xu Ma (2017) Identification of Influencing Factors for supplier selection of prefabricated Residential Components based on Dematel-BP. *Mathematics in practice and cognition* 47 (09): 22–34.

Zhan, Y. (2017) Supplier selection of prefabricated components for prefabricated residential Buildings based on DEMATEL Method.) Chongqing Jiaotong University.

*Civil Engineering and Urban Research – Mohamed & Hou (Eds)*

# Research on 3D building model construction based on UAV oblique photogrammetry

Zengzeng Lian*

*School of Instrument Science and Engineering, Southeast University, Nanjing City, China*
*School of Surveying and Land Information Engineering, Henan Polytechnic University, Jiaozuo City, China*

Jingcheng Xu* & Jiaqi Dong*

*School of Surveying and Land Information Engineering, Henan Polytechnic University, Jiaozuo City, China*

ABSTRACT: Traditional aerial photogrammetry can only obtain the height information and top texture information of buildings, and the extraction of side textures can only be supplemented by manual shooting, which contradicts the demand for 3D data acquisition by the rapid development of 3D digital cities. The Phantom 4 RTK collected the oblique image of the laboratory building and successfully constructed the 3D model of the key laboratory building of the Henan University of Technology through aerial triangulation and other steps. The experimental results show that the use of UAV oblique photogrammetry technology can well realize the construction of the 3D model of buildings.

## 1 INTRODUCTION

Building 3D modeling is a science and technology that uses computer graphics and image processing technology to convert the 2D floor plan of a building into a 3D model and display it in three dimensions (Li et al. 2016; Tatum & Liu 2017). It has a wide range of applications in the fields of urban landscape planning, architectural design, and the protection of ancient buildings. In recent years, with the acceleration of the construction of digital cities, virtual cities, and smart cities in various places, 3D modeling of buildings has become a research hotspot in the fields of surveying and mapping, GIS, and architecture. The core of building 3D modeling technology is to construct its three-dimensional model according to the geometric information of the building, use related modeling software to generate its three-dimensional model, and assign the model surface texture map to display the three-dimensional graphics (Sun & Wang 2018).

In recent years, the rapid development of UAV photogrammetry technology has provided more abundant technical means for the acquisition, processing, and analysis of short-range surface space mapping data, and also greatly improved the accuracy of extracting ground objects, making building refined 3D modeling become a technology trend (Tan et al. 2021). Using oblique photogrammetry technology to construct 3D models can quickly acquire large-area 3D objects on the premise of ensuring object coordinate accuracy and texture quality. Compared with traditional fine monomer modeling, airborne radar technology, and 3D laser scanning technology, it is very convenient to collect data, build a model at a low price, and consume little time. It can be used in building informatization, smart city construction, and other fields. Therefore, it has gradually become the mainstream 3D modeling technology in the market (Yu et al. 2021).

To sum up, for the research and engineering practice in the fields of a digital city, virtual city, and smart city, there is an urgent need for a practical, highly operable, and highly automated modeling

---

*Corresponding Authors: zengzenglian@163.com, xujingcheng1997@163.com and 1191809324@qq.com

DOI 10.1201/9781003372417-92

method. This paper proposes a building method based on UAV oblique photogrammetry. It provides a new idea for the building information modeling management of buildings.

## 2   3D MODELING PRINCIPLE OF OBLIQUE PHOTOGRAMMETRY BUILDINGS

Oblique photogrammetry technology is developed based on traditional measurement technology. It breaks through the limitations of vertical photography and can obtain images of buildings in multiple orientations and angles at the same time, including their top images and rich textures on the sides (Zhao et al. 2021). When the aircraft is sailing, the instantaneous coordinates, flight speed and altitude, image overlap, and other data will be automatically recorded by the system, so that the oblique images can be sorted and analyzed later. Therefore, the same feature can appear on at least 4 images and a maximum of 8 images, so it is convenient to analyze the structure type of the feature in the internal work, and the texture mapping can also select the one with the highest definition and the smallest deformation image. The advantage of this technology is that it can reflect the real situation and shape of ground objects in all directions, capture the high-precision side texture information of ground objects, especially buildings, and can locate and model the ground objects in the image to build a real 3D model of the city, its application fields are becoming more and more extensive (Tan et al. 2022).

The workflow of oblique photogrammetry includes field image acquisition, image preprocessing, automatic aerial triangulation, and 3D triangular mesh construction. Among them, field image acquisition can be divided into survey area observation, survey area route planning, flight plan implementation, and data storage; image preprocessing mainly includes geometric correction and radiation correction of images; 3D triangular mesh construction mainly includes the segmentation and merging method, triangular network Growth algorithm, and point-by-point interpolation (Sun et al. 2022).

The data processing flow of oblique photogrammetry technology includes image preprocessing, joint adjustment of multiview images, intensive matching of multiview images, automatic extraction of high-precision DSM, and urban 3D modeling. The basic flow of its data processing is shown in Figure 1.

## 3   EXPERIMENT AND ANALYSIS

### 3.1   *Overview of the experimental area*

The State Key Laboratory Building of the Henan University of Technology has a construction area of 30,000 square meters, the main body of the building has 13 floors, and the building height is 53.855 m. The seismic fortification intensity of the building is 7 degrees.

### 3.2   *Data collection*

The experiment uses DJI Phantom 4RTK small multirotor high-precision aerial survey drone. Its gimbal system is equipped with a 1-inch 20-megapixel CMOS sensor, intelligent infrared perception, and visual obstacle avoidance system; further, the UAV is equipped with a TimeSync system RTK module and high-sensitivity GNSS system so that it has centimeter-level navigation and positioning accuracy and a high-performance imaging system. The specific UAV parameters are shown in Table 1.

To ensure the modeling accuracy and sufficient data in the overlapping area, the designed flight is about 60 m in altitude, 70% in the sideways overlap, 80% in the course overlap, and has 3 flight sorties, and a total of 685 oblique images were acquired.

Data collection is carried out on the comprehensive experimental building, and image control points are laid around the experimental building, as shown in Figure 2. A total of six image control

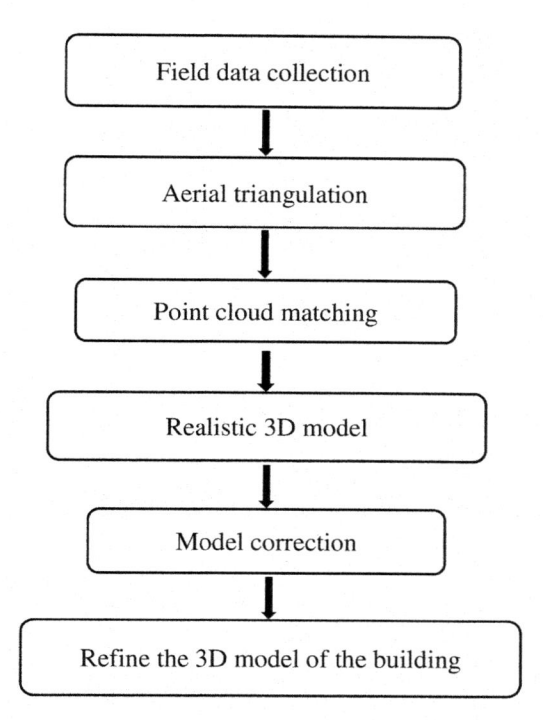

Figure 1.    Flowchart of the data processing of oblique photogrammetry.

Table 1.    UAV technical parameters.

| | |
|---|---|
| Weight | 1,388 g |
| Maximum ascent speed | Automatic flight: 6 m/s |
| | Manual control: 5 m/s |
| Maximum descent speed | 3 m/s |
| Maximum flight speed | Attitude mode: 58 km/h |
| | Positioning mode: 50 km/h |
| Working temperature | to 40°C |
| Maximum flight time | About 30 mins |
| Maximum takeoff altitude | 6,000 m |
| Image sensor | One inch CMOS |
| | 20 million effective pixels |
| Lens | FOV 84° |
| | 8.8 mm/24 mm |
| | f/2.8-f/11 |
| Maximum photo resolution | 5472 × 3648 (3:2) |
| | 4864 × 3648 (4:3) |
| RTK-GNSS positioning accuracy | Vertical 1.5 cm + 1 ppm (RMS) |
| | Horizontal 1 cm + 1 ppm (RMS) |

points is deployed, and RTK collects the coordinates of the image control points, as shown in Table 2. The drone sailed in the shape of a "well" and collected a total of 332 images, as shown in Figure 3. To ensure that the five directions of the experimental building can be collected, the front, rear, left, right, and top of the facility were taken in all directions. For high-definition image data, the image resolution is 5472*3648, so the modeling accuracy is high.

Figure 2. The layout of image control points.

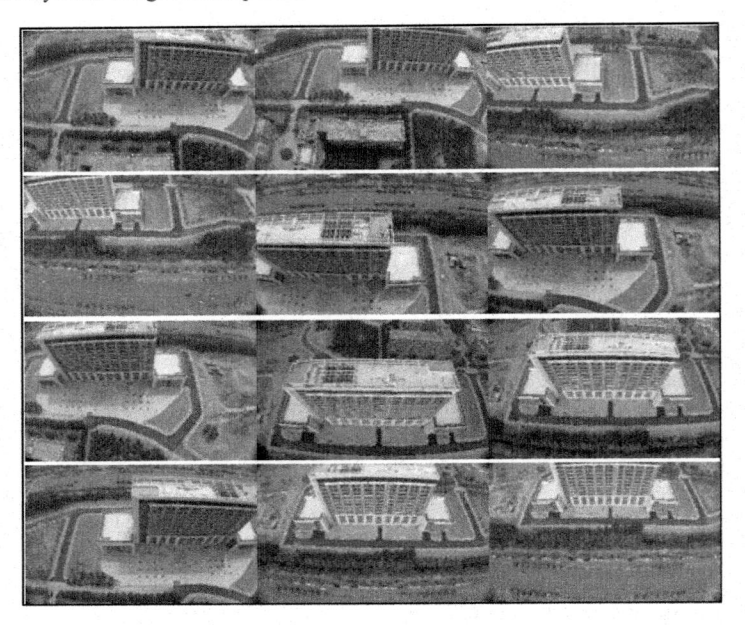

Figure 3. UAV image data.

Table 2. Image control points coordinates.

| Point | X (m) | Y (m) | H (m) |
| --- | --- | --- | --- |
| xk001 | 432019.057 | 3895310.881 | 75.54 |
| xk002 | 432031.83 | 3895251.709 | 75.607 |
| xk003 | 432112.3 | 3895260.836 | 75.292 |
| xk004 | 432191.516 | 3895271.221 | 75.073 |
| xk005 | 432160.945 | 3895329.979 | 75.524 |
| xk006 | 432094.782 | 3895321.05 | 75.278 |

## 4  3D BUILDING MODEL CONSTRUCTION AND CORRECTION

### 4.1  *3D building model construction*

First, a massive point cloud of the model is generated, and then according to the corresponding triangular mesh growth rule, the point cloud data is converted into triangular mesh model data,

654

followed by the transformation of the mesh data to the white model, and then the white model is textured mapping, and then generated the real 3D model of the experimental building, as shown in Figures 4–6. The generation model is generated in sequence according to the grid, which takes a long time.

Figure 4.    Triangular mesh model.

Figure 5.    White molds from different angles.

Figure 6.    Reality model from different angles.

## 4.2   *3D building model correction*

After importing the tiles that need to be repaired, you can first select the analysis option. At this time, the software will automatically detect the position that needs to be repaired, but most of the detected objects are dangling objects, which can be repaired automatically. However, the automatic monitoring is very limited, and most of them still need to be repaired manually. Add the repaired four tiles into the decoration directory, rework and submit the production project, and get the real 3D model of the repaired laboratory building, as shown in Figures 7 and 8.

## 5   CONCLUSION

This paper studies the basic principle, technical composition, and operation flow of oblique photogrammetry technology. Taking the key laboratory building on the campus of the Henan University

a. Before correction b. After correction

Figure 7.    Partially enlarged comparison chart.

Figure 8.    The overall effect of the 3D building after correction.

of Technology as an example, the oblique photogrammetry data of the building was collected by drones. The 3D model of the building is constructed by using oblique photogrammetry technology, and the 3D model of the building is repaired for voids and suspended objects.

## ACKNOWLEDGMENTS

This work was financially supported by the Doctoral Scientific Fund Project of Henan Polytechnic University (Grant No. B2017-10); the Henan Polytechnic University Funding Scheme for Young Backbone Teachers (Grant No. 2022XQG-08); the Natural Science Foundation of Henan Province (Grant No. 202300410180). The authors of this paper are very grateful for their financial support.

## REFERENCES

Li, M., L. Nan, N. Smith and P. Wonka (2016). "Reconstructing building mass models from UAV images." *Computers & Graphics* 54: 84–93.

Sun, J., B. Peng, C. C. Wang, K. Chen, B. Zhong and J. Wu (2022). "Building displacement measurement and analysis based on UAV images." *Automation in Construction* 140: 104367.

Sun, S. and B. Wang (2018). "Low-altitude UAV 3D modeling technology in the application of ancient buildings protection situation assessment." *Energy Procedia* 153: 320–324.

Tan, Y., G. Li, R. Cai, J. Ma and M. Wang (2022). "Mapping and modelling defect data from UAV captured images to BIM for building external wall inspection." *Automation in Construction* 139: 104284.

Tan, Y., S. Li, H. Liu, P. Chen and Z. Zhou (2021). "Automatic inspection data collection of building surface based on BIM and UAV." *Automation in Construction* 131: 103881.

Tatum, M. C. and J. Liu (2017). "Unmanned Aircraft System Applications in Construction." *Procedia Engineering* 196: 167–175.

Yu, D., S. Ji, J. Liu and S. Wei (2021). "Automatic 3D building reconstruction from multi-view aerial images with deep learning." *ISPRS Journal of Photogrammetry and Remote Sensing* 171: 155–170.

Zhao, S., F. Kang, J. Li and C. Ma (2021). "Structural health monitoring and inspection of dams based on UAV photogrammetry with image 3D reconstruction." *Automation in Construction* 130: 103832.

*Civil Engineering and Urban Research – Mohamed & Hou (Eds)*
*© 2023 the Authors, ISBN 978-1-032-44487-1*

# Research on concrete bridge crack safety evaluation system based on particle swarm optimization algorithm

Yifan Gu*
*Wuhan University of Technology, Wuhan, China*

ABSTRACT: Nowadays, automation and intelligence have become more and more popular in recent months. However, compared with other industries, the bridge cracks safety assessment is almost blank. This paper aims to study this aspect and discuss the theoretical and technical basis of this technology.

## 1 INTRODUCTION

The automation of various industries has improved the work efficiency, but the safety assessment of Bridges is still almost entirely dependent on manual. This paper aims to study the construction of a library. The data of this library can be enriched. Based on this library and the data already existed, we can build a database that can continuously improve the numerical accuracy of dangerous membership functions. By these, we can accurately evaluate the safety assessment of diversified bridge cracks.

## 2 DESIGN OF PARTICLE SWARM OPTIMIZATION ALGORITHM

The main parameters of PSO include M (population size), W (inertia weight, which represents the ability to inherit previous particles), C1, C2 (learning parameters, the former is self-learning, the latter is mutual learning between particles), and the maximum number of iterations (Gmax operation end judgment condition).

The particle swarm optimization (PSO) algorithm has the advantage of a complex adaptive system. Particle swarm optimization PSO algorithm has n particles x in the n-dimensional search space; x = (x1, x2, ..., Xn), where the velocity and position of the ith particle can be expressed as v; v = (vi1, vi2, ..., vin). The historical optimal position reached by the particle gtbest and the global optimal position reached by the particle gtbest determine the velocity and position of the particle, and ultimately determine the evolution direction of the population and the convergence speed of the algorithm. The updating speed and position of each particle are

Vijt+1 = wvtij+ c1r1(ptbestij)+ c2r2(gtbestij-xtij )

Xijt+1=xtij+vtij

In the type: (I) = 1, 2, ..., n; T is algebra; C1 and C2 are cognitive learning parameters and social learning parameters respectively. Suganthan's experiment shows that when C and C are constants, a better solution can be obtained, generally C1=C2€ [0.4]. R1 and R2 are random numbers between 0 and 1, and w is inertial weight, which can balance the ability of global search and local development in the algorithm. A large w is conducive to global search, while a small W is conducive to local search.

---

*Corresponding Author: 1549239087@qq.com

DOI 10.1201/9781003372417-93

The particle swarm optimization algorithm not only considers its optimal historical information but also makes full use of the optimal information between different particles to achieve the purpose of fast convergence to the optimal solution. However, its disadvantages are poor local search ability, easy falling into local extremum, low search accuracy, and gradually reduced convergence speed in the late evolution.

## 3 ADAPTIVE SEARCH IMMUNE PARTICLE SWARM OPTIMIZATION

Immune particle swarm optimization (IPSO), which introduces the information processing, immune memory, and self-regulation mechanism of the immune algorithm to the particle swarm algorithm, can improve the optimization of two algorithms. IPSO can complement the advantages of the two algorithms, adopt the method of maintaining population particle diversity based on particle swarm concentration mechanism and guide population renewal through the immune operators of the immune algorithm (vaccination and vaccine selection). In the immune operator, the inoculated vaccine is the extraction of prior knowledge of the problem. Immune selection includes immune detection of vaccinated population individuals and annealing selection of offspring population to select individuals into the parent with the corresponding probability. Through the above algorithm, it can ensure that the offspring with degradation phenomenon in the process of crossover and mutation, that is, the offspring whose fitness is not as good as the parent will be replaced.

Antibody concentration refers to the proportion of similar antibodies in the population.

$$Densty_{xi} = 1 \sum I = 1 n + M \parallel xi f \parallel - f xj \parallel$$

The selection probability of antibody concentration refers to the proportion of the ability of an antibody to respond to antigen and other antibody activations of a single particle in the population to the ability of all particles in the population.

$$Selection\_probability = 1 densty_{xi} \sum I = 1 n + M1densty_{xi}$$

The more particles there are, the lower the probability that a specific particle (I) will be selected, thus ensuring the diversity of the population. When the search speed of particle swarm optimization is reduced in the later stage, the disturbance factor can be added to solve the problem to improve the ability to search on the population based on maintaining the diversity of particles, convergence speed and accuracy:

$$N + 1 = wvtij + c1r1 (pbesttij - xtij) + c2r2 (gbesttij - xtij) + c3 (r3) - 0.1$$
$$xt+1ij=xtij+vtij \text{ (Cheng 2021)}$$

The algorithm flow is shown in the figure below

## 4 SIMULATION EXPERIMENT

Select five common test functions to test. The functions are shown in Table 1.

In Table 1, the first three functions are unimodal, and there is a minimum point of 0. Function 4 is a typical nonlinear multimode function with many local minimum points in its search space. Function 5 has a global minimum point of 0, and reaches a local minimum when $Xi \approx \pm K \pi I$ (i=1, 2, ..., N; k=1, 2, ..., n).

In the experiment, all test functions will be set as 6-dimensional functions. Each test function is tested 100 times with each algorithm. Then, the minimum and average values of the function values are counted, as shown in Table 2 (Lv 2020).

## 5 HAZARD DEGREE MEMBERSHIP FUNCTION MODEL OF CONCRETE BRIDGE CRACK

To establish a quantitative description of cracks and the risk of cracks, a hazard degree membership function model of concrete bridge cracks is established (Weng 2021). Based on the measured data

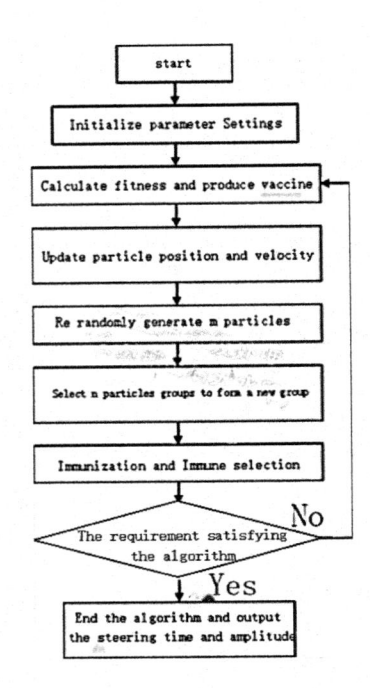

Figure 1. Algorithm Flow Chart.

Table 1. Testes functions.

| Function number | Tested function | Function formula |
|---|---|---|
| 1 | Sphere function | $f(x) = \sum i = 1nxi2$ |
| 2 | Step function | $f(x) = \sum i = \ln|xi|i+1$ |
| 3 | Rosenbrock function | $f(x) = \sum i = \ln[100(xi + 1 - xi2)2+(xi-1)2]$ |
| 4 | Rastrigin function | $f(x) = \sum i = \ln[xi2 - 10 \cos(2\pi xi)+10]$ |
| 5 | Griewank function | $f(x) = 14000 \sum i = \ln xi2 - \Pi i = \ln\cos xii + 1$ |

Table 2. Test results of three algorithms for test functions.

| Algorithm | Parameter | Sphere | Step | Rosenbrock |
|---|---|---|---|---|
| Particle swarm optimization | Average | 2.321 4 E-16 | 2.475 3 E-13 | 0.357 8 |
| | Minimum | 4.584 4 E-16 | 2.868 2 E-19 | 2.078 1 E-11 |
| Immune particle swarm | Average | 1.7203 E-14 | 6.043 3 E-13 | 0.477 2 |
| | Minimum | 1.094 9 E-16 | 8.844 7 E-18 | 2.742 5 E-11 |

| Algorithm | Parameter | Rastrigin | Griewank |
|---|---|---|---|
| Particle swarm optimization | Average | 4.491 8 | 4.158 9 E-15 |
| | Minimum | 1.445 2 E-8 | 0 |
| Immune particle swarm | Average | 4.566 9 | 3.584 9 E-15 |
| | Minimum | 1.771 7 E-8 | 0 |

of bridge cracks, the function is established according to the final estimation of different parameters by the particle swarm optimization algorithm. Hazard degree is the basis for evaluating the damage of cracks to Bridges. The main parameters are the length center length (L), width (W), depth (H), shape (S) and location parameter F (X, Y). Then the fracture hazard model F is defined as:

F=a1L+a2H+a3S+a4f

The specific values of a1, a2, a3 and a4 are obtained by referring to the study of Xu (2020), known data, particle swarm optimization algorithm, and deep learning technology, thereby obtaining the function model of L, W, H, S, and F that tends to be accurate, and the risk membership function model is obtained.

## 6 BRIDGE CRACK EVALUATION RESULTS GRADING STANDARDS

Refer to the research method of safety assessment based on a model test (Liu 2015), which can obtain sub-dimensional values.

Table 3. Rating of bearing capacity.

| D (fractual dimension value) | Bearing capacity condition | Evaluation scale |
| --- | --- | --- |
| [1.00,1.25] | excellent | 1 |
| (1.25,1.50] | good | 2 |
| (1.50,1.75] | commonly | 3 |
| (1.75,1.90] | poor | 4 |
| [1.90,2.00] | danger | 5 |

## 7 CLASSIFICATION OF FRACTURES

(1) Mild crack: box girder web crack width is less than 0.2 mm, allowing a small number of short oblique cracks, and transverse crack; crack width of the roof or bottom plate is less than 0.2 mm, allowing a small number of longitudinal cracks; diaphragm crack width is less than 0.2 mm.

(2) Moderate crack: the web crack width of the box girder is less than 0.2 mm. There are a lot of oblique cracks, and horizontal cracks and the crack length is long. The crack width of the roof or bottom plate is less than 0.2 mm, and the number of longitudinal cracks is more. Local cracks are less than 0.2 mm under anchor or toothed plate.

(3) Severe crack: there are more or longer cracks in the web, roof, and bottom plate of the box girder, and the crack width is more than 0.2 mm. Oblique cracks appear on the web of the box girder and run through the height or thickness of the web. The web crack and the transverse crack of the floor run through to form a connected crack. Many longitudinal cracks appear in the roof of the box girder, and the cracks have a trend of continuous protection. Transverse cracks appear in the tensile position of the roof or bottom plate, and the cracks continue to expand.

## 8 INFLUENCE OF CRACKS ON CONCRETE CONSTRUCTION

In concrete bridges, cracks will damage the surface protective layer of the bridge, and then the interior of the bridge will be in direct contact with the outside world. Therefore, the cement in the bridge will react with some substances. These reactions will generate calcium carbonate, thus reducing the alkalinity of concrete and carbonating. The steel materials will be corroded, which will accelerate the aging of the bridge.

Concrete bridge cracks will not only harm the appearance, but also reduce the strength and stiffness of the bridge, and even affect the safety performance of the bridge.

Bridge cracks will make the internal material of the bridge dissolve away, over time, increase the internal stress of the bridge, and affect the load capacity of the bridge.

Transverse crack generally in the crack and steel bar intersection will have an impact on the steel bar, but the impact is limited. Generally the intersection of cracks and reinforcement will have an impact on reinforcement, but the impact is limited, the general impact range is 2 to 5 cm on both sides of the crack, and when the width of the crack is effectively controlled, the influence of transverse cracks on reinforcement will gradually weaken.

Longitudinal crack on the bridge harm is much larger than the transverse crack, longitudinal crack on the mechanism of reinforcement is also more complex, generally divided into the following two cases.

(1) Due to the influence of construction quality, concrete shrinkage, freeze-thaw, and so on, the structure first produces longitudinal cracks along the direction of reinforcement, and then the existence of longitudinal cracks leads to corrosion of reinforcement along the direction of cracks.

(2) Due to poor construction quality, the insufficient thickness of concrete protective layer or carbonization of concrete, $Cl^-$ and $CO_2$ enter the concrete, causing corrosion of steel bars.

Reinforcement corrosion is caused by the volume expansion after cracking, along the direction of reinforced concrete reinforced more contact with the outside world, which makes cause steel corrosion speed is accelerated, resulting in the cycle of "rust to cracking, corrosion", eventually led to the spalling of the concrete protective layer, the serious influence the durability of a reinforced concrete bridge.

With the continuous expansion of longitudinal cracks, it will greatly weaken the grip force of concrete on steel bars, which will lead to the shedding of the concrete protective layer, so that the bearing capacity of reinforced concrete Bridges will be greatly reduced.

Therefore, the development of the longitudinal crack will affect the safety performance of the structure, so the control of the longitudinal crack width of the bridge has an important role in ensuring the durability and safety of the bridge.


## 9 CONCLUSION

Nowadays, with the trend of automation and intelligence in all occupations, the safety assessment of bridges still relies almost entirely on manual data collection and evaluation.

This paper points out a method based on artificial intelligence particle swarm optimization to automatically evaluate the influence of bridge cracks on the bridge and some relevant theoretical basis. The library automation evaluation system can increase and iteration of the data in the database, based on existing data and the continuous extension of the database to build a can improve data richness, constantly improve the accuracy of subordinate function database, and by using particle swarm optimization (PSO) algorithm to realize the accurate evaluation of diverse bridge crack safety assessment.

## REFERENCES

Cheng Jian, Li Enhua. Bridge Multi - Ting based on particle swarm optimization. Cable tower bridge concrete fracture safety assessment analysis [D]. Chongqing jiaotong university, 2020. The DOI: 10.27671 /, dc nki. GCJT maintenance decision optimization [J]. *Engineering and Construction*, 2020, 34(04):767–769.

Hu Wenjian, Yang Yang, Liu Baoan, Zheng Jian, Chang Shengqiang. 5 g large-scale based on particle swarm algorithm MIMO communication system efficiency evaluation [J]. *Journal of electronic measurement technology abroad*, 2022 9 (02) : 46 to 52. DOI: 10.19652 / j.carol carroll nki femt. 2103363.

Liu Qiujin. Analysis on the cause and type of cracks and their influence on concrete Bridges [J]. *Heilongjiang Science and Technology Information*, 2015(16):208.

Lvjiapeng, shixianjun, Wang Kang Establishment method of Gaussian process fault prediction model based on Improved Particle Swarm Optimization [j] *Electro Optic and Control*, 2020, 27 (11): 75–80+96.

Wang Lei. Influence of Cracks on durability of reinforced Concrete Bridges and Preventive Measures [J]. *Transportation Construction and Management*, 2014(24):138–140.

Weng Jianjun, Hu Chong, Li Longhao, Yan Qingxin, WANG Qunpeng, Zheng Shuaixiang. *Journal of Wuhan University of Technology* (Transportation Science and Engineering), 201, 45(04):799–804.

Wu Xi, Chang Chunguang, Yan Xin. Science technology and engineering, 2019, 19(27):304–310.

Xu Tian. *Research on performance evaluation model of in-service concrete bridge based on apparent crack disease* [D]. Southeast university, 2021. DOI: 10.27014 /, dc nk:. Gdnau. 2021.002741.

Xu Yuec.2020.000521.

Xu Yueting. *Cable tower bridge concrete fracture safety assessment analysis* [D]. Chongqing jiaotong university, 2020. The DOI: 10.27671 /, dc nki. GCJTC. 2020.000521.

*Civil Engineering and Urban Research – Mohamed & Hou (Eds)*
*© 2023 the Authors, ISBN 978-1-032-44487-1*

# Research on monitoring method of time-varying gravity dam deformation based on Bayesian dynamic linear model

Lin Cheng* & Jiamin Chen*
*State Key Laboratory of Eco-hydraulics in Northwest Arid Region, Xi'an, China*

Pengfei Xie*
*Sinohydro Bureau 14 Co Ltd., Yunnan, China*

Chunhui Ma* & Jie Yang*
*State Key Laboratory of Eco-hydraulics in Northwest Arid Region, Xi'an, China*

ABSTRACT: Given the time-dependent nature and uncertainty of concrete gravity dam deformation, the Bayesian Dynamic Linear Model (BDLM) has the advantages of dynamic modeling ability and interpretability. Based on the BDLM framework, this paper introduces the dynamic modeling idea into the regression model to construct the monitoring model for time-varying concrete gravity dam deformation. The proposed method can update the model in real-time, quantify the model error, and predict uncertainty. Finally, combined with the actual project, the physical reason for the state variable mutation is analyzed. For the three mature models, the performance of BDLM is improved by more than 50% with strong generalization ability.

## 1 INTRODUCTION

A concrete gravity dam is a common dam type in water conservancy and hydropower engineering construction. Its high-quality, efficient, and safe construction and long-term efficient and safe operation are critical to the national economy and people's livelihood (Li 2022). The dynamic change process of gravity dam deformation is the comprehensive reflection of the structural state of the dam body and foundation and an important basis for analyzing and evaluating the working condition of the dam (Hu 2005). For one thing, the external environment of the dam, such as the downstream water level and temperature, is constantly changing, especially when the dam experiences the most unfavorable load combination or new load combination forms, which results inevitably in constant change of dam deformation. For another, the material properties of dam and bedrock change slowly with time, generally presenting a process of attenuation. It changes the dam behavior slowly and then causes the time-varying deformation to change. Therefore, to analyze the deformation behavior of concrete dams correctly, the time-varying characteristics and uncertainty of concrete dam deformation should be fully considered.

The statistical, deterministic, and hybrid models are commonly used in analyzing dam prototype monitoring data. With the wide application of artificial intelligence, more and more intelligent algorithms (Hu 2021) such as long-short term memory (LSTM) (Hu 2020), Gaussian process regression (GPR) (Li 2021), and support vector machines (SVM) are used to solve the problem of deformation prediction accuracy. However, most models assign almost the same input weights to different influence factors and do not consider the temporal and spatial effects of each factor on

---

*Corresponding Authors: chenglin@xaut.edu.cn, 2200421257@stu.xaut.edu.cn, 992349972@qq.com, 1137300843@qq.com and 1529759737@qq.com

   DOI 10.1201/9781003372417-94

dam deformation. Based on the time-space model of variable intercept panel data, Gu Chongshi and others (Wang 2020) made full use of the deformation data of multiple measuring points at Jingpin Hydropower Station, solving the shortage of conventional statistical model that only considers the deformation behavior of single point in time sequence. Li Mingchao et al. (Li 2021) elaborated on the three correlations of factor correlation, dynamic causal relationship, and sequence similarity from the five aspects of dimension, correlation, test, measure, and model, proposing a dynamic monitoring model of dam deformation with both correlation and similarity. However, the model does not consider the characteristics of real-time updating of influence factors and deformations, so there are still some defects in reflecting the structure's authenticity. BDLM, a state space model suitable for sequential reasoning, has developed rapidly in recent years. Goulet (Goulet 2017) proposed the BDLM framework. The time-dependent response was modeled and decomposed into corresponding hidden components to construct the structure and external effects. Subsequently, it was applied in real-time structural health monitoring across structural groups (Goulet 2018; Wang 2019), the financial field (West 1989), and the aerospace business (Gibbs 2011). Firstly, the above research shows that BDLM is suitable for stable monitoring data and non-stationary monitoring data, and dam deformation data is a typical non-stationary data. Secondly, BDLM can update the model with the newly acquired observation data to make the prediction more convenient and efficient. Thirdly, BDLM quantifies the uncertainty in deformation prediction. Therefore, it is suitable for analyzing gravity dam safety monitoring data.

In conclusion, considering the time-varying characteristics and uncertainty of concrete gravity dam deformation, the Bayesian inference real-time updating strategy is applied to probability prediction under the given data. In this work, the weight of each influence factor of the statistical model is regarded as dynamic change, and the dynamic regression component of BDLM is used to update the real-time online prediction. The hidden components are estimated by the Kalman filter. Finally, the BDLM analyzes actual engineering monitoring data to verify its performance.

## 2  PRINCIPLE OF BDLM ALGORITHM

In this section, the dynamic prediction strategy of BDLM is first described. Then the application of BDLM in the deformation prediction of a gravity dam is further clarified by combining the physical interpretation principle of each component (water pressure, temperature, and time effect component) of the traditional statistical model for gravity dam deformation.

### 2.1  Bayesian dynamic linear model

In BDLM, the influence factors of dam deformation are regarded as Gaussian random variables, and the state transition is defined by a linear function (Zhang 1992). The BDLM model consists of two linear equations: the observation equation and the state equation. The former describes the relationship between time effect data $y_t$ and state variables $\boldsymbol{\theta}_t$ at time t∈[1:T]. The state equation describes dynamic changes of the hidden state variables $\boldsymbol{\theta}_t$ with time. The mathematical formula of two equations (Zhang 1992) is defined as follows.

$$y_t = \boldsymbol{C}_t\boldsymbol{\theta}_t + V_t, \begin{cases} y_t \sim N(E[y_t], cov[y_t]) \\ \boldsymbol{\theta}_t \sim N(\boldsymbol{m}_t, \boldsymbol{S}_t) \\ V_t \sim N(0, R_t) \end{cases} \tag{1}$$

$$\boldsymbol{\theta}_t = \boldsymbol{A}_t\boldsymbol{\theta}_{t-1} + \boldsymbol{W}_t, \boldsymbol{W}_t \sim N(0, \boldsymbol{Q}_t) \tag{2}$$

Data $y_t$ represents the observed variable at time $t \in[1:T]$. $\boldsymbol{\theta}_t$ reflects the hidden state variables which are not directly observed at time t. Over time, observations are defined as the function of state variables $\boldsymbol{\theta}_t$, observation matrix $\boldsymbol{C}_t$, and Gauss measurement error $V_t$, where the mean value of Gaussian measurement error $V_t$ is zero, and the covariance matrix is $R_t$. $E[y_t]$ and $cov[y_t]$ are the expectation and covariance of observed variables $y_t$. The transformation of hidden state variables between time steps $\theta_t$ is defined by the transformation matrix $A_t$ and the Gaussian model error $W_t$ where the mean value is zero, and the covariance matrix is $Q_t$. $m_t$ and $S_t$ are mean and variance of state variables $\theta_t$.

In this paper, BDLM is combined with the traditional model for time-varying statistics to construct the dynamic prediction model with physical interpretation. The general formula of BDLM provides the prediction of state parameters and observation values at each time, so as to achieve the result of the one-step head prediction. The posterior distribution of state parameters can be obtained by updating and modifying the prior data and observation data of state parameters. Figure 1 shows the flow chart of advanced prediction, and the specific process is as follows.

(1) The distribution of states $\theta$ is estimated as prior information according to the sample information at the time:

$$p(\theta_t|D_t) \sim N(m_t, S_t) \tag{3}$$

(2) The posterior information at time t + 1 is estimated according to the prior information at time t:

$$p(\theta_{t+1}|D_t) \sim N(\mu_{t+1}, \Sigma_{t+1}) \tag{4}$$

In the formula, $\mu_{t+1} = E(\theta_{t+1}|D_t) = A_{t+1}m_t$ and $cov[y_{t+1}] = Var(y_{t+1}|D_t) = C_t\Sigma_{t+1}C'_t + R_{t+1}$.

(3) One-step ahead prediction is conducted:

$$p(y_{t+1}|D_t) \sim N(E[y_{t+1}], cov[y_{t+1}]) \tag{5}$$

In the formula, $E[y_{t+1}] = E(y_{t+1}|D_t) = C_t\mu_{t+1}$ and $cov[y_{t+1}] = Var(y_{t+1}|D_t) = C_t\Sigma_{t+1}C'_t + R_{t+1}$.

(4) The posteriori information is updated at time t + 1:

$$P(\theta_{t+1}|D_{t+1}) \sim N(m_{t+1}, S_{t+1})$$
$$m_{t+1} = \mu_{t+1} + F_{t+1}e_{t+1}, S_{t+1} = \Sigma_{t+1} - F_{t+1}F_{t+1}^T cov[y_{t+1}]$$
$$e_{t+1} = y_{t+1} - E[y_{t+1}], F_{t+1} = \frac{\Sigma_{t+1}C'_{t+1}}{cov[y_{t+1}]} \tag{6}$$

In the formula, $e_{t+1}$ is the error of one-step ahead prediction, and $F_{t+1}$ is the adaptive coefficient.

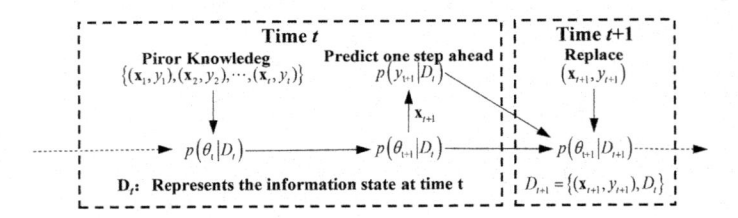

Figure 1.    Flow chart of one-step ahead prediction.

## 2.2    *Statistical regression model*

According to the theoretical analysis of structural mechanics of concrete dam and dam foundation, the deformation of concrete gravity dam can be attributed to an external load (such as water pressure and temperature) and time-varying effect (Wu 2003). Therefore, the deformation of concrete gravity dam can be divided into different influence factors:

$$Y_t = Y_t^H + Y_t^T + Y_t^\theta = a_0 + \sum_{i=1}^{3} a_i H^i + \sum_{i=1}^{2} [b_{1i} \sin(i\alpha) + b_{2i} \cos(i\alpha)] + c_1\tau + c_2 \ln\tau \tag{7}$$

In the formula, $Y_t^{\mathrm{H}}$ is the water pressure component, $Y_t^{\mathrm{T}}$ is the temperature component, and $Y_t^{\theta}$ is the aging component. $a_i$, $b_{1i}$, $b_{2i}$, $c_1$ and $c_2$ represent the regression coefficient of each component. $H$ is the water depth in front of the dam, $\alpha = 2\pi t/365$, and $\tau$ divides the number of days from the monitoring date to the starting date by 100 days.

## 2.3  Dynamic regression modeling based on BDLM

Dynamic regression can describe the dynamic correlation among temperature, water pressure, and deformation of a concrete gravity dam. In this paper, the regression coefficients of each component in the traditional regression model are regarded as state variables, and the influence factors are taken as the observation matrix. Thus, dynamic regression modeling is carried out, and dynamic regression is defined as the following formula.

$$y_t^{\mathrm{Re}} = x_t \boldsymbol{\beta}_t + V_t^{\mathrm{Re}}, V_t^{\mathrm{Re}} \sim N\left(0, R_t^{\mathrm{Re}}\right) \tag{8}$$

Data $y_t^{\mathrm{Re}}$ reflects the dynamic regression measurement, and Re indicates the establishment of regression model. $\boldsymbol{\beta}_t$ is the state variable, namely the dynamic regression coefficient $\boldsymbol{\beta}_t = [a_{0,t}, a_{1,t}, a_{2,t}, a_{3,t}, b_{11,t}, b_{12,t}, b_{21,t}, b_{22,t}, c_{1,t}, c_{2,t}]$. $x_t$ is the time series of each influence factor.

Temporal reasoning is given by the following formula.

$$\boldsymbol{\beta}_t = \boldsymbol{\beta}_{t-1} + W_t^{\mathrm{Re}}, \ W_t^{\mathrm{Re}} \sim N\left(0, \boldsymbol{Q}_t^{\mathrm{Re}}\right) \tag{9}$$

In the BDLM model, dynamic regression can be expressed as follows.

$$\boldsymbol{\theta}_t^{\mathrm{Re}} = \boldsymbol{\beta}_t, \boldsymbol{C}_t^{\mathrm{Re}} = x_t, \boldsymbol{A}_t^{\mathrm{Re}} = I, V_t^{\mathrm{Re}} = \left(v_t^{\mathrm{Re}}\right)^2, \boldsymbol{W}_t^{\mathrm{Re}} = \left(\boldsymbol{w}_t^{\mathrm{Re}}\right)^2 \tag{10}$$

## 3  CONSTRUCTION OF GRAVITY DAM DEFORMATION PREDICTION MODEL BASED ON BDLM

In this paper, the deformation prediction model of a concrete gravity dam is constructed based on BDLM to realize the prediction and analysis of gravity dam deformation. The main task can be divided into estimating the unobserved state parameters and super parameters and predicting the future structural response according to the obtained structural response time series. The flow chart of the model is shown in Figure 2, and the specific steps are as follows.

**Step 1: data preprocessing.** To eliminate the influence of different dimensions in the data, it is necessary to normalize the deformation monitoring data of the gravity dam and scale the data to [0.1, 0.9]. The calculation formula is as follows.

$$X_{std} = (X - X_{\min})(0.9 - 0.1)/(X_{\max} - X_{\min}) \tag{11}$$

$X_{std}$ is the normalized data, $X_{min}$, the minimum value, and $X_{max}$, the maximum value in the formula.

Due to the instability of the monitoring system, most time series are uneven, but the object of time series analysis is usually an equal time step sequence, which needs to be completed by interpolation.

**Step 2: selection of influence factors and determination of prior distribution.** The influence factors of concrete gravity dam deformation selected this time, namely the state variables of BDLM, are defined as $\{H^1, H^2, H^3, \sin(\alpha), \cos(\alpha), \sin(2\alpha), \cos(2\alpha), \tau, \ln\tau\}$. The relatively mature Conjugate Normal Inverse Gamma (CNIG) (Ni 2020) is used to determine the prior distribution of state parameters and hyperparameters. That is, the state parameters obey normal distribution, and the hyperparameters obey Inverse Gamma Distribution (IGD).

**Step 3: confirmation of the main probability parameters of the model.** The main parameters of BDLM include $V_{t+1}$, $W_{t+1}$, $m_t$, and $S_t$, which can be determined by the forward filtering backward sampling (FFBS) (Wang 2020), and $W_{t+1}$ can be approximately obtained by the discount factor (Fan 2019; Zhu 2008).

**Step 4: recursive update of the forecast.** The determined super parameters are obtained from step 3, and then the real-time recursive updating is carried out by using the recurrence formula described in section 2.1, so the state variables at each time point are obtained. Finally, the predicted value of the gravity dam deformation in the future is obtained by the recursive update of the determined observation matrix in each period.

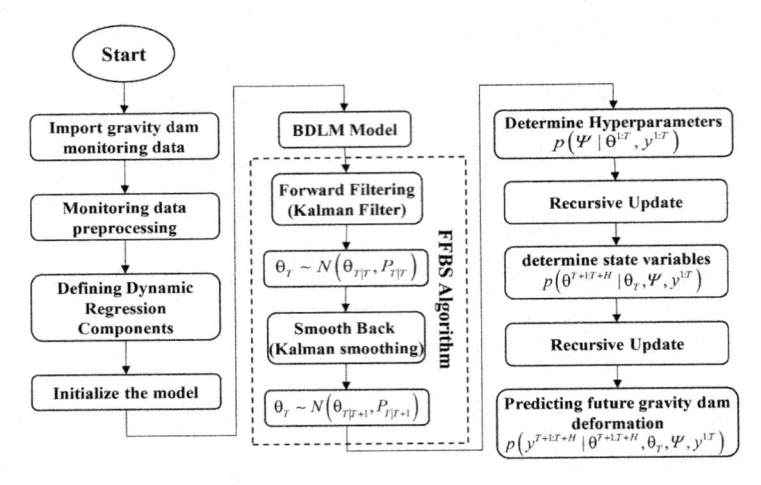

Figure 2. Flow chart of gravity dam deformation prediction based on BDLM.

## 4 ENGINEERING EXAMPLES

### 4.1 *Overview of deformation monitoring data of concrete gravity dam*

A gravity dam is located in Yongding County, Fujian Province, and its upstream elevation is shown in Figure 3. The height of the gravity dam is 113.0 meters, the total length of the dam crest is 308.5 meters, and the elevation of the dam crest is 179.0 meters. The normal water level of the reservoir is 173 meters, the regulated storage capacity is 1.122 billion cubic meters, and the check flood level is 177.80 meters. It is mainly used for power generation and has comprehensive benefits such as flood control, shipping, and aquaculture.

An extension line is arranged at the elevation of 179.00 meters on the dam crest to monitor the horizontal displacement of the dam crest. Twelve measuring points are arranged from EX1 to EX12 from the left to the right of the dam. The location of the measuring points is shown in Figure 3. The data series of the extension line is from January 1, 2003, to December 30, 2008, and its time series is shown in Figure 4. The deformation downstream is negative; otherwise, it is positive. As can be seen from Figure 4, except for the EX1 and EX12 measuring points on the bank side of the dam, the other points show strong similarity.

### 4.2 *Establishment and analysis of the model*

According to the above-mentioned BDLM modeling algorithm, the EX12 measurement point with relatively stable data has been used in this paper as the experimental sample for one-step prediction in advance. A total of 2010 data sets are used in the data time series from January 1, 2003, to December 31, 2008.

The change process line of the state variable is shown in Figure 5. It can be seen from the process line of state variables (dynamic regression coefficient) in Figure 5 that each component

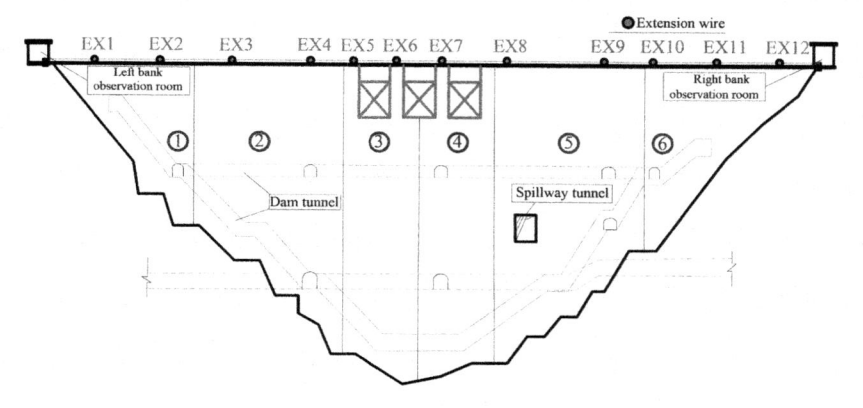

Figure 3.   Elevation view of upstream dam.

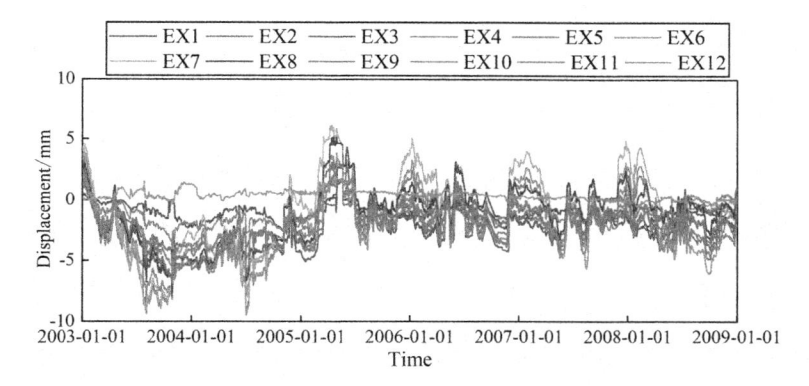

Figure 4.   Process line of each measuring point.

presents an obvious step-type mutation, and the most obvious is the change of regression coefficient of the temperature component. It shows that the influence of temperature change on horizontal displacement is significant during the service period of a gravity dam. Two reasons for the mutation of state variables are internal and external factors. Internal factors refer to the working environment of the dam, such as the downstream water level and temperature, which are constantly changing, so the deformation will be in constant change. The internal cause is that the structural state of the gravity dam changes slowly due to the influence of the environment during its service. That is to say, the structural elastic parameters and compression conditions change with time, which leads to the time-varying deformation behavior.

BDLM model boasts strong interpretability. Figure 6 shows the real-time updating diagram of each component of BDLM. Firstly, the constant term is highly correlated with the trend of the measured value, which can objectively reflect the trend of the measured value. Secondly, the water pressure component strongly correlates with the upstream water level, which reflects the influence of the reservoir water level on horizontal displacement. Thirdly, the temperature component shows strong periodicity. In winter, the displacement value of the temperature component is at the peak value of a year, which shows a trend of deformation upstream of the dam. Fourth, the aging component is gradually increasing and stable, consistent with the dam deformation law of the actual dam operation. So real-time online update of each component by BDLM has strong reliability.

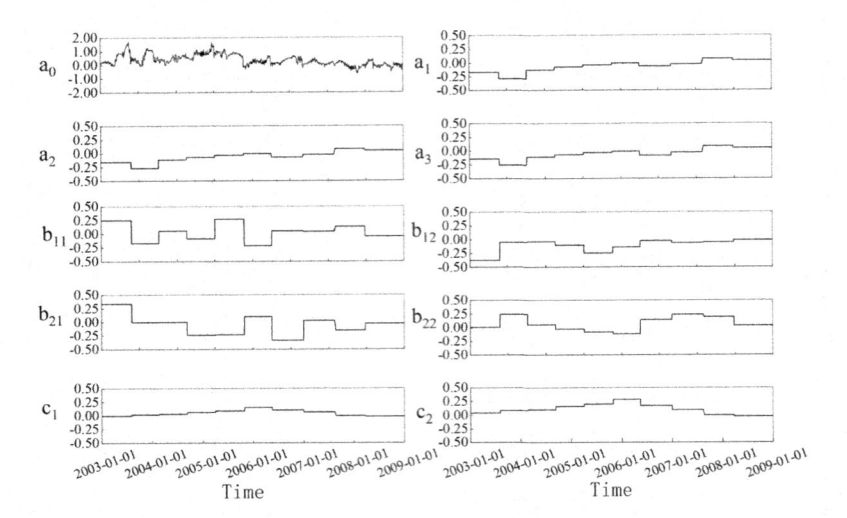

Figure 5. Curve of the regression model coefficients changing with time.

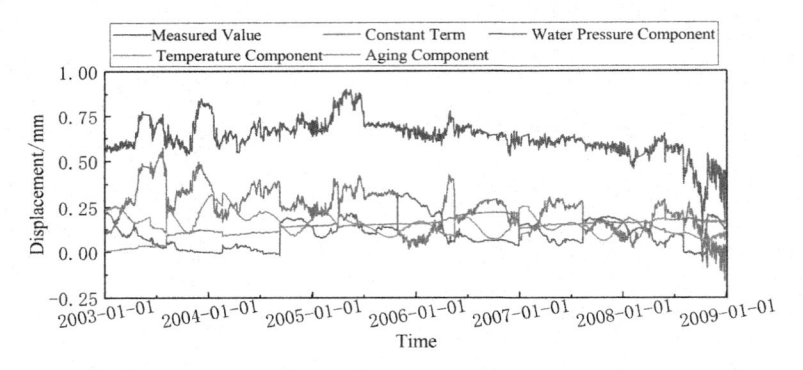

Figure 6. Separation results of each component of displacement.

### 4.3 Analysis of model prediction accuracy

To prove the accuracy of BDLM in the one-step prediction process, we compared and analyzed BDLM with GPR, boosting algorithm, SVM, and stepwise regression-based statistical model (SR-SM) for the last 150 groups of prediction.

The prediction results are shown in Figure 7. The comparison of evaluation indexes of each model is shown in Table 1. First, the measured values fall within the 95% confidence interval. Second, compared with the predicted values of other models, the measured values are in the range of -0.5 millimeters $\sim$ 0.5 millimeters, and the prediction accuracy of BDLM is better than that of other models. Third, in the MSE index, the performance of BDLM is 68.9% higher than that of the GPR model, 84.6% higher than that of the SVM model, and 88.6% higher than that of the SR-SM model. In conclusion, BDLM has obvious advantages in prediction.

To verify the universality of the model, we also apply the model to other measurement points, and the prediction indexes are shown in Table 1. It can be seen that the accuracy of BDLM is also better than that of other models at other measuring points.

670

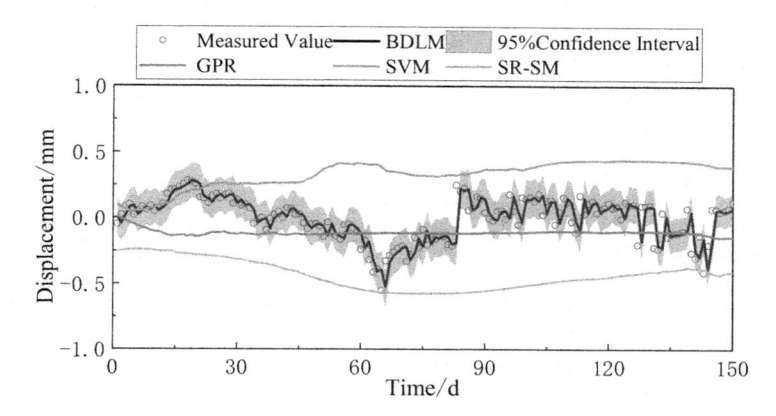

Figure 7.    Comparison of prediction effects of each model.

Table 1.    RMSE Forecast Indicators for Each Mode.

| Model | BDLM/mm | GPR/mm | SVM/mm | SR-SM/mm |
|---|---|---|---|---|
| EX1 | 0.017 | 0.146 | 0.098 | 0.251 |
| EX2 | 0.010 | 0.142 | 0.135 | 0.192 |
| EX3 | 0.010 | 0.051 | 0.039 | 0.079 |
| EX4 | 0.012 | 0.061 | 0.075 | 0.093 |
| EX5 | 0.009 | 0.073 | 0.090 | 0.077 |
| EX6 | 0.010 | 0.103 | 0.121 | 0.062 |
| EX7 | 0.010 | 0.079 | 0.025 | 0.081 |
| EX8 | 0.013 | 0.068 | 0.049 | 0.049 |
| EX9 | 0.017 | 0.098 | 0.079 | 0.062 |
| EX10 | 0.016 | 0.103 | 0.137 | 0.069 |
| EX11 | 0.022 | 0.158 | 0.141 | 0.254 |
| EX12 | 0.061 | 0.196 | 0.397 | 0.456 |

## 5    CONCLUSION AND PROSPECT

In this study, an effective method is proposed to analyze the development trend of the influencing factors on the overall deformation of the dam. The model helps analyze the dynamic effects of environmental evolution and material aging. Through the hidden components of the model, BDLM can provide valuable data support for the intelligent safety monitoring system, which is beneficial to the dam project and can be applied to other fields. The advantages can be concluded by verifying the engineering data, as listed below.

Firstly, the BDLM can well capture the deformation and hidden state of concrete dams, and the deformation behavior of concrete dams is characterized by the superposition of water pressure components, temperature components, and time-dependent components. In addition, model reasoning involves dynamic state updating, which allows the stochastic uncertainty of deformation to be calculated by considering prediction errors. It also reflects the model's advantages in dynamic prediction and uncertainty quantification.

Secondly, by comparing BDLM with other models, the prediction accuracy of BDLM is improved by 50%, which indicates that BDLM can be well applied in dynamic early warning and prediction of dam safety monitoring and provide a timely decision-making basis for practical projects.

Thirdly, BDLM has obvious advantages in the interpretability of the model, and it has good application in real-time prediction. Moreover, the good real-time update strategy of the

BDLM framework makes the real-time updating of each component quickly find out the physical mechanism of deformation mutation.

## ACKNOWLEDGMENT

This work was partly supported by the Joint funds of the natural science fundamental research program of Shaanxi province of China and the Hanjiang-to-Weihe river valley water diversion project under Grant 2019ILM-55.

## REFERENCES

Fan, X.P. & Qu, G. & Liu, Y.F. (2019). Dynamic prediction of bridge non-uniform sampled extreme stress based on the BDLM. *Journal of Fuzhou University* (Natural Science Edition). 47(1):6

Gibbs, B.P. (2011). *Advanced Kalman filtering, least-squares and modeling: a practical handbook*. Wiley.

Goulet, J.A. & Koo, K. (2018). Empirical Validation of Bayesian Dynamic Linear Models in the Context of Structural Health Monitoring. *Journal of bridge engineering*. 23(2): 05017017.

Goulet, J.A. (2017). Bayesian dynamic linear models for structural health monitoring. *Structural Control and Health Monitoring*. 24(12): e2035.

Hu, A.Y. & Bao, T.F. & Yang, C. L. et al. (2020). LSTM-ARIMA-based Prediction of Dam Deformation: Model and Its Application. *Journal of Yangtze River Scientific Research Institute*. 37(10):6.

Hu, J. & Ma, F.H. (2021).Comparison of hierarchical clustering-based deformation prediction models for high arch dams during the initial operation period. *Journal of Civil Structural Health Monitoring*. 2021(12)

Hu, L.Z. (2005). *Study on Safety Monitoring Time-Varying Model and Its Application of Concrete Dam's Deformation*. Hohai University.

Kang, F. & Li, J.J. & Dai, J. (2019).Prediction of long-term temperature effect in structural health monitoring of concrete dams using support vector machines with Jaya optimizer and salp swarm algorithms. *Advances in Engineering Software*. 131(MAY):60–76.

Li, M.C. & Ren, Q.B. & Kong, R. et al. (2021). Dynamic modeling and prediction analysis of dam deformation under multidimensional complex relevance. *SHUILI XUEBAO*. 50(6):12.

Li, Q.B. & Ma, R. & Hu, Y. et al. (2022). *A review of intelligent dam construction techniques*. Tsinghua Univ (Sci& Technol). 1–18.

Li, Y.T. & Bao, T.F. & Shu, X.S. et al. (2021). A Hybrid Model Integrating Principal Component Analysis, Fuzzy C-Means, and Gaussian Process Regression for Dam Deformation Prediction. *Arabian Journal for Science and Engineering*.

Lv, L.T. & Li, J.H. & Lv, H. et al.(2019). Research on Bayesian Dynamic Model and Forecasting Algorithm in Data Mining Application. *Computer Engineering and Applications*. 40(20):4.

Ma, J.J. & Su, H.Z. & Wang, Y.H. et al.(2021). Combinatorial Prediction Model for Dam Deformation Based on EEMD-LSTM-MLR. *Journal of Yangtze River Scientific Research Institute*. 38(05):47–54.

Ni, Y.Q. & Wang, Y.W. & Zhang, C.A.(2020). Bayesian approach for condition assessment and damage alarm of bridge expansion joints using long-term structural health monitoring data. *Eng structs*. 212:110520.

Wang, H. & Zhang, Y.M. & Mao, J.X.et al.(2019). Modeling and forecasting of temperature-induced strain of a long-span bridge using an improved Bayesian dynamic linear model. *Engineering Structures*. 192(AUG. 1):220–232.

Wang, J.M. & Gu, C.S. & Zhang, C. et al. (2020). Deformation behavior analysis of Jinping arch dam based on spatiotemporal model of variable intercept panel data. *Journal of Hydroelectric Engineering*. 39(11):21–30.

Wang, Y.W. & Ni, Y.Q. (2020).Bayesian dynamic forecasting of structural strain response using structural health monitoring data. *Structural Control and Health Monitoring*. 27(8).

WEST, M & Harrison, J.(1989).*Bayesian Forecasting and Dynamic Models*. Springer.

Wu, Z.R.(2003). *Safety Monitoring Theory & Its Application of Hydraulic Structures*. Higher Education Press.

Zhu, W.B. (2008). Application of Bayesian Multivariable Dynamic Linear Model in Dam Monitoring Analysis. *Water Resources and Power*. 2008(02):68–71.

Civil Engineering and Urban Research – Mohamed & Hou (Eds)
© 2023 the Authors, ISBN 978-1-032-44487-1

# Study on trajectory prediction, monitoring and early warning of disaster-causing objects in water intake area of coastal power plant

Yunjia Sun*, Chen Li* & Baisu Zhu*
*CCCC Tianjin Port Engineering Institute Co., Ltd., Key Laboratory of Coastal Engineering Hydrodynamic, CCCC, Tianjin, China*
*CCCC First Harbor Engineering Company Ltd., Tianjin, China*

ABSTRACT: The cold source condition of the coastal power plant is very important for its safe operation. A large number of marine disaster-causing objects' accumulation in the water inlet channel could lead to severe accidents, e.g., unit emergency shutdown. In order to better predict the movement and accumulation characteristics of marine disaster-causing objects, a marine tidal current model in the project area is constructed; the particle tracking model is used to calculate the movement trajectory of pollutants by calculating the pollutants entering the open channel; a new reliable technical support for power plant operators is provided.

## 1 INTRODUCTION

Safety is the bottom line of the development of coastal power plants. However, in recent years, incidents, where marine organisms or foreign objects threaten the safety of water intake of coastal power plants, have occurred many times at home and abroad. As this trend increases, the safety of cold sources has become an important factor in the safety of power plants.

According to the World Association of Nuclear Operators (WANO), about 20 percent of such incidents directly impact safety-related systems. In extreme cases, such as marine organisms surging and sudden influx of pollutants, accidents, including the destruction of pollution retaining structures and nuclear power units' shutdown, seriously affect the water intake safety of the power plant.

Due to insufficient prediction of marine biological outbreaks, the early warning of marine biological pollution remains limited, which is also the focus of the current research. For abiotic pollutant accumulation events, the current research on marine hydrodynamics tends to be mature, and pollutants can be traced by various technical means. Therefore, it is possible to realize early warning and abiotic pollutants monitoring.

The movement of organisms and pollutants at sea is affected by the tidal current and the water intake in the cold source of the power plant. Tidal current in a specific area has periodic characteristics, so under the condition of mastering the characteristics of the regional tidal current and water intake, the motion trajectory of pollutants can be analyzed and predicted. Since the proportion of pollutants entering the open intake channel can be calculated by mathematical analysis method, effective early warning information for power plant operators could be yielded, and technical support for disaster prevention and mitigation of power plants could be provided.

## 2 METHOD

The movement characteristics of disaster-causing objects such as floating garbage and marine organisms without autonomous motion ability (hereinafter referred to as pollutants) in the water body are mainly affected by the tidal current and especially affected by continuous, one-way water

---

*Corresponding Authors: sunyunjia@ccccltd.cn, lichen11@ccccltd.cn and zhubaisu@ccccltd.cn

DOI 10.1201/9781003372417-95

intake near the water intake. The pollutants mainly gather into the intake channel, resulting in blocking the sewage net and reducing the water intake efficiency. Several related research focused on current tidal models (Doeoes 2017; Enright 2002; Li 2019; Qiao 2015; Wang 2017) and obtained important research results.

In this study, through the construction of the marine tidal current model in the project area, the particle tracking model is used to calculate the movement trajectory of pollutants by calculating the pollutants entering the open channel, so as to provide reliable technical support for the power plant operators.

## 2.1 Numerical model

The mathematical model of pollutant transport is based on the tidal current dynamic model. The hydrodynamic model of Mike21 software is used as the basis, and the governing equation of flow movement is a two-dimensional shallow water equation. To verify the accuracy of the tidal current model, it is necessary to verify the measured data in the project area, including tidal level and flow direction data.

The simulation of floating objects is carried out by using the Particle Tracking Module of the Mike21 hydrodynamic model. This method simulates the motion of particles in water according to the Langevin equation. The basic equations are as follows:

$$dX_t = a(t, X_t)dt + b(t, X_t)\xi_t \qquad (1)$$

Where $a=$ floating object; $b=$ diffusion term; and $\xi=$ random number.

In order to simulate the Euler trajectory of a particle, it is necessary to discretize the equation of motion in time steps:

$$Y_{n+1} = Y_n + a(t, X_t)Y_n\Delta_n + b(t, X_t)Y_n\Delta W_n \qquad (2)$$

$$\text{Where} \quad \Delta W_n = W_t - W_s \in N(\mu = 0, \sigma^2 = \Delta_n) \qquad (3)$$

In the model, the particles of floating objects are simulated by source terms, and the number of floating objects is represented by setting different source intensities.

## 2.2 Statistic method

After establishing the tidal current model and particle tracking model in the engineering area, the sea area near the project area is divided as the location of pollutants. Then the proportion of particles in different positions entering the open channel under the action of tidal current and water intake is calculated. The specific methods are as follows:

(1) Taking the water intake as the center, the surrounding sea area is divided into 16 directions, namely E, ENE, …i…, ENE (each 22.5° is set as an azimuth, where E is the starting point, rotating counterclockwise, ENE is the endpoint, and i is numbered from 1 to 16), as the direction where the pollutant appears.
(2) Several source points of pollutants in different positions to the gate are set up from each azimuth, such as 1km, 2km, …, j…, N km. Since the source of pollutants can only be in the water body, it is reasonable to assume that the area n km to the project from all directions are water bodies. Hence, 16n sources of the above pollutants are deployed.

The motion trajectories of pollutants at each position under the action of tidal current and water intake are calculated, respectively. The proportion of pollutants entering the open channel is determined according to the following method:

Assuming that the pollutant under the k-th working condition is located at the position of X (i, j) and Y (i, j), the pollution point continuously releases the amount of pollution for 1 hour, and the total amount released is Tp. Under this condition, the movement trajectory of

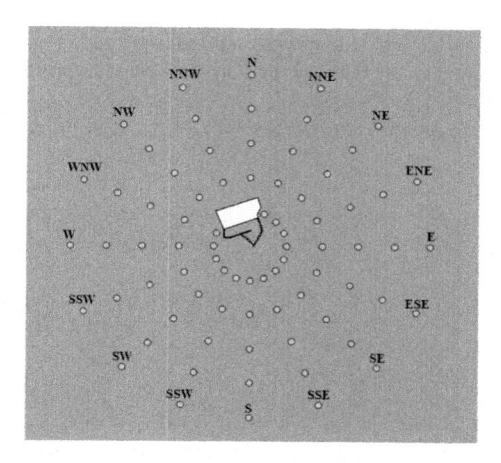

Figure 1. The location where the pollutant is released.

pollutants is calculated, and the number of pollutants entering the open channel mouth Ip is calculated. The proportion of pollutants entering the open channel is as follows:

$$P_p = I_p / T_p \qquad (4)$$

Where $P_p$ = Percentage of pollutant flux through a section; $I_p$ = Pollutant flux through a section; $T_p$ = Total amount of pollutant released.

(3) According to the 16n working conditions calculated above, the proportion of pollutants entering the open channel under different working conditions is counted, and the proportional cloud map is drawn to determine the key monitoring range.

## 3 CASE STUDY

A coastal power plant is located on the island. The project area is affected by the outer sea tide and wave, which shows east-west reciprocal flow. Topography and mountain boundary constraints lead to a chaotic local flow pattern; on the side of the east dyke, the flood current is from north to south, and the ebb current is from south to north; a wide range of clockwise circulation appears during the flood tide near the intake gate, and its circulation scale and center gradually shift to the west in the flood tide process; during the ebb tide, large-scale circulation doesn't exist, and the ebb tide flow flows to the northeast around the head of the east dyke, forming a circulation on the side of the east dyke. The local flow field of the project is shown in Figure 2.

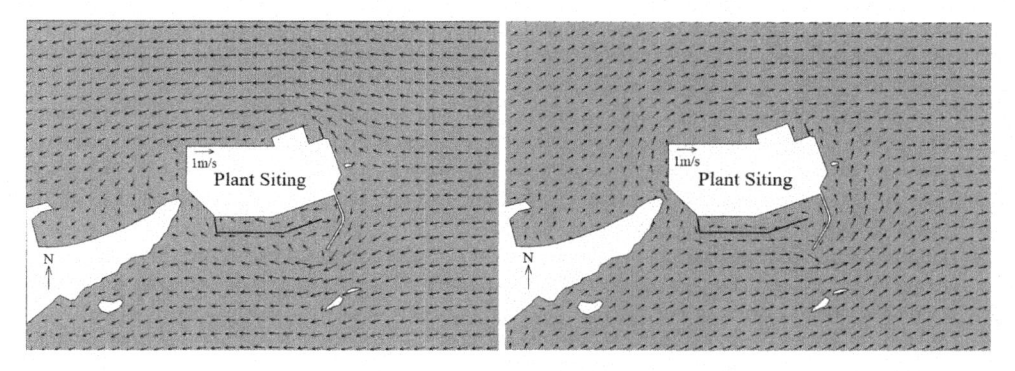

Figure 2. The flow field of the project area (maximum flood tide on the left, maximum ebb tide on the right).

Taking the water intake east dyke as the center and taking the 1~6km as the radius, the space was divided into 16 directions to release pollutants four times: maximum flood tide, high slack, maximum ebb tide, and low slack, respectively. The pollutant release point is shown in Figure 3. The release of pollutants lasted 1 hour, and the statistical time of pollutant flux was within 24 hours after the pollutant release.

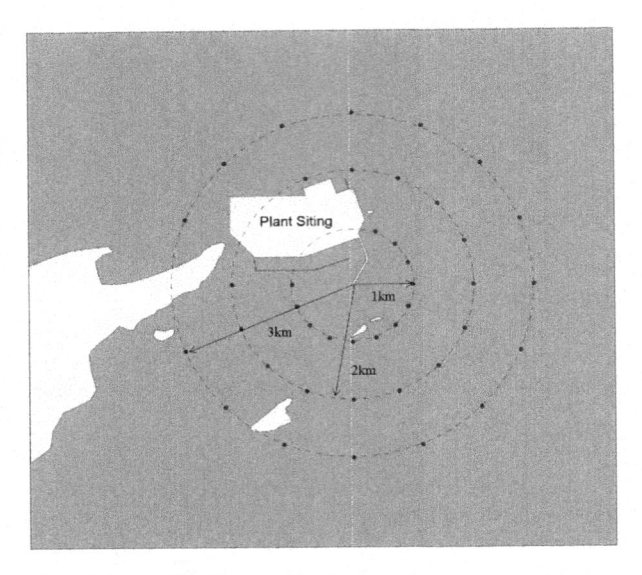

Figure 3.   Pollutant release location (1~3km location is shown).

The numerical simulation results show obvious temporal and spatial distribution characteristics of pollutants released into the open channel in different directions. Generally, the main way for pollutants to enter the open channel is through the open channel from the west dyke during the ebb tide; for the pollutants from SW~W, there are different amounts of pollutants flowing into the channel in a tidal cycle; while for the pollutants from N~E~S, if they cannot move to the west side of the dyke head with the water flowing during the flood tide, there is no 'opportunity' to enter the open channel.

Taking as an example the release of pollutants from the WSW direction and 1km distance to the head of the dyke, the main movement characteristics of pollutants are as follows:

(1) The pollutants that 'appear' at maximum flood tide would move with the tide to the southwest, then move to the northeast with the ebb tide near Island-2, and afterward, move to the east along the south dyke; most of the pollutants at the entrance move with the intake water to the inside of the channel, and only a few pollutants bypass the east dyke and flow out of the project area to the northeast, as shown in Figure 4-a.

(2) For the pollutants that 'appear' at high slack, since the position is in the center of the rotating flow at this time, the velocity is slow, and most of them move northeast around the east dyke with the ebb tide; some of the pollutants are stranded in the rotating flow near Island-4, and some of the rest move with the current to the north side of the plant site and then move to the west side with the flood tide, as seen in Figure 4-b.

(3) The pollutants that 'appear' at maximum ebb tide are brought out of the project area by the strong ebb tide and moved to the northeast side of the plant site; then, no pollutants flow to the open channel, as shown in Figure 4-c.

(4) For the pollutants that 'appear' at low slack, due to the weak power of the tidal current, the pollutants are trapped in the swirling area near the south side of the breakwater at the

beginning, and then move to the west with the flood tide; at this time, some of the pollutants move northward along the waterway between Island-1 and plant siting; some of them move along the southwest direction toward Island-2, and then enter the channel with ebb tide process, as seen in Figure 4-d.

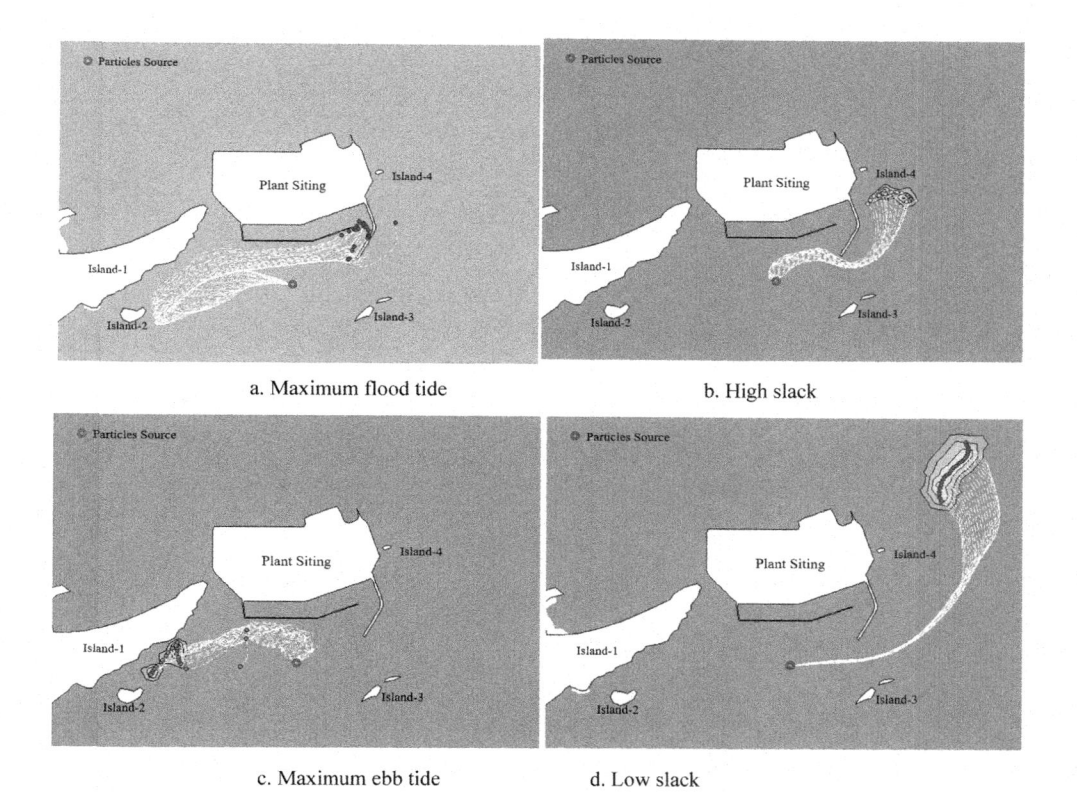

a. Maximum flood tide          b. High slack

c. Maximum ebb tide          d. Low slack

Figure 4. Motion track of pollutants released at different times in the direction of WSW from the position of 1km to the head of the dyke.

Formula (4) is used to calculate the proportion of pollutants entering the open channel under different working conditions, and the cloud maps are shown in Figure 5. As can be seen from the diagram, after the pollutants are released from different directions, at different times, and at different locations, the number of pollutants entering the open channel varies greatly.

Generally, except few conditions of direction, such as E and N, where no pollutants enter the open channel during the whole tide, there are different amounts of pollutants entering the open channel in most directions. Taking 1 km from the dyke head as an example, the number of pollutants released from SSW into the open channel is the largest, reaching 75%, which is followed by WSW - 70% of the pollutants entering the open channel at maximum flood tide.

The pollutants entering the open channel are mainly from the direction of WSW~W, and the flux into the open channel varies greatly at different times. When the pollutant distance from the dyke head exceeds 6km, very few pollutants enter the open channel in a tidal cycle.

Note that the model calculation assumes that the pollutants come from all directions of the dyke head. The solid waste floating at sea mainly comes from the nearshore. It is necessary to observe more pollutants from the coastal area. In particular, observing pollutants from the southwest of the plant site should be strengthened.

According to the on-site real-time monitoring of pollutants, the pollutants that have entered the observation range are marked in time, their position and orientation from the water intake are grasped, and the pollutant information is input into the particle tracking model. The follow-up trajectory of the pollutant is predicted, and the time to the water intake is calculated to give an early warning for the nuclear power operation units or take emergency measures in time.

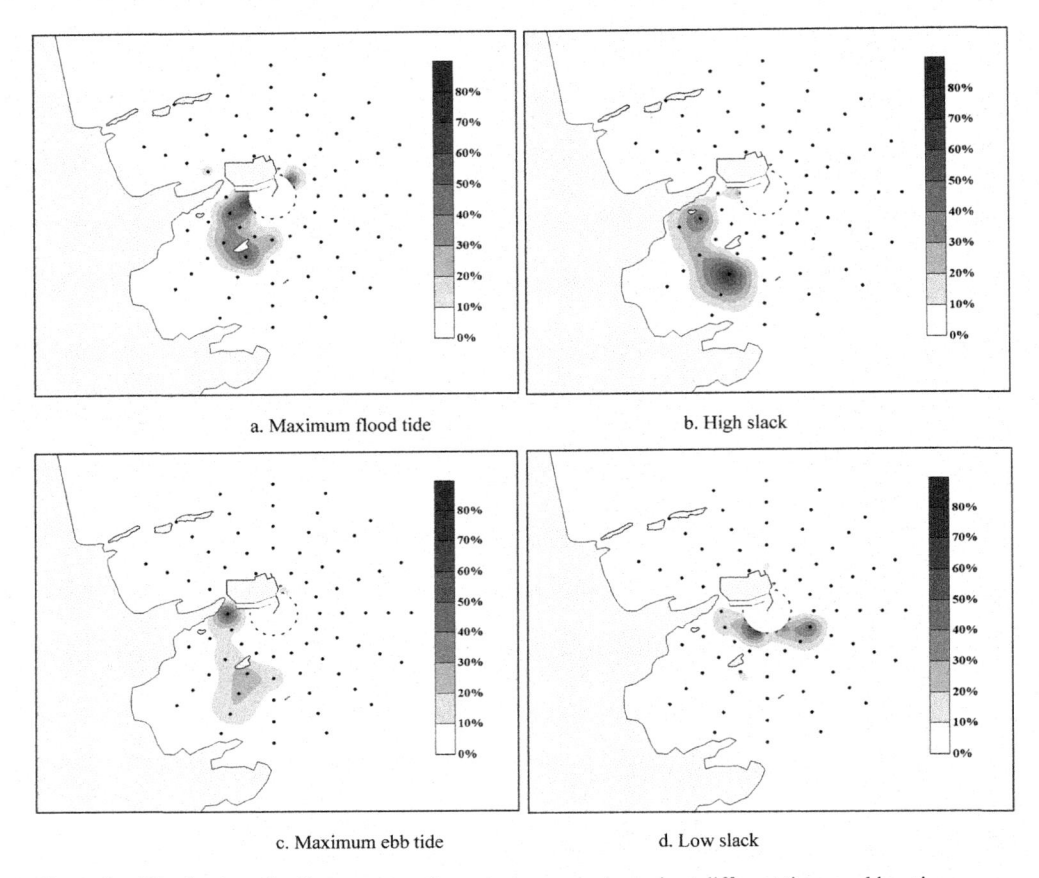

a. Maximum flood tide    b. High slack

c. Maximum ebb tide    d. Low slack

Figure 5.    Distribution of pollutant proportion entering open channels at different times and locations.

## 4    CONCLUSIONS

By establishing the mathematical model of tidal current and pollutant movement in the engineering area, the movement trajectories of pollutants from different directions, positions, and the proportion of pollutants entering the open channel are simulated. Furthermore, the motion characteristics of floating disasters in engineering would be mastered. According to the simulation results, monitoring pollutants in key areas could be reinforced. After marking the pollutants entering the monitoring range, the subsequent movement trajectory is predicted by the particle tracking model, and the proportion of entering the open channel and the arrival time is calculated, which would directly provide early warning for the power plant operation units, and improve the operation efficiency of the power plant.

## REFERENCES

Doeoes, K., & Engqvist, A. (2007). Assessment of water exchange between a discharge region and the open sea – a comparison of different methodological concepts. *Estuarine Coastal & Shelf Science*, 74(4), 709–721.

Enright, D., Fedkiw, R., Ferziger, J., & Mitchell, I. (2002). A hybrid particle level set method for improved interface capturing. *Journal of Computational Physics*, 183 (1), 83–116.

Li, W., & Sun, Z. (2019). Effects of the Liao River discharge on the dispersion of pollutants. *Marin Environ Sci* (2), 9.

Qiao, Y., Xu, Y., & Zhang, M. (2015). Numerical simulation of movement track of pollutant for Zhuanghe breeding sea area based on implicit hydrodynamic model. *Journal of Water Resources and Water Engineering*, 26(04), 73–79.

Wang, K, Lin J., & Du, J. (2017). Numerical simulation of oil spill drift-diffusion in the sea. *Journal of Hydrodynamics: A*, 32(2), 7.

*Civil Engineering and Urban Research – Mohamed & Hou (Eds)*
*© 2023 the Authors, ISBN 978-1-032-44487-1*

# Statistical inference of calculated model uncertainty coefficient in structural performance modeling

Cheng Kaikai*
*Department of Civil Engineering, Xi'an Shiyou University, Xi'an, China*

Hao Yun*
*College of New Energy, Xi'an Shiyou University, Xi'an, China*

Zhang Yumin*
*Department of Civil Engineering, Xi'an Shiyou University, Xi'an, China*

ABSTRACT: Classical statistics method is applied to infer the estimation values of distribution parameters for basic variables. However, this method fails to consider the statistical uncertainty caused by the limited sample numbers, leading to unreasonable inference results. Some scholars have proposed the expressions of the estimation values of distribution parameters for basic variables allowing for statistical uncertainty. However, those expressions are theoretically imperfect because an approximate method is used in the derivation. The statistical uncertainty in structural performance modeling exists mainly in the statistical analysis of the calculated model uncertainty coefficient. The paper proposes the Bayesian method to determine the probability distribution of the calculated model uncertainty coefficient in structural performance modeling. This method has a good theoretical basis, and the determined probability distribution can fully reflect the influence of the statistical uncertainty and obtain more reasonable inference results.

## 1 INTRODUCTION

The international standard General Principles of Structural Reliability (International Organization for Standardization 2015) introduces structural tests into the specific design process and proposes a more empirical design assisted by a testing approach. Structural performance studies refer to the combination of structural tests and theoretical studies to establish analytical models of structural performance, where the most prominent problem is the indiscriminate use of classical statistics in statistical analysis (Housing and Urban-Rural Development of the People's Republic of China 2012). The classical statistical approach does not consider the effect of statistical uncertainty due to insufficient sample size, and in most cases, it will result in over-estimation or under-estimation of structural performance. Statistical uncertainty in structural performance modeling exists mainly in the statistical analysis of the calculated model uncertainty coefficient. Therefore, the statistical extrapolation of calculated model uncertainties in structural performance modeling becomes a necessary task.

Haldar et al. (Haldar et al. 2000) put forward the concept of statistical uncertainty caused by the limited sample size in parameter estimation of basic structural variables and proposed the estimation method of distribution parameters when considering statistical uncertainty for variables subject to normal distribution. But it is limited to the statistical uncertainty of the mean value of the basic variables. On this basis, Chinese scholar Yao Ming Chu (Yao 1992) derived formulas for the statistical uncertainty of the variance of the basic variables, and the obtained results are more comprehensive and safe. For more non-normal random variables, the literature (Yang 2000)

---

*Corresponding Authors: chengkaikai_1990@126.com, 121041760@qq.com and 35813930@qq.com

  DOI 10.1201/9781003372417-96

derived formulas for calculating the mean and variance of log-normally distributed variables and the statistical uncertainty of extreme value type I distributed variables are analyzed using the Monte-Carlo method. However, the approximate calculation methods are all used in the derivation process, which is theoretically flawed and will produce large errors when the sample size is small.

Based on this, this paper firstly introduces the traditional methods for determining the probability distribution of the calculated model uncertainty coefficient and its parameter estimation and briefly describes the estimates of the mean, variance, and coefficient of variation of a normally distributed variable when statistical uncertainty is considered (Yao 1992; Yang 2000). Then according to the joint posterior distribution of the unknown parameters of the fundamental variable $\mu, \sigma$ based on the Bayesian method, the probability distribution of the calculated model uncertainty coefficient $\eta$ determined by the full probability formula is proposed. This corresponds to a weighted average of the joint densities of $\mu, \sigma$, so that the probability distribution of $\eta$ can fully reflect the influence of statistical uncertainties in the distribution parameters and obtain more reasonable inference results.

## 2  ESTIMATION OF DISTRIBUTION PARAMETER WITHOUT CONSIDERATION OF STATISTICAL UNCERTAINTY

The probability model of structural performance is the basis of the structural performance analysis model in design, which can be expressed as (Joint Committee on Structural Safety 2001; Yao 2016b):

$$y = \eta g(x_1, x_2, \ldots, x_m) \tag{1}$$

Where $x_1, x_2, \ldots, x_m$ are the basic variables, $g(\cdot)$ is the fitted structural performance function, and $\eta$ is the calculated model uncertainty coefficient. The probabilistic model of structural performance can be determined according to the probability distribution of the structural performance function and the calculated model uncertainty coefficient. Whereas the statistical uncertainty in structural performance function exists mainly in the statistical analysis of the calculated model uncertainty coefficient $\eta$, its probabilistic properties need to be inferred from the experimental results:

$$\eta_i = \frac{y_i}{g(x_1, x_2, \ldots, x_m)} \tag{2}$$

The probability distribution of $\eta$ is generally taken to be the normal distribution, and its mean $\mu_\eta$, variance $\sigma_\eta^2$ and coefficient of variation $\delta_\eta$ are estimated using classical statistical methods, which can be written as (Yao 2016a):

$$\mu_\eta = \bar{\eta} = \frac{1}{n} \sum_{i=1}^{n} \eta_i \tag{3}$$

$$\sigma_\eta^2 = s_\eta^2 = \frac{1}{n-1} \sum_{i=1}^{n} (\eta_i - \bar{\eta})^2 \tag{4}$$

$$\delta_\eta = \frac{\sigma_\eta}{\mu_\eta} = \frac{s_\eta}{\bar{\eta}} \tag{5}$$

where $\eta_1, \eta_2, \ldots, \eta_n$ are the $n$ sample observations of $\eta$; $\bar{\eta}, s_\eta^2$ are the sample mean and sample variance.

In structural design analysis, the mean $\mu_\eta$ and variance $\sigma_\eta^2$ of the basic variable can be estimated from the mean $\bar{\eta}$ and the unbiased estimates of the variance $s_\eta^2$ of the sample observations of the basic variable. However, when the sample size is insufficient, the value of $\bar{\eta}, s_\eta^2$ obtained from the test may not be the true value even without considering the test error. If the same tests are repeated with multiple groups, the values of $\bar{\eta}, s_\eta^2$ are also different among the groups, and the smaller the sample size, the greater the differences will generally be. This variability in the statistical analysis results due to insufficient sample size is known as the statistical uncertainty, and the larger the sample size, the smaller the statistical uncertainty (Haldar et al. 2000; Yao 1992). The relative error ranges of $s_\eta^2$ at 90%, 95%, and 99% guarantees are shown in Table 1. It can be seen that the larger the sample size, the smaller the relative error ranges of $s_\eta^2$. And the larger the required guarantee rate for the same sample size, the larger the relative error ranges of $s_\eta^2$.

Table 1. The relative error ranges of $s_\eta^2(\%)$.

| $n$ | $1\text{-}\alpha = 0.90$ | $1\text{-}\alpha = 0.95$ | $1\text{-}\alpha = 0.99$ |
|---|---|---|---|
| 30 | $-39 \sim 47$ | $-45 \sim 58$ | $-55 \sim 80$ |
| 40 | $-34 \sim 40$ | $-39 \sim 49$ | $-49 \sim 68$ |
| 50 | $-31 \sim 35$ | $-36 \sim 42$ | $-46 \sim 57$ |
| 60 | $-29 \sim 32$ | $-34 \sim 38$ | $-42 \sim 52$ |
| 70 | $-27 \sim 29$ | $-31 \sim 35$ | $-40 \sim 48$ |
| 80 | $-25 \sim 27$ | $-29 \sim 33$ | $-37 \sim 44$ |
| 90 | $-24 \sim 26$ | $-28 \sim 31$ | $-35 \sim 42$ |
| 100 | $-22 \sim 24$ | $-26 \sim 29$ | $-34 \sim 39$ |

The European Code "Euro code-basis of structural design" recommends that when the sample size is not greater than 100, the small sample extrapolation should be used to extrapolate the design values for resistance (European Committee for Standardization 2002), which represents a reference standard for practical experimental studies that need to take into account the effects of statistical uncertainty. However, regardless of the sample size, the current statistical analysis of test results indiscriminately adopts the classical statistical methods without considering the effects of statistical uncertainty due to inadequate sample size (number of tests) and, in most cases, will result in over-estimation or under-estimation of structural performance, leading to additional construction costs or failure risks of the structure. It is, therefore, necessary to consider the effect of statistical uncertainty on the statistical extrapolation of the calculated model uncertainty coefficient.

## 3 THE TRADITIONAL ESTIMATION METHOD FOR THE CALCULATED MODEL UNCERTAINTY COEFFICIENT

Assuming that the basic variable $\eta$ obeys the normal distribution, its mean and variance are unknown and noted $\mu_\eta, \sigma_\eta^2$. Its sample size is $n$, if the sample size is limited, there is statistical uncertainty in estimating the parameters of the parent distribution $\mu_\eta, \sigma_\eta^2$, which can be expressed in terms of statistical uncertainty variables, that is:

$$\mu_\eta = a\bar{\eta} \tag{6}$$

$$\sigma_\eta^2 = bs_\eta^2 \tag{7}$$

where $a$ and $b$ are the statistical uncertainty variables for the mean and variance of the basic variables $\eta$.

When the standard deviation of the basic variable is unknown, and the effect of statistical uncertainty is taken into account, the estimates of the mean and variance $\hat{\mu}_\eta, \hat{\sigma}_\eta^2$ and coefficient of variation $\hat{\delta}_\eta$ of a normally distributed random variable should be written as:

$$\hat{\mu}_\eta = a\bar{\eta} = \bar{\eta} \tag{8}$$

$$\hat{\sigma}_\eta^2 = \frac{2n^3 - 5n^2 - n + 2}{2n(n-1)(n-3)} \cdot s_\eta^2 \tag{9}$$

$$\hat{\delta}_\eta = \frac{\hat{\sigma}_\eta}{\hat{\mu}_\eta} = \frac{s_\eta}{\bar{\eta}} \sqrt{\frac{2n^3 - 5n^2 - n + 2}{2n(n-1)(n-3)}} \tag{10}$$

When the standard deviation of the basic variable is known to be $\sigma_\eta$, the estimates of the mean and variance $\hat{\mu}_\eta, \hat{\sigma}_\eta^2$ and coefficient of variation $\hat{\delta}_\eta$ can be expressed as:

$$\hat{\mu}_\eta = a\bar{\eta} = \bar{\eta} \tag{11}$$

$$\hat{\sigma}_\eta^2 = \sqrt{\frac{n+1}{n}} \cdot \sigma_\eta^2 \tag{12}$$

$$\hat{\delta}_\eta = \frac{\hat{\sigma}_\eta}{\hat{\mu}_\eta} = \frac{\sigma_\eta}{\bar{\eta}} \sqrt{\frac{n+1}{n}} \tag{13}$$

## 4  BAYESIAN ESTIMATION METHOD FOR THE CALCULATED MODEL UNCERTAINTY COEFFICIENT

To obtain the probability distribution of the calculated model uncertainty coefficient $\eta$, the joint posterior distribution or belief distribution of the distribution parameters is established under small sample conditions using the Bayesian or belief inference method firstly, and then the final probability distribution is determined using conditional probability methods.

When the variance $\sigma^2$ is unknown, the Jeffreys joint prior distribution with parameter $\mu, \sigma$ is taken to be:

$$\pi_{\mu,\sigma}(y_1, y_2) = \frac{1}{y_2} \tag{14}$$

The likelihood function created using the sample observations $x_1, x_2, \cdots, x_n$ is

$$p_{X_1,\cdots,X_n|\mu,\sigma}(x_1,\cdots,x_n|y_1,y_2) = \prod_{i=1}^{n} \frac{1}{\sqrt{2\pi}y_2} e^{-\frac{1}{2}(\frac{x_i-y_1}{y_2})^2} = \left(\frac{1}{\sqrt{2\pi}}\right)^n \left(\frac{1}{y_2}\right)^n e^{-\frac{1}{2}\frac{(n-1)s^2+n(y_1-\bar{x})^2}{y_2^2}} \tag{15}$$

The posterior distribution is

$$p_{\mu,\sigma|X_1,\cdots,X_n}(y_1, y_2|x_1,\cdots,x_n) = \frac{\frac{1}{y_2}\left(\frac{1}{\sqrt{2\pi}}\right)^n \left(\frac{1}{y_2}\right)^n e^{-\frac{1}{2}\frac{(n-1)s^2+n(y_1-\bar{x})^2}{y_2^2}}}{\int_0^\infty \int_{-\infty}^\infty \frac{1}{y_2}\left(\frac{1}{\sqrt{2\pi}}\right)^n \left(\frac{1}{y_2}\right)^n e^{-\frac{1}{2}\frac{(n-1)s^2+n(y_1-\bar{x})^2}{y_2^2}} dy_1 dy_2} \tag{16}$$

The probability distribution of $\eta$ can be written as

$$f_\eta(t) = \int_0^\infty \int_{-\infty}^\infty \frac{1}{\sqrt{2\pi}y_2} e^{-\frac{1}{2}\left(\frac{t-y_1}{y_2}\right)^2} p_{\mu,\sigma|X_1,\cdots,X_n}(y_1, y_2|x_1,\cdots,x_n) dy_1 dy_2$$

$$\propto \left[1 + \frac{\left(\frac{t-\bar{x}}{s\sqrt{1+\frac{1}{n}}}\right)^2}{n-1}\right]^{-\frac{(n-1)+1}{2}} \tag{17}$$

From statistics, $\frac{\eta-\bar{x}}{s\sqrt{1+\frac{1}{n}}}$ follows a $t$-distribution with degree of freedom $n-1$. Based on the nature of the $t$-distribution, the following conclusion can be obtained:

$$\hat{\mu}_\eta = \bar{x} \tag{18}$$

$$\hat{\sigma}_\eta = \sqrt{\frac{n-1}{n-3}\left(1 + \frac{1}{n}\right)} \cdot s \tag{19}$$

$$\hat{\delta}_\eta = \frac{\hat{\sigma}_\eta}{\hat{\mu}_\eta} = \frac{s}{\bar{x}} \sqrt{\frac{n-1}{n-3} \left(1 + \frac{1}{n}\right)} \tag{20}$$

When the variance $\sigma^2$ is known, the Jeffreys joint prior distribution of $\mu$ can be taken as

$$\pi_\mu(y_1) = 1 \tag{21}$$

Follow the similar steps, $\frac{\eta - \bar{x}}{\sqrt{1 + \frac{1}{n}}\sigma}$ follows the standard normal distribution. Therefore, there are

$$\hat{\mu}_\eta = \bar{x} \tag{22}$$

$$\hat{\sigma}_\eta = \sqrt{1 + \frac{1}{n}} \cdot \sigma \tag{23}$$

$$\hat{\delta}_\eta = \frac{\hat{\sigma}_\eta}{\hat{\mu}_\eta} = \frac{\sigma}{\bar{x}} \sqrt{1 + \frac{1}{n}} \tag{24}$$

Where $\bar{x}, s^2$ are the unbiased estimates of the mean and variance of the basic variable $\eta$, respectively, obtained from equations (3) and (4). For cases where the variance $\sigma^2$ is known, the inferred results by the method in the paper are identical to those in the literature (Yao 1992), which also proves the correctness of the method in this paper.

## 5 COMPARATIVE ANALYSIS OF BAYESIAN METHOD AND TRADITIONAL METHOD

As deduced above, it is clear that the main difference in inferring the parameter estimates for the calculated model uncertainty coefficient when considering the effect of statistical uncertainty lies in the case where the standard deviation of the basic variable is unknown. And when the standard deviation of the basic variable is known, the inference results of traditional inference methods and Bayesian methods under small sample conditions are consistent. The relationship between the ratios of the coefficient of variation estimation values to the sample estimation values and different sample sizes $n$ when the standard deviation of the basic variable is unknown are shown in Table 2.

As can be seen from Table 2, the statistical uncertainty decreases when the sample size increases. When the sample sizes are the same, the statistical uncertainty on the coefficient of variation calculated in this paper is greater than that in the literature (Yao 1992). And in this paper, the joint posterior distribution of the unknown parameters $\mu, \sigma$ is determined by the Bayesian method, and then the probability distribution of $\eta$ is determined according to the full probability formula. So the probability distribution of $\eta$ determined in this way can reflect the influence of the statistical uncertainty fully, and the corresponding statistical uncertainty is more fully considered. Then the values of $\hat{\delta}_\eta/\delta_\eta$ obtained from this paper are larger than that from the literature (Yao 1992).

Table 2. Relationship between the coefficient of variation and sample size under statistical uncertainty.

| | $\hat{\delta}_\eta/\delta_\eta$ | | | $\hat{\delta}_\eta/\delta_\eta$ | |
|---|---|---|---|---|---|
| $n$ | Reference (Yao 1992) | This paper | $n$ | Reference (Yao 1992) | This paper |
| 5 | 1.2349 | 1.5492 | 30 | 1.0262 | 1.0535 |
| 6 | 1.1738 | 1.3944 | 40 | 1.0194 | 1.0394 |
| 7 | 1.1391 | 1.3093 | 50 | 1.0154 | 1.0312 |
| 8 | 1.1164 | 1.2550 | 60 | 1.0128 | 1.0258 |
| 9 | 1.1003 | 1.2172 | 70 | 1.0109 | 1.0220 |
| 10 | 1.0882 | 1.1892 | 80 | 1.0095 | 1.0192 |
| 15 | 1.0552 | 1.1155 | 90 | 1.0085 | 1.0170 |
| 20 | 1.0403 | 1.0833 | $\geq 100$ | 1.0076 | 1.0153 |

# 6 CONCLUSION

(1) The classical statistical method is used to infer the estimation value of the distribution parameters of basic variables. This is reasonable in the case of a large sample size. However, when the sample size is small (<100), it is necessary to consider the statistical uncertainty in estimating the distribution parameters due to the limited sample size.

(2) For the case of small samples, the statistical uncertainty in estimating the distribution parameters is expressed in terms of additional random variables, and expressions are derived for estimating the parameters when the standard deviations are known and unknown. The smaller the sample size, the greater the statistical uncertainty. When the sample size is greater than 100, the statistical uncertainty on the coefficient of variation is small and can be disregarded. However, this method is theoretically imperfect because an approximate method is used in the derivation process.

(3) The paper proposed a Bayesian approach for determining the probability distribution of the calculated model uncertainty coefficient in structural performance modeling, which has a good theoretical basis and fully considers the influence of statistical uncertainty, which can obtain more reasonable inference results.

## ACKNOWLEDGMENTS

The authors would like to acknowledge assistance from the Natural Science Foundation of Shaanxi Province, China (2022JM-279; 2021JM-406), the Scientific Research Program Funded by the Shaanxi Provincial Education Department (No.21JK0834).

## REFERENCES

European Committee for Standardization. (2002). *Eurocode basis of structural design*(EN1990-2002). BSI, London, UK.

Haldar A, Mahadevan S. (2000) *Probability, reliability, and statistical methods in engineering design.* New York, Chichester: Wiley.

Housing and Urban-Rural Development of the People's Republic of China. (2012). *Unified standard for reliability design of building structures* (GB/T50152-2012). China Architecture and Building Press, Beijing, China.

International Organization for Standardization. (2015). *International Standard: General Principles on Reliability for Structures* (ISO2394: 2015). International Organization for Standardization, Switzerland.

Joint Committee on Structural Safety (JCSS). (2001). *JCSS probabilistic model code: part 3: resistance models.* JCSS, Copenhagen, Denmark.

Yang, Y. C., Bai, X. L. & Zhang, M. (2000). *Estimation method of distribution parameters with consideration of statistical uncertainty.* Engineering safety and durability-Proceedings of the 9th Annual Meeting of China Civil Engineering Society, Hangzhou, China.

Yao, J. T., Cheng, K. K. & Liu, W. (2016b). Statistical uncertainty and its impact on the establishment of structural performance model by testing. *Journal of Xi'an University of Architecture and Technology* (Natural Science Edition), 48(5), 639–642/675.

Yao, J. T., Cheng, K. K. & SONG, C. (2016a). Small samples method to the establishment of the probability characteristics of structural performance. *Journal of Xi'an University of Architecture and Technology* (Natural Science Edition), 48(2), 155–159/1775.

Yao, M. C. (1992). *Statistical uncertainty for the estimation of parameters of basic variables with small-size samples.* The second national Academic Exchange meeting of "Engineering Structure Reliability" held by the Structural Reliability Committee of Bridge and Structural Engineering Society of China Civil Engineering Society, Nanjing, China.

Civil Engineering and Urban Research – Mohamed & Hou (Eds)
© 2023 the Authors, ISBN 978-1-032-44487-1

# Research on intelligent analysis system for aircraft-loaded bridge health monitoring

Qijie Teng* & Xuekui Gao
*China Airport Planning & Design Institute Co., Ltd, Beijing, China*

Fulai Wang, Shengping Ma & Rui Liu
*Qingdao International Airport Group Co., Ltd, Qingdao, China*

ABSTRACT: Aircraft-loaded bridges have the mechanical characteristics of large load, large width-span ratio, and prominent spatial effect, and their unique characteristics in load effect, structural form, and operation requirements. In this paper, the first intelligent analysis and management system for health monitoring of aircraft-loaded bridges in China is introduced in detail, and the stress characteristics of bridge structure under the load of A380f are analyzed in detail by solid finite element analysis. This paper is of great importance for guiding the establishment of the health monitoring system, structural design, and maintenance management of aircraft-loaded bridges.

## 1 INTRODUCTION

The aircraft-loaded bridge is an effective method in solving the intersection of traffic flows of landside and airside of the airport; that is, the bridge is set up when taxiways or runways need to cross roads, railways, tunnels, rivers, seas, and other obstacles. The aircraft-loaded bridge is an important part of airport engineering. There are more than 70 taxiway bridges in China's airports, such as Guangzhou Baiyun International Airport, Beijing Capital International Airport, Shanghai Pudong International Airport, and Macao International Airport, which have been built with single-span and multi-span layouts. Many airports in the United States have built aircraft-loaded bridges, such as Kennedy International Airport, O'Hare International Airport, Dallas Fort Worth International Airport, and Memphis International Airport. Most aircraft-loaded bridges in China and abroad are built over roads or rivers, and a few taxiway bridges are built over the sea, such as Macao International Airport and Tokyo International Airport.

Taxiway bridges in China can be divided into three types in terms of usage: the first category is the taxiway bridge over the access road. With the expansion of the hub airport in recent years, the number of access road lanes whose capacity matches the passenger throughput has increased correspondingly. The rail transit-led, multimodal integrated transportation to enter and leave the airport has also gained popularity. Therefore, the access road is often accompanied by underground structures such as the subway, high-speed rail, or intercity rail. The structural form of taxiway bridge structures tends to be continuous multispan, and a tendency to increase in the length and span of the bridge; the second category is the bypass taxiway bridge, such as the Shanghai Hongqiao airport taxiway bridge system; the third category is taxiway bridges that cross over the roads or waters in the flight area.

Generally, the taxiway bridge investment accounts for less than 10% of the total investment of the expansion project. However, due to the engineering requirements of the airport project, sometimes the scale of the bridge project is large, which is important to ensure the early opening of the taxiway.

---

*Corresponding Author: Tengqj_2005 @163.com

DOI 10.1201/9781003372417-97

At the same time, the taxiway bridge has the characteristics of large load, large width-span ratio, and prominent spatial effect. It has unique characteristics in load effect, structural form, and demand for operation. As a risk point of important structures in the airfield, the maintenance and management of the taxiway bridge are significant to ensure the structural safety of the built taxiway bridge and the safe operation of the airport.

Figure 1. Taxiway bridge of Qingdao Jiaodong international airport over access road.

## 2 INTELLIGENT ANALYSIS AND MANAGEMENT SYSTEM FOR HEALTH MONITORING

*Outline of Action for The Construction of Four-Type Airports* (2020-2035) sets out the objectives and tasks for constructing a safe airport. To ensure the safety of airport construction, it is necessary to strengthen the monitoring and detection of infrastructure operation, improve the level of facility maintenance, enhance the durability and reliability of facilities, and improve the use efficiency of assets throughout the life cycle of the project.

In November 2019, the information circular *Application of Aircraft-Loaded Bridges in Airport Engineering* issued by the Civil Aviation Administration of China clearly pointed out that more than 30 of the 60+ existing taxiway bridges in China have been in service for more than ten years. In order to ensure the safety of the bridge structure and the safe operation of the airport, attention should be paid to the maintenance and management of existing bridges, regular inspection should be done, and a bridge safety monitoring system can be applied if conditions permit. The draft of the *Guidelines for Design of Aircraft-Loaded Bridges in Civil Airports* under preparation clearly states that the taxiway bridge can be equipped with necessary structural monitoring facilities.

The traditional method for identifying the status of bridge structures is the "Weight Stacking Method," which is used before operation and during routine inspections after five years of operation and special inspections. During routine inspections, the workload in daily maintenance is heavy, and processes such as information sharing have received centralized and traceable management (Mu 2010; Su 2019).

Through the implementation of the *Research on Structural State Perception and Evaluation of Key Technologies for Aircraft Bridges* project and by deploying "Action Effect Test" sensors on representative bridge structures, collecting data and information from various deployed sensors, using data mining and data fusion technology to extract the characteristic parameters of action effects under aircraft load, the problem of lack of calculation and analysis could be made up. Other important aspects such as reliability of assessment conclusions, the accuracy of health monitoring threshold setting, and the improvement of the level of management, maintenance, and the quality of repairing of the aircraft-loaded bridge structures could also be improved (Guo 2021; Li 2019).

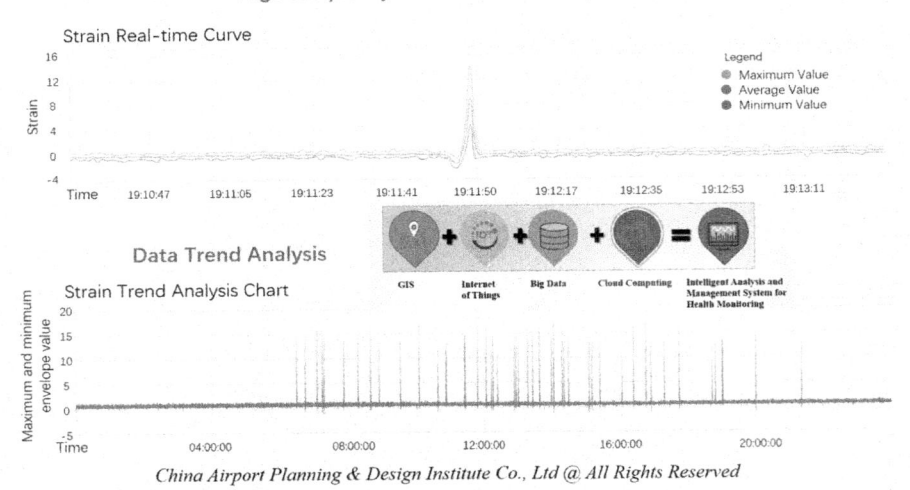

*China Airport Planning & Design Institute Co., Ltd @ All Rights Reserved*

Figure 2. Modular user interface of the health monitoring system.

## 3 SOLID ELEMENT ANALYSIS OF AIRCRAFT-LOADED BRIDGE

### 3.1 *Calculation model set-up*

*Midas FEA* is used to establish the solid element model of the whole bridge, which divide the bridge into 57769 hexahedral elements with a unit size of 0.2*0.2*0.5m (Lu 2016).

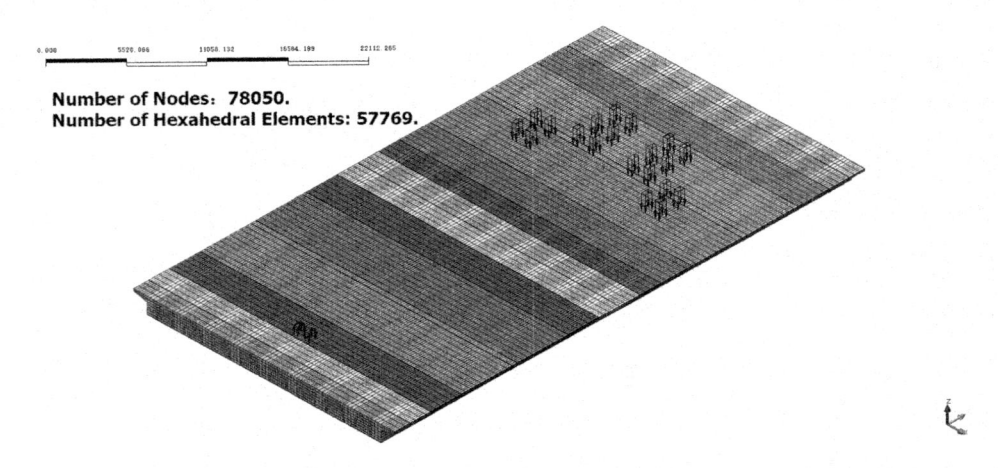

Figure 3. Solid finite element model (Loading Case: A380F).

### 3.2 *Calculation parameters and load cases*

(1) Calculation Parameters

The grade concrete is C50; Volumetric weight = 26 kN/m³; the elastic modulus E = 34.5Gpa; the Poisson's ratio $\gamma = 0.2$.

(2) Load Cases

A380F, 592t in total; The six wheels with the most unfavorable aircraft-loaded are arranged at the mid-span.

(3) Boundary Conditions

Based on the actual constraints of bridge bearings.

## 3.3 *Calculation results and analysis*

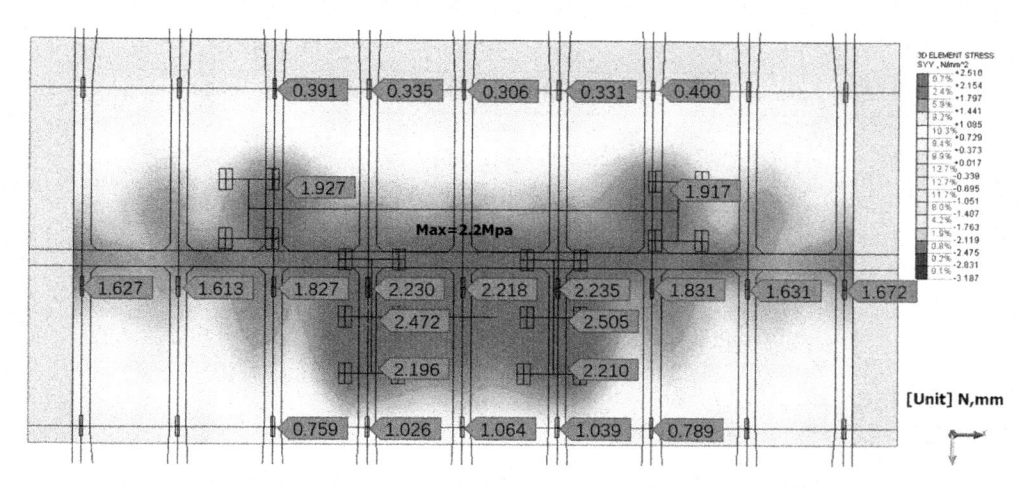

Figure 4. Mid-span longitudinal stress nephogram for ClassF A380F (Unit: Mpa).

The above results show that the overall stress performance of the bridge is good, and all nine webs bear the stress. Under the action of 6 wheel sets, three webs in the middle bear larger stress with the maximum stress of 2.4Mpa, the maximum stress of 2.2MPa occurs at the sensor position, and the stress of 1.62~1.85Mpa for the three webs occurs on both sides of the bridge.

It can be seen from Figure 4 that under the action of 6 wheelsets, the transverse bridge is under stress, and the three webs in the middle are the largest, which is 1.2 MPa. Therefore, the transverse bridge rebar configuration should be properly reinforced for this design.

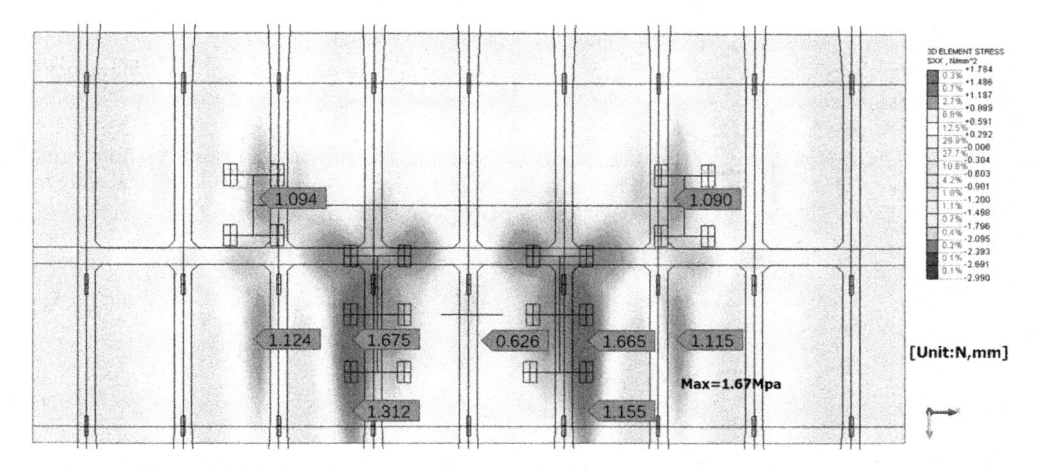

Figure 5. A380 Mid-span transverse nephogram for Class F A380F (Unit: Mpa).

# 4 RESPONSE ANALYSIS OF BRIDGE STRUCTURE UNDER AIRCRAFT LOAD-DATA ANALYSIS

The solid element calculation also calculates the load cases of class C and class E aircraft. According to the measured data of health monitoring, the bridge structure response is within the range of the control value. It shows that the bridge structure has a stable response and good structural performance (JTG 3362-2018).

Table 1. Measured and controlled values under aircraft load (Unit: $\mu\varepsilon$).

| Web No. | L/4 | | L/2 | |
| --- | --- | --- | --- | --- |
| | Measured values | Controlled values | Measured values | Controlled values |
| 1# | 24 | 56 | 40 | 70 |
| 2# | 16 | 57 | 34 | 69 |
| 3# | 36 | 69 | 40 | 82 |
| 4# | 36 | 79 | 34 | 87 |
| 5# | 26 | 82 | 18 | 89 |
| 6# | 32 | 78 | 33 | 86 |
| 7# | 18 | 69 | 19 | 82 |
| 8# | 33 | 57 | 22 | 70 |
| 9# | 30 | 56 | 35 | 69 |

# 5 CONCLUSION

This system has established a top-level monitoring platform for the aircraft-loaded bridge, which is the system of the bridge group. In the future, this system can access the monitoring data of all airport road and bridge infrastructures such as taxiway bridges and airport tunnels. At the same time, the system has realized the collection and transmission of high-precision and high-frequency data, a unique damage assessment method, and powerful computing power. It will better serve industry managers, owner-managers, and engineering technicians.

This system allows industry managers to understand the dynamics of different airport bridges and the diseases of bridges of different ages, the current situation of airport taxiway bridges in the industry, and the number of pre-alarm. Also, it helps industry managers to comprehensively understand and judge the national airport taxiway bridge system and provide transparent and efficient decision-making information through intelligent means.

This system provides detailed, quantitative, and economical maintenance suggestions for owner's managers, which can provide information feedback when making decisions and gradually distribute them to branch managers at all levels.

However, technicians should quantify the bridge condition assessment and formulate the maintenance strategy for the whole life cycle based on the following factors: the calculation results of the refined finite element model, the bridge state perception through a bar graph, and the loading strain curve.

In the future, the data from the health monitoring system will provide technical reserves for the integration of construction and operation according to the needs of the airport runway and bridge projects in mountainous or coastal areas.

ACKNOWLEDGMENTS

Thanks to China Airport Planning & Design Institute Co., Ltd for its generous funding of this research project and the strong support of Qingdao International Airport Group Co., Ltd.

## REFERENCES

Hongwei Lu (2016).Application of MIDAS/FEA in Static Load Test of Concrete Cable-stayed Bridge.*Journal of Highways & Automotive Applications*.173, 195–199.

Jinhai Li (2019). *Design and Implementation of Bridge Health Monitoring Information Management System,Master of Engineering,School of Information and Communication Engineering*, China.

Qiang Su (2019), Bridge Safety State Assessment Based on Load Test and Health Monitoring System. *Journal of China Highway*.2019(17), 100–103.

*Specifications for Design of Highway Reinforced Concrete and Prestressed Concrete Bridges and Culverts* (JTG 3362-2018). Beijing: Issued by the Ministry of Transport of the People's Republic of China (2018) (In Chinese).

Tianhui Guo, Yaoyue Cui (2021). Research on Bridge Health Monitoring System Based on Load Test Modification. *Journal of Highway*. 2021(8), 201–205.

Zhifeng Mu (2010). Static Load Test of Prestressed Concrete Simply Supported T-beam Bridge.*Journal of China & Foreign Highway*. 30 (6), 153–156.

*Civil Engineering and Urban Research – Mohamed & Hou (Eds)*
*© 2023 the Authors, ISBN 978-1-032-44487-1*

# Non-parametric identification method of nonlinear rolling coefficients of floating structure based on Hilbert transform

Minghao Yan
*Center for Marine Civil Engineering Technology, Ludong University, Yantai, China*

Weihao Sun & Qinglai Fan*
*School of Civil Engineering, Ludong University, Yantai, China*

ABSTRACT: Under the interference of external factors such as wind, waves, and currents, the floating structure such as a ship has six degrees of freedom of violent motion. Among them, the rolling motion is the largest swing amplitude, which will seriously threaten the safety of offshore floating structures. Two identification methods of nonlinear rolling hydrodynamic coefficients according to Hilbert theory, i.e., identification based on first-order solution and identification considering high-order super harmonics, are studied to estimate the restoring moment coefficient and damping moment coefficient of nonlinear rolling. The two methods are compared with the traditional identification method based on energy conservation, and the accuracy of the rolling hydrodynamic coefficient identification method based on the Hilbert transform is verified.

## 1 INTRODUCTION

Under the interference of external factors such as wind, waves, and currents, the ship structure on the sea will have six degrees of freedom of violent motion. Among them, the rolling motion is the largest swing amplitude, which will threaten the safety of ships. Therefore, it is of great significance to obtain the hydrodynamic coefficient of floating structures in a rolling motion.

Roll damping studies mainly focus on roll conditions with attenuation or forced stimulus. The damping coefficients obtained from these tests, usually in calm water, are assumed to represent roll damping in waves. However, this assumption may not be reliable, especially given moderate sea conditions. Hou and Zou (Hou 2015; Zou 2016) identified the hydrodynamic coefficients of roll motion in regular and irregular waves by parameter identification technology, but did not compare them with roll attenuation coefficients. Somayajula and Falzarano (Somayajula & Falzarano 2017) obtained the roll damping coefficient of the research ship from irregular waves based on the calibration of the roll RAO curve through advanced system identification techniques. Among semi-empirical methods, the Ikeda method described by Himeno (Himeno 1981) is the most widely used semi-empirical method for predicting roll damping. Kawahara et al. (2012), Irkal et al. (2019), and Kim et al.(2020) have shown that the simplified method cannot predict the experimental attenuation coefficients for some ship types and load conditions.

Based on the Hilbert transform, Feldman (1994) proposes a practical method to identify nonlinear damping and recovery coefficients of a single degree of freedom system with free vibration response. Lewandowski (2011) used the Hilbert transform method to predict the linear and quadratic terms of ship roll damping moment based on the experimental data obtained in the model pool. More recently, Kim and Park (Kim 2015; Park 2015) used free vibration signals based on the Hilbert transform to identify nonlinear roll damping and recovery moments of FPSO. These research

---

*Corresponding Author: 7154@ldu.edu.cn

     DOI 10.1201/9781003372417-98

processes have successfully studied the performance of Hilbert transform in ship rolling motion recognition.

Therefore, the nonlinear roll coefficient identification method based on the Hilbert transform is studied systematically in this paper. One numerical model test compares the nonlinear roll coefficient identification method based on traditional energy with the proposed method.

## 2 IDENTIFICATION METHOD OF HYDRODYNAMIC COEFFICIENTS OF NONLINEAR ROLLING BASED ON HILBERT TRANSFORM

### 2.1 Identification based on first-order solution

The normalized rolling motion equation is rewritten into two typical models of quadratic damping and cubic damping, shown in eq. (1) and (2),

$$\ddot{\varphi} + c_1\dot{\varphi} + c_2|\dot{\varphi}|\dot{\varphi} + k_1\varphi + k_3\varphi^3 + k_5\varphi^5 + \ldots = m(t) \tag{1}$$

$$\ddot{\varphi} + c_1\dot{\varphi} + c_3\dot{\varphi}^3 + k_1\varphi + k_3\varphi^3 + k_5\varphi^5 + \ldots = m(t) \tag{2}$$

Set $m(t) = 0$, the free vibration equation of rolling motion can be obtained by rewriting the normalized rolling motion equation (eq. (1) and (2)) into the general form,

$$\ddot{\varphi} + 2h_0(\dot{\varphi})\dot{\varphi} + \omega_0^2(\varphi)\varphi = 0 \tag{3}$$

In the eq. (3), $h_0(\dot{\varphi})$ and $\omega_0^2(\varphi)$ is defined as the nonlinear damping coefficient and undamped natural frequency of the system. The Hilbert transformation of the signal $\varphi(t)$ is defined by the integral transformation,

$$H[\varphi(t)] = \tilde{\varphi}(t) = \frac{1}{\pi}PV\int_{-\infty}^{\infty}\frac{\varphi(\tau)}{t-\tau}d\tau \tag{4}$$

In equation (4), $P.V.$ represents the Cauchy integral. Using the Hilbert transform, the analytic signal $\Phi(t)$ can be constructed by taking the original signal $\varphi(t)$ as the real part and the projection $\tilde{\varphi}(t)$ of the Hilbert transform of the original signal as the imaginary part.

$$\Phi(t) = \varphi(t) + j\tilde{\varphi}(t) = A(t)e^{j\Psi(t)} \tag{5}$$

In equation (5), the modulus of the analytic signal is to determine the instantaneous amplitude (or the envelope) $A(t)$ of the signal, and $\Psi(t)$ is the instantaneous phase. Both ($\Psi(t)$ and $A(t)$) should be represented using the original signal and its Hilbert transform projection as follows,

$$A(t) = \sqrt{\varphi^2(t) + \tilde{\varphi}^2(t)} \tag{6}$$

$$\Psi(t) = \arctan\left[\tilde{\varphi}(t)/\varphi(t)\right] \tag{7}$$

The instantaneous angular frequency is obtained by differentiating the instantaneous phase angle $\Psi(t)$,

$$\omega(t) = \dot{\Psi}(t) \tag{8}$$

The instantaneous undamped modal frequency and modal damping coefficient can be estimated by using the envelope and instantaneous frequency extracted from the free-decaying signal, as shown in eq. (9) and (10),

$$\omega_0^2(t) = \omega^2(t) - \frac{\ddot{A}(t)}{A(t)} + \frac{2\dot{A}^2(t)}{A^2(t)} + \frac{\dot{A}(t)\dot{\omega}(t)}{A(t)\omega(t)} \tag{9}$$

$$h_0(t) = -\frac{\dot{A}(t)}{A(t)} - \frac{\dot{\omega}(t)}{2\omega(t)} \tag{10}$$

Where $\dot{\omega}(t)$, $\dot{A}(t)$ and $\ddot{A}(t)$ are the first and second derivatives of instantaneous frequency $\omega(t)$ and envelope (amplitude) $A(t)$.

Then the nonlinear recovery moment $k(\phi)$ and damping moment $c(\dot{\phi})$ can be identified from the following expression,

$$k(\phi) = \omega_0^2(\phi)\phi \approx \begin{cases} \omega_0^2(t)A(t), & x > 0 \\ -\omega_0^2(t)A(t), & x < 0 \end{cases} \qquad (11)$$

$$c(\dot{\phi}) = 2h_0(\dot{\phi})\dot{\phi} \approx \begin{cases} 2h_0(t)A_{\dot{\phi}}(t), & x > 0 \\ -2h_0(t)A_{\dot{\phi}}(t), & x < 0 \end{cases} \qquad (12)$$

According to Equations (4) $\sim$ (6), through the analysis of the roll angular velocity signal, it can be known that $A_\varphi(t)$ is the envelope of roll angular velocity.

## 2.2 Identification considering higher order superharmonics (PISH)

In fact, the main components have primary natural frequencies, and some secondary components have higher multiple frequencies. For example, taking the non-stationary roll angle envelope $A(t)$ as a new signal, the signal decomposition process can be used to decompose it into the sum of several components.

$$A(t) = \sum_{i=1}^{N} A_i(t) \cos\left(\int \omega_{Ai}(t)dt + \varphi_{Ai}(t)\right) \qquad (13)$$

Where $A_i(t)$ and $\omega_{Ai}(t)$ are the envelope and instantaneous frequency of the $i$th harmonic of envelope signal $A(t)$ respectively, and $\varphi_{Ai}(t)$ is the phase Angle between the first harmonic and the I th harmonic of roll Angle envelope $A(t)$. These instantaneous parameters ($A_i(t)$, $\omega_{Ai}(t)$ and $\varphi_{Ai}(t)$) for each component can be estimated by equations (5) $\sim$ (8).

By this decomposition process, the envelope of $A(t)$ can be estimated in the following way, and $A_{env}(t)$ is expressed as:

$$A_{env}(t) = \sum_{i=1}^{N} A_i(t) \cos \varphi_{Ai}(t) \qquad (14)$$

Similarly, we use other instantaneous parameters, such as the roll Angle velocity envelope $A_{\dot{\phi}}(t)$, instantaneous modal frequency $\omega_0(t)$, and modal damping $h_0(t)$, as new signals. Through signal decomposition, we can obtain their first and higher order superharmonics. The main result of such signal decomposition is the estimation of its envelope

$$A_{\dot{\phi}_{env}}(t) = \sum_{i=1}^{N} A_{A_{\dot{\phi}}i}(t) \cos \varphi_{A_{\dot{\phi}}i}(t) \qquad (15)$$

$$\omega_{0_{env}}(t) = \sum_{i=1}^{N} a_{\omega_0 i}(t) \cos \varphi_{\omega_0 i}(t) \qquad (16)$$

$$h_{0_{env}}(t) = \sum_{i=1}^{N} a_{h_0 i}(t) \cos \varphi_{h_0 i}(t) \qquad (17)$$

Where $A_{A_{\dot{\phi}}i}(t)$, $a_{\omega_0 i}(t)$ and $a_{h_0 i}(t)$ are the envelopes of the $i$th harmonic of $A_{\dot{\phi}}(t)$, $\omega_0(t)$ and $h_0(t)$ respectively. $\varphi_{A_{\dot{\phi}}i}(t)$, $\varphi_{\omega_0 i}(t)$ and $\varphi_{h_0 i}(t)$ are phase angles corresponding to their first harmonic and the $i$th harmonic.

$A(t)$, $A_{\dot{\phi}}(t)$, $\omega_0(t)$ and $h_0(t)$ are replaced by corresponding envelope lines $A_{env}(t)$, $A_{\dot{\phi}env}(t)$, $\omega_{0_{env}}(t)$ and $h_{0_{env}}(t)$ in Equations (12) and (13), and the static characteristics are obtained as follows,

$$k(\phi) = \begin{cases} \omega_{0env}^2(t)A_{env}(t), & x > 0 \\ -\omega_{0env}^2(t)A_{env}(t), & x < 0 \end{cases} \qquad (18)$$

$$c(\dot{\phi}) = \begin{cases} 2h_{0_{env}}(t)A_{\dot{\phi}env}(t), & x > 0 \\ -2h_{0env}(t)A_{\dot{\phi}env}(t), & x < 0 \end{cases} \tag{19}$$

## 3 NUMERICAL MODEL TEST OF FREE ROLLING

In this section, a numerical model is used to verify the proposed nonlinear roll motion recognition scheme.

The roll motion equation is,

$$\ddot{\phi} + 0.32\dot{\phi} + 0.16\dot{\phi}^3 + 16\phi + 19.2\phi^3 = m(t) \tag{20}$$

As non-parametric identification, this equation and its corresponding parameters are not used in the identification process of IPS and PISH methods but are used to generate roll responses and verify the identification results. First, the accuracy of IPS and PISH methods is studied by using rolling motion responses of the ship model. Then, the two methods based on the Hilbert transform are compared with the traditional method for parameter identification of the correlation model.

### 3.1 *Identification of hydrodynamic coefficients of nonlinear rolling based on Hilbert transform*

The differential equation of free roll was obtained by setting roll moment $m(t)$ to 0 and then numerically resolved by using the fourth order Runge-Kutta method. The initial conditions of the model are $\varphi_0 = 0.5rad$ and $\dot{\varphi}_0 = 0rad/s$, respectively. Free vibration signals and phase planes are shown in Figures 1~2.

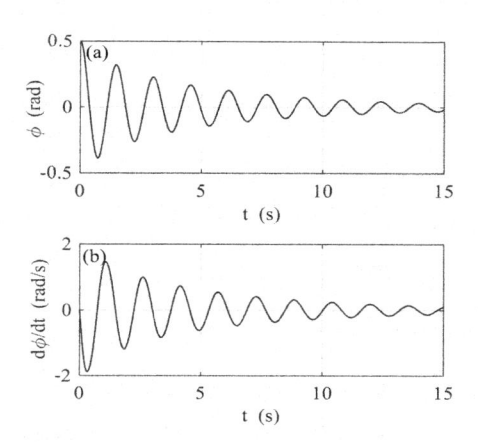

Figure 1. Free vibration signals of model I

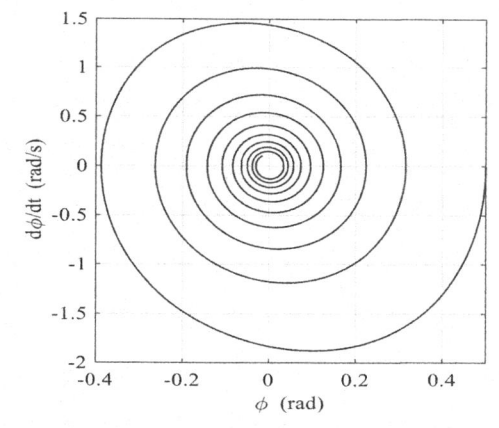

Figure 2. Phase planes of model I.

First, we use the current non-parametric method to identify roll motion. The envelope and instantaneous frequency are obtained for free vibration signals according to equations 5~8. In the initial stage of free roll, the envelope is not a smooth curve, but a non-stationary time series containing fast oscillation and modulation. Then, the instantaneous undamped modal frequency and modal damping coefficient are estimated according to Equations 9 and 10. From these estimated instantaneous parameters, the rolling motion of model I was identified using the IPS method, and the instantaneous characteristics of oscillation were smoothed by the Savitsky-Golay filtering process.

According to Equations 11~12, the nonlinear recovery moment and damping moment are determined, and the results are shown in Figure 3. It can be found that the identified recovery moment and damping moment curves are very close to the actual values, especially for the damping moment.

By comparing the estimated value with the measured value, it can be observed that with the gradual increase of roll angle, the calculated results of the recovery moment have a large deviation.

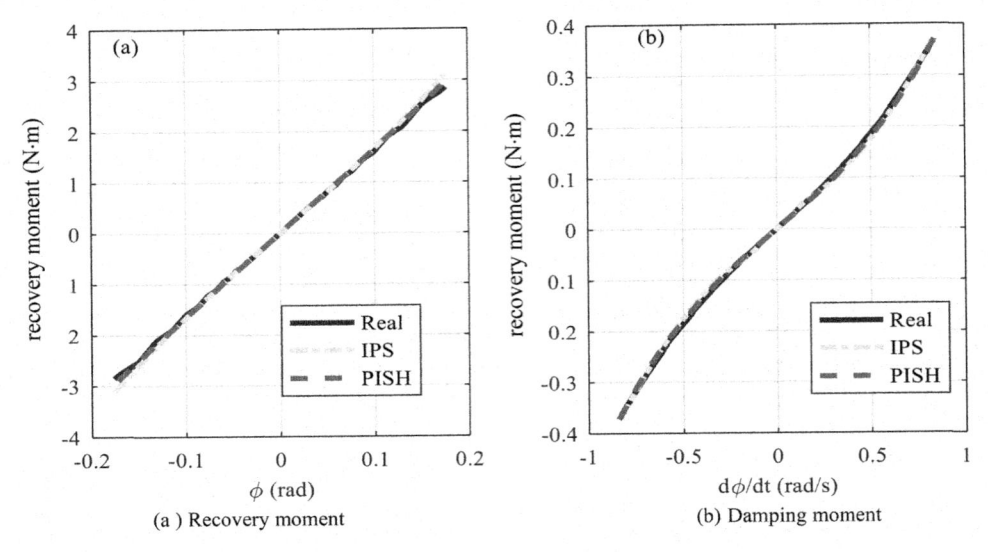

(a ) Recovery moment            (b) Damping moment

Figure 3.   Hydrodynamic coefficient of model I.

### 3.2 *Comparisons of results*

Damping moment coefficient and roll period are the two most important parameters to describe rolling motion. The energy method is one of the most commonly used traditional methods to predict roll damping using free attenuation data. In this method, the recovery torque is assumed to be linear. The dependence on the damping moment model is a major shortcoming of this method.

The recovery moment and damping moment coefficients of rolling motion can be obtained by least square fitting. As shown in Table 1, the identified linear coefficients are close to the actual values, and the errors are all within 15%. Compared with the nonlinear coefficient identification results, PISH results are more accurate, and the identification errors are less than 15%.

Table 1.   Comparisons of results obtained from different methods.

| Model NO. | Parameter | Actual value | Estimated value | | | Error/% | | |
|---|---|---|---|---|---|---|---|---|
| | | | Energy method | IPS | PISH | Energy method | IPS | PISH |
| I | $c_1$ | 0.32 | 0.3215 | 0.3537 | 0.3125 | 0.47 | 10.53 | 2.34 |
| | $c_2$ | 0.16 | 0.1376 | 0.1171 | 0.1820 | −14 | 26.81 | 13.75 |
| | $k_1$ | 16.00 | – | 16.1005 | 16.2971 | – | 0.63 | 1.86 |
| | $k_3$ | 19.20 | – | 50.6202 | 16.3576 | – | 163.65 | 14.80 |

## 4   CONCLUSIONS

This paper studies the identification method based on first order solution (IPS) and the identification method considering higher order superharmonics (PISH). It is compared with the nonlinear roll coefficient identification method based on the traditional method. According to the research results, the following conclusions are drawn:

Two nonlinear system identification methods based on the Hilbert transform can simultaneously estimate the damping moment and recovery moment of ship rolling motion without any prior information about the recovery moment and the damping moment model. In the identification process, only the free roll attenuation data are used, and no special tests are required to determine the transverse metacenter height and natural frequency. In addition, the identification accuracy of nonlinear coefficients is improved because the AM (amplitude modulation) and PM (phase modulation) characteristics of nonlinear rolling motion are considered in the PISH algorithm.

The method can be further applied to the hull structure in irregular waves by combining the Random Decampel Technique (RDT) with the Hilbert transform method. The ship response in irregular waves is converted into a random decrement signal by the random decrement technique. Then, from these random decrement signals, the nonlinear recovery and damping moments are identified using the Hilbert transform recognition method.

## ACKNOWLEDGMENTS

The work reported in this paper is supported by the Open Project Program of Shandong Marine Aerospace Equipment Technological Innovation Center, Ludong University (Grant No. MAETIC2021-07).

## REFERENCES

Feldman, M. (1994). Nonlinear system vibration analysis using Hilbert transform – I. Free vibration analysis method 'FREEVIB.' *Mech. Syst. Signal Process.* 8(3), 309–318.

Himeno, Y. (1981). Prediction of ship roll damping state of the art. The University of Michigan.

Hou, X. R. & Zou, Z. J. (2015). SVR-based identification of nonlinear roll motion equation for FPSOs in regular waves. *Ocean Eng.* 109, 531–538.

Hou, X. R. & Zou, Z. J. (2016). Parameter identification of nonlinear roll motion equation for floating structures in irregular waves. *Appl. Ocean Res.* 55, 66–75.

Irkal, M.A.R., Nallayarasu, S., Bhattacharyya, S.K. (2019). Numerical prediction of roll damping of ships with and without bilge keel. *Ocean Eng.* 179, 226–245.

Kawahara, Y., Maekawa, K., Ikeda, Y. (2012). A simple prediction equation of roll damping of conventional cargo ships on the basis of Ikeda's method and its limitation. Shipp. *Ocean Eng.* 2 (4), 201.

Kim Y, Park M J. (2015) Identification of the nonlinear roll damping and restoring moment of an FPSO using Hilbert transform. *Ocean Eng.* 109, 381–388.

Kim, M., Jung, K.H., Park, S.B., Lee, G.N., Duong, T.T., Suh, S. B., Park, I. R. (2020). Experimental and numerical estimation on roll damping and pressure on a 2-D rectangular structure in free roll decay test. *Ocean Eng.* 196, 106801.

Lewandowski E M. (2011) *Comparison of some analysis methods for ship roll decay data.* In: Proceedings of 12th International Ship Stability Workshop. Washington D.C., USA, June, 326–330.

Somayajula, A. & Falzarano, J. (2016). Application of advanced system identification technique to extract roll damping from model tests in order to accurately predict roll motions. *Appl. Ocean Res.* 67, 125–135.

Civil Engineering and Urban Research – Mohamed & Hou (Eds)
© 2023 the Authors, ISBN 978-1-032-44487-1

# Risk factors identification in old residential areas renovation PPP projects based on hierarchical holographic modeling

Bing Zhao* & Shengyue Hao*

*School of Economics and Management, Beijing Jiaotong University, Beijing, China*

ABSTRACT: This paper constructs an HHM-based framework for PPP projects of old residential area renovation. Based on the economic subsystem, social subsystem, and environmental subsystem, a risk identification model is constructed from the perspective of each stage of the PPP project, and then a list of risks based on sustainability is obtained. The conclusion is that risks of old residential areas renovation PPP projects are concentrated in the project implementation stage, followed by the project identification and preparation stages. The study lays a solid foundation for further assessing and controlling the management risks.

## 1 INSTRUCTIONS

Currently, the urbanization process of China has gradually shifted from the stage of high growth and rapid expansion to the stage of stock quality improvement and renovation, which is led by the enhancement of urban quality. The renewal of old residential areas has become one of the urgent livelihood issues that government needs to solve (Wang & Liu 2020). The government has issued a clear requirement to promote the transformation of old residential areas comprehensively. Extensive social resources and private sectors are encouraged to participate in the transformation projects. Therefore, public-private partnership is considered to be adopted in this field.

However, PPP projects have features of multiple participating parties, long operation periods, and high sensitivity to the environment. And the contract terms of PPP are naturally incomplete contractual, leading to actual risks during the operation (Xiang & Song 2016). While renovating old urban residential areas has become an important national livelihood project, it's important to comprehensively and accurately identify the risk factors. Therefore, the research will discuss the risk identification of the PPP renovation projects of old residential areas based on Hierarchical Holographic Modeling (HHM) from the perspective of the whole life cycle of the project. We hope to contribute to this field.

## 2 THE FEATURES OF OLD RESIDENTIAL AREAS RENOVATION PPP PROJECTS

### 2.1 Complexity

Complexity is one of the features that must be mentioned. The public-private partnership in the renovation projects of old residential areas has the following characteristics:

(1) The complexity of renovation projects: Normally, the building density and population density of old residents and the staff management of the renovation projects are intricate. Besides, each residential area has different transformation needs, which also increases the complexity of the project.

---

*Corresponding Authors: bingzhao@bjtu.edu.cn and haoshyue@bjtu.edu.cn

 DOI 10.1201/9781003372417-99

(2) The complexity of society: The renovation projects of old residential areas are livelihood projects that greatly impact society. The project execution must conform to residents' intentions and be supervised by residents. Not only the internal community, but also the surrounding environment must be coordinated, such as the traffic. On all accounts, the renovation should consider all public opinions (Xiao & Chen 2019).

(3) The complexity of participants: Residents are important participants in the PPP renovation projects. The public authority and the private sectors are also essential. The research on the complexity of the public and private sectors is still enriching in recent years.

## 2.2 *The categories of PPP renovation projects*

According to the construction scope of existing PPP projects of the old residential areas renovation based on the open project database of CPPPC, this thesis summarizes PPP renovation projects into four categories.

(1) Residential buildings renovation: Repair and energy-saving renovation of residential buildings, e.g., reconstruction of electricity, water, and heating pipeline, erection of external elevator and security equipment, and resolution to the aging problem of the residential building structure.

(2) Supporting facilities upgrading: Upgrading water, electricity, gas, communication, roads, and other infrastructure, service facilities, and communal areas within and directly around the district.

(3) Idle real estate revitalization: Renewal of unused estate premises and areas, e.g., parking garages (courts), cultural and sports facilities, etc.

(4) Community service introduction: Perfection of school, health care, housing keeping, and commercial facilities in community endowment, babysitting, and compulsory education.

## 3 MODEL CONSTRUCTION

### 3.1 *HHM-framework constructing*

This paper analyzes the PPP project of old residential areas renovation from five perspectives: project stakeholders (participants), PPP project lifecycle, risk-loss types, project types, and project management.Figure 1 shows the HHM-based risk identification framework for PPP projects in the old residential area renovation.

(1) Project Stakeholder Dimension: The successful implementation and operation of the PPP project cannot be achieved without the assistance and joint efforts of the government, the social capital, the project company(SPV), and the property owners of the community (Yang & Wang 2018). This paper considers the residents as an important part of the stakeholders, which helps to comprehensively consider the risks brought by the residents' behavior to the project.

(2) PPP project lifecycle dimension: The whole life cycle of a PPP project is divided into project identification, project preparation, project procurement, and project implementation. Each of these four phases includes various tasks. Smooth execution of each project stage will lay the foundation for the implementation and operation of the entire PPP project.

(3) Risk loss dimension: The types of risk losses include economic, social, and environmental losses. Project operation does require not only the monitoring of economic losses, but also the consideration of social and environmental losses.

(4) Project type dimension: In the above analysis, this paper classifies the projects into different types based on the focus of the old residential areas renovation projects, which are Residential buildings renovation, supporting facilities upgrading, idle real estate revitalization, and community service introduction.

(5) Project management dimension: The successful implementation and operation of any construction project need a systematic project management system, which requires good schedule control, cost control, quality control, optimized contract management, safety management, and information management.

### 3.2 *Risk identification model of the PPP project of old residential areas renovation*

Based on the HHM structure framework, this paper shows the intrinsic qualities and essence of the old district renovation PPP project from different levels and dimensions. On this basis, through iteration, a system is divided into multiple subsystems, each of which can be decomposed again, and then come to get all the possible risks, dissecting the project's risks layer by layer.

Based on the types of losses in the HHM structure, including economic losses, social losses, and environmental losses, risk factors are identified from the perspective of the whole project life cycle. The risk is studied from three major subsystems, namely the economic, environmental, and social subsystems, to reflect their economic, environmental, and social losses.

Therefore, we will identify possible risks using the three subsystems of loss types and the whole life cycle of a PPP project, based on the "whole life cycle - loss types" sub-framework as a risk identification model. Figure 2 shows the risk identification model for PPP projects renovating old residential areas.

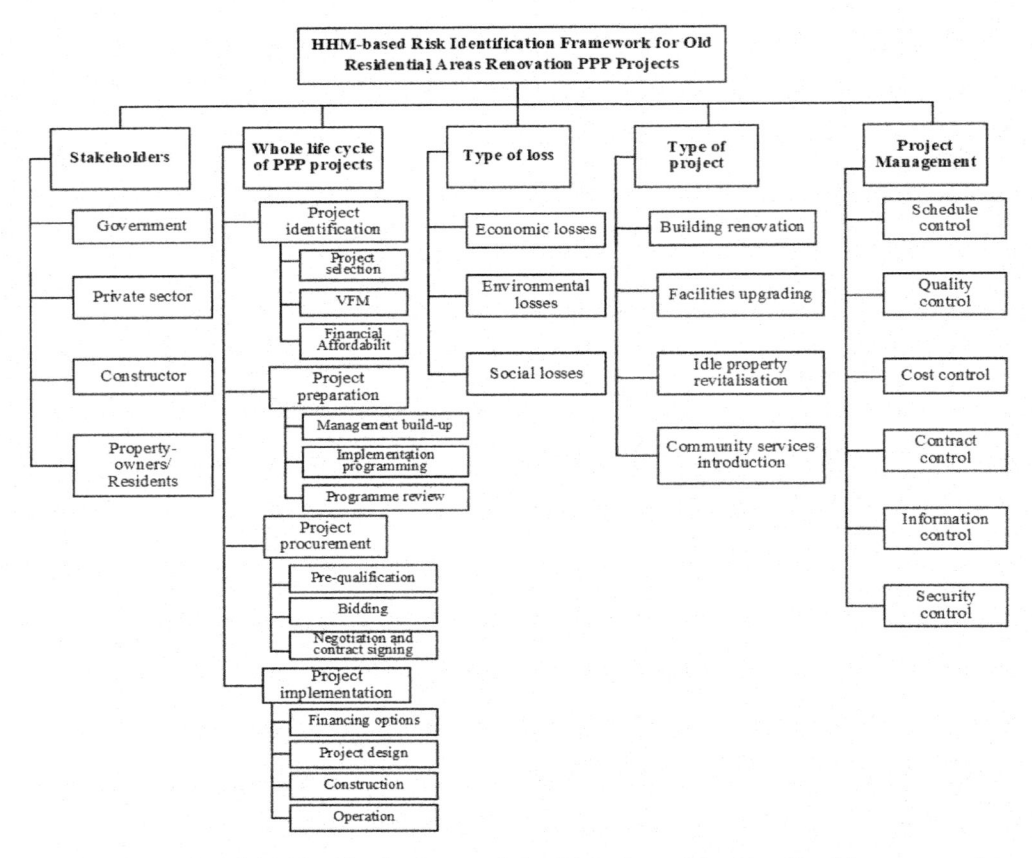

Figure 1. HHM-based risk identification framework for PPP projects of the old residential areas renovation.

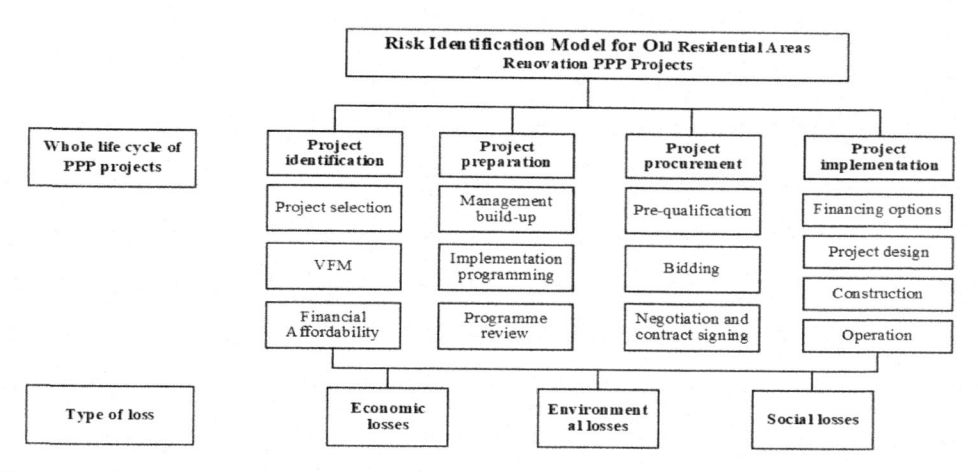

Figure 2. Risk identification model for PPP projects of the old residential areas renovation.

## 4 RISK IDENTIFICATION

Based on the analysis above, the next part of this paper will use the risk identification model that has been constructed based on economic, social, and environmental subsystems to identify the risks that exist in the various stages of the PPP project of old residential areas renovation.

### 4.1 Risk identification in the economic subsystem

The risk identification of the economic subsystem of the PPP project of old residential areas renovation is for the risks that exist in the project itself and in the whole life cycle of the PPP project that may affect the project costs and objectives. The execution phase of a PPP project covers the design, construction, and operation stages of a general project and is the most important part of the PPP project cycle, with more risks than other stages. Specific risk factors and sources of risk are shown in Table 1.

Table 1. Risk factors and sources of economic subsystem.

| Stage | Risk Factors | Risk Sources |
| --- | --- | --- |
| Identifying | Macroeconomic risks | Domestic and international financial market fluctuations, inflation. |
| | Policy risks | Industrial structure adjustment, economic policy numbering, and imperfect laws and regulations. |
| | Project selection risks | Insufficient investigation of the project's investment estimates, financial returns, social benefits, and uncertainty analysis. |
| Preparing | PPP model selection risks | Subjective preferences and information errors lead to the wrong choice of the PPP model. |
| | Programming risks | Inadequate preparation of project brief and design development, division of roles and responsibilities between the parties, project planning, risk sharing, mode of operation, transaction structure, contract system management, and supervision structure. |
| | Delays in program review | Cumbersome approval process and inadequate prep. |

*(continued)*

Table 1. Continued.

| Stage | Risk Factors | Risk Sources |
|---|---|---|
| Procuring | Qualification risks | Bidder qualification risks. |
| | Contract terms define risks | Deviations in the understanding of contract terms between the purchasing parties. |
| | Risks of contract changes | Contract terms need to be changed. |
| Implementing | Financing Risks | Unreasonable financing structure, difficulty in raising funds, poor credit of financiers. |
| | Survey and design risks | Inaccurate geological and hydrological information and limited survey standards; lack of safety and reasonableness in project program design. |
| | Schedule risks | Force majeure, poor construction organization design, and delay in equipment procurement lead to construction delays. |
| | Cost risks | Construction cost overruns due to inflation, rising labor, design changes, etc.; operating cost overruns due to unsound operating model and low operating efficiency. |
| | Quality risks | Project quality does not meet the requirements and standards. |
| | Security Risks | Improper operation of site construction personnel and inadequate safety in the use of machinery and equipment. |
| | Operating income risks | The unstable consumer market, competition from similar products, a downward adjustment of unit price charged, lower than expected consumer demand. |
| | Operating Management Risks | Inadequate capacity and mismanagement of operation managers. |
| | Residual risks | Salvage value from leftover materials and equipment. |

### 4.2 Risk identification for environmental subsystem

The risk identification of the environmental subsystem mainly considers the impact of the entire renovation project on the old district's ecological and living environment and surroundings (Cheng & Liu 2019). The identification and preparation stages of the renovation PPP project should make the basic planning of the whole project according to the renovation needs, which are often inseparable as far as the impact on the environment is concerned, as shown in Table 2.

Table 2. Risk factors and sources of environmental subsystem.

| Stage | Risk Factors | Risk Sources |
|---|---|---|
| Identifying and preparing | Risks of disrupting soil and water balance | An incomplete collection of water and soil information on old residential areas. |
| | Risks of harmful pollution | Underestimate the harmful pollution generated by the project to the neighborhood. |
| | Risks of urban landscape destruction | The project may cause damage to the appearance of the old neighborhood. |
| Procuring | Risks of garbage pollution | Tender documents and other office paper are discarded at will, causing waste pollution. |

*(continued)*

Table 2. Continued.

| Stage | Risk Factors | Risk Sources |
|---|---|---|
| Implementing | Risks of land resource destruction | Damage to nearby land resources from the laying of water, electricity, and communication equipment. |
| | Temporary facility impact risks | The adverse impact of large temporary facilities on land resources, traffic, and residents' life in the area near the project. |
| | Risks of waste contamination | Risks from solid, liquid waste. |
| | Geological hazard risks | Causes ground collapse, slope instability, and other disasters. |
| | Risk of greenery damage | The project harms the greenery and ecology of the old residential area. |
| | Risks of light and groundwater contamination | Construction causes light pollution and groundwater contamination in older neighborhoods. |
| | Noise and dust pollution risks | Construction generates a lot of noise and dust pollution. |
| | Risks of landscape change | The original landscape of the old residential area was changed by the renovation. |
| | Traffic congestion risks | Project construction and operation adversely affect the surrounding traffic. |

## 4.3 Risk identification for social subsystems

As a people's livelihood project, the social benefits and social impacts of the old district renovation project are important measurement factors for the success of the project. Therefore, the risk identification of the social subsystem mainly focuses on the social benefits and impacts, as shown in Table 3.

Table 3. Risk factors and sources of social subsystem.

| Stage | Risk Factors | Risk Sources |
|---|---|---|
| Identifying | Social Development Risks | Unfavorable factors for regional social development. |
| | Inheritance risks | Lack of inheritance of regional history. |
| | Harmonious risks | Disharmony with community culture and poor integration. |
| | Acceptance risks | Residents do not accept or oppose the renovation. |
| Preparing | Risks of legal imperfection | The law is not complete. |
| | Third-party personnel risks | Third-party companies own personnel risk. |
| | Liability allocation risks | No clear division of public and private roles and responsibilities. |
| Procuring | Risks of bidding irregularities | Non-transparent procedures, unfair competition, and improper dealings. |

(continued)

Table 3. Continued.

| Stage | Risk Factors | Risk Sources |
| --- | --- | --- |
| Implementing | Security risks for the population | The construction poses a threat to the safety of the residents. |
| | Risks of the destruction of culture, monuments | Risk of damage to old residential areas and surrounding monuments. |
| | The risks of disrupting the normal life of residents | Disrupting the normal life of the residents. |
| | Risks of temporary possession of public resources | The construction has caused the temporary loss of public resources for the people living in the old neighborhood. |
| | Risks of information leakage | Leakage of residents' information during project operation. |
| | Health and education risks for the residents | Negative impact on the health and education of local residents. |
| | Completion Risks | Project completion does not meet social expectations. |
| | Risks of cooperation | Rational definition of operational responsibilities and roles of public and private institutions. |
| | Expropriation risks | Poor government credibility and changing policies. |
| | Risks of damage to existing facilities | The construction process caused damage to the existing facilities in the old. |

## 5 CONCLUSION

From the list of risks, we obtained a total of 51 risk factors and their corresponding sources of risk. It can be seen that the risks of PPP projects of old residential areas renovation are concentrated in the project implementation stage, followed by the project identification and preparation stages. Therefore, in terms of risk management, emphasis should be placed on risk control in these stages.

The priority is to do a good job of project initiation and identification according to the policy requirements and project characteristics of urban old neighborhood renovation, value for money and financial commitment evaluation according to the project characteristics of urban old neighborhood renovation, fully demonstrate the necessity and feasibility of the public-private partnership, and clarify the allocation of rights and responsibilities and risk sharing mechanism between the public and private.

In addition, during the implementation of the project, the preparation of the implementation plan, especially the construction plan, should fully reflect the characteristics of the urban old district renovation. The formulation of a good implementation plan is the key to guaranteeing the implementation of the project and can reduce the occurrence of risks for all parties. It is also important to consider the impact on residents' lives and the surrounding environment.

## REFERENCES

Cai J. G. & Sai Y. X. (2014). An Analysis of Influencing of Risk Factors of Risk Factors in Shantytowns Renovation Project based on Interpretative Structural Modeling. *Science and Technology Management Research*. 34(06):240–244.

Cheng M. (2019). Risk Analysis of PPP Projects of NIMBY Facility Based on Grounded Theory. *Construction Economy*. 40(10):61–65.

Jiang L. (2021). The Mode and Experience of Promoting Social Capital to Participate in the Renovation of Old Residential Communities in Cities from the Perspective of Transaction Cost. *Price: Theory & Practice*. (06):17–22.

Liu G. W. (2018). Research on the Relationship between the Risk Factors of Urban Renewal Projects Based on ISM. *Construction Economy*. 39(02), 89.

Ma L. Q. (2020). Study on the Risk Assessment of Decision-making in the Early Stage of Urban Shantytown Reconstruction Project. *Journal of Shandong Technology and Business University*. 34(06), 56–65.

Qi X. (2009). Analysis on Critical Risk Factors Causing the Failures of China's PPP Projects. China Soft Science. (05):107–113.

Shao R.W. & Chen Q. S. (2019). Risk assessment on social exclusion in urban renewal. *Urban Problems*. 2019, (07), 77–85.

Wang Z. B. (2022). Study on the Way and Countermeasure of Upgrading and Reconstruction of Old Residential Area in China. *Urban Development Studies*. 27(07): 26–32.

Xiang P. C. & Chang W. (2015). Risk Factors Identification in Interregional Major Engineering Projects Based on HHM. *World Sci-Tech R & D*. 37(01), 67–72.

Xiang P. C. & Song X. P. (2016). The Financing Risk Evaluation of Urban Infrastructure Projects Based on PPP Mode. *Journal of Engineering Management*. 30(01):60–65.

Xiao Y. (2019). Research on the Implementation Effect Evaluation of old Community Upgrading and Renovation. *Construction Economy*. 40(01), 102–106.

Yang Z. (2018). Research on Risk Identification of Infrastructure PPP Project Based on HHM. *Construction Economy*. 39(03): 39–43.

Zhou Z. H. (2021). Research on risk identification model of agricultural infrastructure PPP project based on HHM-DEMATELISM. *Project Management Technology*. 19(09), 13–18.

*Civil Engineering and Urban Research – Mohamed & Hou (Eds)*
*© 2023 the Authors, ISBN 978-1-032-44487-1*

# Research on deformation monitoring data processing of water conservancy and hydropower projects based on comprehensive analysis model of wavelet theory

Huacheng Yang*

*College of Architectural Engineering, Yunnan University of Business Management, Kunming, Yunnan Province, China*

ABSTRACT:   The purpose of water conservancy and hydropower project deformation monitoring data processing is to monitor the long-term operation safety of water conservancy and hydropower projects, continuously feedback on the construction process, improve the construction level, and ensure the safe construction of water conservancy and hydropower projects. It can be seen that the deformation monitoring of water conservancy and hydropower projects plays an important role in ensuring the construction safety of water conservancy and hydropower projects. Based on this, this paper briefly expounds on the basic concepts of wavelet theory and then combines the comprehensive analysis model to analyze and discuss the practical application of water conservancy and hydropower engineering deformation monitoring data processing based on wavelet theory, aiming to provide some useful references and improve the water conservancy and hydropower engineering deformation monitoring data processing.

## 1 INTRODUCTION

The deformation monitoring of water conservancy and hydropower projects mainly refers to the periodic or continuous monitoring, observation, and analysis of the deformation phenomenon of water conservancy and hydropower projects using special instruments and methods (Cui 2020). Deformation monitoring data processing of water conservancy and hydropower projects is help-ful for the timely discovery of potential safety hazards in the construction process and plays an important role in improving the construction safety management level of water conservancy and hydropower projects. Ensuring construction safety is important in the construction process of water conservancy and hydropower projects. Strengthening the deformation monitoring of water conservancy and hydropower projects and improving the data processing level of deformation mon-itoring are important means to ensure the construction safety of water conservancy and hydropower projects. Therefore, it is necessary to apply wavelet theory to explore the practical application of water conservancy and hydropower engineering deformation monitoring data processing based on wavelet theory, which has positive practical significance for improving water conservancy and hydropower engineering deformation monitoring data processing.

## 2 EXPLANATION OF WAVELET THEORY

### 2.1 *Overview of wavelet analysis*

Wavelet analysis is a milestone in the development history of Fourier analysis (Liu et al. 2020). In recent years, it has become a hot spot of common concern in many disciplines in the world (Liu & Zhou 2020). The development process of the wavelet transform can be roughly divided into

---

*Corresponding Author: 1803683519@qq. com

DOI 10.1201/9781003372417-100

three stages: first, the budding and isolated application period of wavelet analysis ideas. The main feature of this stage is the application of some specially constructed wavelets to specific problems in some scientific research fields. For example, French geophysicists J. Morlet and A. Grossman used wavelets to analyze seismic data for the first time and proposed the concept of wavelet analysis. Second, wavelet research boom and unified tectonic period. Many foreign scholars have carried out in-depth research on "wavelet." For example, in 1986, Y. Meyer successfully constructed the first real wavelet basis when he doubted the existence of a wavelet basis. Later, P. Lemarie and GBattle also independently constructed smooth wavelets with exponential decay, and the function system generated by their scaling and translation constitutes the standard orthonormal basis of L2(R). S. Mallat and Y. Meyer proposed the multi-resolution analysis (MRA) theory, which unified various specific wavelet construction methods proposed before. At the same time, S. Mallat also gave a numerical algorithm of discrete wavelets based on multi-resolution analysis, namely the Mallat tower algorithm. Third, the full application period. Since 1992, wavelet analysis has entered the stage of comprehensive application.

The advantage of wavelet analysis over Fourier transform is that it has good localization properties in both the time domain and frequency domain (Lin & Tang 2018) and can focus on any details of the object due to the use of a gradually finer time domain or spatial domain sampling steps for high-frequency components, which poses a new challenge to the traditional Fourier analysis method. It can extract a lot of useful information from the signal and is the processing framework of various signal processing methods (such as the time-frequency analysis method, multi-scale analysis, and sub-band coding). The multi-scale refinement analysis of the signal is carried out through the operation functions such as scaling and translation (Tong et al. 2020). The wavelet analysis developed in recent years is a powerful tool for analyzing non-stationary signals. It provides a new technology for the field of signal processing and promotes the Signal processing has entered a new historical development period.

### 2.2 *Wavelet analysis algorithm*

Wavelet analysis is a time-frequency localized analysis method with a fixed window size but a variable shape and variable time window and frequency window (Tian 2019). The core content of wavelet analysis is wavelet transform. There are many wavelet transform algorithms, each with its characteristics. In this paper, combined with deformation monitoring and analysis, the Mallat algorithm of wavelet transform is mainly used. Generally, the Mallat algorithm performs wavelet transform on the original data, which can quickly and concisely perform wavelet transform and inverse transform on the original sequence. Let the original sequence be f(k) (k = 0, 1, 2, ..., N-1). According to the wavelet decomposition method, there are:

$$\begin{cases} C_{j,k} = \sum_{n \in Z} c_{j-1} h_{n-2k} \\ d_{j,k} = \sum_{n \in Z} c_{j-1} g_{n-2k} \end{cases} \tag{1}$$

In formula (1), h(n) and g(n) are the filter coefficients of the low-pass filter H and the high-pass filter G, respectively, and j is the number of decomposition layers. On this basis, construct the reconstruction algorithm:

$$C_{j-1,n} = \sum_{n \in Z} \left( C_{j,k} h_{n-2k} + d_{j,k} g_{n-2k} \right) \tag{2}$$

## 3   CONSTRUCTION OF A COMPREHENSIVE ANALYSIS MODEL FOR THE EVALUATION OF DATA PROCESSING ACCURACY FOR DEFORMATION MONITORING OF WATER CONSERVANCY AND HYDROPOWER PROJECTS

Using the comprehensive analysis model, the processing accuracy of the deformation monitoring data of the water conservancy and hydropower project is evaluated (Wang et al. 2018).

First, establish a water conservancy and hydropower project deformation monitoring data processing index system, including data indicators such as horizontal displacement, vertical position, inclination, and crack observation.

Second, use AHP and entropy value method to determine the combined weight of index factors. The subjective weight $W_1$ (j) of the index is calculated by the analytic hierarchy process and the objective weight $W_2$ (j) of the index is calculated by the entropy method. The subjective weight and guest official weight of the comprehensive indicators can be combined to obtain the combined weight, W (j) (j = 1 - n). According to the principle of minimum relative information entropy, we get:

$$minF = \sum_{j=1}^{n} w(j)[\ln w(j) - w_1(j)] + \sum_{j=1}^{n} w(j)[\ln w(j) - w_2(j)] \quad (3)$$

Combined with formula (3), the optimization is obtained by using the Lagrange multiplier method:

$$w(j) = \frac{[w_1(j)w_2(j)]^{0.5}}{\sum_{j-1}^{n} [w_1(j)w_2(j)]^{0.5}} \quad (4)$$

After calculating the combined weight of each index, the combined weight of the index and the data processing accuracy index of the next sub-region affected by a single index are combined to obtain the overall accuracy. The comprehensive analysis model for the evaluation of the processing accuracy of the deformation monitoring data of water conservancy and hydropower projects is as follows:

$$Z_i = \sum_{j=1}^{5} w(j) \times F_{ij} (i = 1, 2, \ldots, n) \quad (5)$$

In formula (5), it represents the overall accuracy of water conservancy and hydropower engineering deformation monitoring data processing. The larger the index, the higher the accuracy of water conservancy and hydropower engineering deformation monitoring data processing. The accuracy evaluation criteria are 0.7–1 for high accuracy, 0.4–0.7 means general accuracy, and less than 0.4 means low precision.

## 4 COMPARISON AND ANALYSIS OF WATER CONSERVANCY AND HYDROPOWER ENGINEERING DEFORMATION MONITORING DATA PROCESSING ACCURACY EVALUATION BASED ON COMPREHENSIVE ANALYSIS MODEL OF WAVELET THEORY

A combined model is constructed based on the comprehensive analysis model, combined with the wavelet theory (Xie 2018). Here, we analyze and compare the residual calculation results of different models based on the horizontal radial displacement observation value of a horizontal deformation monitoring point set up in a water conservancy and hydropower project in Phase 4. The steps are as follows: First, perform wavelet denoising data preprocessing on the original data sequence (the denoising method mainly uses the improved threshold method and the translation invariant method for denoising), and obtain the denoised data sequence. Second, the processed data is judged to see if it meets the conditions for combined modeling, and the data sequences that do not meet the requirements need to be processed by data sequence transformation (Zheng et al. 2020). Third, the combined modeling of the data is used to obtain the observation data; fourth, the accuracy of the observation data is evaluated. Fifth, use the comprehensive analysis model to obtain the accuracy evaluation results and compare and analyze the results obtained by the combined model. The details are shown in Table 1 and Table 2.

Table 1. Comparison of residuals of different models.

| Observation period | Comprehensive analysis model | Relative error (%) | Combination model | Relative error (%) |
|---|---|---|---|---|
| 1 | −0.4035 | 0.45 | −0.344 | 0.38 |
| 2 | −0.6704 | 0.75 | −0.384 | 0.43 |
| 3 | −0.6856 | 0.76 | −0.4245 | 0.48 |
| 4 | −0.6392 | 0.72 | −0.4276 | 0.48 |

Table 2. Comparison of the accuracy evaluation of observed data.

| Observation period | Comprehensive analysis model | Combination model |
|---|---|---|
| 1 | 0.64 | 0.72 |
| 2 | 0.55 | 0.68 |
| 3 | 0.56 | 0.65 |
| 4 | 0.51 | 0.58 |

Combining the data in Tables 1 and 2, it can be seen that the observed data under the comprehensive analysis model and the combined model are different. Although the comprehensive analysis model can predict the settlement (Zhao 2020), the residual error is very large. The absolute value of the residual error in the last three periods is larger than 6mm. The combined model predicts the residual error to a certain extent compared with the comprehensive analysis model, and the prediction accuracy is controlled within 5 mm. Moreover, the accuracy of the deformation monitoring data of water conservancy and hydropower projects under the combined model is improved compared with the comprehensive analysis model(Zhang 2019). This shows that under the wavelet theory, using the comprehensive analysis model for modeling can improve the accuracy of the model. In this way, more accurate deformation monitoring data of water conservancy and hydropower projects can be obtained to grasp the construction situation of water conservancy and hydropower projects in time, avoid potential risk factors in time, and ensure the safe construction of water conservancy and hydropower projects.

## 5 CONCLUSION

Compared with the traditional analysis methods, the wavelet theory and the comprehensive analysis model are organically combined to construct a combined model, which is applied to the deformation monitoring data processing of water conservancy and hydropower projects, which can improve the processing accuracy of deformation monitoring data and ensure the water conservancy and hydropower engineering. Safe operation of hydropower projects. Therefore, in the construction process of water conservancy and hydropower projects, in order to strengthen safety management, relevant technical personnel can reasonably apply the comprehensive analysis model to practice on the basis of wavelet theory, improving the accuracy of monitoring data processing results, and ensuring construction safety. Due to the limitation of the article space, the research on the deformation monitoring data processing of water conservancy and hydropower projects based on the comprehensive analysis model based on wavelet theory is not in-depth and perfect. In the future stage, we should continue to pay attention to the comprehensive analysis model based on wavelet theory. The relevant research trends, continuous stages, and learning, enrich the research experience to make up for the lack of research in this paper.

## REFERENCES

Cui Chengeng. Application of wavelet analysis in GPS dynamic monitoring data processing [J]. *Gansu Science Journal*, 2020, 32(02):63–68.

Lin Lisen, Tang Linbo. Application research of comprehensive analysis model based on multi-source hydrogeological information [J]. *Jiangxi Water Conservancy Science and Technology*, 2018, 44(06):456–458+464.

Liu Qianju, Chen Daiming, Chen Shaoyong, Li Yuanjiu. Application of wavelet theory in the detection of gross errors in dam safety monitoring data [J]. *Northwest Hydropower*, 2020, (S1): 129–132.

Liu Zhikun, Zhou Lanting. PCA-IPSO-SVM prediction model of concrete dam deformation based on wavelet theory [J]. *China Rural Water Resources and Hydropower*, 2020, (07):185–189+195.

Tian Maozhen. A brief discussion on the application of multiple linear regression in data processing of dam deformation monitoring [J]. *Small and Medium Enterprises Management and Science* (Mid-School), 2019, (10): 184–185.

Tong Shanshan, Sun Lijuan, Zeng Wen. Discussion on practical teaching of enterprise decision management based on two types of comprehensive analysis models [J]. *Science and Technology Vision*, 2020, (10): 59–60.

Wang Yingjie, Cao Tienan, Xia Chengwen, Pan Rui. Research on data processing countermeasures for power transformer condition monitoring [J]. *Electronic Testing*, 2018, (23):116–117.

Xie Run. Research on data processing and management technology of subway construction safety monitoring [J]. *Jiangxi Building Materials*, 2018, (12):30–31.

Zhang Jiaying. Application of Excel in nonferrous metal construction monitoring data processing [J]. *World Nonferrous Metals*, 2019, (20):202–203.

Zhao Qiang. Dynamic monitoring data processing of large-span bridges based on wavelet analysis [J]. *Journal of Heilongjiang Institute of Engineering*, 2020, 34(02):38–42.

Zheng Qiuyuan, Fu Yun, Chen Dahua, Hu Jiawang. Research on Power System Fault Diagnosis Based on Wavelet Theory [J]. *Mechanical Design and Manufacturing Engineering*, 2020, 49(10):68–71.

*Civil Engineering and Urban Research – Mohamed & Hou (Eds)*
© 2023 the Authors, ISBN 978-1-032-44487-1

# Parameter calculation and structural design of wave-maker in Harbor Basin

Ping He*, Tao Han* & Chen Li*
*CCCC Tianjin Port Engineering Institute Co., Ltd., Key Laboratory of Coastal Engineering Hydrodynamic, CCCC, Tianjin, China*
*CCCC First Harbor Engineering Company Ltd., Tianjin, China*

ABSTRACT: In this article aiming at the current AC servo wave-maker, the working principle and transmission mode are described in detail. Moreover, based on the theory of potential function, the main design parameters of the wave-maker are calculated according to the ability of the proposed wave-maker. According to the parameters, the structure of the wave-maker is designed, and the specific specifications of the ball screw and servo motor are determined. The results show that wave height and period meet the design requirements and that the generated wave with a higher standard of uniformity and repeatability are superior to their counterparts generated according to generated industrial specifications.

## 1 INSTRUCTION

In the design stage of national marine engineering, marine national defense engineering, and military equipment construction, as well as the research stage of marine disaster prevention and treatment, it is necessary to conduct a physical model test and research to obtain the regular pattern between physical model and wave. Therefore, it is necessary to generate wave equipment—wave-maker manually. With the development of computer and control technology, irregular wave-maker manufacturing technology has been improved, perfected, and widely used in various research units. In the design of a wave-maker for a harbor basin, the moving speed and stroke of the wave pusher determine the capacity of the wave-maker, the maximum load force determines the structural stiffness of the wave-maker and the agility and rapidity of response, and the maximum power determines the maximum wave making capacity (Li 2020). The accuracy of these basic parameters relates to the performance and cost of the wave-maker. Based on the previous design and application experience of wave-makers, this article uses the theory of potential function to calculate the main performance parameters of wave-maker, further determine the mechanical structure and key component specifications of wave-maker, and ensure the performance of wave-maker to the greatest extent. The results are applied to the port basin wave-maker construction project of our unit.

## 2 PRELIMINARY SCHEME

### 2.1 *Performance of the wave-maker*

The average-period ($T_a$) waves in China's coastal maritime space are about 2.0s~7.0s, and long-period waves in some areas can reach 15.0s (Zhang 2011). In order to meet the needs of the long-period wave, assume that the model gravity is similar to the prototype gravity, and the scale factor($\lambda = 10$) is allowed, so the period of the wave made by the harbor wave-maker proposed by our company is 0.6s~4.74s. In addition, the width of the harbor wave-maker unit is 3.5m, the

---

*Corresponding Authors: ydheping@163.com, hantao@ccccltd.cn and lichen11@ccccltd.cn

DOI 10.1201/9781003372417-101

height of the wave pushing plate is 1.2m, the maximum water depth is 0.9m, and the maximum wave height is 0.4m.

## 2.2 *Determination of the wave-making type*

In the wave physical model research, the swing-plate wave maker and the push-plate wave maker have been widely used due to their simple wave-making principle, compact structure, and easy maintenance. Among them, the swing-plate wave maker is used for making waves in deep water, which only generates waves within a certain depth range of the water surface, and the wave action at the lower part of the water is extremely small. The push-plate wave maker uses the horizontal reciprocating motion of the push-plate to push the water to form a wave. Along the water depth direction, the velocity and displacement of the water body in front of the push-plate are the same, which is suitable for shallow water wave generation. Therefore, the push-plate wave maker is preferred as the design type for this project.

## 2.3 *Determination of the driving mode*

There are three main driving modes: servo motor-drive, hydraulic-drive, and pneumatic-drive, and the comparison of them is shown in Table 1. Compared with the other two drive modes, servo motor-drive has great advantages in control accuracy, rapidity, and control flexibility, and has the merit of convenient installation and debugging, no noise, no oil pollution, and so on. Therefore, this project will be driven by servo motors, and digital low-inertia AC servo motors will be used to drive the ball screw to rotate directly, and the rotation will be converted into the reciprocating linear motion of the push-plate through the screw nut pair, thereby pushing the body of water to create waves. The transmission principle is shown in Figure 1.

Table 1. Comparison of three driving modes.

| | Items | | | | | |
|---|---|---|---|---|---|---|
| Modes | Precision | Bearing capacity | Structure | Noise | Maintenance | Pollution |
| Motor-drive | High | Secondary | Secondary | Low | Easy | Little |
| Hydraulic-drive | Secondary | High | Complex | Low | Complex | Oil |
| Pneumatic-drive | Low | Low | Easy | Low | Easy | Little |

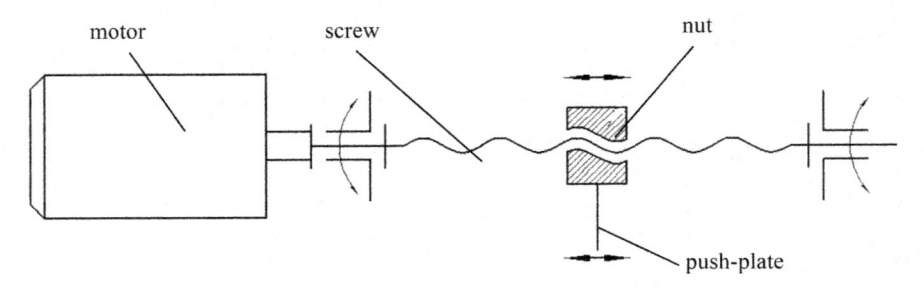

Figure 1. Transmission principle of the wave-maker.

## 3 PARAMETER CALCULATION AND STRUCTURAL DESIGN

### 3.1 *Parameter calculation*

The maximum stroke, maximum speed, unit maximum load force, and load power of the wave-maker are the parameters needed in the design of the wave-maker. Whether these parameters can be given accurately is the key to the performance and cost of the wave-maker. This paper will use the theory of potential function (Yu 2011) to calculate the above parameters.

To use the relationship between wave height and stroke to determine the maximum stroke of the push-plate, we must calculate the maximum wave height under various water depths and periodic conditions first, which can be calculated by Equation 1:

$$H_{max} = 0.142 L \tanh(kh) \tag{1}$$

where $h$ =water depth; $k$ =the number of waves; $L$ =wavelength, which can be calculated by Equation 2:

$$L = (gT^2/2\pi) \cdot \tanh(2\pi h/\lambda) \tag{2}$$

When calculating, $h$=0.3, 0.4, 0.5, 0.6, 0.7, 0.8, 0.9m; $T$ =0.5, 0.75, 1.0, 1.25, 1.5, 1.75, 2.0s. The calculation results are shown in Table 2.

Table 2. The calculation of maximum wave height.

| Wave period (s) | Water depth (m) | | | | | | |
|---|---|---|---|---|---|---|---|
| | 0.3 | 0.4 | 0.5 | 0.6 | 0.7 | 0.8 | 0.9 |
| 0.5 | 0.06 | 0.06 | 0.06 | 0.07 | 0.07 | 0.07 | 0.08 |
| 0.75 | 0.13 | 0.14 | 0.14 | 0.14 | 0.14 | 0.14 | 0.14 |
| 1.0 | 0.17 | 0.19 | 0.21 | 0.22 | 0.22 | 0.22 | 0.22 |
| 1.25 | 0.22 | 0.27 | 0.28 | 0.30 | 0.30 | 0.33 | 0.33 |
| 1.5 | | | | 0.36 | 0.39 | 0.42 | 0.44 |
| 1.75 | | | | | | 0.49 | 0.52 |
| 2.0 | | | | | | | 0.58 |

What is considered is the maximum wave height that the water depth can generate in front of the wave-maker. The actual test section is a certain distance from the wave-maker, and the water depth of the test section is relatively shallow. Considering the needs of the actual test, 0.4m is taken as the maximum wave height.

### 3.1.1 Maximum stroke

The relationship between wave height and itinerary is:

$$S_{max} = H_{max} \cdot \frac{sh^2 kh + 2kh}{4 sh^2 kh} \tag{3}$$

According to Table 2 and Equation (3), calculate the maximum stroke of the wave-maker under various cycles and water depths. The calculation results are shown in Table 3, combined with the specifications and models of the guide rail and the lead screw, which take 60cm as the maximum stroke of the wave-maker.

Table 3. The calculation of maximum stroke.

| Wave period (s) | Water depth (m) | | | | | | |
|---|---|---|---|---|---|---|---|
| | 0.3 | 0.4 | 0.5 | 0.6 | 0.7 | 0.8 | 0.9 |
| 0.5 | 0.03 | 0.03 | 0.03 | 0.03 | 0.03 | 0.03 | 0.03 |
| 0.75 | 0.07 | 0.07 | 0.06 | 0.06 | 0.06 | 0.06 | 0.06 |
| 1.0 | 0.13 | 0.13 | 0.12 | 0.12 | 0.11 | 0.11 | 0.11 |
| 1.25 | 0.19 | 0.19 | 0.18 | 0.18 | 0.18 | 0.18 | 0.2 |
| 1.5 | | | | 0.30 | 0.29 | 0.29 | 0.36 |
| 1.75 | | | | | | 0.52 | 0.47 |
| 2.0 | | | | | | | 0.55 |

### 3.1.2 Maximum speed

The maximum speed of the wave-maker is calculated by Equation (4):

$$\dot{x}\max = \omega S/2 \tag{4}$$

where $\omega = 2\pi/T$. The results are shown in Table 4.

Table 4. The calculation of maximum speed.

| Wave period (s) | Water depth (m) | | | | | | |
|---|---|---|---|---|---|---|---|
| | 0.3 | 0.4 | 0.5 | 0.6 | 0.7 | 0.8 | 0.9 |
| 0.5 | 0.19 | 0.19 | 0.17 | 0.17 | 0.17 | 0.17 | 0.17 |
| 0.75 | 0.26 | 0.26 | 0.26 | 0.26 | 0.26 | 0.26 | 0.26 |
| 1.0 | 0.39 | 0.38 | 0.38 | 0.37 | 0.36 | 0.35 | 0.35 |
| 1.25 | 0.5 | 0.5 | 0.51 | 0.5 | 0.5 | 0.47 | 0.46 |
| 1.5 | | | | 0.63 | 0.62 | 0.58 | 0.55 |
| 1.75 | | | | | | 0.6 | 0.56 |
| 2.0 | | | | | | | 0.58 |

### 3.1.3 Unit maximum load force

Using the theory of potential function, the load force of the wave-maker when working can be calculated according to the following Equation (5):

$$F(t) = (m + 2mAl)\ddot{x} + 2DAl\dot{x} + R_f \tag{5}$$

where $\dot{x} = \frac{1}{2}\omega S\cos(\omega t)$, $\ddot{x} = -\frac{1}{2}\omega S\sin(\omega t)$; $m$= mass of push-plate (960kg), $l$ = width of push-plate (3.5m); $t$ =time; $R_f$ = friction resistance of push-plate structure (300N); $m_A, D_A$ are described in reference 4 (Zhang 1996).

When $dF/dt = 0$, the unit maximum load force can be obtained, and the results are shown in Table 5.

Table 5. The calculation of unit maximum load force.

| Wave period (s) | Water depth (m) | | | | | | |
|---|---|---|---|---|---|---|---|
| | 0.3 | 0.4 | 0.5 | 0.6 | 0.7 | 0.8 | 0.9 |
| 0.5 | 1083 | 1083 | 1109 | 1110 | 1110 | 1110 | 1121 |
| 0.75 | 1588 | 1599 | 1605 | 1607 | 1614 | 1610 | 1608 |
| 1.0 | 2684 | 2754 | 2826 | 2909 | 2946 | 2967 | 2975 |
| 1.25 | 3662 | 4086 | 4547 | 4959 | 5444 | 5534 | 5672 |
| 1.5 | | | | 7358 | 8192 | 8600 | 5774 |
| 1.75 | | | | | | 10542 | 10878 |
| 2.0 | | | | | | | 12869 |

### 3.1.4 Unit maximum load power

The maximum load power of the unit is calculated by Equation (6):

$$P(t) = F(t) \cdot \dot{x}(t) \tag{6}$$

When dP/dt = 0, the maximum load power of the unit can be obtained, and the results are shown in Table 6.

Table 6. The calculation of unit maximum load power.

| Wave period (s) | Water depth (m) | | | | | | |
|---|---|---|---|---|---|---|---|
| | 0.3 | 0.4 | 0.5 | 0.6 | 0.7 | 0.8 | 0.9 |
| 0.5 | 181 | 181 | 181 | 181 | 181 | 181 | 181 |
| 0.75 | 386 | 386 | 395 | 392 | 398 | 393 | 392 |
| 1.0 | 1132 | 1081 | 1005 | 998 | 987 | 981 | 976 |
| 1.25 | 2104 | 2083 | 2185 | 2299 | 2545 | 2449 | 2454 |
| 1.5 | | | | 4294 | 4711 | 5379 | 4469 |
| 1.75 | | | | | | 5918 | 5681 |
| 2.0 | | | | | | | 6919 |

According to the above calculation results, we can obtain the design parameters of the wave-maker: the maximum stroke is 60cm; maximum movement speed is 0.63m/s; maximum load force of the unit is 12869N; maximum load power of the unit is 6919W.

## 3.2 *Structural design*

In this scheme, the servo motor is used to drive, and the rotation of the motor is transformed into the linear motion of the push-plate through the ball screw. According to the calculation of the above design parameters, the key components of the wave-maker (servo motor, ball screw, linear guide rail) can be selected. On this basis, the SolidWorks software is used for modeling, and its overall structure is shown in Figure 2.

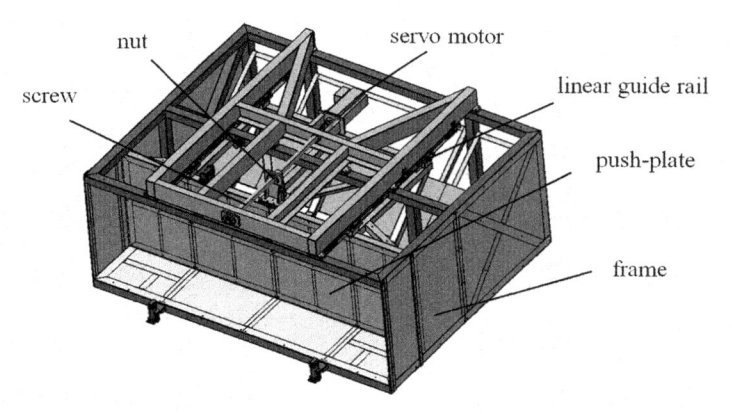

Figure 2. Structure of wave-maker.

The push-plate is the main part of the wave-maker, and the wave generation mainly depends on its reciprocating motion. According to the design requirements, the size of the push-plate is set as 1.2m high, 3.5m wide, and 10mm thick, and the truss structure is adopted. In consideration of the working environment of the push-plate, the material of the push-plate is stainless steel, with a brand of 0cr18, a tensile strength of 520MPa, and a hardness of 187HBS.

## 4 PERFORMANCE VERIFICATION

In order to test the uniformity and repeatability of the generated wave by the wave-maker, the group test is carried out when the maximum water depth is 0.8m. The target value of the group test is shown in Table 7, three groups in total. Each group is tested twice. The capacitive wave height sensor is used to detect the wave data. The average wave height, average period, maximum wave height, and maximum wave period obtained from the analysis of the measured wave data are shown in Table 8. After calculation, the uniformity and repeatability of measured wave data are shown in Table 9, which are better than the requirements in JTJ/T234-2001 code for the wave model test.

Table 7. Test group target value.

| Group | Water depth (m) | Wave height (m) | Wave period (s) |
|---|---|---|---|
| 1 | 0.8 | 0.15 | 2.2 |
| 2 | 0.8 | 0.2 | 3.0 |
| 3 | 0.8 | 0.3 | 4.0 |

Table 8. Group measured value.

| Group | Times ($i$) | Average wave height (m) | Average wave period (s) | Maximum wave height (m) | Maximum wave period (s) |
|---|---|---|---|---|---|
| 1 | 1 | 0.148 | 2.203 | 0.15 | 2.208 |
|   | 2 | 0.149 | 2.203 | 0.151 | 2.208 |
| 2 | 1 | 0.196 | 3.021 | 0.201 | 3.073 |
|   | 2 | 0.199 | 3.018 | 0.202 | 3.066 |
| 3 | 1 | 0.297 | 4.029 | 0.305 | 4.122 |
|   | 2 | 0.293 | 4.025 | 0.302 | 4.1 |

Table 9. Uniformity and repeatability.

| Group | Times (i) | Wave height uniformity (%) | wave period uniformity (%) | Wave height repeatability (%) | Wave period repeatability (%) |
|---|---|---|---|---|---|
| 1 | 1 | 1.35 | 0.23 | 0.68 | 0.00 |
|   | 2 | 1.34 | 0.23 |  |  |
| 2 | 1 | 2.55 | 1.72 | 1.53 | 0.23 |
|   | 2 | 1.51 | 1.59 |  |  |
| 3 | 1 | 2.69 | 2.31 | 1.35 | 0.53 |
|   | 2 | 2.99 | 1.86 |  |  |

## 5 CONCLUSION

This paper explores the working principle and transmission mode of wave-maker in the harbor basin. According to the specific project, the main technical parameters of the wave-maker are calculated using the theory potential function. On this basis, the structure of the unit is designed, and the three-dimensional model is established using SolidWorks software. After the performance test of the wave-maker designed for the project, the uniformity error of wave height and period is less than 3%, and the repeatability error of wave height and period is less than 2%, meeting the requirements of wave model test procedures.

## REFERENCES

Li, X. (2020). Mechanical structure design of push plate wavemaker. *Mechanical engineering and automation* (03): 91–93.

Yu, Y. (2011). *Random waves and their engineering applications*. Dalian University of Technology Press.

Zhang, F. (1996). *Report on the development of harbor snake wave generator*. Tianjin research institute for water transport engineering, M.O.T.

Zhang, R. (2011). *Study on wave simulation equipment of port engineering*. Tianjin University of Technology. Ph.D. thesis, 8–9.

*Civil Engineering and Urban Research – Mohamed & Hou (Eds)*
*© 2023 the Authors, ISBN 978-1-032-44487-1*

# Numerical simulation research on comparison of smoke exhaust means in road tunnels

Mingxuan Li* & Xiujuan Mei*
*Sichuan Fire Research Institute, Ministry of Emergency Management, Sichuan, China*

ABSTRACT: To deal with the problems in road tunnels, which are insufficient evacuation passages, long evacuation routes, and small refuge spaces. The advantages and disadvantages of natural smoke venting, longitudinal mechanical smoke exhaust, and transverse mechanical smoke exhaust are analyzed, the calculation method of smoke production rate in case of fire is proposed, and the effects of various smoke removal means under different fire loads are calculated. The result shows natural smoke venting is only suitable for short tunnels, and the top vents are better than the side wall vents for mechanical smoke exhaust in terms of smoke removal efficiency. This study provides alternatives for the future selection of smoke removal methods for tunnels, which can improve the fire safety performance of tunnels and protect people's lives and property from tunnel fire.

## 1 INTRODUCTION

Due to the narrow and long tunnel space, if a fire breaks out, the fire develops rapidly, the smoke, the poor visibility, and the high temperature pose a huge threat to the drivers and passengers in the tunnels and bring great difficulties to the firefighters, because of the traffic jam, large size firefighting equipment cannot be transported into the tunnels, and it is more difficult to save people and put out fires. At the same time, the tunnel is a kind of underground building with good airtight conditions, and the heat is not easy to dissipate. Therefore, compared with ordinary open-air roads, the number of accidents in tunnels is relatively small, but the consequences and impacts are more serious. Currently, the means for removing the smoke of tunnel fire are mainly divided into two solutions: natural smoke venting and mechanical smoke exhaust.

(1) Natural smoke venting

Natural smoke venting, that is, through the openings of the tunnel and various shafts, the smoke is exhausted by the buoyancy of smoke (Ding 2013). This kind of smoke exhaust method requires relatively low cost and simple operation and maintenance; however, the change of natural wind is uncertain and unstable, so it is not very reliable. This solution removes the smoke through the shaft through the stack effect generated by the shaft. Compared with the traditional natural smoke exhaust method, the pressure difference of the exhaust from the shaft is larger, achieving a good smoke removal effect. At present, it is widely used in urban shallow-buried tunnels.

(2) Mechanical smoke exhaust

Longitudinal mechanical smoke exhaust and transverse mechanical smoke exhaust are two typical mechanical smoke exhaust solutions, and transverse smoke exhaust methods include full transverse smoke exhaust and semi-transverse smoke exhaust (Han 2010). The simple and easy way of tunnel ventilation and smoke exhaust is longitudinal ventilation, as shown in Figure 1.

---

*Corresponding Authors: limingxuan@scfri.cn and 2993350228@qq.com

DOI 10.1201/9781003372417-102

It generates wind pressure in a certain direction when a fire happens by using a jet fan or an axial fan installed on the ceiling of the tunnel and forms a unidirectional airflow with a wind speed not less than the critical wind speed in the tunnel to prevent the spread of smoke, and prevent the movement of smoke to other areas.

Figure 1. Longitudinal smoke mechanical exhaust.

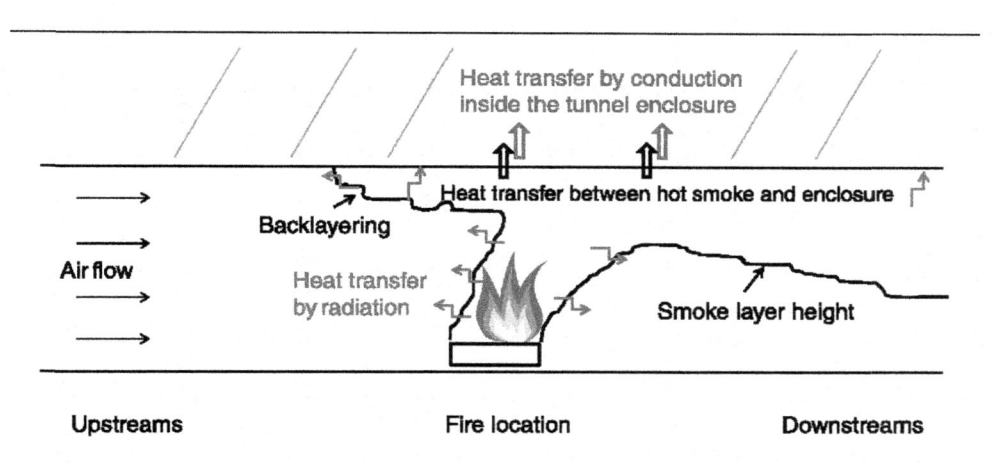

Figure 2. Airflow under the longitudinal smoke mechanical exhaust.

Longitudinal smoke mechanical exhaust is more effective in one-way traffic tunnels (Jiang 2017). When a fire happens, the drivers and passengers on the upstream side can evacuate and escape in a smoke-free environment, as shown in Figure 2. Moreover, the smoke management system does not need to set up an exhaust air duct, which can effectively save construction cost. Although the principle of longitudinal ventilation smoke exhaust is simple and economical, its disadvantages are also obvious. While controlling the flow direction of the smoke, it also supplies a large amount of oxygen for combustion, which will speed up the burning of the fuel, and at the same time, the spread of the smoke downstream is accelerated. As a result, the visibility of the downstream tunnel is reduced, and the air quality deteriorated. It is dangerous to the evacuation of downstream personnel and firefighters. The full transverse smoke exhaust method divides the tunnel into three layers (Mashimo 2002), as shown in Figure 3A. The makeup air supply duct is arranged at the bottom of the tunnel, and the exhaust duct is installed at the ceiling; mechanical makeup air shall be less than the mass flow rate of the mechanical smoke exhaust (Xu 2013). For systems with makeup air supplied by fans, supply fan activation shall be sequenced with exhaust fan activation. The tunnel forms a transverse upward airflow, removing the smoke through the top air duct. During this process, the wind direction is upward, so no longitudinal airflow is generated to greatly reduce the spread of smoke in the tunnel and have good air quality (Zhang 2019). However, the full transverse smoke exhaust system needs to set up mechanical air supply vents at a low position, so it takes more effort to set up a separate air supply duct, which is difficult to construct, and the project cost and maintenance cost is also high; therefore, it is mostly used in large and long road tunnels.

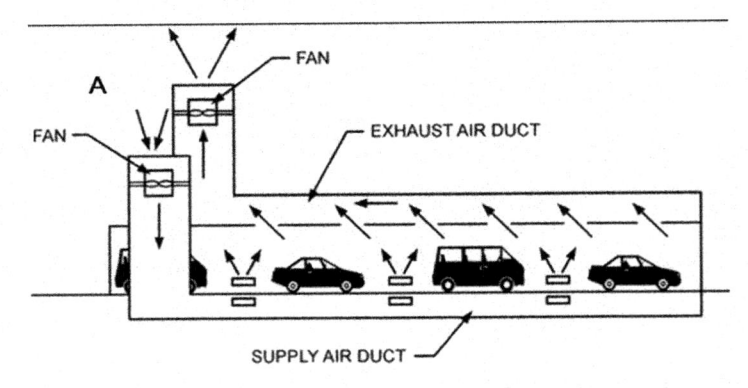

Figure 3. Transverse mechanical smoke exhaust, A) Full transverse mechanical smoke exhaust, B) Semi-transverse mechanical smoke exhaust.

The semi-transverse mechanical smoke exhaust method includes the transverse smoke exhaust and the natural makeup air supply (Tang 2013), as shown in Figure 3. The openings on both sides of the tunnel naturally supply air, and there is no need to set up additional makeup air supply ducts, which can reduce the cross-sectional area of the tunnel, save construction costs, and have a good smoke removal effect (Tong 2014).

## 2   SIMULATION SET UP

In this study, a comparative study on the smoke removal effect under different smoke exhaust means, which is transverse mechanical smoke exhaust and natural smoke venting in the tunnel, was conducted to investigate the fire temperature distribution and smoke concentration distribution under different ventilation conditions in tunnel fires under different fire loads. The ICEM CFD and FLUENT software are employed to conduct this numerical simulation study.

The simulation model of this study is a full-scale experimental tunnel. As shown in Figure 4, the total length of the tunnel is 140m, and the width of the openings at the two ends are 8.9m and 10.9m. The left side of the tunnel is 70m long and 5.4m high. The middle section of the tunnel is 61.1m long, 5.0m high, and 5.8m wide. The right side of the tunnel is 8.9m long and 5.8m high. There are four mechanical smoke exhaust vents in the tunnel, all of which are distributed in the middle section of the tunnel. As shown in Figure 2, mechanical smoke exhaust vents 1 and 2 are located at the top of the tunnel, and mechanical smoke exhaust vents 3 and 4 are located on the tunnel's side walls. As shown in Figure 1(a), the fire fuel is located in the middle of exhaust vent 1 and exhaust vent 2. The distances between fuel and vents 1 or 2 are 13.8m, as shown in Figure 1(b). The distances between the fuel and vents 3 or 4 are 13.8m.

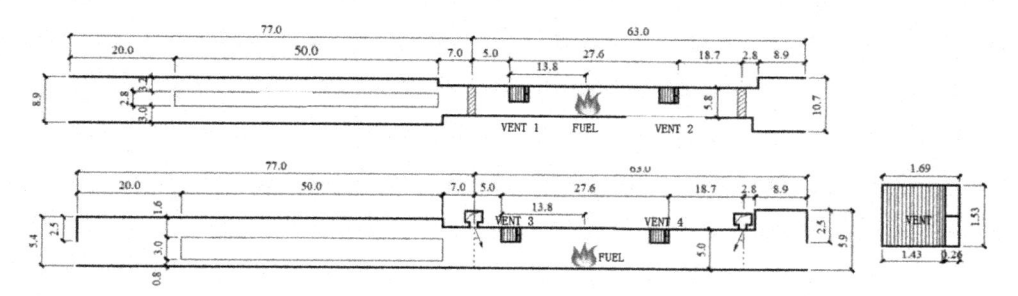

Figure 4.    Tunnel size for the simulations, A) tunnel layout, B) tunnel section view.

The volumetric rate of the smoke exhaust is calculated according to the smoke production of the fire source. Assuming that the type of the fire plume is an axisymmetric plume, the smoke production can be calculated by using the Heskestad equation in NFPA92B. The rate of smoke production shall be calculated using the equations shown below:

$$M = 0.071Q_c^{1/3}z^{5/3} + 0.0018Q_c \quad z > z_1 \tag{1}$$

$$M = 0.032Q_c^{3/5}z \quad z \leq z_1 \tag{2}$$

$$z_1 = 0.166Q_c^{2/5} \tag{3}$$

where $Z_l$ is the limiting elevation (m), $Q_c$ is the convective portion of heat release rate (Btu/sec), Z is the distance above the base of the fire to the smoke layer interface (m), and m is the mass flow rate in plume at height Z (kg/sec). The radiant heat is also a part of the heat release loss of the fire source, and 30% of the heat release rate is lost through radiation; the remaining 70% of the heat release rate enters into a smoke layer to drive the smoke flow. Therefore, the convective portion of the heat release rate of the fire shall be determined from Equation 4.

$$Q_c = 0.7Q \tag{4}$$

where $Q_c$ is the convective portion of the heat release rate of the fire (Btu/s), Q is the heat release rate of the fire (Btu/ft), and 0.7 is the convective fraction (dimensionless).

$$V_e = \frac{MT_s}{\rho_0 T_0} \tag{5}$$

$$T_s = \frac{Q_c}{Mc_p} + T_0 \tag{6}$$

721

Where: $V_e$ is the volumetric rate of smoke production, m³/s; $\rho_0$ is the density of smoke, 1.2kg/m³; $T_s$ is the temperature of the smoke, K; $C_p$ is the specific heat of the ambient air, 1.02kJ/ (kg. K); $T_0$ is the temperature of ambient air, 301K.

Table 1.  Smoke production from fire sources.

| $Q$ (MW) | $Z$ (m) | $Z_1$(m) | $V_e$(m³/s) |
| --- | --- | --- | --- |
| 1 | 1.5 | 2.28 | 4.03 |
| 2 | 2.0 | 3.01 | 8.10 |
| 3 | 2.5 | 3.54 | 12.54 |
| 5 | 3.0 | 4.34 | 20.66 |
| 20 | 4.0 | 7.56 | 72.61 |

In this study, methanol was used as the fuel, the inner wall of the tunnel was considered adiabatic, and the influence of external wind at the ends of the tunnel was not considered. The ambient temperature around the tunnel is 28°C, and the atmospheric pressure is 101.325 Kpa, as shown in Table 2.

Table 2.  Simulation inputs.

| Fuel | Wall condition | (°C) | Atmospheric pressure (Kpa) |
| --- | --- | --- | --- |
| methanol | adiabatic | 28 | 101.325 |

The design fire is 1MW, 2MW, 3MW, and the smoke exhaust is 6.5 times the smoke production, the design fire is 5MW, the smoke exhaust is 5.0 times the smoke production, and the design fire is 20MW, and the smoke exhaust is 2.5 times.

## 3   RESULTS AND DISCUSSION

As shown in Figure 5, in the case of 1MW, the smoke spreads to the left and right sides of the fuel in the two directions under the ceiling. In the case of natural smoke venting, when the simulation time comes to 1000 seconds, the smoke layer on the left side of the fuel has dropped to the ground, and the temperature of the smoke layer on the right side of the train has reached 80°C, which is 60°C higher than the temperature that the human body can tolerate. The $CO_2$ concentration at the head of people is about 0.5%, much higher than the $CO_2$ concentration in the environment, which will affect the normal evacuation of personnel. Therefore, people should evacuate as soon as possible in case of fire before the environment become acceptable. The result also shows natural smoke venting is not fit for long tunnels.

As shown in Figure 5, in the case of 1MW, two smoke exhaust vents on the top and two smoke exhaust vents on the side walls are used for mechanical smoke exhaust. When the numerical simulation comes to 1000 seconds, the smoke does not spread to the openings at both ends of the tunnel, and the temperature at the height of the person's head does not exceed 60°C. And as shown in Figure 5, the $CO_2$ concentration in the tunnel is significantly lower than that in the case of natural smoke exhaust, and the smoke is under control. The mechanical smoke exhaust effect with the top vents is slightly better than that of the mechanical smoke exhaust with the vents on the side walls.

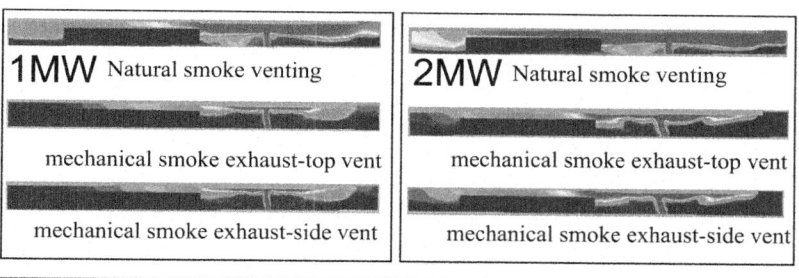

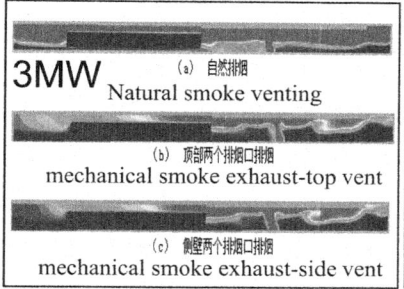
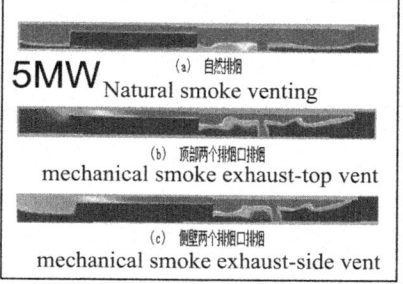

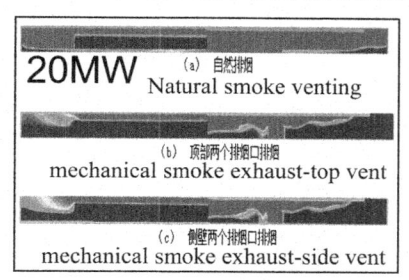

Figure 5. Smoke spread.

## 4 CONCLUSIONS

In order to understand the smoke spread characteristic in road tunnels in case of fire, the smoke under different fire loads and smoke exhaust means (natural smoke exhaust, mechanical smoke exhaust with the vents on the top or side wall of the tunnels) were studied. In this study, methanol was used as the fuel, and the design fire of 1MW, 2MW, 3MW, 5MW, and 20MW were used. The simulation results show that the effect of the mechanical smoke exhaust is better than that of natural smoke exhaust, and the mechanical smoke exhaust solution with the vents on the ceiling of the tunnel is slightly better than the mechanical smoke exhaust solution with the vents on the side wall of the tunnel. When the design fire is 1MW, 2MW, 3MW, and the smoke exhaust is 6.5 times the smoke production, the design fire is 5MW, the smoke exhaust is 5.0 times the smoke production, and the design fire is 20MW, and the smoke exhaust is 2.5 times, the fire smoke cannot be contained between the two smoke exhaust vents, in this case, more smoke exhaust vents should be activated to remove smoke. Otherwise, the personnel cannot evacuate the tunnel in time and would die from the fire smoke.

1. The smoke grows very fast in the tunnels. Because of the structural characteristics of the tunnels, small spaces, and inclined tunnels of the road tunnel are very easy for smoke to spread. If the ventilation and smoke exhaust equipment are not operated in time in case of fire, the spread of

the smoke will be accelerated. Natural smoke venting is fit for short tunnels but not for long tunnels.

2. Dense smoke spread very easily in long tunnels, and it is difficult for people to evacuate if no mechanical smoke exhaust is adapted to handle it. After a fire breaks out, if the smoke cannot be removed, the whole tunnel will be quickly filled with smoke, making it difficult for people to escape.
3. The mechanical smoke exhaust solution with the vents on the tunnel's ceiling is slightly better than the mechanical smoke exhaust solution with the vents on the side wall of the tunnel.

## ACKNOWLEDGMENTS

The authors gratefully acknowledge the financial support of the Sichuan Province Research and Development Program under Grant No. 2016JY0169.

## REFERENCES

Ding Yuan, Wang Qing. (2013) *From entry to mastery of ANSYS ICEM CFD J.* Tsinghua University Press. 3, 43–47.

Han Zhanzhong, Wang Jing, Lan Xiaoping. (2010) *FLUENT - Examples and Applications of Fluid Engineering Simulation J.* Beijing Institute of Technology Press. 5, 63–68.

Jiang Shuping. (2017) Statistics of highway tunnels in China J. Tunnel Construction. 37, 6–14.

Lilly, Quintiere B, James. (2002) Enclosure Fire Dynamics J. *Fire Safety Journal.* 6, 13–17.

Mashino H. (2002) State of the road tunnel safety technology in Japan J. *Tunnelling & Underground Space Technology.* 17, 145–52.

Mingxuan, Li. (2017) Experimental Research on the Smoke Control System in a Complex Road Tunnel Fire. *J. Procedia Engineering.* 211, 379–387.

Tang Jiapeng. (2013) *FLUENT 14.0 Super Learning Manual J.* People's Posts and Telecommunications Press. 37, 643–4.

Tong Y, Wang X, ZHAI J, et al. (2014) Theoretical predictions and field measurements for potential natural ventilation in urban vehicular tunnels with roof openings *J. Building & Environment.* 82, 45–48.

Tong Y, Zhai J, WANG C, et al. (2016) Possibility of using roof openings for natural ventilation in a shallow urban road tunnel *J. Tunnelling & Underground Space Technology.* 54, 92–101.

Xu Zhisheng, Li Weiping, Li Weiping. (2013) *Fire Smoke Control in Highway Tunnels: Research on Centralized Smoke Exhaust System of Independent Exhaust Channels J.* People's Communications Press. 6, 33–38.

Zhang Xingkai. (1997) *The principle and application of fire in underground engineering J.* Capital University of Economics and Business Press. 7, 43–47.

*Civil Engineering and Urban Research – Mohamed & Hou (Eds)*
*© 2023 the Authors, ISBN 978-1-032-44487-1*

# A model test study on vibration construction of U-shaped sheet pile for bank protection reinforcement

Kexiong Wu*, Pengyan Bi, Ling Zhang, Huimei Shi & Yang Ming
*GuangXi Key Laboratory of New Energy and Building Energy Saving, Guilin, Guangxi, China*
*Guilin University of Technology, Guilin, Guangxi, China*

Guoping Xu
*Guilin University of Technology, Guilin, Guangxi, China*

ABSTRACT:    Taking the Wulongqiao Revetment reinforcement Project in the Huzhou section of the Grand Canal as the research background, this paper starts with a test of the vibration construction model test of u-shaped sheet pile and studies the influence mechanism of the driving of u-shaped sheet pile on the soil around the pile to obtain the dynamic response mechanism of u-shaped sheet pile, bank revetment and soil mass under the condition of vibration construction. The results show that the process of vibration driving of the u-shaped sheet pile has a great influence on the deformation and settlement of the original bank revetment, but a relatively small influence on the soil behind the original bank revetment. The attenuation of the old revetment is slower than that of the soil, and the acceleration of the old revetment is significantly greater than that of the rear caused by the vibration wave of vibration construction of the u-shape sheet pile. This paper shows the superiority of this new form of bank protection and greatly reduces the impact on the surrounding environment, which shows the superiority of this new form of bank protection. The paper results can provide support and a basis for the extension of the u-shape sheet pile driving project.

## 1   INTRODUCTION

In recent years, with the continuous development of inland water transport, the contradiction between channel construction, land resource protection, and ecological environment construction has become increasingly prominent. Therefore, the inland waterway gradually adopts the Deepening Bank Protection Foundation to improve the channel standard. For example, the hang-ping-shen line channel reconstruction project has completed the upgrading base on maintaining the channel width and deepening the shore protection foundation. At the same time, because of the limitation of stone resources, the original gravity revetment is gradually being replaced by a new type of revetment structure, construction efficiency, and construction quality. However, the optimization of the Bank revetment structure has gradually become one of the important contents of inland waterway construction research.

The Revetment reinforcement structure is the most representative of Sheet-pile types, such as concrete sheet-pile and steel sheet-pile. Among them, the U-shape sheet pile was more and more used in the field of Slope Control, bank protection, and reinforcement, municipal construction, and other engineering construction (Chen et al. 2020) for the advantages of excellent mechanical performance, obvious economic benefit, wide application in geology and good durability in engineering, etc. Shangguan Jingling and others (ShangGuan et al. 2013; ShangGuan 2016) found that the deformation of U-type prestress concrete retaining sheet piles has certain ductility and elasticity

---

*Corresponding Author: 9820070@glutnn.cn

DOI 10.1201/9781003372417-103

by studying the horizontal bearing capacity, bending bearing capacity and failure characteristics of U-type prestress concrete retaining sheet piles. Liu Zhenyu (Liu 2016) analyzed the reduction problem of the bending rigidity of the u-shaped sheet pile by the locking hole and established the finite element model of the u-shaped sheet pile. Besides, as the reason for the influence of sheet pile driving on surrounding buildings and the environment, the dynamic response characteristic of vibration construction has been paid more and more attention by domestic and foreign scholars. GORBUNOVM P (Gorbunov-Posadov 1968) studied the law of soil displacement caused by pile driving. Grabe (Grabe et al. 2006) summarized the influence of different piling methods on the surrounding soil by monitoring the deformation of soil mass at different distances from the pile driving point. Meanwhile, Lee (Lee et al. 2012) studied the sheet pile characteristics under vibration construction in sandy soil and pointed out that the sheet pile driving process will affect the structural stability. Xu Zhigang (Xu et al. 2018) studied the dynamic response characteristics of the u-shaped sheet pile by field test, which indicated that the ground vibration acceleration decreased exponentially with the increase of pile driving distance. At the same time, by combined with three u-shaped steel sheet pile pressing methods, Li Pengju (Li et al. 2015) found that the pile pressing curve would have sudden changes at the interface of the soil layer, established a relationship between pile pressing curve spectrum and formation strength, and put forward the concept of "Dynamic investigation".

To sum up, most scholars at home and abroad mainly study the mechanical properties, deformation properties, and soil deformation properties caused by vibration construction of u-shape sheet piles through field tests. However, the research on the dynamic response of vibration construction is relatively scarce, and it is difficult to be applied to other projects because of the large discrete type of site Coefficient and soil physical and mechanical parameters.

Based on the research background of the Wulongqiao revetment project in the Huzhou section of Grand Canal, Zhejiang Province, the paper designs a laboratory model test of u-shaped sheet pile driving. By analyzing the vibration and deformation of soil around sheet pile and bank protection, it studies the influence mechanism of u-shape sheet pile driving on pile and soil around the pile, which provides a test basis of field construction of U-shape sheet piles.

## 2 MATERIALS AND METHODS

As China's north-south waterway, Grand Canal (Zhejiang Section) is responsible for the economic development of Zhejiang Province, such as a large number of coal, mining materials, oil, steel, and other cargo transport tasks, and its project of third-level Channel Regulation plays an important role in promoting the construction of comprehensive transport system. Based on the background of the Wulongqiao revetment reinforcement project in the Zhejiang section of the Grand Canal River, this test simulated vibratory piling of a u-shaped concrete sheet pile on the right bank revetment of the Wulong Bridge, and a reinforced revetment pile is 7.5 M350 I u type concrete sheet pile.

### 2.1 *Trial model*

This test attempts an optimal height reduction of Wulong Bridge North side of the right bank reinforcement project site conditions. According to the actual situation, the size and location of the u-shaped concrete sheet pile and the old bank protection are reduced by a 10:1 proportion, and the influence of the boundary effect is taken into account, this trial makes 5 model piles and arranges to monitor instruments in the middle pile body. The way of driving the piles was to simulate the way of hammering by gravity hammer. The materials used in the model are all organic glass, including the model groove, the model pile, and the old revetment, in which the model groove is 80cm high, 60cm long, and 60cm wide, and the model pile is 10cm wide, 3.5cm high and 40cm, 50cm long and 60cm long, respectively. The section of the old revetment model is shown in Figure 2.

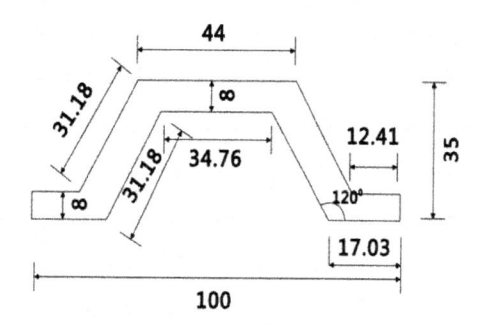

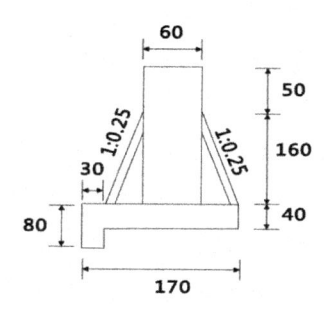

Figure 1.   Section size of model pile.

Figure 2.   Section size of model revetment.

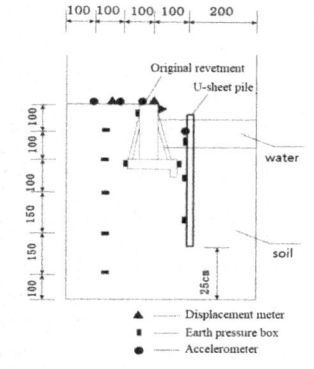

Figure 3.   Data acquisition system.

Figure 4.   Piling system.

Figure 5.   Instrument layout diagram.

## 2.2   *Test apparatus*

This test is mainly divided into two systems, the piling system, and the data acquisition system, as shown in Figure 3 and Figure 4.

The piling system consists of a plexiglass frame, 1 kg weights, and a rope. We can see the layout of the apparatus shown in Figure 5. It needs to ensure the perpendicularity of the weight by drawing the rope from a small hole in the plexiglass frame and controlling the piling force by the high weights lifted by the rope. At the same time, this process must provide a condition of about 1 Hz of pile driving frequency.

## 2.3   *The model pile driving*

There are set up three types of model piles of 40cm, 50cm, and 60cm in length. Before pile driving, the fixed guide groove is installed at first and the pile body is pressed into the soil 10cm by static pressure. During pile driving, the horizontal and vertical displacement of the old revetment is monitored once for every 10 driving of the sheet pile, the earth pressure box and the accelerometer are monitored throughout, and additional monitoring is carried out in the final driving stage. Every process keeps a detailed record.

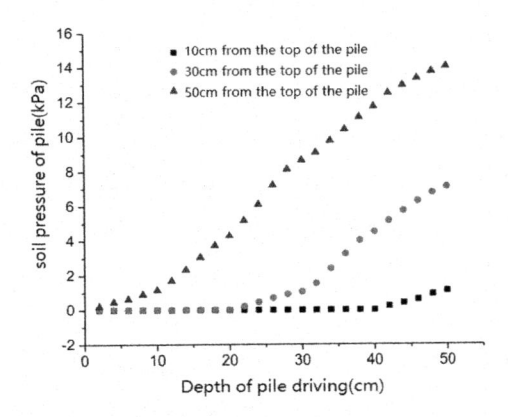

Figure 6. The relationship between the soil pressure and the depth of the pile.

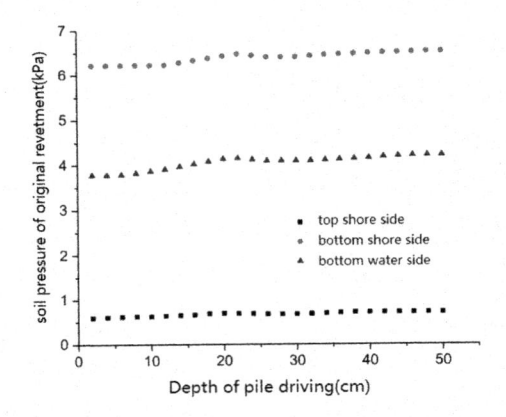

Figure 7. The relationship between the soil pressure of old revetment and the depth of the pile.

## 3 RESULTS & DISCUSSION

### 3.1 *Variation law of earth pressure*

#### 3.1.1 *Earth pressure on pile body*

As we can see in Figure 6, it is a curve diagram of the variation of earth pressure at different positions of the pile body with the depth of penetration. The curve shows that the earth pressure is 0 when the pile body does not enter the water, the earth pressure increases linearly after entering the water. Furthermore, the increased amplitude of earth pressure increases after the pile body enters the soil. There is a little fluctuation of the change value, which indicates that the vibration process causes the disturbance of the surrounding soil body, and the stress state between soil particles is changed. At the same time, the earth pressure on the pile increases with the depth of the pile.

#### 3.1.2 *Earth pressure of old revetment*

Figure 7 shows the relationship between the earth pressure at different positions of the old revetment and the depth of the pile. The soil pressure near the top of the old revetment increases slightly with the increase of the depth of the soil, which shows that the soil near the top of the old revetment is less affected by the pile vibration. The Earth pressure at the bottom of the old revetment increased obviously, especially at the waterside, which indicated that the old revetment was subjected to force from part of the soil and controlled the deformation of the rear soil.

#### 3.1.3 *Earth pressure in the rear of revetment*

Figure 8 illustrates the curves of the Earth's pressure at different depths with the depth of the soil. It can be seen from the diagram that the earth pressure of the soil behind the old bank revetment changes little in the process of pile driving, and there are some differences in the law of the Earth pressure at different depths. With the increase of the depth, the change of the Earth pressure is larger at the deeper depth, which shows that the sheet pile driving will only affect the soil in a certain range of the depth, but has little influence on the soil outside the range.

### 3.2 *Displacement analysis*

Figure 9 illustrates the relationship between the uplift of the old revetment and the number of hammering, based on the analysis of a single pile penetration during the second hammering. As can be seen from the diagram, with the increase of the number of hammering, the uplift of the

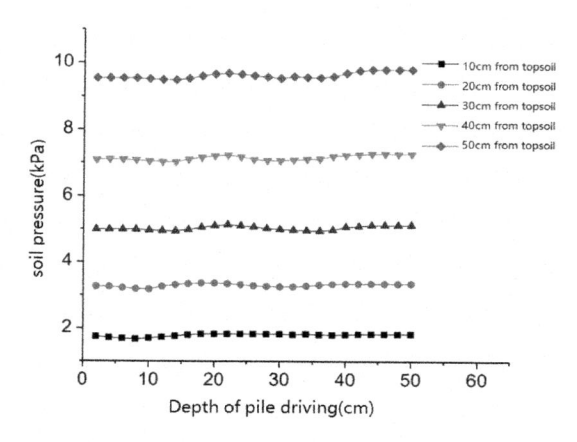

Figure 8. The relationship between the soil pressure at different depths and the depth of the pile.

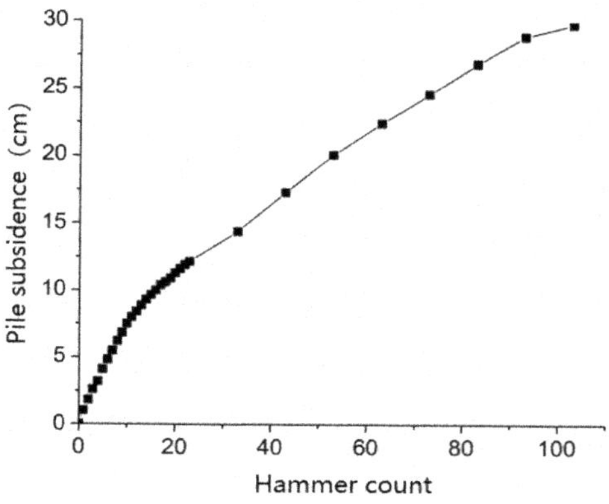

Figure 9. The relationship between hammering number and uplift amount of old revetment.

old revetment continues to increase, because, with the increase of the number of hammering, the pile body rapidly sinks, and part of the soil body is squeezed to the old revetment, causing the old revetment to rise. The soil particles are moving in the course of hammering, making the soil have an upward trend. However, in the later period, the resistance of soil increases, the penetration of piles decreases, and the effects of the soil compaction weaken. Therefore, the growth rate of the uplift of the old revetment gradually decreased.

During the whole penetration process, the displacement of the old revetment and the soil behind it changes as shown in Figure 10. It can be seen that at the initial stage of pile driving, both the vertical displacement of the old revetment and the soil appear negative value, that is, slight settlement occurs. The reason is that there are a few pores in the soil mass, after the vibration, the pores decrease, resulting in soil mass settlement. The effect of Soil compaction is not obvious in this process. Moreover, with the pile body sinking gradually, the displacement of the old bank protection increases gradually, showing the displacement of uplift and deviating pile body, while the soil body settlement lasts longer, and the amount of uplift increases in the later period of pile driving. It is shown that the old revetment effectively blocks the displacement trend of the rear soil mass and limits the deformation of the rear soil mass, which proves the characteristics and advantages of new and old bank revetment.

With the increase of pile length, the slope of the displacement of the old revetment and the soil decreases, as shown in Figure 11. With the increase of the pile length, the depth of the pile gradually exceeds the bottom of the old revetment, and the vibration and compression of the old revetment become smaller and smaller. It is shown that in homogeneous soil, the displacement of the old revetment and the soil behind it occurs mainly near the bottom of the old revetment.

## 3.3 Acceleration analysis

### 3.3.1 Vibration response analysis of old revetment and ground

From Figures 12 and 13, it is shown respectively that the attenuation curves of surface vibration waves of the old revetment and the soil behind the revetment at the same time during pile driving. It can be seen obviously that the amplitude of vibration acceleration is close to each measuring point, and the attenuation time is basically between 0.2 s to 0.6 s. The acceleration attenuation curve shows that the attenuation of the old bank revetment is slower while the soil is faster. For the reason that the attenuation of ground vibration is not only related to the depth of pile, but also to the vibration frequency, i.e., the strength of foundation soil.

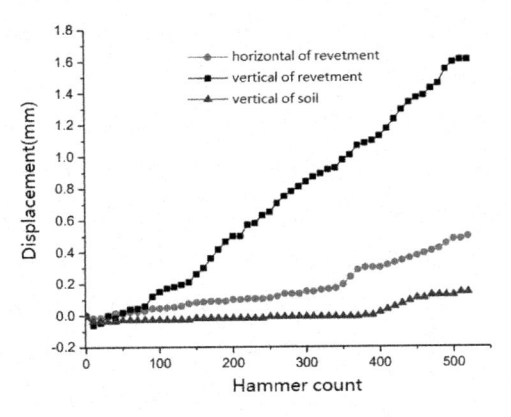
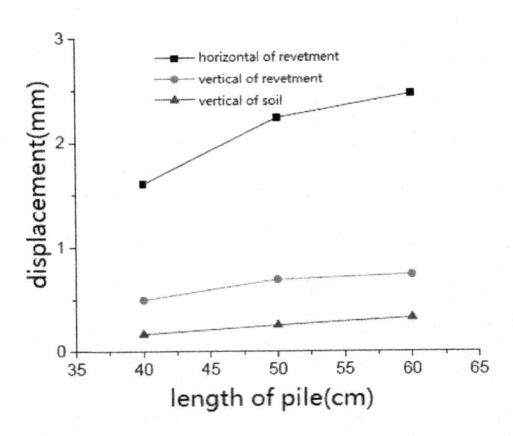

Figure 10.   Displacement variation of old revetment and soil.

Figure 11.   Maximum displacement of old revetment and soil under different pile lengths.

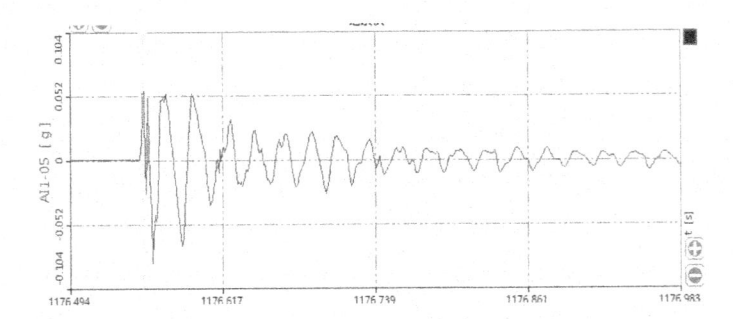

Figure 12.   Acceleration attenuation curve of old revetment.

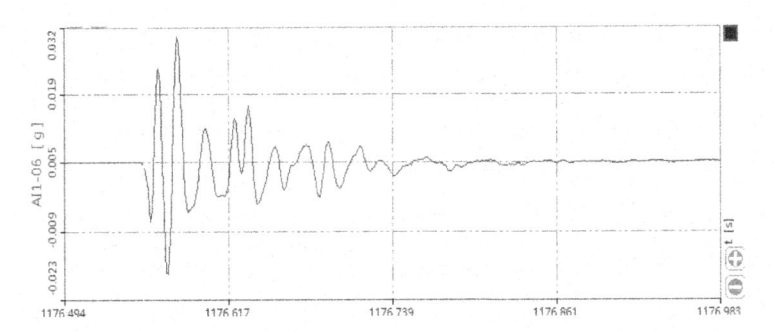

Figure 13.   Acceleration attenuation curve of soil surface.

Figures 14 and 15 are the Spectra corresponding to the acceleration vibration wave in Figure 12 and Figure 13 respectively. The amplitude of acceleration is reached at 43 Hz and 50 Hz, and the peak values are 0.007 m/s 2 and 0.0012 m/s 2 respectively. It can be seen from the vibration spectrum that the vibration frequency of the old revetment is lower than that of the soil, so the vibration attenuation is slower than that of the soil. At the same time, the vibration frequency of the old revetment is mainly concentrated in a small range, and the variation range of the soil vibration frequency is large, which reflects the complexity of the soil particle nature and composition.

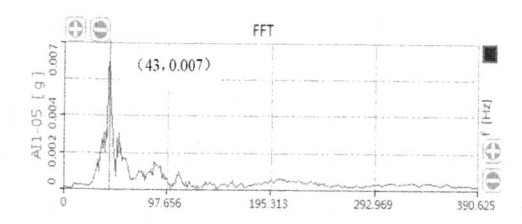

Figure 14.  Vibration spectrum of old revetment.       Figure 15.  Vibration spectrum of soil surface.

### 3.3.2  *Analysis of pile tip penetration depth and amplitude of ground vibration acceleration*

In order to study the change law of the acceleration effectively, the average value of the acceleration at a certain penetration depth is taken as the acceleration of each measuring point when the sheet pile acts on the soil layer, and the relationship between the vertical vibration acceleration amplitude of the old revetment and the soil surface behind the old revetment and the depth of the pile tip is shown. As we can see from the diagram in Figure 16 that when the penetration depth is small (less than the height of the old revetment), the acceleration gradually increases with the penetration depth, and the increased range is larger. By contrast, as the penetration depth continues to increase, the acceleration changes little and presents certain volatility. In addition, the acceleration of the old revetment is significantly greater than that of the soil behind it.

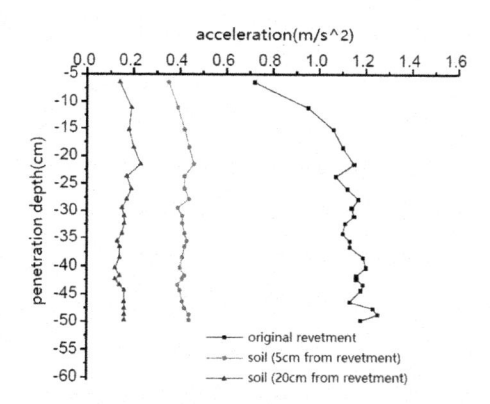

Figure 16.  Relationship between acceleration and pile penetration depth.

## 4  CONCLUSIONS

Based on the design of the u-shape sheet pile driving model test, this paper studies the dynamic response of the u-shape sheet pile. The conclusions are as follows:

(1) Through the variation law of earth pressure and displacement under vibration construction, it can be found that the earth pressure of pile body increases gradually with the increase of the depth of pile, and when the depth of sheet pile reaches the bottom of old bank protection, the earth pressure of the bottom of old bank protection has obviously changed. The soil pressure in the rear of the old revetment increases slightly, but the increase is small, and the vertical displacement is obviously lower than that of the old revetment. The results show that the disturbance of rear soil is less under the cooperation of new and old revetments, which shows the advantage of the combination of new and old revetments.

(2) The vibration wave caused by the vibration construction of the u-shaped sheet pile in the old revetment attenuates more slowly than that in the soil, because the vibration frequency of the old revetment is lower than that of the soil. Besides, the vibration frequency of the old

revetment is mainly concentrated in a small interval, and the variation range of soil vibration frequency is large, which reflects the complexity of soil particle property and composition.

(3) when the penetration depth is small (less than the height of the old revetment), the acceleration increases gradually with the penetration depth increasing. On the other hand, as the penetration depth continues to increase, the acceleration changes little and presents a certain fluctuation. More than that, the acceleration of the old revetment is significantly greater than that of the soil behind it.

## ACKNOWLEDGMENTS

This paper is one of the achievements of Guangxi Key Laboratory of New Energy and Building Energy Saving (Guikeneng 22-J-21-25) and the project of improving the basic scientific research ability of young and middle-aged college teachers in Guangxi (2021KY0253).

## REFERENCES

Chen Y.J., Li J.L., Zou F.J., Gou L.F., etc. (2020) Application of concrete u-sheet pile in bank protection project of Zhengji River. *Technical Supervision in Water Resources*, 05:209–212.

Gorbunov-Posadov M I. (1968) Displacement and compaction of soil by a driven pile. *Soil Mechanics & Foundation Engineering*, 5(5):313–318.

Grabe J, Mardfeldt B, Mahutka K P. (2006) Zur last abtragung vonpfahl konstruktion en im hafenbau[ J]. *HANSA*, 143(2):42–51.

Lee S.H., Kim B.I., Han J.T. (2012) Prediction of the penetration rate of sheet pile installed in the sand by a vibratory pile driver. *KSCE JOURNAL OF CIVIL ENGINEERING*, 12(3):316–324.

Li P.J., Zhou H.Q., Su H., Wu R.Z., Teng Y.L. (2015) Field Experimental study on dynamic investigation technique based on the construction of U-shaped steel sheet pile. *Chinese Journal of Rock Mechanics and Engineering*, 34(S1):3553–3563.

Liu Z.Y. (2016) Research on bending stiffness reduction effect of U-shaped steel sheet pile lock. *China Water Transport*, 16(03):304–306.

ShangGuan J.L., Huang G.L., Geng S.H. (2013) Calculation method of bearing capacity of U-shaped pre-stressed concrete sheet pile. *Journal of Nanjing Tech University* (Natural Science Edition), 35(05):100–104.

ShangGuan J.L. (2016) Experimental study on bending resistance of U-shaped prestressed concrete sheet pile. *Construction Materials & Decoration*, 31:41–43.

Xu Z.G., Chen L., Wang F.X. (2018) Experimental study on dynamic response characteristics of U-shaped plate pile vibration construction. *Water Conservancy Construction and Management*, 38(01):34–40.

*Civil Engineering and Urban Research – Mohamed & Hou (Eds)*
© *2023 the Authors, ISBN 978-1-032-44487-1*

# Two-dimensional numerical simulation analysis of Qiliuqiu-Beitang river water environment improvement

Zhihao Fang
*School of Naval Architecture and Maritime, Zhejiang Ocean University, Zhoushan, China*
*The Key Laboratory for Technology in Rural Water Management of Zhejiang Province, Zhejiang University of Water Resources and Electric Power, Hangzhou, China*

Dongfeng Li*
*The Key Laboratory for Technology in Rural Water Management of Zhejiang Province, Zhejiang University of Water Resources and Electric Power, Hangzhou, China*

ABSTRACT: There are many rivers in Shaoxing Binhai New Area. Water diversion through sluice regulation is an important measure to improve the water environment of plain rivers and lakes. In order to demonstrate the effect of water diversion on the improvement of the river and lake water environment, this paper selects the Qiliuqiu-Beitang River in the new area as the research object, establishes a two-dimensional numerical model for simulation, and analyzes its water level, velocity, BOD concentration and DO concentration through the water diversion scheme. The results show that the BOD concentration of this reach reaches the class III water index after water diversion, and DO concentration reaches the index of class III water or even class II water. The diversion scheme is feasible.

## 1 INTRODUCTION

Shaoxing Binhai New Area is a provincial-level new area under the jurisdiction of Shaoxing City, Zhejiang Province. It is located on the South Bank of Hangzhou Bay and the north of Shaoxing City. It is the bridgehead for Shaoxing to fully integrate into the regional integration development of the Yangtze River Delta and the construction of the Hangzhou Shaoxing Ningbo integration demonstration zone. However, the fluidity of its river network is poor, the water quality is deteriorating, and the eutrophication of rivers and lakes occurs from time to time. Since the opening and closing of different diversion sluices, different drainage sluices, and sluices in the river network have different effects on the changed position, it can guide the sluice dispatching and serve the improvement of the water environment through the opening and closing operation of sluice (Ji 2022; Lu et al. 2013; Wu 2019).

The use of the plane two-dimensional hydrodynamic mathematical model of the water environment is an important tool and means to improve the water environment (Gao et al. 2015; Tong et al. 2016; Yang et al. 2018). Therefore, this paper analyzes the impact of the water diversion scheme on river networks through a two-dimensional mathematical model. The geographical location of the Qiliuqiu-Beitang River selected in this calculation is shown in Figure 1.

---

*Corresponding Author: lidf@zjweu.edu.cn

DOI 10.1201/9781003372417-104

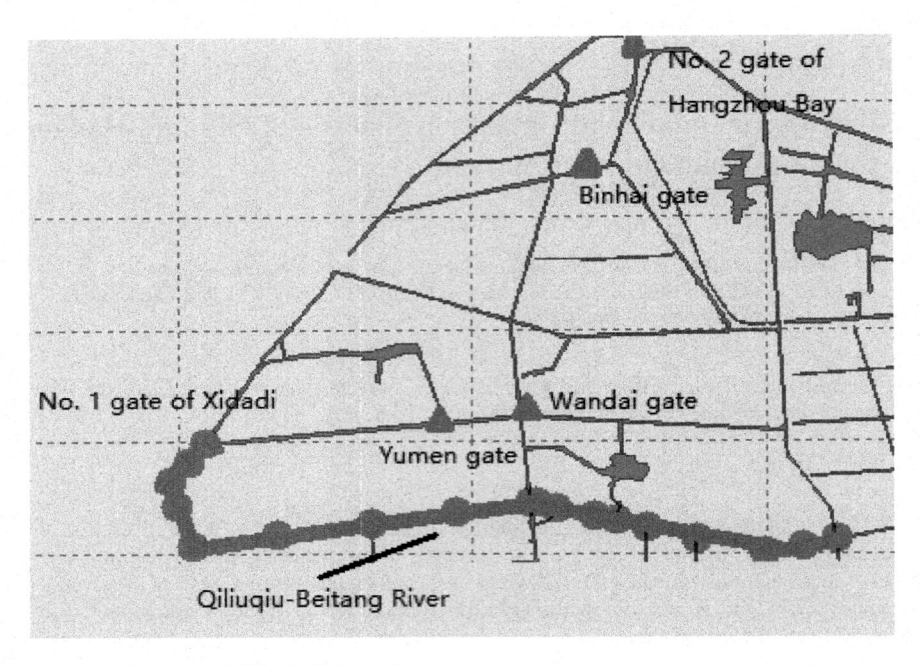

Figure 1.  Location map of Qiliuqiu-Beitang River.

## 2  TWO-DIMENSIONAL HYDRODYNAMIC PLANE MATHEMATICAL MODEL OF PLAIN RIVER NETWORK

### 2.1  *Basic equation*

Flow continuity equation:

$$\frac{\partial z}{\partial t} + \frac{\partial (hu)}{\partial x} + \frac{\partial (hv)}{\partial y} = 0 \tag{1}$$

X. Y-direction flow equation:

$$\frac{\partial u}{\partial t} + u\frac{\partial u}{\partial x} + v\frac{\partial u}{\partial y} - fv + g\frac{\partial z}{\partial x} + g\frac{u\sqrt{u^2 + v^2}}{c^2 h} = \lambda\Delta u \tag{2}$$

$$\frac{\partial u}{\partial t} + u\frac{\partial v}{\partial x} + v\frac{\partial \beta v}{\partial y} + fu + g\frac{\partial z}{\partial y} + g\frac{v\sqrt{u^2 + v^2}}{c^2 h} = \lambda\Delta v \tag{3}$$

Where: x and y are spatial coordinates; z is the water level; t is the time; u and v are the velocity components in the X direction and Y direction respectively; h is the water depth; $f$ is Coriolis force, $f = 2W \sin \Phi$; $W$ is the rotation velocity of the earth; $\Phi$ is the latitude of the earth; c is Xie Cai coefficient; g is the gravitational acceleration.

### 2.2  *Initial conditions*

The initial value of the mathematical model, given the initial time of the river network, the water level is the constant water level of 2.8m and the flow velocity is 0; the initial condition of water quality is class V water, that is, BOD is 10 mg /L and DO is 2mg /L.

In this calculation, the introduced clean water is class III water, that is, BOD is 4 mg /L, DO is 5 mg /L, and the planned and designed flow is given at the diversion outlet.

The specific diversion scheme is shown in Table 1 below. Then the water level change, velocity change, and water quality change of the main river Huantang West River environment are analyzed.

Table 1. calculation conditions of the diversion scheme.

| Gate name | Flow (m³/s) | Opening and closing conditions |
|---|---|---|
| No. 1 gate of Xidadi | 20 | open |
| No. 2 gate of Hangzhou Bay | −60 | open |
| Binhai gate | 0 | close |
| Yumen gate | 0 | close |
| Qianjin gate | 0 | close |
| Wandai gate | 0 | close |

## 3  ANALYSIS OF CALCULATION RESULTS

### 3.1  Analysis of water level along the way

It can be seen from Figure 2 that the water surface elevation of the Qiliuqiu-Beitang River is relatively stable at the same time. With the passage of time, the water surface elevation of each point in the Qiliuqiu-Beitang River gradually decreases with time, and the change rate tends to decrease. From 0:00 on April 7 to 0:00 on April 10, the water level elevation decreased from 2.8m to 2.33m. Secondly, it can be seen from the figure that the water level elevation changed significantly in the first 48 hours and slowly in the last 24 hours.

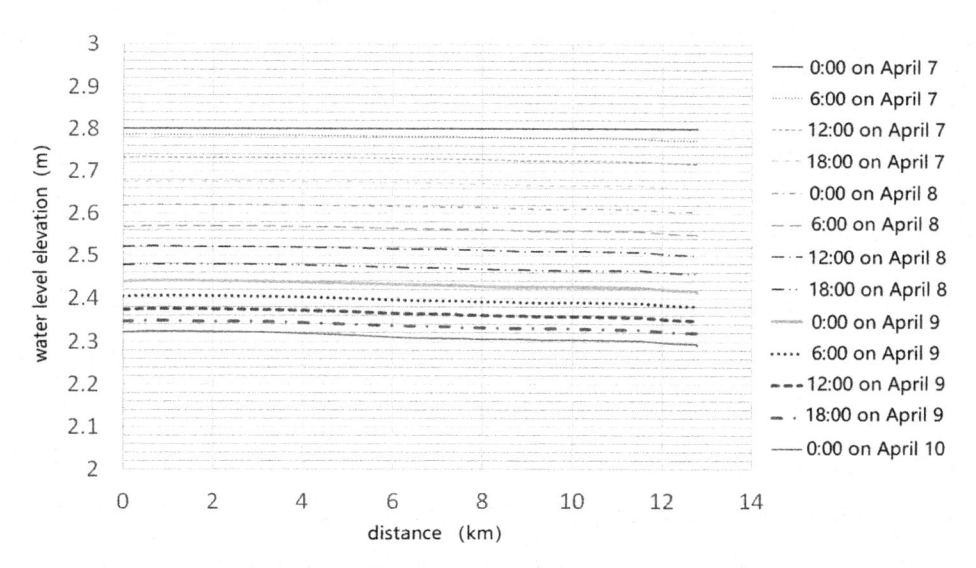

Figure 2.  Variation of water surface elevation of Qiliuqiu-Beitang River with distance at different times.

### 3.2  Analysis of flow velocity along the river

As can be seen from Figure 3, the flow velocity of the Qiliuqiu-Beitang River fluctuates greatly and is not stable enough. At a distance of 8km from the No. 1 gate of Xidadi, the velocity is 0, which is caused by the closure of the Hongqi gate.

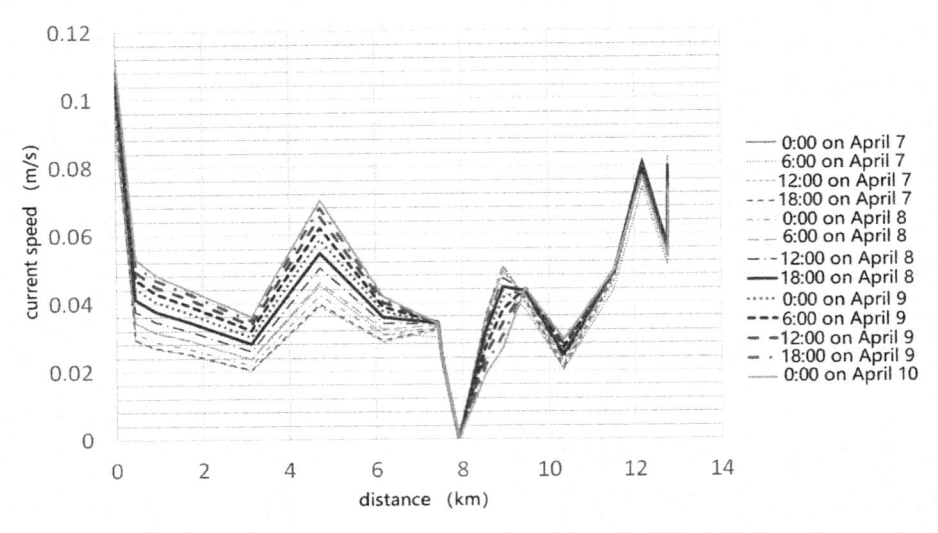

Figure 3.    Variation of velocity with distance at different times along the Qiliuqiu-Beitang river.

### 3.3  *BOD concentration analysis*

It can be seen from Figure 4 that the changing trend of the Qiliuqiu-Beitang River first decreases, then increases sharply, and then tends to be stable at each moment, which shows that with the flow of water, the river section far away from No. 1 gate of Xidadi can also be improved, but the improvement is not very obvious. After 0:00 on April 10, the BOD concentration in this river section is close to meeting the standard of class III water.

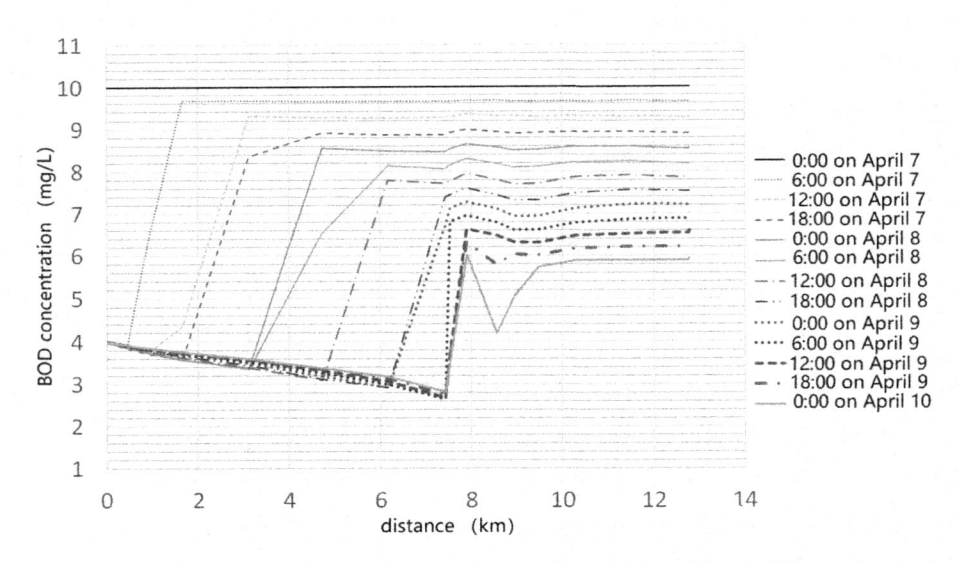

Figure 4.    Variation of BOD concentration with distance in Qiliuqiu-Beitang River at different times.

### 3.4 *Do concentration analysis*

It can be seen from Figure 5 that the DO concentration in the Beitang river of Qiliuqiu is relatively stable before 6:00 on April 10, fluctuates greatly after 6:00 on April 7, and the DO concentration changes most at the point farthest from the origin of the river. With the passage of time, the DO concentration of each point in the river changes greatly with time, and at 6:00 on April 7, after 0.5km from the origin, the DO concentration does not increase but decreases sharply. It is speculated that the closure of the Hongqi gate makes the water flow unable to flow so that the DO concentration in that section cannot be improved. In addition, by the completion of the simulation calculation, the improvement of water quality basically meets the requirements.

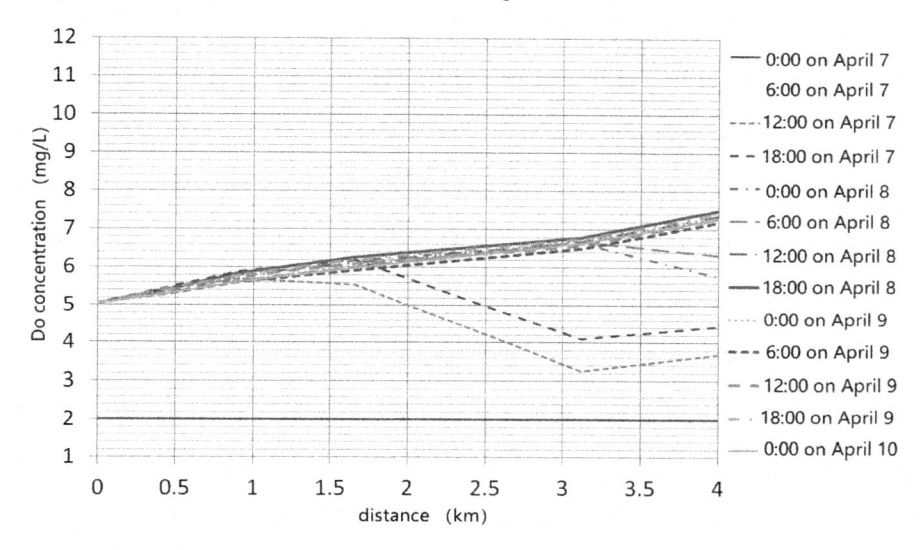

Figure 5.    Variation of DO concentration with distance in Qiliuqiu-Beitang River at different times.

Figure 6 is the variation curve of DO concentration with distance at different times. From the figure, it can be seen that the changes in hydrodynamic and water quality indexes are as follows;

First: as shown in the figure, the flow velocity at the selected point reached the maximum value of 0.000075 m/s 3 hours after the opening of No. 1 gate of Xidadi and Hangzhou Bay gate 2, but dropped sharply to 0.00004 m/s 14 hours after the opening of the gate. When the gate (Hongqi gate) of the Qiliuqiu-Beitang River is closed, the flow velocity in the reach is approximately unchanged, and the change rate is almost 0.

Second: according to the figure, the BOD concentration generally shows a significant reduction trend. Due to the unstable change of flow rate, it sometimes increases and sometimes decreases. Due to gate closing, the proportional relationship between BOD and flow rate cannot be seen temporarily.

## 4    CONCLUSION

By establishing a two-dimensional mathematical model, it is found that the water surface elevation is relatively stable after water diversion, and the water surface elevation of each point decreases gradually with time; the flow velocity changes greatly and fluctuates; the final BOD concentration in this reach is close to the standard of class III water; with the passage of time, the DO concentration at each point in the river does not change greatly with time, and tends to decrease slowly. Finally, all points on the river reach can meet the expected standards and reach the indicators of class III water or even class II water. To sum up, the drinking water scheme can improve the water

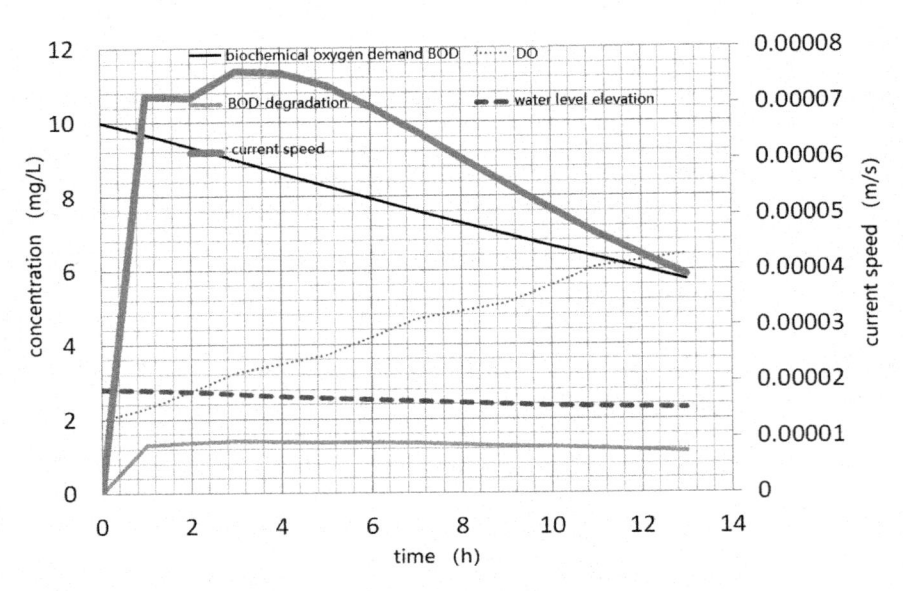

Figure 6. Variation diagram of DO concentration with distance at different times.

environment, and the scheme is feasible. Due to the limited space, numerical simulation analysis of many surrounding rivers should be carried out in the future, so as to make the conclusion more convincing.

## ACKNOWLEDGMENTS

This research was supported by the Funds Key Laboratory for Technology in Rural Water Management of Zhejiang Province(ZJWEU-RWM-202101), the Joint Funds of the Zhejiang Provincial Natural Science Foundation of China (No. LZJWZ22C030001, No. LZJWZ22E090004), the Funds of Water Resources of Science and Technology of Zhejiang Provincial Water Resources Department, China (No.RB2115, No.RC2040), the National Key Research and Development Program of China (No.2016YFC0402502), the National Natural Science Foundation of China (51979249).

## REFERENCES

Gao Q, Tang Q.H, Meng Q.Q. (2015) Evaluation on improvement effect of connecting water environment of the tidal river and lake water system. *People's Yangtze River*, 46(15): 38–40 + 50.

Ji H.T. (2022) Analysis of the impact of sluice operation on water quality based on hydrodynamic model. *Electromechanical technology of hydropower station*, 45 (01): 82–84 + 121.

Lu X.Y, Wang M.M, Huang F. (2013) Influence of Fenghua River and Yongjiang River Regulation on flood control capacity of Yaojiang River and its countermeasures. *Journal of Zhejiang water resources and Hydropower College*, 25 (01): 44–47.

Tong Y.Y, Li D.F, Nie H. (2016) Study on a hydrodynamic and plane two-dimensional mathematical model of the river network in Datian Plain. *Journal of Zhejiang University of water resources and hydropower*, 28 (01): 14–17.

Wu J.J. (2019) Influence of sluice operation on river water quality change. *Engineering technology research*, 4 (04): 213–214.

Yang W, Zhang L.P, Li Z.L, Zhang Y.J, Xiao Y, Xia J. (2018) Study on river lake connection scheme of urban lake group based on water environment improvement. *Journal of Geography*, 73(01):115–128.

*Civil Engineering and Urban Research – Mohamed & Hou (Eds)*

# Numerical simulation on the effects of temperature difference and heat loss on the oscillations of thermo-solutocapillary convection

Jungeng Fan

*Key Laboratory of Electromagnetic Processing of Materials, Ministry of Education, Northeastern University, Shenyang, Liaoning Province, China*

Ruquan Liang*

*School of Mechanical and Vehicle Engineering, Linyi University, Linyi, China*

ABSTRACT: This paper uses a mixed solution of Toluene/n-hexane as a liquid bridge and captures any small changes in the two-phase free interface using the improved Level-Set method to study the effects of the temperature difference and heat loss on the thermo-solute capillary convective flow structure and oscillation characteristics in a microgravity environment. The results show that the interface deformation under different temperature differences has an "S"-type distribution in which the shape of the upper is concave and the under is convex, and the degree of them increases with the growth of the temperature difference. The radial velocity at the upper of the liquid bridge exhibits a single-cycle sinusoidal oscillation and axial velocity exhibits a double-cycle small amplitude pulsating oscillation under small temperature difference conditions. The intermediate position radial velocity is double-cycle small amplitude pulsation oscillation, and the axial velocity is single-cycle small amplitude pulsation oscillation. The upper and the middle position speed oscillation are large amplitude pulsation oscillations of multiple cycles under the condition of large temperature differences.

## 1 INTRODUCTION

The semiconductor material in the melt during the growth of the floating zone is often doped with impurities. The non-uniform distribution of impurity concentrations on the free surface will produce solute capillary convection caused by the concentration gradient. It will couple with hot capillary convection to form Marangoni convection on the free surface. Such flows are very common in many industrial applications, e.g., in crystal growth (Timofeev et al. 2015; Fang et al. 2012), welding (Fujii et al. 2008; Lu et al. 2008; Xu et al. 2007), evaporation (Girard et al. 2008; 2006). The method used to study this phenomenon is to use a liquid bridge model. A so-called liquid bridge is a volume of liquid kept by surface tension between two co-axial solid rods. After the upper disc is heated and a certain amount of solute is introduced, the surface tension based on temperature and concentration changes to form a thermo-solute Marangoni stress at the free interface, ultimately causing the fluid to move at the free interface. When the temperature and concentration difference between the upper and lower discs are small, a stable axisymmetric vortex structure appears inside the liquid bridge, but when the temperature and concentration difference reach a critical value, time-based periodic oscillation occurs inside. It was first experimentally observed by Chun (Chun & Wuest 1979) et al.

In general, people have a very limited understanding of the coupled thermo-solute capillary convection in the floating zone. There was no report on the oscillating coupled thermal-solute

---

*Corresponding Author: 1810567@stu.neu.edu.cn

DOI 10.1201/9781003372417-105

capillary convection study considering the dynamic deformation of the free surface. Therefore, this paper carries out research in this area and considers the effect of free surface dynamic deformation on thermal-solute capillary convection. The purpose is to provide guidance for the crystal growth process using the floating zone method to improve crystal quality. In this paper, the mathematical model of thermal solute coupled capillary convection and the Level Set method used in interface tracking are introduced in the second part, and the accuracy of the model used is verified. In the third part, the effects of four different temperature differences on the thermo-solute capillary convective flow structure and oscillation characteristics are discussed. Then, the influence of heat loss on the thermo-solute capillary convective flow structure and oscillation characteristics is discussed. In the last part, the effects of different variables on the flow and oscillation in the liquid bridge are summarized.

## 2 PROBLEM FORMULATION AND RESULTS

The physical model of the liquid bridge is shown in Figure 1, and the cylindrical liquid bridge with radius R and distance H is suspended initially between two coaxial disks and surrounded by gas in a rectangular container with radius 2R and distance H. Different temperatures and concentrations are applied at the upper disk (T1, C1) and lower disk (T0, C0), where T1 > T0 and C1 > C0. The general governing equations of the problem under microgravity are given by the following non-dimensional mass conservation, Navier-Stokes, energy conservation, and solute diffusion equations.

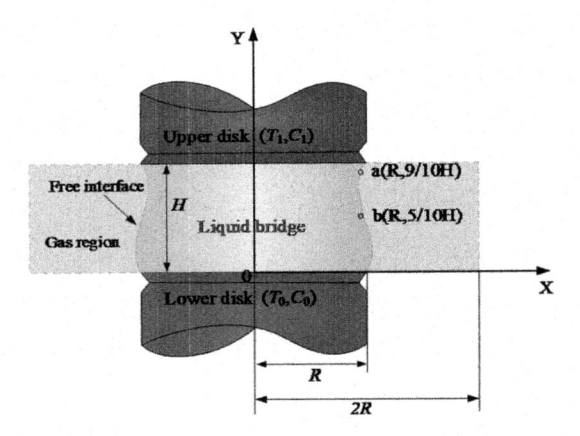

Figure 1.   Schematic of liquid bridge model and layout of monitoring points.

### 2.1  *Effect of temperature difference on thermal-solute capillary convective flow structure*

This paper mainly studies the influence of temperature difference and heat loss on the flow structure of the coupled thermo-solute capillary convection in the floating zone and the influence on the oscillation characteristics of the free interface. The selected liquid bridge working medium is a mixed solution, and the physical properties are shown in Table 1. The calculation conditions are shown in Table 2 and Table 3 respectively, in which the liquid bridge radius is 2.5 mm and the height to diameter ratio is 1. In order to monitor the free interface oscillation characteristics, two fixed monitoring points "a" and "b" are set in the right interface corner area (R, 9/10H) and the intermediate position (R, 5/10H). The physical model is shown in Figure 1.

Figure 2 shows the speed vector under different temperature differences. It can be seen from the figure that the change of the temperature difference has little effect on the velocity vector field in the vicinity of the central axis of the liquid bridge (x = -1.0 mm ∼1.0 mm), but great changes have occurred due to the change of the temperature difference for a velocity vector field of other

Table 1.  Physical properties of the toluene/n-hexane fluid.

| Property | symbol | value |
| --- | --- | --- |
| density | $\rho_l$ | $699 \text{kg/m}^3$ |
| Diffusion coefficient | $\alpha_l$ | $1.88*10^{-7}\text{m}^2/\text{s}$ |
| dynamic viscosity | $\mu_l$ | $3.37*10^{-4}\text{kg/(m·s)}$ |
| Temperature surface tension coefficient | $\sigma_T$ | $9.43*10^{-5}\text{N/(m·K)}$ |
| Concentration surface tension coefficient | $\sigma_C$ | $-8.62*10^{-3}\text{N/m}$ |
| Prandt number | $Pr$ | 5.54 |
| Lewis number | $Le$ | 25.8 |

regions. It can be clearly observed from figure 2(a) to figure 2(d) that the flow velocity of the fluid near the interface on both sides of the liquid bridge increases as the temperature difference increases. When the temperature difference is 1.5 K ∼ 2.5 K, the fluid flow near the interface is relatively stable. When the temperature difference is increased to 6 K, the fluid flow near the interface is no longer stable, and a large degree of fluctuation occurs which promotes the absorption of the surface material of the liquid bridge and the convection capillary convection, thereby causing the interface fluid flow to become disordered. Since the vicinity of the liquid bridge interface is mainly the surface recirculation area, the increase in the flow rate means that the volume reflow and surface reflow conversion speed are accelerated. By comparing the velocity vector diagrams under different temperature differences, it can be seen that as the temperature difference increases, the flow velocity of the liquid bridge interfaces fluid increases. When it is aggravated to a certain extent, the interface fluid flow becomes disordered. The conversion speed of surface recirculation and volume recirculation also increases as the temperature difference increases.

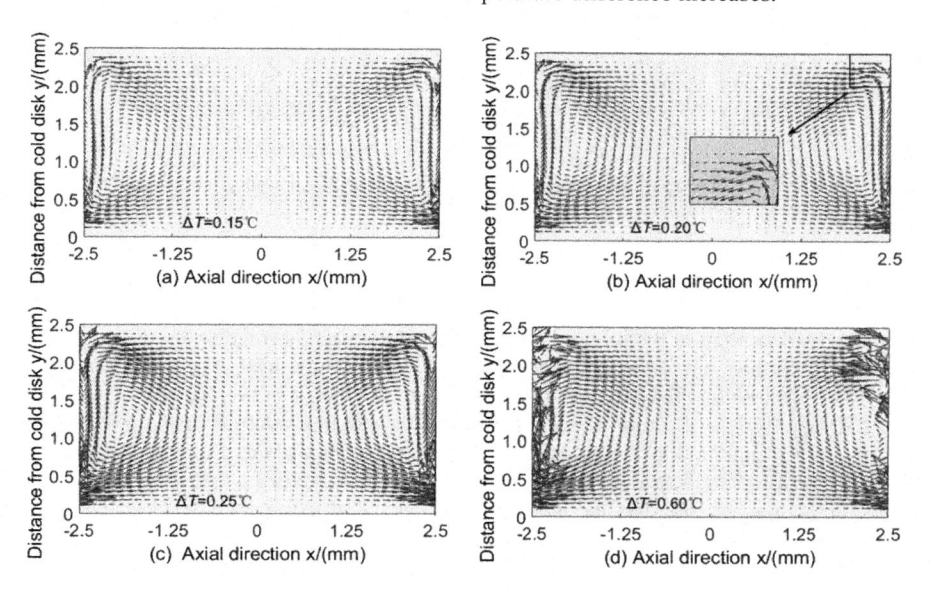

Figure 2.  Velocity vector at different temperatures.

## 2.2  *Effect of temperature difference on temperature and concentration on the free surface*

Figure 3(a) shows the temperature change of the free surface on the right side of the liquid bridge and (b) shows the change in the concentration of the free surface on the right side. it can be seen from Figure 5(a) that the temperature of the liquid bridge interface is distributed as a convex function under different temperature differences and the temperature gradient at the upper end of the liquid bridge is significantly larger than the temperature gradient at the lower end. With the increase in temperature difference, the gradient difference between the upper end and the lower end of the liquid bridge gradually decreases, indicating that the temperature distribution at the liquid bridge interface tends to be evenly distributed with the increase in temperature difference. For the concentration change of the free interface, it can be seen from Figure 3 (b) that although the temperature difference is different the concentration distribution curve is in the shape of "F". The hot end of the liquid bridge (y = 2.25 mm ~ y = 2.50 mm) and the cold end (y = 0.0 mm ~ y = 0.50 mm) region form a large concentration gradient and the concentration of the near-hot end region of the liquid bridge increased sharply from 0.0165g/g to 0.055g/g, while the liquid bridge near the cold end region. The concentration is sharply reduced from 0.012g/g to 0.017g/g to 0. Since the near hot end and the near cold end of the liquid bridge are the areas of volume reflow and surface reflow conversion, a large concentration gradient is formed in the area. At the same time, it can be seen that in the range of liquid bridge height y = 0.50 mm to y = 2.25 mm, the concentration shows a decreasing trend with the increase of temperature difference. The temperature change is more uniform than the concentration change.

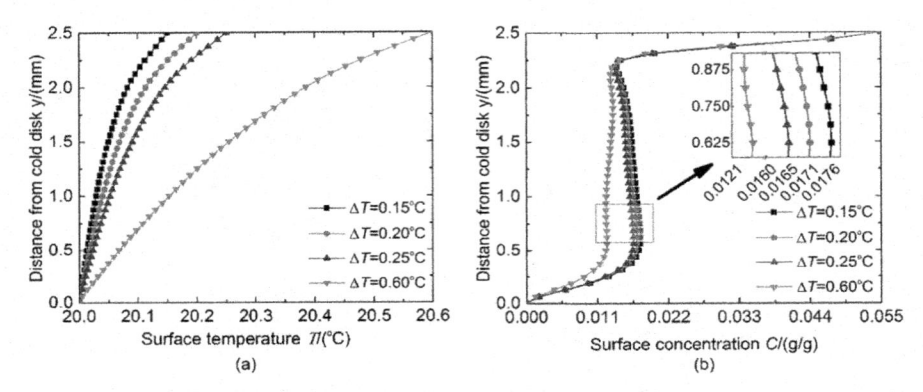

Figure 3.  (a) Variation of temperature on the right free surface, (b) Variation of concentration on the right free surface.

## 2.3  *Effect of heat loss on thermal-solute capillary convective flow structure*

Figure 4 is a contour map of the internal temperature of the liquid bridge under different heat loss conditions. Comparing the temperature contour maps under three heat loss conditions, it can be seen that the heat loss has little effect on the temperature change in the vicinity of the center of the liquid bridge (x = −1.25 mm ~ 1.25 mm). The temperature contour of the area is almost a straight line indicating that the heat transfer pattern in the vicinity of the center of the liquid bridge is mainly in the form of heat conduction. However, for areas close to the free interface, there are some subtle changes in the temperature field of the area due to heat loss. Compared with the temperature in the vicinity of the center, the temperature in the vicinity of the free interface of the liquid bridge becomes no longer uniform. The degree of unevenness increases with the increase of the Bi number. The closer to the free interface, the more obvious this situation is. The heat transfer method near the free interface is mainly convective heat transfer under the influence of the surface reflow of the liquid bridge, which results in a certain unevenness in the temperature distribution of the region compared with the central region. The heat loss of the liquid bridge is increased when the Bi number is increased. The temperature value at the same height position of the free interface is continuously decreased so that the temperature distribution unevenness near the interface is gradually increased. Figure 4 is a contour plot of the internal concentration of the liquid bridge under different heat loss conditions. Concentration contours exhibit large non-uniformities compared to temperature contours. Under the three heat loss conditions, the concentration values ??in most of the middle part of the liquid bridge are almost unchanged. However, for the area close to the free interface, as the Bi number increases, the concentration contour of the area gradually approaches the upper end of the liquid bridge forming a large concentration gradient at the upper end of the liquid bridge.

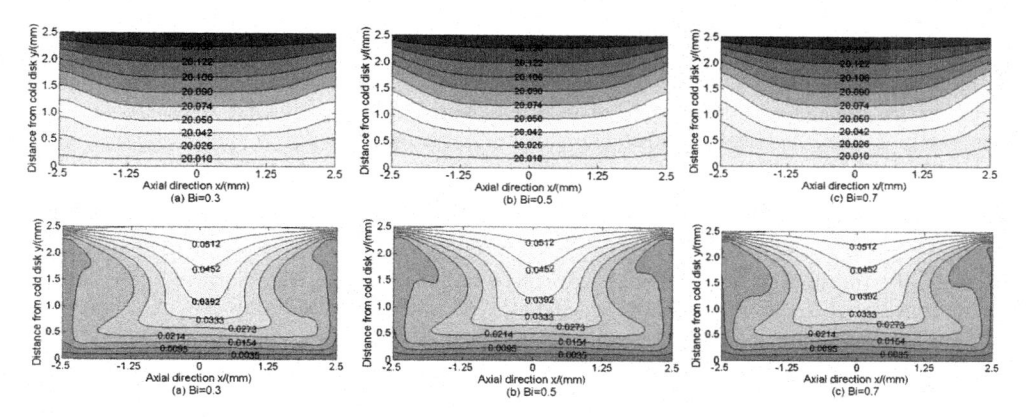

Figure 4. The iso-thermals (up) and iso-concentrations (down) under different heat loss conditions.

## 2.4 *Effect of heat loss on oscillation characteristics*

Figure 5 shows the radial velocity oscillation of the upper detection point "a" under different heat loss conditions. It can be seen from the figure that the radial velocity of the upper monitoring point "a" under different heat loss conditions undergoes a stable development process and then begins to oscillate.

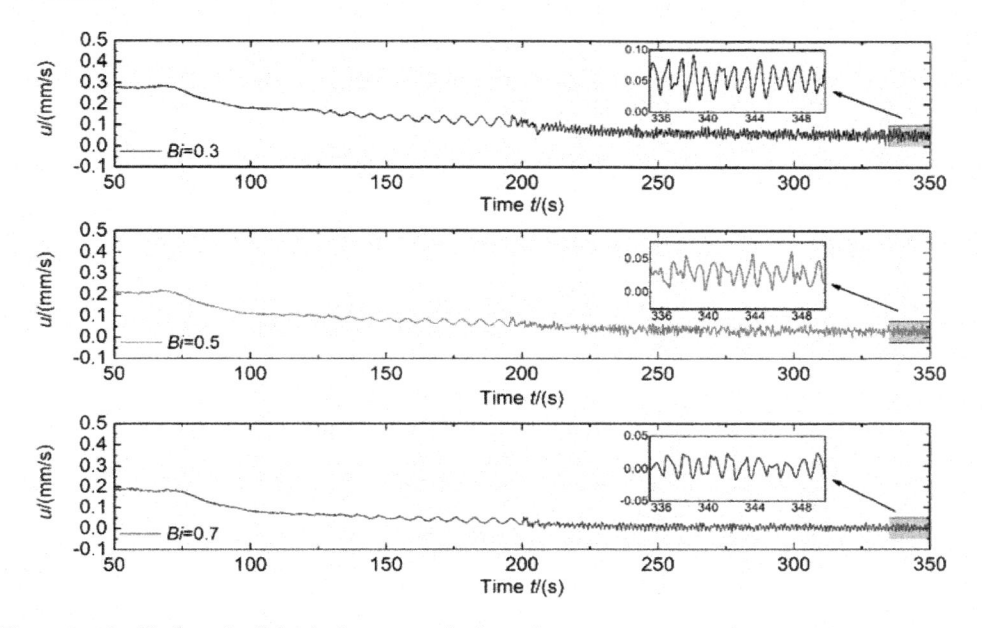

Figure 5. Oscillation of radial velocity at a monitoring point.

The amplitude gradually increases and finally enters the high-frequency stable oscillation phase. For the high-frequency stable oscillation phase, it can be seen from the figure that when Bi=0.3, the average radial velocity is 0.97 mm/s, the average amplitude is 0.055 mm, and the average oscillation period is 1.13 s; when Bi=0.5, the velocity, amplitude and oscillation period are 0.83mm/s, 0.12mm and 1.16s respectively; when Bi=0.043, the velocity, amplitude and oscillation period are 0.83mm/s, 0.036mm, and 1.23s respectively. The radial velocity value and amplitude gradually decrease as the Bi number increases, and the average oscillation period gradually increases. Since the upper monitoring point "a" is in the high concentration region, the fluid flow is greatly affected by the solute capillary convection. As can be seen from Figure 4.35, the concentration of the monitoring point "a" is reduced with the degree of heat loss increases. The influence of solute capillary convection is weakened, which leads to a gradual decrease in flow velocity and amplitude.

## 3 CONCLUSION

In summary, for the first time, the mass conserved level set method is used to study the thermal solute convective flow and oscillation characteristics of the floating zone coupled with temperature difference and heat loss in the microgravity environment. The following conclusions can be drawn:

The flow velocity of the liquid bridge interface fluid increases with the increase of the temperature difference. When it is aggravated to a certain extent, the interface fluid flow becomes disordered, and the conversion speed of surface recirculation and volume recirculation also increases with the increase of the temperature difference. The interface deformation under different temperature differences has an "S"-type distribution. The degree of convexity and depression increases with the increase of the temperature difference. The speeds at the upper and middle positions of the liquid bridge exhibit different forms of oscillation. The radial velocity at the upper end of the liquid bridge exhibits a sinusoidal oscillation under small temperature difference conditions. This sinusoidal oscillation is a single-cycle oscillation, but the axial velocity exhibits a small amplitude pulsation oscillation. This small amplitude oscillation is twice the period; Both the radial velocity and the axial velocity of the intermediate position exhibit small amplitude pulsation oscillation, but the small amplitude pulsation oscillation of the radial velocity is twice the periodic oscillation. The small amplitude pulsation oscillation of the axial velocity is the single periodic oscillation. The upper end and the middle position velocity oscillation are both large amplitude pulsation oscillations under the condition of large temperature differences. This large amplitude pulsation oscillation is a multiple period oscillation.

## ACKNOWLEDGMENTS

This research was funded by the National Natural Science Foundation of China under the grants of 51976087 and 51676031.

## REFERENCES

Chun C H and Wuest W 1979. *Acta Astronaut.* 6 1073–82
Fang H, Tian J, and Zhang Q 2012 *Int. J. Heat Mass Transfer* 55 8003–09
Fujii H, Sato T and Lu S 2008. *Mater. Sci. Eng. A* 495 296–303
Girard F, Antoni M and Faure S 2006. *Langmuir* 22 11085–91
Girard F, Antoni M and Sefiane K 2008. *Langmuir* 24 9207–10
Lu S, Fujii H and Nogi K 2008 *J. Mater. Sci.* 43 4583–91
Timofeev V V, Kalaev V V and Ivanov V G 2015 *Int. J. Heat Mass Transfer* 87 42–48
Xu Y L, Dong Z B, and Wei Y H 2007 *Theor. Appl. Fract. Mech.* 48 178–186

*Civil Engineering and Urban Research – Mohamed & Hou (Eds)*
*© 2023 the Authors, ISBN 978-1-032-44487-1*

# Evaluation of internal environmental quality of civil air defense project based on matter-element extension evaluation model optimized by AHP entropy weight method

Yipeng Ning
*Air Force Engineering University, Xi'an, China*

Dongmei Yao & Xin Luo*
*National Defense Engineering Research Institute, Academy of Military Sciences, Beijing, China*

Yilun Zhang
*Unit of the Chinese People's Liberation Army, Xi'an, China*

Erlei Bai & Ao Yao
*Air Force Engineering University, Xi'an, China*

ABSTRACT: In this paper, aiming at the problem of poor internal environment quality existing in the current underground civil air defense projects, scientific and feasible maintenance and optimization measures are explored. Starting from the internal environmental quality of underground civil air defense projects, the evaluation index system of internal environmental quality of underground civil air defense projects is established, and a matter-element extension evaluation model based on AHP entropy weight method optimization is constructed (Zhang et al. 2005). Combined with the change information of the position of each index measuring point, a comprehensive evaluation of the internal environmental quality of the underground civil air defense project is carried out. The evaluation results show that the internal environmental quality of the underground civil air defense project is generally low, and the two indicators of noise and CO2 concentration need to be further optimized. The evaluation results are scientific and reasonable.

## 1 INTRODUCTION

As an important part of the modern national defense military system, civil air defense engineering(Yang et al. 2013) plays a more important role in the modern high-tech local war than ever before. The civil air defense project is deep underground. The interior of the underground civil air defense project is relatively isolated from the external environment, with strong tightness, poor lighting and ventilation conditions, difficult air flow, large contrast with the natural environment, and a great impact on the psychology and physiology of the stationed personnel. Adopting scientific and effective evaluation methods can not only provide theoretical support for the technical research of the internal environmental quality improvement of civil air defense projects, but also play an important role in solving the health problems of personnel in daily activities.

In this paper, the matter-element extension evaluation model optimized based on the AHP entropy weight method is used to evaluate the internal environment of civil air defense engineering(Chen et al. 1997), which provides a theoretical basis for the subsequent research on the internal environment improvement technology of civil air defense engineering. At the same time, the application prospect of the matter-element extension evaluation model optimized based on the AHP entropy weight method in the internal environment quality evaluation of underground civil air defense engineering is studied.

---

*Corresponding Author: 13630283725@163.com

## 2 CONSTRUCTION OF INTERNAL ENVIRONMENTAL QUALITY EVALUATION MODEL

Firstly, the index weight coefficient is determined by using the analytic hierarchy process and entropy weight method(Chen et al. 2019), and then the internal environment assessment model is established by combining the matter-element analysis method, extension theory, and correlation degree theory. Finally, the single objective evaluation model can be transformed into a multi-objective comprehensive evaluation model.

### 2.1 Construction of main characteristic matter element matrix

In this paper, the research and evaluation object is set as N, its characteristics are recorded as C, and its corresponding characteristic quantity is recorded as V. Let R=[N,C,V], then R represents the multi-dimensional parametric matter element with multi characteristic elements of the evaluation thing N, that is, the main characteristic matter element matrix.

$$R = \begin{bmatrix} N & C_1 & V_1 \\ & C_2 & V_2 \\ & \cdots & \cdots \\ & C_n & V_n \end{bmatrix} = \begin{bmatrix} R_1 \\ R_2 \\ \cdots \\ R_n \end{bmatrix} \tag{1}$$

### 2.2 Construction classical domain matter-element matrix

By combining the code requirements for the construction of underground civil air defense engineering facilities, relevant national standards, and the research results of other scholars in the literature synthesis, and combined with expert opinions, the summary table of internal environmental assessment level and standard range of underground civil air defense engineering is determined, and each level is described as a quantitative classical domain matter-element matrix $R_d$.

$$R_d = (N_d, C, v_{di}) = \begin{bmatrix} N_d & C_1 & v_{d1} \\ & C_2 & v_{d2} \\ & \vdots & \vdots \\ & C_i & v_{di} \end{bmatrix} \tag{2}$$

### 2.3 Construction of nodal matter-element matrix

The nodal matter-element matrix ($R_e$) is composed of the comprehensive allowable value range of the internal environmental quality evaluation index of the underground civil air defense project.

$$R_e = (N_e, C, v_{ei}) = \begin{bmatrix} N_e & c_1 & (y_{e1}, p_{e1}) \\ & c_2 & (y_{e2}, p_{e2}) \\ & \vdots & \vdots \\ & c_i & (y_{ei}, p_{ei}) \end{bmatrix} \tag{3}$$

### 2.4 Construction of the matter-element matrix to be evaluated

$$R_f = \begin{bmatrix} N & c_1 & v_1 \\ & c_2 & v_2 \\ & \vdots & \vdots \\ & c_n & v_n \end{bmatrix} \tag{4}$$

Firstly, it needs to determine the correlation degree function of the matter element to be evaluated for different categories, and assuming that the bounded real domain interval is $D_0$:

$$D_0 = [c, d] \tag{5}$$

The modulus of $D_0$ is defined as:

$$|D_0| = |d\text{-}c| \tag{6}$$

The distance from point $x_0$ to interval $D_0$ is:

$$\rho(x_0, D_0) = \left| x - \frac{(c+d)}{2} \right| - \frac{(d-c)}{2} \tag{7}$$

Therefore, the expression of the correlation function is:

$$k(x) = \begin{cases} -\rho(x, D_0) & (x \in D_0) \\ \dfrac{\rho(x, D_0)}{\rho(x, D) - \rho(x, D_0)} & (x \notin D_0) \end{cases} \tag{8}$$

## 2.6 *Determination of weight by AHP entropy weight method*

The specific steps of determining the weight by analytic hierarchy process are as follows:

The score range is 1-9 points, and then establish a judgment matrix, as shown in Formula 9 below:

$$A = \{A_1, A_2, A_3, \cdots, A_n\} \tag{9}$$

The judgment matrix constructed according to the expert scoring results is shown in the following formula 10:

$$A = \left(a_{ij}\right)_{n \times n} = \begin{bmatrix} a_{11} & a_{12} & \cdots & a_{1n} \\ a_{21} & a_{22} & \cdots & a_{2n} \\ \vdots & \vdots & \vdots & \vdots \\ a_{n1} & a_{n2} & \cdots & a_{nn} \end{bmatrix}_{n \times n} \tag{10}$$

In the actual analysis, the reference factors are compared in pairs each time. For example, $a_{ij}$ is used to represent the comparison results of the importance of indicators $x_i$ and $x_j$. The specific comparison result score value is found and determined by referring to the importance scale table.

Consistency verification for judgment matrix A, when the consistency meets the requirements, the following Formula 11 shall be met:

$$\lambda_{\max} = n \tag{11}$$

In the above formula: n – the order of judgment matrix A;

The consistency index determined by the maximum eigenvalue and order of the matrix is used for consistency verification, and the consistency index CI (consistency index) is defined as:

$$CI = \frac{\lambda_{\max} - n}{n - 1} \tag{12}$$

The closer the CI value is to 0, the better the consistency. However, for the possibility of many inconsistencies in practice, in order to measure the size of CI, the random consistency index R.I. is introduced, as shown in Table 1.

The concept of the ratio of consistency index CI and random consistency index R.I. is introduced, and the allowable range of its error is specified to test whether the consistency of the judgment matrix meets the requirements. The formula of test coefficient CR is shown in formula 13 below:

$$CR = \frac{CI}{RI} \tag{13}$$

Table 1. Standard values of random consistency index R.I.

| Matrix order | 1 | 2 | 3 | 4 | 5 | 6 | 7 | 8 | 9 | 10 |
|---|---|---|---|---|---|---|---|---|---|---|
| R.I. | 0 | 0 | 0.58 | 0.9 | 1.12 | 1.24 | 1.32 | 1.41 | 1.45 | 1.49 |

If CR < 0.1, the consistency of the judgment matrix meets the requirements. If CR ≥ 0.1, the consistency of judgment matrix A does not meet the requirements and needs to be adjusted.

In subjective weight calculation, each vector of judgment matrix A is normalized after geometric average, and the obtained column vector is used as weight vector Q:

$$q_i = \frac{\left(\prod_{j=1}^{n} a_{ij}\right)^{\frac{1}{n}}}{\sum_{k=1}^{n} \left(\prod_{j=1}^{n} a_{kj}\right)^{\frac{1}{n}}} \tag{14}$$

$$Q = (q_1, q_2, q_3, \cdots, q_n)^T \tag{15}$$

The specific steps of determining the weight by entropy weight method are as follows:

(1) Construct the initial matrix hypothesis and select the measured data of m test positions, which are $C_i$ (i=1, 2,..., m); There are n indexes to be evaluated, which are $P_j$ (j = 1, 2,..., n). Each evaluation index in the study is each environmental quality evaluation index. The measured values of each index at different test positions can form an initial matrix X:

$$X = (x_{ij})_{m \times n} \tag{16}$$

In the above formula, $x_{ij}$ – the measured value of the j-th index to be evaluated at the i-th measuring point;

(2) Initial matrix standardization processing the initial matrix X is standardized to obtain the standard matrix V:

$$v_{ij} = \frac{x_{ij} - \min(x_j)}{\max(x_j) - \min(x_j)} \left(0 \leq v_{ij} \leq 1\right) \tag{17}$$

$$V = (v_{ij})_{m \times n} \tag{18}$$

(3) Calculate the characteristic proportion of the index, calculate the proportion of the measured value sample of the j-th index to be evaluated at the i-th measuring point, and regard it as the probability in entropy calculation to measure the amount of information:

$$p_{ij} = \frac{v_{ij}}{\sum_{i=1}^{m} v_{ij}} \left(0 \leq p_{ij} \leq 1\right) \tag{19}$$

(4) Calculate the information entropy of the index. The information entropy of the j-th index to be evaluated is:

$$e_j = -\frac{1}{\ln n} \sum_{i=1}^{n} p_{ij} \ln p_{ij} \quad (j = 1, 2, \cdots, m) \tag{20}$$

(5) Calculate the difference coefficient of the index. The greater the difference coefficient, the greater the degree of variation of the index, the more information it carries, the greater the impact on the final target result, and the higher the proportion of weight.

$$d_j = 1 - e_j \tag{21}$$

748

(6) Calculate the objective weight of the index, which can be expressed as:

$$w_j = \frac{d_j}{\sum\limits_{j=1}^{m} d_j} \quad (j = 1, 2, \cdots, m) \tag{22}$$

(7) Finally determine the combination weight:

$$z_i = \frac{(w_i \cdot q_i)^{1/2}}{\sum\limits_{i=1}^{m} (w_i \cdot q_i)^{1/2}} \tag{23}$$

$$\sum\limits_{i=1}^{m} z_i = 1 \quad (i = 1, 2, \cdots, n) \tag{24}$$

## 2.7 *Calculation of comprehensive correlation degree*

Determine the comprehensive correlation degree according to the relevant formula:

$$K_j(N_v) = \sum\limits_{i=1}^{n} \omega_i k_i(v_i) \tag{25}$$

## 3 CONSTRUCTION OF INTERNAL ENVIRONMENTAL QUALITY EVALUATION MODEL

### 3.1 *Construction of internal environmental quality status evaluation index system*

Establishing a scientific and reasonable index system (Abdul et al. 2015; Lu 2020; Piasecki 2019; Zuhaib et al. 2018) is the premise of internal environmental quality evaluation. This study mainly selects the indexes that can reflect the current situation characteristics of internal environmental quality of civil air defense projects and have a great impact on human health according to relevant literature and code for design of civil air defense basement (GB50038-2005), and finally determines the current situation evaluation indexes of internal environmental quality of 6 civil air defense projects.

The entropy weight method is used to calculate the index weight coefficient, and the calculation results are shown in Table 2.

### 3.2 *Construction of the classical domain and node domain of the evaluation model*

According to the corresponding model construction process rules, the classical domain matter-element matrix R1-4 of the internal environment evaluation of the underground civil air defense project is as follows:

$$R_1 = (N_1, C, v_{1i}) = \begin{bmatrix} N_1 & C_{1-3} & v_{1,1-3} \\ & C_4 & v_{1,4} \\ & C_5 & v_{1,5} \\ & C_6 & v_{1,6} \\ & C_7 & v_{1,7} \\ & C_8 & v_{1,8} \end{bmatrix} = \begin{bmatrix} N_1 & C_{1-3} & [60, 70] \\ & C_4 & [750, 1000] \\ & C_5 & [0, 35] \\ & C_6 & [0, 500] \\ & C_7 & [0, 0.005] \\ & C_8 & [0, 0.05] \end{bmatrix} \tag{26}$$

Table 2. Weight table of internal environmental quality evaluation system of underground civil air defense project based on AHP entropy weight method.

| Target layer A | Criterion layer B | Weight value | Index layer C | Weight value | Subjective weight | Objective weight | Comprehensive weight |
|---|---|---|---|---|---|---|---|
| A: Internal environmental quality of underground civil air defense project | $B_1$: thermal environment | 0.4844 | $C_{1-3}$: Comfort index | 1 | 0.4844 | 0.1549 | 0.4418 |
| | $B_2$: Light environment | 0.1095 | $C_4$: Illumination | 1 | 0.1095 | 0.1736 | 0.1119 |
| | $B_3$: Acoustic environment | 0.0632 | $C_5$: Noise | 1 | 0.0632 | 0.1482 | 0.0551 |
| | $B_4$: Air quality | 0.3430 | $C_6$: $CO_2$ concentration | 0.6929 | 0.2377 | 0.2103 | 0.2943 |
| | | | $C_7$: $PM_{2.5}$ concentration | 0.2308 | 0.0792 | 0.1564 | 0.0729 |
| | | | $C_8$: $PM_{10}$ concentration | 0.0762 | 0.0260 | 0.1565 | 0.024 |

$$R_2 = (N_2, C, v_{2i}) = \begin{bmatrix} N_2 & C_{1-3} & v_{2,1-3} \\ & C_4 & v_{2,4} \\ & C_5 & v_{2,5} \\ & C_6 & v_{2,6} \\ & C_7 & v_{2,7} \\ & C_8 & v_{2,8} \end{bmatrix} = \begin{bmatrix} N_2 & C_{1-3} & [51, 60; 70, 79] \\ & C_4 & [750; 500] \\ & C_5 & [35, 40] \\ & C_6 & [500, 707] \\ & C_7 & [0.005, 0.014] \\ & C_8 & [0.05, 0.16] \end{bmatrix} \quad (27)$$

$$R_3 = (N_3, C, v_{3i}) = \begin{bmatrix} N_3 & C_{1-3} & v_{3,1-3} \\ & C_4 & v_{3,4} \\ & C_5 & v_{3,5} \\ & C_6 & v_{3,6} \\ & C_7 & v_{3,7} \\ & C_8 & v_{3,8} \end{bmatrix} = \begin{bmatrix} N_3 & C_{1-3} & [40, 51; 79, 85] \\ & C_4 & [500, 200] \\ & C_5 & [40, 45] \\ & C_6 & [707, 900] \\ & C_7 & [0.014, 0.038] \\ & C_8 & [0.16, 0.5] \end{bmatrix} \quad (28)$$

$$R_4 = (N_4, C, v_{4i}) = \begin{bmatrix} N_4 & C_{1-3} & v_{4,1-3} \\ & C_4 & v_{4,4} \\ & C_5 & v_{4,5} \\ & C_6 & v_{4,6} \\ & C_7 & v_{4,7} \\ & C_8 & v_{4,8} \end{bmatrix} = \begin{bmatrix} N_4 & C_{1-3} & [0, 40; 85, 100] \\ & C_4 & [200, 0] \\ & C_5 & [45, 60] \\ & C_6 & [900, 1000] \\ & C_7 & [0.038, 0.105] \\ & C_8 & [0.5, 1.58] \end{bmatrix} \quad (29)$$

According to the corresponding model construction process rules, the nodal matter-element matrix re of the internal environment assessment of the underground civil air defense project is as follows:

$$R_e = (N_e, C, v_{ei}) = \begin{bmatrix} N & C_{1-3} & [0, 100] \\ & C_4 & [0, 1000] \\ & C_5 & [0, 60] \\ & C_6 & [0, 1000] \\ & C_7 & [0, 0.105] \\ & C_8 & [0, 1.58] \end{bmatrix} \quad (30)$$

Table 3. Calculation results of internal environment correlation coefficient of underground civil air defense project.

| Test site | $K_i(v_i)$ | Excellent | Good | Qualified | Unqualified | $MAXK_i(v_i)$ | Correlation evaluation grade |
|---|---|---|---|---|---|---|---|
| Test site 1 | $K_{1-3}(v_{1-3})$ | $-0.5000$ | $-1.2571$ | $-1.0629$ | $-1.0692$ | $-0.5000$ | Excellent |
| | $K_4(v_4)$ | $-0.7400$ | $-0.6533$ | $-0.4800$ | $-0.5000$ | $-0.4800$ | Qualified |
| | $K_5(v_5)$ | $-0.5000$ | $-0.3750$ | $-0.1667$ | $0.1667$ | $0.1667$ | Unqualified |
| | $K_6(v_6)$ | $-0.7900$ | $-0.6416$ | $0.0259$ | $-0.0455$ | $0.0259$ | Qualified |
| | $K_7(v_7)$ | $-0.4615$ | $-0.3750$ | $0.1250$ | $-0.0789$ | $0.1250$ | Qualified |
| | $K_8(v_8)$ | $-0.4648$ | $-0.3667$ | $0.3529$ | $-0.2400$ | $0.3529$ | Qualified |
| Test site 2 | $K_{1-3}(v_{1-3})$ | $-0.4074$ | $-0.1111$ | $0.3333$ | $-0.2000$ | $0.3333$ | Qualified |
| | $K_4(v_4)$ | $-0.7000$ | $-0.8500$ | $-0.9063$ | $-0.9250$ | $-0.7000$ | Excellent |
| | $K_5(v_5)$ | $-1.0640$ | $-1.0800$ | $-1.1067$ | $0.0000$ | $0.0000$ | Unqualified |
| | $K_6(v_6)$ | $-0.9300$ | $-0.8805$ | $-0.6500$ | $0.3500$ | $0.3500$ | Unqualified |
| | $K_7(v_7)$ | $-0.4627$ | $-0.3793$ | $0.0833$ | $-0.0526$ | $0.0833$ | Qualified |
| | $K_8(v_8)$ | $-0.4713$ | $-0.3974$ | $0.1176$ | $-0.0800$ | $0.1176$ | Qualified |
| Test site 3 | $K_{1-3}(v_{1-3})$ | $-0.3810$ | $0.1111$ | $-0.0714$ | $-0.3500$ | $0.1111$ | Good |
| | $K_4(v_4)$ | $-0.5000$ | $-0.7500$ | $-0.8438$ | $-0.8750$ | $-0.5000$ | Excellent |
| | $K_5(v_5)$ | $-1.0840$ | $-1.1050$ | $-1.1400$ | $0.0000$ | $0.0000$ | Unqualified |
| | $K_6(v_6)$ | $-1.0380$ | $-1.0648$ | $-1.1900$ | $0.0000$ | $0.0000$ | Unqualified |
| | $K_7(v_7)$ | $-0.4648$ | $-0.3871$ | $0.0000$ | $0.0000$ | $0.0000$ | Unqualified |
| | $K_8(v_8)$ | $-0.4359$ | $-0.2143$ | $0.1765$ | $-0.5600$ | $0.1765$ | Qualified |

Table 4. Calculation results of comprehensive correlation coefficient of the internal environment of underground civil air defense project.

| Test site | Excellent | Good | Qualified | Unqualified | $MAXK_i(B)$ | Correlation evaluation grade |
|---|---|---|---|---|---|---|
| 1 | -0.6086 | -0.8747 | -0.5073 | -0.5440 | -0.5073 | Qualified |
| 2 | -0.6581 | -0.568 | -0.3067 | -0.4885 | -0.3067 | Qualified |
| 3 | -0.6508 | -0.5209 | -0.5071 | -0.2701 | -0.2701 | Unqualified |

## 4 CASE STUDY ON INTERNAL ENVIRONMENTAL QUALITY ASSESSMENT OF A CIVIL AIR DEFENSE PROJECT IN XI'AN

### 4.1 Calculation of single index correlation degree

### 4.2 Calculation of comprehensive correlation degree

The internal environmental quality evaluation grades of an underground civil air defense project in Xi'an are mainly "qualified" and "to be optimized". The internal environmental quality grade of the civil air defense project is generally not high and needs to be improved. On the premise of giving full play to the social and economic benefits of underground civil air defense projects, we need to adhere to the simultaneous development and maintenance, establish a daily maintenance and management mechanism, do a good job in the combination of peacetime and wartime while meeting people's daily production and life, and further optimize the internal environmental quality of underground civil air defense projects.

# 5 CONCLUSION

Based on the results and discussions presented above, the conclusions are obtained as below:

(1) In this paper, the matter-element extension evaluation model optimized based on the AHP entropy weight method is used to evaluate the internal environment of civil air defense projects. The results are basically consistent with the data of previous research results. It shows that the matter-element extension evaluation model optimized based on the AHP entropy weight method is credible for the internal environment evaluation results of civil air defense projects and can truly and accurately reflect the internal environment conditions of civil air defense projects.

(2) The internal environmental quality of underground civil air defense projects detected in this paper is generally low. In the evaluation index system, noise and carbon dioxide concentration need to be optimized and improved. The main reason is that people's daily life and work activities produce noise sources and spread noise, resulting in the increase of carbon dioxide concentration in the internal environment, and the old ventilation and dehumidification equipment can't effectively reduce the carbon dioxide concentration in the internal environment.

(3) This paper constructs the internal environmental quality evaluation index system and evaluation model of civil air defense projects, so as to provide a reference for the internal environmental evaluation of civil air defense projects. In the follow-up research, we can consider changing the number of evaluation indexes and evaluation methods to further improve the accuracy of evaluation results.

## ACKNOWLEDGMENTS

This work was financially supported by Project approved by the State Civil Air Defense Office, grant number RF20SC01J—S0.

## REFERENCES

Abdul-Wahab S A, En S C F, Elkamel A, et al. (2015) A review of standards and guidelines set by international bodies for the parameters of indoor air quality[J]. *Atmospheric Pollution Research*, 6(5): 751–767.

Chen G, Liu Y, Chi Q Y, et al. (1997) Comprehensive evaluation of indoor environmental quality of civil air defense fortifications by fuzzy mathematics [J] *Journal of the Second Military Medical University*, (03): 75–78.

Chen H Y, Lin Q, Liu W W, et al. (2019) Attribute recognition theoretical model based on entropy weight for comprehensive evaluation of indoor air quality[J]. *Environmental Engineering*, 37(09): 205–209.

Chen X, Liu Y. (2019) Subjective multidimensional evaluation model of minimal space based on human perception[J]. *Industrial architecture*, 49(10): 80–84+116.

GB/T17216-2012, *environmental hygiene requirements for civil air defense works in peacetime*[S]

Lu Y H. (2020) A Novel Indoor Air Quality Standards and Design Methods in Environmental Assessment[C]//*Journal of Physics: Conference Series*. IOP Publishing, 1549(2): 022083.

Piasecki M. (2019) Practical Implementation of the Indoor Environmental Quality Model for the Assessment of Nearly Zero Energy Single-Family Building[J]. *Buildings*, 9(10): 214.

Yang X C, Liu M, Uzzal H. (2013) Comparative analysis of indoor environmental indicators of typical green building evaluation system[J]. *Civil Architecture and Environmental Engineering*, 35(S1): 174–176.

Zhang F, Yang Z, Yu W H. (2005) Matter element model and extension evaluation method of indoor environment evaluation[J]. *Journal of Tianjin University*, (04): 307–312.

Zuhaib S, Manton R, Griffin C, et al. (2018) An Indoor Environmental Quality (IEQ) assessment of a partially-retrofitted university building[J]. *Building and Environment*, 139: 69–85.

*Civil Engineering and Urban Research – Mohamed & Hou (Eds)*
© *2023 the Authors, ISBN 978-1-032-44487-1*

# Numerical simulation of size effect of progressive failure of circular tunnel surrounding rock

Hai-feng Li, Biao Wang, Lei Chen, Yang Wang & Shuai Tao*
*POWERCHINA Huadong Engineering Corporation Limited, Hangzhou, Zhejiang, China*
*Zhejiang Huadong Engineering Consulting Corporation Limited, Hangzhou, Zhejiang, China*

ABSTRACT: Tunnel section size is the main parameter of tunnel section design, which is bound to affect the unloading failure process of the brittle surrounding rock, such as the rockburst process. The Gu Ming-cheng and Tao Zhen-yu rockburst criterion is introduced into the numerical calculation to simulate the rock burst process of surrounding rock with different diameters, and the influence of diameter on rockburst grade and circumferential stress distribution of surrounding rock is analyzed. The calculation results show that with the increase in tunnel diameter, the number of rockburst units at all levels increases. The larger diameter of the tunnel, the more obvious the circumferential stress fluctuation of the monitored unit. At the junction of the elastic-plastic zone, the circumferential stress of the monitored element is maximum. The number of failure elements increases monotonously with the diameter of the tunnel. The larger the diameter of the tunnel, the more the number of failure and rockburst elements in the surrounding rock of the tunnel.

## 1 INTRODUCTION

With the continuous and rapid development of the national economy and the further implementation of the energy strategy, China's construction in the field of hydropower and energy such as pumped storage is intensively carried out, which will involve a large number of underground tunnels and other structural forms. Tunnel construction will develop in a deeper, longer, and larger direction. In the design and construction process of deep tunnels, the existence of high ground stress is one of the important factors affecting the stability of the tunnel. With the increase in tunnel depth, the frequency of geological disasters such as rockburst will increase significantly (Gu 2001; Xu et al. 2009; Zhang & Fu 2008). A large number of engineering practices have proved that the stability of deep-buried tunnels is mainly manifested as hard rockburst and large deformation of soft rock. Many hydropower projects face rockburst problems under high ground stress, such as Ertan Hydropower Station and Baihetan Hydropower Station. Rockburst is a strong instability phenomenon of surrounding rock in deep engineering (Feng et al. 2012; Liu et al. 2017; Xu et al. 2009). It is a sudden failure of surrounding rock in rock engineering under high stress, accompanied by the sudden release of strain energy, which seriously threatens the safety of field construction personnel and equipment (Feng et al. 2012; Liu et al. 2016; Patel & Martin 2020; Xu et al. 2009).

Although a large number of studies have been carried out on rockburst, the mechanism of rockburst has been studied based on different means such as field monitoring, laboratory test, theoretical analysis, and numerical calculation, and good results have been achieved in engineering practice, the size effect of tunnel section in the process of rockburst is rarely involved, and the size effect has been a hot topic in the field of rock mechanics. Liu et al. (2017) studied the influence of tunnel size effect on rockburst process under high ground stress and buried depth of Jinping II deep tunnel group. Gong et al. (Gong et al. 2019; Si et al. 2018; Si & Gong 2021) carried out

---

*Corresponding Author: tao_s@hdec.com

DOI 10.1201/9781003372417-107

rockburst simulation tests on cube red sandstone and fine-grained granite samples with circular penetrating holes under three-dimensional high-stress conditions, observed the rock burst failure process and the symmetrical 'V' shape notch formed by the hole wall, and studied the influence of stress environment on the failure of the hole wall. With on-site monitoring and so on, it provides an effective means to analyze the rockburst process, but the numerical calculation method has become a powerful tool for analyzing geotechnical engineering problems due to the advantages of small restrictions by external conditions and low cost. In this paper, Gu Ming-cheng and Tao Zhen-yu's (Gu-Tao) rockburst criterion (Gu 2001; Zhang & Fu 2008) is introduced into the numerical calculation to simulate the rockburst process of surrounding rock of circular tunnels with different diameters. The influence of diameter on rockburst grade and circumferential stress distribution of surrounding rock is analyzed and discussed, which provides a reference for the design and construction of deep-buried hard rock tunnels.

## 2 CALCULATION MODEL, SCHEME AND ROCK BURST CRITERION

### 2.1 *Calculation model, scheme and constitutive parameters*

The length (x direction) and height (y direction) of the model are both 10 m, which are divided into 40000 rectangular elements with the same area, as shown in Figure 1.

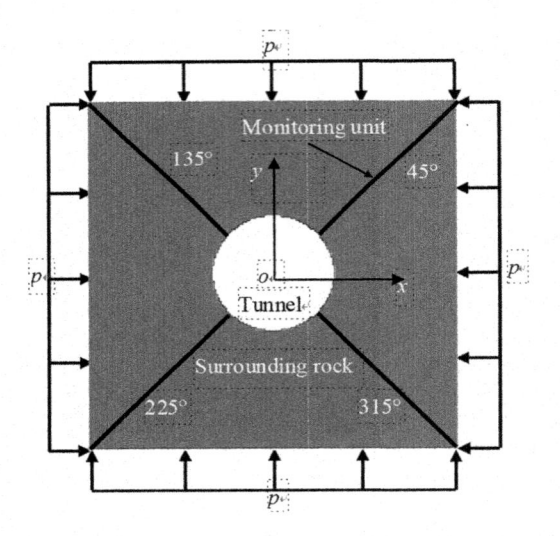

Figure 1.   Model geometry and boundary conditions.

In this paper, three calculation schemes are completed. The excavation radius $R$ of schemes 1 to 3 is 1.11 m, 1.33 m, and 1.66 m, respectively. Except for the excavation radius, the parameters are the same. Other parameters are shown in Table 1. The numerical calculation is carried out under the small deformation mode. Before the tunnel excavation, in order to make the calculation model reach the static equilibrium state as soon as possible, the stress of each small unit is set to the value equal to the confining pressure. After tunnel excavation, the stress of surrounding rock is redistributed. When the calculation reaches 50000-time steps, the model reaches the static equilibrium state again. This chapter only gives the results when the model reaches the static equilibrium state in each scheme.

In order to facilitate the expression, the surrounding rock of the tunnel is divided into four regions, which are the 1st to 4th quadrants. The coordinate origin is taken at the center of the tunnel. The horizontal right is the positive x-axis, and the vertical upward is the positive y-axis.

Table 1. Mechanical parameters of surrounding rock.

| Scheme | $P$(MPa) | $K$(GPa) | $G$(GPa) | $c$(MPa) | $c_r$(MPa) | $\sigma_t$(MPa) | $\sigma_c$(MPa) | R(m) | $\varphi_0$(°) | $\varphi_r$(°) | $\psi$(°) | $\upsilon$ |
|---|---|---|---|---|---|---|---|---|---|---|---|---|
| Scheme 1 | 50.0 | 28.1 | 26.2 | 23.4 | 7.8 | 11.7 | 175 | 1.11 | 60.0 | 40.0 | 0.0 | 0.25 |
| Scheme 2 | 50.0 | 28.1 | 26.2 | 23.4 | 7.8 | 11.7 | 175 | 1.33 | 60.0 | 40.0 | 0.0 | 0.25 |
| Scheme 3 | 50.0 | 28.1 | 26.2 | 23.4 | 7.8 | 11.7 | 175 | 1.66 | 60.0 | 40.0 | 0.0 | 0.25 |

The region of x > 0 and y > 0 is the first quadrant, and the region of x < 0 and y > 0 is the second quadrant.

## 2.2 Rockburst criterion

In this paper, the Gu-Tao rockburst criterion (Gu 2001; Zhang & Fu 2008) is used as the main stress rockburst criterion. The rockburst is divided into three levels (slight rockburst, medium rockburst, and severe rockburst), and the ratio of the maximum principal stress $\sigma_1$ to the uniaxial compressive strength of the rock $\sigma_c$ is used to judge:

$$\sigma_c/\sigma_1 = 5.0 \sim 6.67 \text{ (slight rockburst)} \tag{1}$$

$$\sigma_c/\sigma_1 = 2.5 \sim 5.0 \text{ (medium rockburst)} \tag{2}$$

$$\sigma_c/\sigma_1 < 2.5 \text{ (severe rockburst)} \tag{3}$$

## 3 RESULT ANALYSIS AND DISCUSSION

### 3.1 Influence of different diameters on rockburst grade of tunnel surrounding rock

Figure 2–4 show the calculation results of schemes 1–3 at 50000time steps, respectively. The dark area in the figure is the high-value area of rockburst at all levels and shear strain increment.

Figure 2 (a-c) gives the calculation results of the slight, medium, and severe rockburst in scheme 1 (R = 1.11m). It can be found that the units with slight and medium rockburst are only distributed on the surface of the cavern, while the units with serious rockburst are slightly concentrated on several points inside the surrounding rock. Figure 2 (d) shows the distribution of the high-value area of shear strain increment of tunnel surrounding rock. It can be found that the high-value area of shear strain increment is concentrated in points, distributed on the surface of the cavern, and no shear zone is formed in the surrounding rock.

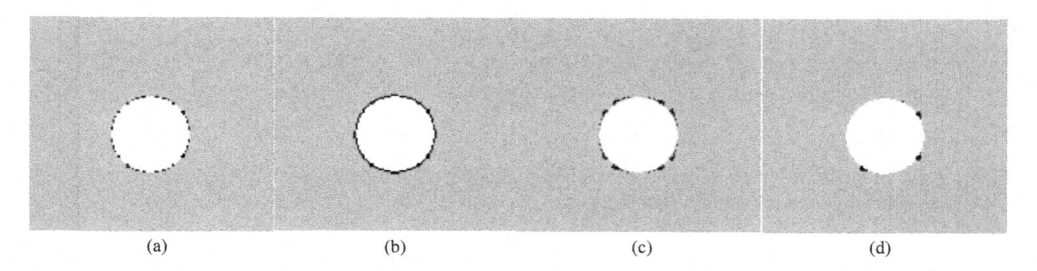

(a)          (b)          (c)          (d)

Figure 2. Numerical results in scheme 1: (a) slight rockburst; (b) medium rockburst; (c) serious rockburst; (d) elements having higher shear strain increments.

Figure 3 (a-c) gives the calculation results of the slight, medium, and severe rockburst in scheme 2 (R = 1.33m). It can be found that the number of units at all levels of rockburst in scheme 2 is significantly higher than in scheme 1. Among them, the unit with medium and severe rockburst formed a rockburst notching the first quadrant. Although the unit with a slight rockburst did not form an obvious rockburst pit, it was not limited to the surface of the cavern, and it also occurred inside the surrounding rock. Figure 3 (d) shows the distribution of the high-value area of shear strain increment of tunnel surrounding rock. It can be found that the high-value area of the shear strain increment is at the position of the rock burst notch formed by the medium and severe rock burst units. The two shear bands converge to form a shape similar to the rock burst pit, and the short shear band is also formed in the second and fourth quadrants. The position of the high-value area of the shear strain increment is basically the same as that of the rock burst unit.

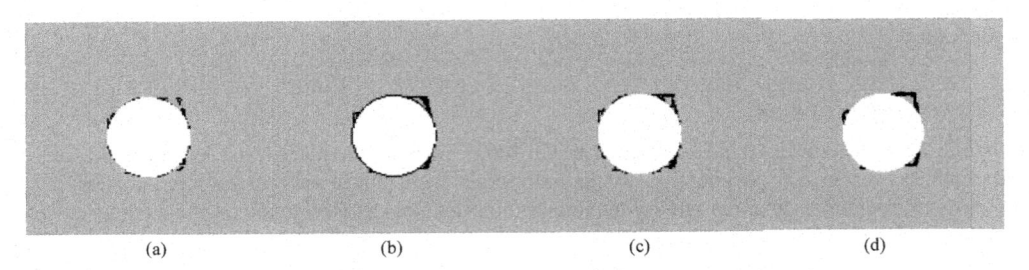

Figure 3.   Numerical results in scheme 2: (a) slight rockburst; (b) moderate rockburst; (c) serious rockburst; (d) elements having higher shear strain increments.

Figure 4 (a-c) gives the calculation results of the slight, medium, and severe rockburst in scheme 3 (R = 1.66m). It can be found that the number of rockburst units at all levels in Scheme 3 is significantly higher than that in the two schemes, and rockburst pits are formed within four quadrants. The most obvious rockburst pits are formed by the units with serious rockburst. Except for the location of rockburst pits, the number of units with serious rockburst on the surface of the cavern is less, while the units with slight and medium rockburst are mainly distributed along the surface of the cavern except for the location of rockburst pits. Figure 4 (d) shows the distribution of the high-value area of shear strain increment of tunnel surrounding rock. It can be found that the high-value area of shear strain increment is at the position of the rockburst notch formed by medium and severe rockburst units. The shear zone converges to form a similar shape to the rockburst pit, and its position is basically the same as that of the rockburst pit.

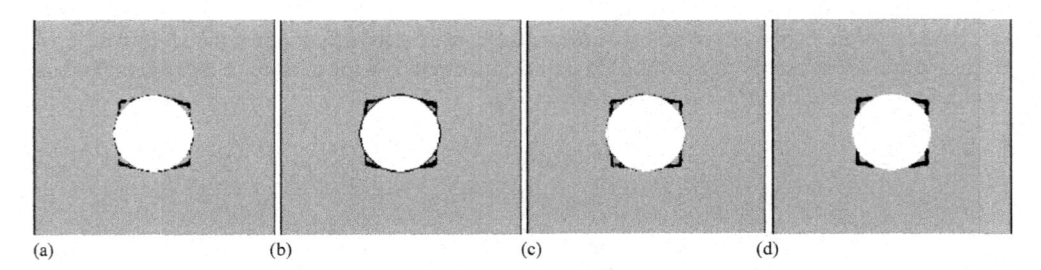

Figure 4.   Numerical results in scheme 3: (a) slight rockburst; (b) moderate rockburst; (c) serious rockburst; (d) elements having higher shear strain increments.

## 3.2   *The influence of different tunnel diameters on the circumferential stress distribution of tunnel surrounding rock*

Four rows of elements whose center is located on the four vertices connecting the center of the tunnel to the model are selected as the monitored elements, namely, the elements in the directions of 45°, 135°, 225° and 315°, as shown in Figure 1. Figure 5 (a-c) show the circumferential stress distribution of the monitored unit in four directions from Scheme 1 to Scheme 3, where the ordinate is the circumferential stress value ($\sigma_\theta$) of the monitored unit, and the abscissa is the distance (R) from the center of the monitored unit to the center of the tunnel.

It can be found from Figure 5 (a) that in Scheme 1 ($R = 1.11$ m), the overall trend of the circumferential stress distribution of the monitored unit is basically the same, and there is no large fluctuation in the distribution curve. At this point, it can be found from Figure 3 (a-c) that the number of units of rock burst at all levels in the surrounding rock of the tunnel is small, and they are distributed within the surface of the tunnel.

It can be found from Figure 5 (b) that in Scheme 2 ($R = 1.33$ m), the circumferential stress distribution curves of the monitored units in the directions of 45° and 315° show obvious fluctuations. The absolute value of circumferential stress increases first, then decreases gradually, and finally tends to a certain value. When the monitored unit in 135° and 225° directions is near the surface of the tunnel, the absolute value of the circumferential stress increases monotonically. With the increase in the distance from the center of the monitored unit to the center of the cavern, the circumferential stress gradually tends to a certain value. It can be found from Figure 3 (a-c) that the units where rockburst occurs are mainly concentrated in the first and fourth quadrants of the model, and the units where moderate and severe rockburst occur in the first quadrant form obvious rockburst pits. In the regions passed by the monitored units in the first and fourth quadrants, there are different levels of rockburst, so the circumferential stress distribution curve fluctuates greatly.

From Figure 5 (c), it can be found that in Scheme 3 ($R = 1.66$ m), the circumferential stress distribution of the monitored unit is basically the same, and the distribution curves have obvious fluctuations. Within 2.5 m from the center of the tunnel, the circumferential stress of the monitored unit has two peaks, that is, the absolute value of the circumferential stress of the unit near the surface of the tunnel first decreases to the minimum (near 0 MPa), then increases to the maximum, and finally decreases and gradually tends to a certain value. It can be found from Figure 4 (a-c) that in the four quadrants of the model, all the units that occur rock burst at all levels form rock burst pits, and the monitored units pass through the area where the rock burst pits are located. The units inside the rock burst pits are not destroyed, while the units that form rock burst pits are destroyed, which makes the undamaged units unloaded, so the absolute value of the circumferential stress is low, while the absolute value of the circumferential stress of the monitored unit at the junction of the elastic-plastic zone can reach the maximum value, which can explain the phenomenon of large fluctuations in the circumferential stress distribution curve of the monitored unit in Figure 5 (c).

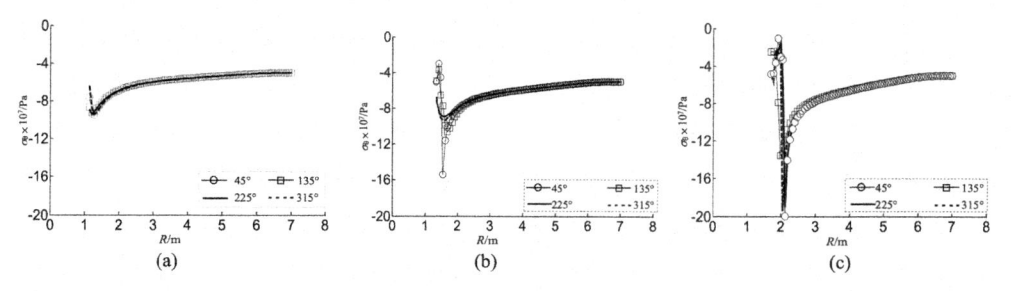

Figure 5. Tangential stress distribution of monitored elements under different schemes: (a) scheme 1; (b) scheme 2; (c) scheme 3.

Figure 6 presents the curves of the number of elements entering the plastic state, slight, moderate, and serious rockburst in the surrounding rock of the tunnel in Schemes 1 to 3. It can be found that the elements in the above four states increase monotonously with the increase of the radius of the tunnel. The larger the diameter of the tunnel is, the more the elements in the surrounding rock of the tunnel are damaged and rockburst. When the hole diameter is the same, the number of medium rockburst units is more, followed by serious rockburst.

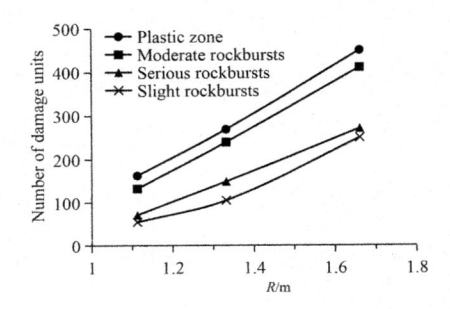

Figure 6. The number of damaged units distributed in all schemes.

## 4 CONCLUSIONS

Gu Ming-cheng and Tao Zhenyu's (Gu-Tao) rockburst criterion is introduced into the numerical calculation to simulate the rock burst process of circular tunnel surrounding rock with different tunnel diameters. The influence of tunnel diameter on the rockburst grade of surrounding rock and the circumferential stress distribution of surrounding rock is analyzed and discussed. The calculation results show that with the increase of the radius of the circular tunnel, the number of units that cause rockburst at all levels in the surrounding rock of the tunnel increases. The larger the tunnel diameter is, the easier it is to form rockburst pits, and the higher the shear strain increment is, the easier it is to form the shear zone that penetrates into the surrounding rock. With the increase of the radius of the tunnel, the fluctuation of the circumferential stress distribution curve of the monitored unit is more obvious, and the circumferential stress is more prone to the peak. Since the monitored unit passes through the area where the rockburst notch is located, the circumferential stress value of the unit near the surface of the tunnel is reduced to close to 0 MPa due to unloading, and the circumferential stress value of the monitored unit at the junction of the elastoplastic zone reaches the maximum. With the increase in the radius of the tunnel, the number of damaged units increases monotonously. The larger the diameter of the tunnel, the more damage, and rockburst units in the surrounding rock of the tunnel. When the diameter is the same, the number of medium rockburst units is more, followed by serious rockburst. In this paper, the principal stress rockburst criterion is introduced into the calculation process. In order to judge the rockburst process of surrounding rock more accurately, the circumferential stress and energy criterion can be introduced into the numerical calculation, and different factors can be used to judge whether the rockburst occurs.

## REFERENCES

Feng X T, Chen B R, Li S J, Zhang C Q, Xiao Y X, Feng G L, Zhou H, Qiu S L, Zhao Z N, Yu Y, Chen D F, Ming H J, 2012. Studies on the evolution process of rockburst in deep tunnels. *Journal of Rock Mechanics and Geotechnical Engineering*, **4(4)**: 289–295.

Gong F Q, Si X F, Li X B, Wang S Y, 2019. Experimental investigation of strain rockburst in circular caverns under deep three-dimensional high-stress conditions [J]. *Rock Mechanics and Rock Engineering*, **52(5)**: 1459–1474.

Gu M C, 2001. Research on rockburst in Qinling railway tunnel[J]. *Research on Water Resources and Hydropower*, **3(4)**: 19–26.

Liu N, Zhang C S, Chu W J, Ni S H, 2017. Discussion on size effect of rock burst risk in deeply buried tunnel[J]. *Chinese Journal of Rock Mechanics and Engineering*, **36(10)**: 698–725.

Meng F Z, Zhou H, Wang Z Q, Zhang L M, Kong L, Li S J, Zhang C Q, 2016. Experimental study on the prediction of rockburst hazards induced by dynamic structural plane shearing in deeply buried hard rock tunnels[J]. *International Journal of Rock Mechanics and Mining Sciences*, **86**: 210–223.

Patel S, Martin C D, 2020. Impact of the initial crack volume on the intact behavior of a bonded particle model[J]. *Computers and Geotechnics*, **127**: 1–10.

Si X F, Gong F Q, 2021. Rockburst simulation test and strength weakening effect of deep high-stress circular tunnel under internal unloading condition[J]. *Chinese Journal of Rock Mechanics and Engineering*, **40(2)**: 276–289.

Si X F, Gong F Q, Luo Y, Li X B, 2018. Experimental simulation on rockburst process of deep three-dimensional circular cavern[J]. *Rock and Soil Mechanics*, **39(2)**: 2514–2521.

Xu N W, Tang C A, Zhou J F, Tang L X, Liang Z Z, 2009. Numerical simulation of rockburst on the drain tunnel in the Jinping Second Level Hydropower Station[J]. *Journal of Shandong University (Engineering Science)*, **39(4)**: 134–139.

Zhang J J, Fu B J, 2008. Rockburst and its criteria and control[J]. *Chinese Journal of Rock Mechanics and Engineering*, **27(10)**: 2034–2042.

*Civil Engineering and Urban Research – Mohamed & Hou (Eds)*
*© 2023 the Authors, ISBN 978-1-032-44487-1*

# Numerical analysis of the probability of blockage of slag discharge wellbore based on the DDA method

Shengquan Yung* & Chengwu Peng*
*School of Civil Engineering, Hunan University of Science and Technology, Xiangtan, China*

Zhongyu Lu
*Guangxi Road and Bridge Engineering Group Co.Ltd, Nanning, Guangxi, China*

Ke Cao
*School of Civil Engineering, Hunan University of Science and Technology, Xiangtan, China*

ABSTRACT:   This paper takes the ventilation shaft project of Ma Luan Mountain tunnel of Ping Shan–Yan tian District Highway in Shenzhen city as the background. We adopted the discontinuous deformation analysis (DDA) method to study and analysed the slag blocking problem of small and medium diameter shafts in the construction of ventilation shafts. Wellbore rock falling is a dynamic process of discontinuous displacement. We numerically modelled the wellbore rock falling process with the ball element which is made up of the block base unit. In this study, the binding algorithm was used to bind the ball particles into the corresponding shape blocks. The collision and friction processes were simulated during the movement of blocks. At the same time, we analyzed the probability of rock falling blockage in a small diameter wellbore. The research results not only reflect the engineering practice of slag discharge shaft working conditions, but also provide theoretical basis and practical guidance for smooth slag discharge shaft. The research results can be used for reference for similar projects.

## 1   INTRODUCTION

The discontinuous deformation analysis (DDA) method proposed by Shi (1997) can be correctly applied (Fang & Zhuo 2005; Gong et al. 2020; Wu et al. 2003; 2006; Wu 2015; Xia & Xu 2010), the key of which lies in the scientific treatment of the contact problem between blocks. When dealing with rock materials or non-rock materials, static problems or dynamic problems (Andersen et al. 2017; Fan et al. 2017; Jiao et al. 2007; Jiang et al. 2016; Liu & Li 2019; Ning et al. 2010; Wang et al. 2014; 2019), DDA method is often the first choice to solve such problems since it owes great advantages in simulating rock block movement with rotating, opening and closing between blocks. Quality and time factors are also involved in the process of calculation. Therefore, in recent years, this method has been widely used in slope and field dam stability analysis, etc. which has been verified, improved, and developed.

The ventilation shaft located in the Ma Luan Mountain tunnel of Shenzhen Ping Shan–Yan tian District Highway Project has a designed excavation section diameter of 16.8m and shaft excavation depth of 170m. As shown in Figure 1, the most design adopts "drilling rig reverse well forward expansion method" construction. Since the difficulty of construction technology and the shortage of construction period, the construction party directly carries out full section blasting excavation instead of drilling and blasting process. The 1.4m small diameter central wellbore is the slag discharge channel. There are many problems with this diameter slagging wellbore. For example, in the slag discharge process, the wellbore cross-section diameter is too small to decrease the risk

---

*Corresponding Authors: sqyung@163.com and 819100910@qq.com

DOI 10.1201/9781003372417-108

of plugging. It is difficult to find the blockage in time and then dredge the blockage as the long length of the wellbore.

Therefore, in this paper, the DDA method is used to simulate the falling process of rock blocks in the wellbore. The relationship between them and the blocking probability of the slagging wellbore is studied by changing the inner wall flatness of the slagging wellbore and the lumpiness of rock blocks thrown into the wellbore. The research results are very necessary to guide the smooth construction of shaft blasting excavation.

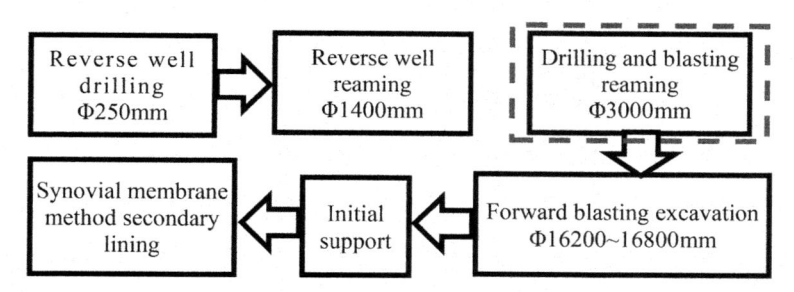

Figure 1. Shaft construction process.

## 2 INFLUENCING FACTORS AND NUMERICAL MODELLING ANALYSIS OF WELLBORE CLOGGING

### 2.1 Influencing factors of wellbore clogging

There are three main factors affecting the blockage of the slag discharge shaft: the flatness of the inner wall of the slagging wellbore, the lumpiness of rock thrown into the slagging wellbore, and the length of the wellbore.

#### 2.1.1 Inner wall flatness of the slagging wellbore
The inner wall flatness of the wellbore can affect slag discharge to a great extent. The worse the flatness is, the greater the probability of rock blocks thrown into the wellbore will be. The inner wall flatness of the wellbore is closely related to well quality. The poor quality well makes the inner wall of the wellbore rough because of complex engineering geological conditions. However, the good quality well makes it different obviously.

#### 2.1.2 The lumpiness of rock thrown into a slagging wellbore.
The larger the rock mass, the higher the probability of blockage of the slag discharge shaft. The smaller the rock, the result will be opposite.

#### 2.1.3 The length of the wellbore
The longer the wellbore length is, the more complex the movement trajectory of rock blocks thrown into the wellbore will be. When a large number of rock blocks are thrown into the wellbore at the same time, the greater the probability of wellbore blockage.

In addition, the shape of the rock block and the friction factor between the rock block and the inner wall of the wellbore also affect the probability of wellbore plugging. However, these factors have a relatively lower impact than the above factors.

Therefore, we used the DDA method to analyze the probability of wellbore clogging with a certain wellbore length. At the same time, we took the influences of the flatness of the wellbore wall and the lumpiness of the thrown rock block into consideration in the simulation results.

### 2.2 *Numerical modelling analysis of wellbore clogging*

In our research, we took the sphere as the basic block unit and then used the binding algorithm to bind the ball particles into blocks with corresponding shapes to simulate the collision and friction process during block movement. The simulation of the rock block is shown in Figure 2.

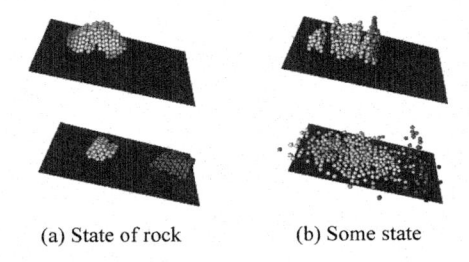

(a) State of rock  (b) Some state

Figure 2.  Simulation of rock block.

We used DDA analysis software to simulate the falling process of the rock block after being thrown into the wellbore under different conditions. Then we tracked the falling state of the rock block. To ensure the authenticity of the results, the rock was rotated randomly at any Angle and thrown into the wellbore. The simulation process of the slagging wellbore is shown in Figure 3.

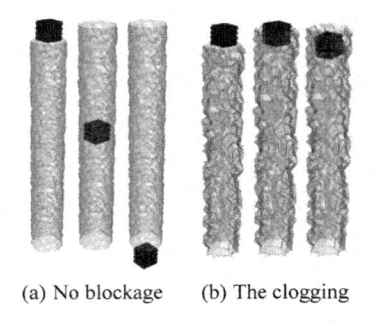

(a) No blockage  (b) The clogging

Figure 3.  The simulation process of slagging wellbore.

## 3  THE ESTABLISHMENT OF THE PROBABILITY ANALYSIS MODEL

### 3.1 *Engineering background*

The depth of the ventilation shaft is 194.0m and the excavation diameter is 16.8m. According to survey data, the blasting excavation height of the ventilation shaft is about 170.0m.

### 3.2 *The height and diameter of the wellbore model*

In our engineering practice, we reasonably simplified the model and selected the height of the wellbore model as 10.0m and the diameter as 1.4m.

### 3.3 *Random simulation of inner wall flatness of slagging wellbore*

The degree of concave and convex in the inner wall of the wellbore has certain randomness. The degree of concave and convex may change in different positions. The random distribution follows probability distribution to some extent, which is related to the process of well formation and engineering geological conditions.

Figure 4 shows the cross-section of the slagging wellbore model. The inner wall radius of the designed borehole is 0.7m. The MC method obtained the random wellbore radius which fluctuated randomly around 0.7m. According to the actual situation and statistical rules of borehole quality, we built four wellbore models that we can see in Figure 5. The degree of overbreak and underbreak gradually increased from 0.01m to 0.15m.

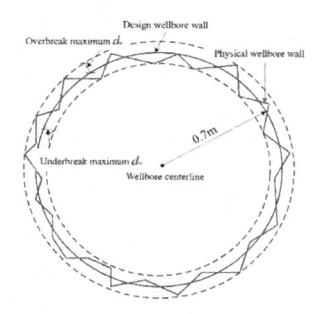

Figure 4.    Random simulation of the wellbore wall.

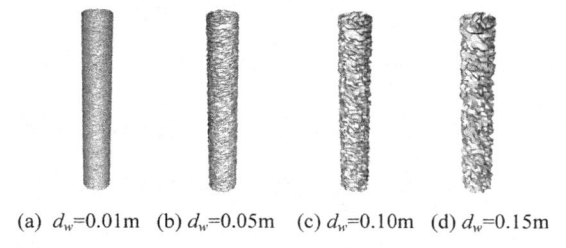

(a) $d_w$=0.01m    (b) $d_w$=0.05m    (c) $d_w$=0.10m    (d) $d_w$=0.15m

Figure 5.    Wellbore model of inner wall with different flatness.

### 3.4   *The blocks simulation of different lumpiness*

According to the actual situation of the construction site, the rock blocks with different lumpiness in the simulation process can be reasonably simplified. Figure 6 shows a partially simulated rock block model. We divided the rock blocks into cuboids of different sizes. The cross-section of the block is square. The side lengths of the square respectively are 0.2m, 0.4m, 0.6m, 0.8m, and 1.0m. According to the field blasting parameters, the rock blocks of various sizes were constructed by changing the ratio of height to width. The longest edge of the rock block is 3.0m. We limited the

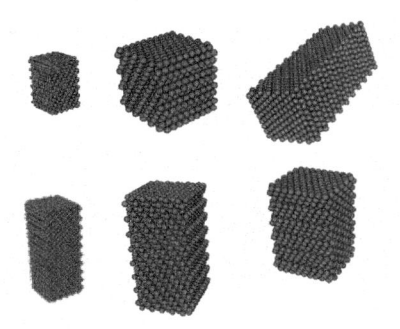

Figure 6.    The partial simulated rock block model.

maximum aspect ratio of constructed blocks to 4.0. The size of the block ranges from (0.2m× 0.2m× 0.2m) to (1.0m× 1.0m× 3.0m). There were 32 types of rock. We listed only 20 larger block models and their corresponding statistics in Table 1. The values of the block parameters are as follows: density $\rho$=2.7×103kg/m³, elastic modulus E=1010 N/m², Poisson's ratio $\mu$=0.28, friction coefficient f=0.22.

Table 1. Blocking probability analysis of different wellbore models.

| Order no | Size:/m (Length*Width*Height) | Aspect ratio | Fragmentation / m | Probability of blockage under overbreak and underbreak conditions /% | | | |
|---|---|---|---|---|---|---|---|
| | | | | 0.01 | 0.05 | 0.1 | 0.15 |
| 1 | 0.4×0.4×1.4 | 3.5 | 1.51 | 0 | 0 | 0 | 0 |
| 2 | 0.4×0.4×1.6 | 4.0 | 1.70 | 10 | 10 | 10 | 20 |
| 3 | 0.6×0.6×0.6 | 1.0 | 1.04 | 0 | 0 | 0 | 0 |
| 4 | 0.6×0.6×0.9 | 1.5 | 1.24 | 0 | 0 | 10 | 15 |
| 5 | 0.6×0.6×1.2 | 2.0 | 1.47 | 0 | 0 | 20 | 30 |
| 6 | 0.6×0.6×1.5 | 2.5 | 1.72 | 20 | 20 | 20 | 20 |
| 7 | 0.6×0.6×1.8 | 3.0 | 1.99 | 30 | 30 | 30 | 30 |
| 8 | 0.6×0.6×2.1 | 3.5 | 2.27 | 95 | 95 | 95 | 95 |
| 9 | 0.6×0.6×2.4 | 4.0 | 2.55 | 100 | 100 | 100 | 100 |
| 10 | 0.8×0.8×0.8 | 1.0 | 1.39 | 0 | 0 | 5 | 25 |
| 11 | 0.8×0.8×1.2 | 1.5 | 1.65 | 0 | 10 | 30 | 75 |
| 12 | 0.8×0.8×1.6 | 2.0 | 1.96 | 35 | 65 | 65 | 85 |
| 13 | 0.8×0.8×2.0 | 2.5 | 2.30 | 100 | 100 | 100 | 100 |
| 14 | 0.8×0.8×2.4 | 3.0 | 2.65 | 100 | 100 | 100 | 100 |
| 15 | 0.8×0.8×2.8 | 3.5 | 3.02 | 100 | 100 | 100 | 100 |
| 16 | 1.0×1.0×1.0 | 1.0 | 1.73 | 100 | 100 | 100 | 100 |
| 17 | 1.0×1.0×1.5 | 1.5 | 2.06 | 100 | 100 | 100 | 100 |
| 18 | 1.0×1.0×2.0 | 2.0 | 2.45 | 100 | 100 | 100 | 100 |
| 19 | 1.0×1.0×2.5 | 2.5 | 2.87 | 100 | 100 | 100 | 100 |
| 20 | 1.0×1.0×3.0 | 3.0 | 3.32 | 100 | 100 | 100 | 100 |

## 4 STATISTICAL RESULTS OF WELLBORE CLOGGING PROBABILITY

In this paper, we constructed 20 rock block models and 4 wellbore models with different degrees of overbreak and underbreak. We conducted more than 2,000 numerical simulations and calculated the corresponding plugging probability.

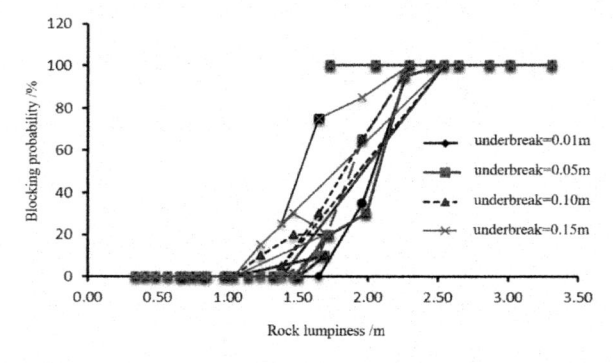

Figure 7. Relationship between the blocking probability and block size.

We can see in Table 1 and Figure 7 that the blocks with lumpiness below 1.00m do not clog in all wellbore models. However, the blocks with lumpiness greater than 2.25m can be blocked. Therefore, any rock blocks with lumpiness less than 1.00m can be thrown into the wellbore. Any rock blocks with lumpiness greater than 2.25m shall not be thrown into the slagging wellbore unless secondary crushing is performed. The block with lumpiness between 1.00m and 2.25m determines whether to throw into the wellbore depending on the block width. The blocks with a width of 1.00m

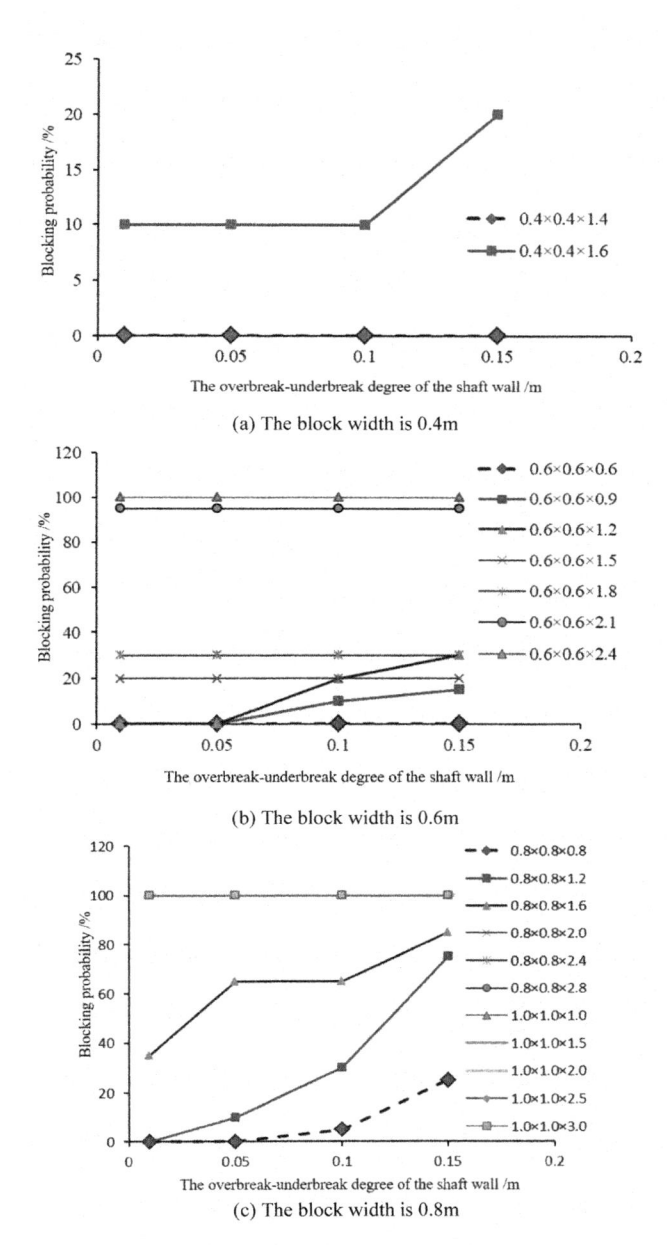

(a) The block width is 0.4m

(b) The block width is 0.6m

(c) The block width is 0.8m

Figure 8. Relationship between the probability of wellbore blockage and the degree of overbreak and underbreak.

can be blocked and cannot be thrown into any wellbore. Figure 7 shows the relationship between plugging probability and lumpiness.

Meanwhile, we can see in Figure 8 that the probability of wellbore blockage increases with the increase of overbreak and underbreak degrees. With the increase in the block width, the trend becomes more obvious. In similar projects, if the slagging wellbore is constructed by the drilling and blasting method, the overbreak and underbreak degree of slagging wellbore should be strictly controlled. Therefore, the recommended control value is 0.15m.

Considering only the blocks in Table 1 with an aspect ratio of 2.0 or less, we can know that: ① the rock blocks with a width less than 0.4m can be thrown into the wellbore. ② When the degree of overbreak and underbreak is less than 0.05m, the rock blocks with a width of less than 0.6m can be thrown into the wellbore. When the degree of overbreak and underbreak is $0.1m{\sim}0.15m$, we know that: ① the rock blocks with a width of less than 0.4m and a length of less than 1.4m can be thrown into the wellbore. ② Nearly cube blocks with a width of less than 0.6m can be thrown into the wellbore.

## 5 CONCLUSIONS AND RECOMMENDATIONS

In this paper, we used the DDA method to build block models with different lumpiness and simulated four wellbore models with different degrees of overbreak and underbreak. We then analyzed the probability of rock fall and blockage in a small diameter wellbore. The research results not only reflect the engineering practice of slag discharge shaft working conditions, but also provide theoretical basis and practical guidance for smooth slag discharge shaft. At the same time, the results also ensure the construction period and safety of the shaft. Specific conclusions and suggestions are as follows:

(1) Small diameter slagging wellbore might be blocked. In order to avoid blocking, we should strictly control the size and lumpiness of the rock blocks.
(2) The small diameter slag discharge wellbore constructed by the drilling and blasting method often appears the phenomenon of overbreak and underbreak. The probability of wellbore blockage will increase with the increase of overbreak and underbreak degrees.
(3) The research results can provide a reference for blasting fragmentation control and small diameter shaft slag discharge control of similar projects.

## REFERENCES

Andersen, O., Nilsen, H.M., Raynaud, X. (2017) Virtual element method for geomechanical simulations of reservoir models. *Comput Geosci.*, 21(3): 877–893.

Fan, H., Zheng, H., Zhao, J.D. (2017) Discontinuous deformation analysis based on strain-rotation decomposition. *Int. J. Rock Mech. Min. Sci.*, 92(2): 19–29.

Fang, C.Y., Zhuo, J.T. (2005) Summarization of numerical simulation for jointed rocks. *Hydro-Science and Engineering*, 25(2): 70–78.

Feng, X.X., Jiang, Q.H., Zhang, H.C. (2020) Study on the dynamic response of dip bedded rock slope using discontinuous deformation analysis (DDA) and shaking table tests. *Int J Numer Anal Methods Geomech.*, 45(3): 411–427.

Gong, S.L., Ling, D.S., Hu, C.B. (2020) Study on large rotation problem of a block in discontinuous deformation analysis. *Rock Soil Mech.*, 41(11): 3810–3822.

Hashimoto, R., Sueoka, T., Koyama, T. (2021) Improvement of Discontinuous Deformation Analysis Incorporating Implicit Updating Scheme of Friction and Joint Strength Degradation. *Rock Mech Rock Eng.*, 2021(54): 4239–4263.

Jiang, W., Chen, W., Sun, G.H. (2016) Research on the failure mode of Outang landslide using DDA method. *JDPME*, 36(04): 551–558.

Jiao, Y.Y., Zhang, X.L., Liu, Q.S. (2007) Simulation of rock crack propagation using discontinuous deformation analysis method. *CJRME*, 26(4): 682–691.

Liu, G.Y., Li, J.J. (2019) Research on the Effect of Tree Barriers on Rockfall Using a Three-Dimensional Discontinuous Deformation Analysis Method. *Int J Comput Methods*, 17(08): 1950046–1950046.

Lu, B., Wu, A.Q., Xu, D.D. (2020) Stiff servo-controlled numerical test method based on mixed higher order discontinuous deformation analysis. *CJRME*, 39(8): 1572–1581.

Ning, Y.J., Yang, J., Chen, P.W. (2010) Numerical simulation of rock blasting in jointed rock mass by DDA method. *Rock Soil Mech.*, 31(7): 2259–2263.

Shi, G.H. (1997) *Numerical manifold method and discontinuous deformation analysis*. Tsinghua University Press, Beijing.

Wang, S.H., Zhu, C.J., Zhang, Z.S. (2019) Stability analysis of the multi-slip surface of slope based on dynamic strength reduction DDA method. *JCCS*, 44(4): 1084–1091.

Wang, W., Zhu, W.S., Chen, Y.J. (2014) Analysis of roadway stability in jointed rock masses. *JCCS*, 39(1): 57–63.

Wu, A.Q., Ding, X.L., Chen, S.H. (2006) Research on deformation and failure characteristic of an underground powerhouse with complicated geological conditions by DDA method. *CJRME*, 25(1): 1–8.

Wu, J.H. (2015) The elastic distortion problem with large rotation in discontinuous deformation analysis. *Comput Geotech.*, 69: 352–364.

Wu, J.H., Ohnishi, Y., Shi, G.H. (2003) Three-dimensional discontinuous deformation analysis (3D DDA) and its application to the rock slope toppling. *CJRME*, 22(6): 937–842.

Xia, C.C., Xu, C.B. (2010) Study of fracturing algorithm of intermitting joint by DDA and experimental validation. *CJRME*, 29(10): 2027–2033.

Yu, P.C., Zhang, Y.B., Peng, X.Y. (2019) Evaluation of impact force of rock landslides acting on structures using discontinuous deformation analysis. *Comput Geotech.*, 114(6): 1–4.

*Civil Engineering and Urban Research – Mohamed & Hou (Eds)*
*© 2023 the Authors, ISBN 978-1-032-44487-1*

# Remote monitoring and early warning system of shield tunneling

Mao Hongmei
*State Key Laboratory of Rail Transit Engineering Informatization (FSDI), Xi'an, China*
*School of Urban Rail Transit Engineering, Shaanxi Railway Institute, Weinan, China*

Nie Hongbin
*School of Urban Rail Transit Engineering, Shaanxi Railway Institute, Weinan, China*

ABSTRACT: Due to the shortcomings of traditional shield tunneling monitoring technology such as complex environment, difficult data collection, large workload, untimely early warning, etc., the requirements of modern enterprises on construction process monitoring cannot be fully satisfied. In consideration of this, a kind of unmanned monitoring technology by erecting a remote monitoring platform adopting big data + early warning technology is established in this article. Functions such as data collecting, transmission, preprocessing, storage, calculation, analysis, and application can be realized using this platform built. With this technology, segment data is collected through a carbon fiber piezoresistive model, connected with equipment data through a dedicated I/O interface, and transmitted through network terminals, routers, and gateway nodes using energy transmission technology. By processing the transmitted data using the ETL method, data cleaning and transformation function can be realized. With transmitted data integrated into the corresponding database cloud platform, unmanned remote monitoring can be realized through analysis and early warning using data extraction and anomaly processing technology based on a proper decision made. By comparing the monitoring data using the platform with manual monitoring data, it shows that the platform monitoring data is close to the manual monitoring data with residual error within 5%, which proves that the accuracy of the platform fully meets the current subway construction requirements and verifies the feasibility of the platform application.

## 1 INTRODUCTION

With the acceleration of China's urbanization process, an efficient, low-carbon, and energy-saving urban rail transit system has developed rapidly. The number and difficulty of tunnel construction in subways, high-speed railways, and other projects have gradually increased (Fu et al. 2022; Li et al. 2021; Loy et al. 2022). The shield tunneling method is often adopted for tunnel construction based on the characteristics of the geological and hydrological environment and stratigraphic structure in China. However, accidents such as cracking, seepage, and water inrush occur during construction due to backward early warning methods and untimely monitoring (Zhou et al. 2020). In view of the current situation that the subway force is difficult to determine, the environment and force coupling mechanism are unclear, and the communication channel is constantly changing, the famous Chinese scholar Ding Lieyun buried fiber Bragg grating (FBG) sensors in the subway crossing the Yangtze River. Taking the sensors as basic sensing units, an integrated early warning system that combines "sensing, transmitting, notifying and controlling was built (Di et al. 2013). Based on the "chain" structure and artificial intelligence technology, the scholar Zhang Zhiming, and others combined regular tunnel monitoring, and measuring, personnel attendance, monitoring and positioning, and realized the organic integration of regular monitoring, early warning and learning (Zhang & Ye 2019). Scholar Tan Wen used the electrode detection method to determine the filling degree of

DOI 10.1201/9781003372417-109

concrete for the lining cavity problem. In order to determine the filling quality of concrete, three cameras were set up inside the lining to monitor directly in the form of video (Tan 2020). Based on the tunnel width, scholar Huang Xiaocheng and others obtained the physical and mechanical parameters of the surrounding rock in front of the tunnel face according to the seismic wave method (TSP). Through a high-order non-collinear response surface model built based on the length change between points taking the expected reliability of the tunnel function as the threshold, the risk of tunnel collapse can be controlled (Huang et al. 2022). Scholars such as Long Bin analyzed the technical difficulties faced by the shield tunnel/TBM construction of the Sichuan-Tibet Railway, and controlled construction risks by accessing monitoring data and building a remote early warning platform (Long et al. 2021). Scholar Yao Zhibin and others adopted the object-oriented B/S+C/S structure to establish a database management system to realize dynamic quantitative and intelligent early warning of rock bursts. With this method, multi-project management, detailed data collection, searching, analysis, result export and other functions can also be realized (Yao et al. 2021). According to the research of the above scholars, the construction methods of the remote monitoring and early warning platform mainly include data intelligent collection, transmission, cloud computing, chain structure analysis, etc. The process of data transmission and analysis is relatively mature. Therefore, the research focus of monitoring and early warning platform components is data collection. At present, the intelligent data acquisition methods include optical grids, electrodes, and seismic wave methods. The optical grating modulates the refractive index of the fiber core through deformation to form a diffraction grating and transmits data through the change in the number of gratings. This method is greatly affected by environmental factors such as temperature and humidity. The electrode method converts the deformation signal into an electrical signal. This method is less affected by the environment but greatly affected by a magnetic field, and the circuit layout is cumbersome. The seismic wave method requires the assistance of vibroseis, and imaging analysis is difficult. Taking the relationship between deformation and resistance (referred to as piezoresistance) as a basis, unmanned monitoring is realized in this paper through a platform built in a combination of the Internet of Things, industrial big data, and other technologies, which provides data support for shield construction safety, risk early warning, quality control, etc. The monitoring results of the platform were compared with the manual monitoring data of Ningbo Metro for reliability verification.

## 2 SEGMENT DATA COLLECTING METHOD

A piezoresistive sensing model is used for segment data monitoring. A certain amount of conductive carbon fibers is added to the segments based on the filed-tested mix ratio. Conductive materials are adopted as segment bolting, and an intelligent conductive early warning ring is formed by segment assembling to test shield tunnel segments settlement, peripheral convergence, and cracking. When segment cracks due to deformation caused by an external force, the carbon fiber mixed into the interior is deformed, resulting in segment resistance change, and equivalent resistance is formed in a specific direction of the force.

Its piezoresistive model is shown in Formula (1) (Tan et al. 2019; Xv et al. 2022),

$$R_{cf} = e^{\left[ 1 - \left( u_{cf} - v_{cf} \right) \right] \left[ \frac{\frac{r_1^2}{r_1^2 - r_2^2} (\sigma_v + \sigma_h) + 2(\sigma_v - \sigma_h) \cos(2\theta)}{E} \right]^2} + C_{cf} \tag{1}$$

In the equation, $v_c$ represents the parameter related to length and section; $u_{cf}$ represents the Poisson ratio; $\varepsilon_{cf}$ represents the deformation per unit length; $C_{cf}$ represents the carbon fiber integral constant; $r_1$ indicates the outer diameter of the segment; $r_2$ indicates the inner diameter of the segment.

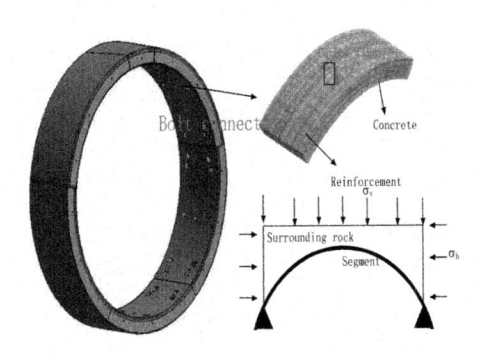

Figure 1. The piezoresistive model of segment intelligent early warning ring.

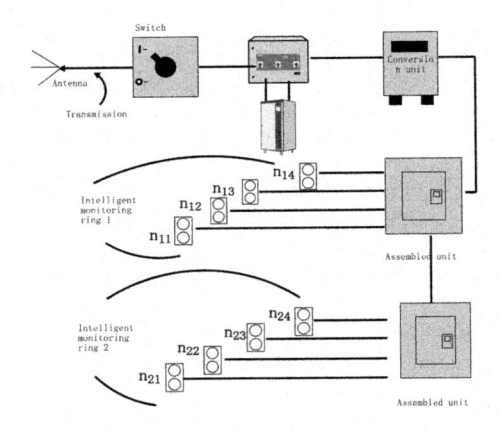

Figure 2. Data transmission.

## 3 DATA TRANSMISSION METHOD

The piezoresistive model described above is mainly responsible for collecting the stress and deformation data of segments, connecting and communicating with the relevant data of the shield equipment layer, and regularly reading various data generated by the equipment during operation (such as analog data, digital data, etc.) through data collector. The data is then cached in the temporary library and uploaded to the corresponding database of the platform through the network in time. And through a connection created by a dedicated data I/O interface between the data collecting layer and equipment layer, device data can be read at a second-level frequency. The system adopts wireless transmission and establishes a transmission system based on the Internet of Things. It is designed according to the organizational structure of network terminals, routing displays, and gateway nodes. Taking the current threshold of the carbon fiber segment as the terminal, an electrical signal is converted into a digital signal through a converter by collecting the data of segment current change under external force. The transformation of the entire shield segment intelligent monitoring ring data and equipment-related data is shown in Figure 2. Wireless transmission. Is made through electromagnetic waves sent by a router in accordance with the standard of the International Telecommunications Union (ITU), and the electromagnetic wave loss propagation model is shown in equation (2) (Xie et al. 2022).

$$L = 32.44 + 20\log_{10}f + 20\log_{10}d \tag{2}$$

In the equation, $L$ represents the amount of waveform loss; $f$ represents the transmission frequency; $d$ represents the transmission distance.

## 4 CONSTRUCTION METHODS OF EARLY WARNING PLATFORM

The shield intelligent monitoring and early warning platform is constructed by taking industrial big data as a foundation and following the construction route of data collection, processing, and analysis. Core data of shield construction such as propulsion system data, cutter head system data, screw conveyor system, data and articulation system data are displayed on the main interface of the platform. The main structure of the platform can be divided into six layers: data collecting layer, data preprocessing layer, data storage and calculation layer, data analysis layer, capability layer, and application layer. The collected and transmitted data are preprocessed by the ETL method. The data in heterogeneous data sources, such as relational data, flat data files, etc., are extracted to the temporary middle layer, cleaned, transformed, integrated, and finally loaded into the data ware-house, the average method of the fractal regression model is adopted by bata difference processing in Equation (3) (Xang 2019). Data collected is stored in Alibaba Cloud, and classified based on shield functions, including four main databases: system database, shield machine database, shield tunneling management database, and intelligent analysis expert knowledge database. Data analysis and early warning are realized by data extraction and exception handling. Analysis of the hidden information in the data serves early warning and decision-making. Detailed information is shown in Figure 3.

$$\text{BIC}^*_{\mathcal{A}} = 2n \log \left[ \sum_{i=1}^{n} \rho_\tau \left\{ Y_i - \mathbf{X}_{i(\mathcal{A})}^\top \widehat{\beta}_{(\mathcal{A})}(\tau) \right\} \right] + |\mathcal{A}| \log(n) \tag{3}$$

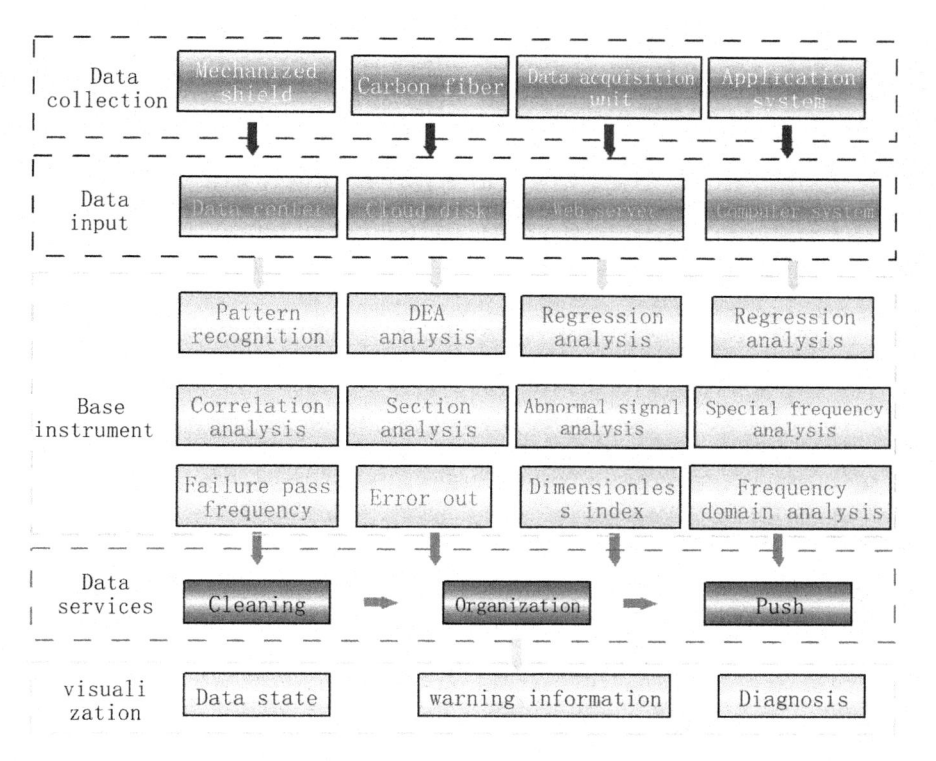

Figure 3.   Diagram of the structure of the early warning platform.

## 5   MONITOR FIRING OF WARNING SYSTEM

Remote monitoring in a shield tunneling project of Ningbo Metro Line 4 was carried out using a shield tunneling big data early warning platform established based on the method shown in Figure 3. This 2.7km tunnel project adopted a slurry shield with an excavation radius of 6.260m. The geological condition encountered during shield excavation involves sand, clay, silt, etc. (Chen 2011; Wang 2020; Wang et al. 2018). Segments of this tunnel are prefabricated with C40 concrete, fiber placed in segment adopts 1mm short-cut high-conductive carbon fiber, and the additive amount of the fiber is 1% of the cement. The copper rod method is used for the current collection. In addition, force sensors are arranged on each scraper of the cutterhead, videos are evenly arranged in the center and surrounding of the cutter head, and a wireless routing device is established for data transmission (Wang 2020; Wang et al. 2018; Xie 2019). The layout of field instruments and test components is shown in Figure 4.

The remote monitoring of the platform during the process of subway shield tunneling is shown in Figure 5. The shield data extracted in Figure 5 are compared with the manual monitoring values, and the residuals are shown in Figure 6. It can be seen from the analysis of Figure 6. that graph (a) shows the change of the difference value with the shield advancing time. The residual error is small at the beginning of the excavation, and as the shield machine continues to advance, the residual error becomes larger, and the maximum value is 5%. During the whole process, the residuals show a Gaussian distribution, see (b), while figures (c) and (d), represent the residuals change over the entire excavation section. And the residuals increase linearly along with the location of the section.

Figure 4.   On-site data collection.        Figure 5.   Early warning platform of shield data.

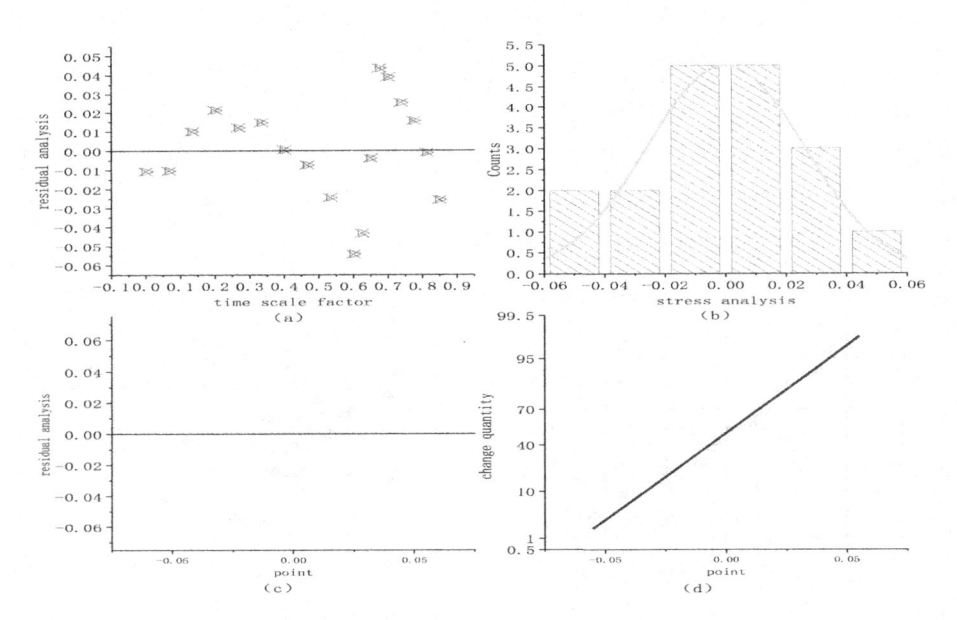

Figure 6.   Residual analysis of warning system.

## 6   CONCLUSIONS AND RECOMMENDATIONS

Through the construction and application of the shield data early warning platform, the following conclusions can be drawn:

1) The remote early warning big data platform realizes unmanned and real-time monitoring. Through industrial big data technology, using its elastic expansion and multiple data copy

methods, non-relational data and relational databases are better connected, effectively solving the needs of real-time data display, historical storage, digging, etc.

2) The integrated collection of shield data is realized through the Internet of Things. According to the stress characteristics of the shield segment structure, the deformation signal is converted into an electrical signal based on the piezoresistive model, and then converted into a digital signal through a data converter, with data sent to the platform cloud processor using wireless energy transmission technology. Thus, an integrated solution of data collection, transmission and storage is realized

3) The self-developed intelligent collection terminal realizes remote early warning and decision-making. The data is cleaned, processed, and sent to the data cloud warehouse through ETL technology. Intelligent early warning is made through data extraction and anomaly analysis and comparison. And decisions can be made using original treatment measures of abnormal data.

Through the comparison of subway shield tunneling and stress monitoring data, it is found that the platform established in this paper has high accuracy, but with a certain difference compared to the actual monitoring value. The next research is suggested to focus on improving the data collection accuracy and optimizing data extraction.

## ACKNOWLEDGMENTS

This work was supported by the National Natural Science Foundation of Shaanxi province (Grant Nos. 2022JQ-420), an Open project of the state key laboratory of rail transit engineering informatization (FSDI)(SKLK20-11), Natural science special fund of Shaanxi education department (21JK0584), Key R&D Programmes of Weinan city (2020ZDYF-JCYJ-187, 2019-ZDYF-JCYJ-129); Young and middle-aged scientific and technological talents program of Shaanxi railway institute(KJRC202004); Scientific research fund of Shaanxi railway institute (KY2020-43).

## REFERENCES

Aiyv Wang. 2020, D. Lanzhou Jiaotong University.

Bin Long, Zhihua Liu, Jing Xiao. 2021, *J. Tunnel construction*, 41:626–633.

Jiaxing Xie, Gaotian Liang, Peng Gao. 2022, *J.Transactions of the Chinese Society for Agricultural Machinery*:1–8.

Jinyang Fu, Ningning Zhao, Yong Qu. 2022, *J. Tunnelling and Underground Space Technology*, 124:44–70.

Lieyun Ding, Cheng Zhou, Xiaowei Ye. 2013, *J. Chain Civil Engineering Journal*, 46:141–150.

Linli Xie. 2019, D. Yunnan University.

Loy J, Tariq S, Nguyen H T. 2022, *J. Building and Environment*, 207:853–877.

Wen Tan.2020, *J. Modern Tunnelling Technology*, 57:199–203.

Wenhui Li, Tanghong Liu, Marinez Pedro. 2021, *J. Tunnelling and Underground Space Technology*, 115:40–55.

Xiaocheng Huang, Deyang Lei, Yunpeng Xie. 2022, *J. Journal of Applied Basic and Engineering Science*, 30:219–235.

Xiaojun Wang, Jiang Yong, Wang Wendo. 2018, *J. Subgrade Engineering*:61–68+74.

Xiaojvan Xv, Jin Luo, Zhaoquan Chen. 2022, *J. Acta Materiae Compositae Sinica*: 1–11.

Yiqiu Tan, Kai Liu, Yingyuan Wang. 2019, *J. Journal of Building Materials*, 22:278–283.

Yuanqing Chen. 2011, *J . Tunnel construction*, 31:162–170.

Zhibin Yao, Wenjing Niu, Yv Zhang. 2021, *J. Chinese Journal of Engineering*, 41:626–633.

Zhiming Zhang, Ying Ye. 2019, *J. Modern Tunnelling Technology*, 56:73–79.

Zhou J, Xiao H, Jiang W. 2020, *J. Measurement*, 15.

*Civil Engineering and Urban Research – Mohamed & Hou (Eds)*
*© 2023 the Authors, ISBN 978-1-032-44487-1*

# Numerical analysis of global-local buckling of sandwich structures

Yongxiang Huang, Shun Liu, Ao Zhao, Di Wang & Minghao Gao
*Shenyang University, Shenyang, P.R. China*

Yanchuan Hui*
*Shenyang University, Shenyang, P.R. China*
*School of Civil Engineering, Wuhan University, Wuhan, P.R. China*
*Shenyang Key Laboratory of Smart Disaster Prevention and Mitigation for Civil Buildings, Shenyang,
P.R. China*

Xiao Liu*
*Shenyang University, Shenyang, P.R. China*
*Shenyang Key Laboratory of Smart Disaster Prevention and Mitigation for Civil Buildings, Shenyang,
P.R. China*

ABSTRACT:  This paper proposes a sandwich beam model. The geometric nonlinear problem of layered beam structure is solved based on Carrera's Unified Formulation (CUF). A special emphasis is put on the interaction of global-local buckling. Sandwich structure is widely used in the construction industry, aerospace, automobile, and other fields. This kind of structure will be unstable. However, instability is not only a single global or local instability, but also a nonlinear problem of global-local coupling. In practical engineering applications, highly soft composite laminated structures are prone to large deflection and posterior buckling. The phenomenon of instability in the sandwich structure is getting increasing attention. In the past three decades, some examples have been studied on the geometric nonlinearities of the sandwich structure. Therefore, this paper is devoted to developing an efficient sandwich structure model to solve complex nonlinear problems.

## 1  INTRODUCTION

This paper proposes a sandwich beam model. In the CUF framework, the unfolding order of beam theory can be freely selected. The accuracy of requirements and calculation results are ensured. Compared with the traditional 2D finite element analysis methods, the proposed model provides accurate predictions. According to several reviews (Carrera and Brischetto (2009), Hu et al. (2008), Sayyad and Ghugal (2017), classical laminate theory, first-order shear deformation theory, high-order theory and zig-zag theory (Ferreira et al. 2011) have been proposed for modeling sandwich composites.

This paper presents a CUF-based sandwich beam model for analyzing the geometric nonlinearity of sandwich beam structures. In the proposed UF, the cross-section and the axial kinematics can be improved by the compact notation of the a priori displacement field approximation. The governing equation is derived from the Virtual Pisplacement principle (PVD) about the basic kernel. For appropriate physical fields, continuous fitting functions can be used, such as the MacLaurin polynomial employed in the text, or a piecewise polynomial, such as the Lagrangian polynomial. When the cross-section of the beam has more non-uniformity, it is difficult to accurately simulate its characteristics. For example, we encounter irregular shapes (I-beams) and significantly different material properties. Because we know that simulating fast-changing curves usually requires high-order polynomials to fit. Therefore, it is safer to use the piecewise function to perform the physical

---

*Corresponding Authors: hui.yanchuan@whu.edu.cn and 489298344@qq.com

  DOI 10.1201/9781003372417-110

field of the cross-section. However, both MacLaurin polynomials and Lagrangian polynomials are viable choices when small cross-section geometry and material changes occur. In this article, the Lagrangian polynomial is taken to describe the mechanical field in sandwich structures. Carrera's Unified Formulation (CUF) has been previously extended for modeling the beam structures on free vibration problems (Hui et al. (2017)) and hygro-thermal problems (Moleiro (Moleiro et al. 2019)). For comparison, the finite element solution results of the commercial code ABAQUS are used.

## 2 PRELIMINARIES

The width, thickness, and length of the sandwich structure are $b$, $h$, and $L$, respectively. The displacement $u$ of the top($\gamma$), bottom($\alpha$), and core($\beta$) layers can be decomposed into a separate representation of the space coordinates:

$$u^{(R)}(\mathrm{x}) = u^{(R)}(\mathrm{I}, \mathrm{II}) = \left\{ u_{\mathrm{I}}^{(R)}(\mathrm{I}, \mathrm{II}) u_{\mathrm{II}}^{(R)}(\mathrm{I}, \mathrm{II}) \right\}^{T}, R = \gamma, \alpha, \beta \tag{1}$$

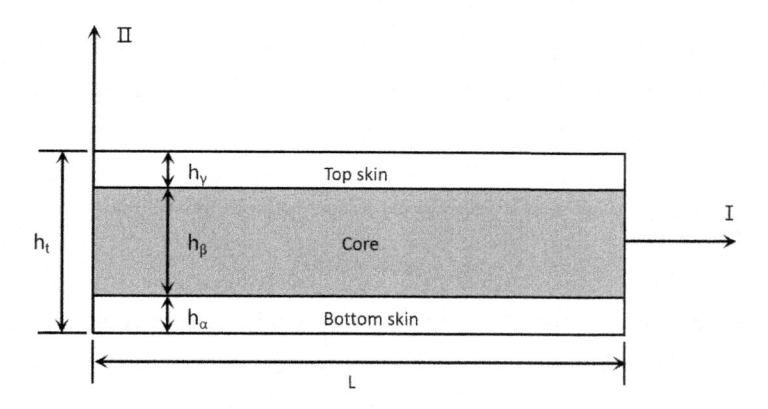

Figure 1.   The sketch of a sandwich beam.

where $T$ stands for the transpose operator. The corresponding first-order derivatives components can be written as:

$$\zeta^{(u)}(u^{(R)}) = \left\{ u_{\mathrm{I,I}}^{(R)} u_{\mathrm{I,II}}^{(R)} u_{\mathrm{II,I}}^{(R)} u_{\mathrm{II,II}}^{(R)} \right\}^{T} \tag{2}$$

Where $\zeta$ is the displacement gradient vector. Considering geometric nonlinearity, the Green-Lagrange strain can be written as follows:

$$D^{(R)} = \begin{bmatrix} D_{\mathrm{I,I}}^{(R)} \\ D_{\mathrm{II,II}}^{(R)} \\ D_{\mathrm{I,II}}^{(R)} \end{bmatrix} = \begin{bmatrix} u_{\mathrm{I,I}}^{(R)} \\ u_{\mathrm{I,II}}^{(R)} \\ u_{\mathrm{II,I}}^{(R)} + u_{\mathrm{II,II}}^{(R)} \end{bmatrix} + \begin{bmatrix} u_{\mathrm{I,I}}^{(R)^2} + u_{\mathrm{II,I}}^{(R)^2} \\ u_{\mathrm{I,II}}^{(R)^2} + u_{\mathrm{II,II}}^{(R)^2} \\ 2u_{\mathrm{I,I}}^{(R)} u_{\mathrm{I,II}}^{(R)} + 2u_{\mathrm{II,I}}^{(R)} u_{\mathrm{II,II}}^{(R)} \end{bmatrix} = \left[ A + \frac{1}{2} B^{(R)}(\zeta^{(R)}) \right] \zeta^{(R)} \tag{3}$$

Where matrices $A$ and $B^{(R)}(\zeta^{(R)}(u^{(R)}))$ are defined as:

$$A = \begin{bmatrix} 1 & 0 & 0 & 0 \\ 0 & 0 & 0 & 1 \\ 0 & 1 & 1 & 0 \end{bmatrix} \quad B^{(R)}(\zeta^{(R)}(u^{(R)})) = \begin{bmatrix} u_{\mathrm{I,I}}^{(R)} & 0 & u_{\mathrm{II,I}}^{(R)} & 0 \\ 0 & u_{\mathrm{I,II}}^{(R)} & 0 & u_{\mathrm{II,II}}^{(R)} \\ u_{\mathrm{I,II}}^{(R)} & u_{\mathrm{I,I}}^{(R)} & u_{\mathrm{II,II}}^{(R)} & u_{\mathrm{II,I}}^{(R)} \end{bmatrix} \tag{4 \& 5}$$

A virtual variation of strain can be written in the following form:

$$\delta D^{(R)} = \delta \left\{ A\zeta^{(R)} + \frac{1}{2}B^{(R)}\zeta^{(R)} \right\} = A\delta\zeta^{(R)} + B^{(R)}\delta\zeta^{(R)} \tag{6}$$

where $\delta$ stands for a virtual variation. For a two-dimensional problem, the material stiffness matrix can be written into the following matrix form:

$$H^{(R)} = \begin{bmatrix} H_{11}^{(R)} & H_{13}^{(R)} & H_{15}^{(R)} \\ H_{31}^{(R)} & H_{33}^{(R)} & H_{35}^{(R)} \\ H_{51}^{(R)} & H_{53}^{(R)} & H_{55}^{(R)} \end{bmatrix} \tag{7}$$

The second Piola-Kirchhoff stress of each layer can be written in the following vector form:

$$E^{(R)} = \left\{ E_{\mathrm{I,I}}^{(R)} \; E_{\mathrm{II,II}}^{(R)} \; E_{\mathrm{I,II}}^{(R)} \right\}^{T} \tag{8}$$

Thus, the constitutive law for each layer is defined by the following expression:

$$E^{(R)} = H^{(R)}D^{(R)} = H^{(R)}A\zeta^{(R)} + \frac{1}{2}H^{(R)}B\zeta^{(R)} \tag{9}$$

The virtual internal $\psi_i$ and external $\psi_e$ virtual work satisfy the following relationship:

$$\delta\psi = \delta\psi_i - \delta\psi_e = 0 \tag{10}$$

The whole sandwich structure's internal work is obtained by summing the top, bottom, and core layers' contributions as follows:

$$\begin{aligned} \delta\psi_i &= \delta\psi_i^{(\gamma)} + \delta\psi_i^{(\alpha)} + \delta\psi_i^{(\beta)} \\ &= \sum_{R=\gamma,\alpha,\beta} \int_{V_0} \delta D^{(R)T} E^{(R)} \mathrm{dv} \\ &= \sum_{R=\gamma,\alpha,\beta} \int_{V_0^R} \delta\zeta^{(R)T} [A + B^{(R)}(\zeta^{(R)}(u^{(R)}))]^T E^{(R)} \mathrm{dv} \end{aligned} \tag{11}$$

where $V_o^R$ is the undeformed volume of a layer. By neglecting the body forces and considering an external force proportional to a scalar parameter $\lambda$, the external work can be written in the following form:

$$\delta\psi_e = \lambda\delta u^T F \tag{12}$$

Where $F$ is an external load force vector. Based on the equations above, a weak formulation of governing equations of the sandwich structure reads as:

$$\sum_{R=\gamma,\alpha,\beta} \int_{v_0} \delta\zeta^{(R)T} [A + B^{(R)}]^T E^{(R)} \mathrm{dv} = \lambda\delta u^T F$$

$$E^{(R)} = H^{(R)}D^{(R)} \qquad D^{(R)} = \left[A + \frac{1}{2}B\right]\zeta^{(R)} \tag{13}$$

These equations are the starting point for developing geometrically non-linear hierarchical finite elements by means of Carrera's Unified Formulation.

# 3 HIERARCHICAL BEAM FINITE ELEMENTS OF SANDWICH STRUCTURES

The transversal displacements of the top, core, and bottom layers are discretized by three separate sets of Lagrange polynomial functions in the framework of beam modeling formulation. Each layer's axial displacement is discretized by the same Lagrange polynomial function in the framework of a finite element solution.

The idea of the proposed CUF-based model is to separate the displacement field of different layers with the corresponding Lagrangian functions $F_\tau^{(\alpha)}$, $F_\tau^{(\gamma)}$ and $F_\tau^{(\beta)}$. It can be noticed that the top, core, and bottom layers share the same shape functions along the axial direction. The maximum number of terms in $F_\tau^{(\alpha)}$, $F_\tau^{(\gamma)}$ and $F_\tau^{(\beta)}$ are $N_m^{(\gamma)}$, $N_m^{(\alpha)}$ and $N_m^{(\beta)}$, respectively. The maximum number of shape functions along the axial direction $N_i$ is $N_n^e$. The displacement approximation can be rewritten as follows for a generic layer:

$$u(\mathrm{I,II}) = F_\tau^{(R)}(\mathrm{II})N_i(\mathrm{I})Q_{\tau i}$$
$$\tau = 1,2,\ldots,N_u^{(R)} \qquad R = \gamma, \alpha, \beta$$
$$i = 1,2,\ldots,N_n^e \tag{14}$$

Where $Q_{\tau i}$ is the generalized nodal displacement vector:

$$Q_{\tau i}^{(R)T} = \left\{ Q_{\tau i}^{(R)u} \quad Q_{\tau i}^{(R)\omega} \right\} \tag{15}$$

Thus, the displacement gradient can be defined as follows:

$$\zeta^{(R)} = \left\{ F_\tau^{(R)}N_{i,I}Q_{\tau i}^u \quad F_{\tau,II}^{(R)}N_iQ_{\tau i}^u \quad F_\tau^{(R)}N_{i,x}Q_{\tau i}^\omega \quad F_{\tau,II}^{(R)}N_iQ_{\tau i}^\omega \right\} = G_{\tau i}^{(R)}Q_{\tau i} \tag{16}$$

Where $G_{\tau i}^{(u)}$ in a matrix form is as follows:

$$G_{\tau i}^{(R)} = \begin{bmatrix} F_\tau^{(R)}N_{i,I} & 0 \\ F_{\tau,II}^{(R)}N_i & 0 \\ 0 & F_\tau^{(R)}N_{i,I} \\ 0 & F_{\tau,II}^{(R)}N_i \end{bmatrix} \tag{17}$$

The following expressions for the virtual variations are derived within the CUF framework:

$$\delta u^{(R)} = F_\tau^{(R)}N_i\delta Q_{\tau i}^{(R)} \quad \delta\zeta^{(R)} = G_{\tau i}^{(R)}\delta Q_{\tau i}^{(R)} \tag{18}$$

The governing equation of the sandwich model then can be written as follows:

$$\sum_{R=\gamma,\lambda,\beta} \delta Q_{\tau i}^{(R)T} \int_{v_0^{(R)}} G_{\tau i}^{(R)T} \left[ A + B^{(R)}(\zeta^{(R)}) \right]^T (E^{(R)})\mathrm{dv} = \delta Q_{\tau i}^T \lambda^1 N_i^T F_\tau^T f$$

$$E^{(R)} = H^{(R)}(D^{(R)}) \qquad D^{(R)} = \left[ A + \frac{1}{2}B(\zeta^{(R)}) \right] G_{E,j}^{(R)} \delta q_{E,j}^{(R)} \tag{19}$$

# 4 NUMERICAL RESULTS: GLOBAL-LOCAL COUPLING BUCKLING ANALYSIS

Two cases of global-local coupling buckling are here presented. In the first case, the global buckling mode occurs first; in the second case, the local buckling mode occurs first. For both cases, the material data of the previous case is used. The ratio of core layer thickness versus skin layer thickness is 50, and the slenderness ratio is 20.

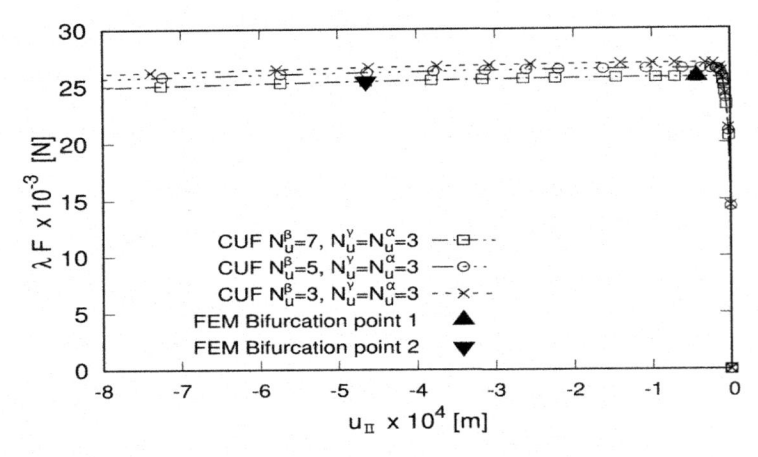

Figure 2.  Bifurcation curves for the local-global-coupling instability of the proposed CUF-based beam model and FEM model (based on Abaqus).

The local buckling instability mode occurs first. The relevant bifurcation curve is shown in Figure 2. The figure also shows the curve corresponding to the model with different nodes in the middle layer. Also, it shows the Bifurcation point 1 corresponding to the local buckling instability and the Bifurcation point 2 corresponding to the local-global coupling buckling instability mode.

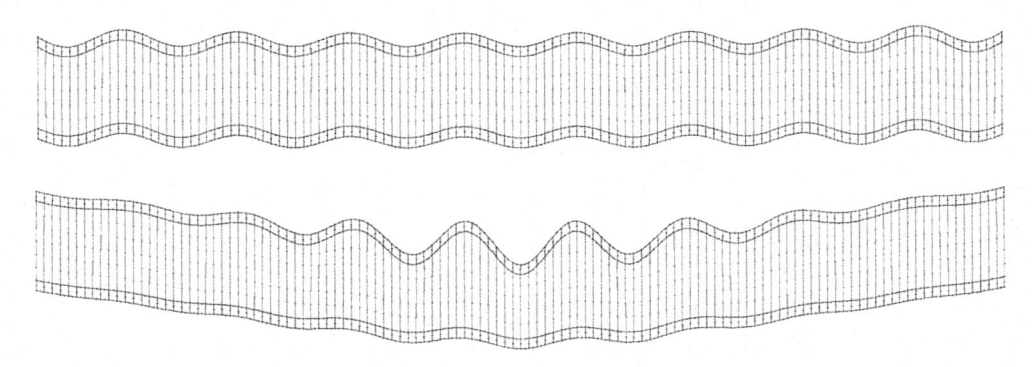

Figure 3.   (a) Local bucking (b) Global-local-coupling buckling.

## 5   CONCLUSIONS

The article focuses on the description of the structural model for geometrically nonlinear analyses. The CUF model was validated by comparison with the finite element model of ABAQUS. The displacement bifurcation curves were analyzed and evaluated. The accuracy is comparable to the reference solution based on 3D FEM and reduces the computational cost. Within the framework of a total Lagrangian approach, a finite element solution based on nonlinear CUF is provided. Furthermore, we show the application of the model to the ensemble-local coupling instability. This model has been verified by us and can solve complex nonlinear problems. But the related parameter analysis has not been carried out. We will analyze the parameters of different surface layers, different thicknesses of middle layers, and different aspect ratios, so as to find the sandwich structure with the best specific strength. In the future, this research can be used to guide the design and optimization of sandwich structures. This sandwich structure has strong scientific and application value.

## ACKNOWLEDGMENTS

This work has been supported by the National Natural Science Foundation of China (Grant Nos.12102269 and 52078306) and the Shenyang Science and Technology Project (21-108-9-21).

## REFERENCES

Carrera, E., Brischetto, S. (1) (2009) A survey with a numerical assessment of classical and refined theories for the analysis of sandwich plates. *Appl. Mech. Rev.*, 62:10803–10803.

Ferreira, A.J.M., Roque, C.M.C., Carrera, E., Cinefra, M., Polit, O. (24) (2011) Two higher-order zig-zag theories for the accurate analysis of bending, vibration, and buckling response of laminated plates by radial basis functions collocation and a unified formulation. *J. Compos. Mater.*, 45:2523–2536.

Hu, H., Belouettar, S., Potier-Ferry, M., Daya, E.M. (2008) Review and assessment of various theories for modeling sandwich composites. *Compos. Struct.*, 84: 282–292.

Hui, Y., Giunta, G., Belouettar, S., Huang, Q., Hu, H., Carrera, E. (2017) A free vibration analysis of three-dimensional sandwich beams using hierarchical one-dimensional finite elements. *Compos. Part. B.-Eng.*, 110: 7–19.

Moleiro, F., Carrera, E., Li, G., Cinefra, M., Reddy, J.N. (2019) Hygro-Thermo mechanical modelling of multilayered plates: Hybrid composite laminates, fiber metal laminates, and sandwich plates. *Compos. Part. B.-Eng.*, 177:107388.

Sayyad, A., Ghugal, Y. (2017) Bending, buckling and free vibration of laminated composite and sandwich beams: A critical review of the literature. *Compos. Struct.*, 171: 486–504.

*Civil Engineering and Urban Research – Mohamed & Hou (Eds)*
© *2023 the Authors, ISBN 978-1-032-44487-1*

# Parametric sensitivity analysis of the impact response of single-layer spherical mesh shells

Ruize Zhong, Chang Lin* & Jun Huang

*Guangzhou Academy of Fine Arts, Guangzhou, China*

ABSTRACT: Large-span spherical shells are vulnerable to solid impact, since they are designed without anti-impact consideration. In order to find the rules of initial conditions on the impact response of single-layer spherical shells, a non-linear finite element computational model of a single-layer spherical shell structure is developed and sensitivity analysis is carried out on parameters such as impact energy, roof mass, and structural damping. Based on the results of a stratified random sampling of the model parameters, the parametric sensitivity analysis method was used to determine the degree of influence of the variation of each parameter of the model on the maximum displacement at the apex of the mesh-shell structure and the maximum axial force response of the bars. The calculation results show that: under the action of concentrated impact, the variation of the Rayleigh damping parameter does not have a significant influence on the impact response of a single-layer spherical mesh shell, and the impact velocity is the most important factor to determine the impact response of mesh shell structure; the variation of the cross-sectional area of various types of mesh shell members does not affect the impact response of mesh shell structure by more than 4%; with the increase of mesh shell span, the influence of roof mass on the impact response decreases.

## 1 INTRODUCTION

Due to natural disasters, terrorist attacks, and other special circumstances, the impact resistance of important structures is gradually attracting the attention of engineers and researchers. Some research results on the impact resistance of large span steel mesh shell structures have been published in China. Li Haiwang and co-workers (Che et al. 2008; Li et al. 2006)were the first to carry out an experimental study and theoretical analysis on the dynamic stability of mesh-shell structures under the impact of blocks, revealing the elastic and elastoplastic dynamic response of the top of the mesh-shell structure to the vertical impact of a block, and finding the critical impact velocity for the instability of the mesh-shell structure through experiments. Shen Shizhao and Wang Duozhi et al. (Wang et al. 2008, 2010; Fan et al. 2010) determined the failure modes and discrimination methods of the mesh-shell structure under impact by means of dynamic non-linear finite element analysis. Wang Xiuli et al. (Ma et al. 2014; Wang et al. 2014) investigated the dynamic response of mesh shell structures subjected to concentrated impacts in different directions at different locations through experimental studies and numerical simulations. The results of their study indicated that the mesh shell structures were most likely to collapse as a whole when subjected to vertical impacts at the apex location, and therefore the impact resistance of mesh shell structures was mainly considered when the apex was subjected to concentrated impacts.

The dynamic response of the top of a mesh-shell structure under the action of a concentrated impact is influenced by the geometrical and physical parameters of the structure, such as bar section size, roof mass, structural damping, etc. Some of the parameters of the mesh shell structure have

---

*Corresponding Author: linlinc@aliyun.com

DOI 10.1201/9781003372417-111

been analyzed in the relevant literature (Ma & Wang 2015), but the sensitivity of the parameters can be quantified to obtain the degree of influence of each parameter on the impact response of the mesh shell structure. This paper introduces the Sobol parameter sensitivity analysis method (Sobol 2001) to carry out a parametric system analysis of the impact response of the actual mesh shell structure. From the results of the study, it can be seen that the impact velocity has the greatest influence on the impact response of the mesh shell structure, the influence of the cross-sectional area of the bars, the roof mass, and the mesh shell defects on the impact response of the mesh shell structure is relatively small, and the influence of the roof mass on the structural response decreases with the increase of the structural span.

## 2 SOBOL SENSITIVITY ANALYSIS METHOD AND ITS CALCULATION

The purpose of sensitivity analysis is to quantitatively describe the extent to which uncertainty in the input parameters of a model affects uncertainty in the output results. Sobol parameter sensitivity analysis is based on the statistical variance of the output results. Suppose a mathematical model is $y = f(x_1, x_2, ..., x_n)$, $x_1 \sim x_n$ are random variables, representing the input parameters; $y$ is a scalar, representing the output outcome. The degree of influence of each input parameter on the output result is obtained by random sampling of the statistical model. For local sensitivity analysis, the degree of influence is generally quantified by fixing a certain random variable $x_i (i=1, 2,..., n)$ and the remaining parameter values are drawn in the sample space with probability, and then substituted into the model to obtain a series of output results, and the variance $V(y)$ of the statistical output results, the smaller $V(y)$, the greater the degree of influence of that input parameter on the output results. However, for global sensitivity analysis, the quantifier is $E(V(y|x_i))$, which is the mean value of $V(y)$ over the range of values of $x_i$. The Sobol method of parametric sensitivity analysis allows the higher order sensitivity of the input parameters to be calculated taking into account the interactions between the parameters. The use of stratified sampling methods in sensitivity analysis alleviates local stacking of the input data and can reduce the computational effort.

The mathematical model targeted by the Sobol parametric sensitivity analysis $y = f(x_1, x_2, ..., x_n)$ can be decomposed as

$$y = f_0 + \sum_{i=1}^{n} f_i(x_i) + \sum_{i=1}^{n-1}\sum_{j>i}^{n} f_{ij}(x_i, x_j) + \cdots + f_{ij...k}(x_i, x_j ... x_k) \tag{1}$$

where $f_0$ is a constant. Sobol proved that this decomposition is unique if the average of the functions on the right-hand side of Equation (1) is 0 and the components can be found in terms of the conditional expectation of the output parameter $y$ as follows.

$$f_0 = E(y) \tag{2}$$

$$f_i(x_i) = E(y|x_i) - E(y) \quad (i = 1,2, \cdots, n) \tag{3}$$

$$f_{ij}(x_i, x_j) = E(y|x_i, x_j) - f_i(x_i) - f_j(x_j) - E(y) \quad (i = 1,2, \cdots, n) \tag{4}$$

where $E(V(y|x_i))$ denotes the expectation of $y$ for random variables other than $x_i$ and $E(V(y|x_i, x_j))$ denotes the expectation of $y$ for random variables other than $x_i$ and $x_j$.

The decomposition Equation (1) can be written in variance form as

$$V(y) = \sum_i V(f_i(x_i)) + \sum_i \sum_{j>i} V(f_{ij}(x_i, x_j)) + \cdots + V(f_{ij...k}(x_i, x_j ... x_k)) \tag{5}$$

Since $V(y) \neq 0$, from Equation (5), it follows that

$$1 = \sum_i S_i + \sum_i \sum_{j>i} S_{ij} + \cdots + S_{ij...k} \tag{6}$$

where $S_i = \frac{V(f_i(x_i))}{V(y)}$ is called the first-order sensitivity of the parameter $x_i$; $S_{ij} = \frac{V(f_{ij}(x_i,x_j))}{V(y)}$ represents the higher-order sensitivity and is influenced by the parameters $x_i$ and $x_j$ together.

The total sensitivity $S_{Ti}$ of a parameter $x_i$ is defined as the sum of the order sensitivities including subscript $i$. For example, when the model encompasses three parameters $x_1$, $x_2$, $x_3$, then $S_{T1} = S_1 + S_{12} + S_{13} + S_{123}$.

Saltelli (Saltelli 2008) implements the Sobol global sensitivity analysis of the mathematical model $y = f(x_1, x_2, ..., x_n)$ using a modified Monte Carlo approach, which is calculated as follows:

I) Generate two matrices $\mathbf{A}$ and $\mathbf{B}$ of m×n by random sampling in the input parameter space.

$$A = \begin{bmatrix} x_1^{(1)} & x_2^{(1)} & \cdots & x_i^{(1)} & \cdots & x_n^{(1)} \\ x_1^{(2)} & x_2^{(2)} & \cdots & x_i^{(2)} & \cdots & x_n^{(2)} \\ \cdots & \cdots & \cdots \cdots & \cdots \cdots \\ x_1^{(m-1)} & x_2^{(m-1)} & \cdots & x_i^{(m-1)} & \cdots & x_n^{(m-1)} \\ x_1^{(m)} & x_2^{(m)} & \cdots & x_i^{(m)} & \cdots & x_n^{(m)} \end{bmatrix} \tag{7}$$

$$B = \begin{bmatrix} x_1^{(m+1)} & x_2^{(m+1)} & \cdots & x_i^{(m+1)} & \cdots & x_n^{(m+1)} \\ x_1^{(m+2)} & x_2^{(m+2)} & \cdots & x_i^{(m+2)} & \cdots & x_n^{(m+2)} \\ \cdots & \cdots & \cdots \cdots & \cdots \cdots \\ x_1^{(2m-1)} & x_2^{(2m-1)} & \cdots & x_i^{(2m-1)} & \cdots & x_n^{(2m-1)} \\ x_1^{(2m)} & x_2^{(2m)} & \cdots & x_i^{(2m)} & \cdots & x_n^{(2m)} \end{bmatrix} \tag{8}$$

II) Define m×n the elements of matrix $\mathbf{C}_i$ the i-th column of which is taken from the i-th column of matrix $\mathbf{B}$ and the remaining columns are taken from the corresponding columns in matrix $\mathbf{A}$.

$$C_i = \begin{bmatrix} x_1^{(1)} & x_2^{(1)} & \cdots & x_i^{(m+1)} & \cdots & x_n^{(1)} \\ x_1^{(2)} & x_2^{(2)} & \cdots & x_i^{(m+2)} & \cdots & x_n^{(2)} \\ \cdots & \cdots & \cdots \cdots & \cdots \cdots \\ x_1^{(m-1)} & x_2^{(m-1)} & \cdots & x_i^{(2m-1)} & \cdots & x_n^{(m-1)} \\ x_1^{(m)} & x_2^{(m)} & \cdots & x_i^{(2m)} & \cdots & x_n^{(m)} \end{bmatrix} \tag{9}$$

III) Calculate the outputs corresponding to the input parameters in the matrices $\mathbf{A}$, $\mathbf{B}$, and $\mathbf{C}_i$, respectively, to obtain three column vectors.

$$y_A = f(\mathbf{A}) \quad y_B = f(\mathbf{B}) \quad y_{Ci} = f(\mathbf{C}_i) \tag{10}$$

The first-order sensitivity of the parameter $x_i$ is approximated as

$$S_i = \frac{V[E(y|x_i)]}{V(y)} = \frac{1/m(y_A \cdot y_{C_i}) - f_0^2}{1/m(y_A \cdot y_A) - f_0^2} = \frac{(1/m)\sum y_A^{(j)} y_{C_i}^{(j)} - f_0^2}{(1/m)\sum (y_A^{(j)})^2 - f_0^2} \tag{11}$$

where $f_0 = E(y)$.

The total sensitivity of the parameter $x_i$ can be approximated as

$$S_{T_i} = 1 - \frac{V[E(y|x_{\sim i})]}{V(y)} = 1 - \frac{1/m(y_B \cdot y_{C_i}) - f_0^2}{1/m(y_A \cdot y_A) - f_0^2} = 1 - \frac{(1/m)\sum y_B^{(j)} y_{C_i}^{(j)} - f_0^2}{(1/m)\sum (y_A^{(j)})^2 - f_0^2} \tag{12}$$

where $x_{\sim i}$ refers to the other input parameters except $x_i$.

# 3 MESH SHELL STRUCTURE MODEL AND ANALYSIS METHODS

The example in this paper is a single-layer Kiewit spherical shell structure subject to impact, with spans of 40 m and 60 m. The rods are Q235 hollow circular tubes. The span of mesh shell structure I is s=40m, the number of ring rod turns m=6, the number of radial rod turns n=6, the sagittal-to-span ratio f/s=1/5, the radial rod section is $\Phi 89 \times 8$, the oblique rod section is $\Phi 89 \times 8$ and the annular rod section is $\Phi 83 \times 8$; the span of mesh shell structure II is s=60m, the number of ring rod turns m=8, the radial rod turns n=8, the sagittal-to-span ratio f/s=1/5, the radial rod section is $\Phi$ The mesh shell structure is as shown in Figure 1.

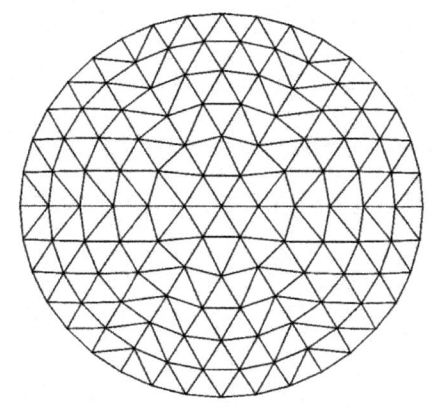 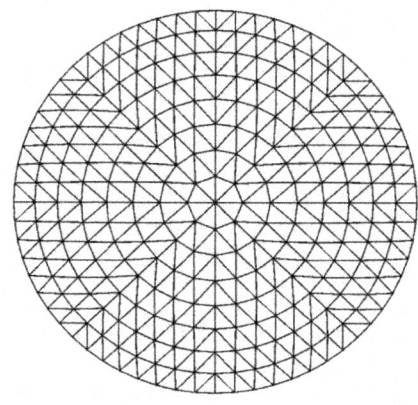

(a) Plan view of the s=40m span mesh shell structure    (b) Plan view of the s=60m span mesh shell structure

Figure 1.    Plan view of the mesh shell structure.

The following input parameters to the mesh structure are treated as random variables and are assumed to be uniformly distributed over a range.

a) Roof mass: Considered as vertical uniform load, according to the loaded code (Ministry of Construction of the People's Republic of China 2010), a reasonable range of 35±5kg/m is taken for the uniform load of the roof. The roof mass is distributed to the nodes when performing the finite element analysis.

b) Rod cross-sectional area: the radial, oblique and circumferential rod cross-sectional diameters vary within ±10 mm of the original diameter, and the wall thickness of the rod remains unchanged. The cross-sectional areas of the radial, oblique and circumferential bars are $A_1$, $A_2$, and $A_3$ respectively.

c) The structure is damped using Rayleigh damping. According to reference (Cao et al. 2008), the damping ratio of the first order mode is 0.02±0.01. The damping ratio of the second order mode is equal to the damping ratio of the first order mode, and the Rayleigh damping equation of the first two orders can be combined to obtain the mass correlation coefficient $\alpha$ and the stiffness correlation coefficient $\beta$. This also ensures that the modal damping ratio increases with the increase of the modal order.

d) Initial geometric defects: two cases of random defects and consistent defects are considered. In the case of random defects, the maximum permissible deviation of local defects is 1/1500 of the span and not greater than 40mm, so the maximum deviation of the mesh shell structure I and II are 26.7mm and 40mm respectively. in the case of uniform defects, the initial defect distribution of the mesh shell structure adopts the first order flexural mode and the maximum defect is 1/300 of the span.

e) Initial impact velocity: The initial impact velocity v can be calculated by taking the mass of the rigid impactor m=400kg for Mesh I and (30±10)m/s for Mesh II. The initial impact velocity can be calculated by taking the mass of the rigid impactor m=500kg for Mesh II and (40±20)m/s.

The output parameters of the mesh-shell structure are the maximum axial force response of the rod $F_{max}$ and the maximum displacement response of the apex $\Delta_{max}$. Due to a large number of input parameters for the mesh-shell structure, it is necessary to first filter some of the parameters that have little effect on the response results and then carry out a parameter sensitivity analysis on this basis.

In this paper, ANSYS/LS-DYNA finite element software was used for the non-linear dynamic response analysis. All mesh shell rods are defined as beam161 units with fixed connections between the rods; impact blocks are defined as solid164 solid units and are treated as rigid bodies; the contact type between the impact blocks and the apex of the mesh shell structure is defined as point contact. The validity of this calculation method is verified in the literature (Wei & Hu 2015).

## 4 PARAMETER FILTERING

The basic steps of parameter screening are: firstly, a certain input parameter $x_i$ is randomly selected, and the rest of the parameters are assumed to remain unchanged as its central value. When the coefficient of variation of the output is large, then the input parameter $x_i$ is more important. Since the sensitivity of an input parameter is calculated with the rest of the input parameters fixed as central values and without considering the whole sample space, this screening process is a local sensitivity analysis method, which is fast and easy compared to global sensitivity analysis.

Tables 1 and 2 give the statistics of the maximum axial force of the bars $F_{max}$ and the maximum displacement at the apex $\Delta_{max}$ calculated for spans 40m and 60m mesh shell structures I and II respectively under the action of concentrated impact.

The input parameters are ranked according to the magnitude of the coefficient of variation of the output results given in Table 1. The ranking of the input parameters corresponding to the coefficient of variation of the maximum displacement at the apex is: impact velocity > cross-sectional area $A_3$ > cross-sectional area A > random defects > cross-sectional area $A_{21}$ > roof mass > structural damping > consistent defects; the ranking of the input parameters corresponding to the coefficient of variation of the maximum axial force of the rod is: roof mass > impact velocity > cross-sectional area $A_1$ > cross-sectional area $A_3$ = random defects > cross-sectional area $A_2$ > consistent defects = structural damping.

The input parameters are ranked according to the magnitude of the coefficient of variation of the output results given in Table 2. The ranking of the input parameters corresponding to the coefficient of variation of the maximum displacement at the apex is: impact velocity > cross-sectional area A > cross-sectional area $A_{32}$ > cross-sectional area $A_1$ > random defects > roof mass = structural damping > consistent defects; the ranking of the input parameters corresponding to the coefficient of variation of the maximum axial force of the rod is: impact velocity > cross-sectional area $A_3$ > roof mass > cross-sectional area $A_2$; random defects > cross-sectional area $A_1$ > structural damping > uniform defects.

A comparison of the two initial geometric defect cases shows that the coefficient of variation is greater when random defects are considered than consistent defects. Furthermore, the nodal position error is closer to the actual situation when considered in terms of random defects, therefore only the random defect case is considered in the following parameter importance analysis. A comprehensive evaluation of the output of the two-span mesh-shell structure shows that the input parameters that affect the degree of structural response are: impact velocity, circumferential rod cross-sectional area, diagonal rod cross-sectional area, radial rod cross-sectional area, and roof mass. In addition, the damping ratio of the structure is less influential, so the following analysis takes the damping ratio of the structure to be 0.02.

## 5 PARAMETER IMPORTANCE

Based on the initial screening of the input parameters, it was found that the six input parameters of radial rod cross-sectional area $A_1$ ($x_1$), oblique rod cross-sectional area $A_2$ ($x_2$), circumferential

Table 1. Statistics of the output of the mesh shell structure I.

| Input parameters | $\Delta_{max}$ (mm) | | | $F_{max}$ (kN) | | |
|---|---|---|---|---|---|---|
| | Average | Standard deviation | Coefficient of variation | Average | Standard deviation | Coefficient of variation |
| Cross-sectional area $A_1$ | 969.93 | 11.08 | 0.011 | 615.10 | 33.74 | 0.055 |
| Cross-sectional area $A_2$ | 976.98 | 23.09 | 0.024 | 621.46 | 10.00 | 0.016 |
| Cross-sectional area $A_3$ | 996.24 | 63.71 | 0.064 | 619.49 | 23.99 | 0.039 |
| Speed of impact | 988.68 | 193.98 | 0.196 | 565.65 | 55.83 | 0.099 |
| Random defects | 968.41 | 13.31 | 0.014 | 634.59 | 24.57 | 0.039 |
| Consistent defects | 973.45 | 1.14 | 0.001 | 619.04 | 8.08 | 0.013 |
| Roof quality | 974.14 | 7.80 | 0.008 | 618.39 | 76.51 | 0.124 |
| Structural damping | 973.84 | 1.86 | 0.002 | 621.75 | 13.25 | 0.021 |

Table 2. Statistics on the output of Mesh Shell Structure II.

| Input parameters | $\Delta_{max}$ (mm) | | | $F_{max}$ (kN) | | |
|---|---|---|---|---|---|---|
| | Average | Standard deviation | Coefficient of variation | Average | Standard deviation | Coefficient of variation |
| Cross-sectional area $A_1$ | 1302.87 | 24.81 | 0.019 | 583.90 | 22.71 | 0.039 |
| Cross-sectional area $A_2$ | 1303.48 | 32.93 | 0.025 | 548.17 | 27.91 | 0.051 |
| Cross-sectional area $A_3$ | 1317.73 | 63.23 | 0.048 | 549.05 | 47.49 | 0.086 |
| Speed of impact | 1293.20 | 469.46 | 0.363 | 620.71 | 135.39 | 0.218 |
| Random defects | 1305.48 | 8.98 | 0.007 | 551.93 | 24.09 | 0.044 |
| Consistent defects | 1303.41 | 0.95 | 0.001 | 597.39 | 6.78 | 0.011 |
| Roof quality | 1302.20 | 2.85 | 0.002 | 548.93 | 43.91 | 0.080 |
| Structural damping | 1303.43 | 1.45 | 0.002 | 597.30 | 14.34 | 0.024 |

rod cross-sectional area $A_3$ ($x_3$), impact velocity ($x_4$), roof mass ($x_5$) and random defects ($x_6$) had a greater influence on the output of the structural impact response (maximum rod axial force $y_1$ and maximum displacement at the vertex $y_2$, and therefore a global sensitivity analysis was performed on these six input parameters. The model outputs are the maximum rod axial force ($y_1$) and the maximum displacement at the apex ($y_2$).

The input matrix **A** and the input matrix **B** were selected by Super Latin sampling, and the matrices $\mathbf{C}_1$ to $\mathbf{C}_6$ were obtained by combining these two matrices. All eight matrices contain 2000 rows of elements, and the total number of model calculations is $2000 \times 8$.

Figure 2 shows the histogram and normal distribution curve of the frequency distribution of the output $y_1$ and $y_2$. From Figure 2, we can see that the output results $y_1$ and $y_2$ are basically normally distributed, indicating that the sampling process is reasonable. During the calculation process, it was found that the mean and variance of the output results stabilized after 200 rounds of calculation.

The results of the Sobol sensitivity analysis of the output $y_1$ and $y_2$ for each input parameter of the mesh shell structure are given in table 3 and table 4 respectively. In the tables, $S_1 \sim S_6$ denote the first order sensitivity corresponding to parameters $x_1 \sim x_6$, and $S_{T1} \sim S_{T6}$ denote the total sensitivity corresponding to parameters $x_1 \sim x_6$.

As can be seen from the results in tables 3 and 4, for the two web shells with different spans, the impact velocity $x_4$ has the greatest influence on the output results $y_1$ versus $y_2$, which is of about 50% importance. This is in line with the reality that the impact force is often an important factor in determining the structural response.

The total sensitivity of the statistical outputs $y_1$ and $y_2$ to the variation of each parameter can be obtained as a percentage of the importance of each parameter, as shown in figure 3. It can be seen

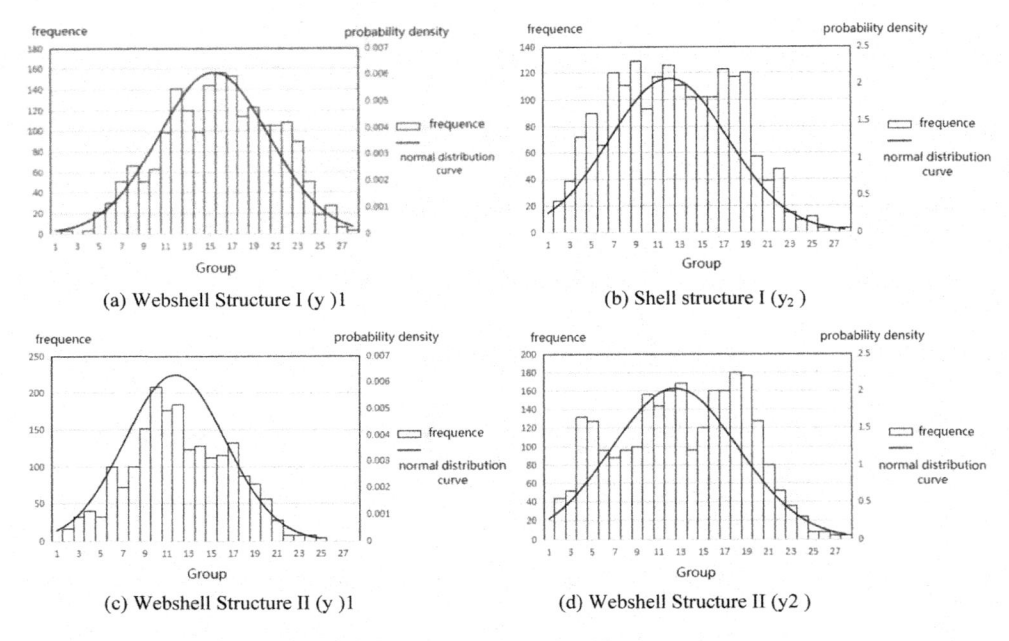

(a) Webshell Structure I (y )1

(b) Shell structure I ($y_2$ )

(c) Webshell Structure II (y )1

(d) Webshell Structure II (y2 )

Figure 2.   Frequency distribution histogram with a normal distribution curve.

from figure 3 that for the mesh-shell structures I and II, the response quantity $F_{max}$ corresponding to the parameter $x_5$ has a higher sensitivity than $\Delta_{max}$ corresponding to the parameter $x_5$. For both spans, the importance ratios of the output parameters $y_1$ corresponding to parameters $x_1$, $x_2$, and $x_3$ do not differ by more than 2%, and the importance ratios of the output parameters $y_2$ corresponding to parameters $x_1$, $x_2$, and $x_3$ do not differ by more than 4%, indicating that the influence of each type of rod on the impact resistance of the mesh shell is equally significant.

Table 3.   Sobol sensitivity corresponding to the output $y_1$.

| Span(m) | 40 | 60 | Span(m) | 40 | 60 |
|---|---|---|---|---|---|
| $S_1$ | 0.18 | 0.09 | $S_{T1}$ | 0.30 | 0.12 |
| $S_2$ | 0.25 | 0.10 | $S_{T2}$ | 0.25 | 0.13 |
| $S_3$ | 0.01 | 0.10 | $S_{T3}$ | 0.26 | 0.11 |
| $S_4$ | 0.27 | 0.48 | $S_{T4}$ | 0.54 | 0.57 |
| $S_5$ | 0.13 | 0.08 | $S_{T5}$ | 0.41 | 0.12 |
| $S_6$ | 0 | 0.07 | $S_{T6}$ | 0.18 | 0.14 |

The importance of the parameters $x_1$, $x_2$, $x_3$, $x_5$, and $x_6$ were recalculated as a percentage of importance if the effect of impact velocity was not considered. Further analysis of the sensitivity of the roof mass $x_5$ reveals that Mesh II is less sensitive than Mesh I to the response output results $y_1$ and $y_2$, due to the increased cross-sectional area of the bars in the 60m span mesh structure and the increased ratio of structural mass to roof mass. Therefore, the larger the span of the single-storey mesh shell structure, the smaller the effect of roof mass on the impact resistance of the structure within the range of parameter variations given in this paper. In addition, the percentage of sensitivity of random defects $x_6$ in the mesh shell structure I to $y_1$, and $y_2$ is smaller than the percentage of sensitivity of this parameter in the mesh shell structure II, which indicates that the larger the span of the structure, the more significant the effect of random defects on the impact response of the structure.

Table 4. Sobol sensitivity corresponding to the output y2.

| Span(m) | 40 | 60 | Span(m) | 40 | 60 |
|---|---|---|---|---|---|
| $S_1$ | 0.10 | 0.09 | $S_{T1}$ | 0.16 | 0.17 |
| $S_2$ | 0.11 | 0.08 | $S_{T2}$ | 0.16 | 0.17 |
| $S_3$ | 0.10 | 0.05 | $S_{T3}$ | 0.12 | 0.21 |
| $S_4$ | 0.38 | 0.52 | $S_{T4}$ | 0.43 | 0.68 |
| $S_5$ | 0.11 | 0.14 | $S_{T5}$ | 0.15 | 0.10 |
| $S_6$ | 0.12 | 0.13 | $S_{T6}$ | 0.13 | 0.15 |

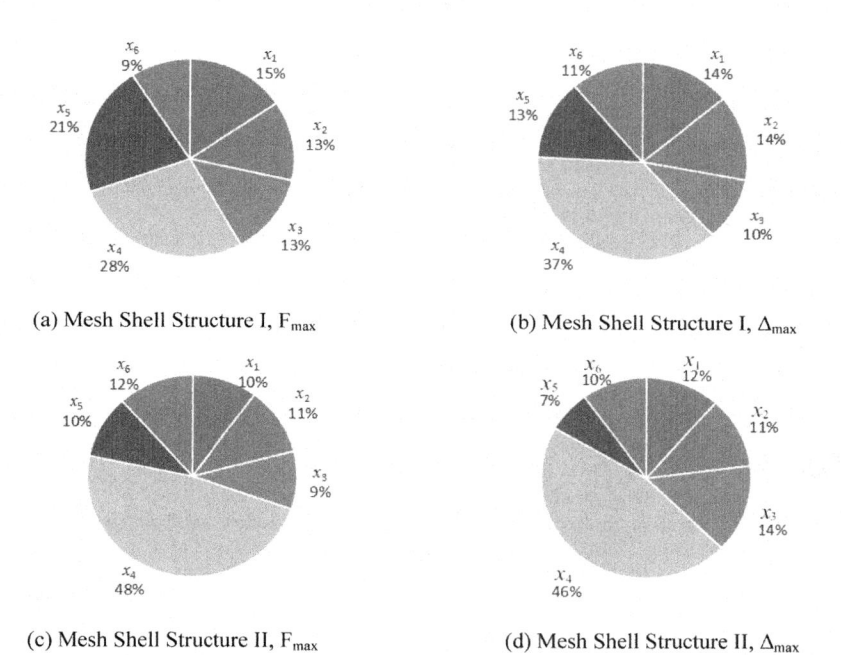

(a) Mesh Shell Structure I, $F_{max}$

(b) Mesh Shell Structure I, $\Delta_{max}$

(c) Mesh Shell Structure II, $F_{max}$

(d) Mesh Shell Structure II, $\Delta_{max}$

Figure 3. Importance of input parameters for the mesh shell structure.

## 6 CONCLUSION

In this paper, Monte Carlo estimates of the sensitivity of the structural response of the mesh shell to the relevant parameters under the action of a concentrated impact are given by means of Sobol sensitivity analysis. The screening and importance analysis of the input parameters leads to the following main conclusions.

(1) Rayleigh damping has little effect on the impact response of a single-layer mesh shell structure and the first order damping ratio of 0.02 can be taken for the dynamic response calculation.

(2) The change in cross-sectional area of each type of rod has an equally significant effect on the impact response of the mesh and shell structure, and the synergistic effect of the mesh and shell structure under impact is more obvious.

(3) As the span of the mesh shell structure increases, the sensitivity of its impact response to changes in roof mass decreases accordingly, while the sensitivity to random defects increases accordingly.

Results reveal that Sobol sensitivity analysis can be very efficient and precise in the evaluation of the parametric importance of impact initial conditions. However, sensitivity analysis should be combined with theoretical inference to reflect the meaning of sensitivity values, and that is continued to be researched.

## ACKNOWLEDGEMENT

This work is supported by Project "Research on Training Mode of Research Ability Involved in the Integration of Arts and Sciences" (Grant No. 2021JGXM077) during phased research results of 2021 Guangdong Academic Degree and Postgraduate Education Reform Research.

## REFERENCES

Cao Z, Xue S and Wang X 2008 Selection of earthquake waves and values of damping ratio for space structures in aseismic analysis *J. Spat. Stru* **14** pp 3–8.

Che W, Li H and Luo Q 2008 Analysis on dynamic buckling of single-layer elliptical paraboloid latticed shells under impact *J. Chi. Quar. Mech* **29** pp 33–39.

Fan F, Wang D, and Zhi X 2010 Failure modes and discrimination method for kiewitt reticulated dome under impact loads *Chi. Civi. Engi. J* **43** pp 56–62.

Li H, Guo K and Wei J 2006 The dynamic response of a single-layer reticulated shell to drop hammer impact *J. Expl. Shoc* **26** pp 39–45.

Ma X and Wang X 2015 Dynamic stability of single-layer reticulated shell structures subjected to impact loads *J. Vib. Shoc* **34** pp 119–124.

Ma X, Wang X, and Wu C 2014 Dynamic response analysis and test research on single-layer reticulated dome under different impact loading points *Proc. Nati. Con. Stru. Eng* Lan Zhou pp 119–124.

Ministry of Construction of the People's Republic of China 2010 GB 50009-2010 Code for structural loading of buildings *Chi. Cons. Indu. Pres*.

Saltelli A 2008 Global Sensitivity Analysis pp The Primer *John Wiley & Sons*.

Sobol I. 2001 Global Sensitivity Indices for Nonlinear Mathematical Models and their Monte Carlo Estimates *J. Math. Com. Simu* **55** pp 271–280.

Wang D, Fan F and Zhi X 2010 Failure mechanism of single-layer reticulated domes subjected to impact loads *J. Expl. Shoc* **30** pp 169–177.

Wang D, Zhi X and Fan F 2008 Failure patterns of kiewitt8 single-layer reticulated domes under impact roads *J. Engi. Mech* **25** pp 144–149.

Wang X, Ma X and Li J 2014 Dynamic Response of single-layer reticulated shells under different points impact loading *J. Spat. Stru* **20** pp 23–28.

Wei D and Hu C 2015 Impact response of single-layer lattice shells *J. Tian. Uni* **S1** pp 127–133.

*Civil Engineering and Urban Research – Mohamed & Hou (Eds)*
© *2023 the Authors, ISBN 978-1-032-44487-1*

# Huantang West River water environment improvement analysis of diversion water based on two-dimensional numerical simulation

Donghui Hu
*Yuyao Water Conservancy Bureau, Yuyao, China*

Haibiao Shen*
*Ningbo Shunnong Group Co., Ltd, Yuyao, China*

Yishan Chen, Zihao Li & Mei Chen
*The Key Laboratory for Technology in Rural Water Management of Zhejiang Province, Zhejiang University of Water Resources and Electric Power, Hangzhou, China*

ABSTRACT: Yubei plain is not only threatened by the flood tide of the outer river, but also has the hidden danger of waterlogging. The waterlogged water in the plain is discharged into the Cao'e River in the west through the No. 1 gate of the west embankment and into the Hangzhou Bay through the new Dongjin gate and the No. 2 gate in the north. In order to improve the water environment of the Yubei plain river network, the influence of the diversion scheme on Huantang West River, the main river in the plain river network, is analyzed by numerical simulation. From the water quality indicators BOD and DO, the water quality after diversion is improved from class V water to class III water, the diversion water quality improvement effect is obvious, and the diversion scheme is feasible. In addition, opening the internal sluice of the river network is conducive to the improvement of water flow and the water environment.

## 1 INTRODUCTION

The riverside area of Shaoxing Binhai New Town starts from the Qiantang River in the north and the Cao'e River in the southwest. The river network in the north of Binhai is shown in Figure 1. Among them, the No. 1 gate of Xidadi has the functions of drainage and water diversion after reconstruction, and the No. 2 gate of Hangzhou Bay mainly undertakes the task of improving the water quality of Yubei plain. The development of the economy and society urgently needs flood control and waterlogging drainage and improvement of the water environment (Lu 2021; Tong et al. 2016; Xie et al. 2010). Due to the lack of measured data on river networks, a large number of hydrodynamic factors such as water level and velocity can be obtained by using a river network mathematical model. However, in the study of the Yubei river network, there are more one-dimensional mathematical models and fewer two-dimensional mathematical models, but two-dimensional mathematical models have many advantages (Chen et al. 2012, 2014, 2015; Lu et al. 2013). Using a two-dimensional mathematical model to study the water environment of the Yubei river network is of great significance. Based on the verification and application of the two-dimensional water quality mathematical model of the Shangyu plain river network, the two-dimensional water quality mathematical model of the Binhai river network is obtained by removing the irrelevant range (Wu & He 2019). The Huantang West River is selected as the research object to analyze the improvement of the water environment through a water diversion scheme.

---

*Corresponding Author: shb198212@163.com

DOI 10.1201/9781003372417-112

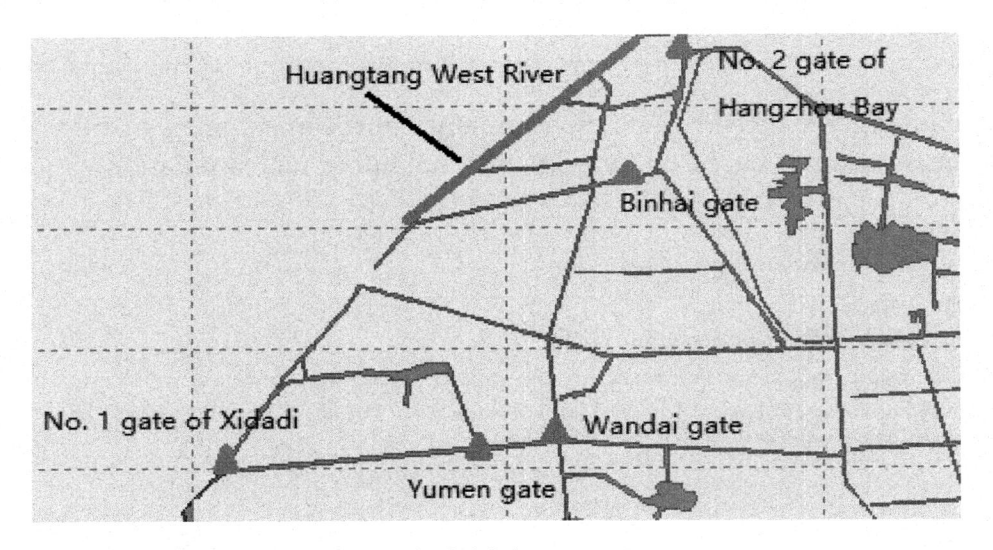

Figure 1.   Location of main gates in the north of Binhai.

## 2   TWO-DIMENSIONAL HYDRODYNAMIC PLANE MATHEMATICAL MODEL OF PLAIN RIVER NETWORK

### 2.1   *Basic equation*

Flow continuity equation:

$$\frac{\partial z}{\partial t} + \frac{\partial (hu)}{\partial x} + \frac{\partial (hv)}{\partial y} = 0 \tag{1}$$

X. Y-direction flow equation:

$$\frac{\partial u}{\partial t} + u\frac{\partial u}{\partial x} + v\frac{\partial u}{\partial y} - fv + g\frac{\partial z}{\partial x} + g\frac{u\sqrt{u^2 + v^2}}{c^2 h} = \lambda\Delta u \tag{2}$$

$$\frac{\partial u}{\partial t} + u\frac{\partial v}{\partial x} + v\frac{\partial \beta v}{\partial y} + fu + g\frac{\partial z}{\partial y} + g\frac{v\sqrt{u^2 + v^2}}{c^2 h} = \lambda\Delta v \tag{3}$$

Where: x and y are spatial coordinates; z is the water level; t is the time; u, v is the velocity component in the X direction and Y direction respectively; h is the water depth; $f$ is Coriolis force, $f = 2W \sin \Phi$; $W$ is the rotation velocity of the earth; $\Phi$ is the latitude of the earth; c is Xie Cai coefficient; g is the gravitational acceleration.

### 2.2   *Initial condition*

The initial value of the mathematical model, given the initial time of the river network, the water level is the constant water level of 2.8m and the flow velocity is 0; the initial condition of water quality is class V water, that is, BOD is 10 mg /L and DO is 2 mg /L; the introduced clean water is class III water, that is, BOD is 4 mg /L, DO is 5 mg /L; the No. 1 gate of Xidadi is opened with a flow of 20 m³ /s, and the No. 2 gate of Hangzhou Bay is opened with a flow of - 60 m³ /s, the direction is opposite to the No. 1 gate; the Binhai gate, Yumen gate, Qianjin gate and Wandai gate are closed.

# 3 ANALYSIS OF CALCULATION RESULTS

## 3.1 *Analysis of water level change*

It can be seen from the variation curve of the water level of Huantang West River with the distance shown in Figure 2 that the water level of Huantang West River decreases continuously with time. It is obvious from the variation curve of the water level difference of Huantang West River with the distance shown in Figure 3 that the water level of the river decreases very fast in the first four hours, and the rate of decline from the 4th to 10th hours is relatively fast, the rate of decline from the 11th to 50th hours is relatively small, and the rate of decline after 50 hours is even smaller.

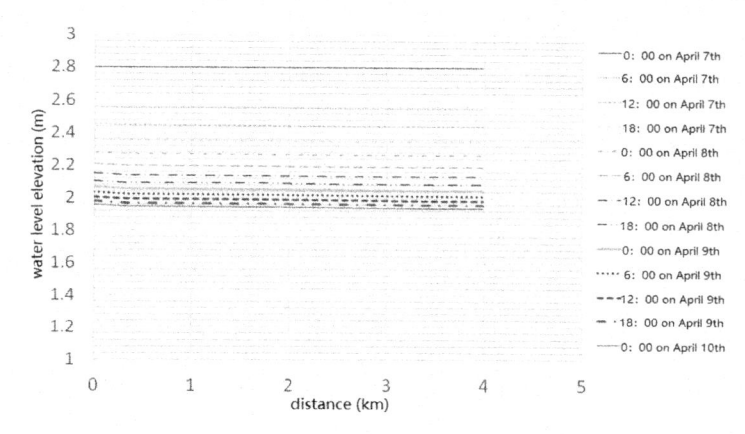

Figure 2. Variation curve of water level of Huantang West River with distance.

Take the typical point at the junction of Huantang West River and Jiuyiqiu-Beitang River as an example. As can be seen from Figure 3 and Figure 4, the difference in water surface elevation at the junction points generally increases in the first 10 hours and decreases in the last 60 hours, that is, the water surface elevation decreases rapidly in the first 10 hours and slowly in the last 60 hours. The rapid decrease of water surface elevation in the first 10 hours is due to the height difference between the flow of the Jiuyiqiu-Beitang River and the Huantang West River. The slow decrease in the last 60 hours is due to the decrease in water surface elevation difference between the two rivers with the water conveyance process.

## 3.2 *Velocity variation analysis*

It can be seen from the variation curve of water quality index and flow rate and water level with time at typical points of Huantang West River in Figure 5 that the longer the improvement time

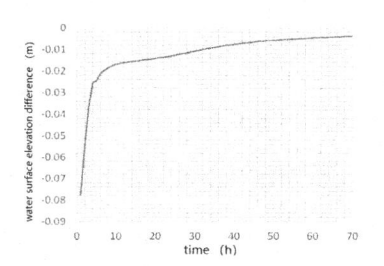

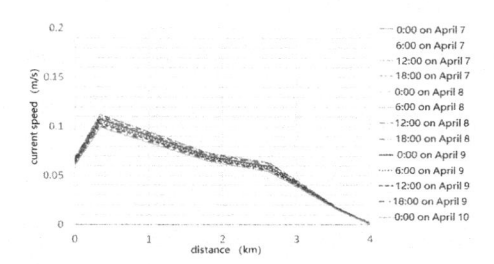

Figure 3. Difference of water surface at points taken by Huantang West River.

Figure 4. Variation curve of the velocity of Huantang West River with distance.

of Huantang West River, the smaller the flow rate of Huantang West River, and at the same time, the greater the distance, the smaller the flow rate. This is because Huantang West River is not connected with the No. 2 gate of Hangzhou Bay, while Jiuyiqiu-Beitang River is affected by the outlet gate, resulting in the continuous accumulation of water flow to the outlet, resulting in the situation shown in the figure.

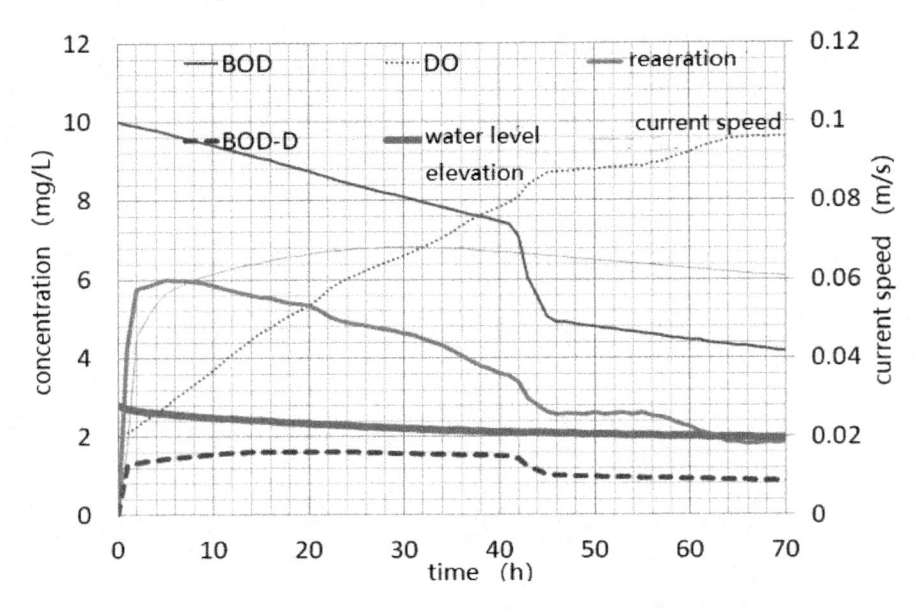

Figure 5. Variation diagram of parameters with time at typical points of Huantang West River.

### 3.3 *Analysis of BOD concentration*

As can be seen from Figure 6, the BOD concentration of Huantang West River is relatively stable before 18:00 on April 8, fluctuates greatly after 0:00 on April 9, and the change of BOD concentration at the point farthest from the river source is the largest. With the passage of time, the BOD concentration at each point in the river changes greatly with time. The farther the distance, the greater the BOD concentration and the worse the water quality. The BOD concentration at each point near Huantang West River generally shows a downward trend, from the initial 10 mg /L to 4.20 mg /L after 70 hours. The water quality index has completed the transformation from five types of water to three types of water.

### 3.4 *Do concentration analysis*

It can be seen from Figure 7 that the DO concentration in Huantang West River has improved significantly from class V water at 0:00 on April 7 to class I water standard at 0:00 on April 10. It can be seen from the figure that the DO concentration in Huantang West River suddenly decreases at 3.5km from the starting point. This is because the river section is not directly connected with the outlet, resulting in almost zero flow rate and no improvement in water quality.

It can be seen from Figure 8 that the DO concentration at the point near Huantang West River generally increases from the initial 2 mg /L to 9.6 mg/L after 70 hours. The water quality at this point belongs to the index of class I water. Through comparative analysis, in the first 40 hours, it can be seen that where the river flow velocity is large, the BOD concentration is small, the DO concentration is large, and the BOD degradation speed is fast. On the contrary, where the flow

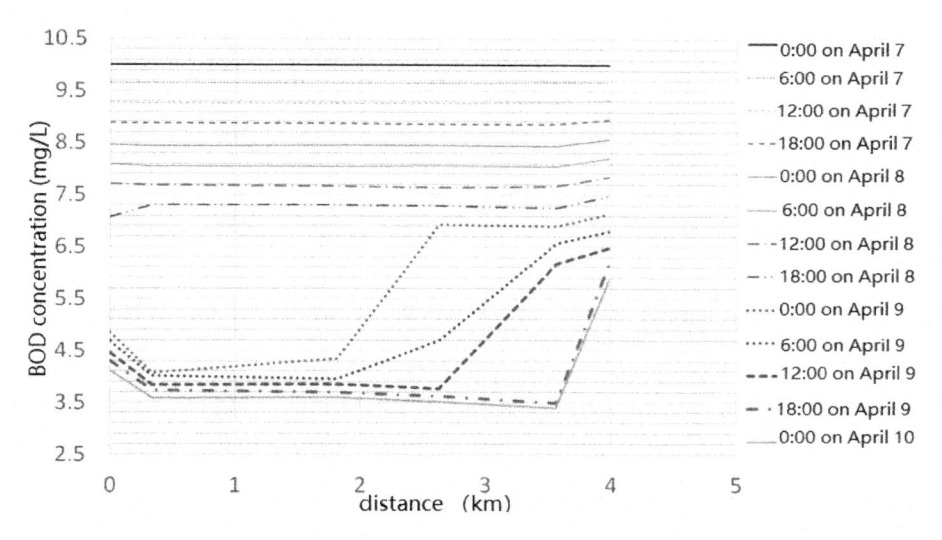

Figure 6. Variation curve of BOD concentration with distance in Huantang West River at different times.

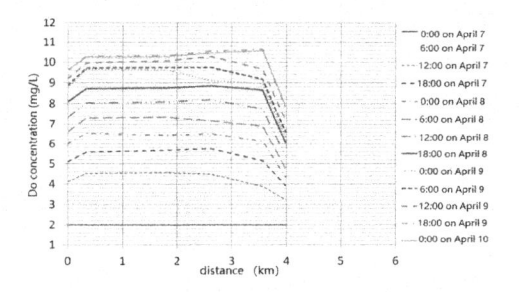

Figure 7. Figure 7 Variation curve of DO concentration with distance in Huantang West River at different times.

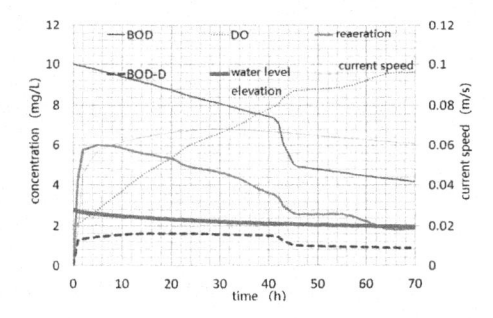

Figure 8. Variation curve of parameters with time at the points taken by Huantang West River.

velocity is small, the BOD concentration is large, the DO concentration is small, and the BOD degradation speed is slow. After 40 hours, the decrease rate of BOD concentration is small, and the decrease rate of DO concentration is small. At this time, the degradation rate of BOD is low.

## 4 CONCLUSION

Based on the establishment of the plane two-dimensional mathematical model of the Yubei plain river network, Huantang West River is selected as the research object. Under the condition of the diversion scheme, the water level, velocity, BOD, and DO are analyzed. After diversion, the water level generally shows a downward trend; the longer the improvement time of Huantang West River, the smaller the velocity of Huantang West River, and at the same time, the greater the distance, the smaller the velocity; BOD concentration showed a decreasing trend as a whole; the concentration of DO is generally increasing, and the scheme improves the water environment. In addition, opening the internal sluice of the river network can promote the flow of water bodies and the improvement of the water environment. The switch of the sluice can be used to adjust and control the diversion of water resources to the required rivers. In the future, the rivers around Huantang West River will be numerically simulated and analyzed to make the scheme more convincing.

## ACKNOWLEDGMENTS

This research was supported by the Funds Key Laboratory for Technology in Rural Water Management of Zhejiang Province(ZJWEU-RWM-202101), the Joint Funds of the Zhejiang Provincial Natural Science Foundation of China (No. LZJWZ22C030001, No. LZJWZ22E090004), the Funds of Water Resources of Science and Technology of Zhejiang Provincial Water Resources Department, China (No.RB2115, No.RC2040), the National Key Research and Development Program of China(No.2016YFC0402502), the National Natural Science Foundation of China (51979249).

## REFERENCES

Chen W.J, Li D.F, Zhang H.W. (2014) Analysis of control conditions of a hydrodynamic model for flood control and drainage of the river network in Shaoxing Plain.*Journal of Zhejiang University of Water Resources and Hydropower.* 26 (03): 38–41.

Lu X.Y, Wang M.M, Huang F. (2013) Influence of Fenghua River and Yongjiang River Regulation on flood control capacity of Yaojiang River and its countermeasures. *Journal of Zhejiang Water Resources and Hydropower College.* 25 (01): 44–47.

Lu Y.J. (2021) Analysis of urban pump gate construction management and river water environment improvement. *Engineering Construction and Design.* (22): 89–91.

Tong Y.Y, Li D.F, Nie H. (2016) Study on the hydrodynamic and plane two-dimensional mathematical model of the river network in Datian Plain. *Journal of Zhejiang University of Water Resources and Hydropower.* 28 (01): 14–17.

Wu Y, He L.Q. (2019) Discussion on constructing a hydrodynamic model of Cao'e River flow area in Shaoxing City. *Zhejiang Water Conservancy Science and Technology.* 47 (03): 21–23+31.

Xie Q.H, Li D.F, Chen D.Y, Fan R.H, Zhang H.W. (2010) Study on water mobility of urban river network based on two-dimensional numerical simulation. *Journal of Zhejiang Water Conservancy and Hydropower College.* 22 (04): 1–6.

Z.H. Chen, X.H. Chen, J. Du, Y.J. Xiong. (2012) Regulation and effect prediction of water environment diversion in the river network area. *Water Resources Protection.* 28(03): 16–21.

Z.T. Chen, L. Hua, Q.N. Jin. (2015) Study on effect evaluation of water diversion to improve the water quality of urban river network. *Journal of Changjiang Academy of Sciences.* 32(07): 45–51.

*Civil Engineering and Urban Research – Mohamed & Hou (Eds)*
© *2023 the Authors, ISBN 978-1-032-44487-1*

# Hydraulic calculation and water hammer analysis of municipal water distribution system with large drop

Zongke Chen, Lishuang Yuan*, Xiaowei Yang, Linyuan Li & Lingxuan Zou
*China Construction Third Engineering Bureau Group Co., Ltd, Wuhan, Hubei, China*

ABSTRACT: When calculating the connection between pipelines, aiming at the problems of large value of loss constraint, a small value of pressure, and inaccurate water hammer effect, the hydraulic calculation, and water hammer analysis process of large drop and long-distance municipal water transmission and distribution system are designed. The triangular curve grid processing pipeline is set as a grid structure from sparse to dense, and the normal imaging numerical relationship is used to control the pipeline constraint conditions Then, a long-distance water transmission and distribution system is set as a grid structure, and the large drop condition is defined. After the elevation value of the pipe centerline is calibrated by using the large drop hydraulic calculation algorithm, the numerical relationship of the water hammer effect is constructed to complete the process analysis. The test results show that the hydraulic calculation error is controlled at about 0.07%, the water level error is controlled at 0.1cm ∼ 2cm, and the difference between the pressure value and the measured standard value is less than 2N.

## 1 INTRODUCTION

The height difference between different riverbeds is very large, and the water transmission and distribution system controls the normal distribution of water flow in the water distribution pipe under the action of gravity. The hydraulic calculation is used to reasonably combine the calculation, scheduling, and optimization of the pipe network (Cong et al. 2020; Han et al. 2020; Wang et al. 2019; Zhang et al. 2019) to transform the pipe network information acquisition process into the digital calculation process of pipe network information. The literature (Ma et al. 2020) constructs a relationship that reflects the mechanical characteristics of the pipeline, but when analyzing the force between the pipe and the water flow, the characteristic relationship parameters are affected by external noise, resulting in a small mechanical value obtained in the final analysis. A multi-level algorithm is used to calculate the force between the pipe and the water flow in the water distribution pipe in Literature (Wang et al. 2020), but the loss constraint set at the connection between the pipes is large, resulting in a small mechanical value obtained by the final analysis. To this end, this paper establishes the hydraulic calculation method of long-distance municipal water transmission and distribution system with a large drop and analyzes the resulting water hammer effect.

## 2 HYDRAULIC CALCULATION AND WATER HAMMER ANALYSIS OF WATER TRANSMISSION AND DISTRIBUTION SYSTEMS

### 2.1 *Divide the pipeline grid structure of long-distance municipal water transmission and distribution system and the judgment of water flow flow*

The long-distance municipal water transmission and distribution system pipeline has an intricate structure, in the division of pipeline structure, the use of a triangular curve grid to treat the pipe

---

*Corresponding Author: wy13156013532@126.com

DOI 10.1201/9781003372417-113

structure as a grid structure from sparse to dense, control the numerical relationship between the dense structure, can be expressed as:

$$\left(g\frac{\partial x}{\partial s}\right) + \frac{\partial}{\partial \alpha}\left(g\frac{\partial x}{\partial l}\right) = 0 \tag{1}$$

where $x$ – the length of the pipe in the plane, m, $s$ – the area enclosed by the pipe, m2, $\alpha$ – the streamline parameter of the pipe, $l$ – the potential line parameter of the pipe, $g$ – gravitational acceleration, m/s2. According to the above formula, the water transmission and distribution system is integrated into one area, and the regional meshing process is shown in Figure 1:

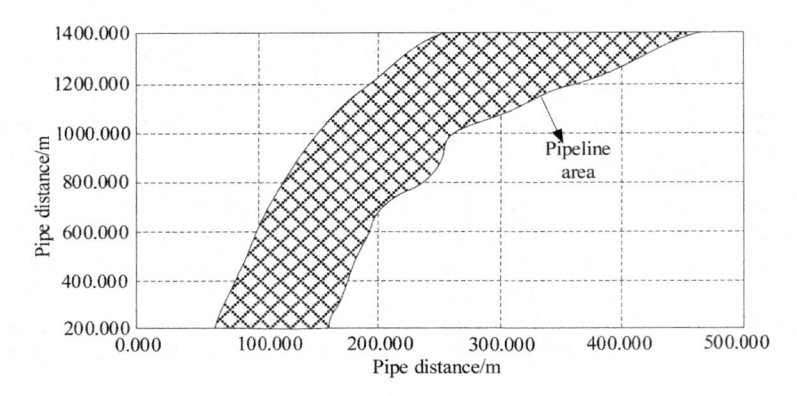

Figure 1. Grid division process of the hydraulic system.

According to the results of the hydraulic system meshing in Figure 1, in the long-distance water distribution system, a numerical relationship of water vector is established, which can be expressed as:

$$\frac{\partial \rho}{\partial t} + \nabla V = 0 \tag{2}$$

Where, $\rho$ – the flowing fluid of the pipeline, $t$ – the water distribution cycle of the water distribution system, the day, $V$ – the volume of water in the water distribution system, m3. Under the control of the vector value, a water flow state judgment process is constructed, which can be expressed as:

$$\mathrm{Re} = \frac{\rho UL}{\mu} \tag{3}$$

where $U$ – characteristic water flow velocity, m/s, $\mu$ – laminar flow state parameters, $L$ – distance parameters, m.

### 2.2 Construct a hydraulic calculation algorithm with large drops

Using the mesh structure obtained by the above division, a hydraulic calculation algorithm with a large drop is constructed, that is, the flow state parameter greater than the value 1 is regarded as the large drop nozzle effect, and the additional water flow generated by the calculation of the large drop can be expressed as:

$$Q_a = \frac{K}{\sqrt{P}} \tag{4}$$

Where $Q_a$ – the additional water flow generated by the large drop, m3/h, $K$ – the flow coefficient of the pipeline, the meaning of the remaining parameters remains unchanged. Corresponding to the transformation of different pipeline branch structures, the simulation obtains the hydraulic line, and the structure is shown in Figure 2:

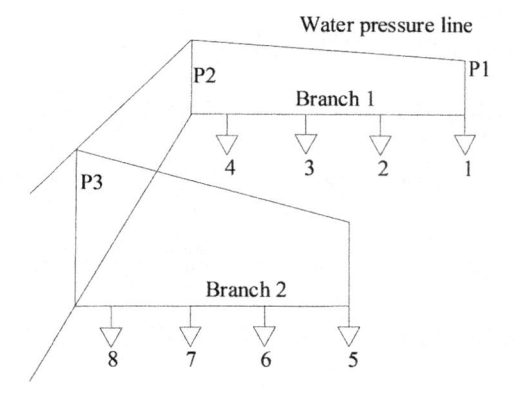

Figure 2. Hydraulic pressure line structure obtained from the simulation.

In Figure 2, P1, P2, and P3 are the potential lines of the pipeline, 1–4 is the sprinkler of branch pipe 1, and 6–9 is the sprinkler of branch pipe 2. Under the hydraulic line structure shown in Figure 2, different structures are affected by the external environment of the municipal water transmission and distribution system, and a certain amount of water head loss is easily generated in the pipeline, which can be expressed as:

$$J = \frac{V^2}{d_i} \tag{5}$$

Where, $J$ – pipe head loss coefficient, $d_i$ – pipe action inner diameter size, m.

### 2.3 Water hammer analysis

The above-obtained water flow values and water pressure parameters are used as indicators of hydraulic drop, and when analyzing the water hammer effect caused by water transmission and distribution, the hydraulic process under different states is calculated, and the analysis of the water hammer process is completed. The process is as follows: Synthesizing the indicators obtained from the above analysis, synthesizing the nodes in each grid structure, and calculating the maximum pressure head value, which can be expressed as:

$$R_{ij} = \frac{H_{ij}^{\max} - H_1}{H_{ij}} \tag{6}$$

Among them, $i$ and $j$ are – the nodes in the water distribution grid structure, $H_{ij}^{\max}$ – the maximum water hammer pressure at the grid nodes, N, $H_1$ – the allowable pressure value at the nodes, N, $H_{ij}$ – the difference in the water hammer pressure values between the two nodes. Synthesizing the pressure generated between the nodes, when encountering the generatable conditions of the water hammer effect, for the relative pressure generated by the water flow in the pipeline, the conditions for judging the water distribution system may produce the water hammer effect can be expressed as:

$$\begin{cases} 1, & H_1 > H_{ij}^{\max} > H_{ij} \\ \dfrac{\alpha - \left| H_{ij}^{\max} \right|}{\alpha}, & -\alpha \le H_{ij}^{\max} < H \end{cases} \tag{7}$$

Where $\alpha$ – the pressure critical point parameter of the pipeline design, the meaning of the remaining parameters remains unchanged. Under the above numerical judgment relationship, the reliability of the two extremum nodes is calculated, and the reliability of the water transmission and distribution system is obtained.

## 3  EXPERIMENT

### 3.1  *Experiment preparation*

It is needed to randomly select a large drop river, place multiple water distribution pipelines at the long-distance trunk road, and work continuously with a 40KW generator and DN100 pump. The layout structure of the water transmission and distribution system is shown in Figure 3:

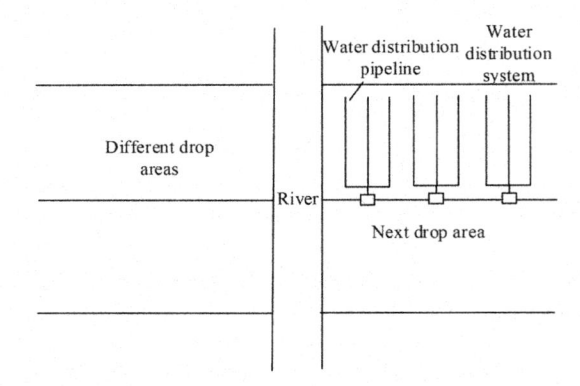

Figure 3.  Layout structure of water distribution system.

Under the system layout structure constructed in Figure 4, according to the layout relationship between each pipeline, the test node of the pipeline is set, the piping parameter values of each water system are measured and counted, and the parameters of the water distribution system are measured as shown in Table 1:

Table 1.  Parameter values of water distribution system measured.

| Distribution pipe nodes | Pipes measure length (m) | Specific resistance (KN/m) | Node pressure (N) |
| --- | --- | --- | --- |
| GD-01-01 | 10.41 | 0.0749 | 21.86 |
| GD-01-02 | 12.41 | 0.0282 | 31.23 |
| GD-01-03 | 16.21 | 0.0728 | 24.84 |
| GD-02-04 | 14.34 | 0.0749 | 18.92 |
| GD-02-05 | 16.21 | 0.0282 | 20.42 |
| GD-02-06 | 15.42 | 0.0728 | 46.88 |
| GD-03-07 | 13.47 | 0.0714 | 43.75 |
| GD-03-08 | 16.21 | 0.0828 | 28.15 |
| GD-03-09 | 15.42 | 0.0542 | 40.62 |
| GD-04-10 | 16.21 | 0.0427 | 34.36 |
| GD-04-11 | 18.61 | 0.0482 | 53.25 |
| GD-04-12 | 17.14 | 0.0714 | 21.86 |

Under the numerical control of the water distribution parameters set in Table 1, the flow value is taken as the flow value of the pipeline, the flow value in the pipeline is added, and the flow control is carried out by two conventional water hammer analysis methods (literature [5] method and literature [6] method) and design water hammer analysis method to compare the performance of the three analysis methods.

## 3.2 Experimental process

Based on the above experimental preparation, the hydraulic calculation method of the three analysis methods is invoked, and the calculation error of the three hydraulic calculations can be expressed as:

$$p = \frac{Q^w}{Cd} \times 100\% \tag{8}$$

where $Q^w$ – corresponding to the actual water flow value of the pipeline, m3/h, $C$ – Hayden coefficient, $d$ – the radius within the pipeline, m.

Count the segments of each pipe node of each water transmission system. When comparing the flow rates of the three analysis methods, the cross-sections at the control tube nodes are equal. After marking the pipe node section, the flow value is used to establish a numerical relationship of the pipe pressure, which can be expressed as:

$$q = \sqrt{\frac{F}{SV}} \tag{9}$$

where $F$ – the pressure value within the pipe, N, $S$ – the cross-sectional area at the pipe node, m2, $V$ – the volume of water flowing through, m3.

## 3.3 Experiment results and analysis

Under the control of the above calculation formula, the calculation error generated by the hydraulic calculation in the three analysis methods is evaluated and compared, and the error result is shown in Table 2.

Table 2. The Calculation Error arising from the hydraulic calculation of the three analysis methods.

| Distribution Pipe node | General Analytical method 1 | Calculation error/% | |
| --- | --- | --- | --- |
| | | General analytical method 2 | Analytical methods for design |
| GD-01-01 | 0.0243 | 0.0136 | 0.0082 |
| GD-01-02 | 0.0277 | 0.0138 | 0.0077 |
| GD-01-03 | 0.0252 | 0.0116 | 0.0088 |
| GD-02-04 | 0.0204 | 0.0141 | 0.0077 |
| GD-02-05 | 0.0224 | 0.0102 | 0.0072 |
| GD-02-06 | 0.0236 | 0.0143 | 0.0097 |
| GD-03-07 | 0.0237 | 0.0111 | 0.0089 |
| GD-03-08 | 0.0255 | 0.0141 | 0.0074 |
| GD-03-09 | 0.0238 | 0.0144 | 0.0075 |
| GD-04-10 | 0.0281 | 0.0146 | 0.0092 |
| GD-04-11 | 0.0215 | 0.01427 | 0.0036 |
| GD-04-12 | 0.0234 | 0.01482 | 0.0025 |

According to the calculation error results shown in Table 2, the calculation error obtained by conventional analysis method 1 is about 0.2%. The calculation error obtained by conventional analysis method 2 is about 0.1%. The calculation error of the designed analysis method is about 0.07%. Compared with the two conventional analysis methods, the designed analysis method produces the smallest error in hydraulic calculation.

Statistically calculate the water level value obtained by the three analysis methods, and calculate the difference value of the water level error generated by the three analysis methods. The error results are shown in Figure 4.

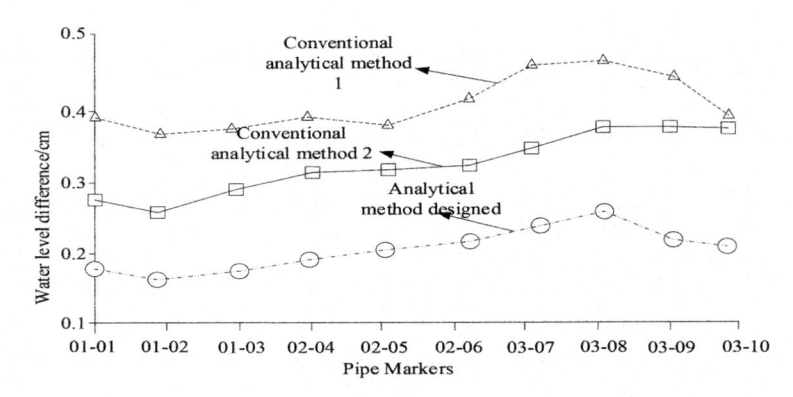

Figure 4. Water level error obtained from three analysis methods.

From the water level error results shown in Figure 5, it can be seen that under the control of the three analysis methods, the water level error generated by the conventional analysis method 1 is between 0.4cm and 0.5cm. The water level error produced by conventional analysis method 2 is between 0.2cm and 0.3cm. However, the water level error produced by the design analysis method is between 0.1C and 2cm, and the numerical error of the designed analysis method is the smallest.

## 4 CONCLUSION

In daily water conservancy projects, affected by external conditions such as large drops and long distances, the water hammer effect often occurs in the process of pipeline water delivery and water distribution, which affects the normal water supply of the city. In this paper, the hydraulic calculation and water hammer analysis method of the large drop and long-distance municipal water transmission and distribution system is designed. After the large drop, the hydraulic calculation algorithm is used to calibrate the elevation value of the pipeline centerline, and the numerical relationship of the water hammer effect is established. The pressure values obtained by measuring 10 pipeline nodes are between 17.86-45.98N, which is close to the measured standard value, and the water level error is between 0.1C and 2cm. It improves the accuracy of hydraulic calculation and water hammer effect analysis of the large drop and long-distance municipal water transmission and distribution system.

## REFERENCES

Cong, X.H., Wang, J.Y., Tang, W.M. (2020) Study on the Factors Affecting Flood Control Hydraulic Calculation for the Trans Boundary River Based on MIKE 21FM Model. *Yellow River*, 42 (8): 45–48 + 55.

Han, K., Ding, F.L., Mao, Z.Y. (2020) Numerical analysis of water hammer in pumping station of long-distance pressurized water transmission project. *Progress in water conservancy and Hydropower Science and Technology*, 40 (2): 69–75.

Ma, K., Gu, Z.J., Bai, C.Q. (2020) An improved MOC and impact analysis of overload on water hammer. *Chinese Journal of Computational Mechanics*, 37 (5): 623–628.

Wang, Q.W., Li, Z.P., Zhu, C.D. (2020) Analysis on Valve Regulation and Water Hammer Control of Gravity Flow Pipeline. *Fluid Machinery*, 48 (6): 38–43 + 50.

Wang, Y.X., Yan, B.K., Tian, Y.L. (2019) Analysis and Protection of Water Hammer for Pump Stopping in Long Distance Water Conveyance Pipeline. *China Water & Wastewater*, 35 (7): 57–61.

Zhang, J.W., Wu, J.H., Gao, J. (2019) Research on Water Hammer Simulation and Protection for Long Distance Gravity Flow Water Conveyance System. *Water Resources and Power*, 37 (5): 57–60.

# Author index

An, Y. 101

Bagen, W. 411
Bai, E. 37, 365, 371, 745
Bai, G. 111
Bao, T. 224
Bi, P. 725
Bu, J.-Q. 18

Cao, K. 63, 760
Cao, S. 207
Chen, F. 284
Chen, H. 345, 588
Chen, J. 664
Chen, L. 101, 292, 753
Chen, M. 789
Chen, S. 613
Chen, Y. 92, 789
Chen, Z. 795
Cheng, C. 340
Cheng, L. 664

Dai, J. 489
Dai, W. 465
Dong, J. 651
Dong, S. 562
Du, J. 3
Du, L. 345
Du, X. 122
Duan, B. 306
Duan, Q. 11

Fan, J. 739
Fan, Q. 692
Fan, Y. 177
Fang, Z. 582, 733
Feng, W. 3

Gao, J. 642
Gao, M. 774
Gao, X. 686
Gu, J. 267
Gu, Y. 549, 658
Guan, S. 484

Guo, L. 28
Guo, S. 516
Guo, Z.-B. 18

Han, T. 11, 711
Hao, H. 495
Hao, J. 574
Hao, S. 698
Hao, Z. 609
He, G. 300
He, P. 711
Hongbin, N. 768
Hongmei, M. 768
Hou, R. 396
Hu, D. 789
Hu, H. 224
Hu, X. 306, 510
Huang, F. 635
Huang, H. 419
Huang, J. 780
Huang, K. 248, 257
Huang, X. 145, 323, 329
Huang, Y. 774
Huaping, Z. 117
Huaqiang, L. 169
Hui, Y. 774

Jia, Y. 145
Jiacheng, Z. 44
Jianfeng, D. 340
Jiang, J. 323, 329
Jiang, Z. 84
Jianhui, X. 335, 340
Jie, W. 335, 340
Jinyan, S. 44
Jiuzeng, S. 436

Kaikai, C. 680
Kang, T. 371
Keran, Y. 411

Lai, X. 207
Lan, C.-H. 233

Li, C. 356, 673, 711
Li, D. 582, 733
Li, H. 431, 593
Li, H.-F. 292, 753
Li, J. 189, 207, 215, 431
Li, K. 377
Li, L. 137, 345, 419, 451, 795
Li, M. 527, 718
Li, Q. 189
Li, R. 549
Li, S. 279, 345
Li, T. 516
Li, W. 201
Li, X. 189
Li, Y. 215, 495, 621
Li, Z. 44, 84, 177, 248, 356, 789
Lian, Z. 651
Liang, F. 403
Liang, R. 739
Lin, C. 780
Lin, H. 311
Liu, D. 549, 601
Liu, J. 151, 356
Liu, L. 163
Liu, Q. 273
Liu, R. 686
Liu, S. 377, 774
Liu, T. 516
Liu, W. 279
Liu, X. 774
Liu, Y. 635
Liu, Z. 627
Long, T.Y. 458
Lou, M. 377
Lu, A. 635
Lu, L. 411
Lu, X. 484, 489
Lu, Y. 75
Lu, Z. 760
Luo, C. 377
Luo, K. 185
Luo, W. 477

Luo, X. 745

Ma, C. 664
Ma, S. 686
Ma, W. 613
Ma, Y. 51
Man, J. 273
Mei, X. 527, 718
Miao, J. 3
Ming, Y. 725
Mu, R. 383
Mu, X. 465

Ning, Y. 37, 365, 745

Pang, J. 224
Peng, C. 760
Peng, J. 345
Peng, X. 442

Qi, Y. 248, 257
Qian, M.A. 44
Qiang, L. 311
Qiao, Y. 329
Qingchen, T. 44

Ren, B. 37, 365
Ren, C. 63
Ren, Q. 279
Ren, X. 207
Ruozhu, W. 169

Shao, J. 273
Shao, Z. 345
Shen, C. 207
Shen, H. 789
Shi, H. 3, 725
Shi, P. 300
Shi, Z. 201
Shoko, K. 411
Shou, L. 224
Shubo, Z. 69
Si, J. 51
Si, X. 549
Si, Z. 306
Sisheng, W. 351
Song, L. 58
Su, J. 543
Sun, G. 318
Sun, H. 371
Sun, J. 145
Sun, L. 601
Sun, Q. 371

Sun, S. 163
Sun, W. 692
Sun, X. 195, 396
Sun, Y. 673

Tan, Y. 383, 557
Tang, J. 92
Tang, R. 306
Tang, S. 510
Tang, X. 489
Tao, B. 215
Tao, G. 613
Tao, S. 292, 753
Tao, W. 117
Teng, Q. 686
Tenzin, S. 51

Wang, B. 111, 292, 753
Wang, C. 151
Wang, D. 774
Wang, F. 686
Wang, H. 11
Wang, K. 130
Wang, L. 177, 224
Wang, N.-X. 233
Wang, Q. 189
Wang, S. 635
Wang, T. 37
Wang, W. 207
Wang, Y. 292, 396, 753
Wang, Z. 273, 323,
    365, 484
Wei, T. 151, 169
Wei, Y. 323
Wu, B. 601
Wu, J. 425
Wu, K. 725
Wu, Q. 207
Wu, Y. 300, 306, 588, 593
Wu, Z. 635

Xiang, H. 329
Xiao, G. 516
Xiao, X. 345
Xiaowen, Y. 335
Xiaozuo, R. 44
Xie, B. 593
Xie, P. 664
Xie, X. 145
Xin, Z. 69
Xing, L. 621
Xiong, J. 185
Xu, G. 725

Xu, J. 37, 651
Xu, K. 84
Xu, Q. 642
Xu, T. 562
Xue, B. 516
Xun, J.-C. 18

Yan, K. 137
Yan, M. 692
Yan, P. 300
Yan, X. 195
Yang, H. 706
Yang, J. 642, 664
Yang, L. 195
Yang, Q. 69
Yang, S. 300
Yang, X. 795
Yang, Y. 390, 495, 549
Yao, A. 37, 365, 745
Yao, D. 745
Yao, Y. 145
Yao, Z. 215
Yi, D. 436
Yifei, W. 311
Yin, Y. 215
Yiqin, Q. 311
Yongmei, Q. 169
Yongyuan, Z. 311
You, C. 284
Yu, H. 157
Yu, Y. 535
Yu, Z. 11
Yuan, L. 795
Yuan, Y. 356
Yugui, L. 335, 340
Yumin, Z. 680
Yun, H. 680
Yun, Z. 335
Yundan, G. 51
Yung, S. 760
Yutao, S. 117

Zang, C. 593
Zeng, J. 177
Zhai, S. 63
Zhan, Y. 257
Zhang, C. 323, 635
Zhang, D. 505
Zhang, E. 101
Zhang, H. 306, 442
Zhang, J.-R. 18
Zhang, L. 101, 215, 613, 725
Zhang, M. 163

Zhang, P. 549
Zhang, T. 621
Zhang, W. 201
Zhang, X. 279, 318, 345, 543, 574
Zhang, Y. 195, 329, 745
Zhang, Z. 356
Zhao, A. 774
Zhao, B. 698

Zhao, H. 151
Zhao, J. 51, 377
Zhao, Y. 390
Zhong, D. 145
Zhong, R. 780
Zhou, C. 442
Zhou, J. 137
Zhou, S. 345, 477
Zhou, X. 345

Zhou, Y. 403
Zhu, B. 673
Zhu, J. 92
Zhu, L. 51
Zhu, N. 356
Zhu, W. 557
Zhu, Y. 51, 122
Zhu, Z. 329
Zou, Q. 588